Geological Nomenclature

ROYAL GEOLOGICAL AND MINING
SOCIETY OF THE NETHERLANDS

GEOLOGICAL NOMENCLATURE

edited by

W. A. Visser

English
•
Dutch
•
French
•
German
•
Spanish

1980
Martinus Nijhoff, The Hague, Boston, London

ISBN 90 247 2403 1

CONTENTS

Preface VII

Authors, Co-authors and Co-operators 'XI

Contents of Thematic Part XXIII

Thematic Part 1

Tables 367

Indices

English 377
Nederlands 417
Français 447
Deutsch 479
Español 513

PREFACE

HISTORICAL NOTES

The compilation of multilingual nomenclatures has become a traditional occupation of the Royal Geological and Mining Society of the Netherlands. Even before the Society's foundation (March 9, 1912) work was done on a geological nomenclature. At the time the Mining Association had entrusted its compilation to Ch. Th. Groothoff, for the mining engineers felt the need of a formal explanatory dictionary of geological terms.

In November 1912 the Board of the Geological Section of the newly founded Geological and Mining Society (the royal title was bestowed in 1952) decided to continue Groothoff's work. A committee was appointed which asked the members of the Society to contribute terms. This, however, appeared to be a not too successful procedure. In 1924 two chapters were completed, Geomorphology and Tectonic Geology. The former was considered far too lenghty, and as a consequence too costly to print. Eventually, in 1928, professor Dr. L. Rutten of the State University in Utrecht was appointed editor. This appointment resulted in the publication of the 'Geologische Nomenclator' in four languages in 1929. It was issued by G. Naeff in The Hague.

Rutten worked along the lines set by a committe appointed by the Board of the Society, *viz.* a thematic arrangement of the terms by chapters subdivided in sections. Dutch was to be the leading language. The lay-out was spacious: a colum of Dutch terms was followed by one for notes and explications, including line drawings, and by three columns for the equivalent terms in respectively German, English and French. There were five indices, one for each of the languages and one for expressions from other languages. It is interesting to note that in the latter nearly all are Italian volcanological terms.

Rutten wrote in his preface that this nomenclature should serve in the first place as a geological dictionary. In the second place, its compilation provided an opportunity to extend the Dutch terminology, at the time in many respects still incomplete. This dual purpose has been attained in eight chapters, each entrusted to a specialist, with a total of about 4000 terms in the leading language. The nomenclature was an immediate success. A few years after its publication it was virtually impossible to obtain a copy. The thematic arrangement of terms was an experiment that succeeded well.

In the meantime the Mining Section of the Society had appointed a committee to compile a mining nomenclature (1928). Work was slow and many mutations in the membership of the committee were unavoidable. In 1946 the text was reported to be completed. The 'Mijnbouwkundige Nomenclator' was printed and issued by J.B. Wolters (Groningen) in 1949. Leading language was Dutch; equivalents were given in German, English, French and Spanish. Although largely a coal-mining nomenclature with almost 3700 terms in the leading language, some 800 terms were included concerning oil drilling and producing techniques, ores and geology. Small print and lay-out, i.e. term and definition in Dutch and the foreign equivalents in one block, resulted in a great reduction of space as compared to the geological nomenclature.

As early as the mid-fifties the feasibility of the issue of a new and revised mining nomenclature was considered. Enthousiastic members of successive Boards of the Mining Section collected about 24000 terms and their translations. However, lack of interest and co-operation outside the Society forced them to abandon the project, obviously to their great disappointment.

In 1959 a new geological nomenclature was published. Its editor, the retired petroleum-exploitation engineer of 'Shell', A.A.G. Schieferdecker, remarked in his preface that the 1929 nomenclature served its purpose adaquately for quite a long period, but scientific progress made adaptation to new geological concepts increasingly necessary. A multitude of new terms were added as well as a number of new chapters. These points are well illustrated by some figures. The nomenclature contains about 5500 terms and their definitions in English, the leading language, and the equivalents in Dutch, French and German, spread over 11 chapters. In the index the terms of all four languages have been entered into one list. The thematic part numbers 350, the index 171 pages (16×24 cm).

The compilation of each chapter or section of a chapter was entrusted to one specialist. However, in the end it appeared that many more had contributed to the final text. Moreover, drafts were sent to specialists abroad for correction and completion. The list of authors and co-operators contains 75 names. The publisher, J. Noorduijn en Zoon at Gorinchem, printed 4400 copies, of which 1000 were at the disposal of the Society to sell to its members at a reduced price and for presentation to the collaborators. In the first years more than half of the edition was sold. In later years there was a slow but steady sale, and in 1977 the book was out of print.

In 1973 it was clear that a new edition would be needed in not too long a future. Preliminary work began the same year; a table of centents was made and authors appointed. Moreover, it was decided that Spanish would be added. The set-up would be the same as in Schieferdecker's. It was realized that Schieferdecker's references to the 1929 nomenclature would also apply to the new one, and as work progressed they proved even more forcibly true. This, the third Geological Nomenclature, is an entirely new book. The number of entries has been more than doubled; several new subjects and chapters have been added. Nevertheless, whether a fully up-to-date piece of work has been produced is open to doubt. Development in the earth sciences and in the areas touching related disciplines is too fast for authors, editor or printer to cope with.

EXPLANATORY NOTES

As mentioned before, the general set-up is similar to the 1959 nomenclature. The English terms and their definitions are followed by the equivalents in Dutch, French, German and Spanish. In several cases, after the English terms, synonyms in languages not belonging to one of the five of the nomenclature have been included. They have attained a certain use in literature and in practice. Normally the leading term is given in the English language. However, occasionally a term typical for one of the other languages and for which no English equivalent is known, is inserted in the leading line.

As a rule, the English term most currently used or preferred is printed in bold-face type. Synonyms, separated by semicolons, are placed in successive order of importance, admittedly a somewhat arbitrary procedure. Obsolete or antiquated terms are indicated as such; inclusion is appropriate, because older literature may be consulted. The terms included in the nomenclature are numbered; these numbers correspond to those in the indices. However, te enhance the thematic arrangement related terms have been listed under the same number. Special care is given to the units of measurement. At present, shortly after the introduction of the 'Système International des Unités', the SI units, also the metric ones are given with the conversions.

The various terms have been entered in five indices, one for each of the languages, while terms not belonging to one of these have been included in the English index. The latter is the most elaborate, in which terms consisting of two coordinate elements can usually be found under the first as well as under the second element. This procedure has been omitted in the other indices. It may cause some inconvenience, but results in a considerable saving of space. In the non-English indices the terms have been listed according to what is considered as their more important element, often a substantive. In all indices terms containing an author's name have been entered under that name only. The terms have been arranged in alphabetical order by electronic means. This has resulted in a few restrictions, which consist of an arrangement based strictly on letters disregarding derivations or commas, and in case of multiple entry repetition of the term in the next line. For instance:

anchor; *anchorage*; *anchor cable*
bed; *bed, bottomset* –; *bedded*; *bedding*; ....; *bed, dirt* –
stream, captor –; *stream, capturing* –; *stream development*; ....; *streamer*; *stream, glacial* –

NOTES OF THANKS

Many people have contributed to this volume. The compilation of each chapter or section was entrusted to a specialist in that particular subject, who was free to seek the participation of other experts. Drafts were sent abroad for correction and completion of French, German and Spanish equivalents. A special expression of thanks is due to

Drs. J. Gravesteijn, head of the documentation department of the 'Bureau de Recherches Géologiques et Minières', Orléans,

Professor Dr. H.-J. Dürbaum, director of the geophysical department of the 'Bundesansalt für Geowissenschaften und Rohstoffe', Hannover,

Professor Dr. E. Perconig, director of the 'Empresa Nacional Adaro de Investigaciones Mineras', Madrid, who kindly disributed the drafs to specialists in their countries and ascertained their timely return, an often tedious and unrewarding task. Although the selection of terms and the definitions are due to the Dutch authors, any comment or remark made by our foreign collegues is gratefully acknowledged. Thanks are due to W.J. van Dalfsen (Ground–Water Survey TNO, Delft) for his work on the units of measurement.

The editor expresses his thanks and appreciation to all those who have contibuted to this nomenclature; their names are entered in the list of authors, co-authors and co-operators.

It goes without saying, that extensive use is made of published literature. For the purpose of checking definitions, the Glossary of Geology and Related Sciences of the American Geological Institute and A Dictionary of Geology by John Challinor (University of Wales Press, Cardiff, 1973) have often been consulted.

Thanks are due to H.G. van Dorssen, bachelor of geology, University of Amsterdam, who rendered great and valuable assistance in the preparation of the indices, the final numbering of the terms and in proof-reading. For the preparation of the typescript the editor is mainly indebted to Miss A. de Kiewit, who is able to produce a flawless copy. She proved also to be an accurate proof-reader. Last but not least, the editor expresses his thanks to his friends in geological and mining diciplines, who helped in proof-reading.

The Board of the Royal Geological and Mining Society of the Netherlands expresses its sincerest thanks to all contributors of this work for their enthousiasm and co-operation.

Nootdorp, March 1980 W.A. Visser

AUTHORS, CO-AUTHORS AND CO-OPERATORS

(institutional affiliations have been mentioned only once)

[0100] **Introduction**

W. A. Visser,
(ret.) Shell Internationale Petroleum Mij.

[0200] **Extraterrestrial Bodies**

W. A. Visser

P. Calas,
Bureau de Recherches Géologiques et Minières, Orléans

R. Mühlfeldt,
Bundesanstalt für Geowissenschaften und Rohstoffe, Hannover

J. M. Torroja Menéndez,
Real Academia de Ciencias, Madrid

[0300] **Geophysics**

[0310] GENERAL TERMS; THE ARCHITECTURE OF THE EARTH

W. A. Visser
A. R. Ritsema

P. Calas

H.-J. Dürbaum,
Bundesanstalt für Geowissenschaften und Rohstoffe, Hannover

E. Perconig,
Empresa Nacional Adaro de Investigaciones Mineras, Madrid

[0320] SEISMOLOGY

A. R. Ritsema,
Kon. Ned. Meteorologisch Instituut, de Bilt

P. Calas
Staff, Département Géophysique,
Bureau de Recherches Géologiques et Minières, Orléans

H. P. Harjes,
Bundesanstalt für Geowissenschaften und Rohstoffe, Hannover

A. Lopez Arroyo,
Inst. Geográfico y Catastral, Madrid

[0330] GEOMAGNETISM AND ROCK MAGNETISM

J. Veldkamp,
(ret.) Rijksuniversiteit te Utrecht

[0330] GEOMAGNETISM AND ROCK MAGNETISM (continued)	A. Gérard, *Bureau de Recherches Géologiques et Minières, Orléans*
	W. Bosun H.-J. Dürbaum, *Bundesanstalt für Geowissenschaften und Rohstoffe, Hannover* S. Plaumann, *Niedersächsisches Landesamt für Bodenforschung, Hannover*
	E. Perconig
[0340] GRAVITY AND ISOSTASY	B. J. Collette, *Vening Meinesz Laboratorium voor Geofysica, Utrecht*
	A. Gérard
[0350] GEOTHERMICS	W. A. Visser W. J. van Dalfsen, *Dienst Grondwaterverkenning TNO, Delft*
	J. Lavigne, *Bureau de Recherches Géologiques et Minières, Orléans*
	L. Jerez Mir, *Empresa Nacional Adaro de Investigaciones Mineras, Madrid*
[0400] **Geochemistry**	by editor assembled from various chapters
	P. Calas H. Gundlach, *Bundesanstalt für Geowissenschaften und Rohstoffe, Hannover*
	A. Alfonso, *Shell España, Madrid*
[0500] **Isotope Geology**	H. J. N. Priem, *ZWO Laboratorium voor Isotopen-Geologie, Amsterdam*
	R. Kreulen, *Vening Meinesz Laboratorium voor Geochemie, Utrecht*
	L. Delbos, *Bureau de Recherches Géologiques et Minières, Orléans*
	E. Perconig

[0600] **Mineralogy and Gemmology**

[0610] MINERALOGY

R. O. Felius,
Geologisch en Mineralogisch Instituut der Rijksuniversiteit te Leiden (thans te Utrecht)

J. Mantienne,
Bureau de Recherches Géologiques et Minières, Orléans

E. Perconig

[0620] GEMMOLOGY

P. C. Zwaan,
Rijksmuseum voor Geologie en Mineralogie, Leiden

[0700] **Crystallography**

P. Hartman,
Geologisch en Mineralogisch Instituut der Rijksuniversiteit te Leiden (thans te Utrecht)

P. Calas

H. Rösch,
Bundesanstalt für Geowissenschaften und Rohstoffe, Hannover

E. Perconig

[0800] **Petrology – Introductory Terms**

[0900] **Petrology of Igneous Rocks**

D. E. Vogel,
Instituut voor Physicochemische Geologie, Katholieke Universiteit, Leuven

[1000] **Petrology of Metamorphic Rocks**

C. Alsac,
Bureau de Recherches Géologiques et Minières, Orléans

F.-J. Eckhardt
U. Vetter,
Bundesanstalt für Geowissenschaften und Rohstoffe, Hannover

Mrs. T. Nodal Ramos
Mrs. A. Argüellos,
Empresa Nacional Adaro de Investigaciones Mineras, Madrid

[1100] **Geotectonics**

H. J. Zwart,
Geologisch en Mineralogisch Instituut der Rijksuniversiteit te Leiden (thans te Utrecht)

P. Calas

F. Kockel,
Bundesanstalt für Geowissenschaften und Rohstoffe, Hannover

E. Perconig

[1200] **Volcanology**	H. J. Wensink, *Geologisch Instituut der Rijksuniversiteit te Utrecht* W. A. Visser
	J. Gravesteijn C. Alsac, *Bureau de Recherches Géologiques et Minières, Orléans*
	K.-P. Burgath, *Bundesanstalt für Geowissenschaften und Rohstoffe, Hannover*
	E. Fernández Vargas, *Empresa Nacional Adaro de Investigaciones Mineras, Madrid*
[1300] **Structural Geology**	H. J. Zwart
	P. Calas
	F. Kockel
	R. Pignatelli, *Empresa Nacional Adaro de Investigaciones Mineras, Madrid*
[1400] **Stratigraphy** [1500] **Palaeontology**	J. P. H. Kaasschieter A. J. Key, *Shell Internationale Petroleum Mij, Rijswijk*
	P. Calas
	R. Wolfahrt, *Bundesanstalt für Geowissenschaften und Rohstoffe, Hannover*
	E. Perconig
[1600] **Geomorphology**	Miss M. I. J. G. Cortenraad (co-ordinator), *Fysisch-Geografisch Instituut der Rijksuniversiteit te Utrecht*
[1610] WEATHERING AND EROSION [1620] KARST AND CAVES	T. W. J. van Asch, *Fysisch-Geografisch Instituut der Rijksuniversiteit te Utrecht*
	P. Calas
	H. Jordan C. Hinze, *Niedersächsisches Landesamt für Bodenforschung, Hannover*
	E. Perconig

[1630] WIND ACTION	Miss M. I. J. G. Cortenraad J. I. S. Zonneveld, *Fysisch-Geografisch Instituut der Rijksuniversiteit te Utrecht* P. Calas K. D. Meyer, *Niedersächsisches Landesamt für Bodenforschung, Hannover* F. Jerez Mir, *Empresa Nacional Adaro de Investigaciones Mineras, Madrid*
[1640] RIVERS [1650] LAKES	Miss M. I. J. G. Cortenraad P. Calas J. Merkt, *Niedersächsisches Landesamt für Bodenforschung, Hannover* E. Perconig
[1660] GLACIERS	Miss M. I. J. G. Cortenraad J. I. S. Zonneveld P. Calas K. D. Meyer E. Perconig
[1670] COASTS AND ISLANDS	P. G. E. F. Augustinus, *Fysisch-Geografisch Instituut der Rijksuniversiteit te Utrecht* Miss M. I. J. G. Cortenraad P. Calas H. Streif, *Niedersächsisches Landesamt für Bodenforschung, Hannover* E. Perconig
[1680] SEQUENCE AND DEVELOPMENT OF LANDFORMS	Miss M. I. J. G. Cortenraad J. I. S. Zonneveld P. Calas H. Jordan E. Perconig

[1700] **Pedology**

G. Steur
H. J. de Bakker,
Stichting voor Bodemkartering, Wageningen

J. Vogt,
Bureau de Recherches Géologiques et Minières, Orléans

G. Roeschmann,
Niedersächsisches Landesamt für Bodenforschung, Hannover

S. Cuadros García,
Empresa Nacional Adaro de Investigaciones Mineras, Madrid

[1800] **Oceanography**

R. Dorrestein,
Kon. Ned. Meteorologisch Instituut, de Bilt

Ph. Bouysse
J. D. Roux,
Bureau de Recherches Géologiques et Minières, Orléans

G. Wissmann,
Bundesanstalt für Geowissenschaften und Rohstoffe, Hannover

E. Perconig

[1840] with collaboration of:
L. M. J. U. van Straaten
J. P. H. Kaasschieter
A. J. Key

[1900] **Sedimentology**

L. M. J. U. van Straaten,
Geologisch Instituut der Rijksuniversiteit te Groningen

[2000] **Clastic Deposits**

P. Nagtegaal
T. J. A. Reijers,
Kon./Shell Exploratie en Productie Laboratorium, Rijswijk

G. Scolari
Staff, Département Documentation,
Bureau de Recherches Géologiques et Minières, Orléans
R. Desbrandes,
Institut Français du Pétrole, Rueil-Malmaison

H. Kudrass,
Bundesanstalt für Geowissenschaften und Rohstoffe, Hannover

Mrs. M. del Carmen Fernández-Luanco Martinez,
Empresa Nacional Adaro de Investigaciones Mineras, Madrid

[2100] **Carbonate Deposits**

T. J. A. Reijers

P. Andreieff

G. Scolari
H. Solari,
Bureau de Recherches Géologiques et Minières, Orléans
R. Desbrandes
Staff, *Shell Française, Paris*

K.-H. Schulze,
Bundesanstalt für Geowissenschaften und Rohstoffe, Hannover

E. Perconig

[2200] **Miscellaneous Deposits**

L. M. J. U. van Straaten

G. Scolari
R. Desbrandes

H. Kudrass

Mrs. M. del Carmen Fernández-Luanco Martinez

[2300] **Diagenesis of Non-Carbonaceous Deposits**

L. M. J. U. van Straaten
P. Nagtegaal
T. J. A. Reijers

G. Dunoyer de Segonzac,
Institut de Géologie, Strasbourg

P. Rothe,
Geographisches Institut der Universität Mannheim

E. Perconig

[2400] **Carbonaceous Deposits**

C. C. M. Gutjahr,
Kon./Shell Exploratie en Productie Laboratorium, Rijswijk
J. Stuffken,
Dutch State Mines, Geleen
P. Eisma†,
(ret.) Kon./Shell Laboratorium, Amsterdam

R. Feys
H. Solari,
Bureau de Recherches Géologiques et Minières, Orléans

H. Schneekloth
H. Jacob
R. Lenz,
Bundesanstalt für Geowissenschaften und Rohstoffe, Hannover

[2400] **Carbonaceous Deposits** (continued)

J. A. Obis Salinas,
Empresa Nacional Adaro de Investigaciones Mineras, Madrid

[2500] **Subsurface Waters, Petroleum and Natural Gas**

[2510] HYDROLOGY; THE HYDROLOGICAL CYCLE

E. Romijn,
Provinciale Waterstaat van Gelderland, Arnhem

[2520] GEOLOGY OF SUBSURFACE WATERS

J. Margat,
Bureau de Recherches Géologiques et Minières, Orléans

[2530] WATER SUPPLY AND WATER QUALITY

H.-J. Dürbaum

E. Perconig

[2540] ROCK PROPERTIES AND FLUID FLOW

E. Romijn
A. J. Dikkers
T. J. A. Reijers

G. Scolari
J. P. Bertrand,
Institut Français du Pétrole, Rueil-Malmaison

H. Beckmann,
Institut für Erdölgeologie der Universität Clausthal

E. Perconig

[2550] GEOLOGY OF PETROLEUM AND NATURAL GAS

[2550] – [2590]:
A. J. Dikkers (co-ordinator),
(ret.) Shell Internationale Petroleum Mij, Den Haag
[2551] – [2554]:
K. de Groot,
Kon./Shell Exploratie en Productie Laboratorium, Rijswijk
E. Eisma†

Staff, *Shell Française, Paris*

E. Perconig

[2560] OIL WELLS AND FORMATION TESTS

W. H. van Eek†,
Instituut voor Mijnbouwkunde, Technische Hogeschool, Delft

[2570] PRODUCTION

A. W. J. Grupping,
Instituut voor Mijnbouwkunde, Technische Hogeschool, Delft

[2580] RESERVOIR ENGINEERING

F. R. van Veen
L. A. Schipper

[2590] OIL ECONOMICS

J. J. Nesselaar,
Shell Internationale Petroleum Mij, Den Haag
K. J. Weber
F. Fernandez Luque

[2590] OIL ECONOMICS (continued)

Miss M. Bakker,
Kon./Shell Exploratie en Productie Laboratorium, Rijswijk

[2560] – [2590]:
Mme M. Moureau,
Institut Français du Pétrole, Rueil-Malmaison

P. Kehrer
H. Jacob
J. Koch
H. Wehner,
Bundesanstalt für Geowissenschaften und Rohstoffe, Hannover

E. Perconig

[2600] **Metallic Mineral Deposits**

G. J. J. Aleva (co-ordinator),
Billiton Geological Consultants, Den Haag

[2610] GENERAL TERMS

C. E. S. Arps,
Rijksmuseum van Geologie en Mineralogie, Leiden

[2620] PROCESSES OF FORMATION

W. Uytenbogaardt,
Instituut voor Mijnbouwkunde, Technische Hogeschool, Delft
M. Zakrewsky,
Instituut voor Aardwetenschappen, Vrije Universiteit, Amsterdam

[2630] CLASSIFICATION

G. J. J. Aleva

[2640] STRUCTURE AND SHAPE OF MINERAL DEPOSITS

I. S. Oen,
Geologisch Instituut der Universiteit van Amsterdam

[2650] TEXTURE AND STRUCTURE OF ORE MINERALS

B. van de Pijpekamp,
Billiton Research, Arnhem

[2660] VARIOUS MINERAL DEPOSITS

G. J. J. Aleva

[2610] – [2660]:
M. Bornuat
J. Gravesteijn,
Bureau de Recherches Géologiques et Minières, Orléans

A. Leube,
Bundesanstalt für Geowissenschaften und Rohstoffe, Hannover

J. Armengot,
Empresa Nacional Adaro de Investigaciones Mineras, Madrid

[2670] ORE DRESSING

C. Blotwijk,
Instituut voor Mijnbouwkunde, Technische Hogeschool, Delft

G. Barbery,
Bureau de Recherches Géologiques et Minières, Orléans

E. Hopmann,
Bundesanstalt für Geowissenschaften und Rohstoffe, Hannover

E. Perconig

[2700] **Mineral Economics**

H. J. de Wijs,
Instituut voor Mijnbouwkunde, Technische Hogeschool, Delft

J. Gravesteijn
M. Jaujou,
Bureau de Recherches Géologiques et Minières, Orléans

H. Walther,
Bundesanstalt für Geowissenschaften und Rohstoffe, Hannover

E. Perconig

[2800] **Exploration and Surveying Methods**

[2810] GEOLOGICAL MAPPING AND RECORDING

by editor assembled from various chapters

P. Calas

R. Vinken,
Bundesanstalt für Geowissenschaften und Rohstoffe, Hannover
. C. Hinze,
Niedersächsisches Landesamt für Bodenforschung, Hannover

E. Perconig

[2820] DRILLING; METHODS AND SOME TECHNICAL TERMS

W. H. van Eek†
and editorial additions from various chapters

Mme M. Moureau

E. Perconig

[2830] REMOTE SENSING

A. A. Kuyp, *Delft*

G. Weecksteen,
Bureau de Recherches Géologiques et Minières, Orléans
B. Rond,
GDTA, Toulouse

[2830] REMOTE SENSING (continued)

R. Mühlfeldt
H. P. Baehr,
Universität, Hannover

E. Perconig

[2840] DEEP SEISMIC METHODS

W. L. Scheen
F. Kalisvaart,
Kon./Shell Exploratie en Productie Laboratorium, Rijswijk

R. Horn,
Bureau de Recherches Géologiques et Minières, Orléans

B. Buttkus,
Bundesanstalt für Geowissenschaften und Rohstoffe, Hannover

E. Perconig

[2850] SHALLOW SEISMIC AND ACOUSTIC METHODS

A. J. A. van Overeem†,
(ret.) CESCO

R. Horn

J. García Rodriguez,
Empresa Nacional Adaro de Investigaciones Mineras, Madrid

[2860] MAGNETOMETRICAL, GRAVIMETRICAL AND RADIOMETRICAL METHODS

R. Smith,
Shell Internationale Petroleum Mij, Den Haag
L. J. Fick,
Billiton Geological Consultants, Den Haag

A. Gérard

E. Perconig

[2870] ELECTRICAL AND ELECTROMAGNETIC METHODS

L. J. Fick,
W. J. van Riel,
(ret.) Ground-Water Survey TNO, Delft

H. Flathe,
Bundesanstalt Für Geowissenschaften und Rohstoffe, Hannover

E. Perconig

[2880] PHYSICAL BOREHOLE MEASUREMENTS

J. Ph. Poley,
Shell Internationale Petroleum Mij, Den Haag
with collaboration of:
A. Brash,
Billiton Geological Consultants, Den Haag

[2880] PHYSICAL BOREHOLE MEASUREMENTS (continued)

F. le Breton
R. Desbrandes,
Institut Français du Pétrole, Rueil-Malmaison

H. Repsold

H. Friedrich,
Niedersächsisches Landesamt für Bodenforschung, Hannover

E. Perconig

[2890] GEOCHEMICAL METHODS

L. J. Fick

E. Wilhelm,
Bureau de Recherches Géologiques et Minières, Orléans

H. Gundlach

E. Perconig

[2900] **Engineering Geology**

N. Rengers,
International Institute for Aerial Survey and Earth Sciences (ITC), Enschedé
with collaboration of:
D. C. Price,
Instituut voor Mijnbouwkunde, Technische Hogeschool, Delft

P. Duffaut,
Bureau de Recherches Géologiques et Minières, Orléans

M. Langer,
Bundesanstalt für Geowissenschaften und Rohstoffe, Hannover

A. Alfonso,
Shell España, Madrid

CONTENTS OF THEMATIC PART

[0100] **Introduction** 0001–0017

[0200] **Extraterrestrial Bodies** 0018–0077
- [0210] THEORIES ON THE ORIGIN OF THE SOLAR SYSTEM 0018–0022
- [0220] THE TERRESTRIAL PLANETS 0023–0044
- [0230] THE MOON 0045–0061
- [0240] METEORITES 0062–0077

[0300] **Geophysics** 0078–0394
- [0310] GENERAL TERMS; THE ARCHITECTURE OF THE EARTH 0078–0093
- [0320] SEISMOLOGY 0094–0275
  - [0321] *General Terms; the Earth's Medium* 0094–
  - [0322] *Stations and Instruments* 0116–
  - [0323] *Earthquakes; Source Parameters* 0138–
  - [0324] *Seismic Waves; Earth Motions* 0161–
  - [0325] *Wave Analysis* 0204–
  - [0326] *Seismicity* 0242–
  - [0327] *Engineering Seismology* 0254–
  - [0328] *Nuclear and Explosion Seismology* 0263–
- [0330] GEOMAGNETISM AND ROCK MAGNETISM 0276–0321
  - [0331] *The Earth's Magnetic Field* 0276–
  - [0332] *Variations of the Magnetic Field* 0288–
  - [0333] *Magnetic Properties of Rocks* 0293–
  - [0334] *Magnetic Measurements* 0316–
- [0340] GRAVITY AND ISOSTASY 0322–0373
  - [0341] *Gravity* 0322–
  - [0342] *Isostasy* 0338–
  - [0343] *Gravimetry* 0349–
- [0350] GEOTHERMICS 0374–0394
  - [0351] *The Earth's Heat* 0374–
  - [0352] *Geothermal Energy* 0388–

[0400] **Geochemistry** 0395–0427

[0500] **Isotope Geology** 0428–0458
- [0510] ISOTOPE GEOCHRONOLOGY 0428–0452
- [0520] STABLE-ISOTOPE GEOLOGY 0453–0458

[0600] **Mineralogy and Gemmology** 0459–0504
- [0610] MINERALOGY 0459–0477
- [0620] GEMMOLOY 0478–0504

[0700] **Crystallography** 0505–0577
- [0710] GEOMETRICAL PROPERTIES OF CRYSTALS 0505–0543
- [0720] OPTICAL PROPERTIES OF CRYSTALS 0544-0577

[0800] **Petrology – Introductory Terms** 0578–0614

[0900] **Petrology of Igneous Rocks** 0615–0878
- [0910] GENERAL TERMS 0615–0638
- [0920] THE MAGMA 0639–0703
  - [0921] *General Terms* 0639–
  - [0922] *Assimilation and Differentiation* 0666–
  - [0923] *Classification of Magmas* 0685–
- [0930] THE INTRUSIVE BODIES 0704–0749
  - [0931] *General Terms* 0704–
  - [0932] *Classification of Intrusive Bodies* 0715–
- [0940] THE CONSTITUENTS OF IGNEOUS ROCKS 0750–0878
  - [0941] *Crystal Forms* 0750–
  - [0942] *Aggregates and Intergrowths* 0771–
  - [0943] *Texture* 0790–
  - [0944] *Structure* 0858–

[1000] **Petrology of Metamorphic Rocks** 0879–1072
- [1010] GENERAL TERMS 0879–0904
- [1020] METAMORPHIC PROCESSES AND PRODUCTS 0905–0944
- [1030] TEXTURE AND STRUCTURE 0945–0975
- [1040] THE CONSTITUENTS 0976–0985
- [1050] CLASSIFICATION 0986–0994
- [1060] STRUCTURAL PETROLOGY 0995–1021
- [1070] MIGMATIZATION 1022–1072

[1100] **Geotectonics** 1073–1145
- [1110] GENERAL TERMS 1073–1079
- [1120] CONTINENTAL STRUCTURE 1080–1112
  - [1121] *Stable Regions* 1080–
  - [1122] *Faulting Tectonics* 1090–
  - [1123] *Mobile Belts* 1093–
- [1130] HYPOTHESES ON MOUNTAIN BUILDING 1113–1145
  - [1131] *Older Hypotheses* 1113–
  - [1132] *Plate Tectonics and Sea-Floor Spreading* 1125–

[1200] **Volcanology** 1146–1477
- [1210] VOLCANIC ACTIVITY 1146–1196
  - [1211] *General Terms* 1146–
  - [1212] *Types of Activity* 1156–
  - [1213] *Mechanism* 1170–
  - [1214] *Paravolcanic and Cryptovolcanic Phenomena* 1180–
  - [1215] *Volcano-Watching* 1190–
- [1220] VOLCANIC PHASES AND PRODUCTS 1197–1408

[1221] *The Intrusive Phase* 1197–
[1222] *The Extrusive Phase: Phenomena* 1207–
[1223] *The Extrusive Phase: Lavas* 1238–
[1224] *The Extrusive Phase: Pyroclastics* 1309–
[1225] *The Extrusive Phase: Miscellaneous Products* 1348–
[1226] *The Late-Volcanic Phase* 1374–
[1230] VOLCANIC LANDFORMS 1409–1477
[1231] *General Terms* 1409–
[1232] *Volcanoes and Craters; Calderas* 1420–
[1233] *Rifts and Subsidence* 1463–
[1234] *Erosional Features* 1471–

[1300] **Structural Geology** 1478–1707
[1310] ATTITUDE OF STRATA 1478–1493
[1320] FRACTURING 1494–1519
[1321] *Joints* 1494–
[1322] *Fissures* 1512–
[1330] FAULTS 1520–1595
[1331] *General Terms* 1520–
[1332] *Orientation and Geometry of Faults* 1544–
[1333] *Displacement along Faults* 1558–
[1334] *Overthrusts* 1568–
[1335] *Fault Systems* 1578–
[1340] CLEAVAGE, SCHISTOSITY AND FOLIATION 1596–1606
[1350] LINEATIONS AND LINEAR STRUCTURES 1607–1614
[1360] FOLDS 1615–1692
[1361] *General Terms* 1615–
[1362] *Fold Elements and Geometry* 1625–
[1363] *Types of Folds* 1642–
[1364] *Fold Systems* 1670–
[1365] *Nappe Tectonics* 1679–
[1370] DIAPIRISM 1693–1701
[1380] GRAVITATIONAL GLIDING 1702–1707

[1400] **Stratigraphy** 1708–1785
[1410] GENERAL TERMS 1708–1738
[1420] STRATIGRAPHICAL UNITS 1739–1785
[1421] *General Terms* 1739–
[1422] *Lithostratigraphical Units* 1747–
[1423] *Chronostratigraphical Units* 1757–
[1424] *Biostratigraphical Units* 1769–

[1500] **Palaeontology** 1786–1986
[1510] GENERAL TERMS 1786–1826
[1520] FLORAL AND FAUNAL PALAEONTOLOGY 1827–1885
[1521] *Palaeobotany* 1827–
[1522] *Palaeozoology* 1847–
[1523] *Ichnology* 1871–
[1530] ACTUOPALAEONTOLOGY 1886–1919
[1540] TAXONOMY 1920–1958
[1550] PHYLOGENY 1959–1986

[1600] **Geomorphology** 1987–2656
[1610] WEATHERING AND EROSION 1987–2062
[1611] *Physical and Chemical Weathering* 1987–
[1612] *Erosion and Denudation* 2012–
[1613] *Removal 2031–*
[1614] *Mass Movement 2037–*
[1620] KARST AND CAVES 2063–2153
[1621] *Karst Processes and Solutional Forms* 2063–
[1622] *Caves* 2094–
[1630] WIND ACTION 2154–2190
[1631] *Erosion and Transport by Wind* 2154–
[1632] *Dunes, Geographical Distribution* 2165–
[1633] *Dunes, Genetic Classification* 2171–
[1640] RIVERS 2191–2329
[1641] *The River Course* 2191–
[1642] *The River System* 2205–
[1643] *River Curves and River Banks* 2216–
[1644] *The River Bed* 2242–
[1645] *Discharge and Drainage* 2249–
[1646] *Flow Characteristics* 2260–
[1647] *Valley Development and Valley Forms* 2270–
[1648] *River Terraces* 2293–
[1649] *Water Divides and River Capture* 2312–
[1650] LAKES 2330–2344
[1651] *Lake Waters* 2330–
[1652] *Morphological Features* 2335–
[1653] *Evolution of Lakes* 2340–
[1660] GLACIERS 2345–2436
[1661] *Snow and Ice* 2345–
[1662] *The Glacier* 2366–
[1663] *Glacial Erosion* 2394–
[1664] *Glacial Transport and Deposits* 2402–
[1665] *Types of Glaciers* 2425–
[1670] COASTS AND ISLANDS 2437–2536
[1671] *General Terms* 2437–
[1672] *Coastal Forms* 2446–
[1673] *Rocky Coasts* 2470–
[1674] *Sandy Shores* 2485–
[1675] *Deltas* 2494–
[1676] *Tidal Flats and Marshes* 2504–
[1677] *Types of Coasts* 2519–
[1678] *Development of Coasts* 2528–
[1680] SEQUENCE AND DEVELOPMENT OF LANDFORMS 2537–2656
[1681] *Tectonic and Structural forms* 2537–
[1682] *Landscape Development, General Terms* 2566–
[1683] *Arid Regions* 2577–
[1684] *Humid Regions* 2588–
[1685] *Glaciated Regions* 2610–
[1686] *Periglacial Regions* 2633–

[1700] **Pedology** 2657–2780
[1710] GENERAL TERMS 2657–2677
[1720] SOIL-FORMING PROCESSES 2678–2685
[1730] SOIL PROFILE AND SOIL HORIZONS 2686–2708
[1740] SOIL TEXTURE 2709–2722
[1750] SOIL STRUCTURE; CEMENTATION 2723–2734
[1760] SOIL CHEMISTRY 2735–2745
[1770] PHYSICAL PROPERTIES 2746–2763

[1780] SOIL MOISTURE 2764–2775
[1790] SOIL EROSION 2776–2780

[1800] **Oceanography** 2781–2848
[1810] GENERAL TERMS 2781–2787
[1820] SEA-BOTTOM TOPOGRAPHY 2788–2806
[1830] WAVES AND TIDES 2807–2826
[1840] MARINE ENVIRONMENTS 2827–2848

[1900] **Sedimentology** 2849–3163
[1910] GENERAL TERMS 2849–2858
[1920] ENVIRONMENTS, FACIES AND DEPOSITS 2859–2927
[1921] *General Terms* 2859–
[1922] *Subaerial Terrestrial Environments* 2865–
[1923] *Subaqueous Terrestrial Environments* 2880–
[1924] *Marine Environments* 2900–
[1930] SEDIMENTARY PROCESSES 2928–2993
[1931] *Provenance of Material* 2928–
[1932] *Transportation* 2950–
[1933] *Mass Movement* 2972–
[1934] *Deposition* 2985–
[1940] SEDIMENTARY STRUCTURE 2994-3076
[1941] *General Terms* 2994–
[1942] *Ripple Marks* 3000–
[1943] *Current Marks* 3017–
[1944] *Imprints and Cracks* 3041–
[1945] *Penecontemporaneous Deformation* 3050–
[1950] SEDIMENTARY TEXTURE 3077-3094
[1960] STRATIFICATION 3095-3151
[1961] *Single Strata* 3095–
[1962] *Interrelation of Strata* 3110–
[1963] *Cross Stratification* 3133–
[1970] GRAIN-SIZE DETERMINATION 3152– 3163

[2000] **Clastic Deposits** 3164-3265
[2010] THE CONSTITUENTS 3164-3179
[2020] THE PARTICLES 3180-3202
[2021] *Shape of Particles* 3180–
[2022] *Size of Particles* 3193–
[2030] THE DEPOSITS 3203-3265
[2031] *Lutites* 3203-
[2032] *Arenites* 3225–
[2033] *Rudites* 3237–

[2100] **Carbonate Deposits** 3266-3364
[2110] GENERAL TERMS 3266-3269
[2120] TEXTURE 3270-3304
[2121] *Carbonate Particles* 3270–
[2122] *Grain-Size Classification* 3274–
[2123] *Depositional Textures* 3294–
[2130] THE DEPOSITS 3305-3321
[2140] CARBONATE SEDIMENT BODIES AND SETTINGS 3322-3364
[2141] *Carbonate Bodies* 3322–
[2142] *Reefs: Morphology and Deposits* 3333–
[2143] *Algal Structures* 3358–

[2200] **Miscellaneous Deposits** 3365-3387
[2210] SILICIOUS DEPOSITS 3365-3374
[2220] EVAPORITES 3375-3387

[2300] **Diagenesis of Non-Carbonaceous Deposits** 3388-3492
[2310] GENERAL TERMS 3388-3397
[2320] ENVIRONMENTAL DIAGENESIS 3398-3416
[2330] BURIAL DIAGENESIS 3417-3471
[2331] *General Terms* 3417–
[2332] *Compaction* 3427–
[2333] *Cementation* 3433–
[2334] *Pressure Solution* 3445–
[2335] *Recrystallisation and Replacement* 3452–
[2340] CONCRETIONS 3472-3492

[2400] **Carbonaceous Deposits** 3493-3609
[2410] GEOLOGY AND OCCURRENCE 3493-3518
[2420] COALIFICATION; CLASSIFICATION OF PEATS AND COALS 3519-3564
[2421] *General Terms* 3519–
[2422] *Biochemical Phase; Environments of Deposition* 3524–
[2423] *Geochemical Phase* 3551–
[2430] PETROGRAPHY AND STRUCTURE OF COAL 3565–3592
[2440] ANALYSIS AND PRODUCTS 3593-3609

[2500] **Subsurface Waters, Petroleum and Natural Gas** 3610-3902
[2510] HYDROLOGY; THE HYDROLOGICAL CYCLE 3610–3633
[2520] GEOLOGY OF SUBSURFACE WATERS 3634-3664
[2521] *Water-bearing Formations* 3634–
[2522] *Ground-Water Level and Head* 3642–
[2523] *The Distribution of Subsurface Waters* 3648–
[2530] WATER SUPPLY AND WATER QUALITY 3665-3678
[2540] ROCK PROPERTIES AND FLUID FLOW 3679-3711
[2550] GEOLOGY OF PETROLEUM AND NATURAL GAS 3712-3781
[2551] *Types of Hydrocarbons; Origin and some Products* 3712–
[2552] *Migration* 3748–
[2553] *Accumulation* 3754–
[2554] *Interrelation of Water, Oil and Gas* 3773–
[2560] OIL WELLS AND FORMATION TESTS 3782–3788
[2570] PRODUCTION 3789-3818
[2580] RESERVOIR ENGINEERING 3819-3892
[2581] *Reservoir Conditions* 3819–
[2582] *Reservoir Processes* 3822–
[2583] *Reservoir Engineering Methods* 3846–
[2584] *Reserves* 3868–
[2590] OIL ECONOMICS 3893-3902

[2600] **Metallic Mineral Deposits** 3903-4166
[2610] GENERAL TERMS 3903-3925
[2620] PROCESSES OF FORMATION 3926-3965
[2621] *Early Theories* 3926–
[2622] *Some Physico-Chemical Concepts* 3929–

[2623] *Petrological and Geological concepts* 3936–
[2630] CLASSIFICATION 3966-4002
[2631] *Related to Time and Depth of Formation* 3966–
[2632] *Related to the Source of the Material* 3968–
[2633] *Various Bases of Classification* 3972–
[2640] STRUCTURE AND SHAPE OF MINERAL DEPOSITS 4003-4079
[2641] *General Terms* 4003–
[2642] *Attitude of Ore Bodies* 4016–
[2643] *Bedded Ore Deposits* 4021–
[2644] *Ore Masses* 4032–
[2646] *Vein Systems* 4061–
[2650] TEXTURE AND STRUCTURE OF ORE MINERALS 4080-4113
[2651] *Crystalline Structures* 4080–
[2652] *Metacolloidal Structures* 4091–
[2653] *Intergrowth Structure* 4098–
[2654] *Stereometric Analysis* 4107–
[2660] VARIOUS MINERAL DEPOSITS 4114-4136
[2670] ORE DRESSING 4137-4166

[2700] **Mineral Economics** 4167-4225
[2710] SAMPLING 4167-4170
[2711] *Sampling of Ore in Place* 4167–
[2712] *Sample Spacing* 4171–
[2713] *Miscellaneous Terms* 4175–
[2720] SAMPLE REDUCTION AND CHEMICAL ANALYSIS 4180- 4184
[2730] STATISTICAL ANALYSIS OF SAMPLES 4185–4192
[2731] *Grade Computation* 4185–
[2732] *Statistics of Grade Distribution* 4187–
[2740] GEOSTATISTICS 4193-4198
[2750] ESTIMATION OF ORE RESERVES 4199-4204
[2760] MINERAL RESOURCES 4205-4215
[2770] MINE VALUTION 4216-4225

[2800] **Exploration and Surveying Methods** 4226-4659
[2810] GEOLOGICAL MAPPING AND RECORDING 4226-4260
[2820] DRILLING; METHODS AND SOME TECHNICAL TERMS 4261-4299
[2830] REMOTE SENSING 4300-4404
[2831] *General Terms* 4300 –
[2832] *Aerial Photography* 4327–
[2833] *Radio Detection and Ranging* 4384–
[2834] *Scanning* 4401–
[2840] DEEP SEISMIC METHODS 4405-4478
[2841] *Field Procedures* 4405–
[2842] *Seismic Recording and Processing* 4420–
[2843] *Interpretation* 4458–
[2850] SHALLOW SEISMIC AND ACOUSTIC METHODS 4479-4520
[2851] *General Terms* 4479–
[2852] *Design and Equipment* 4486–
[2853] *Processing* 4508–
[2854] *General Phenomena and Interpretation* 4512–
[2860] MAGNETOMETRICAL, GRAVIMETRICAL AND RADIOMETRICAL METHODS 4521-4551
[2861] *Magnetometrical Equipment* 4521–
[2862] *Magnetometrical Measurement and Interpretation* 4526–
[2863] *Gravimetrical Methods* 4536–
[2864] *Radiometrical Methods* 4545–
[2870] ELECTRICAL AND ELECTROMAGNETIC METHODS 4552-4577
[2880] PHYSICAL BOREHOLE MEASUREMENTS 4578-4624
[2881] *Logging Devices* 4578–
[2882] *Formation Properties and Log Interpretation* 4604–
[2890] GEOCHEMICAL METHODS 4625-4659

[2900] **Engineering Geology** 4660-4876
[2910] GENERAL TERMS 4660-4665
[2920] BEHAVIOUR AND PROPERTIES OF ROCK AND SOIL 4666-4740
[2921] *General Aspectrs of Material Behaviour* 4666–
[2922] *General Aspects of Material Properties* 4680–
[2923] *Engineering Properties and Behaviour of Soils* 4702–
[2924] *Engineering Properties and Behaviour of Rocks* 4722–
[2925] *Hydraulics* 4735–
[2930] METHODS OF INVESTIGATION 4741-4791
[2931] *Samples and Sampling Procedures* 4741–
[2932] *Laboratory and Field Testing; Instrumentation* 4753–
[2933] *Model Testing and Geotechnical Analysis* 4775–
[2934] *Engineering Geological Mapping* 4784–
[2940] PROJECTS AND ACTIVITIES 4792-4876
[2941] *Soil and Rock Support* 4792–
[2942] *Soil and Rock Qualigy Improvement* 4800–
[2943] *Blasting; Excavation* 4814–
[2944] *Foundations; Urban Geology* 4831–
[2945] *Roads and Bridges* 4840–
[2946] *Underground Excavation; Tunnelling* 4847–
[2947] *Embankments, Dams and Reservoirs* 4859–
[2948] *Construction Materials* 4871–

Thematic Part

[0100] INTRODUCTION

0001
geology, – the science that treats of the history of the earth and its life, the composition of the earth's crust and its structural conditions and the forces that govern the processes taking place in the earth's crust. / geologie; aardkunde / géologie / Geologie / geología

geognosy, antiquated term, synonymous with geology. / geognosie / géognosie / Geognosie / geognosía

geologist, / geoloog / géologue / Geologe / geólogo

0002
earth sciences; geoscience, – the sciences that treat of the solid, liquid and gaseous phases of the earth. / aardwetenschappen / sciences de la terre / Erdwissenschaften; Geowissenschaften / ciencias de la tierra

telluric, – pertaining to the earth. / tellurisch / tellurique / tellurisch / telúrico

0003
cosmogony, – the science that treats of the origin and development of the universe. / cosmogonie / cosmogonie / Kosmogonie / cosmogonía

geogony, – the science that treats of the origin and formation of the earth. / geogonie / géogonie / Geogonie / geogonía

0004
neptunism, – the theory, postulating that virtually all rocks of the earth's crust have been deposited in water. / neptunisme / neptunisme / Neptunismus / neptunismo

plutonism, – the theory, as opposed to neptunism, postulating that many rocks of the earth's crust were formed by solidification from a molten state. / plutonisme / plutonisme / Plutonismus / plutonismo

0005
catastrophism, – the doctrine that changes in the earth's crust have been effected suddenly by successive destructions of violent nature. / catastrofisme / cataclysme; catastrophisme / Katastrofismus / catastrofismo

catastrophist geology, – study of discontinuities in the earth's history; a modernistic, somewhat unusual term. / catastrofengeologie / géologie des catastrophes / Katastrophengeologie / geología catastrófica

0006
uniformitarianism, – the principle that all geological events that took place in the past can be explained by processes and forces active at present / uniformitarianisme; uniformitarisme / uniformitarisme / Uniformitarismus / uniformitarismo

actualism, – a system of geological thought postulating that the same processes and natural laws prevailed in the past as those we can now observe or can infer from observations, although the rate at which they act may be variable. / actualisme / actualisme / Aktualismus / actualismo

0007
palaetiology, – the science that attemps to reconstruct a past state by establishing the causes that led to the present state. / palaetiologie / palétiologie / Palätiologie / paleoetología

comparative ontology, – the science that attempts to infer a process of formation by studying the object that was formed by that process. / vergelijkende ontologie / ontologie comparative / vergleichende Ontologie / ontología comparada

0008
actuogeology – the branch of geology that studies the geological processes at present active on the surface of the earth and their results, in particular those that have been influenced by human activity. / actuogeologie / géologie des phénomènes actuels / Aktuogeologie / actualismo geológico

0009
palinspastic reconstruction, – to conceive the original position of crustal slices from their present relative position according to their inferred geological history. / palinspastische reconstructie / reconstruction palinspastique / palinspastische Rekonstruktion / reconstrucción paleoespástica

0010
physical geology, – the branch of geology that treats of the geological processes, their action and results. / algemene geologie / géologie générale / allgemeine Geologie / geología general

dynamical geology, – an antiquated term for physical geology.

0011
historical geology, – the branch of geology aiming to give a reconstruction and a chronological account of the events during the earth's history. / historische geologie / géologie historique / historische Geologie / geología histórica

palaeogeography, – the branch of geology dealing with the distribution of land and sea during the history of the earth. / palaeogeografie / paléogéographie / Paläogeographie / paleogeografía

palaeoclimatology, – the branch of science dealing with the climatological conditions during the geological history of the earth. / palaeoklimatologie / paléoclimatologie / Paläoklimatologie / paleoclimatología

0012
physical geography, – the branch of geography that treats of the exterior physical features and changes of the earth on land, in water and air. / natuurkundige aardrijkskunde / géographie physique / physische Geographie / geografía física

0013
geodesy, – the determination by observation and measurement of the relative positions of points and areas on the earth's surface and the figure and shape of the earth as a whole. / geodesie / géodésie / Geodäsie / geodesía

0014
experimental geology, – imitation of geological processes in scaled-down models, in order to obtain an insight into the working of those processes or to carry out physical, chemical or mechanical experiments where they apply to geology. / experimentele geologie / géologie expérimentale / experimentelle Geologie / geología experimental

0015
mathematical geology, – the application of mathematical techniques in solving geological problems, especially mathematical statistics and computer technology. / wiskundige geologie / géologie mathématique / mathematische Geologie / geología matemática

0016
economic geology; applied ...., – in its widest sense the application of geology to practical uses, e.g. exploration for, evaluation and exploitation of mineral resources, engineering works, ground water and agriculture. / economische geologie; toegepaste .... / géologie économique; .... appliquée / ekonomische Geologie; angewandte .... / geología económica; .... aplicada

0017
explorationist, – an earth scientist who is versed in exploration methods and their interpretation, and who may have a geological, geophysical or geochemical background. Whatever his background he should be able to comunicate with and understand those who have a background different from his own.

[0200] EXTRATERRESTRIAL BODIES

[0210] THEORIES ON THE ORIGIN OF THE SOLAR SYSTEM

0018
solar system, – the sun and the bodies in orbit around it. / zonnestelsel; planetenstelsel / système solaire / Sonnensystem / sistema solar

0019
dualistic hypothesis, – the hypothesis postulating the bodies revolving around the sun to have originated by close encounter or collision of the primitive sun and another body. / dualistische hypothese / hypothèse dualistique / dualistische Hypothese / hypótesis dualística

0020
monistic hypothesis, – the hypothesis postulating sun and bodies revolving around it to have originated out of a rotating disk of gas and dust. / monistische hypothese / hypothèse monistique / monistische Hypothese / hipótesis monística

0021
solar nebula, – a hypothetical rotating disk from which the sun and the bodies revolving around it originated. / zonnenevel / nébuleuse solaire / Urnebel / nebulosa solar

nebular hypothesis, – / nevelhypothese / hypothèse de la nébuleuse / Urnebel-Hypothese / hipótesis nebular

0022
accretion hypothesis; planetesimal ...., – the now widely accepted hypothesis that postulates the origin of the bodies revolving around the sun by agglomeration from the solar nebula of cold gases, dust and particles of various size, during which process kinetic energy was transferred into heat and at least partial melting took place; otherwise no further physical and chemical differentiation within these bodies can be envisaged. / planetesimaalhypothese / hypothèse de l'accrétion / Zusammenballungstheorie; Planetesimaltheorie / hipótesis de la acreción; ... planetesimal

[0220] THE TERRESTRIAL PLANETS

0023
planetology, – the science that deals with composition, evolution, inner and surficial structure and features of the planets. / planetologie / planétologie / Planetologie / planetología

0024
exogeology, – the "geo"logy of extraterrestrial bodies; the science that deals with evolution, composition, structural and morphological features of their rocky surfaces if present, and that investigates these features by geological methods. / exogeologie / exogéologie / Exogeologie / exogeología

0025
terrestrial planet, – one of the earth-like planets, Mercury, Venus or Mars, that possesses a rocky crust and whose interior may be differentiated in a mantle, either rigid or partly viscous, and a core that may consist of nickel-iron or of silicates with or without some nickel-iron; in these respects the earth's moon and at least some of the moons of the major planets (Jupiter and Saturn) are earth-like / aardse planeet / planète terrestre / terrestrische Planet; erdähnliche Planet / planeta terrestre

0026
impact, – a probably fundamental and universal process in the formation and – in so far not covered by a thick atmosphere – the further development of planets and moons; impacts range from micrometeorites, creating tiny pits, to huge bodies that f.i. sculptered the mare basins on the moon. / inslag / impact / Meteoraufschlag; ....einschlag / impacto

0027
impact crater, – crater produced by the impact of a solid body. / inslagkrater / cratère d'impact / Meteorkrater / cráter meteorítico

primary (impact) crater, – produced by the impact of a foreign body. / primaire (inslag)krater / cratère d'impact primaire / Primärkrater / cráter meteorítico primario

secundary (impact) crater, – produced by material ejected by the primary impact. / secundaire (inslag)krater / cratère d'impact secondaire / Sekundärkrater / cráter meteorítico secundario

0028
cosmic weathering, – physical and chemical changes in rocks by ions from the solar wind impinging on the surface. / kosmische verwering / altération cosmique / kosmische Verwitterung / alteración cósmica

cosmic erosion, – removal of skin of rock by bombardment by (micro-)meteorites and by insolation. / kosmische erosie / érosion cosmique / kosmische Erosion / erosión cósmica

0029
uplands, – on Mars the rugged and densely cratered and presumably the most ancient terrain, possibly comparable to the lunar terrae.

basin; plain, – on Mars and Mercury the relatively smooth and sparsely cratered terrain, presumably produced by large impacts and subsequently filled by successive lava flows, possibly comparable to the lunar maria. / bekken; vlakte / bassin; plaine / Becken; Tiefland / planicie

0030
multi-ringed basin, – on Mars lava-filled circular basins surrounded by concentric mountain ranges, similar to their lunar counterparts. / ringvlakte / bassin annulaire / – – / planicie anulare

0031
wrinkle ridge, – on Mars morphologic features created by lava flows over the plains. / rimpelrug / – – / Lavarücken / – –

0032
lobate scarp, – on Mercury, a shallowly scalloped cliff; on Mars possibly the front of a solidified lava flow. / geschulpte rand / escarpement lobé / sigmoidale Steilrand / escarpado lobulado

0033
elevated plateau, – on Mars regions raised presumably by tectonic activity. / hoog plateau / plateau surélevé / Hochplateau / planicie

0034
fossa (plur. fossae), – on Mars huge canyon-like features due to fracturing, rifting and collapse of the surface related to the elevation of the plateaus. / fossa / fosse / Graben (plur. Grabenbrüche) / fosa

chasma; valle, – on Mars in topographic names denoting the canyons. / chasma; canyon / vallée / Chasma / abismo

0035
shield volcano, – on Mars the huge shield-shaped volcanic mountains. / schildvulkaan / volcan bouclier / Schildvulkan / escudo volcánico

dome, – on Mars the smaller features probably built up from more viscous lavas than the shields. / koepelvulkaan / dôme / Domvulkan / domo

0036
caldera, – on Mars collapse features on top of a shield. / caldera / caldère / Kaldera / caldera

0037
patera, – on Mars in topographic names denoting the most degraded and presumably oldest shields or domes, of which the rim only remains.

tholus, – a less degraded shield or dome.

mons, – the less degraded, relatively fresh and presumably youngest shields or domes.

0038
chaotic terrain, – on Mars a landform consisting of a depressed region with irregularly arranged pieces of broken rock. / chaotisch landschap / terrain chaotique / chaotisches Terrain / terreno caótico.

0039
fretted terrain, – on Mars a landform consisting of a flat lowland bordered by steep cliffs of complex build. / – – / terrain fritté / – – / terreno calcinado

0040
polar blanket, – on Mars wind-blown deposits and layers of ice in the polar areas. / polaire mantel / couverture polaire / Polarablagerungen / casquete polar.

laminated terrain, – on Mars a landform due to the layered texture of the polar blankets. / – – / terrain laminé / Schichtstufenlandschaft / terreno laminado

etched terrain, – on Mars a landform of the southern polar blanket due to erosion by wind action of unbedded deposits. / – – / terrain buriné / äolisch überformte Landschaft / – –

0041
braided terrain, – on Mars a landform similar to deposits in a braided river plain on earth. / – – / terrain anastomosé / "Wildfluss"-ähnliche Struktur / – –

0042
channel; gully, – on Mars morphological features presumably due to fluvial erosion. / ravijn; geul / chenal; ravin / Fliessrinnen / canal

0043
wind action, – on Mars the scouring by wind-blown dust. / winderosie / action du vent / äolische Erosion / acción del viento

0044
creep, – on Venus slow plastic degradation of surface features due to the high prevailing temperatures. / kruip / boursoufline / Kriechen / – –

[0230] THE MOON

0045
selenology; ....graphy; ....physics; ....chemistry, – once proposed but not accepted terms for the moon sciences analogous to the geosciences on earth.

0046
disruption theory, – the now obsolete theory that postulates the moon's origin by disruption from the earth as a tidal resonance effect. / resonantietheorie / théorie de la dislocation / – – / teoría del rompimiento; teoría de resonancía

0047
capture theory, – the theory that postulates the moon having been formed outside the earth's orbit and subsequently captured by gravitational forces. / invangtheorie / théorie de la capture / Einfangtheorie / teoría de la capture

0048
accretion theory, – the theory that postulates the moon's origin in orbit by agglomeration of solid particles of various size. / planetisimaaltheorie / théorie de l'accrétion / Agglomerationstheorie / teoría de la acreción

0049
volcano theory, – the now obsolete theory that postulates the majority of the moon's craters to have been formed by volcanic activity. / vulkaantheorie / théorie volcanique / Vulkantheorie / teoría volcánica

impact theory, – the theory that postulates the majority of the moon's craters to have been formed by the impact of solid bodies of various size. / inslagtheorie / théorie des impacts / Impakttheorie / teoría de los impactos

0050
terra (plur. terrae), – the lunar "continents", – highlands, the bright areas on the moon; densely cratered, rugged and most ancient terrain on the moon. / maancontinent / continent lunaire / lunares Hochland / continente lunar

0051
mare (plur. maria), – the lunar "seas", – plains, the dark areas on the moon; moderately cratered and relatively smooth terrain, younger than the terrae; they are basins formed by collisions with massive bodies that were subsequently flooded by successive and extensive basalt flows. / maanzee / mer lunaire / Mare / mar lunar

sinus, – more or less circular embayment at the edge of a mare. / maanbaai / sinus / Sinus / – –

0052
mascon, – mass concentration, positive gravity anomaly over major maria. / mascon / mascon / Massenkonzentration; Mascon / mascón

0053
multi-ringed basin, – a huge crater formed presumably by an asteroid-sized body causing sufficiently large shock waves to create mountainous swells surrounding it, subsequently flooded or partly overflooded by mare basalt. / ringvlakte / bassin annulaire / – – / planicie anulare

0054
crater, – ring-like walled enclosure, the major morphological features on the moon that originated by impact of major bodies. / krater, ringwalberg / cratère / Impaktkrater / cráter

central mountain; .... peak, – if not flooded by subsequent basalt flows still observable in the central part of a crater. / centrale berg / montagne centrale / Zentralberg / – –

0055
ejecta blanket, – deposit of material ejected from the crater by the force of the impact. / mantel van uitblazingsprodukten / matérial éjecté / Auswurfdecke / – –

0056
ray; ray system, – strip(s) of debris that extend radially from the youngest of the major craters. / straal; stralenkrans / rai / Strahlenkranz / rayo; radiación

0057
rille, – mostly linear, sometimes arcuate or sinuous canyon-like features, presumably due to collapsed lava tunnels or to block faulting. / ril / rainure / Rille / arroyo

0058
wrinkle; .... ridge, – large ridges of sinuous outline commonly parallel to the margins of the major maria. / rimpelrug / croûte plissée / – – / plegamiento

0059
dome; cone, – local arrays of morphological features similar to volcanic features on earth. / dom; kegel / dôme; cône / Dom; Kegel / domo; cono

0060
lunar soil, – a regolith consisting of loose debris. / maanbodem / sol lunaire / Regolithdecke; Regolithschicht / suelo lunar

0061
moonquake, – maanbeving / séisme lunaire / Mondbeben / lunamoto

[0240] METEORITES

0062
meteoritics, – the science of meteorites. / meteorietenkunde / science des météorites / Meteoritenkunde / ciencia de los meteoritos

0063
meteorite, – stony or metallic body, from a few grammes to several tons in mass, roughly equi-dimensional to highly irregular in shape; they reached the earth's surface from inter-

planetary space. / meteoriet, meteoorsteen / météorite / Meteor; Meteorit / meteorito

meteoroid, – extraterrestrial body passing through the earth's atmosphere. / meteoriet / météoroide / Meteor; Meteorit / meteorito

meteor, – transitory luminous streak in the sky produced by the incandescence of a meteoroid. / meteoor / météore / Meteor; Leuchtspur / meteoro

0064
micrometeorite, –
1. the smallest particles, less than I mm in size, that can be recognized as individual meteorites;
2. extraterrestrial particles that pass through the atmosphere without being altered because of their small size.

/ micrometeoriet / micrométéorite / Mikrometeorit / micrometeorito

0065
cosmic dust, – the smallest extraterrestrial particles that pass the earth's atmosphere practically unaltered. / kosmisch stof / poussière cosmique / kosmischer Staub / polvo cósmico

meteoritic dust; meteorite dust, – ablation product from the surface of an extraterrestrial body passing through the atmosphere. / meteorenstof / poussière météorique / kosmischer Staub / polvo meteorítico

0066
iron meteorite; siderite; holosiderite; – "irons", – a group of meteorites consisting of nickeliferous iron with small amounts of silicate minerals, subdivided into three classes according to structure. / sideriet; ijzermeteoriet / sidérite; holosidérite / Siderit; Holosiderit; Eisenmeteorit / meteorito férrico; siderita

0067
Widmanstätten figures; .... structure, – a distinctive network of stripes crossing at certain angles that appears on a polished surface of many iron meteorites upon etching, due to lamellar intergrowth of kamacite (6% Ni) and taenite (26-30% Ni) oriented parallel to the faces of an octahedron. / figuren van Widmanstätten / figures de Widmanstätten / Widmanstätt'sche Figuren / figuras de Widmanstätten

octahedrite, – the class of iron meteorites that display Widmanstätten figures. / octaëdriet / octaédrite / Oktaëdrit / octaedrito

0068
Neumann lines; .... bands, – straight lines tthat appear on a polished surface of some iron meteorites upon etching, due to twinning of cubic crystals of kamacite on a trapezohedron face. / lijnen van Neumann / lignes de Neumann / Neumann-Linien / líneas de Neumann

hexahedrite, – the class of iron meteorites composed mainly of large crystals of kamacite and showing Neumann lines. / hexaëdriet / hexaédrite / Hexaedrit / hexaedrito

0069
araxite, – a class of rare iron meteorites that display neither Neumann lines nor Widmanstätten figures, being composed essentially of a fine-grained mixture of kamacite and taenite (plessite), that is less susceptible to etching. / ataxiet / ataxite / Ataxit / ataxito

0070
stony-iron meteorite; siderolite; "stony-irons", – a group of meteorites consisting of a heterogeneous mixture of nickeliferous iron and silicates subdivided into four classes according to the nature of the silicate minerals. / sideroliet / sidérolite / Stein-Eisen-Meteorit; Siderolit / meteorito férico-rocoso; siderolito

0071
stony meteorite; aerolite; "stones", – a meteorite composed of silicate minerals generally with some nickeliferous iron, subdivided into two groups according to the absence or presence of chondrules. / steenmeteoriet; aeroliet / aérolite; météorite pierreuse / Aerolit; Steinmeteorit / meteorito rocoso; aerolito

0072
chondrule (plur. chondrules, chondri), – globular, radially crystallized aggregates, commonly about 1 mm in diameter. / chondrule / chondrule / Chondrule / condrula

chondrite, – the major group of stony meteorites that contains chondrules, subdivided into five classes according to their mineralogy. / chondriet / chondrite / Chondrit / condrita

achondrite, – the group of stony meteorites that is destitute of chondrules, subdivided according to their mineralogy into at least two and possibly four classes of calcium-poor achondrites and possibly three classes of calcium-rich achondrites. / achondriet / achondrite / Achondrit / acondrita

0073
fusion crust, – the outer layer of many meteorites formed by physical and chemical interaction with the atmosphere during its fall, generally a few tenths of a mm, rarely more than 1 mm thick; within the crust 2 to 4 zones may be distinguished, showing a decrease of alteration inward. The external structure of the crust is characterized by the orientation while moving through the atmosphere, i.e. close-textured or knobby (frontal surfaces), striated, ribbed, porous (lateral surfaces), warty or scoriacious (rear surfaces). / smeltingskorst / croûte de fusion / – – / corteza de fusión

0074
regmaglypt; piezoglypt; "thumb marks", – polygonal or oval, shallow depressions in the surface of a meteorite, from a few mm to many cm in diameter and sometimes with sharp rims separating them.

0075
meteor crater; astrobleme, – the depression in the earth's surface and surrounding swell caused by the impact of a (major) meteorite. / meteoorkrater / cratère de météorite; astroblème / Meteorkrater / cráter meteorítico

0076
impactite, – vesicular, glassy to finely crystalline material, produced where a meteorite has struck the earth's surface. / impactiet / impactite / Impaktit / impactita

coesite; stishovite, – high-pressure forms of quartz (stability fields above respectively 200 and 900 kPa), often found near impact craters, and considered to have been formed by strong shock (impact) metamorphosis.

shatter cone, – conical fracture in rock, up to a metre or more in length, caused by an extremely strong impact.

0077
tektite, – silica-rich (70-80% SiO_2) glass body, generally small (up to 200-300 grammes), rounded (discoid, lensoid) to elongated, with glossy or dull, pitted or corrugated surface. Tektites occur in limited areas in a few regions on the earth's surface, e.g. Bohemia, Moravia (moldavites), Indonesia (billitonites, javaites), Philippines (philippinites or rizalites), Texas (bediasites). They are probably of extraterrestrial origin, presumably splashes from major impacts on the moon. / tektiet / tectite / Tektit / tectita

[0300] GEOPHYSICS

[0310] GENERAL TERMS; THE ARCHITECTURE OF THE EARTH.

0078
geophysics, – the science dealing with the physics of the earth from its innermost parts to the outer fringes of its magnetosphere; also the study of the earth by physical means. / geofysica / géophysique; physique du globe / Geophysik /geofísica.

0079
geonomy, – a somewhat controversial term, denoting the science that treats of the geological and geophysical aspects of the solid earth. / geonomie / géonomie / Geonomie / geonomía

0080
geodynamics, – the science dealing with the forces and processes of the interior of the earth, as affecting the features of the crust. / – – / géodynamique interne / Geodynamik / geodinámica

tectonosphere; tectosphere, – a non-current term denoting the part of the earth, in which the tectonic movements originate, roughly embracing the earth's crust and upper mantle / tectonosfeer / tectonosphère; tectosphère / Tektonosphäre / tectosfera

0081
endogenetic, – a term applied to processes originating within the earth. / endogeen /endogène / endogen; innenbürtig / endogenético; endógeno; endogénico

exogenetic, – a term applied to processes originating at or near the surface of the earth or from outside. / exogeen / exogène / exogen; aussenbürtig / exogenético; exógeno; exogénico

0082
World Data Centre; WDC – one of the worldwide distributed centres for the collection, exchange and general availability of data; for example on geophysics, tsunamis etc.

0083
exosphere (obsolete), – the transition zone between atmosphere and interplanetary / space. / exosfeer / exosphère / Exosphäre / exosfera

0084
atmosphere, – the mass of air / surrounding the earth. / dampkring; atmosfeer / atmosphère / Atmosphäre / atmósfera

0085
hydrosphere, – the aqueous envelope of the earth, consisting of ocean and seas, lakes and streams, snow and ice, and including underground waters. / hydrosfeer /hydrosphère / Hydrosphäre / hidrosfera

0086
earth's crust, – the outer shell of the earth between the surface and the Mohorovičić discontinuity; below the continents to an average depth of 33 km (varying between some 20 and 70 km), below the oceans to a depth around 10 km. / aardkorst / écorce terrestre; croûte .... / Erdkruste / corteza terrestre

Conrad discontinuity, – a poorly defined and not everywhere present seismic discontinuity within the deep part of the continental crust; sometimes considered to mark the transition from felsic to mafic rocks or from amphibolite to granulite facies. / Conrad discontinuïteit / discontinuité de Conrad / Conrad-Diskontinuität / discontinuidad de Conrad

0087
Mohorovičić discontinuity; M-discontinuity; Moho, – the seismic discontinuity across which there is an abrupt increase of the velocities of both longitudinal and transverse waves, defining the base of the crust and the top of the mantle. It is associated with an increase in density and presumably a change in lithology from mafic to ultramafic. /Mohorovičić discontinuïteit / discontinuité de Mohorovičić / Mohorovičić-Diskontinuität / discontinuidad de Mohorivičić

0088
lithosphere, – the riged shell of the earth, considered to embrace the part of the earth between the surface and the asthenosphere, i.e. crust and uppermost part of the mantle. / lithosfeer; steenschaal / lithosphère / Lithosphäre, Steinschale / litosfera

0089
endosphere, – a non-current term denoting the inner parts of the earth below the lithosphere. / endosfeer / endosphère / Endosphäre / endosfera

0090
asthenosphere, – a zone within the upper mantle, in which the propagation velocity of seismic waves is at a minimum. It is present at varying depths below oceans (about 60-100 km) and continents (about 100-300 km), but virtually absent below stable shields. Its temperature is presumably up to the

melting point of rocks, causing a zone of low strength that decouples the overlying mechanically rigid lithospheric plates from the underlying part of the mantle. /asthenosfeer / asthénosphère / Asthenosphäre / astenosfera

0091
earth's mantle, – the part of the earth between Mohorovočić and Gutenberg discontinuities. / aardmantel / manteau terrestre / Erdmantel / manto terrestre

upper mantle, – the part of the mantle to a depth of about 400 km, presumably composed of silicate rocks. / buitenmantel / manteau supérieur / oberer Mantel / parte superior del manto

transition zone; ... layer, – the part of the mantle between depths of about 400 and 1000 km, where abnormally high gradients in seismic wave velocities occur.

lower mantle, – the part of the mantle below the transition zone, where the velocity of seismic waves increases gradually. It is presumably composed of densily packed silicate rocks. / binnenmantel / manteau inférieur / unterer Mantel / parte inferior del manto

0092
Gutenberg discontinuity; Wiechert-Gutenberg ....; Gutenberg-Oldham ...., –the seismic discontinuity between earth's mantle and core at a depth around 2900 km, where there is an abrupt decrease in the velocity of the longitudinal waves and where the transverse waves do not pass. It is associated with an increase in density, a change from solid to liquid phase and a major change in composition. / Gutenberg discontinuïteit / discontinuité de Gutenberg / Gutenberg-Diskontinuität / discontinuidad de Gutenberg

0093
earth's core, – the part of the earth below the Gutenberg discontinuity, probably composed of metallic iron with some nickel and lighter elements, e.g. silicon. /aardkern / noyau terrestre / Erdkern / núcleo terrestre

outer core, – the liquid part of the core through which no transverse seismic waves pass; between depths of about 2900 and 5100 km. /buitenkern / noyau externe / äusserer Kern / parte exterior del núcleo

inner core, – the presumably solid part of the core between a depth of about 5100 km and the centre of the earth at 6371 km. / binnenkern / noyau interne; graine / innerer Kern / parte interior del núcleo

[0320] SEISMOLOGY

[0321] *General Terms; The Earth's Medium*

NOTE:
– for further terms see sections:
[2840] Seismic Exploration Methods
[2850] Shallow Seismic and Acoustic Methods

0094
seismology, – the study of earthquakes, and by extension that of the structure of the interior of the earth through both natural and artificially generated seismic signals. / seismologie; aardbevingsleer / séismologie; sismologie / Seismologie; Erdbebenkunde / sismología

planetary seismology, – the science of seismology applied to the moon and planets. /planetaire seismologie / séismologie planétaire; sismologie ... / Planetär-Seismologie / sismología planetaria

deep seismic sounding, – deep investigation of the lithosphere by means of combined refraction and reflection techniques. / diepe seismische peiling; ... sondering / sondage sismique profond / seismische Tiefensondierung / sondeo sísmico profundo

0095
seismic event, – an earthquake or another transient earth motion, for example an explosion. / seismisch evenement; seismisch gebeuren / événement sismique / seismisches Ereignis / evento sísmico

0096
bradyseism, – a slow upheaval or downward movement of the earth's surface. / bradyseisme / – – / Bradyseismus / bradisismo

0097
microseism; – faint, more or less continuous vibrations of the earth's crust caused by a variety of natural and artificial agents, esp. atmospheric. For the detection of earthquakes it is considered as noise. / microseisme; ruis; microseismische beweging / microséisme; agitation microsismique / mikroseismische Bodenunruhe / microsismo

seismic noise, – any undesired disturbance within a for seismology useful frequency band, e.g. microseisms of 1 sec period relative to the signal of teleseisms. / ruis / bruit / Rauschen / ruido sísmico

noise level, – / ruisniveau / niveau de bruit / Störpegel / nivel de ruido

0098
ellipticity correction, – correction to be applied to distance and angle measurements on the globe because of the ellipsoidal shape of the earth. / ellipticiteitscorrectie / correction d'ellipticité / Elliptizitätskorrektur / corrección de la elipticidad

geocentric latitude, – the celestial latitude of a point at the earth's surface based on or as seen from the centre of the earth. It differs from geodetic latitude because of the ellipsiodal shape of the earth. / geocentrische breedte / latitude géocentrique / geozentrische Breite / latitud geocéntrica

0099
discontinuity; seismic ...., – boundary plane between two zones in the earth with different seismic velocities. / discontinuïteitsvlak / surface de discontinuité / Diskontinuitätsfläche / discontinuidad (sísmica)

0100
low-velocity layer; ... channel; ... zone ;LVL,– any layer or shell of the earth with a seismic wave velocity lower than in the adjacent layers or zones, e.g. Gutenberg low-velocity zone. / lage-snelheidslaag; ....kanaal / couche à faible vitesse / Schicht erniedrigter Geschwindigkeit / capa de baja velocidad

0101
wave guide, – a layer or zone in the solid earth that channels seismic energy. This is also applicable to the atmosphere and the ocean. / golfgeleider / guide d'onde / Wellenleiter / guía de onda

0102
rheology, – the study of deformation and flow of matter / rheologie / rhéologie / Rheologie / reología

0103
stress, – the force per unit area, acting on any surface within the body. / spanning/ contrainte / Spannung / tensión

principal stress, – /hoofdspanning / contrainte principale / Hauptspannung / tensión principal

deviatory stress; differential stress,– the maximum minus the least principal stress /differentiële spanning; verschilspanning; deviatorische spanning / contrainte differentielle / Deviationsspannung / tensión desviatoria

0104
strain, – change in shape or volume of a body as a result of stress. / vervorming / déformation / Zug / deformación

0105
shear; shear strain,– a strain resulting from stresses that cause contiguous parts of a body to slide relative to each other in a direction parallel to their plane of contact. / schuif / cisaillement / Scherung / cizallamiento

0106
stress-strain curve, – plot of the strain in percent change of length against the differential stress. / spanning-vervormingskromme / courbe déformation-contrainte / Spannungs-Deformationskurve / curva de deformación-tensión

0107
strain release, – the strain in the focal rock released by the motion in the focus of an earthquake. /vrijgekomen vervormingsenergie / relâchement des contraintes / Spannungsabfall / liberación de deformación

0108
elastic rebound theory; – mechanism, explaining fault motion as the result of an abrupt release of progressively increasing elastic strain in the rock mass. / theorie der elastische vereffening / théorie du rebondissement élastique / Theorie des elastischen Zurückschnellens / teoría del rebote elástico

0109
dilatancy, – increase of the bulk volume during deformation, caused by a change in structure, and accompanied by an increase of pore volume. / dilatantie; opzwelling / dilatance / Dilatanz / dilatancia

0110
anisotropy; aeolotropy; eolotropy,– said of a medium whose physical proporties, e.g. seismic velocities, vary in different directions. /anisotropie; aeolotropie / anisotropie / Anisotropie; Aeolotropie / anisotropía

0111
anelasticity, – the effect of attenuation of a seismic wave, symbolized by factor Q. / anelasticiteit / anélasticité / Anelastizität / anelasticidad

0112
absorption, – conversion of energy of a seismic wave into other forms of energy, e.g. heating of the medium through which the wave passes. / absorptie / absorption / Absorption / absorción

0113
seismic-electric effect, – variation of electrical resistivity with elastic deformation of rocks (as precursor effect of large earthquakes). / seimisch-electrisch effect / effet sismo-électrique / seismo-elektrischer Effekt / efecto sismo-eléctrico

0114
silent earthquake; creep, – slow deformation that results from long application of a stress below the elastic limit. Part of the creep is permanent deformation, part of it is elastic. / kruip / reptation / Kriechen / reptación

creep recovery; elastic afterworking; elastic aftereffect; transient strain,– gradual recovery of elastic creep strain upon release of applied stress. / kruipherstel; elastische nawerking / rétablissement élastique / elastische Nachwirkung / efecto elástico

0115
rock burst; rockburst; pressure burst; bump; outburst; burst,– a spontaneous rock expansion in a mine or underground cavity. / autoklaas; gesteenteslag / bandon; éclat; autoclase; rupture spontanée / Bergschlag; Gebirgschlag / ruptura violenta de una roca

[0322] *Stations and Instruments*

0116
seismograph, – instrument that records vibrations of the earth, e.g. earthquakes; horizontal, vertical, universal. / seismograaf / séismographe / Seismograph / sismógrafo

broad-band seismograph, – seismograph sensitive for an exceptionally broad period range. / – – / séismographe à large bande; sismographe–.... / Breitband-Seismograph / sismógrafo de banda ancha

vault, seismograph vault, – / kelder; seismograafkelder / cave d'observation séismologique / Seismographenbunker / bóveda

seismogram; earthquake record; seismic record,– record of an earthquake, produced by a seismograph. / seismogram / séismogramme; sismogramme / Seismogramm / sismograma

0117
seismoscope, – instrument indicating the occurrence of an earthquake. / seismoscoop / séismoscope; sismoscope / Seismoskop / sismoscopio

0118
seismometer; geophone,– instrument that detects earth motions. It is the detector of the seismograph system, and does not by itself contain a recording unit. / seismometer / séismomètre; sismomètre / Seismometer / sismómetro; geófono

tromometer; tremometer, – archaic term for seismometer. / tremometer; tromometer / – – / Tremometer; Tromometer / tromómetro

0119
array station, – seismic station consisting of a number of seismometers in an ordered geographical arrangement all recording at a central place. / array station / station composite / Array-Station / – –

0120
accelerometer; strong motion seismometer, – instrument to detect ground acceleration. Seismometer specifically designed to respond to the acceleration of the ground motion by making its natural period much smaller than that of the ground motion. / accelerometer; versnellingsmeter / accéléromètre / Akzelerometer; Beschleunigungsmesser / acelerómetro

accelerogram, – record made by an accelerograph. / accelerogram; versnellingsregistratie / accélérogramme / Akzelerogram; Beschleunigungsregistrierung / acelerograma

peak acceleration, – the maximum acceleration measured in an accelerogram. / piekversnelling / accélération de pointe / maximale Beschleunigung / aceleración máxima

0121
displacement meter, – seismometer designed to respond to the displacement of earth particles, by making its natural period much larger than that of the ground motion. / verplaatsingsmeter / extensomètre / Wegaufnehmer / medidor de desplazamiento

0122
extensometer; strain meter,– instrument designed to detect deformation of the ground by measuring relative displacement of two points. / extensometer; vervormingsmeter / extensomètre / Deformationsmesser / extensómetro

0123
vibrograph; vibration measurer,– apparatus to register rapid vibrations as caused by artificial earthquakes. / vibrograaf / vibrographe / Vibrationsmesser / vibrógrafo

0124
tiltmeter; clinograph, – instrument to measure small changes of inclination of the surface of the earth. / hellingsmeter; tiltmeter; klinograaf / clinomètre / Tiltmeter; Klinograph / clinómetro

tilt, – / kanteling / basculement / Kippung / basculamiento

0125
stationary mass; steady mass, – the inertia mass of a seismometer. / stationaire massa; trage massa / masse stationaire / stationäre Masse; Trägheitsmasse / masa estacionaria; masa inerte

0126
length of the equivalent pendulum, – the length of the mathematical pendulum equivalent to the physical one. / equivalente slingerlengte / longueur équivalent du pendule / äquivalente Pendellänge / longitud equivalente del péndulo

0127
damping, – a resistance decreasing the amplitude of successive vibrations of a seismometer; it results from passing through air, oil or by other mechanical friction or by electromagnetic absorbers. / demping / amortissement / Dämpfung / amortiguación

viscous damping, – seismograph damping in which the damping force is directly proportional to the velocity. / viskeuse demping / amortissement visqueuse / visköse Dämpfung / amortiguación viscosa

dynamic damping, – damping of the motion of a seismometer mass proportional to its velocity. / dynamische demping / amortissement dynamique / dynamische Dämpfung / amortiguación dinámica

0128
damping factor, – ratio of the damped and the undamped frequency of a seismometer. / dempingsfactor / facteur d'amortissement / Dämpfungsfaktor / factor de amortiguación

damping ratio, – ratio between two successive swings of a damped pendulum (seismometer). / dempingsverhouding / rapport d'amortissement / Dämpfungsverhältnis / cociente de amortiguación

0129
critical damping; aperiodical damping, – the minimum damping for which the moving mass after deflection returns to the equilibrium position without oscillating. / kritische demping; aperiodische demping / amortissement critique / Kritische Dämpfung; aperiodische Dämpfung / amortiguamiento crítico

0130
free period, – time for one complete swing when all damping is removed from the seismometer mass. / eigen periode / période d'oscillation libre / Eigenperiode / período libre

0131
static magnification, – ratio between the amplitude of the indicator and that of the gravity centre of the seismometer mass. / statische vergroting / amplification statique /

Indikatorvergrösserung; statische Vergrösserung / amplificación estática

dynamic magnification, – seismograph magnification of ground motion taking into account the free period and damping of the seismometer and the period of ground motion. / dynamische vergrotingsfactor / amplification dynamique / dynamische Verstärkung / amplificación dinámica

0132
dynamic range, – ratio of the maximum to the minimum seismic signal that instrumentally can be handled, specified over a certain frequency bandwidth and measured in decibels. / dynamisch bereik / étendue dynamique / dynamischer Bereich / rango dinámico

0133
response curve; magnification curve, – the curve that gives for each period the ratio of the amplitude of the output signal of the seismograph and the amplitude of the input signal of the ground. This can be displacement, velocity or acceleration. / responsiecurve; vergrotingscurve / courbe de réponse / Responzkurve; Charakteristik / curva de respuesta; curva de amplificación

feedback, – /terugkoppeling / couplage réactif / Rückkopplung / realimentación

0134
phase shift, – the difference in phase between the incoming ground wave and the recorded wave; either phase lag or phase lead. / faseverschuiving / déphasage / Phasenverschiebung / defasaje

phase lag, – / fasevertraging / – – / Phasenverspätung / – –

phase lead, – / faseversnelling / – – / Phasenvorauseilung / – –

0135
phase response, – a graph of the phase shift versus frequency. / faseresponsie; fasecurve / – – / Phasenkarakteristik / respuesta en fase

0136
frequency filtering, – procedure to separate wave trains of different frequency, e.g. in the analysis of a dispersed surface wave and the separation of higher modes from the fundamental mode. / frequentiescheiding / filtrage des fréquences / Frequenzfilterung / filtrado en frecuencia

0137
cut-off filter, – filter shaped to remove unwanted radiation, either above (low pass), or below (high pass) a desired frequency. / afsnijfilter / filtre passe-bas; filtre passe-haut / Abschneide-Filter / filtro de corte

band-pass filter, – filter in which frequency components outside a given range are attenuated. Inverse of band-reject filter. / doorlaat-filter / filtre passe-bande / Durchlass-Filter / filtro de paso-banda

band-reject filter, – filter in which frequency components in a given range are attenuated, all others being passed. Inverse of band-pass filter. / – – / filtre éliminateur de fréquence / Sperrfilter / – –

notch filter, – a filter designed to remove a single frequency. / – – / filtre ponctuel / Kerbfilter / – –

[0323] *Earthquakes; Source Parameters*

0138
earthquake; seism; temblor; tremblor; shock, – a sudden motion or trembling in the earth. / aardbeving; aardschok; aardstoot / tremblement de terre; séisme; / Erdbeben; Erdstoss / terremoto; sismo; temblor; temblor de tierra; sacudida

macroseism; earthquake, – record of an earthquake, as opposed to that of microseism. / macroseisme / macroséisme; tremblement de terre / Makroseisme; Makroseismik / macrosismo

world quake, – earthquake that is recorded all over the world. / wereldbeving / séisme mondial / Weltbeben / sismo mundial

0139
tectonic earthquake, – / tectonische aardbeving; dislocatiebeving / tremblement de terre tectonique / tektonisches Erdbeben; Dislokationsbeben / sismo tectónico

0140
volcanic earthquake, – an earthquake associated with volcanic rather than with tectonic forces. / vulkanische beving / séisme volcanique / Vulkanbeben / sismo volcánico

crypto-volcanic earthquake, – earthquake caused by volcanic phenomena that do not reach the surface. / crypto-vulkanische beving / tremblement cryptovolcanique / Krypto-vulkanisches Erdbeben / terremoto cripto-volcánico

0141
collapse earthquake, – earthquake caused by collapse of natural or artificial cavities. / instortingsaardbeving / tremblement de terre d'effondrement / Einsturzbeben / terremoto de colapso

impact earthquake, – earthquake generated by the fall on earth of a meteorite. / inslagaardbeving / séisme d' impact / Einschlagerdbeben / sismo de impacto

0142
principal earthquake; main shock, – the strongest of a consecutive number of earthquakes. / hoofdbeving / tremblement de terre principal / Hauptbeben / sismo principal

0143
multiple earthquake, – earthquakes of comparable magnitude from the same focus taking place within a period of a few minutes or tens of minutes. / meervoudige aardbeving / tremblement de terre multiple / mehrfacher Erdbeben / terremoto múltiple

0144
swarm earthquakes, – series of earthquakes of long duration not showing a pronounced principal earthquake. /aardbevingszwerm / série de secousses / Erdbebenschwarm / enjambre de sismos

0145
relay earthquake, – shock occurring as an after-effect of an earthquake taking place elsewhere. / relaisbeving / tremblement de terre à relais / Relaisbeben / terremoto de relé

0146
microearthquake, – an earthquake of magnitude two or less on the Richter scale (the limit of two is arbitrary). / microaardbeving / microséisme / Mikrobeben; Mikroerdbeben / microterremoto

tremor; earth tremor; earthquake tremor,– a minor earthquake. / trilling; aardtrilling / séisme mineur / Tremor / temblor

0147
foreshock, – initiating shock preceding the principal earthquake or main shock. / voorschok; voorloper / secousse prémonitoire / Vorbeben / sacudida premonitoria

aftershock, – earthquake following the main shock in an earthquake sequence and originating in the same region. Usually aftershocks decrease in frequency and magnitude with increasing time. / naschok; nabeving / réplique / Nachbeben / réplica

0148
inland earthquake, – earthquake with the origin below a continent. / landbeving / séisme continental / kontinentales Beben / terremoto terrestre; terremoto continental

submarine earthquake; seaquake, – an earthquake occurring beneath the ocean and that, if heavy, can be observed on board a ship in the vicinity of the epicentre. / zeebeving; onderzeese aardbeving / séisme sous-marin / Seebeben; submarines Beben / terremoto (sub)marino

0149
local shock, – /plaatselijke beving / séisme local / lokales Beben / sacudida local; sismo local

teleseism; distant shock, – an earthquake that is distant from the station. / verre beving; teleseisme / téléséisme; séisme lointain / Fernbeben; Teleseisme / telesismo; sismo remoto

0150
precursor, – any natural phenomenon that precedes the occurrence of an earthquake. Also: seismic wave preceding the onset of a much larger seismic wave train, as e.g. in the case of core waves. / voorloper / précurseur / Vorläufer; Vorbote / precursor /

0151
source parameter; focal parameter, – data describing the focus of an earthquake, such as location, depth, time, magnitude, mechanism, etc. / brongegeven; haardparameter / paramètre hypocentral paramètre de la source / Herdparameter / parámetro focal

0152
tripartite method, – method to determine the apparent surface-wave velocity and the direction of propagation of microseisms or earthquake waves by measuring times at three separate station points about 1 km apart. / driepuntsmethode / méthode tripartie / Dreipunktmethode / método tripartito

hyperbole method, – method to determine the location of the earthquake focus. / hyperboolmethode / méthode des hyperboles / Hyperbolmethode / método de las hipérboles.

0153
master event, – well documented earthquake, the travel-times of which are used for a relocation of smaller events in the same region, resulting in an internally coherent picture of the seismicity. / – – / événement majeur / Referenzbeben / – –

0154
epicentre; epicenter (USA), – point on the earth's surface directly above the hypocentre or focus of the earthquake. / epicentrum / épicentre / Epizentrum / epicentro

anticentre; anticenter; antiepicentre, – point at the earth's surface diametrically opposite to the epicentre. / anticentrum / anticentre / Antizentrum / anti-epicentro

0155
epicentral azimuth, – azimuth of the epicentre as seen from the recording station. / epicentraal azimuth / azimut de l'épicentre / Epizenterazimut / azimut epicentral

epicentral distance; angular distance, – distance from the epicentre measured along the earth's surface. / epicentrale afstand / distance épicentrale / Epizentralentfernung; Epizentraldistanz / distancia epicentral

0156
focus; hypocentre; hypocenter (USA); source; seismic focus; centrum, – the point within the earth that is the centre of an earthquake and the origin of its elastic waves. / haard; hypocentrum / hypocentre; foyer / Hypozentrum; Herd / hipocentro; foco

0157
depth of focus; focal depth, – depth of hypocentre below the earth's surface, i.e. shallow (0-60 km), intermediate (60-300 km), deep (over 300 km). / haarddiepte / profondeur hypocentrale; profondeur du foyer; profondeur focale / Herdtiefe / profundidad focal

shallow (0-60 km) / ondiep; normaal / superficiel; normal / flach; normal / superficial

intermediate (60-300 km) / intermediair / intermédiair / intermediär / intermedia

deep (over 300 km) / diep / profond / tief / profunda

high focus, – an earthquake focus calculated to be situated at negative depth. / – – / foyer négatif / negative Herdtiefe / – –

0158
bathyseism, – deep-focus earthquake that is instrumentally detected world wide (obsolete term). / diepe wereldaarbeving / bathyséisme / Bathyseism; Tiefherdbeben / batisismo

0159
origin time; focal time, – time of occurrence of the earthquake. / oorsprongstijd; haardtijd; stoottijd / heure initiale / Herdzeit; Ursprungszeit / tiempo de origen

0160
station azimuth, – azimuth of the station as seen from the focus. / stationsazimuth / azimut d'une station / Stationsazimut / azimut de la estación

station correction, – correction to be applied to the travel time or azimuth of seismic waves as observed in the station. The source for such corrections is to be found in structural features of the underground of the station and in the trajectory of the wave. / stationscorrectie / correction de station / Stationskorrektion / corrección de estación

[0324] *Seismic Waves; Earth Motions*

0161
seismic wave, – elastic wave radiated from an earthquake focus. / aardbevingsgolf / onde sismique / Erdbebenwelle / onda sísmica

0162
wave length (λ), – the distance between successive similar points on the wave cycle. / golflengte / longueur d'onde / Wellenlänge / longitud de onda

wave number (k), – the number of wave cycles per unit distance; reciprocal of wave length. / golfgetal / nombre d'onde / Wellenzahl / número de onda

0163
period (T), – the time for one wave cycle. / periode / période / Periode / período

0164
amplitude (A), – the maximum excursion in a wave cycle measured from the zero position. / amplitude; amplitudo; uitslag / amplitude / Amplitude / amplitud

peak and trough, – the highest and lowest part of a wave form. / top en dal / crête-crête; maximum et minimum / Maximum und Minimum / pico y valle

0165
phase, – the onset of a displacement on a seismogram indicating the arrival of a different type of seismic wave, cf. longitudinal, transverse, surface wave, Airy, T-phases. / fase / phase / Phase / fase

0166
mode, – a stationary vibration of an oscillatory system, more specifically relevant for seismic surface waves. / mode; trillingswijze / mode / Mode / modo

0167
fundamental mode, – the wave with the basic period that is possible in the medium with given structure. / grondtoon; fundamentele mode / mode fondamental / Fundamentalmode / modo fundamental

harmonic, – a frequency which is a simple multiple of a fundamental frequency. / harmonische / harmonique / harmonische / armónico

higher mode; overtone, – higher harmonic surface wave generated by constructive interference during wave propagation in a layer. / hoger harmonische; hogere mode / mode harmonique / Oberwelle / modo superior

0168
wave velocity, – velocity at which a wave train moves forward. / golfsnelheid / vitesse de propagation / Wellengeschwindigkeit / velocidad de onda

wave slowness, – the reverse of wave velocity, mostly measured in seconds per degree: dT/d∧ / – – / lenteur / Wellen-Langsamkeit / lentetud de onda

0169
phase velocity (C), – the velocity of propagation of an individual wave crest, or point of constant phase, through a medium. Especially for seismic surface waves. / fasesnelheid / vitesse de phase / Phasengeschwindigkeit / velocidad de fase

0170
group velocity (U), – the velocity of energy transport of a seismic surface-wave group through the earth. / groepsnelheid / vitesse de groupe / Gruppengeschwindigkeit / velocidad de grupo

0171
travel time; transit time, – time required for a seismic wave to travel from the hypocentre to the recording station. / looptijd / durée de parcours; temps de propagation / Laufzeit / tiempo de propagación; tiempo de recorrido

travel-time curve; time-distance graph; time-distance curve; T. D. curve; T-X graph; hodochrone; hodograph,-graphical presentation of the travel time as a function of the distance from the epicentre measured along the earth's surface. / looptijdkromme / hodochrone; courbe de propagation / Laufzeitkurve; Hodographe / curva de propagación; curva tiempo-distancia; hodócrona

travel-time table, – table showing travel times of seismic waves as a function of epicentral distance. / looptijdtabel / table des temps de propagation / Laufzeittabelle / tabla de tiempos de propagación

0172
arrival; onset, – the appearance of seismic energy on a seismogram. / inzet / arrivée / Einsatz / llegada

0173
impetus, – a sharp onset or arrival of a seismic wave on a seismogram. / scherpe inzet / impetus; début brusque / deutlicher Einsatz; Impetus / impetu; llegada violenta

impulsive, – said of an impetus or sharp arrival of a seismic phase. / – – / impulsion / scharf / impulsiva

0174
emersio, – a gradual onset or arrival of a seismic phase on a seismogram. / geleidelijk beginnende inzet / début peu marqué; émergence / allmähliges Auftreten / emersión; aparición gradual

emergent, – said of an emersio, or gradual arrival of a seismic phase. / emergent / émergent / emergent / emergente

0175
coda; cauda, – end portion of earthquake record. / coda; eindfase / queue; phase finale / Coda; Nachläufer / coda; parte terminal; fase final

0176
transient signal, – a seismic signal with a sudden beginning and of short duration, as opposed to a harmonic oscillation. / voorbijgaand signaal; kortstondig signaal / signal transitoire / transientes Signal / señal transitoria
sıtorıa

0177
signal-generated noise, – disturbance of the seismic record after the arrival of the direct wave signal and caused by secondary waves that are generated at appropriate places in and on the earth by the passing of the front of the original wave. / signaal-opgewekte ruis / – – / signal-erzeugtes Rauschen / ruido generado por la señal

sonic boom, – generated by supersonic flight of airplanes, and giving rise to disturbances of the seismic records in the immediate vicinity. / geluidsbarrièreknal / déflagration sonique / Überschall-Knall / – –

0178
beat, – periodic pulsation caused by the simultaneous occurrence of waves of different frequencies. / – – / battement / Trägerwelle / batido

beat frequency, – the frequency, that is the difference between two given frequencies causing the beat. / fréquence des battements / Trägerfrequenz / frecuencia de batido

0179
body wave; bodily seismic wave, – a seismic wave that travels through the interior of the earth and is not related to any boundary surface: either longitudinal (P-wave) or transverse (S-wave). / lichaamsgolf; ruimtegolf / onde de volume / Raumwelle / onda sísmica interna; onda de volumen

0180
core wave, – seismic wave that traverses the core of the earth. / aardkerngolf / onde du noyeau / Kernwelle / onda del núcleo

0181
preliminary phase, – obsolete term for primary (longitudinal) and secondary (transverse) phases in a seismogram. / voorloper / phase préliminaire / Vorläufer; Vorphase / fase preliminar

0182
longitudinal wave, – body wave in which the direction of vibration coincides with that of the propagation. / longitudinale golf / onde longitudinale / Longitudinalwelle / onda longitudinal

P-wave; primary wave; longitudinal wave; irrotational wave; push wave; pressure wave; push-pull wave, – first arriving longitudinal wave excited by an earthquake. / P-golf / onde P / P-Welle / onda P; onda longitudinal; onda de presión; onda primaria

0183
compressional wave, – longitudinal wave in which the first impulse is in the direction of propagation. / drukgolf; compressiegolf; condensatiegolf / onde de compression / Kompressionswelle; Verdichtungswelle / onda de compresión

dilatational wave, – longitudinal wave in which the first impulse is opposite to the direction of propagation. / rekgolf; dilatatiegolf / onde de dilatation / Dilatationswelle; Verdünnungswelle / onda de dilatación

0184
transverse wave, – body wave in which the direction of vibration is perpendicular to that of the wave propagation. / transversale golf / onde transversale / Transversalwelle / onda transversal

S-wave; secondary wave; transverse wave; shear wave; rotational wave; tangential wave; equivoluminal wave; distortional wave; shake wave, – transverse wave excited by an earthquake. It is the second pronounced wave in the seismogram. / S-golf / onde S / S-Welle / onda S; onda secundaria; onda transversal; onda de cizalla

0185
transformed wave; converted wave, – a reflected wave that has been transformed from a longitudinal to a transverse wave or vice versa. / getransformeerde golf; transformatie golf / onde transformée / Konvertierte Welle / onda transformada

0186
head wave; conical wave, – seismic wave originating along the interface between two media and traveling in the slower medium at a speed greater than the characteristic speed of that medium. / kopgolf / – – / Kopfwelle / onda cónica

0187
anaseism; compression, – earth movement away from the earthquake focus. / anaseisme; compressie / anaséisme; compression / Anaseism; Kompression / compresión; anasismo

kataseism; dilatation; rarefaction, – earth movement toward the earthquake focus. / kataseisme; dilatatie / cataséisme; dilatation / Kataseisme; Dilatation / catasismo; dilatación; rarefacción

0188
surface wave; L-wave,; long wave (obsolete); circumferential wave (obsolete), – a seismic wave that travels along the surface of the earth with an amplitude rapidly decreasing with depth. / oppervlaktegolf; lange golf / onde de surface / Oberflächenwelle / onda de superficie; onda superficial; onda larga

long wave phase; principal wave phase, – / hoofdfase / phase des ondes longues / Hauptphase / fase principal

0189
Rayleigh wave; R-wave, – surface wave with a vertical retrograde elliptical motion at the free surface. / Rayleigh golf / onde de Rayleigh / Rayleigh Welle / onda de Rayleigh

Love wave; Q-wave; querwelle wave; LQ, – a surface wave with a horizontal motion transverse to the direction of propagation. / Love golf / onde de Love / Love Welle; Querwelle; Oberflächenscherungswelle / onda de Love

0190
Airy phase, – predominant surface-wave group corresponding to a maximum or minimum group velocity. / Airy fase; hoofdfase / Airy phase; phase principale / Airy Phase / fase Airy

0191
returns; W-wave, – surface waves that have travelled around the earth by the long arc between epicentre and station or that have passed the station one or more time(s); indicated as R_2, R_3, R_4 etc. for waves of the Rayleigh type. Especially pronounced for mantle waves. / – – / ondes R / Wiederkehr-Wellen / ondas W

0192
visible wave, – seismic wave train reported to have been seen in the landscape during a very heavy earthquake. / zichtbare golf / onde visible / sichtbare Welle / onda visible

0193
coupled wave; C-wave, – a surface wave that is continously generated by another wave with the same phase velocity. / koppelingsgolf / onde couplée / gekoppelte Welle / onda de acoplamiento

0194
mantle wave, – surface wave of very long period largely determined by the properties and structure of the earth's mantle. / mantelgolf / onde de manteau / Mantelwelle / onda del manto

0195
guided wave, – any seismic wave propagating along a surface or discontinuity. / geleidegolf / onde guidée / Leitwelle; geführte Welle / onda guiada

Stoneley wave, – guided wave propagating along the interface of two media. / Stoneley golf / onde de Stoneley / Stoneley Welle / onda de Stoneley

channel wave, – guided wave propagating in a sound channel due to a low-velocity layer in the solid earth. / kanaalgolf / onde guidée / Kanalwelle / onda canalizada

leaking mode; leaky wave, – an imperfectly trapped channel wave of which part of the energy leaks across a layer boundary causing some attenuation. / lekgolf / mode à perte / Leckwelle; Streuungswelle / – –

0196
boundary wave; interface wave, – seismic wave propagating along a free surface or an interface between layers. / grensgolf / onde d'interface / Grenzwelle / onda de frontera

0197
dispersion, – the variation in velocity of seismic surface waves with frequency, due to varying elastic properties of the layers in the earth. / dispersie / dispersion / Dispersion / dispersión

normal dispersion – the recorded wave period decreases with time. / normale / normal / normal / normal

inverse dispersion, – the recorded wave period increases with time. / omgekeerde; inverse / renversé; à l'envers / inverse; umgekehrte / inversa

0198
free oscillation; free vibration, – an independent oscillation of the earth triggered by a major earthquake, that has its own specific natural period dependent on the internal structure of the earth. / eigentrilling / oscillation libre / Eigenschwingung / oscilación libre

spheroidal oscillation, – standing wave in the earth of the free vibration kind arising through mutual interference of propagating Rayleigh waves. / sferoïdale trilling / vibration sphéroïdale / sphäroidische Schwingung / vibración esferoidal

toroidal oscillation ; torsional oscillation, – standing wave in the earth of the free vibration kind arising through mutual interference of propagating Love waves. / toroidale trilling / oscillation toroidale / toroidale Schwingung / oscilación toroidal

0199
earth tide, – tides in the solid earth caused by the gravitational attraction of sun and moon. / aardgetij / marée terrestre / Erdgezeit / marea terrestre

0200
Chandler wobble, – irregular changes in earth rotation due to the angular momentum vector not being precisely co-linear with a principal axis of inertia of the earth. / Chandler schommeling / oscillation Chandler; agitation Chandler / Chandler Schwankung / oscilación de Chandler

0201
T-wave, – a short period (0.5 sec) acoustic wave in the sea, propagating in an oceanic sound channel. / T-golf / onde T / T-Welle / onda T

0202
gravitational wave, – type of atmospheric wave in which the whole atmosphere participates and of which the velocity is determined by the gravity field. It can be generated by atmospheric nuclear explosions or by the impact of a large meteorite. / gravitatiegolf; zwaartekrachtgolf / onde de gravitation / Gravitationswelle / onda de gravedad

0203
earthquake sound; air wave, – an acoustic wave in the air generated by a seismic event. / aardbevingsgeluid / bruit de tremblement de terre / Erdbebengeräusch / ruido del terremoto

[0325] *Wave Analysis*

0204
energy release, – in seismology the energy that is radiated from the earthquake focus in the form of seismic waves. / vrijkomende energie / energie relâché / Energieabstrahlung / liberación de energía

0205
seismic efficiency, – the fraction of the total available strain energy radiated as seismic waves. Also used in the case of underground explosions. / seismisch rendement; nuttig effect / rendement sismique / seismischer Wirkungsgrad / eficiencia sísmica

0206
seismic ray; trajectory; wave ray; orthogonal; path; ray path; wave path, – the line along which a wave or ray travels between two points of disturbance in an elastic medium; orthogonal to the wave front . / pad; golfweg; traject; orthogonaal / trajet; parcours; orthogonal; rayon sismique / seismischer Strahl / rayo sísmico; trayectoria; recorrido

0207
ray parameter, – a function that is constant along a given seismic ray, and that characterizes this ray. / straalparameter / paramètre de rayon sismique / Strahlparameter / parámetro del rayo

0208
apparent velocity, – the travelling speed of a seismic wave along the surface of the earth. It is the reciprocal of the seismic ray parameter. / schijnbare snelheid / vitesse apparente / scheinbare Geschwindigkeit / velocidad aparente

0209
brachistochronic path; minimum time path; least-time path, – the path between two points along which the travel time of a seismic ray is a true minimum. / minimum-tijdweg / temps de propagation minimum / Extremalweg / trayectoria braquistocrónica

0210
isochron(e); isotime curve, – line of equal travel time. / isochroon / isochrone / Isochrone / isócrona

homoseismal line; homoseism; coseismal line; coseismic line, – line on the earth's surface connecting points where a seismic wave arrives at the same time. / homoseiste / homoséiste; ligne homoséiste; isochrone / Homoseiste / línea homosísmica

0211
angle of incidence, – angle between a seismic ray and the vertical. It is the complement of the angle of emergence. / invalshoek / angle d'incidence / Einfallswinkel / ángulo de incidencia

plane of incidence, – plane containing the incident seismic ray and the normal to the earth's surface at the point of incidence. / invalsvlak / plan d'incidence / Einfallsebene / plano de incidencia

0212
grazing incidence, – e.g. of a seismic body wave on the mantle-core boundary. / scherende inval / incidence rasante / streichendes Einfall / incidencia bajo ángulo crítico

0213
total reflection, – reflection in which all of the incident wave energy is returned. / totale reflectie / réflection totale / Total-Reflektion / reflexión total

critical angle, – the smallest angle of incidence at which there is total reflection when the seismic wave passes from one medium to another less refracting medium. / kritische hoek; kritische invalshoek / angle critique / kritischer Winkel / ángulo crítico

0214
scheiteltiefe, – maximum depht of the seismic ray before returning to the surface of the earth. / grootst bereikte diepte / profondeur maximal de propagation des ondes / Scheiteltiefe / profundidad máxima;

0215
angle of emergence; emergence angle, – angle between a seismic ray and the horizontal. It is the complement of the angle of incidence. / emergentiehoek / angle d'émergence / Emergenzwinkel / ángulo de emergencia

0216
shadow zone, – area at the earth's surface, 100-140° from the epicentre of an earthquake, in which, due to refraction by the relatively low velocity inside the earth's core, there is no direct penetration of seismic waves. / seismische schaduwzone / zône d'ombre / seismische Schattenzone / zona de sombra

caustic, – a locus of points at the surface of the earth where at a distance of about 142° from the earthquake focus seismic wave energy is concentrated because of ray refraction on the mantle-core boundary. / caustic / point brillant / Kaustik / cáustica

0217
diffraction, – curving of the wave front around curved discontinuity surfaces, e.g. P-wave diffracted along the core-mantle boundary. / diffractie; straalafbuiging / diffraction / Diffraktion / difracción

0218
scattering, – divergence of seismic energy by inhomogeneities in the medium. / verstrooiing / dispersion / Streuung / dispersión

0219
dissipation, – the loss of energy of a seismic wave along the ray path. / verstrooiing / dispersion; dissipation / Dissipation / dispersión; disipación

attenuation, – decrease of seismic signal strength with distance caused by reflection, absorption and scattering. / demping / atténuation / Dämpfung / atenuación

0220
magnitude; earthquake magnitude, – strength of an earthquake measured on the basis of recorded seismic energy, e.g. body-wave magnitude, surface-wave magnitude. / magnitude / magnitude / Magnitude / magnitud

unified magnitude, – other term for body-wave magnitude. / – – / – – / Raumwellenmagnitude / magnitud unificada.

magnitude scale, – a standard for the measurement of earthquake magnitude. / magnitudeschaal / échelle de magnitude / Magnitudenskala / escala de magnitud

Richter scale, – a magnitude scale. / schaal van Richter / échelle de Richter / Richter Skala / escala de Richter

0221
spectral analysis, – the study of amplitude and phase of a seismic wave as functions of frequency (or period). / spectraalanalyse / analyse spectrale / Spektralanalyse / análisis espectral

0222
power spectrum, – a graph of power density versus frequency. It is the square of the amplitude–frequency responses. / vermogensdichtheid-spectrum / spectre des puissances / Leistungsspektrum / espectro de potencia

0223
frequency domain, – a spectrum shows the seismic wave amplitude as a function of frequency (or period) in the frequency domain. / frequentiedomein / domaine de fréquence / Frequenzbereich / dominio de frecuencia

time domain, – the analysis of an earthquake record with respect to arrival time occurs in the time domain. / tijddomein / domaine de temps / Zeitbereich / dominio de tiempo

0224
cepstrum, – the reciprocal of the wave spectrum.

0225
corner frequency, – the frequency of a seismic wave at which the spectral amplitude begins to decrease. It is related to the dimensions of the seismic source. / hoekfrequentie / – – / Eckfrequenz / frecuencia de codo

0226
dispersion curve, – a plot of seismic wave velocity versus period or frequency. / dispersiecurve / courbe de dispersion / Dispersionskurve / curva de dispersión.

0227
matched filtering, – a filter with the same amplitude- and phase-frequency response as the wave form to be detected.

0228
velocity filtering, – procedure to separate wave trains of different velocity of propagation, e.g. a seismic body wave and microseisms. / snelheidsscheiding / séparation des vitesses / Geschwindigkeitsfilterung / filtrado de velocidades

0229
group-velocity method, – determination of the surface wave dispersion for seismic wave groups to study the average structure along the wave path from epicentre to station. / groepssnelheidmethode / méthode des vitesses de groupe / Gruppengeschwindigkeitmethode / método de velocidad de grupo.

0230
phase-velocity method, – procedure to determine crustal structure by measuring the phase velocity of waves of different periods between two or more seismic stations. / fasesnelheidmethode / méthode de vitesse de phase / Phasengeschwindigkeitsmethode / método de velocidad de fase

0231
source window , – the area at such a distance from the epicent re that the direct wave signals are least distorted by the followed trajectory. / bronvenster / fenètre de la source / teleseismischer Fenster / – –

0232
earthquake mechanism; focal mechanism, – the source motion of an earthquake as determined by recorded characteristics of seismic wave radiation. / aardbevingsmechanisme; haardmechanisme / mécanisme au foyer / Erdbebenmechanismus; Herdmechanismus / mecanismo de los terremotos

0233
focal sphere, – a reference sphere around the focus of an earthquake used for focal mechanism studies. / focale bol / sphère focale / Herdkugel / esfera focal

0234
nodal plane, – a plane through the earthquake focus in which no energy of the longitudinal wave kind is radiated, but where transverse wave energy is at a maximum. / knoopvlak / plan nodal; surface nodale / Knotenfläche / plano nodal; superficie nodal

nodal line, – the projection on a map of the lines of intersection of the two nodal planes with the focal sphere. / knooplijn / ligne nodale / Knotenlinie / línea nodal

0235
null vector; B-axis, – in studies on earthquake mechanism used for the line of intersection of the two nodal planes for longitudinal waves in the earthquake focus. / nulvector; B-as / vecteur nul; axe B / B-Achse / eje B

0236
polarization, – direction of vibration in a transverse wave, used in focal mechanism studies. / polarisatie / polarisation / Polarisation / polarización

0237
Z-phenomenon, – the possible time lag (a few seconds or less) between the generation of P- and S- waves from an earthquake focus, once thought to be significant.

0238
fault plane, – The surface along which fault motion takes place. / breukvlak / plan de faille / Bruchfläche; Verschiebungsebene / plano de falla

deformation plane; displacement plane; plane of action, – the plane perpendicular to the fault plane and parallel to the slip direction. / deformatievlak; vlak van actie / plan de déformation / Deformationsfläche; Formungsebene / plano de deformación

auxiliary plane, – the plane perpendicular to the fault plane and to the slip direction. Its position is determined from the initial motion directions of seismic body waves. / hulpvlak / plan auxiliare / Hilfsebene / plano auxiliar

0239
fault propagation; rupture velocity, – velocity with which the rupture propagates along the fault plane. / breukvoortplanting / propagation de faille; vitesse de rupture / Bruchfortpflanzungsgeschwindigkeit / propagación de ruptura; velocidad de ruptura

0240
duration, – of the duration of the motion in the focus. / duur / durée / Dauer / duración

0241
residual; residual error, – the difference between the measured and the computed value, e.g. travel-time residual (O – C = observed time – computed time). / residu / résidu / Residuum / error residual

[0326] *Seismicity*

0242
seismic activity; seismicity, – measure of frequency and intensity of earthquakes per unit of time and area. / seismiciteit / séismicité / Seismizität / actividad sísmica

specific seismicity, – the square root of the energy per unit area and unit time released by the earthquakes of a given region. / specifieke seismiciteit / séismicité spécifique / spezifische Seismizität / sismicidad específica

0243
intensity; seismic strength, – the strength of an earthquake measured or estimated on the basis of macroseismic effects on structures and human beings. / intensiteit; sterkte / intensité / Intensität / intensidad

intensity scale, – standard of relative measurement of earthquake intensity. / intensiteitsschaal; sterkteschaal / échelle d'intensité/ Intensitätsskala / escala de intensidad

modified Mercalli scale; MM-scale, – a seismic intensity scale.

Medvedev/Sponheuer/Karnik scale; MSK-scale, – a seismic intensity scale.

0244
seismic zoning map, – map showing the maximum intensities that are likely to occur per area. / seismische zoneringskaart / carte des zones sismiques / seismische Zonenkarte / mapa de zonas sísmicas

0245
isoseism; isoseismal line; isoseismal ; isoseist; isoseismic line, – line on the earth's surface connecting points of equal earthquake intensity. / isoseiste; isoseismische lijn / isosiste; ligne isosiste / Isoseiste / línea isosísmica; isosista

0246
pleistoseismal line; innermost isoseismal, – isoseism surrounding the area of maximum strength of the earthquake. / pleistoseiste / pleistosiste; ligne pleistosiste / Pleistoseiste / pleistosista; línea pleistosísmica; línea plistosística

meizoseismal area, – region within the innermost isoseismal where the earthquake had its maximum destructive force. / pleistoseismisch gebied / – – / – – / región pleistosista

0247
seismic region, – area in which earthquakes are frequent. / seismisch gebied / région sismique / seismische Region / región sísmica

0248
macroseismic region, – area in which the earthquake is felt by human beings. / macroseismisch gebied / région macrosismique / makroseismisches Gebiet / región macrosísmica

0249
megaseismic region, – the most heavily disturbed area by an earthquake. / megaseismisch gebied / région mégasismique / megaseismisches Gebiet / región megasísmica

0250
malloseismic region, – an area that is likely to be visited several times in a century by destructive earthquakes. / malloseismisch gebied / région mallosismique / malloseismische Region / región mallosísmica

0251
microseismic region, – area in which the earthquake is recorded by instruments without being felt by human beings. / microseismisch gebied / région microsismique / mikroseismisches Gebiet / región microsísmica

0252
peneseismic region, – area in which earthquakes are not frequent. / peneseismisch gebied / région pénésismique / peneseismische Region / región penesísmica

0253
aseismic region, – area that is not subject to earthquakes. / aseismisch gebied / région asismique / aseismisches Gebiet / región asísmica

[0327] *Engineering Seismology*

0254
engineering seismology; earthquake engineering,– the study of the behaviour of foundations and structures relative to seismic ground motion, and the attempt to mitigate the effect of earthquakes on such structures. / ingenieursseismologie / séismologie de l'ingénieur / Ingenieurseismologie / ingeniería sismológica

0255
earthquake risk; seismic risk, – the risk of damage or loss of lives by earthquakes. / aardbevingsrisico / risque des tremblements de terre / Erdbebenrisiko / riesgo de terremoto

regionalization, – seismic risk classification applied to a certain area. / regionalisatie / régionalisation / Regionalisierung / regionalización

0256
building code, – rules for the way of building structures, houses and bridges, etc. in seismic areas. / bouwcode / code de construction / Baunormen / código de construcción

0257
seismic constant, – the horizontal acceleration in fractions of the accelleration of gravity (g) that a building has to withstand to prevent earthquake hazards. / seismische constante / constante sismique / seismische Konstante / constante sísmica

0258
foundation coefficient, – coefficient expressing the ratio of the effect of an earthquake in a given rock to that in undisturbed crystalline rock under the same conditions. / ondergrondcoëfficient / coefficient de sous-sol / Untergrundskoeffizient / coeficiente de cimentación

0259
ground factor, – a factor depending on the type and structure of the ground and the foundation of a structure that is used in earthquake engineering. / grondfactor / – – / Grundfaktor / factor de suelo

0260
earthquake-resistant design, – the design of earthquake-proof buildings and structures. / aardbevingsbestendig bouwplan / plan de construction antisismique / Erdbebensichere Konstruktion / proyecto antísísmico

earthquake-proof construction, – building of sufficiently strong construction to withstand even heavy shocks. / aardbevingsvrij bouwwerk / construction antisismique / Erdbebensichere Bauwerk / construcción antisísmica

0261
earthquake prediction, – the branch of seismology dealing with the physical conditions or indications that precede earthquakes, in order to extrapolate the size, time and location of the impending shock. / aardbevingsvoorspelling / prévision de séisme / Erdbeben-Vorhersage / predicción de terremotos

0262
earthquake control; earthquake regulation, – stimulation of small-scale fault rupture, e.g. by means of injection of water in the fault plane to prevent an earthquake caused by large-scale fault rupture. / aardbevingsregulatie / séisme contrôlé / Erdbebenkontrolle / control de terremoto

earthquake prevention, – aardbevingpreventie / prévention des séismes / Erdbebenverhütung / prevención de terremotos

[0328] *Nuclear and Explosion Seismology*

0263
nuclear seismology, – seismology of nuclear explosions. / nucleaire seismologie / séismologie nucléaire / Nuklearseismologie / sismología nuclear

0264
forensic seismology, – branch of seismology dedicated to the detection of violations of a test-ban treaty and to the prevention of false alarms. / – – / – – / forensische Seismologie / sismología legal

0265
artificial earthquake, – resulting from industrial or traffic disturbances, explosions, etc. / kunstmatige aardbeving / tremblement de terre artificiel / künstliches Erdbeben / terremoto artificial

0266
test-ban treaty, – treaty to be negotiated between nuclear powers to ban all experimental nuclear explosions. / – – / – – / Kernwaffen-Teststopvertrag / tratado de limitación de pruebras nucleares

threshold treaty, – treaty to ban all nuclear explosions with a yield higher than a stipulated amount. / drempelverdrag / traité limité / Schwellenvertrag / tratado de valor umbral

CTB treaty, – comprehensive test-ban treaty

0267
on-site inspection, – term used in the negotiations on a treaty to ban nuclear test explosions. / plaatselijke inspectie / inspection in situ / Ortsinspektion / inspección in situ

test site, – place for testing nuclear explosions as existing in various countries.

0268
nuclear explosion, – considered to be the source of a seismic signal that should be detected and identified. / kernexplosie / explosion nucléaire / Kernexplosion / explosión nuclear

underground nuclear explosion, – a task for the seismologist is to discriminate between the signal from such an event and the signal from a natural earthquake. / ondergrondse kernexplosie / explosion nucléaire sousterraine / unterirdische Kernexplosion / explosión nuclear subterránea

0269
false alarm, – a seismic event wrongly identified as an underground explosion. / vals alarm / fausse alerte / Falschalarm / falsa alarma

0270
detection capability, – the capability in magnitude of a station or group of stations to discriminate between a natural and an artificial seismic event. / detectievermogen / capacité de détection / Detektionskapazität / capacidad de detección

detection threshold, – the minimum magnitude of a natural or artificial seismic event that can be detected by a station or group of stations. / detectiedrempel / seuil de détection / Detektionsschwelle / umbral de detección

0271
identification capability, – the capability in magnitude of a station or group of stations to discriminate between a natural earthquake and an explosion. / identificatievermogen / capacité d'identification / Identifikationskapazität / capacidad de identificación

identification threshold; threshold of identification,– the minimum magnitude of a seismic event that can be identified as either a natural earthquake or an explosion. / identificatiedrempel / seuil d'identification / Identifikationsschwelle / umbral de identificación

0272
yield, – energy radiation from an explosion, expressed in terms of equivalent, kilotons or megatons of TNT explosive. / opbrengst; rendement / puissance; rendement / Ladungsstärke / rendimiento

0273
coupling, – measure for the transfer of explosive energy into seismic energy, being strongly dependent on the type of the explosion and of that of the medium. / koppeling / couplage / Kopplung / acoplamiento

decoupling, – effort to decrease the fraction of the total explosion energy that is converted into seismic wave energy, e.g. by firing the explosion inside a cavity. / ontkoppeling / découplage / Entkopplung / desacoplamiento

evasion, – procedure to lower the strength of the seismic signal from an explosion by firing in a cavity. / ontduiking / évasion / Umgehung / evasión

0274
shock wave; – supersonic wave in the near field of an explosion. / schokgolf / onde de choc / Stosswelle / onda de choque

near field, – the area in which the shape of the signal is influenced by the dimensions of the source relative to the wave length of the radiated energy.

0275
complexity; wave ...., – measure of the non-simplicity of a seismic signal. Especially used in the field of discrimination between earthquake and explosion signals. / complexiteit / complexité / Komplexität / complejidad

[0330] GEOMAGNETISM AND ROCK MAGNETISM

[0331] *The Earth's Magnetic Field*

NOTE:
– for additional terms see section:
[2860] Magnetometrical and Gravimetrical Exploration Methods

0276
geomagnetism, – magnetism of the earth. / geomagnetisme; aardmagnetisme / géomagnetisme; magnétisme terrestre / Geomagnetismus; Erdmagnetismus / geomagnetismo

geomagnetic field, – magnetic field of the earth. / aardmagneetveld / champ magnétique terrestre / Erdmagnetfeld / campo magnético terrestre

0277
magnetosphere, – space around the earth where the geomagnetic field is present. / magnetosfeer / magnétosphère / Magnetosphäre / magnetosfera

magnetopause, – outer boundary of the magnetosphere. / magnetopause / magnétopause / Magnetopause / magnetopausa

magnetosheath, – zone of transition between the magnetosphere and the interplanetary space. / magnetolaag / – – / – – / – –

0278
dynamo theory, – statement that the geomagnetic field is thought to be sustained by self-exciting dynamo action in the earth's fluid core. / dynamotheorie / théorie dynamomagnétique / Dynamotheorie / teoría dinamomagnética

magnetohydrodynamics, – relationship between magnetic fields and movements in an electrically conducting fluid. / magnetohydrodynamica / théorie hydromagnétique / Magnetohydrodynamik / magnetohidrodinámica

0279
axial dipole field, – dipole field with its axis along the earth's rotation axis. / axiaal dipoolveld / champ d'un dipôle axial / axiales Dipolfeld / campo dipolar axial

eccentric dipole field, – an approximation of the earth's magnetic field by a dipole not coinciding with the centre of the earth. / excentrisch dipoolveld / champ d'un dipôle excentrique / excentrisches Dipolfeld / campo dipolar excéntrico

0280
geomagnetic axis, – axis of the centric dipole field approximating the actual geomagnetic field. / geomagnetische as / axe géomagnétique / geomagnetische Achse / eje géomagnético

geomagnetic pole; axis pole,– one of the two points, where the axis of a centric dipole intersects the surface of the earth. / geomagnetische pool / pôle géomagnetique / geomagnetische Pole / polo geomagnético

geomagnetic latitude, – latitude relative to the geomagnetic equator. / geomagnetische breedte / latitude géomagnetique / geomagnetische Breite / latitud geomagnética

0281
geomagnetic equator, – great circle of the earth whose plane is perpendicular to the geomagnetic axis. / geomagnetische equator / équateur géomagnétique / geomagnetischer Äquator / ecuador geomagnético

0282
magnetic pole; dip-pole, – one of the two points on the earth's surface where its magnetic field is directed vertically downward or upward. / magnetische pool / pôle magnétique / magnetischer Pole / polo magnético

0283
magnetic inclination; magnetic dip, – angle between the magnetic field vector and the horizontal plane. / magnetische inclinatie / inclinaison magnétique / magnetische Inklination / inclinación magnética

isoclinic line, – a contour of equal magnetic dip or inclination. / isocline / isocline / Isokline / isoclina

magnetic equator, – line of zero inclination. / magnetische equator / équateur magnétique / magnetischer Äquator / ecuador magnético

0284
magnetic declination, – angle between geographic meridian and direction of magnetic north. / magnetische declinatie / déclinaison magnétique / magnetische Deklination / declinación magnética

isogonic line, – contour of equal magnetic declination. / isogoon / isogone / Isogon / isogónica

agonic line, – contour connecting points of zero magnetic declination. / agoon / agone / Agon / agónica

0285
isodynamic line; isodynamic, – contour of equal magnetic field strength, either its total value or its horizontal and vertical components. / isodyname / isodyname / Isodyname / isodinámica

0286
magnetic anomaly, – deviation of a magnetic element from normal value. / magnetische anomalie / anomalie magnétique / magnetische Anomalie / anomalía magnética

0287
isoporic line, – contour of equal rate of change of a magnetic element. / isopore / isopore / Isopor / isópora

[0332] *Variations of the Magnetic Field*

0288
variometer, – instument for measuring temporal variations of the geomagnetic field. / variometer / variomètre / Variometer / variómetro

0289
secular variation, – long period changes of the geomagnetic field. / seculaire variatie / variation séculaire / Säkularvariation / variación secular

0290
reversed polarity, – polarity of the geomagnetic field that is opposite to the present direction. / omgekeerde polariteit / polarité inversée / reverse Polarität / polaridad invertida

0291
geomagnetic reversal, – change of the earth's magnetic field between normal and reversed polarity. / geomagnetische ompoling / inversion géomagnétique / geomagnetische Umkehrung / inversión geomagnética

0292
polarity epoch, – geological period of time during which the geomagnetic field was predominantly or entirely of one polarity. / polariteitstijdperk / époque de polarité / Polaritätsepoche / época de polaridad

polarity event, – geologically short period during which the geomagnetic field was opposite to the predominant polarity of that epoch. / polariteitsinterval / événement intercalaire / Polumkehr-Ereignis / suceso de inversión polar

[0333] *Magnetic Properties of Rocks*

0293
rock magnetism, – magnetic properties of rocks. / gesteentemagnetisme / aimantation des roches / Gesteinsmagnetismus / litomagnetismo

0294
palaeomagnetism, – magnetization of rocks acquired in non-recent geological history. / palaeomagnetisme / paléomagnétisme / Paläomagnetismus / paleomagnetismo

archaeomagnetism, – remanent magnetism of rocks acquired in the archaeologic past. / archaeomagnetisme / archéomagnétisme / Archäomagnetismus / arqueomagnetismo

0295
magnetization; intensity of ...., – the magnetizing force that gives rise to the flux density within a magnet or magnetizable body, i.e. the magnetic moment per unit volume. Unit: ampère per metre, A m^{-1}; symbol: I. / magnetisatie / aimantation / Magnetisierung / imanación

pole strength, – the intensity of magnetization at the end faces of a magnet. Unit: magnetization times area, A m; symbol: m.

0296
magnetic (dipole) moment, – of materials, a vector quantity with magnitude of pole strength times length of the magnet or intensity of magnetization times volume of the magnet; the direction is from the negative to the positive pole, i.e. the lines of force emerge at the north pole and reenter at the

south pole, imagining a constant flow (flux) of the magnetic force through the magnet and the area surrounding it. Unit: ampère times metre squared, A m^2; symbol: M. / magnetisch (dipool)moment / moment magnétique / magnetisches (Dipol-)Moment / momento magnético

0297
magnetic polarization, – the magnetization of materials induced by an external field. Unit: ampère per metre, A m^{-1}; symbol: H. / magnetische polarisatie / polarisation magnétique / magnetische Polarisierung / polarización magnética

saturation magnetization, – the maximum value of magnetization. / verzadigingsmagnetisme / saturation magnétique / Sättigungsintensität / imanación de saturación

0298
magnetic susceptibility, – the measure of the ease of magnetization of a body when placed in a magnetic field, i.e. the ratio of magnetization to inducing field strength. Symbol: k. / magnetische susceptibiliteit / susceptibilité magnétique / magnetische Suszeptibilität / susceptibilidad magnética

diamagnetism, – magnetism of material with negative susceptibility, i.e. the south poles are developed in the direction of the inducing field. / diamagnetisme / diamagnétisme / Diamagnetismus / diamagnetismo

paramagnetism, – magnetism of material with positive susceptibility, i.e. the north poles are developed in the direction of the inducing field. / paramagnetisme / paramagnétisme / Paramagnetismus / paramagnetismo

0299
ferromagnetism, – very strong paramagnetism. / ferromagnetisme / ferromagnétisme / Ferromagnetismus / ferromagnetismo

0300
superparamagnetism, – paramagnetism of extremely small particles of ferromagnetic minerals. / superparamagnetisme / superparamagnétisme / Superparamagnetismus / superparamagnetismo

0301
antiferromagnetism, – property of some ferromagnetic materials in which moments of neighbouring magnetic ions are aligned antiparallel. / antiferromagnetisme / antiferromagnétisme / Antiferromagnetismus / antiferromagnetismo

0302
ferrimagnetism, – antiferromagnetism of materials with two kinds of magnetic ions, in which one kind of ions predominates, so that the resulting magnetization resembles ferromagnetism. / ferrimagnetisme / ferrimagnétisme / Ferrimagnetismus / ferrimagnetismo

0303
magnetic anisotropy, – property of materials that causes different susceptibility in different directions. / magnetische anisotropie / anisotropie magnétique / magnetische Anisotropie / anisotropía magnética

0304
magnetic permeability, – of a medium, the ratio of flux density and magnetizing force acting upon the medium. / magnetische permeabiliteit / permeabilité magnétique / magnetische Permeabilität / permeabilidad magnética

0305
magnetic hysteresis, – lagging of magnetization or induction behind a variable magnetic field. / magnetische hysterese / hystérèse magnétique / magnetische Hysterese / histéresis magnética

0306
remanent magnetization; residual ...., – magnetization that remains after removing the inducing magnetic field. / remanent magnetisme / aimantation rémanente / remanenter Magnetismus / imanación remanente

0307
coercivity; coercive force, – magnetic field necessary to remove magnetization. / coërcitiefkracht / coercivité / Koerzitivkraft / coercitividad

0308
Curie point, – temperature at which ferromagnetic material looses its magnetization. / Curiepunt / point de Curie / Curiepunkt / punto de Curie

0309
thermoremanent magnetization, – magnetization acquired by cooling through the Curie point. / thermoremanent magnetisme / aimantation thermorémanente / thermo-remanenter Magnetismus / imanación termoremanente

0310
Koenigsberger ratio, – ratio of remanent magnetization to magnetization induced by the earth's field. / Koenigsberger verhouding / Koenigsberger Q-Faktor / rapport de Koenigsberger / cociente de Koenigsberger / relación Koenigsberger

0311
self-reversal, – natural remanent magnetization opposite to the ambient magnetic field. / zelf-omkering / auto-inversion / Selbstumkehr / auto-inversión

0312
depositional remanent magnetization, – resulting from orientation of ferromagnetic mineral grains along the ambient field during sedimentation. / afzettingsmagnetisatie / aimantation rémanente de dépot / Sedimentations-Remanenz / imanación remanente de deposición

chemical magnetization, – magnetization formed by growth of ferromagnetic grains in a magnetic field. / chemische magnetisatie / aimantation chimique / chemische Magnetisierung / imanación química

0313
viscous magnetization, – semi-stable magnetization that behaves as stable during a time longer than a laboratory measurement. / viskeuze magnetisatie / aimantation visqueuze / viskose Magnetisierung / imanación viscosa

0314
anhysteretic remanent magnetization, – produced by simultaneous application of a constant field and an alternating field whose amplitude decreases to zero. / anhysteretische magnetisatie / aimantation anhysterétique / anhysteretische Remanenz / imanación anhisterética

0315
reversed magnetization, – magnetization that is opposed to the present geomagnetic field. / omgekeerde magnetisatie / aimantation inversée / umgekehrte Magnetisierung / imanación invertida

[0334] *Magnetic Measurements*

0316
magnetometer, – instrument to measure the geomagnetic field or the field of a rock sample. / magnetometer / magnétomètre / Magnetometer / magnetómetro

0317
magnetizing force, – the cause of magnetic induction, i.e. the field of force due to electric currents or magnets. Unit: ampère per metre, A m^{-1}; symbol: H. / magnetiserende kracht / force magnétisante / magnetisierende Kraft / fuerza magnetisanda; fuerza imantanda

magnetic flux, – a measure of the magnetic field. Unit: weber, Wb (= volt sec). / magnetische flux / flux magnétique / magnetischer Fluss / flujo magnético

magnetic flux density; magnetic field strength, – magnetic flux per unit area. Unit: Wb m^{-2} (Oersted, obsolete, in c.g.s.); symbol: B. / magnetische veldsterkte / intensité de champ magnétique / Feldintensität; Feldstärke / densidad de flujo magnético

magnetic induction, – a measure of the field strength within a magnetized body. Unit: tesla, T = Wb m^{-2} (gauss in c.g.s.); symbol: B. / (magnetische) inductie / induction (magnétique) / magnetische Induktion / inducción magnetica

0318
gamma; gauss;– the practical unit of geomagnetic field strength; 10^{-4} gamma = 1 tesla (obsolete).

0319
astatic magnetometer, – instrument to measure magnetic fields; it is almost insensitive for external magnetic fluctuations. / astatische magnetometer / magnétomètre astatique / astatisches Magnetometer / magnetómetro astático

0320
spinner magnetometer, – magnetometer that rotates the specimen whose remanent magnetization is measured, to produce an alternating voltage.

0321
AC demagnetization, – partial demagnetization by an alternating magnetic field. / wisselveld demagnetisatie / désaimantation par champ alternatif / Entmagnetisierung durch Wechselfeld / desimanación por C.A.

thermal demagnetization, – loss of magnetization by a rise of temperature. / thermische demagnetisatie / désaimantation thermique / thermische Entmagnetisierung / desimanación térmica

[0340] GRAVITY AND ISOSTASY

NOTE:
– for additional terms see section:
[2860] Magnetometrical and Gravimetrical Exploration Methods

[0341] *Gravity*

0322
gravitation, – mutual mass attraction of bodies. / gravitatie / gravitation / Gravitation; Anziehung / gravitación.

0323
force of gravity, – / zwaartekracht / pesanteur / Schwerkraft / fuerza de gravedad.

0324
acceleration of gravity, – / versnelling van de zwaartekracht / accélération de la pesanteur / Schwerebeschleunigung / aceleración de la gravedad

gravity, – short for acceleration of gravity. / zwaartekracht / gravité; intensité de la pesanteur / Schwere; Schwerkraft / gravedad

gal, – unit of gravitational field strength (named after Galileo), recognition of which is abrogated after introduction of the SI system; 1 gal = 1 $cm{\cdot}s^{-2}$.

0325
gravitational constant, g. – gravitatieconstante / constante de la gravitation / Gravitationskonstante / constante de gravitación

0326
gravity field, – / zwaartekrachtsveld / champ de pesanteur / Schwerefeld / campo de la gravedad

0327
gravity potential, – potential belonging to the earth's gravity field. / zwaartekrachtspotentiaal / potentiel de la gravité / Schwerepotential / potencial de la gravedad

0328
geoid, – equipotential surface of the earth at mean sea level. / geoïde / géoide / Geoid / geoide

0329
plumbline, – / schietlood / fil à plomb / Lot; Lotlinie / plomada

vertical, – direction of the plumbline. / verticaal / verticale / Lotrichtung / vertical

deflection of the plumbline, – / schietloodafwijking / déviation de la verticale / Lotabweichung / desviación de la plomada; .... de la vertical

0330
spheroid, – used in geodesy for a surface of revolution closely approaching an ellipsoid of revolution of slight flattening; the equilibrium surface of a fluid earth is a spheroid. / sferoïde / sphéroïde / Sphäroid / esferoide

flattening, – of the earth. / afplatting / aplatissement / Abplattung / achatamiento; aplastamiento

0331
international ellipsoid, – ellipsoid of reference accepted in the assembly of the Int. Union of Geodesy and Geophysics in Madrid in 1924 as an adequate approximation of the geoid; it may be considered as a close approximation of the equilibrium figure of a fluid earth. / internationale ellipsoïde / ellipsoïde international / internationales Ellipsoid / ellipsoide internacional

0332
normal value of gravity, – value of gravity upon the international ellipsoid, according to the international formula accepted by the 1930 Assembly of the International Union of Geodesy and Geophysics in Stockholm, i.e. the gravity exerted by a 'normalized' earth that is defined by having the same mass as the real earth and an equipotential surface coinciding with the international ellipsoid which encloses all masses except that of the atmosphere. / normaalwaarde van de zwaartekracht / valeur normale de la gravité / Normalschwere / valor normal de la gravedad

0333
international formula 1930 for normal gravity, – formula giving the normal value of gravity. / internationale formule van 1930 voor de normaalwaarde van de zwaartekracht / formule internationale de 1930 pour la valeur normale de la pesanteur / internationale Formel 1930 für die Normalschwere / formula internacional de 1930 para el valor normal de la gravedad

0334
Potsdam system; .... value,– (system of) gravity values based on the absolute gravity determination at Potsdam ($g = 981\ 274\ cm \cdot s^{-1}$). / Potsdam-systeem; ....-waarde / système de Potsdam; valeur de .... / Potsdamer Schweresystem; .... Schwerewert / sistema Potsdamiano; valor ....

0335
gradient of gravity, – partial derivative, with respect to distance, in a horizontal direction of the acceleration of gravity, and where the acceleration of gravity is considered a scalar. / gradiënt van de zwaartekracht / gradient de la gravité / Schweregradient / gradiente de la gravedad

0336
curvature value, – quantity, determined by the torsion balance, that is related to the second derivative of the gravity potential with respect to the horizontal co-ordinates. / krommingswaarde / valeur du moment / Krümmungsgrösze / valor de curvatura

0337
Eötvös unit (obsolete), – practical unit of second derivative of gravity potential; $1\ E = 10^{-9}\ s^{-2} = 10^{-9}\ gal\ cm^{-1}$.

[0342] *Isostasy*

0338
isostasy, – phenomenon that the masses between the earth's surface and mean sea level in land areas, and the mass deficiencies between sea level and sea floor in water-covered areas, are in general compensated by masses deeper in the earth with reversed sign. /isostasie / isostasie / Isostasie / isostasía.

0339
isostatic compensation, – deeper mass compensating the topographic mass. /isostatische compensatie / compensation isostatique / isostatische Kompensation / compensación isostática

0340
local (isostatic) compensation, – situation in which the isostatic compensation is located vertically below the topographic element. / lokale compensatie / compensation locale / lokale Kompensation / compensación (isostática) local

regional (isostatic) compensation, – situation in which the isostatic compensation of a topographical element is extended in a horizontal sense over a wider area. / regionale compensatie / compensation régionale / regionale Kompensation / compensación (isostática) regional

degree of regionality, – of the regional isostatic compensation; symbol: R. / graad van regionaliteit / degré de régionalité / Grad des Regionalfeldes / grado de regionalidad

0341
compensation mass, – mass forming the isostatic compensation. / compensatiemassa / masse compensatrice / Kompensationsmasse / masa compensadora

0342
isostatic mass compensation, – the condition of topographic and compensation masses being equal. / isostatische massacompensatie / compensation isostatique des masses / isostatischer Massenausgleich / compensación isostática de masas

0343
deficiency of mass, – / massatekort / déficit de masse / Massendefizit / déficit de masa

excess of mass, – massaoverschot / excès de masse / Massenüberschuss / exceso de masa

root, – excess of crustal material below mountains. / wortel / racine / Wurzel / raiz

0344
isostatic pressure compensation, – the condition of pressures at a certain depth in the earth being equal, independent of the overlying topography. / isostatische drukcompensatie / compensation isostatique des pressions / isostatischer Druckausgleich / compensación isostática de presiones

0345
compensation level; .... depth, –
1. in **Pratt's hypothesis**: level from which the compensation is thought to have been realized (in upward sense);
2. in **Airy's hypothesis**: incorrect term for thickness of crust of light material for zero topography.

/ compensatieniveau; ....diepte / niveau de compensation; profondeur de .... / Ausgleichsfläche; ....tiefe / nivel de compensación; profundidad de ....

0346
isostatic equilibrium, – the hydrostatic or floating equilibrium of the rigid crust on the plastic substratum (Airy hypothesis). / isostatisch evenwicht / équilibre isostatique / isostatisches Gleichgewicht / equilibrio isostático

0347
thickness of the M-crust, – the part of the rigid crust above the Mohorovičić discontinuity./ dikte van de korst / épaisseur de l'écorce / Mächtigkeit der Kruste / espesor de la corteza

0348
rigid crust, – lithosphere, as differing from the M-crust. / starre korst / croûte rigide / starre Kruste / corteza rígida

substratum, – the material below the M-crust, not identical to asthenosphere. / substratum / substratum / Substrat / substrato.

[0343] *Gravimetry*

0349
gravimetry, – / gravimetrie / gravimétrie / Gravimetrie / gravimetría

gravity determination; .... measurement,– / zwaartekrachtsbepaling; ....meting / measure de la pesanteur / Schweremessung / determinación de la gravedad; medición de gravedad

0350
torsion balance, – instrument for measuring the gradient and some other second order derivatives of the earth's potential. / torsiebalans / balance de torsion / Drehwaage / balanza de torsión

0351
pendulum, – / slinger / pendule / Pendel / péndulo

reversible pundulum, – pendulum that can swing around two knife edges placed in such a way that their periods are equal; used for absolute gravity determinations. / reversieslinger / pendule réversible / Reversionspendel / péndulo reversible

0352
gravimeter; gravity meter, – instrument for measuring differences in acceleration of gravity. / gravimeter / gravimètre / Gravimeter / gravímetro

null-reading type of gravimeter, – gravimeter in which the beam is always brought back to its initial position by means of a contrivance provided with a dial. / nulmeter / gravimètre basé sur la méthode zéro / Gravimeter mit Nullmethode / gravímetro con indicator cero

0353
gravity station, – / zwaartekrachtstation / station gravimétrique / Schwerestation / estación gravimétrica

base station, – station belonging to a wide grid of stations, where the acceleration of gravity is determined with particular care and to which the field stations are tied in. / basisstation / station de base / Basis-Station / estación de base

0354
level of reference, – / referentieniveau / surface de référence / Bezugsniveau; ....fläche / nivel de referencia

0355
latitude correction; north-south ....; planetary ..., – correction for the effect of latitude on the second order derivative of the potential of gravity of a normalized earth. / breedtecorrectie / correction de latitude / (geographische) Breitenkorrektur / corrección de latitud

0356
topographic correction, – correction for the effect on gravity of the topography above and / or below sea level. / topografische correctie / correction de plateau / topographische Reduktion ; Geländekorrektur / corrección topográfica

0357
elevation correction, – free air correction; with some authors elevation correction ambiguously means free air plus Bouguer correction. / hoogtecorrectie / correction d'altitude / Höhenreduktion / corrección por altitud

datum plane, – plane to which the elevation corrections are referred. / referentieniveau / niveau de référence / Bezugsniveau / plano de referencia

0358
free-air reduction; .... correction, – correction for elevation supposing that there is no mass between station and surface of reference, as if in "free air"; in marine gravimetry the same correction combined with two times the attraction of an infinite slab of water of a height equal to the distance between the apparatus and sea level. / vrijeluchtreductie; ....correctie / correction à l'air libre; reduction .... / Freiluftreduktion / reducción de aire libre; corrección de ....

0359
Bouguer correction; .... reduction, – correction that allows roughly for the effect of the mass between the earth's surface and datum plane (generally the geoid) – or in water-covered areas of the effect of the deficiency of mass between sea level and sea floor. The mass or mass deficiency is assumed to be

identical with an infinite horizontal plate. / Bouguercorrectie; ....reductie / correction de Bouguer / Bouguer-Reduktion; ....-Korrektur / corrección de Bouguer; reducción de ....

Bouguer plate, – the attraction of a Bouguer plate, an infinite plate, is independent of the distance to the gravity station. / Bouguerplaat / couche plane de Bouguer / Bouguerplatte / lámina de Bouguer

0360
terrain correction, – a term commonly used for the effect of topography after application of the Bouguer reduction, sometimes for that part only that is quite near to the station. / terreincorrectie / correction de terrain; .... topographique / Geländereduktion / corrección de terreno

0361
isostatic correction; .... reduction, – correction to allow for the effect of isostasy; in order to apply this reduction it is necessary to make first certain suppositions with regard to the distribution of the compensation masses: the Pratt hypothesis, or nowadays generally the Airy hypothesis (local or regional). / isostatische reductie / réduction isostatique / isostatische Reduktion / reducción isostática; corrección isostática

0362
Hayford zone, – subdivision of the globe in zones for enabling the calculation of the topographic and isostatic reductions around a gravity station (after Hayford). / zone van Hayford / zone de Hayford / Hayfordzone / zona de Hayford

zone, – area between two concentric circles on a sphere. / zone / zone; couronne / Zone / zona

compartment, – radial subdivision of a zone. / compartiment / compartiment / Kompartiment / compartimiento

0363
Bullard's method, – method for computing the effect of topography for Hayford zones A-O, in which first the effect of the spherical cap of height equal to the station height is calculated and after that the effect of the topographical deviations from this cap. / methode van Bullard / méthode de Bullard / Methode von Bullard / método de Bullard

spherical cap, – in gravimetry; part of a spherical shell limited by a circular cone with the apex in the centre of the sphere. / bolkap / calotte sphérique / Kugelkalotte / capa esférica; casquete esférico

0364
topo-isostatic reduction, – combination of topographic and isostatic reduction as usually applied to Hayford's zones 1-18. / topo-isostatische reductie / correction topo-isostatique / topographische-isostatische Reduktion / reducción topoisostática

0365
tidal (force) correction, – correction for the effect on gravity of sun and moon, including the effect of the tides of the solid earth. / getijcorrectie / correction de l'influence luni-solaire / Gezeitenkorrektur / corrección (de la fuerza) luni-solar

0366
indirect effect, – by the application of the gravity corrections the geoid is deformed; the effect of this deformation on gravity is the indirect effect. / indirect effect / effect indirect / indirekter Effekt / efecto indirecto

0367
gravity anomaly, – difference between theoretical (or normal) and observed value of gravity. / zwaartekrachtsanomalie / anomalie de gravité / Schwereanomalie; ....störung / anomalía de la gravedad

milligal; mgal, – the practical unit for expressing gravity anomalies, whose use is abrogated; equal to 10^{-3} cm s^{-1}.

0368
free-air anomaly, – anomaly after application of the free-air reduction. / vrijeluchtanomalie / anomalie à l'air libre / Freiluftanomalie / anomalía de aire libre

Bouguer anomaly, – strictly: anomaly after application of the Bouguer reduction; also used for anomaly after total topographic reduction for Hayford's zones A-O, or after application of the Bouguer reduction and a terrain correction. / Bougueranomalie / anomalie de Bouguer / Bouguer'sche Anomalie; Bouguer.... / anomalía de Bouguer

0369
isostatic anomaly, – anomaly after application of an isostatic reduction. / isostatische anomalie / anomalie isostatique / isostatische Anomalie; (früher auch: Anisostasie) / anomalía isostática

local isostatic anomaly, – anomaly after application of an isostatic reduction in which the compensation is assumed to be local. / lokale isostatische anomalie / anomalie isostatique locale / lokale isostatische Schwerestörung / anomalía isostática local

regional isostatic anomaly, – anomaly after application of an isostatic reduction in which the compensation is assumed to be regional. / regionale isostatische anomalie / anomalie isostatique régionale / regionale isostatische Anomalie / anomalía isostática regional

0370
modified Bouguer anomaly, – anomaly after application of total topographic reduction for Hayford's zones A-O and topographic and isostatic reduction for Hayford's zones 1-18. / gemodificeerde Bougueranomalie / anomalie de Bouguer modifiée / modifizierte Bouguer Anomalie / anomalía de Bouguer modificada

0371
regional gravity map, – gravity map that shows only the gradual changes of gravity. / regionale zwaartekrachtskaart / carte gravimétrique régionale / Karte der regionalen Schwere / mapa regional de la gravedad

0372
isanomalic (contour) map, – map showing lines that connect points of equal anomaly. / isanomalieënkaart / carte isanomale / Isanomalenkarte / mapa isanómalo; mapa de líneas isanómalas

isogal map, – isanomalic contour map. / isogallenkaart / carte isogale / gravimetrische Isanomalenkarte / mapa isogálico

isogam map, – ambiguous term for isanomalic contour map (borrowed from magnetometry). / isogammenkaart / carte isogamme / Isogammenkarte / mapa isógamo; mapa de líneas isógamas

0373
isostatic isocorrection line map, – map showing lines that connect points of equal isostatic reduction. / isostatische isocorrectielijnenkaart / carte d'isocorrection isostatique / – – / mapa de correcciones isostáticas iguales

[0350] GEOTHERMICS

[0351] *The Earth's Heat*

0374
geothermics, – the branch of geophysics that deals with the internal heat and temperatures of the earth. / geothermie / géothermie / Geothermik / geotermismo; geotérmia

earth('s) heat, – the heat contained in the interior of the earth. / aardwarmte / chaleur terrestre; chaleur de la terre / Erdwärme / calor terrestre; calor de la tierra

0375
radiogenic heat; radioactive ...., – heat formed by radioactive decay; the main heat source in the lithosphere. / radiogene warmte / chaleur due à la radioactivité / radioaktive Wärmeproduktion / calor radioactivo; .... radiogénico

0376
thermal; geothermal, – pertaining to surface and subsurface manifestations of earth's heat, where the temperature is higher than ambient temperature in the surroundings, e.g. thermal springs, ... water, volcanism, plutonism. / geothermisch / géothermique; thermique / geothermal; thermal / geotermal

0377
(geo)isotherm, (geo)isothermic plane, – plane of equal temperature (within the earth). / (geo)isotherm, (geo)isothermvlak / géoisotherme; surface isotherme / Geo-Isotherm / (geo)isoterma; plano (geo)isotérmico; superficie (geo)isotérmica

0378
geotherm, – a function relating temperature to depth within the earth. / geotherm / – – / Temperaturtiefenkurve / función geotermal

0379
geothermometry, – measurement of temperatures within the earth. / geothermische metingen / géothermométrie / Geothermometrie / geotermometría

geothermometer; geochemical thermometer, – originally a mercury thermometer based on the overflow principle and that was designed and used in boreholes; in the present usage geochemical thermometer (in literature often abbreviated to geothermometer) means a number of chemical indicators of subsurface temperatures in relation to the origin of the water in a geothermal system. / geo(chemische)thermometer / géothermomêtre / Geothermometer / – –

thermocouple; thermistor, – high-accuracy thermometers, suitable for use in boreholes, based respectively on the thermo-electric effect in a closed circuit composed of two or more electrical conductors and on the relation between electrical resistance and temperature. /thermokoppel / pyromètre à couple / Thermoelement / termopar; par termoeléctrico

0380
geothermal gradient; temperature gradient, – rate of temperature increase with depth. Unit: K m^{-1} (= °C m^{-1}; still in use: °C/100 m or °C/km). / geothermische gradiënt / gradient géothermique / geothermischer Gradient / gradiente geotérmico

geothermal step; the increase of depth related to 1°C increase of temperature. Unit: m K^{-1} (= m/°C). / geothermisch bedrag / degré géothermique / geothermische Tiefenstufe / salto geotérmico; desnivel geotérmico

0381
mass transfer of heat, – by movement of material of higher temperature than that of the surroundings (plutonism, volcanism, thermal springs). / warmteoverdracht door massatransport; convectief warmtetransport / transfert massique de chaleur / konvektiver Wärmetransport / masa transferente de calor

heat convection, – transfer of heat by the motion of deformable media (gases, liquids). / warmteconvectie / convection de chaleur / konvektiver Wärmetransport / convección de calor; convección calorífica; convección

heat conduction, – transfer of heat by molecular interaction; the main way of heat transfer in rocks. / warmtegeleiding / conduction de chaleur / Wärmeleitung / conducción de calor; conducción calorífica; conducción

heat radiation, – heat transfer by the emission of electromagnetic waves. / warmtestraling / rayonnement de chaleur / Wärmestrahlung / radiación de calor; radiación calorífica

0382
thermal conductivity, – the amount of heat transmitted across unit area in unit time as result of unit temperature gradient. Unit: W m^{-1} K^{-1} (= 2.39 mcal.cm^{-1} sec^{-1} $°C^{-1}$); symbol: K. / warmtegeleidingsvermogen / conductivité thermique / Wärmeleitfähigkeit / conductividad térmica

0383
terrestrial heat flow, – the transfer of heat in the earth, in general directed outward and tending to equalize differences of temperature. / terrestrische warmtestroom / flux thermique terrestre; .... de la terre / irdische Wärmestrom; .... Wärmefluss / flujo térmico terrestre; .... de la tierra

0384
heat-current density; heat flow, – the amount of heat flowing through unit area in unit time; in case of heat transfer by conduction it is expressed as the product of temperature gradient and thermal conductivity: W m^{-2} (=0.0239 mcal cm^{-2} sec^{-1}); symbol: q. /warmtestroomdichtheid /densité de flux thermique / Wärmestromdichte; ....flussdichte / flujo térmico

0385
heat-flow unit; HFU, – the practical unit of heat-current density in the lithosphere: 10^{-3} W m^{-2}.

heat-flow rate, – the amount of heat flowing across a certain surface in unit time: W (= 239 mcal sec^{-1}). / snelheid van warmtestroom / flux thermique / Wärmestrom; ....fluss / caudal térmico

0386
heat capacity; thermal ...., – the amount of heat required to raise the temperature of a body by 1 degree. Unit: J K^{-1} (= $239 \cdot 10^{-3}$ °C^{-1}). / warmtecapaciteit / capacité calorifique / Wärmekapazität / capacidad calorífica

specific heat capacity, – the amount of heat required to raise the temperature of unit mass of a material by 1 degree. Unit: J kg^{-1} K^{-1} (= 239 cal g^{-1} sec^{-1}). / soortelijke warmtecapaciteit / capacité calorifique spécifique / spezifische Wärmekapazität / capacidad calorífica específico

0387
thermal diffusivity, – a measure of the penetration rate of heat by conduction into a medium, expressed as the thermal conductivity divided by the product of specific heat capacity and density of the penetrated material. Unit: $m^2 s^{-1}$; symbol: a. / temperatuurvereffeningscoëffici / diffusivité thermique / Temperaturleitfähigkeit / difusividad térmica

[0352] *Geothermal Energy*

0388
geothermal energy, – energy of economic value derived from the natural heat of the earth. / geothermische energie; nuttige aardwarmte / énergie géothermique / geothermale Energie / energía geotérmica

0389
enthalpy; heat content, – a thermodynamic function, expressing the energy contained in a system with certain volume under certain pressure; in literature on geothermal energy it is often used to denote the amount of heat contained in the amount of produced water or steam. Unit: J (= 0.239 cal). / enthalpie; warmte-inhoud / enthalpie / Wärmeinhalt / entalpía; contenido de calor

high-enthalpy system; high-caloric system; high-temperature system, – the temperature of the produced water or steam is sufficiently high for the generation of electricity, i.e. more than 150°C. /hoog-calorisch veld / système à haute enthalpie; .... énergie / System mit hohem Wärmeinhalt / sistema de alta entalpía; sistema calórico alto; sistema calorífico alto

low-enthalpy sytem; low-caloric system; low-temperature system – the temperature is below 150°C. / laag-calorisch veld / système à basse enthalpie; ... énergie / System mit niedrigem Wärmeinhalt / sistema de baja entalpía; sistema calórico bajo; sistema calorífico bajo

0390
geothermal province; heat-flow province, – a geographic area with similar heat-flow characteristics. / geothermische provincie / province géothermique / geothermische Provinz / provincia geotérmica

geothermal anomaly, – an area with heat-flow characteristics different from those in the surrounding province. / geothermische anomalie / anomalie géothermique / geothermische Anomalie / anomalía geotérmica

0391
hyperthermal region, – the geothermal gradient is greater than 80°C.km^{-1}. / hoog-thermisch gebied / région hyperthermale / hyperthermale Region / región hipertermal

semi-thermal region, – the geothermal gradient is between 40 en 80°C.km^{-1}. / midden-thermisch gebied / – – / Region erhöhten Wärmestroms / región mesotermal

thermally normal region, – the geothermal gradient is smaller than 40°C.km^{-1}. / thermisch normaal gebied / – – / Region durchschnittlichen Wärmestroms / región termal normal; región hipotermal

0392
geothermal field, – an installation where geothermal energy of economic value is produced. /geothermisch veld / champ géothermique / geothermisches Kraftwerk / campo geotérmico

0393
vapor-dominated (geothermal) system, – a geothermal field where dry or superheated steam is produced without associated liquid; in the reservoir liquid and vapor coexist with the latter as the continuous pressure-controlling phase. / natuurlijk stoomveld / système à vapeur dominante / geothermisches Dampfkraftwerk / sistema (geotérmica) de vapor dominante

liquid-dominated (geothermal) system; hot-water system, – a geothermal field where hot water is produced; in the reservoir liquid is the continuous pressure-controlling phase; the water temperature is due either to volcanic or magmatic heat or to ahigh geothermal gradient in a sedimentary basin. / warmwaterveld / système à liquide dominante; .... d' eau chaude / geothermisches Heisswasserkraftwerk / sistema (geotérmica) de líquido dominante; sistema de agua caliente

low-temperature (geothermal) system, – a geothermal field where water of a temperature not higher than about 150°C is produced; it is a sub-type of the liquid-dominated system. / warmwaterveld / système à basse température / geothermisches Heisswasserkraftwerk niedriger Temperatur / sistema (geotérmica) de baja temperatura

0394
hot-dry-rock system; magma system, – an as yet hypothetical possibility to produce geothermal energy by circulating water through artificial cracks in deep crystalline basement rocks or in cooling magmatic bodies; in literature often used to denote possible production through artificial fracturing of impervious water-bearing beds in a sedimentary sequence.

[0400] GEOCHEMISTRY

NOTE:
- this chapter is a general introduction to geochemistry; for details see specialised chapters and sections, e.g.
  [0900] Petrology of Igneous Rocks
  [1000] Petrology of Metamorphic rocks
  [1700] Pedology
  [2300] Diagenesis
  [2420] Coalification
  [2551] Types, Origin of Hydrocarbons
  [2620] Processes of Formation of Metallic Mineral Deposits
  [2890] Geochemical Exploration Methods

0395
geochemistry, – the study of the abundances, the distribution and the migration of the chemical elements in lithosphere, hydrosphere, biosphere and atmosphere. / geochemie / géochimie / Geochemie / geoquímica

biogeochemistry, – study of geochemical activity of organisms and of the regulation of geochemical metabolism by living systems. / biogeochemie / biogéochimie / Biogeochemie / biogeoquímica

Gaia theory, – theory postulating that biosphere, lithosphere, hydrosphere and atmosphere are a single, self-regulating cybernatic system, aimed at the maintenance of optimum conditions for life on earth; named after the impersonification of the earth in Greek mythology.

0396
geochemical cycle, – the description of the circulation of elements through different geological reservoirs, and their chemical behaviour and residence time within these reservoirs. The geochemical cycle is commonly divided into an exogenic part (weathering, erosion, transport, sedimentation, early diagenesis), and an endogenic part (late diagenesis, subsidence, metamorphism, anatexis, and tectonic or magmatic upward movement back to the earth's surface). The enrichment of elements into ore deposits can be considered as a special case within the geochemical cycle. Within each major cycle several shorter subcycles may exist. / geochemische kringloop / cycle géochimique / geochemischer Kreislauf / ciclo geoquímico

0397
residence time, – the timespan during which a particle, taking part in a cycle, remains in a phase of that cycle. In general, the smaller the amount of a substance in a part of a cycle, the smaller its residence time and the greater its rate of flow. / verblijftijd / durée de permanence / Verweilzeit; ....dauer / – –

steady state, – the dynamic equilibrium, in which, invariant with time, equal amounts of a substance enter into or disappear from a system or phase of a cycle. / dynamisch evenwicht /équilibre dynamique / dynamisches Gleichgewicht / equilibrio dinámico

0398
differentiation, – the progressive separation of chemical elements, phases, components or minerals. / differentiatie / différenciation / Differentiation / diferenciación

geochemical differentiation, – the processes that led to the present and will govern the future distribution of elements in the earth; three stages can be recognized:
the primary, primordial or pregeological stage producing the core – mantle – crust differentiation;
the second or endogenic stage including the further differentiation of the silicate crust and largely dealing with magmatic and metamorphic differentiation;
the third or exogenic stage dealing with weathering, sedimentation, etc.
/ geochemische differentiatie / différenciation géochimique / geochemische Differentiation / diferenciación geoquímica

0399
Clarke number; .... value, – the average abundance of an element in the earth's lithosphere (to 16 km depth and composed of 95% igneous rocks and 5% sedimentary rocks). Named in honour of F. W. Clarke. /getal van Clarke / (valeur de) Clarke / Clarke-Wert; ....-Zahl / número de Clarke

0400
trace element, – an element present in very small concentrations, conventionally considered to be less than 0.1% (1000 ppm), by some authors, however, less than 1 or 2%. / sporenelement / élément en trace / Spurenelement / elemento traza

minor element, – an element in concentrations smaller than 1%. / – – / élément mineur / – – / elemento menor

Nebenelement, – in German usage, between 0.1% and 1%; also, however, between e.g. 0.01% and 1% or 0.1% and 10%.

0401
geochemical province, – a relatively large segment of the earth's crust with a distinct chemical composition. / geochemische provincie / province géochimique / geochemische Provinz / provincia geoquímica

geochemical relief, the geographical variation in the level of concentration of the elements, including background values and anomaly peaks, and in the homogeneity of their distribution. / – – / – – / Elementverteilung; geochemisches Relief / relieve geoquímoco

0402
primary environment, – in geochemistry, the environment embracing those areas extending downward from the lower levels of circulating meteoric water to those of the deep-seated processes of igneous differentiation and metamorphism. / primair milieu / milieu primaire / primäre Umgebung / medio primario

secondary environment, – in geochemistry, the environment that comprises the surficial processes of weathering, soil formation and sedimentation at the surface of the earth. / secundair milieu / milieu secondaire / sekundäre Umgebung / medio secundario

0403
dispersion, – the process causing the distribution or redistribution of elements by physical and chemical agents. / dispersie / dispersion / Dispersion / dispersión

dispersion pattern, – the spatial distribution of elements resulting from the dispersion process. / dispersiepatroon / figure de dispersion; auréole de .... / Verteilungsmuster / distribución de la dispersión

0404
primary dispersion, – the distribution or redistribution of elements in the primary environment./ primaire dispersie / dispersion primaire / Primärdispersion / dispersión primaria

primary dispersion pattern, –
1. **syngenetic** – formed essentially contemporaneously with the surrounding rock;
2. **epigenetic** – formed by material introduced into a preexisting rock:
   a. **hydrothermal** – formed by movement of hydrothermal solutions
   b. **pressure-temperature related** – formed by the effect of P-T changes.

0405
secondary dispersion, – the dispersion originating in the secondary environment. / secundaire dispersie / dispersion secondaire / Sekundärdispersion / dispersión secundaria

secondary dispersion pattern, –
1. **syngenetic** – formed essentially contemporaneously with the surrounding rock:
   a. **clastic** – related to in situ weathering and the movement of clastics;
   b. **hydromorphic** – related to movement of igneous solutions;
   c. **biogenetic** – related to biological activity;
2. **epigenetic** – formed by material introduced into a preexisting rock: hydromorphic and biogenetic.

0406
absorption, – the incorporation of gases, liquids or dissolved substances by the bulk of a solid. / absorptie / absorption / Absorption / absorción

adsorption, – the condensation of gases, liquids or dissolved substances on the surfaces of solids. / adsorptie / adsorption / Adsorption / adsorción

0407
sorption, – general term – comprising both, absorption and adsorption.

desorption, – the displacement of adsorbed substances from the surface of the solid. / desorptie / désorption / Desorption / desorción

0408
complexing, – the formation of complex coordination compounds by the association of two or more simpler chemical compounds each of which is capable of independent existence; the associated compounds may be of ionic or molecultar nature or they may be oppositely charged compounds. The stable complexes formed can either be solutions or precipitates and may have completely different properties than those of the component parts. / vorming van complexe verbindingen / complexation / Komplexbildung / complejamiento

coordination number, – the number of ions or molecules that in a complex compound surrounds a central atom or ion; it is greater than the valence. / coördinatiegetal / coordinance / Koordinationszahl / número de coordinación

0409
chelation, – the formation of organic metal complex compounds, in which one or more rings are closed by semipolar bonds. / chelatie / chélation / Chelatbildung / quelación

organo-metal complexing, – chelation of metal ions with the nitrogen of organic molecules. / organometallische chelatie / complexation organo-métallique; chélation .... / organometallische Komplexbildung / quelación organo-metálico

0410
organometallic compound, – a compound in which a metal atom is bound to an organic compound directly through a carbon atom. / organometallisch chelaat / chélate organométallique / organometallische Verbindung / complejo organo-metálico

metallo-organic compound, – a compound in which a metal atom is bound to an organic compound through an atom other than carbon, such as oxygen, nitrogen, or sulphur, to form a complex or coordination compound. / metalloörganisch chelaat / chélate métallo-organique / metallorganische Verbindung / complejo metalo-orgánico

0411
solution, – a mixture, in solid, liquid or gas phase, in which the components are homogeneously distributed. / oplossing / solution / Lösung / solución

suspension, – the dispersion of a solid (in particles of 1-0.1 μm) in a liquid. / suspensie / suspension / Suspension / suspensión

precipitation, – the process of separation of a solid substance from a solution when the solubility product is exceeded, i.e. the solution becomes supersaturated in that substance. / neerslaan / précipitation / Fällung / precipitación

0412
evaporation, – the conversion of a liquid into a vapour; dissolved salts and suspended solids remain behind as a precipitate. / verdamping; indamping / évaporation / Eindampfung / evaporación

sublimation, – the conversion of a solid substance into a vapour without passing through a liquid stage. / sublimatie / sublimation / Sublimation / sublimación

condensation, – the process by which a vapour becomes a liquid or solid. / condensatie / condensation / Kondensation / condensación

0413
coagulation, – the aggregation of particles in a colloidal solution to larger particles that may precipitate. / coagulatie / coagulation / Koagulation / coagulación

0414
filtration, – removal of suspended and/or colloidal material from a liquid by passing it through a relatively fine porous medium. / filtratie / filtration / Filtration / filtrado

fractionation, – the differentiation of a mixture by separation of one or more of its phases, component minerals, elements or isotopes. /fractionatie / fractionnement / Fraktionierung / fraccionamiento

0415
desulphurization, – by this process oxysalts, particularly sulphates, are broken down by anaerobic microorganisms to obtain oxygen. A typical reaction is the forming of calcite and hydrogen sulphide from organic matter and calcium sulphate. Common process on the bottom of oceans, seas, salt lakes, marshes, and also active in the early phase of oil genesis. / sulfaatreductie / désulfuration / Desulfurisierung / desulfurización

0416
oxidation-reduction reaction, – a reaction involving a change in the valence of an element, single ion or central atom in a complex ion. / oxydatie-reductie reactie / réaction d'oxydo-réduction / Redox-Reaktion; Oxidations-Reduktions-Reaktion / reacción de oxidación-reducción

oxidation, – the increase in the positive valence or the decrease in the negative valence. / oxydatie / oxydation / Oxidation / oxidación

reduction, – the increase in the negative valence or the decrease in the positive valence. / reductie / réduction; désoxydation / Reduktion / reducción

0417
bacterial reduction, – process caused by sulphur bacteria that obtain the oxygen needed in metabolism by reducing sulphate ions to hydrogen sulphide or elemental sulphur. / bacteriële reductie / réduction bactérienne / bakterielle Reduktion / reducción bacteriana

0418
hydrolysis, – in the usual geological sense, the reaction between water and rock-forming minerals to produce a dilute solution and insoluble hydrous silicates (e.g. clays). / hydrolyse / hydrolyse / Hydrolyse / hidrólisis

0419
hydration, – the combination of an element or compound with water. / hydratie / hydratation / Hydration / hidratación

ion potential, – the attraction of ions in solution to water molecules; the hydration of an ion is proportional to its charge and inversely proportional to its radius. / ionenpotentiaal / potentiel ionique / Ionenpotential / potencial iónica

0420
ion exchange, – the exchange of especially cations between clay minerals and interstitial water. / ionenuitwisseling / échange ionique; .... d'ions / Ionenaustausch / Intercambio iónico

ion-exchange capacity, – sum of exchangeable anions (cations) that a clay or soil can adsorb (milliequivalents per 100 grammes clay or soil at a specific pH). / ionenuitwisselingscapaciteit; adsorptie.... / capacité d'échange ionique; .... d'échange d'ions / Ionenaustausch-Kapazität / capacidad de cambio iónico

0421
redox potential; Eh; oxidation ....; oxidation-reduction ...., – the relative intensity of oxidizing and reducing conditions in solutions, usually expressed in millivolts, whereby a positive sign is assigned to oxidizing, a negative sign to reducing systems. / redox potentiaal / potentiel d'oxydo-réduction; .... redox / Redox-Potential / potencial redox

0422
acidity; pH, – the negative logarithm base 10 of the hydrogen-ion concentration. / zuurgraad / acidité / Azidität / acidez

buffer, – the chemical concept that the acidity of a solution does not change greatly by the addition of moderate amounts of anions or cations, as extended to natural water systems, and especially to the composition of sea water. The dissolved contents of a water body are maintained at a certain level through chemical equilibria between dissolved substances and solids and gases outside the main water body. / buffer / tampon / Puffer / tampón

0423
carbonate compensation depth, – the depth in the sea, at which precipitation and solution of $CaCO_3$ are in equilibrium. / carbonaatcompensatiediepte / profondeur de compensation des carbonates / Karbonat-Kompensationstiefe / profundidad de precipitación de carbonatos

0424
anaerobic; anoxic, – pertaining to conditions without air respectively oxygen. / anaeroob / anaérobie; anoxique / anaerob / anaerobio

aerobic; oxic, – pertaining to conditions with air respectively oxygen. / aeroob / aérobie; oxique / aerob / aerobio

0425
carbon-nitrogen ratio; C/N ....; C-N ..., – ratio of the mass of organic carbon to the mass of total nitrogen in soil or rock. / koolstof-stikstofverhouding; C/N-....; C/N-quotiënt / rapport carbone-azote; .... C/N / Kohlenstoff-Stickstoff-Verhältnis; C/N-....; C:N-.... / relación carbono-nitrógeno; .... C/N

0426
bituminization, – the alteration of vegetal and animal matter leading to an increase of fixed carbon and a decrease of volatiles, as is the case in coalification and the origin of hydrocarbons. / bituminisatie / bituminisation / Bituminisierung / bituminización

0427
caustobiolith, – a combustible rock, consisting wholly or to a large extent of organic matter. / kaustobioliet / caustobiolite / Kaustobiolit / cáustobiolito

NOTE:
– see table 1 – Caustobiolites.

[0500] Isotope Geology

[0510] ISOTOPE GEOCHRONOLOGY

0428
isotope geochronology, – a system of dating based upon naturally occurring radioactive nuclides and the law of radioactive decay, and the interpretation of such ages in terms of geologic events. / isotopengeochronologie / géochronologie isotopique / Isotopengeochronologie / geocronología isotópica

0429
radiometric dating; isotopic ....; radioactive ...., – dating events in the history of the earth, moon and meteorites by means of naturally occurring radioactive nuclides:
1. by measuring the ratio of a radioactive nuclide to its daughter nuclide produced in situ by radioactive decay (**accumulation clocks**),
2. by determining the proportion of a radioactive nuclide that has decayed from a given initial concentration (**decay clocks**),
3. by using deviations of the radioactive equilibrium in the uranium decay chains (**disequilibrium clocks**).
/ radiometrische datering; isotopische .... / datation radiométrique; .... isotopique / radiometrische Datierung; Isotopen.... / datación radiométrica; .... isotópica

absolute dating; .... age, – obsolete and erroneous terms denoting radiometric dating or age; their use should be avoided. / absolute datering; .... ouderdom / datation absolu; âge .... / absolute Datierung; absoluteser / datación absolute; edad ....

0430
concordant age, – when several isotopic dating methods are applied to a mineral, whole-rock or a suite of cogenetic minerals and the resulting ages agree within analytical error, the dates are termed concordant. / concordante ouderdom / âge concordant / konkordantes Alter / edad concordante

discordant age, – when several isotopic dating methods are applied to a mineral, whole-rock or a suite of cogenetic minerals and the resulting ages disagree outside the analytical error, the dates are termed discordant. Discordant age patterns are useful in interpreting the evolution of terrains with a complex geologic history (e.g. polymetamorphic terrains). / discordante ouderdom / âge discordant / diskordantes Alter / edad discordante

0431
half-life, – time required for a given amount of radionuclide to decay to one-half of its initial value. Related to the decay constant λ as $T(^1/_2) = 0.6931/\lambda$. / halveringstijd / demi-période / Halbwertszeit / período de semidesintegración

decay constant; λ, – constant representing the probability that a radioactive nuclide will disintegrate in unit time: $-dN/dt = \lambda N$ where N is the total number of radioactive atoms. / vervalsconstante / constant de désintégration / Zerfallskonstante / constante de desintegración

0432
isotope dilution analysis, – determining the unknown amount of an element of known or normal isotopic composition by mixing it with a known amount of the same element with a different isotopic composition (spike), and measuring the resulting isotopic composition of the mixture. / isotopenverdunningsanalyse / analyse par dilution isotopique / Isotopenverdünnungsanalyse / análisis por dilución isotópica

0433
isochron, – for any radioactive decay system $M \rightarrow D_1$: the linear relationship between a group of samples of equal age in a plot of the ratios radiogenic daughter isotope D_1/stable isotope D_2 (of the same element) versus the ratios radioactive mother M/stable isotope D_2. A powerfull and widely used technique in dating of cogenetic rocks and / or minerals that may facilitate the analysis of:

$^{238}U-^{206}Pb$ data ($^{206}Pb/^{204}Pb$ versus $^{238}U/^{204}Pb$),
$^{235}U-^{207}Pb$ data ($^{207}Pb/^{204}Pb$ versus $^{235}U/^{204}Pb$),
$^{232}Th-^{208}Pb$ data ($^{208}Pb/^{204}Pb$ versus ^{232}T $/^{204}Pb$),
Rb-Sr data ($^{87}Sr/^{86}Sr$ versus $^{87}Rb/^{86}Sr$),
K–Ar data ($^{40}Ar/^{36}Ar$ versus $^{40}K/^{36}Ar$).

/ isochroon / isochrone / Isochrone / isócrona

0434
uranium-lead dating, – age determination on the basis of the accumulation of ^{206}Pb and ^{207}Pb produced in situ by the radioactive decay of ^{238}U and ^{235}U, respectively. (Complex decay chains with many intermediate short-lived radionuclides; half-lives 4.468×10^9 year for ^{238}U and 0.704×10^9 year for ^{235}U; atomic ratio $^{238}U/^{235}U = 137.88$). A widely used technique in dating zircons, titanites and uraninites. Use is made of isochron plots. / uranium-lood datering / datation par les méthodes uranium-plomb / Uran-Blei Datierung / datación por los métodos uranio-plomo

lead-lead dating, – an alternative (but not independent) method of uranium-lead dating by combining the $^{238}U-^{206}Pb$ and $^{235}U-^{207}Pb$ dating methods. Using the present-day $^{238}U/^{235}U$ ratio of 137.88, an age value can be determined by measurement of the radiogenic $^{207}Pb/^{206}Pb$ ratio alone. / lood-lood datering / datation par la méthode plomb-plomb / Blei-Blei Datierung / datación por el método plomo-plomo

0435
Concordia-Discordia diagram, – in a plot of the ratio radiogenic $^{206}Pb/^{238}U$ versus the ratio radiogenic $^{207}Pb/^{235}U$, a single curve ("Concordia") contains all points of concordant $^{238}U-^{206}Pb$ and $^{235}U-^{207}Pb$ ages. Suites of cogenetic minerals having discordant $^{238}U-^{206}Pb$ and $^{235}U-^{207}Pb$ ages usually display a linear relationship ("Discordia"); the upper intersection of this straight line with Concordia yields the age of formation of the suite of samples, the lower intercept may or may not represent a time of episodic lead loss. A powerful and widely used technique in dating suites of cogenetic zircons and/or titanites. / Concordia-Discordia diagram / courbe Concordia-Discordia / Concordia-Discordia-Diagramm / diagrama Concordia-Discordia

0436
thorium-lead dating, – age determination on the basis of the accumulation of ^{208}Pb produced in situ by the radioactive decay of ^{232}Th. (Complex decay chain with many intermediate short-lived radionuclides; half-life 14.01×10^9 year). A technique suitable for dating zircons. Use is made of isochron plots. / thorium-lood datering / datation par la méthode thorium-plomb / Thorium-Blei Datierung / datación por el método thorio-plomo

0437
radiogenic lead, – in its broadest sense: all lead formed by radioactive decay of uranium and thorium since the beginning of the solar system. More specifically: ^{206}Pb, ^{207}Pb and ^{208}Pb accumulated in situ in a rock or mineral by the decay of uranium and thorium. / radiogeen lood / plomb radiogénique / radiogenes Blei; (Zerfallsblei) / plomo radiogénico

primeval lead; primordial ...., – lead with the isotopic composition as existing at the beginning of the solar system, such as preserved in troilitic phases (FeS) of meteorites. It contains four isotopes, ^{204}Pb, ^{206}Pb, ^{207}Pb and ^{208}Pb, with the approximate composition: ^{206}Pb/^{204}Pb = 9.35, ^{207}Pb/^{204}Pb = 10.22, ^{208}Pb/^{204}Pb = 28.96. / oerlood / plomb primitif / Urblei / plomo primordial

common lead, – in its broadest sense: lead in rocks sustaining U-Th-Pb systems that approximate the average crustal ratios of uranium and thorium relative to ^{204}Pb; it may be thought as consisting of primeval lead plus radiogenic lead. More specifically: any lead that in its present position is not associated with a significant amount of uranium and/or thorium (e.g. in feldspars or lead-ore deposits); the isotopic composition may reflect the average lead composition in the source rocks at the time of its separation from the associated uranium and thorium (e.g. generation of the lead-bearing fluids). The term is also used for initial lead, i.e. common lead incorporated in a mineral at the beginning of the time interval dated by the U-Pb and/or Th-Pb methods (the time of formation); the isotopic composition of initial lead varies according to its source. / gewoon lood / plomb commum / gewöhnliches Blei / plomo común

0438
lead-isotope growth curves, – curves illustrating the gradual change with time of the lead-isotopic composition in closed uranium-thorium-lead systems by the continuous addition of radiogenic lead. / lood-ontwikkelingscurven / courbes d'évolution des plombs / Blei-Entwicklungslinien / curvas de evolución del plomo

0439
single-stage lead model, – geochemical model assuming isotopic evolution of lead in a locally closed uranium-thorium-lead system between the time of formation of the earth and the time of extraction of a particular common lead. / enkelvoudig lood-evolutiemodel / modèle d'évolution des plombs à un seul stade / einstufiges Bleimodell / modelo de evolución del plomo a un estadio

multi-stage lead model, – geochemical model assuming isotopic evolution of lead in more than one locally closed uranium-thorium-lead system prior to the time of extraction of a particular common lead. / meervoudig lood-evolutiemodel / modèle d'évolution des plombs à plusieurs stades / mehrstufiges Bleimodell / modelo de evolución del plomo en varios estadios

0440
lead-model age, – the age calculated for the time at which a particular common lead was extracted from the parent uranium-thorium-lead system, assuming a single-stage lead model. / lood-modelouderdom / âge conventionnel du plomb / Blei-Modellalter / edad convencional (modelo) del plomo

0441
anomalous lead, – common lead that has been developed in more than one closed uranium-thorium-lead system prior to the time of extraction of the lead (multi-stage lead model), or that was mixed after its extraction with lead from another environment. The lead-model age of such lead has no geochronological significance:

J-type lead, – model age geologically too young,

B-type lead, – model age geologically too old.

/ anomaal lood / plomb anormal / anomales Blei / plomo anómalo

0442
fission-track dating, – dating by counting under the microscope the number of tracks (about 10^{-3} cm long) created by the spontaneous fission of ^{238}U. (Decay constant for fission still uncertain, probably about $6.9\times10^{-17}a^{-1}$). A technique for dating volcanic or man-made glasses, and minerals such as apatite, zircon, titanite and epidote. Especially of interest for the youngest 10^5 year. / splijtingssporen-datering / datation par la méthode des traces de fission / Spaltspurendatierung / datación por el método de trazas de fisión

0443
uranium-series disequilibrium dating, – dating methods based upon the separation of members of the uranium decay chains from each other by geochemical and/or biochemical processes, due to differences in the chemical properties of the different radionuclides. Two types of processes are utilized for dating:
1. a radionuclide is separated from its parent and subsequently decays at a rate determined by its half-life,
2. a radionuclide is formed by decay of its parent (previously separated from its daughter) until radioactive equilibrium is restored.
The methods are particularly used for dating Pleistocene corals and deep-sea sediments (upper age limit about 300 000 years). / radioactief niet-evenwicht datering / datation par déséquilibre radioactif / radioaktive Ungleichgewichtsdatierung / datación por el desequilibrio radioactivo

0444
ionium-thorium dating, – a uranium-series disequilibrium method for dating young sediments on the bottom of the deep ocean by measuring the ratio ^{230}Th(ionium)/^{232}Th. Ionium (half-life 7.6×10^4 year) is produced in the ^{238}U decay chain and precipitates in sediments intimately mixed with the long-lived ^{232}Th. The ratio ^{230}Th/^{232}Th decreases with depth in the sediments (e.g. in a core sample) and the ages can be calculated (after correction for any uranium-supported ^{230}Th) relative to the surface layer. A useful method for time spans up to several hundred thousand years. / ionium-thorium datering / datation par la méthode ionium-thorium / Ionium-Thorium Datierung / datación por el método ionio-thorio

0445
protactinium-thorium dating, – uranium-series disequilibrium method for dating young sediments on the bottom of the deep ocean by measuring the ratio ^{230}Th/^{231}Pa. Both thorium and protactinium have very short residence times in ocean water and precipitate almost completely. ^{230}Th (half-life 7.6×10^4 year) and ^{231}Pa (half-life 3.4×10^4 year) are produced by the decay of ^{234}U (^{238}U decay chain) and ^{235}U, respectively. By comparing the ratios ^{230}Th/^{231}Pa in a sedimentary sequence (e.g. a core sample) with the ratio in sediments now being deposited, ages up to about 175 000 years can be calculated (after correction for uranium-supported nuclides). / protactinium-thorium datering / datation par la

méthode protactinium-thorium / Protactinium-Thorium Datierung / datación por el método protactinio-thorio

0446
rhenium-osmium dating, – age determination on the basis of the accumulation of ^{187}Os produced by the radioactive decay of ^{187}Re. (Beta emission; half-life still uncertain, possibly around 43×10^9 year). A rarely used dating method with very few applications (molybdenite, iron meteorites). / rhenium-osmium datering / datation par la méthode rhenium-osmium / Rhenium-Osmium Datierung / datación por el método renio-osmio

0447
rubidium-strontium dating, – age determination on the basis of the accumulation of ^{87}Sr produced in situ by the radioactive decay of ^{87}Rb. (Beta emission; for the half-life two values were commonly used until 1977, either 47.2×10^9 year or 50.0×10^9 year; since then the half-life 48.8×10^9 year is recommended; atomic ratio $^{85}Rb/^{87}Rb$ = 2.592 65). A widely used technique for dating biotites, muscovites, feldspars and whole-rocks of igneous and metamorphic rocks. Use is made of isochron plots. / rubidium-strontium datering / datation par la méthode rubidium-strontium / Rubidium-Strontium Datierung / datación por el método rubidio-estroncio

0448
radiogenic strontium, – in its broadest sense: all ^{87}Sr formed by radioactive decay of ^{87}Rb since the beginning of the solar system. More specifically: ^{87}Sr accumulated in situ in a rock or mineral by the decay of rubidium. / radiogeen strontium / strontium radiogénique / radiogenes Strontium / estroncio radiogénico

initial strontium, – strontium incorporated in a rock or mineral at the beginning of the time interval dated by the Rb-Sr method (the time of its formation). The isotopic composition varies according to its source. / initiëel strontium / strontium primaire / Anfangsstrontium / estroncio inicial

common strontium, – strontium has four isotopes, ^{84}Sr, ^{86}Sr, ^{87}Sr and ^{88}Sr. In its broadest sense: any strontium from environments with Rb/Sr ratios approximating the average value of earth, moon and meteorites; its $^{87}Sr/^{86}Sr$ ratio ranges from a primeval value of about 0.699 at the beginning of the solar system to a present-day terrestrial value of around 0.706. More specifically: strontium that in its present location is not associated with a significant amount of rubidium (e.g. strontianite, limestone). / gewoon strontium / strontium commun / gewöhnliches Strontium / estroncio común

0449
potassium-argon dating, – age determination on the basis of the accumulation of ^{40}Ar produced in situ by the radioactive decay of ^{40}K. (^{40}K has a dual decay system; 10.48% decays to ^{40}Ar through electron capture and 89.52% to ^{40}Ca through beta emission; recommended decay constants since 1977: $\lambda_e = 0.58\times10^{-10}a^{-1}$; $\lambda_\beta = 4.962\times10^{-10}a^{-1}$; abundance ^{40}K = 0.011 67 atom % total K; half-life $^{40}K = 1.25\times10^9$ year). A widely used technique for dating biotites, muscovites, hornblendes, calcic plagioclases, sanidines, leucites, glauconites, and whole-rocks of basic igneous rocks and micaceous schists. Use is made of isochron plots. / kalium-argon datering / datation potassium-argon / Kalium-Argon Datierung / datación potasio-argon

potassium-calcium dating, – age determination on the basis of the accumulation of ^{40}Ca produced in situ by the radioactive decay of ^{40}K through beta emission. A rarely utilized method because of the difficulty of measuring the small radiogenic increment of ^{40}Ca against the much larger amounts of calcium commonly present in rocks and minerals (for 96.97% consisting of ^{40}Ca). Occasionally applied to sylvite and ancient micas. / kalium-calcium datering / datation potassium-calcium / Kalium-Kalzium Datierung / datación potasio-calcio

0450
radiogenic argon, – in its broadest sense: all ^{40}Ar formed by radioactive decay of ^{40}K since the formation of the solar system. More specifically: the proportion of ^{40}Ar in argon contained in a rock or mineral that exceeds the abundance of ^{40}Ar in athmospheric argon. / radiogeen argon / argon radiogénique / radiogenes Argon / argón radiogénico

atmospheric argon, – argon in the present earth's atmosphere (0.93% by volume). Isotopic composition: 99.6% ^{40}Ar, 0.063% ^{38}Ar, 0.337% ^{36}Ar, $^{40}Ar/^{36}Ar$ = 295.5. / atmosferisch argon / argon atmosphérique / atmosphärisches Argon; (Luftargon) / argón atmosférico

excess argon, – radiogenic ^{40}Ar present in a rock or mineral but not produced in situ by the radioactive decay of ^{40}K. Presence of excess argon leads to anomalously high K-Ar ages. / overmaat argon / excès d'argon / Überschuss Argon / exceso de argón

0451
carbon-14 dating; radiocarbon ...., – dating the time elapsed since a carbon-containing sample (e.g. a piece of wood) was removed from the carbon dioxide cycle (and thus from the atmospheric exchange reservoir of radiocarbon) by counting the remaining specific ^{14}C activity in the sample. (Beta emission; accepted half-life 5 568 year.) A widely used method for dating biogenic carbonaceous material in archeology, anthropology, pedology, limnology and late Pleistocene geology (upper age limit about 50 000 years or, after enrichment, about 75 000 years). Also applied to non-biogenic material (speleothems, ground water). / koolstof-14 datering; radiokoolstof.... / datation par le carbone-14 / Radiokohlenstoffdatierung; Kohlenstoff-14 .... / datación por el método carbono-14; datación radiocarbono

0452
tritium dating, – dating the time elapsed since a water sample was removed from the atmospheric exchange reservoir of tritium by counting the remaining ^3H activity. (Beta emission; half-life 12.262 year). A useful tool in meteorological and hydrological studies (upper age limit about 100 years). / tritium datering / datation par le tritium / Tritium Datierung / datación por el método tritio

[0520] STABLE ISOTOPE GEOLOGY

0453
stable isotope geology; .... geochemistry, – application of the variation in relative abundances of stable isotopes, particularly D/H, $^{13}C/^{12}C$, $^{18}O/^{16}O$ and $^{34}S/^{32}S$, to the study of geological processes. Stable isotopes provide natural tracers for studying the genesis and/or exchange history of various materials (e.g. water, sediments, biogenic materials, ore deposits, igneous and metamorphic rocks). The partitioning

(fractionation) of stable isotopes between coexisting phases is utilized as a geothermometer. / stabiele isotopen-geologie; ....-geochemie / géologie des isotopes stables; géochimie .... / stabile Isotopengeologie; .... geochemie / geología de los isótopos estables; geoquímica ....

0454
δ-value; delta value, – the isotopic composition of an element is reported as the deviation (δ) from a standard. The δ-value is defined as

$$\delta\ (\text{in}‰) = \frac{R_{sample} - R_{standard}}{R_{standard}} \times 1000$$

where R is the isotope ratio D/H, $^{13}C/^{12}C$, $^{18}O/^{16}O$ or $^{34}S/^{32}S$. / δ-waarde / valeur de δ / δ-Wert / valor de δ

standard, – in the δ-value notation isotopic compositions are reported relative to a standard. The international standards are:
SMOW (standard mean ocean water) for D/H and $^{18}O/^{16}O$;
PDB (belemnite from Pedee Formation) for $^{13}C/^{12}C$, occasionally (e.g. in palaeotemperature studies) for $^{18}O/^{16}O$;
CD (triolote from the Canyon Diablo meteorite) for $^{34}S^{32}S$.
/ standaard / standard; (échantillon de référence) / Standard / muestra de referencia

0455
isotope fractionation factor, – partitioning coefficient of isotope distribution between two coexisting phases A and B, defined as

$$\alpha^A_B = R_A/R_B$$

where R_A and R_B are the isotope ratios in A and B. In many cases it is more convenient to use $10^3 \ln \alpha^A_B$, the value of which is approximated by the difference in isotopic composition between the phases A and B:

$$10^3 \ln \alpha^A_B \simeq \delta_A - \delta_B$$

/ isotopen-fractioneringsfactor / facteur de fractionnement isotopique / Isotopenfraktionierungsfaktor / factor de fraccionamiento isotópico

isotope fractionation curve, – curve defining the equilibrium-isotope fractionation between coexisting phases as a function of the temperature. Two fractionation curves can be combined as, for example:

$$10^3 \ln \alpha^{quartz}_{water} - 10^3 \ln \alpha^{magnetite}_{water} \simeq 10^3 \ln \alpha^{quartz}_{magnetite}$$

/ isotopen-fractioneringscurve / courbe de fractionnement isotopique / Isotopenfraktionierungskurve / curva de fraccionamiento isotópico

0456
isotope thermometry, - utilizing the temperature dependence of isotope fractionation between coexisting phases as a geothermometer. / isotopenthermometrie / géothermometrie isotopique / Isotopenthermometrie / termometría por isótopos

palaeotemperature determination, – determination of ancient sea water temperatures by measuring the oxygen isotopic composition of fossil shells, utilizing the (temperature dependent) fractionation of oxygen isotopes between calcium carbonate and sea water. / palaeotemperatuurbepaling / paléothermometrie / Paläotemperaturbestimmung / paleotermometría

0457
kinetic isotope fractionation, – isotope fractionation due to non-equilibrium processes. The lighter molecules react (or move) faster than the heavier ones, resulting in an enrichment of the light isotope in the reaction product. Kinetic fractionation may play an important role in biological processes and in the diffusion of gases. / kinetische isotopenfractionering / fractionnement cinétique des isotopes / kinetische Isotopenfraktionierung / fraccionamiento cinético de los isótopos

0458
meteoric water line, – relation between δD and $\delta^{18}0$ as generally observed in meteoric waters: $\delta D = 8\ \delta^{18}0 + 10$.
/ meteorisch-waterlijn / ligne des eaux météoriques / meteorische Wasserlinie / línea de aguas meteóricas

[0600] MINERALOGY AND GEMMOLOGY

[0610] MINERALOGY

NOTE:
– this section is a general introduction to the subject only; for further terms see
[0840] The Constituents of Igneous Rocks
[0930] Texture and Structure of Metamorphic Rocks
[0940] The Constituents of Metamorphic Rocks
[2650] Texture and Structure of Ore Minerals

0459
mineralogy, – the study of minerals: their formation and occurrence, composition and properties, classification and practical value to man. / mineralogie / minéralogie / Mineralogie / mineralogía

0460
mineral, –
1. a solid chemical element or compound or a solid solution individualized within a planet or moon by natural proses with a unique physical and chemical character;
2. any naturally formed, inorganic material, i.e. a member of the mineral kingdom as opposed to the plant kingdom and the animal kingdom.

/ mineraal / minéral / Mineral / mineral

mineraloid, – a term to denote an amorphous mineral, used by those American authors who include the requirement of crystalline state in the definition of a mineral. / mineraloide / – – / – – / mineraloide

0461
artificial mineral; synthetic ...., – a man-made product with the composition and crystal structure of a known mineral

and with the same or improved properties. This term can only be used if the requirement of a natural origin is not included in the definition of a mineral. / kunstmatig mineraal / minéral artificiel; .... synthétique; .... de synthèse / künstliches Mineral / mineral artificial; .... técnico

0462
industrial mineral, – any rock or mineral of economic value, exclusive of ores, mineral fuels and gemstones. / industriëel mineraal / minéral industriel; substance minérale industrielle / Industriemineral / mineral industriel

rock-forming mineral, – a mineral that enters into the composition of a rock and determines its classification. / gesteentevormend mineraal / minéral essentiel; .... constituant / gesteinsbildendes Mineral / mineral constituyente

ore mineral, –
1. one of the valuable minerals composing an ore body, from which at least one metal can be economically extracted;
2. sometimes used as a general term to denote an opaque mineral or a mineral with a metallic lustre.

/ ertsmineraal / minerai; minerai métallique; minéral opaque; .... métallique / Erzmineral; (Erzart) / mineral metalífero

0463
isotypism, – the phenomenon that two or more elements or chemically analogous compounds have the same crystal structure. / isotypie / isotypisme / Isotypie / isotipismo

polytypism, – the property of a chemical element or compound to crystallize in different structures due to different modes of packing in one crystallographic direction / polytypie / polytypisme / Polytypie / politipismo

0464
polymorphism, – the characteristic of a chemical element or compound to crystallize in more than one crystalline phase. / polymorfie / polymorphisme / Polymorphie / polimorfismo

isomorphism, – the characteristic of two or more chemically similar and isotypic crystalline substances to form a solid solution series. / isomorfie / isomorphisme / Isomorphie / isomorfismo

0465
pseudomorphism, – the phenomenon that a mineral or mineral aggregate has the outward shape proper to another mineral species, e.g. limonite with the shape of pyrite or limonite pseudomorph after pyrite. / pseudomorfose / pseudomorphisme / Pseudomorphose / (p)seudomorfismo

paramorphism, – the phenomenon that a mineral or monomineral aggregate has the outward shape of another polymorph of that mineral, e.g. kyanite paramorph after andalusite. / paramorfose / paramorphisme / Paramorphose / paramorfismo

0466
streak, – the colour of the mark left behind by rubbing a mineral on a piece of unglazed porcelain. The obtained colour may differ from that of the compact mineral, but is usually constant for the same mineral. / streepkleur / couleur de la trace / Strichfarbe / raya

0467
luminescence, – the emission of light by a mineral with a temperature below that of its incandescence, caused by the absorption of energy from an external source. / luminescentie / luminescence / Lumineszenz / luminescencia

fluorescence, – luminescence during the absorption of energy. / fluorescentie / fluorescence / Fluoreszenz / fluorescencia

phosphorescence, –luminescence some time (from less than one second up to several days) after the absorption of energy. / fosforescentie / phosphorescence / Phosphoreszenz / fosforescencia

0468
striation, – parallel depressions or narrow bands on a crystal face or cleavage plane of a mineral caused by oscillation in growth between differently oriented faces (as in tourmaline) or by repeated twinning lamellae (as in plagioclase). / streping / striation / Streifung / estración

0469
cleavage, – the property of a mineral to split along certain crystallographic planes, reflecting those directions in the crystal structure across which the bonds are weakest. The cleavage of a mineral is described by the cleavage form or the indices of the cleavage plane and its quality in general terms. / splijting / clivage / Spaltbarkeit / crucero

parting, – the breaking of a mineral along smooth or wavy planes of weakness caused by deformation or twinning. Parting resembles cleavage, but it is limited to certain specimens of a mineral species and certain parallel planes. / – – / – – / Absonderung / exfoliación

fracture, – the breaking of a mineral other than along planes of cleavage or parting. The fracture of a mineral can be described in terms of conchoidal (shell-like), fibrous, uneven, etc. / breuk / cassure; fracture / Bruch / fractura

0470
tenacity, – the resistance of a mineral against deformation under stress. / taaiheid / ténacité / – – / tenacidad

0471
brittle, – a mineral is said to be brittle if it can be crushed or powdered readily (e.g. quartz, diamond). / bros / cassant / spröde / quebradizo

malleable, – a mineral is said to be malleable if it can be hammered into thin sheets without breaking (e.g. gold, silver, copper). / pletbaar / malléable; aplatissable / dehnbar / maleable

sectile, – a mineral is said to be sectile if it can be cut with a knife. / snijdbaar / sectile / schneidbar / sectil

0472
flexible, – a mineral is said to be flexible if it can be bent without rupture and without returning to its original form. / buigbaar; flexibel / flexible / biegsam / dúctil

elastic, – a mineral is said to be elastic if it can be bent in such a way that it returns to its original form. / elastisch / élastique / elastisch / elástico

0473
hardness, – by the hardness of a mineral is generally understood its relative resistance to scratching with the scale of Mohs as reference. / hardheid / dureté / Härte / dureza

scale of Mohs, – a scale of ten minerals numbered from 1 to 10 with increasing hardness: 1 talc, 2 gypsum, 3 calcite, 4 fluorite, 5 apatite, 6 orthoclase, 7 quartz, 8 topaz, 9 corundum, 10 diamond. / hardheidsschaal van Mohs / échelle de dureté de Mohs / Mohssche Härteskala / escala (de dureza) de Mohs

0474
zoning, – a variation in the composition of a crystal from core to margin; it is due to separation of the crystal phases during growth by loss of equilibrium in a continuous reaction series. / zonering / zonage / Zonarbau / zonificación

0475
normal zoning, – a zoning from the higher temperature phase in the core towards the lower temperature phase of a solid solution series in the margin of a crystal (especially used with respect to plagioclase). / normale zonering / zonage direct; .... normal / normaler Zonarbau / zonificación normal

inverse zoning, – a zoning from the lower temperature phase in the core towards the higher temperature phase of a solid solution series in the margin of a crystal (especially used with respect to plagioclase). / inverse zonering / zonage inverse / invertierter Zonarbau / zonificación inversa

0476
exsolution, – the process whereby an initially homogeneous solid solution separates into two or more distinct crystalline phases without change in the overall chemical composition, e.g. the perthitic exsolution of Na-K-feldspar into K-feldspar with albite patches. / ontmenging / exsolution / Entmischung / desmezcla

0477
atomic substitution; diadochy, – the partial or total replacement of a chemical element by another one with the same or nearly the same crystallographic and chemical characteristics within a crystal structure. / atomaire vervanging / substitution ionique; diadochie / Diadochie / sustitución atómica

coupled substitution, – more than one simultaneous atomic substitution in a crystal structure. / gekoppelde vervanging / substitution couplée / gekoppelte Diadochie / sustitución acoplada

[0620] GEMMOLOGY

0478
gemmology, – the science that deals with precious minerals and other materials used for personal adornment, and for objets d'art. / edelsteenkunde / gemmologie / Edelsteinkunde / gemología

gem stone, – a mineral with special qualities, i.e. beauty, durability and rarity. / edelsteen / pierre précieuse / Edelstein / piedra preciosa

0479
synthetic stone, – manufactured stone that has essentially the same composition, crystal structure and properties as the natural mineral it represents. / synthetische edelsteen / pierre précieuse synthétique / synthetischer Stein / piedra sintética

curved growth line, – as seen, together with gas bubbles, in synthetic corundum made in an oxy-coal-gas furnace (Verneuil process). / gebogen groeilijn / ligne courbé d'accroissement / gebogene Anwachslinie / curva de crecimiento

0480
composite stone, – a stone consisting of two (**doublet**) or three (**triplet**) parts either of the same or of different materials cemented or otherwise joined together. / samengestelde steen; doublet; triplet / pierre agglomérié; doublet; triplet / zusammengesetzer Stein; Dublette; Triplette / – –

0481
imitation stone, – material such as glass or plastic that may resemble genuine stones in colour and appearance, but differ from them in composition and physical properties. / imitatie / imitation de pierre / Nachahmung / piedra de imitación

swirl mark, – irregular striations in badly annealed glass imitations. / glasslier / rayure de polissage rapide; strie irrégulière / Schlieren / – –

0482
carat weight, – the internationally accepted unit of mass for gem stones, equivalent to one-fifth of a gramme. / karaat / carat / Karat / quilate

0483
idiochromatic, – said of minerals and gem stones whose colour is related to the crystal lattice. / idiochromatisch / idiochromatique / eigenfarbig / idiocromático

allochromatic, – said of minerals and gem stones whose colour is due to the occurrence of microscopic inclusions, impurities or pigments. / allochromatisch / allochromatique / fremdfarbig / alocromático

0484
lustre, – the appearance of a stone's surface (or of a mineral in general) in reflected light. Refractive index and perfection of polish possessed by the stone are the main factors affecting lustre, while hardness is also of some importance. / glans / éclat / Glanz / lustre

vitreous; glassy, – said of the lustre that is possessed by the majority of stones. / glas- / vitreux / Glasglanz / vitreo

adamantine, – the characteristic lustre of diamond. / adamant- / adamantin / Diamantglanz / adamantino

resinous, – as in certain garnets. / hars- / résineux / Harz / resinaso

waxy, – as in turquoise / was- / gras / Wachsglanz / céreo

pearly – as im moonstone. / paarlmoer / perlé / Perlmutterglanz / perlado

silky, – as in fibrous calcite and gypsum. / zijde- / soyeux / Seidenglanz / sedoso

metallic, – as in gold or pyrite. / metaal- / métallique / Metallglanz / metálico

0485
fire, – in gemmology, the flashes of spectrum colours due to the difference in the refractive index of a stone for light of different wavelengths, i.e. dispersion of light waves. / vuur / de feu / Feuer / fuego

0486
iridescence, – the play of rainbow colours seen in cracks and flaws, due to interference of light in thin films of differing refractive index. / iridescentie / iridesence / Schiller / iridiscencia

Some examples are:

opalescence, – iridescence as seen in opals. / opalescentie / opalescence / Opalisieren / opalescencia

labradorescence, – iridescence as seen in labradorite feldspar. / labradorescentie / labradorescence / Labradorisieren / labradorescencia

orient of pearl, – iridescence as seen in pearls. / oriëntparel / orient / Glanz der Perlen / oriente

0487
flaw, – a fracture or a cleavage either on the surface or running partly or completely through the stone. / glis / paille; givre / Fehler / grieta; fractura

0488
chatoyancy, – the 'cat's-eye' effect, due to the reflection of light on fibres or channels arranged in parallel formation within a stone, as seen when the stone is cut en cabochon in the correct direction. / chatoyantie / chatoyant / Chatoyieren / efecto de ojo de gato

cat's eye, – / katteoog / oeuil de chat / Katzenauge / ojo de gato

0489
asterism, – the star effect seen by reflected light in stones containing suitably oriented rod-like inclusions or channels when these are cut en cabochon in the correct direction. / stereffect / astérisme / Asterismus / efecto de estrella; asterismo

0490
sheen, – a shimmering effect due to the reflection of light from structures inside a stone. / glans / éclat / Glanz / brillo

silk, – the fine intersecting rod-like crystals or cavities typically seen in some corundums, which give a silky sheen in reflected light. / zijdeglans / de soie / Seide / seda

0491
fluorescence, – provides in gemmology a useful means of identification; ultraviolet sources of the wavelengths 365 nm (long-wave) and 253.7 nm (short-wave) are used. / fluorescentie / fluorescence / Fluoreszenz / fluorescencia

absorption spectrum, – provides in gemmology a useful means of identification because the absorption bands cross the spectrum in positions characteristic for different gem stones, e.g. ruby and emerald. / absorptiespectrum / spectre d'absorbtion / Absorptionsspektrum / espectro de absorción

0492
artificial treatment, – a technique to alter the colour of a gem stone to a more attractive one. / kunstmatige behandeling / traitement artificiel / künstliche Behandlung / tratamiento artificial

It is done by heating and by:

staining, – with for instance nickel or chromium solutions. / beitsen / coloré artificiel / Färbung / teñido

irradiation, – a bombardment with particles of atomic size. / bestraling / irradiation / Bestrahlung / irradiación

0493
brilliant, – the most important and effective style of cutting for diamond. The cut consists of 58 facets: a table and 32 facets in the crown and 25 in the base. The cut is also used for other stones. / brilliant / brillant / Brilliantschliff / brillante

0494
grading, – the determination of the quality of a diamond by analyzing the 'four C's' – carat weight, clarity, cut and colour. / gradatie / graduation de la qualitĕ / Graduieren / clasificación

point, – term used in describing the mass of diamonds under one carat; the point is 0.01 carat (0.002 grames). / puntje / point / Punktdiamant / punto

0495
ballas, – stones consisting of spherical masses of minute diamond crystals arranged more or less radially. / – – / ballas / Ballas / – –

boart, – a cryptocrystalline form of dark-coloured diamond, sometimes having a radial structure. Unlike ballas it posses-ses a cleavage. / – – / boart / Bort / – –

0496
carbonado, – an opaque, black, tough and compact variety of diamond found in Brazil.

cascalho, – the native name for the diamond or other heavy-mineral bearing gravels of Brazil.

0497
boule, – pear-shaped synthetic corundum or spinel, when produced under the inverted blowpipe of a Verneuil furnace. / smeltpeer / boule synthétique / Schmelzbirne / – –

0498
jade, – in gemmology a term used for both, jadeite and nephrite. / jade / jade / Jade / jade

0499
illam, – the gem-bearing gravel of Sri Lanka.

dullam, – the concentrated gem-bearing gravel of Sri Lanka.

0500
cabochon, – a style of stone cutting with a dome-shaped top. The base may be convex, concave or flat. / cabochon / cabochon / Cabochon / cabujón

0501
table, – the large central facet on the crown of a faceted gem stone. /tafel / table / Tafel / tabla

girdle; setting edge, – the outer edge of a cut stone. / rondist / arête enchassée / Gürtel / filetín

0502
lapidary, – a craftsman who cuts and polishes gem stones other than diamonds. / steensnijder / lapidaire / Edelsteinschneider; Steinschleifer / lapidario

0503
cultured pearl, – a pearl produced by the insertion of an artificial nucleus into the pearl oyster, usually mother-of-pearl, and thereupon the deposition of nacre around it by the mollusc. / gecultiveerde parel / perle de culture / Zuchtperle / perla cultivada

nacre; mother-of-pearl, – / paarlmoer / nacre / Perlenmutter; Mutter der Perle; Perlmutter / nácar; madreperla

endoscope, – an optical instrument used to distinguish between real and cultured, drilled pearls. / endoscoop / endoscope / Endoskop / endoscopio

0504
grain, – the unit of mass commonly used for pearls, equivalent to one fourth of a carat (0.05 grammes). / grein / quart de carat / Grain / grano

[0700] CRYSTALLOGRAPHY

NOTE:

– this chapter is a general introduction only; for further terms see
[0840] The Constituents of Igneous Rocks
[0930] Texture and Structure of Metamorphic Rocks
[0940] The Constituents of Metamorphic Rocks
[2650] Texture and Structure of Ore Minerals

[0710] GEOMETRICAL PROPERTIES OF CRYSTALS

0505
crystallography, – the study of crystals, including their growth, structure, geometrical and physical properties. / kristallografie / cristallographie / Kristallographie / cristalografía

0506
crystal, – a piece of material in the solid state, having an atomic arrangement that is regularly repeated by translation in three dimensions. / kristal / cristal / Kristall / cristal

0507
crystal lattice, – a three-dimensional array of all points in a crystal that are related by translation to an arbitrarily chosen point. /kristalrooster / réseau cristallin / Kristallgitter / red estereocristalina

0508
lattice point, – / roosterpunt / noeud du réseau / Gitterpunkt / nudo de la red

lattice row, – /roosterlijn / rangée du réseau / Gittergerade / línea de la red

lattice plane, – / roostervlak / plan réticulaire / Gitterebene / plano reticular

0509
crystal structure, – the atomic arrangement of a crystal as exemplified by the contents of the unit cell. / kristalstructuur / structure cristalline / Kristallstruktur / estructura cristalina

0510
unit cell, – a cell of the crystal lattice, the edges of which are parallel to the crystallographic axes and have a length equal to the translation period along these axes. / elementaircel / maille élémentaire / Elementarzelle / celdilla cristalina; célula ....

0511
crystallographic axis, – one of the three non-planar translation directions of the crystal lattice used to describe the atomic positions and the crystal geometry. The axes are chosen so as to comply with the crystal symmetry. / kristallografische as / axe cristallographique / kristallographische Achse / eje cristalográfico

0512
point group; crystal class, – one of the 32 ways in which symmetry elements, occurring in a crystal and going through a common point, can be combined. / puntgroep; kristalklasse / groupe ponctuel; classe cristalline / Punktgruppe; Kristallklasse / clase cristalina

0513
space group, – one of the 230 ways in which the symmetry elements that can occur in a crystal structure, can be combined in space. / ruimtegroep / groupe spatial; .... d'espace / Raumgruppe / grupo espacial

0514
crystal system, – one of seven groups, each comprising a number of point groups, allowing the description of a crystal being based on one set of crystallographic axes with specific dimensional properties. / kristalstelsel / système cristallin / Kristallsystem / sistema cristalino

0515
triclinic, – / triklien / triclinique / triklin / triclínico

monoclinic, – / monoklien / monoclinique / monoklin / monoclínico

orthorhombic, – / orthorhombisch / orthorhombique / (ortho)rhombisch / ortorómbico

trigonal, – / trigonaal / trigonal; rhomboédrique / trigonal / trigonal

tetragonal, – / tetragonaal / tétragonal; quadratique / tetragonal / tetragonal

hexagonal, – / hexagonaal / hexagonal / hexagonal / hexagonal

cubic; isometric (obsolete); regular (obsolete), – / kubisch / cubique / kubisch / cúbico

0516
rhombohedral cell, – a unit cell occurring in the trigonal system and having the form of a rhombohedron. / rhomboëdrische cel / maille rhomboédrique / rhomboedrische Zelle / malla romboédrica

0517
face, – a naturally occurring planar surface bounding a crystal. / vlak / face / Fläche / cara

0518
edge, – the line of intersection of two faces. / ribbe / arête; côte / Kante / arista

0519
zone, – a series of three or more faces intersecting along parallel edges. / zone / zone / Zone / zona

0520
crystal form, – a set of crystal faces related to each other by the symmetry elements of the point group. / kristalvorm / forme cristalline / Kristallform / forma cristalina

0521
habit, – the overall shape of a crystal expressed in qualitative terms; often also used to indicate the combination of forms. / habitus / habitus / Habitus / hábito

combination, – the crystal forms present on a crystal. / vormenkombinatie / faciès / Tracht; Flächenkombination / combinación

0522
tabular; platy, – / plaatvormig / tabulaire; aplati / tafelig / tabular

isometric; equidimensional, – / isometrisch / isométrique / isometrisch / isométrico

pyramidal, – / piramidaal / pyramidal / pyramidal / piramidal

prismatic, – / prismatisch / prismatique / prismatisch / prismático

lath-like, – / latvormig / en lattes / leistenförmig / en forma de listón

acicular; needle-like, – / naaldvormig / aciculaire; en forme d'aiguilles / nadelig / acicular

fibrous, – / vezelig / fibreux / faserig / fibroso

0523
euhedral; idiomorphic, – bounded by crystal faces. / euhedrisch; idiomorf / idiomorphe / automorph; idiomorph / euhedral

subhedral, – bounded by partially developed crystal faces.

anhedral, – not bounded by crystal faces, the surface being irregular or curved. / anhedrisch / anhédrique / allotriomorph / anhedral

0524
whisker, – a hair-like crystal (abnormal habit caused by extreme growth in one direction only). / – – / whisker / Haarkristall / triquita

0525
dendrite, – a single crystal that has crystallized in a branching tree-like pattern. / dendriet / dendrite / Dendrit / dendrita

0526
growth sector; .... pyramid, – that part of a crystal that was always bounded by one and the same crystal face during growth. / groeisector / secteur de croissance / Wachstumssektor / sector de crecimiento

0527
step; ledge, – the end of a growth layer on a crystal face. / trede / lisière; marche / Stufe / escalón de crecimiento; .... borde; .... orilla

0528
growth hillock, – a hill with a very small slope (a few degrees or less) on a crystal face, caused by superimposed growth layers, each having a smaller dimension than the layer on which it lies. / groeiterras / figure de croissance / Wachstumshügel / figura de crecimiento

0529
growth spiral, – a step in the form of a spiral caused by its winding up during growth around the point where a screw dislocation emerges on the crystal face. / groeispiraal / spirale de croissance / Wachstumsspirale / espiral de crecimiento

0530
etch pit; .... figure, – a pit in a crystal face produced by the attack of a solvent. / etsput; etsfiguur / figure de corrosion / Ätzfigur / figura de corrosión

0531
crystal defect, – an imperfection in the ideal crystal structure; there are **point, line** and **plane defects.** / kristalfout / défaut cristallin / Fehlordnung / defecto cristalino

dislocation, – a line defect in the crystal structure; there are **edge, screw** and **mixed dislocations.** / dislocatie; rand....; schroef.... / dislocation; ....-coin; ....-vis / Versetzung; Stufen....; Schrauben.... / dislocación

0532
grain boundary, – the contact surface of two grains of a polycrystalline solid, either of different minerals or of the same mineral in different orientation. / korrelgrens / joint de grain / Korngrenze / contorno del grano; límite del ....

0533
zoning, – a compositional variation within a crystal from core to margin. / zonering / structure zonée / Zonarbau / zonificación; estructura zonada

0534
hourglass structure, – a type of zoning in which a prominent growth sector has a cross section resembling that of an hourglass. / zandloperstructuur / structure en sablier / Sanduhraufbau / estructura en reloj de arena

0535
twin, – an intergrowth of two or more crystals of the same mineral in such a way that the lattices have a special mutual orientation. / tweeling / macle / Zwilling / macla

twin law, – the special mutual orientation of a twin crystal. / tweelingswet / loi de macle / Zwillingsgesetz / ley de la macla

0536
composition surface; .... plane, – the surface or plane along which the individuals of a twin are joined. / vergroeiingsvlak / plan d'accolement / Verwachsungsfläche; Zwillingsebene / plano de unión; .... de yuxtaposición

0537
contact twin, – a twin of which the composition surface is a plane. / contacttweeling / macle de contact / Ebenenzwilling / macla de contacto

interpenetration twin; penetration twin, – a twin of which the individuals appear to have grown through one another. / penetratietweeling; doordringings.... / macle de pénétration / Durchdringungszwilling; Durchkreuzungs.... / macla por penetración

0538
polysynthetic twinning, – repeated twinning of three or more individuals according to the same law and with parallel composition planes. / polysynthetische vertweelinging / maclage polysynthétique / polysynthetische Verzwillingung / maclado polisintético

0539
slip plane; glide ...., – a plane in a crystal along which translation (slip) of one part of the crystal may take place due to plastic deformation. / glij(d)vlak / plan de cisaillement / Gleitebene / plano de deslizamiento

slip band; deformation lamella, – a region in a crystal where slip has taken place. / deformatielamelle / bande de déformation / – – / banda de deformación; lamela de ....

0540
cleavage plane, – a plane in a crystal along which it can break when yielding to stress. / splijtvlak / plan de clivage / Spaltfläche / plano de exfoliación; .... de crucero

0541
fracture, – the breaking of a crystal other than along cleavage planes. / breuk / fracture / Bruch / fractura

conchoidal fracture, – a fracture yielding an irregular surface consisting of smooth and curved shell-like areas. / schelpachtige breuk / fracture conchoïdale / muscheliger Bruch / fractura concoidea

parting, – a type of fracture along planes of weakness caused by deformation, twinning or inclusions. / – – / cassure / Teilung / – –

0542
epitaxy, – an intergrowth of two crystals of different minerals in such a way that the lattices have a special mutual orientation (the **epitaxy relationship**), in which usually a lattice plane and a lattice row in that plane of each of the crystals are parallel. / epitaxie / épitaxie / Epitaxie / epitaxia

epitactic, – adjective of epitaxy; (often the incorrect adjective epitaxial is used). / epitactisch / épitactique / epitaktisch / epitáctico

0543
topotaxy, – a strongly preferred orientation of a crystal aggregate, produced by either a transformation fron a polymorph or by a chemical reaction, whereby the lattice orientation is such that a lattice plane and a lattice row in that plane are parallel to a plane and a row of the original crystal. / topotaxie / topotaxie / Topotaxie / topotaxia

topotactic, – adjective of topotaxy. / topotactisch / topotactique / topotaktisch / topotáctico

[0620] OPTICAL PROPERTIES OF CRYSTALS

0544
optical crystallography, – the part of crystallography that deals specifically with optical properties of crystals. / optische kristallografie / optique cristalline / optische Kristallographie / cristalografía óptica

0545
transparent, – capable of transmitting light to such an extent that an object may be seen. / doorzichtig / transparent / durchsichtig / transparente

translucent, – capable of transmitting light but not to such an extent that an object may be seen. / doorschijnend / translucide / durchscheinend / translúcido

opaque, – not capable of transmitting light. / opaak / opaque / opak / opaco

0546
colour, – the visual perception of light transmitted or reflected by a mineral, whereby this light has obtained a spectral composition different from the incident white light. / kleur / couleur / Farbe / color

0547
orthoscopic illumination, – an illumination, usually of a polarising microscope, that produces an image of the object on the retina of the eye, or on a screen or photographic film. / orthoscopische belichting / illumination orthoscopique / orthoskopische Beleuchtung / iluminación ortoscópica

conoscopic illumination, – an illumination, usually of a polarising microscope, that produces an image on the retina of the eye, or on a screen or photographic film, of the light source as seen through the object. / conoscopische belichting / illumination conoscopique / konoskopische Beleuchtung / iluminación conoscopica

0548
light ray, – an infinitely thin beam of light, radiating from a point source. / lichtstraal / rayon lumineux / Lichtstrahl / rayo de luz

0549
wave surface; ray-velocity ...., – the locus of all the points reached after a set time by light rays radiating from a point source in all directions. / golfoppervlak / surface d'onde / Wellenoberfläche / superficie de la onda

wave front, – a plane, associated with a light ray, that is tangent to the wave surface at the point where the light ray pierces the wave surface. / golffront / plan d'onde / Wellenfront / plano de la onda

0550
front normal, – the normal to a wave front. / frontnormaal / normale au plan d'onde / Normale zur Wellenfront / normal al plano de la onda

0551
refractive index; index of refraction, – a number that expresses the ratio of the velocity of light in vacuo to the velocity of a wave front within a crystal. / brekingsindex / indice de réfraction / Brechungsindex / índice de refracción

0552
indicatrix, – a closed surface of which each radius vector represents with its direction a vibration direction of a polarised light ray in the crystal and with its length the magnitude of the corresponding refractive index. / indicatrix / ellipsoïde des indices; indicatrice / Indikatrix / indicatriz

0553
uniaxial, – pertaining to an indicatrix with the shape of an ellipsoid of revolution with one circular section. / eenassig / uniaxe / einachsig / uniaxial

biaxial, – pertaining to an indicatrix with the shape of a triaxial ellipsoid with two circular sections. / tweeassig / biaxe / zweiachsig / biaxial

0554
optic axis, – direction in the crystal perpendicular to a circular section of an indicatrix. / optische as / axe optique / optische Achse / eje óptico

optic axial plane, – the plane containing the two optic axes of a crystal with a biaxial indicatrix. / vlak van de optische assen / plan des axes optiques / optische Achsenebene / plano de los ejes ópticos

0555
optic angle; optical ....; optic axial ...., – the acute angle between the two optic axes of a biaxial crystal. / optische assenhoek / angle des axes optiques / optischer Achsenwinkel / ángulo de los ejes ópticos

0556
bisectrix, – the bisectrix of the angle made by the two optic axes of a biaxial crystal. / bisectrice / bissectrice / Bisektrix / bisectriz

acute bisectrix, – the bisectrix of the acute angle between the optic axes. / scherpe bisectrice / bissectrice aiguë / spitze Bisektrix / bisectriz aguda

obtuse bisectrix, – the bisectrix of the obtuse angle between the optic axes. / stompe bisectrice / bissectrice obtuse / stumpfe Bisektrix / bisectriz obtusa

0557
optic sign; optical sign – a sign, positive or negative, characterizing the shape of an indicatrix. / optisch teken / signe optique / optischer Charakter / signo óptico

positive, – the acute bisectrix or the axis of revolution corresponds to the largest refractive index.

negative, – the obtuse bisectrix or the axis of revolution corresponds to the smallest refractive index.

0558
relief, – according to W. R. Phillips, the degree of visibility of a transparent grain in its surrounding medium. / reliëf / relief / Relief / relieve

0559
Becke line, – in transmitted light microscopy, a bright line that appears near the boundary of a grain when the object is slightly out of focus. / lijn van Becke / frange de Becke; ligeré de Becke / Beckesche Linie / línea de Becke

Kalb line, – in reflected light microscopy, a bright line that appears at the junction of a soft and a hard mineral (in polished section). If a specimen is being lowered away from the objective a light line appears to move into the softer mineral. / lijn van Kalb / frange de Kalb / Kalbsche Linie / línea de Kalb

0560
birefringence, –
1. the ability of a crystal to split a beam of ordinary light into two beams of unequal velocities and (eventually) of unequal absorption, the vibration directions of which are perpendicular to each other;
2. the difference between the refractive indices of these two beams;
3. the difference between the largest and smallest refraction index of a mineral.

/ dubbelbreking / double réfraction; biréfringence / Doppelbrechung / birefringente

0561
optically isotropic, – said of a mineral that shows no birefringence. / optisch isotroop / (optiquement) isotrope / optisch isotrop / opticamente isótropo

optically anisotropic, – said of a mineral that shows birefringence. / optisch anisotroop / (optiquement) anisotrope / optisch anisotrop / opticamente anisótropo

0562
pleochroism, – the property of a birefringent crystal to absorb differentially various wavelengths of transmitted light in various directions, and thus to show different colours in different directions. / pleochroïsme / pléochroisme / Pleochroismus / pleocroísmo

0563
optical path difference; retardation, – the difference of the distances two parallel wave fronts travel in air after their refraction through a crystal. / gangverschil / retard; différence de marche / Gangunterschied / retardo

0564
extinction, – the more or less complete darkness of a birefringent crystal obtained between crossed polarizers at two positions where the vibration directions in the crystal are parallel to those of the polarizers. / uitdoving / extinction / Auslöschung / extinción

Extinction may be:

straight, – / recht / droite / gerade / recta

inclined; oblique, – /scheef / oblique / schief / oblicua

symmetrical, – / symmetrisch / symétrique / symmetrisch / simétrica

0565
undulatory extinction, – a type of extinction that occurs successively in adjacent areas, as the microscope's stage is turned. / onduleuze uitdoving / extinction onduleuse; .... roulante / undulöse Auslöschung / extinción ondulante

0566
extinction angle, – the angle between an extinction position and a known crystallographic direction of a section of a birefringent crystal. / uitdovingshoek / angle d'extinction / Auslöschungswinkel / ángulo de extinción

0567
elongation, – the apparent optic sign of a crystal or a grain with an elongate shape. / elongatie / allongement (optique) / Elongation / elongación

positive, – when the larger axis of the elliptic cross section of the indicatrix makes an angle less than 45° with the elongate direction.

negative, – when the smaller axis makes an angle less than 45° with the elongate direction.

0568
interference colour, – the colour displayed by a birefringent crystal between crossed polarizers. / interferentiekleur; polarisatiekleur / teinte de polarisation / Interferenzfarbe / color de interferencia

anomalous interference colour, – a colour that does not fit into the normal Newtonian scale of interference colours due to dispersion and/or absorption effects. / anomale interferentiekleur / teinte de polarisation anomale / anomale Interferenzfarbe / color de interferencia anómalo

0569
interference figure, – the conoscopic image obtained from a convergent light beam passing through a crystal between crossed polarizers. / interferentiefiguur; assenbeeld / image d'interférence(s); figure de polarisation / Interferenzfigur; Achsenbild / figura de interferencia

0570
isogyre, – the black or shadowy part of an interference figure that is produced by extinction. / isogyre / ligne neutre / Isogyre / línea neutra

melatope, – a point on the isogyre where an optic axis emerges. / melatoop / mélatope; point noir / Melatop / – –

0571
isochromatic curve, – the assembly of points of an interference figure having the same optical path difference. / isochromaat; isofasische lijn / courbe isochromatique / isochromatische Kurve / curva isocromática

0572
dispersion, – the phenomenon that extinction angles and interference figures are different for different wavelengths. / dispersie / dispersion / Dispersion / dispersión

axial dispersion, – / assendispersie / dispersion des axes optiques / Dispersion der optischen Achsen / dispersión axial

crossed dispersion, – / gekruiste dispersie / dispersion croisée; .... droite; .... normale / gekreuzte Dispersion / dispersión cruzada

0573
horizontal dispersion, – / horizontale dispersie / dispersion horizontale / horizontale Dispersion / dispersión horizontal

inclined dispersion, – / hellende dispersie / dispersion oblique; .... diagonale / geneigte Dispersion / dispersión inclinada

0574
reflectance; reflectivity, – in reflected light microscopy, the ratio of the intensity of the reflected beam to that of the incident beam (E) taken as unity; it is expressed either as a fraction (R) or as a percentage (R%); the ratio R/E is also called the **reflecting power**. The phenomenon depends on the optical symmetry of the crystal surface. Generally the special case of normal incidence is considered. / reflectiviteit / réflectance; pouvoir réflecteur; réflectence; réflectivité / Reflexion / reflectividad

0575
bireflection, – a change in the reflectivity, that depends on birefringence and biabsorption. / bireflectie / biréflectance / Bireflexion / bireflexión

0576
internal reflection, – the beam passes through the optically denser medium, n_2, and is reflected from the surface of the less dense medium, n_1. / interne reflectie / réflexion interne / Innenreflexion / reflexión interna

external reflection, – when a beam passing through a medium of refractive index n_1 is reflected from the surface of an optically denser medium n_2, i.e. where $n_2 > n_1$. / externe reflectie / réflexion externe / Reflexion / reflexión externa

0577
glare, – light forming a secondary component to the incident beam, causing errors that have to be taken into account in reflectivity measurements in ore-mineral microscopy. / – – / reflet / – – / deslumbramiento

primary glare, – the effect due to some light from the illuminator being reflected by the objective upwards before it has fallen upon the specimen.

secondary glare, – the effect due to light on its way up from reflection by the specimen being partly reflected down again by the objective.

[0800] PETROLOGY – INTRODUCTORY TERMS

NOTE:
– for further terms see chapter:
[2600] Metallic Mineral Deposits.

0578
petrology, – the science that treats of the natural history of rocks, including their origin, present conditions, alterations and decay. / petrologie / pétrologie / Petrologie / petrología

0579
petrography, – a branch of petrology dealing with the description and systematic classification of rocks. / petrografie / pétrographie / Petrographie; Gesteinskunde / petrografía

0580
petrochemistry, – study of the chemical composition of rocks (not equivalent to petroleum chemistry). / geochemie / pétrochimie / Geochemie / petroquímica

0581
petrogenesis; petrogeny, – a branch of petrology dealing with the origin and formation of rocks; especially igneous rocks. / petrogenese / pétrogenèse / Petrogenese / petrogénesis

0582
petrogenetic cycle, – sequence of rock-forming processes including the magmatic, sedimentary, metamorphic, and anatectic stages, whereby the cycle is closed by palingenesis of a new magma. / petrogenetische kringloop / cycle pétrogénétique / petrogenetischer Kreislauf / ciclo petrogenético

0583
crystalline rock, – rock consisting wholly of crystals or fragments of crystals, but lacking glassy material. Also used collectively for igneous and metamorphic rocks. / kristallijn gesteente / roche cristalline / kristallines Gestein / roca cristalina

0584
infracrustal, – adjective for rocks formed at great depth. It is synonymous with plutonic but also used for metamorphic rocks and processes. / infracrustaal / infracrustal / infrakrustal / infracórtical

supracrustal, – adjective for rocks overlying the basement. Intercalated intrusives are also included. / supracrustaal / supracrustal / suprakrustal / supracórtical

0585
mineral facies, – a term having a wider scope than metamorphic facies. It includes rocks of any origin whether igneous, metamorphic or metasomatic, whose constituents have been formed within the limits of a certain pressure-temperature range which is characterized by the stability of certain index-minerals. / mineraalfacies / faciès minéralogique / Mineralfazies / facies mineralógica

0586
assemblage; mineral ....; mineral association, – a group of minerals found together in a rock. For metamorphic rocks the term **metamorphic assemblage** is used as a synonym. / associatie / association minéralogique / Vergesellschaftung; Mineral-....; Mineral-Assoziation / unión; montaje; asociación mineralógica

paragenesis, – characteristic association or occurrence of minerals: a mineral assemblage representing equilibrium conditions. / paragenese / paragenèse / Paragenese / paragénesis

paragenetic, – pertaining to paragenesis. / paragenetisch / paragénétique / paragenetisch / paragenético

0587
leucocratic; light-coloured, – adjectives for rocks consisting predominantly of light-coloured (felsic) minerals (dark minerals usually less than 30-35%). / leukokraat / leucocrate / leukokrat / leucocrato; leucocrático

hololeucocratic, – adjective for rocks consisting completely or almost completely of felsic minerals. / hololeukokraat / hololeucocrate / hololeukokrat / hololeucocrato; hololeucocrático

mesocratic; mesotype, – adjectives for rocks with about equal amounts of light- and dark-coloured (mafic) minerals (dark minerals usually between 30 and 60%). / mesokraat / mésocrate; mésotype / mesokrat; mesotyp / mesocrato; mesocrático

0588
melanocratic; melanic; dark-coloured; chromocratic, – adjectives for rocks consisting predominantly (at least 50-60%) of dark-coloured minerals. / melanokraat / mélanocrate / melanokrat / melanocrato; melanocrático

holomelanocratic; hypermelanic, – adjectives for rocks consisting completely or almost completely of mafic minerals. / holomelanokraat / holomélanocrate / holomelanokrat / holomelanocrato; holomelanocrático

0589
monomineralic; anchi....; monogene; monogenic, – adjectives for rocks composed essentially of a single mineral, e.g. anorthosite or dunite. / (anchi-)monomineraal / monominéral / (anchi)monomineralisch / monomineral; monomineralógico; monomicta

polymineralic, – adjective for rocks composed of two or more minerals. / polymineraal / polyminéral / polymineralisch / polimineral

0590
polygenetic; polygene; polygenic, – resulting from more than one process of formation (as in polygenetic inclusions), or derived from more than one source or originating or developing at various places and times (as in a mountain range resulting from various orogenic episodes). Also used in the sense of polymineralic. / polygeen / polygénétique / polygen; polygenetisch / poligénico

0591
generation, – all the crystals of the same mineral species that crystallized at the same time. If there are olivine phenocrysts in a groundmass containing olivine there are said to be two 'generations' of olivine. / generatie / génération / Generation / generación

0592
habit, – the sum of the external characteristics of a mineral or rock. In its application to rocks the term implies more than structure or texture, including also other features that control the outward appearance, such as lustre, degree of alteration and fracture. / habitus / habitus / Habitus / hábito

0593
mode; mineralogical composition, – the actual mineral composition of a rock expressed quantitatively in percentages by mass or volume. / modus; modale samenstelling; kwantitatieve mineralogische samenstelling / mode; composition modale / Modus; quantitative mineralogische Zusammensetzung / modo

modal, – pertaining to the mode. / modaal / modal / modal / modal

0594
principal constituent; .... component; .... mineral, – minerals present in major quantities in a rock, making up the bulk of the rock volume. / hoofdbestanddeel / minéral principal / Hauptgemengteil / componente principal

auxiliary constituent; .... component; .... mineral, – minerals present in minor quantities in a rock. / nevenbestanddeel / minéral auxiliaire / Nebengemengteil / componente subordinado; .... auxiliar

accessory constituent; .... component; .... mineral; accessory, – minerals occurring in very small amounts in a rock. / accessorisch bestanddeel; .... mineraal / minéral accessoire / akzessorisches Mineral; .... Gemengteil / componente accesorio; mineral ....

essential constituent; .... component; .... mineral; specific mineral, – minerals occurring in a rock that are necessary to its classification and nomenclature, but that are not necessarily present in large amounts. / kenmerkend bestanddeel; .... mineraal / minéral essentiel / wesentliches Mineral; mitbestimmender Gemengteil / mineral esencial

NOTE:
– in the German language the use of the plural 'Minerale' instead of 'Mineralien' is increasingly gaining ground.

0595
varietal mineral; characterizing accessory; distinctive ...., – minerals that distinguish one variety of rock from another. / specifiek mineraal / minéral caractéristique / charakteristischer Übergemengteil / mineral característico; .... fundamental

vikariierender Übergemengteil; stellvertretender Gemengteil, – German terms for accessory minerals that replace a principal or auxiliary constituent more or less completely, e.g. tourmaline replacing mica in granites.

0596
primary mineral, – those minerals that were formed in the original ore-forming or rock-forming episode. / primair mineraal / minéral primaire / primäres Mineral / mineral primario

secondary mineral, – those minerals that were formed later than the original rock constituents and as a consequence of processes not directly connected with the process of formation of the original rock. / secundair mineraal / minéral secondaire / sekundäres Mineral / mineral secundario

0597
felsic, – adjective for the minerals composed essentially of silicon and aluminium (feldspars, feldspathoids, quartz, corundum) that are actually present in a rock; also for rocks predominantly composed of such minerals. / felsisch / felsique / felsisch / félsico

felsite, – rock composed predominantly of felsic minerals. / felsisch gesteente / roche felsique; felsite / felsisches Gestein; Felsit / felsita

0598
mafelsic, – adjective for a rock in which felsic and mafic minerals are present in equal amounts. / mafelsisch / – – / mafelsisch / mafélsica

0599
mafic; hypermelanic; ferromagnesian, – adjectives for the minerals composed essentially of iron, magnesium and silicon, and that are actually present in a rock; also for rocks composed predominantly of such minerals. / mafisch / mafique; ferromagnésien / mafisch / ferromagnésico; máfico; ferromagnesiano

mafite, – rock composed predominantly of mafic minerals. / mafisch gesteente / roche mafique / mafisches Gestein; Mafitit / mafita

melanes, – mafic minerals.

0600
ultramafic, – adjective for a rock that is predominantly composed of mafic minerals and as such synonymous with mafic; the term is generally accepted, however, and widely used for rocks, usually of ultrabasic chemistry that have a colour index of more than 70. / ultramafisch / ultramafique / ultramafisch / ultramáfica

ultramafite, — ultramafic rock. The ultramafites are separated in two classes: those that have been emplaced as a magma and have crystallized in place and those designated as **alpine-type,** that are generally strongly tectonized and do not show evidence of magmatic emplacement such as contact-metamorphic aureoles. / ultramafisch gesteente / roche ultramafique / ultramafisches Gestein; Ultramafit / ultramafita

alpine-type; alpinotype; alpine, — / alpinotype / type alpin / alpinotyp / tipo alpino

0601
acid; acidic; persilicic, — adjectives for rocks with a higher percentage of silica than orthoclase. In general SiO_2 taken to be above 66%. / zuur / acide / sauer / ácida

acidite, — acid rock. / zuur gesteente / roche acide / saures Gestein / roca ácida

0602
intermediate; mediosilicic; neutral, — adjectives for rocks intermediate in composition between acid and basic, with in general SiO_2 between 54 and 66%. / intermediair / intermédiaire / neutral; intermediär / mediosilícea; intermedia

0603
basic; subsilicic, — adjectives for rocks with an $Si0_2$ content generally taken between 45 and 54%. / basisch / basique / basisch / básica

basite, — basic igneous rock. / basiet / basite / Basit / basita

0604
ultrabasic, — adjective for rocks with an $Si0_2$ content generally taken to be lower than 45%. /ultrabasisch / ultrabasique / ultrabasisch / ultrabásica

ultrabasite, — ultrabasic rock. / ultrabasiet / ultrabasite / Ultrabasit / ultrabasita

0605
oversaturated; silicic, — adjectives for rocks containing free quartz. / oververzadigd / sursaturé / übersättigt / silícica

saturated, — adjective for minerals that can form in the presence of free silica, for rocks containing neither free quartz nor unsaturated minerals or having quartz in the norm, and for magmas from which such rocks can crystallize. / verzadigd / saturé / gesättigt / saturado

unsaturated; undersaturated, — adjectivs for minerals that are unstable in the presence of free silica, for rocks containing such minerals or carrying them in the norm, and for magmas from which such rocks can crystallize. / onverzadigd / soussaturé / ungesättigt / no saturado; subsaturado

critically undersaturated, — adjectives for a rock having feldspathoids and olivine in its norm. / kritisch onderverzadigd / sous-saturé / kritisch untersättigt / infrasaturada

0606
norm; standard mineral composition, — the chemical composition of a rock expressed in terms of standard minerals. It may or may not correspond to the actual mineral composition or mode. According to the way in which the normative composition was calculated, a **Niggli norm,** a **CIPW norm** a **Barth norm** is obtained. The **Rittmann norm** is calculated according to a method devised specifically for volcanic rocks. / norm; normatieve samenstelling / norme; composition (minéralogique) virtuelle / Norm / norma

epinorm, — the chemical composition of a rock expressed in terms of standard metamorphic minerals of the epizone. / epinorm / épinorme / Epinorm / epinorma

mesonorm, — the chemical composition of a rock expressed in terms of standard metamorphic minerals of the mesozone. / mesonorm / mésonorme / Mesonorm / mesonorma

0607
normative mineral; standard ...., — the minerals constituting the norm of a rock. / normatief mineraal; standaard.... / minéral normatif; .... virtuel / normatives Mineral / mineral normativo

femic, — adjective used for the CIPW normative minerals rich in iron, magnesium and calcium. Also used in Niggli's chemical classification to designate fm-rich al-poor magma types. / femisch / fémique / femisch / fémico

salic, — adjective used for the CIPW normative minerals rich in silicon and aluminium. Also used in Niggli's chemical classification to designate al-rich fm-poor magma types. / salisch / salique / salisch / sálico

0608
megacryst, — any crystal or grain in an igneous or metamorphic rock that is significantly larger than the surrounding matrix. / megakrist / mégacristal / Megakristall / macrocristal; megacristal

0609
anhydrous, — adjective for a magma, a rock or a mineral that is completely or almost completely without water. / droog / anhydre / trocken; wasserfrei / anhidro

0610
opaque constituent; opacite, — / opaak bestanddeel / minéral opaque / opaker Bestandteil / componente opaco; opacita

0611
polarizing microscope; petrographic ...., — / polarisatiemicroscoop / microscope polarisant / Polarisationsmikroskop / microscopio de polarización; .... petrográfico

0612
Chayes point counter, — a mechanical device that allows traversion of a microscopic slide in a series of regularly spaced points. It tabulates the minerals encountered in order to enable estimation of the model composition of the investigated rock. / puntenteller / compteur de points / Punktzähler / contador de puntos

0613
Rosiwall method, — method by which the quantitative mineralogical composition of a rock is estimated from the total length of linear intercepts of each mineral determined with the aid of a micrometer-fitted stage. / methode van Rosiwall / méthode de Rosiwall / Rosiwall-Methode / Método micrométrico de Rosiwall

0614
universal stage; U-stage, – accessory for the polarizing microscope that allows the investigated thin section or grain mount to be rotated relative to the microscope stage. Used for the measurement of optical characteristics and for the determination of the spatial orientation of individual grains. / tafel van Fedorov; universele draaitafel / platine universelle / Universaldrehtisch / platina universal; .... de Fedorow

[0900] PETROLOGY OF IGNEOUS ROCKS

[0910] GENERAL TERMS

0615
plutonism, – a general term for the phenomena associated with the formation of igneous bodies. / plutonisme / plutonisme / Plutonismus / plutonismo

0616
igneous; magmatic; pyrogenous; pyrogenic; eruptive; orthotectic; (obsolete terms: anogenic; anogene), – of, pertaining to, or derived from magma; as in igneous rocks. / magmatisch / igné; magmatique; éruptif / magmatisch / ígnea; eruptiva; magmática

magmatite, – igneous rock. / stollingsgesteente / roche ignée / Eruptivgestein; Erstarrungs....; Massen....; Magma....; Magmatit / roca magmática

pyrogenesis, – the intrusion and extrusion of magmas and their derivatives. / pyrogenese / – – / Pyrogenese / pirogénesis

0617
plutonic; plutonian; abyssal; deep-seated; hypogene (obsolete), – adjectives for processes taking place at great depth and for rocks formed at great depths./ plutonisch; abyssaal / plutonique; abyssal / plutonisch; abyssisch; abyssal / plutónica; abisal; profunda

plutonite; irruptive rock, – plutonic rock. / dieptegesteente; plutoniet / roche plutonique / Tiefengestein; Plutonit / roca plutónica

hypabyssal; subvolcanic, – adjectives for rocks and processes intermediate between plutonic and volcanic. / hypabyssaal / hypabyssique; hypoabyssal; de semi-profondeur / hypabyssisch; hypabyssal / hipabisal; subvolcánica; somera

0618
intrusion; emplacement; invasion; irruption; injection, – the process of emplacement of magma in pre-existing rock; magmatic activity; giving rise to plutons. / plaatsname; intrusie / intrusion / Intrusion / intrusión; erupción; inyección; emplazamiento

0619
permissive; passive; suctive, – adjectives for a magma or intrusion if space is not forcibly created by the magma itself. / passief / – – / passiv / pasivo

intrusive; irruptive, – of and pertaining to intrusion and pluton. / intrusief / intrusif / intrusiv / intrusivo

aggressive; invasive, – adjectives for a magma or intrusion that created space by forcible intrusion. / agressief; actief / – – / aggressiv; aktiv / activo

0620
injection complex, – an assemblage or association of rocks consisting of igneous intrusions in intricate relationship to sedimentary and metamorphic rocks. / intrusief complex / complexe intrusif / Intrusiv-Komplex / complejo intrusivo; .... plutónico

0621
autointrusion; autoinjection, – intrusion, along rifts, of the rest magma into its own crystal mush. / autointrusie / autoinjection / Autointrusion / autointrusión; autoinyección

0622
magmatic pressure; volcanic ...., – the pressure exerted by a magma on its surroundings; the pushing of hypomagma toward the surface may be due to gravity, to the internal gas pressure or, as thought by others, to an external cause like folding or faulting movements. / magmatische druk / pression magmatique / magmatischer Druck / presión magmática

0623
magmatic stoping; stoping, – emplacement of magma through detachment of pieces of country rock that sink or float in the magma chamber or are assimilated. / magmatische opslokking / stoping / Aufschmelzung / excavación magmática

overhead stoping; overhand ...., – the overlying beds have been broken out. / opslokking van het dak / – – / Dachaufschmelzung / arranque por realce

underhand stoping, – stoping of the underlying beds. / opslokking van de vloer / – – / Aufschmelzung des Liegenden / arranque por rebaje

arrested stoping, – congelation of apophyses radiating from the roof of a batholith. / gestuite opslokking / – – / – – / arranque interrumpido

0624
deroofing, – fusing of the roof of a batholith leading to a deroofing extrusion or an areal eruption. / doorsmelting van het dak / fusion du toit / Durchschmelzung des Daches / – –

0625
zone melting; zone refining, – process through which magma migrates upward through a solid wall rock. As a result of thermal convection within the 'magma chamber' crystallization will take place in the cooler bottom part of the chamber, combined with fusion of overlying wall rock near the hotter top part. / – – / fusion par zone / zonares Aufschmelzen / zona de fusión

0626
groundmass; matrix, – the material between the phenocrysts in a porphyritic rock. It includes the basis as well as the smaller crystals of the second generation. / grondmassa; matrix / pâte / Grundmasse; Matrix / pasta; matriz

mesostasis; basis; base, – the last-formed interstitial material, either glassy or aphanitic of an igneous rock. / mesostasis; basis / mésostase / Mesostasis; Basis / mesostasis

0627
glass, – a general term for the amorphous consolidation products of magmas, whether forming the whole of a rock, as in obsidian or pumice, or only a groundmass or mesostasis. / glas / verre / Glas; Gesteinsglas / vidrio

0628
petrographic province; magma ....; igneous ....; comagmatic region, – a natural region in which the magmatic rocks, erupted during the same period of igneous activity, are regarded to have been derived from a common parent magma; or are thought to be the expression of a common endogenic process. / petrografische provincie; comagmatisch gebied / province magmatique; .... pétrographique / komagmatische Region; petrographische Provinz / provincia petrográfica

0629
consanguinity, – the relationship between igneous rocks that are thought to have the same parent magma. / (bloed)verwantschap / consanguinité / Verwandschaft / parentesco; consanguinidad

consanguineous; comagmatic – / comagmatisch / consanguin; comagmatique / verwandt; comagmatisch / comagmático; consanguíneo

0630
rock association; rockkindred; suite, – group of chemically and petrographically related igneous rocks within a petrographic province. / (gesteente)associatie / association magmatique / (Gesteins)vergesellschaftung / asociación (de rocas)

rock series; igneous rock series, – an assemblage of related igneous rocks of the same general form of occurrence (plutonic, hypabyssal, volcanic), that occurs in one district and is characterized by continuous variation of the properties (composition, texture, etc.) of its members. / gesteentereeks / série de roches ignées / Gesteinsserie / serie de rocas ígneas

0631
primary rock, – a rock whose constituents are newly formed particles that have never been constituents of previously formed rocks and that are not the products of alteration or replacement. / primair gesteente / roche primaire / Primärgestein / roca primaria

primary gneiss. – a rock that exhibits planar or linear structures characteristic of metamorphic rocks but that lacks observable granulation or recrystallization and is therefore considered to be of igneous origin. / primaire gneis / – – / Primärgneis / gneis primario

primary flowage, – movement within an igneous rock that is still partly fluid. / primaire vloei / fluidalité primaire / primärer Fluss / flujo primario

0632
crystallization, – the process of matter becoming crystalline from a gaseous, fluid or dispersed state. / kristallisatie / cristallisation / Kristallisation / cristalización

0633
piezocrystallization, – crystallization of a magma under the influence of directed pressure. / piëzokristallisatie / piézocristallisation / Piezokristallisation / piezocristalización

0634
anchieutectic, – adjective for rocks composed almost exclusively of two or more minerals which are thought to represent a eutectic mixture. / anchieutektisch / – – / anchieutektisch / anchieutectico

0635
heteromorphism, – rocks with identical or similar chemical composition and a different mineralogical composition. / heteromorfie / hétéromorphisme / Heteromorphismus; Heteromorphie / heteromorfismo

heteromorphic, – / heteromorf / hétéromorphe / heteromorph / heteromorfo; heteromórfico

0636
cenotypal; cenotype; kainotype, – obsolete adjectives for fresh or nearly fresh volcanic and hypabyssal rocks, irrespective of real age, as contrasted with altered volcanic and hypabyssal rocks to which the now obsolete adjective **palaeotypal** was applied. / neovulkanisch / cénotypique / cenotyp; jungvulkanisch / cenotipo; (cenotipalo)

palaeotypal, – / palaeovulkanisch / paléovolcanique / paläotyp; altvulkanisch / paleotipo

0637
granide, – granitic rock s.l.

0638
trap; trapp; trap rock; trappide, – field name for any dark-coloured, fine-grained non-granitic, hypabyssal or extrusive rock. Also used in a geomorphologic sense, as in trap basalts, Deccan traps. / trap / trapp / Trapp / trapp

[0920] THE MAGMA

[0921] *General Terms*

0639
magma, – naturally occurring mobile rock material, generated within the earth and capable of intrusion and extrusion, from which igneous rocks are thought to have been derived through solidification and related processes. Magma may or may not contain suspended solids (such as crystals and rock fragments) and/or a gas phase. / magma / magma / Magma / magma

0640
crystal mush; .... mesh; .... mash, – partially crystallized magma: an aggregate of solid crystals lubricated by compressed water vapour or residual melt. / kristalbrei / – – / Kristallbrei / semicristalizada

0641
dissolved gas; gas in solution, – molecularly dispersed gases in the magma. / opgelost gas / gaz dissous / gelöstes Gas / gas disuelto

0642
resorption, – the partial or complete refusion or solution by a magma of crystals with which it is not, or no longer in equilibrium. / resorptie / résorption / Resorption / resorción

corrosion; magmatic ...., – partial resorption. / corrosie / corrosion / Korrosion; magmatische Korrosion / corrosión

0643
latent magma, – subcrustal, extremely viscous magma under high pressure that reacts on short-lasting stresses like a solid body. With release of pressure the viscosity decreases and the magma becomes sufficiently fluid to flow. / latent magma / – – / latentes Magma / magma latente

0644
primary magma, –
1. (obsolete) subcrustal magma of pregeological origin;
2. magma directly derived from subcrustal material.
/ oermagma; magma sensu stricto / magma primaire / Urmagma; Primärmagma / magma primario

prototectite (obsolete), – rock formed by crystallization of primary magma.

palingenetic magma; palingenic ....; reborn ....; neomagma, – magma formed in situ by ultrametamorphism. / palingeen magma; herboren .... / magma palingénétique / palingenes Magma / magma palingenético

0645
parental magma, – the magma from which a particular igneous rock solidified or from which another magma was derived. / moedermagma / magma parental / Stammagma; Mutter.... / magma primario; .... original

0646
rest magma; residual liquor, – that part of the magma that during differentiation has not yet crystallized. / restmagma / magma résiduel / Restmagma / magma residual

0647
residual liquid, – the volatile components of a magma remaining in the magma chamber after much crystallization has taken place. / restvloeistof / liquide résiduel / Restschmelze / líquido residual

0648
magma blister, – a pocket of magma whose formation has raised the overlying land surface.

0649
macula, – a local pocket of magma that is formed by the fusion of shale and that acts as a type of magma chamber.

0650
volatile component; gaseous ....; fugitive constituent; volatile; volatile flux, – components with a sufficiently high vapour tension to be concentrated in the gas phase. / vluchtig bestanddeel; gasvormig .... / composant volatil; élément gazeux / flüchtiges Bestandteil / componente volátile; elemento gaseoso;.... volátil

gas content, – / gasgehalte / teneur en gaz / Gasgehalt / contenido en gas

connate fluid, – fluids derived from the same magma.

0651
orthomagmatic stage; orthotectic ...., – the main stage in the crystallization of silicates from a typical magma. / orthomagmatisch stadium / stade orthomagmatique / liquidmagmatisches Stadium / estadio ortomagmático

0652
emulsion stage, – the stage in the crystallization of a magma in which the concentration of water exceeds the solubility, and a new water-rich phase is formed. / emulsief stadium / stade émulsif / emulsives Stadium / estadio de emulsión

0653
pneumatolytic stage, – the stage in the crystallization of a magma during which the solid and gaseous phases are in equilibrium. / pneumatolytisch stadium / stade pneumatolytique / pneumatolytisches Stadium / estadio (p)neumatolítico

pegmatitic stage, – the stage in the crystallization of a magma in which the concentration of volatiles has become sufficient to permit the formation of coarse-grained rocks. / pegmatitisch stadium / stade pegmatitique / pegmatitisches Stadium / estadio pegmatítico

0654
hydrothermal stage, – the stage in the crystallization of a magma during which the solid phases are in equilibrium with an aqueous residual solution. / hydrothermaal stadium / stade hydrothermal / hydrothermales Stadium / estadio hidrotermal

0655
pyrogenic, – formed in the orthomagmatic stage in a volatile-poor melt. / pyrogeen / pyrogène / pyrogen / pirogénico

pneumatogenic, – formed by the action of gases. / pneumatogeen / pneumatogène / pneumatogen / (p)neumatogénico

0656
aqueo-igneous; hydatopyrogenic; hydroplutonic, – formed from a hydrous magma; used in those cases where the presence of water is essential. / hydatopyrogeen / – – / hydatopyrogen; hydroplutonisch / hidatopyrógeno; hidatopirogénico; hidroplutonismo

0657
endoblastesis, – late or postmagmatic crystallization of minerals in igneous rocks from hydrothermal fluids. / endoblastese / – – / Endoblastese / endoblastesis

endoblastic, – formed by endoblastesis. /endoblastisch / – – / endoblastisch / endoblástico

0658
deuteric; epimagmatic; paulopost, — relating to the changes taking place in an igneous rock during the latest stages of its consolidation and in direct continuation of the consolidation of the magma itself. / deuterisch / deutérique / deuterisch / deutérico

0659
autometamorphism, — the process of chemical adjustment of an igneous assemblage to falling temperature, attributed to the action of its own volatile fluxes. / autometamorfose / autométamorphisme / Autometamorphose / autometamorfosis; autometamorfismo

autopneumatolysis, — the interaction takes place with gases at high temperatures. / autopneumatolyse / autopneumatolyse / Autopneumatolyse / auto(p)neumatolisis

autometasomatism, — the interaction takes place with the last water-rich liquid fraction. / autometasomatose / autométasomatose / Autometasomatose / autometasomatismo

0660
allogene; allogenic; allothigene; allothigenic; allothigenous, — terms meaning generated elsewhere, applied to those constituents that came into existence outside of, and previously to, the rock of which they now constitute a part. / allogeen / allogène / allothigen / alotígeno; alotigénico

0661
raft; stoped block, — country rock swallowed by the magma. / opgeslokt blok / bloc digéré / verschluckte Scholle / — —

0662
inclusion; enclosure,— nongenetic term for a fragment of mineral or rock included in igneous rock. They may be related in origin to the evolution of the magma or they may be incorporated from wall or country rocks. / insluitsel / enclave / Einschluss / inclusión

On a genetic basis the inclusions can be subdivided as follows:

0663
xenolith; accidental inclusion; noncognate ....; exogenous ...., — rock fragment foreign to the igneous rock in which it occurs. / xenoliet; vreemd insluitsel / enclave enallogéne; xénolithe / Xenolith; exogener Einschluss; fremder ....; enallogener .... / xenolito

epixenolith, — xenolith derived from the adjacent wall rock.

xenocryst; disomatic crystal, — enclosed crystal, foreign to the rock in which it occurs. / xenokrist / xénocrystal / Fremdmineral; Xenokristall / xenocristal

0664
autolith; cognate inclusion; endogenous ...., — inclusion genetically related to the rock in which it occurs. / autoliet; verwant insluitsel / enclave homéogène; autolite / Autolith; endogener Einschluss; homögener ....; protogene Bildung / — —

0665
schlieren; schliere; swirls, — in plutonic rock, small masses or streaks that have the same general mineralogy as the main body but that, because of some differences in the ratios of the minerals, are darker or lighter; the boundaries tend to be transitional. Some schlieren are modified inclusions, others may be segregations of minerals. / slieren / schlieren / Schlieren / schlieren

schlieric,— / streperig / — — / schlierig / — —

[0922] *Assimilation and Differentiation*

0666
magmatic evolution, — the continuing change in composition of a magma as a result of magmatic differentiation, assimilation or mixing of magmas. / magmatische evolutie / évolution magmatique / magmatische Evolution / evolución magmática

0667
differentiation; magmatic ...., — the process of developing more than one rock type, in situ, from a common magma. / differentiatie; splitsing / différenciation / Differentiation; Spaltung / diferenciación magmática

differentiate, — product of differentiation. / differentiaat / différenciat / Differentiat / diferenciado

differentiated, — adjective for a pluton in which there is more than one rock type due to differentiation. / gedifferentieerd / différencié / differentiiert / diferenciado

0668
diaschistite (obsolete), — a rock containing minor intrusions that differ in composition from the parent magma out of which they formed by differentiation. / diaschistiet; schizoliet / — — / Schizolith / roca diasquística

aschistite (obsolete), — a non-differentiated rock. / aschistiet / — — / aschistes Ganggestein / roca asquística

0669
complementary (obsolete), — adjective for a group of rocks differentiated from the same magma and having an average composition equal to that of the parent magma. / complementair / — — / komplementär / complementario

0670
fractional crystallization; fractionation; fractionalization; crystallization differentiation, — separation of a cooling magma into parts by the successive crystallization of different minerals at progressively lower temperatures. / gefractioneerde kristallisatie / cristallisation fractionnée / fraktionierte Kristallisation; Kristallisationsdifferentiation / cristalización fraccionada

crystal fractionation, — magmatic differentiation resulting from fractional crystallization. / differentiatie door gefractioneerde kristallisatie / différenciation par cristallisation fractionnée / Kristallisationsfraktionierung / cristal formado por cristalización fraccionada

0671
crystal sorting, — the separation by any process of crystals

from a magma, or of one crystal phase from another during crystallization of a magma. / kristalafscheiding / ségrégation cristalline / Differentiationstrennung / segregación cristalina

0672
gravitational differentiation; gravitative ...., — differentiation accomplished through the existence of a gravitational field, as by sinking of a heavy phase (liquid or crystals), or the rising of a light phase (liquid, crystals or gases) through the magma. / gravitatieve differentiatie / différenciation par gravité; .... par densité / gravitative Differentiation / diferenciación gravitatoria

0673
crystal settling; .... sedimentation, — sinking of heavier crystals in a body of magma. / kristalsedimentatie / décantation cristalline / Differentiation durch die Schwere / decantación

crystal flotation; flotation, — upward migration of lighter crystals in a body of magma. / flotatie / flottation / Flotation / flotación

0674
crystal accumulation, — the development of layering by the process of crystal settling. / — — / accumulation cristalline / Kristall-Akkumulation / acumulación

0675
filter pressing; filtration differentiation, — a differentiation process whereby a residual liquid is squeezed from a partly crystallized magma. / filterperseffect / effet de filtre-presse / Filterpressung; Abpressungsfiltration; Auspressungsdifferentiation / diferenciación por filtración

0676
gas streaming, — process of differentiation in a magma accomplished by bubbles of gas that may attach themselves to crystals and buoy them upward or that may cause interstitial liquids to move. / — — / entrainement gazeux / — — / — —

0677
flowage differentiation, — concentration in a vertical conduct of early formed crystals that were detached from the wall and were moved to a zone of more rapid flow. / vloeidifferentiatie / — — / Flieszdifferentiation / diferenciación por flujo

0678
differentiation by liquid immiscibility; liquation (differentiation), — a process of magmatic differentiation involving division of the magma into two or more separate liquid phases that become separated from each other by gravity or other processes. / liquatie; vloeibare ontmenging / liquation; différenciation par immiscibilité / Liquation; liquide Entmischung / licuación; diferenciación por inmiscibilidad

0679
pneumatolitic differentiation; — gaseous transfer ....; volatile transfer ...., — separation in a magma of a gaseous phase that moves relative to the magma and releases dissolved substances in areas of reduced pressure. / pneumatolytische differentiatie / différenciation par transfert gazeux; .... pneumatolitique / pneumatolytische Differentiation / diferenciación (p)neumatolítica

0680
liquid fractionation, — differentiation due to vertical diffusion in response to pressure-temperature gradients. / diffusieve differentiatie / différenciation par diffusion / Diffusionsdifferentiation / diferenciación por difusión

0681
assimilation; digestion; dissolution; magmatic solution; .... digestion; .... dissolution, — incorporation of solid or fluid foreign material into a magma. / assimilatie; magmatische vertering / assimilation / Assimilation / asimilación

hybrid, — the magma or rock resulting from assimilation / hybried; bastaard / hybride / hybrid / híbrido

hybridization; hybridism; contamination, — the process of creating new magma through assimilation. / hybridisatie; contaminatie / hybridation; contamination / Hybridisierung / hibridisación; hibridación

0682
syntexis, — modification of a magma as a result of assimilation of incorporated country rock; also anatexis of more than one kind of rock and mingling of the magmas so formed. / syntexis / contamination / Syntexis / síntesis

syntectite, — rock formed by syntexis. / syntectiet / syntectite / Syntektit / roca formada por síntesis

0683
marginal assimilation, — assimilation along the main contact of the invading magma. / randassimilatie / assimilation marginale; .... de bordure / Randassimilation / asimilación marginal

0684
basification, — development of a more basic rock; e.g. contamination of granite by assimilation of country rock. / basificatie / basification / Basifizierung / basificación

[0923] *Classification of Magmas*

0685
class, — a subdivision that is based on the relative proportions of salic and femic standard minerals in the CIPW system and on the modal amount of dark constituents in Johannsen's classification. / klasse / classe / Klasse / clase

order, — the basic unit of the class based on proportions of mineral molecules in the CIPW system and on the modal amount of leucocratic constituents in Johannsen's classification. / orde / ordre / Ordnung / orden

section, — subdivision of the order in the CIPW system. / sectie / section / Sektion / sección

rang, — alternative subdivision of the class in the CIPW system, based on relative proportions of certain oxide molecules. / rang / rang / Rang / categoría; rango

0686
CIPW norm, — a norm system that offers a taxonomic system,

published about 75 years ago by Cross, Iddings, Pirsson and Washington.

0687
clan, — a group composed of igneous rocks that are closely related in chemical composition. / stam / clan / Stamm / grupo

family, — subdivision of a clan. / familie / famille / Familie / familia

0688
colour index; colour ratio, — a number indicating the percentage of dark minerals in an igneous rock. / kleurindex / indice de coloration / Farbzahl / índice de color

0689
diagnostic mineral; symptomatic mineral, — a mineral such as olivine or quartz, whose presence, in an igneous rock, indicates whether the rock is undersaturated or oversaturated with respect to silica. / karakteristiek mineraal / minéral caractéristique / symptomatisches Mineral / mineral característico; .... índice

0690
Atlantic suite; intrapacific province, — the assemblage of alkalic and alkali-calcic rocks characteristic for the volcanic areas of the central Atlantic and Pacific oceanic areas. / atlantische serie / série atlantique / atlantische Sippe; .... Reihe / serie atlántica

Pacific suite; circumpacific province; anapeirean rocks, — the assemblage of calc-alkali rocks, represented by andesites, granodiorites and associated rocks characteristic for the circumpacific area and similar macro-tectonic environments. / pacifische serie / série pacifique / pazifische Sippe; .... Reihe / serie pacífica

Mediterranean suite, — the assemblage of potassium-rich rocks characteristic for the volcanism in the Mediterranean area. / mediterrane serie / série méditerranéenne / mediterrane Sippe / serie mediterránea

0691
andesite line; Marshall line, — the geographic-petrographic boundary around the Pacific Ocean that separates the basalts of the Atlantic suite from the predominantly andesitic rocks of the Pacific suite. / andesietlijn / ligne andésitique / Andesit-Linie / línea andesítica

0692
alkali-lime index; alkali-calc ...., — the mass percentage of silica in a sequence of igneous rocks on a variation diagram, where the mass percentages of CaO and of (K_2O + Na_2O) are equal, i.e. the crossing point of the curves for CaO and (K_2O + Na_2O). Superior results are claimed by Kuno for an alkali-lime index similarly derived in a plot against the solidification index. / alkalikalk index / indice alcalis-chaux (obsolete) / Alkali-Kalkindex / índice alcalis-calcio.

On the basis of their alkali-lime index the rocks can be subdivided into four classes for which Peacock gives the following limiting values:

alkalic series, — alkali-lime index less than 51. / alkalische reeks / série alcaline / Alkalireihe / serie alcalina

alkali-calcic series, — alkali-lime index between 51 and 56. / alkalikalkreeks / série alcalino-calcique / Alkalikalkreihe / serie alcalino-cálcica

calc-alkalic series, — alkali-lime index between 56 and 61. / kalkalkalische reeks / série calco-alcaline / Kalkalkalireihe / serie cálco-alcalina

calcic series, — alkali-lime index greater than 61. / kalkreeks / série calcique / Kalkreihe / serie cálcica

0693
peralkaline, — said of an igneous rock with Al_2O_3 less than (Na_2O, + K_2O), as expressed in molecular proportions. / peralkalisch / hyperalcalin / peralkalisch / peralcalino

agpaitic, — is used in the same sense as peralkaline, but defined as al less than na +k, where al, na, and k, stand for relative atomic abundances of Al, Na, and K. / agpaitisch / agpaitique / agpaitisch / agpaítico

peraluminous, — said of an igneous rock with Al_2O_3 greater than (Na_2O + K_2O), as expressed in molecular proportions. / peralumineus / hyperalumineux / peraluminisch / peraluminoso

metaluminous, — said of an igneous rock with (Na_2O + K_2O + CaO), greater than (Al_2O_3) greater than (Na_2O + K_2O) as expressed in molecular proportions. / metalumineus / métalumineux / metaluminisch / metaluminoso

0694
alkalic; alkaline, — adjectives for igneous rocks with such a relative abundance of alkali metals that it is mineralogically expressed by the presence of sodium pyroxenes, sodium amphiboles and (or) feldspathoids. / alkali-; alkalisch / alcalin / Alkali-; alkalisch / alcalino

subalkalic, — adjective for igneous rocks that contain no alkali minerals other than feldspars. / subalkalisch; subalkali- / subalcalin / Subalkali-; subalkalisch / subalcalino

subaluminous, — adjective for igneous rocks with no excess of Al_2O_3 over that required to form feldspars or feldspathoids. / subalumineus / subalumineux / subaluminisch / subaluminoso

calc-alkalic, — adjective for igneous rocks in which the proportions of lime and alkalis in relation to other constituents are such that plagioclase feldspar is formed. / kalkalkali-; kalkalkalisch / calco-alcalin / Kalkalkali-; Alkalikalk- / calco-alcalino

0695
alkali basalt; alcalic ....; alkali-olivine ...., — class of basalts characterized by the absence of a reaction relation between olivine and pyroxene, when defined on modal composition. When defined on the basis of the CIPW norm it is a basalt with normative olivine and nepheline with nepheline less than 5%. / alkalibazalt / basalte alcalin / Alkalibasalt / basalto alcalino

tholeiite, — class of basalts characterized by the absence of olivine in the matrix when defined on modal composition. When defined on the basis of the CIPW norm, it is a basalt carrying normative hypersthene. /tholeïet / tholéite / Tholeit; Tholeiit / tholeiita; toleita

high-alumina basalt, — class of basalts intermediate between alkali basalt and tholeiite and without definite mineralogical Characteristics. Chemically it is characterized by Al_2O_3 greater than 16.5% and greater than $Na_2O + K_2O$ greater than for any given SiO_2 content intermediate between that of tholeiite and alkali basalt.

0696
ophiolite; ophiolitic suite; non-sequence type ophiolite, – consanguineous association of mafic and ultramafic rocks consisting predominantly of peridotites or serpentinites with subordinate gabbro, diabase, norite, and basalt or spilite; occurring without a definite stratigraphic sequence. They are considered characteristic for the geosynclinal stage of the eugeosyncline. / ofioliet; ofiolietserie/ ophiolite; série ophiolitique / Ophiolit; Ophiolitserie / ofiolita

ophiolite; sequence type ...., – consanguineous association of mafic and ultramafic rocks, usually forming layered masses showing a more or less regular stratigraphic sequence from underlying ultrabasic rocks through gabbroic rocks to overlying basalts or spilites. Some of these are considered to represent oceanic upper crust and mantle.

Steinmann's trinity, – the association of ophiolite with abyssal sediments and radiolarian chert, considered characteristic for a deep-sea depositional environment. / triniteit van Steinmann / trilogie de Steinmann / Steinmann-Trilogie / triología de Steinmann

0697
variation diagram,
1. **Harker variation diagram,** – plot of SiO_2 against the other major oxides showing the genetic relationships within a rock series and the nature of processes that may have affected such a series.
2. **Larsen variation diagram,** – plot of the mass percentages of the constituent oxides against $\frac{1}{3}SiO_2 + K_2O + FeO + MgO + CaO$.
/ Harker variatiediagram; Larsen .... / diagramme de variation selon Harker; .... selon Larsen / Harker Variationsdiagramm; Larsen .... / diagrama de variación de Harker; .... Larsen

saturation line, – the line on a variation diagram of an igneous rock series that represents saturation with respect to silica. / verzadigingslijn / ligne de saturation / Sättigungslinie / línea de saturación

0698
AFM diagram, – triangular diagram based on the molecular quantities of alumina, total iron and magnesia. Used to characterize differentiation series. / AFM diagram / diagramme AFM / AFM-Diagramm / diagrama AFM

ferro-femic index, – index number derived from the AFM diagram; it is a measure for the degree of enrichment in iron for a particular differentiation series.

0699
Zavaritsky diagram, – diagram representing the principal chemical characteristics of a rock by two adjoined projections of a vector in space. / diagram van Zavaritsky / diagramme de Zavaritsky / Zavaritsky-Diagramm / diagrama de Zavaritsky

0700
accumulative phase, – that portion of a frequency distribution curve for a magmatic series that lies toward the low-silica side of the modal peak, presumably representing accumulation of solid crystals by gravitational differentiation. / – – / domaine d'accumulation / – – / fase acumulativa

0701
differentiation index, – a numerical expression of the extent of differentiation in a magma; given by the sum of normative quartz, orthoclase, albite, nepheline, leucite and kalsilite percentages. / differentiatie-index / indice de différenciation / Differentiationsindex / índice de diferenciación

0702
crystallization index, – a number calculated from normative minerals belonging to the system anorthite-diopside-forsterite. It is a measure for the progression of partial magmas or igneous rocks away from the primitive system anorthite-diopside-forsterite. / kristallisatie-index / indice de cristallisation / Kristallisationsindex / índice de cristalización

0703
solidification index; SI, – index used for the separation of rock types in a series, calculated from the mass percentages of the oxides according to:

$$\frac{100\,MgO}{MgO + FeO + Fe_2O_3 + Na_2O + K_2O}$$

/ stollingsindex / indice de solidification / Erstarrungsindex / índice de solidificación

[0930] THE INTRUSIVE BODIES

[0931] *General Terms*

0704
exogene effect, – the influence of an igneous mass on the rocks it invades. / extern effect / effet exogène / exogener Effekt / efecto exógeno

endogene effect, – the influence of an igneous intrusion on its margin. / intern effect / effet endogène / endogener Effekt / efecto endógeno

NOTE:
– these terms are not to be confused with adjectives as used e.g. in exogenetic and endogenetic processes.

0705
roof, – the country rock bordering the upper surface of an igneous intrusion. / dak / toit / Dach / roca del techo

crown; top, – highest part of an intrusive mass. / rug; nok / sommet / Scheitel / cumbre; cima; vértice; cúpula

floor, – the country rock bordering the lower surface of an igneous intrusion. / bodem / mur / Boden / suelo; piso; fondo; plataforma; muro

wall, – the side of an igneous intrusion. / wand / parois; éponte / Wand / pared

0706
roof pendant; pendant, — country rock projecting from the roof of a batholith, giving it a very irregular form. / dakfranje / lambeau de toit; .... témoin / — — / — —

0707
selvage; selvedge; salband, — marginal zone of a rock mass characterized by a distinctive feature of fabric or composition; for instance chilled margins, or glassy inner margins of pillows in pillow lava. / flank / salbande / Salband / salbanda

chill zone; chilled margin, ... zone; .... border, — marginal area of an igneous intrusion characterized by a smaller grain size than the interior of the pluton due to rapid cooling. / afkoelingsrand / bordure figée / Kontakthof / aureola de contacto; zona de ....

basic border; mafic margin, — the marginal area of an igneous intrusion, characterized by a relatively more basic composition than the interior of the pluton. / basische randzone / bordure basique / basische Randzone / borde básico

0708
vitrification; vitrifaction, — the superficial melting (fusion) of country rock by nearby magma, mostly due to blowpipe action of volcanic gases, and consecutive glassy solidification of the molten skin. / verglazing / vitrification / Frittung; Verglasung; oberflächliche Anschmelzung; Vitrifizierung / vitrificación

0709
country rock; wall rock, — the rock intruded by, and surrounding an igneous intrusion. / nevengesteente / roche encaissante / Nebengestein; Nachbar....; Alt.... / roca encajante

0710
contact breccia; intrusion breccia, — a breccia around an igneous intrusion, caused by wall-rock fragmentation, and consisting of both intrusive material and wall rock. / intrusiebreccie / brèche intrusive / Intrusivbreccie / brecha intrusiva; brecha de contacto

0711
injection breccia, — fragmental rock formed by the introduction of largely foreign rock fragments along veins and fractures in the host rock. Some are thought to have been formed as the result of meteorite impact. / — — / brêche d'injection / Injektionsbreccie / brecha de inyección

0712
granite tectonics, — the study of structural features of plutonic rocks and the processes that created them. / intrusietektoniek / — — / Granittektonik / tectónica del granito

0713
cooling crack; contraction fissure, — a joint or fissure formed as a result of cooling of an igneous rock. / afkoelingsbarst / fissure de contraction; .... de retrait / Kontraktionsspalte; Absonderung / grieta de contracción; fisura de ....; fractura de ....

0714
flow joint, — a joint parallel to the flow layers of a plutonic rock.

[0932] *Classification of Intrusive Bodies*

0715
pluton; irruptive; intrusive; .... body; injected body; .... mass; intrusion, — an igneous body formed by penetration of magma; or a body of rock formed by metasomatic replacement. / plutoon; intrusief / intrusif; venue intrusive profonde / Pluton; Intrusivkörper / plutón; eruptivo; intrusivo; energio inyectado; masa; intrusión

intrusive rock, — intrusiegesteente; intrusief....; intrusiva / roche intrusive / Intrusivgestein / roca intrusiva

0716
homogeneous, — adjective for a pluton that does not show magmatic differentiation. / homogeen / homogène / homogen / homogéneo

0717
concordant, — adjective for plutons with contacts parallel to the bedding. / concordant / concordant / konkordant / concordante

interjection, — concordant pluton, concordant injection.

0718
discordant; transgressive, — adjective for plutons with contacts that cut across bedding or foliation. / discordant / discordant / diskordant / discordante

0719
conformable, — adjective used for the contact of a pluton to indicate that it is parallel to the internal structure of the pluton. / conform; aangepast / conforme / gleichförmig / concordante; conforme

0720
simple intrusive, — formed by a single intrusion. / enkelvoudig intrusief / intrusion simple / einfaches Intrusiv / intrusivo simple; .... sencillo

0721
composite intrusive; multiple ...., — pluton composed of several parts that intruded sequentially, separated by periods of crystallization (e.g. a composite batholith). / gemengd intrusief; meervoudig .... / intrusion composée / zusammengesetztes Intrusiv; mehrfaches .... / intrusivo compuesto; .... múltiple

partial pluton, — one of the several parts of a composite intrusive. / deelintrusief / — — / Teilintrusiv / intrusivo parcial

composite intrusion; multiple ...., — the process forming a composite intrusive. / gemengde intrusie; meervoudige .... / intrusion composée / zusammengesetzte Intrusion; mehrfache .... / intrusión compuesta; .... múltiple

0722
compound intrusive, — formed by a single intrusion but having

several parts separated from each other by country rock. / samengesteld intrusief / groupe d'intrusions / zusammengesetztes Intrusiv / intrusivo mezclo

0723
feeder, — supply channel of an igneous body. / toevoerkanaal; ....pijp / cheminée d'alimentation / Zufuhrkanal; ....spalte / alimentador; conducto de alimentación

run, — branching or fingerlike extension of the feeder of an igneous intrusion. Runs usually spread laterally along several stratigraphic levels. / — — / ramification / — — / ramificación

0724
cupola, — an upward projection of an igneous intrusion into its roof. / koepel / coupole / Kuppe / cúpula

0725
apophysis; tongue, — a branch or offshoot of a larger intrusive body, generally visibly connected with it. / apofyse; tong / apophyse / Apophyse; Nebengang / apófisis

0726
synorogenic; syntectonic; synkinematic,— adjectives for a pluton emplaced during orogenesis, also for the intrusive process itself. / synorogeen; syntectonisch; synkinematisch / synorogénique; syntectonique; syncinématique / synorogen; synkinematisch / sintectónico; sinorogénico

postorogenic; posttectonic; postkinematic; postkinetic, — adjectives for a pluton emplaced after an orogeny, also for the intrusive process itself. / postorogeen; posttectonisch; postkinematisch / postorogénique; posttectonique / postkinematisch; postorogen / posttectónico; postorogénico

anorogenic, — lacking in, or unrelated to tectonic disturbance. / anorogeen / anorogénique; atectonique / anorogen / anorogénico

0727
autochthonous, — adjective used for magma produced by liquefaction in place, or for the rock crystallized from such a magma. / autochtoon / autochtone / autochthon / autóctono

allochthonous, — adjective used for magma emplaced at a distance from its source, or for the rock crystallized from such a magma. / allochtoon / allochtone / allochthon / alóctono; extraño

parautochthonous, — adjective for the mobilized part of autochthonous magma or pluton that has the characteristics of an allochthonous body but is part of an autochthonous one. / parautochtoon / parautochtone / parautochthon / parautóctono

0728
subjacent body; .... mass; dome, — a generally discordant pluton without known floor, that is presumed to widen downward to an unknown depth. / onderliggend massief / corps sous-jacent; massif .... / Liegendkörper / masa subyacente

0729
batholith; bathylith; batholite; bathylite; batholyte; batholyth; subtuberant mountain; intrusive ....; central granite, — a large irregularly shaped plutonic mass with an area of more than 100 km^2 in surface exposure. Its diameter is assumed to increase with depth, and a visible floor has never been observed. / batholiet / batholite / Batholith; Zentralgranit / batolito

0730
stock; boss, — a small batholith with a surface exposure of less than 100 km^2. / kern / stock; amas / Stock / macizo magmático; cuerpo ....

0731
globulith, — an intrusive body, or a group of closely associated bodies, having a globular or botryoidal shape and almost concordant contacts resulting from the effect of the intrusion(s) on the immediate surroundings.

0732
chonolith; chonolite, — intrusion of such an irregular shape that it cannot be classified into any of the other categories. / chonoliet / chonolite / Chonolith / conolito

0733
phacolith; phacolite, — a minor, concordant, concave-convex usually granitic intrusion into folded strata. / fakoliet / phacolite / Phakolith / facolito

0734
laccolith; laccolite; cistern rock; dyke with expanded summit, — a lens-shaped body of swelled out magma, having a flat base and vaulted covering strata. / laccoliet / laccolite / Lakkolith / lacolito

bysmalith; plutonic plug, — a conical or cylindrical laccolith of which the vertical dimensions greatly exceed the horizontal ones. / bysmaliet; intrusieplug / — — / Bysmalith; vulkanischer Pflock / bismalito

bell-jar intrusion, — a type of bysmalith where the overlying sediments have become domed and severely fractured.

0735
lopolith, — a large dish- or spoon-shaped intrusive. / lopoliet / lopolite / Lopolith / lopolito

0736
sphenolith, — a wedgelike igneous intrusion, partly concordant and partly discordant. / sfenoliet / — — / Sphenolith / esfenolito

0737
ethmolith, — a discordant pluton with a funnel-shaped cross section. / ethmoliet / — — / Ethmolith / ethmolito

0738
sole injection, — an igneous intrusion that was emplaced along a thrust plane.

0739
sheet, — general term for a tabular igneous intrusion irrespective of its attitude. / plaat / feuillet / Blatt / plancha; hoja

0740
sill, – a tabular igneous intrusion that is parallel to the planar structure of the surrounding rock. / intrusieplaat / filon couche; sill / Lagergang; Instrusivlager; Sill; Intrusivgang / capa intrusiva; sill

0741
dyke; dike,– a tabular body of igneous rock that cuts across the structure of adjacent rocks. / gang / filon; dyke / Gang; Lavagang / dique

dyke rock; dykite, – the rock composing a dyke. / ganggesteente / roche filonienne / Ganggestein / roca filoniana

dyke wall; dyke ridge; hogback, – the outcrop of a dyke, standing out as a conspicuous wall, laid bare by erosion and denudation. / lavamuur / – – / Lavamauer; Gang.... / dique muro; .... cresta

0742
ring dyke; ring-fracture intrusion; ring boss, – dyke that is arcuate or roughly circular in plan and is vertical or dips toward the centre of the arc. / ringintrusie / ring dyke / Ringintrusion; Ringgang / dique anular; .... circular

cone sheet, – a dyke that is arcuate in plan and dips toward the centre of the arc. It occurs in concentric sets that are assumed to converge in the magmatic centre. / trechtergang / cone sheet / Trichtergang; Kegelgang / – –

ring complex, – ring dykes associated with cone sheets. / – – / complexe annulaire / – – / – –

0743
ring fault; ring fracture, – a steep-sided fault pattern that is circular in outcrop and that is associated with cauldron subsidence. / ringvormige verschuiving / fracture annulaire / kreisförmige Verwerfung; Kreisbruch / dislocación anular

cauldron subsidence, – depression caused by the sinking of an approximately cylindrical part of the roof of a magma chamber along ring faults, sometimes accompanied by magmatic intrusion along these faults to form ring dykes. / breukkuil / subsidence en chaudron / Kesseleinbruch; ....bruch / fosa tectónica circular

0744
secundine dyke, – a dyke that has intruded hot country rock.

welded dyke, – a secundine dyke with boundaries obscured by continued mineral growth of the (granitic) country rock into the intrusion.

0745
replacement dyke, – a dyke formed by gradual transformation of wall rock by solutions along fractures or permeable zones. / vervangingsgang / dyke de remplacement / Verdrängungsgang / dique de substitución

0746
headed dyke; ductolith, – horizontal dyke with teardrop-like terminal expansion.

0747
roofed dyke, – a dyke that has an upward termination.

0748
intrusive vein; eruptive vein, – sheet-like intrusion of magma, rich in gases and of irregular form. / intrusieve ader / filon intrusif / Intrusivgang; Eruptivader / veta intrusiva

0749
dyke swarm; dyke complex; dyke set, – a group of dykes without a directly observable relationship to a plutonic body. They may radiate from a single point or lie in a parallel linear arrangement. / gangencomplex / réseau de dykes; essaim de .... / Gangkomplex; Gangschwarm / sistema de diques

cluster; dyke system, – a group of criss-crossing dykes, demonstrably related to an exposed pluton. / gangsysteem / faisceau de dykes / Gangsystem / grupo de diques; sistema de ....

[0940] THE CONSTITUENTS OF IGNEOUS ROCKS

[0941] *Crystal Forms*

0750
euhedral; idiomorphic; automorphic, – adjectives for those minerals in a rock that are bounded by their own crystal faces. The terms idiomorphic and automorphic are also used to indicate rocks or rock textures characterized by predominance of euhedral minerals. / idiomorf; automorf / idiomorphe; automorphe / idiomorph; automorph / euhedral; idiomórfico; automórfico

euhedron; idiomorph, – euhedral crystal.

0751
subhedral; hypidiomorphic; hypautomorphic; subidiomorphic, – adjectives for those minerals in a rock that are only partly bounded by their own crystal faces. The terms hypidiomorphic, hypautomorphic and subidiomorphic are also used to indicate rocks or rock textures characterized by predominance of subhedral minerals. / hypidiomorf; hypautomorf / hypidiomorphe / hypidiomorph; hypautomorph / subhedral; hipidiomórfico

subhedron, – subhedral crystal.

0752
anhedral; xenomorphic; allotriomorphic, – adjectives for those minerals in a rock that are not bounded by crystal faces. The terms xenomorphic and allotriomorphic are also used to indicate rocks or rock textures characterized by predominance of anhedral minerals. / xenomorf; allotriomorf / xénomorphe / xenomorph; allotriomorph / anhedral; alotriomórfico; xenomórfico

anhedron, – anhedral crystal.

0753
crystal form, – the geometric shape of a crystal. / kristalvorm / forme cristalline / Kristallform / forma cristalina

0754
equant; equidimensional, – having approximately the same dimensions in all directions. / equidimensionaal / équidimensionnel / isometrisch / equidimensional

equiform, – adjective for crystals having approximately the same shape. / gelijkvormig / équiforme / gleichförmig / equiformo

0755
tabular; platy, – flattened in one plane; also used for geological bodies. / plat; plaatvormig / aplati; tabulaire / tafelig; plattig; tafelförmig / tabular

acicular, – needle-like in form. / naaldvormig / aciculaire / nadelig; faserig; nadelförmig / acicular

columnar; prismatic, – rod-like in form. / stengelig; prismatisch / prismatique / prismatisch; stengelig; säulig / columnar

0756
embayment, – an irregular corrosion or modification of the outline of a crystal by the magma from which it previously crystallized, or in which it occurs as a foreign inclusion.

embayed crystal; corroded ....; brotocrystal, – / gecorrodeerd kristal / cristal corrodé / korrodierter Kristall; teilweise resorbierter .... / cristal corroido

0757
occult mineral, – a mineral that, according to the chemical analysis, should be present in an igneous rock, but is not found because it is submicroscopic, or because it is held in solid solution, or because it has not yet crystallized and forms part of the glassy matrix. / occult mineraal / minéral occulte / okkultes Mineral / mineral oculto

released mineral, – a mineral formed during the crystallization of a magma due to failure during an earlier phase to react with the liquid. Thus failure of olivine to react with the melt to form pyroxene may enrich the melt in silica, so that ultimately crystallization of quartz as a released mineral may result.

0758
phenocryst; inset; phanerocryst; porphyrocryst, – the large, conspicuous crystals in porphyritic rocks. / fenokrist; eersteling / phénocristal / Einsprengling; Phänokristall; Porphyr.... / fenocristal

0759
microlite; microlith, – a microscopic crystal that polarizes light and has some determinable optical properties. / microliet / microlite / Mikrolith / microlito

0760
crystallite, – a general term for minute bodies occurring in glassy igneous rocks. They are not referable to any definite mineral species or crystal form. They are undeveloped, embryonic forms and represent early efforts toward crystallization. / kristalliet / cristallite / Kristallit / cristalito

crystallitic, – of, pertaining to, or composed of crystallites. / kristallitisch / cristallitique / kristallitisch / cristalítico

0761
belonite, – a crystallite of elongated form with rounded or pointed ends. Longulites, clavalites and spiculites are included under this term. / beloniet / bélonite / Belonit / belonito

0762
longulite, – a crystallite of elongated, rod-like form. / longuliet / longulite / Longulit / longulito

bacilite, – a rod-like form made up of a number of parallel longulites. / bacilliet / bacillite / Bazillit / bacilito

0763
clavalite, – a rod-like crystallite clubbed at the ends. / clavaliet / clavalite / Clavalit / clavalito

0764
spiculite, – a spindle-shaped crystallite considered to represent the coalescence of a linear series of globulites. / spiculiet / spiculite / Spiculit / espiculito

0765
globulite, – a spherical crystallite. / globuliet / globulite / Globulit / globulito

globosphaerite, – a closely aggregated mass of globulites. / globosferiet / globosphérite / Globosphärit / globosferito

cumulite, – globulites, loosely aggregated into irregular cloudy masses. / cumuliet / cumulite / Cumulit / cumulito

0766
margarite, – a linear string of globulites. / margariet / margarite / Margarit / margarita

0767
axiolite; axiolith, – an elongated spherulite. / axioliet / axiolite / Axiolit / axiolito

axiolitic, – adjective for a rock or a structure characterized by the presence of axiolites. / axiolithisch / axiolithique / axiolithisch / axiolítico

0768
arculite, – crystallites grouped in a bow-shaped aggregate. / arculiet / arculite / Arkulit / arculito

0769
scopulite, – a crystallite in the form of a rod or a stem terminated by divergent brushes or plumes or a number of lateral brushes. / scopuliet / scopulite / Skopulit / escopulito

0770
trichite, – a hairlike crystallite, usually black and opaque and either straight or curved, or in groups radiating from a point. / trichiet / trichite / Trichit / triquito

[0942] *Aggregates and Intergrowths*

0771
crystalline aggregate, – / kristallijn aggregaat / aggrégat cristallin / kristallinisches Aggregat / agregado cristalino

glomerocryst, – an aggregate of crystals of the same mineral. / glomerokrist / agglomération monominéral / Agglomeration / glomerocristal

nodule, – a rounded mass of crystals, either of the same or different minerals, e.g. pyroxene -, olivine -, peridotite nodule. / knol / nodule / Knolle / nódulo

0772
columnar, – an aggregate made up of column-like individuals. / stengelig / prismatique / säulig; stengelig / columnar

bladed, – the columns are flattened like a knife blade. / platstengelig / prismatique aplati / plattstengelig / hojoso

0773
fascicular; fibrous, – an aggregate consisting of an aggregate of acicular (needle-shaped) crystals or fibers. / vezelig / fibreux / faserig / fibroso

0774
capillary; filiform, – an aggregate in hair-like or thread-like crystals. / haarvormig / capillaire / haarförmig / capilar

vermiform, – having the shape of a worm. / wormvormig / vermiculaire / wurmförmig / vermicular

0775
lamellar; lamellate, – made up of plates or leaves. / bladerig / feuilleté; lamellaire / blätterig; lamellar / laminar; lamelar

foliaceous; foliated, – the laminae are thin and separable. / schubbig; schilferig / folié / schuppig / foliáceo; foliado

micaceous, – the foliaceous structure of mica. / glimmerachtig / micacé / glimmerartig / micáceo

0776
radiated; divergent, – the crystals radiate from a centre. / radiaalstralig; straalgewijs / radié / radialstrahlig / radiado

reniform, – consisting of radiating individuals terminating in rounded masses resembling a kidney in shape. / niervormig / réniforme / nierenförmig / reniforme

0777
botryoidal, – consisting of groups of globular forms resembling clusters of grapes. The globular forms consist of radiating individuals. / trosvormig; botryoïdaal / botryoïdal / traubig / botroidal

0778
spherulite; sphaerolite; radiolite, – radial aggregate of acicular and fibrous minerals occurring in silicious lavas and shallow intrusive rocks, particularly those rich in glass. / sferoliet / sphérolite / Sphärolith / esferulito

In German literature distinction is made between:
1. **Sphärokristall;** homogener Belonosphärit, – spherulite composed of a single mineral species;
2. **Pseudosphärolith,** – spherulite consisting of more than one mineral, often quartz and feldspar;
3. **Sphärolith;** Felsosphärit, – spherulite composed of a mixture of crystalline fibres and glass or microfelsite, of glass and microfelsite or entirely of microfelsite.

0779
granospherite, – spherulite composed of radially or concentrically arranged grains. / granosferiet / granosphérite / Granosphärit / granosferito

compound spherulite, – consisting of a radial plumose core surrounded by a denser outer spherical shell with radial structure. May be up to ten metres in diameter. / samengestelde sferoliet / sphérolite composée / zusammengesetzter Sphärolith / esferulito compuesto

0780
lithophysa, – granospherite in a glassy rock. / lithofyse / lithophyse / Lithophyse / litofiso

0781
variole, – a pea-size spherulite, usually composed of radiating crystals of plagioclase or pyroxene. Varioles generally occur in basic volcanic or hypabyssal rocks. / variole / variole / Variol / variola

variolite, – rock containing numerous varioles. / varioliet / variolite / Variolit / variolita

0782
amygdale, – the filling of a cavity or vesicle formed by expanding gases in lavas or shallow intrusions. Usually they are spherical or ovoid but many are irregular. The filling consists of deuteric or secondary minerals such as opal, chalcedony, chlorite, calcite and zeolites. The form **amygdule** is a diminutive of amygdale. / amandel / amygdale / Mandel / amígdala

pipe amygdule, – elongate amygdule found at the base of lava flows to which it is approximately normal. It may branch out upward and is thought to have been formed due to gases or vapour generated in the underlying material by the heat of the lava.

0783
intergrowth, – the state of interlocking crystals of two different minerals as a result of their simultaneous crystallization. / vergroeiing / texture symplectique / Verwachsung; Durchwachsung / intercrecimiento; textura simplectítica

0784
synantetic, – adjective for minerals formed between certain pairs of other minerals by the interaction between the latter. Well-known examples are those that form the coronas around olivine in gabbros and diabases. / synantetisch / synantétique / synantetisch / sinantético

0785
myrmekite, – an intergrowth of plagioclase, commonly of the series oligoclase-albite, and vermicular quartz. Myrmekite is usually synantetic between crystals of plagioclase and potash feldspar forming fringes around the former at the con-

tact with the latter. / myrmekiet / myrmékite / Myrmekit / mirmequita

perthite, – an intergrowth of alkali feldspars. / perthiet / perthite / Pertit / pertita

0786
corona; reaction rim; celyphitic rim; kelyphitic border; kelyphite, – a peripheral zone around a mineral composed of a different mineral species or several species and that represents the reaction of earlier crystallized material with the surrounding magma, or the reaction of a mineral with its neighbours due to changed physical conditions. The terms **corrosion border, corrosion rim, corrosion zone,** and **resorption border** specifically apply to a reaction between crystals and magma. / mineraalkrans; reactieschaal; ....rand; kelyfitische rand / bordure réactionnelle; couronne ....; auróle de réaction; couronne kélyphitique; auréole .... / Korona; Reaktionsrand; ....rinde; ....hof; Kelyphitrinde; kelyphitische Rinde; Kelyphit / corona; aureola quelifítica; .... kelifítica; borde de reacción; zona de ....; zona de corrosión

coronite, – a rock containing mineral grains surrounded by coronas. / coroniet / coronite / Koronit / coronita

0787
armoured relic, – an unstable relic that is prevented from further reaction by a rim of reaction products. / gepantserd relikt / rélique cuirassé / gepanzertes Relikt / relicto armado

armouring, – formation of a reaction rim resulting from a change in physical conditions, or a loss of equilibrium in a discontinuous reaction series.

0788
refractory, – non reactive. / niet reagerend / refractaire / – – / refractario

0789
inclusion; endomorph, – any foreign body enclosed by a mineral; not to be confused with an inclusion in igneous rock. / insluitsel / inclusion / Einschluss / inclusión

[0943] *Texture*

0790
texture, – a term indicating the intimate mutual relations of the mineral constituents and glassy matter in a rock made up of a uniform aggregate. Texture is defined by crystallinity, granularity and fabric. /textuur / texture / Textur / textura

micro-, – prefix used to indicate that a particular texture or feature is only visible at a microscopic scale; for instance microgranitic, micropoikilitic. / micro- / micro- / mikro- / micro-

0791
crystallinity, – one of the component factors of texture. It refers to the degree to which the rock is crystalline (e.g. holocrystalline, hypocrystalline or holohyaline), and to the degree to which the crystalline character of an igneous rock is developed (e.g. macrocrystalline) or apparent (e.g. phaneritic, aphanitic). / kristalliniteit / cristallinité / Kristallinität / cristalinidad

0792
granularity; grain size, – one of the component factors of texture. On the basis of the size of the constituent crystals the crystalline rocks can be subdivided into phanerocrystalline, macrocrystalline, mesocrystalline, microcrystalline and cryptocrystalline rocks. / korrelgrootte / grain / Korngrösse / granularidad; granulosidad

NOTE:
– see table 3a – Grain-Size Classes of Igneous Rocks

0793
fabric, – one of the component factors of texture. It represents the appearance or pattern defined by the shape and arrangement of crystalline and noncrystalline parts in a rock. / maaksel; weefsel; patroon / fabrique / Gefüge / fábrica

0794
monoschematic, – adjective for a uniform or homogeneous fabric. / monoschematisch / – – / monoschematisch / monoesquemático

polyschematic, – adjective for a fabric with more than one component element. / polyschematisch / – – / polyschematisch / poliesquemático

macropolyschematic; chorismitic, – adjectives for a macroscopically distinguishable polyschematic fabric; a coarse mixed fabric. / macropolyschematisch / chorismitique / makropolyschematisch; chorismatisch / macropoliesquemático

chorismite, – a macropolyschematic rock, for instance a breccia, conglomerate or heterogeneous migmatite. / chorismiet / chorismite / Chorismit / – –

0795
phanerocrystalline; phaneritic; eucrystalline, – adjectives for holocrystalline rocks in which the crystals of the essential minerals can be distinguished with the naked eye. / fanerokristallijn; zichtbaar .... / phanérogène; eucristallin / phanerokristallin; holo....; phaneride / fanerocristalina; fanerítica

phanerite, – a phanerocrystalline rock. / faneriet / – – / Phanerit / fanerita

The phanerocrystalline rocks can be subdivided according to grain size. The following subdivision is the most widely used one.

0796
very coarse-grained, – grain size larger than 3 cm. / zeer grofkorrelig / à grain très grossier / sehr grobkörnig / de grano muy grueso

coarse-grained, – grain size between 0.5 and 3 cm. / grofkorrelig / à grain grossier / grobkörnig / de grano grueso; gruesogranular; gruesocristalina

medium-grained, – grain size between 1 and 5 mm. / middelkorrelig / à grain moyen / mittelkörnig / de grano medio; mediogranular

fine-grained, – grain size between 0.1 and 1 mm. / fijnkorrelig / à grain fin / feinkörnig / de grano fino; finogranular; finocristalina

0797
macrocrystalline; mega....; macromeritic; macroscopic; megasopic, – adjectives for rocks with crystals large enough to be distinctly visible with the use of a simple handlens, including the fine-grained rocks with grain sizes between 0.5 and 1 mm. / macrokristallijn / macrocristallin / makrokristallin / macrocristalina; mega....

mesocrystalline, – adjective for a rock with crystals intermediate in size between micro- and macrocrystalline, including the fine-grained rocks with grain sizes between 0.1 and 0.5 mm. / mesokristallijn / mésocristallin / mesokristallin / mesocristalina

microcrystalline; micromeritic, – adjectives for a crystalline rock or a matrix in which the individual crystals can only be distinguished under the microscope. / microkristallijn / microcristallin / mikrokristallin / microcristalina; microgranular

dyscrystalline, – partly synonymous with microcrystalline but also used for incompletely crystallized rocks. / – – / mal cristallisé / schlecht kristallisiert / discristalino; mal cristalizada

0798
aphanitic; felsitoid, – synonymous with microcrystalline but including cryptocrystalline. / afanitisch / aphanitique / aphanitisch; aphanide / afanítica

aphyric, – an aphanitic texture without phenocrysts. / afirisch / aphyrique / aphyrisch / afírica; afídica

aphanite; felsite, – aphanitic rock. / afaniet / roche aphanitique / Aphanit / afanita

The microcrystalline rocks can be subdivided according to grain size in very fine-grained and dense.

0799
very fine-grained, – grain size between 0.1 and 0.01 mm. / zeer fijnkorrelig / à grain très fin / sehr feinkörnig / de grano muy fino

dense, – grain size between 0.01 and 0.001 mm. / dicht / dense / dicht / denso

0800
(micro)cryptocrystalline; phanerocrystalline-adiagnostic; microaphanitic; microfelsitic (obsolete); felsitic (obsolete), – adjectives for a rock or groundmass in which the crystals are too small to be separately distinguished under a polarizing microscope. Grain size less than 0.001 mm. / cryptokristallijn / cryptocristallin; / kryptokristallin / criptocristalina

0801
dubiocrystalline, – adjective for rocks or rock textures whose crystallinity cannot be determined with certainty under the microscope. / dubiokristallijn / – – / dubiokristallin / de cristalinidad dudosa

0802
ghost; phantom, – a visible outline of a former crystal shape or other rock structure that since has been partly or completely obliterated. / fantoom / fantôme / Phantom / borroso

0803
holocrystalline; pleo...., – adjectives for a rock consisting only of crystalline material and no glass. / holokristallijn / holocristallin / holokristallin / holocristalino

0804
hypcrystalline; hemi....; semi....; mio....; mero....; hypohyaline, – adjectives for a rock that is partly glassy, partly crystalline. / hypokristallijn; ....hyalien; halfglazig / hypocristallin; semi.... / hypokristallin; halbkristallinisch / hipocristalino

In the CIPW system the hypocrystalline rocks were subdivided on the basis of the ratio of crystals to glass (obsolete terms):

percrystalline, –	crystals:glass >7:1
docrystalline, –	7:1>crystals:glass >5:3
hyalocrystalline, –	5:3>crystals:glass >3:5
dohyaline, –	3:5>crystals:glass >1:7
perhyaline, –	1:7>crystals:glass

0805
holohyaline, – adjective for a rock entirely consisting of glass. / holohyalien / holohyalin / holohyalin / holohialino

glassy; hyaline; vitreous, – having glass-like properties; or being composed predominantly of glass. / glazig; glasachtig / vitreux; hyalin / glasartig; hyalin / vitreo; hialino

0806
granular, – a textural term applied to holocrystalline rocks in which most of the mineral grains are equidimensional. / korrelig / grenu; équigranulaire / körnig / granular; granuda; equigranular

0807
equigranular; even grained; homeocrystalline, – composed predominantly of crystals of (nearly) the same size. Used for textures and for rocks. / gelijkkorrelig / équigranulaire / gleichmässig körnig / equigranular

inequigranular, – composed of crystals that are unequal in size. Used for textures and for rocks. / ongelijkkorrelig / hétérogranulaire / ungleichmässig körnig / inequigranular; heterogranular

0808
panidiomorphic-granular; panautomorphic-granular; panidiomorphic; panautomorphic; idiomorphic-granular; automorphic-granular, – consisting predominantly of euhedral minerals. Used for textures and for rocks. / panidiomorf; panautomorf / panidiomorphe / panidiomorph(körnig) / panidiomórfico; idiomórfica granular; automórfica ....

hypidiomorphic-granular; hypautomorphic-granular, – consisting of euhedral, subhedral and anhedral crystals. Used for textures and for rocks. / hypidiomorf; hypautomorf / hypidiomorphe / hypidiomorph(körnig); hypautomorph / hipidiomorfa; hipautomórfica; hipidiomórfica granular

xenomorphic-granular; allotriomorphic-granular, – consisting predominantly of anhedral minerals. Used for textures and for rocks. / panxenomorf; panallotriomorf / xénomorphe / allotriomorph(körnig); panallotriomorph; panxenomorph / panalotriomórfica; xenomórfica

0809
lamprophyric, – resembling lamprophyre. Also adjective for a texture that is inequigranular-panidiomorphic. / lamprofierisch / lamprophyrique / lamprophyrisch / lamprofídica

0810
granitic; granitoid; euganitic, – resembling granite; also adjective for a texture that is hypidiomorphic-granular. / granitisch / granitoïde / eugranitisch / granítica

granolite, – rock with a granitic texture. / – – / – – / Granolit / granolita

0811
aplitic; saccharoidal; sugary; sucrosic; autallotriomorphic, – resembling aplite; also adjective for a texture that is fine-grained panxenomorphic. / aplitisch / aplitique; saccaroïde / autallotriomorph / aplítica; sacaroidea

0812
gabbroid, – resembling gabbro; also adjective for a medium- or coarse-grained panxenomorphic texture. / – – / gabbroïque / Gabbro; gabbroid / gabroidea

0813
monzonitic, – adjective for a texture characterized by euhedral plagioclase crystals and some interstitial orthoclase. / monzonitisch / monzonitique / monzonitisch / monzonítica

0814
pegmatoid, – adjective for a texture of a pegmatitic rock that lacks graphic intergrowths, also name for a rock with such a texture. / pegmatoïde / pegmatoïdique / pegmatoid; pegmatitähnlich / pegmatoidea

0815
fibrocrystalline, – characterized by the presence of fibrous crystals. / fibrokristallijn / fibrocristallin / fibrokristallinisch / fibrocristalina

0816
interlocking texture; crystalline ....; serrate ...., – texture in which grains with irregular boundaries interlock by mutual penetration. / vertande textuur / texture entrelacée / runitische Textur / textura entrelazada; .... interlazada

0817
porphyritic, – adjective for the texture of an igneous rock in which phenocrysts are set in a finer groundmass, which may be crystalline and/or glassy. Sometimes restricted to cases where phenocrysts and matrix formed during two different generations of crystallization. Also used for rocks having such a texture and for the mineral forming the phenocrysts. For rocks with megascopic phenocrysts in an aphanitic matrix the term **porphyro-aphanitic** may be used. / porfirisch; porfiritisch / porphyrique / porphyrisch; porphyritisch / porfirítica; porfírica; porfídica

porphyry, – rock with a porphyritic texture. In European usage a distinction is made between porphyries with a surplus of alkalifeldspar over plagioclase; / porfier / porphyre / Porphyr / pórfiro; pórfido,
and those with a surplus of plagioclase over alkalifeldspar. / porfiriet / porphyrite / Porphyrit / porfirito

0818
feldsparphyric, – containing feldspar phenocrysts.

In the CIPW system the porphyritic rocks were subdivided on the basis of the ratio of groundmass to phenocrysts (obsolete terms):

perpatic, –		groundmass:phenocrysts >7:1
dopatic, –	7:1>	groundmass:phenocrysts >5:3
sempatic, –	5:3>	groundmass:phenocrysts >3:5
dosemic, –	3:5>	groundmass:phenocrysts >1:7
persemic, –	1:7>	groundmass:phenocrysts

0819
pseudoporphyritic, – adjective for the texture of plutonic rocks, especially granites, containing relatively large crystals of feldspar that have grown, at least in part, after the rock solidified, thus giving the rocks a porphyritic appearance. / pseudoporfirisch; porfierachtig / porphyroïde / pseudoporphyritisch; porphyrartig / (p)seudoporfídica

0820
seriate, – adjective for the texture of inequigranular rocks in which the sizes of the crystals vary gradually or in a continuous series. / continu variabel / à grain hétérogène / serial-porphyrkörnig; serial-porphyrisch / seriada

hiatal, – adjective for the texture of inequigranular rocks where the sizes of the crystals do not vary gradually but form a broken series or where two or more sizes are notably different from one another as in porphyritic rocks. / discontinu variabel / à grain hétérogène / porphyrkörnig; wechsel....; mittelgleich.... / con grano heterométrico

0821
cumulophyric; glomeroporphyritic; glomerophyric, – adjectives for porphyritic rocks or their textures in which the phenocrysts are arranged in clusters or irregular groups. / cumulofirisch; glomeroporfirisch / gloméroporphyrique / glomeroporphyritisch / glomerofídico

cumulophyre, – rock with cumulophyric texture. / cumulofier / – – / – – / cumulofídico

synneusis, – said of the texture of a rock, not necessarily a porphyritic one, in which some of the crystal components are aggregated in clusters. / – – / synneusis / – – / sinneusis

0822
gregaritic, – adjective for a texture of porphyritic rocks in which independently oriented grains of the groundmass (especially of augite) occur in clusters.

0823
vitroporphyric: hyalinocrystalline, – adjectives for the texture of porphyritic rocks with a glassy groundmass. A distinction is sometimes made between vitrophyric rocks with a megascopically glassy matrix and **vitriphyric** rocks where the matrix is glassy under the microscope also. / vitrofirisch / hyaloporphyrique / vitrophyrisch; vitroporphyrisch / vitroporfídica

vitrophyre; glass porphyry, – rock with vitrophyric texture. / vitrofier / vitrophyre / Vitrophyr / vitrófiro; vitrofídica

0824
felsophyric; aphanophyric, – adjectives for porphyritic rocks

with a cryptocrystalline groundmass. / felsofirisch / felsophyrique / felsophyrisch / felsofídico; afanítica

felsophyre; aphanophyre, – rock with a felsophyric texture. / felsofier / felsophyre / Felsophyr / felsófiro; felsofídica

0825
pilotaxitic, – adjective for the texture of holocrystalline porphyritic rocks in which the groundmass consists essentially of lath-shaped feldspar microlites, often showing a more or less noticeable flow arrangement. / pilotaxitisch / pilotaxitique / pilotaxitisch / pilotáxica

felty, – a texture of feldspar microlites with random orientation and a matrix containing glass. / viltachtig / – – / filzig; filzartig / – –

0826
orthophyric, – adjective for a texture of holocrystalline porphyritic rocks such as some trachytes, in which the feldspars of the groundmass show square or shortly rectangular cross sections. / orthofirisch / orthophyrique / orthophyrisch / ortofídica

0827
arabesquitic, – adjective for a texture in porphyritic rocks in which an apparently homogeneous matrix shows arabesque-like patterns in polarized light. / arabeskachtig / en arabesque / – – / arabesquéitica

0828
rapakivi, – adjective for a porphyritic texture characterized by rounded alkalifeldspar; also used for rocks with such texture. / rapakivi / rapakivi / rapakivi; Rapakivi-Gefüge / en rapaquivi

0829
skedophyric, – adjective for a texture of porphyritic rocks in which the phenocrysts are scattered throughout the matrix.

skedophyre, – rock with skedophyric texture.

planophyric, – adjective for a texture of porphyritic rocks in which the phenocrysts are arranged in layers.

planophyre, – rock with planophyric texture.

linophyric, – adjective for a texture of porphyritic rocks in which the phenocrysts are arranged in lines or streaks.

linophyre, – rock with linophyric texture.

(obsolete terms)

0830
cumulus; primary precipitate, – the accumulation of crystals that precipitated from a magma without having been modified by later crystallization; also adjective for the texture of an accumulative rock. / – – / cumulus / Kumulus / cúmulo

cumulate; accumulative rock, – an igneous rock formed by the accumulation of crystals that settled out of a magma by the action of gravity. / – – / cumulat / Kumulat / acumulado

cumulus crystal; primary precipitate ...., – a unit of the cumulus. / – – / cristal cumulus / Kristall-kumulus / cristal cúmulo

0831
intercumulus, – the space between crystals of a cumulus. / – – / intercumulus / Interkumulus / intercúmulo

intercumulus liquid; .... material; interprecipitate liquid; .... material, – magmatic liquid or material present in the intercumulus. / – – / liquide intercumulus / Interkumulus-Schmelze / líquido intercúmulo

0832
orthocumulate, – a cumulate composed chiefly of one or more cumulus minerals plus the crystallization products of the intercumulus liquid. / – – / orthocumulat / Orthokumulat / ortoacumulado

mesocumulate, – cumulate with a small amount of intercumulus material; intermediate between an orthocumulate and an adcumulate. / – – / mésocumulat / Mesokumulat / mesocumulado

adcumulate, – cumulate in which the growth of unzoned homogeneous crystals (**adcumulus growth**) has forced the intercumulus liquid out of the intercumulus space. Intercumulus material comprises less than five percent of the rock. / – – / adcumulat / Adkumulat / adcumulado

0833
heterad crystallization, – adcumulus growth in which cumulus crystals and poikilitic crystals of the same composition continue to develop until little or no interstitial liquid remains. / – – / cristallisation hétéradcumulaire / – – / cristalización heteracumular

heteradcumulate, – cumulate formed by heterad crystallization; also adjective for the texture of such a rock. / – – / hétéradcumulat / – – / heteracumulado

0834
crescumulate, – cumulate where large elongated crystals are oriented roughly at right angles to the cumulate layering in the rock; also adjective for a rock with such a texture. / – – / crescumulat / – – / cresacumulado

0835
lithoidal; stony, – a term applied to dense, fine-grained, crystalline igneous rocks, to devitrified glasses, or to a dense groundmass of crystalline material as distinguished from glassy varieties. / lithoidaal / – – / lithoidisch; steinig / litoideo; litoidal

0836
microlitic, – a general term relating to the texture of the groundmass of volcanic rocks in which there is a felty aggregate of randomly oriented or parallel microlites in an often glassy base. It includes hyalopilitic, pilotaxitic and trachytic textures. / microlitisch / microlitique / mikrolitisch / microlítica

0837
hyalopilitic, – adjective for a texture of volcanic rocks, especially andesites, of which the groundmass consists essentially of glass and lath-shaped or needlelike feldspar microlites. Often the microlites show a more or less noticeable flow arrangement. / hyalopilitisch / hyalopilitique / hyalopilitisch / hialopilítica

0838
trachytic, – adjective for a texture characteristic for trachytes. The groundmass is commonly holocrystalline or with a small amount of glassy residue. It consists essentially of lath-shaped feldspar microlites with some degree of parallel disposition in consequence of flow (**pilotaxitic** texture). / trachitisch / trachytique / trachytisch / traquítica

0839
trachytoid, – adjective for a texture in phaneritic extrusive igneous rocks in which the minerals in the matrix have a subparallel or randomly divergent alignment. / trachitoïde / – – / trachytoid / traquitoidea

bostonitic, – adjective for a texture typical of bostonites with clusters of divergent feldspar laths in a trachytoid matrix. / bostonitisch / bostonitique / bostonitisch / bostonítica

0840
intersertal, – adjective for a texture common in basaltic rocks in which the hypocristalline or glassy groundmass occurs only as angular patches in the interstices of the abundant crystals. / intersertaal / intersertal / intersertal / intersertal

hyalo-ophitic, – adjective for a texture resembling ophitic texture, in which the spaces of an open network of feldspar laths are occupied by glass, it also resembles intersertal texture, however the glassy groundmass is more abundant and continuous. / hyalo-ofitisch / hyalo-ophitique / hyalo-ophitisch / hialofítica

0841
intergranular, – adjective for a texture in which the interstitial augite between crystals of plagioclase occurs in an aggregate of grains and not in large crystals in optical continuity. / intergranular / intergranulaire / intergranular / intergranular

0842
diktytaxitic, – adjective for a rock texture characterized by numerous jagged irregular vesicles bounded by crystals, some of which protrude into the cavities.

0843
radiolitic, – adjective for a texture characterized by radial, fan-like aggregates of acicular crystals, resembling sectors of spherulites. / radiair / radiée / radialstrahlig / radiolítica

0844
poikilitic; poecilitic (obsolete); semipegmatitic (not recommended), – adjectives for a texture in igneous rocks in which small crystals are irregularly scattered without common orientation in a larger crystal of another mineral. / poikilitisch / poecilitique / poikilitisch / poiquilítica

oikocryst; host, – the enclosing crystal in a poikilitic fabric. / oikokrist; gastheer / hôte / Wirtkristall / mineral englobante; .... posterior

chadacryst; cadacryst; guest, – the enclosed crystal in a poikilitic fabric. / chadakrist; gast / – – / Chedakristall; Gast .... / mineral huésped

0845
ophitic; doleritic; diabasic, – adjectives for a texture, or a rock with such a texture, characterized by plagioclase crystals wholly or partly enclosed by augite. Exclusive use of the term ophitic for textures where augite predominates over plagioclase and of the term diabasic for textures where plagioclase predominates over augite is not recommended. / ofitisch / ophitique; doléritique / ophitisch / ofítica

subophitic, – plagioclase and augite are of about equal size, resulting in incomplete enclosure of the plagioclase. / subofitisch / subophitique / subophitisch / subofítica

Several varieties of ophitic texture have been recognized:

0846
poikilophitic, – lath-shaped feldspars are poikilitically enclosed by large anhedral pyroxene crystals. / poikilofitisch / poecilophitique / poikilophitisch / poiquilofítica

sporophitic, – large pyroxene crystals enclose much smaller, widely separated, plagioclase crystals.

trachyophitic, – the enclosed feldspar crystals have a parallel orientation.

basiophitic, – the ophitic rock has a mesostasis composed of augite.

nesophitic, – the pyroxene is interstitial to plagioclase and occurs in isolated areas.

oxyophitic, – the ophitic rock has a mesostasis composed of quartz, orthoclase, or both.

basioxyophitic,, – the ophitic rock has a mesostatis intermediate between that of an oxyophitic and a basiophitic rock.

oxymesostasis, – mesostasis of an oxyophitic rock.

porphyrogranulitic, – large phenocrysts of feldspar and augite or olivine in a groundmass of smaller lath-shaped feldspar crystals and irregular augite grains.

0847
graphic intergrowth, – an intergrowth of crystals, commonly feldspar and quartz, that produces a type of poikilitic texture in which the larger crystals have a fairly regular geometric outline and orientation. / grafische vergroeiing / texture graphique / graphische Verwachsung; schriftgranitische ....; runitische .... / textura gráfica; intercrecimiento gráfico

graphic; runic, – adjectives for a texture characterized by graphic intergrowhts. / schriftgranitisch / graphique; pegmatitique / graphische Impliktions- ; Schrift- / gráfico

graphic granite, – / schriftgraniet / granite graphique / Schriftgranit / granito gráfico; .... granofídico

0848
granophyric; pegmatophyric; pegmatitic, – commonly used as an adjective for a texture characterized by irregular intergrowths of quartz and feldspar and where phenocrysts and groundmass mutually penetrate each other indicating simultaneous crystallization. Less common is its use as an adjective for porphyritic rocks with a crystalline-granular groundmass. / granofirisch / granophyrique; micropegmatitique; micropegmatoïde / granophyrisch; pegmatophyrisch / granofídica

cryptographic, – adjective for a cryptocrystalline granophyric texture. / cryptografisch / cryptographique / kryptopegmatitisch / criptográfica

graphophyre, – in the CIPW classification porphyritic rocks with a megascopic granophyric groundmass

graphiphyre, – in the CIPW classification rocks with a microscopic granophyric groundmass.

0849
symplectitic; symplektitic; symplektic, – characterized by the occurrence of symplectite, the complex intergrowth of two different minerals; also used for the intergrowth itself. / symplektitisch / symplectique / symplektisch / simplectítica

dactylitic; dactylotype, – adjectives for a texture characterized by symplectitic intergrowth wherein one mineral shows fingerlike projections into the other mineral; also used for the intergrowth itself. / dactylitisch / – – / daktylitisch / dactilítica

dactylite, – rock with dactylitic texture. / dactiliet / – – / Daktylit / dactilita

0850
spinifex texture, – intergrowths of acicular skeletal olivine and pyroxene grouped in intersecting sheaf-like forms. / spinifex textuur / texture spinifex / – – / textura espinifex

0851
perthitic, – characterized by the presence of perthite, an intergrowth of alkali feldspars. / perthitisch / perthitique / perthitisch / pertítica

0852
myrmekitic, – characterized by intergrowths of vermicular quartz and feldspar; also used for the intergrowth itself. / myrmekitisch / myrmékitique / myrmekitisch / mirmequítica

0853
hyperite texture, – texture in which a fibrous amphibole-reaction is formed at the contact between olivine and plagioclase crystals. / hyperitische structuur / structure hypéritique / hyperitische Struktur / estructura hiperítica

0854
clathrate, – adjective for a texture found chiefly in leucite rocks, in which leucite crystals are surrounded by tangential augites in such a way that it resembles a net or section of a sponge, the felted mass of augite prismoids representing the threads or walls, and the clear, round leucite crystals the holes.

0855
ocellar texture, – occurs in nepheline- or leucite-bearing rocks and is characterized by phenocrysts, which are aggregates of smaller crystals (biotite or acmite) arranged radially or tangentially around larger euhedral crystals of nepheline or leucite, or that form rounded eye-like branching forms. / – – / texture ocellaire / Ozellarstruktur / estructura ocelar

ocellus, – phenocryst characterizing ocellar texture.

0856
reticulate texture; mesh ....; reticulated ....; reticular ...., – a texture resembling a network; caused by the alteration of certain minerals e.g. by the serpentinization of olivine. / maastextuur; netvormige textuur / texture maillée; .... réticulée / Maschentextur; Netz.... / textura reticular; .... en mallas

0857
protoclastic, – a term for igneous rocks or their texture in which the earlier formed minerals show granulation and deformation, the result of differential flow of the magma before complete solidification. / protoklastisch / protoclastique / protoklastisch / protoclástica

Protoklase, – German term for the process causing protoclastic texture.

[0944] *Structure*

0858
structure, – a term used for the larger features of rocks, e.g. columnar -, pillow -, bedded structure. The term also denotes some small-scale features of heterogeneous rocks due to the juxtaposition of aggregates differing in texture or composition, e.g. amygdaloidal -, spherulitic -, orbicular -, miarolitic -, schistose -, gneissose structure. / structuur / structure / Struktur / estructura

0859
massive; compact, – adjectives for a structure (or a rock having such a structure) without stratification, flow banding, foliation, schistosity, or other similar features. / massief / massif / massig; richtungslos / masivo

0860
banded structure; banding, – structure of igneous and metamorphic rocks caused by an alternation of layers, streaks or flat lenses of different mineralogical composition and/or structure. In igneous rocks it is thought to be caused by flow of heterogenous material (**flow banding, fluxion banding,** primary gneissic banding) or by successive deposition of layers of different composition. / bandenstructuur; banding / structure rubanée; structure protogneissique; rubanement / primäre Bänderung; gebänderte Struktur / estructura bandeada

0861
eutaxitic, – adjective for a banded structure in extrusive rocks caused by alternating layers of different structure, composition or colour. / eutaxitisch / eutaxitique / eutaxitisch / eutaxítica

eutaxite, – rock with eutaxitic structure. / eutaxiet / eutaxite / Eutaxit / eutaxita

0862
crystallization banding, – banding developed in an igneous rock by successive deposition of crystals, such as those settling at the bottom of a pool of magma. / – – / rubanement par cristallisation / Kristallisationsbänderung / cristalización bandeada

0863
rhythmic layering, – macroscopically visible layering in intru-

sives with a repetition of zones of varying composition according to a set pattern. For types with alternating dark and white bands the term **zebra layering** is sometimes used. / ritmische gelaagdheid / litage rhythmique / rhythmische Bänderung / bandeado rítmico

harrisitic layering, – rhythmic layering in which a crescumulate texture has been developed. / harrisitische gelaagdheid / – – / harrisitische Bänderung / – –

Willow-Lake type layering, – rhythmic layering with crescumulate textures, developed along the margin of cooling plutonic bodies.

0864
rhythmic crystallization, – crystallization of different minerals in parallel or concentric layers; sometimes leading to orbicular structure. / ritmische kristallisatie / cristallisation rhythmique / rhythmische Kristallisation / cristalización rítmica

0865
cryptic layering, – macroscopically less obvious type of layering characterized by a vertical change in chemical composition. / cryptische gelaagdheid / – – / kryptische Bänderung / bandeado críptico

0866
mineral-graded layer, – cumulate sheet characterized by a gradual stratigraphic change in proportions of two or more minerals.

0867
orbicular; spheroidal; nodular, – adjectives for the structure of a rock characterized by the occurrence of orbicules. / orbiculair; kogel- / orbiculaire / kugelig; sphäroidisch; sphärisch / orbicular

orbicule, – spherical to subspherical body from microscopic size to several feet across whose components are arranged in concentric layers. / orbicule; kogel / orbicule / Kugel; Sphäroid / orbícula

The orbicular rocks can be subdivided into the following categories:

isothrausmatic, – the nuclei of the orbicules have the same composition as the matrix.

homeothrausmatic, – orbicules and matrix are comagmatic.

heterothrausmatic, – the nuclei of the orbicules are of various kinds.

allothrausmatic, – the orbicules are formed around xenoliths.

crystallothrausmatic, – the orbicules are formed around phenocrysts.

0868
flow structure; (flow texture); fluidal ....; flowage ....; fluxion ....; rhyotaxitic ...., – structure (texture) caused by flow in a crystallizing magma. / vloeistructuur; fluïdale .... / structure fluidale / Fluidaltextur; Fluktuations....; Fliessgefüge / textura fluidal

igneous lamination; flow layering, – flow structure, manifest in the form of compositional or textural banding. / vloeilaminatie / fluidalité magmatique / magmatische Lamination / laminación ígnea; bandeado fluidal

flow line, – alignment of platy minerals in streaks or swirls along planes of lamellar flowage. / vloeilineatie / ligne d'écoulement / Fliessrichtung / línea de flujo

0869
platy (flow) structure; linear flow ....; planar flow ....; pseudostratification, – structure of tabular sheets in igneous rocks, suggesting stratification. It is formed by contraction during cooling, and the structure is parallel to the cooling surface. / plutonische gelaagdheid; pseudo.... / – – / – – / (p)seudoestratificación

0870
trough banding, – a crescent-shaped mineralogic layering or alignment parallel to the floor of channels running across the bottom of a magma chamber.

0871
gneissoid, – gneiss-like in structure or texture, but of non-metamorphic origin, e.g. viscous magmatic flow forming a gneissoid granite. / gneisachtig / gneissique / gneisähnlich / gneisoide

0872
amygdaloidal; amygdaloid, – adjectives for a structure or a rock characterized by the presence of numerous amygdales. / amygdaloïdaal / amygdalaire / amygdaloïdisch / amigdaloide; amigdalar

amygdaloid, – a rock that has an amygdaloidal structure. / amandelsteen / roche amygdaloïde; .... amygdalaire / Mandelstein / roca amigdaloide; .... amigdalar

0873
miarolitic; drusy, – adjectives for the structure of plutonic rocks containing miarolitic cavities. / miarolitisch / miarolitique / miarolitisch; kleindrusig / miarolítica

miarolitic cavity; druse, – cavities of irregular shape in certain plutonic rocks, into which crystals of the rock constituents project. / miarolitische holte / cavité miarolitique / miarolitischer Hohlraum; .... Drusenraum / cavidad miarolítica

0874
vesicular; cellular, – adjectives for a rock or its structure characterized by numerous vesicles. / caverneus; cellig / vacuolaire; vésiculaire; celluleux / blasig; zellig / vesicular

vesicle; vacuole, – cavity formed in lavas and shallow intrusives by gases. They are spherical, ovoid or irregular in shape, and may subsequently be filled with deuteric or secondary minerals to form amygdules. / blaasvormige holte / vacuole; vésicule / Blasenraum; blasenförmiger Hohlraum / vesícula; vacuola

0875
pumiceous; spongy; foamy; frothy; spumous, – adjectives for the structure of a vesicular rock in which the partitions between the vesicles form a fine network. / puimsteenachtig; sponzig; schuimig / ponceux / bimssteinartig; schwammig; schaumig / pumítica; espumosa; esponjosa

0876
spherulitic; sphaerolitic; lithophysal, – terms applied to the structure of rocks that are characterized by the presence of spherulites. / sferolitisch / sphérolitique / sphärolitisch / esferulítica

pyromeride, – a devitrified rhyolite characterized by spherulitic structure, and thus having a nodular appearance. / pyromeride / pyroméride / Pyromeride / piromérido

0877
variolitic, – adjective for the structure of rocks characterized by the presence of varioles. / variolitisch / variolitique / variolitisch / variolítica

0878
perlitic, – adjective for the structure in glassy rocks that is caused by the cracking of the glass, due to contraction on cooling, into small spheroids with imbricated shells like the parts of an onion. / perlitisch / perlitique / perlitisch / perlítica

[1000] PETROLOGY OF METAMORPHIC ROCKS

[1010] GENERAL TERMS

0879
metamorphism, – the mineralogical and structural adjustment of hard rocks to physical or chemical conditions that have been imposed at depths below the zones of weathering, cementation and diagenesis, and that differ from the conditions under which the rocks in question originated. / metamorfose / métamorphisme / Metamorphose / metamorfismo

NOTE;
– there is a gradual transition form diagenesis to metamorphism without clearly defined limit.

0880
metamorphic rock; metamorphite, – / metamorf gesteente; metamorfiet / roche métamorphique; .... crystallophyllienne / metamorphes Gestein; Metamorphit / roca metamórfica

0881
isochemical metamorphism; treptomorphism, – metamorphism without bulk change in chemical composition. / isochemische metamorfose / métamorphisme isochimique; .... topochimique / isochemische Metamorphose / metamorfismo isoquímico

allochemical metamorphism, – metamorphism accompanied by addition or removal of material. / allochemische metamorfose / métamorphisme allochimique / allochemische Metamorphose / metamorfismo aloquímico

0882
ultrametamorphism, – metamorphic processes under extreme conditions, where partial or complete anatexis of the affected rocks occurs, e.g. migmatization. / ultrametamorfose / ultramétamorphisme / Ultrametamorphose; Ultrametamorphismus / ultrametamorfismo

0883
meta-, – prefix used in combination with a rock name to indicate the parentage of a metamorphic rock; e.g. metabasalt, metasediment. / meta- / méta- / Meta- / meta-

protolith; parent rock, – the unmetamorphosed rock out of which a given metamorphic rock originated. / moedergesteente / roche originelle / Muttergestein / roca original; roca madre

metabasite, – metamorphic rock of basic composition in which all traces of original texture have disappeared making for instance differentiation between metabasalt and metagabbro impossible. / metabasiet / métabasite / Metabasit / metabasita

0884
metablastesis, – preferred crystallization of a mineral or group of minerals by isochemical recrystallization or metasomatism without evidence of a separate mobile phase. / metablastese / métablastèse / Metablastese / metablastesis

metablastic, – formed by metablastesis. / metablastisch / métablastique / metablastisch / metablástico

0885
metamorphic assemblage; .... complex, – a group of associated metamorphic rocks. / metamorfe associatie / assemblage métamorphique / metamorphe Vergesellschaftung / complejo metamórfico

0886
tectonic overpressure, – contribution of orogenic forces to formation pressure during metamorphism. This contribution can at most be equal to the strength of the rock, 10 to 20 kPa (1000 to 2000 bars). / tektonische overdruk / surpression tectonique / tektonischer Druck / sobrepresión tectónica

0887
regional metamorphism; dynamothermal ...., – terms, in contrast to local metamorphism, used when the processes of metamorphism have affected the rocks over extensive areas. Although within the limits of a natural region different grades of metamorphism may be exhibited, the variation is of a gradual and orderly kind. / regionale metamorfose / métamorphisme régional; .... général / Regionalmetamorphose / metamorfismo regional

burial metamorphism; load ....; static ....; constructive ....; geothermal ...., – a type of regional metamorphism unaccompanied by orogenesis or plutonism. Due to the absence of deformation original rock textures are largely preserved. / begravingsmetamorfose; statische ....; belastings.... / métamorphisme d'enfouissement / Versenkungsmetamorphose; Belastungs....; statische .... / metamorfismo estático

0888
anchimetamorphism, – mineralogical and structural adjustment of solid rocks at physical conditions intermediate between those of diagenesis and those of metamorphism. Therefore anchimetamorphic rocks have some diagenetic characteristics, but also show neomineralization without, however, development of equilibrium associations. / anchimetamorfose / anchimétamorphisme / Anchimetamorphose / anchimetamorfismo

0889
injection metamorphism, – metamorphism accompanied by intimate injection of sheets and streaks of liquid magma in zones near deep-seated intrusive contacts, often related to migmatization. / injectiemetamorfose / – – / Injektionsmetamorphose / metamorfismo de inyección

0890
mantled gneiss dome, – a dome in metamorphic terrain with a core of gneiss. The gneiss was remobilized from an original basement and rose through its cover of younger, also metamorphosed rocks. The gneiss is surrounded by a concordant sheath of the basal part of the overlying metamorphic sequence. / omhulde gneiskoepel / dome gneissique / ummantelte Gneiskuppe / domo gnéisico

0891
local metamorphism, – a term, in contrast to regional metamorphism, used when the metamorphism is caused by a local process. / lokale metamorfose / – – / lokale Metamorphose / metamorfismo local

0892
dynamometamorphism; dynamic metamorphism; dislocation ....; cataclastic ....; kinetic ....; mechanical ...., – local metamorphism concentrated along narrow belts of shearing or crushing. Since elevated temperature is not a prerequisite for this form of metamorphism, its products may range from incoherent cataclasites to coherent tectonites with considerable recrystallization, depending on the depth at which the crushing process took place. / dynamometamorfose; dislocatiemetamorfose / dynamométamorphisme / Dynamometamorphose; Dislokations....; kinetische Metamorphose; mechanische ....; Druckmetamorphose; Stauungs .... / dinamometamorfismo; metamorfismo cataclástico; .... cinético; .... dinámico; .... de dislocación; .... mecánico

0893
shock metamorphism, – the totality of observed permanent physical, chemical, mineralogical and morphological changes produced in rocks and minerals by the passage of transient high-pressure shock waves acting over short time intervals ranging form a few microseconds to a fraction of a minute. / schokmetamorfose / métamorphisme de choc / Schockmetamorphose / metamorfismo de choque

impact metamorphism, – shock metamorphism caused by hypervelocity impact of a meteorite body. / inslagmetamorfose / métamorphisme d'impacte / Impaktmetamorphose / metamorfismo de impacto

0894
shock zone, – a volume of rock in or around an impact or explosion crater in which a distinctive shock-metamorphic deformation or transformation effect is present. / schokzone / zone de choc / Shockzone / zona de choque

shock lamellae; planar features, – distinctive, multiple, closely spaced, parallel, plate-like microscopic planes, distinct from cleavage planes, occurring in shock-metamorphosed minerals, and regarded as unique and important indicators of shock metamorphism. / schoklamellen / lamelles de choc / Schocklamellen / lamelas de presión

0895
diaplectic, – adjective for glass-like mineralogic features caused by shock waves without melting. The stage of structural order of diaplectic materials is intermediate between that of crystalline phases and glassy matter. / diaplectisch / diaplectique / diaplektisch / diapléctico

0896
thermal metamorphism; thermometamorphism; pyromorphism (obsolete), – metamorphism controlled predominantly by temperature and to a less extent by confining pressure (as a function of depth); there is no simultaneous deformation. / thermische metamorfose / métamorphisme thermique / thermische Metamorphose / metamorfismo térmico; .... termal

contact metamorphism, – local thermal metamorphism connected with the intrusion or extrusion of magma. Here, not only heat, but also material emanating from the magma effects the metamorphic changes. / contactmetamorfose / métamorphisme de contact / Kontaktmetamorphose / metamorfismo de contacto

pyrometamorphism; igneous metamorphism (obsolete), – contact-metamorphic alterations at extremely high temperatures in xenoliths and along the immediate contacts of some plutons. / pyrometamorfose / pyrométamorphisme / Pyrometamorphose / pirometamorfismo

0897
baking; optalic metamorphism; caustic ...., – the indurating, burning and fritting effects produced by lavas and small dykes on the rocks with which they come into contact. / kaustische metamorfose / cuisson; métamorphisme optalique / kaustische Kontaktmetamorphose / metamorfismo cáustico

fritting, – the partial melting of mineral grains so that now each of them is surrounded by a zone of glass. Fritting is due to contact action of basalt and other lavas on neighbouring rocks. / verglazing / frittage / Frittung / vitrificación

0898
exomorphism; exometamorphism; exomorphic metamorphism, – the modifications effected in the country rock by intrusion of magma (contact metamorphism in the usual sense). / exomorfose / exométamorphisme / Exomorphose; exomorphe Kontaktwirkung / exomorfismo

endomorphism; endometamorphism; endomorphic metamorphism, – the changes produced within the intrusive rock itself by the complete or partial assimilation of fragments of the invaded formation or by reaction with the invaded formation. /endomorfose / endométamorphisme / Endomorphose; endomorphe Kontaktwirkung / endomorfismo

0899
contact zone; .... aureole; aureole; thermal ....; metamorphic ....; metamorphic zone; exomorphic ...., – a zone surrounding an igneous intrusion in which contact metamorphism of the country rock has taken place. / contactzone; .... aureool; ....hof / zone de contact; auréole de .... / Kontakthof; ....aureole / zona de contacto; aureola metamórfica

0900
contact metasomatism; pyrometasomatism, – contact metamorphism accompanied by addition of material from emanations issuing form the invading magma. / contactmetasomatose; pyro.... / pyrométasomatose / Kontaktmetasomatose; metasomatische Kontaktmetamorphose; pneumatolyti-

sche .... / metasomatismo de contacto; pirometasomatismo; pirometasomatosis; (p)neumatolisis de contacto

0901
polymetamorphism; superimposed metamorphism, – multiple-phase or multiple metamorphism, whereby two or more successive metamorphic events have left their imprint on the same rocks. / polymetamorfose; gesuperponeerde metamorfose / polymétamorphisme / Polymetamorphose / polimetamorfismo; metamorfismo múltiple; .... superpuesto

plurimetamorphism, – a term proposed for those cases where more than one metamorphic event is part of the same orogeny. / plurimetamorfose / – – / Plurimetamorphose / plurimetamorfismo

0902
eometamorphism, – early or initial metamorphism. / eometamorfose / – – / Eometamorphose / metamorfismo inicial

0903
prograde metamorphism, – metamorphic changes in response to higher-grade metamorphic conditions (pressure, temperature) than those to which the rock was adjusted formerly. / progressieve metamorfose / métamorphisme prograde / progressive Metamorphose / metamorfismo progresivo

progressive metamorphism, – progressive change in the degree of metamorphism across a metamorphic terrain. Since in most languages this term is used as a synonym for prograde metamorphism its use is not recommended.

0904
retrograde metamorphism; retrogressive ....; diaphthoresis; polymetamorphic ...., – metamorphic changes in response to lower-grade metamorphic conditions than those to which the rock was adjusted formerly. / retrograde metamorfose / métamorphisme régressif; rétrométamorphisme / rückschreitende Metamorphose; Retrometamorphose; Retromorphose; retrograde Metamorphose / metamorfismo retrogrado; retromorfismo; retrometamorfismo

diaphthorite, – rock formed by retrograde metamorphism. / diaftoriet / – – / Diaphtorit / diaftorita

diaphthoresis, – term sometimes used exclusively for retrograde metamorphic reactions taking place during the cooling phase of the same metamorphic cycle that produced the assemblage that is being replaced. / diaftorese / diapthorèse / Diapthorese / diaftoresis

[1020] METAMORPHIC PROCESSES AND PRODUCTS

0905
recrystallization, – the formation of new crystals in a rock, essentially in the solid state. The newly formed minerals may have either the same composition as their precursor minerals or a different one. / rekristallisatie / recristallisation / Neukristallisation; Um....; Re.... / recristalización

neomineralization, – the newly formed minerals have a composition different from their precursors. / neomineralisatie / (néo)formation (de minéraux) / Mineralneubildung / neomineralización

0906
pretectonic; prekinematic, – adjectives used to indicate recrystallization that preceded tectonic activity. / pretektonisch; prekinematisch / antétectonique / prätektonisch; präkinematisch / pretectónico; precinemático

paratectonic; synkinematic, – adjectives used to indicate recrystallization that proceeded contemporaneously with tectonic activity. / paratektonisch; synkinematisch / syntectonique; syncinématique / paratektonisch; parakinematisch / sintectónico; sincinemático

posttectonic; postkinematic, – adjectives used to indicate recrystallization that postdated tectonic activity. / posttektonisch; postkinematisch / post-tectonique / posttektonisch; postkinematisch / posttectónico; postcinemático

0907
Deformationskristalloblastese, – German term for paratectonic recrystallization.

0908
crystalloblastesis, – metamorphic recrystallization. / kristalloblastese / cristalloblastèse / Kristalloblastese / cristaloblástesis

crystalloblast, – crystal formed by crystalloblastesis. / kristalloblast / – – / Kristalloblast / cristaloblasto

0909
crystallizing force; force of crystallization; crystalloblastic strength; form energy,– the expansive force of a crystal that is growing within a solid medium. The force varies according to crystallographic direction. / kristallisatiekracht / force de cristallisation; pression de croissance / Kristallisationskraft; Formenergie / fuerza cristaloblástica; .... de cristalización

crystalloblastic series; idioblastic series; crystalloblastic order, – an arrangement of metamorphic minerals in order of decreasing force of crystallization, so that crystals of any of the listed minerals tend to assume idioblastic outlines at surfaces of contact with simultaneously developed crystals of all minerals occupying lower positions in the series. / kristalloblastische reeks / série cristalloblastique / idioblastische Reihe / serie cristaloblástica; orden cristaloblástico

0910
integration; sammelkristallization, – the action, depending on surface tension or on total free surface energy, by which smaller grains become unstable in relation to larger grains and will eventually be incorporated into the latter. / collectieve kristallisatie / – – / Sammelkristallisation / cristalización integral

regenerated crystal, – a crystal newly formed by integration. / geregenereerd kristal / cristal régénéré / regeneriertes Kristall / cristal regenerado

0911
metamorphic differentiation, – a collective term for the various processes by which contrasted mineral assemblages develop from an initially uniform parent rock during metamorphism. / metamorfe differentiatie / différenciation métamorphique / metamorphe Differentiation / diferenciación metamórfica

0912
metamorphic diffusion, – the migration by diffusion (in solution or in solid state) of material from one part of a rock to another during its recrystallization. / metamorfe diffusie / diffusion métamorphique / metamorphe Diffusion / difusión metamórfica

0913
metamorphic convergence; mesitis, – the phenomenon that the same end product of metamorphism may be formed by different metamorphic processes and/or from different parent rocks. / metamorfe convergentie / convergence métamorphique / metamorphe Konvergenz; Mesitis / convergencia metamórfica

NOTE:
– use of the term **convergence** (convergency) in the same sense for crystalline rocks in general is not recommended since the same term is also used with different meanings in oceanography, stratigraphy and palaeontology.

0914
cataclasis, – deformation of a rock by crushing without chemical reconstitution. The cataclastic texture thus produced may be developed to a greater or lesser extent, and some or all of the mineral constituents of a rock may be affected. / kataklase / cataclase / Kataklase / cataclasis

cataclasite, – rock formed by cataclasis. / kataklasiet / cataclasite / Kataklasit / roca cataclástica; cataclasita

0915
mylonitization; mylonization, – deformation of a rock by extreme microbrecciation, due to mechanical forces with a definite direction in shear zones, without noteworthy chemical reconstitution of granulated minerals. Characteristic features of the rocks produced by this process are a flinty, banded or streaked appearance and undestroyed 'eyes' and lenses of the parent rock in a granulated matrix. / mylonitisatie / mylonitisation / Mylonitisierung / milonitización

mylonite, – microbreccia with flow structure, formed by mylonitization. / myloniet / mylonite / Mylonit / milonita

0916
protomylonite, – coherent crush breccia representing an early stage in the mylonitization process. Primary structures may have been retained in larger lenticular remnants of the parent rock. The term is also used for contact-metamorphic rock mylonitized along the contact with the pluton as a result of overthrust. / protomyloniet / protomylonite / Protomylonit / protomilonita

ultramylonite; flinty crush rock; hyalomylonite, – homogeneous aphanitic microbreccia representing a late stage in the mylonitization process. Porphyroclasts have been obliterated; fluxion banding is absent or only faintly visible. / ultramyloniet / ultramylonite / Ultramylonit / ultramilonita

blastomylonite, – a recrystallized mylonite. / blastomyloniet / – – / Blastomylonit / – –

0917
pseudotachylite, – ultramylonite-like rock that is not closely associated with shear planes and may fill irregular fractures near a fault or otherwise simulate a dyke of igneous rock. / pseudotachyliet / pseudotachylite / Pseudotachylit / (p)seudotaquilita

0918
phyllonite; phyllite-mylonite, – intensely deformed rock resembling a phyllite; whereas, however, evolution of phyllite from slate involves progressive increase of grain size, the fine-grained structure of a phyllonite is the result of reduction of grain size during deformation of originally coarser rocks. Recrystallization and neomineralization have usually been active in their development. / fylloniet / phyllonite / Phyllonit; Phyllitmylonit / filonita

phyllonitization, – the formation of rocks of phyllitic appearance by mylonitization of coarser-grained rocks, accompanied by varying amounts of new crystallization. / fyllonitisatie / phyllonitisation / – – / filonitización

0919
devitrification, – conversion of the glassy texture of a rock into a crystalline texture after its solidification. / ontglazing; devitrificatie / dévitrification / Entglasung / desvitrificación

0920
marmorization; marmarization; marmorosis; marmarosis, – the conversion of limestone into marble. / marmorisatie / marmorisation / Marmorisierung / marmorización

marble, – a usually granoblastic metamorphic rock composed of calcite or dolomite. / marmer / marbre / Marmor / mármol

0921
propylitization, – the hydrothermal alteration of andesitic and related rocks into a greenstone-like rock **(propylite)** composed essentially of chlorite, sericite, epidote, carbonates and quartz with disseminated pyrite. / propylitisatie / propylitisation / Propylitisierung; Propylitbildung; Propylitisation / propilitización

propylite, – / propyliet / propylite / Propylit / propilita

0922
spilitization, – hydrothermal albitization of a basalt, forming a spilite. / spilitisatie / spilitisation / Spilitisierung / espilitización

spilite, – / spiliet / spilite / Spilit / espilita

0923
serpentinization, – the process whereby ferromagnesian minerals are replaced by serpentine. / serpentinisatie / serpentinisation / Serpentinisierung; Serpentinisation / serpentinización

serpentinite, – rock formed by serpentinization and consisting predominantly of serpentine-group minerals. / serpentiniet / serpentinite / Serpentinit / serpentinita

0924
greisenization; greisening; greisenisation, – hydrothermal or pneumatolitic alteration by fluorine-bearing vapours whereby feldspar is being replaced by mica, resulting in a quartz-mica rock or greisen. Topaz, fluorite, lepidolite, tourmaline, cassiterite and wolframite are possible additional products of greisenization. / greisenvorming / greisenification / Greisenbildung / greisenización

greisen, – / greisen / greisen / Greisen / greisen

0925
fenitization, – alkalimetasomatism of quartz-feldspathic country rock near carbonatite complexes. / fenitisatie / fénitisation / Fenitisation / fenitización

fenite, – end product of fenitization. / feniet / fénite / Fenit / fenita

0926
kaolinization; kaolinisation; kaolisation, – replacement of aluminium-silicate minerals by kaolinite; also in general, the formation of kaolinite by hydrothermal processes. / kaolinisatie / kaolinisation / Kaolinisierung; Kaolinisation / caolinización

0927
saussuritization, – the replacement of anorthite-bearing plagioclase by saussurite. Saussuritization is often accompanied by chloritization of the ferromagnesian minerals. / saussuritisatie / saussuritisation / Saussuritisierung; Saussuritisation; Saussuritbildung / Sausuritización

saussurite, – a fine-grained aggregate of zoisite, epidote, albite, calcite, sericite and zeolites. / saussuriet / saussurite / Saussurit / sausurita

0928
uralitization, – the conversion of pyroxene into hornblende, usually as a finely-fibrous aggregate. / oeralitisatie / ouralitisation / Uralitisierung; Uralitisation / uralitización

0929
silicification; silification,– the entire or partial replacement of rocks and fossils by silica, either as quartz, chalcedony or opal. / verkiezeling; silicificatie / silicification / Verkieselung; Silifizierung; Silizifikation / silicificación

desilication,– the removal of silica; from a magma by reaction with the wall rock or from a rock by the breakdown of silicates followed by the removal of silica. / desilicificatie; ontkiezeling / désilicification / Desilifizierung; Entsilifizierung / desilificación

0930
dedolomitization,– a process whereby, during metamorphism, part or all of the magnesium in a dolomite or dolomitic limestone is used for the formation of magnesium minerals (e.g. forsterite, brucite) and calcium-magnesium minerals (e.g. tremolite), resulting in the enrichment in calcite. / dedolomitisatie / dédolomitisation / Dedolomitisation; Dedolomitisierung / desdolomitización

ophicalcite,– a rock composed of calcite and serpentine, formed by recrystallization of a dedolomitized siliceous dolomite. / ophicalciet / ophicalcite / Ophicalcit / oficalcita

0931
replacement,– the process of practically simultaneous capillary solution and deposition by which a new mineral of partly or wholly differing chemical composition may grow within the body of an old mineral or mineral aggregate. / vervanging / substitution; replacement / Verdrängung / reemplazamiento; sustitución

0932
carbonatization; carbonation, – introduction of, or replacement by carbonates. / carbonatisatie / carbonatisation / Karbonatisierung; Karbonatisation / carbonatización

alunitization,– introduction of, or replacement by alunite. / alunitisatie / alunitisation / Alunitisierung; Alunitisation / alunitización

adularization, – introduction of, or replacement by adularia. / adularisatie / adularisation / Adularisierung; Adularisation / adularización

albitization, – introduction of, or replacement by albite. / albitisatie / albitisation / Albitisierung; Albitisation / albitización

scapolitization; dipyrization, – introduction of, or replacement by scapolite. / skapolitisatie / scapolitisation / Skapolitisierung; Skapolitisation / escapolitización

amphibolization, – introduction of, or replacement by amphibole. / amfibolisatie / amphibolitisation / Amphibolisierung; Amphibolisation / amfibolitización

sericitization, – introduction of, or replacement by sericite. / sericitisatie / séricitisation / Sericitisierung; Sericitisation / sericitización

biotitization, – introduction of, or replacement by biotite. / biotitisatie / biotitisation / Biotitisierung; Biotitisation / biotitización

chloritization, – introduction of, or replacement by chlorite. / chloritisatie / chloritisation / Chloritisierung; Chloritisation / chloritización

tourmalinization, – introduction of, or replacement by tourmaline. / toermalinisatie / tourmalinisation / Turmalinisierung; Turmalinisation / turmalinización

zeolitization, – introduction of, or replacement by zeolite. / zeolitisatie / zéolitisation / Zeolithisierung; Zeolithisation / ceolitización

analcimization; analcitization, – introduction of, or replacement by analcime. / analcimisatie / analcimisation / Analcimisation / analcimización

0933
adinole, – quartz-albite rock formed by sodium metasomatism; often with argillaceous sediments as parent rocks. / adinole / adinole / Adinol / adinola

spilosite, – spotted adinole formed at an early stage of its development. / spilosiet / spilosite / Spilosit / espilosita

desmosite, – banded adinole formed at an early stage of its development. / desmosiet / desmosite / Desmosit / desmosita

0934
amphibolite, – a metamorphic rock composed mainly of hornblende and plagioclase. / amfiboliet / amphibolite / Amphibolit / anfibolita

feather amphibolite, – amphibolite where porphyroblastic amphibole crystals form stellate or sheaf-like groups on the planes of foliation or schistosity. / garvenschist / – – / Garbenschiefer /– –

0935
fels; granofels, – granoblastic rock, lacking schistosity or foliation. / – – / corne / Fels / roca corneánica

0936
gneiss,– an irregularly banded metamorphic rock, in which the schistosity is rather poorly developed because in general less than 50% of the minerals show preferred orientation. Quartz and feldspar are usually principal constituents but not necessarily so. Mineral content (garnet-biotite gneiss) parentage **(orthogneiss)** or texture **(augengneiss)** may be indicated with a prefix. / gneis / gneiss / Gneis / gneis; neis

pencil gneiss; stengel gneiss, – a gneiss that splits into prismatic fragments. / stengelgneis / gneiss en crayon / Stengelgneis / gneis apizarrado; .... con disyunción prismática

0937
granulite,–
1. relatively coarse, granular rock formed at the high pressures and temperatures of the granulite facies and exhibiting a crude gneissic structure due to the parallelism of flat lenses of quartz and/or feldspar, e.g. leptite;
2. in Anglo-Saxon usage, a sedimentary rock composed of sand-sized aggregates of constructional origin, e.g. an oolite. Syn.: granulyte;
3. in French usage, a muscovite granite.

/ granuliet / leptynite / Granulit / granulita

leptite,– granular, quartz-feldspathic metamorphic rock formed by regional metamorphism of the highest grade. / leptiet / leptite / Leptit / liptita

0938
greenschist,– schistose metamorphic rock whose green colour is due to the abundance of chlorite, epidote or actinolite it contains. / groenschist / schiste vert / Grünschiefer / esquisto verde

greenstone,– term applied to non-schistose varieties of greenschist.

0939
hälleflinta,– a Swedish term for granulose rocks of horny aspect that are compact or porphyritic and sometimes banded. The mineral composition, quartz, feldspars, micas, chlorite, etc., indicates a metamorphic origin from rhyolite or corresponding volcanic tuffs.

0940
hornfels, – a contact-metamorphic rock composed of a mosaic of equidimensional grains without preferred orientation (granoblastic or hornfelsic texture); porphyroblasts may be present. / hoornrots / cornéenne / Hornfels / corneana; cornubianita

leptynolite, – a fissile or schistose variety of hornfels containing mica, quartz and feldspar with or without minerals such as andalusite and cordierite. / leptinoliet / leptynolite / Leptinolith / leptinolita

0941
maculose rock; spotted slate; .... schist; knotted slate; .... schist, – rock in which porphyroblasts of minerals such as andalusite, cordierite, chloritoid, ottrelite, biotite, etc., are well developed, or in which spotting appears as the result of incipient crystallization of these minerals and of the segregation of carbonaceous matter. / vleklei; nopjeslei; ....schist / schiste tacheté; .... noduleux / Fleckenschiefer; Frucht....; Knoten.... / roca mosqueada; .... moteada; esquisto mosqueado; pizarra mosqueada

The German terms, also adopted in English usage, are of a descriptive character:

Fleckenschiefer,– a schist showing minute flecks or spots of indeterminate material;
Fruchtschiefer,– a schist having concretionary spots suggestive of grains of wheat;
Knotenschiefer,– a schist showing conspicuous subspherical or polyhedral clots often composed of definitely individualized minerals.

0942
slate, – a fine-grained metamorphic rock with perfect fissility (slaty cleavage). The minerals usually cannot be determined megascopically. A product of dynamo- or regional metamorphism of mudstones, siltstones and other fine-grained clastic sediments. / lei; leisteen / ardoise / Schiefer; Tonschiefer / pizarra

pencil slate, – a slate that splits into roughly prismatic pencil-like fragments due to intersecting parallel surfaces (S-planes). / griffellei / phyllade en crayon / Griffelschiefer / – –

0943
phyllite, – a fine-grained schistose rock, sometimes with incipient segregation banding; schistosity surfaces have a lustrous sheen, given off by colourless mica and chlorite. Phyllites have the same origin as slates, but their grain size has coarsened as a result of somewhat more advanced metamorphism. / fylliet / phyllade / Phyllit; Tonglimmerschiefer / filita

0944
schist, – phanerocrystalline metamorphic rock with a strongly developed parallel fabric and excellent fissility. A common characteristic of schists is that they split into foliae, which are mineralogically similar, whereas in gneisses such foliae are usually dissimilar. The mineralogy of the schist may be indicated by a prefix. Contrary to gneisses quartz is also mentioned in the name if it is present in sufficient quantities (quartz-mica schist). / schist / schiste / Schiefer / esquisto

[1030] TEXTURE AND STRUCTURE

0945
banded structure; banding, – structure of igneous and metamorphic rocks caused by an alternation of layers, streaks or flat lenses of different mineralogical composition and/or texture. In metamorphic rocks it may have been caused by segregation during metamorphism or represent a feature inherited from a sedimentary or magmatic parent rock. For a very fine banding the terms **ribbon** or **stripe** may be used as in ribbon jasper or ribbon slate. / bandenstructuur; banding / structure litée / Lagentextur / estructura bandeada

0946
segregation banding, – banding in metamorphic rocks caused

by metamorphic differentiation. / segregatiebanding / rubanement de ségrégation / Segregationsbänderung / segregación en bandas; .... bandeada

0947
tectonic unmixing, – mechanical segregation of minerals due to recrystallization in a shear zone. / tektonische ontmenging / ségrégation tectonique / tektonische Segregation / segregación tectónica

0948
foliation, – a laminated structure resulting from a more or less pronounced aggregation of particular constituent minerals of the metamorphosed rock into layers, lenticles, streaks or discountinuous bands, sometimes very rich in one mineral, and contrasting with contiguous lenticles or streaks, rich in other minerals. All these show at any one place a common parallel orientation. / foliatie / foliation / Schieferung; Kristallisationsschieferung / foliación

NOTE:
– many, especially American, authors use the terms foliation and schistosity synonymously to cover all those megascopically conspicuous, parallel fabrics of metamorphic origin that impart a definite fissility to the rocks in which they occur.

Distinction is made between:

closed foliation, – the foliae appear megascopically as a continuous felt of flakes or rods. / aaneengesloten foliatie / foliation continue / – – / foliación continua

open foliation; interrupted ...., – the dominant minerals are megascopically discontinuous. / onderbroken foliatie / foliation discontinue / – – / foliación discontinua

foliate; foliated, – / bladerig / folié / blätterig; schuppig; geblättert / foliado

0949
decussate, – adjective for a texture of metamorphic rocks, in which axes of contiguous crystals lie in diverse, criss-cross directions that are not random but rather are part of a definite mechanical expedient for minimizing internal stress. It is most noticeable in rocks composed largely of minerals with a flaky or columnar habit. / ongericht / non-orienté / dekussat / divergente

0950
diablastic, – adjective for a texture in metamorphic rocks of which the constituents interpenetrate and are intricately intergrown. / diablastisch / diablastique / diablastisch / diablástica

0951
gneissose structure; gneissic ....; gneissosity, – a composite structure due to the alternation of schistose and granulose bands and lenticles, that are dissimilar both in mineral composition and in texture. The foliation is interrupted; and while a gneiss may split along a plane of schistosity, it does so less readily than a schist, and exposes a much rougher fracture surface. There can be all transitions between schistose, granulose and gneissose structure. / gneistextuur; ....structuur / texture gneissique / Gneistextur / estructura gnéisica; .... néisica

0952
augen structure; flaser ....; eyed .....; phacoidal ...., – structure in metamorphic rocks characterized by augen enveloped in layers of laminar or linear constituents. Use of the term flaser structure is sometimes limited to cases where the matrix surrounding the flaser is cataclastic. / lenticulaire structuur; ogen.... / structure oeilée / Flasertextur; flaserige Textur; lentikulare ....; Augenstruktur / textura ojosa; .... en flaser; .... en augen

augen; flaser, – coarse, lenticular or roughly lenticular minerals, grains or aggregates in foliate metamorphic rocks. Also used as a prefix, as in augen gneis, flaser gabbro, etc. / ogen; lenzen / yeux / Flaser; Augen / flaser; ojos; augen

0953
granoschistose, – adjective for a texture in a monomineralic metamorphic rock produced by the parallel elongation of mineral grains of a normally isodiametric mineral. / – – / – – / – – / granoesquistosa

0954
granulitic, –
1. adjective for a texture characterized by greatly flattened lenses of rather coarse quartz that occurs in some rocks of the granulite facies. / granulitisch / leptynique / granulitisch / granulítica
2. (obsolete) adjective for a structure characterized by a mesh of plagioclase crystals with interstitial olivine and augite; occurring in magmatic rocks. Use of **intergranular** for this type of structure was recommended in 1920 by the Committee on British Petrographic Nomenclature;
3. adjective for a structure resulting from the production of flattened or granular fragments in a rock by crushing;
4. (obsolete) adjective for the structure of granular holocrystalline igneous rocks. Instead, use of the synonyms **panidiomorphic-, hypidiomorphic-** or **xenomorphic-granular** is recommended.

0955
helicitic, – adjective for a texture consisting of bands or streaks of inclusions; they indicate the original bedding or schistosity of the parent rock and continue through later formed metamorphic crystals. The relict inclusions may occur in curved, contorted or sigmoidal patterns in which case the contorted pattern is also a pre- or paracrystalline feature. / helicitisch / hélicitique / helizitisch / helicítica

0956
heteroblastic, – adjective for the texture of crystalloblastic rocks in which the essential constituents are of two or more distinct orders of magnitude. / heteroblastisch / hétéroblastique / heteroblastisch / heteroblástica

porphyroblastic; pseudoporphyritic, – the texture of a heteroblastic rock in which large blastic crystals (porphyroblasts) of one or more minerals are embedded in a matrix made up of smaller elements. / porfiroblastisch / porphyroblastique / porphyroblastisch / porfidoblástica

0957
porphyroid, – non-magmatic rock having the appearance of a porphyry. / porfieroïd /porphyroïde / Porphyroid / porfiroida

0958
pressure shadow; strain ....; stress ....; pressure fringe, – an area adjoining a porphyroclast, characterized by a growth fabric rather than a deformation fabric. / drukschaduw / – – / Druckschatten / sombra de presión; cola de ....

0959
pseudoporphyroblastic, – resembling a porphyroblastic texture but formed by a process other than growth, for instance differential granulation, e.g. mortar structure. / pseudoporfiroblastisch / pseudoporphyroblastique / pseudoporphyroblastisch / (p)seudoporfidoblástica

0960
homeoblastic, – adjective for the texture of crystalloblastic rocks in which the essential constituents are approximately of equal size. Dependent on the habit of the essential minerals homeoblastic rocks may be called: granoblastic, lepidoblastic, nematoblastic or fibroblastic. / homeoblastisch / homéoblastique / homöoblastisch / homoblástica

granoblastic; granulose, – adjectives for the texture of non-schistose metamorphic rocks that are composed of equidimensional grains. / granoblastisch / granoblastique / granoblastisch / granoblástica

mosaic; cyclopean; granuloblastic, – adjectives for a granoblastic texture with smooth grain boundaries. / mozaïektextuur; plaveisel / en mosaïque / Pflastertextur; Mosaik....; Hornfels.... / en mosaico

sutured, – adjective for a granoblastic texture in which the grains have irregular interlocking contacts. / vertand- / engrené / Sutur- / – –

0961
lepidoblastic, – adjective for the texture of schistose metamorphic rocks characterized by the presence of an abundance of phyllosilicates. / lepidoblastisch / lépidoblastique / lepidoblastisch / lepidoblástica

0962
nematoblastic, – adjective for the texture of schistose metamorphic rocks characterized by the presence of an abundance of columnar or prismatic crystals. / nematoblastisch / nématoblastique / nematoblastisch / nematoblástica

0963
fibroblastic, – adjective for the texture of schistose metamorphic rocks characterized by the presence of an abundance of fibrous crystals. / fibroblastisch / fibroblastique / fibroblastisch / fibroblástica

0964
knitted, – adjective for a texture developed by serpentine when replacing clinopyroxene. / verweven / fenestré; maillé / gestrickt / en malla

lattice, – adjective for a texture developed by serpentine when replacing amphibole. / tralie / maillé / Gitter- / reticular

0965
maculose; knotted; spotted, – adjectives for a structure typically developed in thermally metamorphosed argillaceous rocks. / nopjestextuur / tacheté / Flecktextur / mosqueada; moteada

0966
mortar; murbruk; porphyroclastic, – adjectives for a cataclastic structure in which granulated fragments of quartz and feldspar occur interstitial to, or as borders on the edges of much larger rounded relicts which were more resistant to deformation. / mortel / en mortier; porphyroclastique / Mörtel-; Murbruk- / en mortero

0967
strain recrystallization, – recrystallization in which a deformed mineral changes to a mosaic of undeformed crystals of the same mineral. / dynamische rekristallisatie / – – / Strain-Rekristallisation / recristalización dinámica; granulación

0968
undulatory extinction; wavy .....; oscillatory ....., – a type of extinction that occurs successively in areas adjacent to a single mineral grain on rotating the microscope stage. / onduleuze uitdoving / extinction ondulée; .... roulante / undolöse Auslöschung; undulierende ..... / extinción ondulosa; .... ondulante

0969
mylonitic, – adjective for a structure characteristic of mylonites; caused by microbrecciation, and having the appearance of a flow structure. / mylonitisch / mylonitique / mylonitisch / milonítica

0970
overprint; superprint; imprint; metamorphic overprint, – the development or superposition of metamorphic structures on original structures; the evidence of deformation in a rock fabric. / opdruk / surimpression / Überprägung / sobrepuesto; superpuesto

0971
palimpsest; relic structure, – adjective for a structure in metamorphic rocks in which remnants of the original texture are preserved. / palimpsest; relict / rélique / palimpsest; relikt / residual; relictica

blasto-, – prefix used to indicate a palimpsest structure in metamorphic rocks, e.g. blastogranitic, blastophitic, blastoporphyritic. / blasto-; relict / blasto- / blasto-; relikt / blasto-

0972
plane parallel, – adjective for a texture in crystalline schists characterized by the arrangement of tabular and prismatic crystals in parallel planes. / planair / planaire; laminaire / schieferig; lamellar / laminar; planar

linear parallel; parallel, – adjectives for a texture in crystalline schists characterized by parallel alignment of linear constituents. / gestrekt; stengelig / linéaire / linear; gestreckt; stengelig / linear

0973
lineation; secondary ...., – general terms for any linear structure in a rock, irrespective of scale or origin; e.g. prismatic crystals (**mineral streaking; streaking**), rod-like aggregates, axes of folds or microfolds, or lines of intersection of different schistosities. / lineatie / linéation / Linearstreckung / lineación

dynamofluidal, – linear texture in dynamometamorphic rocks.

0974
poikiloblastic; sieve; poeciloblastic, – adjectives for a metamorphic texture due to the development, during recrystallization, of a new mineral around numerous relicts of the original minerals, thus simulating the poikilitic texture of igneous rocks. Also used for large xenoblasts enclosing numerous small idioblasts. / poikiloblastisch; zeef- / poeciloblastique / poikiloblastisch; Sieb- / poiquiloblástica; poeciloblástica

0975
schistosity, – a structure in metamorphic rocks due to the arrangement of the lamellar, tabular, rod-like and highly cleavable minerals such as mica, chlorite, talc and amphibole in parallel planes or in parallel directions. This parallel arrangement imparts to the crystalline schists a characteristic fissility (not to be confused with foliation). / schistositeit / schistosité / Schieferung; Schieferigkeit / esquistosidad

schistose; schistic, – adjectives for rocks displaying schistosity. / schisteus / schisteux / schieferig / esquistosa

In German literature the following descriptive terms are used:

eigentliche Schieferigkeit, – the tabular constituents of the crystalline schist show a perfect parallel arrangement;
schuppige Schieferigkeit, – the tabular constituents of the schist show imperfect parallel arrangement;
gewundene Schieferigkeit; gefältete ...., – the cleavage planes are gently undulating;
Zickzackschieferigkeit, – the cleavage planes are strongly crumpled.

schistoid, – resembling schist. / schistachtig / – – / – – / esquistoida

[1040] THE CONSTITUENTS

0976
metacryst; metacrystal, – crystal formed by metamorphic processes. / metakrist / – – / Metakristall / metacristal

metamorphic water; regenerated ...., – water involved in, or driven out, by the process of metamorphism. / metamorf water / eau métamorphique / metamorphes Wasser / agua metamórfica

0977
piezogene, – pertaining to the formation of minerals primarily under the influence of pressure. / piëzogeen / – – / piezogen / piezógeno

thermogene, – pertaining to the formation of minerals primarily under the influence of temperature. / thermogeen / – – / thermogen / termógeno

0978
contact mineral, – mineral formed by contact metamorphism. / contactmineraal / minéral de métamorphisme de contact / Kontaktmineral / mineral de contacto

0979
stress mineral (obsolete), – mineral for whose formation stress was thought to be an essential factor.

antistress mineral (obsolete), – mineral for whose formation the absence of stress was thought to be an essential factor.

0980
idioblast, – a constituent of metamorphic rocks formed by recrystallization or neomineralization and that is bounded by its own crystal faces. / idioblast / idioblaste / Idioblast / idioblasto

hypidioblast; subidioblast, – metamorphic mineral only partly bounded by crystal faces. / hypidioblast / hypidioblaste / Hypidioblast / hipidioblasto

xenoblast; allotriblast, – terms applied to crystals that have grown during metamorphism without developing their crystal faces. / xenoblast / xénoblaste / Xenoblast / xenoblasto

xenoblastic, – / xenoblastisch / xénoblastique / xenoblastisch / xenoblástico

0981
porphyroblast; pseudophenocryst; phenoblast, – pseudoporphyritic crystal in a metamorphic rock. / porfiroblast / porphyroblaste; phénoblaste / Porphyroblast / porfidoblasto; fenoblasto

porphyroclast; – pseudoporphyritic crystal in a cataclastic rock. / porfiroklast / porphyroclaste / Porphyroklast / porfidoclasto; fenoclasto

poikiloblast, – large xenoblast enclosing numerous small idioblasts. / poikiloblast / crystal poecilitique / Poikiloblast / poiquiloblasto

0982
pseudomorph; false form, – a mineral that formed by alteration, substitution, incrustation, or paramorphism from another mineral species whose outward form it maintained. Such a mineral is said to be **pseudomorphous after** its predecessor. / pseudomorf / pseudomorphe / pseudomorph / (p)seudomórfo

pseudomorphous after, – / pseudomorf naar / (pseudomorphe) d'après / pseudomorph nach / (p)seudomorfo según;

pseudomorphism, – the process of becoming and the condition of being a pseudomorph. / pseudomorfose / pseudomorphose / Pseudomorphose / (p)seudomorfismo

0983
allothimorphic (obsolete), – adjective for a constituent of a metamorphic rock, that has retained the shape it had in the parent rock.

eleutheromorphic (obsolete), – adjective for a new mineral in a metamorphic rock that has developed freely and whose form is independent of pre-existing minerals, such in contrast to a pseudomorph.

0984
holoblast, – crystal completely formed during metamorphism and therefore not possessing a pre-existing core. / holoblast / holoblaste / Holoblast / holoblasto

0985
relic, – mineral, remains of a mineral, texture, structure or other feature in a metamorphic rock that was a part of the

parent rock or was formed during an earlier phase of metamorphism and survived later stages of metamorphism. Also used as an adjective, as in: relic texture. / relict / relique / Relikt / mineral relíctico; relicto

NOTE:
- in German usage a distinction is made between: **primäres Relikt,** – a relict constitent of the parent rock, **sekundäres Relikt,** – a relict of an earlier phase of metamorphism.

stable relic, – a relic that was not only stable under the conditions prevailing while it was formed, but also under the newly imposed conditions. / stabiel relict / relique stable / stabiles Relikt / relicto estable

unstable relic; metastable ...., – a relic that is unstable under newly imposed metamorphic conditions but, nevertheless, persists on account of low velocity of transformation. / instabiel relict / relique instable / instabiles Relikt / relicto inestable

[1050] CLASSIFICATION

0986
para, – prefix used with a metamorphic rock name to indicate a sedimentary parentage, e.g. paraamphibolite. / para / para / Para / para

ortho, – prefix used with a metamorphic rock name to indicate an igneous parentage, e.g. orthoamphibolite. / ortho / ortho / Ortho / orto

0987
metamorphic facies; densofacies, – a term comprising all the rocks that are metamorphosed within the limits of a certain pressure-temperature range, which is characterized by the stability of certain characteristic mineral parageneses. / metamorfe facies / facies métamorphique / metamorphe Fazies; Metamorphose-Fazies / facies metamórfica

metamorphic subfacies, – a subdivision of a metamorphic facies based on mineralogical transitions that take place, within the limits of this facies, in mineral associations other than those used for the definition of the facies itself. / metamorfe subfacies / sousfacies métamorphique / metamorphe Subfazies / subfacies metamórfica

NOTE:
- See table 2 – Metamorphic Facies

0988
critical mineral; typomorphic ....; index ....; critical .... association, – minerals or mineral associations that are stable only under the conditions of one metamorphic facies and will change upon change of conditions. / kritisch mineraal; typomorf ....; kritische mineraalassociatie / minéral typomorphe; association typomorphe / kritisches Mineralgemengteil; kritische Mineralassoziation; typomorphes Mineral / mineral índice; .... guía; .... tipomorfo

0989
ACF diagram, – triangular plot devised by Eskola, in which for silica-saturated rocks the relationship between the chemical (reduced to three parameters: Al_2O_3, CaO, FeO + MgO) and the mineralogical composition is visualized. / ACF diagram / diagramme ACF / ACF-Diagramm / diagrama-ACF

A'KF diagram, – a plot with a similar purpose as the ACF diagram, but based on different parameters (Al, K_2O, FeO + MgO). / A'KF diagram / diagramme A'KF / A'KF-Diagramm / diagrama A'KF

AFM diagram, – triangular plot devised by Thompson to visualize the effect of partitioning of FeO and MgO over the ferromagnesian phases on the mineralogical composition of metamorphic rocks. / AFM diagram / diagramme AFM / AFM-Diagramm / diagrama-AFM

0990
charnockite series, – a series of rocks of plutonic appearance, compositionally similar to the granitic rock series but characterized by the presence of orthopyroxene. A completely separate nomenclature exists for the rocks of this series, e.g. m-charnockite. / charnockitische reeks / série charnockitique / charnockitische Reihe / serie charnockítica

m-charnockite, – member of the charnockite series containing mesoperthite as the only feldspar. / m-charnockiet / – – / M-Charnockit / m-charnockita

0991
metamorphic facies series, – a group of metamorphic facies characteristic for a particular area or terrain. The series is represented by a curve, or a group of curves, in a pressure-temperature diagram illustrating the ranges of the different types of metamorphism and metamorphic facies. / metamorfe faciesreeks / série de facies métamorphiques / metamorphe Faziesserie / serie de facies metamórfica

petrogenetic grid, – a P-T diagram in which the equilibrium curves of the most important metamorphic reactions are indicated. / petrogenetisch raster / – – / petrogenetisches Raster / – –

0992
metamorphic grade; .... rank, – the extent or rank of metamorphism, measured by the amount of degree of difference between the original parent rock and the metamorphic rock. / graad van metamorfose / degré métamorphique / Metamorphosegrad / grado metamórfico

isofacial; isograde; isogradal; isophysical, – pertaining to rocks belonging to the same metamorphic (sub)facies. / isograad / de même facies métamorphique / isofaziell / isográdica

isograd, – a line on a map joining points in which metamorphism proceeded at similar values of pressure and temperature. / isograad / isograde / Isograd / isograda

0993
depth zone, – characteristic physico-chemical environments at various depths in the earth's crust that give rise to different metamorphic phenomena. Grubenmann recognized three such depth zones; called the **epizone**, the **mesozone** and the **catazone** in increasing order of depth. This concept was later modified, whereby temperature rather than depth was stressed as the most important factor contributing to this zonation. / dieptezone / zone de profondeur / Tiefenzone; ....stufe / zona (metamórfica) de profundidad

epizone, – uppermost zone of metamorphism including the products of low-temperature contact metamorphism and low-temperature hydrothermal metasomatism. / epizone / épizone / Epizone / epizona

mesozone, − middle zone of metamorphism. / mesozone / mésozone / Mesozone / mesozona

catazone; katazone, − deepest zone of metamorphism including the products of high-temperature contact metamorphism as well as perimagmatic (pneumatolytic) metasomatism and injection metamorphism. / katazone / katazone / Katazone / catazona

plutonic metamorphism, − a term sometimes used as a synonym for catazonal metamorphism.

0994
geological thermometry; geothermometry, − any method by which the prevailing temperature at the time of formation of a rock or a mineral can be determined. / geothermometrie / géothermométrie / geologische Thermometrie; Geothermometrie / geotermometría

geological thermometer; geothermometer, − mineral or mineral assemblage that can be used for geothermometry. / geologische thermometer / géothermomètre; thermomètre géologique / geologisches Thermometer / termómetro geológico; geotermómetro

geobarometry; geopiezometry, − any method by which the prevailing pressure at the time of formation of a rock or mineral can be determined. / geobarometrie / géobarométrie / Geobarometrie / geobarometría

[1060] STRUCTURAL PETROLOGY

0995
structural petrology; petrofabrics; petrofabric analysis; structural....; petrogeometry; petromorphology; fabric analysis; microtectonics, − the branch of petrology dealing with all textural and structural features of rocks and with the interpretation of the spatial relations of these textures and structures. It deals with the interpretation of chemical and mineralogical facts. / structurele petrologie; maakselanalyse / pétrofabrique / Gefügekunde / petrología estructural

0996
petrotectonics; tectonic analysis, − structural petrology including the analysis of the movements that produced the rocks' fabric. / petrotektoniek / − − / Petrotektonik / petrotectónica

tectonic facies, − a collective term for rocks that owe their present characteristics mainly to specific tectonic movements.

0997
precrystalline, − adjective used to indicate that the deformation preceded the recrystallization of the rock. / prekristallijn / − − / präkristallin / precristalino

paracrystalline, − adjective used to indicate that deformation and recrystallization proceeded simultaneously. / parakristallijn / − − / parakristallin / paracristalino

postcrystalline, − adjective used to indicate that the deformation postdated the crystallization or recrystallization of the rock. / postkristallijn / − − / postkristallin / postcristalino

0998
tectonic transport; .... flow, − direct (gliding, rotation) or indirect (recrystallization) componental movement of a rock mass during deformation. / tektonisch transport / − − / tektonischer Transport / flujo tectónico; transporte ....

intragranular, − tectonic transport taking place along glide planes. / intragranulair / − − / intragranular / intragranular

intergranular, − tectonic transport taking place by rotation and displacement of grains relative to each other. / intergranulair / − − / intergranular / intergranular

0999
componental movement, − the relative movement of fabric elements. / − − / − − / Teilbewegung / movimiento composicionaI

deformation lamella, − a type of slip zone that is produced by active slip within a mineral grain during tectonic deformation or shock. / deformatielamel / déformation en lamelles / Deformationslamelle / deformación lamelar

1000
internal rotation, − rotation with respect to internal coordinates within a grain, producing distortion of the crystal lattice; e.g. calcite **twinning.** / interne rotatie / rotation interne / Internrotation / rotación interna

external rotation, − rotation with respect to external coordinates of grains or aggregates, regardless of what internal rotation there may be. / externe rotatie / rotation externe / Externrotation / rotación externa

tectonic rotation, − internal rotation of a tectonite in the direction of tectonic transport. / tektonische rotatie / rotation tectonique / tektonische Rotation / rotación tectónica

1001
tectonite; tektonite, − a rock formed or deformed by interrelated componental movements. / tektoniet / tectonite / Tektonit / tectonita

primary tectonite − a tectonite formed by componental movements in a fluid medium; e.g. depositional fabrics formed in air, water or melt. / primaire tektoniet / − − / Primärtektonit / tectonita primaria

fusion tectonite, − a primary tectonite with a fabric caused by movement within the melt. / smelttektoniet / − − / Schmelztektonit / tectonita fundida

secondary tectonite, − a tectonite with a fabric formed by componental movement in an essentially solid medium (solid flow). / secundaire tektoniet / − − / Sekundärtektonit / tectonita secundaria

1002
S-tectonite; slip tectonite, − tectonite with a fabric dominated by one set of visible S-planes (expressed megascopically as a single well-defined planar fabric component), linear parallelism of fabric elements being inconspicuous or absent. / S-tektoniet / − − / S-Tektonit / tectonita-S

S-plane; S-surface, – non-genetic term for one of a set of parallel surfaces of discontinuity or mechanical inhomogeneity such as foliation, bedding, schistosity, etc. / S-vlak / surface S / S-Fläche / plano-S

slip surface; flow ...., – plane separating adjacent layers of slip or flow. / – – / surface de glissement / Gleitfläche / superficie de deslizamiento

1003
R-tectonite, – tectonite with a linear fabric caused by rotation of fabric elements. / R-tektoniet / – – / R-Tektonit / tectonita-R

B-tectonite, – tectonite with a planolinear fabric, defined by two or more sets of S-planes intersecting along the b(B) kinematic axis. / B-tektoniet / – – / B-Tektonit / tectonita-B

$B \perp B'$-tectonite, – tectonite in which two kinematic B axes are perpendicular to each other. / $B \perp B'$-tektoniet / – – / $B \perp B'$-Tektonit / tectonita-$B \perp B'$

$B \wedge B'$-tectonite, – tectonite in which two kinematic B axes lie at an angle of less than 90° to each other. / $B \wedge B'$-tektoniet / – – / $B \wedge B'$ Tektonit / tectonita-$B \wedge B'$

1004
fabric; rock ....; structural ....; petro...., – the sum of all the textural and structural features of a rock. / maaksel / fabrique / Gefüge / fábrica

se; se-fabric, – a fabric external to an inclusion (such as a porphyroblast). / extern maaksel / – – / Externgefüge; se / S_e; esquistosidad exterior

si; si-fabric, – the fabric within an inclusion. / intern maaksel / – – / Interngefüge; si / S_i; esquistosidad interior

1005
megafabric; megascopic fabric; macrofabric, – the fabric of a rock as seen in outcrop. / megamaaksel / – – / Megagefüge / megafábrica

mesofabric; mesoscopic fabric, – the fabric of a rock as seen in hand specimen. / mesomaaksel / – – / Mesofefüge / mesofábrica

microfabric; microscopic fabric, – the fabric of a rock as seen under the microscope. / micromaaksel / – – / Mikrogefüge / microfábrica

subfabric, – any part of the fabric of a rock; the spatial array of any particular fabric element. / – – / – – / Teilgefüge / subfábrica

1006
fabric domain; domain, – a three-dimensional area or volume, defined by boundaries such as structural or compositional discontinuities, within which the rock fabric is uniform. / maakseldomein / – – / – – / dominio de fábrica

1007
deformation fabric; tectonic ...., – fabric in which the fabric elements owe their orientation to deformation. / deformatiemaaksel / – – / Deformationsgefüge / fábrica deformada

homotactic, – a deformation fabric where the subfabrics are all identical and conformable in symmetry. / homotaktisch / homotactique / homotaktisch / homotáctica

heterotactic, – a deformation fabric where the subfabrics do not conform to a common symmetry. / heterotaktisch / hétérotactique / heterotaktisch / heterotáctica

1008
mimetic, – adjective for a tectonite fabric formed by recrystallization or neomineralization, reflecting a pre-existing anisotropic fabric. Also used as an adjective for the crystallization process itself. / mimetisch / mimétique / mimetisch / mimética

mimetic crystallization, – / mimetische kristallisatie / cristallisation mimétique / mimetische Kristallisation; Abbildungskristallisation / cristalización mimética

1009
belteroporic, – adjective for fabric in deformed rocks that bears no direct relation to deformation. The development of preferred orientation in such a fabric is controlled entirely by the anisotropic nature of the medium in which the crystals have grown. / – – / – – / belteropor / – –

1010
growth fabric; growth-zone ...., – fabric formed by growth of grains in place by crystallization, independent of stress and resultant movement, e.g. growth from the walls of a fissure, growth in a pressure shadow, etc. / groeimaaksel / – – / Wachstumsgefüge / fábrica de crecimiento

1011
depositional fabric, – fabric formed by mechanical deposition from a moving or stagnant fluid medium, e.g. sedimentation or crystal settling. / sedimentatiemaaksel / – – / Absatzgefüge; Geopetalgefüge / fábrica deposicional

1012
fabric element, – any aggregate, crystal, or part thereof that behaves as a unit with respect to the forces applied to it. The shape of fabric elements may be described as equant, planar, planolinear, linear or triaxial. / maakselelement / – – / Gefügeelement / elemento de fábrica

superindividual, – a fabric element consisting of a monomineralic aggregate of crystals, e.g. a homogeneous aggregate of grains produced by granulation of a single large crystal. / superindividu / – – / Überindividuum; Gefügeelement höherer Ordnung / superindividual

1013
preferred orientation, – non-random orientation of fabric elements. / voorkeursoriëntatie / orientation préférentielle / Regelung; Regel; Einregelung / orientación preferente; .... preferencial

NOTE:
– in German usage the term **Regelung** refers to the orientation process, **Regel** to the accomplished product.

lattice orientation, – a preferred orientation of the space lattices of crystals. / roosteroriëntatie / orientation crystallographique / Gitterregelung; Gitterregel; Regelung nach Kornbau; Regel nach .... / orientación cristalográfica

dimensional orientation, – a preferred orientation of inequidimensional grains or aggregates according to their external form. / vormoriëntatie / – – / Formregelung; Formregel; Regelung nach Korngestalt; Regel nach .... / orientación dimensional

1014
blastetrix, – in an anisotropic medium, any surface to which a direction of greatest ease of growth is perpendicular. / – – / – – / Blastetrix / – –

1015
fabric diagram; petrofabric ....; orientation ...., – graphic representation of the data of fabric elements. / maakseldiagram / diagramme structurologique / Gefügediagramm / diagrama estructural; .... de fábrica

Wulff net; stereonet, – meridional or polar plotting device used for stereographic projection with conservation of true angular relationships. / net van Wulff / canevas de Wulff / Wulff'sches Netz / red de Wulff; falsilla de ....

Schmidt net; equal-area net, – meridional or polar plotting device used to plot vectoral data with conservation of true areal relationships. / net van Schmidt / canevas de Schmidt / Schmidt'sches Netz / red de Schmidt; falsilla de ....

1016
fabric axis; reference axis, – one of the three mutually perpendicular axes used in structural petrology as reference in the orientation of fabric elements, and in establishing the geometrical relationship between fabric and the overall structure of the rocks. / maakselas / axe de la déformation / Gefügeachse / eje de deformación; eje de fábrica

b-axis; b-direction, – fabric axis lying parallel to the most prominent lineation and/or normal to the most obvious symmetry plane of the fabric. / b-as / axe b / b-Achse / eje-b

c-axis; c-direction, – fabric axis perpendicular to a cleavage or other structural surface. / c-as / axe c / c-Achse / eje-c

a-axis; a-direction, – fabric axis normal to the b-c plane. / a- as / axe a / a-Achse / eje-a

1017
t-direction; translation glide line, – the direction of transportation in a glide plane. / t-richting / – – / T-Richtung / dirección-t

T-plane; translation glide plane, – plane along which parts of a crystal have been displaced. / T-vlak / – – / T-Fläche / plano-T

f-axis, – the fabric axis of rotation, around which a gliding or twinning plane may be bent, normal to the t-direction. / f-as / – – / f-Achse / eje-f

1018
pole, – the point on the surface of a reference sphere where a line, perpendicular to a plane passing through the centre of the sphere, penetrates. Also the stereographic projection of such a point. / pool / pôle / Pol / polo

pi-diagram; π....; pole ....; point ....; scatter .... , – a plot of poles of planar elements in a fabric. / pooldiagram / diagramme de pôles / Poldiagramm / diagrama-π

1019
maximum, – a single area of concentration of points representing orientation of fabric elements, e.g. a point maximum or a girdle maximum. / maximum / maximum / Maximum / máximo

1020
girdle, – belt or concentration of points approximately coincident with a great or a small circle of a stereographic net. / gordel / guirlande / Gürtel / cintura; corona

pi-circle; π-circle, – the girdle that represents poles to folded surfaces. / poolgordel / courbe π / Polgürtel / círculo-π

girdle axis, – the normal to a girdle. / gordelas / pôle de la guirlande / Gürtelachse / eje de la corona; polo de la ....

pi-axis; pi-pole; π-axis; beta-....; β-axis, – the normal to the great circle approximated by segments of a folded surface. / β-as / axe π / β-Achse / eje-β

1021
beta-diagram; β-diagram, – constructed by determination of the intersections of a number of cyclographically projected S-planes. It is used to determine a regional fold axis. / β-diagram / diagramme-β / β-Diagramm / diagrama-β

cyclographic projection, – the projection of the intersection of the S-plane with the sphere of reference in a stereographic projection. / cyclografische projectie / projection planosphérique / zyklographische Projektion / proyección plano esférico

[1070] MIGMATIZATION

1022
migmatization; migmatitization, – the process by which rock is converted into migmatite. / migmatisatie / migmatisation; migmatitisation / Migmatisierung; Migmatitisierung / migmatitización

migma, – mobile mixture of solid rock material and magma, formed by palingenesis. / migma / migma / Migma / migma

migmatite, – mixed rock (chorismite) consisting of two or more petrographically different parts; one is the country rock in a more or less metamorphic state, the other one is of pegmatitic, aplitic, granitic or of a generally plutonic appearance. / migmatiet / migmatite / Migmatit / migmatita

1023
mictite, – coarse, composite rock formed as the result of incorporation and partial or complete assimilation of country-rock fragments in a magma under conditions of relatively low temperature and probably at relatively high levels in the crust. / – – / mictite / Mictit; Miktit / mictita

1024
palingenesis, – the in situ formation of magma by anatexis and/or metasomatism. / palingenese / palingenèse / Palingenese / palingénesis

palingenetic, – / palingeen / palingénétique / palingenetisch / palingenético

1025
anatexis, – melting of pre-existing rock. / anatexis / anatexie / Anatexis / anatexia

anamigmatization, – high-temperature, high-pressure anatexis resulting in the formation of magma.

1026
anatexite; anatectite, – rock formed by anatexis. / anatexiet / anatexite / Anatexit / anatexita

For anatectic granite the synonym **ultrametagranite** is sometimes used.

1027
diatexis, – high-grade anatexis that includes the mafic minerals. / diatexis / – – / Diatexis / diatexia

diatexite; diatectite, – rock formed by diatexis. / diatexiet / – – / Diatexit / diatexita

1028
diachyte, – rock formed by marked mechanical and/or chemical contamination of anatectic magma by cognate basic material.

1029
differential anatexis; .... melting; .... fusion; partial melting; .... anatexis; selective fusion, – partial melting of a rock. / differentiële anatexis; selectieve .... / anatexie différentielle; fusion partielle / selektive Anatexis; differenzielle ....; partielle .... / anatexia parcial; .... diferencial; fusión selectiva

1030
metatexis, – selective anatexis of the low-melting components of a rock. / metatexis / – – / Metatexis / metatexia

metatexite, – rock formed by metatexis. / metatexiet / – – / Metatexit / metatexita

1031
lipotexite; lipotectite, – non-liquefied basic material within anatectic magma. / – – / – – / Lipotexit / lipotexita

metatectite, – a lipotexite with a mineralogy and a texture that have been affected by metasomatism accompanying anatexis. / – – / – – / Metatektit / metatectita

1032
restite, – geochemically immobile part of a rock during partial mobilization of rock components. / restiet / restite / Restit / restita

atexite; atectite, – basic material that is not changed during anatexis. / atexiet / atexite / Atexit; Atectit; Atektit / atexita; atectita

1033
mianthite, – dark-coloured enclosures, patches, streaks, etc. in an anatexite.

1034
ectexis, – migmatization by formation of a melt in situ. / ectexis / – – / Ektexis / ectexia

ectexite; ectectite, – rock formed by ectexis. / ectexiet / – – / Ektekt / ectexita

1035
aoritic (obsolete), – adjective for the stages of migmatization, through partial melting in situ, that preceded the magmatic stage.

1036
entexis, – migmatization by introduction of neosome from an outside source. / entexis / entexie / Entexis / entexia

entexite; entectite, – rock formed by entexis. / entexiet / matériel entectique / Entekt / entexita

1037
neosome, – newly formed part of a migmatite. / neosoom / néosome / Neosom / neosoma

palaeosome, – part or host rock of a migmatite. / palaeosoom / paléosome / Paläosom; Palasom / paleosoma

1038
leucosome, – leucocratic part of a migmatite, generally rich in quartz and feldspar. / leukosoom / leucosome / Leukosom / leucosoma

melanosome, – melanocratic part of a migmatite, rich in mafic minerals. / melanosoom / mélanosome / Melanosom / melanosoma

1039
kyriosome, – fundamental mass or matrix of a migmatite. / – – / – – / Kyriosom / quiriosoma

akyrosome, – subsidiary mass (bands, lenses, veins, etc.) of a migmatite. / – – / – – / Akyrosom / aquirosoma

1040
metatect, – the fluid or more mobile part of a migmatite, formed as a result of anatexis. / metatekt / matériel métatectique / Metatekt / metatecto

chymogenetic; chymogenic, – adjectives for metatect. / – – / – – / chymogen / quimogenético

1041
metaster, – the portion of a migmatite that remained solid (immobile or less mobile) during migmatization. / – – / – – / Metaster / metastero

stereosome, – as metaster, but used in the same sense for chorismites in general. / – – / – – / Stereosom / estereosoma

stereogenetic; stereogenic, – adjectives for stereosome. / – – / – – / stereogen / estereógeno

1042
agmatite; plutonic breccia, – breccia-like migmatite. / agmatiet / agmatite / Agmatit / agmatita

1043
dictyonite; diktyonite, – magmatite with a net-like structure. / diktyoniet / – – / Diktyonit; Netzgestein / dictionita

1044
phlebite; composite gneiss; veined ...., – veined migmatite. / flebiet / phlébite / Phlebit; Adergneis / flebita

1045
arterite, – the phlebitic material is presumed to have been injected. / arteriet / artérite / Arterit / arterita

venite, – the phlebitic material is derived from the palaeosome by exudation. / veniet / vénite / Venit / venita

1046
mixed gneiss, – term used in cases where the palaeosome is of sedimentary parentage. / – – / – – / Mischgneis / – –

diadysite, – a French term referring to a phlebite with granitic veins cross-cutting an older schistosity. / – – / diadysite / Diadisit / diadisita

1047
crocydite, – phlebite with a fluff- or flake-like texture caused by the distribution pattern of the vein material. / – – / – – / Crocydit; Krokydit / – –

1048
merismite, – a mixed rock (chorismite) characterized by the irregular penetration of the units of the fabric. This term includes such migmatites as agmatites, dictyonites and phlebites. / merismiet / – – / Merismit / – –

1049
stromatite; stromatholith (obsolete), – chorismite with layered, banded or stratified structure. / stromatiet / – – / Stromatit / estromatita

epibolite, – migmatite in which granitic layers are concordant with the gneissosity of the metamorphic palaeosome. / – – / épibolite / Epibolit / epibolita

lit-par-lit, – adjective for the injection mechanism that causes a stromatic structure in migmatites; also used for the structure itself. / lit-par-lit / lit par lit / lit-par-lit / lit-par-lit

1050
surreitic; dilation, – adjectives for a structure in migmatites where the neosome is concentrated in zones of tension (**heterokinetic** spaces).

1051
ptygmatic, – adjective for a structure or a type of folding common in migmatites and characterized by strongly contorted, closely folded pegmatitic or aplitic veins (**ptygmatic fold** or **ptygma**). / ptygmatisch; / ptygmatique / ptygmatisch; Torsionstextur; Injektionsfältelung / ptigmática

1052
ophthalmite; opthalmite, – chorismite characterized by augen and/or lenticular aggregates of minerals. / ofthalmiet / – – / Ophthalmit / oftalmita

1053
stictolith; stictolite, – migmatite with spotted appearance. / stiktoliet / – – / Stictolith / estictolita

1054
nebulite, – migmatite with ghost-like relics of pre-existing rock, lacking distinct boundaries between the various textural elements. / nebuliet / nébulite / Nebulit / nebulita

1055
embrechite, – French term, more or less synonymous with migmatite. Used to indicate regionally metamorphosed rocks containing a granitic component beside the normal metamorphic assemblage. / embrechiet / embréchite / Embrechit / embrechita

ectinite, – in contrast to embrechite, a regionally metamorphosed rock formed without migmatization or metasomatism. / ectiniet / ectinite / Ectinit / ectinita

1056
regional metasomatism, – metasomatic processes affecting extensive areas whereby the introduced material may be derived from partial fusion of rocks at deeper levels. / regionale metasomatose / métasomatose régionale / Regionalmetasomatose / metasomatismo regional

1057
mobilization, – petrogenetically neutral term pertaining to the formation of minerals or groups of minerals in rocks by geochemical migration of mobile components, e.g., in melts, solutions or in a gas phase. / mobilisatie / mobilisation / Mobilisierung / movilización

mobilizate; mobilisat, – geochemically mobile phase formed by mobilization. / mobilisaat / mobilisat / Mobilisat / movilizado

rheomorphism, term restricted to those cases of mobilization, where mobility is, at least in part, caused by fusion or partial fusion commonly accompanied, if not promoted, by addition of new material by diffusion. / rheomorfose / rhéomorphisme / Rheomorphose / rheomorfismo

1058
mobile component, an element or group of elements that can migrate beyond the limits of a single mineral (or any other given system.) / mobiel bestanddeel / composant mobile / – – / componente movil

1059
permeation, – passage of geochemically mobile components through a rock. / permeatie / – – / Permeation / – –

permeation gneiss, – gneiss formed or modified by permeation.

petroblastesis, – formation of rocks chiefly through permeation. / petroblastese / – – / Petroblastese / petroblastesis

1060
imbibition, – penetration of rock matter by solutions or absorption of solutions by rocks. / – – / impregnation / Imbibition / imbibición

1061
ichor, – fluid phase considered responsible for such processes as granitization and migmatization. / ichoor / ichor / Ichor / ichor

1062
granitization; granitisation; granitification; transformation, – the process by which solid rocks are converted into rocks of granitic character without passing through a magmatic stage; according to others, a group of processes by which a solid rock (without enough liquidity at any time to make it mobile or rheomorphic) is made more like granite than it was before, in minerals, or in texture and structure, or in both. Some authors include anatexis in the process of granitization. / granitisatie / granitisation / Granitisierung; Granitisation / granitización

1063
pegmatitization, – formation of, or replacement by, a pegmatite. / pegmatitisatie / pegmatitisation / Pegmatitisierung / pegmatitización

1064
feldspathization, – secondary growth of feldspar in a rock, usually as a result of metasomatism. One of the initial stages of granitization. / veldspatisatie / feldspathisation / Feldspatisierung / feldespatización

1065
degranitization, – partial or complete removal, for instance by partial anatexis or metasomatism, of the granitic components of a rock leading to its basification.

1066
basic front; mafic ....; magnesium ...., – a zone enriched in basic constituents that are expelled from country rocks undergoing granitization. The introduced elements (mainly Fe and Mg but commonly including Al, Ca, K, H, Ti, P and Mn) are those that are either incompatible with a granite composition or in excess of the appropriate amounts. / basisch front / front basique; front mafique / basische Front / frente básico

basic behind, – a zone in which residual mafic minerals are concentrated as a result of degranitization. / restiet / restite / Restit / restita

1067
metasome; metasom, – metasomatically formed new part of a migmatite. Also the replacing mineral in a case of metasomatic replacement. / metasoom / – – / Metasom / metasoma

1068
diabrochite, – metamorphic rock that owes its composition to intensive penetration by ichor, or to in situ partial fusion, without injection of a visible granitic component, as in a migmatite.

1069
transfusion, – the process whereby potassic ultrabasic magmas are formed by reaction of emanations from deep-seated sources with crustal rocks. Also used in a broader sense for the production of any igneous rock through the intermediary of gaseous or hydrothermal fluids.

1070
envelope, – the migmatized and/or metamorphosed part of a regional body of rock undergoing granitization. / – – / – – / migmatische Hülle; .... Umhüllung / zona de borde

core, – the innermost, completely granitized part of a regional body of rock undergoing granitization. / – – / – – / granitischer Kern / núcleo

1071
skialith, – a vague remnant of country rock in granite, obscured by the process of granitization. / skialiet / – – / mixogener Einschluss / – –

1072
surmicaceous inclusion, – mica-rich inclusion, commonly a degranitized remnant, in granite or migmatite. / biotietnest / enclave surmicacée / Biotitputzen / – –

[1100] GEOTECTONICS

[1110] GENERAL TERMS

1073
(geo)tectonics, – the study of the structure on a large scale of mountain chains, continents, oceans or the whole lithosphere, and the origin of that structure. / (geo)tektoniek / (géo)tectonique / (Geo)tektonik / (geo)tectónica

1074
diastrophism, – movements of the solid crust that result in a relative change of position, i.e. orogeny or epeirogeny. / diastrofisme / diastrophisme / Diastrophismus / diastrofismo

1075
deformation, – change on any scale in position, orientation andz/ or change in shape of a rock mass. / deformatie; vervorming / déformation / Deformation; Verformung / deformación

1076
morphotectonics, – the tectonical interpretation of geomorphological features. / morfotektoniek / morphotectonique / Morphotektonik / morfotectónica

1077
suprastructure, – the deformed uppermost part of the earth's crust, made up of unmetamorphosed, diagenetic or low-grade metamorphosed rocks. / bovenstructuur / superstructure / Oberbau; oberes Stockwerk / supraestructura

infrastructure, – the deformed lower part of the crust, consisting of medium to high-grade metamorphosed rocks. / onderstructuur / infra-structure / Unterbau; tiefes Stockwerk; Sockel / infraestructura

1078
alpine-type, – refers to an orogeny, when it involves intense rock deformation including the formation of large nappes. / alpine gebergtevorming / orogenèse de type alpin / alpinotyp; Deckengebirgs- / tipo alpino

germanotype, – refers to very gentle folding and fault-block formation outside orogenic regions. / saxonische orogenese / orogenèse de type saxon / germanotyp; Bruchfaltungs-; saxonisch / tipo germánico

1079
prekinematic, – a term applied to processes occurring before rock deformation. / prekinematisch / précinématique / präkinematisch / precinemático

synkinematic, – a term applied to processes occurring simultaneously with orogeny. / synkinematisch / syncinématique / synkinematisch / sincinemático

interkinematic, – a term applied to processes between two phases of rock deformation. / interkinematisch / intercinématique / interkinematisch / intercinemático

late kinematic, – a term applied to processes occurring in a late stage of rock deformation. / laat-kinematisch / tardicinématique / spätkinematisch / cinemático tardío

postkinematic, – a term applied to processes occurring after rock deformation. / postkinematisch / postcinématique / postkinematisch / postcinemático

[1120] CONTINENTAL STRUCTURE

[1121] *Stable Regions*

1080
craton, – large region consisting of exposed or covered Precambrian rocks, that behaved rigidly during Phanerozoic orogenic events. / kraton / craton / Kraton / cratón

1081
shield, – part of a craton where Precambrian rocks are exposed. / schild / bouclier / Schild / escudo

1082
greenstone belt, – usually strongly folded sequence of mafic and ultramafic volcanic rocks of Archaean age, forming synclines in surrounding granitic or gneissose rock. / – – / zone de roches verdes / greenstone belt / facies 'esquistos verdes'

1083
platform, – part of a craton where the Precambrian is covered by flat-lying, slightly deformed or undeformed and unmetamorphosed sediments and volcanics. / platform / plate-forme / Plattform; Tafel / plataforma (cratónica)

1084
anteclise, – part of a platform with shallow basement, whose top possesses the shape of an open antiform. / – – / antéclise / Anteklise / anteclise

syneclise, – part of a platform with shallow basement, whose top has the shape of an open synform. / – – / synéclise / Syneklise / sineclise

1085
aulacogen, – elongated basin in cratonic regions and filled with a thick sequence of sediments, often accompanied by volcanites. / aulacogeen / aulacogène / Aulakogen / aulacogen

1086
basin, – any depression in the earth's crust formed by subsidence and filled with sediments and some volcanics. / bekken / bassin / Becken / cuenca

epicontinental basin, – a basin within a continent and filled with marine as well as continental deposits. / epicontinentaal bekken / bassin épicontinental / Epikontinentalbecken / cuenca epicontinental

1087
block, – a rigid, relatively stable unit. / blok; massief / bloc / Block; Masse; Massiv / bloque

1088
inversion, – the interruption of the subsidence of a basin or part of it by temporary uplift and erosion; in grabens faults may change direction of throw and appear reverse in the inversion stratigraphic level. / omkering; inversie / inversion / Inversion / inversión

1089
stable shelf, – the part of the continent, adjoining a geosyncline and covered by a thin sequence of undeformed sediments. / – – / plate-forme stable / stabiler Shelf; .... Schelf / – –

[1122] *Faulting Tectonics*

1090
epeirogeny; epiro .... ; .... genesis, – in distinction to orogenesis, the relatively slow and wide-reaching movements, resulting in the rise or subsidence of large crustal blocks without significant internal deformation. / epeirogenese / épirogénie; épirogenèse / Epirogenese; Epeirogenese / epirogénesis

1091
taphrogenesis; taphrogeny, – the forming of rifts or grabens. / tafrogenese / taphrogenèse / Taphrogenese / tafrogénesis

1092
rift, – a major, elongated and relatively narrow, deep depression in the earth's crust, either continental or oceanic, and bounded by normal faults. / rift / rift / Graben; Rift / surco; grieta; fractura

[1123] *Mobile Belts*

1093
mobile belt, – elongated zone where strong subsidence, sedimentation and folding, usually accompanied by regional metamorphism, have taken place. More or less synonymous with **orogenic belt.** / – – / zone mobile; ceinture .... / mobiler Gürtel; orogener .... / cinturón orogénico; faja .... ; movil ....

1094
geosyncline, – according to the original definition (J.D. Dana), an elongated mobile, gradually subsiding downwarp in the earth's crust. It is being filled with sediments, which may attain a considerable thickness and that subsequently are intensively folded. / geosynclinaal / géosynclinal / Geosynklinale / geosinclinal

NOTE:
– the term 'geosyncline' has been used indiscriminately for any sedimentary basin regardless of size and development and of position on a cratonic region. This use gave rise to a classification, which has been omitted here; use of it is not recommended.

1095
orthogeosyncline, – a compound geosyncline along a continental margin; the later site of mountain building. / orthogeosynclinaal / orthogéosynclinal / Orthogeosynklinale / ortogeosinclinal

1096
miogeosyncline, – a marginal part of an orthogeosyncline at the continental side without ophiolites, but with shallow-marine sediments and often containing significant amounts of carbonates. / miogeosynclinaal / miogéosynclinal / Miogeosynklinale / miogeosinclinal

1097
eugeosyncline, – a part of an orthogeosyncline at the ocean side, characterized by the occurrence of ophiolites and sometimes by deep-marine sediments. / eugeosynclinaal / eugéosynclinal / Eugeosynklinale / eugeosinclinal

1098
ophiolite, – a sequence of mafic and ultramafic rocks, occurring in orogenic belts, usually strongly deformed and metamorphosed, and consisting of peridotites, gabbros, sheeted-dyke complexes and pillow lavas. Ophiolites are at present being considered as pieces of ocean floor, generated at an oceanic ridge and thrust up during mountain building. / ofioliet / ophiolite / Ophiolit / ofiolita

Steinmann trinity, – the association of serpentinized ultramafic rocks, pillow lavas and radiolarian cherts. / triniteit van Steinmann / triade de Steinmann / Steinmann Trinität / trilogía de Steinmann

1099
geanticline, – an elongated region or rise in a geosyncline, where subsidence is considerably less than on either side, resulting in a reduced thickness of sediments. / geanticlinaal / géanticlinal / Geantiklinale / geanticlinal

1100
orogeny; orogenesis; mountain building; tectogenesis, – the process of deformation, metamorphism and igneous activity, taking place during a specific period in a restricted, often elongate region of former strong subsidence and sedimentation. / orogenese; orogenetische beweging; gebergtevorming / orogenèse / Orogenese; Gebirgsbildung / orogénesis; tectogénesis; movimiento orogénico

1101
orogen; orogene; orogenic belt; tectogen, – a region on earth where orogeny has taken place. / orogeen / orogène; orogénie / Orogen / orógeno

1102
mountain chain; .... system, – often synonymous with orogen. / gebergteketen / chaîne de montagnes / Gebirgskette / cadena de montañas

1103
cordillera, – generally a whole mountain province including all the subordinate mountain ranges and groups and the interior plateaus and basins. / ketengebergte / cordillère / Kettengebirge; Kordillere / cordillera

1104
root, – a downward bulge of continental crust in the mantle. / wortel / racine / Gebirgswurzel; Tiefenwulst / raíz

NOTE:
– not to be confused with the root of a nappe.

1105
orogenic cycle; orogenetic .... , – a presumed cyclic repetition in the orogenic activities during the history of the earth, consisting of geosynclinal subsidence, an orogenic phase and uplift. / orogenetische kringloop / cycle orogénique / orogener Zyklus / ciclo orogénico

1106
orogenic period, – one of the several periods in the geological history of the earth during which major orogenic movements took place. / orogenetische periode / période orogénique / orogene Periode / período orogénico

anorogenic period, – a period of relative crustal quiescence. / anorogene tijd / période anorogénique / anorogene Zeit; .... Periode / período anorogénico

1107
orogenic phase, – one of the various stages of an orogenic period. / orogenetische fase / phase orogénique / orogene Phase / fase orogénica

folding phase, – a short period of fold formation, evidenced by an unconformity. / plooiingsfase / phase de plissement / Faltungsphase / fase de plegamiento

1108
magmatic phase, – a period of intensified magmatic activity often related to an orogenic cycle. / magmatische fase / phase magmatique / magmatische Phase / fase magmática

Distinction is made between:

initial, – related to the intrusive or volcanic activity during the geosynclinal phase. / begin- / initiale / initial- / inicial

synorogenic, – related to the presumably intrusive and migmatizing granitic phase during the orogenic phase. / synorogenetisch / synorogénique / synorogen / sinorogénico

late orogenic, – related to the presumably intrusive granites or granodiorites forming batholiths in a late stage of the orogeny. / laatorogenetisch / tardi-orogénique / spätorogen; subsequent / orogénico tardío

postorogenic, – related to the mostly volcanic activity of the later or postorogenic stage. / nafase / post-orogénique / postorogen; final / postorogénico

1109
foreland, – the undeformed, stable area toward which a mountain-building movement is mainly directed. / voorland / avant-pays / Vorland / antepaís

foredeep; foretrough, – a marginal depression at the foreland side of a mountain system. / voordiep / avant-fosse / Vortiefe; Vorsenke; Saumtiefe / prefosa

1110
hinterland, – the undeformed stable area, away from which the mountain-building movement is mainly directed. / achterland / arrière-pays / Rückland / hinterland; retropaís

backdeep; backtrough, – a marginal depression at the hinterland side of a mountain system. / achterdiep / arrière-fosse / Rücktiefe; .... senke / retrofosa

1111
marginal trough; .... deep, – a long trench-like, late-mountain system, filled or being filled with erosional products from the orogen. / randdiep / fosse marginale / Saumtiefe; Randsenke / fosa marginal

1112
intramontane basin, – a more or less saucer-shaped, late-orogenic depression surrounded by mountains, filled or being filled with erosional products from the mountains. / intramontaan bekken / bassin intramontagneux / intramontanes Becken / cuenca intramontañosa

[1130] HYPOTHESES ON MOUNTAIN BUILDING

[1131] *Older Hypotheses*

1113
fixism, – the doctrine, assuming that the division in space of the continental shields has remained the same throughout geological history. / fixisme / fixisme / Fixismus / fixismo; (doctrina fixista)

mobilism, – the doctrine, assuming that during geological time the continents moved relative to each other. / mobilisme / mobilisme / Mobilismus / mobilismo; (doctrina mobilista)

1114
contraction hypothesis; shrinkage ...., – the folding phenomena in the earth's crust are due to the cooling and consequent shrinkage of the globe. / contractiehypothese / hypothèse de la contraction / Kontraktionshypothese / hipótesis de la contracción

1115
continental drift, – (A. Wegener, 1915), the continents are made up of light and rigid material and float in heavier and more plastic material underlying the oceans; they may drift for considerable distances. / continentale verschuiving / dérive des continents; translations continentales / Kontinentalverschiebung / deriva continental; traslación ....

1116
Pangaea, – the reassembly of all present continents in one supercontinent in late Palaeozoic time. / Pangaea / Pangée / Pangaea / Pangea

1117
hypothesis of radioactivity, – (J. Joley, 1924), tensional phenomena in the earth's crust caused by expansion due to radioactive heat, followed by magmatic activity, subsequent cooling and contraction causing compression. / radioactiviteitshypothese / hypothèse du réchauffement radioactif / Radioacktivitätshypothese / hipótesis de radioactividad

1118
buckling hypothesis, – (F.A. Vening Meinesz, 1930), the deep-sea trenches and the observed negative gravity anomalies represent geosynclinal belts, caused by a downbuckling of the crust into the substratum due to lateral compressive forces. / inknikkingshypothese / hypothèse du flambement; .... du flambage / Einsenkungshypothese / hipótesis del hundimiento; .... de los bucles

1119
oscillation hypothesis, – (E. Haarmann, 1930), primary tectonic features, such as geotumors and geodepressions in gravitational gliding of masses from these geotumors into the geodepressions. / oscillatiehypothese / hypothèse de l'oscillation / Oszillationshypothese / hipótesis de la oscilación

1120
geotumor, – a major updoming of the surface of the earth's crust. / geotumor / géotumeur; intumescence profonde; pli de fond / Geotumor; Grundfalte / pliegue de fondo; intumescencia profunda

geodepression, – a major downbending of the earth's crust. / geodepressie / géodépression / Geodepression / geodepresión

1121
undation hypothesis, – (R.W. van Bemmelen, 1931), primary orogens are formed by differential movements operating in a vertical sense in the earth's crust as a result of changes in density in the tectonosphere caused by physicochemical processes. / undatiehypothese / hypothèse de l'ondation / Undationshypothese / hipótesis de la ondación

1122
undation, – differential movements in a vertical sense result-

ing from mass displacements in the tectonosphere, that are caused by changes in density of matter at depth by physicochemical processes. / undatie / ondation / undation / ondación

asthenolith, – assumed magma and migma accumulation at the base of the crust, from which the energy may be derived for mountain-building processes. / asthenoliet / asthénolite / Asthenolith / astenolito

1123
expansion hypothesis, – (L. Egyed, 1957, 1960), due to the expanding earth, the geosynclinal phase is considered the result of tensional stress, and the orogeny that of magma intrusion, causing compression and gliding tectonics. / expansiehypothese / hypothèse de l'expansion / thermische Expansionshypothese / hipótesis de la expansión

1124
mantle convection, – a variety of hypotheses proposed by many authors from 1839 up to recent years, assuming some form of convective overturn in the earth's mantle due to thermal disequilibrium. Recently convection in the asthenosphere has been proposed as the motor of plate tectonics. / mantelconvectie / convection du manteau / Mantelkonvektion / convección del manto

[1132] *Plate Tectonics and Sea-Floor Spreading*

1125
plate tectonics; new global ...., – theory stating that the lithosphere is divided into a number of plates that move independently with regard to one another. Plates are bounded by zones of seismic activity, and movements can be of three kinds, divergent, convergent or transcurrent. / platentektoniek / tectonique de plaques; .... globale / Plattentektonik; Global.... / tectónica de placas

1126
pole of rotation, – point on the surface of the earth, the centre of the spherical arc as described by an object moving along the surface of a sphere, e.g. lithospheric plates. / rotatiepool / pôle de rotation / Rotationspol / polo de rotación

angle of rotation, – the spherical angle over which a lithospheric plate has moved. / rotatiehoek / angle de rotation / Rotationswinkel / ángulo de rotación

1127
Tethys, – former ocean between the Eurasian and Gondwana plates. / Tethys / téthys / Tethys / Tethys; Mar de .....

1128
sea-floor spreading, – process whereby oceanic crust is continuously formed in the central rift of oceanic ridges, and the two, thus newly formed plates diverge, move laterally away. / spreiding van de zeebodem / expansion du fond océanique / Neubildung ozeanischer Kruste / formación de suelo oceánico

1129
(mid-)oceanic ridge system, – earth-encircling system of ridges occurring in all oceans, often with a rift on the axis, where new ocean floor through intrusion and extrusion of basalt is supposed to be formed. / (midden)oceanische rug / système de dorsales (médio-océaniques) / System zentralozeanischer Rücken / sistema de cadenas (medio-)oceánicas; .... axial de cadenas oceánicas

1130
triple junction, – place where three zones of seismic activity or three plates meet. / – – / jonction triple / – – / junción triple

1131
continental collision, – event when two continental plates converge and collide, causing strong compression and mountain building. / continentbotsing / collision de plaques continentales / Kontinent-Zusammenstoss; Kontinentprall / colisión continental

1132
suture (zone), – important fault in an orogenic belt, often supposed to mark the boundary between two former plates in a collision orogen. / sutuur / (zone de) suture; zone d'affrontement / Sutur; Narbe / zona de sutura; sutura

1133
obduction, – the process whereby a continental plate overrides another one, causing an extensive orogenic region. / opslokking / obduction / Obduktion / obducción

1134
subduction zone, – locality on earth where convergent movement takes place, and one, the subducting (oceanic) plate, is thrust under the overriding (continental) plate; it results in orogeny on the continental side and in the formation of island arcs and deep-sea trenches on the oceanic side. / opslokkingszone / zone de subduction / Subduktionszone / zona de subducción

Benioff zone, – a zone of earthquake hypocentres from shallow below deep-sea trench and outer arc to deep (up to 700 km) below the continent. / zone van Benioff / zone de Benioff / Benioff-Zone / zona de Benioff

1135
island arc, – row of islands arranged in an arcuate shape, considered to mark, with the trench on its convex side, the surface expression of a subduction zone. / eilandenboog / arc insulaire / Inselbogen / arco de islas

1136
back-arc basin, – usually a marine basin behind an island arc, with thin continental or oceanic crust, and sometimes filled with shallow-marine to terrestrial sediments. / – – / bassin marginal / Becken im Rücken des Inselbogens / – –

1137
volcanic inner arc, – of a double island arc, the one at the continental side characterized by strong acid and andesitic magmatism and volcanism. / vulkanische binnenboog / arc interne volcanique / vulkanischer Innenbogen / arco interior volcánico

intermediate depression, – an often marine depression between the arcs. / intermediaire depressie / dépression entre arcs / intermediäre Depression / depresión intermedia

non-volcanic outer arc, – of a double island arc, the one at the ocean side with strongly folded, sometimes metamorphic sediments and characterized by an axis of strong negative gravity anomalies, the **axis of Vening Meinesz.** / niet-vulkanische buitenboog / arc externe non volcanique / nichtvulkanischer Aussenbogen / arco exterior no volcánico

axis of Vening Meinesz, – / as van Vening Meinesz / axe de Vening Meinesz / Achse von Vening Meinesz / eje de Vening Meinesz

1138
(deep-sea) trench, – elongated and narrow, deep trough on the convex side of island arcs or along continents in the eastern Pacific Ocean, supposed to be formed by subduction of the oceanic plate. It also marks the site of intersection of a Benioff zone with the surface of the solid earth. / diepzeetrog / fosse océanique / Tiefseegraben / fosa oceánica

1139
paired metamorphic belt, – a double metamorphic belt in an orogen, consisting of a high pressure-temperature zone on the ocean side, and a low pressure-temperature zone on the continental side. / dubbele metamorfe zone / couple de zones métamorphiques; zones métamorphiques couplées / paariger metamorpher Gürtel / doble zona de metamorfísmo

1140
orocline, – an orogenic belt, bent or flexed in plan to a horseshoe or elbow shape as a result of a tangential impressed strain. / orocline / orocline / Orogenschlinge; .... bogen / oroclino

1141
transform fault, – a type of transcurrent fault, where two plates move past each other. Transform faults in mid-oceanic ridges show reverse directions of apparent offset and plate motion, while seismic activity is confined to the stretch of offset. In other cases, in their apparent termination they may be transformed into another structural feature, e.g. the transcurrent movement may be transformed into compression. / transformatiebreuk / faille transformante; .... de raccord(ement); .... de transformation / Transform-Störung / falle transformante

1142
oceanization, – the hypothetic conversion of continental into oceanic crust. / oceanizatie / océanisation / Ozeanisation / oceanización

1143
sphenochasm, – a triangular gap of the oceanic crust separating two continental blocks, bounded by faults that converge towards a point. / sfenochasme / sphénochasme / Sphenochasmus / sfenochasma

rhombochasm, – a parallel-sided gap in the continental crust, occupied by oceanic crustal material. / rhombochasme / rhombochasme / Rhombochasmus / rombochasma

1144
hot spot, – more or less fixed place in the mantle with high heat flow, resulting in volcanic activity at the surface above it. / warmtegebied / point chaud / hot spot / punta caliente

1145
mantle plume, – diapiric rise of hot material from the lower part of the earth's mantle into the upper mantle. / mantelpluim / panache mantellique / Mantelaufwölbung / pluma del manto

[1200] VOLCANOLOGY

[1210] VOLCANIC ACTIVITY

[1211] *General Terms*

1146
volcanology; vulcanology, – the branch of geology that deals with volcanism, its causes and its phenomena both at the earth's surface and at deeper levels. / vulkanologie / volcanologie / Vulkanologie / vulcanología

1147
volcanism; vulcanism, – the aggregate of processes associated with the surface phenomena involved in the transfer of materials from the earth's interior to or immediately below its surface. / vulkanisme / volcanisme / Vulkanismus / vulcanismo

1148
volcanogenic, – of volcanic origin. / vulkanogenetisch / volcanogène / vulkanogenetisch / vulcanogenético

volcanic; vulcanic, – includes anything to do with igneous action at or immediately below the earth's surface, whether or not a volcano is formed. / vulkanisch / volcanique / vulkanisch / volcánico

1149
volcanic activity; .... action, – the visible phenomena pertaining to a volcano, such as explosions, lava effusions and fumarolic activity. / vulkanische werking / activité volcanique / vulkanische Tätigkeit / actividad volcánica

1150
palaeovolcanism, – an obsolete term, originally denoting volcanic activity of pre-Tertiary age; now sometimes used to indicate volcanic activity of the geological past. / palaeovulkanisme / paléovolcanisme / Paläovulkanismus / paleovulcanismo

neovolcanism, – an obsolete term, originally denoting of Tertiary and Quaternary age; now sometimes used to indicate volcanic activity of recent times. / neovulkanisme / néovolcanisme / Neovulkanismus / neovulcanismo

subrecent volcanism, – volcanism of prehistoric times. / subrecent vulkanisme / volcanisme subactuel / subrezenter Vulkanismus / vulcanismo subreciente

1151
orogenic volcanism, – volcanism related to and contemporaneous with mountain building. / orogeen vulkanisme / volcanisme d'orogène; .... orogénique / orogener Vulkanismus / vulcanismo orogénico

1152
epigenetic volcanism, – younger than the main period of activity. / epigenetisch vulkanisme / volcanisme épigène / epigenetischer Vulkanismus / vulcanismo epigenético

1153
active volcano, – a volcano that erupts from time to time or shows continuous activity; also a volcano in the fumarole or solfatara stage. / werkende vulkaan / volcan actif / tätiger Vulkan; aktiver .... / volcán en actividad

subactive volcano, – a volcano that has been active in recent times, but of which no historical eruptions are known. / subactieve vulkaan / volcan subactif / subaktiver Vulkan / volcán en subactividad

1154
ephemeral activity, – eruptive activity lasting a few weeks or a few months only. / kortstondige werking / activité éphémère / einmalige Ausbruchstätigkeit; kurzfristige Tätigkeit / actividad efímera

1155
dormant volcano, – a volcano that at present is inactive, but later may recur to activity; also the state between periods of activity. / vulkaan in ruststadium / volcan endormi; .... dormant; .... assoupi; .... sommeillant; .... en repos; cratère endormi; .... en phase de sommeil / derzeit ruhender Vulkan; schlummernder ....; untätiger ....; Vulkan in derzeitigem Ruhezustand / volcán latente

extinct volcano, – a volcano that does not show any activity and of which no activity is expected in the future. / uitgedoofde vulkaan; uitgestorven .... / volcan éteint / erloschener Vulkan / volcán apagado; .... extinto

[1212] *Types of Activity*

1156
volcanic cycle, – regular repetition of similar activities during an eruption of a volcano. Also the repetition of periods of repose and periods of activity. / vulkanische cyclus / cycle volcanique / vulkanischer Zyklus / ciclo volcánico

1157
preliminary stage; fore-phase, – the period of minor eruptions preceding the main or paroxysmal state. / inleidende fase; voorfase / état initial / Vorphase / etapa preliminar

1158
Hawaiian-type eruption, – volcanic activity that is characterized by low gas pressure and very fluid, effusive basaltic lava, sometimes forming an incandescent lake. It may cause lava flows and fountains but explosive phenomena are rare. / Hawaii-werking / activité hawaïenne; projection hawaïenne; éruption hawaïenne / hawaianische Tätigkeit; Hawaiitätigkeit / actividad de tipo hawaiano

1159
Strombolian-type eruption, – volcanic activity that is characterized by fire fountains of fluid, basaltic lava from a central crater and medium gas pressure. / Stromboli-type eruptie / activité strombolienne; éruption .... / strombolianische Tätigkeit; Strombolitätigkeit / actividad de tipo estromboliano

1160
Vulcanian-type eruption, – volcanic activity that is characterized by very viscous lava that quickly crusts over after an eruption. Gases accumulate beneath the congealed cover and blow off at longer and longer intervals, with correspondingly greater violence, resulting into explosions emitting dark 'cauliflower' clouds with volcanic fragments of many dimensions. / Vulcano-type eruptie / éruption vulcanienne; activité .... / vulcanianische Tätigkeit; Vulcanotätigkeit / actividad de tipo vulcaniano

Vesuvian-type eruption, – usually applied as synonymous to Vulcanian-type eruption.

1161
Plinian-type eruption, – a modification of the Vulcanian-type eruption and characterized by a gas-blast of enormous power from an eruptive magma of low or medium viscosity and high gas pressure, generally taking place after a long period of quiescence. Also a violent eruption during the course of which the material changes greatly in composition. / Plinisch eruptietype / éruption plinienne; activité plinienne / plinianische Tätigkeit / actividad de tipo pliniano

1162
gas-emission phase; intermediate gas phase; gas phase, – this stage is characterized by the paroxysmal emission of highly compressed vapours and gases with a preponderant ejection of incandescent liquid lava with a high proportion of ash. / (intermediaire) gasfase / phase gazeuse (intermédiaire) / (intermediäre) Gasphase / fase de gas

1163
Peléan-type eruption, – volcanic activity that is characterized by explosions of extreme violence and the formation of nuées ardentes. The material usually is very silicic and very viscous. / Pelée-type eruptie / éruption péléene; activité péléene / peleanische Tätigkeit; Peleantätigkeit / actividad de tipo peleeano

1164
paroxysmal eruption, – a most violent explosive action, usually after a period of quiescence of the volcano, during which subterranean pressures accumulate. It sometimes leads to the destruction of the volcano and is generally preceded and followed by smaller explosions. / paroxysmale eruptie / éruption paroxysmale / paroxysmale Eruption / erupción paroxísmica

1165
Mount Katmai activity, – volcanic eruption that produces floating glowing clouds. The clouds spread out and move; they form streams of incandescent sand and pumice, in which every particle is surrounded by gas and thus partially held in suspension. / Mount-Katmai werking / activité de

type Mont Katmai / Mount Katmai-Tätigkeit / actividad de tipo Monte Katmai

1166
ultravolcanian, – pertaining to volcanic explosions in which the ejected material is solidified and nearly cold. The material consists of angular blocks with lapilli and ashes; luminous phenomena and real bombs are missing. These eruptions may occur in the beginning when the crater vent is being opened. / ultravulkanisch / ultravulcanien / ultravulkanisch / ultravolcánico

1167
after-phase; concluding stage, – the period of minor eruptions after the main phase. / slotfase; nawerking / phase finale; .... terminale; manifestations ultimes / Endphase; Schluss....; Ausklingen des Ausbruchs / fase final

1168
fumarolic stage, – a late or decadent type of volcanic activity characterized by the emission of gases and vapours. / fumarolenstadium; ....toestand / phase fumerollienne; .... des fumerolles; stade fumerollien / Fumarolenstadium / fase de fumarola

solfataric stage, – a late or decadent type of volcanic activity characterized by the emission of sulphurous gases. / solfatarenstadium; ....toestand / phase solfatarienne; .... des solfatares; stade solfatarien / Solfatarenstadium / fase de solfatara

1169
repose period; .... phase; dormancy, – a quiet period between successive volcanic outbursts. / rustperiode / phase de repos; .... de tranquillité; période de sommeil / Ruheperiode / período de reposo

[1213] *Mechanism*

1170
mechanism of the eruption, – / mechanisme van de uitbarsting / mécanisme de l'éruption / Ausbruchsmechanismus / mecanismo de erupción

1171
volcanic energy; potential eruptive ....; potential eruptive capacity, – the internal power becoming gradually available in the magma in the form of vapour tension and leading to volcanic outbursts. / vulkanische energie / énergie volcanique / vulkanische Energie; potentielle Ausbruchsenergie / energía volcánica

kinetic eruptive energy; .... eruptive capacity, – the energy available in the form of vapour tension at a given moment. / kinetische uitbarstingsenergie / énergie éruptive cinétique / kinetische Ausbruchsenergie; Ausbruchskapazität zu einem bestimmten Zeitpunkt / energía cinética eruptiva.

1172
dissolved gas; gas in solution, – molecularly dispersed gases in the magma, i.e. gases in true solution. / opgelost gas / gaz dissous / gelöstes Gas / gas disuelto

volatile component; gaseous ....; fugitive constituent, – materials in a magma whose vapour pressures are sufficiently high to be present in the gaseous phase. / vluchtig bestanddeel; gasvormig .... / composant volatil; élément gazeux / flüchtiges Bestandteil; gasförmiges .... / componente volátil

resurgent (magmatic) gas, – gases in magma, possibly derived from dissolved or assimilated country rock. / resurgent (magmatisch) gas / gaz résurgent / resurgentes (magmatisches) Gas / gas resurgente

1173
gas pressure, – the expansive force of gases in the magma or of gases escaping from a solfatara or fumarole. / gasdruk / pression de gaz; .... gazeuse / Gasdruck / presión de gas

pressure release, – the reduction of confining pressure; when this happens in the magma chamber, the magma may become fluid and possibly active. / drukontlasting; drukafname / baisse de pression / Druckentlastung / descarga de presión

1174
retrograde increase of vapour tension, – increase of gas pressure in a magma chamber during the process of crystallization, by which the gas content per unit volume of the residual melt increases. / retrograde dampdrukverhoging / augmentation rétrograde de pression gazeuse / thermisch rückläufige Dampfdrucksteigerung; Dampfdrucksteigerung mit fortschreitender Kristallisation / aumento retrógrado de presión de gases

retrograde boiling, – separation of the gas phase from a cooling magma as a result of the residual enrichment of the dissolved gases by progressive crystallization of the magma. / retrograad koken / – – / retrogrades Sieden / cocción retrógrada

1175
explosive phenomena, – all phenomena connected with the sudden expansion of the gases in the magma. / explosieve verschijnselen / phénomènes explosifs / explosive Erscheinungen / fenómenos explosivos

explosivity index, – percentage of the pyroclastics in the total of products produced:

$$E = 100 \times \frac{\text{pyroclastics}}{\text{total products}}.$$

/ indexcijfer voor de explosiviteit / indice d'activité explosive / Explosivitätsgrad / índice de actividad explosiva

1176
explosion point, – the physico-chemical stage at which the gas content of the magma escapes explosively. / explosiepunt / point d'explosion; stade explosif / Explosionspunkt / estado explosivo

1177
level of explosion; explosion level, – the horizon in the upper part of a magma column where gases nucleate; it fluctuates as a function of confining pressure and gravity. / explosieniveau / niveau d'explosion / Explosionsniveau / nivel de explosión

explosion focus; eruptive ....; volcanic ...., – a point near the top of the magma column, situated directly underneath the solidified plug; it acts as a choke for the accumulating volcanic gases. / explosiehaard / centre d'explosion / Explosionsherd / foco de explosión

1178
gas fluxing; gaseous excavation; volcanic blowpiping, – the wearing out of the walls of a magma chamber by the rapid rise of free, juvenile gas through a column of molten magma, followed by the rising of the liquid into the cleared space. / uitholling door hete gassen / ouverture causée par des gaz sous pression / Aushöhlung durch heisse Gasströme / excavación por gases ardientes

1179
relaxation, – the release of stress in a magma, e.g. after the initial explosions. / ontspanning / relaxation; détente / Entspannung / laxitud; relajación de esfuerzo

[1214] *Paravolcanic and Cryptovolcanic Phenomena*

1180
paravolcanic phenomena, – all secondary manifestations of volcanic activity, e.g. earth tremors, atmospherics, structural deformation. / vulkanische bijverschijnselen / manifestations paravolcaniques / vulkanische Begleiterscheinungen / fenómenos paravolcánicos

1181
cryptovolcanism, – volcanic phenomena possibly in crust or upper mantle, the effects of which do not necessarily bring volcanic material to the surface, but do cause structural deformation. / cryptovulkanisme / cryptovolcanisme; volcanisme souterrain / Kryptovulkanismus / criptovulcanismo

1182
volcano-tectonic structure, – a geological structure controlled by both, volcanic and tectonic processes. / vulkano-tektonische struktuur / structure volcano-tectonique / vulkantektonische Struktur / estructura volcano-tectónica.

Distinction is made between:
volcano-tectonic horsts
volcano-tectonic troughs
volcano-tectonic fractures

volcano-tectonic depression, – a large-scale depression, usually linear in shape, due to collapse of the roofs of one or more magma chambers by rapid evacuation of the magma; its formation is primarily controlled by volcanic processes, but structurally it resembles a graben. / vulkano-tektonische depressie; .... verzakking; tektono-vulkanische depressie / dépression volcano-tectonique / vulkano-tektonische Senke; .... Depression / depresión volcáno-tectónico

1183
cryptovolcanic caldera, – a large depression of complex structure with walls consisting mainly of sediments. The subsidence or collapse at the earth's surface may be a result of the injection of a laccolith or sheets of igneous magma at shallow depth. Little if any magma escaped at the surface. / cryptovulkanische depressie / dépression cryptovolcanique / kryptovulkanische Depression; Einsenkung; Einbruchskessel / caldera criptovolcánica

1184
cryptovolcanic structure, – highly deformed, strongly brecciated, generally circular structure previously believed to have been produced by volcanic explosions, but lacking any direct evidence of volcanic activity; now often believed to have been formed by meteorite impact. / cryptovulkanische struktuur / structure cryptovolcanique / kryptovulkanische Struktur / estructura criptovolcánica

cryptoexplosion structure, – a non-genetic, descriptive term to designate a roughly circular structure formed by a sudden, explosive release of energy and exhibiting intense, often localized rock deformation with no obvious relation to volcanic or tectonic activity. They are believed to be the result either of the impact of meteorites or of obscure volcanic activity. / cryptoëxplosie-struktuur / structure cryptoexplosive / Kryptoexplosionsstruktur / estructura criptoexplosiva

1185
ring depression, – the annular, structurally depressed area surrounding the central uplift of a cryptoexplosion structure. / ringvormige verzakking / dépression annulaire / kreisförmige Senke / depresión anular

1186
cauldron subsidence, –
1. the subsidence of an approximately cylindrical portion of the roof of a magma chamber along ring faults, sometimes accompanied by the intrusion of magma to form ring dykes;
2. the structure thus formed.

/ ketelvormige inzakking / subsidence en chaudron / Kesseleinbruch / hundimiento en caldera

subterranian cauldron subsidence; underground ...., – a cauldron subsidence in which the ring faults do not penetrate to the surface. / ondergronds instortingsbekken / subsidence en chaudron souterraine / unterirdischer Kesseleinbruch / hundimiento en caldera subterránea

1187
ring-fracture stoping, – large-scale magmatic stoping that is associated with cauldron subsidence.

1188
roof foundering; roof collapse, – subsidence of the roof of a magma chamber after the extrusion of a large quantity of magma. / instorten van het dak / écroulement du toit / Zusammenbruch des Herddaches / desplome de techo

1189
pseudovolcano, – term applied to large circular hollows generally not associated with any positive indications of recent volcanic activity, e.g. craters of doubtful meteoritic origin, or those which are thought to be the result of phreatic explosions or cauldron subsidence. / pseudovulkaan / pseudo-volcan / Pseudovulkan / (p)seudovolcán

[1215] *Volcano-Watching*

1190
volcano-watching; volcanological observations, – the observation and recording of premonitory events and of volcanic activity in general, in order to be able to give warning of impeding, possibly disastrous events of both volcanic and paravolcanic nature. / vulkaanbewaking / surveillance des volcans / Überwachung vulkanischer Vorgänge / observación vulcanológica

1191
premonitory events; .... symptoms, – precursory warnings of a nearing eruption, such as increase of volcanic tremors, change of inclination of the volcano flank, change of temperature in the crater. / voorboden / signes précurseurs; phénomènes prémonitoires; .... annonciateurs; manifestations prévolcaniques / Eruptions-Vorzeichen; ....-Vorboten / signos precursores

1192
transient magnetic anomaly, – temporary deviation of the local magnetic field that may precede earthquakes and volcanic eruptions. / magnetische storing / perturbation magnétique / erdmagnetische Störung / alteración magnética

1193
volcanic earthquake; .... tremor, – an earthquake or tremor associated with volcanic rather than tectonic forces. It is caused by a pushing up of magma shortly before the eruption, or by explosions, and it is generally of local significance only. / vulkanische beving / secousse volcanique; tremblement de terre ....; séisme .... / vulkanische Beben; Ausbruchsbeben / terremoto volcánico

1194
bradyseism, – a slow upheaval or downward movement of the earth's surface, occasionally caused by magmatic or volcanic activity. / bradiseisme; bodemschommelingen / bradyséisme / Bradyseismus / bradisismo

tumescence, – a slight uparching of a volcanic edifice caused by the accumulating magma in the reservoir. It may or may not be followed by an eruption. / opwelving / intumescence / Aufwölbung / intumescencia

1195
pyrometer, – instrument for measuring high temperatures, e.g. of molten lavas, by electrical or optical means; non-optical instruments make use of the increase of electric resistance in a metal when heated. / pyrometer / pyromètre / Pyrometer / pirómetro

optical pyrometer, – instrument for measuring temperature by means of determining the intensity of the light of a particular wavelength emitted by a hot body. / optische pyrometer / pyromètre optique / optisches Pyrometer / pirómetro óptico

1196
Seger cone; pyrometric cone, – made of a mixture of clay and salt, that softens at a definite, known temperature; used to determine high temperatures, e.g. of volcanic gases. / kegel van Seger / cône de Seger / Segerkegel / cono de Seger

[1220] VOLCANIC PHASES AND PRODUCTS

[1221] *The Intrusive Phase*

1197
volcanic phase, – a particular stage in the sequence of volcanic events. / vulkanische fase / phase volcanique / vulkanische Phase / fase volcánica

Distinguished are:

intrusive phase, – injection of magma in the crust. / inpersingsfase / phase d'intrusion / Intrusionsphase / fase intrusiva

abyssal phase, – injection in the deeper parts. / abyssische fase / phase abyssale / abyssische Phase / fase abisal

hypabyssal phase, – injection in the shallower parts. / hypoabyssische fase / phase hypabyssale / hypoabyssische Phase / fase hipabisal

extrusive phase, – extrusion of magmatic material at the surface. / extrusiefase / phase d'extrusion / Extrusionsphase / fase extrusiva

late-volcanic phase, – with fumarolic activity only. / postvulkanische fase / phase post-volcanique / postvulkanische Phase / fase postvolcánica

1198
magma; igneous ....; internal molten ...., – a completely or partially molten mass of silicates within or beneath the earth's crust, containing molecularly dissolved gases, being capable to intrude as such into fissures and to erupt at the earth's surface. / magma / magma / Magma / magma

melt; fused rock; molten ...., – liquid and fused rock. / smelt / bain magmatique / Schmelze / fusión

1199
hypomagma, – relatively immobile magma, undersaturated with gases, and that can only exist under pressures greater than the vapour pressure of the dissolved gases. / dieptemagma; hypomagma / hypomagma / Hypomagma / hipomagma

1200
pyromagma, – a highly mobile magma, oversaturated with gases and therefore vesiculate or frothy; it exists at smaller depth than hypomagma. / pyromagma / magma neuf; pyromagma / Pyromagma / piromagma

dermolithic solidification, – congealing of the surface of a pyromagma. / oppervlakkige verstarring / solidification superficielle / oberflächliche Erstarrung / solidificación dermolítica

1201
epimagma, – a gas-poor, vesicular magmatic residue of a semisolid, pasty consistency, due to depletion in gas content

and cooling. / epimagma / épimagma / Epimagma / epimagma

clastolithic sedimentation, – specifically heavier, semisolid, pasty epimagma sinking through the frothy pyromagma. / klastolithische sedimentatie / décantation clastolitique / klastolitische Anhäufung / acumulación clásticolítica

1202
magma chamber; volcanic ....; pocket of magma, – a closed and restricted accumulation of magma in the crust at shallow depth (up to a few kilometres), from which volcanic materials are derived. / magmakamer; ....haard / chambre magmatique; réservoir .... / Magmakammer; ....herd; Schmelzherd / cámara magmática

central hearth, – / centrale haard / foyer central / Zentralherd / hogar central

1203
satellitic injection, – formed as a branch of the magma chamber, i.e. a small laccolith or a sill that may feed a volcanic centre. / perifere haard / foyer périphérique / Zweigintrusion; sekundärer Magmaherd / inyección satélite

1204
dyke chamber; fissure ...., – a magma chamber whose width is relatively small in comparison to its length and its height. / gangvormige magmakamer / chambre magmatique en forme de dyke / gangförmige Magmakammer / cámara magmática en forma de dique

1205
cupola, –
1. a dome-shaped vault, i.e. the upward projection of an igneous intrusion into its roof;
2. an indentation or cavity in the ceiling of some lava tube caverns.

/ koepel / coupole / Kuppel / cúpula

1206
hemidiatreme; laccolithic dome, – a magma column that, not reaching the surface, domed the overlying beds. / hemidiatrema / hémidiatrème / Hemidiatrema; Reztypus-Neck; lakkolithische Aufwölbung / hemidiatrema

subvolcano, – a small intrusive body, situated at little depth and becoming smaller in downward direction. It may or may not have been in connection with a volcano. / subvulkaan / subvolcan / Subvulkan / subvolcán

[1222] *The Extrusive Phase: Phenomena*

1207
extrusion; external volcanism, – the emission of magmatic material onto the earth's surface; also the accumulation around and above the vent, such as a lava flow, a layer of pyroclastic rocks, or a volcanic dome. / extrusie / extrusion; volcanisme extrusive / Extrusion; Oberflächenvulkanismus / vulcanismo extrusivo; extrusión

1208
eruptive phase; period of activity, – the period of activity of a volcano or a volcanic structure. / eruptieve fase; periode van werking / phase éruptive; .... d'activité; .... de paroxysmes / eruptive Phase; Tätigkeitsperiode / fase eruptiva

explosive phase, – the period of explosive activity of a volcano or a volcanic structure. / explosieve fase / phase explosive / explosive Phase / fase explosiva

1209
element of eruptive activity, – a single eruptive phenomenon; the sequence of such phenomena characterizes the eruptions, or also the persistent activity of a volcano, as e.g. jets of ash, scoriae, lava rags, breccia, lava fountains, blasts of gas, effusion of lava. / uitbarstingsbestanddeel / élément d'activité éruptive / Tätigkeitselement der Eruption / elemento de actividad eruptiva

1210
emission, – the discharge of material. / lozing; uitwerpen / émission / Förderung / emisión; descarga

1211
eruption; outburst, – the process by which solid, liquid or gaseous material is extruded or emitted onto the earth's surface or/and into the atmosphere as a result of volcanic activity. / uitbarsting; eruptie / éruption / Eruption; Ausbruch / erupción

active crater, – a crater in eruption. / werkende krater / cratère actif; bouche active / tätiger Krater / cráter en actividad

1212
intermittent eruption, – eruption during which the discharges occur at intervals. / intermitterende eruptie / projection intermittente / intermittierende Eruption / erupción intermitente

1213
effusion, – the outflow of lava. / effusie / effusion / Effusion / efusión

effusive activity, – / effusieve werking / activité effusive / effusive Tätigkeit / actividad efusiva

1214
lava eruption, – volcanic eruption emitting lava flows without or with minor explosive phenomena. / lava-uitbarsting / éruption de lave / Lavaausbruch / erupción de lava

detritus eruption, – an eruption producing much pyroclastic debris. / uitbarsting van los materiaal / éruption de fragments pyroclastiques / Lockermassenausbruch / erupción de piroclastos

1215
emission of gas; .... of vapour, – the process of release or emission of gases from magma or from volcanic substances. / uitstroming van gas; ontgassing / émission gazeuse; dégagement de gaz / Entgasung / emanación de gas

volcanic emanation, – volatile and non-volatile material emitted at the earth's surface from volcanoes, fumaroles or lavas, usually consisting of water vapour and volcanic gas(es). / vulkanische emanatie / émanation gazeuse; .... volcanique / vulkanische Emanation / emanación volcánica

exhalation, – the emission of volcanic vapours and gases. / exhalatie / exhalaison; .... volcanique / Gasförderung; Dampfförderung / exhalación

1216
explosive eruption; .... emission, – an outburst characterized by an explosion associated with the emission of gases and volcanic and non-volcanic fragmental products. / explosieve uitbarsting / éruption explosive / Explosivausbruch / erupción explosiva

explosive activity, – / explosieve werking / activité explosive / explosive Tätigkeit / actividad explosiva

1217
explosion; discharge, – a sudden and violent outburst accompanied by a loud sound as a result of a forcible expansion of the gases in the magma. / explosie; losbarsting / explosion; dégagement gazeux explosif / Explosion / explosión

detonation, – an explosion caused by the combustion of gases, or by the abrupt release of gases from a vent. / ontploffing; detonatie / détonation / Detonation / detonación

1218
indirect eruption, – an eruption resulting from a sudden release of entrained magmatic gases by heating of congealed gas-charged glassy rocks in a volcanic vent. / indirecte eruptie / éruption indirecte / indirekte Eruption / erupción indirecta

1219
mixed eruption, – an eruption during which explosions are accompanied by lava flows. Also an eruption producing both, new and old material. / gemengde uitbarsting / éruption mixte; explosion .... / gemischter Ausbruch; gemischte Ausbruchstätigkeit / erupción mixta

phreatomagmatic explosion, – a volcanic explosion producing fragmental material as well as great amounts of steam. It is caused by the contact of the rising magma with either surface or shallow subsurface water. Magmatic gases may also be released. / freatomagmatische explosie / explosion phréatomagmatique / phreatomagmatischer Ausbruch / explosión freáticomagmática

magmatic eruption, – an eruption originating in the magma itself and not being caused by influences from outside. / magmatische eruptie / éruption magmatique / magmatischer Ausbruch / erupción magmática

embryonic eruption, – an eruption without surface expression, and whose energy produces volcanic tremors only. / embryonale eruptie / éruption embryonnaire / nichtextrusive Eruption / erupción embrionaria

1220
eruption point; .... centre; volcanic centre, – a place on the earth's surface where solid, liquid and gaseous materials are ejected. / eruptiepunt / centre d'éruption / Eruptionspunkt / centro de erupción

1221
volcanic vent; .... chimney; .... orifice; .... channel; channel of ascent; conduit; pipe; diatreme; feeding vent; eruption channel, – the opening in a volcanic structure through which volcanic materials are extruded; also, the channel through which they pass. / kraterpijp; explosiepijp; eruptiekanaal / évent; cheminée; diatrème / Vulkanschlot; Förderkanal; Schusskanal; Diatrema / conducto volcánico; canal de erupción; diatrema; chimenea

diatreme, – strictly a more or less vertical and cylindrical channel in the earth's crust through which magmatic materials have passed; also the contents of such a channel. / diatrema; eruptiefzuil / diatrème; cheminée / Diatrema; Schusskanal / diatrema

1222
embryonic volcano; abortive ...., – a breccia-filled volcanic pipe without surface expression and regarded to be produced by phreatic explosions. Examples of Permian age occur in Scotland; of Miocene age in Swabia, Germany. / vulkaanembryo / embryon de volcan; volcan embryonnaire; .... avorté / Vulkanembryo(n); Breccienschlot / volcán embrionario

1223
rootless vent, – a vent not directly connected to the conduit that brought the volcanic material to the surface; it may be an overflow or an outflow from an otherwise solidified lava flow. / wortelloze pijp / centre émissif sans racine / wurzelloser Förderkanal / conducto volcánico sin raíz

1224
eruption cylinder; cylindrical vent, – a channel created by the reaming of a conduit under high pressure and, as a consequence, high velocity of a flow of gas that is charged with ash and lava blocks, torn out of the conduit walls. / uitblazingscylinder; blaasschacht / centre émissif cylindrique / zylinderförmiger Schlot / conducto volcánico en forma de cilindro

1225
fissure vent, – / spleetvormig eruptiekanaal / fissure d'éruption / Eruptionsspalte / conducto volcánico en forma de fisura

1226
magma column; magmatic ...., – fluid magma filling the volcanic vent. / magmazuil; ....kolom / colonne magmatique / Magmasäule / columna magmática

plastic lining, – a layer of viscous, semisolid material formed when hot lava congeals on the cooler walls of the vent. / bepleistering / enduit de lave; paroi tapissée de lave; placage de lave / Lavabelag / revoque de lava

1227
radiating dyke; radial dyke, – dyke radiating from a volcanic centre. / radiaal uitstralende gang; radiale gang / dyke rayonnant; / Radialgang / dique radial

1228
central eruption, – the outflow of lavas or the ejection of pyroclastics or gases from a vent, located in the central part of a volcanic structure. / centrale eruptie / éruption centrale; .... normale / Zentraleruption; Zentralschlot.... / erupción central

1229
summit eruption; top ...., – eruption at the top of a volcano,

either explosive or effusive. / toperuptie / éruption sommitale / Gipfelausbruch / erupción apical

eccentric eruption, – an eruption emanating from an adventive crater. / excentrische eruptie / éruption excentrique / exzentrischer Ausbruch / erupción excéntrica

1230
flank eruption, – an eruption on the flank of a volcano or a caldera. / flankeruptie / éruption latérale; .... de flanc / Flankeneruption / erupción lateral

1231
shifting of the volcanic vent, – a displacement of the volcanic outlet, usually as a result of the obstruction of the earlier eruption channel. / kraterpijpverplaatsing / déplacement du centre d'émission / Schlotverlagerung / desplazamiento de la chimenea volcánica

1232
areal eruption; extrusion by deroofing, – a larger volcanic unit embracing independent volcanoes that clearly are fed from the same magma reservoir. Also the overflow of magma welling up as a result of the deroofing of a magma chamber, and producing lava floods around its periphery. / areaaleruptie / éruption aréale / Flächeneruption / erupción areal

1233
fissure eruption, – a linear volcanic eruption, in which lava, generally mafic and of low viscosity, comes to the surface along a line of crustal weakness, usually without explosive activity. The result may be an extensive basalt plateau. / spleeteruptie; lineaire eruptie / éruption linéaire; .... fissurale / Spalteneruption / erupción fisural

fissure effusion, – when along the entire length of a fissure lava wells up onto the surface. / spleeteffusie / effusion fissurale / Spaltenerguss / efusión fisural

1234
gjá (Icelandic, 'chasm'; plur. gjár); gaping volcanic fissure, – a dilatation fissure, from which lava flows quietly without explosive activity. / open vulkaanspleet; gapende .... / fente volcanique béante; fissure .... béante / klaffende Vulkanspalte; Gjaspalte / fisura volcánica entreabierta

1235
explosion graben, – a long and narrow depression due to the activity of a number of closely-spaced volcanic explosions located along a line. / explosieslenk / fossé d'explosion / Explosionsgraben / fosa de explosión

1236
subaerial eruption, – eruption occurring in the open air. / subaerische eruptie / éruption subaérienne / subaerischer Ausbruch / erupción subaérea

subaquatic eruption, – eruption occurring under water. / onderwatereruptie / éruption subaquatique / Unterwasserausbruch / erupción subacuática

subglacial eruption, – eruption occurring below the surface of ice cap or glacier. / subglaciale uitbarsting / éruption subglaciale / subglazialer Ausbruch / erupción subglaciar

submarine eruption, – eruption occurring below the surface of the sea. / onderzeese uitbarsting / éruption sous-marine / untermeerischer Ausbruch / erupción submarina

1237
fire spitting, – / vuurspuwend / crachant du feu / feuerspeiend / ignívomo

volcanic sounds, – the noises caused by volcanic activity, as explosions, avalanches of rock, etc. and of subterranean disturbances. / vulkanische geluiden / bruits volcaniques / vulkanische Geräuscherscheinungen / ruidos volcánicos

drum fire, – a prolonged series of continuous explosions. / trommelvuur / feu roulant / Trommelfeuer / explosiones en cadena

subterranean rumbling; .... noise, – a continuous low, heavy subterranean sound as a result of magmatic activity. / onderaards gerommel / bruits souterrains; grondements .... / unterirdische Geräusche / retumbo

[1223] *The Extrusive Phase: Lavas*

1238
volcanic rock; eruptive ....; volcanite, – rock that has issued at the earth's surface either ejected explosively or extruded as lava. The term includes near-surface intrusions that form part of the volcanic structure. / vulkanisch gesteente; eruptief....; vulkaniet / roche volcanique / vulkanisches Gestein; Eruptivgestein; Vulkanit / roca volcánica

volcanic facies, – deposits entirely or partly composed of volcanic material. / vulkanische facies / faciès volcanique / vulkanische Fazies / facies volcánica

volcanic sequence, – a succession of volcanic rocks, the composition of which has undergone change during the active period of a volcano. / vulkanische opeenvolging / succession volcanique / vulkanische Schichtfolge / secuencia volcánica

1239
extrusive body, – a mass of volcanic material formed by emissions. / extrusielichaam / masse extrusive; venue de surface; .... éruptive externe / Extrusivmasse; ....körper / masa extrusiva

eruptive body, – a mass of volcanic material formed either as an injection or as an extrusion. / eruptielichaam / massif éruptif; appareil .... / Eruptivkörper / cuerpo eruptivo; macizo ....

1240
extrusive rock, – igneous products that have been ejected onto the earth's surface, including both, lava flows and fragmental material. / extrusiegesteente / roche extrusive / Extrusivgestein / roca extrusiva

effusive rock, – solidified from melts or emulsions that have been poured out upon the earth's surface. / effusiegesteente; uitvloeiings.... / roche effusive; .... d'épanchement / Effusivgestein; Erguss.... / roca efusiva

1241
centroclinal bedding, – layered volcanic deposits, that dip towards the eruption centre. / centroclinale gelaagdheid / – – / trichterförmiger Schichtenbau; zentroklinaler .... / estratificación centroclinal

periclinal bedding, – layered volcanic deposits, that dip radially outward from the eruption centre. / periclinale gelaagdheid / stratification périclinale / mantelförmiger Schichtenbau; periklinaler .... / estratificación periclinal

quaquaversal bedding, – somewhat antiquated term, denoting layered volcanic deposits; they dip outward on the flanks and inward toward the axis of the cone at and around the top of volcanoes.

1242
lava, – extruded igneous material and the rock formed upon consolidation. / lava / lave / Lava / lava

congealed lava; solidified lava, – lava in a solid state. / gestolde lava / lave à l'état solide; lave figée / erstarrte Lava / lava solidificada

1243
taxite, – a general term for volcanic rocks having a clastic appearance, due to a mixture of materials of varying texture and structure in the same flow. / taxiet / taxite / Taxit / taxita

eutaxite, – a taxite of which the consolidation products are disposed in parallel arrangement and in an alternation of layers of different texture, mineral composition or colour. / eutaxiet / eutaxite / Eutaxit / eutaxita

ataxite, – a brecciated or irregularly mottled taxite. / ataxiet / ataxite / Ataxit / ataxita

1244
crateral magma, – igneous material that appears at the surface during an eruption; at depth phenocrysts have formed and gases have escaped. / oppervlaktemagma / magma superficiel / Oberflächenmagma / magma superficial

1245
hollow dyke, – a fissure only partly filled with lava, as found in the crater of Vesuvius after the eruption of 1906. / holle gang / filon creux / hohler Gang / dique hueco

1246
lava column, – column of fluid or solidified lava in the volcanic vent. / lavakolom / colonne de lave; .... lavique / Lavasäule / columna de lava

1247
protrusion; upheaval (of plug), – a massive column of very viscous lava that may be forced upwards from the vent by the pressure of the rising magma. / uitstoting / ascension (du bouchon) / Stosskuppe / ascenso (del tapón)

1248
vent agglomerate, – an agglomerate localized within or surrounding a volcanic vent. / kraterpijpagglomeraat / agglomérat de cratère / Schlotagglomerat / aglomerado de chimenea

vent breccia, – / kraterpijpbreccie / brèche de cratère / Schlotbreccie / chimenea con brechas

volcanic friction breccia; autobrecciated lava, – a breccia composed of broken or crushed rock fragments formed in a volcanic vent where the rising column of nearly congealed lava was shattered against the walls. The fragments may later be cemented by newly rising magma. / kraterpijpwrijvingsbreccie / brèche de friction / vulkanische Reibungsbreccie / brecha volcánica de fricción

1249
crumble breccia, – an incoherent accumulation of lava blocks of different dimensions, crumbled from a lava dome or spine. The large openings between the blocks may gradually be filled with finer material. / afbrokkelingsbreccie / brèche d'écroulement / Abbröckelungsbreccie; Schutt.... / brecha de desmoronamiento

1250
pahoehoe lava, – a lava that in solidified form is characterized by a smooth, billowy or ropy surface, and that has a skin of glass a fraction of a cm to several cm thick. It is usually of a basaltic composition. / pahoehoe lava / lave pahoehoe / Pahoehoelava / lava pahoehoe; .... cordada; .... rugosa

slab pahoehoe, – a lava flow the surface of which has been fractured and broken up and consists of innumerable slabs, sometimes arranged in a fairly regular imbricated fashion, but more commonly lacking any regular arrangement. It appears to result from an increase in viscosity of the liquid flowing beneath the solidified crust, with resultant increase in the amount of drag on the crust. / schervenlava / lave en tessons / Scherbenlava / lava en losas

1251
shark-tooth projection, – a lava structure with fine, sharp points commonly less than 5 cm long, formed where two masses of viscous lava are pulled apart; the tips of the projections point away from the direction of relative movement of the lava masses to which they are attached. Such projections may occur at breaks or in cracks in the pahoehoe crust or along the edge of a flow.

lava toes, – small bulbous protrusions with distinct concentric arrangement that develop at the front of a moving pahoehoe lava, usually less than 1 m in diameter and only a few dm long. They are due to bursts in the solidified crust and the subsequent emergence of fluid lava. / lava-uitlopers / coulants de lave / Lavazehen; Lavaausläufer / retoños de lava

1252
ropy lava; corded lava; helluhraun (Icelandic), – from origin a rather fluid lava that is a variation of pahoehoe. Its surface features resemble ropes coiled on the deck of a vessel. The 'coils' consist of lines of glassy scoriae, so arranged by surface currents in the lava stream. / touwlava / lave cordée; (lave en boyaux); (.... en bourrelets) / Stricklava; Gekröselava; Seillava; Wulst(en)lava / lava cordada

1253
pillow lava, – a lava consisting of a mass of more or less ellipsoidal bodies each with a billowy surface. The pillows range from 10 cm to 7 m in diameter; they lie merged in one another, not unlike an irregular collection of sofa pillows. Radial joints are conspicuous in cross sections. Such lavas are known from submarine outflows; they are usually of basaltic or andesitic composition. / koeklava / lave en galettes; .... en coussins; .... en oreillers / Kissenlava / lava almohadillada

1254
palagonite, – a hydrated basaltic glass, usually brown to yellow or orange, formed when basalts are erupted beneath water or ice. It is usually found in pillow lavas as interstitial material or amygdules. / palagoniet / palagonite / Palagonit / palagonita

palagonitization, – the formation of palagonite by hydration of volcanic glass of a basaltic composition. / palagonitisatie / palagonitisation / Palagonitbildung / palagonitización

1255
pseudopalagonite, – well bedded, recemented palagonite; reworked scree material, containing only shattered fragments of basalt globes. / pseudopalagoniet / pseudopalagonite / Pseudopalagonit / (p)seudopalagonita

1256
basalt globe; basaltic lava nodule, – friable but quite dense or slightly porous lava nodule with thick tachylitic (sideromelan) cooling rims, fracturing along radial and tangential planes. / bazaltbol / (globe de basalte) / Basaltkugel / globo de basalto

1257
plateau basalt; flood ...., – single or successive, evenly extended basaltic lava flow(s) that build up a plateau. The high-temperature and very fluid lavas reach the surface through fissure eruptions. / plateaubazalt / basalte des plateaux / Deckenbasalt; Plateau.... / basalto de meseta

basaltic plateau; basalt ....; basalt plain, – / bazaltplateau / plateau basaltique / Basaltplateau; ....tafelland / meseta basáltica

1258
cone basalt, – differs from a plateau basalt by a low iron content and a higher viscosity, and hence rather forming a shield volcano than a plateau. / kegelbazalt / basalte des cônes / Schildvulkan-Basalt / basalto de cono

1259
block lava; apalhraun (Icelandic), – lava with a surface composed of loose, angular blocks with sharp edges, but fairly smooth faces. It is not synonymous with aa lava. / bloklava; schol.... / lave en blocs; .... à blocaux / Blocklava; Schollen.... / lava en bloques

1260
aa lava; aphrolithic ....; aphrolith, – a lava with an exceedingly rough fragmental top consisting of sharp, angular fragments of lava and clinkers cemented together. / aa-lava / lave aa; .... en gratons; .... à blocs scoriacés; .... scoriacée / Aa-Lava; Zackenlava / lava aa; lava afrolítica

sprouts; spinose ...., – protuberances causing the rough, jagged or spinous surface of aa lava. They are very irregular in form and range from a few millimetres to more than 20 cm. / stekels; stekelige uitwassen; pieken / épines / Stacheln / espinas

clinker field, – a large area composed of aa lava with its typical rough and jagged surface, resembling the clinker or slag of a furnace. / slakkentop van aa-lava / cheire scoriacée / Zackenoberfläche einer Aa-Lava / terreno con escorias

1261
dermolith; fluent lava, – obsolete terms, denoting lava with a smooth, coherent surface, varnished and vitrified to a depth of a few mm. / huidlava / lave lisse / Lava mit geglätteter Oberfläche / dermolítica

1262
volcanic glass, – a usually homogeneous amorphous rock, formed by cooling of a molten lava too rapidly to permit crystallization. / vulkanisch glas / verre volcanique / vulkanisches Glas; Gesteinsglas / vidrio volcánico

obsidian, – a black or dark-coloured volcanic glass, sometimes banded or microspherulitic, and usually of rhyolitic to andesitic composition. It has a glassy or satiny luster and a conchoidal fracture. / obsidiaan / obsidienne / Obsidian / obsidiana

pitchstone; fluolite, – obsidian containing more than 4% water. / peksteen / fluolite / Pechstein / pecilita; piedra pez

tachylite; hyalo-basalt; basalt obsidian; .... glass, – a volcanic glass, that may be black, green, or brown and of basaltic composition. It is commonly found in quickly chilled margins of dykes, sills, or flows. / tachyliet / tachylite / Tachylyt / taquilita

1263
lava flow; .... stream; .... river, – a subaerial or subaqueous outpouring of lava in molten state. The lava usually has a medium or a low viscosity forming a sheet of variable length, depth, and relatively small width. Also the rock body formed upon solidification. / lavastroom / coulée de lave; torrent de ....; fleuve de ....; coulée volcanique; torrent de feu / Lavastrom / colada de lava

farinaceous lava, – the particles of a white-hot lava stream seeming to be in a state of mutual repulsion and flowing like meal as at Helemaumau. / – – / lave farineuse / weissglühende mehlige Lava / lava farinácea

1264
lava tongue; lobe of lava, – an offshoot or a ramification of a larger lava flow; it may have a length of several km. / lavatong / langue de lave; lobe de .... / Lavazunge / lengua de lava

1265
parasitic lava flow, – a subsidiary lava flow issuing from a fissure on the slope or at the foot of a volcanic structure. / parasitische lava-uitvloeiing / coulée de lave parasite / parasitärer Lavaerguss / colada de lava parásita

1266
superfluent lava flow; summit overflow, – a flow of lava issuing from a summit crater and streaming down the flanks of a volcano. / terminale uitvloeiing van lava / effusion terminale / terminaler Lavaausfluss / efusión terminal

effluent lava flow, – a lava flow that is discharged from a volcano by way of a lateral fissure. / uitvloeiing uit de glooiing / effusion de flanc / Lateralerguss / efusión de flanco

flank outflow, – a lava flowing out of the flank of a volcanic structure. / – – / – – / Flankenerguss / – –

1267
interfluent lava flow, – a lava flow that is discharged into and through subterranean fissures and cavities in a volcano and that may never reach the surface. / onderaardse uitvloeiing van lava / coulée de lave souterraine; .... sousjacente / unterirdischer Lavaerguss / colada de lava subterránea

1268
flow breccia; clinker ...., – a volcanic breccia that is formed contemporaneously with the movement of a lava flow. / stroombreccie / brèche d'épanchement / Strombreccie; vulkanische Fliessbreccie / brecha de flujo

welded flow breccia; .... clinker, – that part of the aa or the block-lava flows, which is below the surface, and where the clinker or block fragments are stuck firmly together by adherence of still plastic lava. / aaneengelaste stroombreccie / brèche d'épanchement unie / verschweisste Strombreccie; .... vulkanische Fliessbreccie / brecha de flujo compactada

1269
agglomerate lava, – a lava that during its flow collected bombs, slags and ash; also the blocks on the surface of a block lava. / agglomeraatlava / lave d'agglomérat / Agglomeratlava / aglomerado de lava

1270
lava scratches; grooving marks, – parallel scratches on lava surfaces produced by sliding of lava flows over each other shortly before consolidation, and indicating the direction of the movement. / lavakrassen / stries; striage de la lave; rayures sur lave / Lavaschrammen / estrías

grooved lava, – / gekraste lava / lave striée / geschrammte Lava / lava estriada

1271
chilling effect, – the cooling influence of already solidified lava on liquid lava, e.g. in a lava lake. It may, among other effects, temporarily terminate gas emanations. / afkoelingseffect / effet de refroidissement / abkühlende Wirkung / efecto de enfriamiento

1272
intumescence of lava; tumefaction of ....; lava upheaval, – a vaulted or inflated part in the molten or solidified crust of a lava flow or of lava on the bottom of a crater, developed by a localized pressure rise. / zwelkoepel / intumescence de lave; tuméfaction / Lavadom; Schollendom / intumescencia de lava

tumulus, – a domical upheaval on the surface of a lava flow whose forward motion was obstructed by some barrier; also caused by pressure due to the difference in rate of flow between the cooler crust and the more fluid lava below. / tumulus / tumule / Tumulus / túmulo

1273
carapace, – the solidified surface of a lava dome. / lavakorst / carapace; calotte / Lavadomkruste / costra

1274
blow hole, – a minute crater on the surface of a viscous lava flow. / spuitgat / puits; bouche / Durchschlagloch; Blasloch / abertura de deflación

1275
lava sheet; extruded ...., – lava flow, in parts very thick, covering a large area. / lavadek; lavaplaat / nappe de lave / Lavadecke; Eruptivgesteinsdecke; Pedionit / manto de lava; colada

1276
lava flood, – enormous emission of lava, that flowed out rapidly with an output of millions of tons hourly, leading to the accumulation of many lava sheets to form a basalt plateau. / lavavloed / déluge de lave / Lavaflut; Flächenerguss / diluvio de lava; inundación de lava

1277
lava bed, – a single lava flow; a unit flow of lava. / lavabank / couche de lave / Lavabank / capa de lava; colada

lamination, – the planar array of constituents of lava parallel to the surface of the flow, or to other laminations. / laminatie / lamination / Lamination / laminación

1278
lava field; lava plain; lava plateau; lava desert, – a broad area of lava flows or lava sheets, characterized by low relief. Vent structures, eroded or non-eroded, and also some pyroclastics may occur. / lavaveld; lavavlakte / planèze; champ de lave / Lavafeld; Deckenerguss / campo de lava

pressure plateau, – an uplifted area of lava, the elevation of which is due to the intrusion of new lava from below that does not reach the surface. The lower part of such a flow may remain fluid for weeks. / – – / – – / Lavafeld-Aufpressung; ....Aufwölbung / – –

1279
malpais; scorched area, – a region with rough and barren lava flows. / lava-steenwoestijn / désert de lave / Lavawüste / desierto de lava; malpais

1280
lava delta, – a delta-like body of lava formed where a lava flow enters the sea. / lavadelta / delta de lave / Lavadelta / delta de lava

lava-flow coast, – a convex shoreline, formed by recent lava flows. / kust met lavastromen / côte de lave / Lavastromküste / costa de coladas de lava

1281
lava lake, – liquid lava, usually of basaltic composition, filling or partly filling a crater or a depression. The term also refers to the solidified and partly solidified stages; then, one of the adjectives solidified or congealed should be added. / lavameer / lac de lave / Lavasee / lago de lava

1282
lava fountain, – a gas-charged fluid emitted from a lava lake, volcanic crater vent or fissure, shooting up to heights ranging between a few metres and more than 500 m above the surface. / lavafontein / fontaine de lave; jets de lave liquide / Lavafontäne; Schlacken.... / surtidor de lava

clastogenetic lava flow; rootless lava flow, – a flow originating from lava fountains through accumulation of liquid lava spatters to such a degree, that they weld together and flow downhill in a compact mass. / wortelloze lavastroom / coulée de lave sans racine / klastogener Lavastrom; wurzelloser .... / colada de lava sin raíz

1283
fire fountain; spouts of fire, – a rhythmic, rise-and-fall eruption of incandescent lava fragments either from a central volcanic vent or along a fissure. / vuurwerk / fontaines de feu; gerbes enflammées; .... de feu / Feuerfontäne; ....säule; Glutsäule / fuente de fuego

1284
lava pit; fire pit; lava cauldron, – a circular volcanic crater or depression with steep walls that contains an active lava lake. / lavameerkrater / puits de lave / Lavaseekrater / hoya de lava

lava ring, – a wall of congealed lava rising above the average level of a lava-filled pit, formed by a fluctuating level of the lava lake, lava waves, subsequent lava drainage or spatter. / lavaring / anneau de lave / Lavaring / anillo de lava

1285
platform of solid lava, – a patch of solidified lava raised above the general level of the fluid lava of the lake. / plaat van gestolde lava/ terrasse de lave figée; plaque de lave .... / Lava(ufer)terrasse / terraza de lava

bench lava, – solidified lava in a crater, forming terraces. / terrassenlava / lave morte / Uferterrassenlava / bancal de lava

1286
lava scarp, – a line of cliffs across a lava flow or an assemblage of lava flows produced by faulting or erosion; also its steep face. / lavahelling / escarpement de lave / Lavahang / escarpe de lava

1287
lava cascade, – a cascade of lava formed when during its course the flow passes a precipitous part. / lavaval / cascade de lave; cataracte de feu / Lavafall; ....kaskade / cascada de lava

fan of lava, – a mass of congealed lava formed on a steep slope, due to the continually changing direction of effusions from a single source. / lavawaaier / éventail / Lavafächer / abanico de lava

1288
squeeze-up; squeezed-up lava toadstool; lava push, – a small extrusion of rather viscous lava, commonly injected into fractures of the solidifying crust. Its form may be bulbous or linear, and its height may reach from a few centimetres to almost a metre. / lavakurk / protubérance; amas boudiné; meringuée / Lavapilz; Lavapfropfen / extumescencia

1289
steptoe; volcanic nunatak; dagala (Etna area); kipuka (Hawaii),– an isolated rock mass surrounded by one or more lava flows. / lavavenster / fenêtre / Lavafenster / nunatak volcánico

1290
lava tunnel; .... tube; supply channel; feeding tube, – a hollow space in the shape of a tubular tunnel inside an otherwise solidified lava flow, formed by the withdrawal of the fluid lava after the formation of a superficial crust. Also the tunnel, wholly or partly filled with the still flowing, fluid lava. / lavatunnel / tunnel de lave / Lavatunnel / tunél de lava

collapse canal, – a sinuous depression in the surface of a lava flow due to collapse of the roof of a lava tunnel. / instortingskanaal / canal de lave / Lavamulde; eingestürzter Lavatunnel / canal de hundimiento

1291
lava cavern; lava cave, – subterranean cavity in a lava, usually within a fracture, a lava upheaval, or a collapsed lava tunnel. / lavagrot; ....hol; grotte de lave / Lavahöhle; Lavagrotte / caverna de lava

1292
lava tree-mold; lava tree-cast, – a cylindrical hollow in a lava formed when a flow surrounds the trunk of a tree; the tree may be burned to charcoal and ash, that in turn may be removed by wind and water, leaving an empty tube. Its inner wall may preserve in detail a mold of the surface of the bark. / lavaholte met stamafdruk / moule de tronc d'arbre; périmorphose d'un tronc; voûte de lave avex soupirail / Verdrängungsmetamorphose eines Baumstammes; Baumstamnegativ / molde del tronco de un árbol por lava

lava sheeth, – lining of lava that envelops an empty tube. / lavakoker / gaine de lave; tourbillon figée de lave / Lavahülle / forro de lava

1293
lava stalactite, – a pendent cone formed by dripping of remelted lava from the roof of an empty lava tunnel. / lavastalaktiet / stalactite de lave / Lavastalaktit / estalactita de lava

lava stalagmite, – an erect cone of lava on the floor of an empty lava tunnel built up by lava dripping from above, usually from a stalactite. / lavastalagmiet / stalagmite de lave / Lavastalagmit / estalagmita de lava

1294
scoria (plur . scoriae); cinder; slag, – rough, cinder-like, dark-coloured, often vesicular crust on the surface of a lava. Also the fragments of lava thrown out by an explosive eruption. The term is usually restricted to basaltic and closely allied lavas. / slak / scorie / Schlacke / escoria

scoriaceous; slaggy, – / slakkig / scoriacé / schlackig / escoriáceo

1295
scoria moraine; pseudomoraine, – volcanic debris of vesicular and cindery material from the surface of a lava flow, now accumulated in front of, and along the sides of the flow. / lavamorene / moraine de lave / Lavamoräne / morrena de escoria

medial moraine, – a narrow strip of volcanic debris, formed by the union of the scoria moraines when two lava flows join. / middenmorene / moraine médiane / Mittelmoräne / morrena intermedia

lava levee, – solidified, usually scoriaceous wall of lava, formed upon overflow of the natural channel. / oeverwal van lava / levée de lave / vulkanischer Randschutt / albardón de lava

1296
lava ball; pseudobomb; flotation bomb; volcanic dumpling, – a spherical or ellipsoidal mass of lava with a scoriaceous core and a compact shell. The diameter may range from a few mm to a few m. It is formed by the coating of a scoria fragment in a fluid lava. Layered balls also occur, consisting of an alternation of scoriaceous and compact zones. / lavabal / boule de lave / Lavakugel / bola de lava

accretionary lava ball, – a rounded mass of lava formed by the accretion of viscous lava around a rolling solid fragment. They range in diameter from a few cm to 5 m or more. / aangegroeide lavabal / boule de lave accrûe / angewachsene Lavakugel / bola de lava acrecional

1297
lava rag; lava clot; slag lump, – a small mass of scoriaceous lava of indefinite shape, thrown out during an eruption. / lavaklodder / motte de lave; motte de magma / Schlackenklumpen; Wurfschlacke / mancha de lava

1298
hornito, – a spatter cone built up around a rootless vent on a lava flow. / hornito / hornito / Hornito / hornito

driblet spire; spire, – a type of hornito characterized by extreme height relative to its diameter; some straight columns are less than a metre in diameter and as much as 4 m high.

spatter cone; driblet cone, – a low, steep-sided cone built up of clots of lava, dragged along by escaping gases, and deposited around the opening of emission. / kegel met lavaspetters; slakkenschoorsteen; lava.... / cône d'agglutinat; jets de lave fondue / Schweisschlackenkegel; Hornito; Lavaschornstein; Schlacken....; Lavaturm; Lavaorgel; Lavakegel / cono de salpicaduras de lava

1299
lava tower, – ragged pillar, predominantly consisting of scoriaceous material in the core, with a coating of denser lava. These pillars, having diameters of several metres and heights which may attain more than 10 m, are formed by the scoriae that consolidated at the surface of a lava flow or lake, and concentrated in eddies. After the drainage of the molten lava these scoriaceous accumulations were left behind as pillars. / lavatoren / tour de lave / Lavaturm / torre de lava

1300
littoral cone, – cone built up of ejected debris from a rootless vent, and formed at the shoreline where aa flows reach the sea. Generally they are better bedded than typical cinder cones and they may reach a height of 100 m and a basal diameter of more than 1 km. / aan de kust gevormde kegel / cône littoral / an der Küste gebildeter Kegel; littoraler Spratzkegel / cono litoral

1301
lava colonnade, – an exposure of columns in a mass of volcanic material, whether lava or tuff. / lavaorgel / orgue de lave; colonnade volcanique / Lavaorgel / columnata volcánica

giant's causeway, – a more or less horizontal surface made of prismatic columns due to contraction by cooling, resembling a kind of platform. / bazalt plaveisel / chaussée de géants / Basaltsäulengepflasterte Verebnung / pavimento de basalto

1302
cylindrical jointing, – occasionally found in andesite. / cylindrische klieving / joints de retrait cylindriques; fracturation cylindrique / zylindrische Absonderung / diaclasamiento cilíndrico

1303
polyhedric parting, – a jointing in 4-, 5-, or 6-sided columns as sometimes found in porphyries. / polyëdrische klieving / division polyédrique; .... en polygones / polyedrische Absonderung / disyunción poligonal

1304
spheroidal jointing, – a series of concentric and spheroidal or ellipsoidal cracks produced about compact nuclei of igneous rock, e.g. in basalts. / sferoïdale klieving; kogelvormige .... / désagrégation en boules / kugelige Absonderung / diaclasamiento esferoidal

1305
cavernous; cellular, – porous texture due to weathering out of certain constituents rather than to expansion of gases. / caverneus; cellig / caverneux; vacuolaire / porig; zellig / cavernosa

vesicular, – porous texture of a rock, especially a lava, due to the expansion of gases during the fluid stage of the lava. / vesiculair / vésiculaire; vésiculé / blasig; vesikular / vesicular; con huecos

1306
vesicle cylinder, – a cylindrical zone in a lava that contains abundant vesicles with diameters from 1 to 10 cm or more and lengths up to 1 m or more. The vesicles probably are formed by the generation of steam from underlying water-bearing strata. / cylindervormige blaasjes / vacuoles cylindriques / zylindrische Blasen / huecos cilíndricos

1307
foamy lava; pumiceous ....; thread-lace ...., – brittle lava containing many large and irregular vesicles, the separating walls of which are very thin and more fragile than those of pumice. It breaks by the pressure of the fingers. The crust of the lava is rough, showing holes and peaks of glass. / schuimlava / lave en échaudé / Schaumlava / lava espumosa

1308
pumice; pumice stone; foam; volcanic foam; pumicites, – a light-coloured, rough and cinder-like, glassy rock, usually of rhyolitic composition. It may float on water and is economically useful as a light-weight aggregate and as an abrasive. / puimsteen / ponce / Bimsstein / pumita

pumiceous, – spongy, with numerous cavities, and sometimes fibrous to thready; the adjective is usually applied to pyroclastic ejectamenta. / puimsteenachtig / ponceux / bimssteinartig / pumítica

[1224] *The Extrusive Phase: Pyroclastics*

1309
pyroclastics, – fragmental material projected in the air, issued from volcanic vents and produced by explosions. It includes particles of variable size of both volcanic and non-volcanic – usually older – rocks. / pyroklastica / produits pyroclastiques; (matériel pyroclastique) / pyroklastische Produkte; Pyroklastite / piroclastos

pyroclast, – any particle or fragment that has been ejected by an erupting volcano. / pyroklast / pyroclaste / Pyroklast / piroclasto

pyroclastic activity, – eruptions during which pyroclastic material is thrown out. / pyroklastische werking / activité pyroclastique / pyroklastische Tätigkeit / actividad piroclástica

1310
pyroclastic deposit; tephra, – collective term for deposits composed of all rock material fragmented by a volcanic eruption, such as pyroclastics, accidental ejecta, mudflow deposits, etc., that have not yet consolidated. / tefra; klastisch-vulkanische afzetting / téphra; dépôt volcanique clastique / pyroklastische Ablagerung; Tephra / tefra; depósito volcánico clástico

volcanoclastic deposit, – any deposit rich in fragments or grains of volcanic rocks, formed by volcanism, weathering and erosion of volcanic products. / vulkanoklastische afzetting / dépôt volcanoclastique / vulkanoklastische Ablagerung / depósito volcanoclástico

1311
pyroclastic rock, – consolidated deposit of fragmental volcanic material. / pyroklastisch gesteente / roche pyroclastique / pyroklastisches Gestein / roca piroclástica

1312
tephrochronology, – chronology based on the study of the successive deposits of fragmental volcanic products. / tefrochronologie / téphrochronologie / Tephrochronologie / tefracronología

1313
cognate ejecta, – pertaining to ejected material directly derived from or related to the vent magma, i.e. pyroclasts that are essential or accessory. / autolieten / éjecta consanguins / magmaeigene Auswürflinge; authigene .... / eyecciones consanguíneas

ejecta(menta); clastic ...., – a general term for material thrown out by a volcanic eruption, including fragmental material, derived from the eruptive magma as well as re-ejected portions of the volcanic cone and derived from older rocks, e.g. country rock. / (gruisvormige) uitwerpselen; klastische ....; efflata; ejectamenta / projections; éjecta clastiques / Auswürflinge; lockere Eruptionsprodukte / eyecciones

1314
essential ejecta; juvenile ....; authigenous ...., – pyroclastic material of consanguineous magmatic origin. / autogene efflata; juveniele pyroklastica / éjecta authigènes / juvenil-authigene Auswürflinge / eyecciones esenciales

accessory ejecta; resurgent ...., – material derived from previously solidified rocks of the eruption centre and of consanguineous origin. / accessorische pyroklastica / éjecta consanguins remaniés / resurgent-authigene Produkte / eyecciones accesorias

accidental ejecta; allothigenous ....; noncognate, ..., – fragments of non-volcanic or volcanic rock from the country rock or from the walls of the vent. / allogene uitwerpselen; niet-juveniele .... / éjecta énallogènes / allothigene Auswürflinge / eyecciones accidentales

1315
incandescent detritus, – ejecta thrown out in a glowing state. / gloeiende uitwerpselen / projections incandescentes / glühende Auswürflinge / emisiones incandescentes

nonluminous detritus, – ejecta of the same magma and of the same eruption, but already cooled down. / niet-lichtende detritus / projections refroidies / abgekühlte Auswürflinge / emisiones no luminosas

1316
volcanic dust, – unconsolidated, fine-grained material thrown out during volcanic eruptions and having a grain size of less than 0.05 mm. Dust can remain in suspension in the air for a long time, up to several years. / vulkanisch stof / poussière volcanique / vulkanischer Staub / polvo volcánico

NOTE:
– see table 3b – The overlapping Grain-Size Classes of Volcanic Ejecta

1317
volcanic ash, – unconsolidated, fine-grained material thrown out during volcanic eruptions; grain size between 0.05 and 4 mm. The term is sometimes also used for its consolidated counterpart or tuff. / vulkanische as / cendre volcanique / vulkanische Asche / ceniza volcánica

1318
volcanic quicksand, – a deposit, mainly composed of fine ash at an elevated temperature, in consequence of which the gas-filled interstitial volume is sufficiently great to give the mass an extraordinary degree of potential mobility. / vulkanisch drijfzand / sable mouvant volcanique / vulkanischer Treibsand / arena movediza volcánica

1319
trituration; comminution; pulverization, – the reduction of older rock into a fine powder, as has been suggested for certain ashes of Mt. Etna. The term usually applies to the reduction of a rock to progressively smaller particles by weathering, erosion and tectonic movements, and by mechanical breaking. / verpulvering / trituration; pulvérisation / Zermalmung; Pulverisierung; Aufbereitung / pulverización

1320
mantle of ash, – a new deposit of fine-grained volcanic material covering the landscape. / asdek / manteau de cendre / Aschendecke / manto de ceniza

1321
(volcanic) pisolite; chalazoidite, – pellet with an accretionary

structure formed by rain falling through dense volcanic dust and ash clouds. They may grow to pea size, and roll downhill on the surface of loose ash, thus acquiring a spiral structure. / (vulkanisch) pisoliet / pisolite (volcanique) / (vulkanisch) Pisolith; Schlammkügelchen / pisolita; chalazoidita

(volcanic) mud ball; .... pellet; accretionary lapilli, – spherical or somewhat flattened, sometimes silicified mass of mud, with a diameter from 2.5 to 85 mm, composed of concentric shells of fine or coarse volcanic ash, formed around grains of sand or lapilli. They are formed during prolonged rotation in whirling eruption clouds. / schalenkogel; modder....; sferoliet / sphérolite / Aschen-Sphärolit / esferolito; lapilli acrecional

sulphur (mud) ball, – small spherical skin of sulphurous mud that formed on a bubble of hot volcanic gas and became firm on contact with the air. / zwavel(modder)kogeltje / sphérule (de boue) de soufre; perle de .... / Schwefel(schlamm)kügelchen / bolita de fango de azufre

1322
volcanic sand, – a deposit of unconsolidated pyroclasts with a diameter from 0.05 to 2 mm, hence falling within the ash-size category. / vulkanisch zand / sable volcanique / vulkanischer Sand / arena volcánica

1323
scoriaceous sand, – / slakkenzand / sable scoriacé / Schlackensand / arena escoriácea

1324
lapilli (plur.; sing. lapillo); scoriaceous ....; porous ....; lava pellet, – fragmental material issued by volcanic eruptions of a size range that has been variously defined between the limits of 1 to 64 mm. The fragments may consist either of juvenile lava, still liquid or plastic when ejected, or of broken rock of any sort from the walls of the vent or from the bed rock; that is, they may be essential, accessory or accidental ejecta. / lapilli / lapilli / Lapilli / lapilli

volcanic gravel, – an unconsolidated deposit of pyroclasts with a size range – although variously defined – between 2 and 75 mm. / vulkaangrind / gravier volcanique / Vulkanschotter / grava volcánica

1325
crystal lapilli, – more or less well-shaped crystals already formed in the magma and explosively ejected in great quantities. / kristallapilli / lapillis cristallins / Kristallapilli / lapilli de cristales

1326
Pele's hair; filiform lapilli; capillary lava, – filaments of lava glass; natural spun glass formed by blowing out of molten lava during quiet fountaining of the fluid. Thus long threads are formed that may be more than 2 m in length and range in diameter from about 0.01 to 0.5 mm. / glasdraden; Pelea's haar / fils étirés; cheveux de Pélé; filaments vitreux / Fadenlapilli; Glasfäden; Lava....; Peles Haar / pelo de Pele; filamentos vítreos

Pele's tear; tear-shaped bomb, – small, solidified drops of volcanic glass formed by quiet fountaining of fluid lava, sometimes in association with Pele's hair. The drops may be tear-shaped, spherical, or nearly cylindrical, measuring from 1 mm to more than 1 cm in length. / glasdruppel; (Pelea's) traan / larme de lave; .... de Pélé; goutte; gouttelette / Tränenbombe; Lavaträne; Peleträne / lágrima de Pele; gota de lava

1327
block; boulder, – a usually angular fragment that has been erupted in a solid or nearly solid state. Its size ranges from 5 cm in diameter to several cubic metres. It may be essential, accessory, or accidental. / blok; lavabrok / bloc (volcanique) / ausgeworfener Block / bloque volcánico

1328
bomb, – a mass of lava, generally larger than lapilli, issued by explosion from a volcanic vent. Bombs are viscous when thrown into the air, receiving their particular shapes during the flight. The shape is used in the descriptive classification. / bom / bombe / vulkanische Bombe / bomba

1329
rotational bomb, – showing a spiral or spherical form resulting from rotation during the flight. The term includes spindle-shaped, spheroidal, and tear-shaped bombs. / rotatiebom / bombe ayant subi une rotation / Rotationsbombe / bomba de rotación

spindle-shaped bomb; almond-shaped bomb; bipolar fusiform bomb, – a rotational bomb spun into spindle shape by rapid rotation during its passage through the air and generally flattened as a result of impact with the ground. / bipolaire bom; spindelvormige bom / bombe en fuseau; bombe fusiforme / spindelförmige Bombe; bipolare Bombe / bomba fusiforme

spheroidal bomb; spherical bomb; globular bomb, – / bolvormige bom / bombe sphéroïdale / kugelige Bombe / bomba esférica

tear-shaped bomb, – a rotational bomb having the shape of a tear drop, with an ear at its constricted end. / druppelvormige bom / bombe en forme de goutte / Tropfenförmige Bombe / bomba en forma de gota

unipolar bomb, – a bomb lengthened and protracted on one end. / unipolaire bom; peervormige bom / bombe unipolaire / unipolare Bombe; ausgeschwänzte Bombe / bomba unipolar; .... piriforme

twisted bomb, – / gedraaide bom / bombe tordue / gedrehte Bombe / bomba torneada

pancake bomb, – / pannekoekbom / bombe en forme de galette / Pfannkuchenbombe / bomba en forma de tortilla

ribbon bomb, – an elongated and flattened bomb, derived from ropes of lava. / lintvormige bom; bandvormige bom / bombe rubanée / bandförmige Bombe / bomba en forma de cinta

cored bomb; perilith, – a bomb composed of a core of older volcanic or non-volcanic rock and a skin of solidified younger lava. / samengestelde bom / bombe allogène / perilithische Bombe; Bombe mit Kern / bomba con núcleo

olivine bomb; peridotite bomb, – cored bombs, the cores of which consist of a large nodule of respectively olivine or mixed olivine and pyroxene (and spinel) crystals. / olivijnbom; peridotietbom / bombe dòlivine; bombe de péridotite / Olivinbombe; Peridotitbombe / bomba de olivina; bomba de peridotita

bread-crust bomb, – a bomb showing a network of opened

cracks on its surface as a result of continued expansion of the interior after solidification of the crust. / broodkorstbom / bombe en croûte de pain / Brotkrustenbombe / bomba en forma de corteza de pan

explosive bomb; exploding bomb, – a bread-crust bomb that has thrown off violently fragments of its crust during its flight or shortly afterwards. This is due to the tension caused by continued expansion of its interior after solidification of the bomb's crust. / exploderende bom / bombe explosive / explodierende Bombe / bomba de explosión

cow-dung bomb, – a bomb whose flattened shape is due to its impact while still viscous. It usually has a scoriaceous surface. / slakkenkoek / bombe en bouse de vache / Lavaflatschen; Lavafladen / bomba de boñiga de vaca

slag bomb, – a bomb of a very scoriaceous texture. / slakkenbom / bombe scoriacée / Schlackenbombe / bomba escoriácea

pumaceous bomb, – a bomb consisting of irregularly elongated pieces of pumice with glassy bread-crust skin. / puimsteenbom / bombe de ponce / Bimssteinbombe / bomba de pumita

1330
mantle of scoriae; scoriaceous cover, – coarsely vesicular pyroclastic rock, usually of basaltic or andesitic composition, and lying on top of a different rock, e.g. a lava. / slakkenschede / gaine de scories / Schlackenmantel / manto de escoria

1331
reticulite; thread-lace scoria, – an extremely brittle pyroclastic rock, consisting of numerous spherical vesicles of which the walls have collapsed, leaving a network of threads. / – – / réticulite / Retikulit; schwammiger Bimsstein / reticulita

1332
tuff, – an indurated, pyroclastic deposit, predominantly consisting of fine-grained volcanic fragments; sedimentary particles may be present. The deposit may or may not be laid down in water, and it may be well sorted or heterogeneous. / vulkanische tuf / tuf volcanique / vulkanischer Tuff / toba

tuffite; tuffogenic rock, – an indurated clastic rock consisting both of pyroclastic and non-volcanic detrital material with a predominance of pyroclastic fragments. / tuffiet / tuffite / Tuffit / tufita

tuffaceous, – pertaining to sediments containing up to 50% tuff. / tuffeus / tufacé / tuffitisch / tobáceo; tufítico

1333
dust tuff, – a tuff mainly consisting of very fine-grained material with a grain size less than 0.05 mm. / stoftuf / tuf en poussière / Staubtuff / toba de polvo; toba pelítica

ash tuff, – a tuff mainly consisting of fine-grained material with a grain size between 0.05 and 4 mm. / astuf / cinérite / Aschentuff / toba de ceniza

lapilli tuff, – a tuff consisting of fragments with a grain size between the limits 1 to 64 mm in a fine-grained matrix. / lapillituf / tuf à lapilli / Lapillituff / toba de lapilli

pumice tuff, – a tuff mainly consisting of pumiceous pyroclasts. / puimsteentuf / tuf à ponces / Bimssteintuff / toba de pumita

scoria tuff, – a tuff consisting of fragments of scoria in a fine-grained matrix. / slakkentuf / tuf à scories / Schlackentuff / toba de escorias

lithic tuff, – a tuff predominantly consisting of fragments other than pyroclasts. / litische tuf / tuf lithique / lithischer Tuff / toba lítica

vitric tuff, – a tuff predominantly consisting of fragments of volcanic glass. / glastuf / cinérite / Glastuff / toba vítrea

crystal tuff, – a tuff predominantly consisting of crystals and fragments of crystals blown out during the eruption. / kristallentuf / tuf à cristaux / Kristalltuff / toba cristalina

crystal-lithic tuff, – a tuff in which both crystal and lithic fragments are present, but where crystal fragments predominate.

lithic-crystal tuff, – a tuff in which both lithic and crystal fragments are present, but where lithic fragments predominate.

crystal-vitric tuff, – a tuff consisting of 50-75% crystal fragments and 25-50% glass fragments.

chaotic tuff, – a massive, unstratified tuff consisting of a mixture of equally distributed fine and coarse material. / chaotische tuf / tuf chaotique / chaotischer Tuff / toba caótica

pisolitic tuff, – a tuff consisting of accumulations of pisolits, i.e. pellets with an accretionary structure. / pisoliettuf / tuf pisolitique / Pisolithtuff / toba pisolítica

1334
sedimentary tuff, – any deposit of tuff deposited either subaerial or subaquatic. The term is sometimes applied to subaerial tuffs deposited in lakes and seas. / sedimentaire tuf / tuf sédimentaire / Sedimenttuff / toba sedimentaria

1335
subaeric tuff; explosion tuff, – a tuff, the constituents of which after ejection have been laid down directly without reworking. / explosietuf / tuf d'explosion / Explosionstuff; atmosphärisch extrudierter Tuff / toba de explosión

subaqueous tuff, – formed by subaquatic eruptions. Near the eruption point the tuffs are unstratified, at greater distances stratified. / in water afgezette tuf / tuf subaquatique / subaquatisch extrudierter Tuff / toba subacuática

1336
stratified dry tuff, – a subaerial, graded tuff formed by sorting of the particles in the air, the heavy fragments falling first. / gelaagde droge tuf / tuf stratifié / geschichteter Trockentuff / toba seca estratificada

1337
reworked tuff, – a tuff carried by flowing water and redeposited in another locality. / verplaatste tuf / tuf transporté; tuf remanié / umgelagerter Tuff; aufgearbeiteter Tuff / toba redepositada

1338
puzzolana; pozzuolana, – a siliceous tuff or ash of trachytic

composition. It hardens under water when mixed with water and lime, therefore used in cement. Named after the city of Pozzuoli, Italy. / puzzolaan(aarde); pozzolaan(....) / pouzzolane / Puzzolan; Puzzolanerde / puzolana

1339
tuffisite, – term used for the filling of Swabian volcanic pipes, and predominantly consisting of fragmented country rock. / tuffisiet / tuffisite; tuf intrusif / Tuffisit / tufisita

1340
ignimbrite; ash flow sheet; ash flow tuff; ash flow cooling unit; flood tuff; sillar (Peru), – an extrusive rock unit made up of glass shards, crystals and lithic fragments, commonly in that order of decreasing abundance. It is formed by the deposition and consolidation of incandescent ash flows and nuées ardentes. The deposit is generally but not exclusively silicic and often but not necessary welded. Ignimbrites are frequently characterized by a) a considerable thickness, a great areal extent and a flat plateau-like surface, b) the development of a columnar structure as a result of shrinkage during cooling, c) secondary crystallizations of e.g. feldspar and tridymite, formed by deuteric pneumatolytic action, d) glass shards bending around the crystal fragments of e.g. quartz and feldspar, e) the existence in some cases of a kind of flow structure resulting from the parallel arrangement of the larger glass particles. / ignimbriet; smelttuf / ignimbrite; tuf soudé / Gluttuff; Ignimbrit; Schmelztuff / ignimbrita

1341
welded tuff; tuff lava, – a tuff in which the individual fragments have remained sufficiently plastic to become partly or wholly welded; a tuff that has been indurated by the combined action of the heat retained by particles, the weight of overlying material, and hot gases. The term includes both air-fall and ignimbrite deposits. / gewelde tuf / tuf soudé / Schweistuff / toba soldada

1342
piperno, – a welded, trachytic tuff, characterized by conspicuous lenses of glass ('**fiamme**'). Type locality is Soccavo near Naples. / piperno / piperno / Piperno / piperno

pipernoid, – pertaining to the macro- or microscopic structure of a volcanic rock showing dark patches and stringers in a light-coloured groundmass.

1343
volcanic conglomerate, – indurated coarse deposits, largely consisting of rounded and subangular water-worn volcanic boulders and gravels, possibly mixed with sedimentary or other country rocks. They may be formed by mudflows. / vulkanisch conglomeraat / conglomérat volcanique / vulkanisches Konglomerat / conglomerado volcánico

1344
agglomerate, – a non-stratiform rather chaotic assemblage of unsorted predominantly angular volcanic fragments, either loose or consolidated, in a matrix of finer volcanic material. The deposit is localized near to its centre of eruption; the shapes of the fragments are not determined by the action of running water. / agglomeraat / agglomérat / Agglomerat / aglomerado

1345
volcanic breccia; eruptive ....; lava ...., – a rock assemblage that consists of angular volcanic fragments, larger than 2 mm in diameter, with or without a matrix. / vulkanische breccie / brèche volcanique / vulkanische Breccie / brecha volcánica

1346
explosion breccia, – a breccia composed of rock fragments of many dimensions and produced by volcanic explosions. / uitbarstingsbreccie / brèche de projection / Explosivausbruchsbreccie / brecha de explosión

agglutinate, – a deposit of pyroclasts of which the scoriaceous fragments were erupted in a rather liquid state, and have been cemented by vitric material. There is no matrix of tuff. / aaneengebakken slakken; gekoekte .... / scories agglomérées; .... agglutinées / Schweisschlacken; Schlackenagglomerat / escorias aglutinadas

lava of tuff agglomerate, – a composite lava formed as a result of lava breaking through masses of tuff, picking them up and welding them together. / tufagglomeraatlava / lave de tufs agglomérés / Tuffagglomeratlava / lava de tobas aglomeradas

1347
palagonite tuff; .... breccia; móberg (Icelandic), – tuff-like material consisting of angular fragments of palagonite (hydrated basaltic glass). It may contain also fragments of augite and olivine, microlites of plagioclase, and broken pieces of basalt. The term is also used for volcanic breccias and tuffs, formed by subglacial eruptions and characterized by the presence of basalt globes (lava nodules) in a matrix, largely consisting of basaltic glassy grains. / palagoniettuf; ....breccie / tuf palagonitique; brèche .... / Palagonittuff; ....breccie; ....formation / toba de palagonita

hyaloclastite, – a breccia-like material consisting of aphanitic or glassy fragments formed by scattering (granulation) of a (rising) basic magma in contact with sea water. The glassy fragments may or may not show a palagonitic alteration. / – – / – – / Hyaloklastit / – –

[1225] *The Extrusive Phase: Miscellaneous Products*

1348
juvenile water; maqmatic ...., – derived directly from magma and coming to the earth's surface for the first time. / juveniel water; magmatisch .... / eau juvénile / juveniles Wasser; magmatisches .... / agua juvenil; .... magmática

juvenile gas; magmatic gas, – gas derived directly from magma and coming to the earth's surface for the first time. / juveniel gas; magmatisch gas / gaz juvénile / juveniles Gas; magmatisches Gas / gas juvenil

1349
volcanic gas, – volatiles released during volcanic eruptions or escaping from fumaroles, previously dissolved in the magma. Water vapour (probably totally derived from non-magmatic subsurface waters or nearly so) is the dominant gas, amounting up to 90% of the total; other constituents are carbon dioxide, carbon monoxide, sulphur dioxide, hydrogen sulphide, sulphur trioxide. / vulkanisch gas / gaz volcanique / vulkanisches Gas / gas volcánico

(volcanic) fume, – gaseous emission, which may contain dust, rising from the crater, from fumaroles or solfatares, from lava flowing into water (sea or lake), or from a body of molten lava. / vulkaandamp / fumée volcanique; vapeurs volcaniques / vulkanischer Dampf / humo volcánico

1350
blowpipe flame, – volcanic flames with a temperature up to 1300°C, as e.g. occurring in gas vents of Kilauea. / steekvlam / flamme d'émission volcanique / Stichflamme / llama de soplete

1351
hydroexplosion, – explosion caused by contact of hot lava with water. / hydroexplosie / explosion phréatique / Hydroexplosion / hidroexplosión

phreatic gas, – formed by the contact of atmospheric or subsurface water with ascending magma. / freatisch gas / gaz phréatique / phreatisches Gas / gas freático

phreatic eruption; semivolcanic ....; pseudovolcanic ....; secondary ...., – steam and mud eruptions caused by the expansion of volatile matter such as water, sulphuric gases, etc. above the roof of a hot igneous body. The ejecta, blown out at relatively low temperatures, contain no fresh incandescent matter. / freatische eruptie; semivulkanische .... / éruption phréatique / phreatischer Ausbruch; semi-vulkanische Eruption / erupción freática

1352
lahar; volcanic mudflow, – a flow of water mixed with chiefly volcanic material composed of fragments of all sizes. The resulting deposits are usually unsorted; they may attain thicknesses of tens of metres and may reach distances up to 40 km. / lahar; vulkanische modderstroom / coulée de boue / vulkanischer Schlammstrom; .... Murstrom; Lahar / lahar; corriente de barro volcánico

1353
hot lahar; hot (volcanic) mudflow; a flow of water mixed with hot volcanic material, formed by eruption through a crater lake. Also a flow produced by heavy rains on freshly deposited hot volcanic fragments. / warme lahar / lahar chaud; courant boueux chaud / heisser Schlammstrom; .... Lahar / lahar caliente

1354
outbreak lahar, – a volcanic mudflow that originates when the wall of a crater lake collapses. / doorbraaklahar / lahar de rupture / Durchbruchlahar / lahar de descarga

1355
cold lahar; volcanic mudflow, – a flow of water mixed with chiefly volcanic material formed by the mobilization of rain-soaked debris on volcanic slopes. It may also originate by the collapse of the wall of a crater lake or by melting ice caps. / koude lahar; gewone .... / lahar froid; torrent boueux froid / kalter Schlammstrom; .... Lahar / lahar frío; corriente de barro volcánico

rain lahar; mud lava, – a lahar caused by heavy rains on the slopes of a volcano, removing loose material. They may either be hot or cold, depending on the moment of origin, i.e. soon after an eruption, or at a later date, when the material has cooled down. / regenlahar / lahar causé par des pluies / Regenlahar / lahar de lluvia

1356
lahar deposit; mudflow ...., – settled lahar or mudflow material. / laharafzetting; modderstroomconglomeraat; dépôt de lahar / Laharablagerung; Schlammtuff / depósito de lahar

1357
glacier burst; jökulhlaup (Iceland), –
1. a water flood caused by subglacial volcanic outbursts;
2. a flood due to a catastrophic draining of an ice-dammed lake.

/ gletsjerlahar / lahar de fonte glaciaire / Schmelzwasser-Lahar; ....-Schlammstrom / lahar de fundición glacial

1358
ebullition; ebullitim, – agitation of the water of a crater lake or of the sea by volcanic heat. The formed gases rise in bubbles or may even throw up the water in jets. / opborreling / ébullition / Aufsprudeln; Aufkochen; Kochen / borboteo

upwelling, – a rise of water that originates when volcanic gases escape with little force, driving up the water to a height of a few metres. Also a relatively quiet eruption of lava and volcanic gases without much force. / opwelling / jaillissement / Aufwallung / surgencia

1359
water cupola; .... dome, – a vaulted uprising of the surface of the sea above a submarine volcanic explosion. It immediately precedes the ejection of gases and tephra. In 1928 a water cupola has been observed of a height of 26 m and a width of 100-200 m at its base. / waterkoepel; ....dom / eau soulevée en coupole / Wasserkuppel; ....dom / cúpula de agua

water fountain, – the initial effect of an explosion, visible above the surface of the water, in which case no great height is reached. Also the natural sequel from upwelling water or a water cupola to the escape of gases without ejection of solid material. The water column rises high and losing width. Mud generally appears at the base of the fountain. / waterfontein / fontaine; jet d'eau / Wasserfontäne / surtidor de agua

1360
eruption cloud; explosion ...., – a gaseous cloud containing fragmental – predominantly volcanic – material, emitted from a crater. As the cloud may rise to great heights, its appearance and shape is largely determined by condensation of atmospheric water vapour. / eruptiewolk / nuée volcanique / Eruptionswolke; Ausbruchs.... / nube volcánica

nucleation, – pertaining to volcanic clouds, i.e. the phenomena of condensation of vapour from volcanic vents and from the atmosphere, thus giving the illusion of a copious emission from the crater. / kernvorming / condensation nucléaire / Nukleation; Bildung von Kondensationskernen / nucleación; condensación de núcleos

1361
eruption column, – the lower portion of an eruption cloud, having a vertical, columnar shape. In this portion the ash and the debris-laden gases are still rising with great speed. / eruptiekolom; rook.... / gerbe éruptive; colonne de fumée / Eruptionssäule / columna eruptiva

1362
pine-tree (shaped) cloud, – an eruption cloud with a vertical shaft (eruption column) terminating upwards in a canopy of

clouds. / pijnboomvormige wolk / gerbe en panache; fumée en forme de pin-parasol / Pinienwolke / nube volcánica en forma de pino

cauliflower cloud, – an eruption cloud with a short vertical column and whirling border parts, caused by the rolling friction with the air. / bloemkoolvormige wolk / nuée en forme de chou-fleur / blumenkohlförmige Wolke / nube volcánica en forma de coliflor

1363
ash cloud; dust ...., – an eruption cloud consisting of volcanic ash and gases. / aswolk / nuage de poussières; .... de cendres / Aschenwolke / nube de ceniza

ash fall; ash shower; rain of ash, – the descent of volcanic ash directly from the air. / asregen / pluie de cendres / Aschenregen / lluvia de cenizas

1364
smoke ring, – ring-shaped whirls, i.e. circular vortices of ash and vapour. / rookring / tourbillon / Rauchkringel / torbellino de humo

1365
electrified ash cloud, – ash clouds with an electric charge, due to friction of the particles. / electrisch geladen aswolk / nuée de cendres électrisées / elektrisch geladene Aschenwolke / nube de ceniza electrizada

1366
nuée ardente; glowing cloud, – a gas-generating, turbulent cloud, erupted from a volcano, of which the heavy fractions of incandescent debris avalanche and follow the depressions on the flank of the volcano covering the adjacent landscape, while the lighter fractions of volcanic gases, ash and dust cauliflower upwards. They are highly mobile, travelling over 100 km per hour and are essentially gas-glass-dust emulsions. The lower part of the nuée ardente is comparable to an ash flow, and the terms are sometimes used synonymously in this sense. / gloedwolk / nuée ardente / Glutwolke / nube ardiente

1367
glowing avalanche; fire ....; hot ....; volcanic ...., –
1. an avalanche of incandescent blocks, lapilli, ash, and dust;
2. the debris of a disintegrating lava mass;
3. the lower layer of a glowing cloud (nuée ardente) with fragmental debris that follows the depressions in the landscape.

/ gloedlawine; gloeiende lavalawine; ladu / avalanche incandescente / Glutlawine; glühende Schuttlawine; .... Detrituslawine / alud ardiente

1368
mixed avalanche, – an avalanche consisting of either a mixture of incandescent and old rocks of the volcano or a mixture of dust, ash and coarse fragments. / gemengde afstortingslawine / avalanche mixte / gemischte Lawine / alud mixto

debris avalanche, – an avalanche of existing rock, sometimes caused by an eruption. / steenlawine / avalanche de pierres / Schuttlawine / alud de derrubios

1369
(hot) ash avalanche; ash slide; ash flow, – freshly fallen hot sand and ash, flowing down the flank of a volcano with the velocity and aspect of a snow avalanche, but generally in an impressive silence. The accumulation of ejected material on the upper portions of the cone, consisting principally of fine hot ash, is unstable and in a state of potential mobility. / (hete) aslawine; .... asstroom / avalanche sèche / (heisse) Aschenlawine / alud de ceniza ardiente

incandescent ash flow; .... pumice flow; .... tuff flow; hot tuff flow; hot sand flow, – a fiery gas-saturated flow, consisting mainly of fine material. / hete tufstroom / coulée de cendres incandescentes / Gluttuffstrom; heisser Aschenstrom; .... Sandstrom / corriente de ceniza incandescente

1370
fluidization, – the mixing process of gas and loose, fine-grained material so that the mass flows like a liquid, e.g. the formation of a nuée ardente during a volcanic eruption. / fluïdisatie / fluidisation / Fluidisation / fluidización

gas skin; gas film, – the envelope of hot gas surrounding each of the particles in an ash flow or nuée ardente. / gasomhulling / pellicule gazeuse / Gashülle / película gaseosa

1371
flashing arc, – a phenomenon that may occur during powerful explosions. The arcs originate in the crater and are due to sound waves, whose spherical fronts of compression and dilatation, spreading out in all directions, refract light to different extents and are thus made visible. / flikkerboog / arc lumineux / Explosionslichtbogen; sichtbare Schallwelle / arco luminoso

1372
globe lightning; fire globe, – a ball of fire, which may be up to one metre in diameter and may remain steady for some tens of seconds. Although not an exclusive paravolcanic phenomenon, it is sometimes seen near the base of a crater cloud. / bolbliksem / éclair en boule; globe / Kugelblitz / relámpago esférico

1373
volcanic thunderstorm, – electric discharges in steam clouds that accompany volcanic eruptions; they are possibly caused by friction of ash particles in the vent. / vulkanisch onweer / orage volcanique / vulkanisches Gewitter / tormenta volcánica

eruption rain; volcanic rain, – rain induced by a volcanic eruption, and that would not have fallen otherwise. / eruptieregen / pluie d'éruption / Eruptionsregen / lluvia de erupción

[1226] *The Late-Volcanic Phase*

1374
postvolcanic phenomena; posteruption ...., – occurring after the main extrusive phase. They represent a late- rather than a postvolcanic phase. / postvulkanische verschijnselen; vul-

kanische nawerking / phénomènes post-volcaniques; séquelles du volcanisme / postvulkanische Erscheinungen; Nachklangerscheinungen; Nachhall.... / fenómenos postvolcánicos

1375
fumarolic activity, – this term is applied when gases only escape more or less violently from the crater, the slope or the foot of the volcano. / fumarolenwerking / activité fumerollienne / Fumarolentätigkeit / actividad fumarolica

1376
volcanic emanation, – volatile and non-volatile material emitted at the earth's surface from volcanoes, fumaroles or lavas, usually consisting of water vapour and volcanic gas(es). / vulkanische emanatie / émanation gazeuse; .... volcanique / vulkanische Emanation / emanación volcánica

exhalation, – the emission of volcanic vapours and gases. / exhalatie / exhalaison (volcanique) / Aushauchung; Exhalation / exhalación

1377
(volcanic) fume, – gaseous emission, which may contain dust, rising from the crater, from fumaroles or solfataras, from lava flowing into water (sea or lake), or from a body of molten lava. / vulkaandamp / fumée volcanique; vapeurs volcaniques / vulkanischer Dampf / humo volcánico

1378
gas-barren ground, – an area devoid of vegetation as a result of volcanic emanations. / – – / – – / gasverätztes Ödland / yermo volcánico

1379
fumarole, – violent or passive emission of gases and vapours, escaping from fissures in craters, on the slope or at the foot of a volcano; fumaroles may also occur in apparently chaotic clusters or fields. The chemically active fluids frequently alter the character of the surrounding rocks. Fumaroles are sometimes classified by the composition and the temperature of the gases. The term is also applied to the hole or vent from which the gases and vapours are emitted. / fumarole / fumerolle / Fumarole / fumarola

1380
fumarole field; .... area, – an areal grouping of fumaroles that occur in rows on fissures or in an apparently random pattern. / fumarolenveld / champ de fumerolles / Fumarolenfeld / campo de fumarolas

1381
permanent fumarole, – a fumarole emitting continuously from the same vent or field during several centuries. / bestendige fumarole / fumerolle permanente / ständige Fumarole / fumarola perenne

temporary fumarole, – a fumarole, whose activity lasts up to a few years at most. / tijdelijke fumarole / fumerolle temporaire / vergängliche Fumarole / fumarola temporal

1382
primary fumarole, – a fumarole situated at a fissure in or on a volcano and fed directly from the main source of activity, thus giving a true index of internal conditions. / primaire fumarole / fumerolle primaire / primäre Fumarole / fumarola primaria

secondary fumarole; rootless ...., – a gas vent developed on lava flows at a distance from their source; or on flowing cloud deposits and hot lahars. / wortelloze fumarole; secundaire .... / fumerolle sans racine; .... (d'origine) secondaire / wurzellose Fumarole / fumarola sin raíz; .... secundaria

1383
crater fumarole, – a fumarole within a volcanic crater. / kraterfumarole / fumerolle de cratère / Kraterfumarole / fumarola de cráter

fissure fumarole, – a fracture, the entire length of which serves as a gaseous vent. / spleetfumarole / fissure fumerollienne; alignement de fumerolles; fumerolle de fissure / Spaltenfumarole / fumarola fisural

tunnel fumarole, – a fissure fumarole of which the opening is almost shut off by a roof of incrustation products. / tunnelfumarole / tunnel fumerollien / Tunnelfumarole / fumarola de túnel

1384
leucolitic fumarole, – a fumarole, arising from a red-hot lava. / leucolitische fumarole / fumerolle leucolitique / leucolitische Fumarole / fumarola leucolítica

croicolitic fumarole, – a fumarole, arising from a partly cooled-down lava. / kroikolitische fumarole / fumerolle croicolitique / kroikolitische Fumarole / fumarola croicolítica

1385
eruptive fumarole, – a fumarole, where the gases escape explosively from an opening in a lava flow or from fractures in the volcano flank, dragging with them clots of lava. / eruptieve fumarole / fumerolle éruptive / eruptive Fumarole / fumarola eruptiva

roaring fumarole, – a fumarole where the gases escape vigorously, producing more than normal noise. / blazer / soffine / Blaser; brausende Fumarole / soplo; fumarola bufante

1386
flaming orifice, – a fumarole emitting burning gases, bluish when containing hydrogen or hydrogen sulphide, and yellow when containing sodium. / vlammende fumarole / fumerolle en flammes / flammende Fumarole; brennender Gasaustritt / fumarola llameante

1387
chlorine fumarole; dry ...., – a fumarole emitting anhydrous chlorides at temperatures above 330°C. / chloride fumarole / fumerolle d'acide chlorhydrique / Chloridfumarole; Salz.... / fumarola anhidra

1388
acid fumarole, – a fumarole emitting hydrochloric acid and sulphur dioxide at temperature above 80°C. / zure fumarole / fumerolle acide; .... sulfurique / saure Fumarole / fumarola ácida

alkalic fumarole, – a fumarole emitting mainly ammonium chloride and steam at temperatures above 80°C. / alkalische (salmiak) fumarole / fumerolle alcaline; .... ammoniacale / alkalische Salmiakfumarole / fumarola alcalina

1389
steam fumarole; .... vent, – a fumarole emitting water vapour that contains various volcanic gases. / stoomfumarole / soufflard / gasführende Wasserdampffumarole / fumarola de vapor

cold fumarole, – a fumarole emitting nearly pure water vapour at a temperature below 100°C. / koude fumarole; waterdamp.... / fumerolle aqueuse; .... froide / reine Wasserdampffumarole / fumarola fría

1390
fumarole mound; .... pimple, – a low, rounded natural hill on which a fumarole occurs. It may be formed by differential deflation through wind action and/or by accumulation from small eruptions. / fumarolenheuvel; ....puist/ pustule fumerollienne; verrue .... / Fumarolenhügel / montículo de fumarola

mud field, – an area where the soil is saturated with ground water due to the presence of fumaroles. / slijkveld / champ de boue / Schlammfeld / campo de barro

1391
solfatara, – a type of fumarole characterized by the emission of sulphurous gases (sulphur dioxide, sulphur trioxide, hydrogen sulphide), and associated with the approaching extinction of volcanic activity. / solfatare / solfatare; soufrière / Solfatare / solfatara

1392
solfatara field; solfataric area, – an areal grouping of solfataras occurring in rows on fissures or apparently in a random pattern. / solfatarenveld / champ de solfatares / Solfatarenfeld / campo de solfataras

1393
acid solfatara, – a solfatara emitting sulphur dioxide, sulphur trioxide, water vapour and carbon dioxide. /zure solfatare / solfatare acide / saure Solfatare / solfatara ácida

1394
sublimate; sublimate product, – a solid deposit, precipitated from a volcanic gas or vapour, such as sulphur, sal ammoniac, hematite. / sublimatie-product / produit de sublimation; sel sublimé; sublimé de sel / Sublimationsprodukt / producto de sublimación; sublimado

1395
mud volcano; macaluba; hervidero, –
1. an accumulation, usually forming a small flat cone, of mud and rock ejected by volcanic gases. Mud volcanoes are found in regions where volcanism is in a fumarolic stage;
2. similar accumulations may be formed by diapirism or by escaping petroliferous or other non-volcanic gases.

/ moddervulkaan; slijk....; slik.... / volcan de boue / Schlammvulkan; Salse; Makkalube; Maccaluba / volcán de barro; macaluba; hervidero

mud eruption, – / moddereruptie / éruption boueuse / Schlammausbruch / erupción de barro

1396
mud cone; puff cone, – a small cone of mud built around the opening of a mud volcano or mud geyser. Also, a similar accumulation on the surface of a mud flow or a lahar, produced by the escaping gases from beneath. / modderkegel / cône de boue / Schlammkegel / cono de barro

1397
mud lava, – a water-saturated muddy flow containing volcanic debris. / modderlava / lave boueuse; coulée de boue / Schlammlava / lava de barro

1398
mud pot; sulphur-mud pool, – a type of hot spring containing boiling mud, usually sulphurous and often multicoloured. / slijkpoel; modderwel; slikbron / source de boue sulfureuse; fumerolle boueuse / Schlammpfuhl; Schwefelschlammkrater / barrizal; cráter de barro

mud geyser, – an intermittent spring that erupts sulphurous mud; a type of mud volcano. / modderfontein / jet de boue / Schlammsprudel / geíser de barro

1399
sulphur mud, – a mixture of water and dark-coloured rock dust with yellow sulphur. / zwavelmodder / boue sulfureuse / Schwefelschlamm / barro de azufre

sulphur-mud flow, – a flow of sulphurous mud issued from a mud volcano or mud pool. / zwavelmodderstroom / coulée de boue sulfureuse / Schwefelschlammstrom / colada de barro sulfuroso

1400
sulphur-mud column; .... tree, – clots of sulphur mud that stick together and form a column projecting above the surface. Lateral openings in the column permit growth in various directions whereby capricious ramifications are formed. / zwavelmodderzuil; ....schoorsteen; ....boom / colonne de soufre; .... de boue sulfureuse / Schwefelschlammsäule; Schwefelschornstein / columna de barro sulfuroso

1401
gas seepage, – the more or less diffuse emanation of gas from the surface, i.e. in a dispersed manner and not from one or more distinct openings. The term pertains to both, volcanic and petroliferous phenomena. / doorsijpeling van gas / source de gaz diffuse / diffuse Gasförderung / escapes de gas

1402
mofette, – the exhalation of carbon dioxide with some oxygen and nitrogen, and occasionally with water vapour, usually in an area of late-stage volcanic activity. Also, the opening from which the gas is emitted. / mofette / mofette / Mofette / mofeta

death valley; poison ...., – a depression (crater floor or valley) filled with heavy toxic volcanic gases such as carbon dioxide, carbon monoxide, hydrogen sulphide and sulphur dioxide. / dodendal; stikvallei / vallée de la mort; gouffre de la mort / Totental / valle de la muerte

1403
secondary steam eruption, – an emission of steam formed by

the contact of water (rain, river, sea) with hot volcanic material. / lahareruptie; secundaire stoom.... / éruption secondaire / sekundäre Dampferuption / erupción secundaria de vapor

1404
geyser; pulsating spring, – named after Stora Geysir in Iceland, a type of intermittent hot spring that more or less regularly erupts a column of hot water and steam into the air, as a result of ground water coming into contact with rock or steam hot enough to create steam under conditions preventing free circulation. / geyser; intermitterend spuitende bron; pulserende bron / geyser; source jaillissante intermittente / Geysir; intermittierende Springquelle / géiser

geyser basin, – a depression that contains many springs, geysers and steaming fissures, fed by the same ground-water flow. / geyserbekken / bassin de geysers / Geysirbecken / cuenca de géiseres

1405
geyser jet, – the plume of hot water and steam emitted during the eruption of a geyser. / waterstraal van een geyser / jet de geyser / Wasserstrahl eines Geysirs / surtidor de géiser

geyser crater, – the bowl- or funnel-shaped opening of the geyser pipe that often contains a pool of water. / krater; ketel (van de geyser) / cratère de geyser / Geysirkrater; Quellbecken / cráter de géiser

geyser pipe; .... shaft, – the narrow tube or well of a geyser extending downward from the geyser pool. / toevoerbuis / cheminée de geyser / Geysirschlot; Quellrohr / chimenea de géiser

1406
geyserite; siliceous sinter, – incrustations of opaline silica deposited by precipitation from the waters of a geyser. / geyseriet; kiezelige sinter / geysérite / Geyserit; kieselsäurehaltiger Sinter / geiserita; sintersilíceo

siliceous geyser, – a geyser, the water of which contains silicic acid. / kiezelzuurhoudende geyser / geyser siliceux / kieselsäurehaltiger Geysir / géiser silíceo

1407
thermal spring; laug (plur. laugar) (Icelandic), – a pool of water or a spring whose temperature is higher than the local annual mean atmospheric temperature. It is not necessarily associated with volcanic activity. / thermale bron; therme / source thermale / Therme; Thermalquelle / fuente termal

boiling spring; hver (plur. hverar) (Icelandic), – a natural pool of hot water through which bubbles of steam or volcanic gas rise to the surface, often with much force. / kokende bron / source bouillante / kochende Quelle; heisser Sprudel / manantial hirviente

1408
hypogene spring, – a spring, sometimes pulsating, in which the water comes from a great depth, and having a distinct relation to volcanic phenomena. / hypogene bron / source hypogène / hypogene Quelle / manantial hipogénico

[1230] VOLCANIC LANDFORMS

[1231] *General Terms*

1409
positive volcanic form; protuberance, – structure formed due to an excess of supply of volcanic material in comparison to the withdrawal of material.

Distinguished are:
strictly volcanic forms in which the volume of ascending magma is equal to the volume of the volcanic structure above the surface:
shield volcano
lava dome
cinder cone
strato volcano
volcano-tectonic forms in which the volume of ascending magma is larger than the volume of the volcano above the surface:
updoming above a laccolith
regional updoming above a batholith
/ positieve vulkaanvorm; vulkanische verheffing / soulèvement volcanique / Vulkanbaute; Vulkanberg; positives vulkanisches Relief / forma volcánica positiva; levantamiento volcánico

1410
negative volcanic form; hollow, – structure formed due to a lack of supply of volcanic material in comparison to the withdrawal by eruptions, erosion or volcano-tectonic processes.

Distinguished are:
strictly volcanic forms:
explosion forms: more material is removed by the explosion than is supplied by the ascending magma,
crater
maar
collapse forms: the removal of magma from the chamber by an eruption, causing a caving-in of the roof,
caldera
volcano-tectonic forms:
rift structures
subsidence structures.
/ negatieve vulkaanvorm; vulkanische inzinking / affaissement volcanique / negative Vulkanform; vulkanische Depression; negatives vulkanisches Relief / forma volcánica negativa; hundimiento volcánico

1411
volcano, – a hill or mountain built up from the accumulation of volcanic products around a crater, i.e. the accumulation of lavas and/or pyroclastics. / vulkaan; vuurspuwende berg / volcan / Vulkan; feuerspeiender Berg / volcán

1412
crater, – a roughly circular depression that is usually located at the summit of a volcanic structure. It is formed by an explosive eruption or by the gradual accumulation of lava and/or pyroclastics around a vent. / krater / cratère; boucle volcanique / Krater; Sprengtrichter / cráter

crater basin, – / kraterbekken / bassin du cratère / Kraterbecken / cuenca del cráter

1413
volcanic edifice; .... pile; .... structure; .... mass, – a structure built by a volcano and rising above the earth's surface. / vulkaanlichaam; ....massief / édifice volcanique / Vulkangebäude; ....massiv / edificio volcánico

supravolcano, – volcanic body forming part of the landscape. / supravulkaan / volcan superficiel / aufgesetzter Vulkan / supravolcán; volcán superficial

1414
superstructure, – of the volcano, formed by successive eruptions and by intrusions within the accumulated material on top either of older volcanic or of non-volcanic rocks. / bovenbouw / édifice externe; suprastructure / Oberbau / edificio externo; superestructura

substratum of a volcano, – / onderbouw / substratum d'un volcan / Untergrund des Vulkans; Sockel des .... / estructura subvolcánica

1415
subaerial volcano, – / subaerische vulkaan / volcan subaérien / Festlandsvulkan / volcán subaéreo

subaquatic volcano, – / onderwatervulkaan / volcan subaquatique / Unterwasservulkan / volcán subacuático

1416
submarine volcano, – a volcano on the sea floor, usually erupting tholeiitic basalt. / onderzeese vulkaan / volcan sousmarin / untermeerischer Vulkan / volcán submarino

volcanic island; island volcano, – a submarine volcano that has been sufficiently built up to be exposed above the level of the sea. / vulkaaneiland / île volcanique / Vulkaninsel / isla volcánica

1417
crater island, – a submarine volcano, of which the crater rim only is exposed above sea level. / kratereiland / – – / Kraterinsel / isla de cráter

caldera island, – a submarine caldera, of which the rim only is exposed partly or entirely above sea level. / caldera-eiland / – – / Kalderainsel / isla de caldera

1418
bomb pit, – a small craterlet formed by the impact of volcanic bombs or blocks falling on sandy or ashy deposits on the slope or foot of the volcano. / inslagtrechter / trou d'impact / Einschlagtrichter / embudo de impacto

bomb sag, – depression and/or disturbance in stratified material produced by the impact or weight of a volcanic bomb and subsequent burial. / – – / – – / Eindellung / abolladura

1419
volcanic barrier lake, – a lake whose waters are impounded by the formation of a natural dam or barrier by lava or volcanoclastics. / vulkanisch stuwmeer / lac de barrage volcanique / vulkanisch aufgestauter See / embalse volcánico

[1232] *Volcanoes and Craters; Calderas*

1420
shield volcano; basaltic dome, – a low, broad volcano in the shape of a flattened dome built up almost entirely of very fluid, basaltic lava; its slopes seldom exceed 10° at the summit and about 2° at the base. / schildvulkaan / volcan en bouclier; cône de lave / Schildvulkan; Aspit(e) / volcán en escudo; .... de lava

dyngja, – the Icelandic term for a shield volcano of moderate dimensions.

1421
imbricated volcano, – a volcano built up of lava sheets overlapping each other like tiles of a roof, younger layers being intruded between older ones. / dakpansgewijs gelaagde vulkaan / volcan à structure imbriquée / Vulkanbau mit Schuppenstruktur / volcán de estructura imbricada.

1422
lava dome; volcanic dome; plug dome; **tholoid;** cumulo volcano; .... dome; lava cupula, – protrusion of highly viscous lava forming a steep mound above its vent. Its form is different whether it originates in a crater, on the flank of a volcano or isolated on a fault. The term should not be applied to shield volcanoes or tumuli. / lavadom; tholoïde; cumulovulkaan / mamelon; tapon; dôme éruptif; dôme volcanique; coupole; tholoide / Lavadom; Lavakuppel; (Quellkuppe; Staukuppe); Tholoid(e) / domo de lava; cúpula de lava

1423
lava dome with tongue-like offshoot, – highly viscous lava, poured out over an inclined surface. The lava first tends to build up a dome, but soon begins moving downward in a tongue-like offshoot, that may pass into a flow. / lavadom met tongvormige uitloper / dôme éruptif avec lobe de lave / Lavadom mit zungenförmigem Auswuchs; Staukuppenstrom / domo eruptivo con lóbulo de lava

1424
spine; obelisk; spire; monolith; belonite, – a pointed mass of viscous or solidified lava that sometimes protrudes from the throat of a volcano, as at Mount Pélée, where the spine was 300 m high. / (rots)naald / aiguille; monolithe / Lavadorn; Lavanadel; Belonit / aguja; pitón

1425
puy, – dome of lava that was highly viscous during extrusion; named after the Chaîne des Puys in central France. / puy / puy / Puy / puy

1426
ubehebe, – a low cone composed of pyroclastics and rock fragments from the underlying terrain; named after the Ubehebe craters of the northern end of Death Valley, California.

1427
lava cone, – an accumulation of viscous lava in the shape of a cone displaying concave flanks at the base. / lavakegel / cône de lave / Lavakegel / cono de lava

cinder cone; scoria ....; slag ...., – cone formed by the accumulation of cinders and other scoriaceous ejecta, usually as a result of explosive eruptions or lava fountains, and normally of andesitic or basaltic composition. The cone may be beautifully symmetrical, having rather steep slopes and a nearly circular ground plane. / slakkenkegel / cône de scories; .... de débris; garde / Schlackenkegel / cono de escorias

ash cone, – / askegel / cône de cendre / Aschenkegel / cono de ceniza

1428
tuff cone; tephra cone, – formed of fine-grained volcanic material, such as ashes and sand. / tufkegel / cône de tufs / Tuffkegel; Tuffvulkan / cono tobáceo

1429
stratovolcano; stratified volcano; bedded volcano; composite cone, – a volcano that is composed of alternating layers of lavas and pyroclasts. / stratovulkaan; gemengde vulkaan / stratovolcan; volcan mixte / Stratovulkan; Schicht....; gemischter ....; Vesuvtypus....; rheuklastischer .... / estratovolcán; volcán compuesto; cono ....

1430
pseudoaspite; pseudo shield volcano, – a stratovolcano mainly built up by lavas and having the shape of a shield volcano. / pseudoaspite / (pseudoaspite) / Pseudoaspite(e); schildförmiger Stratovulkan / (p)seudovolcán en escudo

1431
explosion crater; .... vent, – depression formed chiefly, or entirely, by explosion. / explosiekrater / cratère d'explosion / Explosionskrater / cráter de explosión

1432
explosion funnel, – the conical depression in, or forming, a crater. / explosietrechter / entonnoir / Kratertrichter / embudo de explosión

1433
maar; explosion lake, – a low-relief crater, less than 10 km in diameter and commonly filled with water. It is surrounded by an inconspicuous wall of ejectamenta consisting of blasted older rock and/or fresh volcanic material. Its type occurrence is in the Eifel, Germany. / maar / maar / Maar / maar

1434
central volcano, – a volcano with a crater in the central part of the structure. / vulkaan met centrale krater / volcan à cratère central / Vulkan mit zentralem Krater / volcán con cráter central

volcanic cone, – a cone-shaped hill or mountain of lava and/or pyroclastics that is built up around a central vent. / vulkaankegel / cône volcanique / Vulkankegel / cono volcánico

1435
normal cone, – a cone-shaped volcano with a relatively small crater. / kegelberg / cône / Kegelberg / cono

Distinction can be made into (German terms):
Konide, – a normal cone
Homate, – a relatively low cone and wide crater
Homakonide, – a small cone inside a greater crater
Konihomate, – a cone with a wide flat top

1436
summit crater, – situated at the top of a volcano. / topkrater / cratère terminal; .... sommital / Gipfelkrater / cráter de cumbre

1437
adventive cone; parasitic cone; lateral cone; minor cone; subsidiary cone; (subordinate cone), – a tuff or lava cone on the flank or foot of the major volcano. / adventiefkegel; parasitische kegel / cône adventif / Adventivkegel; parasitärer Kegel; Lateralkegel / cono adventicio; cono parasítico; cono lateral

1438
subterminal crater, – situated in the outer slope of the volcano but near the top. / subterminale krater / cratère subterminal / Subterminalkrater; Hang.... / cráter subterminal

1439
adventive crater; parasitic ....; lateral ....; subcrater, – a small crater on the flank of a larger volcano. / adventiefkrater; neven....; parasitaire .... / cratère adventif / Adventivkrater; Neben....; Seiten....; Nebenausbruchstelle; Lateralkrater; parasitärer Krater / cráter adventicio; .... parasítico; .... lateral

1440
bocca; aperture, – a small opening in any part of a volcano from which volcanic products escape. / bocca; explosiegat / bouche / Bocca; Ausbruchsöffnung / abertura; hoyo de explosión

1441
simple volcano, – a regularly built volcano with a single crater. / enkelvoudige vulkaan; eenassige ....; monoconische .... / cône simple / einfacher Vulkan; einachsiger .... / volcán simple

compound volcano; composite ...., – a volcanic structure having more than one cone. / meerassige vulkaan; samengestelde ....; polyconide / volcan complexe / mehrachsiger Vulkan; polyzentrischer .... / volcán compuesto

1442
twin volcano; double cone; twin dome, – two volcanic cones, situated near to each other, having the same base. / tweelingvulkaan; dubbel.... / volcans jumelés / Zwillingsvulkan; Doppel....; Geschwister.... / volcán maclado

twin crater; double ...., – two craters of more or less the same dimensions, situated very near to each other or intersecting. / tweelingkrater / cratères jumelés; cratère double / Doppelkrater; Zwillings.... / cráter maclado

1443
central cone, – cone within the main crater at the top of a volcano or in a caldera. / centrale kegel / cône intérieur; .... central / zentraler Kegel / cono central

1444
nested, – pertaining to cones, craters or calderas, in a complex volcanic structure with one or more younger volcanoes inside an older one.

1445
somma volcano; cone-in-cone structure, – a volcano with a large crater or caldera, containing a more or less centrally located younger cone; named after Monte Somma (with the cone of Vesuvius). / sommavulkaan; omwalde vulkaan; dubbelvulkaan / cône (double) emboité; sommavolcan / Sommavulkan; Vulkan mit Ringwall / volcán con cono en cono

cone-in-cone crater, – a large crater or caldera, containing one or more small craters or boccas at its bottom. / krater met centrale boccas / cratère à bouches multiples / Krater mit zentralen Bocchen / cráter de bocas múltiples

1446
solitary volcano, – a single and isolated volcanic edifice. / alleenstaande vulkaan; vrijstaande .... / volcan isolé / alleinstehender Vulkan / volcán aislado

1447
volcano belt; .... zone; ....girdle, – a distribution of volcanoes in definite zones or belts. / vulkaangordel; ....zone / zone volcanique / Vulkangürtel; ....zone / zona volcánica; cinturón volcánico; franja volcánica

1448
chain of volcanoes; row of ....; volcanic chain, – a more or less linear arrangement of volcanoes. / vulkaanrij; ....reeks / chaîne de volcans; volcans alignés / Vulkankette; ....reihe / cadena volcánica

1449
row of craters; chain of ...., – usually located along a fissure. / kraterrij / chaîne de cratéres; alignement de .... / Kraterreihe / hilera de cráteres

1450
volcanic cluster; cluster of volcanoes; .... of cones, – a group of volcanic vents in a restricted area and apparently in random arrangement without identifiable structural control. / opeenhoping van vulkanen / groupe de volcans / Vulkangruppe / grupo volcánico

volcanic shield cluster, – a group of volcanic shield structures as on Hawaii. / schildvulkaangroep / groupe de volcans en bouclier / Schildvulkangruppe; ....gebirge / grupo de volcanes en escudo

1451
crater ring, – a rim of fragmental material surrounding the crater. / kraterwal / enceinte du cratère; mur; cloison; corniche circulaire / Kraterwall; ....ring / borde del cráter

1452
rampart, – a ring-like or crescentic ridge of pyroclastics at the top part of a volcano. / ringwal / rempart / Ringwall / muro anular

spatter rampart, – / slakkenwal / rempart de scories / Schlackenwall / muro de salpicaduras

1453
wall, – of the crater, the inner side, inclined towards the bottom. / kraterwand / paroi / Kraterwand / pared del cráter

moat; fosse, – a valley-like depression at the inner side of a crater, between the wall and a lava dome. / ringdal / fossé / Ringtal / fosa

1454
crater lake, – an accumulation of rain and/or ground water in a crater. / kratermeer / lac de cratère / Kratersee / ibón de cráter

1455
pit crater; inbreak ....; sink, – circular or elliptical collapse structure with steep walls, partly filled with lava and without accumulated volcanic material surrounding it (Hawaii). / instortingskrater; put....; smeltschacht / cuve d'effondrement; cratère ....; cratère citerne / Einbruchskrater; Einsturz....; Pit.... / cráter en hoya

1456
caldera; volcanic sink, – a large volcanic depression, more or less circular in form and with a diameter many times that of the included vents. / caldera; caldeira / caldeira / Kaldera; Caldera / caldera

1457
simple caldera, – formed by one paroxysmal outburst only. / enkelvoudige caldera / caldeira simple / einfache Kaldera / caldera sencilla

caldera complex, – a group of associated calderas with a discrete magmatic assemblage. / calderacomplex / ensemble de caldeiras / Kalderakomplex / complejo de calderas

1458
explosion caldera, – formed chiefly, or entirely, by explosive eruptions that have removed large masses of rock. / explosiecaldera / caldeira d'explosion / Explosionskaldera / caldera de explosión

phreatic caldera, – an explosion crater of caldera size, formed by superheating of near-surface water-rich sediments and volcanic ash by a magma. No liquid magma is erupted. / caldera gevormd door een freatische uitbarsting / caldeira formée par une éruption phréatique / phreatisch gebildete Kaldera / caldera freática

1459
collapse caldera; sunken ...., – formed primarily by subsidence of the roof of a magma chamber due to the removal of pyroclastics or by subterranean withdrawal of magma. / instortingscaldera / caldeira d'effondrement / Einbruchskaldera; Einsturz.... / caldera de desplome

conca (caldera), – a conchoidal collapse structure, bounded by gradually sloping walls, due to repeated volcanic activity in the region. / conca / conca / Conca / conca

1460
summit caldera, – a caldera, formed at the top of a large volcanic complex. / topcaldera / caldeira sommitale / Gipfelkaldera / caldera en cima

1461
somma, – a more or less complete, circular or crescentic

ridge, surrounding a crater or caldera. / somma / somma / Somma / somma

caldera ring, – a ridge surrounding a caldera with a very steep inner wall. / calderawal / rempart / Kalderawall; Ringgebirge / anillo de caldera

atrio, – a crescent-shaped valley between central cone and caldera ring or somma. / atrio / enclos; atrio / Atrio; interkolliner Raum; hufeisenförmiges Krater-Zwischental / atrio

1462
sandsea, – a flat plain in a caldera formed by ashes and other pyroclasts, and washed by rain. / zandzee / plaine de sable; .... de cendres / Sandsee / mar de arena

[1233] *Rifts and Subsidence*

1463
cauldron; caldron; volcanic basin, – a subsided area, due to volcanic activity, regardless of shape, size or depth. / inzinking; keteldal / chaudron / Kessel / caldero; olla

1464
external collapse; .... subsidence, – a collapse area on the flank of a volcano as the result of a paroxysmal effusion of lava. / hellingverzakking / effondrement latéral / Hangversenkung; Einbruch am Aussenhang / desplome lateral

volcanic sector graben, – a collapse area on the flank of a volcano with walls converging towards the top. / vulkaansectorslenk / secteur de subsidence volcanique / Sektorgraben (im Vulkanmantel) / sector de hundimiento volcánico

1465
(volcanic) fissure, – a fracture in a volcano or in a lava dome, traceable in the field as a deep and nearly straight cleft, usually no more than a few metres in width, and that may extend for many kilometres. / spleet / fissure; fente; crevasse / Spalte / fisura; grieta

1466
(volcanic) rent, – a violent spreading of the crater floor or of the flank of the volcano, caused by the pressure of outbreaking lava. / eruptiespleet; vulkaan....; scheur; breuk / fente / Riss; Eruptionsspalte / grieta

1467
rift zone, – a zone marked at the surface by open fissures and rows of cinder and spatter cones, and associated with underlying dike complexes (Hawaii). / breukzone / zone de failles / Bruchzone / zona de rift

1468
flank fissure, – a cleft in the side of a volcano. / vulkaanmantelspleet / fissure de flanc / Vulkanmantelspalte / fisura lateral

radial rift; .... fissure; radiating crack, – a cleft extending from the centre of the volcano or the volcanic complex. / radiale spleet / fissure radiale; fente .... / radiale Spalte / fisura radial; rift ....

1469
concentric rift; .... fissure, – a cleft concentric with the rim of either crater or caldera. / concentrische spleet / fissure concentrique; fente .... / konzentrische Spalte; Ringspalte / fisura concéntrica; rift concéntrico

1470
transverse rift, – a cleft that strikes obliquely to any other structural trend in a volcanic complex. / dwarslopende spleet / fente transversale; fissure excentrique / Querspalte; Transversal....; querverlaufende Spalte / fisura transversal; rift ....

[1234] *Erosional Features*

1471
breached crater; .... cone; horseshoe-shaped crater, – a crater or cone, the rim of which has been broken through by erosion, by lava overflow eroding the weak scoria-built wall, or by a volcano-tectonic disturbance. / hoefijzervormige krater / cratère égueulé / hufeisenförmiger Kraterwall; Halbmondkrater / cráter en herradura; .... fracturado

notch; cleft, – a breached opening in the crater rim of a volcano. / bres; inkeping / brèche; échancrure / Bresche; Einschnitt; Scharte / brecha

1472
intercrateral downfall; internal avalanche, – a slump of parts of the crater wall. / afstorting in de krater / avalanche intercratérique / Absturz im Krater / derrumbe intercrateral

1473
erosion caldera, – developed from any kind of caldera through erosional processes, sometimes resulting in a very large depression. / erosiecaldera / caldeira d'érosion; cirque .... / Erosionskaldera; ....kessel / caldera de erosión

1474
furrowed volcano, – the first stage of destruction by erosion, the flanks being cut by valleys that do not yet effect the shape of the volcano. / gegroefde vulkaan / volcan raviné / radial gefurchter Vulkanberg / volcán surcado

barranco; barranca; furrow, – a deep, steep-sided drainage valley on the slope of a volcano resulting from erosion, e.g. by mudflows, and/or from coalescence of smaller channels. / barranco / barranco; barranca / Barranco; Radialschlucht / barranca

1475
cleaved volcano; dissected ...., – representing the second stage of destruction by erosion, deep and wide valleys having destroyed the regular form of the volcano and its crater rim. / gekloofde vulkaan; gekerfde .... / volcan disséqué; .... démantelé / Radialschluchtenberg / volcán encañado

1476
volcanic skeleton; .... wreck, – the last stage of destruction by erosion, the cone or dome shape being destroyed. An arrangement of radial ridges or a centrifugal pattern of con-

sequent valleys may survive. / vulkaanruïne / ruine volcanique; vestige de volcan / Vulkanruine / esqueleto volcánico; volcán en ruinas

1477
neck; volcanic plug; .... butte; lava plug, – a mass of solidified lava or fragmental igneous rock filling the conduit or vent of an extinct or inactive volcano; the term is usually applied to the erosional remnant after removal of the surrounding rock mass. / neck; eruptiefzuil / neck; diatrème; suc / Halskuppe; Schlotpfropfen / cuello

[1300] STRUCTURAL GEOLOGY

[1310] ATTITUDE OF STRATA

1478
structural geology, – the branch of geology that deals with the ways rocks respond to the application of deforming stress and with the structures that result from deformation. Generally the term is used for features of microscopic, mesoscopic and macroscopic size, but usually not exceeding that of the geological map. There is a certain overlap with the term tectonics. / structurele geologie / géologie structurale / strukturelle Geologie / geología estructural

NOTE:
– for terms on stress, see section:
[2922] General Aspects of Material Properties

1479
microtectonics, – a comprehensive term used for structures on the scale of hand specimen or thin section. The term is somewhat ambiguous; the use of **microstructure** is to be preferred. / microtektoniek; microstructuur / microtectonique; microstructure / Mikrotektonik; Klein.... / microtectónica

NOTE:
– see also section:
[1060] Structural Petrology

1480
structural position, – / tektonische ligging / disposition structurale / tektonische Lage / posición estructural

attitude, – a genaral term to describe the relation of some directional feature in a rock to a horizontal plane. / stand / allure / Anordnung; Schicht(en)stellung / disposición

1481
upturned strata, – / opgerichte lagen / couches redressées / aufgerichtete Schichten / capas levantadas

1482
tilted strata, – / gekantelde lagen / couches basculées / gekippte Schichten; schräggestellte .... / capas basculadas

inclined strata; dipping beds, – / hellende lagen / couches inclinées geneigte Schichten / capas inclinadas

inclined bedding, – / hellende gelaagdheid / stratification inclinée / geneigte Schichtung / estratificación inclinada

1483
strike; strike line, – the line formed by the intersection of a planar feature with a horizontal plane. / strekking / direction / Streichen / dirección

direction of the strike; – bearing of the ....; azimuth of the ...., – / strekkingsrichting; azimuth van de strekking / sens de la direction; azimut de la .... / Streichrichtung / rumbo

1484
dip; true dip, – the angle which a planar feature makes with the horizontal plane, as measured in a plane normal to the strike. / helling / pendage; inclinaison / Fallen; Einfallen / inclinación; buzamiento

apparent dip, – the angle in a vertical plane that a planar feature makes with the horizontal plane as seen in a plane not perpendicular to the strike. / schijnbare helling / pendage apparent / scheinbares Einfallen / inclinación aparente

1485
angle of dip; – dip angle; angle of inclination, – / hellingshoek / angle de pendage; angle d'inclinaison / Fallwinkel; Einfallswinkel / ángulo de buzamiento

1486
original dip; primary dip, – the dip of beds immediately after deposition. / oorspronkelijke helling; primaire .... / inclinaison originelle / ursprüngliche Neigung; primäre .... / inclinación primitiva

1487
dip flattens out; slackening of the dip, – / afnemen van de helling; geringer worden .... / l'inclinaison devient faible; .... s'adoucit; .... s'atténue / Neigung wird schwächer; .... verflacht sich / atenuación del buzamiento

1488
gradient of the bed, – the rate of regular descent of a bed. / gradiënt van de laag / – – / Gradient der Schicht / gradiente de la capa

1489
down-dip, – / hellingafwaarts; in de richting van de helling / en aval-pendage / in der Richtung des Einfallens; in der Fallrichtung / en el sentido de la inclinación

up-dip, – / hellingopwaarts; in de richting tegengesteld aan de helling / en amont pendage / entgegen dem Einfallen; entgegengesetzt der Fallrichtung / en sentido contrario a la inclinación

1490
direction of dip, – azimuth at right angles to the strike in the direction of descent of the dipping planar feature. / hellingsrichting / azimut du pendage / Richtung des Einfallens; Fallrichtung / azimut del buzamiento

1491
overturned; inverted, – / overhellend; overkipt / renversé; retourné / überkippt / invertido

1492
trace, – the intersection of a geological surface or planar fabric element with a given reference plane or surface. If the element is linear, the trace is its orthogonal projection upon the reference plane. / spoor / trace / Spur / traza

1493
trend, – the general azimuth of the trace (usually on a horizontal or topographic surface) of a planar or linear feature. / richting / direction générale / Spurrichtung / dirección

[1320] FRACTURING

[1321] *Joints*

1494
fracture, – general term for a surface along which loss of cohesion has taken place. / breuk / fracture / Bruch / fractura

fracturing, – / breukvorming / fracturation / Bruchbildung / fracturación

fractured, – / met breuken doorsneden / fracturé / zerlegt durch Brüche; zerbrochen / fracturado

1495
joint, – fracture along which there has been little or no movement. / diaklaas / diaclase / Kluft / diaclasa

jointed, – / gekliefd; doorkliefd; gespleten / diaclasé / zerklüftet; ge.... / diaclasado

jointing; jointage, – / klieving; splijting / – – / Klüftung; Zerklüftung; Klüftigkeit; Kluftbildung / diaclasamiento

1496
joint surface; .... face, – / diaklaasvlak; splijtings.... / plan de diaclase / Kluftfläche / plano de diaclasa

1497
spacing, – of joints. / spatiëring / espacement / Maschenweite; Kluftabstand / espaciado (de diaclasas)

1498
joint set, – a group of more or less parallel joints. / groep diaklazen; groepering van .... / groupe de diaclases parallèles / Kluftschar / grupo de diaclasas

joint system, – a system consisting of two or more joint sets, or any group of joints with a characteristic pattern. / diaklaassysteem; ....netwerk / système de diaclases; réseau de .... / Kluftsystem; ....netz / sistema de diaclasas

1499
conjugate joint system; system of conjugate joints, – a system of joints consisting of two sets that are symmetrically disposed about some structural feature or about an inferred stress axis. / systeem van conjugate diaklazen / système de diaclases conjuguées / System sich kreuzender Kluftscharen / sistema de diaclasas conjugadas

1500
slab jointing; sheet ...., – a structure produced in rock by the formation of numerous closely spaced parallel joints, the rock being thereby divided into slabs. / plaatvormige klieving / débit en plaques; .... en dalles / Absonderung in plattigen Kluftkörpern / división en placas

1501
columnar jointing; prismatic ...., – in basalts, etc. as a result of contraction. / zuilvormige klieving / débit columnaire; .... colonnaire; .... prismatique / säulenförmige Absonderung; prismatische .... / disyunción columnar

columnar structure; prismatic ...., – / zuilstructuur / structure prismée / Säulentextur / estructura columnar

1502
rhomboidal jointing, – / romboëdervormige klieving / débit rhomboédrique / Absonderung in rhomboidalen Kluftkörpern / disyunción romboédrica

1503
tension joint, – joints produced by tensional forces. / rekdiaklaas / diaclase d'extension / Zerrkluft / diaclasa de tensión

extension joint; .... fracture, – a joint or fracture formed by separation of a body across a surface normal to the direction of least principal stress. / – – / – – / – – / diaclasa (de extensión)

1504
shear joint, – joint caused by failure in shear. / schuifdiaklaas / diaclase de cisaillement / Scherkluft / diaclasa de cizallamiento

1505
feather joint; pinnate shear ...., – pertaining to closely-spaced joints arranged in a way similar to the barbs of a feather, sometimes appearing along faults. / – – / diaclase satellite de cisaillement / Fiederkluft / diaclasa plumosa de cizallamiento

1506
release joint, – joint perpendicular to a former compression direction and produced by expansion of rock upon release of load. / ontspanningsdiaklaas / diaclase de décompression / Entlastungskluft / diaclasa de distensión

strike joint, – a joint that strikes parallel to the strike of strata or layering. / langsdiaklaas / diaclase directionelle / streichende Kluft / diaclasa direccional; .... según la dirección

dip joint; cross ...., – a joint that strikes approximately perpendicular to the strike of bedding or layering. / dwarsdiaklaas / diaclase croisée / Querkluft / diaclasa cruzada

1507
longitudinal joint, – a joint that strikes parallel to the trend of the fold axis. / longitudinale diaklaas; overlangse .... / diacla-

se longitudinale / Längskluft; L-Kluft; bc-.... / diaclasa longitudinal

transverse joint, – a joint that strikes perpendicular to the trend of the fold axis. / transversale diaklaas / diaclase transversale / Transversalkluft; ac-Kluft; Q-.... / diaclasa transversal

oblique joint; diagonal ...., – a joint that strikes oblique to the strike of the associated strata or to the trend of the fold axis. / diagonale diaklaas / diaclase diagonale; .... oblique / Diagonalkluft; D-Kluft; hko-.... / diaclasa oblicua; .... diagonal

1508
joint pattern, – / diaklaaspatroon / dessin des diaclases / Kluftmuster; .... netz / patrón de diaclasas

1509
main joint; master ....; major ...., – a more persistent member of a joint set. / hoofddiaklaas / diaclase principale / Haupkluft / diaclasa principal

minor joint; – secondary...., – secundaire diaklaas / diaclase secondaire / untergeordnete Kluft / diaclasa secundaria

1510
systematic joint, – pertaining to joints that are planar and parallel, or sub-parallel, so that they form sets. / – – / – – / regelmässige Kluftschaar / diaclasa sistemática

non- systematic joint, – irregular, curved or conchoidal joints. / – – / – – / unregelmässige Kluft / diaclasa no sistemáticas

1511
plumose structure; .... markings, – fine structure resembling a feather defined by differential relief on a joint face. / pluimstructuur / – – / federförmige Struktur / estructura plumosa

[1322] *Fissures*

1512
fissure; gash, – a fracture with a component of displacement normal to the fracture surface. / spleet; (open scheur) / fissure; fente / Spalte / fisura

1513
joint fissure, – / diaklaasspleet / fissure / Spalte; Kluftspalte / fisura

open joint, – / ongevulde diaklaas; open .... / fissure vide / klaffende Spalte; offene .... / fisura abierta

1514
fissure vein; infilled fissure, – a fissure filled with mineral matter different from the walls. / gang; gevulde spleet / filon; (fissure filonienne); (fente ....) / Gang; Spaltengang; Gangspalte; Lagergang / filón fisural

1515
vein filling; fissure ...., – minerals without economic value or occasionally ore minerals. / gangvulling; spleet.... / remplissage filonien; (.... de fissure) / Gangfüllung; Spalten.... / releno filoniano; .... de fisura

1516
dyke; dike, – a discordant wall-like fissure filling. / gesteentegang / dyke / Gesteinsgang; Eruptivgang / dique

clastic dyke, – sedimentary dyke consisting of a variety of broken rocks derived from underlying or overlying material. / klastische gang / filon clastique / sedimentärer Gang / dique clástico

1517
cooling fissure, – / afkoelingsspleet / fissure de refroidissement / Abkühlungskluft / fisura de enfriamiento

1518
desiccation fissure, – / uitdrogingsspleet / fente de dessiccation / Trockenriss / grieta; fisura de desecación

1519
tension fissure, – / rekspleet / fissure d'extension / Zerrspalte; Spannungsrisz; Dehnungsriss / fisura de tensión

tension gash, – a filled tension joint often lensoid in shape. / – – / diaclase de tension rempli / Dehnungsfuge / – –

[1330] FAULTS

[1331] *General Terms*

1520
fault tectonics, – / breuktektoniek / tectonique de failles; style tectonique à ....; tectonique cassante / Verwerfungstektonik; Bruch.... / tectónica de fallas

1521
fault, – a fracture surface along which appreciable displacement has taken place. / breuk / faille / Verwerfung; Verschiebung; Sprung; Bruch / falla

1522
dislocation (obsolete), – displacement of rocks by movement along a fracture or fault. / dislocatie / dislocation / Dislokation; Verwerfung; Störung / dislocación

1523
displaced strata, – / verplaatste lagen / couches disloquées / versetzte Schichten; verworfene .... / capas dislocadas

offsetting of strata, – by faulting. / verspringen van lagen / décalage de couches / Absetzen von Schichten / desviación ortogonal

1524
lateral extent; extent in lateral direction; length, – of a fault. / laterale uitgestrektheid; lengte in de strekking / extension latérale / streichende Erstreckung; seitliche .... / extensión lateral

1525
profound fault; deep-reaching ...., – / diep-doordringende breuk / faille profonde / tiefgreifende Verwerfung / falla profunda

shallow fault; near-surface ...., – / oppervlakkige breuk / faille superficielle / flachgründige Verwerfung; oberflächennahe .... / falla superficial

1526
vertical extent; downward ....; depth, – of a fault. / uitgestrektheid naar de diepte / extension en profondeur / Tiefenerstreckung; Teufen.... / extensión en profundidad

to die out downward; to terminate at depth; to end with ...., – / verdwijnen naar de diepte / disparaître en profondeur; s'amortir en .... / nach unten aufhören; .... unten enden; gegen die Teufe hin .... /desaparecer en profundidad; terminar en ....

to die out upward, – / verdwijnen naar boven / disparaître en haut; s'amortir vers la surface / nach oben aufhören; nach oben enden; zum Hangenden hin .... / desaparecer hacia arriba; terminar hacia ....

1527
active fault, – / werkende breuk / faille vivante / lebende Verwerfung; aktive .... / falla activa

dead fault, – / uitgewerkte breuk / faille morte / inaktive Verwerfung / falla inactiva; .... muerta

rejuvenated fault; revived ....; renewed ....,– / herleefde breuk / faille ayant rejoué; .... rajeunie / wiederbelebte Verwerfung; wiederaufgelebter Bruch / falla rejuvenecida

1528
fault surface; .... plane, – a surface or plane along which faulting has taken place. / breukvlak / plan de faille / Verwerfungsfläche; ....ebene; Störungsfläche / superficie de falla

1529
wall; face, – of a fault. / breukwand / éponte; paroi / Wand; Nebengestein / pared

hanging wall, – the rock mass above a fault wall. / bovenliggende breukwand / paroi supérieure / Hangendes / labio levantado; .... techo

footwall, – the rock mass at the lower side of a fault wall. / benedenliggende breukwand; onderliggende .... / paroi inférieure / Liegendes / labio hundido; .... muro; .... yacente

1530
wall rock; country ...., – / wandgesteente; neven.... / roche encaissante; .... bordière / Nebengestein; Salband / roca encajante

1531
fault block, – rock mass of which the opposite boundaries are fault planes. / breukschol / bloc faillé /Bruchscholle / bloque fallado

1532
fault zone; zone of faulting, – / breukzone / zone de faille; .... broyée; .... de broyage / Störungszone; Bruch....; Verwerfungs.... / zona de falla

1533
shear zone, – a zone in which shearing has occurred, resulting in crushing and brecciation or ductile deformation of the rock. / schuifzone / zone de cisaillement / Scherzone / zona de cizallamiento

1534
closed fault, – a fault in which the two walls are in contact. / gesloten breuk / faille fermée / geschlossene Verwerfung / falla cerrada

open fault, – a fault in which the two walls are separated. / open breuk / faille ouverte; .... disjonctive / klaffende Verwerfung / falla abierta

1535
plane fault, – a fault whose surface is planar. / vlakke breuk / faille plane; .... plate / glatte Verwerfung; ebene .... / falla plana

curved fault; arcuate ...., – a fault whose surface is curved or arcuate. / gebogen breuk / faille courbée / gebogene Verwerfung / falla arqueada; .... plegada

1536
trailed fault, – an earlier fault that is deflected by a younger fault. / meegesleepte breuk / faille décalée / abgelenkte Störung / falla deflectada; .... desviada

fault deflection; .... deviation, – / afbuiging van een breuk / déviation d'une faille / Verwerfungsablenkung / desviación de la falla

1537
drag; dragging; flexing, – of the beds by a fault. / sleuring / retroussement; rebroussement / Schleppung / arrastre

1538
slickenside, – usually striated surface of rock produced by friction. / wrijfspiegel / miroir de faille / Harnischfläche; Rutsch....; Gleit.... / espejo de falla

polished surface; polish, – / spiegel / miroir / Harnisch; Spiegel / superficie pulimentada

stria (plur. striae); striation, – / streep / strie / Rutschstreife; Strieme; Schramme / estría

scratch, – / kras / rayure / Rutschschramme; Friktionsstreife / raya

groove; furrow, – / groeve / cannelure / Furche / ranura; acanaladura

scale, – / schub / écaille / Ruschel; Schuppe / escama

step, – / trede / ressant; échelon / Stufe / escalón; grada

1539
fault breccia; tectonic ...., – broken rock, filling a fault zone, that results from friction between the walls of a fault. / breukbreccie / brèche de faille; .... de friction / Verwerfungsbreccie; tektonische Breccie / brecha de falla

friction breccia, — / wrijvingsbreccie / brèche de friction / Reibungsbreccie / brecha de fricción

crush breccia, — / vergruizingsbreccie / brèche de broyage / Reibungsbreccie / brecha de trituración

1540
fault gouge; clay ...., — pulverized country rock resulting from friction between the walls of a fault. / breukklei / argile de frottement; .... de friction; .... de trituration / Störungsletten / pulverización por falla

1541
mylonite, — strongly deformed, fine-grained, laminated rock, usually occurring in fault zones. / myloniet / mylonite / Mylonit / milonita

ultramylonite; flinty crush rock, — strongly deformed, completely structureless rock. / ultramyloniet / ultramylonite / Ultramylonit / ultramilonita

blastomylonite, — a recrystallized mylonite. / blastomyloniet / blastomylonite / Blastomylonit / blastomilonita

phyllonite, — rock resembling phyllite formed by strong deformation in fault zones. / phylloniet / phyllonite / Phylonit / filonita

NOTE:
- for further details, see:
  [1020] Metamorphic Processes and Products

1542
pseudotachylite, — a glassy rock formed by partial fusion of a mylonite. / pseudotachyliet / pseudotachylite / Pseudotachylit / (p)seudotaquilita

1543
hartschiefer, — a strongly banded and partly schistose rock, resulting from intense dynamic metamorphism, and associated with other rocks of mylonitic habit. The alternating bands have been formed from ultramylonite by recrystallization and metamorphic differentiation. / hartschiefer / hartschiefer / Hartschiefer; Schiefermylonit / esquisto duro

[1332] *Orientation and Geometry of Faults*

1544
strike fault, — a fault of which the trend coincides approximately with the strike of the associated strata or layering. / strekkingsbreuk / fail le directionnelle / streichende Verwerfung / falla direccional

1545
oblique fault; diagonal ...., — a fault of which the trend is oblique to the strike of the associated strata or layering. / diagonale breuk / faille oblique; .... diagonale / diagonale Verwerfung; Diagonalverwerfung / falla oblicua

cross fault; dip ...., — a fault whose trend is at a right angle to the strike of the associated strata or layering. / dwarsbreuk / faille perpendiculaire; .... orthogonale / Querverwerfung / falla perpendicular; .... ortogonal

1546
longitudinal fault, — a fault whose strike is parallel with the general structure. / longitudinale breuk / faille longitudinale / streichende Verwerfung / falla longitudinal

transverse fault, — a fault whose strike is transverse to the general structure. / transversale breuk / faille transversale / Transversalverwerfung; ....störung / falla transversal

1547
bedding fault, — a fault along or parallel to a bedding plane. / breuk in het laagvlak / faille dans le plan des couches / schichtenparallele Verwerfung; .... Störung / falla paralela al plano de estratificación

1548
hade, — the angle made by a fault plane with the vertical, i.e. the complement of the dip. / afwijking uit het lood / écart avec la verticale / Fallwinkel einer Verwerfung / complemento del buzamiento

1549
steep(—dipping) fault; high-angle ...., — fault dipping at an angle of 45 degrees or more, i.e. the angle with the horizontal. / steilstaande breuk / faille à pente raide / steile Verwerfung; .... Störung / falla muy inclinada

low-dipping fault; low-angle ...., — fault dipping at an angle of less than 45 degrees. / zwakhellende breuk / faille faiblement inclinée / schwach geneigte Verwerfung; flache .... / falla poco inclinada

1550
dip separation, — the distance between two parts of a disrupted bedding plane, measured in the fault plane parallel to its inclination. / — — / rejet incliné / Versatz an einer Verwerfung / separación de inclinación

1551
throw; vertical separation; normal ...., — the vertical component of the dip separation. / verticale spronghoogte / rejet vertical / vertikale Sprunghöhe / rechazo vertical

heave, — the horizontal component of the dip separation.
1. for normal faults. / horizontale gaping / rejet horizontal / Horizontalversatz / rechazo horizontal
2. for reverse faults. / horizontale verdubbeling; .... bedekking / recouvrement horizontal / horizontale Deckung; .... Schubweite / rechazo horizontal; solapado ....

1552
stratigraphic throw; .... separation, — the distance between the two displaced parts of a faulted bedding plane, measured perpendicular to that bedding plane. / stratigrafische spronghoogte / rejet stratigraphique; .... perpendiculaire aux couches / stratigraphische Sprunghöhe / rechazo vertical estratigráfico

stratigraphic heave; .... gap, —
1. for normal faults, the width of the gap between two parts of a disrupted bed, measured in the direction of the faulted bedding plane. / stratigrafische gaping / rejet parallèle aux couches / stratigraphischer Horizontalversatz / rechazo horizontal estratigráfico

stratigraphic heave; .... overlap, –
2. for reverse faults, the width of the overlap between two parts of a disrupted bed, measured in the direction of the faulted bedding plane. / stratigrafische verdubbeling; .... bedekking / recouvrement stratigraphique / stratigraphische Deckung; .... Schubweite / solapado estratigráfico

1553
apparent heave, –
1. for normal faults, the gap between the horizontal projections of the fault traces, measured perpendicular to the strike of the disrupted bedding plane. / schijnbare gaping / rejet horizontal apparent / scheinbare horizontale Sprungweite / rechazo horizontal aparente
2. for reverse faults. / schijnbare verdubbeling; .... bedekking / recouvrement horizontal apparent / scheinbare horizontale Deckung; .... horizontale Schubweite / solapado aparente

1554
strike separation, – the distance measured in the strike of the fault between the two outcropping parts of the faulted planar feature. / – – / rejet en direction / streichender Versatz / separación de rumbo

1555
offset; normal horizontal separation, – the distance between two outcropping parts of a faulted bedding plane, measured in a direction normal to the strike of that bedding plane. / verzet; kortste dagzoomafstand / rejet transversal; .... horizontal normal / Horizontaldistanz / separación normal; .... ortogonal

1556
gap (of the outcrop of the bed), – a term applied to normal faults for the component of the strike separation measured in the strike plane. / dagzoomhiaat / rejet longitudinal / – – / gap

overlap (of the outcrop of the bed), – a term applied to reverse faults for the component of the strike separation measured parallel to the strike of the displaced strata. / dagzoomverdubbeling / recouvrement longitudinal / – – / solapado

1557
omission of beds; elimination of ....; cutting out of ...., – the beds being faulted out. / uitsnijden van lagen / supression des couches / Schichtenunterdrückung; Ausfall von Schichten / omisión de capas; .... estratos

repetition of beds, – / herhaling van lagen / répétition de couches / Schichtenwiederholung / repetición de capas; .... de estratos

[1333] *Displacement along Faults*

1558
normal fault, – a fault in which the hanging wall has moved downward relative to the footwall. / afschuiving / faille normale; .... directe / Abschiebung / falla normal

reverse fault, – a fault in which the footwall has moved downward relative to the hanging wall. / opschuiving / faille inverse; .... anormale / Aufschiebung; Wechsel / falla inversa

1559
downthrow; downthrown side; .... block, – / afgeschoven vleugel; .... blok / pan effondré; compartiment ....; lèvre affaissée; .... abaissée; côté affaissé / abgesunkene Scholle / labio hundido

faulted down; down-faulted, – / afgeschoven / déplacé en bas; rejeté en profondeur / abgeschoben / desplazado en profundidad

upthrow; upthrown side; .... block, – / opgeschoven vleugel; .... blok / côté relevé; lèvre relevée lèvre soulevée; / gehobene Scholle / labio levantado

1560
strike-slip fault; transcurrent ....; tear ...., – a fault in which the net slip is practically in the direction of the fault strike. / horizontaalverschuiving / faille à rejet horizontal; faille à rejet en direction; faille de décrochement / Blattverschiebung; Horizontal.... / falla horizontal de desgarze

wrench fault, – a vertical or almost vertical strike-slip fault. / – – / faille verticale de décrochement / steile Blattverschiebung / falla vertical de desgarze

transform fault, – a strike-slip fault, first recognized by J. Tuzo Wilson in mid-oceanic ridges, along which ridge segments are offset. / transformatiebreuk / faille transformante; .... de raccord(ement); .... de transformation / Transform-Störung / falla transformante

1561
sinistral fault; left lateral ...., – a strike-slip fault in which the net slip is such that to an observer looking towards the fault the opposite block appears to have moved to the left. / – – / faille sénestre / linkshändige Blattverschiebung; .... Transversalstörung / falla sinistral

dextral fault; right lateral ...., – a strike-slip fault where the opposite block appears to have moved to the right. / – – / faille dextre / rechtshändige Blattverschiebung; .... Transversalstörung / falla dextral

1562
net slip, – the shortest distance measured in the fault plane betweeen two formerly adjacent points. / ware verplaatsing / rejet net; déplacement véritable / wahre Sprunghöhe / desplazamiento neto

direction of the net slip, – / richting van de ware verplaatsing / sens de rejet net / wahre Verschiebungsrichtung; wahre Bewegungsrichtung / dirección del desplazamiento neto

1563
vertical slip; .... component, – the vertical component of the net slip; this is the same as the vertical component of the dip slip. / verticale verplaatsing; .... component / composante verticale (du rejet) / vertikale Komponente / componente vertical (del desplazamiento neto)

1564
dip slip, – the component of the slip in the direction of the true

dip of the fault plane. / hellende verplaatsing / rejet incliné / desplazamiento de inclinación

1565
strike slip, – the horizontal component of the slip parallel to the fault strike. / horizontale verplaatsing / – – / – – / desplazamiento de rumbo

1566
translational fault, – a fault with translational movement. / translatiebreuk / faille à déplacement rectiligne / Translationsstörung / falla de traslación

translational movement, – movement in such a way that straight lines on opposite sides of the fault that were parallel before faulting remain parallel after faulting. / translatie / – – / Translation / traslación

1567
rotational fault; hinge ...., – a fault whose displacement has a rotational component around a pivotal axis perpendicular or parallel to the fault plane. / draaibreuk / faille rotatoire / Verwerfung mit Rotation / falla rotacional

curvilinear fault, – a fault with a rotational displacement in the fault plane. / breuk met kromlijnige bewegingsrichting / faille à déplacement curviligne / Kippschalenverwerfung / falla curvilínea

[1334] *Overthrusts*

1568
overthrust, – a reverse fault with a low-angle thrust plane and large net slip. / overschuiving / chevauchement / Überschiebung / cabalgamiento; cobijadura; corrimiento

distance of thrust, – / overschuivingsafstand / distance de charriage / Förderweite; Schub.... / recorrido; distancia de cabalgamiento; .... de corrimiento

overthrust plane; thrust ...., – / overschuivingsvlak / surface de charriage; .... de chevauchement / Überschiebungsfläche; Schub.... / plano de cabalgamiento; .... de corrimiento

1569
overthrust block; .... mass, – the block above the overthrust plane. / overschuivende schol / block charrié; masse charriée; .... chevauchante / überschiebende Scholle / bloque corrido

stationary block, – the block of relatively undeformed rocks beneath the overthrust plane. / overschoven blok / masse chevauchée / überschobene Scholle / masa cabalgada

1570
drag fold, – minor folds produced by shearing stresses set up by the overthrust movement. / sleepplooi / pli d'entraînement / Schleppfalte / pliegue de arrastre

1571
imbricate structure; shingle-block ...., – a structure formed by a repetition of overthrust wedges being arranged so as to overlap like tiles. / schubstructuur / structure en écailles; régime imbriqué / Schuppenstruktur / estructura imbricada

imbrication, – / schubvorming / écaillage / Verschuppung / imbricación

wedge; shingle, – / schub / écaille / Schuppe / escama

1572
basal thrust plane; sole, – the basal plane underlying a pile of overthrusts. / basaal overschuivingsvlak / plan basal des chevauchements / basale Schubfläche / zócalo; (plano basal de los corrimientos)

1573
listric surface, – moderately inclined overthrust plane concavely flexed towards the surface. / listrisch vlak / surface listrique / listrische Fläche; Schaufelfläche / superficie lístrica

1574
underthrust, – the fault plane along which a rock mass has been thrusted under another one. / onderschuiving / sous-charriage / Unterschiebung / bajocorrimiento

1575
(overthrust) nappe; .... sheet, – an overthrust mass that has been displaced over a great distance. / (overschuivings)dekblad / nappe de chevauchement; .... de charriage / Überschiebungsdecke / manto de cabalgamiento; corrimiento arrastre

1576
overthrust fold; fold overthrust, – a recumbent fold of which the reversed middle limb has been considerably sheared out as a result of the great horizontal translation. / plooioverschuiving / pli-faille couché; pli écaille; faille d'étirement; charriage du premier genre / überschobene Falte / pliegue cobijante; .... recumbente

Distinguished are:

upper limb, – / bovenvleugel / flanc supérieur; série chevauchante / Hangendschenkel / flanco superior; miembro ....; (serie cabalgante)

middle limb, – / middenvleugel / flanc médian / Mittelschenkel / miembro mediano

lower limb; floor limb, – / ondervleugel / flanc inférieur; série chevauchée / Liegendschenkel / flanco inferior; miembro ....; (serie cabalgada)

1577
break thrust, – overthrust along a fault plane cutting through the crest of a fold. / – – / pli-faille / Scheitelüberschiebung / pliegue-falla

[1335] *Fault Systems*

1578
fault system; framework of faults, – a system of parallel or nearly parallel faults. / breuksysteem / système de failles / Verwerfungssystem / sistema de fallas

network of faults, – a not too irregular system of faults striking in different directions. / netwerk van breuken / réseau de failles; failles en réseau / Verwerfungsnetz / red de fallas

1579
intersecting faults, – / snijdende breuken / failles croisées / sich kreuzende Verwerfungen / fallas cruzadas

1580
major fault; main ....; dominant ...., – / hoofdbreuk / faille principale; .... majeure; .... maîtresse / Hauptverwerfung / falla principal

minor fault; secondary ...., – / secundaire breuk / faille secondaire / sekundäre Verwerfung / falla secundaria

1581
auxiliary fault; subsidiary ....; associated ...., – / nevenbreuk / faille accessoire; .... d'accompagnement; .... satellite / Nebenverwerfung; Begleit....; Fiederstörung / falla auxiliar

1582
branching fault; ramifying ...., – / vertakte breuk / faille ramifiée / verzweigte Verwerfung / falla ramificada

branch; offshoot, – / aftakking / branche / Abzweigung / rama

1583
splay faulting, – minor faulting that diverges from a longer fault at an acute angle. / vertakking / – – / Fiederstörung / divergencia de fallas

splay, – divergent small fault at the extremities of large normal faults, especially rifts. / vertakking / – – / divergierende Störung / falla en lisel; chaflán

1584
fault set; group of parallel faults, – / groep van evenwijdige breuken / système de failles parallèles; faisceau de failles / Verwerfungsschar / grupo de fallas paralelas

conjugate fault system, – a system of faults consisting of two sets that are symmetrically disposed about some other structural feature or about an inferred stress axis. / conjugaat breuksysteem / système de failles conjuguées / System sich kreuzender Verwerfungen / sistema de fallas conjugadas

1585
en échelon faults, – / en échelon breuken / failles en échelons; relais de failles / gestaffelte Verwerfungen / fallas en escalonamiento; .... en relevo

1586
block-faulted area, – / schollengebied / région compartimentée en blocs; .... désintégrée en blocs / Bruchschollengebiet / área fallada en bloques

1587
step faults, – closely associated parallel faults with the same direction of displacement. / trapbreuken / failles à gradins; .... en escalier; .... étagées / Staffelverwerfungen; ....störungen / fallas en gradería

1588
synthetic fault, – auxiliary fault that dips in the same direction as the major fault. / synthetische breuk / faille synthétique / synthetische Verwerfung / falla homotética; .... sintética

antithetic fault, – auxiliary fault that dips against the dip of the major fault. / antithetische breuk / faille en retour; .... antithétique / antithetische Verwerfung / falla antitética

1589
horst; fault ridge; upthrown block, – a block, structurally in a higher position than the surrounding area and bounded by normal faults. / horst / horst; bloc surélevé; bloc soulevé; môle tectonique / Horst; Hochscholle / meseta tectónica; pilar tectónico; horst

1590
graben; fault trough; .... trench; downthrown block, – a relatively depressed fault block lying between normal faults with roughly parallel strikes. / slenk / fossé (d'effondrement) / Graben; Tiefscholle / fosa de hundimiento; .... tectónica; (graben)

rift valley; rift, – a depression or valley in a graben. / rift / – – / Graben; ....senke; Rift / valle tectónico; rift

1591
fault basin, – a region depressed relative to the surrounding regions and separated from them by faults. / breukbekken / cuvette d'effondrement; bassin .... / Kesselbruch; tektonisches Becken / cuenca de hundimiento tectónico

1592
peripheral fault, – fault along the periphery of a structurally elevated or depressed region. / perifere breuk / faille périphérique / Randstörung / falla periférica

1593
boundary fault; marginal ....; border ...., – / randbreuk / faille bordière; .... marginale; .... limite / Randverwerfung; Grenz.... / falla marginal

1594
radial faults; radiating ...., – a group of faults that radiate from a common centre. / radiale breuken; radiaire .... / failles radiales; .... rayonnantes / radiale Verwerfungen; Radialverwerfungen / fallas radiales

1595
fracture zone, – a fault or fault zone in the ocean floor. / breukzone / zone de fracture / Bruchzone / zona de fractura

[1340] CLEAVAGE, SCHISTOSITY AND FOLIATION

1596
cleavage; rock cleavage, – tendency of rock to break along parallel planes yielding smooth surfaces. / druksplijting / clivage / Schieferung / clivaje; crucero; exfoliación

cleavable, – / splijtbaar / clivable / spaltbar / exfoliable

cleavability, – / splijtbaarheid / clivabilité / Spaltbarkeit / exfoliabilidad

1597
fissility, – the property of rocks to split along roughly planar and approximately parallel surfaces. / deelbaarheid / fissilité / Teilbarkeit / fisibilidad

fissile, – tending to split along the bedding or cleavage planes. / deelbaar / fissile / teilbar / físil

1598
cleavage plane, – / splijtvlak / plan de clivage / Schieferungsebene; ....fläche / plano de exfoliación

1599
shaly cleavage; .... parting, – cleavage parallel to the bedding planes. / schaliesplijting / délit schisteux / Schieferung in Ton / exfoliación paralela

slaty cleavage; flow ...., – quarrymen's term, denoting the ability to split in thin laminae, due to the parallel arrangement of small platy minerals, often phyllosilicates and quartz. / leisplijting / clivage ardoisier / Dachschieferung / exfoliación pizarrosa; pizarrosidad

1600
crenulation cleavage; strain-slip ....; shear ....; herringbone ....; false ...., – closely spaced cleavage associated with limbs of microfolds that usually represent crenulations of an older cleavage. / – – / schistosité de crénelure / Runzelschieferung / exfoliación de crenulación; clivaje de ....

1601
fracture cleavage, – spaced cleavage in deformed but only slightly metamorphosed competent rocks that are usually poor in platy minerals. / breuksplijting / schistosité de fracture; (clivage cassant) / Bruchschieferung / exfoliación por fractura

1602
cleavage fan, – a family of diverging cleavage planes. / (splijtvlak)waaier / – – / Schieferungsfächer / abanico de planos de exfoliación

divergent cleavage fan, – cleavage planes diverge towards the core of the fold.

convergent cleavage fan, – cleavage planes converge towards the core of the fold.

1603
schistosity; schistose structure, – foliation in schist or coarse-grained crystalline rock due to parallel arrangement of mineral grains. / schistositeit; schiststructuur / schistosité cristallophyllienne / Schieferung (in Glimmerschiefer und Gneis) / esquistosidad

schistose, – / schisteus / schisteux / schieferig / esquistoso

linear schistosity, – a schist possessing also a lineair element. / stengelige schistositeit / schistosité linéaire / Griffelschieferung / esquistosidad lineal

NOTE:
– for further details, see
[1030] Texture and Structure (of metamorphic rocks)

1604
foliation, – a fabric surface of unspecified origin in metamorphic rocks. / foliatie / foliation / Kristallisationsschieferung; Schieferung / foliación; hojosidad

foliated rock, – said of a rock with a foliation. / bladerig gesteente / roche à structure feuilletée; .... à structure foliacée / kristalliner Schiefer / roca foliácea; .... hojosa

1605
tectonite, – a rock that owes its fabric to tectonic deformation. / tektoniet / tectonite / Tektonit / tectonita

fabric, – a rock fabric is the manner of mutual arrangement in space of the components of a rock body and of the boundaries between these components. / maaksel / pétrofabrique / Gefüge / textura; fábrica

1606
differentiated layering, – a planar structure consisting of layers of different mineralogical composition, that is produced by metamorphic differentiation, often parallel to the axial planes of the associated folds. / – – / litage par différenciation / – – / diferenciación metamórfica estratificada

[1350] LINEATIONS AND LINEAR STRUCTURES

1607
lineation, – includes all linear structures in rocks without regard to origin. / lineatie / linéation / Lineation / lineación

linear structure, – / lineaire structuur / structure d'étirement / lineares Gefüge; Lineartextur / estructura linear

1608
rodding structure, – a thin rod-shaped monomineralic family of linear bodies in folded metamorphic rocks. / – – / – – / Mullionsstruktur / estructura calamiforme

mullion, – a linear structure occurring on a surface or on opposed surfaces of a layer. / – – / meneau / Mullionsstruktur / – –

1609
boudinage, – a structure produced by the stretching of a competent layer, resulting in a series of block-like bodies that are often elongate in one direction. In cross section they show a sausage shape, and commonly exudation of quartz or calcite occurs between the individual boudins. / boudinage / boudinage /Boudinage / budinaje

pinch-and-swell structure, – is related to boudinage, but the layer has not been separated into individual boudins. / – – / structure à renflements et étranglements / An- und Abschwellen / budinaje imperfecto

1610
pencil cleavage, – a property of certain metamorphic rocks to break into long pencil-like fragments as a result of two intersecting cleavages. /griffelsplijting / bébit en bâtons de craie; clivage ardoisier / Griffelschieferung / disyunción en lápices

1611
intersection lineation, – is defined by the intersection of planar features. / – – / linéation d'intersection / – – / alineación de intersección

1612
crenulation lineation, – is defined by the axes of crenulation (small-scale) folds. / – – / linéation de crénelure / Kleinfältelungsachsen; Runzelungsachsen / alineación de crenulación

1613
stretching lineation, – is due to the stretching of rock particles, e.g. pebbles. / – – / linéation d'étirement / Streckungslineation / lineación de estiramiento

flattening, – / – – / aplatissement / Plättungslineation / aplastamiento

1614
plunge, – the vertical angle between any linear feature and the horizontal. / onderduiking / ennoyage sous; disparition sous / Abtauchen / inmersión

pitch, – the angle between a linear feature, lying in some specified plane, and the horizontal, measured in the specified plane. / – – / – – / Fallen; Einfallen; Abtauchen / sesgo

[1360] FOLDS

[1361] *General Terms*

1615
fold tectonics, – / plooiingstektoniek / tectonique de plissement / Faltentektonik; Faltungs.... / tectónica de plegamiento

1616
fold, – a bend or flexure produced in rock. / plooi / pli / Falte / pliegue

flexure, – usally synonymous with fold; sometimes used to indicate a double bend without reversal of dip. / flexuur / flexure / Flexur / flexura

1617
monocline, – local steepening of an otherwise uniformly and gently dipping or horizontal sequence of strata. / monoclinaal; monocline / monoclinal / Monoklinale; Monokline / monoclinal

1618
antiform, – a fold in which the beds dip in opposite directions from the crest, i.ie. from the axial plane. / antiform / – – / Antiform / antiforma

anticline; anticlinal fold, – a fold with a core of stratigraphically older rocks. / anticlinaal; anticline / anticlinal; pli .... / Antiklinale; Antikline; Sattel / anticlinal

1619
synform, – a fold in which the beds dip from opposite directions toward a trough, i.e. toward the axial plane. / synform / – – / Synform / sinforma

syncline; synclinal fold, – a fold with a core of stratigraphically younger rocks. / synclinaal; syncline / synclinal; pli .... / Synklinale; Synkline; Mulde / sinclinal

1620
antiformal syncline; pseudoanticline, – antiform with a core of younger rocks. / pseudo-anticlinaal / faux anticlinal / Pseudosattel; Pseudo-Antiklinale; Antiform-Mulde / sinclinal antiforme

1621
synformal anticline; pseudosyncline, – synform with a core of older rocks. / pseudosynclinaal / faux synclinal / Pseudomulde; Pseudo-Synklinale; Synform-Sattel / anticlinal sinforme

1622
face, – folds are said to face in the direction of the stratigraphically younger rocks along their axial surface. / – – / – – / – – / encararse

facing, – / – – / vergence / – – / encarado

1623
bedding-plane slip, – slip of beds along the bedding planes.

1624
competent, – describes the property of rocks that are relatively more resistant to deformation than adjacent rocks, which are then considered incompetent. / competent / compétent / kompetent / competente

incompetent, – / incompetent / incompétent / inkompetent / incompetente

[1362] *Fold Elements and Geometry*

1625
limb, – the less curved sections of the folded surface. / flank / flanc; aile; côté / Flanke; Schenkel; Flügel / flanco

1626
inter-limb angle, – the angle between the tangents to the

limbs at the inflexion points. / plooiingshoek / angle du pli / Öffnungswinkel / ángulo del pliegue

Descriptive terms:

gentle, – > 120° / zwak / modéré / weit / suave

open, – 70° – 120° / open; matig / ouvert / offen / abierto

close, – 30° – 70° / dicht / fermé / geschlossen; eng / cerrado

tight, – < 30° / zeer dicht / serré / sehr eng / estrecho

isoclinal, – pertaining to a fold with parallel or almost parallel limbs. / isoclinaal / isoclinal / isoklinal / isoclinal

1627
inflection point; inflexion ...., – point where, seen in section, the limbt is intersected by its tangent, and coincides with the reversal in curvature. / inflectiepunt / point d'inflexion / Wendepunkt / punto de inflexión

inflection line; inflexion ...., – locus of reversal of curvature. / inflectielijn / ligne d'inflexion / Wendeachse / línea de inflexión

1628
core, – of an anticline or syncline. / kern / noyau; coeur / Kern / núcleo; corazón

detached core, – / afgeknepen kern / noyau étranglé / abgequetschter Kern / núcleo desprendido; .... separado

1629
hinge point, – point of maximum curvature of a folded surface, as seen in profile. / scharnierpunt / point charnière / Scheitel; Scharnier / punto de charnela

hinge line, – locus of maximum curvature in the folded surface. / scharnierlijn / ligne charniére / Scheitellinie; Scharnier / línea de charnela

hinge zone, – part of the folded surface that shows the sharpest curvature. / ombuiging / zone charnière / Scheitelbereich; Scharnierzone / zona de charnela

1630
fold axis, – any line parallel to the hinge line in a cyclindrical fold. / plooi-as / axe de pli / Faltenachse / eje del pliegue

1631
axial plane; .... surface, – the surface that contains all the hinge lines of a fold. / assenvlak / plan axial; surface axiale / Achsenebene / superficie axial ....; plano ....

axial surface trace, – the intersection of the axial surface with the topographic surface or any other specified surface. / spoor van het assenvlak / trace du plan axial / Spur der Achsenebene / traza de la superficie axial

1632
crestal surface, – surface formed by joining the crest lines of all folded surfaces in an antiform. / kruinvlak / plan de crête / Faltenspiegel; Sattel.... / superficie de cresta

crest line, – the line following the highest part of the fold. / kruinlijn / crête anticlinale / Scheitellinie / línea de cresta

1633
crest point, – the highest point of the fold seen in profile. / kruin / crête; toit / Scheitelpunkt / punto de cresta

1634
amplitude, – the vertical distance between inflection and crest lines, as seen in section. / amplitude / amplitude / Amplitude / amplitud

1635
culmination; apex; crest-line apex; .... culmination, – the highest point on the crest line. / culminatie; apex / sommet / Kulmination; Apex; Scheitel / culminación; vértice

saddle, – the depression between culminations along the crest of an anticline. / zadel; depressie / ensellement; abaissement axial; zone d'ennoyage; dépression / Achsendepression; Quereinsattelung / depresión; plegamiento en silla de montar

1636
trough surface, – the surface that joins the trough lines of all folded surfaces in a synform. / trogvlak / enveloppe des arêtes synclinales / Muldenspiegel / plano del seno

trough line, – the line following the lowest parts of a folded surface. / troglijn / arête synclinale / Muldenlinie / seno

1637
plunging fold, – a fold whose hinge line is not horizontal. / duikende plooi / pli plongeant / abtauchende Falte / pliegue buzante

1638
doubly plunging anticline, – an anticline of which the hinge line plunges in opposite directions away from a culmination. / tweezijdig duikende anticlinaal / anticlinal à double plongement / beidseitig abtauchender Sattel / anticlinal doblemente buzante

doubly plunging syncline, – a syncline of which the hingle line plunges in opposite directions towards a depression. / tweezijdig duikende synclinaal / synclinal à double plongement / konvergent eintauchende Mulde / sinclinal doblemente buzante

1639
structural bulge; .... nose; .... promontory, – the termination of a plunging anticline without individual closure. / anticlinale neus / saillant anticlinal / antiklinale Nase / promontorio estructural

1640
brachyanticline, – a doubly plunging anticline of which the ratio of width to length is greater than 1:2 and less than 1:5. / brachyanticlinaal / brachyanticlinal; (coupole ellipsoïdale) / Brachyantikline / braquianticlinal

brachysyncline, – a doubly plunging syncline of which the ratio of width to length is greater than 1:2 and less than 1:5. / brachysynclinaal / brachysynclinal / Brachysynkline / braquisinclinal

1641
dome; periclinal structure, – anticlinal structure, of which the

ratio width to length is approximately 1:1. / koepel / dôme; (coupole) / Dom; Kuppel / domo; cúpula

basin, – synclinal structure, of which the ratio width to length is approximately 1:1. / schoteltrog / cuvette / Trog; Becken / cuenca

[1363] *Types of Folds*

1642
major fold; main fold, – / hoofdplooi / pli principal; pli majeur / Hauptfalte / pliegue principal

minor fold; small fold. / secundaire plooi / pli annexe; pli secondaire; pli mineur / Nebenfalte; untergeordnete Falte / pliegue secundario

1643
crenulation; plication; puckering, – small-scale fold. / rimpel / plissotement; crénelure / Runzelung; Fältelung / crenulación

microfold, – a fold visible on the scale of hand specimen or thin section. /microplooi / micropli / Kleinfalte / micropliegue

1644
symmetrical fold, – a fold with identical limbs. / symmetrische plooi / pli symétrique / symmetrische Falte / pliegue simétrico

asymmetric fold; unsymmetric fold, – / asymmetrische plooi / pli dissymétrique; pli asymétrique / asymmetrische Falte; unsymmetrische .... / pliegue asimétrico

1645
upright fold, – a fold whose axial plane is vertical or very steep. / rechte plooi / pli droit / aufrechte Falte; stehende .... / pliegue simétrico

1646
inclined fold, – a fold of which the axial plane is inclined. / hellende plooi / pli déjeté / vergente Falte; schiefe ....; geneigte .... / pliegue inclinado

1647
overturned fold; overfold, – a fold in which one limb has passed the vertical, so that the beds in that limb are inverted in relation to their depositional position. / overhellende plooi; / pli renversé; pli déversé / überkippte Falte / pliegue invertido

Terms relating to overturned folds:

normal limb, – the limb in which the beds are not inverted. / normale flank / flanc normal / normaler Schenkel / flanco normal

inverted limb; reversed limb, oversteepened limb, – the overturned limb of a fold. / abnormale flank; omgekeerde vleugel / flanc renversé; .... inverse / überkippter Mittelschenkel; inverser .... / flanco invertido

1648
knee fold, – a fold whose limbs are at a nearly right angle to each other. / knieplooi / pli en genou / Kniefalte / pliegue en rodilla

1649
box fold, – a double-hinged fold with subrectangular shape in profile. / kofferplooi / pli coffré / Kofferfalte / pliegue en cofre

1650
fan(-shaped) fold, – a fold with limbs converging toward the core. / paddenstoelplooi / pli en éventail / Fächerfalte / pliegue en abanico

1651
chevron fold; zigzag fold; concertina fold, – fold with narrow hinge zone and straight limbs. / zigzagplooi / pli en chevron / Zickzackfalte; Harmonika.... / pliegue en zigzag; .... cabrío

1652
parallel fold, – a fold, in which each bed maintains the same thickness throughout the fold. / afstandsgetrouwe plooi; evenwijdige ....; parallelle .... / pli parallèle / parallele Falte / pliegue paralelo

concentric fold, – parallel fold, in which the various surfaces have the same centre of curvature. / concentrische plooi / pli concentrique / konzentrische Falte / pliegue concéntrico

1653
similar fold, – a type of fold, in which each successively lower folded surface shows the same geometrical form as the surface above. Similar folding implies thinning on the limbs of the folds and thickening at the hinge. / vormgetrouwe plooi; congruente .... / pli semblable; pli similaire / kongruente Falte / pliegue similar

1654
cylindrical fold, – a folded surface that can be generated by a straight line, moved parallel to itself. Sections perpendicular to the hinge line yield identical profiles. / cylindrische plooi / pli cylindrique / zylindrische Falte / pliegue cilíndrico

non-cylindrical fold, – a fold, of which profiles perpendicular to the hinge line are not identical. / niet-cylindrische plooi / pli non-cylindrique / nicht-zylindrische Falte / pliegue nocilíndrico

1655
conical fold, – a non-cylindrical fold that can be generated by a straight line revolving around a fixed point. / kegelvormige plooi / pli conique / kegelförmige Falte / pliegue cónico

1656
reclined fold; neutral fold, – a fold whose hinge line is parallel to the dip direction of the axial surface, and that closes sideways.

1657
polyclinal fold, – a fold with more than one axial surface. / – – / pli polyclinal / trikline Falte / pliegue policlinal

1658
ptygmatic fold, – closely folded veins or dykes, commonly found in migmatites; the inter-limb angle is often less than 0°, i.e. limbs and tangents intersect at opposite sides of the median surface. / ptygmatische plooi / pli ptygmatique / ptygmatische Falte / pliegue ptigmático

1659
rootless fold, – an isolated fold hinge, usually in strongly foliated rocks. / – – / – – / allochthone Falte / pliegue desenraizado

1660
intrafolial fold, – a fold that is restricted to a narrow zone parallel to its axial surface and usually occurring in strongly foliated rocks. / – – / – – / – – / pliegue intrafoliar

1661
kink, – an open, asymmetric, sharply hinged, straight-limbed fold of small dimensions, occurring in thinly laminated rocks or in crystals. / knikplooi / – – / Zickzackfalte; Harmonika.... / – –

kink band, – the short limb of a kink fold. / knikzone / – – / Knickzone; ....band / – –

1662
shear fold, – fold apparently formed by shear in a direction parallel to the fold's axial plane. / schuifplooi / pli de cisaillement / Scherfalte / pliegue de cizallamiento; .... de estiramiento

1663
disharmonic folding, – folding in which individual folds change in form or magnitude when traced along the axial surface. / disharmonische plooiing / plissement disharmonique; .... dysharmonique / disharmonische Faltung / pliegue disarmónico

décollement; detachment; shearing off, – the disruption and slip of one rock mass over another one. / wegschuiving / décollement; (déclenchement) / Abscherung / despegadura

1664
supratenuous fold, – a major fold in which the beds thicken towards the syncline. / – – / – – / Beule; Kompaktionsbeule / pliegue supratenuo

draping, – a term applied to a form of anticlinal flexure in strata laid down over some positive relief feature, and of which the individual beds, due to differential compaction, are thinner on the axis than on the limbs. / – – / – – / Drapierfalte; ....struktur / – –

1665
buckling; buckle fold; flexural slip; .... flow, – the wave-like deflection of a rock layer resulting from a compressive stress acting along its length. It is a mechanism by which folding can develop. / – – / flambement; flambage / Biegegleitfaltung / pliegues en bucle

1666
incipient folding, – / plooiing in wording; beginnende plooiing / plissement embryonnaire / Embryonalfaltung / plegamiento incipiente; .... embrional

incipient stage of growth, – of a fold. / eerste aanleg van een plooi / amorce du pli; embryon-pli / Anfangsstadium der Faltung / estadio incipiente de desarrollo; esbozo de ....; estadio embrional de un pliegue

1667
precursory folding, – / vóórplooiing / plissement précurseur / Vorfaltung / plegamiento precursor

penecontemporaneous folding; – progressive folding during deposition. / synsedimentaire plooiing / plissement contemporain de la sédimentation / synsedimentäre Faltung / plegamiento penecontemporáneo

1668
refolded fold, – / meervoudige plooi; geplooide .... / pli replié; pli replissé; pli reployé / wiedergefaltete Falte / pliegue replegado

super(im)posed folding; multiple ....; double ....; refolding, – / herplooiing / replissement; reploiement / wiederholte Faltung; überprägende .... / plegamiento sobreimpuesto

1669
folding phase, –
1. in the sense of H. Stille – Faltungsphase –, a short period of fold formation, evidenced by an unconformity, e.g. Sudetic folding plase, Pyrenean ....;
2. a term used to describe the relative timing of fold formation in multiply folded rocks, where dating of folding events is not possible.

/ plooiingsfase / phase de plissement / Faltungsphase / fase de plegamiento

[1364] *Fold Systems*

1670
fold system, – a group of folds showing common characteristics and trends, and presumably being of common origin. / plooisysteem / système de plis; groupement de plis / Faltensystem / sistema de pliegues

1671
anticlinorium, – a composite major anticline consisting of several subsidiary folds. The enveloping surface of the subsidiary folds is antiformal in shape. / anticlinorium / anticlinorium / Antiklinorium / anticlinorio

synclinorium, – a composite major syncline consisting of several subsidiary folds. The enveloping surface of the subsidiary folds is synformal in shape. / synclinorium / synclinorium / Synklinorium / sinclinorio

1672
enveloping surface, – a surface, tangential to a folded layer. / omhullend oppervlak / (surface) enveloppe / Faltenspiegel / superficie de envolvimiento

1673
median surface, – surface passing through the inflection lines

of adjacent folds. / mediaan oppervlak / surface médiane / Mittelebene / superficie mediana

1674
fold bundle, – / plooibundel / faisceau de plis / Faltenstrang; ....schar; ....bündel / fajo de pliegues

fold belt; folded zone, – / plooiingszone / zone de plissement; ceinture plissée; / Faltungszone; Faltengürtel / zona de plegamiento; faja plegada

1675
virgation, – sheaf-like divergence of folds in an orogenic belt. / virgatie; uitwaaiering / virgation / Virgation / virgación

1676
en échelon folds, – / plooien en échelon / plis en échelons; plis en coulisses; plis à relais / gestaffelte Falten; versetzte .... / pliegues en relevo

1677
parasitic fold, – small fold on a larger fold and sharing the geometrical elements. / parasitaire plooi / pli parasite / parasitäre Falte; Adventiv-Falte / pliegue parásito

1678
cross fold, – fold with its axis perpendicular to, or oblique to, the main folds. / dwarsplooi / pli transverse / Querfalte; Faltendurchkreuzung / pliegue cruzado

[1365] *Nappe Tectonics*

1679
nappe tectonics, – / dekbladentektoniek / tectonique en nappes / Deckentektonik; ....bau / tectónica de mantos de corrimiento

1680
recumbent fold, – a fold with a more or less horizontal axial plane. / liggende plooi / pli couché / liegende Falte / pliegue recumbente

Terminology concerning recumbent folds:

normal limb, – / bovenvleugel / flanc normal; .... supérieur / Hangend-Schenkel / flanco normal; .... superior

reversed limb; inverted ...., – / omgekeerde vleugel / flanc renversé; .... inverse / inverser Mittelschenkel / flanco invertido

stretched-out middle limb; drawn-out middle ...., – / uitgerekte middenvleugel / flanc médian étiré / ausgedünnter Mittelschenkel / parte central estirada

1681
piled-up recumbent folds; packet of recumbent .... / stapel(ing) van liggende plooien; pakket van liggende .... / plis couchés empilés; (paquet de plis superposés) / Paket liegender Falten; Stapel liegender .... / pliegues recumbentes apilados

1682
fold nappe, – recumbent fold with great amplitude. / plooidekblad / pli-nappe; nappe de recouvrement / Deckfalte; Überfaltungsdecke; Faltendecke / manto de recubrimiento

1683
nappe, – a rock mass of large dimensions that has been displaced over other rocks by means of a thrust or large-scale recumbent folding. / dekblad / nappe / Decke; Überschiebungsmasse; ....decke; Schubmasse; Deckschale; Schubbrett; Gleit.... / manto

1684
root zone, – region of origin of one or more nappes. / wortelgebied / zone de racines; région des .... / Wurzelzone; Deckenwurzel; Narbenzone / zona de raices; región de ....

1685
allochthonous, – large rock masses that have been displaced over considerable horizontal distances. / allochthoon / allochtone / allochthon / alóctono

autochthonous, – large rock masses that are more or less in place. / autochthoon / autochtone / autochthon / autóctono

1686
front; head; edge, – the most forward part of a nappe. / front; kop / tête; front / Deckenstirn / fronte

frontal region; front zone, – / frontzone / région des plis frontaux / Stirnzone; ....region / región frontal

1687
frontal lobe; .... pucker; leading edge, – anticlinal hinge in the frontal zone of a nappe. / frontscharnier / charnière frontale / Stirnscharnier; ....einrollung / charnela frontal

digitation, – an offshoot from the frontal lobe. / vertakking / digitation / Stirnlappe; ....schuppe; Deckenaufspaltung; Digitation / digitación

1688
plunging nappe, – a nappe, of which the front plunges downward. / duikend dekblad / nappe plongeante / Tauchdecke / manto buzante

1689
backfolding; backward ...., – a type of folding in which the folds are overturned in a direction opposite to the main displacement of the nappes. / terugplooiing / plissement en retour; .... à rebours / Rückfaltung; Rücküberkippung / retroplegamiento

back fold; backward fold, – / – – / pli en retour; pli à rebours / Rückfalte / retropliegue

1690
involution, – the refolding together of two nappes, or the wrapping of an older nappe around a younger one. / inkapseling / encapuchonnement / Einwicklung / involución

1691
pile of nappes; packet of superimposed ...., – / dekbladensta-

pel(ing) / pile de nappes / Deckenpaket; ....stapel; ....gruppe; ....system / paqueto de mantos; pila de ....; paqueto de mantos superpuestos; pila de mantos ....

overlapping nappes, – / overgrijpende dekbladen / nappes chevauchantes / übergreifende Decken; überschiebende .... / mantos cabalgantes

1692
klippe, – the erosional remnant of a nappe. / klippe / – – / Klippe / – –

[1370] DIAPIRISM

1693
diapirism, – the piercement of relatively light and mobile rock material through or into an overlying sequence. Diapirism may involve rock salt and associated evaporites, subcompacted clays and magmas. / diapirisme / diapirisme / Diapirismus / diapirismo

diapir; diapiric structure; pircement ...., – / diapier; diapiere structuur / (pli) diapir; pli à noyau de percement / Diapir / diapiro

1694
salt tectonics, – the structural deformation of salt formations and adjacent rocks. / zouttektoniek / tectonique salifère / Salztektonik; Halokinese; Halotektonik / tectónica salina

halokinesis, – according to F. Trusheim, the proposer of the term, a collective term for all processes connected causally with the autonomous, isostatic movement of salt. / halokinese / halocinèse / Halokinese / halocinesis

halokinetic structure, – a structure due to halokinesis; its type depends on thickness of the salt formation and its depth. / halokinetische structuur / structure halocinétique / halokinetische Struktur / estructura halocinética

halotectonic structure, – a salt structure that originated predominantly as a result of compressive tectonic forces. / halotektonische structuur / structure halotectonique / halotektonische Struktur / estructura halotectónica

1695
salt dome; salt pillow, – a non-piercement halokinetic structure of the shallower and thinner parts of an evaporitic sequence; it possesses a rather flat base and arched top. / zoutkussen / dôme de sel; coupole salifère / Salzkissen; Salzbeule / domo de sal; eczema salino (no perforante)

salt plug; salt stock; piercement dome, – a more or less cylindrical, often mushroom-shaped or overhanging diapiric salt structure. / zoutpijler / noyau de sel / Salzstock; Diapir / bloque de sal; núcleo de sal; (domo perforante)

salt wall, – a dyke-shaped piercement structure of salt. / zoutmuur / mur de sel / Salzmauer / dique salino

1696
cap; cap rock, – a cap of limestone, anhydrite or other rock of secondary origin immediately overlying the salt in a salt structure. / kap; hoed / capuchon; chapeau / Hut; Hutgestein / capuchón

1697
rim syncline; peripheral sink, – ring-shaped depression caused by displacement of the underlying salt into a growing salt structure. / randsynclinaal; randinzinking / synclinal bordier; dépression périphérique / Randmulde; Randsenke / sinclinal periférico

residual anticline, – a local structural high between two adjacent rim synclines, formed by beds overarching nondisplaced salt masses. / – – / pseudo-anticlinal / Restsalzkissen / (anticlinal residual)

1698
diapiric fold, – fold structure whose more mobile core pierces overlying, less mobile rock. / diapierplooi / pli diapir; pli à noyau de percement; pli à noyau percant / Diapirfalte; Injektiv.... / pliegue diapírico

1699
salt glacier, – outflow of rock salt at the surface, as preserved in arid climates, in diapiric folds. / zoutgletsjer / glacier de sel / Salzgletscher / glaciar de sal

1700
mud volcano, – a conical or dome-shaped outflow of waterbearing clay, often accompanied by some gas, due to diapirism of a high-pressure clay at depth along some line of weakness. / moddervulkaan / volcan de boue / Schlammvulkan / volcán de fango

1701
flow folding; flowage, – folding by the flowage of rocks, especially in very mobile rocks like salt, clay or migmatites. / vloeiplooiing / plissement par fluage / Fliessfaltung / plegamiento fluidal

[1380] GRAVITATIONAL GLIDING

1702
gliding tectonics; gravitational ...., – deformational features in large masses of rocks, either near the surface or at depth, due to gliding down. / gravitatietektoniek / tectonique de glissement; .... d'écoulement / Gleittektonik; Schweregleitung / tectónica de deslizamiento (por gravedad)

1703
gravitational gliding, – / afglijding door de zwaartekracht / écoulement par gravité / Abgleitung durch die Schwerkraft bedingt; Schweregleitung / deslizamiento por gravedad

1704
(gravity-)collapse structure, – structure resulting from gliding of rocks under the influence of gravity. / afglijdingsstructuur / déformation par glissement / Einbruchsstruktur; Kollaps....; Rutsch.... / estructura de colapso

1705
gliding plane, – / glijvlak / surface de glissement; plan d'écoulement / Schubbahn; Gleitfläche / superficie de deslizamiento

1706
cascading folds, – system of folds with subhorizontal and subvertical limbs, thought to have originated by gravitational gliding. / cascadeplooien / plis en cascades; plis en chaise / Kaskadenfalten / pliegues en cascada

1707
gravitational fault, – a fault that originated by gliding of a sedimentary mass; it may be preserved by subsequent burial. / afglijdingsbreuk / faille gravitationnelle / synsedimentäre Störung; Abrutsch-.... / falla gravitacional

growth fault, – a synsedimentary gravitational fault occurring in areas of fast sedimentation, e.g. deltas of major rivers, and striking parallel to the coast. At its downthrown, seaward, side sedimentation rate greatly exceeds that on its landward side. The lower part of the fault plane is curved, and it loses its identity in a bedding plane. Compressional folding may occur in the downslided block.

[1400] STRATIGRAPHY

[1410] GENERAL TERMS

1708
stratigraphy, – the branch of geology that deals with the formation, composition, sequence, and correlation of the stratified rocks as part of the earth's crust. / stratigrafie / stratigraphie / Stratigraphie / estratigrafía

1709
lithostratigraphy; rock stratigraphy, – the branch of stratigraphy that deals with the lithology of strata and their lithologic characteristics. / lithostratigrafie / lithostratigraphie / Lithostratigraphie / litoestratigrafía

lithostratigraphic(al) classification, – the organization of rock strata into units on the basis of their lithologic characteristics. / lithostratigrafische classificatie / classification lithostratigraphique / lithostratigraphiische Klassifikation / clasificación litoestratigráfica

1710
biostratigraphy, – the branch of stratigraphy that deals with remains or evidence of former life in sediments and with the organization of such strata into units based on fossil content. / biostratigrafie / biostratigraphie / Biostratigraphie / bioestratigrafía

biostratigraphic(al) classification, – the organization of sedimentary strata into units based on their fossil content. / biostratigrafische klassificatie / classification biostratigraphique / biostratigraphische Klassifikation / clasificación bioestratigráfica

1711
ecostratigraphy, – the study and classification of stratified rocks with respect to their origin and depostional environment. / ecostratigrafie / écostratigraphie / Ökostratigraphie / ecoestratigrafía

1712
chronostratigraphy, – the branch of stratigraphy that deals with the age of strata and their time relationships. / tijdstratigrafie; chronostratigrafie / chronostratigraphie / Chronostratigraphie / cronoestratigrafía

chronostratigraphic(al) classification, – the organization of strata into units on the basis of their age or time of origin. / chronostratigrafische klassificatie / classification chronostratigraphique / chronostratigraphische Klassifikation / clasificación cronoestratigráfrica

1713
geochronology, – the science of dating and determining the time sequence of events in the history of the earth's crust. It has two branches: dating based on stratigraphy and palaeontology and on radiometry; the former is expressed in relative relationships, the latter usually in years. /geologische tijdrekening; geochronologie / chronologie géologique; géochronologie / geologische Zeitrechnung; Geochronologie / geocronología

1714
superposition, – the sequence in which rocks are placed above one another. / superpositie; opeenvolging / superposition / Überlagerung / superposición

law of superposition, – the law that underlying strata must be older than overlying strata where there has been neither inversion nor overthrust. / principe van superpositie / loi de superposition; critère de .... / Lagerungsgesetz / principio de superposición

1715
facies, – the sum of the lithological and palaeontological characteristics exhibited by a sedimentary deposit in a particular place. / facies / faciès / Fazies / facies

NOTE:
– for further terms see section:
[1920] Sedimentary Environments, Facies and Deposits.

1716
correlation, – the determination on palaeontological or physical evidence of the equivalence in geologic age and/or stratigraphic position of two or more stratigraphic units in separate locations. / correlatie / corrélation / Korrelation / correlación

1717
zone, – used for a minor stratigraphic interval in any category of stratigraphic classification. / zone / zone / Zone / zona

1718
interval, – the body of strata between two stratigraphic markers. / interval / intervalle / Interval / intervalo

1719
conformity, – true stratigraphic continuity in the sequence of strata, parallel and unbroken, marking a continuous period of uninterrupted deposition. / conformiteit; concordantie / concordance stratigraphique; continuité .... / stratigraphische Konkordanz; gleichförmige Übereinanderfolge / concordancia estratigráfica

conformable, – pertaining to strata or groups of strata lying above another in parallel and unbroken sequence. /conform; concordant / concordant; en concordance; en continuité / konkordant / concordante

1720
unconformity, – a surface of erosion or non-deposition that separates younger strata from older rocks. / nonconformiteit; discordantie / discordance stratigraphique; discontinuité .... / stratigraphische Diskordanz / disconformidad; discordancia estratigráfica

unconformable, – having the relation of unconformity to the underlying rocks. / nonconform; discordant / discordant; en discordance / diskordant / discordante

1721
angular unconformity; clino-nonconformity, – an unconformity in which the beds below the plane of erosion or non-deposition dip at a different angle, mostly steeper, than the beds above the unconformity. / hoekdiscordantie; klino-nonconformiteit / discordance angulaire / Winkeldiskordanz; ungleichförmige Überlagerung / discordancia angular

nonangular unconformity; parallel unconformity; disconformity, – an unconformity in which the beds on the opposite sides of the unconformable contact are parallel. / parallele nonconformiteit; disconformiteit / transgressivité parallèle; discordance plate / Erosionsdiskordanz / discordancia no angular

1722
erosional unconformity, – a surface separating older strata, which have been subjected to erosion, from covering strata. / erosie-nonconformiteit / discordance de ravinement; .... érosionelle; disconformité / Erosionsdiskordanz / discordancia por erosión

1723
hiatus, – a break in a rock-stratigraphic sequence, either as a result of erosion or non-deposition. It represents the period of time missing between beds above and below an unconformity. / hiaat / lacune; hiatus / Schichtlücke; Hiatus / hiato; lacune; lacuna

1724
diastem, – a depositional break or hiatus of minor duration. / diasteem / diastème / Sedimentationsunterbrechung; kleine Sedimentationslücke / diastema

1725
convergence; convergency, – the reduction in thickness of a sedimentary interval in a specific direction. / convergentie / convergence / Konvergenz / convergencia

1726
transgression, – the flooding of a land area by a positive movement of base level, resulting in onshore migration of the sea. / transgressie / transgression / Transgression / transgresión

regression, – the retreat of the sea from a land area. / regressie / régression / Regression / regresión

1727
onlap; transgressive overlap, – the disposition of sediments, lying above an unconformity, where they were deposited in a transgressing sea; the progressively younger sediments extend further onto the previous land surface than their predecessors. / dakpansgewijze ligging / débordement transgressief / übergreifende Auflagerung / desborde transgresivo

offlap; regressive overlap, – the disposition of sediments laid down as a sea regresses, the progressively younger sediments being deposited seaward of the shore line that marked the previous maximum extent of the sea. / terugwijkende bedekking / position en retrait / regressive Lagerung / retiro regresivo

1728
emergence, – a term that signifies that part of the sea floor became land, but does not imply whether the sea level dropped or the land rose. / emersie / émersion / Auftauchen / emersión

submergence, – a term that signifies that part of a land area became inundated by the sea, but that does not imply whether the land sank or the sea level rose. / onderdompeling / submersion; ennoyage / Überfluting; Untertauchen / sumersión

1729
synchronous; –contemporaneous; simultaneous, – pertaining to coincidence of the time of sedimentation; occurring at the same time. / synchroon; gelijktijdig / synchrone / synchron; gleichalt / contemporáneo; sincrónico

1730
coevel, – contemporary, of the same age. / gelijktijdig / contemporain / gleichalt / coevo; coetáneo

1731
heterotopical, – pertaining to synchronous deposits of different geological provinces. / heterotoop / hétérotope / heterotopisch / heterotópico

1732
chronotaxial, – pertaining to strata of equivalent age in separate successions. / chronotax / chronotaxe / chronotax / cronotaxial

1733
homotaxial, – pertaining to strata, in separate successions, with apparently the same position (order of deposition) in the stratigraphic sequence. / homotax, equivalent / homotaxe équivalent / homotax äquivalent / homotaxial

1734
heterochronous, – pertaining to a sequence of sediments representing a later development of a similar lithofacies but of successively younger (or older) geological age. / heterochroon / hétérochrone / heterochron / heterócrono

1735
isomesical, – pertaining to deposits, not necessarily synchronous, laid down in identical media, e.g. two or more river deposits. / isomesisch / isomésique / isomésique / isomesisch / isomésico

1736
diachronism, – the condition of a lithological unit whose base is not a time plane, i.e. whose age is different in different successions. / diachronisme / diachronisme / Diachronismus / diacronismo

diachronous, – differing in geological age. / diachroon / diachrone / diachron / diácrono

1737
dendrochronology, – study and matching of the width of annual growth rings of a certain tree species in order to date (climatic) events in the recent past. / dendrochronologie / dendrochronologie / Dendrochronologie / dendrocronología

1738
azoic, –
1. without life, often pertaining to the earliest part of earth history;
2. loosely, pertaining to rocks without regonizable fossils or fossil fragments (= barren).

/ zonder spoor van leven, azoïsch / azoïque / azoisch / azóico

1739
type locality, – the specific geographic locality in which a stratotype is situated or, in case of absence of a designated stratotype, the locality where the stratigraphic unit or boundary was originallu defined or named. / typelokaliteit / localité type / Typuslokalität / localidad tipo

type area; ..... region, – the geographic territory in which a type locality or a stratotype is situated. / typegebied / région type / Typusgebiet / región tipo; área .....

[1420] STRATIGRAPHICAL UNITS

[1421] *General Terms*

1740
stratotype; type section, – the originally or subsequently designated type of a formal stratigraphic unit or of a stratigraphic boundary, identified as a specific interval or a specific level in a specific sequence of rock strata. / stratotype, typesectie / stratotype / Stratotypus / estratotipo

holostratotype, –
1. the original stratotype selected by the original author when establishing a stratigraphic unit or boundary;
2. the component stratotype of a composite stratotype selected as the prime section.

/ holostratotype / holostratotype / Holostratotypus / holoestratotipo

1741
unit stratotype, – the type section of strata serving as the standard for the definition of a stratigraphic unit. The upper and lower limits of a unit-stratotype are its boundary stratotypes.

1742
composite stratotype, – a unit stratotype composed of several specific type intervals of strata known as component stratotypes. / samengesteld stratotype / zusammengesetzter Stratotypus / estratotipo compuesto

component stratotype, – a specific lithostratigraphic interal that, in conbination with other component stratotypes, forms a composite stratotype. / componentstratotype / Teil-Stratotypus / componente del estratotipo

1743
parastratotype, –
1. a supplementary stratotype used in the original description for elucidating the holostratotype
2. one of the component-stratotypes of a composite stratotype, in addition to the holostratotype.

/ parastratotype / parastratotype / Parastratotypus / paraestratotipo

1744
boundary stratotype, – a specific level in a specific rock sequence that serves as the standard for the definition of a stratigraphic boundary. / grens-stratotype / Grenz-Stratotypus / estratotipo del límite

1745
neostratotype, – a new stratotype designated to replace an older or the original one that has been destroyed or otherwise nullified. / neostratotype / néostratotype / Neostratotypus / neoestratotipo

lectostratotype, – a stratotype that is selected subsequently to the original description of the unit or boundary, in absence of an adequately designated original stratotype. / lektostratotype / lectostratotype / Lectostratotypus / lectoestratotipo

1746
hypostratotype; reference section; auxiliary reference section, – a stratotype selected to extend the knowledge of the unit or boundary, established already by a stratotype, to other geographical areas. It is always subordinate to the holostratotype. / hypostratotype / – – / Hypostratotypus / hipoestratotipo

[1422] *Lithostratigraphical Units*

NOTE:
– see table 4 – The Stratigraphical Hierarchies

1747
lithostratigraphic(al) zone; lithozone, -an informal lithostratigraphical unit that, although unified in a general way by lithological features, is insufficiently known or for which there is insufficient need to designate it as a formal unit. / lithostratigrafische zone; lithozone / zone lithostratigraphique / lithostratigraphische Zone / zona litoestratigráfica

1748
lithostratigraphic(al) unit; rock unit, – a body of strata characterized by consisting dominantly of a certain lithologic type or combination of lithologic types, or by having other unifying lithologic features. /lithostratigrafische eenheid / unité lithostratigraphique / lithostratigraphische Einheit / unidad litoestratigráfica

1749
lithostratigraphic(al) horizon; lithohorizon, – a surface of distinctive lithostratigraphic change. / lithostratigrafische horizon / horizon lithostratigraphique; niveau .... / lithostratigraphischer Horizont / horizonte litoestratigráfico

horizon, – the plane that separates two beds and, hence, without thickness. / horizon / horizon / Horizont / horizonte

1750
bed; stratum, –
1. unit layer in a stratified rock sequence that is lithologically distinguishable from other layers above and below; smallest formal unit in the hierarchy of lithostratigraphic units;
2. (loosely) smallest division in a stratified succession, marked by more or less well-defined division planes (called bedding planes) from the other beds above and below.

/ laag / couche; strate / Bank; Lage / estrato; capa

1751
key bed; marker bed, – a bed with sufficiently distinctive characteristics to make it clearly identifiable in correlation. / gidslaag / couche guide; couche repère / Leitbank / estrato guía; capa guía

1752
lentil; lens; lenticle, – a lens-shaped body of rock; when enclosed by lithologies different from the body itself and when of some geographic extent it is commonly, though not necessarily named. / lens, lensvormig lichaam / lentille; amas lenticulaire / Linse; linsenförmiger Körper / lentejón

tongue, – a part of a lithostratigraphic unit extending out beyond the main body. / tong / tongue / zungenförmiger Körper / lengua

1753
member, – formal lithostratigraphic unit next in rank below a formation; a named entity because of its lithologic characteristics that distinguish it from the adjacent parts of the formation. / laagpakket; afzetting / – – / Formationsglied / miembro

1754
formation, – any sedimentary bed or consecutive series of beds sufficiently homogeneous or distinctive to be regarded as a unit; practicability of mapping and of delineation in cross sections is an important consideration in the establishment of formations. / formatie / formation / Formation / formación

1755
group, – a sequence of two or more adjacent formations that have significant unifying lithological properties in common; the formal lithostratigraphic unit next in rank above a formation. / groep / groupe / Gruppe / grupo

1756
complex, –lithostratigraphic unit composed of diverse types of rock (sedimentary, metamorphic, igneous) and characterized by a highly complex structure to such an extent that the original succession of the component strata can be obscured partly or completely. / complex / complexe / Komplex / complejo

[1423] *Chronostratigraphical Units*

NOTE:
– see table 4.

1757
chronostratigraphic(al) unit; time-stratigraphic(al) ...., – a specific interval of relative geological time. / chronostratigrafische eenheid; tijdstratigrafische .... / unité chronostratigraphique / chronostratigraphische Einheit /unidad cronoestratigráfica

time-rock unit, – a body of strata deposited during a specific interval of relative geological time.

1758
chronostratigraphic(al) horizon; chronohorizon, – a stratigraphic level that is everywhere of the same age (isochronous). / chronostratigrafische horizon / horizon chronostratigraphique / chronostratigraphischer Horizont / horizonte cronoestratigráfico

chronozone, – the division of lowest rank in the hierarchy of time-rock units; a zonal unit embracing all rocks formed anywhere during the time span of some geological feature or some specific interval of strata. / ouderdomszone; chronozone / chronozone; phase / Chronozone / cronozona

1759
moment; instant, – the geochronological equivalent of a chronostratigraphic horizon. / moment / moment / Moment / momento

1760
chronomere, – in stratigraphy, any interval of geological time. / chronomeer / chronomère / Chronomere / – –

1761
chron, –the smallest interval of geological time, geochronological equivalent of a chronozone; chrons may be grouped together to form an age. / chron / chrone / Chron / crono

1762
age, –
1. geochronological equivalent of stage;
2. (loosely) any great period in earth history, often characterized by its dominant type of life, e.g. age of reptiles.

/ tijd, tijdsnede / âge / Alter / edad

stage, – time-rock unit of relatively minor rank in the hierarchy of formal chronostratigraphic terms, representing a relatively minor period of geological time. / etage / étage / Stufe / piso

1763
subage, – geochronical equivalent of a substage. / ondertijd / subâge / Subalter / subedad

substage, – subdivision of a stage. / onder-etage / sous-étage / Unterstufe / subpiso

superstage, – grouping of two or more adjacent stages. / super-etage / super-étage / Superstufe / superpiso

1764
epoch, – geochronical equivalent of a series, subdivision of a period. / tijdvak / époque / Epoche / época

series, – time-rock unit in the chronostratigraphic hierarchy, ranking below a system and above a stage. / serie; afdeling / série / Serie / serie

1765
period, – geochronical equivalent of a system. / periode / période / Periode / período

system, – time-rock unit of major rank, above series and below an erathem, in the formal chronostratigraphic hierarchy. / systeem / système / System /sistema

1766
era, – geochronological equivalent of an erathem. / era / ère / Ära / era

erathem, – largest formal time-rock unit recognized in the chronostratigraphic hierarchy and usually consisting of several adjacent systems. / eratheem / érathème / Ärathem / eratema

1767
eon; aeon, – the largest interval of relative geological time; geochronologic unit greater than an era. / eon; aeon / éon / Äon / eón

eonothem, – time-rock unit equivalent of an eon. / eonotheem; aeonotheem / éonothème / Äonothem / eonotema

1768
biochron, – total time represented by a biostratigraphic zone. / biochron / biochrone / Biochron / – –

[1424] *Biostratigraphical Units*

NOTE:

– see table 4.

1769
biostratigraphic(al) zone; biozone, – general term for any kind of biostratigraphic unit. / biozone; biostratigrafische zone / biozone / Biozone / zona bioestratigráfica; biozona

1770
biostratigraphic(al) unit, – a body of rock distinguished by its fossil content. / biostratigrafische eenheid / unité biostratigraphique / biostratigraphische Einheit / unidad bioestratigráfica

1771
biohorizon, – level of biostratigraphic change or of distinct biostratigraphic character. / biohorizon / – – / – – / biohorizonte

1772
interval zone; bio-interval zone; interbiohorizon zone, – the stratigraphic interval between two biohorizons. / intervalzone / – – / – – / zona de intervalo

1773
acme zone; epibole; flood zone; peak zone, – a group of strata representing the maximum development of some species, genus, or other taxon, but not its total range. / acme-zone / épibole / Acme-Zone / zona de apogeo

1774
assemblage zone; cenozone, – a group of strata characterized by a distinctive natural assemblage of all or some of the fossil forms present in that group of strata. / assemblagezone; cenozone / cénozone; coenozone / Faunenzone / zona de asociación; cenozona

1775
barren interzone, – interval without fossils between successive biozones. / steriele interzone / intervalle stérile interzonal / fossilleere Zwischenzone / interzona esteril

barren intrazone, – interval without fossils within a biozone. / steriele intrazone / intervalle stérile intrazonal / – – / – –

1776
biostratigraphic(al) interval-zone; bio-interval zone; interbiohorizon zone, – interval between two biostratigraphic horizons, that is not the range zone of any taxon or concurrence of taxons and that does not necessarily contain a particularly distinctive fossil assemblage or feature. / biostratigrafische intervalzone / – – / – – / zona de intervalo bioestratigráfico

1777
concurrent-range zone; overlap zone; range-overlap zone, – the coincident or concurrent parts of the range zones of two or more specific taxons selected from all taxons contained in a sequence of strata. / overlappende-looptijdzone / zone de cooccurrence / Überlappungszone / zona de distribución concomitante

1778
genus zone, – a type of taxon-range zone, in which the taxon that defines the zone is a genus. / genus-zone / – – / Gattungszone / género-zona

1779
lineage zone; phylozone; evolutionary zone; morphogenetic zone; phylogenetic zone, – the body of strata containing specimens of a segment of an evolutionary line or trend, defined at top and bottom by changes in the features of the line or trend. / phylozone; morphogenetische zone; phylogenetische zone / – – / phylogenetische Zone / filozona; zona filogenética

1780
Oppel zone, – a zone characterized by an aggregation of selected taxons of restricted and largely concurrent range and selected as indicative of approximate contemporaneity. / Oppel-zone / Oppel-zone / Oppel-Zone / Oppel-zona

1781
range, –
1. in stratigraphy, the distribution of any given taxon throughout geological time, i.e. in vertical direction. / looptijd / série / Zeitspanne; Reichweite / serie
2. in biology, the geographical area in which a particular taxon occurs, i.e. in horizontal direction. / verspreiding / extension / geografische Verbreitung / distribución

range zone; acrozone,– a group of strata representing the stratigraphic range of one or more taxa of the total assemblage of fossil forms present. / looptijdzone; acrozone; verspreidingszone / zone d'extension / Verbreitungszone / acrozona

1782
subbiozone; subzone, – finer biostratigraphic division than biozone. / onderbiozone; subzone / – – / Subbiozone / subbiozona

zonule, – finer biostratigraphic division than subbiozone (subzone). / zonula / zonule / – – / zónula

superbiozone; superzone, – grouping of two or more biozones on ground of common biostratigraphic features. / superbiozone; superzone / superbiozone; superzone / Superbiozone; Superzone / super-biozona

1783
taxon-range zone, – the body of strata representing the total range, both horizontal and vertical, of specimens of a particular taxon, e.g. species zone, genus zone. / taxon-looptijdzone / zone d'extension d'un taxon / – – / zona de distribución de taxón

1784
teilzone; local-range zone; topozone, – the body of strata representing the local range of a specific taxon in a given area. / teilzone / topozone / Topozone / zona de distribución local; topozona

1785
guide fossil; key fossil; index fossil, – any fossil type that is sufficiently wide-spread and abundant in a more or less restricted thickness of sediments to have value for an indication of geological horizon and relative age. / gidsfossiel / fossile indicateur; fossile caractéristique; fossile repère / Leitfossil / fósil guía; fósil característico

zonal guide fossil, – species, or other taxon, of known, limited vertical range in a local rock succession. / zonaal gidsfossiel / fossile caractéristique de zone / Zonen-Leitfossil / fósil guía zonal

[1500] PALAEONTOLOGY

[1510] GENERAL TERMS

1786
palaeontology; paleontology, – the science of the life of past geological ages; based on the study of the fossil remains of organisms. / palaeontologie; paleontologie / paléontologie / Paläontologie / paleontología

micropalaeontology, – the study of microfossils. / micropalaeontologie / micropaléontologie / Mikropaläontologie / micropaleontología

1787
neontology, – the study of existing life. / neontologie / néontologie / Neontologie / neontología

1788
biosphere, – the part of the uppermost crust of the earth, its surface and the lower portion of the atmosphere that is inhabited by (living) organisms. / biosfeer / biosphère / Biosphäre / bioesfera

1789
fossil, – remain or trace of animal or plant that has been preserved in the earth's crust. / fossiel / fossile / Fossil / fósil

fossiliferous; fossil-bearing, – fossielhoudend / fossilifère / fossilführend / fosilífero

pseudofossil, – object such as a concretion that may be mistaken for a fossil. / pseudofossiel; schijnfossiel / reste problématique; pseudofossil / Problematikum; Pseudofossil; vermeintliches Fossil / (p)seudofósil

1790
microfossil, – fossil remain of organisms whose average representatives are of microscopic size, or whose average nepionic and neanic stages are microscopic in size, or of dwarfed forms of megafossils. / microfossiel / microfossile / Mikrofossil / microfósil

nannofossil, – a fossil remain of less than 0.030 mm, predominantly derived from planktonic organisms. /nannofossiel / nannofossile / Nannofossil / nanofósil

1791
biostratonomy, – the branch of stratigraphy that deals with the processes by which the remains of organisms become embedded in rock. / biostratonomie / biostratinomie / Biostratinomie / bioestratonomía

1792
biogenesis, – formation by the action of organisms. / biogenese / biogenèse / Biogenesis / biogénesis

biogenetic; biogenic, – formed by the action of organisms. / biogenetisch / biogénétique / biogenetisch / biogenético

1793
morphology, – in palaeontology and biology, the form of an organism considered either as a whole or in its gross aspects. / morfologie / morphologie / Morphologie / morfología

1794
ontogeny, – developmental history of an individual organism from egg to adult. / ontogenese; ontogenie / ontogénèse; ontogénie / Ontogenese; Ontogenie / ontogenía; ontogénesis

metamorphosis, – a change of form during the ontogeny of an animal. / metamorfose; gedaanteverwisseling / métamorfose / Metamorphose; Umgestaltung; Verwandlung / metamorfosis

1795
fauna, – collectively, the animal life of any given region or given age. / fauna / faune / Fauna / fauna

faunal, – pertaining to a natural assemblage of animals. / fauna- / faunique; faunistique / Faunen- / faunístico

1796
faunula, – in palaeontology, an association of animals occurring in a single stratum or a limited succession of strata. / faunula / faunule / Faunula / fáunula

1797
flora, – collectively, the plants of any given age or region. / flora / flore / Flora / flora

floral, – pertaining to a natural assemblage of plants. / flora- / floral / floral / floral

1798
nekton, – collectively, the swimming animals in aquatic environments that can move independently of currents. / nekton / necton; nekton / Nekton / necton

nektonic, – pertaining to the swimming forms of pelagic life. / nektonisch / nectonique / nektonisch / nectónico

1799
plankton, – collectively, all the floating or drifting plants and animals suspended in the water of aquatic habitats, and not independent of currents and other water movements. / plankton / plancton; plankton / Plankton / plancton

planktonic, – pertaining to the floating and drifting forms of pelagic life. / planktonisch / planctonique / planktonisch / planctónico

nannoplankton, – planktonic organisms with maximum dimensions of less than 0.030 mm. / nannoplankton / nannoplancton / Nannoplankton / nanoplancton

1800
zooplankton, – collectively, all those animals that live suspended in aquatic habitats, and can not move independently of currents and other water movements. / zoöplankton / zooplancton / Zooplankton / zooplancton

phytoplankton, – collectively, all plants that live suspended in aquatic habitats such as diatoms, dinoflagellates and coccolithophores. / fytoplankton / phytoplancton / Phytoplankton / fitoplancton

1801
meroplankton, – the planktonic, larval stages of animals that inhabit the sea bottom during most of their life. / meroplankton / meroplancton / Meroplankton / meroplancton

limnoplankton, – lake plankton. / limnoplankton / limnoplancton / Limnoplankton / limnoplancton

1802
geobios, – life of the terrestrial environment. / geobios / géobios / Geobios / geóbios

hydrobios, – life of the aquatic environment. / hydrobios / hydrobios / Hydrobios / hidróbios

limnobios, – life of the fresh-water environment. / limnobios / – – / Limnobios / limnóbios

1803
metabolism, – the sum of the processes or chemical changes in an organism or a single cell by which food is built up into living protoplasm and by which protoplasm is broken down into simpler compounds with the exchange of energy. / stofwisseling; metabolisme / métabolisme / Stoffwechsel; Metabolismus / metabolismo

photosynthesis, – synthesis of chemical compounds with help of radiant energy, especially light. Commonly used for the synthesis of carbohydrates from water and carbon dioxide in the chlorophyll-containing tissues of plants exposed to sunlight. / fotosynthese / photosynthèse / Fotosynthese / fotosíntesis

1804
biota, – the flora and fauna of a region or a period. / biota / bios / Bios; Lebewelt / – –

1805
assemblage, – collectively, a community of living or fossil organisms in space and in time. / gemeenschap / assemblage; ensemble / Gemeinschaft; Gesellschaft / asociación; conjunto

1806
colony, – a group of specimens of a given plant or animal species that is organized together in such a way that the independence of the individual members is partly or completely lost. / kolonie / colonie / Ansiedlung; Kolonie / colonia

1807
facies fossil, – a fossil type, usually a species, adapted to life in a special environment. / faciesfossiel / fossile de faciès / Faziesfossil / fósil de facies

1808
circulus, – a fossil community of homoeomorphic forms for which a genetic relationship cannot be claimed. / circulus / – – / Plete / – –

1809
cosmopolitan species, – a species that lives or has lived all over the earth or a large part of it. / cosmopolitische soort / espèce cosmopolite / kosmopolitische Art / especie cosmopolita

1810
continuum, – pattern of gradually overlapping populations of different subspecies (or species) along a gradient.

1811
polymorphism; pleomorphism, – the existence of a species in several morphological forms independent of the variations of sex. / polymorfie / polymorphisme / Polymorphismus / polimorfismo

dimorphism, – occurrence of two distinct morphological forms in a single population. / dimorphisme / dimorphisme / Zweigestaltigkeit; Dimorphismus / dimorfismo

1812
dispersion, – spatial distribution pattern of specimens within a population; one distinguishes random, uniform, clumped. / dispersie; spreiding / – – / – – / dispersión

1813
autochthonous; indigenous, – pertaining to an organism or a taxon living or having lived permanently in a particular habitat. / inheems; autochtoon / autochtone / heimisch; autochtoon / autóctono

1814
endemic, – pertaining to any plant or animal taxon that is restricted to a particular geographical area. / endemisch / endémique / endemisch / endémico

pandemic, – widely distributed, as opposed to endemic. / pandemisch / pandémique / pandemisch; kosmopolitisch / pandémico

1815
allochthonous, – pertaining to an organism or a taxon that has immigrated from another habitat. / allochtoon / allochtone / allochthon / alóctono

1816
condominant, –
1. pertaining to two or more dominant species in a habitat;
2. pertaining to one of the several animal of plant species dominating a community.

/ condominant / condominant / – – / condominante

dominant, – pertaining to species that is or are numerically the most abundant in a particular environment or fossil assemblage. / dominant; overheersend / dominant; prépondérant / vorherrschend; überwiegend / dominante; preponderante

1817
sympatric, – pertaining to two or more populations of closely related species that occupy identical or broadly overlapping geographical areas. / sympatrisch / sympatrique / sympatrisch / – –

allopatric, – pertaining to two or more related populations that live in mutually exclusive, although usually adjacent, geographical areas. / allopatrisch / allopatrique / allopatrisch / – –

1818
synchronic, synchronous; contemporaneous, – pertaining to species that occur at the same time level. / synchroon; gelijktijdig / synchronique; synchrone; contemporain / synchron; gleichalt / sincrónico; contemporáneo

allochronic; allochronous, – pertaining to species that do not occur at the same time level. / allochroon / allochronique; allochrone / allochron / alocrónico; no contemporáneo

1819
polytopic, – occurring in different places, e.g. a subspecies composed of widely separated populations. / polytoop / polytopique; polytope / polytopisch / politópico

1820
homeomorphism; homomorphy; homeomorphy, – similarity of form, especially resemblance in external characteristics, in different groups of organisms. / homeomorfie / homéomorphie / Homoömorphie / homeomorfismo

1821
dendroid; arborescent, – having a many-branched, tree-like mode of growth, e.g. in certain colonial corals and graptolites. / dendritisch; boomvormig / dendritique; dendroïde / dendritisch / dendrítico; dendroide

coralline, – pertaining to any non-coral organism that bears a superficial resemblance to a mass of coral, e.g. some Hydrozoa, Bryozoa, Algae. / koraalachtig / – – / korallenähnlich / coraliforme

1822
accretion, – growth by simple external addition, as opposed to incorporation. / accretie / accrétion / Akkreszenz / acreción

1823
pseudomorph, – in palaeontology, a fossil in which the original skeletal substance of the organism has been replaced by a secondary material in the course of fossilization, thus preserving its shape. / pseudomorf / pseudomorphe / pseudomorph / (p)seudomorfo; (p)seudomórfico

1824
mold; mould, – a fossil in which the original skeletal parts have been dissolved, leaving a space that preserves their shape. / afdruk; steenkern / empreinte / Abdruck; Ausguss / molde

internal mold; .... mould, – a mold that shows the form and markings of the inner surfaces of skeletal parts, such as shells. / steenkern / moule interne / Steinkern; innerer Ausguss / molde interno

external mold; .... mould, – a mold that shows the form and markings of the outer surface of skeletal parts, such as shells. / afdruk / moule externe / Skulptursteinkern / molde externo

1825
cast, – a type of fossil consisting of a pseudomorph in which

the skeletal parts of the organism have been dissolved and the resulting space replaced by a secondary material, producing a replica of the original form. / afgietsel / moulage / Ausguss; Abguss / réplica

internal cast; endocast, – cast of an internal cavity. / afgietsel / teste de substitution / Steinkern / réplica interna

external cast; exocast, – a cast that preserves the outer surface of skeletal parts, such as a shell. / uitwendig afgietsel / moulage externe / Abguss; Skulptursteinkern / réplica externa

1826
imprint, – the impression made on a soft surface of sediment by organisms. / afdruk / empreinte / Abdruck / impronta

[1520] FLORAL AND FAUNAL PALAEONTOLOGY

[1521] *Palaeobotany*

1827
palaeobotany; phytopalaeontology, – the study of fossil plants and their distribution in time. / palaeobotanie; fytopalaeontologie / paléobotanique; paléontologie végétale / Paläobotanik; Pflanzenpaläontologie; Phytopaläontologie / paleobotánica

1828
palynology, – study of fossil sporomorphs and other acid-resistant microfossils, such as dinoflagellates, acritarchs and silicoflagellates. / palynologie / palynologie / Palynologie / palinología

1829
palynomorph, – all acid-resistant microfossils that are object of palynological studies. / palynomorf / palynomorphe / Palynomorph / palinomorfo

1830
microplankton, – loosely, all acid-resistant planktonic microfossils studied in palynology such as dinoflagellates, acritarchs and silicoflagellates. / microplankton / microplancton; microplankton / Mikroplankton / microplancton

1831
sporomorph, – collectively, spores and pollen grains. / sporomorf / sporomorphe / Sporomorph / esporomorfo

1832
spore, –
1. in botany, an asexual reproductive structure, commonly unicellular, usually produced in a sporangium;
2. in zoology, but in this sense not used in palaeontology, in Protozoa a caselike structure in which a few sporozoites are formed by multiple fission.

/ spore / spore / Spore / espora

sporangium, – a spore case / sporangium / sporangium; sporange / Sporangium / esporangio

1833
miospore, – all fossil spores and sporelike bodies smaller than 0.20 mm. / miospore / miospore / Miospore / miospora

1834
homospory; isospory, – the production of spores of a single type from an individual plant. / homosporie; isosporie / homosporie; isosporie / Homosporie; Isosporie / isospórea

heterospory, – the production of spores of different sizes (mega- and microspores) from one individual plant. / heterosporie / hétérosporie / Heterosporie / heterospórea

1835
isospore; homospore, – a spore, derived from a plant that produces only one kind of spores serving both male and female gametophytic functions. / isospore; homospore / isospore; homospore / Homospore; Isospore / isospora

1836
microspore, – a spore, produced within a microsporangium and forming a microgametophyte; generally of much smaller size than the megaspore of the same species. / microspore / microspore / Mikrospore / microespora; microspora

microsporangium, – a sporangium that produces microspores. / microsporangium / microsporange / Mikrosporangium / microesporangio

1837
megaspore; macrospore, – a spore, produced within a megasporangium and forming a megagametophyte; generally recognizable on account of their relatively large dimensions. / megaspore; macrospore / mégaspore; macrospore / Megaspore; Makrospore / megaspora; macroespora; macrospora

megasporangium, – a sporangium that produces megaspores. / megasporangium / mégasporange / Megasporangium / megaesporangio

1838
microgametophyte, – a male gametophyte that results from the growth of a microspore and that produces male gametes. / microgametofiet / microgamétophyte / Mikrogametophyt / microgametófito

megagametophyte, – a female gametophyte that results from the growth of a megaspore and that produces female gametes. / megagametofiet / mégagamétophyte / Megagametophyt / megagametófito

1839
pollen, – microspores of the higher plants (gymnosperms and angiosperms). / stuifmeel; pollen / pollen / Blütenstaub; Pollen / polen

1840
coccolith, – minute plates of low-magnesium calcite that are attached to the cell membrane of the coccolithophorids, a group of unicellular, mainly marine planktonic algae. / coccoliet / coccolithe / Coccolith / cocolito; coccolito

1841
cyst, – the spore-like cell, commonly a resting stage of unicellular algal organisms. / cyste; kiste / cyste / Zyst / quiste

1842
frustule, – the siliceous shell of a diatom, consisting of two valves, one overlapping the other. / frustula / frustule / Schale; Frustel / frústula

1843
gyrogonite; gyrolith, – fossilized sporangium of charophytes. / gyrogoniet / oogone; gyrogonite (obsolete) / Gyrogonites / girogonio

1844
cryptozoon, – problematic Cambrian and Ordovician fossils, problably algae; reef forming, hemispherical, irregular, or spreading structures made of concentric laminae of calcite. / cryptozoön / cryptozoon / Cryptozoon / cryptozoon

1845
zoochlorellae, – small green algae that occur as intracellular symbionts in certain protozoa, sponges, coelenterates, etc. / zoöchlorellae / zoochlorelles / Zoochlorellen / zooclorelas

zooxanthellae, – small yellow or brown flagellates that often occur as intracellular symbionts in certain foraminifera and radiolaria. / zoöxanthellae / zooxanthelles / Zooxanthellen / zooxantelas

1846
halophyte, – a plant that can tolerate a considerable amount of salt in the soil in which it grows. / halofiet / halophyte / Halophyt / halófita

[1522] *Palaeozoology*

1847
palaeozoology, – the science of fossil animals and their distribution in time. / palaeozoölogie / paléozoologie / Paläozoologie / paleozoología

1848
palichthyology, – that branch of palaeozoology that deals with fossil fish remains. / palichthyologie / palichthyologie / Palichthyologie / paleoictiología

1849
faunal province, – a topographically defined region characterized by a specified assemblage of animals. / fauna provincie / province faunistique / Faunenprovinz / provincia faunística

1850
facies fauna, – a group of animals adapted to a life in a special type of environment. / faciesfaune / faune de faciès / Faziesfauna / fauna de facies

1851
dwarf fauna; depauperate fauna, – a fauna characterized by fossils of stunted size. / dwergfauna / faune naine / Zwergfauna / fauna enana

1852
zoobenthos, – collectively, the animals living on or in the bottom of aquatic environments. / zoöbenthos / zoobenthos / Zoobenthos / zoobentos

1853
infauna, – comprising all fauna inhabiting the sandy or muddy surface layers of the sea bottom. / infauna / endofaune / in Sediment lebende Fauna / – –

epifauna, – comprising all fauna living upon or associated with more or less firm substrates on the sea bottom (e.g., rocks, stones, shells, plants). / epifauna / épifaune / Epifauna / epifauna

meiofauna, – collectively, the microscopic and small macroscopic metazoan fauna on the surface of the sea bottom. / meiofauna / méiofaune / Meiofauna / meiofauna

1854
endobiont, – animal living in burrow or moving through the sediment. / endobiont / endobionte / Endobiont / endobionte

1855
epibiont, – an animal living on the sea floor on hard substrates (such as rocks, plants, animal shells); one recognizes vagile and sessile epibionts. / epibiont / épibionte / Epibiont / epibionte

1856
epiphyte, – any organism living attached to a plant; often used for plants living attached to another plant or object, not growing from the soil. / epifiet / épiphyte / Epiphyt / epífito

1857
geophagous, – pertaining to animals that obtain their food by ingesting sediment or soil. / geophaag / géophage / geophag / geófago

lithophagous,
1. rock-boring, as done by certain molluscs and echinoids;
2. feeding on living coral;
3. ingestion of gravel, as done by certain birds.
/ lithophaag / lithophage / lithophag / litófago

xylophagous, – eating of, or boring in wood, thereby destroying it, such as done by some marine annelids and pelecypods. / xylophaag / xylophage / xylophag / xilófago

1858
amphibious, – pertaining to an animal that is capable to live on land as well as in water. / tweeslachtig; amfibisch / amphibie / amphibisch / anfibío

1859
bivalve, – a mollusk having two shells hinged together. / tweekleppig / bivalve / Bivalve; Muschel / bivalvo

1860
conodont, – minute, toothlike microfossils that may be cone, bar, blade or platform shaped; are composed of concentric layers or longitudinal fibrous bundles of calcium metaphosphate; and range in size from less than 0.1 mm to more than 4 mm. / conodont / conodonte / Conodont / conodonto

1861
skeleton, – collectively, the hard supporting structure of an animal; it is either internal (endoskeleton) or external (exoskeleton). / skelet; geraamte / squelette; carcasse / Skelett; Gerippe; Gerüst / esqueleto; armazón

endoskeleton, – any animal skeleton of various composition that is situated within the body and provides support and permanent shape to the body and its parts, present in echinoderms and vertebrates. / endoskelet; inwendig skelet / endosquelette / Innenskelett / endoesqueleto

exoskeleton, – any animal skeleton of various composition that forms the outer covering of the body, providing support and permanent shape to the body and its parts such as in arthropods, mollusks. / exoskelet; uitwendig skelet / exosquelette / Aussenskelett / exoesqueleto; exosqueleto

1862
spicule, – a minute needle-like carbonate or silicon dioxide body that forms part of the supporting skeleton of many animals, such as sponges, octocorals and radiolarians. / spicula / spicule / Nadel / espícula

1863
coprolith, – fossilized excrement of an animal. / coproliet / coprolithe; excrément fossile / Koprolith / coprolito

faecal pellet, – small coprolith (mostly 1 mm or less in size), mainly of invertebrates, present especially in marine deposits. / microcoproliet / pelote fécale; microcoprolithe / Kotpille / pellet fecal; pelet ....

1864
gastrolith, – gizzard stone; stomach stone, – pebble ingested by reptiles and birds to assist in breaking down food in the stomach. / gastroliet; maagsteen / gastrolithe / Gastrolith / gastrolito

1865
otolith, – calcareous concretion in the membranous labyrinth of lower vertebrates, especially fishes. / otoliet / otolithe / Otholith / otolito

1866
nepionic, – pertaining to the infantile post-embryonic stage in the development of an organism before the neanic stage, i.e. before the appearance of distinctive specific characteristics. / nepionisch / népionique / nepionisch / nepiónico

neanic; nealogic, – pertaining to the stage of development of an organism in which progressive development of the distinctive specific characteristics occurs; adolescent. / neanisch / néanique / neanisch / neánico

gerontic, – pertaining to the old-age stage in the life history of an organism. / gerontisch / gérontique / gerontisch / geróntico

1867
gigantism, – abnormally large development in size of the whole body or parts of the body of an organism. / gigantisme; reuzengroei / gigantisme / Gigantismus; Riesenwuchs / gigantismo

1868
megalospheric; macrosperic, – one of two different kinds of test that occur in many species of foraminifera because of the alternation of sexual and asexual modes of reproduction; the test is usually small in over-all size but has a large proloculus as compared to the microspheric form. / megalosferisch; macrosferisch / mégalosphérique; mégasphérique / megalosphärisch / megasférico; macrosférico

microspheric, – one of the two different kinds of test that occur in many species of foraminifera because of the alternation of sexual and asexual modes of reproduction; the test is usually of much larger size than the megalospheric form, the proculus is much smaller. / microsferisch / microsphérique / mikrosphärisch / microsférico

1869
trimorphism, – the concept of the presence of three forms in every foraminiferal species: two representing the megalospheric generations, one the microspheric phase. / trimorfisme / trimorphisme / Trimorphismus / trimorfismo

1870
multilocular; multichambered, – composed of many chambers, as in foraminifera. / veel kamers bevattende / multiloculaire; pluriloculaire / mehrkammerig; vielkammerig / multilocular

[1523] *Ichnology*

1871
ichnology, – study of traces made by organisms, including their description, classification and interpretation. / ichnologie / ichnologie / Ichnologie; Spurenkunde / icnología

1872
palichnology, – the study of ichnofossils. / palichnologie / palichnologie / Palichnologie / paleoicnología

1873
ichnofossil; trace fossil, – a general term for fossil tracks, trails, burrows and borings, produced by organisms. / ichnofossiel; sporenfossiel / ichnofossile; trace fossile; fossile de trace / Spurenfossil; Ichnofossil / icnofósil

1874
ichnocoenose; ichnocenosis; ichnocenose, – assemblage of traces or trace fossils; components include the ichnofauna or animal and the ichnoflora or plant traces (e.g. algal borings). / ichnocoenose; ichnocenose / ichnocénose / Ichnozönose / icnocenosis

1875
lebensspur, – sedimentary structure produced by a living animal within the sediment or on the sediment surface. One recognizes resting, crawling, grazing, feeding and dwelling structures. / levensspoor / trace de vie; trace d'activité / biogene Spur; organogene Spur; Lebensspur / huella animal

surface lebensspur, – structure formed on the sediment surface as a result of animal activity. / oppervlaktelevensspoor / trace de vie superficielle / – – / huella superficial

1876
trace; spoor, – an individually distinctive biogenic structure, especially when related more or less directly to the morphology of the organism that produced it. / spoor / trace / Spur / traza

1877
track; footprint, – impression left on the sediment surface by an individual foot or podium of a living animal. / voetindruk / empreinte / Trittsiegel; Fussspur / pisada

trackway, – a succession of tracks reflecting directed locomotion. / voetspoor / piste; galerie / Fährte / pista

1878
trail; trailway, – continuous groove produced during locomotion by an animal having part of its body in contact with the substrate surface, or a continuous subsurface trace made by an animal travelling from one point to another. / kruipspoor; sleepspoor / piste; galerie de reptation / Kriechspur / pista; rastro

1879
repichnia; crawling trace, – trackways and trails (epistratal or intrastratal) made by animals treveling from one place to another. / kruipspoor / trace de locomotion / Kriechspur / traza de reptación

1880
cubichnia; resting trace, – shallow depressions made by animals that temporarily settle onto, or dig into, the substrate surface. / rustspoor / trace de station / Ruhespur / traza de estación

domichnia; dwelling structure, – burrows or dwelling tubes providing more or less permanent domiciles, mostly for semisessile, suspension-feeding animals. / woongang / terrier d'habitation / Wohnbaute / estructura de habitación

1881
fodinichnia; feeding structure, – more or less temporary burrows constructed by deposit-feeding animals; typically these structures also provide shelter for the animals. / voedingsgang / structure de nutrition / Fressspur / estructura de nutrición

pascichnia; grazing trace, – grooves, pits and furrows made mostly by mobile deposit-feeding animals, at or near the substrate surface. / graasspoor / trace de pacage / Weidespur / traza de pastos

1882
boring, – a biogenic structure excavated by an organism into a rigid substrate. / boorgang / perforation / Bohrgang / perforación

burrow, – excavation made by a living animal within unconsolidated sediment. / graafgang / terrier / Grabgang / conducto

1883
bioerosion structure, – a biogenic structure excavated mechanically or biochemically by an organism into a rigid substrate. / bio-erosie structuur / structure de bioérosion / Bioerosionsstruktuur / estructura de bioerosión

1884
biostratification structure, – a biogenic sedimentary structure consisting of stratification features imparted by the activity of organisms. / biostratificatiestructuur / structure de biostratification / biogenes Schichtgefüge / estructura bioestratigráfica

1885
bioturbation, – reworking and further degradation of sediment by living organisms. / bioturbatie / bioturbation / Verwühlung, Bioturbation / bioturbación

bioturbation structure, – a biogenic sedimentary structure reflecting the disruption of stratification features or sediment fabrics by the activity of an organism. / bioturbatiestructuur / structure de bioturbation / Wühlgefüge; Bioturbationsgefüge / estructura de bioturbación

[1530] ACTUOPALAEONTOLOGY

1886
actuopalaeontology, – the branch of palaeontology in which studies of modern organisms and their relations to the environment are directed towards the understanding of fossil analogues. / aktuopalaeontologie / actuopaléontologie / Aktuopaläontologie / actualismo paleontológico

1887
ecology; bionomics, – the branch of biology that deals with the interrelationships between organisms and their animate and inanimate surroundings. / oecologie; ecologie; bionomie / écologie; bionomie / Ökologie; Bionomie / ecología

1888
autoecology, – econology of a single specimen or of all specimens in one species. / autoëcologie / auto-écologie / – – / autoecología

synecology, – study of the ecology of communities. / synecologie / synécologie / – – / sinecología

1889
palaeoecology, – the branch of palaeontology that deals with the relationship between extinct organisms and their environment. / palaeoëcologie / paléoécologie / Paläoökologie / paleoecología

1890
community, – collectively, all of the organisms inhabiting a common environment and interacting freely with one another. / levensgemeenschap / société; communauté; groupement / Lebensgemeinschaft / comunidad

1891
climax community; climax formation, – final or stable community in a successional series; self perpetuating, in equilibrium with the physical and chemical factors of the environment, composed of a definite group of plant and animal species. / climax-levensgemeenschap / formation climax / – – / climax de comunidad

climax species, – particular species of organisms characteristic for a climax community. / climax-soort / espèce climax / – – / climax de especies

1892
association, –
1. one of the major climax community types; the largest subdivision of a biome;
2. loosely, all those organisms living together in any given combination of environmental conditions.

/ gemeenschap; associatie / association / Gemeinschaft; Vergesellschaftung; Assoziation / asociación

1893
ecosystem, – collectively, all animals and plants in a community in combination with the associated animate and inanimate environmental factors. / ecosysteem / écosystème / Ökosystem / ecosistema

1894
sere, – series of communities that follow one another in slow but definite sequence, ending in a climax typical of a particular climate and geography. / sere / série / – – / – –

1895
biome, – world-wide complex of communities characterized by prevailing climatic and soil conditions, e.g. the tropical rain forsts everywhere form the tropical rain-forest biome. / bioom / biome / Biom / bioma

1896
biocoenosis; biocenosis; biocenose; life assemblage, – collectively, assemblage of organisms living together as a community. / biocoenose ; levensgemeenschap / biocénose; biocoenose / Biozönose; Lebensgemeinschaft / biocenosis

thanatocoenosis, – an assemblage of dead organisms formed in a place where they once lived in a biocoenosis. / thanatocoenose / thanatocénose; thanatocoenose / Thanatozönose / tanatocenosis

taphocoenosis, – a group of organisms brought together after death. / taphocoenose; taphocenose / taphocénose; taphocoenose / Taphozönose / tafocenosis

1897
taphonomy, – the study of the processes affecting an organism from its death to its possible fossilization. / taphonomie / taphonomie / Taphonomie; Biostratinomie / tafonomía

1898
symbiosis, – the condition in which two different species live in intimate association, whereby for at least part of the time each one derives a certain advantage from the metabolic partnership. / symbiose / symbiose / Symbiose / simbiosis

1899
commensalism, – growth together of different organisms or species in such a way that no harm is done to any member. / commensalisme / commensalisme / Kommensalismus / comensalismo

1900
environment, – total of physical, chemical and biological factors to which an organism is subjected. / milieu; omgeving / milieu; environment / Milieu; Umgebung / ambiente; medio ambiente

NOTE:
– further environmental terms have been assembled in sections
[1651] Lake Waters
[1840] Marine Environments
[1920] Sedimentary Environments
[2422] Biochemical Phase, Environments of Deposition

1901
habitat, – the specific place where a particular plant or animal normally lives. / woonplaats; verblijfplaats; standplaats / habitat / Wohnort; Standort / habitat

1902
niche, – ecological role of an organism with reference to its special place in its inanimate environment and with reference to other species (plur.) associated with it; of primary importance are food and nutrition relationships. / niche / niche / – – / nicho

biotope, – niche, especially in the geographical or spatial sense. / biotoop / biotope / Biotop / biotopo

1903
biotic, – pertaining to those environmental factors that are the result of living organisms and their activities. / biotisch / biotique / biotisch / biótico

1904
photic, – pertaining to the uppermost layer of a lake or sea extending to the maximum depth of sunlight penetration. / fotisch / photique / photisch / fótico

euphotic, – pertaining to the uppermost portion of a body of water in which there is sufficient light for photosynthesis; in the sea it can be up to 100 m thick, in certain lakes possibly 5 m or less. / eufotisch / euphotique / euphotisch / eufótico

dysphotic; disphotic, – pertaining to that part of a body of water, below the euphotic zone, in which there is insufficient light for photosynthesis but sufficient light for certain animal responses, in clear sea water usually between depths of 100 and 800 m. / dysfotisch / dysphotique / dysphotisch / disfótico

1905
aphotic, – pertaining to that part of a body of water that receives no sunlicht of biological significance, usually in the sea below a depth of 800 m. / afotisch / aphotique / aphotisch / afótico

1906
aerobic, – pertaining to organisms capable to live in the presence of molecular oxygen. / aëroob / aérobie / aerob / aerobio

anaerobic, – pertaining to organisms capable to live in the absence of molecular oxygen. / anaëroob / anaérobie / anaerob / anaerobio

1907
benthos, – collectively, all those organisms living on or in the bottom of lake or sea (the term is often used with reference

to animals only, i.e. bottom fauna). / benthos; bodembewoners / benthos / Benthos / bentos

benthonic; benthic, – pertaining to organisms living on or in the bottom of lake or sea. / benthonisch / benthonique; benthique / benthonisch; benthal / bentónico

1908
epibenthos, – the organisms living on the bottom of a large body of water between low tide level and a depth of 200 m. / epibenthos / – – / epibenthos / epibentos

archibenthic, – pertaining to organisms living in a body of water between the sublittoral zone and depths of 800 to 1100 m, i.e. at base of the dysphotic zone. / archibenthisch / – – / archibenthisch / archibentónico

1909
eurybenthic, – pertaining to species of bottom organisms that are able to live in a wide depth range in a body of water. / eurybenthisch / – – / eurybenthisch / euribentónico

stenobenthic, – living on the bottom of a body of water over a narrow depth range only. / stenobenthisch / sténobenthique / stenobenthisch / stenobentónico

1910
pelagic, – pertaining to communities of organisms that live in the open waters of sea and lakes, in direct independence from bottom or shore. / pelagisch / pélagique / pelagisch / pelágico

epipelagic, – pertaining to a community of suspended organisms inhabiting a body of water between surface and a depth of 200 m. / epipelagisch / – – / epipelagisch / epipelágico

mesopelagic, – pertaining to a community of suspended organisms inhabiting a body of water between 200 and 1000 m depth. / mesopelagisch / – – / mesopelagisch / mesopelágico

bathypelagic, – pertaining to a community of suspended organisms inhabiting a body of water between 1000 and 4000 m depth. / bathypelagisch / bathypélagique / bathypelagisch / batipelágico

abyssopelagic, – pertaining to a community of suspended organisms inhabiting a body of water below 4000 m depth. / abyssopelagisch / – – / abyssopelagisch / pelágico abisal

1911
eurybathic, – pertaining to species of aqueous organisms that are able to live in a wide depth range. / eurybathisch / eurybathe / eurybathisch / euribatial; euribático

stenobathic, – pertaining to aqueous organisms restricted to a narrow depth range only. / stenobaat / sténobathe / stenobathisch / estenobatial; estenobático

1912
euryhaline, – pertaining to organisms that are able to withstand a wide range of salinity in their aqueous environment. / euryhalien / euryhalin / euryhalin / eurihalino

stenohaline, – pertaining to organisms capable to live under a narrow range of salinity only. / stenohalien / sténohalin / stenohalin / estenohalino

1913
eurythermal, – pertaining to organisms that are able to live under widely varying temperatures. /eurytherm / eurytherme / eurytherm / euritermo

stenothermal, – pertaining to organisms able to exist within a limited temperature range only. / stenotherm / sténotherme / stenotherm / estenotermo

1914
eurytopic, – pertaining to species of organisms that have a wide range of suitable ecological conditions. / eurytoop / eurytope / eurytop / euritópico

stenotopic, – pertaining to organisms with a narrow range of suitable ecological conditions. / stenotoop / sténotope / stenotop / estenotópico

1915
halolimnic, – pertaining to marine organisms that are able to survive in fresh water also. / halolimnisch / – – / – – / halolímnico

1916
barrier, – any factor that inhibits further spread of a taxon (species, genus etc.); such barrier can be biological, climatic, chemical or physical, e.g. predators, desert, salinity, lack of proper food etc. / barrière; hindernis / barrière / Barriere / barrera

geographic(al) barrier, – topographic feature that prevents the further spread of a taxon and/or prevents interbreeding of two closely related populations separated by such feature. / geografische barrière / barrière géographique / geographische Barriere / barrera geográfica

1917
vagility, – relative power of dispersal or ability to cross ecological barriers. / vagiliteit / – – / Vagilität / vagilidad

1918
vagrant, – pertaining to benthos that is capable of active movement on, in or above the substrate. / vagrant; zwervend / vagile / – – / bentos errante

1919
sedentary, – organisms that move about, but very little, or are permanently attached to a substrate. /sedentair / sédentaire / sitzend; festsitzend; sesshaft / sedentario

sessile, – organisms that are permanently attached to a substrate. / sessiel / sessile / sessil; festgeheftet / sésil

[1540] TAXONOMY

1920
taxonomy, –
1. the systematic classification of organisms;
2. the science of classification of organisms, including principles, procedures and rules.

/ taxonomie / taxonomie; taxinomie / Taxonomie / taxonomía

1921
classification, – that branch of taxonomy that deals with de definition, ranking and arrangement of the various categories to which organisms are assigned. / klassificatie / classification / Klassifikation; Einteilung / clasificación

natural classification, – taxonomic classification based on the phylogenetic relationships of the organisms classified. / natuurlijke klassificatie / classification naturelle / natürliche Klassifikation / clasificación natural

1922
uninominal nomenclature, – the designation of a taxonomic category by a scientific name consisting of a single word; required for the categories above the species. / uninomische nomenclatuur / nomenclature uninominale / uninominale Nomenklatur / nomenclatura monómica

binominal nomenclature; binomial ....; binary ...., – the system of nomenclature, first standardized by Linaeus and adopted by the International Congresses of Zoology and Botany, by which the scientific name of an organism is designated by both a generic and specific trivial name. / binomische nomenclatuur / nomenclature binominale / binominale Nomenklatur / nomenclatura binómica

trinominal nomenclature, – an extension of the binominal system of nomenclature to permit the designation of subspecies by a three-word name. / trinomische nomenclatuur / nomenclature trinominale / trinominale Nomenklatur / nomenclatura trinómica

1923
validity, – conformity or non-conformity of a taxonomic name to all the rules of nomenclature; a name in harmony with these rules is valid and the correct name for use; if it fails to conform to any one or more rules, it is invalid or nonvalid. / validiteit; geldigheid / validité / Gültigkeit / validez

1924
law of priority, – the valid name of a genus or a species can only be that name under which it was first designated in conformance with the requirements of the International Rules of Nomenclature. / prioriteitswet / loi du priorité / Prioritätsgesetz / ley de prioridad

1925
hierarchy, – in classification, the system of ranks that indicates the taxonomic level of various taxonomic categories. / hierarchie / hiérarchie / Hierarchie / jerarquía

1926
taxon (plur. taxa), – formally named group of organisms, e.g. species, genus, family. / taxon / taxon / Taxon / taxon

1927
category, – term used to designate a group of any definite rank to which organisms are assigned. / kategorie / catégorie / Kategorie / categoriá

1928
complex, – a neutral term for a number of related taxonomic units, most commonly involving units in which the taxonomy is difficult or confusing. / complex / complexe / Komplex / complejo

1929
group, – a neutral term for a number of related taxonomic units, especially an assemblage of closely related species within a genus. / groep / groupe / Gruppe / grupo

phylum, – a primary division of the animal and plant kingdoms, containing a group of closely related classes. / fylum; stam / phylum; embranchement / Phylum; Hauptabteilung; Stamm / filum

class, – a major subdivision of a phylum; in a few phyla some biologists recognize certain superclasses and/or subclasses. / klasse / classe / Klasse / clase

order, – a taxonomic category in the classification of plants or animals, ranking between a class and a family; an order exists of one or more families. / orde / ordre / Ordnung / orden

1930
family, – a taxonomic category including one or more genera that have certain phylogenetic characters in common; a category ranking above a genus and below an order. / familie / famille / Familie / familia

superfamily, – a taxonomic category below an order and including a number of families that have certain features in common. / superfamilie / superfamille; super-famille / Superfamilie / superfamilia

subfamily, – a taxonomic category intermediate between a family and a tribe. / subfamilie /sous-famille / Unterfamilie / subfamilia

1931
tribe; tribus, – a taxonomic category inserted between the genus and the subfamily. / tribus / tribu / Tribus / tribu

1932
form genus, –
1. a series of related genera resulting from the splitting up of an old familiar genus;
2. in palaeobotany used when the genus name refers to only parts of the plant, e.g. leafs, seeds.

/ vormgenus / genre de forme / Formgattung / forma del género

1933
genus, – a taxonomic category that consists of one or more species of organisms, presumably of common phylogenetic origin. / genus; geslacht / genre / Genus; Gattung / género

subgenus, – a group of species that have special characteristics in common and that is distinct from other such groups or subgenera. / ondergeslacht; subgenus / sous-genre / Untergattung; Subgenus / subgénero

1934
gens, – a species group considered to be the aggregate of all those species that possess in common a large number of essential properties and are continuously related either in space or in time. / gens; phylogenetische reeks / gens / Gens / gens

1935
superspecies; artenkreis, – a monophyletic group of entirely or largely allopatric species. / superspecies / – – / Artenkreis / superespecie

1936
formenkreis, –
1. in biology, a collective category of allopatric subspecies or species;
2. in palaeontology, a group of related species or variants.
/ – – / cercle de formes / Formenkreis / – –

1937
species, – group of natural populations that actually (or potentially) interbreed and that are reproductively isolated from all other such groups; a taxonomic category lower than a (sub)genus and above a sub-species. / soort; species / espèce / Art; Spezies / especie

1938
rassenkreis, – a species consisting of several intergrading subspecies that inhabit imperfectly separated adjacent biotopes. / – – / – – / Rassenkreis / – –

1939
subspecies, – a geographically defined aggregate of local populations that differs taxonomically from other such subdivisions of species. / ondersoort; subspecies / sous-espèce / Unterart; Subspezies / subespecie

race, – commonly applied as synonym of a subspecies, sometimes applied to a local population within a subspecies. / ras / race / Rasse / raza

1940
variety, – term often used loosely for various intraspecific forms; more precisely the term is restricted to discontinuous variants within a single interbreeding population. / variëteit / variété / Varietät, Spielart; Abart / variedad

1941
population, – the individuals of a given locality that potentially form a single interbreeding community. / populatie / population / Population / populación; población

1942
ecotype, – a descriptive term used for plant subspecies that owe their more conspicuous characteristics to the selective effects of local environments. / ecotype / écotype / Ökotyp / ecotipo

1943
congeneric, – pertaining to a group of species in the same genus. / congenerisch / congénérique / kongenerisch / congenérico

conspecific; cospecific, – pertaining to specimens or populations in one species. / gelijksoortig / conspécifique; co-spécifique / konspezifisch / coespecífico

1944
infraspecific, – pertaining to categories within a species, e.g. subspecies. / infraspecifisch / infraspécifique / infraspezifisch / infraespecífico

1945
monotypic; monotypical, – pertaining to a taxon that contains only one immediately subordinate zoological or botanical unit, e.g. an order containing only one family, a genus containing only one species. / monotypisch / monotypique / monotypisch / monotípico

polytypic, – pertaining to a taxon that contains two or more immediately subordinate categories, e.g. a species consisting of a group of subspecies that replace each other geographically. / polytypisch / polytypique / polytypisch / politípico

1946
type locality, – the locality at which a holotype, lectotype, or neotype was collected. / typelokaliteit / localité type / Typlokalität / localidad tipo

1947
type genus, – the genus selected by the original author, when establishing a new family, and that he considers as most appropriately representative, whose name serves as the base for the name of the family. / typegeslacht / genre type / Typus-Gattung / género tipo

1948
generitype; generotype, – the type species of a genus. / generitype; generotype / généro type / – – / Genotypus / – –

type species, – a particular species in a genus that was originally designated as having the characteristics of the genus in which it occurs. / typesoort / espèce type / Typus-Art / especie tipo

1949
holotype, – the single specimen designated as the type by the original author at the time of the publication of the original description. / holotype / holotype / Holotypus / holotipo

allotype, – a paratype of the opposite sex to the holotype which is designated as such. / allotype / allotype / Allotypus / alotipo

1950
paratype, – a specimen other than the holotype considered by the author in the original description and that was designated as such. / paratype / paratype / Paratypus / paratipo

1951
syntype; cotype, – any specimen of the author's original material of a particular taxon when no holotype was designat-

ed. / syntype; cotype / syntype; cotype / Syntypus; Cotypus / sintipo; cotipo

lectotype, – one of a series of syntypes that, subsequent to the publication of the original description of a taxon, is selected (by publication) to serve as the type specimen of that particular taxon. / lectotype / lectotype / Lectotypus / lectotipo

neotype, – a specimen selected as type subsequent to the original description in cases when the primary types are definitely known to be destroyed. / neotype / néotype / Neotypus / neotipo

1952
plesiotype, – a specimen or specimens upon which subsequent descriptions or figures are based. / plesiotype / plésiotype / Plesiotypus / plesiotipo

1953
topotype, – a specimen, not of the original type series, collected at the type locality. / topotype / topotype / Topotypus / topotipo

1954
metatype, – a specimen compared by the author of the species with the holotype and determined by him as conspecific with it. / metatype / métatype / Metatypus / metatipo

homotype, – a specimen compared by another than the author of a species with the type and determined by him to be conspecific with it. / homotype / homotype / Homotypus / homotipo

1955
synonymy, – a chronological list of all scientific names that have been applied, correctly or incorrectly, to a particular taxon. / synonymie / synonymie / Synonymie-Liste / sinonimía

synonym, – one or more different names applied to the same taxon. / synoniem / synonyme / Synonym / sinónimo

homonym, – one of two or more identical but independently proposed names for the same or different taxa. / homoniem / homonyme / Homonym / homónimo

1956
cheironym, – a manuscript name. / cheironiem / chironyme / – – / queirónimo

1957
tautonym, – a name in which the specific name is merely a repetition of the generic name. / tautoniem / tautonyme / Tautonym / tautónimo

1958
nomen conservandum, – scientific name whose usage has been preserved by decision of the International Commission on Zoological Nomenclature, even though its use would be in conflict with the rules of nomenclature.

nomen nudum, – a published scientific name that does not meet the requirements of the International Rules of Nomenclature.

nomen rejectum, – published scientific name that is rejected for any valid reason and should not be used for that particular organism.

[1550] PHYLOGENY

1959
phylogeny; phylogenesis, – the line, or lines, of direct descent in a given group of organisms; also the study of such relationships. / fylogenese; fylogenie; afstammingsleer / phylogenèse; phylogénie / Phylogenese; Phylogenie; Stammesgeschichte / filogenía; filogénesis

1960
phylogenetic tree; dendrogram, – a diagrammatic presentation of assumed lines of descent, based on palaeontological, morphological or other evidence. / fylogenetische stamboom / arbre phylogénique / Ahnenreihe; phylogenetische Reihe / árbol filogenético

phyletic, – pertaining to an evolutionary line of descent. / fyletisch / phylétique / phyletisch / filético

monophylectic, – pertaining to a taxon whose units are all derived from a single immediate line of descent. / monofyletisch / monophylétique / monophyletisch / monofilético

polyphylectic, – having evolved along several lines of descent and so alike as to be included in one taxon, as with some genera of Foraminifera. / polyfyletisch / polyphylétique / polyphyletisch / polifilético

1961
plexus, – the multiple lines of descent found in the phylogeny of many species. / plexus / – – / Plexus / plexo

1962
cladogenesis, – phyletic splitting or branching. / cladogenese / cladogenèse / Kladogenese / cladogénesis

1963
taxogenesis, – origin and history of a taxon. / taxogenese / – – / Taxogenese / taxogénesis

1964
evolution, – origin, ancestry and differentiation of organisms; includes the slow, long-time continuous adaptation of organisms to their ever-changing environment through such agencies as variation, mutation and selection. / evolutie; afstamming / évolution; descendance / Evolution; Abstammung / evolución

1965
lineage; branch; line of evolution, – a series of species (and genera) that form an evolutionary series, each one being ancestral to its successor in the geological sequence. / tak; afstammingslijn; afstammingsreeks / branche; rameau; lignée / Ast; Zweig / línea de evolución; línea evolutiva

1966
Darwinism, – concept of organic evolution, developed by Charles Darwin, resulting from variation and the selection

of favoured individuals through natural selection. / Darwinisme / Darwinisme / Darwinismus / darwinismo

neo-Darwinism, – concept of evolution that combines the original Darwinian ideas with modern concepts of genetics and speciation. / neo-Darwinisme / Néo-darwinisme / Neo-Darwinismus / neo-darwinismo

natural selection, – complex of processes by which the collective factors of the environment eliminate those individuals of a population that are least fitted to that particular environment; 'survival of the fittest'. / natuurlijke teeltkeus / sélection naturelle / natürliche Zuchtwahl; .... Auslese / selección natural

1967
Lamarckism; Lamarckianism, – the theory, advocated by Jean Baptiste de Lamarck, to explain evolution on the basis of inheritance of acquired characteristics and habits. / Lamarckisme / Larmarckisme / Lamarckismus / lamarckismo

neo-Lamarckism, – the theory that evolutionary changes are caused directly by environmental influences on the organism as well as by the reaction of the organism to the environment. / neo-Lamarckisme / Néo-lamarckisme / Neo-Lamarckismus / neo-lamarckismo

1968
palingenesis, – the recapitulation in the young developmental stages of an organism of characteristics of their phylogenetic ancestors. / palingenese / palingenèse / Palingenese / palingénesis

biogenetic law; recapitulation theory, – the theory (now rejected) that the individual development of an organism passes through stages resembling the adult conditions of its successive ancestors. / biogenetische grondwet / loi biogénétique / biogenetisches Grundgesetz / ley de recapitulación

1969
speciation, – process of formation of new species through evolutionary development. / speciatie / spéciation / – – / especiación

centre of origin, – area in which a species first appeared and from which it subsequently spread. / ontstaansgebied / centre d'origine / Entstehungsgebiet / centro de origen

divergence, – the spreading or phyletic ramification of a group of organisms from a common ancestral type into divergent descendant types, each with a distinct and characteristic adaptive status and ecological relationship. / divergentie / divergence / Divergenz / divergencia

1970
microevolution, – evolution based on the gradual accumulation of small genetic changes within a species, leading to the formation of a slightly differing new subspecies. / microëvolutie / microévolution / Mikroevolution / microevolución

macroevolution, – evolution of categories higher than infraspecific ones. / macroëvolutie / macroévolution / Makroevolution / macroevolución

1971
quantum evolution, – relatively rapid evolution, under strong selective pressure from one type of adaptation to a distinctly different one.

iterative evolution, – repeated development of new forms from a conservative stock. / iteratieve evolutie / évolution répétitive / iterative Evolution / evolución iterativa

parallel evolution, – the phenomenon whereby related but distinct phyletic stocks develop similar forms by nearly identical evolutionary changes through time. / parallele evolutie / développement parallèle / parallele Entwicklung / evolución paralela

1972
homologous, – pertaining to similarity of structures in different groups of organisms, assumed to be due to inheritance of these structures from a common ancestry. / homoloog / homologue / homolog / homólogo

1973
convergent evolution; adaptive convergence, – evolutionary processes that result in superficial morphological similarity among distantly related taxons. / convergente evolutie / évolution convergente; convergence adaptive / konvergente Entwicklung / evolución convergente

convergence; convergency, – the morphological similarity among distantly related animals or plants. / convergentie / convergence / Konvergenz / convergencia

1974
isomorph, – pertaining to an organism, similar to another, although phylogenetically unrelated to it. / isomorf / isomorphe / isomorph / isomorfo

1975
orthogenesis, – evolution continuously in a single direction over a considerable period of time; the term usually carries the implication that the direction is determined, irrespective of the effects of natural selection, by some factor internal to the organism. / orthogenese / orthogenèse / Orthogenese / ortogénesis

rectilinear evolution, – continued evolutionary change in the same direction and over a long period of geological time; similar to orthogenesis but without implication concerning the factors that determine and maintain this direction of evolution. / gerichte evolutie / évolution rectiligne / gerichtete Evolution / evolución recto-lineal; .... rectilínea

orthoselection, – natural selection causing evolution to progress continuously in a single direction over a considerable period of time. / orthoselectie / – – / – – / ortoselección

1976
cryptogenic, – pertaining to a type or species of unknown descent. / cryptogeen / cryptogène / kryptogen / criptógeno

1977
katagenesis, – a decrease or apparent regression in the evolutionary rate of change of an organism. / katagenese / catagenèse / Katagenese / katagénesis; catagénesis

bradytely, – persistence of a group of organisms for a very long period of time with practically no evolutionary advancement. / bradytelie / bradytélie / – – / braditelía

1978
rudimentary; vestigial, – said of organic structures that are

not as highly developed in the descendants as they were in the ancestors. / rudimentair / rudimentaire / rudimentär; verkümmert / rudimentario

1979
genotype, – in genetics, the category to which an individual belongs on the basis of its genetic constitution, without regard to its external appearance. / genotype / génotype / Genotypus / genotipo

phenotype; phaenotype, – the category to which an organism belongs on the basis of visible characteristics as the result of an interaction between genotype and environment. / fenotype / phénotype / Phänotypus / fenotipo

archetype; prototype, – a hypothetical ancestral type arrived at by elimination of specialized characteristics. / archetype; prototype; oertype / archétype / Prototyp / arquetipo

1980
adaptation, – peculiarity of structure, physiology, or behaviour of an organism that assists in fitting the organism into a particular environment; also the process by which such fitness is acquired. / aanpassing; adaptatie / adaptation / Anpassung; Adaptation / adaptación

adaptive, – fitted for a particular environment. / adaptief / adaptive / adaptiv / adaptable

homoplastic, – pertaining to a similar adaptation through parallel evolution. / homoplastisch / homéomorphe / – – / homoplástico

preadaptation, – fitness for an environment which the organism does not occupy, at the time when the adaptation appears; usually applied to an advantageous mutation that will permit the invasion of a new habitat. / pre-adaptatie / préadaptation / Präadaptation / preadaptación

1981
isolation, – effective separation and the prevention of interbreeding of a group of specimens of any species from another group of the same species or different species. / isolement; afzondering / isolement / Isolation; Absonderung / aislamiento

1982
degeneration, – loss of function or structure as an evolutionary process over many generations. / degeneratie / dégénérescence / Degeneration / degeneración

mutation, – a spontaneous change in the genetic constituents of an organism. / mutatie / mutation / Mutation / mutación

saltation, – discontinuous variation in a phylectic series produced at a single step by mutation. / saltatie / saltation / Saltation / saltación; salto

1983
neomorph, – pertains to a characteristic that is newly formed by mutation and is not present in an ancestor. / neomorf / néomorphe / neomorph / neomorfo

1984
cline, – a gradual and nearly continuous change of a characteristic in a series of continuous populations.

1985
acme, – maximum development, usually in abundance or frequency, of a species, genus or other taxon. / acme; bloeitijd / acme; épanouissement / Acme; Blütezeit / apogéo

paracme, – the state of decline in the phylogeny of a taxon, following the acme of its development. / paracme / paracme / Paracme / para-apogéo

relic, – any species or other taxon that is a survivor of a nearly extinct group. / relict / relique / Relikt / relicto

1986
extinction, – the disappearance of a species or other group of organisms representing a particular evolutionary line. / uitsterven / extinction / Aussterben; Artentod / extinción

[1600] GEOMORPHOLOGY

NOTE:
– for terms on volcanoes, sea floor, reefs and fresh-water marshes see sections
[1230] Volcanic Landforms,
[1820] Sea-Bottom Topography
[2140] Carbonate Sediment Bodies and Settings,
[2422] Biochemical Phase; Environments of Deposition

[1610] WEATHERING AND EROSION

NOTE:
– see also
[1700] Pedology
[1930] Sedimentary Processes

[1611] *Physical and Chemical Weathering*

1987
geomorphology, – the description of nature and natural phenomena and the investigation of the history of geologic changes through the interpretation of topographic forms. / geomorfologie / géomorphologie / Geomorphologie / geomorfología

1988
weathering, – disintegration and decomposition of rock, i.e. destruction by physical, chemical and/or biological processes. / verwering / altération (météorique); désagrégation et décomposition; désagrégation physico-chimique / Verwitterung; Zersetzung; Zerfall / meteorización

1989
physical weathering; mechanical ....; disintegration; disaggregation; rock breaking, – / fysische verwering; mechanische .... / désagrégation physique; .... mécanique / physikalische Verwitterung; mechanische .... / meteorización física

1990
block disintegration, – rock separation into blocks. / uiteenvallen in blokken / désagrégation en blocs / Zerstückelung; Zerfall / desintegración en bloques

mineral disintegration; granular ....; granulation, – the final separation of rocks into the individual, composing mineral parts. / verkorreling / désagrégation granulaire; décomposition arénacée / Zerfall in Mineralkörner / desintegración mineral

1991
crumbling; breaking down; reduction, – of rocks. / vergruizing; verbrokkeling / effritement; émiettement / Zerbröckelung; Vergrusung / desagregación

1992
(destruction by) insolation, – disintegration of rocks due to changes in temperature. / droge verwering; verwering door temperatuurwisseling / désagrégation thermique / Temperatursprengung / destrucción por insolación; desagregación térmica

1993
sun crack; shrinkage ...., – / krimpscheur / fente de contraction; fissure de retrait; forme de .... / Schrumpfungsriss; Trockenriss / rotura (por insolación); fisura de retracción (por insolación)

heat crack, – crack due to vigorous changes in temperature, causing a rock to fall apart into two or more pieces. / hittebarst; ....scheur / craquelure thermique / Kernsprung / desintegración por calor; rotura por ....

1994
exfoliation; scaling; desquamation; onion-skin weathering, – a weathering process, consisting in the peeling off of thin slabs from the rock surface. / bolschaalverwering; afbladdering; afschilfering / exfoliation; desquamation / Abblätterung; Auf....; schalenförmige Abschuppung / exfoliación; descamación

1995
sheeting; splitting; topographic jointing; unloading, – division of rock into sheets or beds by joint-like fractures, that are generally parallel to the ground surface. / splijting / exfoliation / – – / hojosidad por fracturas

Kernsprung, – a German term, denoting boulders that have been fractured into a small number of large fragments, due to expansion of a kind similar to unloading.

1996
frost weathering; .... wedging; .... splitting; .... shattering; .... work; congelifraction, – disintegration and mechanical breakdown of rock by freezing of water, present in the pores and along joints and bedding planes. / gelivatie; vorstverwering / gélivation / Frostverwitterung; ....sprengung; Spaltenfrost; Spaltenfrostverwitterung / meteorización por helada; desagregación por ....

gelivity, – sensitiveness of rock to frost weathering. / gevoeligheid voor vorstverwering / gélivité / Empfindlichkeit für Frostsprengung / gelividad

1997
salt wedging, – weathering due to the dilatation of crystals by absorption of water / zoutwigwerking / – – / Salzsprengung / hidratación de sales; meteorización por hidratación de sales

1998
chemical weathering; decomposition; decay, – chemical destruction and alteration of rock, resulting in the formation of new stable chemical compounds. / chemische verwering / décomposition (chimique); altération chimique / chemische Verwitterung / meteorización química

1999
humic decomposition, – decomposition by the chemical action of vegetal acids. / biologische verwering / décomposition humique / biologische Verwitterung / descomposición húmica

2000
carbonation; carbonatization,– a decomposition, by which carbon dioxide is added to the minerals of a rock, forming carbonates or bicarbonates. / (verwering door) koolzuurinwerking; carbonatisatie / carbonatation / Karbonatisierung / carbonatización

2001
honeycomb, – pertaining to a type of solution weathering causing facets in limestone resembling honeycombs. / honingraat / alvéole / Waben / alveola

2002
tafone (Corsica), – cavernous weathering pit.

2003
spheroidal weathering, – weathering of blocks, formed by jointing, to spheroidal masses. / wolzakverwering / décomposition en boules / Wollsackverwitterung / meteorización esferoidal

2004
selective weathering; differential ...., – / selectieve verwering / décomposition sélective; désagrégation .... / selektive Verwitterung; Verwitterungsauslese / meteorización selectiva

2005
detritus; detrital material; waste; weathering waste; rock ....; debris; rubble, – any loose material that results directly from rock disintegration. / verweringspuin; gesteentegruis; detritus; detritisch materiaal / éboulis; matériel détritique;

débris / Verwitterungsschutt; Gesteins....; Verwitterungstrümmer; Gesteinsgrus / detritus; material detrítico

detrition, – / puinvorming / formation d'éboulis / Trümmerbildung; Gesteinszerfall / – –

2006
debris plain, – plain covered with rock waste. / puinveld; ....vlakte / champ de débris / Trümmerfeld; Schuttfläche / llanura de derrubios

2007
talus slope; scree, – slope consisting of fallen-down rock waste. / puinhelling / talus d'éboulis / Schutthang; ....halde / glacis / coluviones de ladera; talus de derrubios

talus cone, – a cone of debris at the base of a cliff or slope or at the end of a rockfall track. / puinkegel / cône d'éboulis / Schuttkegel / cono de deyección; .... de derrubios

talus fan; apron of rubble; alluvial fan, – an extensive deposit of detrital material, with in general, a low outward slope. / puinwaaier / cône d'alluvions / Schuttfächer / cono de aluviones; cono aluvial

2008
regolith; mantle rock, – the cover of loose material consisting of soils, sediments and broken rock overlying the bedrock. / regoliet; los oppervlaktemateriaal / manteau de débris superficiel; régolite / Lockerboden; Schuttdecke / regolita; manto rocoso; .... superficial de derrubios

2009
residual soil; waste mantle, – the more or less thick cover of weathered material that has not undergone transportation other than creep. / verweringsbodem; ....laag / sol résiduel / Verwitterungskrume; ....boden / suelo residual; covertura residual superficial

saprolite, – weathered residual rock in place. / saproliet / – – / – – / saprolita

2010
weathering crust, – crust formed by superficial weathering of a rock. / verweringskorst / manteau d'altération (superficielle); croûte .... / Verwitterungskruste; ....rinde / costra de meteorización; .... de alteración superficial

2011
bedrock, – the solid rock underlying unconsolidated superficial formations and weathered zone. / vast gesteente / roche solide / anstehendes Gestein / roca viva; roca sólida

[1612] *Erosion and Denudation*

2012
erosion, – in parts of Europe this term is being used predominantly for the wear of solid rock by the impact of detrital fragments and particles carried by rivers, wind, glaciers or the sea, including the removal of the loosened material. / erosie / érosion; .... linéaire / Erosion; linienhafte ....; lineäre .... / erosión

erosion; denudation; degradation, – in the USA and Germany the meaning predominantly attached to these terms is the combined effect of all processes of terrestrial degradation, including weathering, transportation, chemical and mechanical action of running water and the mechanical action of wind, ice etc. / denudatie / abaissement; érosion; dénudation / Erosion; Abtragung; Denudation / erosión; denudación; degradación

2013
denudation; ablation; surface wastage, – by European authors this term is used predominantly in the limited sense of transportation of rock fragments; according to some (German) authors the term is restricted to subaerial processes. / denudatie; ablatie / dénudation; ablation / Denudation; Abtragung; Ablation / denudación; ablación

2014
corrasion, – mechanical wear of rocks by the effect of material being transported by wind, ice or running water. / corrasie / corrasion / Korrasion / corrasión

2015
stream erosion; .... corrasion, – mechanical wear by rivers. / fluviatiele erosie; .... corrasie / érosion fluviatile / Flusserosion; fluviatile Erosion / erosión fluvial

glacial erosion; .... abrasion; .... corrasion, – mechanical wear by glaciers. / glaciale erosie / érosion glaciaire / Gletschererosion / erosión glacial; abrasión ....

wind erosion; .... corrasion; .... abrasion; eolian erosion; .... corrasion; .... abrasion; wind scour, – mechanical wear by wind. / winderosie; ....corrasie / érosion éolienne; corrasion / Korrasion; Sandschliff / erosión eólica; abrasión ....

2016
marine abrasion; .... erosion; wave erosion, – the erosive action of breakers on the coast. / abrasie; mariene erosie / abrasion; érosion marine / Abrasion; Brandungserosion; marine Erosion; .... Abrasion / abrasión marina; erosión ....

2017
corrosion; chemical erosion, – erosion by solution. / corrosie; chemische erosie / corrosion; érosion chimique / Korrosion; chemische Verwitterung / corrosión; erosión química

2018
gully erosion; gullying; ravinement, – erosion by which a great number of consequent ravines are or gullies developed on slopes. / geulerosie / érosion en ravins; ravinement / dichte Zerschneidung durch Erosionsrinnen / erosión superficial por numerosos abarrancamientos; .... en barrancos; abarrancamiento

2019
dissection, – erosion resulting in destroying the continuity of a relatively even surface by the cutting of ravines or valleys. / doorsnijden met dalen / dissection / Zertalung; Zerschneidung / disección

2020
incision, –
1. the down-cutting action of a stream. / insnijding / creusement; incision / Einschneidung / incisión
2. notch; cut, – the cutting made by a stream. / insnijding / incision / Einschnitt; Backriss; Schlucht / incisión

2021
headward erosion; headward migrating ....; headwater ....; retrogressive ...., – erosion extending in upstream direction. / terugschrijdende erosie / érosion régressive; .... remontante / rückschreitende Erosion; rückläufige .... / erosión remontante

2022
lateral erosion, – / zijwaartse erosie / érosion latérale / Seitenerosion; Lateral....; Breiten.... / erosión lateral

vertical erosion; grooving ....; downward ...., – / diepte-erosie; verticale erosie / érosion verticale; creusement / Tiefenerosion; Vertikal.... / erosión vertical

2023
undercutting action; undermining ...., – of running water. / ondergravende werking / sapement / Unterwaschung; Unterschneidung / – –

2024
differential erosion, – / selectieve erosie / érosion différentielle / selektive Erosion / erosión diferencial

2025
base level of erosion; general base level of ....; permanent base level of ...., – the theoretical limit towards which erosion constantly tends to reduce the land. / erosiebasis; algemene .... / niveau de base de l'érosion; .... de base général / Erosionsbasis; allgemeine .... / nivel de base de erosión; .... de base general

local base level of erosion; temporary base level of ...., – e.g. at the confluence of a tributary with the main stream, or at the top of a hard ridge. / locale erosiebasis; plaatselijke ....; tijdelijke .... / niveau de base partiel; .... temporaire; .... local / örtliche Erosionsbasis; lokale .... / nivel de base parcial; .... de base temporal

2026
rill; wet-weather ....; gully; wash-out channel, – / geultje; regengeul / rigole de ruissellement / Regenfurche; ....rille; Spülrinne; Wasserriss; Regen....; Erosionsrinne; Runse; Rachel / cárcava

ravine; gully; gulch, – / stortbeekkloof; ravijn / gorge de torrent / Schlucht; Runse; Tobel / barranco

2027
pothole, – an excavation in the bed of a stream, due to the erosive action of whirling water. / kolkgat / marmite torrentielle / Kolk; Kolkloch; Strudeltopf; ....loch; ....kessel; Erosionskessel / marmita torrencíal

2028
plunge pool, – a relatively large erosion hollow, e.g. below a waterfall. / kolkgat / marmite de géant; chaudière / grosser Erosionskessel / marmita de gigante

2029
earth pyramid; .... pillar; demoiselle, – boulder-capped pinnacle consisting of easily eroding material. / aardpyramide / pyramide de terre; demoiselle; cheminée de fée; pilier de terre / Erdpyramide; Erdpfeiler / pirámide de tierra; pilar de ....

2030
badland, – an area without or with very scarce vegetation, dissected by innumerable deep gullies. This severe gully erosion mostly occurs in arid and semi-arid areas by the action of occasional torrential rains. / – – / territoire raviné; badland / stark zertaltes Ödland / tierras malas

[1613] *Removal*

2031
surface wash, – the downslope transport of regolith material over the surface through moving water. / afspoeling / transport par ruissellement / Abspülung / lavado superficial

2032
sheet wash; .... flood; sheet-flood erosion; sheet ...., – the removal of finely weathered particles in sluggishly off-flowing films of water or in a network of shallow rills not yet concentrated or collected into definite channels. / vlakte-erosie; vlakteafspoeling / ruissellement en nappe; .... en surface; érosion en nappe; / Schichtfluterosion; Schichtabspülung; Flächenspülung; ....erosion; flächenhafte Erosion; .... Abtragung; .... Abspülung / – –

2033
rill wash; rill erosion, – the removal of material by way of extremely fine gullies. / – – / érosion en rigoles; ruissellement en filets / Rinnenerosion; Rillenabspülung / – –

2034
subsurface wash, – the transport of small soil grains, clay particles, clay and other colloids and material in ionic solution by water flowing within the regolith. / – – / lessivage latéral; transport par l'eau interstitielle / Tonverlagerung / – –

2035
rain splash, – transport of soil particles downslope due to the impact of raindrops. / spaterosie / érosion par impact de la pluie / – – / – –

2036
soil erosion; accelerated ...., – transport and removal of soil as a result of human activity. / bodemerosie / érosion du sol / Bodenerosion / erosión del suelo

[1614] *Mass Movement*

2037
mass movement; .... transport; .... wasting, – the gravitative transfer of loosened material where large quantities of debris move together as a coherent whole in close grain to grain contact. / massabeweging; ....transport / movement en masse; transport en .... / Rutschung / movimiento en masa; transporte en ....

2038
soil creep, – the slow downslope movement of soil and rock debris; usually it is not perceptible except to observations of long duration. / kruip; kruipen / reptation / Bodenkriechen; Gekriech / desplazamiento lento de restos superficiales

2039
seasonal creep, – the slow downslope movement of the regolith as a result of the net effect of movements of its individual particles. These movements are caused by systematic reworking of the surface soil layers due to soil moisture and temperature variations, by random movements due to organisms and by the action of downhill shear stress. / – – / reptation saisonnière / periodisches Bodenkriechen / – –

continuous creep, – a slow but continuously distributed strain (deformation) of the soil resulting from stress that is either external to the material affected or is caused by the weight of a considerable thickness of such material. / – – / reptation permanente / permanentes Bodenkriechen / – –

2040
talus creep; rock ...., – the downslope movement of debris on talus slopes; the movement is especially rapid where frequent freeze and thaw occurs. / kruipen van puin / reptation des éboulis; .... de débris / Schuttkriechen / desplazamiento de talud

2041
frost creep; solifluction, – the downward movement of material due to expansion and contraction by alternate freezing and thawing. The term solifluction is confined mostly to phenomena of cold climates. / solifluctie; bodemvloeiing / solifluction; solifluxion / Solifluktion; Bodenfliessen / soliflucción; solifluxión

2042
gelifluction, – flowage of the material when the soil becomes thoroughly water-logged.

2043
slide; sliding, – a type of mass movement in which the material is slipping along a well defined surface; the mass above the slide surface moves as a unit without internal shear. / afglijding / glissement / Rutsch; Rutschung im Verband / deslizamiento

2044
landslide; landslip, – collective term for several types of mass movement of rock debris, characterized by a relatively low to a high speed and a small water content of the moving mass. Landslides may cause obstruction in a valley and the formation of a lake behind it. / aardverschuiving / glissement de terrain / Bergsturz; ....rutsch / deslizamiento de tierras

2045
slump; slumping, – the downward slipping of one or several units of earth or rock debris along curved slip planes, taking place as an intermittent movement over short distances; usually a backward tilting of the surface of the units results. / afschuiving; aardverschuiving / glissement en forme d'écailles; .... de loupes / schuppenförmige Abrutschung / deslizamiento en forma de escamas

2046
debris slide; earth ....; soil slip, – the rapid rolling or sliding of unconsolidated earth debris, moving largely along a single slip plane; usually the amount of water is small. / aardstorting; puin....; puinafglijding / glissement de débris / Erdrutsch; Schuttrutschung / deslizamiento de derrubios; .... de suelos

2047
debris avalanche, – a flowing movement of rock debris in more or less narrow tracks down steep slopes. The debris mass disintegrates substantially during movement. / puinlawine; steen.... / avalanche de pierres / Steinlawine; Schutt.... / avalancha de piedras

2048
rock slide, – the movement of masses of loosened blocks of fresh or only slightly weathered rock, often mixed with the debris of rock decay, sliding down bedding, joint or fault surfaces, which sometimes have a lubricating function. / bergafglijding / glissement de montagne; .... en masse; éboulement / Bergsturz; Felsrutsch; ....rutschung / deslizamiento de montaña; .... en masa

2049
rock fall, – movement of masses of fresh or only slightly weathered rock, precipitating from cliffs, mountain or valley sides when these are undercut at the base; the movement is not guided by slip planes and is independent of lubricating surfaces. / bergstorting / éboulement; écroulement / Bergsturz; Fels....; Fall.... / deslizamiento de masas rocosas

2050
flow, – a type of mass movement in which there is no sharply defined failure surface, but instead shear is distributed throughout the moving mass. / – – / coulée / Rutschung mit gestörtem Verband / flujo

2051
earth flow, – the movement of water-saturated clayey or silty earth material down slopes, at a rate quicker than solifluction, but slower than mud flows; the time occupied in attaining a new equilibrium may range from hours to years. / aardstroom / – – / Erdrutsch; Erdschlipf / desprendimiento por deslizamiento

mud flow, – slow to very rapid movement of water-saturated rock and earth debris down distinct channels. / modderstroom / coulée de boue; .... boueuse / Schlammstrom; Mure / colada de lodo; .... de barro

2052
landslide scar, – the niche left at the place where a landslide was detached. / puinafglijdingsnis / niche de départ; .... d'ar-

rachement / Abrissnische; ....narbe / cicatriz del deslizamiento; nicho de partida; .... de arranque

2053
scarp, – upper part of a slip surface that becomes visible after downward slipping of a unit of earth or rock debris along a curved slip plane. / – – / escarpement / Abriszkante / – –

2054
landslide track, – the path followed by the landslide. / puinafglijdingsbaan / chemin d'éboulement; trajectoire ....; glissoir / Bahn des Erdrutsches; Bahn des Bergsturzes / trayectoria del deslizamiento

2055
slip plane; slip surface; shear plane, – the plane over which a unit of rock or earth or rock debris slips downwards. / glijvlak; afglijdingsvlak / surface de glissement / Gleitfläche / superficie de deslizamiento

2056
landslide topography, – characteristic, irregular topography of a region in which landslide material came to rest. / puinafglijdingsreliëf / relief d'éboulement / Bergsturzrelief; Rutschungsmorphologie / topografía de deslizamientos; relieve de ....

2057
block field; stone ....; felsenmeer, – a field of blocks that may be produced by several causes, the most important of which is frost shattering under the influence of a cold climate. / blokkenveld / mer de rochers; chaos de rocs; champ de pierres / Blockmeer; Felsen.... / campo de piedras; mar de ....

stone river; rock stream; block ....; boulder ...., – a block field concentrated into stream-like masses, moving downslope or along the axis of a valley. / steenstroom / coulée pierreuse; .... de pierres; .... de blocaille; .... de blocs / Blockstrom / colada pedregosa; .... de piedras; .... de bloques

2058
outcrop bending; .... curvature, – the phenomenon of turned-down strata in the weathered, superficial zone of the rocks in a slope; this phenomenon, often to be seen in quarries, is caused by creep. / – – / phénomène de fauchage / Hakenschlagen / cabeceo

2059
terracette; bench, – narrow ledge in a slope, formed by backward tilting of the surface of the units of a slump movement or by other slow, downslope movements. / – – / surface d'arrachement; banquette; rideau / – – / grada

2060
sheep track; .... path; pasture track, – narrow bench in a slope, serving as path for cattle; these benches can have the same origin as terracettes, but they also may be caused simply by the walking of the cattle. / – – / sentier de vaches; .... de moutons / Viehgangel; Schafstieg; Kuhtritt / asenderado; senderos de cabras

2061
field terrace, – terrace not caused by mass movements, but thrown up by men in behalf of agriculture. / – – / terrasse de culture / Kulturterrasse; Anbau....; Terrassenfeld / terraza artificial

2062
lynchet; reen, – the scarp of a field terrace or sheep track. / graft / rideau / – – / lindero

[1620] KARST AND CAVES

[1621] *Karst Processes and Solutional Forms*

2063
karst forming, – the development of forms caused predominantly by solution in regions with easily leachable rocks. / karstvorming / karstification / Karstbildung; Verkarstung / formación de karst; .... de carst

karstification, – the processes of solution and infiltration whereby the surface features and subterranean drainage network of a karst landscape are developed. / verkarsting / karstification / Verkarstung; Karstbildung / carstificación; karstificación

2064
karst; .... region, – limestone or dolomite region having a subterranean drainage due to rock solution, and showing typical karst phenomena. / karstgebied; karst / région karstique; Karst / Karstgebiet; Karst / carst; karst; región kárstica

karst landscape, – landscape showing a karst topography. / karstlandschap / paysage karstique / Karstlandschaft / paisaje cárstico; .... kárstico

NOTE:
– in Spanish both forms karst and carst are in use; in the following the latter is omitted.

2065
karst phenomena, – the typical phenomena caused by predominant action of corrosion. / karstverschijnselen / phénomènes karstiques / Karsterscheinungen; ....phänomene / fenómenos kársticos

2066
base level of karst erosion, – level on which the processes of corrosional landscape development conclude their action, e.g. on the top of non-soluble strata or at the base level of erosion. / corrosiebasis van de karst / niveau de base karstique / Korrosionsbasis / nivel de base kárstico

2067
deep karst, – karst phenomena in regions with limestone strata of very great thickness. / diepe karst / karst profond; .... complet; .... parfait / tiefer Karst; Tiefkarst; Holo.... / karst profundo

shallow karst, – karst phenomena in regions with comparatively thin limestone strata underlain by non-soluble strata. / oppervlaktekarst; ondiepe karst / karst superficiel; .... partiel; .... imparfait / seichter Karst; flacher ....; Halbkarst; Meso.... / karst somero; .... superficial

2068
naked karst, – karst topography not covered by any rock or weathering residue. / onbedekte karst / karst nu / nackter Karst / karst desnudo; .... descubierto

2069
covered karst, – a fossil or currently developing karst in limestone, which underlies superficial deposits or other rock, and that may produce landforms at the surface reflecting subsurface karstification. / bedekte karst / karst couvert / bedeckter Karst / karst cubierto; .... fósil

2070
subterranean karst, – karst topography developed under non-soluble and pervious strata (karst phenomena are younger than the covering layers). / onderaardse karst / karst sous-jacent / unterirdischer Karst / karst subterráneo

2071
karst depression, – depression caused by karst processes, e.g. doline, uvala, polje. / karstlaagte / dépression karstique / Karstwanne / depresión kárstica; dolina

2072
doline; dolina; sinkhole; sink; karst funnel, – an enclosed depression formed by the solutional enlargement of joints and subsequent settling of the surface and/or by subsidence caused by roof collapse of shallow caverns; sometimes, if choked, a small lake is present. / doline / doline / Doline; Erdfall; Karsttrichter / dolina

solution sink; funnel sink, – doline caused by solution of the bedrock. / oplossingsdoline; corrosie.... / doline de dissolution / Trichterdoline; Erosions....; Korrosions....; Doline / dolina de disolución; hundimiento por ....

collapse doline; .... sink, – doline formed by the collapse of the roof of an underground cavern. / instortingsdoline / doline d'effondrement / Einsturzdoline; Erdfall; / dolina por colapso; hundimiento por ....; dolina de hundimiento

NOTE:
- '**Erdfall**' is used often in the sense of '**Nachsackungserscheinung**' (the sinking away of covering material in subterranean karst hollows, and the forms caused by this process; **Nachsackungsdoline; Pinge**).

2073
uvala (Yugoslavia), – complex closed depression with more than one hollow, formed by the coalescence of neighbouring dolines. / uvala / ouvala / Uvala; Schüsseldoline / uvala

2074
polje (Yugoslavia), – a large karst basin with flat bottom, karstic drainage and steep peripheral slopes. The polje floor is in most cases covered by more or less impermeable residual material or alluvia, causing lake formation. / polje / polje / Karstwanne / polje

hum (Yugoslavia); pepino-hill; haystack hill (Puerto Rico); mogote (Cuba), – a residual hill of karst limestone, lying within a polje. / hum; karstrestberg / hum; butte-témoin karstique / Hum; Karstrestberg; Mosor; Karstinselberg / hum

ponor (Yugoslavia), – a swallow hole in a polje

2075
swallow hole; sink; swallet; streamsink, – the location in a karst limestone at which a surface stream wholly or partially commences an underground course. / verdwijngat / aiguigeois; chantoir; entonnoir; gouffre absorbant / Schluckloch; Saugloch; Schlundloch; Schwinde / – –

coastal swallow hole, – swallow hole in which sea water disappears, probably in consequence of the sucking action of ground-water currents. / zeeponoor / perte marine / Meeresschwinde / – –

2076
jama (Yugoslavia); **karst pit,** – a chimney-like shaft in limestone, some 100 m deep or more, that usually ends in underlying caverns; most jamas are formed by corrosion. / karstpijp / aven; abîme; jama / Karstschlot; ....brunnen; Naturschacht; Erdloch; Jama / – –

NOTE:
- some French regional terms are used for both, swallow hole and jama, e.g. **igue** (Petits Causses); **embut** (Jura); **bétoire; mardelle** (Bassin Parisien).

light hole, – a karst pit seen from the cave on which the pit debouches. / lichtschacht / cheminée / Lichtschacht / – –

2077
sand pipe; gravel ....; solution ...., – a subterranean karst pit filled by material from the overlying strata. / geologische orgelpijp; aardpijp / puits naturel; abannet / Erdpfeife / – –

2078
karst window, – a section of a subterranean river exposed by roof collapse. / karstvenster / fenêtre karstique / Karstfenster / ventana kárstica

2079
natural bridge; .... arch; karst bridge, –
1. a part of the roof of a former cavern, now forming a bridge over a collapse valley;
2. the not-collapsed part of the roof of a subterranean river between windows.

/ natuurlijke brug / pont naturel; arche naturelle / Naturbrücke / puente kárstico; .... natural

2080
cockpit karst, – a tropical type of karst landscape consisting of irregular solution hollows (the cockpits) separated by residual steep-sided conical hills. (The original meaning of 'cockpit' is a pit used during cockfights.) / cockpit-landschap / karst à dolines / Kugelkarst; Dolinen.... / karst con dolinas irregulares; paisaje de cockpits

2081
cone karst, – tropical karst in which the relief is dominated by projecting residual forms. / kegelkarst / – – / Kegelkarst / – –

2082
tower karst, – tropical karst with steeply sloping residual hills (sometimes vertical or overhanging) and swampy alluvial plains around the towers. / torenkarst / karst à tourelles / Turmkarst / karst con torretas

2083
blind valley, – a karst valley abruptly terminated by the passage underground of the water course, which has hitherto remained at the surface. / blind dal / vallée aveugle / blindes Tal; Blindtal / valle ciego

steephead; cul-de-sac valley, – form due to spring sapping or to collapse of substantial lengths of cave roof. / – – / vallée reculée / Sacktal / – –

2084
dry valley, – valley that has lost its water courses to underground drainage. / droog dal / vallée sèche; .... morte / Trokkental / valle seco; .... muerto

2085
collapse valley, – a valley caused by the collapse of the roof of an extensive cave system. / instortingsdal / vallée d'effondrement / Einsturztal / valle de hundimiento; .... colapsado

2086
karren, – minor solution channels on a karst surface caused by rain or soil water before it infiltrates underground. / lapies; karren / lapiés; lapiez / Karren; Schratten / lapiez

scallop; flute (USA); facet; pocketing; solution ripple; flow marking; current mark, – small, rather regularly spaced, asymmetrical corrosion pockets on limestone surfaces caused by a turbulent flow of water. The steep slope is at the upstream side. / uitschulping / cupule / Fliessfazette / golpe de gobia

2087
solution flute; scallop (USA), – longitudinal groove on fairly steep to vertical surface. Flutes run in sets straight down the steepest inclination with sharp ribs between. / oplossingsgroef / – – / Rillenkarren / estría de solución

solution runnel, – longitudinal groove in a limestone surface that increases in width and depth downstream through gathering volumes of run-off water. / geultjeskarren / lapié à rigoles / Rinnenkarren / canales de solución

solution slot; grike, – joint or cleavage plane, widened by solution. / spleetkarren / lapié de diaclases / Kluftkarren / pistas de solución

2088
Karrenfeld, – a German term, denoting a limestone surface of somewhat greater extent, intensively carved by minor solution channels. / karrenveld / champ de lapiés / Karrenfeld / – –

2089
karst spring; resurgence; rise – a spring in which original surface waters reappear after having been diverted to underground routes. / karstbron / source karstique; résurgence / Karstquelle / – –

2090
exsurgence, – the re-emergence at the surface, as a stream, of meteoric water that has fallen upon and percolated through a calcareous mass. / – – / (source d')exsurgence / – – / resurgencia

2091
perched karst spring, – emergence of underground water some way above the basement of a calcareous mass caused by the interbedding of an impermeable or less permeable stratum. / – – / source d'affleurement / Schichtquelle im Karst / fuente kárstica colgada

2092
estavelle, – a French term denoting a stream sink that in time of flood may reverse its function and acts like a spring. / estavelle / estavelle / Wechselschlund; Estavelle; Speiloch / – –

2093
vauclusian spring, – a spring whose water flows from large solution channels in limestone. / vauclusebron / source vauclusienne / Vauclusequelle; Riesen....; starke Karstquelle / fuente vauclusiana

[1622] *Caves*

2094
speleology; caveology, – / grottenkunde; speleologie / spéléologie / Höhlenkunde; Speläologie / espeleología

2095
cave; cavern; grotto; hollow; shake, – / grot; hol; spelonk / caverne; cavité; grotte / Höhle; Grotte; Spelunke / caverna; gruta; cavidad

2096
bedding cave, – a passage, usually wide and low, formed along a bedding plane in horizontal or slightly tilted rocks. / laaggrot / caverne suivant la stratification / Schichtfugenhöhle; Schichtgrenz.... / caverna según la estratificación

fault cave, – a cave developed along a fault or fault system. / breukgrot / caverne suivant une faille / Störungshöhle / caverna según fractura

fissure cave, – a cave developed along a fissure. / spleetgrot / caverne suivant une diaclase / Kluft(fugen)höhle / caverna según fisura

2097
shaft cave, – a cave consisting mainly of more or less cylindrical and vertical tube(s). / schachtgrot / gouffre / Schachthöhle / sima

2098
stoopway, – a passage passable only by stooping. / bukgang / passage surbaissé / Kriechgang / laminador

crawl passage; .... way; rabbit run; cat run; drainpipe, – / kruipgang / chatière / – – / gatera; galería en gatera

squeeze, – a narrow passage only just passable with effort. / – – / chatière / – – / – –

2099
tube, – a cave passage with a nearly circular or elliptic cross

section. / buis / boyau; méat / – – / tubo; paso en tubo

2100
meander channel, – a narrow erosion channel in the floor of a passage. / canyongeul / – – / Schluchtgang / meandro; galería meandriforme

2101
master cave, – a hall situated upstream of an actual or former sump; it is wide and high, and its floor is not or slightly inclined. / hoofdzaal / salle / – – / vestíbulo

2102
bridge, – in a cave, a rock slab in its natural position spanning a passage from wall to wall and inclining less than 45 degrees to the horizontal (not a fallen block). / brug / pont / – – / puente

partition, – in a cave, a rock in its natural position spanning a passage from floor to ceiling and inclining more than 45 degrees to the horizontal (not a fallen block). / schot / cloison / – – / partición

2103
blade, – a thin, more or less sharp projection jutting out from the roof, the wall or the floor, and of which it constitutes an integral part. / kling / lame tranchante / – – / estalactita de roca; estalagmita de roca

2104
abyss; pit; pot; pothole; chasm; shaft; ghyll, – / afgrond; put; schacht / abîme; gouffre; puits; aven; scialet / Schacht / sima

2105
pothole; pot, –
1. a vertical slope open to the surface;
2. a cave system containing slopes requiring the use of a tackle.

/ – – / marmite / – – / pozo

well; cenote (Yucatan); open pit, – / put / doline en puits / Brunnen / dolina en pozo

2106
ceiling pocket, – a down-facing solution cavity in a cave ceiling unrelated to joints. / – – / cloche de dissolution; coupole de .... / – – / – –

ceiling cavity, – solution cavity opening in a cave ceiling; orientation determined by joints. / – – / – – / – – / cavidad ascendente

ceiling tube, – a half tube developed in a cave ceiling, elongated along a joint. / – – / – – / – – / tubo ascendente

2107
ice cave, –
1. a pit at the bottom of which snow accumulates and forms ice. / ijskelder / glacière / Eiskeller / gruta helada
2. a cavern in which underground glaciers are formed. / ijsgrot / grotte glacée / Eishöhle / gruta glaciar

2108
blowing cave, – a cave from whose entrance air blows out. / – – / trou souffleur; soufflard / – – / cueva sopladora

breathing cave, – a cave at whose entrance air is alternately blown out and sucked in. / – – / – – / – – / caverna respirante

2109
blowhole, – a hole in the roof of a sea cave from which air escapes by the action of the waves. / – – / trou souffleur; soufflard / – – / bufador; bufadero

2110
cavern breakdown, – the collapse of the roof or the walls of a cave. / grotverval / effondrement / Höhlenverfall / hundimiento de caverna

roof fall; breakdown; block ...., – / loslaten van het dak / décollement; délitement; effondrement du toit; chute de bancs / Deckenbruch; ....einsturz / bloque caido

2111
wall fall; flake ....; flake, – large, thin fragments or plates of rock, standing almost on end in the passage section. / schilfer / – – / – – / escama de pared

2112
rock burst; bump, – spontaneous rock expansion. / autoklaas / rupture spontanée / Bergschlag; schlagendes Gebirge / estallido de roca

2113
pressure arch, – an arch developed above a gallery by crumbling of the roof due to rock pressure. / drukgewelf / – – / Bruchgewölbe / arco de presión

pressure dome, – a result of rock pressure above a gallery. / drukkoepel / cloche / Bruchdom / bóveda de presión

2114
boulder choke, – a tumble of rocks blocking the passage from floor to roof. / puinverstopping / trémie / Versturzkegel; abschliessender Trümmerkegel / derrumbe

2115
spongework; anastomosis, – an entanglement of interconnected tubular channels of various size developed at random in limestone and not restricted to a plane. / anastomose / cavité anastomosée / – – / laberinto

2116
bedding-plane spongework, – an intricate network of tubular channels of various size along a bedding plane. / laagvoeg-anastomose / – – / – – / laberinto según plano de estratificación

joint-plane spongework, – an intricate network of tubular channels of various size along a joint plane. / breukvlak-anastomose / – – / – – / laberinto según plano de diaclase

2117
roof spongework, – a spongework in the roof of a passage formed at some time, when the passage was filled with clay. / dak-anastomose / – – / – – / pendantes

2118
palette; shield, – a more or less flat, protruding ledge of calcite; it is a remnant of a calcite lentil or vein, formed by solution of the country rock on both sides of it. / – – / – – / – – / escudo

boxwork, – a protruding network of calcite veins formed by solution of the country rock. / – – / boxwork / Netzleiste / boxwork; panal

2119
Efforation, – a German term for erosion caused by flowing water under pressure with the aid of solid particles.

2120
cavitation, – a kind of erosion by rapidly running water in closed conduits whereby vacua are formed that quickly collapse; an impingement attack without the aid of solid particles. / cavitatie / cavitation / Kavitation / cavitación

2121
cave deposit, – a deposit on the floor of a cave, other than chemical precipitates; e.g. clay, sand, gravel. / grotsediment / dépôt souterrain / Höhlensediment / deposito de cueva

2122
cave loam; residual ...., – / grotklei / argile résiduelle ; .... de décalcification; résidu argileux / Höhlenlehm / arcilla de decalcificación

clay filling, – / opvulling met grotklei / colmatage argileux / Ausfüllung mit Höhlenlehm / colmatación por arcilla

2123
petre dirt; saltpetre ....; nitrous ...., – an earthy mass containing calcium nitrate. / – – / – – / Salpeter / salitre

2124
speleothem; speleolite; cave formation, – terms applicable to all secondary mineral deposits in caves precipitated from aqueous solutions. / speleoliet / concrétion; spéléolite / Speleolit; Tropfsteinbildungen; Sintergebilde / espeleotema; concreción

2125
speleogen, – formed by removal of material due to solution or corrasion of the bedrock. / – – / – – / – – / espeleogeno

2126
petromorph, – secondary mineral deposits formed within the bedrock and incidentally exposed within a cave, as e.g. palette, boxwork. / – – / – – / – – / petromorfo

2127
dripstone, – a term used for all speleothems attributable to dripping water. / druipsteen / spéléothème / Tropfstein / – –

2128
flowstone, – a coating of the floor or the wall of a cave, consisting of a sheet of calcium carbonate precipitated from slowly flowing water. / vlietsteen / plancher stalagmitque / Sintertapete / colada estalagmitica

2129
sheet, – a thin coating of calcium carbonate formed on walls, shelves, benches, and ledges by trickling water. / huidje / – – / – – / lámina

2130
pendent formation, – a term applicable to all formations hanging from the ceiling of a cavern. / hanger / pendant; pendeloque / Gehänge / pendante

2131
stalactite, – a cave formation hanging from the roof, usually composed of calcium carbonate. / stalactiet / stalactite / Stalaktit / estalactita

straw stalactite; tubular ....; soda straw (USA), – a hollow, thin-walled stalactite, uniform in diameter throughout its length. / strohalmstalactiet / stalactite fistuleuse; ..... en tuyau de plume; cristal creux en forme de tube; macaroni / Federkielstalaktit / estalactita tubular; macarrón

2132
helictite, – a slender formation of calcium carbonate growing from the roof, the walls or the floor of a cave, twisting and ramifying in every direction. / helictiet / excentrique / – – / excéntrica

2133
cave coral; coral formation, – a very small, stalked formation of calcium carbonate on the floor, the wall or the ceiling of a cave. / – – / stalactite coralliforme; stalagmite .... / – – / estalactita coraliforme; estalagmita ....

2134
sheet stalactite, – a stalactite, thin in one horizontal direction and broad in the other one, thus forming a pendent triangle. / – – / – – / – – / estalactita hojosa

2135
curtain, – a thin sheet of dripstone, hanging from the ceiling or projecting from the wall of a cave. / gordijn / rideau / Vorhang / bandera

2136
botryoidal stalactite; botryoid; clusterite; grape formation, – round or semi-round smooth nodular growths of calcite, usually occurring in clusters on cavern walls. / druiventrosstalactiet / – – / – – / estalactita botroidal

2137
stalagmite, – a cave formation projecting up from the floor. / stalagmiet / stalagmite / Stalagmit / estalagmita

2138
column; stalacto-stalagmite (USA), – a formation due to the union of a stalactite with a stalagmite. / zuil; pilaar / colonne; pilier / Säule / columna

2139
splash cup, – a cup at the top of some stalagmites formed by falling drops. / spatkom / – – / – – / pocillo de goteo

2140
pillar, – a thin stalagmite, not reaching the roof. / kaars / cierge; chandelle / Kerze / vela

2141
terraced stalagmite, – a stalagmite having more or less horizontal ledges. / – – / – – / – – / estalagmita en terraza

splash stalagmite, – a stalagmite having whorls of upstanding petal-shaped protuberances at many levels. / spatstalagmiet / stalagmite d'éclaboussement; .... d'éclaboussure; .... en pile d'assiettes / – – / estalagmita en pila de platos

stool stalagmite; mushroom ....; lily pad; platform, – a calcite formation shaped like a plate and grown around a stalagmite whose top coincided with the surface of a pool with saturated water. / sinterbord / plateau; (stalactite à plateau) / – – / estalagmita en seta

Buckelsinter, – a German term, denoting a stalagmitic formation consisting of innumerable irregular bosses. / bultstalagmiet / – – / Buckelsinter / – –

2142
calcite skin, – a thin coating of calcite on clay walls, formed by ascending solutions. / calcietvlies / dépôt pelliculaire; vernis calcaire / Kalzittapete / pelicula calcítica

sels grimpants, – a French term for ascending solutions from which a thin coating precipitated.

2143
hoarfrost formation, – a thin plastering with needle-shaped crystals of calcite. / calcietrijp / givre calcaire / Faserkalzit / escarcha calcárea

2144
floating calcite scale; calcite ice; lime ice – floating scales of small calcite rhombohedrons. / drijvende calcietschub / calcite flottante / – – / calcita flotante

2145
crust stone, – a fragile, flaky crust of calcite or other minerals covering portions of cave walls. / – – / – – / – – / costra

2146
calcite bubble, – a hollow sphere formed by deposition of calcite around a gas bubble. / calcietblaasje / – – / – – / burbuja de calcita

2147
rimstone pool; gour, – a basin formed within a calcite rim precipitated from slowly overflowing water. / gour / gour / Sinterbecken; ....schale; ....wanne / gour

rimstone bar, – the rim of a rimstone pool. / sinterdam / barre de travertin / Sinterdamm / barra de gour

2148
calcite ledge; .... edge; shelfstone, – a ledge of calcite projecting horizontally into a pool at its water level. / sinterzoom / – – / – – / festón de calcita

2149
cave pearl; .... pisolite, – a small, polished, unattached concretion, formed by successive coatings of calcite deposited round a nucleus. Unpolished forms are included in the general term: cave pisolite and in the French term: dragée. / grotparel; grotpisoliet / perle de caverne; dragée; pisolite / Kalziterbse; Pisolit / 'perla' de caverna

mûre de caverne, – a French term for a cave pisolite having an exterior like a mulberry. / – – / – – / – – / 'mora' de caverna

2150
anthodite, –
1. a cluster formed of radiating, long, needle-like crystals of calcite, sometimes of aragonite or gypsum.
2. a cluster of small tubular stalactites.

/ – – / – – / Kalzitrosette; Gips.... / anthodita; erizo

2151
oulopholite (USA); gypsum flower, – radiating gypsum crystals simulating outward curving flower petals. / gipsbloem / fleur de gypse / – – / flor de yeso; oulopholita

2152
pegostylite, – a pillar, cone, or dome-shaped formation of calcite or aragonite precipitated in ascending water. / pegostyliet / pégostylite / Quellstalagmit; Pegostylit / pegostylita

2153
lublinite; mountain milk; moon ....; rock ...., – a soft cheesy mixture of calcite and water. / lubliniet / lait de lune; lublinite / Bergmilch; Lublinit; Bergmehl; Mondmilch / lublinita; montmilch

[1630] WIND ACTION

[1631] *Erosion and Transport by Wind*

2154
eolian transport; wind-borne ...., – / windtransport / transport éolien / Windverfrachtung / transporte eólico

wind corrasion; eolian ....; wind abrasion; .... attrition; .... grinding; .... scour; .... wear, – the natural sandblast action of wind-blown sand / eolische corrasie; windcorrasie / corrasion éolienne; érosion ....; émoulage ....; attrition .... / Windkorrasion; ....erosion; Sandschliff; Wind....; äolische Erosion / erosión eólica; abrasión

2155
deflation, – the lifting and removal of sand and dust by wind generally after it has already been loosened by some means. / deflatie; eolische ablatie; uitblazing / déflation; ablation éolienne / Deflation; äolische Ablation; .... Denudation; Ausblasung; Ab....; Auswehung; Windablation / deflación; denudación

2156
drift sand; blowing ....; flying ....; wind-driven ....; drifting

...., – / stuivend zand; stuifzand / sable éolien / Flugsand; Staub....; Trieb....; Treib....; Wander.... / arena móvil

sand drift, – / zandverstuiving / vent de sable / Sandtreiben; ....wehen / tempestad de arena

2157
sand plain, – formed by deflation down to ground-water level. / (stuif)zandvlakte / plaine de sable / Sandfeld / llanura de arena

2158
smoking crest, – of a dune, resulting from the removal of finer sand grains by wind (winnowing) at its top. / rokende duinkruin / crête fumante / rauchender Dünenkamm / cresta fumadora

2159
blood rain; dust fall; .... shower, – rain coloured red by dust from the air. / bloedregen; stof.... / pluie de sang; .... de poussière / Blutregen; Staub....; Staubfall / lluvia de sangre

2160
desert pavement; .... armor (USA); .... armour; boulder pavement; lag deposit; .... gravel; deflation residue, – pavement of pebbles or fragments of rock, formed by the action of the wind, which removes the fine material so that the coarser material is left. / woestijnpantser / pavé désertique; pavage ....; résidu de déflation / Wüstenpflaster; Stein....; Steinpanzer; Lesedecke; Auslese....; Deflationsrückstand / residuo de denudación; .... de deflación

2161
ventifact, – general term for any stone, faceted, worn or polished by wind-blown sand. / door zandwind bewerkte steen / roche éolisée / äolisch bearbeiteter Felsen; windpolierter .... / roca eolizada

2162
windkanter; wind-faceted stone, – a stone on which abrasion by wind-blown sand has produced faces that intersect in one or more sharp edges. / windkei / caillou à facettes / Windkanter / roca facetada por el viento

einkanter, – a ventifact having two wind-abraded faces intersecting each other in a single sharp edge. / tweekanter / roche éolisée à deux facettes / Einkanter / canto de una arista

dreikanter; pyramid pebble, – a ventifact with wind-abraded faces that intersect in three sharp edges coming together in one point. / driekanter / roche éolisée en pyramide / Dreikanter / canto de tres aristas

2163
pedestal rock; rock pedestal; mushroom rock, – attached residual column of rock, showing a narrowed foot, caused by eolian abrasion or differential weathering. / paddestoelrots / rocher en forme de champignon; rocher-champignon / Pilzfels; Tisch.... / roca en forma de hongo

balanced rock; rocking stone; perched block, – a loose pedestal rock, having such a narrow foot, that its position is unstable. / schommelsteen / rocher branlant; .... perché / Wackelstein / roca en equilibrio inestable

2164
yardang (Turkey), – sharp-edged ridges, separated by elongated U-profile furrows, originated by wind scour. / yardang / yardang; jardang / Yardang / – –

[1632] *Dunes, Geographical Distribution*

2165
dune, – a hill or ridge of sand piled up by the wind. / duin / dune / Düne / duna

2166
littoral dune; coastal dune, – dunes or ridges along the coast, formed by wind prevailing from sea. / kustduin; zee.... / dune littorale; .... côtière / Küstendüne / duna litoral

2167
exterior dune; outer dune, – coastal dune ridge at the seaside. / buitenduin / dune extérieure; .... externe / Aussendüne / duna exterior; .... externa

interior dune; inner dune, – coastal dune ridge at the landside. / binnenduin / dune interne / Innendüne / duna interior; .... interna

2168
beach dune, – strandduin / dune d'estran / Stranddüne / duna de playa

foredune, – first continuous range of dunes along the landward limit of highest tides. / zeereep; strandloper; zee....; voorduin / avant-dune / Vordüne; Sekundär.... / frente dunar

2169
cliff dune, – dune formed on the edge of a cliff. / klifduin / dune chevauchant une falaise / Kliffdüne / duna sobre el borde de un escarpe

2170
inland dune; continental dune, – dunes or ridges of dunes formed inland without connection to the coast. / landduin / dune intérieure; .... continentale / Binnendüne; Kontinental.... / duna continental

[1633] *Dunes, Genetical Classification*

2171
longitudinal dune; linear ....; seif ...., – a dune or ridge of dunes oriented in the direction of the prevailing wind. / lengteduin / dune longitudinale / Längsdüne; Strich....; Reihen.... / duna longitudinal

transverse dune, – a dune or ridge of dunes perpendicular to the direction of the prevailing wind. / dwarsduin / dune transversale / Querdüne; Wall.... / duna transversal

2172
psammogenic dune, – caused by the effect of sand surfaces in trapping more sand. / – – / dune psammogène / – – / duna psamogenética

phytogenic dune, – caused by the effect of vegetation in arresting sand. / organogeen duin / dune phytogène / organogene Düne / duna fitogenética

2173
streamer, – elongated accumulation formed behind an obstacle. / windvaan / traîne / – – / – –

2174
lee dune; wind-shadow ...., – caused by various types of natural or artificial obstacles. / hindernisduin; windschaduw.... / dune d'obstacle / Hindernisdüne / duna de obstáculo

2175
wandering dune; migrating ....; moving ....; mobile ....; shifting ...., – a dune, the sand of which is blown by the prevailing wind from the windward slope to the leeward slope, so that the dune as a whole is moving in the same direction as the wind. / trekduin; zwerf....; loper / dune migrante; .... mobile; .... mouvante / Wanderdüne / duna viva; .... móvil

fixed dune; stabilized ....; anchored ....; stationary ...., – / vastgelegd duin / dune fixée; .... stabilisée; .... stationnaire / festliegende Düne; befestigte ....; stillstehende ....; ruhende .... / duna fija; .... estacionaria

2176
embryonic dune, – / oerduintje / embryon de dune; dune embryonnaire / Embryonaldüne; Primär....; Ur.... / duna embrionaria

2177
barchan(e); barkhan (Turkestan); crescentic dune; horseshoe ...., – a dune showing a crescentic form, the horns of it pointing down-wind. / barchaan; sikkelduin / barchane; barkhane; dune en forme de croissant; .... en croissant; .... en forme arquée / Barchan; Sicheldüne; Bogen....; Halbmond....; Hufeisen.... / duna creciente en forma arqueada

parabolic dune, – a parabola-shaped dune with points tapering to windward; sometimes proceeding from destroyed transverse dunes. / paraboolduin / dune parabolique / Parabeldüne / duna parabólica

2178
sif, – a Saharan Arabic term for a short longitudinal dune element with a short, often slightly curved crest.

silk, – a Saharan Arabic term for a longitudinal dune consisting of a row of associated sifs.

2179
ghourd, – an Arabic term for a star-shaped pyramidical dune. / sterduin / – – / – – / – –

2180
draa, – an Arabic term for a very large sand dune, carrying dunes of normal size on its gently sloping side, and forming part of a regular series, the wavelength of wich lies between some 300 and 6000 m. Most draa have their crests transverse to the dominant wind direction. / draa / draa; chaîne de dune / – – / dunar

2181
zibar, – an Arabic term for low dunes in flat areas, often in corridors between higher dunes, the sands of which have a slightly less prominent coarse mode (often many coarse grains).

2182
whale back; sand levee, – longitudinal sand ridges broad (1-3 km) and low (about 50 m) and continuing for distances of the order of 300 km, forming a pedestal for one or more longitudinal dune chains.

2183
erg (NW Sahara); sand sea, – and extensive sand region composed of sand complexes without isolated dunes. / – – / – – / – – / erg; mar de arena

2184
aklé, – a Saharan Arabic term for a network-dune pattern whose formation seems to require relatively unidirectional winds and a considerable quantity of sand.

2185
slip face; sandfall ...., – of a dune, the leeward face of a dune on which the sand, which has been blown over it, is avalanching. / storthelling / talus croulant / Sturzseite / cara de sotavento

2186
interdune passage, – wind-swept passage between longitudinal dunes. / – – / couloir (interdunaire) / Windgasse; Dünental / pasadizo interdunal

gassi (Sahara), – interdune passage with a rocky floor.

feidsh (Sahara), – interdune passage with a sandy floor.

2187
blowout; deflation hole; .... basin, – a small basin in an area of sand accumulation; formed by deflation. / windkuil; waaigat / cuvette de déflation; creux de ....; caoudeyre (Gascony) / Windmulde; ....kuhle; Deflationskessel; ....wanne / cubeta de denudación; .... de deflación

blowout dune, – an accumulation of sand derived from a blowout basin. / windkuilduin; waaiduin / pourrière / Haldendüne / duna de denudación; .... de deflación

2188
dune valley; .... slack; .... low, – an elongated depression in a dune terrain. / duinvallei; duindel / couloir interdunaire / Dünental / valle interdunal

2189
duinpan, – a Dutch term for a flat-bottomed depression amidst dunes. / duinpan / dépression / Dünenwanne / depresión

2190
hummocky dune, – a relatively high dune protected from erosion by a dense vegetal cover. / pollenduin; kopje; duinrest / croc; dune confuse; .... morcelée; .... résiduelle / Kupste; Kupstendüne; Reliktendüne; Pflanzen-Sandkuppe / duna en forma de mogote

[1640] RIVERS

[1641] *The River Course*

2191
river; stream, – / rivier; stroom / rivière; fleuve / Fluss; Strom / río

channel; runway, – the actual stretch occupied by water. / watergang; geul / chenal / Rinne; (Rinnsal; Gerinne) / canal

2192
drainage basin; catchment area; river basin; hydrographical ...., – the entire tract of country drained by a river and its tributaries. / stroomgebied; afwaterings....; rivierbekken / aire d'alimentation; .... de drainage; bassin fluvial; .... hydrographique; .... versant; impluvium / Einzugsgebiet; Ernährungs....; Entwässerungs....; Flussbecken; Strom.... / cuenca hidrográfica

2193
river head, – the source or beginning of a stream. / bron; oorsprong / origine / Flusshaupt / nacimiento

headwater region, – / brongebied / – – / Quellgebiet / cuenca de recepción

2194
runnel; rill, – a small, freely flowing water course, normally smaller than a brook. / stroompje / filet / Rinnsal / arroyo

2195
upper course, – the mountainous stretch of a river with predominantly vertical and subordinate lateral erosion. / bovenloop / cours supérieur / Oberlauf / corso superior; .... alto

middle course, – the stretch of a river with predominantly lateral erosion and local and temporary deposition; many, often large tributaries; river is fed by ground water also. / middenloop / cours moyen / Mittellauf / corso medio

lower course, – the lowland stretch of a river with predominantly deposition; diversion into distributaries; river feeds ground water. / benedenloop / cours inférieur / Unterlauf / corso bajo

2196
mouth; debouchure, – / mond; monding / embouchure; bouche / Mündung / desembocadora

2197
stream development, – the ratio of the total length of the river course with all its curves to the straight distance from the source to the mouth of the river. / stroomontwikkeling / développement du cours d'eau / Stromentwicklung / desarrollo del río

2198
ungraded profile; irregular ....; interrupted ...., – longitudinal stream or valley profile, showing no smoothed out course; in the Davisian concept a characteristic of the stage of youth. / onvereffend profiel / profil non régularisé / unausgeglichene Gefällskurve / perfil no equilibrado

profile of equilibrium; graded profile, – longitudinal river or valley profile showing a smoothed-out course; in the Davisian concept a characteristic of the stages of maturity and old age. / evenwichtsprofiel; ....verhang / profil régularisé; pente limite; profil d'équilibre; pente .... / ausgeglichenes Gefälle; normales ....; Normalgefällskurve; Normalgefälle; Ausgleichs....; Gleichgewichts....; Gleichgewichtsprofil; ....kurve / perfil de equilibrio

terminal profile, – a theoretical graded profile, having become so flat, that the stream can no longer transport any load. / erosieterminante; eindverhang / profil terminal d'équilibre; .... définitif; .... d'équilibre définitif / Erosionsterminante; Endkurve der Erosion; Endprofil / perfil de equilibrio definitivo

2199
(k)nick; (k)nickpoint, –
1. terms applicable to a break in the longitudinal profile of a stream or of a valley. / profielknik; verhang.... / ressaut du profil longitudinal; rupture de pente; brisure / Knickpunkt; Gefällsbruch; ....stufe; ....knick / ruptura de pendiente
2. terms applicable only to a nickpoint in the longitudinal profile of a valley. / daltrede; daltrap / gradin de vallée; ressaut de .... / Talstufe / – –

cyclic nick; .... nickpoint, – a break in the longitudinal stream or valley profile at the place of intersection of a newly graded profile with an older one. / cyclische profielknik; verjongingsprofielknik / ressaut cyclique / zyklische Gefällsstufe; zyklischer Gefällsbruch; Erosionshaltstufe / ruptura cíclica

2200
rock step, – nickpoint due to the outcrop of a resistant rock. / structuurtrede / gradin structural / Felsstufe; Härtestufe / gradiente estructural

2201
bar; barrier; hard band, – / drempel / barre; seuil / Schwelle; Barre; Riegel; (Felsband) / barra

2202
rapid, – a river stretch where the current moves with great velocity, the surface usually being broken by obstructions, but without actual waterfall. / stroomversnelling / rapide / Stromschnelle / rápida

2203
waterfall; fall, – water dropping down a precipice, usually caused by a bed of hard rock. / waterval / chute d'eau; chute / Wasserfall / caída de agua

cascade, – a small waterfall or a series of falls resembling steps. / cascade / cascade / Kaskade / cascada

2204
cataract, –
1. a great waterfall, or a series of falls.
2. sometimes used in the sense of rapids.
/ cataract / cataracte / Katarakt / catarata

[1642] *The River System*

2205
river system; stream ...., – / rivierstelsel / système fluvial; réseau .... / Flusssystem; Stromnetz / red fluvial

2206
main stream; master ...., – / hoofdrivier / rivière principale; .... maîtresse / Hauptfluss / corriente principal

trunk stream; stem ...., – a main stream having an axial or central position in a drainage area. / – – / axe fluvial / – – / eje fluvial

2207
side stream, – a river that joins a main stream but that has received its water from a separate drainage area. / nevenrivier / cours d'eau secondaire; affluent / Nebenfluss / río confluente

tributary; affluent; feeder, – a stream flowing into a larger one. / zijrivier / affluent; tributaire / Zufluss; Zubringerfluss / afluente; tributario

2208
junction; confluence, –
1. the meeting or junction of two or more streams;
2. the place of meeting.
/ samenvloeiing; inmonding / confluent / Zusammenfluss; Konfluenz / confluencia

accordant junction, – the junction of two valleys or two rivers on the same level. / gelijkvloerse monding; .... samenvloeiing / raccordement parfait; .... normal; embouchure en raccord / Flachmündung; gleichsohlige Mündung / confluencia a nivel

discordant junction, – a valley or river junction on unequal level. / zwevende monding; trapmonding / confluence à gradin; embouchure à .... / Stufenmündung; Hänge....; ungleichsohlige Mündung / confluencia con desnivel

2209
shifted junction; deferred ...., – / verplaatste samenvloeiing; verschoven .... / confluent entraîné / verlegte Konfluenz; verschleppte .... / confluencia diferida

barbed junction; boathook bend, – a sharp bend formed by a tributary entering the main stream in an upstream direction. / weerhaakvormige samenvloeiing / confluent en crochet; .... en hameçon / – – / confluencia en horqueta

2210
splitting; branching; ramification, – a division of a stream into branches. / splitsing; vertakking / ramification; diramation / Teilung; Spaltung; Verzweigung / ramificación

forking; bifurcation; divarication, – a division into two more or less equal river branches. / bifurcatie; gaffelverdeling / bifurcation; divarication / Gabelung; Bifurkation / bifurcación

2211
distributary, – a river branch flowing away from the main stream and not rejoining it. / aftakking / défluent / Abzweigung / defluente

2212
false channel; .... river, – / dode arm / faux bras; bras mort / blinder Arm / canal abandonado

2213
river diversion, – a passive or spontaneous change in the direction of a stream. / rivieromlegging / déversement fluvial; détournement .... / Flussverlegung / cambio del curso

river deflection; .... deviation, – a more or less spontaneous river diversion, e.g. by warping, alluviation, glaciation, volcanic action, lateral erosion. / rivierafleiding; spontane rivieromlegging / diversion fluviale / Flussablenkung / desviación fluvial

2214
river inversion, – / rivieromkering / inversion d'une rivière / Flussumkehr / inversión de una corriente

2215
Durchbruch(s)berg; Sehnen...., – German terms for a hill, separated from the surrounding mountains by diversion of a tributary to a master stream, at a certain distance upstream of the original point of junction.

[1643] *River Curves and River Banks*

2216
sinuosity, – ratio of channel length to down-valley distance. / – – / sinuosité / – – / sinuosidad

2217
meandering channel, – channel with a sinuosity equal to or above 1.5. / – – / chenal à méandres / – – / canal meandriforme

sinuous channel; straight ...., – channel with a sinuosity less than 1.5. / – – / chenal sinueux / – – / canal sinuoso

2218
braided; anastomosing; anastomosed, – said of a channel with a sinuosity less than 1.5, marked by successive divisions and rejoinings of the flow around alluvial islands. / verwilderd; vlechtend / anastomosé / verwildert / anastomosado

2219
meander, – a more or less regular curve of a river or valley. / meander; rivierkronkel / méandre / Mäander; Serpentine / meandro

2220
free-swinging meander; free ...., – river meander being displaced easily by lateral corrasion, meander cut-off, etc.; vertical corrasion is of no influence. / vrije meander; vlaktemeander / méandre divagant; .... libre / freie Mäander; bewegliche ....; Wiesenmäander; Aufschüttungs....; Fluss.... / meandro libre; .... divagante

2221
valley meander; incised ....; inclosed ....; enclosed ...., – a curve of a meandering valley. / dalmeander; insnijdings....; ingesneden meander / méandre de vallée; .... encaissé / eingesenkter Mäander; eingeschnittener ....; Talmäander; Erosions.... / meandro de valle

intrenched meander; entrenched ....; inherited ...., – a valley meander caused by the incision of a free meander in a former plain or valley floor, so that the upper side of the meander spur is formed by the surface on which the original free meander was situated. / overerfde meander / méandre encaissé; .... imprimé; .... hérité / Zwangsmäander; ererbter Mäander; vererbter .... / meandro heredado

2222
ingrown meander, – a valley meander formed by the incision of a smaller free meander in a former plain or valley floor and having typical slip-off and undercut slopes due to lateral corrasion. / afglijdingsmeander / méandre encaissé; .... sculpté / Gleitmäander / meandro encajado

2223
terrace meander, – a valley meander formed by the incision of a free meander that belonged to a former valley floor; the remnants of the valley floor now form a terrace. / terrasmeander / mándre de terrasse / Terrassenmäander / meandro de terraza

2224
meander belt, – the strip within which the meanders in a river valley are confined. / meandergordel / zone des méandres; lit des .... / Mäandergürtel; ....streifen / zona de meandros

2225
point bar; flood-plain lobe, – part of the flood plain, enclosed by a meander bend. / kronkelwaard / lobe de méandre / Mäanderzunge; ....lobus / – –

2226
meander neck, – / meanderhals; kronkel.... / pédoncule; isthme; col; racine / Mäanderhals / cuello del meandro

2227
meander spur, – / meanderspoor / éperon / Sporn; Talsporn / espolón del meandro

meander core; rock island; cut-off meander spur, – / omloopberg; kronkelberg / mamelon central / Umlaufberg / núcleo del meandro

2228
meander scar, – small bluff caused by the lateral expansion of a meander curve. / meanderklif / échancrure de méandre / Mäanderkerbe / escollo de meandro

valley spur, – / dalspoor / éperon / Talvorsprung; ....sporn / – –

little-trimmed spur, – a valley spur, the upstream side of which is eroded by the downstream migrating meander. / aangevreten spoor / éperon façonné / angenagter Sporn / – –

2229
meander cut-off, – / meanderdoorbraak; ....afsnijding / recoupement d'un méandre / Mäanderdurchbruch; ....abschneidung / estrangulación del meandro

2230
chute cut-off, – cutting of a new channel by a stream in a meander loop along a swale of a point bar. / – – / recoupement par court-circuit / – – / – –

neck cut-off, – cutting of a new channel through the narrow neck between two meander loops or through the capture of a loop by the next one upstream. / – – / recoupement par tangence / – – / – –

2231
avulsion, – with chute cut-off and neck cut-off one of the three modes of channel shifting; the sudden abandonment of a part of the whole of the meander belt by a stream for some new course at a lower level on the flood plain. / – – / déplacement subit du lit / – – / avulsión

2232
meander loop; oxbow, – / meanderbocht / boucle; anse / Mäanderschlinge / arco del meandro

cut-off meander; oxbow lake; mort lake, – abandoned meander loop. / afgesneden meander; hoefijzermeer / méandre recoupé; .... abandonné; lac en forme de croissant; lac-croissant / abgeschnittener Mäander; Altwasser / meandro abandonado; lago semilunar

2233
scroll bar, – arcuate ridge at the inner side of a meander loop. / – – / bourrelet arqué; croissant / – – / barra arqueada

2234
swale, – narrow intervening depression between scroll bars. / kronkelwaardgeul / gouttière / – – / – –

flood-plain scroll, – isolated patch of deposited material along the inside curve of the bend of an incised meander. / – – / croissant de lit majeur / Talschnörkel / – –

2235
convex bank; inner ...., – of a river, generally with deposition. / convexe oever; binnenbocht / rive convexe; .... intérieure; .... plate / konvexe Ufer; innere ....; Flachufer / ribera convexa; .... interior

concave bank; outer ...., – of a river, generally with lateral erosion. / concave oever; buitenbocht / rive concave; ....

extérieure / konkave Ufer; äussere .... / ribera cóncava; .... exterior

2236
undercut bank; cut ....; river cliff; .... bluff, – at the concave bank of a river. / stootoever; holle oever; steile .... / rive raide; .... concave / Prallufer; Stosz....; Steil.... / ribera escarpada

undercut slope; cut-bank ...., – / stootoeverhelling / pente de rebondissement; versant concave / Prallhang; Schnitt....; Unterschneidungs.... / pendiente socavada

2237
slip-off slope, – / afglijdingshelling / pente de glissement; versant convexe / Gleithang / pendiente de deslizamiento

2238
natural levee, – the natural, built-up bank of a river formed during flooding by the deposition of material in suspension. / oeverwal / levée naturelle; bourrelet de rive / natürlicher Damm; levée / dique natural

2239
backswamp; flood basin, – the lowlands at both sides of a river beyond the natural levees. / kom / dépression latérale / Hinterwasser / depresión lateral

flood plain, – / overstromingsgebied / plaine d'inondation / Überschwemmungsgebiet; Inundations.... / llanura de inondación

2240
crevasse, – water-cut channel through the natural levee deposits. / doorbraakgeul; crevasse.... / percée de levée naturelle / Durchbruch / paso; abertura

crevassing, – the flow of excess water through isolated low sections or breaks in the natural levees.

2241
crevasse splay, – a tongue-shaped mass of sediment, deposited by a crevasse. / crevassecomplex / – – / Durchbruchfächer / – –

[1644] *The River Bed*

2242
river bed, – / rivierbedding / lit de (la) rivière / Flussbett / lecho del río

mean-water bed, – / zomerbed / lit ordinaire; lit mineur / Normallbett / lecho ordinario; .... menor

high-water bed, – / winterbed / lit de hautes-eaux; lit majeur / Hochfluttbett; Überschwemmungsbett / lecho mayor

2243
interlaced channels; intertwined ....; tangled ...., – / vlechtwerk van geulen / réseau de rigoles; lacis de ....; tresse de .... / verflochtene Rinnen / canales entrelazados

2244
sand bar; gravel bar; pebble bar; shingle bar, – / zandbank; grind / banc de sable; .... de gravier; .... de galets / Sandbank; Kies....; Geröll.... / banco de arena; .... de grava

2245
aggradation, – the progressive building up of a channel bed. / opvullen / aggradation; alluvionnement; comblement / Zuschüttung; Verfüllung / agradación

2246
bed form, – structure formed at the interface of water (or air) currents and sediment bed, e.g. ripples, sand waves. / bodemstructuur / – – / Sohlform / forma del lecho

2247
riffle; shoal; ford, – shallow part in a channel. / ondiepte; voorde / haut-fond; seuil; gué / Untiefe; Schwelle; Furt / alto fondo

pool; deep; hole, – deep part in a channel. / diepte; kom / mouille; creux; fosse; trou d'eau / Loch; Kolk / fosa

2248
rock mill; churn hole; eddy ....; pothole; pot, – a hole closed at the bottom bored in the rock bed of a stream, with the aid of sand and stones moved by whirling action of the current. / kolkgat / marmite de géant; cuve; oull; tine / Strudeltopf; ....loch; ....kessel / marmita de gigante

[1645] *Discharge and Drainage*

2249
discharge; flow; volume, – the quantity of water passing through unit cross section in unit time. / debiet; afvoer; vermogen; capaciteit / débit; portée; abondance / Wasserführung; ....spende; ....menge; ....ertrag / caudal

2250
drainage ratio; run-off coefficient; discharge efficiency, – the ratio of the run-off in a certain area during a certain period to the precipitation in that area during the same period. / afvloeiingscoëfficient; specifieke afvoer / coefficient d'écoulement; fraction .... / Abflusskoeffizient; ....faktor / coeficiente de escorrentía

2251
flood water, – / wassend water; opkomend .... / eau en crue / steigendes Wasser / agua en crecida

bankfull stage, – the stage, above which banks are flooded. / – – / niveau de pleins-bords / ufervoller Stand / nivel de llenado

overflow, – flooding of the river banks. / buiten de oevers treden / débordement / Überlaufen; Übertreten / desbordamiento

overflow level, – / overloopniveau / niveau de débordement / Überlaufniveau / nivel de desbordamiento

2252
freshet, – a river flood caused by heavy rains or rapidly melting snow. / rivierzwelling / avalaison; avalasse / Sprungschwall / avalancha

flash flood; spate, – a sudden river flood caused by torrential rains. / stortvloed / courant torrentueux / Sturzflut / corriente torrencial

2253
drainage, – / ontwatering / drainage / Entwässerung / drenaje

drainage may be:

ephemeral, – / kortstondig / éphémère / kurzfristig / efímersa

episodical, – / episodisch; toevallig / accidentel; occasionnel / episodisch / episódica

periodical, – / periodiek / périodique / periodisch / periódica

seasonal, – / seizoenmatig / saisonnier / jahreszeitlich / estacional

intermittent, – / intermitterend / intermittent / intermittierend / intermitente

perennial, – / permanent / permanent / ständig; dauernd / permanente

2254
drainage pattern; .... net, – pattern of tributaries and master streams in a drainage basin. / drainagepatroon; ontwaterings .... / configuration du réseau; arrangement de .... / Entwässerungsanordnung; ....muster; Muster des Entwässerungsnetzes / red de drenaje

Types of patterns:

dendritic; tree-like, – / dendritisch; boomvormig / dendritique; arborescent / dendritisch; baumartig / dendrítica

digitate, – / vingervormig vertakt / digité / fingerförmig / digital

craw-foot; bird-foot, – / vogelpootvormig / en patte d'oie / vogelfuss; krähen.... / pata de ave

rectangular; trellised, – / rechthoekig; tralie- / rectangulaire; en espalier / rechtwinklig; gitterförmig / rectangular

pennate, – / veervormig / penné / fiederförmig / pennada

radial, – / radiaal / rayonnant / radial; radiär / radial

annular, – / ringvormig / annulaire / ringförmig / anular

2255
drainage density; .... texture, – the ratio of the cumulative length of a stream and its tributaries to the total drainage area. / rivierdichtheid / densité de réseau / Flussdichte / densidad de la red

2256
external drainage; through-flowing ....; out-flowing ....; exoreic ...., – a drainage whereby the water reaches the open sea directly or indirectly. / externe ontwatering / réseau exoréique; écoulement .... / exoreïsche Entwässerung / drenaje exorreico

internal drainage; endoreic ...., – a drainage whereby the water does not reach the open sea. / interne ontwatering / réseau endoréique; écoulement .... / Binnenentwässerung; endoreïsche Entwässerung / drenaje endorreico

2257
areic; riverless, – relates to areas that almost completely lack superficial drainage. / afvloeiingsloos; zonder bovengrondse afwatering / aréique / flusslos; areïsch / falta de red fluvial; red hidrográfica ausente

2258
dissipating area, – an area in which the surface water is spreading and seeping into the earth. / – – / champ d'épandage / Schwundgebiet / área de expansión; .... de disipación

2259
allogenous stream; allochthonous ...., – a stream that originates in a humid or in a niveoglacial area and then flows through an arid region, in that region is considered to be allogenous. / allochtone rivier / rivière allogène / landfremder Fluss; Fremdlingsfluss / corriente alóctono

[1646] *Flow Characteristics*

2260
current velocity, – / stroomsnelheid / vitesse du courant / Strömungsgeschwindigkeit / velocidad de corriente

2261
wetted cross section, – the area occupied by water in a cross section of a channel normal to the direction of flow. / natte doorsnede / section mouillée / benetzter Querschnitt / sección mojada

wetted perimeter, – of the cross section of a channel normal to the direction of flow, the sum of the widths of bottom and water level and the lengths of the talus slopes below that level. / natte omtrek / périmètre mouillé / benetztes Perimeter; benetzter Umfang / perímetro mojado

2262
hydraulic radius; mean hydraulic depth, – the ratio of the wetted cross section to the wetted perimeter. / hydraulische straal; gemiddelde .... / rayon hydraulique; .... moyen / mittlerer Profilradius / radio hidráulico; .... medio

2263
hydraulic theory; .... geometry, – analysis of the relationships between stream discharge, channel shape, sediment load and slope.

2264
laminar flow, – flow in which the streamlines remain parallel to the axis of flow. / laminaire stroming / courant laminaire; écoulement .... / laminäre Strömung / régimen laminar; corriente ....

turbulent flow, – flow in which the streamlines do not remain parallel to the axis of flow; the usual case in river flow. / turbulente stroming / courant turbulent; écoulement .... / turbulente Strömung / régimen turbulento; corriente turbulenta

helical flow; spiral flow, – three-dimensional flow, e.g. below a river bend. / helicoïdale beweging; schroefdraadvormige .... / écoulement hélicoidal; courant .... / – – / régimen helicoidal

2265
first critical velocity, – in a stream channel, the velocity at which the movement of the water current changes from laminar to turbulent. / eerste kritische snelheid / vélocité critique primaire / erste kritische Geschwindigkeit / velocidad critical primaria

Reynolds number; Re, – expresses the first critical velocity:

$$Re = \frac{\bar{v}R}{\nu}$$

where ν is the kinematic viscosity of the medium, v̄ average velocity and R the hydraulic radius.

2266
second critical velocity, – in a stream channel, the velocity at which the turbulent movement of the water current changes from tranquil to rapid, i.e. when velocity becomes greater than that with which waves can propagate. / tweede kritische snelheid / vélocité critique secundaire / zweite kritische Geschwindigkeit / velocidad critical secondaria

Froude number; Fr; F, – expresses the maximum velocity with which waves can propagate:

$$Fr = \frac{\bar{v}}{\sqrt{gh}}$$

where v̄ is average velocity and h depth

2267
hydraulic regime, – three kinds of flow, often defined by the Froude number:
$F < 1$ tranquil flow
$F = 1$ second critical velocity
$F > 1$ rapid flow
/ – – / régime hydraulique / – – / régimen hidráulico

2268
thread of maximum velocity, – line (often seen at the surface of a stream by floating debris) connecting points of maximum current velocity. / stroomdraad / ligne de courant / Stromstrich / línea de corriente

2269
whirl; swirl, – an eddying motion caused by turbulent flow. / neer; maalstroom / tourbillon / Wirbel; Mahlstrom / torbellino

eddy; whirlpool; vortex, – / draaikolk; neer / tourbillon mobile / Eddy, Strudel / remolino

[1647] *Valley Development and Valley Forms*

2270
valley development; formation of valleys; excavation of ...., – / dalvorming / formation des vallées; creusement des .... / Talbildung / formación de los valles

2271
longitudinal valley profile, – / lengteprofiel van een dal / profil longitudinal d'une vallée / Tallängsschnitt; Längsprofil eines Tales / perfil longitudinal de un valle

transverse valley profile, – the profile across a valley. / dwarsprofiel van een dal / profil transversal d'une vallée; section transversale d'une .... / Talquerschnitt; Querprofil eines Tales / perfil transversal de un valle

2272
V-shaped valley, – / V-vormig dal / vallée (à profil) en V; .... entaillée / Kerbtal; V-förmiges Tal; V-Tal / valle con perfil en V; .... en forma de V

2273
U-shaped valley, – besides by glacial erosion this valley form can originate through fluvial erosion in rocks with a tendency to develop perpendicular walls (e.g. sandstone, limestone) and through fluvial erosion in arid climates, in which lateral denudation is largely absent. / U-vormig dal; trogdal / vallée (à profil) en U / U-förmiges Tal; Trogtal / valle con perfil en U; .... en forma de U

trough-shaped valley, – a valley with very flat concave slopes that pass gradually into the broad valley bottom. / – – / vallée en auge / Muldental / artesa

Kastental, – German term, denoting a flat-bottomed valley with steep walls that originated under periglacial circumstances.

2274
asymmetric valley, – / asymmetrisch dal / vallée dissymétrique; .... asymétrique / asymmetrisches Tal / valle asimétrico

2275
two-storey valley; two-cycle ...., – a valley-in-valley form caused by rejuvenation. / tweecyclisch dal / vallée en vallée; .... emboîtée / zweizyklisches Tal; Tal-in-Tal / valle de dos ciclos

multiple-cycle valley, – a valley formed during several cycles of erosion. / polycyclisch dal; meercyclisch dal / vallée polycyclique / mehrzyklisches Tal / valle policíclico

2276
reception basin; headwater ....; watershed (USA), – / verzamelgebied; bron.... / bassin de réception / Quelltrichter; Erosions....; Quellmulde / cuenca de recepción

valley head, – / dalbegin; daleinde / tête de vallon; fin de vallée / Talschluss; Talbeginn / cabecera de valle

open valley, – a valley lacking a closed valley head, but joining another valley across a low valley divide. / – – / vallée ouverte / offenes Tal / valle abierto

2277
slope crack, – break in slope in the valley wall. / hellingknik / rupture de versant / Talkante / ruptura en la ladera

valley wall; .... side; valley-side slope, – / dalwand / versant d'une vallée / Talwand; Talhang; Talgehänge; Tallehne / vertiente de un valle

2278
valley floor; .... bottom; .... plain, – a flat floor built up by alluvial deposits. / dalbodem / fond de (la) vallée / Talboden; Talsohle; Talaue; Talgrund; Sohlenfläche / fondo del valle

flood-plain valley, – a valley with a valley floor. / – – / vallée à fond / Sohlental / valle de fondo

valley fill, – / dalopvulling / alluvions; remblaiement de vallée / Alluvionen; Talaufschüttung / aluviones

2279
thalweg; valley line, – the line joining the lowest points along a valley, from the source of a river to its mouth. / – – / thalweg / Talweg; Tallinie / thalweg

2280
valley mouth, – / dalmonding / embouchure de vallée / Talausgang; Talmündung / desembocadora de un valle; entrada de un ....

2281
dell, – a smooth, oblong and often branched depression, showing a regular grade and having slopes, that pass into each other in smooth, rounded-off forms; a valley floor and a continuously flowing brook are lacking. / delle / vallon / Delle / vallecito

2282
drowned valley, – a valley partly submerged by a transgression of the sea. / verdronken dal / vallée submergée / ertrunkenes Tal / valle sumergido

extended valley, – a valley extended downstream by a regression of the sea. / verlengd dal / vallée prolongée / verlängertes Tal / valle suspendido

2283
engrafted valley system, - a valley system consisting of originally independent rivers, contracted to one system in consequence of negative movements of the sea level. / geënt dalsysteem; samengetrokken .... / réseau de vallées intégrées / aufgepropftes Talsystem / red de valles integrados

2284
consequent, – pertaining to valley or river running or flowing in the direction of the original slope of the land. / consequent / conséquent / konsequent; ursprünglich; Abdachungs-; Folge- / consecuente

resequent, – pertaining to a valley or river that is not a survival of an originally consequent one, but is the result of successive new adjustments to structure at deeper levels. / resequent / réséquent / resequent / resecuente

2285
obsequent, – pertaining to a valley or river that drains off in a direction opposite to that of a consequent valley. / obsequent / obséquent / obsequent / obsecuente

2286
subsequent; longitudinal; strike, – pertaining to valley or river running or flowing parallel to the strike of the rock strata, following the softer beds. / subsequent / subséquent / subsequent; Nachfolge-; Schicht- / subsecuente

2287
insequent, – pertaining to valley or river running or flowing in a direction that is not explainable by determinable factors. / insequent / inséquent / insequent / no consecuente

2288
cataclinal, – pertaining to valley or river running or flowing in the direction of the dip of the strata. / cataclinaal / cataclinale / Kataklinal- / cataclinal

anaclinal, – pertaining to valley or river running or flowing in the direction opposite to the dip of the strata. / anaclinaal / anaclinale / Anaklinal- / anaclinal

2289
longitudinal, – pertaining to valley or river running or flowing parallel to conspicuous mountain chains. / lengte- / longitudinale / Längs- / longitudinal

transverse; cross, – pertaining to valley or river that cuts across a mountain range at a more or less right angle. / doorbraak-; dwars- / transversale / Quer-; Durchbruchs-; Bresche- / transversal

2290
antecedent, – pertaining to valley or river when, during elevation of the land, the river was able to incise its channel as fast as the land was rising. / antecedent; doorbraak- / antécédent; percé / antezedent; Durchbruchs- / antecedente

superimposed; superposed; epigenic, – pertaining to valley or river that originally developed on less resistant, subsequently removed strata, so that now in the underlying, more resistant rock, no adaptation is shown to surface and structure of that formation. / epigenetisch / épigénique; surimposé; épigénétique; percé / epigenetisch; Durchbruchs- / sobreimpuesto

2291
radial consequent, – pertaining to valley system or drainage patterns. / radiaal consequent / système en réseau radial / zentrifugales System / sistema en red radial

2292
dendritic, – pertaining to valley system or drainage patterns. / vertakt / système en réseau dendritique / verzweigtes System / sistema en red dendrítica

rectangular; trellis, – pertaining to valley system or drainage patterns. / rechthoekig / système en réseau orthogonal; .... en réseau en treillis / Kammersystem; gitterförmiges Netz / sistema en red ortogonal

[1648] *River Terraces*

2293
river terrace; stream ...., – a former valley floor, dissected by the river. / rivierterras; fluviatiel terras / terrasse fluviale / Flussterrasse; Tal....; Talleiste / terraza fluvial

2294
terrace level, – the surface of a terrace corresponding with the former river profile. / terrasniveau / niveau de terrasse / Terrassenzug / nivel de terraza

2295
terrace edge, – the outer edge of the terrace surface. / terrasrand / rebord de terrasse / Terrassenkante / borde de terraza

terrace slope; .... scarp; .... bluff; .... front; .... face, – the escarpment below the terrace edge. / terraswand / talus de terrasse; abrupt de ....; escarpement de .... / Terrassenböschung; ....steilrand; ....steilhang / escarpe de terraza; talud de ....

2296
terrace flight; flight of terraces; stepping ....; stepped ...., – / terrassentrap / terrasses en gradins; .... étagées / Terrassentreppe; ....stockwerk / terrazas en escalón

2297
bank; bench; berm, – narrow terrace. / bank; berm / berme; vire (Alpes); banquette / Leiste; Gesims; Barre; Bank; Bühne / banco

2298
rock terrace; cut ....; stream-cut ....; erosion ...., – terrace left when a stream incises a new valley below the level of the original floor, which was cut in bedrock. / erosieterras; rots.... / terrasse d'érosion; .... rocheuse; .... de stabilité; .... de profil d'équilibre / Felsterrasse; Erosions.... / terraza de erosión

alluvial terrace; built ....; built-up ....; stream-built ....; drift ....; constructional ....; fill ...., – a terrace formed when a river incises into its own valley fill. / accumulatieterras / terrasse de remblaiement; .... d'accumulation; .... de dépôt; .... alluviale (de graviers); .... construite / Aufschüttungsterrasse; Aufschotterungs....; Akkumulations....; Schotter.... / terraza aluvial

2299
cyclic terrace; valley-plain ....; continuous ....; main ...., – a terrace caused by a geomorphical cycle. / cyclisch terras; doorlopend .... / terrasse cyclique; .... principale; .... continue; .... régionale / durchlaufende Terrasse / terraza cíclica

noncyclic terrace; local ...., – a river terrace caused by other influences than the sequence of geomorphical cycles. / nietcyclisch terras; plaatselijk .... / terrasse non-cyclique; .... locale / lokale Terrasse; örtliche .... / terraza no cíclica

2300
inset terraces, – valley-plain terraces formed during successive periods of vertical and lateral erosion, so that on both sides of the valley remnants of former valley bottoms are left. / ineengeschakelde terrassen / terrasses emboîtées; .... étagées; .... polycycliques / ineinandergeschachtelte Terrassen / terrazas policíclicas

matched terraces; matching ....; paired ...., – valley-plain terraces, matching in level on opposite sides of a valley, and being the remnants of the same valley bottom. / in hoogte overeenkomende terrassen / terrasses couplées; .... appariées / nach Höhe übereinstimmende Terrassen / terrazas emparejadas

2301
fill-in fill terrace, – terrace left when a river, after having incised its valley fill, partly fills up the new valley and incises anew. / ineengeschakeld accumulatieterras / terrasse de remblaiement emboîtée / eingeschaltete Aufschüttungsterrasse / terraza sucesíva

2302
slip-off slope terrace, – local terrace formed on the inner side of an extending meander curve, which is being incised slowly at an irregular rate. / afglijdingsterras / terrasse de glissement / Gleithangterrasse / terraza de deslizamiento

polygenic terrace, – a gently sloping terrace, formed by the smoothing of slip-off slope terraces. / afglijdingsterras / fausse terrasse; terrasse polygénique; .... de glissement latéral / Gleithangterrasse / terraza poligénica

2303
meander-spur terrace, – a terrace on a meander spur. / meanderspoorterras / terrasse d'éperon; .... de lobe / Mäanderspornterrasse / terraza de espolón

2304
meander-scar terrace; alternate ...., – a local terrace formed by the shifting of the meander curves during the process of continuous excavation of a valley at a slow rate. / meanderterras / terrasse de méandres / Mäanderterrasse; Mäanderkerbe.... / terraza de meandros

cusp; salient, – jutting-out point of a meander-scar terrace, caused by the meeting of two meander scars. / terraspunt tussen twee meanderkliffen / pointe entre deux échancrures; saillant / Spitze zwischen zwei Mäanderkerben / cúspide

2305
rock-perched terrace; rock-defined ....; defended-cusps ...., – meander terrace, protected by the outcrop of bedrock at the foot of the cusps. / beschermd meanderterras / terrasse de méandre à armature rocheuse; .... armée / Mäanderterrasse mit geschützten Spitzen / terraza de armadura rocosa

2306
Ecke; Eckflur, – German terms denoting the steps on ridges between tributary valleys, being dissected terraces of the main valley.

2307
structural terrace; rock bench, – a terrace caused by the more rapid weathering and denudation of soft strata overlying more resistant rocks. / denudatieterras; structuur.... / replat structural / Schichtterrasse; Denudations....; Verwitterungs.... / terraza estructural; rellano ....

diastrophic terrace, – a terrace caused by movements of the earth's crust. / tektonisch terras / terrasse diastrophique / diastrophische Terrasse / terraza diastrófica

2308
climatic terrace, – a terrace caused by erosive river action controlled by climatic circumstances. / – – / terrasse climatique / klimatisch bedingte Terrasse / terraza climática

eustatic terrace, – a terrace due to eustatic movement of sea level. / – – / terrasse eustatique / eustatisch bedingte Terrasse / terraza eustática

2309
fluvioglacial terrace, – a fill terrace consisting of fluvioglacial material. / fluvioglaciaal terras / terrasse fluvioglaciaire / Fluvioglazialterrasse / terraza fluvioglacial

congelifractate terrace, – a terrace left when a stream erodes below the level of a valley fill consiting of periglacial solifluction material. / – – / terrasse de matériel congélifracté / – – / terraza formada por material de solifluxión

2310
Stauterrasse, – a German term denoting a terrace left when a temporary dam in a valley, causing an upstream accumulation of alluvial material, is removed and the surface of the valley fill is incised. / opstuwingsterras / terrasse de barrage / Stauterrasse / terraza de ostrución

2311
terrace intersection, – the intersection between a plunging and an emerging terrace. / terrassenkruising / entrecroisement de terrasses / Terrassenkreuzung / entrecruzamiento de terrazas

plunging terrace; dipping ...., – a terrace dipping in downstream direction below the surface of the valley fill or below another terrace. / onderduikend terras / terrasse plongeante / untertauchende Terrasse / terraza buzante

emerging terrace, – a terrace emerging in upstream direction from below the valley fill or another terrace. / opduikend terras / terrasse émergeante / auftauchende Terrasse / terraza emergente

[1649] *Water Divides and River Capture*

2312
interfluve; interstream area; divide, – the higher terrain between two rivers. / waterscheidingsrug; gebied tussen twee rivieren / interfluve / Wasserscheide; Riedel / interfluvio

doab, – an Indian term for a flat interfluve.

2313
Talzwischenscheide, – German term for a divide between two valleys.

2314
watershed; waterparting; divide, – the boundary line between drainage basins or valleys. / waterscheiding; ....slijn / ligne de partage des eaux / Wasserscheide / línea divisoria de cuencas hidrográficas; divisoria de aguas

2315
valley-floor divide, –
1. a watershed situated in a valley. / dalwaterscheiding / ligne de partage des eaux en vallée / Talwasserscheide / divisoria del valle
2. a divide situated between two parts of the same valley, which however drain to two different river basins. / dalwaterscheidingshoogte / col de partage / Talwasserscheide / divisoria del valle

2316
water divide, – a divide on a mountain crest, on hills, etc. / kamwaterscheiding / ligne de faîte / Kammwasserscheide / divisoria de aguas

dividing ridge; .... crest, – / waterscheidingskam / crête de partage / Wasserscheidekamm / cresta divisoria

2317
consequent divide, – a divide between consequent rivers. / consequente waterscheiding / ligne de partage conséquente / konsequente Wasserscheide / divisoria consecuente

2318
subsequent divide; longitudinal ...., – a divide between subsequent rivers. / subsequente waterscheiding / ligne de partage subséquente / subsequente Wasserscheide / divisoria subsecuente

2319
shifting divide; migrating ...., – / zich verplaatsende waterscheiding / ligne de partage migrante / verlegende Wasserscheide; wandernde .... / divisoria emigrante

creeping divide, – a slowly migrating divide. / verschuivende waterscheiding / ligne de partage se déplaçant lentement / sich langsam verschiebende Wasserscheide / divisoria lenta

2320
leaping divide, – a migrating divide in the case of river capture. / verspringende waterscheiding / ligne de partage se déplaçant brusquement / sich schnell verlegende Wasserscheide / divisoria brusca

2321
zig-zag watershed, – a watershed formed by rivers that have broken through the original dividing crest by retrogressive erosion; between the river basins the watershed retained its original position. / zig-zaggende waterscheiding / ligne de partage en zigzag / durchgreifende Wasserscheide; zickzackförmige .... / divisoria en zig-zag

2322
river capture; .... abstraction; .... piracy, – a river diversion by retrogressive erosion of a neighbouring river. / rivieraantapping; ....roof / capture (fluviale); rapine .... / Anzapfung / captura fluvial

capture by lateral erosion; .... by intercision, – / aantapping door zijdelingse erosie / capture par osculation; .... par tangence / Anzapfung durch Seitenerosion / captura por erosión lateral

2323
river beheading, – the loss of a part of a river course to another river by capture or spontaneous diversion. / rivieronthoofding / décapitation fluviale / Köpfen eines Flusses / decapitación fluvial

2324
beheaded valley, – a valley that lost its upstream part by river capture. / onthoofd dal / vallée décapitée; .... troncconnée / geköpftes Tal; gekapptes Tal / valle decapitado

2325
diverter; pirate river; stream robber; captor stream; capturing ...., – / roofrivier; aanvallende rivier / cours d'eau capteur; rivière conquérante / Räuberfluss; Ablenker / río pirata

2326
Ablenkungsschlucht, – a German term for a gorge on both sides of the elbow of capture. / aantappingskloof; afleidingskloof / gorge de capture / Ablenkungsschlucht / desfiladero de captura

elbow of capture, – / onthoofdingsknie / coude de capture / Ablenkungsknie; Anzapfungscnic .... / codo de captura

2327
water gap, – / kort doorbraakdal / percée; cluse vive / kurzes Durchbruchstal / cluse viva

wind gap; air ...., – / daltorso; verlaten doorbraakdal; droog .... / cluse morte; .... sèche; tronccon de vallée / Taltorso; Strunkpass / cluse muerta

2328
dead valley; dry ...., – / droog dal; verlaten dal / vallée morte; .... sèche / Trockental / valle muerto

2329
misfit river; underfit ...., – a river flowing in a valley that has been beheaded by a larger stream. / hongerrivier, verarmde rivier; verkommerde .... / rivière dépérie; .... incompétente; .... sous-adaptée / Hungerfluss; Kümmer....; unterfähiger Fluss / río incompetente

competent river, – a river flowing in a valley, the morphological character of which corresponds with the present hydrological condition of the river. / passende rivier / rivière compétente / fähiger Fluss / río competente

[1650] LAKES

NOTE:
– terms on types of lakes, e.g. proglacial lake, oxbow lake, have been assembled in the pertaining sections.

[1651] *Lake Waters*

NOTE:
– for further terms see section:
[1530] Actuopalaeontology

2330
lake, – a considerable body of iniand water. / meer / lac / See / lago

pond, – a shallow lake usually of small dimensions; often artificial. / vijver; plas / étang / Teich; Weiher / estanque

pool, – a small and rather deep body of standing water. / kom / mare / Kolk / pantano

2331
limnology, – the study of fresh waters, especially that of ponds and lakes, dealing with all physical, chemical, meteorological and biological conditions in such a body of water. / limnologie / limnologie / Limnologie / limnología

2332
epilimnion, – the upper water layers of sea or lake, that are subject to relatively rapid temperature variations, and that contain more oxygen and less carbon dioxide and have a lower pH than the bottom waters, due to the mixing currents derived from wave action. / epilimnion / épilimnion / Epilimnion / epilimnion

2333
mesolimnion, – the zone in a body of water, e.g. in lakes or sea, between the epilimnion and the hypolimnion, in which the change of temperature with depth exceeds 1°C per metre. / discontinuïteitsvlak / mésolimnion / Sprungschicht / – –

thermocline; metalimnion; discontinuity layer, – the layer of water in a lake in which the temperature exhibits the greatest difference in vertical direction, viz. the transition between epilimnion and hypolimnion. / spronglaag; discontinuïteitsvlak / métalimnion / Sprungschicht; Thermokline / metalimnion

2334
hypolimnion, – the lower layer of water in a sea or lake. / hypolimnion / hypolimnion / Hypolimnion / hipolimnion

clinolimnion, – upper part of the hypolimnion where the rate of heating falls almost exponentially with depth. / clinolimnion / clinolimnion / – – / – –

bathylimnion, – lower part of hypolimnion where rate of heating approaches a constant independent of depth. / bathylimnion / bathylimnion / – – / batilimnion

[1652] *Morphological Features*

2335
lake rim, – a narrow ridge of hard rock or loose material damming off a lake. / drempel; afsluitingsrichel / seuil; verrou / Riegel / umbral del lago

2336
lee shore; sheltered ..., – / lij-oever / côte abritée / Leeufer; Gelege....; Windschatten.... / ribera abrigada

weather shore; exposed ...., – / windoever / côte exposée / Windufer; exponiertes Ufer; Brandungsufer / ribera expuesta

2337
shore terrace; lake platform, – a terrace along a lake shore, cut in rock or built up by detrital material. / oeverterras; ....bank / beine lacustre; terrasse .... / Uferterrasse; ....bank; Schar / terraza del lago

erosional terrace; wave-cut ...., – a usually narrow terrace cut by wave action along the shore of a lake. / erosieterras / terrasse d'érosion / Uferterrasse; ausgewaschene Uferbank / terraza de erosión

depositional terrace; wave-built ...., – a terrace composed of lake sediments, of which the surface is controlled by wave movements. / sedimentatieterras / terrasse construite; .... d'accumulation; beine (Lac Léman) / angeschwemmte Uferbank / terraza de deposición

2338
inlet, – the place where the water enters a lake. / toelaat / entrée / Zuflusz / entrada

outlet, – the place where the water leaves a lake. / uitloop / issue / Ausflusz / desague

lake without outlet, – / afvoerloos meer / lac sans issue; lac sans écoulement / abflussloser See / lago sin desague

2339
spillway, – an outlet over a rim damming off a lake. / overlaat / déversoir / Überlauf / rebosadero

[1653] *Evolution of Lakes*

2340
damming; blocking, – the process of obstructing a valley, either by natural or artificial means. / afdamming; versperring / barrage; blocage / Abdämmung; Absperrung / bloqueo

ponding; impounding, – of river water, resulting in the formation of a lake, e.g. by a lava flow, a glacier, a moraine, a landslide. / stuwing / retenue / Stauung / estancamiento; embalsamiento

2341
filling up; silting up, – of a lake by sedimentation. / opvulling; dichtslibbing / comblement; colmatage / Auffüllung; Zuschüttung / colmatación

filling by plant growth, – / dichtgroeiing; verlanding / comblement par végétation; enherbement / Zuwachsen; Verlandung / colmatación por vegetación

2342
drying up; evaporation, – / opdroging; indamping / assèchement par évaporation / Austrocknung; Eindunstung / desecamiento por evaporación

2343
playa lake, – a shallow lake occupying a desert plain after a rain. / playameer / lac de playa / Playasee / lago de playa

2344
relic lake; residual lake, – a lake that has become separated from the parent body of water by diastrophic or other causes. / restmeer; reliktmeer / lac résiduel; lac relique / Reliktsee; Restsee / lago residual

[1660] GLACIERS

[1661] *Snow and Ice*

2345
glaciology, – collectively the branches of science concerned with the causes and modes of ice accumulation and movement and with ice action on the earth's surface. / gletsjerkunde; glaciologie / glaciologie / Gletscherkunde; Glaziologie / glaciología

2346
dust snow; wild ...., – very light snow, density 0.01-0.015 g.cm^{-3}, easily moved by the wind, formed when falling at a temperature below -10°C. / stofsneeuw / neige folle / Wildschnee; Flaum ....; wilder Schnee / nieve blanda

powder snow; sand ...., – dry, loose and glittering snow, density 0.05-0.1 g.cm^{-3}, formed during heavy snowfalls followed by dry cold weather. / poedersneeuw / neige poudreuse / Pulverschnee / nieve en polvo

2347
drifted snow; wind-packed ...., – snow compacted by wind action and luted by frozen moisture, so that solid masses are formed, density 0.2-0.3 g.cm^{-3} / door wind samengepakte sneeuw / neige entassée par le vent / windgepackter Schnee; Pressschnee / nieve aptastada por el viento

Packschnee, – a German term for snow transported by wind action and deposited as a compact, but not luted layer.

2348
(snow) slush, –
1. watery new snow, density 0.6-0.8 g.cm^{-3}, fallen at temperatures above freezing point;
2. old snow drenched with rain or melt water.

/ papsneeuw / neige flasque / Pappschnee / agua-nieve

2349
dry snow, – snow falling at temperatures below freezing point. / droge sneeuw / neige sèche / Trockenschnee / nieve seca

sleet; wet snow, – snow falling at temperatures above freezing point. / natte sneeuw / neige humide / Feuchtschnee / nieve húmida

2350
granular snow; corn ....; sugar ....; buckshot ...., – dry snow composed of fine-grained ice crystals, formed by repeated thawing and refreezing of old snow. / korrelsneeuw / – – / Kornschnee; Gries....; Mehl.... / nieve granular

spring snow, – a very coarse-grained damp snow formed during spring from old snow, which has been subjected to frequent thawing and refreezing. / – – / neige pourrie / Sulzschnee; Salz....; Faul.... / nieve de primavera

2351
depth hoar, – coarse, granular snow at the bottom of a thick layer of snow, formed by evaporation as a result of heat radiating from the underlying rock whereby water has condensed on particles located higher up. Thus an aerated, unstable layer is formed over which the top layers of snow may slide down-slope. / loopsneeuw / neige coulante / Schwimmschnee / – –

2352
ice crust, –
1. a superficial layer on snow, formed by periodic thawing and refreezing;
2. on a rock surface formed by freezing of water condensed from the air.

/ ijskorst; harst / verglas / Harsch; Harscht; Harst / costra de hielo

2353
snowdrift; .... wreath, – a mass of snow piled together by wind. / sneeuwopwaaiing / neige soufflée; congère / Schneewehe; ....verwehung / ventisquero

2354
snow line; .... limit, – the lowest limit of perpetual snow. / sneeuwgrens / limite des neiges éternelles / Schneegrenze / nivel de las nieves perpetuas

2355
snowbreak, – / afsmelten van de sneeuw / fonte des neiges; fusion des .... / Schneeschmelze / fusión de las nieves

2356
(snow) cornice; overhanging snow, – / sneeuwluifel / corniche (de neige) / (Schnee)wächte / cornisa de nieve

2357
nivation, – the action of alternate freezing and thawing around the fluctuating margins of snow accumulations. / nivatie / nivation / Schnee-Erosion; Nivation / nivación

2358
avalanche; snowslide; .... slip, – a large mass of snow in swift motion down a mountain side. / (sneeuw)lawine; sneeuwafglijding / avalanche / (Schnee)lawine / alud de nieve; avalancha de ....

2359
drift avalanche; powdery ....; dry ...., – an avalanche of dry powdery snow initially set in motion by wind. / stoflawine; stuif ....; droge lawine / avalanche poudreuse; .... volante; .... sèche; .... froidie / Staublawine; Pulverschnee....; trockene Lawine; kalte .... / alud en polvo (de nieve)

ground avalanche; wet ...., – snow masses drenched by spring thaw and thus rendered incoherent to the surface of the underlying rock, moving down-slope as an avalanche. / grondlawine; natte lawine / avalanche de fond; .... terrière; .... humide; .... chaudie / Grundlawine; Feuchtschnee....; Schlag....; nasse Lawine; warme .... / alud de fondo

2360
wind-slab avalanche, – an avalanche of sheets of wind-packed snow. / sneeuwschollenlawine / avalanche de plaques de neige / Schneebrettlawine / alud de placas de nieve

2361
ice avalanche, – avalanche formed by the breaking off of parts of a steeply inclined ice sheet or glacier. / ijslawine / avalanche de glace / Eislawine; Gletscher.... / alud de hielo

2362
avalanche chute; .... trench, – a gully formed by an avalanche. / lawinegeul / couloir d'avalanche; sillon ....; cannelure .... / Lawinengraben; ....trog / canal de alud

avalanche track, – the path along which an avalanche has moved on a mountain side. / lawinebaan / glissoir d'avalanche; couloir .... / Lawinenbahn; ....zug; ....gang / camino del alud

avalanche cone, – the mass of material deposited where an avalanche has fallen; it includes snow, ice, rock and other objects which have been carried away by it. / lawinekegel / cône d'avalanche / Lawinenkegel / cono de alud

2363
firn; .... snow; .... ice; névé, – the granular substance, partly snow and partly ice, formed when the snow, that has accumulated in the catchment basin of a glacier, is being transformed into glacier ice. / firnsneeuw; firn / névé / Firnschnee / firn

firnification; snow recrystallization, – conversion of snow into firn. / firnvorming / – – / Verfirnung; Firneisbildung / – –

2364
névé; .... basin; .... field, – / firnbekken; ....kom; ....veld; ....gebied / névé / Firnmulde; ....feld; ....gebiet / chinarro; congesta

névé line; firn ...., – the line on the glacier surface that separates the accumulation and the ablation area of a glacier. / firngrens; ....lijn / limite du névé / Firnlinie; ....grenze / límite del chinarro

2365
penitent snow, – surface melting forms of firn ice or snow,

formed as a result of (all year round) strong insolation – especially in tropical regions – consisting of peculiarly shaped pillars in East-West rows, sometimes resembling statues. / boetelingensneeuw / neige à pénitents / Büsserschnee; Zackenfirn / nieve penitente

[1662] *The Glacier*

2366
glacier; jökull (Iceland); isbrae (Norway), – a field or body of ice, formed in a region where snowfall exceeds melting, and moving slowly down a mountain slope or valley. / gletsjer; (gletscher) / glacier / Gletscher; Ferner; Kees (eastern Alps) / glaciar

2367
glaciation, – the forming of glaciers or the covering with an ice cap. / vergletsjering; glaciatie / glaciation / Vereisung; Vergletscherung / glaciación; helada

deglaciation; deglacieration, – the (gradual) disappearance of an ice cap or ice sheet. / deglaciatie / déglaciation / Deglaziation / deshielo

2368
warm glacier; temperate ...., – a glacier in which the temperature is not far from the melting point at prevailing pressure. / warme gletsjer; gematigde .... / glacier tempéré / temperierte Gletscher / glaciar templado

cold glacier, – a glacier in which the temperatures are (much) lower than the freezing point. / koude gletsjer / glacier polaire / Hochpolar-Gletscher / glaciar polar

2369
dead ice; stagnant ice, –
1. ice belonging to a glacier that lost its motion;
2. small remnants of glacier ice, carried along and stranded in an outwash plain.

/ dood ijs / glace morte / Toteis / glaciar muerto

fossil ice, – ice of the Pleistocene as preserved in frigid regions. / fossiel ijs / glace fossile / Steineis; fossiles Eis / glaciar fósil

2370
translation, – the property of ice grains, when subjected to pressure, to deform by shearing along planes perpendicular to the crystallographic main axis, causing the plastic flow of glacier ice. / translatie / translation / Translation / translación

2371
intergranular film, – of water between ice grains due to pressure melting at granular contacts. It is considered to facilitate flow of glacier ice.

2372
stream line, – of an ice particle in a glacier; trajectory covered by an ice particle during the glacier movement. / stroombaan / trajectoire / Stromlinie / trayectoria

line of motion, – the projection of a stream line on a horizontal plane. / bewegingslijn / ligne de mouvement / Bewegungslinie / línea de movimiento; .... de corriente

2373
slip along shear planes, – / laminaire beweging; beweging langs glijvlakken / glissement entre les feuillets / Laminarbewegung; Bewegung an den Scherflächen / deslizamiento entre planos de cizalla

2374
blue-bands structure; lamination, – an alternation of blue and white layers in the ice of a glacier tongue, caused by ice of low and high air content respectively; explained by some authors as the result of cleavage by pressure, by others as the result of sliding of the ice along shearing planes. / blauwebanden-structuur; blauwe-bladen-.... / structure rubanée; .... en feuillets bleus; .... en bandes bleues / Blätterung; Bänderung; Blaublätterstruktur / estructura laminada; .... en bandas

2375
crevasse; fissure, – in a glacier, due to strain set up by movement over an uneven surface, by taking sharp bends or by differential movements due to other causes. / gletsjerspleet / crevasse / Gletscherspalte / grieta; fisura

2376
marginal crevasse, – crevasse running from the margin of the glacier in a direction upwards and towards the middle of the glacier. They are due to a differential movement of the ice at the sides and the centre of the glacier. / randspleet / crevasse marginale; .... latérale / Randspalte / grieta marginal; .... lateral

transverse crevasse, – crevasse running in a direction perpendicular to the axis of the glacier, at places where the glacier flows over an elevation in its bed. / dwarsspleet / crevasse transversale / Querspalte; Transversal.... / grieta transversal

longitudinal crevasse, – crevasse originating at a point where the glacier enters a wider valley or a plain, allowing the ice to flow sideways. They run in a direction parallel to the stream lines of the glacier. / lengtespleet / crevasse longitudinale / Längsspalte / grieta longitudinal

2377
icefall; glacier cataract, – the passage of a glacier over a steep declivity in its bed. / ijsval; gletsjerval / cataracte de glace / Gletscherfall; ....sturz; ....bruch / catarata de hielo

sérac; ice pyramid, – the blocks into which a glacier breaks on a steep grade / ijspyramide; serac / sérac / Eispfeiler; Serac / pirámide de hielo

2378
ogive, – low ridges or alternating light and dark coloured bands (in the form of upturned pointed arches) on the surface of a glacier; caused by differential melting of the laminated firn and glacier ice or by the lamination itself; the ogive form is caused by the greater velocity of the ice flow in the middle of the glacier and by the intersection of the lamination planes and the convex surface of the glacier. / ogive / ogive / Ogive / ojiva

false ogive, – ridges on a glacier probably caused by the flowing of the ice. / onechte ogive / chevron / Sparre / cabrilla

2379
dirt band; dust ...., – dark bands caused by the accumulation of dirt and dust in the depressions between ogives. / vuile band; puinband; stof....; slijk.... / bande boueuse; .... sale / Schmutzband; Staub.... / banda polvorienta

2380
catchment basin; zone of accumulation; accumulation area; region of alimentation, – of a glacier. / verzamelbekken; voedingsgebied / bassin d'alimentation; aire .... / Nährgebiet; Speise....; Sammel.... / cuenca de alimentación; zona de acumulación

ablation, – the loss of ice due to melting and evaporation. / ablatie / ablation / Ablation / ablación

2381
ice shed, – glacial divide from which ice moves in opposite directions. / ijsscheiding / ligne de partage des glaces / Eisscheide; Eisscheitel / divisoria de hielos

2382
bergschrund, – large deep crevasse often visible during summer at the head of a mountain glacier between the stationary and the mobile part of the névé; in winter the bergschrund is concealed by snow. / randkloof / rimaye / Bergschrund / rimaya

2383
transfluence, – the flowing of glacier ice through a depression in the mountain ridges, that partly enclose the accumulation area of a glacier. / transfluentie / transfluence / Transfluenz / transfluencia

2384
lateral canyon, – the open space between the margin of a glacier and the rock wall, formed by melting of glacier ice by heat radiation of the rock. / ablatiedal; ....kloof / fossé latéral / Ablationstal; Randschlucht / fosa lateral

2385
riding glacier, – after the confluence of two glaciers, the one that flows on top of the other one. / rijdende gletsjer / glacier chevauchant / sich überschiebender Gletscher / glaciar cabalgante

carrying glacier, – a glacier that, after confluence, carries the riding one; the latter is pulled into pieces when the carrying glacier moves at a greater speed. / dragende gletsjer / glacier porteur / tragender Gletscher / glaciar cabalgado

2386
glacier table, – a block of stone supported above the surface of a glacier on a column of ice, the stone having protected the ice underneath it from melting. / gletsjertafel / table de glacier / Gletschertisch / mesa de glaciar

2387
cryoconite, – terrigenous wind-borne dust on glacier ice. / kryokoniet / cryoconite / Kryokonit / kryokonita

cryoconite hole; dust well, – hole formed in the surface of glacier ice as a result of heat absorption by dust deposits. / kryokonietgat; smeltschaal / trou de cryoconite; .... de midi; baignoir / Kryokonitloch; ....schal; Schmelzloch; Mittags.... / hoyo de kryokonita

2388
supraglacial stream, – melt-water stream on the surface of a glacier. / smeltwaterbeek; supraglaciale beek / cours d'eau de fonte / Schmelzwasserbach / torrente supraglaciar; .... superficial

subglacial stream, – melt-water stream in the lowermost part of a glacier, i.e. incised in the bedrock of the glacier. / subglaciale beek / cours d'eau sous-glaciaire / subglazialer Bach / torrente subglaciar; .... de fondo

englacial stream, – melt-water stream traversing the body of a glacier. / inglaciale beek / cours d'eau intérieur / inglazialer Bach / torrente intraglaciar; .... interno

2389
glacier mill; moulin, – a cylindrical shaft in the ice of a glacier in which a melt-water stream sets up a swirling motion so that it may bore out a deep pothole in the rock beneath. / smeltwatergat; ....pijp / moulin / Gletschertrichter; ....mühle; ....topf / molino glaciar

moulin pothole; glacial ....; giant's kettle, – a cylindrical hole bored out in the rock beneath a glacier by boulders rotated in a swirling melt-water stream; often one or more rounded boulders are found at the bottom. / gletsjermolen / marmite de géant / Gletschermühle; ....topf; Riesentopf / marmita de gigante

2390
glacier tongue, – / gletsjertong / lobe du glacier; langue glaciaire / Gletscherzunge / lengua glaciar

entrenched part, – part of the glacier tongue that is enclosed between steep rock slopes. / – – / partie encaissée / eingeengter Teil / lengua de valle; parte encajada

glacier terminus; .... snout, – terminal face of a glacier, where, over a period of years, accumulation and ablation are balanced. / gletsjerfront / front du glacier / Gletscherstirn / frente del glaciar

2391
glacier cave, – the tunnel-shaped outlet of the subglacial stream in the lower end of the glacier. / gletsjerpoort / arche du glacier / Gletschertor / cueva glaciar; arco del ....

2392
glacial stream, – a water course emerging from a glacier cave. / gletsjerbeek / torrent glaciaire / Gletscherbach / torrente glaciar

glacier milk, – turbid glacial stream water carrying fine powder of triturated rock in suspension. / gletsjermelk / lait de glacier / Gletschermilch / leche de glaciar

ice-marginal valley; pradolina (Poland), – valley parallel to the ice front, draining away the melt water. / oerstroomdal / vallée marginale proglaciaire / Urstromtal / valle marginal anteglaciar

2393
glacial spillway; proglacial valley, – the outlet of a promarginal lake; generally perpendicular on the ice front. / smeltwaterkanaal; ....dal / chenal d'eau de fonte / Schmelzwassertal / canal de desague anteglaciar

proglacial lake; ice-marginal lake, – lake situated in front of the end moraine of a glacier or between the end moraine and the withdrawing ice front. / proglaciaal meer / lac proglaciaire / glazialer Stausee / lago anteglaciar

[1663] *Glacial Erosion*

2394
frost work, – the action of glacier ice at its contact with the bedrock. / – – / gélivation / Frostwirkung / – –

2395
glacial scouring, – subglacial erosion by water under pressure, transporting much fine morainec material. / glaciale uitschuring; .... afschuring / marque glaciaire / – – / abrasión glaciar

planation, – the regional result of the scouring action of glaciers. / afschaving / rabotage / Abhobelung / ablación

ice-scour notch, – notch formed on the ice-scour limit. / – – / niche de rongement glaciaire / Schliffkehle / muesca del estregamiento del hielo

2396
plucking; quarrying; exaration, – / losrukken; losbreken / arrachement; exaration / Herausheben; Ausschürfen; Herausreiszen; Exaration; Detraktion / arrancamiento

2397
glacial striae, – scratches on rock surfaces due to the abrasive action of rock fragments held in a glacier. / gletsjerkrassen / stries glaciaires / Gletscherkritzen; ....kratzen; ....schrammen / estrías glaciares

striation; scratching, – / bekrassing / striage; burinage / Kritzung / estriación; estriaje

2398
chattermark, – crescentic mark, caused by large boulders carried in the glacier. / – – / coup de gouge / Sichelwanne / marca falciforme

glacial groove, – / gletsjergroef / cannelure glaciaire / Gletscherfurche / estría glaciar

2399
polishing, – / polijsting / polissage / Schleifung; Glättung; glättende Abschleifung; Detersion / pulimentación

2400
roche moutonnée; hump, – protruding rock that has been smoothed and striated by the glacier on the upstream side, and left rough and rugged, with steeper slopes, on the downstream side, where the glacier pulled away loosened blocks. / bultrots / roche moutonnée / Rundhöcker / roca aborregada

2401
ice push, – the pressure exerted by expanding ice. / ijsdruk / poussée de la glace / Eisdruck; Eispressung / presión de congelación

ice shove; ice thrust; ice crowding, – the movement of expanding or drifting ice. / ijsstuwing / poussée des glaces / Eisschub; Eisschiebung / dilatación por congelación

[1664] *Glacial Transport and Deposits*

2402
glacial deposit; .... drift, – a deposit of material transported by a glacier. / glaciale afzetting / dépôt glaciaire / glaziale Ablagerung / depósito glaciar

fluvioglacial deposit; glacioaqueous ....; aquoglacial ....; glaciofluvial drift; (glacial) outwash, – melt-water deposits laid down over a considerable area in front of the end moraine of a glacier. / fluvioglaciale afzetting; smeltwaterafzetting / dépôt fluvioglaciaire; .... d'eau de fonte / glazifluviatile Ablagerung; Schmelzwasserablagerung; ....absatz / depósito fluvioglaciar

glaciolacustrine deposit, – / glaciale meerafzetting / dépôt glacio-lacustre / glazialer Seeabsatz; glazilimnischer Absatz / depósito glaciolacustre

2403
glacial drift, –
1. all rocky material in transport by a glacier;
2. all deposits made by glaciers;
3. all deposits of glacial origin.

moraine, –
1. all rock debris incorporated in, in transit on or in, or carried and eventually deposited by glaciers;
2. ridge consisting of that debris.

/ morene / moraine / Moräne / morrena

2404
till, – non-stratified drift; moraine deposit. / keileem / – – / – – / till

ablation till, – glacial drift accumulated on top of the glacier as a result of ablation of the (stagnant) ice. / – – / moraine d'ablation / – – / – –

lodgment till, – basal glacial drift lodged over bedrock. / – – / moraine de fond / – – / – –

2405
moraine in transit; carried moraine; moving ...., – in distinction to deposited moraine, all rock debris that is being transported by a glacier. / voortbewogen morene / moraine mouvante / bewegte Moräne; Wandermoräne / morrena móvil

2406
surface moraine; superglacial drift, – a moraine in transit, car-

ried on the surface of a glacier, i.e. a lateral, a medial or an ablation moraine. / oppervlaksmorene / moraine superficielle / Obermoräne / morrena superficial

2407
medial moraine; median ...., – a moraine in transit, carried in the middle of the glacier; formed by:
1. the union of the lateral moraines when two glaciers coalesce;
2. a rocky protuberance in the bed of the glacier, being abraded by the glacier ice, so that the debris formed in this way appears at the glacier surface in the ablation area of the glacier.

/ middenmorene / moraine médiane / Mittelmoräne; Zwischen....; Gufferlinie (Swiss Alps) / morrena central

interlobate moraine, – a medial moraine between two lobes of a continental ice sheet.

2408
englacial moraine; .... drift, – a moraine in transit, embedded in a glacier, formed as the result of:
1. the union of two glaciers; vertically when the glaciers flow alongside of each other (the lateral moraines fuse); more or less horizontally in the case of a carrying or riding glacier (the ground moraine of the riding glacier is taken up in the composite ice body);
2. the abrasion of a protruding rock mass in the bed of the glacier.

/ binnenmorene; inwendige morene / moraine interne; .... intérieure / Innenmoräne / morrena interna

2409
ground moraine; subglacial ...., – moraine in transit, formed beneath the ice by attrition along the bed of a glacier and augmented by the debris falling down crevasses. / grondmorene / moraine inférieure; .... profonde; .... de fond / Grundmoräne; Unter.... / morrena subglaciar; .... inferior

fluted moraine, – ground moraine with long grooves. / – – / moraine cannelée / – – / morrena estriada

2410
deposited moraine; stranded ...., – in distinction to moraine in transit, all rock debris that has been deposited in the zone of ablation of a glacier, where, after retreat of the glacier it may form an obstruction behind which a lake is formed. / afgezette morene / moraine déposée / abgelagerte Moräne; Stapelmoräne / depósito de morrena

2411
lateral moraine, – the moraine material carried on or deposited at the lateral margins of the glacier tongue; composed of small quantities of marginal ablation moraine, but mainly of debris loosened from the valley walls by glacial abrasion and quarrying or fallen on the glacier from the bordering mountain slopes. / zijmorene / moraine latérale / Seitenmoräne / morrena lateral

2412
ablation moraine; .... drift, – a more or less continuous layer of moraine material melted from the glacier ice in the lower reaches of glacier tongue or ice cap. / ablatiemorene / moraine d'ablation / Ablationsmoräne / morrena de ablación

2413
end moraine; frontal ....; terminal ...., – the mass of debris, consisting of surface and ground moraine material and often non-morainic material, piled up in front of a glacier during a stationary stage. (The term terminal moraine often is used also in the meaning of the end moraine of a glacier, which has or had reached its largest expansion.) / eindmorene; front.... / moraine frontale / Endmoräne; Stirn.... / morrena terminal

2414
moraine rampart, – elongated ridge or row of hills of end and lateral moraine material deposited by a retreated glacier. The term is applicable for longitudinal as well as border moraines. / walmorene; morenewal / moraine rempart; rempart morainigue / Wallmoräne; Moränewall / terraplén morrénico

2415
longitudinal moraine, – a moraine rampart formed by deposits of the medial and englacial moraines of a former glacier. / lengtemorene / moraine longitudinale / Längsmoräne / morrena longitudinal

2416
border moraine, – this term includes the flank and frontal moraines. / randmorene / moraine marginale / Randmoräne / morrena de borde

flank moraine, – a moraine rampart at the flank of a shrunken glacier or a former glacier. / oevermorene / moraine riveraine / Ufermoräne / morrena de flanco

frontal moraine, – a moraine rampart at the front of a former glacier. / frontmorene; verlaten morene / moraine frontale; vallum morainique / Stirnmoräne / morrena frontal

2417
recessional moraine; stadial ...., – end moraine formed during a considerable pause in the final retreat of a glacier. / terugtrekkingsmorene / moraine de retrait / Rückzugsmoräne / morrena de retroceso

2418
moraine amphitheatre, – an amphitheatrical arrangement of moraine ramparts, often occurring in areas of former glaciation. / moreneamfitheater / amphithéâtre morainique / Moränenamphitheater / anfiteatro morrénico

2419
upsetted moraine; push ....; thrust ...., – an accumulation of glacial drift, pushed or plowed up into ridges by the thrust of advancing ice. / stuwmorene / moraine de poussée / Stauchmoräne / morrena de empuje

NOTE:
– these terms have also been used incorrectly in the sense of ice-pushed ridge.

ice-pushed ridge; ice-contorted ....; glacial pressure ...., – a ridge of non-morainic material, pressed up by a glacier (usually inland ice). / stuwwal / crête de poussée; bourrelet de .... / Stauwall; Stauch.... / cresta de empuje

2420
drumlin, – elongated flat mound composed of moraine material.
1. **true drumlin,** – built under the glacier ice not far from its margin or formed from an older ground moraine by a readvancing glacier.
2. **rock drumlin,** – sometimes composed of a core of bedrock, modelled by glacial erosion and veneered with a relatively thin layer of till.

/ drumlin / drumlin / Drumlin / drumlin

esker; ås (plur. åsar) (Sweden), – a long often winding ridge of gravelly and sandy drift, in regions of former glaciation, deposited by subglacial streams. / smeltwaterrug / esker / Wallberg; Os / esker

kame, –
1. fan-like melt-water deposits laid down on the stoss slope of an obstructing moraine;
2. mounds of fluvioglacial material left behind by dead ice that melted away.

/ kame / kame / Kame / kame

kame terrace; ice-marginal ...., – fluvioglacial deposits formed between the somewhat retreated glacier front and the moraine rampart or glacial pressure ridge belonging to it, and left behind by the ice in the shape of a terrace at the flank of the moraine or the pressure ridge. / kameterras / terrasse de kames / Kamesterrasse / terraza glaciar marginal

2421
crag and tail form, – topography of ice-smoothed knobs of solid rock that give place leeward to tapering streamlined tails of ground moraine.

2422
boulder clay, – unstratified, unsorted and unwashed glacial deposits containing a mixture of rock of all dimensions and all degrees of angularity. / keileem / argile à blocaux / Blocklehm; Geschiebemergel; ....lehm / arcilla en bloques

erratic boulder; erratic, – boulder transported from its original resting place by ice. / zwerfsteen; .... kei / galet erratique; bloc .... / erratisches Geschiebe; Findling / bloque errático; canto ....

2423
transition cone, – transitional zone between the end moraine and the glacial outwash, where moraine and fluvioglacial deposits interfinger due to oscillations of the glacier front. / overgangskegel / cône de transition / Übergangskegel / cono de transición

2424
fluvioglacial drift plain; outwash plain; frontal apron; sandur (plur. sandar) (Iceland), – plain formed by fluvioglacial deposits. / fluvioglaciale vlakte; spoelzandwaaier / plaine alluviale fluvio-glaciaire; .... alluviale pro-glaciaire / glaziale Aufschüttungsbene; Sanderebene / llanura aluvial fluvioglaciar; .... de lavado

[1665] *Types of Glaciers*

2425
ice cap; ice carapace; jökull (Iceland), – a cover of perennial ice and snow, moving in all directions from the centre. / ijskap / calotte glaciaire locale / Eiskappe / manto de hielo

ice sheet; ice mantle, – a very large ice cap. / ijsdek / calotte glaciaire / Eisdecke; Deckgletscher / masa de hielo

inland ice; continental ice sheet; .... glacier, – an ice sheet of continental dimensions as that of Antarctica. / landijs / inlandsis; calotte continentale; glaciation régionale / Inlandeis; Binneneis / inlandsis

2426
plateau glacier, – glacier formed on a névé-covered mountain plateau, usually flowing over its edges in hanging glaciers (Scandinavia, Alaska). / plateaugletsjer / glacier de plateau / Plateaugletscher; (Hochlandeis) / glaciar de meseta

mountain glacier, – in distinction to a plateau glacier, a glacier formed on a mountain slope. / gebergtegletsjer / glacier de montagne / Gebirgsgletscher / glaciar de montaña

2427
valley glacier, – a glacier following the course of a valley. / dalgletsjer / glacier de vallée / Talgletscher / glaciar de valle

2428
alpine-type glacier, – a valley glacier being fed by one or more névé basin(s). / gletsjer van het alpine type; .... van het firnbekken-type / glacier du type alpin; .... de vallée simple / alpiner Gletscher; Firnfeldgletscher; Firnmulde.... / glaciar de tipo alpino

2429
Turkestan-type glacier, – a valley glacier not having a névé basin but being fed by snow avalanches exclusively. / gletsjer van het Turkestan type; .... van het lawine type; lawinevergletsjeringtype / glacier de type Turkestan / turkestanischer Gletschertypus; Firnkesselgletscher / glaciar de tipo Turkestan

2430
cirque glacier; Pyrenean-type ...., – a glacier situated in a cirque. / kaargletsjer / glacier de cirque / Kargletscher / circo glaciárico

2431
Mustag-type glacier, – glacier without névé basins; fed by the snow-drowned valleys of a region which has risen to a considerable height. / gletsjer van het Mustag type / glacier de type Mustag / Mustaggletschertypus; Firnstromgletscher / glaciar de tipo Mustag

2432
trunk glacier; dendritic ....; Himalayan-type ...., – a glacier having one tongue formed by the union of several glaciers. / gletsjer van het dendritische type; .... Himalaya type / glacier polysynthétique / Eisstromnetz / glaciar dendrítico

2433
hanging glacier; wall-sided ....; glacierit, – glacier on a steep declivity, usually a valley wall and often occupying a steep gorge in this escarpment. / hellinggletsjer / glacier suspendu / Hängegletscher / glaciar suspendido

2434
reconstructed glacier; remanié ...., – a glacier that has been formed anew at the foot of an icefall, as may happen in an area located above the snow line. / geregenereerde gletsjer; herstelde .... / glacier régénéré / regenerierter Gletscher / glaciar regenerado

2435
piedmont glacier; Malaspina-type ...., – glacier ice, spreading in cake-like sheets over level ground at the foot of glaciated regions and formed by fusion of the lower parts of two or more independent valley or wall-sided glaciers. / voorland-gletsjer; piedmont.... / glacier de piedmont / Vorlandgletscher; ....vergletscherungstypus; Malaspinatypusgletscher / glaciar de pie de monte

foot glacier, – piedmont glacier formed by one valley or wall-sided glacier.

2436
tidewater glacier; tide ....; tidal ...., – a glacier whose foot is situated in the tidal zone; this type often produces icebergs. / getijdengletsjer / glacier de marée; .... débouchant en mer / Gezeitengletscher / glaciar de marea

[1670] COASTS AND ISLANDS

[1671] *General Terms*

2437
coast, – a strip of land of indefinite width extending along the sea shore. / kust / côte / Küste / costa

coast line; shore ...., – the line that forms the boundary between land and water; the outline of the shore. / kustlijn / ligne de côte; trait de ....; ligne de rivage / Küstenlinie / línea de costa

2438
shore, – the coastal strip extending from the low-water line to the landward limit of effective wave action. / oever / rivage / Uferzone; Strand.... / litoral

2439
surf zone; breaker ...., – area between the outermost breakers and the limit of wave uprush. / brandingszone / zone des brisants; .... de déferlement / Brecherzone; Brandungs.... / zona de rompiente

wash zone, – the area that undergoes the erosive action by lapping and breaking of waves. / spoelzone / zone lavée par le jet de rive / Spülzone / – –

splash zone, – the area that undergoes the splashing effect of the breakers. / spatzone / zone des paquets de mer / Spritzzone; Spritzer.... / – –

spray zone, – the area affected by the spray of the breakers. / sproeizone; zone van brandingsnevels / zone des embruns / Sprühzone / – –

2440
coastal plain, – a plain of low elevation along the coast, usually composed of fluviatile or marine sediments, but occasionally of a rock bottom. / kustvlakte / plain littorale; .... côtière / Küstenebene / llanura litoral; .... costera

2441
shore profile, – profile of the sea bottom immediately in front of and perpendicular to the shore line. / kustprofiel / profil littoral / Küstenprofil / perfil litoral

graded shore profile, – shore profile that has reached its profile of equilibrium, typical of the stages of maturity and old age of the shore-line cycle. / vereffend kustprofiel / profil littoral à forme d'équilibre / ausgeglichenes Küstenprofil / perfil litoral en equilibrio

2442
wave base, – the ultimate limit of marine abrasion. / golfbasis; abrasieterminante / niveau de base des vagues; .... limite d'abrasion / Wellenbasis; Abrasionsterminante / nivel de base (de las olas)

2443
island, – a body of land extending above and completely surrounded by water, excepted the continents. / eiland / île / Insel / isla

reef; stack, – a rocky islet lying near or at the surface of the water. / klip; rif / écueil; récif / Klippe; Riff / arrecife

sand cay; .... key, – a comparatively small and low island of sand. / zandeilandje / caye de sable / Sandinselchen / cayo

2444
continental island, – an island situated on the continental shelf. / continentaal eiland / île continentale / Kontinentalinsel / isla continental

oceanic island, – an island rising from the deep sea floor off the shelf. / oceanisch eiland / île océanique / ozeanische Insel / isla oceánica

2445
water-faceted stone; aquafact, – boulder or cobble, faceted under influence of water movement. Most examples are formed on beaches, by wave action, and display two faces coming together in a sharp edge parallel to the trend of the beach. / – – / caillou façonnée par l'eau / – – / aguafacto; roca facetada por el agua

[1672] *Coastal Forms*

2446
cape, – a point or extension of land jutting out into the sea, either in the form of a peninsula or merely as an angle or projecting point on a coast. / kaap / cap / Kap / cabo

promontory; headland; head, – a cape of comparatively high land. / hoge kaap / promontoire / Vorgebirge / promontorio

winged headland, – a promontory having two spits extending in opposite directions. / – – / promontoire ailé / Vorgebirge mit zwei Nehrungsspitzen / promontorio alado

2447
foreland, – a more or less extensive area of land, usually of low elevation, protruding into the sea. / voorland / saillant; cap / Vorland / – –

cuspate foreland, – a triangular foreland of alluvial material. / – – / saillant triangulaire / – – / – –

2448
tongue, – a long narrow strip of land, projecting into a body of water. / landtong / langue de terre / Landzunge / lengua (de tierra)

2449
point, – the extreme end of a cape, or the outer end of any land area protruding into the sea. / punt / pointe / Spitze; Landspitze / punta

2450
re-entrant; indentation, – / inham / rentrant; indentation / Einbuchtung / indentación

bight, – a slight indentation in a coast forming an open bay, often crescent-shaped. / bocht / anse / Bucht / ancón

2451
bay; embayment, – a recess or inlet in the shore between two headlands or capes. / baai / baie / Bucht; Meeresbucht; ....busen / bahía

2452
ria, – a Spanish term for a long and narrow, often branched inlet, whose depth diminishes gradually inward; formed by drowning of a river valley in a rocky coast, where the strike of the formations is normal or at an angle to the shore. / ria / ria / Ria / ría

cala, – a short and narrow ria formed in a limestone coast. / cala / calanque / Cala / cala

2453
fjord; fiord; (sea) loch; firth (Scotland), – a narrow inlet of the sea, with steep sides plunging far below the level of the water, formed by glacial scour and filled by sea water after deglaciation. / fjord / fjord / Fjord / fiordo

2454
förde, – a Danish term for an elongated, comparatively narrow inlet typically formed by submergence of a subglacial channel in a formerly glaciated area. / förde / förde / Förde / fiordo

bodden, – a German term for a broad inlet typically formed by submergence of a former glacial tongue basin (south Baltic coast).

fiard, – a Swedish term for an irregular inlet, formed by glacial scour in a low rocky coast.

2455
skerries (Orkneys), skär (Sweden), – a great number of small islands and peninsulas caused by a partial submergence of a glacially modelled bedrock landscape; usually situated in front of a fiard zone. / scheren / écueils / Schären / – –

2456
estuary, – a more or less funnel-shaped river mouth, affected by the tides. / estuarium; trechtermonding / estuaire / Ästuar; Trichtermündung / estuario

2457
liman, – a drowned river valley on a low coast, protected from the open sea by a barrier beach. / liman / liman / Liman / limán; bañado

2458
lagoon, – the expanse of water separating an offshore bar from the shore. / lagune; haf / lagune / Lagune; Haff / lagoon; laguna

tidal lagoon, – / getijdenlagune / lagune à marée / Gezeitenlagune / laguna de marea

2459
bar, – a submerged or emerged embankment of sand, gravel or mud built on the sea floor in shallow water by waves and currents; it may also be composed of shells. / strandwal / barre / Strandwall; Barre; Sandbank / barra

offshore bar; .... ball; longshore bar, – an offshore ridge of unconsolidated material, more or less parallel to the coast, and submerged at high tide. / zandbank; (strandbank) / barre sous-marine littorale / (Sand)riff; Unterwasserriff / cordón litoral

2460
offshore trough; low, – a depression in the offshore sea floor, parallel to the trend of the coast. / – – / sillon d'avant-côte / (Rifftal) / surco litoral

2461
barrier; .... beach, – an offshore ridge of unconsolidated material (not necessarily attached to a headland), emerging (at least partly) at high tide. / (vrije) strandwal; lido / cordon littoral; lido / (freier) Strandwall; Lido / cordón litoral

outer shore line, – the shore line at the open sea side of the barrier. / buitenkustlijn / rivage extérieur / Aussenküstenlinie / ribera externa

inner shore line, – the mainland shore line. / binnenkustlijn / rivage intérieur / Innenküstenlinie / ribera interna

2462
fixed coastal barrier, – a barrier that has become partly or wholly attached to the mainland by landward migration or by silting-up of the enclosed lagoon or tidal-flat area. / vaste strandwal / cordon appuyé / angelehnter Strandwall / restinga

lunate bar, – a crescentic bar commonly found off the entrance to a harbour. / – – / barre en croissant / sickelförmige Barre / – –

2463
bay-mouth bar; .... barrier, – a narrow and elongated em-

bankment of sedimentary material connecting the headlands of a bay or nearly so. / strandwal (p.p.); schoorwal (p.p.) / cordon littoral (p.p.); flèche (littorale) (p.p.) / Nehrung (p.p.) / – –

2464
spit, – narrow embankment formed primarily by longshore currents and beach drift, attached at one end to a headland and terminating at the other end in open water. / haakwal; schoorwal / flèche; .... à pointe libre; poulier / Haken; (Nehrung) / flecha

recurved spit; hook, – a spit turned landward at the outer end. / weerhaakwal / flèche à crochet / gekrümmter Haken / – –

2465
cuspate bar; .... barrier, – a V-shaped bar or barrier formed by the growing together of two oblique spits. / V-haak / cordon en V / V-förmiger Haken / cordón en V

loop; looped bar, – / lushaakwal / cordon en boucle / Hakenschlinge / – –

2466
flying bar; .... spit, – formed behind an island which has subsequently disappeared. / relicthaak / flèche relicte; .... témoin / Hakenrelikt / flecha testigo

2467
tombolo, – a bar or barrier of unconsolidated material connecting an island with the mainland. / tombolo / tombolo; flèche de jonction; flèche-isthme / Tombolo; Inselnehrung / tÓmbolo

2468
barrier island, – a detached portion of a barrier beach between two inlets. / strandwaleiland / île de cordon libre / Strandwallinsel; Dünen.... / isla de cordón libre

2469
shoal; bank p.p., – a shallow part of sea, lake or river that is always covered by water, though not deeply. / bank / banc; haut-fond / Untiefe; Bank; Riff (p.p.) / – –

[1673] *Rocky Coasts*

2470
cliff; sea ....,– a high, steep face of rock, formed at the coast. / klif; kust.... / falaise; .... littorale / Kliff; Küstenkliff / acantilado

nip, – a small cliff. / klifje / microfalaise; ressaut / kleines Kliff; Absatz / microacantilado; resalte; escarpe

2471
undercutting; undermining, – of a cliff by wave action and corrosion. / ondermijning / sapement; excavation / Unterhöhlung; ....spülung; ....schneidung / socavado; socavamiento

2472
notch, – groove at the foot of a cliff formed by undermining. / brandingsnis / rainure; cannelure creusée par les vagues / Brandungskehle; Hohl....; Hohlkerbe / gruta (de acantilado)

overhang, – the overhanging edge of an undermined cliff. / brandingsluifel / encorbellement / Überhang; überhängende Wand / cornisa

sea cave; sea cavern, – cave formed by wave action. / brandingsgrot / grotte formée par les brisants / Brandungshöhle; Kliff.... / caverna marina

2473
cliff fall, – / afstorting; instorting / éboulement / Nachstürzen; Nachbrechen / desplome de acantilado

cliff landslide; .... landslip,– / afschuiving van het klif / glissement; décollement / Kliffrutschung / deslizamiento de acantilado

2474
gully; cleft, – a wave-cut chasm in a cliff. / brandingskloof / couloir / Brandungsgasse / hendidura

cove, – a small sheltered recess in a (cliff) coast. / inham / crique / Nische / cueva (de acantilado)

2475
plunging cliff, – a cliff bordering deep water directly; a shallow sea on a wave-cut platform is lacking as a result of a rapid relative subsidence of the land. / deels verdronken klif / falaise plongeante / untertauchendes Kliff / acantilado buzante

2476
undercliff, – a secondary cliff, formed in a landslide or cliff-fall deposit of a major cliff. / secundair klif / falaise d'éboulement; fausse falaise / (Sekundärkliff) / falso acantilado

2477
abandoned cliff; ancient ...., – cliff abandoned by the sea in consequence of negative (relative) movement. / verlaten klif; dood .... / falaise morte / totes Kliff / acantilado abandonado

2478
line of cliffs, – cliffs on a graded shore profile. / vereffende klifkust / falaises alignées / Klifflinie / línea de acantilados

2479
wave-cut bench; .... rock bench, – a narrow platform cut by the waves. / – – / banquette (rocheuse); trottoir / schmale Brandungsplatte / rasa

wave-cut platform; marine-cut ....; wave-eroded ....; wave-beveled ....; abrasion ....; marine-erosion ....; wave-cut bench; platform terrace, – the flat rocky zone along the coast that is produced by the combined action of the direct attack on the cliff base, the to-and-fro motion at the water/rock interface, and the undertow. / marien erosieterras; abrasieplat; ....vlak; brandingsvlakte / plate-forme d'abrasion; .... d'érosion marine; .... d'aplanissement / Abrasionsplatte; Brandungs....; Brandungsplatform; ....fläche; marine Erosionsterrasse / plataforma de abrasión

2480
wave-built platform; marine-built ....; built ....; wave-built terrace, – seaward continuation of the wave-cut platform, formed by sediments that accumulated in the deeper water beyond, after having been transported across it. / litoraal accumulatievlak; marien accumulatieterras / plate-forme d'accumulation; terrasse construite; .... d'accumulation / marine Aufschüttungsterrasse; Akkumulations.... / plataforma de acumulación

offshore slope; frontal ...., – the slope below the outer edge of the wave-built platform. / buitenste helling; onderzeese puinhelling / talus frontal / meeresseitige Böschung; Vorstrand-Abhang / talud frontal

2481
betrunked valley, – valley that has lost its lower course by retreat of a cliff, often resulting in the formation of a hanging valley. / afgeknot dal; ontworteld dal / vallée tronquée / abgeschnittenes Tal; litorales Hängetal / valle truncado

2482
dismembered valley system, – valley system, the main valley of which has been betrunked so far by cliff retreat, that now the side valleys debouch independently. / verbrokkeld dalnet; uiteengevallen .... / réseau de vallées démembré / aufgelöstes Talsystem / red de valles desmembrada

2483
hanging valley, – a valley debouching at a considerable height above sea level; a phenomenon caused by cliff retreat. / zwevend dal / vallée suspendue; .... perchée; valleuse; cran / hängendes Tal; Hängetal / valle suspendido; .... colgado

2484
marine terrace; coastal ....; shore ...., – strip of a fossil wave-cut platform, at the landward side of which the old cliff is situated and at the seaward side of which a new cliff has been formed. / kustterras; marien terras / terrasse littorale / Küstenterrasse; marine Terrasse / terraza marina

[1674] *Sandy Shores*

2485
beach; strand, – a shore consisting, at least partly, of unconsolidated material, e.g. sand, shingle, cobbles. / strand / plage; (grève); estran / Strand / playa

shore face, – the zone seaward from the low-tide shore line, permanently covered with water, over which the sand and gravel particles oscillate actively with changing wave conditions. / oeverfront / avant-côte / (Vorstrand) / – –

2486
ridge; full; ball, – an elevation of the foreshore parallel to the trend of the beach. / strandrug / (crête); banc / Strandriff / cresta

runnel; low; swale, – a depression of the foreshore parallel to the trend of the beach. / zwin / bâche; baïne (Gascogne); lague / Strandpriel; ....mulde; ....rinne / corredor

rip channel; cross ...., – a channel connecting two successive runnels, running transverse to the beach. / mui / chenal (transversal) de bas de plage / Rippstromrinne / canal transversal

2487
foreshore; beach face, – the part of the beach lying between the upper limit of the wave swash at high tide (usually the berm crest) and low-tide shore line. / nat strand / bas de plage; basse plage / nasser Strand / playa baja

backshore, – that part of the beach, lying landward from the upper limit of the wave swash at high tide. / stormstrand; droog strand / haute plage; arrière-plage / trockener Strand / playa alta

2488
berm, – a more or less terrace-shaped feature on the backshore formed by the deposition of sediment by wave action. / – – / gradin de plage / Berme / escalón de playa; berma

berm crest; .... edge, – the seaward limit of a berm. / – – / – / Bermenkante / – –

2489
beach scarp, – a low, steep escarpment cut into the beach profile by wave action. / vloedklif / microfalaise de plage / Strandböschung / escarpe de playa; microacantilado de ....

2490
beach ridge, – rigde of beach material formed along the shore by wave action. / strandrug / crête de plage; levée de .... / Strandwall; Uferwall / cresta de playa

normal beach ridge, – a ridge having its base well below low-tide level; along medium to high-energy shores. / normale strandrug / crête; levée de plage / Uferwall / cresta de playa normal

2491
chenier, – a shallow-based beach ridge composed of sands and shells and resting on clay. / chenier; rits / chenière / Chenier / – –

chenier plain, – a plain of tidal marshes or swamps along an open sea shore, bordered by cheniers. / cheniervlakte / plaine à chenières / Chenierebene / – –

2492
strand plain; foreland, – a shore that is advancing seaward and extends some distance along the coast. / strandvlakte / plage progradée / Strandfläche; ....ebene / playa de acreción

2493
beach cusp, – cusp-shaped, rhythmically interspaced features on the beach, with seaward pointing ends. / strandhoorn / croissant de plage / Strandhorn / cuspilito

NOTE:
– for further terms see section
[1942] Ripple Marks

[1675] *Deltas*

2494
delta, – a deposit of sediments made by a stream at the place of its entrance into an open body of water, resulting in progradation of the shore line. / delta / delta / Delta / delta

The following morphological types, as resulting from local circumstances, such as offshore currents, are distinguished:

2495
cuspate delta, – the 'classical' delta of triangular shape. / puntdelta; driehoeks.... / delta en pointe / – – / delta en punta

arcuate delta, – / boogdelta / delta arqué / Bogendelta / – –

lobate delta; digitate ....; bird-foot ...., – / vogelpootdelta / delta lobé; .... digité; .... en patte d'oie / Vogelfussdelta / delta en pata de pájaro

2496
subdelta, – a smaller delta forming part of a larger, lobate or bird-foot delta. / subdelta / sous-delta / Subdelta / subdelta

2497
protruding delta, – delta extending offshore. / open kustdelta / delta saillant / vorgeschobenes Delta / delta saliente

2498
bay delta, – / opvullingsdelta / delta de fond de baie / Ausfüllungsdelta / delta de fondo de bahía

2499
inner tidal delta, – / binnendelta / delta de marée interne / inneres Gezeitendelta / delta de marea interna

outer tidal delta, – / buitendelta / delta de marée externe / äusseres Gezeitendelta / delta de marea externa

2500
wave delta; washover, – a delta-like deposit, formed by waves breaking over a barrier or any other elongated obstruction. / overslagdelta / delta construit par les vagues débordantes / Sturmdelta / delta de oleaje

storm delta; washover fan, – a delta-like deposit formed by the breaching of a barrier or barrier spit, usually during a storm. / inbraakdelta / delta de tempête; (delta par rupture de cordon) / Sturmdelta / delta de tormento

2501
anastomosing, –
1. branching of delta channels due to overloading. / vlechting / anastomose / Verästelung / anastomosamiento
2. branching of delta channels due to capture; both courses remain active. / anastomosering / anastomose par captures / – – / anastomosamiento por capturas

2502
bayou (Louisiana), – in general a secondary water course characterized by a slow or imperceptible current.

2503
pass, – the mouth of a delta-stream channel where it debouches into the open sea. / mondingsgeul / passe / Mündungsrinne / boca

[1676] *Tidal Flats and Marshes*

2504
tidal river, – a river in which the tide is noticeable for a considerable distance upstream. / getijstroom / rivière à marées / Tideflusz / río de mareas; río mareal

tidal section, – of a river, the part affected by the tide. / getijdentraject / – – / Tidegebiet; ....zone / zona mareal

2505
tidal-flat area, – a relatively extensive area of unconsolidated sediments, the greater part of which alternately emerges and submerges with the tide. / waddengebied / vey / Wattengebiet / zona intermareal; estero

tidal flat, – part of a tidel-flat area lying between the levels of mean high and mean low tide. / wad / vey; slikke / Watt / estero

2506
flat, – a tidal flat enclosed between major tidal channels or between major tidal channels and the land. / plaat; bank / banc / Bank; Plate; Platte; Riff / bajo fondo; bajío

tidal pond; .... lake, – irregularly shaped lake in the channelled zone of a tidal-flat area, adjacent to the channels and their accompanying levees. / getijdenmeer / – – / Gezeitentümpel / – –

2507
waddeneiland, – Dutch term for the islands along the Frisian and Danish coasts, forming part of a partly dune-covered barrier separating the tidal-flat area from the open sea. / waddeneiland / île dunaire; (île de type frison) / – – / isla dunar

marsh island, – / kweldereiland / butte de schorre; talard (Brittany) / Marschinsel; Hallig / – –

2508
tidal basin, – portion of a tidal-flat area, drained at ebb tides by the channel system of a single tidal inlet. / wadkom / bassin d'un système de chenaux de marée / Gezeitenbecken / cuenca mareal; .... de un sistema de canales de marea

tidal inlet, – / zeegat / passe / Tief; Seegat; Gatt / – –

2509
tidal channel, – a major channel followed by the tidal currents. / getijgeul / chenal de marée / Wattrinne; Hauptstrom....; Tief; Balje; Balge / canal de marea

tidal wedge, – tidal channel, narrowing or shallowing at the downstream end, and in which either the ebb or the flood current dominates. / getijschaar / – – / Gezeitentrichter / – –

gully, – a minor channel incised in mud or sand flats below the mean high-tide level. / priel / chenal de marée; étier; marigot / Priel; Gezeitenrinne / embocadero

2510
flood channel, – tidal channel in which the flood currents are stronger than the ebb currents. / vloedschaar; vloedgeul / chenal de flot / Flutrinne / canal de creciente

ebb channel, – tidal channel in which the ebb currents are stronger than the flood currents. / ebschaar; ebgeul / chenal de jusant / Ebbrinne / canal de vaciante

2511
tidal divide, – divide between two adjacent channel systems. / wantij / ligne de partage des eaux / Wasserscheide / divisoria de marea

2512
belt of marshes, – of a tidal-flat area. / kwelderzoom / bordure de marais / Marschensaum; ....gürtel / cinturón de marismas

2513
warping, – conversion of a lagoon or tidal-flat area into a marsh. / verlanding / colmatage / Verlandung / colmatación

vertical accretion; upward growth; rise in level, – / opslibbing / exhaussement / Aufschlickung; Aufwuchs / acreción vertical

lateral spread; outward growth; horizontal ...., – / aanslibbing; aangroei / extension latérale; croissance / Anlandung; Anwuchs / extensión lateral

2514
salt marsh; (high marsh); salting, – the area above mean high-tide level, covered by sea water during the highest tides only; usually covered with a thick mat of halophytic vegetation. / kwelder (Netherlands); gors (province of South Holland); schor; schorre (province of Zeeland) / schorre; herbu; marais salé / Salzmarsch; Aussengroden; (Heller) / marisma; saladar

salt pasture, – / kwelder (p.p.); buitendijkse weide / pré-salé / Groden / salobral

marine marsh; coastal swamp; paralic ...., – / kustmoeras / marais côtier / Küstensumpf / marisma litoral

2515
algal marsh, – salt marsh occurring most landwards in tidal-flat areas; composed of 'meadows' of blue-green algae; they might also form fringing algal mats immediately down the backslope of tidal channel levees, transitional to tidal ponds. / algenmoeras / trottoirs à algues / Algenmarsch / – –

mangrove marsh; .... swamp, – a saline to brackish coastal swamp of the tropics, covered by a typical halophytical vegetation. / mangrovemoeras / marais à mangrove / Mangrovesumpf / marisma de manglares

2516
marsh bar, – the elevated outer margin of a salt marsh, genetically partly a natural levee, partly formed by waves. / kwelderwal / lavée externe de marais / erhöhter Aussenrand des Grodens / – –

bluff, – miniature cliff, bordering the salt marsh on the seaward edge. / kwelderklif / microfalaise / Abbruchkante; ....wand; Grodensteilkante / bancal

2517
marsh creek, – a drainage channel in a salt marsh. / kwelderkreek; schor.... / chenal de marais; étier; kreek; creek; marigot / Priel; Grodenpriel / caño

pan; salt pan; marsh pan, – shallow depression in a salt marsh. / kwelderpan / cuvette à sel / – – / – –

marsh basin, – depression between raised banks of creeks in a salt marsh. / kwelderkom / cuvette / – – / cubeta

2518
upland, – dry land at the land side of salt marshes, not reached by storm tides. / (hoger gelegen) achterland / terre ferme au bord du marais / Hochland / tierra firme

[1677] *Types of Coasts*

2519
monoclinal coast, – coast formed by a monoclinal flexure. / flexuurkust / côte de flexure / Flexurküste / costa de flexura

2520
fault coast, – a coast formed directly by faulting. / breukkust / côte de faille / Bruchküste; Verwerfungs.... / costa de falla

fault-scarp coast; fault-line ...., – a coast formed by erosion along a pre-existing fault scarp. / breuklijnkust / côte de ligne de faille / Bruchlinienküste / costa de escarpe de falla; .... de línea de falla

2521
longitudinal coast; conformable ....; Pacific-type coast, – coast parallel to the trend of a coastal orogenic belt. / lengtekust; concordante kust; kust van het Pacifische type / côte concordante; .... à structure longitudinale; .... longitudinale; .... du type pacifique / Längsküste; Konkordanz....; konkordante Küste; pazifischer Küstentypus / costa longitudinal; .... de tipo pacífico

Dalmatian-type coast, – a drowned longitudinal coast, the mountain ranges of which have become islands, and the valleys have become straits. / Dalmatisch kusttype; kanalenkust / côte à type dalmate / dalmatischer Küstentypus / costa de tipo dálmata

2522
transverse coast; unconformable ....; Atlantic-type ...., – coast transverse to the strike of mountain ranges or rocks. / dwarskust; discordante kust; kust van het Atlantische type / côte discordante; .... à structure transversale; .... transversale; .... du type atlantique / Querküste; Diskordanz....; diskordante Küste; transversale ....; atlantischer Küstentypus / costa transversal; .... de tipo atlántico

diagonal coast, – coast oblique to the direction of the strike of mountain ranges or rocks. / diagonaalkust / côte oblique / Diagonalküste; Schräg.... / costa oblicua

2523
neutrale Küste; indifferente ...., – German terms for a coast in which the rocks do not show a distinct direction of the strike, e.g. in the case of horizontal bedding.

neutral coast, – a coast whose essential features are not influenced by relative or absolute changes of sea level, e.g. delta, alluvial plain, volcanic or fault coasts.

2524
raised coast, – a coast having undergone a raise. / opgeheven kust; rijzingskust / côte soulevée; .... de soulèvement / Hebungsküste; gehobene Küste / costa alzada; .... levantada; .... de levantamiento; .... de alzamiento; .... de elevación

depressed coast, – a coast having undergone subsidence. / dalingskust / côte affaissée; .... d'affaissement / Senkungsküste; gesunkene Küste / costa hundida; .... de hundimiento

2525
emerged coast, – a coast having undergone a relative raise. / opduikingskust / côte d'émersion; .... à mouvement négatif / aufgetauchte Küste; Auftauchküste / costa emergida; .... de emersión

submerged coast; drowned ...., – a coast having undergone a relative lowering. / verdrinkingskust / côte d'immersion; .... de submersion; .... à mouvement positif / untergetauchte Küste; Untertauchküste; Senkungs.... / costa sumergida; .... de inmersión

2526
stable coast; stationary ...., – a coast showing no signs of a relative raise or subsidence. / stabiele kust; onbewegelijke .... / côte stable / Stillstandsküste; stabile Küste / costa estable

compound coast, – a coast, the features of which combine elements of submerged and emerged coasts as a result of submergence followed by emergence. / gemengd kusttype / côte du type mixte / gemischter Küstentypus / costa de tipo mixto

2527
coast of transverse deformation; composite coast, – a coast consisting of alternating zones of submergence and emergence, connected with zones of down- and upwarping transverse to the coast. / samengestelde kust / côte composée / zusammengesetzte Küste / costa compuesta

contraposed coast, – a coast in which, in the course of time, a cover of loose marine deposits has been removed exposing the bedrock, and resulting in rejuvenation of the coastal forms. / epigenetische kust / côte contraposée; ....'épigénique' / epigenetische Küste / costa contrapuesta; .... epigénica

[1678] *Development of Coasts*

2528
coast of retrogradation; abrasion coast, – coast produced by the cutting back of the shore in a line of cliffs. / abrasiekust / côte d'abrasion; .... d'érosion marine / Abrasionsküste; Angriffs.... / costa de abrasión; .... de erosión marina

coast of progradation; .... of advance; prograded coast; accretion ...., – coast showing a regular seaward movement in consequence of sedimentation. / sedimentatiekust; aangroeiende kust / côte d'accumulation / Anwachsküste; Anschwemmungs.... / costa de acumulación

2529
ingression coast, – coast, the lower parts of which have been invaded by the sea, resulting e.g. in drowned valleys, bays. / ingressiekust / côte d'ingression / Ingressionsküste / – –

2530
multicycle coast, – emergent coast formed during several interrupted cycles so that two or more elevated marine terraces have been formed. / polycyclische kust; meercyclische .... / côte polycyclique / mehrzyklische Küste / costa policíclica

2531
ungraded shore line, – a shore line showing many promontories or bays, typical of an early stage of development.

graded shore line; smooth shore ....; straight shore ...., – a shore line showing no promontories or bays; typical of advanced development. / vereffende kustlijn; gladde kust; rechtlijnige .... / rivage régularisé; côte rectiligne / glatte Küstenlinie; Ausgleichsküste; geradlinige Küste / costa abierta

2532
crenulate shore line; indented shore ...., – ungraded shore line, showing many sharp headlands. / getande kust / côte indentée; .... crénelée / gezahnte Küste / costa indentada

ragged coast, – a sharply indented coast with numerous peninsulas and islands. / gerafelde kust / côte déchiquetée / zerrissene Küste / costa partida

2533
embayed coast, – ungraded coast showing many bays. / baaienkust / côte à baies; .... échancrée / gebuchtete Küste; Buchtenküste / costa de bahías

lobate coast, – an ungraded coast showing a shore line with lobes. / gelobde kust / côte lobée / gelappte Küste / costa lobulada

2534
attached island; land-tied ...., – an island that has become attached to the mainland by tectonic movements or by sedimentation. / aangehecht eiland; vastgeslibd .... / île rattachée / Angliederungsinsel; landfest gewordene Insel / isla unida; .... ligada

detached island, – / afgesnoerd eiland / île détachée / Abgliederungsinsel / isla desunida; .... separada

2535
upheaved island; uplifted ...., – rijzingseiland / île d'émersion / Hebungsinsel / isla de emersión

island formed by aggradation, – / opwaseiland / île d'accumulation / Aufschüttungsinsel / isla de acumulación

2536
plain of marine erosion, – an emerged sea bottom beveled by marine erosion. / abrasievlakte / plaine d'abrasion / Abrasionsfläche; ....ebene; Brandungsplattform; Schürre / llanura de abrasión marina; .... de erosión marina

[1680] SEQUENCE AND DEVELOPMENT OF LANDFORMS

[1681] *Tectonic and Structural Forms*

2537
tectonic landform, – landform determined by dislocations such as faults or folds. / tektonische vorm / forme tectonique / tektonische Form / forma tectónica

2538
dome mountain, – mountain formed in consequence of uplift by doming. / koepelgebergte / montagne à structure en dôme; .... à structure de voûtement / Kuppelgebirge; Wölbungs....; Dom.... / montaña en domo

fold mountain; folded ...., – mountain formed by folding and uplift. / plooiingsgebergte / montagne à structure plissée; .... de plissement / Faltengebirge / montaña de plegamiento

2539
fault-block mountain; block ...., – mountain formed by differential, vertical movements or by tilting of a number of adjacent fault blocks. / schollengebergte / montagne-blocs; montagne fracturée; .... à structure faillée / Schollengebirge; Bruchschollen.... / montaña en bloques; .... con estructura fallada

fault-wedge mountain, – mountain consisting of a tilted fault block and as a consequence showing asymmetrical slopes; the steeper slope is determined by a fault. / scheefgesteld schollengebergte / montagne fracturée penchée / Keilschollengebirge / – –

Bruchfaltengebirge, – German term for fault-block mountains, built up of folded rocks. / – – / montagne plissée-faillée / Bruchfaltengebirge / montaña de pliegues-falla

2540
horst mountain, – mountain that came into existence by the formation of a horst. / horstgebergte / massif surélevé / Horstscholle; Hoch....; Horstgebirge / montaña en horst

2541
monoclinal mountain; homoclinal ridge, – a mountain range or ridge, formed by a belt of resistant strata dipping uniformly in one direction. / monoclinale keten / chaîne monoclinale; crête .... / Monoklinalkamm; Schichtkamm / cresta monoclinal

anticlinal mountain; .... ridge, – a mountain range developed along the axis of an anticline. / anticlinale keten / crête anticlinale / Antiklinalkamm / cresta anticlinal

synclinal mountain; .... ridge, – a mountain range developed along the axis of a syncline. / synclinale keten / crête synclinale / Synklinalkamm / cresta sinclinal

2542
bordering mountain chain, – a mountain chain at the border of a continent or at the outer side of an orogenic belt. / randgebergte / montagne marginale / Randgebirge; ....kette / cadena de montaña marginal

2543
tectonic valley; structural ...., – a valley mainly determined by dislocations. / tektonisch dal; .... bepaald dal / vallée tectonique; .... structurale / tektonisches Tal; tektonisch angelegtes .... / valle tectónico

warped valley, – a valley whose longitudinal profile is bent by tectonic forces, often causing lake development behind the warp.

2544
monoclinal valley; homoclinal ...., – a valley formed in a belt of weak rocks situated between resistant beds, that dip uniformly in one direction. / monoclinaal dal / vallée monoclinale; combe (Swiss Jura) / Monoklinaltal; Isoklinaltal / valle monoclinal

anticlinal valley, – a valley that follows the axis of an anticline. / anticlinaal dal / vallée anticlinale / Antiklinaltal; Scheiteltal; Satteltal / valle anticlinal

synclinal valley, – a valley that follows the axis of a syncline. / synclinaal dal / vallée synclinale; val; vallon (Swiss Jura) / Synklinaltal; tektonisches Muldental / valle sinclinal

2545
Walmtal, – a German term for a cross valley developed in a depression of the longitudinal axis of an anticline. / – – / vallée en abaissement d'axe des plis; .... de dépression axiale / Walmtal / valle de depresión axial

cluse, – in the Swiss Jura cross valleys in anticlinal ridges.

2546
rift valley; fault trough, – a valley resulting from the formation of a graben, often partly occupied by long and narrow, deep lakes. / slenkdal; ....laagte; breuktrog / fossé d'effondrement; vallée ....; fossé; vallée en fosse (d'effondrement) / Grabensenke; Bruch....; Grabental / fosa tectónica

graben, – a structural term, often used in the morphological sense of fault trough.

2547
fault pit, – a depression surrounded by a system of concentric, arcuate faults. / breukbekken / cuvette de failles / Kesselbruchfeld / cubeta de fallas

fault valley, – a valley resulting from the formation of a fault, e.g. between tilted block mountains. / breukdal / vallée de faille / Bruchtal; Verwerfungstal; Spaltental / valle de falla

fault-line valley; fault-zone ...., – a valley developed in the relatively weak shatter belt of a fault. / breuklijndal / vallée de ligne de faille; .... de zone de faille / Bruchlinietal; Bruchzonental / valle de línea de falla

2548
fault scarp; .... escarpment; .... ledge, – an abrupt cliff or

scarp produced directly by faulting; also used when more or less affected by erosion and denudation. / breuktrede; verschuivingsklif / escarpement de faille; gradin de ....; ressaut de .... / Bruchstufe; Verwerfungs....; Verwerfungsabsturz / escarpe de falla

rejuvenated fault scarp; revived fault ...., – a fault scarp that has been redeveloped by renewed movement after some dissection had taken place in a period of rest. / verjongde breuktrede / escarpement de faille rajeuni / wiederbelebte Bruchstufe / escarpe de falla rejuvenecido

resurrected fault scarp; exhumed fault ...., a fault scarp that has been re-exposed by erosion long after having been buried. / blootgelegde breuktrede / escarpement de faille exhumé; .... de faille rajeuni par l'érosion / aufgedeckte Bruchstufe / escarpe de falla exhumado; .... de falla rejuvenecido por erosión

2549
levelled fault, – a fault not showing any differences in topographical height on either side of it, as a result of the action of exogenic levelling forces. / vereffende breuk; genivelleerde verschuiving / faille nivelée; .... rasée; .... sans relief / geebnete Verwerfung; abgehobelte .... / falla nivelada; .... enrasada

2550
fault-line scarp, – a scarp produced by the differential erosion on either side of a fault, after levelling of the original fault scarp by erosion or sedimentation. / breuklijntrede / escarpement de ligne de faille / Bruchlinienstufe / escarpe de línea de falla

resequent, – pertaining to a fault-line scarp, facing in the same direction as the original fault scarp.

obsequent, – pertaining to a fault-line scarp, facing in the direction opposite to the original fault scarp.

2551
monoclinal scarp; .... fold scarp; downwarping scarp, – a scarp caused by a monoclinal flexure. / flexuurtrede / gradin de flexure / Flexurstufe / escarpe monoclinal

2552
earthquake scarplet, – a low scarp or step caused by the dislocation of the surface as the result of an earthquake. / aardbevingstrede / gradin sismique; .... de tremblement de terre / Erdbebenstufe; rezente Verwerfung / escalón (de terremoto); .... sísmico; desnivel sísmico

earthquake rift, – a rupture in the earth's surface caused by an earthquake. / aardbevingsspleet / crevasse sismique; .... de tremblement de terre / Erdbebenspalte / grieta sísmica

2553
fault facet; triangular fault ....; spur-end ...., – triangular facet on the end of a spur of a dissected fault scarp; the facet is a remnant of the surface of the fault plane, roughened by weathering. / breukvlakfacet / facette indicatrice d'une faille; .... de faille / Bruchstufenfacette / faceta de falla; .... triangular

2554
structural landform, – landform caused by the structure of the rock formations. / structuurvorm / forme structurale / Strukturform / forma estructural

original; primary, – pertaining to land forms caused by rock formations that have not yet been affected by erosion, thus showing their original morphology, e.g. a lava tableland. / oorspronkelijk / primitive / ursprünglich / forma primaria

secondary, – pertaining to land forms resulting from differential erosion of rock formations with a complex structure. / afgeleid / dérivée / abgeleitet / forma secundaria

2555
stripped structural plateau; .... structural table land, – a plateau developed on horizontal strata when relatively weak materials are removed by erosion so as to expose the surface of a resistant stratum. / – – / plate-forme structurale / Schichttafelland / meseta estructural; plataforma ....

structural terrace; .... bench, – a more or less extensive ledge, developed on a resistant stratum in a formation with horizontal bedding, as a result of the removal of the less resistant strata. / structureel terras; denudatieterras / replat structural; terrasse structurale / Schichtterrasse; Denudations....; Destruktions-Längsstufe / terraza estructural

2556
mesa (Spanish for table); table mountain, – an isolated, flat-topped mountain, capped by an essentially horizontal protective cover. / tafelberg; mesa / montagne tabulaire / Tafelberg; Mesa / mesa

2557
stripped bedding plane, – the exposed top of a (usually resistant) stratum from which less resistant beds have been removed by erosion. If extending over a considerable area it can form an extensive, flat surface. The surface of a mesa or a structural plateau may be a bedding plane. / blootgelegd laagvlak / plan de stratification découvert / blossgelegte Schichtfläche / plano de intersección

2558
cuesta (Spanish for slope), – an asymmetrical, more or less lobed ridge, showing an escarpment on one side and a gentler slope on the other one. / cuesta / côte; cuesta / Schichtstufe; Abtragungs....; Destruktions....; Cuesta / cuesta

cuesta scarp; front slope, – the steep side of a cuesta formed by the outcrop of a resistant layer. / cuestafront / front de côte; .... de la cuesta / Stufenstirn; ....hang; ....abfall; Stirnhang; Steilabfall einer Stufe; Trauf (Swabia) / frente de cuesta

cuesta backslope, – the gentler slope on the opposite side of the cuesta scarp, dipping in the direction of the strata, but at a smaller gradient. / cuestavlakte / revers de la cuesta / Landterrasse; Stufenlehne; ....fläche; obere Plattform einer Schichtstufe / espalda de la cuesta

2559
dip slope, – slope in the topography corresponding approximately with the dip of the underlying strata. / – – / pente structurale / – – / pendiente de buzamiento

face slope, – steep slope facing in the direction opposite the dip slope.

2560
inner vale; .... lowland, – a depression formed in the less resistant rocks that separate an 'oldland' from a cuesta landscape. / randlaagte; perifere laagte / dépression périphérique / peripherische Senke / depresión periférica

oldland, – a region, usually composed of folded or crystalline rocks bordering or emerging out of a cuesta landscape; the cuesta backslopes are dipping away from it. / – – / massif ancien / – – / zócalo

2561
outlier; butte, –
1. an erosional remnant detached from a cuesta- or tableland scarp by erosion; usually showing a flat top, caused by the same resistant stratum that protects the adjacent tableland or cuesta. / getuigeberg / butte témoin / Auslieger; Zeugenberg / cerro testigo
2. a mountain or hill situated in front of a mountain range, or hill range, but separated from the latter by a depression.

/ voorberg / avant-butte / Auslieger; Vorberg / – –

2562
hogback, – a ridge formed on the outcrop of steeply inclined resistant strata, the dip slope approaching the face slope in steepness. / – – / hogback; crête monoclinal / Schichtrippe / crestón

flatiron, – remnant of a hogback ridge with triangular dip slopes as a result of erosion. / – – / chevron; facette / – – / crestón aserrado

2563
relief inversion, – the development of depressions on the places of former heights, and of hills or mountain chains on the places of former valleys; a process caused by differential erosion or, in unconsolidated strata, by differential settling. / omkering van het reliëf / inversion du relief / Reliefinversion; Inversion des Reliefs; Reliefumkehrung; ....umkehr / inversión del relieve

2564
Appalachian relief, – a type of relief to be found in old fold mountains consisting of a great number of anticlines and synclines and in which the forms have adapted themselves to the structure and the differential resistance of the formations; secondary structural forms predominate. / appalachisch reliëf / relief appalachien / appalachisches Relief / relieve apalachiano

2565
Jurassian relief, – a type of relief found in young fold mountains consisting of a great number of parallel anticlines and synclines and in which the majority of the main features are primary structural forms, erosion having had only relatively little influence. / jurassisch reliëf / relief jurassien / jurassisches Relief / relieve jurasiano

[1682] *Landscape Development; General Terms*

2566
cycle of erosion, –
1. the period, according to the concept of Davis, needed for the effective wearing down of highlands to featureless lowlands by the action of denudative forces;
2. the sequence of forms that are displayed on the earth's surface during that period.

/ erosiecyclus / cycle d'érosion / Erosionszyklus / ciclo de erosión

In Davis' school the following erosion cycles are distinguished:

2567
arid cycle; desert ...., – the sequence of land forms that develop under arid circumstances. / aride cyclus / cycle aride / arider Zyklus / ciclo árido

fluvial cycle, – the sequential landscape forms that are characteristic for humid regions and that are produced by fluvial processes (described by Davis as the **geographical, normal** or **fluvial cycle**). / fluviatiele cyclus / cycle fluviatile / fluviatiler Zyklus / ciclo fluvial

karst cycle, – the sequence of forms that result from the dominance of solution processes and underground drainage. / karstcyclus / cycle (d'érosion) karstique / Karstzyklus / ciclo kárstico

glacial cycle, – the sequence of land forms that develop through glacial erosion. / glaciale cyclus / cycle d'érosion glaciaire / glazialer Zyklus / ciclo de erosión glacial

periglacial cycle, – the sequence of land forms that develop under periglacial circumstances. / periglaciale cyclus / cycle d'érosion périglacial / periglazialer Zyklus / ciclo de erosión periglacial

marine cycle; shore-line ...., – coastal forms that develop under the influence of marine erosion and other coastal processes. / mariene erosiecyclus / cycle d'érosion littorale; .... littoral; .... marin / mariner Zyklus; litoraler .... / ciclo de erosión litoral

2568
structure; process; stage, – the three factors that determine, according to Davis, the development of forms in the cycles of erosion. / structuur; proces; stadium / structure; processus; stade / Struktur; Vorgang; Stadium / estructura; proceso; estadio

2569
initial surface, – the not eroded surface of an area before the cycle of erosion has commenced. / oorspronkelijk oppervlak; oeroppervlak / surface primitive / Uroberfläche / superficie inicial

initial relief, – the relief of the initial surface. / oorspronkelijk reliëf; oerreliëf / relief primitif / Urrelief / relieve inicial

2570
stage of youth, – the first principal stage of a cycle of erosion. / jong stadium; jeugdstadium; jeugdig stadium / stade de jeunesse; .... initial / Jugendstadium; Früh....; Stadium der Unreife / estadio de juventud

stage of maturity, – the second principal stage of a cycle of erosion. / stadium van rijpheid; rijp stadium / stade de maturité / Reifestadium / estadio de madurez

stage of old age; .... of senility, – the third principal stage of a cycle of erosion, characterized by the development of a peneplain. / oud stadium; ouderdomsstadium / stade de sénilité; .... pénultième; .... final; .... de vieillesse; .... de postmaturité / Stadium des Greisenalters; greisenhaftes Stadium; Spätstadium / estadio de senilidad

2571
interrupted cycle of erosion, – a cycle interrupted by rejuvenation, volcanic action, etc. / onderbroken erosiecyclus / cycle d'érosion interrompu / unterbrochener Erosionszyklus / ciclo de erosión interrumpido

2572
rejuvenation; rejuvenescence, – development of young forms in a mature or old-aged landscape. / verjonging / rajeunissement / Verjüngung / rejuvenecimiento

2573
erosion system, – the total of processes and climatic conditions resulting in the geomorphological development in a certain climate. / denudatiesysteem / système d'érosion / – – / sistema de erosión

generation, – of denudation, the land forms generated during a period in which the (climatic) conditions remained unchanged. / reliëf-generatie / génération de formes de relief / Reliefgeneration / – –

sequence, – of erosion systems, land forms generated by a sequence of erosion systems. / reliëfsequentie; ....successie / séquence morphogénique / Reliefsequenz / secuencia de relieves

2574
accelerated development; waxing ....; ascending ...., – according to W. Penck, development in a region, where the rate of uplift is more rapid than the rate of downward erosion, so that the relative relief is increased and convex slopes are formed. / opgaande ontwikkeling / évolution ascendante / aufsteigende Entwicklung / evolución ascendente

declining development; waning ....; descending ...., – according to W. Penck, development in a region, where, because of a diminishing rate of uplift, erosion is able to decrease the relative relief and form concave slopes. / neergaande ontwikkeling / évolution descendante / absteigende Entwicklung / evolución descendente

uniform development, – according to W. Penck, development in a region, where the rate of uplift is equal to the rate of degradation, resulting in a constant relief and straight slopes. / vormconstante ontwikkeling / évolution uniforme / gleichförmige Entwicklung / evolución uniforme

2575
relief intensity, – the average difference in altitude between the mountain or hill tops and the valley bottoms in unit area, e.g. 1 km^2. / reliëfintensiteit / vigueur du relief / Reliefenergie / fortaleza del relieve; magnitud del ....

Mittelgebirge, – a German term for mountains having a relief intensity less than 1000 $m.km^{-2}$. / middelgebergte / moyenne montagne / Mittelgebirge / montaña media

high mountains, – mountains having a relief intensity more than 1000 $m.km^{-2}$. / hooggebergte / haute montagne / Hochgebirge / alta montaña

2576
alpine relief; steep ....; bold ...., – / alpien reliëf; hooggebergtereliëf / relief escarpé; .... fort / Hochgebirgsrelief; alpines Relief; Steilrelief / relieve alpino; .... escarpado; .... fuerte

[1683] *Arid Regions*

2577
desert, – a land region, not covered by permanent ice or snow, that possesses a very sparse vegetation or none at all, due either to aridity of climate or low air temperatures, or due exclusively to aridity of climate. / woestijn / désert / Wüste / desierto

2578
sandy desert; erg (plur. areg) (Sahara); koum (Central Asia), – area of sand accumulation in arid regions. / zandwoestijn / désert de sable / Sandwüste; Dünenwüste / desierto de arena; .... arenoso

stony desert, – desert showing a surface covered with desert armour. / steenwoestijn; kiezel.... /désert de pierres / Geröllwüste; Schotter.... / desierto de piedras; .... pedregoso

reg (Sahara), – stony desert with armour on fine pebbles.

serir; sserir (Sahara), – stony desert with armour on coarser pebbles.

rocky desert; hamada (Sahara), – rocky uplands of a desert, that have been swept clear of sand and dust by the wind; the rocky surface usually is covered by coarse rock fragments. / rotswoestijn / désert rocheux / Felsenwüste; Schutt....; Stein.... / desierto rocoso

2579
closed basin, – a more or less circular or elliptic basin in arid regions, caused by tectonic movements and/or by eolian corrasion and deflation. / gesloten bekken / bassin fermé / Wanne; geschlossenes Becken / cuenca cerrada

2580
bolson, – basin of interior drainage in an arid or semi-arid region. / bolson / bolson / Bolson / – –

playa; sebkha, – plain in the centre of a bolson, or along the coast, formed by a surface of extraordinary flatness; usually composed of clay encrusted by or mixed with fine salt crystals, which have been left behind by evaporated lakes formed after rains. / playa; sebkha / playa; sebkha / Playa; Sebkha / playa

alkali flat; salt plain; salina, – a playa showing a salt-covered surface or having a soil strongly charged with salt or soluble soil alkalies. / zoutvlakte / salina; salar / Salina; Salar; Saltztonebene / salina

2581
bahada; bajada, – coalescing alluvial fans, forming a continuous waste slope around the centre of a desert basin or, in the case of an isolated mountain massif, at the outer margin

of the pediment encircling the massif. / bahada / bahada / Bahada / bajada

2582
pediment; rock ....; desert ....; conoplain, – panplane (mountain-fringing rocky plain) in a mountainous desert; an eroded bedrock surface, commonly veneered with alluvium and caused by the coalescing of the flood plains of intermittent rivers by lateral planation, or by back-wearing, sheet-flood transportation and rock-floor robbing. / pediment / pédiment; glacis rocheux désertique / Pediment; aride Felsebene; .... Felsfussfläche / pedimento; glacis rocoso desértico

glacis, – pediment-like form, developed in less resistant rocks. / glacis / glacis / – – / glacis

2583
piedmont slope, – gentle slope, composed of a pediment and a bahada; situated at the foot of a mountain. / piedmonthelling / glacis de piémont / Bergfussfläche; ....ebene / glacis de piedemonte

2584
pediplain; pediplane; panfan; desert plain, – widely extending, rock-cut and usually alluviated surface, formed by the coalescence of a number of pediments and occasional desert domes (the 'peneplain' of the arid cycle). / pediplain / pédiplaine; pédiments soudés; .... coalescents / verwachsene aride Felsebenen / pedillanura; pedimentos soldades; .... coalescentes

2585
rock fan, – fan-shaped rock surface, having its apex where intermittent mountain streams debouch upon a piedmont slope; rock fans are caused by lateral planation by the rivers and form, after coalescence with neighbouring fans, a pediment. / – – / cône rocheux / Felskegel / cono rocoso

2586
pediment pass, – pass caused by the meeting of the apices of two rock fans, penetrating, from opposite directions, into a complex of desert mountains. / pedimentpas / col de pédiment / Pedimentpass / – –

2587
wadi, – ravine or river bed in the desert, occupied by water during or after scarce rain showers only. / wadi / oued; wadi / Wadi / guadi

[1684] *Humid Regions*

2588
truncated upland, – old-stage, planed-down mountain massif. / rompgebergte / massif arasé; carcasse / Rumpfgebirge / macizo truncado; .... peneplanizado

2589
peneplain, –
1. in the sense of Davis, a degradational surface, having lost nearly all its relief by passing through the cycle of erosion;
2. sometimes the term is used in the wider sense of degradational surface in general.

/ schiervlakte; peneplain / pénéplaine / Fastebene; Peneplain / penillanura

peneplanation, – the development of a peneplain. / peneplainisatie; schiervlaktevorming / pénéplanation / Einrumpfung; Bildung einer Fastebene / peneplanización

degradation; leveling, – / vereffening; vervlakking / nivelage; aplanissement / Einebnung / nivelación; nivelamiento

2590
altiplanation, – the development of more or less horizontal degradational surfaces at high altitudes. / altiplanatie / altiplanation / Hochebenenbildung / altiplanización

2591
surface of denudation; .... of truncation; platform of level of erosion, – these terms for a degradational surface are noncommittal in regard to the genesis of such a surface. / vereffeningsvlakte; denudatie.... / surface d'érosion; plaine ....; surface d'aplanissement; plaine de troncature / Rumpffläche; Einebnungs....; Verebnungs....; Abtragungs....; Wellungsebene; Rumpf.... / superficie de erosión; .... de denudación

peneplane, – a term that has been proposed by the English translators of Penck's Morphological Analysis of Land Forms.

2592
etchplain, – a surface of denudation caused by deep weathering and subsequent denudational removal of the weathered material.

2593
primary peneplain, – degradational surface, developed in a region where neither altitude nor relief have ever been great because denudation has kept pace with long-continued slow uplift. / oorspronkelijke schiervlakte / pénéplaine d'emblée / Primärrumpf; Trug.... / penillanura primaria

incipient peneplain; partial ....; local ...., – landscape consisting of a broad alluvial valley-floor and extensive valley-floor side strips, caused by degradational processes that are directed toward the valley floors. / vereffeningsvlakte in wording; embryonale vereffeningsvlakte / pénéplaine inachevée; .... naissante; .... embryonnaire; .... locale; .... partielle / Vorrumpf / penillanura embrionaria

2594
peneplain torso, – / schiervlakterest / pénéplaine-reliquat / Altrumpf / penillanura relicto

2595
buried; fossil, – pertaining to peneplains or surfaces of truncation, that are covered by later deposits. / bedekt; fossiel; begraven / fossile; enfouie; ensevelie / zugedeckt; fossil; begraben / fósil

resurrected; exhumed; revealed; stripped, – pertaining to a degradational surface denuded of its younger cover. / blootgelegd / exhumée / abgedeckt; entblösst; wiederaufgedeckt; ausgegraben; exhumiert / exhumada

2596
intersecting, – pertaining to peneplains or surfaces of truncation; an old, often fossil and exhumed surface of denudation and a younger one, intersecting each other. / elkaar snijdend / à facettes / sich kreuzend / solapada

morvan landscape, – in the sense of Davis, a landscape of intersecting peneplains.

2597
summit level; crowning ...., – level formed by (accordant) mountain summits. / topvlakte; topniveau / niveau des crêtes / Gipfelflur; Scheitel.... / nivel de cumbres; .... de crestas

Grenzgipfelflur, – German term for the summit level in the top level of denudation.

2598
summit concordance; accordance of summit levels, – / gelijkheid in tophoogte / uniformité du niveau des crêtes / Konstanz der Gipfelhöhen; Gipfelhöhenkonstanz; Gipfelniveau / uniformidad del nivel de crestas

top level of denudation, – the level above which mountains, given a certain rate of tectonic uplift, cannot rise in consequence of the increase of degradational forces at higher altitudes. / bovenste denudatieniveau / niveau supérieur de dénudation; .... limite de soulèvement / oberes Denudationsniveau; .... Abtragungsniveau; obere Erhebungsgrenze / techo de denudación; nivel superior de ....

2599
storeyed summit level; stairway of summit levels, – two or more adjacent summit levels of different altitudes, due to neighbouring rocks of different resistance or due to dissection of peneplains of different heights. / trapsgewijs gelegen topniveaus / escalier de sommets / Gipfelstockwerk; Gipfelflurtreppe / nivel de cumbres escalonado

storeyed landscape, – landscape in which mountain summits or land surfaces show two or more levels. / trapsgewijs gelegen landschap / paysage étagé / Stockwerklandschaft; Landschaft mit Stockwerkbau / paisaje escalonado

2600
multiconvex relief, – a relief composed of a multitude of convex hills of approximately the same height. / multiconvex reliëf / relief multiconvexe / – – / relieve multiconvexo

2601
faceted surface of degradation; intersecting surfaces of ...., – system of degradational surfaces, formed during several periods of marginal levelling and land subsidence. / polygenetische denudatievlakte / surface dénudative à facettes; .... dénudative à méplats / polygenetische Abtragungsfläche / superficie de denudación con facetas

inset surfaces of denudation, – / in-elkaar-geschakelde denudatievlakten / surfaces emboîtées / ineinandergeschachtelte Abtragungsflächen; Schachtelrelief / superficies de denudación encajonadas

stepping surfaces of denudation; stepped surfaces of ....; step-like surfaces of ...., – / trapvormige denudatievlakten / surfaces dénudatives étagées / Flächentreppe; Rumpf.... / superficies de denudación escalonadas

2602
slope retreat; .... recession, – / hellingterugtrekking; ....terugwijking / recul des versants / zurückfliehen eines Wandes; zurückweichen eines ....; zurückziehen eines .... / retroceso de vertientes

waxing slope, – upper part of the ideal slope profile. / – – / versant convexe; partie convexe du versant / – – / – –

constant slope, – steepest part of the ideal slope profile. / – – / inflexion du versant / – – / falda

waning slope, – the part of the ideal slope situated between the constant slope and the valley floor. / – – / versant concave; partie concave du versant / – – / repié

2603
wash slope; basal ...., – the less steep slope found at the foot of the gravity slope. / – – / versant colluvial / Haldenhang / recuesto

valley-floor side strip, – nearly level surface between wash slope and valley floor, caused by degradational processes, directed toward the valley floor. / – – / niveau d'aplanissement partiel / Subhaldenhang / nivel de nivelamiento parcial; .... de aplanamiento parcial

2604
piedmont flat; .... bench, – an extensive terrace-like bench sloping outward and surrounding a rising and expanding dome in a series of levels. / piedmont-vlakte / plate-forme de piedmont / Piedmontfläche / plataforma de piedemonte

piedmont benchland; .... stairway, – a system of piedmont flats. / piedmont-trap / gradin marginal; .... de piedmont / Piedmonttreppe; Rumpf....; Rumpfflächen....; Gebirgs....; Bergfuss....; Randstufe / piedemonte abancalada

plain of lateral planation; zone of ....; **panplane,** – mountain-fringing, rocky plain caused by the coalescing of the widening flood plains of adjacent streams leaving the mountain range. / zone met zijdelingse vereffening / zone d'aplanissement latéral / Zone lateraler Verebnung / zona de nivelamiento; .... de planización lateral; .... de aplanamiento lateral; pedimento

2605
mountain knot, – place where several mountain chains meet. / gebergteknoop / noeud de montagnes / Gebirgsknoten / nudo de montañas

mountain spur; spur ridge; offshoot, – / gebergte-uitloper / éperon de montagne; contrefort / Gebirgsausläufer / espolón de montaña

front range, – / voorgebergte / avant-montagne / Vorgebirge / precordillera

2606
buttress; rocky spur; rock salient, – / vooruitspringende rotsmassa / saillie de roche; saillant de ....; avancée de ....; éperon; bastion; contrefort / Felsvorsprung; ....sporn; ....ausläufer; Bastion; Strebepfeiler / contrafuerte

2607
residual mountain; remainder ....; erosion residual, – a mountain rising above a peneplain or another surface of de-

nudation. / restberg / montagne résiduelle / Restberg; Restling / montaña residual

2608
monadnock, – a residual mountain preserved in consequence of the greater resistance of the rocks composing it. Sometimes this term is used in the neutral sense of residual mountain / monadnock; hardkop / monadnock (de résistance) / Härtling / cabezo; morro; peñon

mosor, – a residual mountain that is preserved because of its remoteness from the main drainage lines. / mosoor / monadnock de position / Fernling; Fernberg; Mosor / – –

inselberg; bornhardt, – a steep-sided residual mountain, rising abruptly from a surface of denudation and consisting of the same rock as this surface; mainly occurring in the less humid tropical and subtropical regions. Often this term is used for monadnock. / inselberg; bornhardt / morne; mont insule / Inselberg / monte isla

2609
tor, – small inselberg-like from, mostly consisting of a pile of exhumed core stones. / tor / chiest; bosse ruiniforme / – – / – –

core stone, – rounded remnants of massive rocks in which weathering proceeds along diaclase zones. / kernblok / boule / – – / cebollón

[1685] *Glaciated Regions*

2610
cirque; corrie (Scotland); cwm (Wales); botn (Norway), – a deep, steep-walled recess in a mountain, carved by glacial erosion that, after melting of the ice, may be occupied partly by a lake. / kaar; ijsnis / cirque (glaciaire); oule (Pyrenées); van (French Alps) / Kar / circo (glaciar)

cirque threshold, – rock bar at the outlet of the cirque. / kaardrempel / seuil de cirque / Karschwelle; ....riegel / nivel de fondos de circos; .... de circo

2611
valley-head cirque, – cirque form at the head of a valley. / dalkaar; onecht kaar / cirque (de tête de vallée) / Talkar; unechtes Kar / circo de cabecera de valle

2612
nivation cirque, – basin formed by nivation. / nivatienis; ....bekken / niche de nivation / Nivationsnische / nicho de nivación

2613
compound cirque, – cirque consisting of several individual cirques, that grew together by back-wearing of the side walls. / samengestelde kaar / cirque composé; grand-cirque / Grosskar / circo compuesto

cirque platform, – platform caused by the growing together of several cirques. / kaarterras / plate-forme de cirques / Karplatte; ....terrasse / plataforma de circos

2614
cirque stairway; glacial ....; cirque steps, – staircase form, caused by several cirques, situated in a row, above each other. / kaartrap / cirque en gradins; .... en escalier; .... étagé; escalier en cirques / Stufenkar; Kartreppe / circo en escalera; .... escalonado

2615
cirque-floor level; cirque niveau; level of cirque excavation, – the level at which most cirques in a region have their floors. / kaarbodemniveau; kaarniveau / niveau de fonds de cirques; .... de cirques / Karsohlenniveau; Karniveau / nivel de fondos de circos; nivel de circo

2616
fretted upland, – mountains entirely affected by cirque formation. / kaargebergte / montagne déchiquetée par des cirques; .... rongée par des cirques / Kargebirge / cirras recortadas

2617
serrate ridge; arête; sharp crest, – a sharp-crested divide, often caused by cirque cutting on either side. / scherpe bergkam; graat / arête / Grat / arista; cresta cortante

2618
glacial horn; cirque mountain; Matterhorn peak; tind (Norway), – sharp mountain summit, modelled from a rounded preglacial mountain by all-sided cirque excavation. / kaarling; nissenberg / montagne à cirques; aiguille glaciaire / Karling / montaña de circos; aguja glaciar

nunatak, – mountain peak projecting above a glaciated area. / nunatak / nunatak / Nunatak / nunatak

2619
trough end; .... headwall, – the hemicircular upstream end of a glacial U-shaped valley trough, showing steep walls, between which a broad valley bottom is situated. / trogeinde / fin d'auge / Trogschlusz / extremo de la artesa

trough shoulder; .... bench, – a bench in the transverse profile of a former glacial valley trough, whose floor is incised. / trogschouder / épaulement de l'auge / Trogschulter / hombrera

2620
glacial basin; .... rock basin, – basin caused by local widening and deepening of a valley by glacial erosion and often occupied by elongated lakes, e.g. the alpine border lakes. / glaciaal bekken / bassin de vallée glaciaire; .... de surcreusement; cuvette de .... / glaziales Talbecken / cubeta de valle glaciar

rock-basin lake; paternoster lake, – lake in a glacial basin of a trough valley. / trogbekkenmeertje / bassin lacustre glaciaire / Talbeckensee / cubeta lacustre glaciar

2621
glacial diffluence pass, – lower part in the trough wall, where a distributary ice stream has left the main valley. / glaciale diffluentiepas / col de diffluence glaciaire / glazialer Diffluenzpass / pasa de difluencia glaciar

2622
giant stairway; glacier ....; glacial ...., – the floor of a glacial valley, shaped as a broad staircase, composed of a series of treads of various lengths, separated by steep risers of different heights. / glaciaal trapdal / vallée à gradins glaciaires / glaziales Stufental / valle glaciar escalonado

2623
glacial step, – sudden descent in the overall smooth longitudinal profile of a glacial valley. / glaciale daltrede / rupture de pente glaciaire; gradin ....; ressaut .... / glaziale Talstufe / ruptura de pendeiente de origen glaciar; escalón glaciar

2624
diffluence step, – valley step, rising downstream, caused by the weakening of the glacial action on the place of glacial diffluence. / diffluentietrede / gradin de diffluence / Diffluenzstufe / escalón de difluencia

confluence step, – valley step, rising upstream, caused by the strengthening of the glacial action on the place of glacial confluence. / confluentietrap / gradin de confluence / Konfluenzstufe / escalón de confluencia

2625
glacial threshold; .... rock bar; riegel, – glacially modelled rock bar, usually forming a part of a glacial step. / glaciale drempel / verrou; barre glaciaire; seuil rocheux .... / Riegel; Felsriegel; Gesteins....; Gesteinsschwelle / umbral glaciar

Riegelberg, – German term for the higher part of a glacial threshold, isolated by the formation of subglacial erosion channels. / drempelberg / bosse glaciaire / Riegelberg; glazialer Inselberg / monte isla glaciar

2626
reversed gradient, – local valley gradient, opposite to the general valley gradient; a phenomenon often seen at the downstream side of glacially overdeepened valley parts. / tegengesteld verhang / contrepente / rückläufiges Gefälle; widersinniges .... / contrapendiente

2627
hanging glacial valley; perched ...., – tributary valley showing a discordant junction with the main valley in consequence of the more vigorous glacial action in the main valley. / zwevend glaciaal dal / vallée glaciaire suspendue; .... glaciaire perché / glaziales Hängetal; hangendes Glazialtal / valle glaciar suspendido

2628
truncated spur, – valley spur, situated between two side valleys and originating from a period of fluvial erosion, but truncated by the glacial widening and deepening of the main valley during a period of glacial erosion. / afgeknotte mondingsspoor / éperon tronqué / abgestutzter Sporn; abgestumpfter .... / espolón truncado

2629
terminal basin; tongue ...., – depression formed by glacial erosion and marginal accumulation in the area of the end of the glacier tongue. / tongbekken / bassin terminal; cuvette terminale / Zungenbecken / cubeta terminal

2630
morainic-belt topography, – / eindmorenereliëf / paysage de moraines terminales / Endmoränenlandschaft / paisaje de morrenas terminales

till-plains topography; knob and kettle ....; sag and swell surface, – / onregelmatig grondmorenereliëf / paysage de moraine de fond / Grundmoränenlandschaft / paisaje de morrena de fondo

2631
limit of glacial polish; .... of glacial striation, – the limit of the scouring action of ice in a formerly glaciated region. / slijpgrens / limite du polissage glaciaire; .... de la striation glaciaire / Schliffgrenze / límite del pulimento glaciar

2632
kettle hole; kettle, – basin created by melting of blocks of (dead) ice, buried by fluvioglacial sediments, sometimes occupied by a lake. / dood-ijsgat; soll / mare; mardelle; trou d'eau / Soll; Toteispinge; Toteisloch / marmita glaciar; solle

[1686] *Periglacial Regions*

2633
periglacial, – pertaining to
1. the area adjacent to the border of the Pleistocene ice sheets;
2. the climate;
3. by extension, the phenomena induced by this climate even if located outside the main periglacial zone.

/ periglaciaal / périglaciaire / Periglazial / periglaciar (area); periglacial (climate)

2634
cryopedology, – the science that treats of frozen soil and ground. / kryopedologie / cryopédologie / Kryopedologie / criopedología

2635
frozen ground; tjäle (Sweden); frost ground; gelisol, – / hal; ijsbodem / sol gelé / Frostboden; Dauerfrost.... / suelo helado

2636
tundra, – treeless plains in arctic regions characterized by a permanently frozen subsoil and a seasonally thawn surface zone. / toendra / toundra / Tundra / tundra

stringbog, – treeless organic terrain, characterized by ridges of peat and low vegetation, interspersed with depressions that often contain shallow ponds. / strengenveen / tourbière réticulée; .... cordée / Aapamoor; Strang.... / turbera reticulada

2637
permafrost; pergelisol; perenne tjäle (Sweden); merzlota (Russia), – the ground that remains in a permanently frozen

state in areas of frigid climate. / pergelisol; permafrost / pergélisol / Pergelisol; Dauerfrostboden / permafrost; suelo permanente helado

2638
talik, – an unfrozen zone, either overlying, within or underlying the permafrost.

open talik; suprapermafrost ...., – an unfrozen zone between the base of the seasonal frost and the permafrost table.

intrapermafrost talik, – an unfrozen zone within the permafrost.

subpermafrost talik; subgelisol, – the unfrozen zone below the permafrost.

2639
needle ice; pipkrake; mush frost; efflorescent ice, – tufts of ice needles, formed by ice segregation at or just beneath the ground surface as a result of direct cooling of the surface. / naaldijs / aiguilles de glace; pipkrakes / Kammeis; Nadeleis....; Haareis; Pipkrake / agujas de hielo; pipkrake

ground ice; anchor ice, – ice formed to considerable depths in perennially frozen ground. / grondijs / glace profonde / Bodeneis; Aufeis / hielo profundo

2640
frost crack; .... fissure, – fissure originated by thermal contraction of the ground due to the lowering of the temperature. / vorstspleet; ....scheur / crevasse de gel; fente de gel / Frostspalte; ....risz / grieta de enfriamiento

ice wedge, – wedge-shaped body of vertically foliated ice; it builds up over a number of years. / ijswig / coin de glace / Eiskeil / cuña de hielo

2641
sand wedge; tesselon, – frost crack filled with vertically stratified wind-blown sediments or other material, mainly of loess-grain size. It originates under circumstances of extreme aridity. / opgevulde ijswig / fente en coin comblée; coin de sable; .... de loess / aufgefüllte Eiskeilspalte / cuña de arena

soil wedge; ground vein; seasonal frost crack, – frost crack that develops primarily in the seasonally frozen layer; it is filled with mineral soil / – – / – – / – – /cuña de tierra

2642
frost heaving, – the predominantly upward movement of mineral soil during freezing caused by the migration of water to the freezing plane and its expansion upon freezing. / opwelving (door vorst) / bosse; ondulation; boursouflure / Frosthebung; ....beule; ....blähung; ....auffaltung; ....auftreibung / elevación por helada

2643
pingo; frost mound; bulgannyakh (Russia), – an ice-cored hill, that has been domed up from beneath by either the intrusion under pressure of water, which freezes, or by the growth of segregated ice lenses; a perennial intra-permafrost feature. / pingo; hydrolaccoliet / butte à lentille de glace; hydrolaccolite; pingo / Aufeishügel; Schwellungs....; Hydrolakkolith / hidrolacolito

tundra hummock; thufur (Iceland), – about 0.5 m high, vegetation-covered frost mound, initially formed by tussock-forming grasses and sedges in wet, meadow tundra environments; composed predominantly of mineral soil **(earth hummock)** or of organic material **(peat hummock)**. / begroeide toendra-vorstbulten / buttes gazonnées / Rasenhügel; Erdbülte / césped almohadillado

2644
palsa, – a 1 to 7 m high mound covered by peat and composed predominantly of mineral soil, usually occuring in bogs. They form by the combined action of peat accumulation and ice segregation in the underlying mineral soil.

2645
thermokarst, –
1. (obsolete) irregular, hummocky terrain due to melting of ground ice;
2. the process of ground-ice melting accompanied by local collapse or subsidence of the ground surface, often as a result of human activity.

/ thermokarst / thermokarst / Thermokarst / termokarst

2646
thaw depression, – shallow, rounded depression caused by the melting of ground ice, causing the subsidence of the surface. / thermokarstkom / cuvette thermokarstique / Tauniederung; Thermokarst.... / cubeta termokárstica

thaw lake; thermokarst ....; tundra pond, – thaw depression with a (semi)circular pond or lake within it. / thermokarstmeer / lac thermokarstique / Thermokarstsee / lago termokárstica

alas, – circular or oval depression with steep sides and a flat floor, covered with grass or by a thermokarst lake.

thaw sink; cave-in lake, – depression, originated by thawing of a very thin permafrost and subterranean drainage into the underlying thawed sediments. / dooizakking / effondrement de dégel / Tausenke / criokarst

2647
cryoturbation, – the churning process in the active layer, caused by alternation of freeze and thaw. / kryoturbatie / cryoturbation / Kryoturbation; Froststauchung; ....zerrung; ....verbrodelung / crioturbación

active layer; suprapermafrost ...., – zone of seasonal freezing and thawing, overlying the permafrost. / actieve laag; opdooilaag / mollisol; couche active / Mollisol; Fliessschicht / capa activa

2648
involution, – pertaining to periglacial structures produced by freezing-induced pressure in the active layer, leading to deformation and displacement of sediments; more regular in form and distribution than cryoturbation structures. / verplooiing; involutie / involution; interpénétration / Brodel; Verknetung; Verstümmelung / involución; interpenetración

2649
kneaded texture, – / kneedstructuur / structure d'involution; .... chiffonée; .... tourmentée / Brodeltextur; Würge....; Walz.... / textura retorcida; .... amasada

festoon, – upfolded or dragged-up, pointed parts of a layer as occurring in a cryoturbate formation. / flard / feston; corne / Stiche; Lappen / festón.

pocket; sag, – downfolded or dragged-down parts of a layer as occurring in a cryoturbate formation. / zak; instulping / poche; incurvation / Tasche; Einstülpung / bolsada de crioturbación

2650
cryoplanation, – general reduction of the land surface by periglacial weathering, mass movements and the, only seasonally, flowing rivers. / vereffening door vorst / cryonivellement / Frostglättung / crionivelamiento; crioplanización

2651
gravelly layer, – a thin layer of gravel, between the surficial cover of sliding and solifluction material and the underlying beds. / keienvloer / base graveleuse / Steinsohle / capa gravelosa; base ....

2652
patterned ground, – group term for the more or less symmetrical forms, that are characteristic of - but not necessarily confined to - material subject to intensive frost action. Classified on the basis of geometric forms and the presence of absence of sorting. / structuurbodem / sol structuré; sol figuré / Strukturboden; Frostmuster....; Frostgefüge.... / suelo estructurado; .... geométrico

2653
polygonal ground, – a patterned ground with polygonal structures. / polygoonbodem / sol polygonal / Polygonboden / suelo poligonal

striped ground; striated ...., – patterned ground with stone or soil stripes. / geribde bodem / sol strié; sol rayé / Streifenboden / suelo estriado; .... rayado

stone ring; .... circle; .... polygon, – three-dimensional polygonal structures, each composed of a centre of fine material bounded by a border of coarse rock debris. / steenkrans; ....polygoon / cercle de pierres; polygone de .... / Steinring; ....kranz; ....polygon / círculo de piedras; polígono de ....

2654
stone stripe; sorted ...., – bands of fine rock debris alternating with channels filled with coarse rock fragments, orientated parallel to the direction of steepest slope. / steenrichel / bande pierreuse / Steinstreif; ....band / banda pedregosa

soil stripe, – similar to stone stripe except that the texture of both coarse and fine material is considerably finer than in stone stripes. / – – / bande terreuse / Erdstreif / banda terrosa

2655
tundra polygon; ice-wedge ....; Tamyr ...., – large polygons that often possess ice wedges along their borders. / toendrapolygoon / sol à polygones de toundra / Tetragonalboden; Schachbrett....; Eiskeilspalten.... / suelo de polígonos de tundra; polígono de tundra

2656
stone pavement; lag gravel, –
1. veneer of coarse particles resting on the surface, due to wind action (in areas without vegetation);
2. the vertical sorting and upthrusting of larger stones by frost action.

/ plaveisel van stenen / pavage de pierres; .... de blocs / Pflasterboden; Steinplatten.... / pavimento de piedras

congelifract; block field, – angular rock fragments and boulders, produced by frost action. / vorstverweringspuin / éboulis de gélivation; matériel cryoclastique / Frostverwitterungsschutt; Felsenmeer / campo de piedras

[1700] PEDOLOGY

[1710] GENERAL TERMS

2657
soil, –
1. the unconsolidated material on the earth's surface that serves as a medium for the growth of plants;
2. the unconsolidated or secondarily indurated or cemented material on the earth's surface that has been influenced by genetic and environmental factors, such as parent material, climate, living organisms and topography over a period of time, resulting in a product that differs from the original material in many properties and characteristics.

/ bodem; grond / sol / Boden / suelo

parent material; .... rock, – more or less weathered mineral or organic matter from which the overlying soil developed. / moedermateriaal / matériel-parental; ....-originel; .... de départ; roche-mère / Ausgangsgestein; Mutter.... / material de partida

2658
soil science; pedology, – the science dealing with soils as a natural resource on the surface of the earth, including soil formation, classification, mapping and their physical, chemical and biological properties and fertility. / bodemkunde / science du sol; pédologie / Bodenkunde / ciencia del suelo; edafología

2659
soil genesis; pedogenesis, –
1. mode of origin of soils, referring to the processes or factors of soil formation from the parent material;
2. the division of soil science that treats of soil formation.

/ bodemvorming; pedogenese / pédogenèse; évolution des sols / Bodenbildung; ....entwicklung / génesis del suelo; edafogénesis

soil-forming factor, – the variable and inter-related agents active in and causing soil formation, usually grouped into: parent material, climate (including water in soils), organisms (including man), topography (relief), time. / bodemvormende factor; bodemfactor / facteur de pédogenèse / Faktor der Bodenbildung / factor de formación del suelo

2660
soil suitability; land classification, – the aptitude of (a group

of) soil(s) with respect to some particular purpose or use. / bodemgeschiktheid; landclassificatie / aptitude des sols / Bodeneignung / aptitud del suelo; clasificación de tierra

2661
soil classification, – systematic arrangement of soils into categories based upon their characteristics and/or properties. / bodemclassificatie / classification des sols / Bodensystematik; ....klassifikation / clasificación de suelos

NOTE:
– in the past decade soil classification has been and still is being strongly developed. The number of classes increased considerably. In some systems a completely new nomenclature was introduced, often applicable to a particular country only. Correlation keys between some old and some new systems are presented in the FAO-Unesco, the USA and the Canadian systems:
FAO-Unesco Soil Map of the World 1 : 5 000 000, Volume I, Legend, Paris 1974,
Soil taxonomy, U.S. Dep. Agr. Handbook, Washington, D.C., 1975,
The Canadian System of Soil Classification, Can. Dep. Agr. Publ. 1646, Ottawa, 1978.

2662
soil taxonomy, – the systematic classification of soils. / bodemtaxonomie / taxonomie des sols / Bodentaxonomie / taxonomía de suelos

taxon (plur. taxa); soil ...., – a group of soils or a group of soil classes that, according to natural relationships, is similar in selected properties and distinguished from other soils by differences in these properties. / taxon / taxon / Taxon / taxon

2663
soil series, – basic unit of many soil-classification sytems (e.g. USA, UK), a subdivision of a family, consisting of soils, that are alike in all major characteristics except the texture of the A-horizon. Soil series are often named after places in the area where the soil is first defined. / bodemserie / série de sols / – – / serie de suelos

2664
soil family, – in categorical systems of soil classification an intermediate class between the great group and the soil series. / bodemfamilie / famille de sols / – – / familia de suelos

2665
great group (obsolete), – a higher category in the USA 1938 and 1949 soil-classification systems.

2666
soil order, – highest categorical level in e.g. USA, UK, Canadian and Netherlands' systems of soil classification. / orde / ordre / – – / orden

classe de sols, – highest categorical level in the French sytem of soil classification.

Bodentyp, – highest categorical level in the German Federal soil-classification system.

2667
zonal soil, –
1. soil characteristic of a large climatic zone;
2. (obsolete) one of the three orders of the USA (1938, 1949) soil-classification systems.

/ zonale grond; .... bodem / sol zonal / zonaler Boden / suelo zonal

2668
intrazonal soil, – high category in various soil-classification systems, showing characteristics that reflect a dominating influence of some local factor as relief, parent material or age over the effects of climate and vegetation. / intrazonale grond; .... bodem / sol intrazonal / intrazonaler Boden / suelo intrazonal

2669
azonal soil, –
1. (obsolete) soil lacking a profile, predominantly due to climatic influence;
2. soil without distinct horizons.

/ azonale grond / sol azonal / azonaler Boden / suelo azonal

2670
soil survey; soil mapping, – the systematic examination, description, classification and mapping of soils, classfied according to kind and intensity of field examinations, i.e. detailed, reconnaissance or general surveys. / bodemkartering / cartographie des sols / Bodenkartierung / cartografía de suelos

2671
soil map, – map showing the distribution of mapping units, e.g. soil series. / bodemkaart / carte des sols; .... pédologique / Bodenkarte / mapa de suelos

detailed soil map, – map showing soil units in sufficient detail to indicate differences significant to potential use and that are genetically homogeneous within the delineated units. Scale mostly 1:25 000 or larger. / bodemkundige detailkaart / carte des sols détaillée; .... des sols à grande échelle / Bodenspezialkarte / mapa detallado de suelos

reconnaissance soil map, – map showing more widely defined soil units than soil series. These maps show wide variety in cartographic detail and in scale (mostly 1:50 000 or less). / bodemkundige overzichtskaart / carte pédologique de synthèse; .... des sols à petite échelle / Bodenübersichtskarte / mapa de suelos a nivel de reconocimiento

2672
mapping unit; map ...., – unit of soil, defined in terms of one or more taxonomic units (e.g. soil series), used in practical soil mapping. It describes the soil bodies delineated on the map, but includes small proportions of impurities that cannot be excluded in mapping procedures. / kaarteenheid / unité cartographique / Bodeneinheit; Kartier.... / unidad cartográfica (de suelos)

2673
soil association, –
1. a group of taxonomic soil units occurring together in a characteristic pattern over a geographic region;
2. a mapping unit on small-scale maps, in which two or more taxonomic units (occurring together in a characteristic pattern) are combined because the map scale does not permit individual delineation.

/ bodemassociatie; samengestelde kaarteenheid / association des sols / Bodengesellschaft / asociación de suelos

soil complex, – mapping unit used in detailed surveys where two or more taxonomic units are so initimately intermixed that it is impossible or impractical to separate them; there is more intimate mixing in small areas than in the case of soil associations. / bodemcomplex / complexe des sols / Komplex; Bodeneinheit / complejo de suelos

2674
catena, – soil sequences of the same age and from similar parent material, formed under similar climatic conditions, but different in characteristics due to relief and/or drainage conditions. / catena / catena; séquence; chaîne de sols / Bodentypenreliefsequenz; Bodencatena / catena

2675
density of observation, – the number of observations in pits or auger holes per unit of area (mostly per hectare, $10^4 m^2$). / waarnemingsdichtheid; borings.... / généralité des observations / Bohrabstand; Bohrungsdichte / densidad de observaciones

2676
Munsell colour system; .... soil-colour charts, – widely used colour-designation system developed by Munsell Color Company, Baltimore, USA. Colour notation consists of digit(s) and capitals for hue and a digit for each value and chroma, e.g. 10YR6/4 is a colour with hue 10YR (yellow-red), value 6 chroma 4, designated as light yellowish brown. Soil colours are measured by comparison with a soil-colour chart. / Munsell kleursysteem; .... kleurenkaart / code Munsell / Munsell Farbtafeln / código de colores Munsell; tabla de colores

2677
hue, – characteristic of colour, depending on the wavelength of the reflected light. / kleurtoon / teinte; tonalité / Farbe / tinte

value, – the relative lightness or intensity of colour. / helderheid (van kleur) / luminosité; intensité; brillance / Helligkeit; Farbtiefe / claridad

chroma, – relative saturation of a colour; inversely related to greyness. / chroma; kleurverzadiging / chroma; saturation; pureté / Chroma; Farbintensität / saturación cromática

[1720] SOIL-FORMING PROCESSES

NOTES:
- see also sections
  [1686] Periglacial Regions
  [2422] Biochemical Phase; Environments of Deposition,
- there is no clear limit between soil formation and diagenesis; for further terms see section
  [2320] Environmental Diagenesis,
- several soil-forming processes lead to the formation of mineral deposits; for further terms see section
  [2660] Various Metallic Mineral Deposits.

2678
eluviation; leaching, – removal of soil material in suspension. / uitspoeling; eluviatie / éluviation; lessivage / Eluviation; Verlagerung / eluviación

illuviation, – process of deposition of soil material removed from one horizon to another. / inspoeling / illuviation / Illuviation; Anreicherung; Einwaschung; Einlagerung / illuviación

2679
decalcification, – removal of $CaCO_3$ from the soil by leaching. / ontkalking / décalcification; décarbonatation / Entkalkung / decalcificación

2680
podzolisation, – process resulting in the genesis of podzol soils by destruction of clay in the topsoil and leaching of sesquioxides and mobile organic matter that precipitate in the subsoil and thus form an illuvial horizon. / podzolering / podzolisation / Podsolierung / podsolización

2681
lateritization – the leaching of iron and aluminium (hydr)oxides and subsequent deposition at a deeper level; in deeply weathered zonal soils of humid tropics with low silica-sesquioxide ratios in the clay fractions. / lateritisatie / latéritisation / Lateritisierung / lateritización

2682
mineralization, – the conversion of an element from an organic compound into an inorganic compound as a result of microbial decomposition in a soil. / mineralisatie / minéralisation / Mineralisierung / mineralización

2683
humification, – decomposition of organic matter in soils leading to the formation of humus, in which the original plant structure is not observable. / humificatie / humification / Humifizierung / humificación

2684
gleysation, – process developing in badly drained soils that results in reduction of iron and other compounds, producing mottled grey colours. / vergleying / gleyification / Vergleyung / gleyificación

2685
ripening, – processes in alluvial material leading to the transformation of weak mud into a firm soil. / rijping / maturation / – – / maduración edáfica

physical ripening, irreversible loss of water, decrease of pore space, shrinkage, cracking and increase of permeability.

chemical ripening, – transformation of organic matter, exchange of cations (especially Na and Mg in saline mud), loss of carbonates, oxidation of sulphides and of iron.

[1730] SOIL PROFILE AND SOIL HORIZONS

2686
soil profile, – vertical section of the soil through all its horizons and extending into the parent material. / bodemprofiel / profil / Bodenprofil / perfil de suelo

2687
soil horizon, – layer of soil material approximately parellel to the surface and differing from adjacent, genetically related layers in physical, chemical and/or biological characteristics, or in colour, structure, texture, consistency etc. / bodemhorizon; ....horizont / horizon / Bodenhorizont / horizonte

2688
soil layer, – any stratum in the soil, approximately parallel to the surface, without characteristics as produced by soil formation. / bodemlaag; grond....; horizon(t) / couche / Bodenschicht / capa de suelo

2689
pedon, – the (artificial) unit of soil-horizon classification; it is a three-dimensional body of soil with lateral dimensions large enough to study the horizon's shape and relations. Its area ranges from about 1 to 10 m^2. / pedon / pedon / Pedon / pedon

2690
hydromorphic characteristic, – the fossil or active phenomena in the soil indicating soil formation under (strong) action of water, such as mottling, blue and grey colours. / hydromorf kenmerk / caractéristique d'hydromorphie / hydromorphes Bodenmerkmal / característica hidromorfa

2691
cutan; skin, – coatings or concentrations of clay, humus or iron on ped (unit of soil structure) surfaces, sand grains or in pores and ooids. / cutan; huidje / cutane; enduit / Cutan; Häutchen / cután

examples: argillocutan, ferricutan, clay skin, iron skin, humus skin

2692
mottle, – spots or streaks of a colour differing from that of the soil matrix; described in terms of colour, contrast, abundance and size. / vlek / tache; bariolage / Fleck / mota

mottled, – / gevlekt / tacheté; panaché / gefleckt / motado

rust; .... mottle, – patches or streaks of reddish or brownish colours, caused by inhomogeneous distribution of ferric iron; sometimes concretional. / roestvlek / rouille / Rostfleck / mancha rojiza

2693
buried profile; .... soil; .... horizon, – profile or horizon covered by an alluvial, colluvial or peat deposit. / begraven profiel; .... grond; .... horizon(t) / sol enfoui; horizon .... / begrabener Boden; .... Horizont / suelo enterrado; perfil ....

2694
solum, – the upper and most weathered part of the soil profile, including the A- and B-horizons. / solum / solum / Solum / solum

2695
ABC soil, – soil with a profile having the A-, B- and C-horizons. / ABC-profiel; ....-grond / sol ABC / A-B-C-Boden; Boden mit A-B-C-Profil / suelo ABC; perfil ABC

AC soil, – soil with a profile having the A- and C-horizons only. / AC-profiel; ....-grond / sol AC / A-C-Boden; Boden mit A-C-Profil / suelo AC; perfil AC

2696
subsoil; substratum, – the layer beneath the solum, either conform or unconform to the overlying material. / substraat / substratum / Substrat / substrato

2697
epipedon, – a soil horizon that forms or has been formed at the surface. / epipedon / épipedon / Epipedon / epipedon; horizonte a capa superficial

2698
A-horizon, – mineral soil horizon having:
1. organic matter accumulation, formed or forming at or adjacent to the surface,
2. lost clay, iron or aluminium, resulting in concentration of quartz or other resistant minerals or
3. characteristics of 1 or 2, but transitional to a B- or C-horizon.

/ A-horizon(t) / horizon A / A-Horizont / horizonte A

A1-horizon, – mineral soil horizon, formed or forming at or adjacent to the surface, in which humified organic matter is intimately associated with mineral matter. / A1-horizon(t) / horizon A1 / Ah-Horizont (if containing organic matter); Ai-Horizont (if not containing visible organic matter) / horizonte A1

sod; turf; sward, – the uppermost part of the soil, filled with roots, belonging to A1. / heideplag; grasplag / plaque de bruyère; .... de gazon / Heideplagge; Gras....; Sode; Grasnarbe / cesped

2699
Ap-horizon, – A-horizon disturbed by ploughing or other soil cultivations. / Ap-horizon(t) / horizon Ap / Ap-Horizont / horizonte Ap

ploughed layer; tilled ....; plowed .... (USA), – surface layer of the soil moved by ploughing or other superficial cultivation. / bouwvoor / couche arable / Krume; Ackerkrume / capa de laboreo

2700
A2-horizon; E-.... (UK), – a mineral horizon showing a concentration of sand and silt fractions high in resistant minerals, resulting from a loss of silicate clay, iron or aluminium or some combination of these. / A2-horizon(t) / horizon A2 / Al-Horizont (partly, loss of clay); Ae-.... (partly, loss of sesquioxides) / horizonte A2

bleicherde; bleached layer; .... horizon, – the light-coloured, leached A2-horizon of podzol soils. / loodzand / horizon cendreux / Bleicherde; ....horizont / horizonte blanqueado

2701
eluvial horizon, – horizon from which material has been re-

moved either in solution or in suspension; such as the A2- and E-horizons. / eluviale horizon(t); uitspoelings.... / horizon éluvial / Eluvialhorizont; Auslaugungs....; Auswaschungs.... / horizonte eluvial; .... eluviado

2702
B-horizon, – horizon of altered soil material, below the A-horizon, showing:
1. accumulation of clay, iron, aluminium, together with organic matter, or organic matter only and/or
2. blocky or prismatic structures, together with stronger colours than the A-horizon.

/ B-horizon(t) / horizon B / B-Horizont / horizonte B

illuvial horizon, – general term for a soil layer in which material (e.g. organic matter, clay, sesquioxides) from an overlying horizon has been precipitated from solution or deposited from suspension. / inspoelingshorizon(t); ....laag / horizon illuvial / Illuvialhorizont; Einlagerungs....; Anreicherungs....; Einwaschungs.... / horizonte iluviado

2703
colour-B-horizon, – a soil horizon having a stronger colour and a higher clay content than the overlying A- and the underlying C-horizon, caused by weathering. / kleur-B-horizon(t) / horizon-B de couleur / verbraunter Horizont; verlehmter ....; (B)-Horizont; Bv-.... / horizonte B coloreado

textural-B-horizon, – horizon with accumulation of illuviated clay. / briklaag; textuur-B-horizon(t) / horizon B2 de texture; horizon argillique; .... argilo-illuvial / Bt-Horizont; Bt-Illuvialhorizont; Tonanreicherungs.... / horizonte B textural

2704
orterde, – common name for a not indurated podzol-B-horizon. / oerlaag / – – / Orterde / – –

ortstein, – indurated part of a podzol-B-horizon. / oerbank / alios / Ortstein / alios

2705
textural subsoil lamella, – thin layer(s) in the sandy subsoil of some podzols, with illuviated clay and iron. / banden-B-horizon(t) / horizon B en bandes / gebänderter B-Horizont / horizonte en bandas

lamella; fiber, – thin illuvial horizon(s), ranging from a few mm to some cm in thickness below the main B-horizon. / fiber / bande / Bändchen / banda

2706
C-horizon; C-layer, – a layer of unconsolidated material, similar in composition to the material from which the overlying solum has developed. / C-horizon(t) / horizon C / C-Horizont / horizonte C

2707
Cr-horizon; Cg-...., – horizon showing strong reduction as a result of ground-water action. / G-horizon(t) / horizon C_1 / Gr-Horizont / horizonte Cg

2708
D-layer, – stratum that underlies the solum and is different from the material from which the solum has been formed. / D-laag; D-horizon(t) / horizon D / D-Horizont; D-Schicht / horizonte D

[1740] SOIL TEXTURE

NOTE:
– see table 5 – Soil-textural Classes according to various Usages

2709
soil texture; particle-size distribution; grain-size ...., – the relative proportions of the various size groups of individual grains in a mass of soil; specifically the proportions of clay, silt and sand below 2 mm in equivalent diameter. Usually expressed as a mass percentage. / textuur; granulaire samenstelling; korrelgrootteverdeling / texture du sol; composition granulométrique / Korngrössenverteilung; Körnung; Korngrössenzusammensetzung; Textur / textura del suelo

2710
particle size; grain size, – the equivalent diameter of a soil particle, as measured by particle-size analysis. / korrelgrootte / fraction granulométrique / Korngrösse / tamaño de partículas

2711
particle-size analysis; mechanical .... (obsolete), – determination of the amounts of the various fractions in a soil sample, e.g. by sedimentation, sieving, etc. / korrelgrootte analyse; mechanische ....; granulaire .... / analyse granulométrique; analyse mécanique / Bestimmung der Körnung; Korngrössenanalyse; mechanische Analyse / análisis granulométrico

2712
soil fraction; particle-size ....; soil separate (USA), – size groups of mineral particles. / fractie; korrelklasse; korrelgrootte.... / fraction granulométrique / Fraktion; Kornfraktion / fracción del suelo

2713
textural class; .... classification, – arbitrary grouping of particle-size classes. / textuurklasse; ....indeling; korrelgrootte indeling / classe granulométrique; classification .... / Bodenart; Bodenarteneinteilung; Korngrössenklasse / clase textural

NOTE:
– differences in soil-texture classification exist between countries; moreover, classification is different from that used in sedimentology; compare tables 5 and 6.

2714
clay; clay fraction, – soil fraction and textural class consisting of particles smaller than 0.002 mm. / klei; lutumfractie / fraction argileuse / Ton; Tonfraktion / arcilla; fracción arcilla

2715
silt, – soil fraction and textural class consisting of particles

between 0.002 and 0.050 (0.063) mm in diameter. / silt / limon / Schluff / limo

2716
slibfractie (Netherlands), – soil fraction consisting of particles smaller than 0.016 mm. Use of this term is not recommended. / afslibbaar; afslibbare delen; slib / – – / Abschlämmbares (smaller than 0.001 mm, obsolete) / – –

sloef (Netherlands), – soil fraction consisting of particles between 0.002 and 0.016 mm.

zand (Netherlands), – in some classifications this textural class is considered to comprise particle sizes between 0.016 and 2 mm. Use in this sense is not recommended.

2717
leem (Netherlands), – soil fraction or textural class consisting of particles smaller than 0.05 mm (used for aeolian sediments).

2718
sand, –
1. any soil particle between 0.05 (0.063) and 2.0 mm in diameter;
2. soil fraction consisting of mineral particles between 0.05 (0.063) and 2.0 mm in equivalent diameter;
3. a soil textural class.

/ zand / sable / Sand / arena

median of the sand fraction, – the grain size above and below which half of the mass of the sand fraction (0.05 - 2 mm) occurs. / mediaan van de zandfractie; M50 / médiane du sable / Median der Sandfraktion / mediano de la fracción arena

2719
fine earth, – organic-free part of a soil sample, consisting of particles smaller than 2 mm in diameter. / minerale (bodem)delen / terre fine; particules élémentaires (minérales) / Feinboden; Feinerde / tierra fina

2720
loam, – mixtures of clay, silt and sand, i.e. consisting of particles smaller than 2 mm. / leem / limon / Lehm / limo

2721
fine texture(d), –
1. texture of all clay loams and clays (USA);
2. soil containing large quantities of 1.

/ kleiig; ....e textuur; zwaar / texture argileuse / tonreich; schwer(er Boden) / textura fina

medium texture(d), –
1. texture of very fine sandy loams, loams, silty loams and silts (USA);
2. soil containing large quantities of 1.

/ lemig; ....e textuur / texture limoneuse / – – / textura media

2722
stone (UK); **coarse fragment** (USA), – particles and fragments larger than 2 mm in diameter; in pedology subdivided according to shape and size, varying in different countries. / grind / élément grossier / Bodenskelett / grava

[1750] SOIL STRUCTURE; CEMENTATION

2723
soil structure, – the spatial arrangement of primary soil particles and their aggregates. / structuur; bodem.... / structure / Gefüge; Bodengefüge / estructura (del suelo)

2724
ped; aggregate, – unit of soil structure formed by natural processes, e.g. crumb, block, prism. / structuurelement / agrégat; élément structural / Aggregat; Bodenaggregat / agregado

ped surface; ped face, – natural boundary of a ped. / wand van structuurelement / face / Fläche eines Aggregats; Aggregatoberfläche / cara de agregado

2725
clod, – coherent mass of soil produced artificially, e.g. by ploughing, digging, etc. / kluit / motte / Klumpen; Bröckel / terrón; agregado

2726
grade of structure, – a classification of soil structure on the basis of ped adhesion, cohesion or stability within the soil profile. Expressed as weak, moderate, strong. / structuurgraad / degré d'agrégation / Zusammenhalt (der Aggregate); Aggregierungsstufe / grado de estructura

structure type, – a classification of structure based on the shape of peds and their arrangement in the soil profile. / structuurvorm; ....type / type de structure / Gefügeform; ....typ / tipo de estructura

2727
crumb, – soil-structure type, consisting mainly of spheroidal porous peds, whose faces possess slight or no accommodation to the adjoining aggregates. / kruimel / grumeau; granule; grenue; grumuleuse / Krümelgefüge / grumo; migajosa

granular, – soil-structure type, consisting mainly of spheroidal, relatively non-porous peds, whose faces possess slight or no accommodation to the adjoining aggregates. / granulair / granulaire / Korngefüge / granular

2728
platy, – soil-structure type, consisting of peds with horizontal dimensions greatly in excess over the vertical dimension; their faces are mostly horizontal. / platig / lamellaire; feuilletée; plate / Plattengefüge / laminar

2729
blocky; angular ....; nut (obsolete), – soil-structure type, consisting of sharply angular peds, whose three dimensions are of the same order of magnitude. / blokkig / polyédrique; angulaire / Polyedergefüge / estructura en bloques angulares

subangular blocky; subangular; nut (obsolete); nuciform (obsolete), – soil-structure type, consisting of peds with rounded faces, whose three dimensions are of the same order of magnitude. / afgerond-blokkig / subangulaire / Subpolyedergefüge / estructura en bloques subangulares!

2730
prismatic; prism, – soil-structure type, consisting of peds with a considerably greater vertical than horizontal dimension; they possess well-defined vertical faces without rounded caps. / prismatisch; prisma / prismatique / Prismengefüge / prismática

columnar, – soil-structure type, consisting of peds with a considerably greater vertical than horizontal dimension; they possess well-defined vertical faces with rounded caps. / kolom / structure en colonnes; structure colonnaire / Säulengefüge / columnar

2731
slickenside; slicken, – polished and grooved surface on peds, produced by a soil mass sliding past another one. Typical for fine-textured montmorillonitic (smectitic) soils (Vertisols). / wrijfspiegel / surface de glissement / Harnisch; Glattfläche / cara de desligamiento

2732
concretion, – local concentration of material (such as $CaCO_3$ or Fe_2O_3) in the form of a grain or nodule of varying size, shape, hardness and colour, different in composition from the surrounding material. / concretie / concrétion / Konkretion / concreción

NOTE:
– for further terms see section
[2340] Concretions

2733
cemented, – pertaining to a layer of soil with a brittle to hard consistency, consolidated by substances other than clay, such as silica, carbonates, iron or aluminium oxides, etc. / verkit / cimenté / verfestigt / cementado

hardpan, – similar to cemented layer; hardened soil layer caused by cementation of particles, e.g. by organic matter, silica, sesquioxides, calcium carbonate. / bank; oerbank / horizon induré / Ortstein; Raseneisenstein / capa endurecida

NOTE:
– for further terms see
[2320] Environmental Diagenesis
[2333] Cementation
[2660] Various Metallic Mineral Deposits

2734
fragipan, – subsoil layer with relative high bulk density, hard when dry, brittle when moist. / fragipan / fragipan; couche dure; .... durcie / – – / fragipan

clay pan, – compact subsurface layer with a much higher clay content than the overlying material, formed by downward movement of clay or by clay formation in place. / briklaag; kleiband / horizon argileux / Tonanreicherungshorizont / capa de arcilla

iron pan, – indurated layer in which iron oxide is the cementing agent. / ijzerband; ....fiber / alios ferrugineux / Raseneisenstein / capa cementada

duripan, – mineral subsoil layer cemented by silica; air-dry fragments do not slake in water or hydrochloric acid.

[1760] SOIL CHEMISTRY

2735
(calcium) carbonate content; free-carbonate ...., – content of not adsorbed (free) carbonates in the soil; determined by measuring the volume of CO_2 and calculated as $CaCO_3$. / (koolzure) kalkgehalte / teneur en calcaire / Carbonatgehalt; Kalk.... / contenido en carbonato (de calcio)

2736
calcareous, – pertaining to soils or other material that contain sufficient $CaCO_3$ to effervesce with 0.1N HC1. / kalkhoudend; ....rijk / calcaire; carbonaté / carbonathaltig; kalk....; carbonatreich; kalk.... / calcáreo

non-calcareous, – pertaining to soils or other material that contain no or very little $CaCO_3$; not effervescing with HC1. / kalkarm; kalkloos / non calcaire / carbonatarm; kalkarm; carbonatfrei; kalkfrei / no calcáreo

2737
alkaline soil, – soil with a pH greater than 7.3. / basische grond / sol alcalin / alkalischer Boden / suelo alcalino

acid soil, – soil with a pH less than 7.0 (for practical purposes mostly less than 6.6) in the surface layer or the root zone. / zure grond / sol acide / saurer Boden / suelo ácido

2738
acidification, – removal of non-hydrogen cations (especially of calcium ions) from the soil by leaching. / verzuring / – – / Entbasung; Versauerung / acidificación

2739
acid sulphate soil; catclay soil, –
1. soil or soil layer containing soluble acid aluminium and ferric sulphates, high in exchangeable aluminium; pH of water content less than 4; often with yellow mottles of the mineral jarosite;
2. soil with a sulphuric mineral or organic horizon, pH of water content less than 3.5; with yellow mottles.

/ katteklei / sol sulphaté acide / sulphatsaurer Boden; Maibolt / suelo ácido de sulfatos

2740
soil-organic matter, – plant and animal residues in the soil at various stages of decompositon, as determined after sieving through a 2 mm sieve. / organische stof; .... bestanddelen / matière organique / organische Substanz / materia orgánica

NOTE:
– for further terms see section
[2423] – Biochemical Phase; Environments of Deposition.

2741
decomposition, – transformation and alteration of organic debris, i.e. by oxidation and/or activity of (micro)organisms. / afbraak / décomposition / Abbau; Zersetzung; Verwesung / descomposición

2742
litter, – fresh and partly decomposed organic residues, as

leaves, lying on the soil surface. / strooisel / litière / Streu; Förna / hojarasca

2743
humus, –
1. the more or less stable and decomposed part of soil-organic matter;
2. incorrectly used as a synonym of soil-organic matter.
/ humus / humus / Humus / humus

2744
mor (USA); raw humus, – not incorporated, not or slightly decomposed organic remains that rest sharply demarcated on the underlying mineral soil. / ruwe humus / humus brut / Rohhumus / humus bruto

moder, – terrestric form of humus consisting of small excrements (0.05-0.1 mm) without intricate mixing with mineral particles; mesotrophic to oligotrophic conditions. / moder / moder / Moder / moder

mull; mild humus, – terrestric form of humus consisting of an intricate and homogeneous mixture of organic matter and mineral particles (especially clay); formed by decomposition of excrements of small soil animals (e.g. earthworms); mostly eutrophic. / mull / mull; humus doux / Mull / humus dulce

2745
crotivina; crotovine; krotovina (USSR), – an animal burrow in a soil horizon, that is filled with organic matter and/or material from another soil horizon. / crotovina; krotovina / crotovine; crotovina / Krotowine / crotovina

[1770] PHYSICAL PROPERTIES

2746
consistency; consistence, –
1. resistance of material to deformation or rupture;
2. degree of cohesion or adhesion of a soil mass.
/ consistentie / consistance / Konsistenz / consistencia

2747
plasticity; plastic, – quality of wet soil to change shape under stress and to retain that shape. / plasticiteit; plastisch / plasticité; plastique / Plastizität; plastisch / plasticidad

2748
plasticity constants, – the liquid and plastic limits to plastic deformation. / plasticiteitsconstanten; waarden van Atterberg / limites de plasticité / Plastizitätsgrenzen / índices de plasticidad

2749
upper liquid limit, – minimum percentage by mass of moisture at which a soil sample will barely flow under standard treatment. / vloeigrens / limite supérieure de plasticité; .... de liquidité / obere Plastizitätsgrenze; Zerfliess.... / límite superior de plasticidad

lower plastic limit, – minimum percentage by mass of moisture at which a soil sample can be deformed without rupture. / uitrolgrens / limite (inférieure) de plasticité / untere Plastizitätsgrenze / límite inferior de plasticidad

2750
stickiness; sticky, – the quality of wet soil to adhere to other objects. / kleverigheid; kleverig / adhérence; collant / Haftfestigkeit des Bodens / adherencia

sticky point, – the moisture percentage by mass of a mixed, kneaded soil that barely fails to adhere to a standard steel surface when the shearing speed equals 5 $cm.s^{-1}$. / kleefgrens / point d'adhésivité / Klebepunkt / punto de adherencia

2751
bulk density; apparent .... (obsolete), – in pedology, the mass of dried soil at 105°C per unit bulk volume. / volumemassa; schijnbare dichtheid (obsolete) / densité apparente; masse volumique / Volumengewicht; Raum....; scheinbare Dichte (obsolete) / densidad aparente

2752
(soil) porosity, – volume percentage of the total bulk (of soil) not occupied by solid particles. / poriënvolume / porosité / Gesamtporenvolumen / porosidad

pore space, – total space not occupied by particles in a volume (of soil). / poriënvolume / volume des vides / Porenvolumen / porosidad

2753
(soil) permeability, – in pedology, the ease with which gases, liquids or roots can penetrate a soil or a soil layer. / doorlatendheid / perméabilité / Durchlässigkeit / permeabilidad

2754
specific surface; U-figure, – the total surface of all the soil particles, or of the clay, silt or sand fractions, in an amount of soil, mostly expressed in m^2g^{-1}. / specifiek oppervlak; soortelijk .... / surface spécifique (du sol) / spezifische Oberfläche / superficie específica

2755
adsorption complex, – substances (mainly organic and inorganic colloids) in soil capable of adsorbing other materials especially ions. / adsorptiecomplex / complex adsorbant / Sorptionskomplex / complejo adsorbente

exchangeable cations, – cations that are held by the adsorption complex of the soil and are easily exchanged with other cations. / uitwisselbare kationen; (.... basen); omwisselbare kationen (.... basen) / cations échangeables / austauschbare Kationen / cationes cambiables

2756
base saturation; .... percentage, – percentage of total cation exchange capacity of soil saturated with cations other than hydrogen. / basenverzadiging / saturation en bases / Basensättigung / saturación de bases

2757
tilth, – the physical condition of a soil as related to its ease of tillage and its impedance to seedling emergence and root penetration. / – – / état d'ameublissement / Bodengare / sazon; laboreo

2758
slaking; surface capping, – unfavourable property of some soils to collapse under influence of (heavy) rainfall or pressure. / slempigheid; slemp / battance / Verschlämmung / disgregación

2759
crust, – thin surface layer, much more compact, hard and brittle when dry than the material immediately below. / korst / croûte / Kruste / costra

2760
surface sealing, – the packing of dispersed particles in the surface layer of a soil, rendering it relatively impermeable to water. / korstvorming; oppervlakkige slemp; oppervlakteslemp; oppervlakkige verslemping / scellement superficiel / Dichtlagerung; Dichtschlämmung / impermeabilización superficial

2761
ploughsole; plowsole (USA); plowpan (USA); pressure pan; induced pan; traffic pan, – subsurface layer with relatively high bulk density and low porosity as a result of pressure applied by e.g. tillage or other artificial means. / ploegzool / semelle de labour / Pflugsohle / piso de arado

2762
self-mulching, – ability of montmorillonitic soils (Vertisols) to form a surface-layer mulch (a root-protecting layer) consisting of granules. / – – / automulchage / Selbstmulcheffekt / acolchamiento

2763
gilgai, – microrelief of soils, consisting of basins and knolls or of valleys and ridges, all of tiny size (about 25 to 40 cm), and caused by expansion and shrinkage with change of moisture. It occurs in soils with large quantities of expanding clay minerals (smectite, montmorillonite). / gilgai reliëf / gilgai / Gilgai / relieve gilgai

[1780] SOIL MOISTURE

NOTE:
– for further terms see section
[2520] Geology of Subsurface Waters

2764
soil moisture, – water contained in the soil. / bodemvocht / eau du sol / Bodenwasser; ....feuchte / humedad del suelo

2765
phreatic ground water, – water that fills all the voids in the soil and the underlying strata and that is free to move under influence of gravity. / freatisch grondwater / nappe phrétique / freatisches Grundwasser / agua freática

ground-water table; phreatic level, – the level to which the ground water rises in a well or borehole, and at which level the hydraulic pressure equals atmospheric pressure. / grondwaterspiegel; freatisch vlak; grondwatertafel (Flemish) / surface de la nappe phréatique / Grundwasserspiegel / capa freática

2766
water-table class, – classification of average seasonal fluctuations of the ground-water table, in the Netherlands defined by the mean lowest water tables or by a combination of mean highest and mean lowest water tables. / grondwatertrap / – – / Grundwasserstufe / clase de capa freática

near-equivalent:
drainage class (USA)

2767
capillary water, – water retained in the soil by capillary forces. / hangwater / eau capillaire / Haftwasser; Kapillar.... / agua capilar

capillary zone; .... fringe, – zone above the phreatic water table that is saturated with water by capillary forces. / capillaire zone; vol-capillaire zone / frange capillaire / Kapillar-Saum / franja capilar

2768
infiltration, – downward movement of water from the surface into the soil. / infiltratie / infiltration / Versickerung / infiltración

soil drainage, –
1. property of soils allowing the removal of water from them;
2. the degree of removal of water from the soil;
3. methods of removal of water from the soil;
4. removal of an excess of water from an area.

/ drainage (1, 2, 3); ontwatering (2, 3); afwatering (4) / drainage / Entwässerung (2, 3); Drainage (1, 4) / drenaje

2769
water content; moisture ...., – of soils,
1. moisture content expressed as a percentage of the oven-dry mass of soil;
2. ratio of the volume of water in a soil to the total bulk volume.

/ vochtgehalte; water.... / teneur en eau / Wassergehalt; Feuchte / contenido de humedad

2770
moisture pressure; soil-water ...., – the positive or negative pressure (relative to the external gas pressure) to which a solution, identical in composition to a soil water, must be subjected in order to be in equilibrium with that soil water through a porous membrane. / vochtspanning / potentiel capillaire / Kapillarpotential / – –

moisture suction, – the negative pressure in soil water. / zuigspanning / pression de succion de l'eau / – – / – –

2771
moisture tension; .... pression, – the equivalent pressures in soil water; the following are determined: wilting point, field moisture capacity and the moisture equivalent at about pF 2.5, or at a tension of $\frac{1}{3}$ atm. / vochtspanning / tension de l'eau / (Boden)wasserspannung / tensión de humedad

pF, – negative logarithm of the soil-moisture tension, expres-

sed in centimetres height of water column.
Units: 100 cm water column = 9.810 kPa
1 atm = 101.325 kPa
1 bar = 100 kPa
1 kgf/cm^2 = 98.070 kPa

2772
moisture-retention curve; moisture-release culve, – graphs showing the soil moisture percentage (by mass or volume) against applied tension resp. pressure. / pF -kromme; vochtkarakteristiek / courbe de rétention d'eau / pF-WG-Kurve; Wasserspannungskurve / curva de retención de humedad

2773
wilting point; (permanent) .... percentage; 15-atmosphere ....; 15-bar ...., – the (usually mass) percentage of water held by a soil, that has been saturated, after application of a pressure of 15 atm (resp. about 15 bar). / (vochtgehalte bij) verwelkingspunt / point de flétrissement / Welkepunkt / índice de marchitez; porcentaje de 15 atm

available water; .... moisture, – portion of soil moisture that can be absorbed by plant roots; especially that portion held in soils up to approximately 15 atm or about 15 bar pressure, i.e. at about pF 4.2. / beschikbaar water; .... vocht / eau utile; capacité de rétention utile; eau disponible / nutzbares Wasser; nutzbare Feldkapazität / agua útil

2774
field (moisture) capacity, – mass (or volume) percentage of water remaining in a soil after having been saturated with water and after free drainage has ceased, i.e. at about pF 2 to 2.2 or about 60 cm moisture tension percentage. / (vochtgehalte bij) veldcapaciteit / capacité (de rétention) au champ / Feldkapazität / capacidad de campo

2775
water control, – complex of measures to ensure optimal water and air conditions in the soil for a given form of land use in a specific area. / waterbeheersing / contrôle du régime hydrique / Wasserversorgung / control del régimen hídrico

[1790] SOIL EROSION

NOTE:
– for further terms see sections
[1610] Weathering and Erosion,
[1930] Sedimentary Processes.

2776
soil erosion, – wearing away of the land surface or soil and rock particles by water, wind, ice or by gravity. / bodemerosie / érosion / Bodenerosion / erosión

2777
accelerated erosion, – increase in the rate of erosion, caused by activities of man or animals. / versnelde erosie / érosion accélérée / beschleunigte Erosion / erosión acelerada

2778
rill erosion, – formation of numerous small and shallow channels by unequal removal of surface soil by running water. / prielerosie; geul.... / érosion en rigoles / Rinnenerosion / erosión en surcos

gully erosion, – erosion that cuts deep and narrow channels into the land surface. / geulerosie / érosion en ravins; .... en ravelines; ravinement / Grabenerosion; Tiefen.... / erosión en cárcava

2779
sheet erosion, – accelerated removal of a uniform surface layer by running water. / vlakte-erosie; ....afspoeling / érosion en nappe / Flächenerosion / erosión laminar

2780
soil creep; solifluction, – slow mass movement of soil material downslope, primarily under the influence of gravity. / solifluctie / reptation; solifluxion / Bodenkriechen; Solifluktion / deslizamiento del suelo

[1800] OCEANOGRAPHY

[1810] GENERAL TERMS

2781
oceanography, – the study of the sea, embracing and integrating all knowledge pertaining to the sea's physical boundaries, the chemistry and physics of sea water, and marine biology; strictly the description of the marine environment. / oceanografie / océanographie / Ozeanographie / oceanografíia

oceanology, – the study of the oceans and related sciences. / oceanologie / océanologie / Meereskunde; ....forschung / oceanología

2782
ocean, – specifically, the large coherent part of the world ocean separating two or more continents. / oceaan / océan / Ozean / océano

sea, – specifically, a body of salt water with horizontal dimensions of order 10-2000 km, mostly partly land-locked and in connection with an ocean; in some cases completely landlocked. / zee / mer / Meer / mar

2783
adjacent sea; marginal sea, – / randzee / mer adjacente; mer bordière / Randmeer / mar adyacente

2784
estuary, – semi-enclosed coastal body of water having a free

connection with a sea or ocean and within which sea water is measurably diluted with fresh water derived from land drainage; horizontal dimensions mostly of order 1 to 200 km. / estuarium / estuaire / Ästuar; Fluss-Mündungsgebiet / estuario

2785
sea current; ocean ...., – / zeestroming / courant marin / Meeresströmung / corriente marina

surface current, – / oppervlaktestroom / courant de surface / Oberflächenstrom / corriente superficial

2786
upwelling, – upward flow of water from a few hunderds of metres depth to the surrace, associated with divergence of surface currents. / opwelling / upwelling / Aufrieb / revesa de fondo

2787
convergence, – area or zone in which the water converges at the surface and slowly sinks down. / convergentie / convergence / Konvergenz / convergencia

[1820] SEA-BOTTOM TOPOGRAPHY

2788
continental shelf; shelf, – the submerged part of the continents, extending from the low-water line to the depth at which there is a marked increase of slope toward greater depths. / continentaal plat; vastelandsplat / plate-forme continentale; plateau continental / Kontinentalschelf / plataforma continental

shelf edge, – the line along which the above-mentioned increase in slope occurs. Conventionally this edge is taken at 200 m. / continentale rand / rebord continental / Schelfrand / borde del talud continental

2789
continental borderland, – this term is appropriate when the zone below the low-water line is highly irregular, and includes depths well in excess of those typical for continental shelves. / continentaal randgebied / bordure continentale / randlicher Kontinentalabfall / orla continental

2790
continental terrace, – the submerged part of the continents, extending from the low-water line, to the base of the continental slope. / vastelandsterras; continentaal terras / terrasse continentale / Kontinentalterrasse / terraza continental

continental slope; .... talus, – declivity from the offshore border of the continental shelf, or continental borderland, to oceanic depths at about 4000 m. / continentale helling; vastelandsglooiing / pente continentale / Kontinentalhang / talud continental

2791
insular shelf; island ...., – zone around an island or island group, extending from the low-water line to the depth where there is a marked descent towards oceanic depths. Conventionally its edge is taken at 200 m. / eilandsplat / socle; plateau insulaire; plate-forme .... / Inselschelf / plataforma insular

insular slope; island ....; insular talus, – declivity from the offshore border of an island shelf to oceanic depths. / eilandsglooiing / pente insulaire / Inselhang / talud insular

2792
abyssal plain, – an area of the ocean floor below 4000 m depth with a slope of less than 1 in 1000. / abyssale vlakte / plaine abyssale / Tiefsee-Ebene / llanura abisal

2793
basin, – depression of the deep-sea floor more or less equidimensional in form, but not necessarily large and pronounced. / bekken / bassin / Becken / cuenca

cauldron, – small depression of a more or less circular, elliptical or oval shape; use of this term is not recommended. / ketel / trou; ombilic / Kessel / hoya

trough, – long and broad depression of the sea floor with gently sloping sides. / inzinking / auge / Mulde / hondonada; depresión

trench, – long but narrow depression of the sea floor having relatively steep sides, conventionally applied where soundings exceed 6000 m. / trog / fossé; fosse / Graben; Rinne / fosa

2794
deep, – well defined deepest area of a depression of the deep-sea floor, conventionally applied where soundings exceed 6000 m; use of this term is not recommended. / diep / fosse / Tief / fosa

2795
submarine canyon, – elongated steep-welled cleft running across or partially across the continental shelf, the continental borderland and/or slope, and progressively deepening in a direction away from the shore. / onderzeese canyon / canyon sous-marin / – – / cañón submarino

submarine valley, – this term is appropriate when the sides have a more gentle slope. / onderzeese vallei / vallée sous-marine / Canyon / valle submarino

2796
rise, – long and broad elevation of the deep-sea floor that rises gently and smoothly. / verheffing / élévation; seuil / Erhebung; Aufwölbung / elevación; loma

ridge, – long elevation of the deep-sea floor having steeper sides and less regular topography than a rise. / rug / dorsale; ride / Rücken / dorsal

gap, – steep-sided furrow that cuts transversely across a ridge or rise. / groef / sillon; couloir / – – / surco

sill, – submarine ridge or rise separating partially closed basins from one another or from the adjacent ocean. / drempel / seuil / Schwelle / umbral

2797
plateau, – a comparatively flat-topped elevation of the sea floor of considerable extent across the summit and usually

rising more than 200 m on all sides. / plateau / plateau / Plateau / planicie

2798
seamount; seapeak, – elevation rising 1000 m or more from the sea floor and of limited extent across the summit. / onderzeese berg / montagne sous-marine; mont sous-marin / unterseeischer Berg / monte submarino

guyot, – obsolete term for seamount with flat top.

tablemount, – a seamount, the top of which is comparatively smooth. / tafelberg / guyot / Tafelberg / meseta

2799
seascarp; escarpment, – an elongated and comparatively steep slope of the sea floor, separating flat or gently sloping areas. / onderzees klif / escarpement sous-marin / Abbruch / escarpe

2800
seaknoll; knoll, – an elevation rising less than 100 m above the sea floor, and of limited extent across the summit. / – – / colline sous-marine / Aufragung / – –

2801
terrace, – a bench-like structure bordering an undersea feature. / terras / replat sous-marin; terrasse / Terrasse / rellano submarino

2802
bank, – an elevation of the sea floor located on a continental (or island) shelf and over which the water is shallow. / bank / banc / Bank / banco

2803
shoal, – a submerged ridge, bank or bar consisting of, or covered by unconsolidated sediments (mud, sand, gravel) with depths less than 20 m. / ondiepte / haut-fond; banc / Untiefe / bajo; bajío

2804
reef, – ridge, bank or bar composed of rock or coral. / rif / récif / Rif / arrecife

2805
pinnacle, – a sharp pyramidal or cone-shaped rock partly or completely covered by water. / klip / aiguille; pinacle / Spitze / aguja; escollo

2806
spur, – prolongation of a mountain range on or across the continental or insular shelf. / uitloper / éperon / Ausläufer / espolón

[1830] WAVES AND TIDES

2807
wind wave; 'sea', – wave caused by the wind and still being activated. / zeegang / mer du vent / Windsee / mar

swell, – waves generated by winds earlier and elsewhere. / deining / houle / Dünung / oleaje

2808
wave height, – the vertical distance between a crest and the preceding trough. / golfhoogte / hauteur; creux / Wellenhöhe / altura de ola

wave length, – the horizontal distance between similar points on two successive waves measured perpendicular to the crests. / golflengte / longueur d'onde / Wellenlänge / longitud de ola

wave period, – the time for a wave crest to traverse a distance equal to one wavelength. The time for two successive wave crests to pass a fixed point. / golfperiode / période d'onde / Wellenperiode / período de ola

2809
shallow water; wave in ...., – commonly water of such a depth that surface waves are noticeably affected by bottom topography. It is customary to consider water of depths less than one-half the surface wave length as shallow water. / ondiep water / eau peu profonde / seichtes Wasser / aguas someras

deep water; wave in ...., – water of a depth at least one-half the wavelength. / ondiep water / eau profonde / Tiefwasser / aguas profundas

2810
(wave) refraction, –

1. the process by which the direction of a wave moving in shallow water at an angle to the contours is changed. The part of the wave advancing in shallower water moves more slowly than that part still advancing in deeper water, causing the wave crest to bend toward alignment with the underwater contours;
2. the bending of wave crests by currents.

/ (golf)refractie / réfraction / Refraktion; Brechung / refracción

2811
breaker, – a wave breaking on a shore, over a reef, etc. / breker / brisant / Brecher / rompiente

Breakers may be classified into three types:

spilling, – bubbles and turbulent water spill down front face of wave, breaking generally across over quite a distance. / overschuimend / déferlement glissant / Schwalbbrecher / roción

plunging, – crest curls over air pocket; breaking is usually with a crash. / overstortend / déferlement plongeant / Sturzbrecher / vuelco; .... volcada

surging, – wave peaks up, but bottom rushes forward from under wave, and wave slides up beach face with little or no bubble production. / deinend / déferlement frontal / Reflexionsbrecher / ampollada

2812
surf, – the wave activity in the area between the shoreline and the outermost limit of breakers. / branding / déferlement / Brandung / batiente

surf zone, – the area between the outermost breaker and the limit of wave uprush. / brandingsstrook / zone de déferlement / Brandungszone / batidero

2813
uprush; swash; runup, – the rush of water up the beach following the breaking of the wave. / golfoploop / jet de rive / Auflaufen der Wellen / – –

backwash; backrush, – the seaward return of the water following the uprush of the waves. / golfterugloop / retrait / zurückziehende Welle / – –

2814
ripple, –
1. the ruffling of the surface of water, hence a little curling wave or undulation;
2. a wave less than 5 cm long controlled to a significant degree by both, surface tension and gravity.

/ rimpel / ride / Riffel; Rippel / rizadura

2815
solitary wave; wave of translation, – a wave consisting of a single elevation and neither followed or preceded by another elevation or depression of the water surface. / eenlinggolf; translatiegolf / onde solitaire / Einzelwelle / ola de traslación

2816
tsunami, – a Japanese term for a long-period wave caused by any large-scale disturbance on the sea floor of short duration, such as volcanic eruptions or earthquakes. Commonly miscalled **'tidal wave'.** / tsoenami; vloedgolf / tsunami; raz de marée / Tsunami; Flutwelle / maremoto

seiche, – a standing-wave oscillation of an enclosed water body that continues after the cessation of the originating force, which may have been either seismic or atmospheric. / seiche; haling / seiche / Seiche / seca

2817
mean sea level, – the avergage height of the surface of the sea for all stages of the tide over a 19-year period. / gemiddeld zeeniveau / niveau moyen de la mer / mittleres Meeresniveau / nivel medio del mar

2818
tide, – the periodic rise and fall of the sea that results from gravitational attraction of moon and sun acting upon the rotating earth. It is preferable to reserve this term for the vertical movement of the water. / getij / marée / Gezeit(en) / marea

high tide; high water, – the maximum height reached by each rising tide. / hoogwater; hoog tij / marée haute; haute mer; pleine mer / Hochwasser / pleamar

low tide; low water, – the minimum height reached by each falling tide. / laagwater; laag tij / marée basse; basse mer / Niedrigwasser; Niederwasser / bajamar

2819
tidal range; range of the tide, – the difference in height between consecutive high and low waters. / tijverschil / amplitude de marée; marnage / Tidenhub / amplitud de marea

2820
mean high water; .... high tide, – the average height of the high waters over a 19-year period. / gemiddeld hoogwater / haute mer moyenne / mittleres Hochwasser / pleamar media

mean low water;.... low tide, – the average height of the low waters over a 19-year period. / gemiddeld laagwater / basse mer moyenne / mittleres Niedrigwasser / bajamer media

2821
tidal current, – the alternating horizontal movement of water associated with the rise and fall of the tide caused by the tide-producing forces. / getijstroom / courant de marée / Gezeitenstrom; ....strömung / corriente de marea

2822
flood current, – the tidal current toward shore or up a tidal stream, usually associated with the increase in the height of the tide. / vloedstroom / courant de flot / Flutstrom / flujo

ebb current, – the tidal current away from shore or down a tidal stream; usually associated with the decrease in the height of the tide. / ebstroom / (courant de) jusant / Ebbestrom / reflujo

2823
bore; eagre, – a very rapid rise of the tide in shallow estuaries when the tidal range is large, and in which rise the advancing water presents an abrupt front of considerable height. / vloedbranding; stuwvloed / mascaret / Flutbrandung / barra; macareo

2824
storm surge, – a rise above normal water level on the open coast due to the action of wind stress on the water surface. / stormvloed / raz de marée de tempête / Sturmflut / mar ampollada

2825
longshore current, – the current in the breaker zone moving essentially parallel to the shore, usually generated by waves breaking at an angle to the shoreline. / kuststroming / courant littoral de houle; dérive littorale / Küstenströmung; küstenparallele Meeresströmung / corriente costera

2826
rip current, – a strong surface current flowing seaward from the shore. It usually appears as a visible band of agitated water and is the return movement of water piled up on the shore by incoming waves and wind. / – – / courant de retour; rip current / Abschleifungsstrom / resaca

undertow, – a seaward current near the bottom on a sloping inshore zone. It is caused by the return, under the action of gravity, of the water carried up on the shore by waves. Often a misnomer for rip current. / onderstroom / flot de fond / Unterstömung / resaca de fondo

[1840] MARINE ENVIRONMENTS

NOTE:
– see also sections
[1530] Actuopalaeontology
[1670] Coasts and Islands,
[1924] Sedimentary Marine Environments

2827
salinity, – a measure of the quantity of dissolved solids in water. In oceanography it is formally defined as the total amount of dissolved solids in parts per thousand ($^0/_{00}$) by mass when all the carbonate has been converted to oxide, the bromide and iodide to chloride, and all organic matter is completely oxidized. / zoutgehalte / salinité / Salzgehalt / salinidad

sigma-t, (symbol σ_t), – a conventienly abbreviated value of the density of a sea-water sample of temperature t and salinity S:
$\sigma_t = \varrho_{(S,t)} - 1000$,
where $\varrho_{(S,t)}$ is the value of the sea-water density in S.I. units at standard atmospheric pressure.

2828
water type, – coherent body of water within the ocean characterized by nearly uniform temperature and salinity. / watertype / eau type / Wasserkörper / agua tipo

water mass, – coherent body of water within the ocean formed by mixing of two water types. / watermassa / masse d'eau / Wassermasse / masa de agua

NOTE:
– above terms have been interchanged and used loosely.

2829
thermohaline, – pertaining to both, temperature and salinity acting together. / thermohalien / thermohalin / thermohalin / termohalino

thermohaline circulation, – vertical circulation induced by surface cooling. It causes convective overturning and mixing of water types. / thermohaliene circulatie / circulation thermohaline / – – / circulación termohalina

2830
thermocline, – laywer with a vertical negative (downward) temperature gradient that is appreciably greater than the gradients above and below it; the principal thermoclines in the ocean are either seasonal, due to heating of the surface water in summer, or – at greater depths – permanent. / thermocline / thermocline / Temperatursprungschicht / termoclina

halocline, – layer with a vertical positive (downward) salinity gradient that is appreciably greater than the gradients above and below it; often occurring in estuaries. / halocline / halocline / (Salzgehalt-)Sprungschicht / haloclina

pycnocline, – layer with a vertical positive (downward) density gradient that is appreciably greater than the gradients above and below it; may coincide with thermocline and/or halocline. / pycnocline / pycnocline / – – / picnoclina

2831
homoiohaline, – pertaining to marine waters of stable or constant salinity. / homoiohalien / homohalin / – – / homohalino

poikilohaline, – pertaining to marine waters of variable salinity. / poikilohalien / poikilohalin / – – / poikilohalino

2832
oligohaline, – pertaining to marine waters with salinities between 0.5 and 5 $^0/_{00}$. / oligohalien / oligohalin / oligohalin / oligohalino

mesohaline, – pertaining to salinities of marine waters between 5 and 18 $^0/_{00}$ / mesohalien / mésohalin / mesohalin / mesohalino

polyhaline; brachyhaline, – pertaining to salinities of marine waters between 18 and 30 $^0/_{00}$. / polyhalien / polyhalin / polyhalin / polihalino

2833
mixohaline, – pertaining to a salinity of marine waters between 0.5 and 30 $^0/_{00}$. / mixohalien / mixohalin / mixohalin / mixohalino

2834
euhaline, – pertaining to a salinity of marine waters between 30 and 40 $^0/_{00}$; normally used for the salinity of open ocean water. / euhalien / euhalin / – – / euhalino

mixoeuhaline, – pertaining to marine waters with salinities above 30 $^0/_{00}$ but less than the salinity of the adjacent euhaline sea. / mixoeuhalien / mixohalin / – – / mixoeuhalino

2835
hypersaline, – pertaining to a salinity of marine waters greater than 40 $^0/_{00}$. / hypersalien / hypersalin / – – / hipersalino

2836
littoral, –
1. pertaining to, or inhabiting or taking place on or near the shore;
2. the portion of the sea floor between the extremes of high and low tide (intertidal zone).

/ litoraal; tot de kust behorend / littoral / litoral; küsten- / litoral

intertidal zone, – / getijdenzone / zone intertidale / Intertidenbereich / intermareal; estero

2837
supralittoral, – pertaining to that part of the coastal realm that is situated above high tide level but that may be flooded by spring tides or lies within the splash zone of the waves. / supralitoraal / supralittoral / supralitoral / supralitoral

2838
eulittoral, – pertaining to the zone in a body of water that ex-

tends from high-tide level tot the limit of attached plants, usually at a depth between 40 and 60 m. / eulitoraal / eulittoral / — — / eulitoral

2839
sublittoral, — pertaining to the benthic zone in a body of water extending from low-tide level to a depth of approximately 200 m, or to the edge of the continental shelf. / sublitoraal / pré-littoral; sublittoral / sublitoral / sublitoral

infralittoral; inner sublittoral, — the shallow part of the sublittoral zone corresponding to that part of the bottom of a body of water, below the littoral zone, that is compatible with the existence of phanerogams or photophilous algae (euphotic zone); generally taken between low-tide level and 100 m water depth. / infralitoraal / infralittoral / infralitoral / infralitoral

circalittoral; outer sublittoral, — pertaining to the deeper part of the sublittoral zone, extending from the infralittoral to the edge of the continental shelf; generally taken between 100 and 200 m water depth. The term is badly chosen because this zone has few relations, and often not at all, with the littoral zone. / circalitoraal / circalittoral / circalitoral / circalitoral

2840
back shore; back beach, — the coastal zone between mean high-water level and storm tide. / — — / arrière-plage / — — / — —

fore shore; beach face, — the coastal zone between mean low-water level and the upper limit of wave wash at high tide. / — — / avant-plage; bas de plage / — — / — —

inshore; shore face, — the coastal zone between the breaker limit and mean low tide.

offshore, — the zone extending from the outer limit of the breaker zone to the depth at which waves cease to influence the seabed, effectively to the shelf edge. / buitengaats; uit de kust gelegen / au large de la côte; avant-côte; sous-marin / küstenfern / alta mar; mar abierta

2841
neritic, — the portion of the sea situated above the continental shelf; also both bottom and water between surface and bottom; commonly taken from low-tide level to approximately 200 m. / neritisch / néritique / neritisch / nerítico

epineritic; inner neritic, — that portion of the marine benthic environment extending from low tide to a depth of water of about 40 m. / epineritisch / épinéritique / epineritisch / epinerítico

infraneritic; outer neritic, — that portion of the marine benthic environment between 40 and 200 m depth. / infraneritisch / infranéritique / infraneritisch / infranerítico

2842
oceanic, —
1. pertaining to the sea, especially where the water is more than 200 m deep, i.e. off the continental shelf;
2. pertaining to the aquatic environment of the deep parts of the sea from the surface to close to the bottom.

/ oceanisch / océanique / ozeanisch / oceánico

2843
waterbloom; plankton bloom, — explosive multiplication of plankton, generally phytoplankton, e.g. **red tide**; it is often caused by upwelling of water. / waterbloei / eaux rouges / Wasserblüte / marea roja

2844
pelagic, —
1. pertaining to the whole of the marine environment, comprising the neritic zone and the deeper parts of the ocean;
2. the sediments of the deep ocean deposited far from land.

/ pelagisch / pélagique / pelagisch / pelágico

benthonic; benthic, — pertaining to organisms living on or in the sea floor. / benthonisch / benthonique / benthonisch / bentónico

2845
epipelagic, — pertaining to the waters of the ocean between surface and 200 m depth. / epipelagisch / épipélagique / epipelagisch / epipelágico

mesopelagic, — pertaining to the waters of the ocean between 200 and 1000 m depth. / mesopelagisch / mésopélagique / mesopelagisch / mesopelágico

bathypelagic, — pertaining to the waters of the ocean between 1000 and 4000 m depth. / bathypelagisch / bathypélagique / bathypelagisch / batipelágico

2846
bathyal, — pertaining to marine benthic environments and organisms on the continental slope and to the sediments formed there; classically between 200 and 2000 m, now commonly taken between 200 and 4000 m. / bathyaal / bathyal / bathyal / batial

2847
abyssal, —
1. pertaining to the bottom of the deep sea beyond the continental slope (classically beyond 2000 m, now commonly taken below 4000 m) and bounded at its deep side by the zone of oceanic trenches and deeps (below 5000 or 6500 m, depending the authority);
2. pertaining to those oceanic depths where no sunlight penetrates, below 800-1100 m.

/ abyssaal / abyssal / abyssal / abisal

abyssopelagic, — pertaining to that portion of the deep waters of the ocean that corresponds with the abyssal environment. / abyssopelagisch / abyssopélagique / abyssopelagisch / pelágico-abisal

2848
hadal, — pertaining to the sea-floor environment in deep-sea trenches at depths exceeding 6000 m. / hadaal / hadal / hadal / hadal

[1900] SEDIMENTOLOGY

[1910] GENERAL TERMS

2849
sedimentology, – the science of the sedimentary deposits and their formation. / sedimentologie / sédimentologie / Sedimentologie / sedimentología

2850
sediment; sedimentary deposit, – a body of solid material, unconsolidated or consolidated, formed at or near the surface of the earth, at low temperatures (below 200°C, and including diagnetic rocks). Low-temperature hydrothermal deposits (e.g. many kaolin bodies) are excluded, but pyroclastic deposits are normally included. Though it is sometimes proposed to limit the term to unconsolidated material, most authors do not make this distinction. /sediment; afzetting; afzettingsgesteente / sédiment / Sediment; Ablagerung; Schichtgestein / sedimento

sedimentary rock, – strictly: a consolidated sedimentary deposit; in practice the term is almost always applied to both consolidated and unconsolidated material. / sedimentgesteente; sedimentair gesteente; afzettingsgesteente / roche sédimentaire / Sedimentgestein / roca sedimentaria

2851
sedimentary petrology, – the petrology of sedimentary rocks. / sedimentpetrologie / pétrologie sédimentaire / Sedimentpetrologie / petrología sedimentaria

sedimentary petrography, – the petrography of sedimentary rocks. / sedimentpetrografie / pétrographie sédementaire / Sedimentpetrographie / petrografía sedimentaria

2852
sediment-petrological province, – a complex of sediments forming a unit as to source, age and regional distribution. / sediment-petrologische provincie / province pétrographique sédimentaire / sedimentpetrologische Provinz / provincia petrológica sedimentaria

2853
lithology, – the science of the rocks, especially of the sedimentary rocks. / lithologie / lithologie / Lithologie / litología

lithogenesis, – the origin and formation of rocks, especially sedimentary rocks, unconsolidated and consolidated. In Russian literature synonymous with diagenesis. / lithogenese / lithogenèse / Lithogenese / litogénesis

2854
syngenetic, – said of a primary sedimentary feature (e.g. ripple marks) formed contemporaneously with the deposition of the sediment. / syngenetisch /syngénétique / syngenetisch / singenético

syndepositional; synsedimentary, – occurring or formed during deposition. / synsedimentair / synsédimentaire / synsedimentär / sinsedimentario

2855
sedentary deposit, – a sediment or soil that is formed in place, without transportation, by the disintegration of the underlying rock or by the accumulation of organic material. / sedentaat / eluvions; sédiment résiduel; sol .... / sedentäre Ablagerung / material residual

2856
relic sediment, – a sediment having been formed under conditions differing from the present ones, but still lying at the surface, e.g. terrestrial or littoral deposits at the surface of the outer shelf. / relictsediment / relique sédimentaire / Reliktsediment / sedimento relicto

2857
subsidence, – local or regional sinking of the surface of the earth. / daling / subsidence / Senkung; Absenkung; Subsidenz / subsidencia

submergence, – the inundation of parts of the land area, either by subsidence or by a rise of water level. / verdrinking; onderdompeling; bedekking met water; onderlopen / inondation; submergence; ennoiement / Überflutung; Untertauchen / submersión

emergence, – the appearance of parts of the sea bottom above water, either by uplift of the sea floor or by fall of the water level. / opduiking; droogvalling; drooglopen / émersion; émergence; exhaussement / Auftauchen; Trockenfallen / emersión

2858
positive movement, – a relative rise of sea level with regard to the land. / positieve beweging / mouvement positif / positive Bewegung / movimiento positivo

negative movement, – a relative lowering of sea level with regard to the land. / negatieve beweging / mouvement négatif / negative Bewegung / movimiento nagativo

eustatic movement; eustatic change of level, – a world-wide change in sea level. / eustatische beweging / mouvement eustatique / eustatische Bewegung / movemiento eustático

[1920] ENVIRONMENTS, FACIES AND DEPOSITS

[1921] *General Terms*

2859
environment; sedimentary ...., – a spatially restricted complex at the surface of the earth characterized by certain physical, chemical and biological conditions. / milieu; omgeving / environnement / Milieu; Ablagerungsmilieu / ambiente; medio

environmental, – pertaining to (an) environment(s). / milieu- / d'environnement / Milieu- / ambiental

2860
facies; sedimentary ...., – the total of palaeontological and li-

thological properties of a sediment, as they reflect the conditions under which it was formed. / facies / faciès / Fazies; Sedimentsfazies / facies

2861
lithofacies, – the sum of the lithological characteristics exhibited by a sedimentary deposit in a particular place. / lithofacies / lithofaciès / Lithofazies / litofacies

biofacies, – the sum of the biological characteristics exhibited by a sedimentary deposit in a particular place. / biofacies / biofaciès / Biofazies / biofacies

tectofacies, – the sum of primary tectonic characteristics of a sedimentary deposit in a particular place. / tectofacies / tectofaciès / Tektofazies / tectofacies

2862
lithotope, – the area of a (uniform) sedimentary environment. / lithotoop / lithotope / Lithotop / litotopo

tectotope, – a stratum or succession of strata with features characteristic for accumulation in a common tectonic environment, e.g. flysch. / tektotoop / – – / Tektotop / tectotopo

2863
isopic, – relates to deposits that exhibit the same facies. / isopisch / isofaciès; isopique / isopisch; isotopisch / isópico

2864
palaeotemperature, – the mean air or water temperature at a given place and time in geological history. / palaeotemperatuur / paléotempérature / Paläotemperatur / paleotemperatura

palaeocurrent, – a current (generally of water) that existed in the geological past, and whose direction is inferred from the properties of the sediment formed at that time. / vroegere stroom / paléocourant / Paläoströmung / paleocorriente

[1922] *Subaerial Terrestrial Environments*

NOTE:
– see also sections
[1630] Wind Action
[1664] Glacial Transport and Deposits

2865
continental, – said of primary deposits and their environments of formation that are neither marine nor marginal-marine, such as littoral, lagoonal, deltaic or estuarine. / continentaal / continental / kontinental / continental

terrestrial, – said of deposits and their environments.
1. continental (i.e. including lakes, rivers, etc.);
2. subaerial and subglacial.
/ terrestrisch / terrestre / terrestrisch / terrestre

2866
subaerial, – formed, situated or occurring in contact with the open air. / subaërisch / subaérien / subaerisch / subaéreo

2867
talus, –
1. accumulation of rock fragments (usually coarse and angular) derived from and lying at the base of a cliff or a steep slope;
2. scree.
/ hellingpuin / talus; cailloutis / Geröllhalde; Hangschutt / depósito de talud

scree, – steeply sloping accumulations of coarse, little weathered rock fragments, deposited by falling and rolling under the influence of gravity. / hellingpuin / cailloutis / Geröllhalde / depósito de gravedad

2868
colluvium; colluvial deposit, – rock waste consisting for a considerable part of fine-grained weathering products, accumulated by soil creep, mud flows, rainwash, etc., at the base of a slope. / colluvium; colluviale afzetting / colluvions / Kolluvium; kolluviale Ablagerung / coluvión

2869
wash, – loose surface material deposited by running water, as on the lower slope of a mountain. / samenspoelsel / – – / Schwemm- / – –

2870
aeolian; eolian, – pertaining to, formed or deposited by the wind. / aeolisch / éolien / äolisch / eólico

niveo-aeolian; niveolian, – pertaining to windborne clastic material that was originally deposited together with snow. / niveo-aeolisch / niveo-éolien / – – / niveo-eólico

aeolianite; eolianite, – a consolidated sedimentary rock consisting of clastic material deposited by the wind. / aeolianiet / éolianite / Äolianit; Eolianit / eolianita

2871
wind-blow sand; blown sand; aeolian sand, – sand that is transported or that has been deposited by the wind. / door de wind aangevoerd zand; stuifzand; stuivend zand; aeolisch zand / sable éolien / äolischer Sand; Flugsand / arena eólica

cover sand, – an aeolian deposit of fine to very fine sand, formed under glacial climate conditions, and possibly in part the product of heavy snowstorms, / dekzand / sable de couverture / – – / – –

loess, – an unconsolidated aeolian deposit, predominantly composed of silt. / loess / loess / Löss / – –

2872
dune, –
1. a hill or ridge of sand piled up by the wind;
2. megaripple.
/ duin / dune / Düne / duna

sand dune, – a hill of sand, accumulated by the wind. / (zand)duin / dune de sable / Sanddüne / duna de arena

clay dune, – a small and local accumulation of dried clay flakes, formed by the wind. / kleiduin / dune d'argile / Düne aus Tonscherben / duna de arcilla

2873
glacial, – formed by or related to (a) glacier(s) or ice sheet(s). Also pertaining to an ice age or a region of glaciation. / glaciaal / glaciaire / glazial / glacial

subglacial, – situated, occurring or formed below a glacier or an ice sheet. / subglaciaal / subglaciaire / subglazial; subglaziär / subglacial

2874
till, – unsorted, normally unstratified deposit, laid down directly underneath a glacier or ice sheet. / keileem; keienleem / moraine de fond; argile à blocaux / Geschiebelehm; Grundmoräne / – –

tillite, – indurated till. / tilliet / tillite / Tillit / tilita

2875
glacial meal; glacier meal, – rock flour produced by glacier abrasion. / glaciaal gesteentemeel / roche finement triturée / glaziales Gesteinsmehl / suelo glacial

2876
subterranean, – situated, formed or occurring below the earth's surface. / ondergronds / sousterrain / unterirdisch / subterráneo

2877
spelean, – formed in or pertaining to a cave or caves. / grotten- / spéléo / Höhlen- / – –

speleothem, – any deposit that is formed in a cave by precipitation from a watery solution. / speleotheem / spéléothème / Speleothem / espeleotema

2878
guano, – a phosphate-rich deposit, formed by accumulation of excrements of birds or bats. / guano / guano / Guano / guano

2879
redbed; red bed, – mostly clastic sediments, deposited on land and/or along coasts, with a red colour due to finely divided hematite, and forming together a geological formation, e.g. the Old Red Sandstone. / – – / couche rouge / Rotschicht / capa roja

[1923] *Subaqueous Terrestrial Environments and Deposits*

NOTES:
- for terms concerning carbonaceous depositional environments see section [2422] Biochemical Phase; Environments of Deposition.
- see also sections
  [1646] Flow Characteristics
  [1664] Glacial Transport and Deposition.

2880
subaquous; aqueous, – situated, occurring or formed below water. / subaquatisch; .... gevormd / subaquatique; immergé / subaquatisch / subacuático

2881
high-energy environment, – an aqueous sedimentary environment not protected from and characterized by strong actions of waves and/or currents. / milieu van hoge energie / environnement à haute énergie / – – / medio de alta energía

low-energy environment, – an aqueous sedimentary environment not protected from and characterized by weak actions of waves and/or currents. / milieu van lage energie / environnement à basse énergie / – – / medio de baja energía

2882
density stratification, – stratification in a body of water with layers of downward increasing density. / dichtheidsgelaagdheid / stratification par densité / Dichteschichtung / – –

2883
fluvial; fluviatile, – pertaining to a river or to rivers, e.g. fluvial environment, fluvial deposit. / fluvatiel, rivier- / fluviatile; fluvial / fluviatil; Fluss- / fluvial

fluvioglacial; glaciofluvial, – pertaining to the meltwater streams, flowing from wasting glacier ice, and especially to the deposits and landforms produced by such streams. / fluvioglaciaal / fluvioglaciaire / fluvioglazial / fluvioglacial

2884
torrential, – pertaining to or formed by a torrent. / stortbeek- / torrentiel / Sturzbach- / torrencial

flash-flood deposit, – a deposit from a sudden flood of relatively great volume and short duration, carrying a very large load of sedimentary material. / stortvloedafzetting / coulée de boue / Schichtflutablagerung / depósito torrencial

2885
flood-plain deposit,
1. a fluvial deposit formed during flood(s) outside the river channel;
2. a fluvial deposit in a flood plain (including older river channels).

/ riviervlakte-afzetting / dépôt de plaine d'inondation; remblaiement fluviatile / Überflutungssediment / depósito de llanura de inundación

fanglomerate, – the lithified rudaceous deposit of an alluvial fan. / puinwaaier-conglomeraat / fanconglomérat; fanglomérat / Fanglomerat / fanglomerado

2886
alluvial deposit; alluvium, – detrital deposits of modern rivers, especially those whose top parts, after formation, lie emerged for prolonged periods. / alluvium; alluviale afzetting / alluvions / Alluvion; alluviale Ablagerung / depósito aluvial

alluvation, – the formation of alluvial deposits. / aanslibbing; alluviatie / alluvionnement / Ablagerung / aluvión

2887
paludal; palustral, – pertaining to or formed in marshes or bogs. / moeras- / palatre; de marais / palustrisch / palúdico; palustre

2888
lacustrine, – pertaining to, formed in or characteristic of a lake or lakes. / lacustrien; lacustrisch; meer- / lacustre / lakustrisch; See- / lacustre

limnic, – formed in, or pertaining to a body of fresh water, especially a lake or pond. / limnisch / limnique / limnisch; See- / límnico

2889
nearshore, – extending lakeward or seaward a short distance from the shoreline. / oever-nabij; kust-nabij / près du rivage / Küstennah / litoral

2890
limnetic, – relating to the pelagic zone of lakes, from the surface down to the lower limit of effective light penetration (photic zone). / limnetisch / limnique; limnétique / limnetisch / – –

2891
lake ball, – a more or less spherical mass of tangled fibers of living or dead vegetation, formed by wave action on the bottom of a lake, and often thrown up on the beach. / meerbal / boule d'algues; .... lacustre / Seeball; Meerball / – –

2892
playa; sebkha, – relatively large, flat area of clay, silt or sand, characteristically with saline incrustations, and occassionally covered with water. Playas are formed in regions with arid climates, both in inland basins, where the inundations are the result of rainstorms, and along the coast, where they are inundated mostly by sea water. / playa; sebkha / playa; sebkha / Salzsumpf; Salzpfanne; Playa; Sebkha / playa

coastal playa; .... sebkha, – / kust-playa, kust-sebkha / sebkha côtière / Küstenplaya; Küstensebkha / playa costera

2893
salt lake, – lake containing (hyper)saline water. / zoutmeer / lac salé / Salzsee / lago salado

2894
salina, – a Spanish term for
1. a playa or sebkha;
2. a salt lake or hypersaline lagoon;
3. a salt marsh;
4. salt works.

salar, – a term used in the SW USA and in Chile for a salt flat or salt-encrusted depression.

2895
delta, – a clastic deposit built forward by a river where it enters a body of standing water, i.e. lake or sea. Not to be used in a geomorphological sense. / delta / delta / Delta / delta

deltaic, – pertaining to or characterized by a delta. / deltaïsch; delta- / deltaïque / deltaisch; Delta- / deltáico

2896
topset bed, – one of the more or less horizontal layers deposited on top of the foreset beds. / toplaag / couche sommitale / 'Topset'Lage / – –

foreset bed, – one of the inclined layers deposited on the advancing frontal slope of a delta. / frontlaag / couche frontale / 'Foreset'Lage / – –

bottomset bed, – one of the horizontal or gently inclined layers, deposited in water in front of the advancing margin of a delta, and progressively buried by foreset beds. / bodemlaag / couche basale / 'Bottomset'Lage / – –

2897
bar finger, – a narrow, elongated, lenticular body of sand underlying, but several times wider than, a distributary channel in a bird-foot delta. It is produced and lengthened by the seaward advance of the lunate bar at the distributary mouth. / – – / bar digitée / – – / barras digitadas

2898
paralic, – pertaining to the marine border; deposits formed on the landward side of a coastal area with occasional invasions of the sea, e.g. coal-bearing formations with marine intercalations. / paralisch / paralique / paralisch / parálico

2899
shoestring sand(stone), – a large-scale, very long and relatively narrow body of sand(stone), e.g. a fossil coastal barrier or river deposit.

[1924] *Marine Environments*

NOTES:
– see also sections
  [1675] Deltas
  [1676] Tidal Flats and Marshes
  [1840] Marine Environments
– for terms concerning carbonate depositional environments see section
  [2130] Carbonate Sediment Bodies and Settings.

2900
marine, – pertaining to, formed in or by the sea. / marien; zee- / marin / marin / marino

holomarine, – pertaining to deposits laid down in a normal marine environment with absence of recognizable organic remains of land and/or fresh-water / origin. / holomarien / holomarin / holomarin / holomarino

submarine, – situated, occurring or formed at the bottom of the sea. / submarien / sous-marin / submarin / submarino

2901
normal marine, – environment, having water with a normal, open oceanic salinity range (33-40 ‰) as evidenced by its effects on organisms. / normaal marien / – – / – – / – –

restricted marine, – environment having water with a markedly differing salinity range from open oceanic water as evidenced by its effects on organisms, and the precipitation of certain minerals (e.g. gypsum). / beperkt marien / – – / – – / – –

2902
fluviomarine, – pertaining to deposits brought into the sea from the land and there rearranged by current systems, and thus being characterized by the co-occurrence of land, fresh water and marine animal and/or plant remains. / fluviomarien / fluvio-marin / fluviomarin / fluviomarino

terrigenous, – derived from land, or consisting of material that has been derived from land. / terrigeen / terrigène / terrigenisch / terígeno

2903
glacial marine; glaciomarine, – said of marine sediments that contain clastic material supplied by glaciers, icebergs or other drift ice. / glaciomarien / glacio-marin / glaziomarin / glaciomarino

marine till, – a deposit formed on the sea floor from floating glacier ice or shelf ice, or produced by melting of icebergs or floes. / mariene till / tillite / glaziomarine Ablagerung / 'till' marino

2904
estuarine, – pertaining to or characteristic of or formed in an estuary or estuaries. / estuarien / estuarien; estuaire / Ästuar- / de estuario

2905
lagoonal, – pertaining to, formed in or characteristic of a lagoon or lagoons. / lagunair / lagunaire / lagunär / de albufera

2906
littoral, –
1. pertaining to the coast;
2. situated, living, formed or occurring at the shore (either of sea or lake);
3. pertaining to the intertidal zone.

/ littoraal; kust-; oever- / littoral / litoral; Küsten- / litoral

2907
tidal deposit, – deposit formed under influence of the tides, both above and below tide level, e.g. tidal-flat sediments and deposits formed in tide-swept parts of the open sea. The term **'tidalite'** seems superfluous and etymologically unsound. / getijdeafzetting (s.l.) / dépôt tidal / Gezietenablagerung; Wattsediment / depósito de marea

intertidal deposit, – deposit formed between the levels of extreme high and low tides. / getijdeafzetting (s.s.) / dépôt intertidal / Intertidal-Sediment / depósito intermareal

2908
supratidal; supralittoral, – pertaining to the littoral zone above the level of the highest tides. Also used for the zone above mean high-tide level. / stormtij- / supratidal / supratidal / supramareal

subtidal, – pertaining to the littoral environment below lowest low-tide level. Since the sea bottom in this environment may be strongly influenced by tidal currents, the term is somewhat confusing. / – – / subtidal / subtidal / submareal

2909
salt marsh, – plant-covered area along a shore, usually situated for the greater part above mean high-tide level, and flooded regularly or from time to time by saline or brackish water. / schor; kwelder / marais salé; .... littoral / Salzmarsch / marisma

salt pan, – a small depression in the surface of a salt marsh, normally without (larger) vegetation, flooded from time to time by salt water, that through evaporation may cause hypersaline conditions and/or precipitation of a salt crust. / – – / pan salé / Salzpfanne; Salzsumpf / salitral

2910
sand flat, – a sandy tidal flat, normally barren of vegetation. / zandwad; zandplaat / replat sableuse; vey; vase sableuse / Sandwatt / llanura de arena

mud flat, – a muddy tidal flat. / slikwad; slikplaat / replat boueuse; slikke / Schlickwatt / llanura de barro

2911
wash-over deposit; wash-over, – material deposited by water flowing over a normally submerged elevation, specifically a more or less delta-like deposit formed on the edge of a lagoon by overflow of a coastal barrier.

2912
shingle, – coarse water-worn gravel, especially on beaches. / (grof) (stand)grind / galets de plage / Geröll / – –

beach concentrate, – an accumulation of heavy-mineral grains on a beach, mostly formed by wave action.

beach placer, – a beach concentrate, rich in ore-mineral grains. / strandplacer / minéralisation littorale; placer marin / Strandseife / placer de playa

2913
unda; undaform zone, – the part of the sea floor that is situated within the zone of wave action and in which, therefore, the bottom is repeatedly stirred and reworked.

undathem, – a rock unit formed in an unda.

2914
shallow marine, – the zone between low tide and/or wave base or the base of photic zone. / ondiep marien / – – / – – / – –

2915
shelf facies, – the facies characteristic for the continental shelf, i.e. to a depth of approximately 200 m. / shelf-facies / faciès de plateforme / Schelffazies / facies de plataforma

epicontinental, – pertaining to shelf seas. / epicontinentaal / épicontinental / epikontinental / epicontinental

neritic, – orginally used (Haeckel) for the aquatic environment overlying the shallow parts of the sea floor; now often considered to pertain to both, bottom and waters of the continental shelf. / nerietisch / néritique / neritisch / nerítico

2916
clinoform, – the subaqueous land form that extends from the wave base to the more or less level, deeper parts of the shelf.

clinothem, – a deposit formed in the clinoform.

2917
fondoform, – all the floor in a body of water, usually oceanic below a clinoform.

fondothem, – a deposit formed in a fondoform.

2918
oceanic; off-shelf, –
1. pertaining to the aquatic environment of the deep parts of the ocean, from the water surface down to close to the bottom;
2. pertaining to the ocean, especially the parts of it beyond the continental shelf.

/ oceanisch / océanique / ozeanisch / oceánico

deep marine, – the zone below the wave base or the photic zone; based on the presence or absence of carbonate dissolution effects (with depth) it might be possible to further subdivide this environment into a deeper and a deepest zone. / diep marien / – – / – – / – –

2919
pelagic, –
1. pertaining to the whole of the marine environment, inhabited by plankton and nekton and to the communities or organisms that live in direct independence from bottom or shore. The environment comprises both the neritic zone and the deeper parts of the ocean;
2. pertaining to sediments of the deep ocean, formed far from land, and consisting mostly of remains of pelagic organisms and / or very fine lutaceous material (predominantly clay) of terrigenous, volcanic or other origin.

/ pelagisch / pélagique / pelagisch / pelágico

2920
hemipelagic mud, – terrigenous mud deposited in the sea at depths greater than about 200 m, such as 'blue mud' and 'green mud'. The term was given by Krümmel who wanted to convey that the material, though deposited in deep 'pelagic' environment, was terrigenous like many 'non-pelagic' shelf sediments. Since it has appeared that much of the 'pelagic' red clay is also of terrigenous origin, the term has lost its significance. / hemipelagisch slik / boue hémipélagique / hemipelagischer Schlick / fango hemipelágico

ooze; pelagic ooze, – fine-grained pelagic deposit which contains at least 30% of material of organic origin. / slib / vase / Schlamm / barro

2921
blue mud, – black or dark grey, locally somewhat bluish terrigenous mud, formed in the sea at depths greater than about 200 m. Its dark colour (below the oxidized surface layer) is mainly due to iron sulphides. / blauw slik / boue bleue / Blauschlick / fango azul

green mud, – greenish terrigenous mud, formed in the sea at depths greater that about 200 m. Its green colour may be due to glauconite or green-coloured clay minerals. / groen slik / boue verte / Grünschlick / fango verde

2922
deep-sea deposit, – comprehensive term for bathyal and abyssal sediments, i.e. deposited beyond the shelf edge; formerly also used as a synonym for pelagic deposits. / diepzee-afzetting / dépôt de mer profonde / Tiefsee-Ablagerung / depósito de mar profundo

2923
brown clay, – an argillaceous deep-sea deposit, containing less than 30% $CaCO_3$ and less than 30% siliceous organic remains, mostly formed at depths greater than 4400 m. / bruine diepzeeklei / boue brune / brauner Tiefseeton / arcilla oscura abisal

red clay, – an argillaceous deep-sea deposit, containing less than 30% $CaCO_3$ and less than 30% siliceous organic remains, mostly formed at depths greater than 4400 m. / rode diepzeeklei / boue rouge / roter Tiefseeton / arcilla roja abisal

2924
hadal, – pertaining to the sea floor environment in deep-sea trenches at depths exceeding 6000 m. / hadaal / hadal / hadal / hadal

2925
euxinic, – pertaining to an environment of restricted water circulation and anaerobic conditions, as in the Black Sea (Pontus Euxinus). Also pertaining to the black or dark-coloured bottom sediments formed in such an environment. / euxinisch / euxinique / euxinisch / euxínico

2926
flysch, – a sedimentary facies of mountain belts. The deposits were formed during the early stages of orogenesis, and are characterized by a thick series of thinly bedded, mainly marly and argillaceous beds, usually with inttercalations of arenaceous turbidites. / flysch / flysch / Flysch / – –

wildflysch, – a variety of flysch containing numerous exotic elements, emplaced by slumping (as in olistostromes), by tectonism (in tectonic melanges) or both. / wildflysch / 'wildflysch' / Wildflysch / – –

molasse, a sedimentary facies forming thick series, resulting from the denudation and erosion of young, mostly still actively rising mountain ranges, and deposited in shallow marine, lacustrine and terrestrial environments in subsiding areas directly in front of the mountain ranges. / molasse / molasse / Molasse / molasa

2927
melange, – a mappable sedimentary body, consisting of fragments, blocks and slabs of all sizes (from less than 1 cm to

more than 1 km) of various composition and origin, resting in a generally pelitic, often sheared matrix. / mélange / mélange / Mélange / – –

[1930] SEDIMENTARY PROCESSES

[1931] *Provenance of Sedimentary Material*

NOTE:
– see also
[1700] Pedology
[2422] Biochemical Phase; Environments of Deposition

2928
weathering, – the physical disintegration and chemical and biogenic decomposition of rock that produce an in situ mantle of waste and prepare sediments for transportation. Most weathering occurs at the surface, but may reach considerable depths. / verwering / altération / Verwitterung; Zerlegung / meteorización; (eflorescencia)

chemical weathering, – the processes of weathering by which chemical reactions transform rocks and minerals into new, stable chemical combinations. / chemische verwering / altération chimique / chemische Verwitterung; .... Zerlegung / meteorización químiqua

mechanical weathering, – the processes of weathering by which physical forces break down or reduce a rock to smaller fragments, not involving chemical change. / mechanische verwering / altération mécanique / mechanische Verwitterung; .... Zerlegung / meteorización mecánica

2929
decomposition, – chemical processes whereby complex compounds are broken down into simpler ones. / ontleding / décomposition / Zersetzung; Auflösung / descomposición

disintegration, – the breaking down by physical processes of solid material into small particles. / desintegratie / désagrégation; désintégration / Desintegration; Zerstörung / desintegración

2930
eluvium, – an accumulation of rock debris produced in place by the decomposition or disintegration of rock. / eluvium / éluvions / Eluvium; Eluvialboden / eluvión

residual deposit, – the weathering residue left in situ after leaching and eluviation have removed the more mobile material. / verweringsresidu / formation résiduel; dépôt .... / Residuallagerstätte / depósito residual

insoluble residue, – / onoplosbaar residu; oplossingsresidu / résidu non-soluble; .... insoluble / unlöslicher Rest; Rückstand / residuo insoluble

2931
clay with flints, – a residual deposit formed on chalk or other carbonate rocks containing flints. / vuursteen-eluvium / argile à silex / – – / arcilla con pedernal

2932
aerobic decay; oxic ...., – decomposition of organic matter primarily by micro-organisms in the presence of free oxygen. / aërobe ontbinding; .... ontleding / décomposition aérobié / aerobe Zersetzung / descomposición aerobia

anaerobic decay; anoxic ...., – decomposition of organic matter primarily by micro-organisms in the absence of free oxygen. / anaërobe ontbinding; .... ontleding / décomposition anaérobié / anaerobe Zersetzung / descomposición anaerobia

2933
brecciation, – the breaking up of rocks, leading to the formation of a breccia. /brecciatie / bréchification / Brecciierung / brechificacion

brecciated structure, – structure of rocks, consisting for an essential part of angular rock fragments. / breccie-structuur / structure brêchique / brecciöses Gefüge / estructura brechada

2934
granulation, – the formation of grains or granules; also the state of being granulated. / verkorreling; korreling; gekorreldheid / désagrégation en grains / Granulation; Granulierung / granulación

2935
splitting, –
1. the tendency of a layered rock to separate along a surface of parting;
2. breaking of a rock fragment along bedding or cleavage planes resulting in the production of two or three subequal parts.

/ splijting / desquamation; fente / Spaltung / partición

2936
erosion, – the mechanical wear of solid rock material by the impact of detrital fragments and particles that move against or along it by flowing water, ice or air and the removal of the detached or originally loose particles by the same agencies. / erosie / érosion / Erosion; Abtragung / erosión

scour(ing), – / uitschuring; afschuring / décapage; érosion / Einschneiden / erosión

erodibility, – the capability of being eroded. / erodeerbaarheid / capacité à l'érosion; érodibilité / Erodierbarkeit / capacidad de erosión

2937
winnowing, – the selective sorting or removal of fine particles by wind action, leaving the coarser grains behind. The term is often applied to removal by or sorting in water. / ziften / vannage / Ausblasen durch den Wind; Windsichtung / – –

washing, – a term more appropriate than winnowing for removal by or sorting in water.

2938
corrasion, – mechanical wear of rocks by particles (rock fragments) transported by moving water, ice or air. / corrasie / corrasion / Korrasion / corrasión

abrasion, – mechanical wear of rocks or particles; sometimes used for surf action only. / afslijping; abrasie / abrasion / Abrasion; Abrieb / abrasión

2939
sand blasting, – mechanical wear by a powerful, sand-laden current (normally of air). / zandstraling / – – / Sandstrahlen / – –

wet blasting, – mechanical wear by particles moved by a powerful water current.

2940
grinding, – the process whereby rock material is worn down, whetted, and/or crushed by or against other rock material. / vermaling; slijping / attrition / Mahlen; Schleifen / amolar

2941
chipping, – the abrasion process whereby chips are split off from rock, pebbles or sand grains. / – – / débit d'esquilles / Absplittern / – –

2942
corrosion, –
1. erosion of rocks by solution;
2. partial resorption (solution) of crystals or mineral grains by diagenetic processes.

/ corrosie / corrosion / Korrosion / corrosión

2943
source area, – the area from which the clastic particles of a given sediment are derived. / herkomstgebied; oorsprongsgebied / région d'origine / Liefergebiet; Herkunftsgebiet / área fuente

provenance; area of ...., – / herkomst(gebied) / provenance; région de .... / Herkunft / origen

2944
source rock; mother rock; parent rock, – the rock from which the clastic particles of a given sediment are derived. / oorsprongsgesteente; moedergesteente / roche mère / Muttergestein / roca madre

2945
autochthonous, – said of sedimentary rocks chiefly composed of material, e.g. **interclasts**, that was originally formed at the place where it is now found. / autochtoon / autochtone / autochthon / autóctono

allochthonous, – said of sediments chiefly composed of material, e.g. **extraclasts**, that was originally formed or deposited elsewhere. / allochtoon / allochtone / allochthon / alóctono

2946
allogenic; allothigenic; allothigenetic; allothigenous, – said of constituents of (sedimentary) rocks that were originally formed or deposited elsewhere. / allogeen; allothigeen / allogène; authigène / allothigen / alógeno

authigenic, – said of constituents of (sedimentary) rocks that were formed at the place of deposition. / authigeen / autigène / authigen / autígeno

2947
derived; reworked, – said of constituents that have been eroded from an older formation and incorporated in a younger one, e.g. mineral grains, fossils. / verspoeld; geremaniëerd / dérivé; remanié / aufgearbeitet / derivado; retrabajado; reelaborado

2948
dispersal fan; dispersal shadow, – the area downstream of a parent rock or formation where loose fragments or grains (e.g. glacial erratics, specific mineral grains), derived from it, are found. / strooiwaaier / aire de dispersion / Schwemmfächer; Schwemmkegel / abanico de dispersión

2949
exotic constituent; .... boulder; .... block, – fragments of rock varying in size from less than 1 cm to hundreds of metres, unrelated to the surrounding rock mass, especially those that are emplaced by sliding, e.g. in Wildflysch and Argille Scagliose. / exotisch bestanddeel / constituant allogène; .... allochtone / exotischer Bestandteil / constituyente exótico

erratic; .... block, – a relatively large, isolated rock fragment found in or on a formation of different composition. / erraticum; zwerfsteen / erratique / erratischer Block / errático

glacial erratic, – a rock fragment, found in formerly glaciated country, that has been transported over great distances by glacier ice or polar ice cap. / erraticum; zwerfsteen / erratique / erratischer Block / canto errático

[1932] *Transportation*

2950
load, – the total material that is moved by a natural transporting agent, such as a stream, a glacier, wind. / last / charge / Fracht / carga

2951
stream load, – the total material other than water that is transported by a stream. It consists of bed load, suspended load and dissolved load. / stroomlast / charge solide; débit .... / Flussfracht / carga de corriente

sediment load, – the total of bed load and suspended load, transported by wind, water or ice. / sedimentlast / charge solide / Sedimentfracht / carga de sedimento

2952
bed load; bottom ....; traction ...., – the sedimentary material transported by water, on or close to the bottom by rolling and sliding and by saltation. / beddingmateriaal; bodemlast / lit de fond; charge de .... / bodennahe Fracht; Bodenfracht / carga de tracción

2953
traction, – the process of transportation of bed load. / tractie; bodemtractie / traction / Fracht / tracción

2954
saltation, – 'jumping' transport of bottom particles whereby

they are intermittently thrown upward into flowing air or water and move downcurrent a short distance until they land again on the bottom. / saltatie / saltation / Saltation / saltación

saltation load, – the part of the bed load that moves by saltation. / saltatielast / charge / Saltationsfracht / carga de saltación

2955
suspended load; suspension ...., – the sedimentary material that during transport by a stream is kept in suspension continuously or most of the time. / suspensiemateriaal; suspensielast / charge en suspension / Suspensionsfracht; Schwebgut / carga en suspensión

2956
dissolved load; solution load, – the dissolved material transported by a stream. / opgelost materiaal / charge dissoute / Lösungsfracht / carga en solución

2957
sediment charge, – the ratio of the mass or volume of load to the mass or volume of water in a stream passing a given cross section per unit time. / sedimentlast; sediment-waterverhouding / débit solide / – – / carga sedimentaria

sediment discharge; .... transport rate, – the amount of sedimentary material moved by a stream through a given cross section per unit time. / sedimentafvoersnelheid / taux de transport / Rate des Sedimenttransports / descarga sedimentaria

2958
capacity, – the total mass of detrital load that is transported by a stream per unit time through a given cross section. / capaciteit / capacité / Kapazität / capacidad

competence, – the ability of a stream to transport detrital material as measured by the maximum particle size that can just be moved. / competentie / compétence / Kompetenz / competencia

2959
threshold velocity, – the minimum velocity of wind or water required to start movement of bottom material. / drempelwaarde van de stroomsnelheid / vitesse d'entrainement / – – / limen de velocidad

2960
flow regime, – one of several states of flow, essentially characterized by bed form. / stroomregime / régime d'écoulement / Stromfeld; Strömungsregime / régimen de flujo

NOTE:
– see also section
[1646] Flow Characteristics

2961
drift, – a general term, used especially in Great Britain for all surficial, unconsolidated rock debris transported from one place and deposited in another, and used in distinction to the solid bedrock. The term is often used specifically for glacial deposits. / – – / dérive / – – / deriva

sand drift, –
1. eolian transport of sand. / eolisch zandtransport / transport éolien / äolischer Sandtransport / – –
2. accumulation of drifted sand. / stuifzand / accumulation éolienne / Flugsand / – –

2962
littoral drift; .... drifting, – the displacement of loose material along a coast. / – – / dérive littorale / Küstenlängstransport; Küstendrift / deriva litoral

beach drift; .... drifting, – the movement of loose material on a beach under the influence of waves and currents, displacing it residually and parallel to the shore. / – – / dérive côtière / Stranddrift / deriva de playa

drift line, – a line of drifted material washed ashore and left stranded. / lijn van aanspoelsel / – – / Spülsaum / línea de desplazamiento

2963
upstream side; stoss ...., – the side of an elevation facing into a current of water or air, or glacier movement. / loefzijde / face amont / Luvseite / corriente arribo

windward (side), – / loefzijde / face au vent / Luvseite / barlovento

2964
downstream side, – the side of an elevation facing in het direction of a current. / stroomafwaartse zijde / face aval / stromabgewandte Seite; Leeseite / corriente abajo

leeward (side), – / lij; lijzijde / (face) sous le vent / Leeseite; leewärts / sotavento

2965
lee slope, – the slope, e.g. of a ripple or dune, on the leeward side. / lijhelling / pente sous le vent / Leehang / pendiente de sotavento

2966
slip face, – the lee slope of a dune or ripple, standing at or near the angle of repose of loose sand, and advancing downstream by a succession of slides whenever that angle is exceeded. / glijhelling / talus croulant / – – / talud dunar

2967
lag concentrate, – residual accumulation of clastic or other fragmentary material formed in moving water or air due to coarseness, high density or roundness of particles. / residuaire concentratie / concentration résiduaire par déflation / Residualkonzentrat / concentrado residual

lag gravel, – a lag concentrate consisting of gravel. / residuair grind / résidu de déflation / Residualkies; Lesedecke / gravera

2968
rafting, – the transportation over sea of land-derived material, bottom sediment and organisms by floating ice, wood, algae etc. / – – / – – / Verdriften / acarreo

2969
ice rafting, – the transportation of rock and other material on or within icebergs or ice floes. / – – / – – / Eisdrift / acarreo

ice-rafted, – said of material, such as boulders or sand, deposited by the melting of floating ice, containing this material. / – – / tillites / vom Eis verdriftet / acarreado

2970
drift-wood; driftwood, – wood floating on or cast ashore by water. / drijfhout / bois flottant / Treibholz / bosque a la deriva

2971
ponding, – the impoundment of stream water to form pond or lake; also used for the impounding of turbidity currents by pre-existing subaqueous obstructions. / opstuwing / – – / – – / estaucamiento

[1933] *Mass Movement*

NOTE:
– see also section
[1614] Mass Movement

2972
avalanching, – the sudden sliding down of loose material (e.g. sand) along a subaerial or subaqueous slope, such as the lee side of a large wind dune, or of a current ripple. / afglijding / avalanche / Abgleiten; Abrutschen / avalancha

2973
sand avalanche, – short-lasting, sliding movement of smaller or larger masses of sand down a slope. / zandafglijding / avalanche de sable / Agbleiten von Sand / avalancha de arena

sand flow, – a more or less sliding, relatively long lasting mass movement of sand on sloping parts of the sea floor or of lake bottoms, especially in submarine canyons. / zandstroom / coulée de sable / Sandstrom / flujo de arena

sand fall, – a more or less free fall of sand 'clouds', originating where a sand flow passes over a precipice. / zandval / chute de sable / Sandfall / pared de arena

2974
mud flow, – mass movement of a muddy deposit. / modderstroom / coulée de boue; .... boueuse / Schlammstrom / flujo de barro; corriente de fango

2975
slump(ing), – the relatively rapid downward movement of a mass of unconsolidated and incoherent material along one or more, usually curved slip planes. The term applies especially to subaerial conditions. / afschuiving / slumping / Abrutschen / – –

2976
slide; sliding, – movement of a coherent mass of (sedimentary) material along a slope, due to gravity, having a 'visible' velocity. / afglijding; afglijden / glissement / abgleiten; Rutschung / deslizamiento

2977
olistostrome, – large mass of intimately mixed heterogeneous material, emplaced by subaqueous slumping, e.g. in Wildflysch and Argille Scagliose. / olistostroom / olistostrome / Olistostrom / olistostroma

olistolith, – exotic block or rock mass. / olistoliet / olistolite / Olistholith / olistolito

2978
tilloid, – pebbly mud or mudstone, resembling a till or tillite, possibly originated through gravity movements, and of non-glacial origin. / tilloïed / tilloïde / Tilloid / tilloide

2979
density current, – a gravity-induced flow due to differences of density in a water body. / dichtheidsstroming / courant de densité; .... dense / Dichtestrom / corriente de densidad

2980
suspension current, – a current in water, other fluids or air moving in consequence of its relatively high density due to great quantities of suspended particles. / suspensiestroom / – – / Suspensionsstrom / corriente de suspensión

2981
turbidity current, – a current in water, moving in consequence of its relatively high density due to great quantities of suspended particles. / troebelingsstroom / courant de turbidité / Trübestrom; Turbiditstrom / corriente de turbidez

steady turbidity current, – a persistent turbidity current such as one produced where a stream heavily laden with sediment flows into a body of deep standing water. / duurzame troebelingsstroom / courant continu de turbidité / stetiger Trübestrom / corriente de turbidez constante

suspended current, – a turbidity current that is not in contact with the bottom, but flows over an underlying layer of denser, standing water. / – – / courant turbide stratifié / – – / corriente suspendida

2982
nose of a turbidity current, – the forward part of a turbidity current. / vooreinde van een troebelingsstroom / partie frontale d'un courant de trubidité / Anfang eines Trübestroms / frente de corriente de turbidez

tail of a turbidity current, – the rear part of a turbidity current. / achtereinde van een troebelingsstroom / queue d'un courant de turbidité / Ende eines Trübestroms / extremo de una corriente de turbidez

2983
turbidite, – a stratum deposited by a turbidity current. / turbidiet / turbidite / Turbidit / turbidita

2984
fluxoturbidite, – a deposit transitional between a slump deposit and a turbidite, characterized by poor development of grading and of sole marks. / fluxoturbidiet / fluxoturbidite / Fluxoturbidit / fluxoturbidita

[1934] *Deposition*

2985
deposition, – any process yielding a deposit. / afzetting / processus de dépôt / ablagern; Ablagerung / deposición

deposit, – material laid down by settling from a fluid or gaseous medium, by chemical precipitation or by organisms. / afzetting / dépôt / Ablagerung / depósito

2986
depocentre, – an area or site of maximum deposition. The thickest part of any specified stratigraphic unit in a depositional basin. / depocentrum / dépocentre / Depozentrum / depocentro

2987
accumulation, – deposition of rock material by which the surface is appreciably raised, either locally or regionally. Also: the deposit resulting from such a process. / op(een)hoping; accumulatie / accumulation / Akkumulation / acumulación

2988
accretion, – a process by which a body of rock material increases in size by the external addition of fresh particles. Said of minor bodies (e.g. stromatolites) as well as of large ones (e.g. beaches). / aangroei; aanwas / accrétion / Anwachs; Zuwachs / acreción

2989
sedimentation, – any process (sedimentary or magmatic) whereby particles of solid material settle from a fluid medium onto a non-fluid substrate. / sedimentatie / sédimentation / Sedimentation / sedimentación

sedimentation rate; rate of sedimentation, – the amount of sediment accumulated at a given place per unit time. / sedimentatiesnelheid / taux de sédimentation / Sedimentationsrate / grado de sedimentación; coeficiente de ....

2990
sanding up, – filling with sand. / opvulling met zand; verzanding / comblement sableux; remplissage avec du sable; ensablement / Versanden / – –

silting up, – the partial or complete filling with silt or other fine-grained material. / opslibbing; dichtslibbing / comblement silteux / Verschlicken / – –

2991
settling, – the sinking of solid particles through a fluid. / bezinking / décantation / Sink-; absinken / asentamiento

settling velocity, – / bezinkingssnelheid / vitesse de décantation / Sinkgeschwindigkeit / velocidad de asentamiento

2992
flocculation, – the aggregation of colloidal or other very small particles, e.g. minute clay flakes, by electrolytes. / flocculatie; uitvlokking / floculation / ausflocken / floculación

floccule, – small aggregate formed by flocculation of material suspended or dissolved in water. / vlok / floculat / Flocke / floculo

2993
redeposition; re-deposition, – reworking of clastic material into a new deposit, such as a turbidite, or reprecipitation of mineral matter after temporary solution. / hernieuwde afzetting / redéposition / erneute Ablagerung / redeposición

resedimentation; re-sedimentation, – sedimentation of reworked sedimentary material, e.g. as a turbidite. / hersedimentatie / resédimentation / Resedimentation / resedimentación

secondary deposit, –
1. deposit formed by redeposition;
2. intergranular deposit formed by diagenesis, e.g. cement in pores.

/ secundaire afzetting / dépôt secondaire / sekundäre Ablagerung / depósito secundario

[1940] SEDIMENTARY STRUCTURE

[1941] *General Terms*

2994
sedimentary structure, – shape and relations of the component parts of a rock formed by sedimentary (depositional, deformational or diagenetic) processes. / sedimentaire structuur / structure sédimentaire / Sedimentgefüge; ....textur / estructura sedimentaria

2995
primary sedimentary structure, –
1. structure formed during the deposition of the sediment, especially those produced by mechanical processes, such as ripple structures and bedding;
2. sedimentary structure formed before diagenesis, e.g. those caused by deposition, bioturbation or intraformational corrugation.

/ primaire sedimentaire structuur / structure sédimentaire primaire / primäre(s) Sedimentgefüge; .... textur / estructura sedimentaria primaria

secondary sedimentary structure, –
1. structure formed after deposition, e.g. by bioturbation;
2. structure formed after burial of the sediment, e.g. by diagenesis.

/ secundaire sedimentaire structuur / structure sédimentaire secondaire / sekundäre(s) Sedimentgefüge; ....textur / estructura sedimentaria secundaria

2996
geopetal structure, – any structure that indicates either top or bottom of a layer at the time of deposition; more in particular a structure where an originally horizontal surface is preserved, as e.g. in the mud, filling the lower part of a hollow fossil. / geopetale structuur / structure centripète / geopetales Gefüge / estructura geopetal

2997
proximal, – pertaining to the part of a deposit that lies closest to the source area or to the point where deposition began. / proximaal / proximal / proximal / proximal

distal, – pertaining to the part of a deposit that lies farthest from the source area or from the point where deposition began. / distaal / distal / distal / distal

2998
directional structure, – any structure that indicates the direction of the current that produced it, e.g. cross bedding, current marks and ripple marks. / – – / structure directionnelle / gerichtete Textur / estructura directional

2999
striation, –
1. surface condition, characterized by fine, narrow, curved or straight, parallel grooves;
2. one such groove.

/ striatie / striation / Striemung; Riefung / estriación

lineation, – linear structure. / lineatie / linéation / Lineation / lineación

[1942] *Ripple Marks*

3000
ripple; ....mark, – small-scale rhythmic relief patterns, formed on the surfaca of un consolidated sediment under the influence of water or air moving over it or of wave motion. / ribbel / ride / Rippel / rizadura; 'ripple'

ripple marked, – said of a surface showing ripple marks. / geribbeld / ridé / gerippelt / – –

3001
ripple crest, – the ridge-shaped, elevated part of a ripple. / ribbelkam / crête de ride / Rippelkamm / – –

ripple trough, – the low part of a ripple, situated between successive ripple crests. / ribbeltrog / creux de ride; intercrête / Rippeltal / – –

3002
ripple height; .... amplitude, – the elevation of a ripple crest above the nearest trough(s). / ribbelhoogte / hauteur de ride / Rippelhöhe / amplitud de la rizadura; ....del 'ripple'

ripple length; .... wavelength; chord, – the horizontal distance between successive crests or troughs. / ribbellengte; ....golflengte / longueur d'onde de ride / Wellenlänge / longitud de la rizadura; .... del 'ripple'

ripple index, – the ripple length divided by the ripple height. / ribbelindex / indice d'ondulation / Rippelindex / índice de la rizadura; .... del 'ripple'

3003
transverse ripple mark, – ripple mark having its crest transverse to the generating current(s) or wave motion. / transversale ribbel / ride transversale / transversale Rippel / rizadura transversal; 'ripple mark' ....

3004
current ripple mark, – asymmetric ripple mark, normally with wave lengths less than 60 cm, formed by relatively low-velocity water currents, the ripple crest being transverse to the current direction. / stroomribbel / ride de courant / Strömungsrippel / rizadura de corriente; 'ripple' de ....

3005
lunate ripple mark; crescentic ...., – asymmetric current ripple, the crest of which consists of regular series of crescents with cusps pointing downstream. / sikkelribbel / ride en croissant / sichelförmige Rippel / rizadura lunada; 'ripple' lunado

3006
linguoid ripple mark, – more or less tongue-shaped ripple mark, formed by low-velocity currents in very shallow water. / linguoïdale ribbel; tongribbel / ride linguoïde / Zungenrippel / rizadura linguoida; 'ripple' linguoido

3007
megaripple; 'dune'; giant ripple, – normally asymmetric ripple of large dimensions, with wave lengths between some 60 cm and many metres. It is formed in sand by relatively rapid, but still tranquil flow (Froude number less than 1). / megaribbel / Mégaride; dune / Megarippel / rizadura gigante; 'ripple' ....

sand wave, – term mostly used for regular series of large-scale submarine sand accumulations, having their crests transverse to the currents that formed them. The distance between the crests is usually several tens to several hundreds of metres. / zandgolf / vague de sable / Sandwelle; Riesenrippel / ola de arena

3008
antidune, – ripples of relatively low elevation, more or less sinusoidal in cross section, formed in water flowing with approximately (or more than) the second critical velocity (Froude number greater than 1). The ripples are broadly in phase with waves at the water surface. They are either stationary or migrate slowly upstream or downstream. Fossil antidunes are especially common in turbidites, where they were probably formed in phase with waves at the interface between turbid and clear water. / antiduin / antidune /Gegenrippel / antiduna

3009
chute and pool structure, – structure in sands and sandstones, etc., developed by water flowing with regularly spaced hydraulic jumps. / hydraulische sprongribbel / – – / – – / – –

3010
rhomboid ripple mark, – very low, asymmetric ripples, arranged in patterns of diamonds, and formed in very thin

sheets of rapidly flowing water, the long axes of the diamonds being parallel to the current direction. / ruitribbel; rhomboïdale ribbel / ride losangique / Rhomboidrippel / rizadura romboide; 'ripple mark' ....

3011
wave-current ripple, – ripple mark produced by the combined action of waves and currents. / stroom-golfribbel / ride de vagues et courants / – – / – –

longitudinal ripple mark, – ripple mark with long, straight crest, formed parallel to the direction of a water current. Most of the well-developed ripples of this kind are generated by the combined action of currents and waves. / longitudinale ribbel / ride longitudinale / longitudinale Rippel / rizadura longitudinal; 'ripple mark'....

3012
wave ripple mark, – ripple mark produced by waves. / golfribbel / ride de vagues / Seegangsrippel; Wellenrippel / – –

oscillation ripple mark, – ripple mark formed by oscillating water motion, mostly by normal waves, sometimes, on flat parts of beaches, by regular alternation of swash and backwash. / oscillatieribbel / ride d'oscillation / Oszillationsrippel / rizadura de oscillación; 'ripple mark' de ....

3013
interference ripple mark, – ripple mark formed by interfering waves. Similar patterns are produced as transition stages of wave ripples that are re-oriented by change of wave direction. / interferentieribbel / ride d'interférence / Interferenz-Rippel; komplexe Rippel / rizadura de interferencia; 'ripple mark' de ....

rectangular cross-ripple mark, – interference ripples forming a rectangular pattern.

3014
metaripple, –
1. a term used by Bucher for megaripples. Bucher wrongly supposed that the large megaripples were formed by transformation of antidunes owing to decrease of flow velocity;
2. a term used by Van Straaten for ripples the shape of which has essentially changed as a result of changing water movement produced by decreasing amplitude of wave motion.

/ metaribbel / métaride / Metarippel / – –

3015
flat-topped ripple mark, – ripple mark with flat, wide crest between narrow troughs, mostly formed by planing off during a stage of lowered water level. / ribbel met afgeplatte kam / ride aplanie; ride aplatie / Rippel mit gekappter Kamme / – –

starved ripple; incomplete ...., – ripple mark consisting of an isolated ripple crest of sand, resting on a smooth surface of mud, clay or hard rock. / – – / ride incomplète / unvollständige Rippel / rizadura incompleta; 'ripple' incompleto

3016
wind ripple, – small-scale ripple mark in sand or snow, produced by the wind, and characterized by plane stoss sides and a notable asymmetry. / windribbel / ride éolienne / Windrippel / – –

wind granule ripple, – relatively large ripple (wavelength often more than 25 cm) with asymmetrical or almost symmetrical cross section, produced by wind in coarse sand and granules.

adhesion ripple; wind anti-ripple(t), – small aeolian ripple mark, formed by adhesion of sand grains to moist sand surfaces. It grows in up-wind direction. Similar ripples are found in snow. / adhaesieribbel; kleef .... / antiripplet / Adhäsionsrippel; Haft ..... / rizadura de adhesión; 'ripple' de ....

[1943] *Current Marks*

3017
plane bed, – condition of flatness, specifically of sand bottoms planed off by water flowing with velocities at or near the second critical value (Froude number = 1). / vlakke bodem / lit plan / Horizontalschichtung; Parallelschichtung / – –

3018
mark; marking, – any relief feature on either top or bottom surface of a layer. / merk; teken; sculptuur / marque; figure / Marke / marca

3019
sole mark, – relief feature on the bottom surface of a layer. Used by Kuenen for features in the sole of turbidite beds. According to Dzulynski and Sanders, who consider it preferable to restrict the term mark(ing) to original features, most of the sole features of turbidites, that are due to current action on the underlying bed, should rather be called sole casts. Exceptions are e.g. relief features due to load structures. / zoolmerk; basisvlak-sculptuur / figure basale; structure .... / Sohlmarke / – –

sole cast, – a relief feature in the sole of a bed, representing the negative of a relief feature in the surface of the underlying bed. / afgietsel op basisvlak / moulage basal / Ausguss einer Sohlmarke / – –

3020
current mark; flow mark, – any structure formed by the action of a water current on the surface of a rock or of unconsolidated sediment. / stroomspoor / trace de courant / Strömungsmarke / – –

3021
scour mark, – a depression in the surface of a sediment (rock), formed by current scour, e.g. a flute mark, but not a tool mark. / erosiesculptuur / trace d'affouillement / Erosionsmarke / – –

transverse scour mark, – more or less ripple-shaped mark produced by current scour of soft sediment surfaces, the mark having its crests transverse to the current direction. / transversale erosiesculptuur / trace d'affouillement transversale / transversale Erosionsmarke / – –

3022
channel – a long, narrow erosional feature, straight or curved. / geul / chenal / Rinne / canal

channel fill, – the sediments deposited in a channel, filling it completely or for a considerable part. / geulopvulling / remblaiement de chenal; remblayage / Rinnenfüllung / releno de canal

3023
wash-out, – an elongate interruption in the continuity of a layer, or layers, formed by scour due to flowing water, and filled with younger sediment(s). / uitspoeling / ravinement / – – / – –

3024
scour and fill structure; cut and fill ...., – a small-scale, elongate erosion structure, formed in the direction of the current, and subsequently filled. / uitschuring- en opvulling-structuur / structure de creusement et de remblaiement / – – / – –

3025
flute, – a depression eroded by water currents in the surface of a rock, mostly limestone or dolomite. The upstream side is usually steeper than the downstream side. Flutes often form reticulate patterns and are then termed **scallops** by some authors. Sometimes they are elongated transverse to the stream and resemble sharp-crested current ripples.

flute mark, – an erosion hollow in mud, scooped out by a water current flowing over it. Flutes are mostly elongated parallel to the current, and are deepest on their upstream side, becoming flatter and flaring out in downstream direction. They are sometimes arranged in series diagonal to the current. Fossil flutes are hardly ever preserved, because lutaceous material having been weathered away, the arenaceous fills are exposed. / stroom-uitschulping / 'flute' / Kolkmarke / – –

fluting, – the process that produces flutes. / uitschulpen / formation de 'flutes' / auskolken / – –

3026
flute cast, – the preserved fill of a flute. / afgietsel van stroom-uitschulping / moulage de 'flute' / Ausguss einer Kolkmarke / – –

corkscrew flute cast, – a flute cast with a spiral well developed generally near or on the narrow, deep, upstream end. / – – / – – / Korkenzieherkolk / – –

3027
parting lineation; current lineation, – a relief of very small, straight ridges and grooves, not more than a few mm wide, but relatively long, formed parallel to a current, usually in very thin sheets of water, and characteristically found on bedding planes of thin-bedded sandstones. / stroomlineatie / cannelure / Strömungsstreifung / – –

3028
longitudinal furrow and ridge, – long parallel furrows, rounded in cross section, mostly a few mm to circa 1 cm wide, separated by narrow, sharp-crested ridges. The furrows are formed in water, parallel to a current, and are especially known as negatives (casts) from turbidites.

3029
rill mark; rill marking, – dendritic system of very small erosion furrows (mostly not deeper than 1 cm), formed in sand, e.g. by drainage of wet beaches. The term is also used for markings of approximately similar shape at the base of turbidites. / – – / marque de ruissellement / Rieselmarke / – –

3030
swash mark, – a delicate arcuate ridge on a beach, marking the farthest advance of wave uprush, and consisting of sand, various light debris etc. / golfoploopricheltje / lèche de vagues; marque d'atterrissement de .... / Schwappmarke / – –

3031
water-level mark, – a small incision or terrace-shaped feature, formed at a former, usually higher position of the water level. / waterstandsmerk / marque de niveau (d'eau) / Wasserstandsmarke / – –

3032
tool mark, – a mark, produced in a soft sediment surface, by an object carried along by a current. / stroomspoor / marque d'objet / – – / – –

drag mark, – a continuous tool mark: a long groove or system of grooves scoured out by an object (e.g. a plant remnant) along the bottom by a water current. / sleepspoor / trace de traînage / Schleifmarke / – –

3033
groove mark, – long, narrow, remarkably straight trough, usually without appreciable change in depth or width, and produced by an object dragged along the bottom by a water current. / groef / rainure; cannelure / Rillenmarke; Schleifmarke / – –

groove cast, – relief feature in the base of a sedimentary layer, formed by filling up of a groove mark in the surface of the underlying layer. / groefafgietsel / – – / Ausguss einer Schleifmarke / – –

3034
cross groove, – a groove in two or more sets of intersecting grooves. / kruisende groef / cannelures divergentes / sich kreuzende Schleifmarken / – –

3035
ruffled groove mark, – a groove mark, bordered on both sides by small ruffles, pointing downstream. It is a transition between a groove mark and a chevron mark. / rimpelgroef / – – / – – / – –

3036
chevron mark, – a linear row of chevron-shaped features (with more or less rounded apices), pointing downstream, and produced by rucking up of a coherent upper layer of mud by an object carried along by a current. / chevronspoor / trace en chevron; marque en .... / Fiedermarke / – –

3037
impact mark, – short, discrete mark, either single or one of a set, resulting from an object, that, being transported by a current, has impinged on a soft sedimentary surface, e.g. a prod mark, bounce mark or skip and roll mark. / impactspoor / marque d'impact / Stossmarke / marca de choque

prod mark, – impact mark that is deepest on the downstream end. / stootspoor; prik.... / marque en coin / Stechmarke / – –

3038
roll mark, – a mark made in unconsolidated sediment by an object rolling along the bottom, e.g. an ammonite shell that was carried along by a water current and rotated like a wheel. / rolspoor / marque de roulage / Rollmarke / marca de rodadura

bounce mark, – a tool mark, produced by an object that, as it was carried along by flowing water, struck against the bottom, and rebounded immediately. It is widest and deepest in the middle and fades out in both upstream and downstream direction. / stuitspoor / marque de ricochet / Prallmarke / – –

3039
brush mark, – an elongate depression with one or more crescentic wrinkles at the downstream end, produced by an object that, as it was carried along by flowing water, brushed for a few moments against a muddy bottom. / – – / marque de brossage / Quastenmarke / – –

saltation mark; skip ...., – mark produced in a soft sediment surface by a hard object, carried along by a current, and touching the bottom at regular intervals. / saltatiespoor / marque de saltation / Hüpfmarke / marca de saltación

3040
imbricate(d) structure, – the slanting, overlapping arrangement of flat or elongate clastic elements, mostly due to current action, in which case the elements dip upstream. / imbricatie; dakpansgewijze ligging / structure imbriquée / Dachziegellagerung / estructura imbricada

imbrication, – the production of an imbricated arrangement. Also an imbricated structure. / imbricatie / imbrication / Dachziegellagerung / imbricación

[1944] *Imprints and Cracks*

3041
rain print; rain drop imprint, ... impression, – small, basin-shaped pits, surrounded by circular ridges, formed by impact of rain drops on mud or sand surfaces. / regendruppelindruk / empreinte de gouttes de pluie / Regentropfenmarke; ....eindruck / impresión de Iluvia; huella de ....

hailstone imprint; .... impression, – imprint of a hailstone fallen on a soft sediment surface. / hagelsteenindruk / empreinte(s) de grêlon(s) / Hagelkornmarke; ....eindruck / huella de granizo; impresión de ....

3042
salt-crystal cast, – the filling of the hollow left by solution of a salt crystal in a sediment or at its surface. / zoutkristal-pseudomorfose / pseudomorphose de cristal de sel / Salzkristall-Pseudomorphose / molde de cristal de sal

3043
ice-crystal imprint, – an imprint formed in a soft sediment surface by the growth of an ice crystal. /ijskristalafdruk / empreinte de cristal de glace / Eisnadelabdruck / huella de cristal de hielo

3044
shrinkage crack, – crack formed by shrinkage of material, e.g. owing to dessication (mud) or syneresis (gels). / krimpscheur / fente de retrait / Schrumpfrisz / grieta de contracción.

dessication crack, – a crack in originally water-rich material produced by drying. / uitdrogingsscheur / craquelure de dessication / Trockenriss / grieta de desecación

syneresis crack, – shrinkage crack due to spontaneous water expulsion of gels (or gel admixtures). / synaerese-scheur / fissure de synérèse; craquelure de .... / Synâreseriss / grieta de sinéresis

3045
sun crack, – a mud crack produced by insolation. / uitdrogingsscheur / craquelure d'insolation / Trockenriss / grieta de desecación

3046
mud crack, – a crack in lutaceous sediment, usually the result of subaerial dessication, rarely due to subaqueous syneresis. It may be either hollow or filled with other material. / krimpscheur in slik; .... in klei / boue craquelée / Schrumpfriss / grieta de desecación

incomplete mud crack, – single, bifid or trifid cracks that do not (yet) abut on each other. Thus the mud surface is not yet split up into separate polygons. / beginnende krimpscheur / amorce de polygone de dessication / unvollständige Schrumpfriss / – –

mud-crack polygon, – a polygonal mass of dried mud, bounded by shrinkage cracks. / krimpscheur-polygoon / polygone de dessication / Schrumpfrisspolygon / – –

3047
mud-crack cast, – the fill of a mud crack. / krimpscheur-opvulling / remplissage de fente de retrait / Netzleisten / – –

crumpled mud-crack cast, – a mud-crack cast that displays tortuous folding due to adjustment of the sand fill to compaction of the enclosing mud matrix.

3048
dessication breccia, – a breccia formed where irregular dried-out and mud-cracked polygons have broken into angular fragments, which have been deposited with other sedimentary material. / uitdrogingsbreccie / brèche de dessication / Austrocknungsbreccie / brecha de desecación

3049
salt polygon, – a polygonal part of a salt crust bound by ridges that were formed by expansion of the crust. / zoutpolygoon / polygone de sel / Salzpolygon / polígono de sal

[1945] *Penecontemporaneous Deformation Structures*

NOTE:
– see also section
[2320] Environmental Diagenesis

3050
deformation structure, – structure resulting from any change in shape. / deformatiestructuur / structure de déformation / Deformationsgefüge; ....struktur / estructura deformacional

penecontemporaneous deformation, – deformation occurring in soft rock soon after its deposition. / deformatie direct na afzetting / déformation pénécontemporaine / Deformation kurz nach Ablagerung / deformación penecontemporanea

3051
corrugation, – the process of corrugating, or the state of being corrugated, i.e. wrinkled, or contracted in numerous, closely spaced minor folds. / rimpeling; fijne plooiing / corrugation / Runzelung / corrugación

contortion, – the intricate folding or twisting together of laminated or bedded sediments, whereby the strata may be drawn out or compressed in such a manner as to suggest kneading more than simple folding. Also the state of being contorted. / sterke dooreenplooiing; dooreenkronkeling / contortion / Fältelung / contorción

intraformational contortion; .... corrugation, – contortion or corrugation of separate layers, not affecting the underlying or overlying layers, due to processes taking place before deposition of the sedimentary series was completed. / intra-formationele dooreenplooiing / convolution intraformationelle / intraformationelle Fältelung; .... Runzelung / corrugación intraformacional; contorsión ....

contorted bedding, – bedding that has undergone contortion. / dooreengeplooide gelaagdheid / lit contourné / Gekröseschichtung / estratificación contorsionada

3052
convolution, – the process producing convolute lamination. Also the state of being convoluted, or a separate convolution fold. / convolutie / convolution / Konvolution / convolución

convolute lamination, – an internal structure of sedimentary layers, charactirized by folded and distorted laminations, the deformations usually dying out within the layer, both upward and downward. / convolute laminatie / lamination contournée / Wickelstruktur / laminación convoluta

intrastratal flow structure, – a variety of convolute bedding formed by flowage.

3053
décollement structure, – a structure showing separate folding (intraformational or tectonical) of one or more sets of strata, due to slip along the older or/and younger layers. / décollement structuur / structure de décollement / Abscherungsstruktur / estructura de despegue

3054
overturned foreset bedding, – foreset bedding in which the upper parts of the foresets are bent over owing to current drag. / overkiepte foreset gelaagdheid / stratification à feuillet frontal déversé; .... à élément frontal déversé / – – / estratificación 'foreset' sobreplegada

3055
slump; .... mass, – a mass of unconsolidated, down-slided material. / afglijdingsmassa / glissement / Abrutsch; ....masse / desprendimiento

slump fold, – a fold produced by slumping. / afglijdingsplooi / pli de glissement / Rutschfalte / pliegue producido por desprendimiento

slump ball, – ball-shaped mass of arenaceous material, embedded in lutaceous sediment, and crumpled up by subaqueous sliding. / afglijdingsbal / – – / – – / – –

phacoid, – slump ball, a few cm to a few dm in diameter, shaped rather like the stone of a prune, and probably formed by sliding of sediment that first underwent sigmoidal fracturing. / phacoïd / amygdale; ocelle; lentille / Phakoid / facoide

3056
crinkle mark, – minute subparallel ridges on bedding planes, usually asymmetric in cross section, and probably formed by subaqueous slumping of sediments. The wrinkles on successive bedding planes are often congruent with one another. / – – / – – / Knittermarke / – –

3057
frondescent mark; cabbage-leaf marking, – structures in the sandy base of some turbidity-current deposits, elongated and branching downstream, and with crenulated edges. Possibly they are formed by downsinking of sand into soft mud, the sand flow being directed downstream under the influence of the current passing overhead. / koolblad-structuur / structure frondescente / – – / – –

3058
slide mark, – system of parallel grooves and smoothened strips, produced by sliding, either in unconsolidated or in hard rocks. / afglijdingsspoor / structure de glissement / Harnisch; Striemung / marca de deslizamiento

3059
load structure, – structure due to sinking down of material (mostly sand) into underlying material of lower bulk density (usually mud). Load structures may consist of broad downward bulges of sandy composition with internally deformed laminations, separated by upward pointing ridges of lutaceous composition, or of isolated bodies of sunken arenaceous material. / – – / structure de charge / Belastungsmarke / estructura de carga

sag structure, – a general term for load structures and other related sedimentary structures. / inzakkingsstructuur / structure de charge; figures de .... / Sackungsgefüge / – –

load cast, – this term, introduced by Kuenen, is incorrect, since no cavities or depressions were originally present, and no casting has occurred. / inzakkingsstructuur; structuur met inzakkingsballen / structure de charge / Sackungsgefüge / estructura de carga

3060
pseudonodule, – non-genetic term for more or less roll- or pillow-shaped bodies of sand or sandstone embedded in lutaceous material. Most pseudonodules originate as load structures or slump balls, or by the combined effect of loading and sliding. / pseudonodule / pseudonodule / Pseudonodule / (p)seudonódula

3061
flame structure, – intraformational structure, that, in cross section perpendicular to the bedding, shows tongues of lutaceous material, squeezed up between broad, downward bulges of coarser material, formed by sagging down of the latter into the underlying layer of more mobile mud. / – – / structure en flamme / – – / estructura en llama

3062
bubble mark; .... impression, – small depression, formed by development of a gas bubble at the sediment-water interface. / gasbelafdruk / marque de bulle gazeuse / Blasenabdruck / marca de barbuja

3063
bubble sand; cavernous sand; vesicular sand, – sand with a spongy structure, due to the presence of (smooth-walled) cavities filled with air (entrapped by re-submergence, especially on beaches) or gas (mostly the product of decay of enclosed organic matter). / blaasjes-zand; cavenreus zand / sable vésiculaire; .... à structure spongieuse / Blasensand / arena vesicular

3064
cavernous, – rich in cavities, other than primary pores between clastic particles. / caverneus / caverneux / kavernös / cavernoso

cellular, – rich in cavities larger than primary pores, especially those that are bounded by plane surfaces, / e.g. cellular dolomite. / cellen- / cellulaire / zellulär / celular

void, – open space in solid material of small to moderate size. / holte / vide / Loch / vacio; hueco

3065
sand dome; air dome, – dome-shaped elevation, usually a few cm in diameter, of the surface layer of beach sand, pushed up by air (gas) that was under pressure as a result of flooding. / zand-opwelving / dôme de sable / Gasbeulen / domo de arena

air-heave structure, – structure supposed to be due to pushing up of sediment by enclosed, compressed air. / – – / trace d'éruption; .... de gaz / Gasbeulen / – –

3066
gas pit, – conical pit, usually a few cm to a few dm in diameter, formed in the surface of mud deposits by the escape of gas. / ontgassingskuiltje; ....kuil / microcratère / Gaskrater; Gastrichter / – –

pit and mound structure, – relief of small pits and mounds on soft sediment surfaces due to escape of gas or water.

3067
bioturbation, – disturbance of original depositional structures by movements of (mostly) benthic organisms. / bioturbatie / bioturbation / Bioturbation / bioturbación

ichnofossil; trace fossil, – a sedimentary structure produced by the activity of living organisms. / ichnofossiel; sporen .... / ichnofossile / Spurenfossil; Ichno.... / icnofósil

NOTE:
– for further terms, see section [1523] Ichnology

3068
fluid-escape structure, – diapiric structure due to the disruption of layers or laminae by fluids (mixtures of water and sedimentary material, sometimes also gas), escaping in upward direction.

water-escape structure, – structure, normally consisting of disruptions of layers or laminae due to upward escape of pore water.

3069
dish structure, – a sedimentary structure, occurring in sandstones (mostly turbidites) and consisting of numerous upward concave, more or less dish-shaped bodies, each showing an upward increase of grain size of the sand. The upturned edges of the 'dishes' often appear to be the result of upward escape of pore fluid. / schotelstructuur / structure en coupelle / – – / – –

3070
sedimentary injection; injection, – the squeezing, under abnormally high pressure, of sedimentary material, downward, upward, or laterally, into a pre-existing deposit or rock, either along some plane of weakness or into a pre-existing crack. / sedimentaire injectie / intrusion sédimentaire / sedimentäre Injektion / inyección sedimentaria

sedimentary intrusion; intrusion, – a sedimentary injection on a relatively large scale. / sedimentaire intrustie / intrusion sédimentaire / sedimentäre Intrusion / intrusión sedimentaria

3071
sand volcano, – a small conical accumulation of sand, formed by upward expulsion of sand-laden water. / zandvulkaan / volcan de sable / Sandvulkan / volcán de arena

mud volcano, – a small (up to 2 m) conical accumulation around an orifice, formed by expulsion of gas-rich mud and water. / slikvulkaan / volcan de boue / Schlammvulkan / volcán de barro

3072
mud lump, – a mud diapir forming a small island near the mouth of a major river distributary.

3073
clastic dyke; neptunean dyke; .... dike, – a tabular body of clastic material, cutting across other formations (not necessarily sedimentary) and emplaced either by intrusion or by sinking down of sediment into a fissure under the influence of gravity. / klastische gang / filon clastique / klastischer Gang / dique clástico

3074
clastic pipe, – a body of clastic material of columnar shape, cutting across other structures and measuring from 1 tot 60 m in height. / klastische pijp / – – / – – / – –

3075
tepee structure, – a sharp bend of centimetre to metre scale in a sedimentary layer, in cross section resembling an inverted depressed V, and mostly due to expansion.

3076
enterolithic folding, – intraformational structure of layers of gypsum or anhydrite, that show in cross section perpendicular to the general stratification closely packed folds, somewhat resembling intestinal convolutions. In many cases the folding appears to be due to the force exerted by crystallization of the gypsum or anhydrite precipitated from ground water below the sediment surface. / enterolietische plooiing / plissement entérolithique; structure .... / enterolithische Faltung; Gekrösefaltung / plegamiento enterolítico

[1950] SEDIMENTARY TEXTURE

3077
texture, – the 'microgeometry' of a (sedimentary) rock, determined by size, shape and arrangement of the component crystals or particles. / textuur / texture / Struktur / textura

fabric, – the spatial arrangement and orientation of the elements of a (sedimentary) rock. / maaksel / structure / Textur / fábrica

primary fabric; depositional ...., – fabric formed during the deposition of a sediment. / primaire structuur; afzettingsstructuur / structure primaire / Primärgefüge; Ablagerungs.... / fábrica primaria

3078
particle, – a small, separate piece of matter, e.g. a water particle; for a sediment it is the smallest separable or distinct unit in the composition of the rock. / deeltje / particule / Partikel; Teilchen / partícula

grain, –
1. a more or less equidimensional mineral or rock particle, smaller than a few mm, and generally lacking well-developed crystal faces; in a clastic sediment a small hard, more or less rounded particle;
2. granulometrical texture.

/ korrel / grain / Korn / grano

3079
framework, – primarily coherent part of a sediment, consisting of particles, crystals or crystal aggregates that support each other or are intergrown at their points of contact. / korrelskelet / charpente / Gerüst / trama

3080
matrix; groundmass, – the fine-grained material in which the coarser particles in a heterogeneous mass are embedded. / matrix; grondmassa / matrice / Matrix; Grundmasse / matriz

3081
granular structure; .... texture, – said of rocks that consist of grains (granules), especially when these are of about equal size grade. / korrelig /grenue / körnig / granular

granularity, – / korreligheid / grain / Körnigkeit / granularidad

equigranular, – said of a crystalline rock (igneous, sedimentary or metamorphic) whose constituent crystals are of approximately the same size. / equigranulair / équigranulaire / gleichkörnig / equigranular

3082
packing, – the arrangement of the particles in a sedimentary deposit. / pakking; korrelpakking / arrangement des particules / Packung / empaquetamiento

packing density, – the degree to which the grains of a sediment fill its gross volume. / pakkingsdichtheid / compacité / Packungsdichte / densidad de empaquetamiento

3083
pore; interstice, – a small to minute opening or passageway in a rock or soil, not occupied by solid matter. / porie / pore; interstice / Pore / poro; intersticio

interstitial, – pertaining to an interstice or to its fill, e.g. interstitial water. / interstitiëel; poriën- / interstitiel / interstitial; Poren- / intersticial

3084
liquefaction, – the transformation of a loosely packed sediment, the pores of which are filled with water, into a fluid mass. / vervloeiing; liquefactie / liquéfaction / Liquefaktion; Verflüssigung / licuefacción

3085
quicksand, – water-saturated sand in which the sand grains are loosely packed, so that it may easily flow, and heavy objects placed on it sink rapidly. Sometimes a distinction is made between quicksand proper, in which the grains are at rest, and running sand, in which they are moving. / drijfzand / sable mouvant / Schwimmsand; Treib....; Fliess .... / arena movediza

running sand, – sand, the grains of which are nearly or completely surrounded by interstitial water, and that is moving like a liquid. / loopzand / sable boulant; .... fluent / Fliesssand; Schwimm.... / arena movediza; .... voladora

3086
earth, – said of some very fine-grained, unconsolidated, loosely aggregated materials, e.g. diatomaceous earth, radiolarian earth, Fullers earth. / aarde / terre à - / Erde / tierra

earthy, – / aardachtig / terreux / erdig / terroso

3087
friable, – said of a loosely consolidated rock, the particles of which can easily be rubbed away, or that can be crumbled easily. /kruimelig; gemakkelijk verbrokkelend / friable / krümmelig / friable

3088
sounding sand; squeaking ....; whistling ....; singing ....; roaring ....; musical ...., – sand (usually dry and clean) that,

when mechanically disturbed, emits a distinct sound. /zingend zand / sable musical / tönender Sand / arena musical

3089
mosaic structure, – a structure of more or less equigranular crystalline sedimentary rocks, characterized by relatively simple, straight grain boundaries. / mozaïekstructuur / texture en mosaïque / Mosaikstruktur; Pflasterstruktur / estructura en mosaico

3090
saccharoidal; sugary, – having a finely granular or crystalline, 'sugary' texture, e.g. some sandstones, limestones, dolomites. / suikerachtig / saccharoide / Zucker- / sacaroideo

3091
microcrystalline, – said of crystal aggregates in which no separate crystals can be distinguished with the naked eye. / microkristallijn / microcristallin / mikrokristallin / microcristalino

3092
rose; rosette, – a more or less rose-shaped aggregate of crystals, e.g. of sand-filled barite. / kristalroos; ....rozet / rose / Rosette / rosa

3093
colloform, – said of the rounded, globular texture of a colloidal, or originally colloidal mineral deposit. / colloform / colloforme / kolloform / coloforme

3094
pellet, – a small (mostly less than a few mm diameter) round particle, not formed by rounding of a detrital mineral grain, e.g. a faecal pellet of an invertebrate animal. / propje; bolletje / boulette / Pille / pelet

faecal pellet; fecal ...., – a small excrement, recent or fossil. Most faecal pellets are products of invertebrate organisms. / faecespropje / coprolite / Kotpille / pelet fecal

[1960] STRATIFICATION

[1961] *Single Strata*

3095
stratification; bedding; layering, – a structure produced by deposition of sediments in beds, layers, wedges or other essentially tabular units. Collectively, the presence of beds or laminae in rock sequences, obvious by planes separating rocks of the same or different lithologies. / gelaagdheid / stratification; litage; rubannement / Schichtung / estratificación

stratified; bedded; layered, – formed or lying in beds, layers or strata. / gelaagd / stratifié; lité; rubanné / geschichtet / estratidicado

3096
bedding plane; plane of bedding; stratification plane, – in stratified rocks the division plane that separates adjacent layers, beds or strata. / laagvlak / plan de stratification / Schichtfläche; Schichtebene; Schichtungsfläche / plano de estratificación

interface, – boundary surface between two bodies, especially between bodies of different physical state (solid, liquid, gaseous). / scheidingsvlak / surface de séparation; interface / Grenzfläche / intercara

3097
layer, – a tabular unit of igneous, sedimentary or metamorphic origin, of comparatively homogeneous composition and separated from the material above and below by well defined boundary planes. / laag / couche / Lage / lecho; capa

stratum, –
1. a tabular or sheet-like body, of any thickness;
2. a layer thicker than 1 cm.
/ laag / strate; banc; assise / Lage; Bank; Schicht / estrato

3098
bed, – a tabular unit of sedimentary origin, of comparatively homogeneous composition and separated from the material above and below by well defined boundary planes. Some authors restrict the term to layers thicker than 1 cm. / laag / lit / Schicht; Bank; Lage / lecho

3099
bedding joint, – thin layer differing in composition from the beds between which it occurs. / laagvoeg / joint de stratification / Schichtfuge / – –

3100
film; skin, – thin covering or layer, less than 1 mm thick. / huidje; film / pellicule / Häutchen; Film / pelicula

3101
lamination, – bedding or layering in a sedimentary rock in which the stratification planes are 1 cm or less apart. / laminatie / lamination; structure en lamines; .... en feuillets / Blätterung; Bänderung / laminación

lamina; streak; band, – the thinnest recognizable unit layer of original compsition in a sediment or sedimentary rock, thinner than 1 cm. / lamina / lamine; feuillet / Lamina / lámina

laminated, – consisting of laminae. / gelamineerd / en feuillets; en lamines / laminiert / laminado

3102
lentil; lens, – a body of rock, bounded by converging surfaces, thinning and wedging out towards the edges. / lens / lentille / Linse / lentejón

lenticle, – small lens. / lens(je) / – – / kleine Linse / lentícula

lenticular, – lens-shaped. / lensvormig / lenticulaire / linsig; lenticulär / lenticular

3103
flaser, – originally a German term for a streak or vein, especially a (relatively small) elongate body, pinching out rapidly at right angles to the elongation. / – – / – – / Flaser / – –

3104
wedge out; thin out; pinch out, – said of strata and veins, that grow progressively thinner to disappearance. / uitwiggen / terminaison en biseau / auskeilen; ausdünnen / acuñado

3105
sedimentary prism, – a long, narrow, wedge-shaped sedimentary body.

fill, – any sediment, deposited by any agent, so as to fill a valley, sink or other depression. / opvulling / remblaiement / Füllung / relleno

plug, – a mass of sediment filling partly or wholly a cut-off river meander. / prop / remplissage / Propfen / tapón

nest, – a local concentration of some relatively conspicuous elements, such as a 'nest of pebbles'. / nest / poche / Nest / nido

3106
top, – the upper surface of a layer or geological formation. / bovenkant; ....zijde / sommet / Topkante; Ober....; Oberfläche; Dach.... / techo

bottom; sole, – the lower surface of a layer or geological formation. / onderkant; ....zijde / base / Unterkante; ....fläche; Sohlfläche / muro

3107
geopetal, – pertains to top and bottom relations in rocks at time of formation. / geopetaal / centripète / geopetal / geopetal

substrate; substratum; underlayer, – the stratum that lies directly below another one. / substraat / substratum / Substrat / substrato; sustrato

3108
massive bedding, – said of very thick, homogeneous beds. / – – / litage massif; stratification massive / Bankung / estratificación masiva

thick-bedded, – / dik-gelaagd / en lits épais / grobgeschichtet / estratificación gruesa

thin-bedded, – / dun-gelaagd / en lits minces / feingeschichtet / estratificación fina

3109
blanket deposit, – a sedimentary deposit of great lateral extent and of relatively uniform thickness. / – – / dépôt en nappe / – – / manto

sheet sand(stone), – a blanket deposit of sand(stone). / – – / couverture de grès; nappe / Deck(en)sand(stein) / manto de arena

[1962] *Interrelation of Strata*

3110
sequence, – the succession of sediment(ary rock) layers, lying above one another in a particular area. / sequentie; laagopeenvolging / séquence / Sequenz; Abfolge; Schichtfolge / secuencia

3111
set, – a group of essentially conformable strata or cross strata, separated from other sedimentary units by surfaces of erosion, non-deposition, or abrupt change in character. / groep / faisceau de – / – – / – –

coset, – a sedimentary unit made up of two or more sets, either of strata or cross strata, separated from other strata or cross strata by original flat surfaces of erosion, non-deposition or abrupt change in character. / meervoudige groep / unité sédimentaire composite / – – / – –

3112
regressive deposit, – sediments deposited during the retreat of the sea from a land area, and characterized by an offlap arrangement. / regressieve afzetting / dépôt regressif / regressive Ablagerung / depósito regresivo

offlap, – a situation in a sequence of strata in which each unit is deposited seaward of the shore line that marked the extent of the underlying unit. It may arise from retreat of the sea. / terugwijkende bedekking / position en retrait / regressive Lagerung / retiro regresivo

3113
transgressive deposit, – deposit formed during the encroachment of the sea over a land area and characterized by an onlap arrangement. / transgressieve afzetting / dépôt transgressif / transgressive Ablagerung / depósito transgresivo

onlap, – a situation in a sequence of strata in which, in one direction, each unit extends onto the previous land surface, often as a result of transgression of the sea over the land. / dakpansgewijze ligging / débordement transgressif position débordée / übergreifen; überlappen / desborde transgresivo

3114
basal conglomerate, – coarse, often conglomeratic, usually well-sorted deposit directly upon an erosional break, forming the bottom member of a sedimentary sequence laid down in an encroaching body of water. / basisconglomeraat / conglomérat de base / Basalkonglomerat; Transgressions.... / conglomerado basal

3115
concordant bedding; parallel bedding, – stratification characterized by beds that are parallel and without angular junctions. / concordante gelaagdheid; evenwijdige .... / stratification parallèle / konkordante Lagerung / estratificación concordante

3116
interstratification; interlayering, – the formation or occurrence of layers of a given composition and/or structure between layers of a different composition and/or structure. / vorming van tussenlagen; aanwezigheid van .... / interstratification / Wechsellagerung / interstratificación

interstratified; interbedded, – said of layers lying between, or alternating with layers of different composition. / met tussenlagen / interstratifié / – – / interstratificado

3117
intercalation, – a layer or lens lying between layers of differ-

ent composition or structure. Also the presence or the formation of such bodies. / intercalatie; tussenschakeling; inschakeling; tussenlaag / intercalation / Einlagerung; Einschaltung; Zwischenlage / intercalación

intercalated, – pertaining to an intercalation. / tussenliggend; ingeschakeld / intercalé; interstratifié / zwischengelagert; eingebettet; eingeschaltet / intercalado

3118
interfingering; intertonguing, – lateral intergrading of units; a series of interlocking or overlapping wedge-shaped layers. / vingervormig in elkaar grijpen / interdigitation; imbrication / wechselseitige Verzahnung / interdigitada

3119
cycle of sedimentation, – recurrent sequence of strata, each consisting of several similar lithologically distinctive members arranged in the same order. / sedimentatiecyclus / cycle de sédimentation / Sedimentationszyklus / ciclo de sedimentación

cyclothem, – a symmetrical sedimentary cycle, mostly coal-bearing. / cyclotheem / cyclothème; alternance répétée / Zyklothem / ciclotema

3120
hemicyclothem, – half of a cyclothem, usually referring to either the lower non-marine or the upper marine part of the Pennsylvanian types of cyclothems. / hemicyclotheem / hemicyclothème / Hemizyklothem / hemiciclotema

megacyclothem, – a rhythm of larger order than a single cyclothem, including minor cyclothems; a cycle of cyclothems. / megacyclotheem / mégacyclothème / Megazyklothem; Grosszyklus / megaciclotema

hypercyclothem, – unit, repeated rhythmically, consisting of several megacyclothems. / hypercyclotheem / hypercyclothème / Hyperzyklothem / hiperciclotema

3121
rhythmic bedding, – bedding in which two or more types of sediment regularly alternate with each other. / rhythmische gelaagdheid / litage en alternance répétée; stratification en alternance ré-pétée / rhythmische Schichtung / estratificación rítmica

rhythmic sedimentation; cyclic ...., – the formation of a long series of rhythmically bedded deposits. / rhythmische sedimentatie / sédimentation rythmique; .... en alternance répétée / rhythmische Sedimentation / sedimentación rítmica

rhythmite, – an individual unit of a rhythmical succession, e.g. a cyclothem, a graded layer in a turbidite sequence, or a glacial varve. / – – / rythme élémentaire / Rhythmit / ritmita

3122
banding, – alternation of comparatively thin layers, characterized by a sometimes minor, but usually conspicuous difference in composition. / bandering / rubannement / Bänderung / bandeado

band, – a thin layer with a distinctive lithology, colour or fossil content. / zeer dunne laag; band; snoer / lit; passée / Horizont / banda

banded, – layers displaying differences in colour, texture or composition. / gebandeerd / zoné / gebändert / – –

3123
seam; measure, – a comparatively thin layer occurring in a series of strata of different composition, especially a coal or mineral layer. / laag; delfstoflaag / veine de charbon; couche de ....; couche minéralisée / Flöz / capa

3124
symmict; diamict, – said of layers, especially glacial varves, in which, owing to flocculation, much of the clay has been deposited together with coarser material (mostly silt and sand). / symmict / – – / symmiktisch; diamiktisch / diamíctico

diamictite; mixtite, – a purely descriptive term for a non-sorted sedimentary deposit, with or without rudaceous constituents, such as tillite or tilloid. / – – / diamictite / Diamiktit / diamictica

3125
diastem, – a hiatus due to a comparatively short break in sedimentation. / diaste(e)m / diastème / Diastem; Omissionsfläche / diastema

3126
hard ground, – bed of calcareous or marly composition that has become hardened before it was covered by new sediment, i.e. generally during a break in the deposition in that area. / – – / fond induré; .... durci / Hartgrund / – –

veneer, – a thin but extensive layer of sediment or weathered material at the surface of an older formation. / deklaag(je) / pellicule / Decke / costra

palaeosol, – a fossil soil, or part of it. / palaeosol / paléosol / Paläosol; ....boden; fossiler Boden / paleosuelo

3127
annual layering, – stratification, consisting of annual layers. / jaargelaagdheid / stratification de rythme annuel; litage périodique; .... annuel / Jahresschichtung / estratificación anual

annual layer, – sedimentary layer of a regular, multilayered series, deposited in the course of one year, and usually consisting of two different parts (often grading into each other), as a consequence of seasonal differences in deposition, e.g. a glacial varve. / jaarlaag(je) / couche annuelle / Jahreslage / lecho anual

3128
varve, – one of a series of annual layers, specifically a glacial varve, formed in still water in front of a glacier or ice cap, and normally built up of a fine-sandy or silty summer layer and a very fine-grained, clayey winter layer. / warve / varve / Warve / varva

varved clay, – a dominantly clayey deposit, composed of varves. / warvenklei / argile à varves / Bänderton / arcilla varvada

varvite, – an indurated rock, consisting of ancient varves. / warvengesteente / varvite; laminite / Warvengestein / varvita

3129
fining-upward sequence, – sequence of sediments the grain sizes of which decrease in (originally) upward direction. / – – / – – / – – / secuencia positiva

3130
grading, – the gradual reduction of grain sizes in a given direction within a layer of granular material. / gradering / granoclassement / Gradierung / graduación

graded bedding, – stratification in which each bed is characterized by a gradual change in grain size, normally from coarse at the base to fine at the top; usually such beds have a sharp base and an undefinite top. / gegradeerde gelaagdheid / couche granoclassée / gradierte Schichtung / estratificación gradada

reversed grading, – the decrease in grain size is from top to bottom.

3131
sorted bedding, – graded bedding in which only one size grade is present at each level. / gesorteerde gelaagdheid / – – / – – / – –

3132
flag, – plate of a thin-bedded arenaceous or other fine-grained clastic rock. / – – / dalle; lanze; plaque / Platte; Plattensandstein / placa

flagstone, – a rock consisting of flags. / – – / roche en dalles; .... en plaques / – – / roca laminar

flaggy, – having the character of flags or flagstone, or containing flags. / platig / en dalles; en plaques / plattig / laminar

[1963] *Cross Stratification*

3133
cross stratification, –
1. the arrangement of beds (thicker than 1 cm) or laminae (thinner than 1 cm) at one or more angles to the general dip of the formation;
2. cross bedding.

/ kriskrasgelaagdheid; diagonale .... / stratification oblique / Schrägschichtung / estratificación cruzada

cross bedding, –
1. the arrangement of beds (thicker than 1 cm) at one or more angles to the general dip of the formation;
2. cross stratification.

/ kriskrasgelaagdheid / litage oblique; .... entrecroisée / Schrägschichtung / estratificación cruzada

cross lamination, – the arrangement of laminae at one or more angles to the general dip of the formation. / kriskraslaminatie / lamination oblique / Schrägschichtung / laminación cruzada

3134
current bedding, – sedimentary structure resulting from current action, either by wind or by water; it includes cross stratification and ripple bedding. / stromingsgelaagdheid / stratification de courant; lamination de .... / Schrägschichtung / estratificación entrecruzada

tidal bedding, – bedding formed under influence of the tides, reflecting regular oscillations of water level, or regular reversal of current direction or alternation of currents and slack water stages. / getijdengelaagdheid / stratification tidale; .... de marée / Gezeitenschichtung / estratificación producida por la acción de las mareas

3135
false bedding, – bedding affected by currents, generally erratic and with common changes in directions. / diagonale gelaagdheid / fausse stratification / Schrägschichtung / falsa estratificación

3136
inclined bedding, – bedding laid down with primary dip. / hellende gelaagdheid / stratification oblique / Schrägschichtung / estratificación inclinada

avalanche bedding, – steeply inclined bedding, produced by avalanching of sand down the slip face (lee slope) of aeolian or subaqueous dunes or ripples. / afstortingsgelaagdheid / lit d'avalanche / Leeblattschichtung / – –

3137
angle of repose; angle of rest, – the maximum possible inclination of slopes in accumulations of loose material. / maximale hellingshoek / angle d'équilibre; pente .... / Schüttungswinkel / ángulo de reposo

3138
wavy bedding; .... lamination, – bedding or lamination characterized by undulatory bounding surfaces. / golvende gelaagdheid; .... laminatie / stratification ondulée / wellige Wechselschichtung / estratificación ondulada

3139
delta bedding, – the bedding characteristic of a delta, consisting of comparatively flat topset and bottomset beds, between which more steeply dipping foreset beds are intercalated. / delta-gelaagdheid / stratification deltaïque / Deltaschichtung / estratificación deltaica

3140
foreset bedding; .... lamination, – structure characterized by one or more series of foreset beds or laminae, deposited by water currents, wave action or wind, the layers or laminae dipping in the direction of flow or resultant flow. / foresetgelaagdheid; ....laminatie / stratification oblique tabulaire; litage incliné; stratification à lits frontaire; lamination à feuillets ....; lamination inclinée / 'Foreset'-Schichtung; Vorsetzschichtung / laminación 'foreset'; estratificación ....

foreset bed; .... lamina, – one of a series of parallel, inclined beds or laminae of a cross-bedded unit, deposited by water currents, wave action or wind, the layers or laminae dipping in the direction of flow or resultant flow. / foreset laag; ....lamina / lit frontal; feuillet frontal / 'Foreset'-Schicht; Vorsetzschicht / – –

3141
backset bed; .... lamina, – a cross bed or lamina dipping against the direction of the depositing current (wind, water).

/ backset laag, ...lamina / lit inverse; feuillet .... / 'Backset'-Schicht / – –

3142
planar cross bedding, –
1. cross bedding in which the lower bounding surfaces are planar surfaces of erosion;
2. cross bedding characterized by planar forseset beds.

/ – – / stratification oblique plane / ebene Schrägschichtung / estratificación cruzada planar

angular cross bedding, – cross bedding in which plane (or almost plane) forsets meet the underlying surface at sharp discordant angles. / hoekige kriskrasgelaagdheid / stratification oblique angulaire / winkelige Schrägschichtung / estratificación cruzada angular

tangential cross bedding, – cross bedding in which the foreset beds curve round at the base, meeting the underlying surface at low angles. / tangentiale kriskrasgelaagdheid / stratification oblique tangentielle / bogige Schrägschichtung / estratificación cruzada tangencial

3143
low-angle cross bedding, – cross bedding in which the cross beds have an average maximum inclination of less than 20°.

3144
trough-cross bedding; crescent-type cross bedding, – cross bedding due to infilling of scour throughs having their long axes parallel with or at a relatively small angle to the current. / geultjes-kriskrasgelaagdheid / stratification entrecroisée; .... en gouttière / trogförmige Schrägschichtung / estratificación cruzada en artesa

3145
festoon cross bedding, – cross bedding that in horizontal sections shows subparallel series of more or less semi-elliptical cross sections of beds or laminae, deposited by water currents, the series cutting through one another, and representing the fillings of subsequent, narrow, elongate troughs in the stream bed. / guirlande-kriskrasgelaagdheid / stratification oblique en feston / – – / estratificación cruzada festoneada

3146
herringbone cross bedding; chevron ....; zigzag-...., – cross bedding in which series of foreset beds or laminae in successive layers dip alternately in opposite directions. The structure mostly points to deposition by tidal currents. / zigzag-kriskrasgelaagdheid / stratification oblique en arêtes de hareng; .... oblique en chevrons / Schrägschichtung; Kreuz.... / estratificación cruzada en zigzag

3147
ripple cross bedding; .... lamination, – small-scale cross bedding or lamination due to formation and/or migration of ripples. / ribbel-kriskrasgelaagdheid; ....-laminatie / stratification oblique de rides / Rippelschrägschichtung / estratificación entrecruzada de rizadura

3148
flaser cross bedding; flaser bedding, –
1. ripple cross bedding characterized by the intercalation of numerous flasers of mud (mostly formed in the ripple troughs);
2. cross bedding with lenticles of sand or silt, commonly aligned and usually cross laminated, lying in a matrix of lutaceous sediment.

/ – – / petites stratifications obliques combés / Flaserschichtung / – –

3149
climbing ripple lamination; ripple-drift (cross) lamination, – a structure with series of lee side laminae of current ripples that rise in downstream direction, owing to absence of erosion, or even deposition, on upstream sides of ripples. / klimmende ribbel-laminatie / lamination de rides migrantes; .... vers l'amont / ansteigende Schrägschichtung / – –

3150
lenticular cross bedding, –
1. bedding in which the strata, within the confines of an outcrop, hand specimen or core sample, pinch out;
2. ripple-cross bedding in which isolated (incomplete) or interconnected ripple bodies of sand or silt are embedded in lutaceous sediment.

/ – – / stratification lenticulaire / Linsenschichtung / estratificación lenticular

3151
rib and furrow structure, – a pattern, seen in sections parallel to the general stratification, consisting of small-scale crescentic cross sections of foreset laminae, concave downstream, such as are formed by migration of lunate (crescentic) current ripples. / – – / structure en arrêtes et creux / Schrägschichtungsbogen / – –

[1970] SAMPLING AND GRAIN-SIZE DETERMINATION

NOTE:
– see also sections
[2022] Size of Particles
[2123] Grain-Size Classification

3152
sand trap, – an apparatus to separate and sample sand (and other particles) from flowing water. / zandval / séparateur de sable / Sandfalle / criba

NOTE:
– for further terms on samples and sampling, see sections
[2710] Sampling (of ore)
[2729] Sample Reduction and Chemical Analysis
[2730] Statistical Analysis of Samples
[2820] Drilling; Methods and some technical Terms
[2931] Samples and sampling Procedures

3153
grab sample, – sample obtained by grab sampler, i.e. appara-

tus for sampling subaqueous sediments between two or more 'jaws'. / grijpmonster / échantillon / Greiferprobe / – –

3154
dredge, – a bottom sampler that scoops sediment and benthic organisms as it is dragged behind a moving ship. / dreg / drague / Dredsche; Dredge / draga

3155
corer, – an apparatus retrieving undisturbed cylindrical or box-shaped columnar samples from unconsolidated sediments or soils. / kernapparaat / carottier / Kerngerät / testiguero

core, – a cylindrical or columnar sample, retrieved within the tubular or box-shaped part of a coring apparatus. / kern; boor....; steek.... / carotte / Kern / testigo

3156
gravity corer, – an apparatus for taking core samples from the bottom of a water body, penetrating into it by its own weight. / – – / carottier par gravité / Lot; Schwerelot / testiguero de gravedad

piston corer, – an apparatus for taking core samples from the bottom of a water body, having a piston to ensure full core recovery to the depth of penetration in the sediment. / – – / carottier à piston / Kolbenlot / testiguero de pistón

vibro-piston corer, – a piston corer that penetrates into the bottom by means of vibrations. / – – / carottier à vibropiston / Vibrationskolbenlot / testiguero de pistón vibratil

3157
sieve analysis, a granulometrical analysis carried out by means of sieves. / zeefanalyse / granulométrie par tamisage / Siebanalyse / tamizado

mesh, – one of the openings between the wires of a sieve. / zeefopening / maille / Maschenweite / malla

mesh number, – the number of openings per linear inch.

3158
sedimentation balance, – an apparatus to measure the settling rates of small particles dispersed in a (stagnant) liquid. / sedimentatiebalans / balance de sédimentation / Sedimentationswaage / balanza de sedimentación

3159
elutriation, – literally: washing out, specifically used for:
1. purification of material by washing and decanting;
2. separation of particles of different size or weight by means of a slowly rising current of water or air.

/ elutriatie; afslibbing / élutriation / Elutriation; Auswaschen; Ausspülen / decantación

splitting, – the sampling of a large mass of loose material by dividing it into one or more parts. / splitsen / échantillonnage par la méthode des carrés / splitten; Proben teilen / cuarteo

3160
mechanical analysis, – determination of the particle-size distribution of a rock by sieving or other means of mechanical separation. / mechanische korrelgrootte-analyse / analyse mécanique / mechnische Korngrössenanalyse / análisis mecánico

granulometry, – the measurement of grain sizes. / granulometrie / granulométrie / Granulometrie / granulometría

3161
grade scale, – a system of successive grain-size classes. / korrelgrootten-indeling / échelle des classes granulométriques / Einteilung der Korngrössen / escala de grado

3162
grain-size grade; size grade, – a restricted size range of particles of a sediment or other rock. / korrelgroootte-klasse; korrelgrootte / classe granulométrique / Korngrössenklasse / grado de tamaño de grano

3163
Wentworth scale, – a scale of grain size, the subdivisions of which are based on the powers of 2 expressed in mm. / Wentworth-schaal / échelle de Wentworth / Wentworth-Skala / escala de Wentworth

phi, – particle size expressed as the negative logarithm at the base 2 of the diameter in mm. / phi / Phi / Phi / phi

phi-grade scale; phi scale, – a scale of phi units. / phi-schaal / échelle en Phi / Phi-Skala / escala 'phi'

NOTE:
– see table 6 – Grades of Clastic Sediments

[2000] CLASTIC DEPOSITS

[2010] THE CONSTITUENTS

3164
clastic deposit; fragmental ...., –
1. a deposit, regardless of chemical composition, mainly composed of fragments of pre-existing rocks, crystals or organic remains and of particles newly formed during weathering and erosion, e.g. a conglomerate, a calcarenite, a residual clay;
2. as 1, but excluding bioclastic rocks, e.g. carbonates;
3. a deposit mainly composed of solid products of weathering and erosion that have been transported for some distance, e.g. siliciclastic rocks as conglomerates, sandstones, shales.

/ klastische afzetting / dépôt clastique / klastisches Sediment / depósito clástico

3165
detrital, – said of grains or in general comparatively small fragments loosened from a parent material by weathering, erosion, or activities of organisms and transported for some distance; also said of deposits consisting of detrital particles. / detritisch / détritique / detritisch / detrítico

detritus, – detrital material. / detritus / matériel détritique / Detritus / detrito

3166
clast, – any mineral or rock fragment of any composition, enclosed in sedimentary material. / klastisch fragment; klastische korrel; klast / débris / Klast; Bruchstück / clasto

cryptoclastic, – composed of extremely fine, broken or fragmented particles, barely visible under the microscope. / kryptoklastisch / cryptoclastique / kryptoklastisch / criptoclástica

clastic ratio, – the ratio of the thickness of beds composed of clastics or of clastic material to respectively the thickness of non-clastic beds or clastic material in a sequence of beds.

3167
limeclast; doloclast, – rock fragments composed respectively of limestone or dolomite. / kalksteenfragment; dolomiet .... / débris calcaire; .... dolomitique / Kalkbruchstück; Dolomit.... / clasto calcáreo; .... dolomítico

NOTE:
– for details, see chapter [2100] Carbonate Deposits

3168
bioclastic, – said of a sedimentary rock consisting of the fragmental remains of organisms, such as a limestone composed of shell fragments. / bioklastisch / bioclastique / bioklastisch / bioclástica

bioclast, – fossil fragment forming part of a sediment. / bioklast / bioclaste / Bioklast / bioclasto

phytoclast, – a remnant of a plant or tree that has been transported through water and forms part of a sediment. / fytoklast / phytoclaste; débris de plantes / Phytoklast / fitoclasto

3169
skeletal, – pertaining to hard remains of organisms. Also said of rocks or rock materials that contain or consist of such remains. / skelet- / squelettique / Skelett- / esquelético

3170
pyroclastic deposit; tephra, – a deposit that is composed of fragmental materials, produced and fragmented by volcanism. / pyroklastische afzetting / dépôt pyroclastique / pyroklastische Ablagerung / depósito piroclástico

pyroclast, – any particle or fragment that has been ejected by an erupting volcano. / pyroklast / pyroclaste; débris volcanique / Pyroklast / piroclasto

NOTE:
– for details, see section [1224] The Extrusive Phase: Pyroclastics

3171
tuff, – consolidated pyroclastic deposit. / tuf / tuff volcanique / Tuff / toba

tuffaceous, – pertaining to or having the character of a tuff. / tuf-; tufachtig / tuffacé / Tuff- / tobáceo

tuffite, – a tuff containing a considerable amount of non-volcanic (e.g. detrital) material. / tuffiet / tuffite / Tuffit / tufita

3172
volcanoclastic, – pertaining to any deposit rich in fragments or grains of volcanic rocks, formed by volcanism, weathering and erosion of volcanic products. / vulkanoklastisch / volcanoclastique / vulkanoklastisch / volcanoclástico

3173
heavy mineral, – mineral with a density greater than that of bromoform (2.89). / zwaar mineraal / minéral lourd / Schwermineral / mineral pesado

light mineral, – mineral with a density less than that of bromoform. / licht mineraal / minéral léger / Leichtmineral / mineral ligero

3174
accessory mineral, – mineral that is represented by (very) small percentages of the total detrital grains in sedimentary deposits. / accessorisch mineraal / minéral accessoire / akzessorisches Mineral / mineral accesorio

3175
stable mineral; resister, – a mineral (especially a heavy mineral) that as a sand grain survived many successive cycles of deposition, weathering and erosion, and hence occurs in most, even highly mature arenaceous sediments, e.g. tourmaline, zircon. / doorloper / minéral stable; .... résistant / resistentes Mineral / mineral estable

3176
immature, – said of a clastic sediment, containing much easily weatherable material, and showing as a rule poor sorting and rounding of the particles. / onrijp / peu évolué; jeune / unreif / inmaduro

submature, – said of a clastic sediment characterized by little or no clayey material and by poorly sorted and angular grains.

mature, – said of clastic sediments that contain no easily weatherable minerals, that are well sorted, and the sand grains of which are often well-rounded. / rijp / évolué / reif / maduro

supermature, – said of a clastic sediment whose well-sorted grains are subrounded to well-rounded.

3177
micaceous, – containing mica. / glimmerhoudend / micacé / glimmerig / micáceo

glauconitic, – containing, resembling to or consisting of glauconite. / glauconietisch / glauconieux / glaukonitisch / glauconítico

3178
lithic, – said of medium-grained sedimentary rocks (e.g. sandstone) and pyroclastic deposits containing abundant rock fragments. Also said of such fragments. / met (veel) gesteentebrokjes / lithique / lithisch / lítico

3179
association; assemblage, – the minerals composing a sedimentary rock, especially the minerals occurring as detrital grains, e.g. heavy mineral association. / (mineralen)associatie / association; paragenèse / Assoziation / asociación

[2020] THE PARTICLES

[2021] *Shape of particles*

3180
sphaeroid; spheroid, – a body that in shape approaches a sphere. / sferoïde / sphéroide / Sphäroid / esferoide

spheroidal,
1. shaped like a spheriod;
2. containing, or characterized by spheroids.
/ sferoïdaal / sphéroidal / kugelig; sphäroidisch / esferiodal

sphericity, – the degree to which the shape of a particle approaches that of a sphere. / bolvormigheid; sfericiteit / sphéricité / Kugeligkeit / esfericidad

3181
equidimensional; equant, – said of particles or crystals having roughly the same diameters in all directions (e.g. between $\frac{2}{3}$ and $\frac{3}{2}$ of the mean). / isometrisch / équidimensional / isometrisch / equidimensional

3182
flatness, – / platheid; afplatting / aplatissement / Plattigkeit; Abplattung / aplanamiento

3183
platy, –
1. flat, tabular. / plat; afgeplat / plat / plattig / tabular; planar
2. said of sedimentary rocks splitting into thin layers. / platig; dunsplijtend / tabulaire / plattig / tabular; planar

3184
disk-shaped; discoidal, – / schijfvormig / discoidal / scheibenförmig; diskoidal / discoidal

3185
acicular, – needle-shaped. / naaldvormig / aciculaire / nadelförmig / acicular

rod-shaped, – / staafvormig / prismatique / stabförmig / – –

3186
chip, – a small, flat fragment split off from a crystal or rock. / schilfer / esquille / Splitter / esquirla

flake, – a small, flat, thin piece of material. / schub; schilfer / écaille; esquille / Schuppe / laminilla

3187
roundness, – the degree of rounding of edges and corners of a clastic fragment, irrespective of its general shape. / afronding / arrondissement / Rundung / redondamiento

roundness index, – ratio of average radius of curvature of the several edges or corners of the particles to the radius of curvature of the maximum inscribed sphere. / afrondingsindex / indice d'arrondissement / Rundungsindex / índice de redondamiento

roundness class, – an arbitrarily defined range of roundness values for the classification of sedimentary particles.

3188
well-rounded, – said of clastic elements, the entire surface of which consists of broad convexities. /sterk afgerond / bien arrondi / gut gerundet; stark gerundet / bien redondeado

rounded, – / afgerond / arrondi / gerundet / redondeado

subrounded, – /matig afgerond / subarrondi / angerundet / subredondeado

3189
subangular, – said of clastic elements showing beginning effects of wear. / zwak afgerond / subangulaire / subangular; Kanten gerundet / subangular

angular, – said of clastic elements having sharp edges and corners. / hoekig; niet afgerond / angulaire / eckig; angular / angular

3190
sharp-edged, – / scherpkantig / à arêtes vives / scharfkantig / de aristas angudas

rough; rugged, – / ruw; oneffen / rugueux / rauh; uneben / rugoso

3191
pivotability; rollability, – the capability of a clastic particle to be rolled. / rolbaarheid / aptitude à la rotation / pivotability; Rollfähigkeit / capacidad de rodadura

3192
surface texture, – the microrelief of the surface of detrital particles, not determining the general shape of the particles. / oppervlaktetextuur / état de surface / Oberflächenrelief / textura superficial

[2022] *Size of Particles*

NOTE
– see also section
[1970] Grain-Size Determination.

3193
particle size, – various definitions; in practice (for gravel, sand and silt) usually the width of the minimum square (sieve mesh) aperture through which the particle will pass. / deeltjesgrootte / dimension des particules / Partikelgrösse / tamaño de particula

particle diameter, – / deeltjesdoorsnede / diamètre des particules / Partikeldurchmesser / diámetro de la particula

grain size, – / korrelgrootte / taille du grain / Korngrösse / tamaño de grano

3194
coarse-grained, – / grofkorrelig / à grain grossier / grobkörnig / de grano grueso

medium-grained, – / middelkorrelig / à grain moyen / mittelkörnig / de grano medio

fine-grained, – / fijnkorrelig / à grain fin / feinkörnig / de grano medio

NOTE:
– in the phi scale the divisions have been assumed at: -1 and -2 respectively; see table 6.

3195
equivalent diameter, – the diameter of a quartz sphere having the same settling velocity as the particle under consideration. / equivalente diameter / diamètre équivalent / Äquivalentdurchmesser / diámetro equivalente

nominal diameter, – the computed diameter of a hypothetical sphere having the same volume as that calculated for a given sedimentary particle; it is a true measure of particle size independent of either the shape or the density of the particle.

sedimentation diameter, – measure of particle size, equal to the diameter of a hypothetical sphere of the same specific density and the same settling velocity as those of a given sedimentary particle settling in the same fluid.

3196
effective diameter, –
1. the diameter of the particles in an assumed material that would transmit water at the same rate as the material under consideration and that is composed of spherical particles of equal size and arranged in a specified manner,
2. the maximum diameter of the smallest 10% (by mass) of the particles of a sediment.

/ effectieve diameter / diamètre efficace / effektiver Durchmesser / diámetro efectivo

3197
grain-size frequency distribution, – the proportion of the successive grain-size classes in a particular granular rock. / korrelgrootteverdeling / distribution statistique granulométrique / Korngrössen-(häufigkeits)verteilung / distribución de frecuencias de tamaño de grano

polymodal distribution, – a grain-size frequency distribution having two or more modes. / polymodale korrelgrootteverdeling / distribution polymodale / polymodale (Korngrössen)verteilung / distribución polimodal

3198
skewness, – statistical measure, indicating the relative degree of spread of a cumulative size distribution curve out from the median towards the finer or coarser grades: $\frac{Q_1 Q_3}{Md^2}$

kurtosis, – measure of the peakedness of a frequency distribution: e.g. a measure of concentration of sediment particles about the median diameter: $\frac{Q_3 - Q_1}{2(P_{90} - P_{10})}$

P_{90}, P_{10} – particle diameters such that 90%, resp 10% of the particles are larger than the other 10%, resp. 90%,
Q_3, Q_1 – the quartiles, i.e. the diameter of a grain coarser than 75% of the material and finer than the other 25% (Q_3), respectively the diameter of a grain coarser than 25% of the material and finer than the other 75% (Q_1),
Md – median diameter.

3199
mean diameter, – arithmetic average diameter. / gemiddelde diameter / diamètre moyen / mittlerer Durchmesser / diámetro medio

median diameter; .... grain size, – the diameter of a grain that is coarser than 50% (by mass) of the material, and finer than the other 50%. / mediaan(diameter); ....korrelgrootte / médiane granulométrique / Median(durchmesser) / diámetro mediano

modal diameter, – the diameter that is the most frequent one in a particle-size distribution. / modale diameter; ..... korrelgrootte / diamètre modal / modaler Durchmesser / diámetro modal

3200
sorting, –
1. the process that selects sedimentary particles according to a property or combination of properties;
2. the process that increases the sorting of the grain sizes of a clastic deposit;
3. the degree of grain-size sorting of a clastic deposit.

/ sortering; gesorteerdheid / classement / Sortierung / clasificación

3201
sorting coefficient, – a numerical index for the degree of grain-size sorting of a clastic sediment, e.g. that of Trask:

$$S = \sqrt{\frac{Q_3}{Q_1}}$$

/ sorteringscoëfficiënt / coefficient de classement / Sortierungskoeffizient / coeficiente de clasificación

3202
sorted, – said of clastic deposits, the particles of which are mostly of the same size. / gesorteerd / classé / sortiert / clasificado

well-sorted, – having a sorting coefficient less than 2.5. / goed gesorteerd / bien classé / gut sortiert / bien clasificado

[2030] THE DEPOSITS

NOTE:
– see table 6 – Grades of Clastic Sediments

[2031] *Lutites*

3203
lutaceous, – consisting of detrital material with particle sizes less than $\frac{1}{16}$ mm. / lutietisch / limoneux / lutitisch / lutitica

lutite, – consolidated lutaceous rock. / lutiet / lutite / Lutit / lutita

lutum, – loose material with particle sizes smaller than $\frac{1}{16}$ mm.

3204
silt, – unconsolidated sedimentary material with grain sizes between $\frac{1}{16}$ and $\frac{1}{256}$ mm. / silt / silt / Silt; Schluff / limo

silty, – pertaining to, composed of or containing silt. / siltig; siltachtig / limoneux / siltig; schluffig / limoso

siltstone, – consolidated silt. / siltsteen / siltstone; grès très fin / Siltstein / limolita

3205
aleurite, – a term, originally used by Russian authors for unconsolidated sedimentary deposits with grain sizes (according to Strakhov) between 0.050 and 0.005 mm. / aleurite / aleuriet / Aleurit / aleurita

aleurolite, – a term originally used by Russian authors for a consolidated aleurite. / aleuroliet / aleurolite / Aleurolit / aleurolita

3206
dust, – fine-grained, mostly silt-sized, clastic material that is or has been transported by the wind, e.g. the material composing loess. / stof / poussière / Staub / polvo

3207
fines, – general term for fine-grained material, such as silt or clay particles. / fijnkorrelig materiaal / particules fines; les fines / feinkörniges Material; Feinkorn / finos

3208
pelite, – a fine-grained sediment, consolidated or unconsolidated. The upper particle-size limit is taken by many authors at 0.02 mm. The term is mainly used in Europe. / peliet / pelite / Pelit / pelita

pelitic, – having the granular composition of a pelite. / pelietisch / pelitique / pelitisch / pelítico

3209
rock flour, – very fine-grained (silt or clay size), not weathered mechanical detritus, formed e.g. by glacial abrasion. / gesteentemeel; gesteenteslijpsel / poussière d'abrasion; .... glaciaire / Gesteinsmehl / polvo de abrasión

3210
sludge; slurry, – a semifluid mass of sedimentary material on the bottom of a lake, a sea, etc. / blubber; slappe modder / slush / Schlick; Modder / fango

ooze, –
1. a very soft, often semifluid mud, e.g. in rivers, estuaries and lakes;
2. a deep-sea deposit consisting for more than 30% of the remains of siliceous or calcareous pelagic organisms.

/ slik / boue / Schlamm / barro

mud, – a soft, plastic or semifluid deposit, rich in pore water, partly or entirely consisting of fine-grained material (silt or clay size), sometimes containing much sand, scattered shells etc. / slik; modder / boue / Schlick; Schlamm / fango

muddy, – having the character of mud. / slikachtig; modderig / boueux / schlickig; schlammig / fangoso

3211
argillaceous, – largely composed of, or containing clay minerals (e.g. argillaceous deposits, argillaceous sandstone). / kleiig / argileux / tonig / arcilloso

argillite, – hard argillaceous rock (harder than mudstone) with a conchoidal fracture. / argilliet / argilite / Argillit; Tongestein / argilita

3212
clay, – unconsolidated deposit of very fine particle size (the upper size limit is generally taken at $\frac{1}{256}$ mm) and mainly composed of clay minerals. / klei / argile / Ton / arcilla

3213
clay mineral, – one of a loosely defined group of finely crystalline or colloidal hydrous aluminium silicates; they are the stable secondary products formed by the decomposition of other aluminium silicates (e.g. feldspars). Structurally they consist of stacked sheets (layers) of SiO_4 groups and of Al-hydroxyl or oxygen groups, linked by oxygens common to both. Chemistry and structure of the clay minerals tend to give clayey deposits their specific chemical and physical properties (e.g. plasticity, imperviousness, semi-permeable membrane and ion-exchange capacity, swelling). / kleimineraal / minéral argileux; argile / Tonmineral / mineral de arcilla

3214
two-layers clay mineral, – of which **kaolinite** is the most common representative; structurally composed essentially of one sheet of SiO_4 groups linked with one Al-hydroxyl sheet. No substitution of Al or Si by other elements occurs.

three-layers clay mineral, – of which the **montmorillonite** group possesses one sheet of Al-hydroxyl groups between two sheets of SiO_4 groups, stacked with water molecules in between, which gives the montmorillonites their swelling properties. Substitution of part of the Si by Al and introduction of K in between the layers to neutralize the electric charge results in the **illite** structure. In some minerals Al is entirely substituted by Mg and Fe; apart from K, also Na, Ca or Mg can be introduced between the layers.

mixed-layer clay mineral, – layers of more than one clay mineral are interleaved in a single crystal.

3215
loam, – a sandy and silty clay. The term is normally used only for terrestrial deposits. Various definitions: e.g. a mixture of clay, silt and sand, in which none of these constituents makes up more than 50%; a mixture of clay, silt and sand, in which silt strongly predominates. / leem / limon / Lehm / greda

3216
lean clay, – a clay of low to medium plasticity. / magere klei / argile maigre / magerer Ton / – –

fat clay, – a clay of high plasticity. / vette klei / argile grasse; .... plastique / fetter Ton / arcilla plástica

3217
fire clay, – a clay, usually rich in kaolinite, that on heating changes into hard, fire-resisting material. / vuurvaste klei / argile réfractaire / Feuerton / arcilla refractaria

kaolin; china clay, – a normally soft rock, either hydrothermal or sedimentary, mainly composed of kaolinite. / kaolien / kaolin / Kaolin / caolín

3218
bentonite, – a soft rock, composed of clayey material, usually montmorillonite, and formed by devitrification and accompanying chemical alteration of volcanic rocks, rich in glass, mostly layers of ash or tuff. / bentoniet / bentonite / Bentonit / bentonita

metabentonite, – bentonite that owing to recrystallization has become incapable of absorbing or adsorbing large quantities of water. / metabentoniet / 'metabentonite' / Metabentonit / metabentonita

Fuller's earth, – montmorillonite-rich clay, used as adsorbent of grease etc. / voldersaarde / terre à foulon / Fullererde; Walkerde / tierra de batanero

3219
mudstone, – an indurated mud, lacking the fine lamination or fissility of shale. / – – / – – / Schlammstein / fangolita

pebbly mudstone, –
1. a delicately laminated mudstone, in which thinly scattered pebbles are embedded along somewhat distorted bedding planes;
2. a poorly sorted till-like rock composed of dispersed pebbles in an abundant mudstone matrix.

claystone, – an indurated clay, lacking the fine lamination or fissility of shale. / kleisteen / argilite / Tonstein / argilita

3220
shale, – an indurated, fine-grained detrital rock, characterized by a finely laminated structure and/or high fissility. / schalie / argilite / Schiefer / pizarra

clay shale, – a shale, also characterized by a mainly clayey mineral composition. / kleischalie / argile schisteuse / Tonschiefer / pizarra

shaly, – composed of, containing or having the character of a shale. / schalie-achtig; schalie-houdend / schisteux / schiefrig / pizarroso

paper shale, – a shale, often highly carbonaceous, that easily separates into very thin laminae, suggesting sheets of paper. / papier-schalie / schiste carton / Blätterschiefer / pizarra hojosa

3221
marl, – a consolidated or semiconsolidated mixture of clay and (generally fine-grained) carbonate material (terrigenous or produced locally by organisms or biochemical processes), formed in the sea or in fresh water. According to one definition it contains between 25 and 75% clay. / mergel / marne / Mergel(stein) / marga

marlstone; argillaceous limestone, – a hard, strongly indurated marl. / mergel; kleiige kalksteen / marne / Mergelstein / marga; caliza arcillosa

3222
marly mud; marl mud, – an unindurated sediment, rich in interstitial water, but otherwise having the composition of a marl. / mergelslik / boue marneuse / mergeliger Schlick / fango margoso

3223
marly clay, – a clay having a certain content of fine-grained carbonate material, according to one definition between 10 and 25%. / mergelige klei / argile marneuse / mergeliger Ton / arcilla margosa

clayey marl, – a marl containing more clay than carbonates. / kleiige mergel / marne argileuse / toniger Mergel / marga arcillosa

3224
limy marl, – a marl containing more carbonate than clay. / kalkige mergel / marne calcaire / kalkiger Mergel / marga carbonatada

[2032] *Arenites*

3225
areanceous; arenarious; **psammitic,** – said of deposits, unconsolidated, consolidated or lithified, that are essentially composed of detrital grains of sand size, i.e. according to the Wentworth scale between 2 and $\frac{1}{16}$ mm in diameter. / zandig / sableux / psammitisch; sandig / arenoso; psammítico

arenite, – lithified arenaceous rock. / areniet / arénite / Arenit / arenita

3226
psammite, – term used mainly in Europe, for sandy deposits,

both consolidated and unconsolidated. The size limits are often taken at $\frac{1}{16}$ and 2 mm. / psammiet / psammite / Psammit / psammita

psammitic, – having a psammitic texture. / psammietisch / psammitique / psammitisch / psammítico

3227
sand, – unconsolidated arenaceous material. / zand / sable / Sand / arena

sand size, – having diameters between 2 and $\frac{1}{16}$ mm on the Wentworth scale. / zandkorrelgrootte / sableux; de la taille des sables / Sandkorngrösse / tamaño arena

glass sand, – sand, suitable for glass making, containing more than 98% silica and less than 1% iron oxide. / glaszand / sable pour verrerie / Glassand / arena de vidrio

loamy sand, – various definitions: e.g. a clayey and silty soil or terrestrial deposit containing more than 70% sand. / lemig zand / sable limoneux / lehmiger Sand / arena margosa

3228
sandstone, – a consolidated arenaceous rock, having a dominantly siliceous composition. / zandsteen / grès / Sandstein / arenisca

clean sandstone, – a sandstone containing less than 10 or 15% argillaceous matrix. / schone zandsteen / grès propre / reiner Sandstein / – –

3229
grit, – a coarse-grained sand or sandstone, later specified as being characterized by angularity of grains. Use of term chiefly limited to Great Britain.

3230
orthoquartzite; orthoquartzitic sandstone, –
1. quarz arenite;
2. quartzitic sandstone.

/ orthokwartsiet / orthoquartzite / Quarzsandstein / ortocuarcita

quartz arenite, – an arenite in which quartz forms 95% or more of the (arenaceous) framework. / kwartsareniet / arénite quartzique / Quarzarenit / cuarzarenita

quartzitic sandstone; quartzite (p.p.), – a so strongly cemented quartz arenite that the rock breaks across rather than around the grains. / kwartsietische zandsteen; kwartsiet / grès quartzitique; quartzite / quarzitischer Sandstein; Quarzit / arenisca cuarcítica; cuarcita

3231
greensand, – glauconitic sand. / groenzand / sable vert / Grünsand / arena verde

3232
feldspathic sandstone, –
1. any sandstone containing an appreciable amount of feldspar;
2. a sandstone with 10 to 25% feldspar, but without appreciable matrix.

/ veldspaathoudende zandsteen; veldspaatrijke .... / grès feldspathique / Feldspatführender Sandstein / arenisca feldespática

3233
subarkose, – a rock transitional between a quartz sandstone and an arkose, e.g. feldspathic sandstone. / subarkose / grès subarkosique / Subarkose / subarcosa; sobarkosa

arkose, – an indurated arenaceous deposit consisting chiefly of quartz and feldspar grains, the latter making up more than 25%. The grains are generally poorly rounded, being freshly derived from granitic or granodioritic rocks (or gneisses). The term is also applied to rocks of the same composition, containing a large proportion of granules. / arkose / arkose / Arkose / arcosa; arkosa

3234
subgreywacke; subgraywacke, – a rock transitional between a quartz arenite and a greywacke. / subgrauwacke / subgrauwacke / Subgrauwacke / subgrauvaca.

greywacke; graywacke, – various definitions. According to Füchtbauer and Müller: a field term for dark, (greenish-) grey sandstones, rich in clayey matrix containing mica and chlorite, usually with abundant rock particles and with a varying feldspar content; the arenaceous particles are poorly rounded. / grauwacke / grauwacke / Grauwacke / grauvaca

3235
wacke, – arenaceous rock rich in fine-grained matrix (e.g. from 10 or 15 to 75% of the total material)

3236
placer, – a mineral deposit formed by mechanical concentration of clastic particles. / placer / placer / Seife / placer

[2033] *Rudites*

3237
rudaceous, – said of deposits, unconsolidated, consolidated or lithified, that are essentially composed of rock fragments (rounded or angular) with diameters larger than sand size, i.e. according to the Wentworth scale larger than 2 mm. / rudietisch; psefietisch / pséphitique / ruditisch; psephitisch / rudítico; psefítico

rudite, – a consolidated rudaceous rock. / rudiet / rudite / Rudit / rudita

3238
psephite, – term mainly used in Europe for clastic deposits, consolidated or unconsolidated, with an essential amount of fragments (rounded or angular) larger than 2 mm. / psefiet / pséphite / Psephit / psefita

psephitic, – having the texture of a psephite. / psefietisch / pséphitique / psephitisch / psefítico

3239
phenoclast, – a clastic element larger than a sand grain, i.e. more than 2 mm. / fenoklast / phénoclaste / Phenoklast / fenoclasto

3240
granule, –
1. a grain;
2. a relatively coarse clastic grain, having, according to the Wentworth scale, a diameter between 2 and 4 mm.
/korrel / granule / Korn (1) / gravilla

3241
pebble, – a relatively small rudaceous rock fragment showing signs of abrasion. The diameter, according to the Wentworth scale, is between 4 and 64 mm. / steen; kiezel(steen); rolsteen / gravier; caillou / Geröll / grava

3242
cobble; cobblestone, – a relatively large, more or less rounded rock fragment. According to the Wentworth scale it has diameters between 64 and 256 mm. / kei / galet; pierre / gerundeter Stein / canto

roundstone, – a rounded rudaceous rock fragment. / rolsteen; kei / galet roulé / Geröll / roca redondeada

stone, – a rock fragment larger than a few mm. / steen / pierre / Stein / piedra

3243
block, – an angular rock fragment, larger than 256 mm (Wentworth scale) or 20 cm (many European authors). / blok / bloc anguleux / Block / bloque

boulder, – a rock fragment with a more or less rounded shape, larger than 256 mm (Wentworth scale) or 20 cm (many European authors). / kei / bloc arrondi / gerundeter Block / bloque

3244
gravel, – unconsolidated deposit of coarse clastics. The diameters, according to the Wentworth scale, are larger than 2 mm. Some authors restrict the term to coarse material with diameters less than 5 or less than 20 cm. / grind / gravier / Kies; Schotter / gravera

gravelly, – containing gravel-sized material. / grindhoudend / graveleux / kiesig / semejante a grava

3245
bone bed, – a rudaceous, sometines arenaceous layer containing may fossil bones or bone fragments, usually together with teeth, scales and coprolites. / beenderenlaag; bottenlaag / couche d'ossements; bone bed / – – / – –

3246
clay gall, – gravel-sized, usually flat clast, composed of clay. / kleibal / galet mou / Tongalle; Tongeröll / canto blando

3247
mud ball, – a more or less spherical coherent mass of mud. / slikbal / galet de boue / Schlickgeröll / – –

armoured mud ball, – a subspherical mass of silt or clay, often some 5 to 10 cm in diameter, that has become studded with sand and fine gravel when rolling downstream. / gepantserde kleibal / galet de boue; hérisson de .... / gepanzertes Schlickgeröll / canto armado

3248
mud flake, – clast of mud, mostly very flat and of gravel size, usually incorporated in sand beds, especially near the base. Lithologically the clasts may consist of clay, with or without admixtures of silt and sand. They are produced by reworking of sun-cracked mud layers by wind or water, and by undercutting of mud layers or laminae by water (e.g. in tidal deposits and turbidites). / slikbrok; slikplak / galet plat argileux / Tongalle; Tongeröll / – –

3249
granule gravel, – an unconsolidated deposit consisting mainly of granules (in practice usually called **very coarse sand**). / – – / gravier fin / Grobsand / gravera de gránulos

boulder gravel, – a gravel consisting mainly of boulders. / blokkenafzetting; blokkengrind / gravier grossier / Blockwerk / gravera de bloques

3250
boulder clay, – term used in Great Britain for glacial deposits, consisting of boulders of various size, embedded in a clayey matrix, which usually contains also silt, sand and gravel. **Till** is a more general term for unsorted glacial deposits (with or without boulders). / keileem; keienleem / argile à blocaux / Geschiebelehm / – –

3251
conglomerate, – an indurated sediment containing an essential quantity of rounded rock fragments larger than 2 mm, and normally having a matrix composed of sand. / conglomeraat / conglomérat / Konglomerat / conglomerado

conglomeratic, – pertaining to, or having the texture of a conglomerate. / conglomeratisch / conglomératique / konglomeratisch / conglomerático

3252
mud-flake conglomerate, – conglomerate in which the phenoclasts were originally mostly mud flakes. The latter may have changed by lithification into shale or mudstone pebbles. / slikpakjesconglomeraat / brèche; conglomérat à galets d'argile / Tongallenkonglomerat / – –

3253
puddingstone, – in Great Britain, a conglomerate with well-rounded pebbles, strongly contrasting with the light-coloured, fine-grained matrix or cement. / puddingsteen / – – / – – / pudinga

3254
boulder conglomerate, – a conglomerate consisting mainly of boulders. / blokkenconglomeraat / conglomérat grossier / Block-Konglomerat / conglomerado de bloques

3255
orthoconglomerate, – a conglomerate having an intact framework of pebbles and coarse sand, bound together by a mineral cement. / orthoconglomeraat / orthoconglomérat / Orthokonglomerat / ortoconglomerado

3256
conglomeratic mudstone; paraconglomerate; pebbly mud-

stone, – a conglomerate having more lutaceous matrix than rudaceous clasts, e.g. a tillite, or a lithified slide mass. / paraconglomeraat / paraconglomérat / Parakonglomerat; konglomeratischer Schlammstein / paraconglomerado; fangolita conglomerática

3257
intraformational conglomerate, – a conglomerate formed by penecontemporaneous fragmentation and redeposition, e.g. a mud-flake conglomerate. / intraformationeel conglomeraat / conglomérat intraformationnel / intraformationelles Konglomerat / conglomerado intraformacional

3258
breccia, – a lithified rock, rich in angular rock fragments larger than 2 mm. / breccie / brèche / Breccie; (Brekzie) / brecha

sedimentary breccia, – a breccia formed by sedimentary processes. / sedimentaire breccie / brèche sédimentaire / sedimentäre Breccie / brecha sedimentaria

3259
rubble, – a loose mass, layer or accumulation of rough, irregular, or angular rock fragments broken from larger masses usually by physical (natural or artificial) forces, coarser than sand (diameter greater than 2 mm) and commonly but not necessarily poorly sorted; the unconsolidated equivalent of a breccia.

3260
mud breccia, – dessication breccia containing angular or slightly rounded fragments of fine-grained argillite embedded in somewhat coarser grained and more arenaceous . material.

3261
bone breccia, – a breccia rich in bones and (angular) bone fragments, especially such a deposit formed in a cave or at the bottom of a karst pit, pothole or fissure. / bottenbreccie / brèche ossifère / Knochenbreccie / – –

3262
edgewise breccia; edgewise conglomerate, – breccia or conglomerate in which the flat sides of the phenoclasts have a high inclination to the bedding plane.

3263
collapse breccia, – a breccia formed by the collapse of rock overlying a cavity, e.g. a cave. / instortingsbreccie; samenzakkingsbreccie / brèche d'effondrement / Einsturzbreccie / brecha de colapsamiento

3264
intraformational breccia, – a breccia formed by penecontemporaneous fragmentation and redeposition, often of calcareous or dolomitic composition. / intraformationele breccie / brèche intraformationelle / intraformationelle Breccie / brecha intraformacional

3265
polymict; polymictic, – said of conglomerates and gravels the coarse constituents of which are derived from many different parent rocks / polymict / polygénique / polymikt / polimicto

oligomict;, oligomictic – said of conglomerates and gravels, the constituents of wich are derived from a few different parent rocks only. / oligomict / oligogénique / oligomikt / oligomicto

monomict;, monomictic – said of breccias and conglomerates (gravels), the (coarse) constituents of which are derived essentially from the same parent rock. / monomict / monogénique / monomikt / monomicto

[2100] CARBONATE DEPOSITS

[2110] GENERAL TERMS

3266
carbonate deposit; .... sediment, – deposits formed by the organic or inorganic precipitation from aqueous solutions or composed of detrital carbonate particles. The precipitates consist of carbonates of calcium, magnesium or iron. / sedimentair karbonaatgesteente / dépôt carbonaté; sédiment .... / Karbonatablagerung; ....sediment / depósito carbonatado; sedimento ....

carbonate rock, – rock, either sedimentary or precipitate, composed for more than 50% by mass of carbonate minerals, e.g. $CaCO_3$, $CaMg(CO_3)_2$. / karbonaatgesteente / roche carbonatée / Karbonatgestein / roca carbonatada

3267
calcareous deposit, – said of a deposit that contains up to 50% calcium carbonate. / kalkafzetting / dépôt calcaire / Kalkablagerung / depósito calcareo

3268
orthochemical, – said of a carbonate deposit, formed through direct chemical precipitation. / orthochemisch / de précipitation chimique / orthochemisch / ortoquímica

3269
biolith, – a deposit directly due to the physiological activities of organisms, e.g. coral or reef limestone, diatomite, etc. / biogene afzetting; organogene ....; bioliet / biolite / Biolith / biolito

biolithite, – a limestone constructed by organisms that grew and remained in place, and characterized by a rigid framework of carbonate material that binds allochems and bioclasts (e.g. in the reef core); the major organism should be specified when using the term, e.g. coral biolithite. / organogeen gesteente; biogeen .... / biolithite / Biolithit; Fossilfestkalk / biolitito

[2120] TEXTURE

[2121] *Carbonate Particles*

3270
limeclast, – a fragment of limestone that behaves as a detrital particle, and that has been produced by physical disintegration of a larger mass either within or outside the depocentre of accumulation. / kalksteenfragment / fragment calcaire; débris .... / Kalkbruchstück; Kalkfragment / clasto calcáreo

doloclast, –
1. a fragment derived by erosion from an older dolomite;
2. an intraclast disrupted from partly consolidated dolomitic mud on the bottom of sea or lake.

/ dolomietfragment / fragment dolomitique; intraclaste .... / Dolomitbruchstück; ....fragment / clasto dolomítico

3271
allochem, – a collective term for one of several varieties of discrete and organized carbonate aggregates that serve as the coarser framework grains in most mechanically deposited carbonates; they include intra- and extraclasts, ooids, pellets, lumps and bioclasts. / allochemisch deeltje / allochème / Allochem / aloquímico

orthochem, – particle of carbonate that is formed by direct chemical precipitation within the depositional or the diagenetic environment. / orthochemisch deeltje / orthochème / Orthochem / ortoquímico

3272
intraclast, – fragment of penecontemporaneously, generally weakly consolidated sediment that has been eroded from adjoining parts of the sea bottom and redeposited within the area of original deposition. / autochthoon deeltje / intraclaste / Intraklast / intraclasto

extraclast, – a fragment of sedimentary material, produced by erosion of an older rock outside the area in which it is accumulated. / allochthoon deeltje / extraclaste / Extraklast / extraclasto

3273
bioclastic, – said of a detrital deposit (carbonate) composed of skelets or skeletal fragments. / bioklastisch / bioclastique / bioklastisch / bioclástica

[2122] *Grain-Size Classification*

3274
calcilutite; calcipelite, –
1. a limestone consisting predominantly of calcite particles of silt and clay size, i.e. smaller than $\frac{1}{16}$ mm;
2. the term is also used in a broadened sense to include chemically precipitated crystalline components.

/ kalkslibsteen; calcilutiet / calcilutite; calcisiltite / Kalklutit; Kalkpelit / calcilutita

NOTES:
- in the grain-size classification of limestones and dolomites, the Wentworth/phi scales are applied (table 6).
- the prefix calci refers to limestones; the prefix dolo to dolomites.
  e.g. **calcilutite**
  dololutite
- if the constituents are composed mainly of bioclasts, the prefix bio may be added.
  e.g. **biocalcilutite**
  biodololutite
- if the deposit contains abundant clasts, the prefix litho may be added.
  e.g. **lithocalcilutite**
  lithodololutite

3275
calcisiltite, – a limestone consisting predominantly of detrital carbonate particles of silt size, i.e. between $\frac{1}{16}$ and $\frac{1}{256}$ mm. / kalkslibsteen / aleurolite calcaire; calcarénite fine / Kalksiltit / calcisiltita

3276
calcarenite, –
1. a limestone consisting predominantly (more than 50%) of detrital calcite particles of sand size, i.e. between $\frac{1}{16}$ and 2 mm.
2. a consolidated calcareous sand.

/ kalkzandsteen / calcarénite / Kalkarenit / calcarenita

3277
calcarenitic limestone, – a limestone composed for more than 10% of original calcareous mud matrix (particles with diameters less than $\frac{1}{16}$ mm) accompanied by more than 10% carbonate grains of sand and gravel size. / kalkzandsteenachtige kalk / – – / kalkarenitischer Kalkstein / caliza calcarenítica

calcarenaceous, – said of a sandstone containing an abundant amount (up to 50%) of calcium-carbonate detritus. / kalkzandig / – – / kalksandig / calcarenítica

3278
calcirudite, – a limestone consisting predominantly (more than 50%) of detrital calcite particles larger than 2 mm in diameter, and often also cemented with calcareous material. / calcirudiet / calcirudite / Kalkrudit; Kalksteinkies / calcirudita

calcigravel, – an unconsolidated deposit consisting predominantly of detrital calcite; particles larger than 2 mm. / kalkkiezel / – – / Kalkkies / gravera calcárea

3279
calcibreccia, – calcirudite, rich in angular fragments larger than 2 mm. / kalkbreccie / brèche calcaire / Kalkbreccie / calcibrecha; brecha calcárea

3280
sparite, – contraction of sparry calcite; a descriptive term for the crystalline, clear interstitial component of carbonates, consisting of clear calcite or aragonite; it is coarser than micrite, having a particle size larger than 0.02 mm. / spariet / sparite / Sparit / sparita

spar, – in the present context a microscopic crystal of calcite that forms part of the cementing matrix. / spaat / – – / Spat / – –

dolosparite, – a sparry dolomite crystal, i.e. larger than 0.02 mm. / dolomietspariet / – – / – – / – –

3281
micrite, –
1. an inorganic or biochemical precipitate, formed within the basin of deposition and showing little or no evidence of significant transport;
2. a microcrystalline limestone with less than 1% allochems;
3. a microcrystalline limestone with 1 to 10% allochems, in which case a prefix should be added, e.g. oolite-bearing, fossiliferous.

/ micriet / vase micritique; micrite; calcaire microcristallin; .... micritique / Mikrit / micrita

3282
calcimicrite; carbonate mudstone, – a limestone with particle diameters less than 0.02 mm, i.e. a finely silty and clayey size. / calcimicriet / – – / Kalzimikrit / calcimicrita

dolomicrite; dolomite mudstone, – a sedimentary rock consisting of clay-sized dolomite crystals, interpreted as a lithified dolomite mud and containing less than 1% allochems. / dolomietslibsteen; dolomicriet / dolomicrite; micrite dolomitique / Dolomikrit / dolomicrita

3283
intrasparite, – carbonate with more than 10% allochems of which more than 25% are intraclasts, and in which the sparry-calcite cement is more abundant than the micrite matrix. It commonly characterizes carbonate depositional environments of high physical energy; the spar is usually intergranular cement. / intraspariet / intrasparite / Intrasparit / intraesparita

intrasparenite, – an intrasparite containing sand-sized intraclasts. / intrasparzandsteen / – – / Intrasparalrenit / intraespararenita

intrasparrudite, – an intrasparite containing gravel-sized intraclasts, i.e. size larger than 2 mm. / intrasparrudiet / – – / Intrasparrudit / intraesparrudita

NOTE:
– see table 7 – Particle Size and Composition of Limestones.

3284
intramicrite, – carbonate with more than 10% allochems of which more than 25% are intraclasts in a micrite matrix. / intramicriet / – – / Intramikrit / intramicrita

intramicarenite, – an intramicrite containing sand-sized intraclasts. / intramicareniet / – – / Intramikarenit / intramicarenita

intramicrudite, – an intramicrite containing gravel-sized intraclasts, i.e. median size larger than 1 mm. / intramicrudiet / – – / Intramikrudit / intramicrudita

3285
oosparite, – a carbonate containing at least 25% ooids and no more than 25% intraclasts, and in which the sparite cement is more abundant than the micrite matrix. It is common in depositional environments of high wave or current energy. The spar is usually an intergranular cement. / oöspariet / – – / Oosparit / ooesparita

oospar(ar)enite, – an oosparite containing sand-sized ooids. / oösparenieet / grès oolitique / Oospararenit / ooespar(ar)enita

oosparrudite, – an oosparite containing ooids that are more than 2 mm in diameter. / oösparrudiet / – – / Oosparrudit / ooesparrudita

3286
oomicrite, – a carbonate containing at least 25% ooids and no more than 25% intraclasts, and in which the micrite is more abundant than the sparite cement. / oömicriet / – – / Oomikrit / oomicrita

oomicrudite, – an oomicrite containing ooids, whose median size is larger than 1 mm. / oömicrudiet / – – / Oomikrudit / oomicrudita

3287
biosparite, – carbonate containing less than 25% intraclasts and less than 25% ooids with a volume ratio of bioclasts to pellets greater than 3:1; the sparite cement is more abundant than the micrite matrix. It characterizes high-energy depositional environments with the spar being normally an intergranular cement. Sorting of bioclasts allows further textural subdivison. The major organism should be specified when using the term, e.g. pelecypod biosparite. / biospariet / – – / Biosparit / bioesparita

biospararenite, – a biosparite containing sand-sized bioclasts. / biosparzandsteen / – – / Biospararenit / bioespararenita

biosparrudite, – a biosparite containing rudite-sized bioclasts. / biosparrudiet / – – / Biosparrudit / bioesparrudita

3288
biomicrite, – a carbonate containing less than 25% intraclasts and less than 25% ooids with a volume ratio of bioclasts to pellets greater than 3:1; the micrite matrix is more abundant than the sparite cement. It characterizes environments with relatively low physical energy. The major organism should be specified when using the term, e.g. crinoid biomicrite. / biomicriet / – – / Biomikrit / biomicrita

sparse biomicrite, – a biomicrite in which the bioclasts make up 10-50% of the rock. / bioklast-arme biomicriet / micrite à bioclastes épars; .... peu fossilifère / locker gepackter Biomikrit / – –

packed biomicrite, – a biomicrite in which the bioclasts make up over 50% of the rock. / bioklast-rijke biomicriet / micrite à bioclastes abondants; biomicrite richement fossilifère / dicht gepackter Biomikrit / – –

3289
biomicrosparite, –
1. a biomicrite in which the micrite matrix has recrystallized to microspar;
2. a microsparite containing fossils or fossil fragments.

/ biomicrospariet / – – / Biomikrosparit / biomicroesparita

biomic(ro)rudite, – a biomicrite containing rudite-sized fossils or fossil fragments. / biomic(ro)rudiet / – – / Biomikrudit / biomicrudita

3290
biopelsparite, – carbonate containing less than 25% intraclasts and less than 25% ooids with a volume ratio of fossils

to pellets ranging from 3:1 to 1:3; the sparite cement is more abundant than the micrite matrix. / biopelspariet / — — / Biopelsparit / biopelsparita

3291
biopelmicrite, — carbonate containing less than 25% intraclasts and less than 25% ooids with a volume ratio of fossils to pellets ranging from 3:1 to 1:3; the micrite matrix is more abundant than the sparite cement. / biopelmicriet / — — / Biopelmikrit / biopelmicrita

3292
pelsparite, — a carbonate containing less than 25% intraclasts and less than 25% ooids with a volume ratio of pellets to bioclasts greater than 3:1; the sparite cement is more abundant than the micrite matrix. / pelspariet / — — / Pelsparit / pelesparita

3293
pelmicrite, — a carbonate containing less than 25% intraclasts and less than 25% ooids with a volume ratio of pellets to bioclasts greater than 3:1; the micrite matrix is more abundant than the sparite cement. / pelmicriet / — — / Pelmikrit / pelmicrita

[2123] *Depositional Textures*

NOTE:
— see table 8 — Depositional Texture of Limestones.

3294
mud supported, — term used in carbonate rock classification for muddy carbonate rock that contains more than 10% grains, but not in sufficient amounts to be able to support one another. / slibondersteuning / à structure empâtée / schlammgestützt / — —

grain supported, — term used in carbonate classification for carbonate sedimentary rock in which grains touch each other and form the framework of the rock. / korrelondersteuning / à structure jointive / komponentengestützt; korngestützt / — —

3295
lime mud, — unconsolidated calcareous mud with particles of clay and silt size. / kalkslib / boue calcaire / Kalkschlamm / fango carbonatado; barro ....

3296
lime mudstone, — muddy carbonate rock containing less than 10% grains. / kalkslibsteen / micrite / Kalk-mudstone / caliza mudstone

lime wackestone, — mud-supported carbonate rock containing more than 10% grains, i.e. particles smaller than 2 mm.

floatstone, — mud-supported carbonate rock with more than 10% particles larger than 2 mm.

3297
lime packstone, — grain-supported carbonate rock with a matrix of mud, i.e. particles smaller than 0.02 mm.

grainstone, — grain-supported mud-free carbonate rock with particles smaller than 2 mm.

rudstone, — grain-supported carbonate rock with more than 10% particles larger than 2 mm.

3298
boundstone, — autochthonous limestone whose original components are organically bound during deposition; boundstone is sometimes subdivided into framestone, bindstone and bafflestone.

3299
framestone, — specific term, denoting boundstone, bound by organisms forming a rigid framework (e.g. branching corals).

bindstone, — specific term, denoting boundstone, bound by encrusting and binding organisms.

bafflestone, — specific term, denoting boundstone, bound by in situ, stalk-shaped organisms that reduced the rate of flow of water in the depositional environment.

3300
oncolite, — growth of stacked hemispheroidal algae laminae around a free-moving flake of mud. / oncoliet / oncolithe / Onkolith / oncolito

3301
lump, — composite grain with lobate outline composed of grains texturally similar to the material in which lumps occur. / klont / agrégat / — — / — —

3302
grapestone, — compound grain made up of various carbonate-sand grains.

3303
bahamite, — early-cemented particles, forming rocks that resemble the predominant deposits in the Bahamas. They vary from calcisiltite to calcirudite; the grains are accretionary and commonly composite, existing of smaller particles bound together by precipitated material. / bahamiet / bahamite / Bahamit / bahamita

3304
bird's eye structure, — a structure of patches of finely crystalline calcite in carbonates, indicating supratidal environmental conditions.

[2130] THE DEPOSITS

3305
limestone, – indurated rock consisting almost entirely of calcium carbonate; specifically, a sedimentary rock containing more than 95% calcite and less than 5% dolomite. / kalksteen / calcaire / Kalkstein / caliza

3306
lithographic limestone, – aphanitic compact carbonate rock composed of uniformly sized grains or crystals. / lithografische kalksteen / calcaire lithographique / lithographischer Kalkstein / caliza litográfica

3307
calcareous ooze, – fine-grained pelagic deposits consisting mainly of material of calcareous organic origin, to be defined by the predominant organic constituents, e.g. pteropod ooze, globigerina ooze. / kalkslib / vase / Schlamm / barro

pelleted limestone, – generally calcareous ooze, transformed into a limestone, and characterized by abundant pellets. / pelletkalksteen / calcaire à pelotes / (kot)pillenführender Kalk(stein) / caliza peletífera

3308
coquina, – a limestone composed of loosely aggregated shells and shell fragments that are mechanically sorted. / schelpenkalk / lumachelle / mässig fester Schillkalk; .... fester Fossilkalk / coquina

microcoquina, – fine-grained detrital limestone made up of silt-sized shell fragments. / microcoquina / microlumachelle / Mikroschillkalk / microcoquina

coquinite, – compact, indurated and firmly cemented equivalent of coquina. / schelpenkalksteen / lumachelle (consolidée) / stark verfestigter Schillkalk; .... verfestigter Fossilkalk / coquinita

3309
coquinoid, – said of a deposit consisting of coarse, unsorted and often whole shells and fragments that have accumulated in place; generally with a fine-grained matrix. / – – / lumachellique / – – / lumaquélica

3310
coral limestone, – a limestone predominantly composed of corals or coral fragments. / koralenkalk / calcaire corallien / Korallenkalkstein; Korallenkalk / caliza de corales

reef limestone, – a limestone consisting of the remains of reef-building organisms, such as corals, sponges and bryozoans, and of sediment-binding organic constituents, such as calcareous algae. / rifkalk / calcaire récifal / Riffkalk; Riffkalkstein / caliza arrecifal

3311
coral mud, – lime mud in which corals or coral fragments occur. / koraalhoudend slib / boue corallienne / korallenführender Schlamm / fango de coral

coral sand, – loose sediment consisting of sand-sized fragments of corals. / koraalzand / sable corallien / Korallensand / arena de coral

3312
reef tufa, – drusy, prismatic, fibrous calcite deposited directly from supersaturated water upon the void-filling internal sediment of the mudstone of a reef knoll. / riftuf / – – / – – / – –

3313
crinoidal, – said of a limestone containing abundant crinoid fragments. / crinoiden- / à entroques / Crinoiden- / crinoidea

encrinitic; encrinite, – limestone with more than 50% crinoidal fragments. / crinoidenrijke kalksteen / encrinite / Crinoidenkalk; .... kalkstein / caliza encrinitica

encrinal, – said of a limestone with more than 10% and less than 50% crinoidal fragments. / crinoidenarm / à encrines / crinoidenführend / de crinoides

3314
algal; – said of a limestone predominantly composed of algal fragments. / algen- / à algues / Algen- / de algas

3315
chalk, – fine-grained limestone composed to a large extent of coccolith fragments. / krijtkalk / craie / Kreide; Schreibkreide / – –

3316
dolomitic limestone, – a limestone containing 10-50% dolomite and 50-90% calcite and having an approximate magnesium carbonate equivalent of 4.4-22.7%. / dolomitische kalksteen / calcaire dolomitique / dolomitischer Kalk(stein) / caliza dolomítica

3317
sandy limestone, – limestone homogeneously mixed with 10 to 50% of quartz grains. / zandige kalk / calcaire gréseux / sandiger Kalkstein / caliza arenosa

NOTES:
- other indications are in use, e.g. silty, marly.
- for other, fine-grained, mixtures see section [2031] Lutites.

3318
dolomite, – a common rock-forming rhombohedral mineral, $CaMg(CO_3)_2$, / dolomiet / dolomite / Dolomit / dolomita

dolomite; dolostone (not recommended); dolomitite (obsolete), – a sedimentary rock consisting predominantly (more than 50% by mass) of the mineral dolomite; specifically, a sedimentary rock containing more than 90% dolomite and less than 10% calcite. / dolomiet / dolomie / Dolomitstein / dolomía

3319
calcitic dolomite; calcareous ...., – a dolomite containing 10-50% calcite and 50-90% dolomite and having an approximate magnesium carbonate equivalent of 22.7-41.0%. / kalkige dolomiet / dolomie calcaire / kalzitischer Dolomit(stein); kalkiger .... / dolomía calcarea

3320
orthodolomite; primary dolomite, – dolomite considered to be formed by chemical precipitation, e.g. as an evaporite. / orthochemische dolomiet; primaire .... / dolomie de précipitation primaire; .... orthochimique / Orthodolomit; primärer Dolomit; frühdiagenetischer .... / ortodolomía; dolomía primaria

3321
protodolomite, –
1. a crystalline calcium-magnesium carbonate with a disordered lattice in which the metallic ions occur in the same crystallographic layers instead of in alternate layers as in the dolomite mineral;
2. an imperfectly crystallised artificial material of composition near $CaMg(CO_3)_2$.

/ protodolomiet / protodolomite / Protodolomit; Ca-Dolomit / protodolomita

[2140] CARBONATE SEDIMENT BODIES AND SETTINGS

[2141] *Carbonate Bodies*

3322
carbonate ramp, – huge carbonate bodies built away from positive areas and down gentle regional palaeoslopes. No striking break in slope exists and facies patterns are apt to be wide and irregular. Belts with the highest energy zone are relatively close to the shore. / carbonaathelling / pente à sédimentation carbonatée / flach einfallende Karbonatplattform / talud carbonatado

carbonate platform, – huge carbonate buildup with a more or less horizontal and abrupt shelf margin where high-energy sediments occur. / carbonaatvlakte / plateforme carbonatée / Karbonatplattform / plataforma carbonatada

NOTE:
– The normal process of carbonate sedimentation effectively and rapidly turns ramps into platforms and creates narrow steep ridges at shelf margins. Slopes on some ramps may be so gentle as to make them commonly indistinguishable from platforms. Therefore, these terms are often used interchangeable.

3323
offshore carbonate bank, – complex carbonate buildup often of great size and thickness, e.g. Bahama Bank, well offshore from the coastal carbonate ramps or carbonate platforms. / carbonaatbank in open zee / plateforme carbonatée isolée (du continent) / Karbonatbank im Schelfbereich / banco carbonatado

3324
carbonate buildup, – a body of locally formed, laterally restricted carbonate sediment that possesses topographic relief. / carbonaatopbouwsel / complexe carbonaté / Karbonat-Hügelstruktur / cuerpo carbonatado

carbonate mass, – a carbonate localisation developed with only slight relief caused by facies changes from compactible argillaceous strata to noncompactible carbonates. / carbonaatmassa / – – / Karbonatkörper; Karbonatmasse / masa carbonatada

3325
lime-mud mound; linear mud accumulation, – buildup accumulated both through hydrodynamic processes and in situ organic production. As a consequence a lime-mud matrix dominates such constituents as organic boundstone and bioclasts. / – – / – – / Kalkschlamm-Bioherm / acumulación lineal de fango

3326
reef tract, – a general term indicating the zone with reef-building organisms in the shelf area. / rifzone / zone récifale / Riffbereich / – –

3327
regressive reef, – one of a series of nearshore reefs or bioherms superimposed on basinal deposits during the rising of a landmass or the lowering of the sea level, and developed more or less parallel to the shore. / regressief rif / récif régressif / regressives Riff / arrecife regresivo

built-out carbonate body, – carbonate mass formed as a response to negative sea level changes or uplift; the optimum carbonate production zone shifts outward as a response to the movements. / uitgebouwd carbonaatlichaam / – – / – – / – –

3328
transgressive reef, – one of a series of nearshore reefs or bioherms superimposed on back-reef deposits of older reefs during the sinking of a landmass or a rise of the sea level, and developed more or less parallel to the shore. / transgressief rif / récif transgressif / transgressives Riff / arrecife transgresivo

built-up carbonate body, – carbonate mass formed as a response to positive sea level changes or subsidence; the carbonate production should keep pace with the movements. / opgebouwd carbonaatlichaam / – – / Hügelstruktur-Karbonatkörper; hügelstrukturierter .... / – –

3329
reef cap; coral ...., – a deposit of fossil reef material overlying or covering an island or mountain. / rifkap / chapeau récifal / Rifkappe; Korallenkappe / – –

knoll reef, – a bioherm of organic frame-built growth, represented by a small, isolated, prominent, rounded hill up to 100 m high, consisting of resistant reef material; also used for recent reefs which are isolated, more or less circular in shelf margins or basinal settings. / rifknobbel; knollenrif / intumescence récifal / Knollenriff / pináculo arrecifal

3330
thrombolite, – non-laminated, crypt-algal mound without recent analogs, probably developed in shallow-marine to low-intertidal depositional environments. / tromboliet / thrombolite / Thrombolith / trombolito

3331
bioherm, – a reef-, mound-, lens-like or otherwise circumscribed structure of strictly organic origin, essentially

composed of the remains of sessile organisms, and embedded in rocks of different lithology. / bioherm / bioherme / Bioherm / bioherma

biostrome, – a distinctly bedded and widely extensive or broadly lenticular plate-like mass of rock built by and composed mainly of the remains of sedentary organisms, and not possessing a mound- or lens-like form. / biostroom / biostrome / Biostrom / biostroma

3332
stromatolite, – a calcareous body, attached to the substrate, showing finely laminated accretion structures produced by sediment binding and/or precipitation as a result of the growth and metabolic activity of micro-organisms, principally cyanophytes (blue-green algae). / stromatoliet / stromatolite / Stromatolith / estromatolito

protostromatolite, – unlithified blue-green algal laminated structures. / protostromatoliet / proto-stromatolite / unverfestigte Algenmatte / proto-estromalito

[2142] *Reefs: Morphology and Deposits*

3333
reef, – a product of the actively building and sediment-binding biotic constituents that, because of their potential wave resistance, have the ability to erect, in combination with the associated detrital and bioeroded sediments, rigid, wave-resistant topographic structures. / rif / récif / Riff / arrecife

3334
reefal, – pertaining to a reef and its integral parts, especially to the carbonate deposits in and adjacent to a reef. / van een rif / récifal / zum Riff gehörend / arrecifal

reefoid, – resembling a reef. / rifachtig / – – / riffähnlich / – –

reefy, –
1. containing reefs;
2. containing sedimentary material that resembles the material of a sedimentary reef.

/ rifachtig / d'origine récifale / riff-führend; riffenthaltend; riffartig / (p)seudoarrecifal

3335
hermatypic coral, – reef-building coral, living or having lived in comparatively shallow water within the euphotic zone. Recent hermatypic corals are characterized by the presence of Zooxanthellae. / hermatypisch koraal / coral hermatypique / hermatypische Koralle / coral hermatípico

ahermatypic coral, – coral that does not (or did not) build reefs. Recent examples lack Zooxanthellae, and are found at all depths, from shallow to very deep water. / ahermatypisch koraal / coral ahermatypique / ahermatypische Koralle / coral ahermatípico

3336
reef wall, – a wall-like upgrowth of living reef-building organisms and the skeletal remains of dead coral and other sessile organisms, reaching intertidal level where it acts as a partial barrier between adjacent environments. / rifmuur / paroi récifale / Riffgerüst; Riffmauer / pared arrecifal

3337
reef complex, – a cluster of individual reefs. / rifcomplex / complexe récifal / Riffkomplex / complejo arrecifal

3338
fringing reef, – a curvilinear belt of organic accumulation built directly out from the coast. / franjerif / récif frangeant / Saumriff / orla arrecifal

barrier reef, – a long, relatively narrow coral reef roughly parallel to the shore and separated from it at some distance by a lagoon of considerable depth and width. / barrièrerif / récif-barrière / Barriereriff; Barriere-Riff; Wallriff / barrera arrecifal

3339
faro, – ring-like organic accumulation with a shallow central lagoon located shelfward of a barrier-reef trend in off-reef position

3340
atoll; ring reef, – a ring-shaped coral reef appearing as a low, roughly circular coral island, or a ring of closely spaced coral islets, encircling a shallow lagoon in which there is no pre-existing land or islands of non-coral origin, and surrounded by deep water of the open sea. / atol / atoll / Atoll; Ringriff; Insellagune / atolón

3341
reef core, – the rigid frame of a reef together with the (originally) loose material contained therein. / rifkern / noyau récifal / Riffkern / corteza arrecifal

3342
reef rock, – hard, unstratified rock, cemented by $CaCO_3$, and formed as part of a reef, consisting of sand, shingle and fragments of reef-building organisms. / rifgesteente / roche récifale / Riffgestein / roca arrecifal

reef-rock breccia, – a rock in which masses of coral retain the attitude and position of growth and to which the varied animal and vegetal life of the reef has contributed. / rifsteenbreccie / brèche récifale / Riffgesteins-Breccie / brecha de roca arrecifal

3343
reef flat, – the relatively flat upper part of a reef, composed of a stony platform of coral fragments and coral sand, and generally dry at low tide. / rifplat / platier récifal / Riffplatte / llanura arrecifal

3344
back reef, – the landward side of a reef including the area and the contained deposits (partly or even mainly terrigenous) between the reef and the mainland. / binnenrif / arrière récif; récif interne / – – / – –

3345
lagoon; lagune; laguna, –
1. a shallow stretch of sea water near, or communicating with the sea and partly or completely separated from it by

a low, narrow, elongate strip of land such as a barrier reef or an island;
2. a shallow body of water enclosed or nearly enclosed within an atoll.
/ lagune / lagon / Lagune / laguna

lagoon floor, — the undulating to nearly level bottom of a lagoon. / lagunebodem / plancher du lagon / Lagunenboden / — —

3346
lagoon flat, — a nearly horizontal part of the reef flat located lagoonward from the reef. / lagunevlakte / replat / Lagunen-Gezeitenfläche; Lagunenwatt / — —

lagoon shelf, — the part of the reef that borders the lagoon side of a reef island; the sand-covered, lagoonward-sloping shelf is commonly found where sedimentation conspicuously exceeds organic growth. / laguneplat / — — / Lagunen-Schelf; Lagunenschelf / — —

3347
patch reef, — a growth of coral formed independently on a shelf of less than 70 m depth in the lagoon of a barrier reef or of an atoll; it varies in extent from several kilometres across to a mushroom-shaped growth of a single large colony. / — — / pâté de corail / Fleckenriff; Kuppenriff / mancha arrecifal

pinnacle reef, — a small patch reef, consisting of coral growing sharply upward (slopes from 45° to nearly vertical), often within an atoll lagoon, usually rising close to the water surface. / — — / pinacle corallien / Krustenriff; Säulenriff; Turmriff / — —

3348
lagoon channel, —
1. a stretch of deep water separating a reef from the neighbouring land;
2. a pass through a reef and into and through a lagoon.
/ lagunekanaal / chenal du lagon; goulet / Lagunenkanal / canal de laguna

3349
fore reef, — the seaward side of a reef, commonly a steeply dipping slope with deposits of reef talus. / rifhelling / récif externe / seewärtige Riffseite / — —

3350
reef edge, — the seaward margin of the reef flat, commonly marked by surge channels. / rifrand / — — / Riffrand / — —

3351
reef front, — the upper part of the outer or seaward slope of a reef with abundant living corals and other reef-building organisms. / riffront / front récifal / Riff-Front / frente arrecifal

spur-and-groove structure, — a comb-tooth structure common to all reef fronts, best developed on the windward side, consisting of grooves separated by seaward-extending spurs or ridges. / kamstruktuur / — — / Sporn-und-Rinnen-Struktur / — —

3352
reef-front terrace, — a shelf, or bench-like eroded surface, sometimes veneered with organic growth, sloping seawards to a depth of 15 to 30 m. / riffrontterras / terrasse récifale / Riffront-Terrasse / — —

3353
reef flank, — the part of a reef resting against or surrounding the reef core, and consisting of originally loose and fragmental material, partly or wholly derived from the rigid reef framework. It represents the relatively narrow, transitional zone where biological forces of reef growth compete with physical and biological forces of reef destruction. / rifflank / flanc récifal / Riff-Flanke / flanco arrecifal

reef talus, — reef flank deposits. / rifflankpuin / talus récifal / — — / depósito de talud arrecifal

3354
reef slope, — the face of a reef, rising from the off-reef sea floor. / rifhelling / pente récifale / Riffhang / talud arrecifal

3355
reef detritus, — fragmental material, mainly of sand size or smaller, formed by disintegration of reef rock. / rifdetritus; rifpuin / débris récifaux; éboulis .... / Riffdetritus / material detritico arrecifal

reef breccia, — a sedimentary breccia, formed by consolidation of reef-rock fragments broken by the action of waves and tide. / rifbreccie / brèche récifale / Riffbreccie / brecha arrecifal

3356
off-reef, — pertaining to the sea floor around or seaward of a reef, beyond the base of the subaqueous reef slope. / buiten het rif / avant récif / — — / — —

3357
inter-reef, — area within the reef zone with its contained deposits accumulated under the direct influence of the reef but different from the reef deposits. / tussenrif / interrécifal / — — / interarrecifal

[2143] *Algal Structures*

3358
algal structure, — a sedimentary structure of definite form and usually of calcareous composition, resulting from secretion and precipitation by algal colonies; it includes crusts, small (pseudo)pisolitic and pseudo-concretionary forms, biscuit- and cabbage-like heads of considerable size, laminated structures such as stromatolites, or bedding modified by blue-green algal mats. / algenstruktuur / structure algaire / Algentextur / estructura de algas

3359
calcareous algae, — taxonomically heterogeneous group of predominantly marine algae that secrete calcium carbonate around their tissues. / kalkalgen / algues calcaires / Kalkalgen / algas calcareas

3360
algal reef, — an organic reef largely composed of algal remains

and in which algae are the principal organisms secreting calcium carbonate. / algenrif / récif algaire / Algenriff / arrecife de algas

3361
algal ridge, – a low ridge or elevated margin at the seaward (outer) edge of a reef flat, composed of the secretions of the actively growing calcareous algae. / algendijk / marge algaire; trottoir à algues / Algenbarre; Algenrücken / – –

lithothamnion ridge, – an algal ridge built by Lithothamnion and other red calcareous algae, rising about one metre above the surrounding reef, and extending to depths of 6 to 7 m below sea level. / lithohamniumdijk / trottoir à lithothamniées / Lithothamnien-Rücken / – –

3362
algal rim, – a low, slight rim, built by actively growing calcareous algae on the lagoonal side of a leeward reef or on the windward side of a reef patch in a lagoon. / algenrand / – – / Algenrand / – –

3363
algal head, – a 10-12 cm in diameter bulbous, dome-shaped or columnar mass of mechanically transported, laminated sediments collected by algae (especially blue-green algae) on a tidal flat or in a lake, and bound together by algal filaments. / algendom; algenknobbel / – – / Algendom; Algenkuppel / – –

algal ball, ... biscuit, – hemispherical or disk-shaped, often concentrically laminated, calcareous masses up to 20 cm in diameter. / algenbal / pelote algaire / Onkoid / – –

3364
algal mat, – organic film of algal tissue covering sediment. / algenmat / matte algaire / Algenmatte / – –

[2200] MISCELLAEOUS DEPOSITS

NOTE:
– ferruginous, manganese and phosphoritic deposits have been compiled in chapters
[2300] Diagenesis
[2400] Carbonaceous Deposits
[2660] Various Metallic Mineral Deposits

[2210] SILICEOUS DEPOSITS

3365
siliceous, – composed of or containing silica. / kiezelig / siliceux / kieselig / silíceo

3366
chert, – siliceous sedimentary rock, mostly consisting of finely crystalline aggregates of quartz and fibrous silica, occurring as nodules (mostly in carbonate rocks) and as separate beds (e.g. radiolarian cherts). The fracture is conchoidal or splintery. In folded strata it usually shows several systems of planar micro-joints, lying close together. / silex; hoornsteen / silex / Hornstein / pedernal; silex

novaculite, – a variety (western USA) of bedded chert, light-coloured and essentially composed of cryptocrystalline quartz. / novaculiet / novaculite / Novaculit / novaculita

3367
siliceous shale, – argillaceous chert containing often more than 70% quartz. / kiezellei / schistes siliceux; .... silicifiés / Kieselschiefer / pizarra silícea

siliceous ooze, – an ooze containing more than 30% of siliceous skeletal material produced by planktonic plants and animals. / kiezelslib / boue silicieuse / Kieselschlamm / barro silíceo.

siliceous earth, – general term that includes diatomaceous and radiolarian earths. / kiezelaarde / terre silicieuse / Kieselerde / tierra silícea

3368
radiolarian ooze, – unconsolidated, siliceous, deep-sea deposit (generally over about 4000 m), consisting for more than 30% of radiolarian remains. / radiolariënslik / boue à Radiolaires / Radiolarienschlamm / barro de radiolarios

radiolarian earth, – an unconsolidated deposit, mainly composed of radiolarian remains, raised above sea level. / radiolariënaarde / terre à Radiolaires / Radiolarienerde / tierra de radiolarios

radiolarite, – an indurated deposit, rich in, or mainly composed of radiolarian remains. / radiolariet / radiolarite / Radiolarit / radiolarita

3369
radiolarian chert, – a chert, rich in or mainly composed of radiolarian remains; a radiolarite. / radiolariën-hoornsteen; radiolariet / silex à Radiolaires / Radiolarit / radiolarita

lydite; lydian stone; touchstone, – black or dark grey (carbonaceous) radiolarian chert. / lydiet; toetssteen / phtanite / Lydit / lidita

3370
diatom ooze, – an unconsolidated, siliceous deep-sea deposit beyond the continental shelf edge to depths of about 2000 m, and mainly consisting of diatom frustules. / diatomeeënslik / boue à diatomées / Diatomeenschlamm / barro de diatomeas

frustule, – the siliceous shell of a diatom, consisting of two valves. / diatomeeën-skeletje / frustule de diatomée / Diatomeenschale / frústula

diatomaceous earth; diatom ...., – an unconsolidated deposit mainly consisting of diatom frustules, and lying above sea level. / diatomeeënaarde / terre à diatomées / Diatomeenerde / tierra de diatomeas

diatomite, – a deposit, consolidated or not, mainly composed of diatom frustules. / diatomiet / diatomite / Diatomit / diatomita

3371
diatomaceous chert, – a chert rich in diatom frustules. / diatomeeën-hoorsteen / silex à diatomées / Diatomeen-Hornstein / pedernal de diatomeas

tripoli, – consolidated, normally thin-bedded rock, mainly consisting of diatom frustules. The name is derived from Tripoli (Libya). In America the term is also applied to other porous, siliceous rocks, macroscopically resembling in some degree to the tripoli diatomite. / tripoli / tripoli / tripel; Polierschiefer; Tripoli / trípoli

diatomaceous shale, – a shale rich in diatom frustules. / diatomeeënschalie / schiste à diatomées / Diatomeenschiefer / pizarra de diatomeas

3372
spongolite; spiculite; spicularite, – a siliceous rock, mainly composed of sponge spicules. / spongoliet / spongolite / Spiculit / espongolita

3373
chert ironstone; cherty ...., – sedimentary rock, normally consisting of alternating layers or laminae of chert and iron, mostly siderite or haematite. / kiezel-ijzersteen / roche ferrugineuse silicifiée / – – / – –

3374
porcellanite, – term used for various sedimentary rocks, especially cherts and silicified tuffs, that possess the texture and fracture of porcelain. / porcellaniet / porcelanite / Porzellanit / porcelanita

[2220] EVAPORITES

3375
evaporite, – a rock formed by precipitation of solids dissolved in an evaporating body of open water, either sea, lake, or spring. / evaporiet / évaporite / Evaporit / evaporita

3376
salinity, – the total content of solids dissolved in water, expressed in mol·m^{-3} or in ppm (mg/l). / saliniteit; zoutgehalte / salinité / Salinität / salinidad

saliferous, – said of sediments containing a considerable amount of salts, particularly NaCl. / zouthoudend / salifère / salzführend / salino

NOTE:
- for details see sections
  [1840] Marine Environments
  [2530] Water Supply and Water Quality

3377
hypersaline, – a loose term, denoting salinity far above that of sea water. / hypersalien / hypersalé / hypersalin / hipersalina

bittern, – the concentrated solution that remains after precipitation of most of the dissolved NaCl. It often has a bitter taste, owing to a high content of Mg-ions. / moederloog (p.p.) / liqueur mère / Mutterlauge / agua madre

mother liquor; .... liquid, – the highly concentrated solution, in which the most highly soluble salts of K and Mg have remained. / moederloog / liqueur mère / Mutterlauge / líquido madre; .... residual

3378
salt paragenesis, – the characteristic sequence of deposition of rocks in an evaporating body of water, e.g. from sea water: limestone/dolomite-gypsum/anhydrite-rocksalt-bitter salts, and the from that sequence derived diagenetic minerals. / zoutparagenese / – – / Salzparagenese / – –

3379
whiting, – area of milky-white water in seas in arid climates, containing a dense suspension of aragonite and calcite needles, possibly indicating incipient precipitation.

drewite (obsolete), – mud composed of aragonite needles. / aragonietnaaldenslib / – – / Drewit; Aragonitnadel-Schlamm / – –

3380
primary dolomite; orthodolomite, – dolomite considered to be formed by precipitation from a body of open water, e.g. by evaporation, and not as replacement material. / primaire dolomiet / dolomie primaire / primärer Dolomit / dolomita primaria

3381
gypsum, –
1. the mineral gypsum;
2 a rock composed of gypsum. It seems preferable to use the term gypsite for such rocks.

/ gips; gipsiet / gypse / Gips; Gipsgestein / yeso

gypsite, – an earthy variety of gypsum, containing dirt and sand, formed as efflorescences on the ground in arid regions. It seems advisable to redefine gypsite as any rock, mainly composed of gypsum. / gipsiet / gypsite / Gipsgestein / yeso

anhydrite, – the mineral anhydrite or a rock composed of anhydrite. / anhydriet / anhydrite / Anhydrit / anhidrita

3382
rock salt, – a rock composed mainly of halite; it occupies the largest volume in any salt paragenesis from sea water, Na and Cl forming the dominant ions. / steenzout / sel gemme / Steinsalz / sal de roca

hopper crystal, – a cubic crystal of halite that possesses the shape of an inverted, hollow, stepped pyramid, and formed at the surface of an evaporating brine.

3383
bitter salt, – the K and Mg salts precipitated during the last stage of evaporating sea water. / bitterzout / epsomite / Bittersalz / epsomita

3384
chile saltpetre, – sodium nitrate as found in Caliche in northern Chile. / chilisalpeter / salpètre naturel du Chili / Chile Salzpeter / nitrato de Chile

3385
sinter, – a hard sedimentary rock (silica or lime) formed by precipitation from hot or cold springs, lakes or rivers. / sinter / tuf / Sinter / sinter

3386
trona, – a water-containing sodium carbonate precipitated from ground-water seepages in arid climates.

3387
incrustation, – a crust or hard coating formed by precipitation of one or more substances from a solution. Also the process by which such a crust is formed. / incrustatie / encroûtement / Inkrustierung / incrustación

[2300] DIAGENESIS OF NON-CARBONACEOUS DEPOSITS

[2310] GENERAL TERMS

3388
diagenesis, – all mechanical, chemical, mineralogical and biogenical processes and changes in a sediment and in its interstitial water after its deposition, e.g. compaction, cementation, pressure solution, recrystallization and replacement, at temperatures below those leading to metamorphosis, generally not exceeding 300°. Excluded are the changes brought about by tectonism, metamorphism, volcanism and weathering. / diagenese / diagenèse / Diagenese / diagénesis

3389
rock, – the solid material of which the earth's crust is built. The geologist includes all soft, loose and unconsolidated masses of sediments in their prediagenetic state, that in engineer's and pedologist's terminology are usually not designated as such. / gesteente / roche / Gestein / roca

3390
field of diagenesis, – that part of the realm of lithogenesis which is bounded by three overlapping, not clearly definable interfaces; the depositional, the weathering and the metamorphism interface. It can be subdivided into three diagenetic environments; that of near-surface, shallow burial and deep burial. /diagenetisch veld / domaine diagénétique / Diagenesebereich / campo de diagénesis

3391
diagenetic environment, – a part of the field of diagenesis that is typified by assemblages of characteristic diagenetic features. / diagenetische omgeving / milieu diagénétique / diagenetisches Milieu / ambiente diagenético

3392
near-surface diagenetic environment, – diagenetic environment in which processes act that are related to the depositional or the weathering interfaces. / oppervlakte-nabije diagenetische omgeving / milieu diagénétique superficiel / diagenetisches Milieu nahe der Oberfläche / ambiente diagenético cerca de la superficie

3393
shallow-burial diagenetic environment, – diagenetic environment in which processes act that are not entirely related to the depositional environment and predate stylolite formation. / ondiepe diagenetische omgeving / milieu diagénétique précoce / diagenetisches Milieu mit geringmächtiger sedimentbedeckung / ambiente diagenético de poca profundidad

deep-burial diagenetic environment, – diagenetic environment in which processes act that postdate stylolitisation; this feature therefore characterizes its upper limit; the metamorphism interface is its lower limit. / diepe diagenetische omgeving / milieu diagénétique profond / diagenetisches Milieu mit mächtiger sedimentbedeckung / ambiente diagenético profundo

3394
diagenetic paragenesis, –
1. the sequential order of formation of diagenetic products;
2. a characteristic association of diagenetic products.
/ diagenetisch gezelschap; ....e paragenese / paragenèse diagénétique / diagenetische Paragenese / paragénesis diagenética

3395
segregation, – secondary separation of material by diagenesis. / segregatie / ségrégation / Segregation / segregación

3396
infiltration, – the penetration of a liquid into a rock. / infiltratie / infiltration / Infiltration / infiltración

impregnation, – the filling of porous material with another substance. / impregnatie / impregnation / Imprägnation / impregnación

3397
semi-permeable membrane, – clay layers acting as such; probably (with differential solution and precipitation) the main cause of diagenetic changes in connate (ground- and formation) waters. / semipermeabel membraan / membrane semipermeable / semipermeabele Membran / membrana semipermeable

[2320] ENVIRONMENTAL DIAGENESIS

3398
environmental diagenesis, – changes related to factors of the depositional and weathering environments, such as redox potential, pH value, temperature, climate and drainage. / milieu-gerelateerde diagenese / diagenèse liée au milieu / milieu-bezogene Diagenese / diagénesis ambiental

syngenesis; syndiagenesis, – broad term covering features formed during the accumulation of a sedimentary layer, such as ripple marks, and also penecontemporaneous deformation, early cementation and early concretions. / syngenese / syngenèse / Syngenese / singénesis

NOTE:
– both terms pertain to phenomena of the near-surface diagenetic environment; there is neither a clearly defined limit with soil-forming processes nor with burial diagenesis. For further terms, see sections:
[1730] Soil Profile and Soil Horizons
[1930] Sedimentary Processes
[1940] Sedimentary Structure
[2620] Processes of Formation (of Metallic Mineral Deposits)

3399
vadose diagenesis, – all post-depositional changes in sediments, whose pore space is partly filled with water, and partly with air. / vadose diagenese / diagenèse de percolation / vadose Diagenese / diagénesis vadosa

phreatic diagenesis, – all post-depositional changes in sediments whose pore space is entirely filled with ground water. / freatische diagenese / diagenèse de nappe phréatique / phreatische Diagenese / diagénesis freática

3400
dendrite, –
1. a regularly branched crystal of a type occurring e.g. in many metals and ice;
2. a branched colloidal mineral deposit (mostly Mn- or Fe-(hydr)oxide), formed by ground-water seepage within a porous rock, especially along a joint or bedding plane.

/ dendriet / dendrite / Dendrit / dendrita

3401
horsetail structure, – wispy or ramifying seams in which insoluble residues are accumulated as a result of rock-volume reduction due to differences in solubility. / oplossings paardestaart / structure fibreuse de remplissage / Pferdeschwanzstruktur / estructura en cola de caballo

3402
coating, – thin film of precipitated solid material covering the surface of a rock, or of rock or mineral grains. Also the formation of such a film or films. / film; huidje; omhulling / revêtement / Überzug; Film / corteza

coated, – covered by a film or veneer of secondary material. / – – / revêtu / überzogen / revestido

3403
patina, – a thin outer layer of a rock or rock fragment (pebble, etc.), produced by weathering and/or other soil processes, and differing by its colour and composition from the original material. Also a thin film deposited on a rock or rock fragment that has been exposed to the air, or that has been lying in the soil. / patina / patine / Patina / pátina

cortex, – the covering layer or the veneer of secondary material around a grain.

3404
polish, – the shining appearance of a surface, produced secondarily. / polijsting / poli / Politur / pulido; pulimento

3405
desert pavement; .... armour; .... armor, – a (mainly) residual concentration of stones at the surface in deserts, formed by wind action or (and) sheet wash. / keienvloer in een woestijn / pavage désertique / Wüstenpflaster / cantal de desierto

desert varnish; .... lacquer; .... patina; .... polish, – a thin, shining coating on pebbles and rocks, also filling the pores of their exterior, formed in deserts and consisting of substances rich in Si and Al, in hot deserts also Fe and Mn, the latter elements accounting for dark red to black colours. / woestijnlak / patine désertique / Wüstenlack / baruiz de desierto; pátina de desierto

3406
desert rose, – a more or less flower-shaped aggregate of sand-filled crystals, mostly of calcite, gypsum or barite, formed in desert environment. / woestijnroos / rose des sables / Wüstenrose / rosa del desierto

3407
gloss; luster, – shining appearance of a surface, whether primary or secondary. / glans / brillance / Glanz / lustre

3408
frosting, – the etching or pitting process rendering the surface of detrital grains (usually consisting of quartz) dull and lusterless. / mattering / ternissement / Mattierung / – –

frosted, – said of grains that have undergone frosting. / gematteerd; mat / mat; terni / mattiert / – –

3409
etch figure, – pattern of depressions produced in the surface of crystals or crystal grains by a solvent. / etsfiguur / figure de corrosion / Ätzfigur / figura grabada

pit, – a small indentation or depression left on the surface of a rock particle (especially a clastic one) as a result of some eroding or corrosive process, such as etching or differential solution.

3410
pitted pebble, – pebble with concavities formed by pressure-induced solution at points of contact with other pebbles, that normally have at the contact point a smaller radius of curvature.

percussion mark; .... figure, – a circular, elliptical or crescentic crack; the upper part, or cross section (due to later wear) of a cone of percussion, produced by collision in the surface of a pebble, cobble or rock. / botsfiguur / marque de percussion; figure de .... / Schlagspur / marca de percusión

3411
case hardening, – the lithification of the exterior of loose rock material, e.g. a sand surface, duricrusts or the exterior of a septarian nodule in statu nascendi. / verharding van de buitenkant / durcissement externe / Panzerbildung / litificación periférica

3412
beach rock, – a rock formed by precipitation of carbonate cement in recent beach deposits, mostly sand (calcareous or non-calcareous), locally gravel. / – – / grès de plage / – – / roca de playa.

3413
intrastratal solution, – solution of material from within a sedimentary bed following deposition. / oplossing in de laag / dissolution intraformationnelle / diagenetische Auflösung / solución intrastratal

3414
frayed end, – of mineral particles attributed to intrastratal solution. / gerafeld uiteinde / – – / ausgefranst / – –

hacksaw termination; saw-toothed ....; serrate ...., – said of serrate end of clastic mineral grains attributed to intrastratal solution. / getand uiteinde; gekorven .... / terminaison en dent de scie / gezähnt / borde dentado

3415
chalkification, – the process whereby carbonate sequences, which have diagenetically been stabilised, are brought to exposure where they react to meteoric conditions forming extensive secondary porosity and consequently a pseudo-chalky rock. / verkrijting / crayification / Verkreidung / – –

3416
vadose compaction, – the process whereby grains are denser packed (with fitted grain contacts) than is usually observed after normal deposition in recent surface beach rock. This denser packing results normally from selective solution at grain contacts as a result of undersaturated capillary water. / vadose samendrukking / compaction par percolation / vadose Kompaktion / compactación vadosa

[2330] BURIAL DIAGENESIS

[2331] *General Terms*

3417
burial diagenesis, – diagenesis related to depth of burial, i.e. to overburden pressure and temperature. / begravingsdiepte-gerelateerde diagenese / diagenèse d'enfouissement / Versenkungsdiagenese / diagénesis por soterramiento

3418
early diagenesis, – the first stage of burial diagenesis (in the shallow-burial diagenetic environment), mainly characterized by mechanical compaction, incipient transformation of minerals (e.g. kaolinite formation from feldspars, montmorillonite from volcanic components) and incipient cementation. / vroege diagenese / diagenèse précoce / frühe Diagenese / diagénesis inicial

advanced diagenesis, – the intermediate stage of burial diagenesis whereby montmorillonite has been transformed to mixed layers and illite, kaolinite is generally stable, and whereby pressure solution is becoming an important factor in the reduction of the porosity of sandstones. / gevorderde diagenese / diagenèse avancée / fortgeschrittene Diagenese / diagénesis avanzada

late diagenesis, – the latest stage of burial diagenesis (in the deep-burial diagenetic environment), whereby montmorillonite and kaolinite have been largely transformed, pressure solution is well developed, and whereby the porosities of sandstones have been reduced generally to values of 10% or less. / late diagenese / diagenèse très avancée / späte Diagenese / diagénesis terminal

3419
epigenesis, – late diagenesis; specifically: diagenesis taking place after burial and compaction of the layer under consideration. / epigenese / epigenèse / Epigenese / epigénesis

katagenesis; catagenesis, – those epigenetic processes that took place after lithification of the sediment and prior to its metamorphism. / katagenese / catagenèse; katagenèse / Katagenese / katagénesis

3420
anchimetamorphism, – the transitional stage between late diagenesis and (greenschist) metamorphism. It is characterized by the loss of clastic texture and by fully crystallized illite and chlorite. / anchimetamorfose / anchimétamorphisme / Anchimetamorphose / anchimetamorfismo

3421
lithification, – the conversion of unconsolidated material into coherent and solid rock. / lithificatie; steenwording / lithification / Lithifikation / litificación

3422
consolidation, – any process that causes loosely aggregated earth materials to become firm and coherent rock. / verharding / consolidation / Konsolidierung; Verfestigung / consolidación

consolidated, – said of a rock that has become firm and coherent. / verhard / consolidé / konsolidiert; verfestigt / consolidado

3423
induration, – the process of hardening of unconsolidated or moderately consolidated rocks. / verharding / induration / Verfestigung; Härtung / endurecimiento

3424
framework stability, – used with reference to the composition of clastic frameworks in sandstones that on burial may become mechanically unstable (containing soft lithic grains such as phyllites) and/or chemically unstable (containing chemically unstable grains such as tuff fragments). / bouwselstabiliteit / stabilité structurale / Gerüststabilität / estabilidad de la estructura

3425
alteration, – any change in the mineralogic composition of a rock under the influence of physical or chemical processes, e.g. the generation of an authigenic matrix from unstable lithic components in sandstones. / alteratie / épigénie / Umwandlung / alteración

3426
transformation, – said of the diagenetic, mineralogical change of clay minerals, such as the transformation of montmorillonite to illite. / transformatie / transformation / Transformation / transformación

[2332] *Compaction*

3427
anadiagenesis; anagenesis, — diagenetic processes occurring during the compaction phase of diagenesis. / anadiagenese; anagenese / diagenèse de compaction / Anadiagenese / anadiagénesis

3428
compaction, — the reduction in thickness of a sedimentary body owing to the weight of the overlying material. It generally takes place through reduction of the pore volume and concomitant expulsion of pore liquid. / compactie / compaction / Kompaktion / compactación

differential compaction, — compaction, the total amount of which varies laterally as a result of lateral differences in thickness or lithology. / differentiële compactie / compaction différentielle / differentielle Kompaktion / compactación diferencial

3429
mechanical compaction, — rearrangement of clastic components to a denser packing. / mechanische compactie / compaction mécanique / mechanische Kompaktion / compactación mecánica

3430
repacking, — mechanical grain rearrangement in response to compressive stresses. / korrelverspreidingsverandering / rearrangement mécanique / — — / sobre-empacado

3431
breakage, — brittle fracturing of grains in response to compressive stresses. / barstvorming / fracturation granulaire / — — / arranque; fractura; rotura

3432
chemical compaction, — volume reduction of a sediment or rock that is entirely or partly controlled by dissolution processes (e.g. vadose compaction; pressure solution). / chemische compactie / compaction chimique / chemische Kompaktion / compactación química

[2333] *Cementation*

3433
cementation, — the lithification of loose sedimentary rocks through precipitation of a binding material around rock particles. / cementatie; verkitting / cimentation / Zementation; Verkittung / cementación

cement, — chemically precipitated material, such as authigenous carbonate or silica, that occurs in the pores of sedimentary rocks, binding together the originally loose particles, grains, fragments or pebbles. / cement / ciment / Zement / cemento

3434
authigenesis, — the process that produces new minerals in place within a sediment or sedimentary rock. / authigenese / authigenèse / Authigenese / autigénesis

authigenic; authigenetic; authigenous, — said of constituents (especially mineral constituents) of sedimentary rocks that are formed in place. / authigeen / authigénétique / authigen / autigénico

3435
diagenetic facies, — used in reference to an authigenic mineral assemblage formed in response to a certain stage of diagenesis. / diagenetische facies / faciès diagénétique / diagenetische Facies / facies diagenético

3436
internal sedimentation, — deposition of clastic or authigenic material in pores or other cavities in a rock. / interne sedimentatie / sédimentation interne / interne Sedimentation / sedimentación interna

internal sediment, — sediment, deposited in vugs, body cavities or fractures in (semi-)consolidated rocks. / intern sediment / sediment interne / internes Sediment / sedimento interno

3437
rim cementation; overgrowth ....; syntaxial ...., — growth of crystals in optical continuity to monocrystalline skeletal fragments. / randcement; overgroeiingscement; syntaxiaal cement / cimentation par nourrissage des grains / syntaxiale Zement / cemento syntaxial

granular cementation, — formation of cement in interstices between sedimentary grains, resulting in outward growth of crystalline material adhering to that surface; e.g. growth of calcite in the pores of an unconsolidated sand. / korrelige cementatie / cimentation intergranulaire / granulare Zementation / cemento granular

stromatactis, — open-space sedimentary structures characterized by horizontal or nearly flat bottom and by irregular upper surfaces, formed by the filling of original cavities with internal sediments and/or sparry-calcite cement, as in the central part of a reef core. / stromataktis / stromatactis / Stromatactis / — —

3438
drusy cement; druse ...., — any pore-lining cement often consisting of radially arranged elongated crystals which do not fill the entire pore. / korstcement / ciment de type géode / Drusenzement / cemento de tipo drusa

3439
blocky cement, — cement consisting of crystals characterised by their large approximal equidimensional size. / blokcement / ciment grenu / Blockzement / cemento en bloque

3440
pore filling, — said of cement filling the pores of a sandstone as opposed to authigenic precipitates replacing clastic grains. / poriënvullend / remplissage de pore / Porenfüllung / relleno de poros

3441
bridging, – used to describe the growth habit of authigenic clay minerals such as illite and chlorite where these minerals cross pores in sandstones. / overbruggend / pont argileux / – – / – –

3442
minus-cement porosity, – the sum of porosity and the total of authigenic pore-filling minerals (cement) in a sandstone, expressed in percentage of bulk volume. This value may yield information on the time of cementation (high minus-cement porosity: early cementation) and may be used in analysing the burial history of a sandstone. / minus cement porositeit / porosité initiale / minus-zement Porosität; Porosität vor der Zementbildung / porosidad sin cemento

3443
sand crystal, – a euhedral, sand-filled crystal, formed by cementation of sand. / zandkristal / cristal de sable / Sandkristall / – –

3444
decementation, – removal of cement by solution processes. / decementatie / dissolution du ciment / Lösung des Zements / decementación

[2334] *Pressure Solution*

3445
pressure solution, – solution occurring preferentially at the grain or crystal contacts where overburden or tectonic pressure exceeds the pressure of the interstitial fluid. / oplossing onder druk / dissolution par pression / Drucklösung / solución por presión

3446
pressure molding, – the plastic deformation of soft lithic grains, such as those comprising phyllites, whereby these grains mushroom into and partly fill initially open pore spaces. / korreldeformatie / déformation de compaction / – – / – –

pressure welding, – applied to the process whereby a quartz or other sandstone through the processes of pressure solution, plastic deformation and cementation is altered into a tightly fitting, often oriented, mosaic. / – – / soudure par compaction / – – / – –

3447
shallow-burial plastic deformation, – volume reduction resulting in fitted contacts and distorted grains in comparatively soft pellets and ooids by ruptureless distortion in response to mild compressive stress. / ondiepe plastische vervorming / déformation plastique précoce / plastische Deformation unter geringmächtiger Sedimentbedeckung / deformación plástica de poca profundidad

deep-burial plastic deformation, – volume reduction, resulting in lineations, fitted contacts, elongated grains and twinned calcite in compacted sediments, in response to high pressure-temperature conditions. / diepe plastische vervorming / déformation plastique d'enfouissement profond / plastische Deformation unter mächtiger Sedimentbedeckung / deformación plástica profunda

3448
grain interpenetration, – collective term for faced, fitted and sutured contacts. / korrelineengroeiing / interpénétration granulaire / Korndurchdringung / interpenetración de los granos

3449
faced contact, – straight interface between two grains. / vlak contact / contact rectiligne euhédral / – – / contacto recto

fitted grain contact, – straight or curved interface between two grains. / in elkaar passend korrelcontact / contact linéaire grain à grain / – – / contacto ajustado

sutured contact, – interpenetrating interface between two grains. / sutuurcontact / contact suturé / suturierter Kontakt / contacto por sutura

3450
suturic, – used for the undulated or plicated contact of interpenetrating quartz or other mineral grains and of beds (stylolites) as resulting from pressure solution. / suturisch / suturé / suturisch; suturiert / sutural

3451
stylolite, – a small, more or less columnar, lengthwise striated extension of a rock body that fits into a corresponding socket. Stylolites occur mostly in limestones and dolomites, often in sheets, along bedding planes, joints and other fractures. Their pattern in sections transverse to these boundaries often resemble sutures of skull bones. / styloliet / stylolite / Stylolith / estilolito

[2335] *Recrystallisation and Replacement*

3452
recrystallisation, – pertaining to any change in the fabric of a mineral without change in composition. / rekristallisatie / récristallisation / Rekristallisation / recristalización

3453
replacement; metasomatose, – the simultaneous dissolution of one mineral and precipitation of another, such as the replacement of quartz by calcite. The process of replacement may lead to pseudomorphism. / vervanging; metasomatose / épigénie; remplacement / Verdrängung; Metasomatose / reemplazamiento

3454
grain growth, – refers to unstrained crystals that grow at the expense of their neighbours. / korrelvergroting / – – / Kornwachstum / – –

3455
overgrowth, – secondary enlargement of a clastic grain. Generally but not necessarily developed in crystallographic (lattice) continuity. / overgroei / élargissement; croissance / – – / crecimiento

3456
syntaxial overgrowth, – overgrowth developed in lattice continuity with that of the clastic core, e.g. authigenic quartz on quartz grains. / syntaxiale rand / élargissement syntaxiale / – – / sobrecrecimiento sintaxial

epitaxial overgrowth, – overgrowth on a clastic grain with a different composition, e.g. calcite on quartz. / epitaxiale rand / croissance épitaxique / – – / sobrecrecimiento epitaxial

3457
petrifaction, – the replacement of organic tissue (e.g. shells, bones, wood) by mineral matter, usually silica or lime, so that often the delicate structure is preserved. / verstening / pétrification / Versteinerung / petrificación

beekite, – silica (quartz, quartzite) occurring as aggregates of irregular, concentric rings, rosettes etc. and formed by silicification of calcareous fossils. / beekiet; ringenkiezel / beekite / Beekit / – –

3458
silicification, – replacement by, or introduction of appreciable quantities of silica. / silicificatie; verkiezeling / silicification / Verkieselung; Silifizierung / silificación

calcification, – replacement by, or introduction of calcium carbonate, e.g. dedolomitisation. / calcificatie; verkalking / calcification / Verkalkung / calcificación

pyritization, – replacement by, or introduction of pyrite. / pyritisering; pyritisatie / pyritisation / Pyritisierung / piritización

3459
phosphatization, – conversion into phosphate(s). / fosfatisering / phosphatisation / Phosphatisierung / fosfatización

glauconitization, – formation of glauconite by transformation of pre-existing materials, such as clay minerals or biotites. / glauconitisatie / glauconitisation / Glaukonitisierung / glauconitización

gypsification, – formation of gypsum by transformation from other substances, especially anhydrite. / gipsvorming; vergipsing / transformation en gypse / Bildung von Gips / yesificación

3460
dolomitization, – the replacement of calcium carbonate by calcium-magnesium carbonate in a limestone. / dolomitisatie / dolomitisation / Dolomitisierung / dolomitización

overdolomitization, continued growth of dolomite crystals after complete replacement of the original sediment that ultimately plugs all intercrystalline porosity. / voortgaande dolomitisatie / holodolomitisation / – – / supradolomitización

3461
dolomite mottling, – a textural feature resulting from incipient or arrested dolomitization of limestones, characterized by preferential alteration that leaves patches, blotches, bird eyes, laminae, allochems and/or other structures unaffected. Also a similar phenomenon resulting from arrested or incomplete dedolomitization. / dolomiet-vlekkig / mouchetage dolomitique / Dolomitfleckigkeit / manchas de dolomitización

3462
variegated, – said of a sediment(ary rock) showing variations of colours or tints in irregular spots, streaks, stripes etc. / bont; geschakeerd / bariolé / gescheckt / abigarrado

mottled, – having spots of different colour. / gevlekt / tacheté; moucheté / gefleckt / moteado

3463
reduction sphere, – a more or less spheroidal zone in a sedimentary rock where, owing to reduction of trivalent iron, the red or brown colours have been replaced by white, grey, yellow or green, etc. / bolvormige reductiezone / zone sphérique de reduction chimique / Reduktionshof / esfera de reducción

3464
ferruginous, – containing appreciable quantities of iron. Said especially of sediments with a rusty colour due to the presence of ferric oxide or hydroxide. / ijzerhoudend / ferrugineux / eisenschüssig / ferruginoso

ironstone, – any rock rich in iron. / ijzersteen / roche ferrugineuse / Eisenstein / roca ferruginosa

3465
crystal sandstone, –
1. a sandstone in which secondary calcite, gypsum, barite or other minerals have formed as large poikilitic crystals;
2. a sandstone in which the quartz grains have become enlarged by secondary, largely euhedral quartz growths, thus giving it a 'sparkling' appearance.

/ kristalzandsteen / – – / Kristallsandstein / – –

3466
beef, – term used originally in Purbeck (England) for thin layers or lenses of secondary, fibrous calcite (or other fibrous material, e.g. gypsum), lying between layers of shale. At some places they contain cone-in-cone structures.

3467
vermicular structure, – said of a worm-like crystallization-habit like that of quartz in feldspar (myrmekite) or authigenic kaolinite in soils and sandstones. / wormvormige structuur / structure vermiculaire / wurmförmige Textur / estructura vermicular

ruin marble structure, – structure typically found in micritic limestone with very thin calcite veins, separating blocks with different patterns of iron (hydr)oxide infiltration, that in cross section sometimes gives the impression of a picture of ruins. The structure is also found in other rocks, e.g. in fine-grained sandstones. / ruïnen-marmer-structuur / structure ruiniforme / – – / estructura ruinosa

3468
chickenwire structure, – structure consisting of irregular masses of anhydrite or gypsum, usually more than 1 cm in diameter, separated by thin stringers of dark carbonate, clay or silt. In cross section it forms net-like patterns. / kippengaasstructuur / structure en cage à poulets / Maschendrahttextur / estructura 'chickenwire'

3469
cone-in-cone structure, – a structure of cones, usually consist-

ing of calcite, fitting into one another, the apices pointing downward (when occurring in layers) or inward (when formed in the exterior shell of concretions). / – – / – – / Tütenmergel; Nagelkalk / – –

3470
pegmatitic texture, – coarsely crystalline texture, often present in evaporites. / pegmatietische textuur / texture pegmatitique / Pegmatitstruktur / textura pegmatítica

porphyroblastic structure, – structure with relatively large crystals formed by diagenesis in a finer crystalline matrix, especially in evaporites. / porfieroblastisch / structure porphyroblastique / porphyroblastisch / porfiroblástico

3471
idiotopic, – said of the fabric of a crystalline rock in which the majority of the constituent crystals are euhedral. /idiotoop / idiomorphe / idiotopisch / idiotópico

hypidiotopic, – intermediate between idiotopic and xenotopic; said of the fabric of a crystalline rock in which the majority of the constituent crystals are subhedral. / hypidiotoop / sub-idiomorphe / hypidiotopisch / hipidiotópico

xenotopic, – said of the fabric of a crystalline rock in which the majority of the constituent crystals are anhedral. / xenotoop / xénomorphe / xenotopisch / xenotópico

[2340] CONCRETIONS

3472
concretion, – an indurated concentration of mineral matter formed secondarily, usually within a sedimentary rock, having a globular, disc-like, rod-like or irregularly nodular shape, but always showing rounded surfaces. / concretie / concrétion / Konkretion / concreción

concretionary, – characterized by, consisting of, or producing concretions. / concretie-houdend; concretionnair / concrétionnaire / konkretionär / concrecionario

3473
nodule, – a relatively small, rounded, often irregular body of a mineral or mineral aggregate, differing in composition from the host rock. / knol / nodule / Knolle / nódulo

nodular, – containing, or having the form of nodules. / knollig / nodulaire / knollig / nodular

3474
syngenetic; contemporaneous; primary, – pertaining to concretions formed at the same time as the enclosing rock. / syngenetisch; primair / syngénétique; primair / syngenetisch / singenético

penecontemporaneous, – pertaining to concretions formed close to the surface of a recently deposited sediment. / kort na afzetting / pénécontemporain / kurz nach Ablagerung / penecontemporanea

epigenetic; subsequent; secundary, – pertaining to concretions formed after deposition of the enclosing rock. / epigenetisch; secundair / épigénétique; secondair / epigenetisch; / epigenético

3475
diagenetic differentiation, – the redistribution of material in a sediment by solution and diffusion to centres or nuclei where precipitation occurs. The process may result in the formation of nodules and concretions. / diagenetische differentiatie / différenciation diagénétique / diagenetische Differentiation / diferenciación diagenética

3476
halmyrolysis; submarine weathering, – chemical reactions in sediments on the sea floor. / halmyrolyse / halmyrolyse; altération sous-marine / Halmyrolyse / halmirolisis

3477
manganese nodule, – nodule, rich in (hydr)oxides of manganese and iron and including those of copper, nickel, cobalt and cadmium, up to more than 10 cm in diameter. They occur on (and less abundantly in) the sea bottom, mainly in the deep ocean over extensive areas , and besides on the bottom of some lakes, and in fossil deposits. / mangaanknol / nodule de manganèse / Manganknolle / nódulo de manganeso

3478
pisolith, – a (sub)spherical body, usually consisting of calcium carbonate, and larger than a few mm. / pisolietbolletje / pisolithe / Pisolith / pisolito

pisolite, – a sedimentary rock mainly consisting of pisoliths. / pisoliet / roche pisolithique / Pisolit / pisolita

pisolitic, – pertaining to a pisolite. / pisolietisch / pisolithique / pisolithisch / pisolítico

3479
oolith; ooid, – a small, round, concentrically laminated accretionary body. / oöiede; oölietbolletje / oolithe / Ooid / oolito

oolite, – a rock, mainly composed of ooliths. / oöliet / roche oolithique / Oolith / colita

oolitic, – mainly composed of, or containing ooliths, e.g. oolitic sand or oolitic iron ore. / oölietisch / oolithique / ooidisch; oolithisch / oolítico

3480
phosphorite, – a sedimentary rock or concretion, rich in, or consisting of phosphate(s), often associated with precipitated uranium oxide.- / fosforiet / phosphorite / Phosphorit / fosforita

3481
geode, – round, often subspherical, hollow nodule of at least a few cm diameter, lined inside with crystals projecting inward. / geode / géode / Geode / géoda

3482
spherulite; sphaerolite, – a more or less spherical, crystalline body with a radial structure, formed by diagenesis. The diameter of spherulites may vary from less than 1 mm to more than 1 cm. / sferoliet / sphérolite / Sphärolith / esferulito

spherulitic, – containing spherulites. / sferolietisch / sphérolitique / sphärolitisch / esferulitico

3483
spherosiderite; sphaerosiderite, – siderite concretion having the structure of a spherulite. / sferosideriet / nodule de sidérite / Sphärosiderit / esferosiderita

3484
framboid, – a crystal aggregate, usually microscopic, having a specific, raspberry-like aspect, common in pyrite. / framboïde / framboide / Framboid / framboide

framboidal, – containing framboids. / framboïdaal / framboidal / Framboidal; Himbeer- / framboidal

3485
rattle stone; rattlestone, – a usually ferruginous concretion, whose central kernel is detached from its shell. / klappersteen / aétite / Klapperstein / – –

3486
imatra stone, – a certain type of calcareous concretion found in varved glacial (silty) clays. / imatra-steen / concrétion marneuse; imatra / Imatra-Stein / piedra de imatra

3487
septarian nodule (plur. .... nodules; also: septaria), – a concretion in clayey or marly sediment, mostly calcareous or sideritic, characterized by internal shrinkage cracks, partly or entirely filled with secondary carbonates with or without other minerals. / septarie / septaria / Septarie / nódulo de septarias

melikaria, – boxwork of crack fillings weathered out from a septarian nodule. / melikaria / melikaria / Melikaria / – –

3488
clay ironstone, – clay cemented by siderite, occurring as nodules (often septarian nodules) or as separate beds. / kleiijzersteen / argile à passées ferrugineuses / Toneisenstein / arcilla ferruginosa

3489
loess nodule; .... doll; .... kindchen; .... puppet, – a calcareous concretion formed in loess, mostly of irregular shape. / loesspoppetje / poupée du loess / Lösskindl / muñeca de loess

3490
rhizoconcretion, – a small concretion formed around a root, or part of a root of a plant or tree. / rhizoconcretie / rhizoconcrétion / – – / rizoconcreción

3491
flint, – chert occurring in chalks and other calcareous rocks of Cretaceous and Tertiary age, mostly in the shape of nodules, but also as sheet-like bodies (e.g. along fractures and bedding planes) or branched cylindrical or tubular structures. It lacks the micro-joints found in many older cherts. It is almost exclusively formed by replacement of calcareous sediment and does not form layers like e.g. radiolarian chert. / vuursteen / silex / Flintstein; Kiesel / pedernal

flint curtain, – a plate-shaped 'vein' of flint, normally formed along a joint in a carbonate rock. / vuursteengordijn / veine siliceuse / – – / – –

3492
jasper, – cryptocrystalline quartz aggregate or rock, usually red, due to hematite, also green, yellow, brown etc. Often associated with pillow lavas and formed as primary submarine volcanic deposits or by replacement. Also formed on land in connection with soil processes. The term is sometimes also used for radiolarian cherts. / ijzerkiezel; jaspis / jaspe / Jaspis / jaspe

jaspilite, – ferruginous, bedded chert, normally composed of interbedded hematite and chert or jasper. / jaspiliet / jaspilite / Jaspilit / jaspilita

[2400] CARBONACEOUS DEPOSITS

[2410] GEOLOGY AND OCCURRENCE

3493
carbonaceous deposit, – a deposit composed of organic, in general vegetal, matter or its subsequently produced derivatives, often with mineral and volatile admixtures. The term is usually restricted to peats, brown coals and coals; if rich in substances that upon distillation produce hydrocarbons it is called a kerogenite. / vaste-koolwaterstofhoudende afzetting / dépôt houiller / Kohlenlagerstätte / depósito carbonáceo

NOTE:
– see table 1 – Caustobiolites

3494
peat, – the precursor of humic coal. It consists of fragments of decaying wood embedded in a matrix of miscellaneous disintegrated plant debris, formed in marshes and swamps, under not entirely anaerobic conditions, from the imcompletely decomposed remains of the dead marsh vegetation attacked by bacteria, fungi and other organisms. Stagnant ground water is necessary for its formation to protect the residual plant matter from complete decay. It is distinguished from the lowest rank of brown coal by the presence of free cellulose and contains more than 70% of water. / veen / tourbe / Torf / turba

3495
brown coal, – low-rank humic coal, intermediate between peat and hard coal, contains about 10-75% water. In the English-speaking countries it is synonymous with lignite. / bruinkool / lignite / Weichbraunkohle / lignito pardo

lignite, – in a broad sense, this term is synonymous with brown coal; strictly, in the English nomenclature, it means hard or consolidated brown coal. The term brown coal is preferable. / ligniet / lignite / Hartbraunkohle / lignito

xylite, – whole pieces of fossil wood in brown coal with a well preserved structure. / xyliet / lignite xyloide / Xylit / xilita; lignito xiloide

3496
coal; hard ...., – a combustible organic sedimentary rock with less than 40% of mineral matter (based on dry material), composed of polymers of cyclic hydrocarbons. Coal is formed by the accumulation, decomposition and transformation of plant material. It can be divided into two groups: humic coals and sapropelic coal. If no adjective is added, the term coal is used in the restricted sense of humic coal. / steenkool; kool / charbon; .... de terre; .... minéral / Steinkohle / carbón; .... piedra; .... mineral

3497
humic coal; humolite; banded coal (USA), – coals derived from original organic matter (mostly land plants) that underwent change chiefly through the process of peat formation in the presence of some oxygen. It has a banded appearance. / (humeuze) kool / charbon humique / Humuskohle; Kohle s.s. / carbón húmico

sapropelic coal; sapropelite; lipid coal; non-banded .... (USA), – in contrast to humic coal, it is a coal of which the original material was mostly transformed under complete exclusion of air. It is rich in lipid organic matter with less than 33% of mineral matter, and has a non-banded appearance. Actually it should be classified with the kerogenous deposits. / sapropeliet / charbon sapropélique; gayet / Faulschlamkohle / carbón sapropélico

mixed coal, – simultaneous occurrence of humic and sapropelic coal derived from banded peat. / – – / charbon mixte / – – / – –

3498
coal measures, – strata or stratigraphic units containing coal beds. Formerly used mainly for the British Carboniferous, it is now more widely applied. / steenkoolhoudende formaties / couches à charbon; terrain houiller / kohleführende Schichten; Kohlengebirge; Kohleformation / paquete productivo; .... de capas

coal series, – incorrect use of a chronostratigraphic term for a coal-bearing formation. / steenkoolhoudende formatie / formation carbonifère / Kohleformation; Kohlenserie / formación carbonífera

3499
coal bed, – the formal lithostratigraphic term for the more commonly used term coal seam. / koollaag / couche de charbon / Flöz; Kohlenflöz / capa de carbón; manto de .... (South America)

coal seam, – a bed, stratum or vein of coal. / koollaag / veine de charbon; couche de .... / Flöz / veta; capa de carbón; manto (South America)

coal streak, – a thin band or often a lens only of coal. / koollaagje; koollens; riffel / filet charbonneux; griffe / Kohlenschmitz / lentejón de carbón

3500
guide seam; marker ...., – a coal seam with characteristics that correlate over long distances. / gidslaag / couche repère; .... caractéristique / Leitflöz / capa guía; nivel ....

3501
country rock; wall ...., – the rock enclosing a coal seam. / nevengesteente / stériles; stampes / Nebengestein; Berge / salvandas; hastiales; rocas de caja

overburden, – the non-coal-bearing formation(s) overlying coal measures. / dekterrein / morts terrains; recouvrement / Deckschichten / recubrimiento

roof, – the bed directly overlying a coal seam. / dak / toit / Dachfläche; Hangendes; Firste / techo

floor, – the bed directly underlying a coal seam. / vloer / mur / Sohle; Liegendes / muro

3502
bind, – a tough shale forming the roof of a coal seam. / – – / – – / – – / pizarra fuerte; techo de .... fuerte (miner's slang)

clunch, –
1. a soft shale forming the roof of a coal seam that breaks readily into irregular layers.
2. a stiff, tough or indurated clay forming the floor of a coal seam.

3503
coal pipe; cauldron bottom; horseback, – a cylindrical extension of a coal seam into the overlying rock, representing a rapidly buried tree stump or a mud-filled hole in the roof of a coal seam caused by weathered prostrate tree trunks.

pothole; baumpot, – a cavity left in the roof of a coal seam caused by the dropping of a cast of a fossil tree stump after removal of the coal.

3504
clay ironstone, – a compact, hard, dark grey or brown, fine-grained sedimentary rock consisting of a mixture of argillaceous material (up to 30%) and iron carbonate (siderite), occurring either in layers of nodules or concretions or as relatively continuous but irregular thin beds, usually associated with carbonaceous beds overlying a coal seam. / klei-ijzersteen / argile à passées ferrugineuses; clayat / Toneisenstein; Toneisensteingeoden / arcilla ferruginosa

blackband ironstone, – a dark variety of clay ironstone containing sufficient carbonaceous matter (10-20%) to make it self-calcining (without addition of extra fuel). / koolrijke klei-ijzersteen / – – / Kohleneisenstein / – –

3505
tonstein; kaolin-coal ...., – a German term designating hard argillaceous bands in coal seams, often forming beds of considerable extension. The characteristic and predominant mineral is kaolinite or a mineral belonging to the kaolinite group. Unmistakable identification is only possible by use of the microscope and by X-ray. / – – / gore (blanc) / Tonstein / tonstein

carbonaceous shale; carbargillite, – a coal containing 20-60% mineral matter. / koolhoudende kleisteen / schiste charbonneux / Brandschiefer; Carbargillit / pizarra carbonosa; carbagilita

3506
parting, – a small joint in coal, or a thin layer or streak of

country rock in a coal seam. / – – / nerf stérile; barre; joint / Bergemittel / banda

dirt band; .... bed; .... parting, – a thin stratum of shale, mudstone or soft earthy material in a coal seam or interbedded with coal seams. / steenmiddel; steriel tussenlaagje / passée stérile; nerf / Zwischenmittel; Mittel; Berge; taube Lage / banda estéril; esterilidad

coal split, – a coal seam that is divided by a parting or sedimentary rock and cannot be mined as a single unit. / – – / couche en deux passées / – – / carbón digitado; capa digitada; digitación

3507
coal ball; plant bullion; peat dolomite; seam nodule, – nodules of spheroidal, lenticular or irregular shape consisting mainly of calcareous, dolomitic, sideritic, pyritic or silicious material surrounding or impregnating petrified plant remains and more rarely animal remains. It occurs in coal seams or in adjacent rocks connected with marine incursions. / – – / 'coal ball' / Torfdolomit / concreción dolomítica; ankerita carbonosa

3508
cleat, – a crack or a system of vertical joints in a coal seam, probably originating as desiccation cracks or at least initiated by shrinkage. It is most distinctive in bituminous coal and is not comparable with cleavage in other rocks. Sometimes it is called also **endogenous cracking**, and usually exhibits two systems developed at right angles: face cleat and end cleat. / splijtvlak / fissure endogène / Schwundriss; Schlechte / fisura endógena

end cleat; butt ...., – the minor cleat system in a coal seam.

face cleat, – the major cleat system in a coal seam.

3509
wedge out, – the progressive thinning of a coal seam until it disappears. / uitwiggen / terminaison en biseau; serrage; schistification / auskeilen; vertauben / acuñamiento progresivo; estrechamiento ....

nip, – the pinch, thinning or disappearance of a coal seam, especially as a result of tectonic deformation. / – – / étranglement; pincement; serrée / – – / acuñamiento tectónico; estrechamiento ....

leaf, – the irregular lateral end of a coal seam.

3510
wash-out, – a channel or channel-like feature produced in a coal seam by the scouring action of flowing water, later filled with a younger deposit. / – – / chenal d'érosion / Auswaschung; Auswaschungsrinne / canal erosivo; .... de erosión

rock roll, – a type of wash-out where localized currents have cut into the peat surface, forming elongated ridges on the underside of the roof that penetrate the coal seam. They often have the shape of boat bottoms and sometimes they reach the floor.

want, – the portion of a coal seam that is missing due to a wash-out. / insnoering / étreinte; serrée / Flözverdrückung / – –

3511
coal scare, – lenticular pocket of clean coal in sandstone, usually found in the region of a wash-out.

3512
float coal, – isolated bodies of coal in sandstone or shale, probably redeposited fragments from an eroded deposit. / – – / charbon remanié / – – / carbón flotado

coal gravel, – a secondary coal deposit consisting of transported and redeposited coal fragments. / – – / galets de charbon / Kohlengeröll / fragmentos de carbón arrastrado

pelagochthonous, – a term applied to coal deposits formed from submerged forests and from drift wood.

3513
slack, – coal in small particles disintegrated by weathering. / kolengruis / charbon menu / Kohlengrus; Kohlenklein / finos de carbón; .... de hulla

3514
pebble coal, – coal composed of rounded masses cemented by coaly material as a result of pressure. / – – / charbon en galets / Kugelkohle; Mugelkohle / – –

3515
seat earth, – a bed, usually of clay, underlying a coal seam. It represents the soil that supported the vegetation from which the coal was formed and where the plants were rooted. It commonly contains fossil root casts. / bodemlaag / mur / – – / muro del carbón; muro; piso (South America)

underclay, – a seat earth consisting of clay only, but often containing rootlets. / bodemklei / mur argileux / – – / arcilla de muro

root clay; rootlet bed – a stratum characterized by the occurrence of fossil rootlets of plants, especially the floor of a coal seam. / stigmaria-vloer; wortelvloer / mur à radicelles; sol de végétation / Wurzelboden / arcilla con suelos

3516
seat rock; seat stone, – the nearest bed of clunch, grit or sandstone below a coal seam. / – – / mur / – – / roca de muro

3517
fire clay; sagger; seggar; sagre, – a silicious clay rich in aluminium silicates capable of withstanding high temperatures without deforming. It is deficient in iron, calcium and alkalis and approaches kaolin in composition. It is often found as underclay or seat earth of coal seams and is sometimes extracted for use as refractory clay. / keramische klei / terre réfractaire; argile .... / feuerfester Ton; Fireclay / arcilla refractaria

3518
ganister; galliard, – a hard, compact, even-grained, fine-grained or silty, very pure silica rock (up to 97% SiO_2) composed of sub-angular quartz particles, 0.05-0.5 mm in diameter, cemented with secondary silica. It has a characteristic splintery fracture and is often found as part of the seat earth beneath a coal seam. / – – / – – / Ganister / – –

[2420] COALIFICATION; CLASSIFICATION OF PEATS AND COALS

[2421] *General Terms*

3519
coalification, – the natural, diagenetic process of transformation of plant remains into coal, resulting in a relative increase in carbon content and a relative decrease in the content of hydrogen, oxygen and other elements. / inkoling / carbonification; houillification / Inkohlung / carbonificación; carbonización natural

carbonification, – a once proposed term to replace coalification; it has not found acceptance.

biochemical phase; biogenetic ...., – the first stage in the transformation of organic matter, mostly under the influence of bacteria and fungi. Its main product is peat and it ceases in the lowermost brown-coal ranks. / biochemische fase / carbonification biochimique / biochemische Phase / fase bioquímica

geochemical phase, – the second stage in the transformation of organic matter under the influence of increased temperature and, to an unknown degree, of time. / geochemische fase / carbonification géochimique / geochemische Phase / fase geoquímica

3520
devolatilization; debitumenization, – the loss of volatile constituents and of extractable hydrocarbons, resulting in a proportional increase in carbon content during coalification. / ontgassing / perte des matières volatiles / Entgasung / desvolatiozación

3521
organic metamorphism, – comprises all physical and chemical transformation processes of dead organic matter under the influence of first biochemical, later geochemical processes, generally at temperatures below 250°C. / organische metamorfose / diagenèse et métamorphisme de la matière organique / organische Metamorphose / metamorfismo orgánico; diagénesis orgánica

degree of organic metamorphism; DOM; degree of coalification, – a measure for the stage of coalification in humic coals.

NOTE:
– formally these terms are misnomers, because coalification and the formation of hydrocarbons are diagenetic processes. Not to be confused with 'mineral metamorphism', which belongs to a far higher temperature realm. The term **'organic diagnesis'** has been proposed.

3522
coalification gradient, – the vertical distance or the depth interval in a coalfield that marks the increase by one DOM unit. It depends on the thermal gradient and is different for each coal rank. / inkolingsgradiënt / gradient de carbonification / Inkohlungsgradient / gradiente de carbonificación

Schürmann's law, – a theorem stating that, in a given vertical profile of brown coal, the content of water decreases with depth. / wet van Schürmann / loi de Schürmann / Schürmannsche Regel / ley de Schürmann

Hilt's law, – a theorem stating that in a given vertical profile (or a limited, undisturbed area) the deeper coals are always of a higher rank than those above them. / wet van Hilt / loi de Hilt / Hiltsche Regel / ley de Hilt

3523
isocal, – in maps and sections, lines connecting points of equal calorific value. / – – / – – / Isokalore / isocalorífica

isocarb, – on coalfield maps and geological basin maps, lines connecting pionts of equal coalification. / – – / – – / Isokarbe / isocarbonificada

isovol, – line connecting points of equal content of volatile matter. / isovole / – – / Isovole / isovolatil

[2422] *Biochemical Phase; Environments of Deposition*

3524
peat formation; peatification; ulmification; paludification, – the transformation of vegetable substance in stagnant waters under the influence of small amounts of oxygen. Peats are distinguished according to the types of vegetation or plant genera and species from which they derive. This means largely according to the environments of deposition in relation to the ground-water table. / veenvorming / tourbification / Vertorfung; (Torfbildung) / turbogénesis; turbonificación

3525
humification, – the process of transformation of plant remains into humus or humic acids, essentially by slow oxidation under strictly aerobic conditions. / humificatie / humification / Humifizierung / humificación

humus; soil ulmin, – the generally dark, more or less stable part of the organic matter of the soil, so well decomposed that the original sources cannot be identified. The term is sometimes incorrectly used for the total organic matter of the soil including relatively undecomposed material. / humus / terre végétale; terreau; humus / Humus / humus

humic acid, – acidic, black, organic matter extracted from soils, low-rank brown coal and other decayed plant substances by alkalis. It is insoluble in acids and organic solvents. / humuszuur / acide humique; acide ulmique / Huminsäure / ácido humico

dopplerite; pitch peat; peat gel; brown-coal gel, – a black gelatinous material solidifying as a result of loss of water to a black, lustrous solid, consisting mainly of free humic acids or humic acid salts, such as calcium humate. The term is applicable only to material occuring in peat or lowest-rank brown coal. Collinite would be the equivalent maceral in coals of higher rank. / doppleriet / dopplérite / Dopplerit / dopplerita

3526
disintegration (obsolete), – the transformation of organic residues under aerobic conditions into carbon dioxide, water, etc. / – – / décomposition / Mineralisierung / desintegración

mouldering; rotting, – the process of transformation of organic substance into a residue (black mould) under conditions of air supply inadequate for complete disintegration. / veraarding; vermoddering; vermolming / pourriture / Vermoderung; Vermodern / pulverización

putrefaction, — the process of transformation of vegetable matter in the presence of water and complete exclusion of air by anoxic micro-organisms. It is decomposition of organic matter by slow distillation, the residue becoming enriched in carbon and methane and other gaseous products being formed (H_2, NH_3, H_2S, etc.). The result is a mainly sapropelic product. / rotting / putréfaction; pourrissement / Verfaulung; Verfaulen; Fäulnis / putrefacción

3527
muck, — a dark-coloured soil found in bogs and swamps. It consists of impure, well decomposed vegetable matter, containing a high percentage of mineral matter (usually silt) and not burning readily. It forms surface deposits in some poorly drained areas, e.g. permafrost and lake bottoms. / rottingsslik / sol organique / Anmoor / cieno

dy, — a dark, jelly-like fresh-water ooze composed of organic matter carried into a nutrient-deficient lake in colloidal form and precipitated there. / dy / dy / Dy / — —

3528
gyttja, — a black aquatic ooze, in which the organic matter is more or less determinable, laid down under oxic to anoxic conditions and characteristic of eutrophic and mesotrophic lakes. / gyttja / — — / Gyttja / — —

sapropel, — an aquatic ooze that contains varying amounts of unrecognizable organic debris laid down under anoxic conditions. / sapropeel; rottingsslik / sapropèle; vase putride / Sapropel; Faulschlamm / sapropel

3529
humocoll, — humic material of peat rank.

marsh gas, — volatile decomposition product of peat formation, mainly methane. / moerasgas / gaz des marais / Sumpfgas / gas de pantano

3530
quaking bog; quagmire, — a carpet of bog vegetation that is floating and quivers when walked on. / trilveen / tourbière branlante; branloire / Schwingmoor; ....rasen / pantano oscilante

3531
limnic peat, — peat formed below a body of standing water. Its organic material is often mostly planktonic. / limnisch veen / tourbe lacustre / — — / turba lacustre; .... límnica

3532
moor, — an extensive tract of open, usually more or less wet wasteland overlain by peat. / veen / tourbière; marais; ..... à tourbe; marécage; palus / Moor / turbera

bog; marsh, — a swamp or an area of wet land, often covered with peat. / moeras; broek / marécage; fondrière / Moor / pantano; marisma

paludal, — pertaining to marsh, bog or swamp. / moerassig / marécageux; paludéen; palustre / sumpfig; moorig / palustre

3533
lowmoor bog, — a peat bog that is at or only slightly below the ground-water table and is dependent on it for the accumulation and preservation of peat. / laagliggend veenmoeras / tourbière basse; marais plat / Niedermoor / turbera baja

3534
valley bog, — a bog developed in a valley bottom. / dalveen / tourbière de vallée / Talmoor / pantano de valle

hanging bog, — a bog developed on a moist slope. / hellingveen / tourbière de pente / Gehängemoor / pantano de pendiente

3535
lowmoor peat, — peat that originated in low moors or swamps and contains little or no sphagnum. / waterveen / tourbe de tourbière basse / Niedermoortorf / turba de turbera baja

3536
transition moor, — a moor with peat formation slightly above the ground-water table. / overgangsveen / — — / Übergangsmoor / turbera de transición

meadow peat, — peat derived from stems and roots of grasses. / — — / tourbe de prairie / — — / turba de pradera

wood peat, — / bosveen / tourbe à bois / Bruchwaldtorf / turba de bosque

3537
lagg; bog moat, — marshy border of a raised bog, where drainage water of the raised bog is mixed with water of the surrounding area. The lagg therefore carries a mesotrophic vegetation. / lagg; veenrandmoeras / lagg; marécage bordier / Lagg; Randsumpf / marjal

3538
regeneration sequence; .... complex, — typical succession of peat-forming plant communities in raised bogs. / regeneratiecomplex / série turficale; série de peuplements turficales / Regenerationskomplex / secuencia de regeneración

3539
hollow, — a low part of the regeneration complex. / slenk / — — / Schlenke / parte baja

hummock, — a high part of the regeneration complex. / bult / bosse / Bult / parte alta

3540
bog pool; moor pond, — a shallow pool on a raised bog. / veenmeer / mare de tourbière / Moorsee; Kolk; Blänke / charca (de pantano)

3541
bog burst, — the destruction of the sloping surface of a raised bog by excess of water embodied in the bog after periods of prolonged rainfall. / veendoorbraak / débâcle de tourbière / Moorausbruch / destrucción de la turbera

3542
highmoor bog, — a bog, often in the uplands, whose surface is largely covered by sphagnum mosses which, because of their

high degree of water retention, make the bog more dependent on relatively high rainfall than on the ground-water table for the accumulation of sphagnum peat. / hoogliggend veenmoeras / tourbière haute; marais élevé; fagne / Hochmoor / turbera alta

highmoor peat; moorland peat; moor peat, — peat occurring on highmoor bogs and formed from mosses. / landveen / tourbe de haute tourbière / Hochmoortorf / turba de turbera alta

3543
sphagnum bog, — an acid, very wet bog containing abundant sphagnum moss. / — — / marais à sphaignes / Sphagnum-Moor / pantano de esfagno

sphagnum peat; moss peat, — a highmoor peat composed mainly of bog moss (sphagnum). / mosveen / tourbe à sphagnum; .... à sphaignes / Moostorf; Sphagnum-Torf / turba de esfagno

3544
recurrence horizon, — a level that forms the limit between a strongly decomposed sphagnum peat and a superimposed, only slightly decomposed sphagnum peat. Well known is Weber's **'Grenzhorizont'** occurring in most NW-European raised bogs. / grenslaag / horizon limite / Schwarz-Weisstorf-kontakt; SWK / horizonte de recurrencia

3545
calluna peat; heath peat, — a highmoor peat derived from various heaths. / heideveen / tourbe de bruyère / Calluna-Torf; Heidetorf / turba de matorral

3546
terrestrial peat, — peat formed above the ground-water table. / terrestrisch veen / tourbe terrestre / terrestrischer Torf / turbera terrestre

NOTE:
- in Dutch nomenclature **'laagveen'** (literal translation: 'low peat') and **'hoogveen'** ('high peat') are no genetic terms; they indicate present topographic elevation and position relative to the ground-water level.

3547
tundra peat, — peat formed under sub-arctic conditions by slow accumulation of remains of mosses, heaths, birch and willow. / toendra-veen / tourbe de toundra / Tundra-Torf / turba de tundra

3548
marsh peat; banded peat, — peat derived from a mixture of land-plant debris and sapropelic matter. / — — / tourbe humosapropélique / Mudde / turba humo-sapropélica

3549
oligotrophic peat, mesotrophic peat, eutrophic peat, — classification based on the nutritional requirements of plants in the environment, where the peat was formed.

oligotrophic, — water poor in plant nutrients. / oligotroof / oligotrophe / oligotroph / oligotrófica

mesotrophic, — intermediate between eutrophic and oligotrophic. / mesotroof / mésotrophe / mesotroph / mesotrófica

eutrophic, — water containing abundant plant nutrients. / eutroof / eutrophe / eutroph / eutrófica

3550
ombrogenous, — said of a peat deposit where the moisture is supplied mostly by oligotrophic rainwater and moisture from the air. / ombrogeen / ombrogène / ombrogen / de turbera alta; ombrógena

topogenous, — said of a peat deposit whose moisture content is supplied mostly by (eutrophic) river water. / topogeen / topogène / topogen / de turbera baja; topógena

soligenous, — said of a peat deposit whose moisture content is determined by both rainfall and surface water. / soligeen / soligène / soligen / solígena

[2423] *Geochemical Phase*

3551
coal quality, — the sum of the characteristics of coal type, coal rank and coal grade. / — — / qualité du charbon / Kohlen-Eigenschaft / calidad del carbón

3552
coal type, — in European nomenclature sunonym for microlithotype, the microscopic coal structure. / — — / — — / Microlithotyp / clase de carbón; tipo de ....

coal rank, — one of the various stages in natural coalification indicating the degree of organic metamorphism of humic coal. / inkolingsgraad / rang du charbon / Inkohlungsgrad; Rang / rango de carbón

coal grade, — a grading of coal according to the admixture of mineral matter.

NOTE:
- see table 9 — Coal-Rank Classification.

3553
mineral matter, — the inorganic constituents of humic coal. / minerale bestanddelen / matière minérale / Minerelbestandteile; Mineralgehalt / materia mineral

syngenetic; authigenic; inherent, — terms pertaining to mineral matter inherent in the original plant material and not removable by physical means. / syngenetisch / cendres de constitution / syngenetisch / singenética

epigenetic; extraneous; mineral inclusion, — terms pertaining to mineral matter formed at varous stages of coalification after peat formation. / epigenetisch / cendres secondaires; .... épigénétiques / epigenetisch / epigenética

3554
black damp;choke ...., — a coalmine gas that is non-explosive and consists mainly of CO_2 and N_2 with little O_2. / — — / air grisouteux; mofette / matte Wetter / aire viciado

fire damp, — an explosive coalmine gas, mainly composed of methane and air. / mijngas / grisou / Grubengas; schlagende Wetter; Grubenwetter / grisú

white damp, – a term for carbon monoxide in coal mines. / – – / – – / – – / maleza; ácido (miner's slang)

3555
soft brown coal, – an informal term used in some places for the lowest rank brown coal. / – – / lignite brun; .... tendre / Weichbraunkohle / lignito pardo

earth coal; smut ....; smuth, – an earthy variety of soft brown coal. Also an obsolete name for hard coal. / aardachtige kool / lignite brun terreux / erdige Weichbraunkohle; Schmierkohle / lignito pardo terroso

3556
crumble coal, – an incoherent brown coal that lacks cementing material.

leaf coal, – a variety of brown coal mainly made up of leaves. / bladerkool / charbon feuilleté / Blätterkohle; Laubkohle / carbón hojoso

needle coal, – a variety of brown coal composed of a fibrous needle-like mass of vascular bundles of palm stems. / – – / – – / Nadelkohle / – –

moor coal, – an American term for a variety of friable brown coal characterized by good cleavage: it breaks into cubic or trapezoidal fragments. / – – / lignite clivé; lignite friable / Moorkohle / lignito friable

3557
pitch coal; picurite, – a variety of black, brittle, compact brown coal of a jet-black colour, pitchy lustre and conchoidal fracture resembling jet. It is sometimes wrongly called bituminous brown coal. / – – / jais; charbon piciforme / Pechkohle; Glanzbraunkohle / carbón piciforme

jet, – a hard, lustrous, pure, black variety of brown coal with a conchoidal fracture and taking a high polish. It occurs as isolated masses in kerogenous shale and is probably derived from water-logged pieces of coniferous driftwood. / git / jais; jayet / Gagat; Pechkohle; Jett / carbón azabacheado; azabache

3558
lustrous brown coal; bright brown coal, – an informal term for hard, lustrous coal covering the transition from brown coal to hard coal; it consists predominantly of low-rank hard coal and part of the American sub-bitumous coal. / – – / lignite noir brillant / Glanzbraunkohle / lignito negro brillante

3559
bituminous coal, –

1. orginally a humic coal yielding tar on heating;
2. in the American classification a humic coal group subdivided into three sections: high, medium and low-volatile bituminous coal.

/ bitumineuze kool / houille bitumineuse; charbon .... / 'bituminous coal' / carbón bituminoso

This group includes:

flame coal, – a high-volatile coal with more than 40% volatile matter. / vlamkool / charbon flambant; .... flambant sec / Flammkohle / hulla seca de llama larga; .... lignitosa; .... flamanta

long-flame coal; gas-flame ...., – a high-volatile bituminous coal in the approximate range of gas coal and flame coal with about 35-40% volatile matter. / gasvlamkool / houille sèche à longue flamme / Flammkohle; Gasflammkohle / hulla grasa de llama larga; .... de gas

gas coal, – a high-volatile coal in the approximate fixed carbon range 65-72 with about 28-35% volatile matter. / gaskool / charbon à gaz / Gaskohle / hulla grasa; .... de forja

fat coal, – a humic coal approximately equivalent to medium-volatile bituminous coal with about 19-28% volatile matter. / vetkool / charbon gras; houille gras / Fettkohle / hulla de coque; .... grasa de llama corta

steam coal, – a coal in the approximate fixed carbon range 80-90 with about 12-19% volatile matter. / esskool / charbon demi-gras / Esskohle / carbón de vapor; hulla seca de llama corta; hulla magra antracitosa

NOTE:
– above classification of bitumenous coals is not exact; use of these terms is not recommended.

3560
semi-anthracite, – the coal-rank group comprising coals in the fixed carbon range 86-92. / magerkool / semianthracite; demianthracite; charbon maigre / Semianthrazit; Halbanthrazit; Magerkohle / semi-antracita

lean coal; dry ...., – coal, analogous to semi-anthracite and partly analogous to short-flame coal or dry coal. / magerkool / charbon maigre / Magerkohle; Esskohle / carbón magro

3561
anthracite, – the coal-rank group comprising coals in the fixed carbon range 92-98. It ignites at a high temperature and burns slowly with a smokeless flame. It has a high lustre and can be handled without soiling the fingers. / anthraciet / anthracite / Anthrazit / antracita

3562
meta-anthracite, – the coal-rank group comprising coal with fixed carbon range 98-100, i.e. the highest rank of humic coal. / meta-anthraciet / méta-anthracite / Meta-Anthrazit / meta-antracita

3563
paper coal; dysodil, – low-rank, thinly laminated sapropelic coal, some varieties of which are composed of cuticles of leaves, others of diatoms. / bladerige kool / charbon feulleté; houille papyracée / Papierkohle; Blätterkohle / – –

cannel coal; channel coal, – the non-banded sapropelic coal chiefly consisting of the maceral sporinite. It shows transitions to boghead coal, has a satin sheen or waxy lustre and conchoidal fracture. / kandelkool / gayet / Kennelkohle; (kandelit) / – –

boghead coal; algal coal; torbanite, – a sapropelic coal consisting chiefly of the maceral alginite. It is compact, brownish black to black, very tough and difficult to break, with conchoidal to sub-conchoidal fracture. / algenkool / boghead / Bogheadkohle / carbón boghead

3564
finger coal, – coal altered by intrusions to a coke-like material that is divided into small hexagonal columns.

cinder coal, – coal enriched by heat associated with igneous activity. / veredelde kool / – – / – – / – –

natural coke; carbonite, – coal that has been naturally carbonized to coke by contact with an igneous intrusion or by natural combustion. / natuurlijke cokes / coke naturel / Naturkoks / coque natural

[2430] PETROGRAPHY AND STRUCTURE OF COALS

3565
coal petrography; anthracology; anthracography, – strictly the macroscopic and especially the microscopic composition of the organic constituents of coal ; in general also including the study of the mineral inclusions. / koolpetrografie / pétrographie du charbon / Kohlenpetrographie / petrografía de carbón

coal structure, – humic coal, being an association of various coal macerals, has been subdivided into various structural types. The macroscopic structures are called lithotypes, the microscopic ones microlithotypes. / koolstructuur / texture du charbon / Kohlestruktur / estructura del carbón

3566
maceral, – based on the analogy with minerals of the inorganic rocks, macerals are the more or less homogeneous microscopically recognizable individual organic constituents of coal that, depending on their quantitative participation and their associations, control the chemical, physical and technological properties of a coal of a given rank. They evolve from the different tissues of the initial coal-forming plant during coalification. / maceraal / macéral / Maceral; Gefügebestandteil; Gemengteil / maceral

maceral group, – a group of macerals with comparable properties. / maceraal-groep / – – / Maceralgruppe / grupo maceral

maceral suite, – in the American classification according to Spackman, analogous to the European maceral group.

3567
submaceral, – a subdivision of certain macerals according to particle size of fragments or to the state of preservation. / submaceraal / submacéral / Submaceral / – –

maceral variety, – a subdivision of certain macerals according to recognizable plants and plant tissues. / – – / variété de macéral / Maceral-Varietät / variedad maceral

cryptomaceral, – a subdivision of certain macerals according to special structure that can be recognized by using supplementary methods (e.g. etching). / cryptomaceraal / cryptomacéral / Kryptomaceral / – –

NOTES:
- see table 10 – Coal Macerals,
- suffixes to maceral names

in English	–	inite
Dutch	–	iniet
French	–	inite
German	–	init
Spanisch	–	inita

3568
huminite, – the group of brown-coal macerals analogous to the vitrinite maceral group in hard coal. It contains three sub-groups and six macerals.

humotelinite, – within the maceral group of huminite the subgroup consisting of intact cell walls of tissues or isolated single cells in a humic state of preservation. It contains the macerals textinite and ulminite.

humodetrinite, –
1. within the maceral group of huminite, this is the subgroup consisting of the finest humic fragments with a finely distributed humic gel between them. It contains the macerals attrinite and densinite;
2. in the USA humodetrinite (analogous to translucent humic degradation matter) is partly analogous to attrinite.

humocollinite, – within the maceral group of humite, this is the sub-group that consists of an amorphous humic gel or of intensely gelified plant tissues and humic detritus. It contains the macerals gelinite and corpohuminite.

3569
vitrinite, – the hard-coal maceral group (analogous to the huminite group in brown coal) that is the primary constituent of 'bright coal', and is derived from wood tissue and bark. Vitrinite is analogous to the American anthraxylon if bands are more than 14 microns thick; if less, it is analogous to translucent humic matter.

telinite, –
1. consisting of the cell walls of recognizable plant tissues;
2. in the Russian standard, a maceral analogous to the international terms textinite and texto-ulminite.

vitrodetrinite, – consisting of small fragments of more or less angular and variable outlines originating from the extensive comminution of gelified tissues.

collinite, –
1. hard-coal maceral consisting of a precipitated gel formed from ulmin. It comprises a range of constituents, all of which display an absence of cell structure under normal microscopic conditions;
2. in the Russion standard a brown-coal maceral analogous to the international term gelinite.

3570
liptinite, –
1. a brown-coal maceral group analogous to the exinite group in hard coal;
2. in the Russian standard, liptinite is a synonym of exinite.

exinoid, – an American term denoting, in Spackman's classification, a maceral group of the liptinite (= exinite) suite.

resinoid, – an American term denoting, in Spackman's classification, a maceral group of the liptinite suite (= exinite), consisting of coalified resins.

3571
exinite, – a hard-coal maceral group (analogous to the liptinite group in brown coal) containing the macerals sporinite, cutinite, alginite, resinite and liptodetrinite.

3572
sporinite, – a maceral of the exinite and liptinite groups resulting from pollen and spore exines flattened parallel to stratification.

cutinite, – a maceral consisting of cuticular substance (the waxy layer formed on outer walls of epidermal plant cells) of leaves and stems.

resinite, – a maceral that consists mainly of resinous substance.

alginite, –
1. a maceral formed from algal remains. It is best recognized by luminescent microscopy. It is rare in humic coal but is the characteristic maceral of sapropelic boghead coal.
2. in the Russian standard it is a maceral group.

liptodetrinite, – a maceral consisting of extremely small particles that can not be assigned with certainty to one or other maceral of the liptinite and exinite groups.

3573
inertinite, – a maceral group of both brown and hard coals that contains six macerals, based on reflectance and certain similarities in their technological properties. The material has a very low hydrogen content, and is hardly susceptible to change through diagenesis.

3574
micrinite, – an opaque maceral of the inertinite group showing variable form, finely granular but not cellular structure. It corresponds partly to the American opaque attritus.

macrinite, – a maceral that shows no cell structure, It occurs either as a groundmass or isolated in size, mostly above 10 microns.

semifusinite, – a constituent intermediate between vitrinite, and fusinite, showing a well-defined structure of wood and sclerenchyma. It is also called intermediates and is partly analogous to the American term brown matter.

fusinite, – a maceral that has a charcoal-like appearance, still showing a well defined cellular structure of wood or sclerenchyma.

sclerotinite, – a maceral consisting of strongly reflecting fungal remains, chiefly sclerotia.

inertodetrinite, – a finely detrital maceral comprising all particles in coal with higher reflectivity than vitrinite that cannot be classified as one of the other inertinite macerals.

3575
bituminite, – an ambiguous term, sometimes used for a maceral often present in oil-source rocks deposited in a reducing subaquatic environment. Also denoting structureless organic matter or amorph organic matter. Use is not recommended.

3576
exsudatinite, – a maceral or maceral group, formed by other macerals or solid bitumen as an exudation or sweat in rock pores, fissures etc.

3577
perhydrous maceral, – a maceral having a high hydrogen content, such as the exinites. / – – / – – / – – / perhidromaceral

subhydrous maceral, – a maceral having a low hydrogen content, such as the inertinites. / – – / – – / – – / subhidromaceral

3578
active maceral, – vitrinites and exinites are called active, in contrast to inertinites which show little or no reaction in the coking process. / actief maceraal / macéral actif / – – / maceral activo

3579
lithotype; banded constituent; banded ingredient, – the macrosopic structure of humic coals. / lithotype / structure type / Lithotyp; Kohlentyp / litotipo

NOTE:
– see table 11 – Coal Structure

3580
vitrain, – the lithotype designating macroscopically recognizable bands several mm in width of very bright coal. Microscopically it consists of lithotypes particularly rich in vitrinite. / glanskool / vitrain; charbon brillant; houille claire; lames claires / Vitrain; Glanzkohle / vitreno

clarain, – the lithotype consisting of finely striated bands of coal less than 3 mm thick that have an appearance between that of vitrain and durain. It consists of a variable composition of all maceral groups. / gestreepte glanskool / clarain; houille semi-brillante; .... foliaire; .... moyenne; charbon semi-brillant / Clarain; Halbglanzkohle / clareno

durain, – the lithotype comprising the dull bands in coals. It is analogous to the American splint coal and consists of microlithotypes rich in exinite and inertinite. / doffe kool / durain; charbon mat; houille mate; .... grenue; .... terne / Durain; Mattkohle / dureno

fusain; mineral charcoal; mother of coal, – the extremely friable, charcoal-like lithotype of coal consisting mainly of fusinite. In American nomenclature it is the opaque microscopic component consisting of fusinite and semifusinite more than 0.037 mm in width. / fossiele houtskool / fusain; charbon fibreux; houille mate fibreuse / Fusain; Faserkohle; fossile Holzkohle / fuseno

3581
microlithotype; maceral association; coal type, – a typical maceral association whose minimum band width is fixed at 59 microns. It is the microscopic fine structure of humic coal. / microlithotype / microlithotype / Mikrolithotyp; Streifenart / microlitotipo

NOTES:
– see table 11 – Coal Structure
– suffixes to microlithotype names
  in English – ite
  Dutch – iet
  French – ite
  German – it
  Spanisch – ita

3582
vitrite, – the monomaceral microlithotype consisting of at least 95% vitrinite or huminite. It is partly analogous to the American anthraxylon.

liptite, – the monomaceral microlithotype containing more than 95% of exinite or liptinite. The term sporite is sometimes used if sporinite is the predominant maceral.

inertite, – the monomaceral microlithotype consisting of more than 95% of inertinite.

3583
clarite, – the bimaceral microlithotype consisting principally of vitrinite and exinite.

vitrinertite, – the bimaceral microlithotype consisting principally of the maceral groups vitrinite and inertinite.

durite, – the bimaceral microlithotype consisting principally of inertinites and exinites (particularly sporinite).

3584
trimacerite, – the microlithotype group to which all three maceral groups constribute more than 5%.

3585
component of coal, – in the American nomenclature coal-petrographic entities recognizable visually as bands of layers of coal that have distinctive physical appearance and characteristic microstructural features. It is roughly comparable to the international term microlithotype.

3586
attritus, – this American term designates the thin-banded components of dull coal. It is a collective term not directly comparable with any one of the European microlithotypes and consists of an intimate association of varying proportions of macerals of the vitrinite, exinite and inertinite groups. It is subdivided into translucent and opaque attritus, each having several constituents.

translucent attritus, – in the American nomenclature, the banded component of coal consisting of complex residual organic matter, exclusive of anthraxylon, that transmits light in thin sections.

opaque attritus; .... matrix, – in the American nomenclature, coal material of which the most prominent constituent is opaque matter, chiefly macerals of the European inertinite group.

3587
anthraxylon, – the American term for the translucent banded component of coal, synonymus with vitrinite more than 0.015 mm in width. It is derived from woody tissues of plants and forms lustrous bands together with dull attritus in banded coal.

xylith, – the American term for a brown coal that consists entirely of anthraxylon.

3588
opaque matter, – part of the American opaque attritus consisting of fusinite less than 0.03 mm thick.

3589
granular opaque matter, – part of the American opaque matter consisting of discrete granules 0.05-0.15 mm in diameter. Analogous to the European micrinite.

amorphous opaque matter; massive opaque matter, – part of the American opaque matter, generally equivalent to dispersed humic degradation matter in form, but suggesting a fusinized mode of preservation.

3590
hydrite, – a very common microlithotype in Japanese Tertiary coals. It is analogous to clarite and consists of the macerals vitrinite, degradinite and exinite in considerably varying proportions.

3591
carbominerite, – association of intimately intergrown coal microlithotypes with a special mineral or mineral group.

3592
phyteral, – in contrast to macerals, which represent a purely petrographical concept, phyterals are strictly correlated with certain organs of the initial plant material. Phyterals are always expressed in general botanical terms but rarely used in coal petrography. / phyteraal / – – / Phyteral / fiteral

[2440] ANALYSIS AND PRODUCTS

3593
channel sample, – a special type of coal sampling giving a true average of the characteristics, especially of the rank and chemistry of a coal seam. / sleufmonster / échantillon de saignée / Schlitzprobe / muestra en roza

column sample; pillar ...., – an uninterrupted sample of a coal seam, corresponding to a continuous core sample. / – – / pilier prismatique complet / Säulenprobe; Profilsäule / muestra de testigo

3594
maceration, – in coal petrography and palynology, the separation of single macerals from the coal by various chemical and physical techniques. / maceratie / – – / Mazeration / maceración

combination analysis, – combined maceral and microlithotype analysis according to Mackowsky and Hevía.

ultimate analysis, – elementary chemical analysis; in coal, the determination of carbon, hydrogen, oxygen, sulphur, nitrogen and ash contents. / elementaire analyse / analyse élémentaire / Elementaranalyse / análisis elemental

proximate analysis, – the determination by standard methods of the content of moisture, volatile matter, fixed carbon (by difference) and ash in coal. For all low-rank coal, it includes determination of the calorific value. / – – / analyse immédiate / Kurzanalyse; Immediatanalyse / análisis inmediato

3595
caking power; caking property, – indicates the agglutinating

power of coal compared with inert substances during rapid heating. It is a parameter in the international classification of hard coals. / — — / pouvoir agglutinant / Backfähigkeit / poder aglutinante

Roga test, – a method for the determination of the caking power of coal by means of a shatter test of the coke obtained by a special method. It is not recommended for coals with fixed carbon range higher than 72.

free swelling index; crucible swelling number, – a method of determining the caking power of coal by comparison of the coke button with a standard after very rapid heating in a crucible. / — — / indice de gonflement / Blähgrad; Blähungsgrad / índice de hinchamiento al crisol

3596
coking capacity, – depends on the rate of degasification and on the plastic properties of coal during slow heating. It is a parameter of the international classification of hard coal by type. / — — / pouvoir cokéfiant / Kokungsvermögen / poder coquizante; .... de coquización

dilatometer test, – the determination of the percentage of maximum dilatation of coal on slow heating as an indicator of coking capacity / — — / essai de dilatation / Dilatometer-Test; Kokungsgradbestimmung / prueba del dilatómetro

Gray-King assay, – a method of determining the coking capacity of a coal by comparing the resulting coke against standard buttons after heating slowly with an electrode-carbon mixture.

3597
volatile matter, – those substances in coal other than moisture that are given off as gas and vapour during combustion. / gasvormige bestanddelen / matières volatiles / fluchtige Bestandteile / materia volátil

3598
inherent moisture, – the moisture retained by a coal sample after it has attained approximate equilibrium with the atmosphere to which it is exposed, or that part of the total moisture content of coal that is structurally contained in the material and cannot be removed by natural drying, but requires heating above 100^{0}C. / gebonden vocht / humidité de constitution / hygroskopischer Wassergehalt / humedad intrínseca

free moisture; pit water, – that part of the total moisture that is lost by a coal in attaining approximate equilibrium with the atmosphere to which it is exposed. / vrij vocht / humidité libre / grobe Feuchtigkeit / humedad libre

total moisture; bed ...., – the moisture of coal freshly sampled in situ. / totaal vocht / humidité totale / Gesamtwassergehalt; Rohkohlenwassergehalt / humedad total

3599
ash, – the inorganic residue obtained from coal after burning. Combustion usually alters both the weight and the composition of the original mineral matter. / as / cendres / Asche / ceniza

ash number, – a system of grading coal by numbers 0 – 8 (used in Australia, for example). (0 = 0 to 4% ash on dry basis; 1 = 4 to 8% etc.; 8 = 32% ash). / asgetal / — — / — — / contenido en cenizas

3600
inherent ash; intrinsic ash, – ash arising from mineral matter inherent in the original vegetable material. / — — / cendre intrinsèque / Pflanzenasche / cenizas intrínsecas

extraneous ash; secondary ash; sedimentary ash, – ash arising from mineral matter associated with, but not inherent in humic coal. / — — / cendre extrinsèque; .... sédimentaire / Fremdasche; epigenetische Asche; syngenetische .... / cenizas secundarias

pit ash, – ash in mined coal derived from dirt bands, adjoining shale or cleat minerals that can be removed partly by washing. / — — / cendre de salissure / Waschberge / cenizas brutas; .... en bocamina

potential ash, – mineral matter in coal before incineration. / — — / — — / — — / cenizas del lavado

3601
fixed carbon, – the remaining solid, combustible matter, expressed as a percentage, after the removal of moisture, ash and volatile matter from coal by pyrolysis. It is complementary to volatile matter. / gebonden koolstof / carbone fixe / fixer Kohlenstoff / carbono fijo

carbon ratio, – a technical term denoting the percentage of fixed carbon in pure coal. / — — / pourcentage de carbone fixe / Carbon-Ratio / proporción del carbono; relación del ....

fuel ratio, – a rarely used factor in the analysis and classification of coal, denoting the dry ash-free ratio. / — — / — — / — — / proporción combustible; .... de carbono y volátiles

3602
apparent density; liquid apparent density, – the determination of density by using liquid displacement media that do not penetrate the ultrafine structure of coal. / schijnbare dichtheid / densité apparante / scheinbare Dichte / densidad aparente

true density; helium density, – the density of coal determined by using helium as displacement medium. / werkelijke dichtheid / densité vraie / wahre Dichte / densidad real

3604
calorific value, – the quantity of heat produced by a substance (e.g. coal) upon complete combustion at constant volume. / verbrandingswaarde / pouvoir calorifique / Heizwert / poder calorífico

gross calorific value, – the amount of heat liberated by the complete combustion of a unit mass of coal under specified conditions. The water vapour produced during combustion is assumed to be completely condensed. / bruto verbrandingswaarde / chaleur de combustion / oberer Heizwert; Verbrennungswärme; Brennwert / poder calorífico superior

net calorific value, – the amount of heat liberated by the complete combustion of unit mass of coal under specified conditions. The water vapour produced during combustion is assumed to persist as such. / netto verbrandingswaarde / puissance calorifique / unterer Heizwert / poder calorífico inferior

3605
evaporative unit, – an antiquated unit of calorific value defi-

ned as the weight of steam in pounds that can be generated from one pound of coal at 100°C.

British thermal unit; Btu, – an archaic unit, defined as the amount of heat required to increase the temperature of one pound of water by one degree Fahrenheit; 1 Btu = 252 cal = 1055 J.

quad, – a peculiar unit, equal to 10^{15}Btu.

3606
standard coal equivalent; SCE, – the heat energy of other fuels equal to that of one metric ton of hard coal of a calorific value of 8000 kWh (= 6870 kcal.kg^{-1}). / standaard kolenequivalent / équivalent charbon / Steinkohleneinheit / equivalente standard del carbón

3607
carbonization, – the artificial conversion into carbon of a carbonaceous substance such as coal, by driving off the other components by heat under laboratory conditions. The term is often wrongly used instead of carbonification. / verkoling / carbonisation / Verkohlung / carbonización

charcoal, – coal made by charring wood in a kiln or retort from which air is excluded. It is also called **xylanthrax** to distinguish it from normal mineral coal. / houtskool / charbon de bois / Holzkohle / carbón de madera

3608
coking coal, – a coal that can be usefully converted into coke, i.e. giving a coke sufficiently strong to resist pressure and breakage. There is no direct relationship between the elementary composition of coal and its coking capacity but, generally speaking, coking coals are vitrite-rich coals in the fixed carbon range of about 72-82. / cokeskool / charbon cokéfiable / Kokskohle / carbón de coque

coke, – the combustible, porous melting residue consisting of the fused ash and fixed carbon of a mainly bituminous coal formed by driving off the original volatiles by heat in the absence of air. / cokes / coke / Koks / coque

3609
coal gasification, – not yet economically viable, technical processes to convert coal into combustible, gaseous products, either in a factory or in situ. / kolenvergassing / gazéification du charbon / Kohlevergasung / gasificación del carbón

substitute natural gas; SNG, – combustible gases derived from coal, i.e. CO, H_2, CH_4. The crude gas is commonly accompanied by CO_2, H_2O and N_2 and by sulphur compounds, carbonization products and other trace components.

coal liquefaction, – a not yet economically viable, although in some countries applied, technical processes to convert coal into liquid hydrocarbons. / kolenliquefactie / liquéfaction du charbon / Kohleverlüssigung; Hydrierung / liquefacción del carbón

[2500] SUBSURFACE WATERS, PETROLEUM AND NATURAL GAS

[2510] HYDROLOGY; THE HYDROLOGICAL CYCLE

3610
hydrology, – the science dealing with terrestrial waters, their occurrence, circulation and distribution, their physical and chemical properties and their interaction with the physical and biological environment, including the effects of the activity of man. / hydrologie / hydrologie / Hydrologie; Gewässerkunde / hidrología

3611
geohydrology, – the branch of hydrology that deals with subsurface waters, their occurrence and movements and their replenishment and depletion, the properties of rocks that control water movement and storage and the methods of investigation and utilization. / geohydrologie / géohydrologie; hydrologie souterraine / Geohydrologie; Hydrologie des Grundwassers / geohidrología

hydrogeology, – although often used as a synonym for geohydrology, this science should be considered to deal only with the geological aspects of subsurface waters and water-containing rocks and with these waters as a geological agent. / hydrogeologie / hydrogéologie / Hydrogeologie / hidrogeología

hydrogeochemistry, – the branch of geohydrology that deals with the chemistry of subsurface waters and the diagenesis of their dissolved solids and gases in interaction with the enclosing rocks. / hydrogeochemie / hydrogéochimie / Hydrogeochemie / hidrogeoquímica

3612
hydrological cycle, – everlasting circulation of water: evaporation into the atmosphere, precipitation, movement below and above the surface of the earth, possibly freezing and thawing, evaporation, etc. / hydrologische kringloop; kringloop van het water / cycle naturel de l'eau; cycle hydrologique / Wasserkreislauf; hydrologischer Kreislauf / ciclo hidrológico

3613
atmospheric water, – water in the atmosphere in either gaseous, liquid, or solid state. / atmosferisch water; dampkringwater / eau atmosphérique / atmosphärisches Wasser / agua atmosférica

3614
precipitation, – water supply from the atmosphere to the earth's surface in liquid or solid form. / neerslag / précipitation (atmosphérique) / Niederschlag / precipitación

meteoric water, – water derived from precipitation. / meteorisch water / eau météorique / meteorisches Wasser / agua meteórica

3615
interception, – precipitation that is stored on the vegetal cover and subsequently evaporates. / interceptie / interception / Interzeption; Abfangen / intercepción

3616
runoff, – the part of the precipitation that flows into natural or artificial channels. / afvoer / écoulement / Abfluss / escorrentía

3617
surface runoff, – the water flowing over the ground surface and that finds its way by means of small depressions and rivulets into a channel. / oppervlakte-afvoer / écoulement de surface / Oberflächenabfluss, oberirdischer Abfluss / escorrentía superficial

overland flow, – film of water flowing over impervious soil. / – – / ruisellement de surface / – – / – –

3618
surface retention, – loss in the runoff process as a result of interception and depression storage. / oppervlakteretentie / rétention initiale; .... superficielle / Oberflächenrückhaltung / retención superficial; .... inicial

3619
infiltration, – the downward movement of water from the surface into the soil. / infiltratie / infiltration / Versickerung; Einsickerung / infiltración

3620
infiltration rate, – actual rate of water entering a soil, expressed in mm water column per unit time ($mm.s^{-1}$). / infiltratiesnelheid / taux d'infiltration / Versickerungsgeswindigkeit / grado de infiltración

infiltration capacity, – maximum rate at which water can pass into a given soil under given conditions. / infiltratiecapaciteit / capacité d'infiltration / Versickerungskapazität; Einsickerfähigkeit / capacidad de infiltración

3621
evapotranspiration, – the withdrawal of water from the soil by evaporation and transpiration of plants. / evapotranspiratie / évapotranspiration / Evapo-Transpiration / evapotranspiración

3622
percolation, – movement of water through the soil. / percolatie / percolation / Durchfluss / percolación

3623
subsurface flow, interflow, – the water that infiltrates the soil surface and moves laterally through the upper soil layers until it enters a channel. / oppervlakkige afstroming / écoulement hypodermique / Zwischenabfluss; unechter Grundwasserabfluss / escorrentía subsuperficial

3624
capilary migration, – movement of water in soils or rocks due to molecular attraction of the material for water. It occurs in the unsaturated zone. / capillaire stroming / diffusion capillaire / kapillare Strömung / migración capilar

3625
storage capacity, – of soils or rocks, quantity of water that at a certain moment can be stored above the saturated zone, expressed in mm water column. / waterbergend vermogen / capacité d'emmagasinement / Speicherfähigkeit; Wasseraufnahmefähigkeit / capacidad de almacenamiento

3626
underflow, – a water flow in a permeable stream bed below the channel. / onderstroom / écoulement sous-fluviale / Talgrundwasser / flujo subterráneo

3627
detention time; residence time, – travel time of ground water along a path line. / verblijftijd / durée de rétention / Verweildauer / tiempo de retención

3628
ground-water runoff; base flow; dry-weather flow, – groundwater discharge into a stream at places where the water table intersects the stream channel. / grondwater afvoer; basis afvoer / écoulement souterrain; écoulement de base / Grundwasserabfluss; Trockenwetterabfluss / escorrentía de aguas subterraneas

3629
seepage, –
1. influent seepage from surface water; movement of water towards the ground-water table. / inzijging / suintement / Sickerung / filtración
2. effluent seepage; movement from the ground-water table towards surface water or open air. / kwel / suintement / Sickerung / filtración

3630
spring, – a place where, without the agency of man, water issues at the earth's surface from rock or soil into a body of surface water or the air. / bron / source / Quelle / fuente; manantial

3631
vaucluse spring, – a spring whose water flows from large solution channels in limestone. / vauclusebron / source vauclusienne / Karstquelle / fuente vauclusiana

artesian spring, – a spring whose water issues at the surface under pressure through the confining formation which overlies the aquifer. / artesische bron / source artésienne / artesische Quelle / fuente artesiana

3632
mineral spring, – a spring whose water contains a certain amount of solids in solution, e.g. more than one gramme per litre. / minerale bron / source minérale / Mineralquelle / fuente mineral

thermal spring, – a spring whose water has a temperature higher than the ambient air temperature in the vicinity. / thermale bron / source thermale / Thermalquelle / fuente termal

Distinguished are:
warm spring, – temperature higher than local mean annual temperature, but lower than 37°C.

hot spring, – temperature higher than 37°C.

3633
ground-water balance, – a detailed estimate of the quantities of water added to the ground-water reservoir in a given area during a given period **(ground-water increment)** balanced against estimates of quantities withdrawn or lost from that

reservoir during that period (**ground-water decrement**) together with an estimate of the change in storage. / grondwaterbalans / bilan d'eau (d'une nappe) / Grundwasserhaushalt / balance de agua subterránea

[2520] GEOLOGY OF SUBSURFACE WATERS

[2521] *Water-Bearing Formations*

3634
aquifer; water-bearing stratum; .... bed, – pervious formation or layer holding water and permitting its movement. / watervoerende laag / aquifère / Grundwasserleiter; Aquifer / acuífero

aquifer, – in oilfield practice, the water-saturated part of the reservoir rock adjacent to an oil or gas accumulation.

3635
perched aquifer, – limited ground-water body resting on a layer of less permeable material; it is shallower than and without contact with the phreatic water table. / grondwater met schijnspiegel / nappe perchée / Grundwasserleiter mit schwebendem Wasser / acuífero colgado

3636
confined aquifer, – an aquifer that is overlain by a formation or layer of less permeable or impermeable material. / afgesloten watervoerende laag / nappe captive / Grundwasserleiter mit gespanntem Wasser / acuífero confinado

3637
artesian aquifer, – an aquifer under sufficient head to cause the ground water to rise above the surface of the earth, if opportunity were afforded to do so. / artesische watervoerende laag / nappe artésienne / Grundwasserleiter mit artesischem Wasser / acuífero artesiano

3638
ground-water artery, – a more or less tubular body of permeable material encased in less permeable material and saturated with water. A term especially applicable to deposits of gravel along ancient stream channels that have become buried in less permeable material (shoestring sands) or to deposits in e.g. offshore bars. / grondwaterader / veine de au; conduit naturel (in karst) / Grundwasserader / vena de agua subterránea

3639
aquitard; – a geological formation or layer that transmits water at a very slow rate compared to the aquifer. / slecht doorlatende laag / couche semi-perméable / schwerdurchlässige Schicht / capa semipermeable

3640
aquiclude; impervious bed, – a geological formation or layer that for practical purposes obstructs the flow of ground water. / ondoorlatende laag / couche imperméable / undurchlässige Schicht; Grundwasserstauer / impermeable

aquifugue, – a geological formation or layer that has no (interconnected) interstices and that neither contains nor conducts free water.

3641
ground-water storeys, – superimposed water-bearing strata, separated by layers of less permeable or impermeable material. / grondwaterverdiepingen / système aquifère multicouche / Grundwasserstockwerke / capas acuíferas subterráneas superpuestas

[2522] *Ground-Water Level and Head*

3642
phreatic level, – level where the pressure in the ground water is atmospheric (water in an open observation well would rise to this level). / freatisch niveau / niveau d'une nappe libre / freier Grundwasserstand / nivel freático

3643
water table, – phreatic level, as observed in observation wells. / grondwaterspiegel / surface libre / Grundwasseroberfläche / superficie del agua freática

apparent water table, – free water table of limited dimensions, found at a higher level than the continuous phreatic level, usually involving a perched aquifer. / schijnspiegel / surface libre / Nebengrundwasserspiegel; schwebender Grundwasserspiegel / superficie del agua de una capa colgada

3644
water-table isohyps, – a line on a map connecting all points of the water table that are the same vertical distance relative to a datum level. / hoogtelijn van de grondwaterspiegel; isohypse .... / hydroisohypse; courbe équipotential / Isohyps des Grundwasserspiegels / contorno freático; isoipsa freática

3645
potential, – free energy per unit mass, usually expressed as height above datum divided by the acceleration of gravity. For subsurface water under hydrostatic conditions it is equal to the piezometric head and for phreatic ground water equal to the height of the phreatic level. / potentiaal / potentiel / Potential / potencial

3646
piezometric head; piezometric level; hydrostatic ...., – level relative to a datum plane, to where the water would rise in an observation well, if necessary extended above the surface of the earth. / stijghoogte; piëzometrisch niveau / niveau piézométrique / piëzometrisches Niveau / nivel piezométrico

piezometer, – observation well, or open tube placed in the ground, below piezometric level to measure pore pressures in rock or soil. / piëzometer / piézomètre / Piezometer / piezómetro

3647
ground-water divide, – line of maximum head on the water table, so that water at either side will flow away from it to different drainage systems. / grondwaterscheiding / ligne de partage des eaux souterraines / Grundwasserscheide / divisoria de aguas subterráneas

[2523] *The Distribution of Subsurface Waters*

3648
drainage area; basin; watershed (USA), – a space with natural boundaries that is capable of holding and conducting water, both at the surface and underground. / bekken; stroomgebied / bassin versant / Becken / cuenca

3649
surface water, – water at the surface of the earth, e.g. in depressions, rivers, lakes, channels. / oppervlaktewater / eau de surface / Oberflächenwasser / agua superficial

subsurface water, – water in the earth's crust, including soil water and ground water. / ondergronds water / eau souterraine / unterirdisches Wasser / agua subterránea

3650
saturation, – of water in soils or rocks, the degree of filling of the pore space, expressed as a fraction or a percentage of the pore-space volume. / verzadigingsgraad / saturation / Sättigung; Sättigungsgrad / grado de saturación

3651
zone of aeration; unsaturated zone, – the usually very shallow part of the earth's crust, in which the functional interstices contain both water and air. / onverzadigde zone / zone d'aération; zone non-saturée / ungesättigte Zone / zona de aireación; zona no saturada

3652
soil moisture; soil water, – water in the zone of aeration between the surface of the earth and the ground water. / bodemvocht / eau du sol / Bodenwasser; Bodenfeuchte / agua del suelo

3653
pendular water, – condition of soil moisture in which skins of water are confined to the contact points of soil particles. / pendulair water / eau cunéiforme (obsolete) / Haftwasser; Porenwinkelwasser; Pendulärwasser / – – /

3654
funicular water, – intermediate condition of soil moisture between the pendular and true capillary (saturated) stages. / funiculair water / eau funiculaire (obsolete) / offenes Kapillarwasser / – – /

3655
vadose water, – an antiquated term for water temporarily held and seeping down from the surface to the saturated zone. / vadoos water / eau suspendue; (eau vadose) / vadoses Wasser / agua vadosa

3656
zone of saturation; saturated zone,– the portion of the earth's crust in which the interstices are completely filled with water. / verzadigde zone / zone de saturation / Sättigungsbereich / zona de saturación

3657
capillary zone; capillary fringe, – ground-water belt above the phreatic level, where even the larger pores are filled with water by capillary force. / capillaire zone; vol-capillaire zone / frange capillaire / Kapillarsaum; geschlossenes Kapillarwasser / franja capilar; zona capilar

3658
ground water, – subsurface water in the zone of saturation. / grondwater / eau souterraine / Grundwasser / agua subterránea

3659
phreatic water, – ground water in the zone of saturation having a free water table, in distinction to confined ground water. / freatisch water / eau d'une nappe libre / freies Grundwasser; ungespanntes Grundwasser / agua freática; agua subterránea no confinada

confined ground water, – water in an aquifer that is overlain by a semi-pervious or impervious bed. / afgesloten grondwater / eau d'une nappe captive / gespanntes Grundwasser / agua subterránea confinada

3660
water of compaction, – water driven from a formation by reduction of pore space owing to compaction and diagenesis of the sediments. / compactiewater / eau de compression / ausgepresstes Wasser / agua de compactación

3661
connate water, –

1. the retained part of the water that was entrapped in sediments during their deposition. / connaat water; fossiel .... / eau connée; eau de constitution; eau innée / fossiles Wasser; ursprüngliches .... / agua fósil; egua innata
2. in oilfield practice, the water that was retained around the sand grains by capillary forces during the accumulation of oil or gas. It constitutes the irreducible water saturation. / hechtwater / – – / ursprüngliches Porenwasser / agua intersticial irreducible

3662
formation water; native ...., – stagnant or virtually stagnant water filling the interstices of rocks; it is a type of connate water that does not take part any more in the hydrological cycle. / formatiewater / eau de gisement / Formationswasser / agua de formación

3663
regenerated water, – water released during the metamorphosis of rocks and the related dehydration of minerals. / geregenereerd water / – – / – – / agua regenerada

3664
juvenile water, – water derived from magma, either within the crust of the earth or by volcanic activity, and released for the first time. / juveniel water / eau juvénile / juveniles Wasser / agua juvenil; agua plutónica

[2530] WATER SUPPLY AND WATER QUALITY

3665
well, –
1. artificial, open excavation of more or less permanent nature to produce ground water or to carry out observations. / put / puits / Brunnen / pozo
2. a borehole, – /boorgat; boring / forage; puits / Bohrloch; Bohrung / perforación; sondeo

3666
standing level, – of the water in a non-pumping well. The term is used regardless whether the well is influenced by other wells or not. / niveau in rusttoestand / niveau statique / Ruhewasserspiegel / nivel estático

3667
drawdown, – lowering of the ground-water table caused by pumping or other means of withdrawal. / grondwaterstandsverlaging / rabattement de nappe / Grundwasserabsenkung / descenso del nivel hidrostático subterráneo

cone of depression, – depression of the water table surrounded by a closed water divide that is caused by withdrawal of ground water from a well. / onttrekkingskegel / cône de dépression / Senkungstrichter / cono de depresión

3668
safe yield, – the maximum permanent withdrawal that can be made from an aquifer continuously, including dry periods, without producing an undesired result. / toelaatbare onttrekking; veilige winning / débit de production assuré; débit de securité / zulässige Entnahmemenge / explotación de seguridad

3669
flow meter, – device to measure flow velocities in a groundwater producing or shut-in well, in order to determine yield per unit vertical interval (m^3/h per metre) and thus to trace streaks of lesser permeability in an aquifer or flow from one aquifer into another. In the mechanical flow meters, currently in use, the revolutions of a rotor or impeller are transduced into pulses that are transmitted to the surface and recorded continuously as a function of depth. The number of revolutions per unit time is approximately proportional to vertical flow in the well. / stromingsmeter / moulinet hydrométrique; débimètre / Strömungsmesser; Durchfliessmessgerät / micromolinete de sondeo

3670
catchment area; intake area, – of an aquifer; infiltration area from where water eventually reaches an aquifer. / voedingsgebied / aire d'alimentation / Einzugsgebiet / área de alimentación

3671
induced recharge, – withdrawal of ground water near surface water causing replenishment of the ground water by this surface water. / infiltratie in oevergrondwater / alimentation induite / Uferfiltration / alimentación inducida

artificial recharge, – replenishment of a water-bearing stratum that would otherwise be depleted. / kunstmatige infiltratie / alimentation artificielle / künstliche Grundwasseranreicherung / recarga artificial

3672
contamination, – introduction into water of any undesirable substance not normally present. / verontreiniging / contamination / Verunreinigung / contaminación

3673
permanent electrode system, – various electrode arrays placed permanently in a producing or shut-in ground-water well in order to detect by resistivity measurements up-coning brackish water or contamination from the surface. / vaste electrodenopstelling / – – / – – / sistema de eléctrodes permanentes

3674
water quality, – criteria and specifications of chemical, physical and bacterial constituents of water supply, depending on its use for domestic, industrial or agricultural purposes; determination of organic and inorganic constituents, acidity, specific electrical conductance, temperature, colour, turbidity, odour, taste, coliform organisms. / waterkwaliteit / qualité de l'eau / Wasserqualität / calidad del agua

3675
salinity, – a measure of the quantity of dissolved solids in water. In geohydrology the amounts of the various ions are expressed either in mass units per volume (mg/l) or as a percentage (ppm – parts per million), or in chemical units per volume (e.g. meq/l – milliequivalents per litre). SI units: mole per cubic metre ($mol.m^{-3}$), $kg.m^{-3}$. / saliniteit; zoutgehalte / salinité / Salzgehalt / salinidad

mole, – SI unit of the amount of substance, i.e. a measure of the number of entities in that amount. One mole equals the atomic number expressed in grammes. / mol / mol / Mol / mol

equivalent value, – the chemically equivalent amounts of elementary particles of two or more substances. Unit: milliequivalent, symbol: meq; one meq equals atomic mass divided by valence. / reactiewaarde / équivalent / chemisches Äquivalent / valor equivalente

TDS, – total dissolved solids.

3676
fresh water, – a vague term, denoting water with a very small amount of dissolved solids, in which no taste is discernable; i.e. less than about 150 to 200 ppm of Cl^-. Above that amount taste deteriorates, and in natural waters the ions of sodium and chlorine become dominant. It has been proposed to consider a concentration of 1000 ppm of TDS as the upper limit of the salinity of fresh water (or about 10 $mol.m^{-3}$ of Cl^-). / zoetwater / eau douce / Süsswasser / agua dulce

brackish water, – a vague term, denoting water with a moderate amount of dissolved solids, whose salinity may be found to vary between the limits of 150-200 ppm of Cl^- and the salinity of sea water (around 33 000 ppm of TDS). It has been proposed to consider the limits of salinity of brackish water between 1000 and 10 000 ppm of TDS (or about 100 $mol.m^{-3}$ of Cl^-). / brakwater / eau saumâtre / Brackwasser / agua salobre

saline water, – a vague term denoting water with a considerable amount of dissolved solids, often considered to possess a salinity higher than that of sea water. It has been proposed to consider the limits of saline water between 10 000 and 100 000 ppm of TDS (or between about 100 and 1000 mol.m^{-3} of Cl^-. / zoutwater / eau salée / Salzwasser / agua salada

brine a vague term, denoting water with a very large amount of dissolved solids; in oilfield practice often used for formation water. It has been proposed to consider brine to possess a salinity greater than 100 000 ppm of TDS. / pekel / saumure / Sole; Salzlauge / salmuera

3677
(total) hardness, – of water, a measure of the calcium and magnesium concentration, generally expressed as the equivalent of $CaCO_3$ or CaO in degrees of hardness or in parts per million:
Germany, the Netherlands: 1 mg CaO in 100 ml water
France: 1 mg $CaCO_3$ in 100 ml water
UK: 1 mg $CaCO_3$ in 70 ml water
USA: 1 mg $CaCO_3$ in 1000 ml water, in ppm
1° German = 1.79° French = 1.25° British = 17,9 ppm / hardheid / dureté / Härte / dureza

carbonate hardness; alkalinity, – a measure of the carbonate and hydrocarbonate concentration, expressed as the amount of a standard concentration of sulphuric acid required to titrate a water sample to an end point of pH 4.5. / alkaliniteit; tijdelijke hardheid / alcalinité / Alkalinität / alcalinidad

3678
oxygen demand; biochemical ....; BOD; chemical ....; COD, – estimates of the amount of oxygen required to reestablish the dissolved-oxygen concentration of polluted water, and thus to evaluate the pollution load, either by diluting samples with oxygenated water and measuring the residual oxygen after an incubation period (BOD) or by oxidation with strong oxidizing agents (COD). / zuurstofbehoefte / DBO; DCO / Sauerstoffbedarf / demanda de oxígeno

[2540] ROCK PROPERTIES AND FLUID FLOW

NOTE:
– see also section
[2580] *Reservoir Engineering*

3679
reservoir rock; .... formation, – a rock or formation with porosity and permeability sufficient to allow storage and movement of fluids and gases. / reservoirgesteente; ....formatie / roche réservoir; .... magasin; formation réservoir; .... magasin / Speichergestein; ....formation / roca almacén

3680
porous medium, – a multiphase system whose solid phase is continuous. / poreus medium / milieu poreux / poröses Medium / medio poroso

framework, – the solid phase of a porous medium. / skelet / squelette / Gerüst / – –

3681
porosity, – fraction or percentage of rock or soil volume, occupied by interstices. Symbol: ∅. / porositeit; poriëngehalte / porosité / Porosität; Porengehalt / porosidad

total porosity; bulk ...., – the total of void space. / totale porositeit / porosité totale / totale Porosität / porosidad total

pore volume, – volume, not occupied by solid material. / poriënvolume / volume des pores / Porenvolumen / volumen de los poros

3682
void ratio, – pore volume divided by the volume of solid matter per unit bulk volume of rock or soil. / poriëngetal / taux de porosité / – – / coeficiente de porosidad

3683
interstice; pore; voids (USA), – space in soil, rock or granular material not occupied by solid matter. / tussenruimte; poriën / interstice; vide / Zwischenraum; Hohlraum / intersticio

3684
primary porosity, sedimentary ..., – porosity that developed during sedimentation or that was present within sedimentary particles at the time of deposition. / oorspronkelijke porositeit; primaire .... / porosité primaire / primärer Porosität / porosidad primari

primary interstice, – / primaire tussenruimte / – – / primärer Hohlraum / intersticio primario

3685
secondary porosity – porosity developed in a rock after its depositon or emplacement. / secundaire porositeit; gevormde .... / porosité secondaire / sekundäre Porosität / porosidad secundaria

secondary interstice, – / secundaire tussenruimte / – – / sekundärer Hohlraum / intersticio secundario

secondary porosity may be due to fracturing by (tectonic) stress:
fracture porosity;
or to solution and leaching:
solution porosity
which is a property of especially carbonate rocks;
or to diagenetic processes:
diagenetic porosity,
that tend to reduce porosity.

3686
mud-supported porosity, – homogeneous distribution of pores and framework; the dominant pore size is less than 0.02 mm.

grain-supported porosity, – homogeneous distribution of pores and framework, the dominant pore size is larger than 0.02 mm.

3687
matrix porosity, – the porosity of the finer portion of a rock, in contrast to porosity associated with the coarser particles or constituents. /grondmassaporositeit / porosité matricielle / Matrix-Porosität / porosidad de la matriz

3688
intergranular porosity; interparticle ....; intercrystalline ...., – porosity between grains, particles or crystals. / tussen-korrelige porositeit / porosité intergranulaire / Zwickel-Poren; Interpartikel-Porosität / porosidad intergranular

intragranular porosity; intraskeletal .... (restricted to carbonate rocks) – porosity within individual grains or skeletal parts / inkorrelige porositeit / porosité intragranulaire / Hohlformporosität; Partikel-Porosität; Intra-Partikel-Porosität / porosidad intragranular

3689
shelter porosity, – a type of primary interparticle porosity created by the sheltering effect of relatively large sedimentary particles that (partly) prevent the infilling of pore space beneath them by finer clastic particles. / honingraatporositeit / – – / Sackporosität; Sackgassenporosität / – –

3690
solution opening; .... channel, – secondary interstice created by dissolution of the rock material. / oplossingsruimte / vide de dissolution / Lösungs-Porosität; Lösungs-Hohlraum; Fossil-Kammern / intersticio por disolución

3691
moldic porosity, – texture-related porosity, formed by the selective removal, usually by dissolution or leaching, of a former individual constituent of the sedimentary texture; especially in carbonate rocks. / afdrukporositeit / – – / Partikellösungs-Porosität / porosidad por molde; .... debida a moldes

3692
vug porosity, – porosity as a result of openings in the rock that are not fabric selective; often solution enlarged moulds, sometimes solution channels; especially in carbonate rocks. / – – / porosité vacuolaire / Gesteinslösungs-Porosität / porosidad debida a cavidades

3693
texture-related porosity; texture-selective ....; fabric-selective ...., – secundary porosity that is selectively created in discrete textural elements by diagenetic processes. / maakselporositeit / porosité texturale / – – / porosidad relacionada con la textura

3694
fracture porosity, – porosity formed by fractures; not texture-related. / breukporositeit / porosité fissurale / Kluft-Porosität / porosidad por fractura

3695
functional porosity; interconnected .... , – amount of interconnected pores available for the transmission of fluids and gases. / beschikbare poriënruimte / porosité efficace / nutzbarer Porosität / porosidad por interconexión

pore throat, – the interconnection between individual pores, usually smaller than the pores themselves, rendering porosity effective or, if absent, due e.g. to water held by capillary forces or to diagenetic processes, ineffective to transmit fluids and gases.

3696
Darcy's law, – named after the French hydraulic engineer P. Darcy, who concluded from experiments that the flow rate of water through sands is proportional to loss of energy, inversely so to the length of the flow path and proportional to a coefficient depending on the nature of the sand. / wet van Darcy / loi de Darcy / Darcysches Gesetz / ley de Darcy

3697
permeability; intrinsic .... , – a measure of the resistance to the flow of liquids in a porous medium, defined as the volume of fluid of unit viscosity passing through unit cross section in unit time under the action of unit pressure gradient. It depends on the interconnection of the pores and on pore-surface wetting conditions, but it is independent of the nature of the fluid. / permeabiliteit; doorlatendheid / permeabilité / Permeabilität; Durchlässigkeit / permeabilidad

darcy, – unit of permeability, defined by expressing flow rate in $cm.s^{-1}$, viscosity in centipoise and pressure gradient in $atm.cm^{-1}$. Symbol: k; $1 D = 0.996 \cdot 10^{-8} cm^2$. In oilfield practice the millidarcy, mD, is generally used. / darcy / darcy / Darcy /darcy

3698
effective permeability, – the resistence to flow of each liquid in a two-phase system in a porous medium, e.g. oil and water flowing simultaneously. These permeabilities are dependent on the saturation of each fluid and their sum is less than the permeability for a one-phase system. / effectieve doorlatendheid; .... permeabiliteit / perméabilité effective / effektive Permeabilität; .... Durchlässigkeit / permeabilidad efectiva; .... relativa

relative permeability, – the ratio of effective permeability to that of a one-phase system. / relatieve doorlatendheid; .... permeabiliteit / perméabilité rélative / relative Permeabilität; .... Durchlässigkeit / permeabilidad relativa

3699
mobility, – a measure of the velocity with which a fluid will flow through a given porous rock under a given pressure gradient. / mobiliteit / mobilité / Mobilität / movilidad

mobility ratio, – the relation between the mobilities of two fluids present in the pores of a rock. / mobiliteitsverhouding / taux de mobilité / Mobilitätsverhältnis / relación de movilidad

3700
hydraulic conductivity; .... permeability, – in ground-water flow calculations measured as the volume of fluid passing through unit cross section in unit time under the action of unit pressure gradient. It is a function of fluid properties (viscosity, density), of the degree of saturation and of the medium. Symbol: k; unit $m.s^{-1}$; in practice $cm.s^{-1}$ or metre per day are used. For pure water of 20°C: $1 D = 0.996 \cdot 10^{-3} cm.s^{-1}$. / waterdoorlatendheid /conductivité hydraulique / hydraulische Leitfähigkeit / conductividad hidráulica

3701
primary permeability, – the permeability that developed during sedimentation. / primaire doorlatendheid; .... permeabiliteit / perméabilité primaire / primäre Durchlässigkeit / permeabilidad primaria

secondary permeability, – the permeability that developed in a rock after its deposition or emplacement secundaire doorlatendheid / perméabilité secondaire / sekundäre Durchlässigkeit / permeabilidad secundaria

Distinguished are:

solution permeability,
fracture permeability,
both enhancing permeability,
and
diagenetic permeability,
tending to reduce permeability

3702
transmissivity; transmissibility, – product of permeability or hydraulic conductivity and thickness of the reservoir rock. / doorlaatvermogen / transmissivité / Einheitsergiebigkeit / transmisividad

3703
pore velocity; velocity of ground-water encroachment; – velocity with which a water particle moves through a medium. / werkelijke grondwatersnelheid; opdringingssnelheid van het grondwater / vitesse d'écoulement souterrain / Fliessgeschwindigkeit des Grundwassers / velocidad de corrimiento

seepage velocity; specific discharge, – quantity of water passing in unit time through unit area which is at right angles to the direction of flow. It is a fictitious flow rather than the true velocity through the pores. / filtersnelheid / vitesse de filtration / Filtergeschwindigkeit / velocidad de filtración

3704
specific yield; effective porosity, –
1. the ratio of the volume of water a saturated rock (or soil after having been saturated) will yield by gravity drainage to the gross volume of rock or soil;
2. in oilfield practice, the effective or interconnected porosity is that fraction of bulk porosity from which hydrocarbons can be produced.

/ effectieve porositeit / porosité effective / nutzbarer Porengehalt / porosidad efectiva

specific retention, – the ratio of the volume of water held by molecular and capillary forces at the surface of the grains, and thus retained against gravity drainage, to the total volume of the pores. Values, which are a fraction of the pore volume, depend on grain size, shape and distribution of pores and on wetting conditions. / – – / rétention volumique / – – / retención efectiva

3705
wetting fluid, – a fluid whose molecules, due to capillary effects, adhere to a substance in preference to other fluids; e.g. usually a film of water molecules on sand grains, thus adversely affecting permeability. / bevochtigende vloeistof / fluide mouillant / – – / liquido que moja

non-wetting fluid, – a fluid whose molecules do not adhere to a substance; e.g. usually oil molecules in a sandy or calcareous reservoir. / niet-bevochtigende vloeistof / fluide non-mouillant / – – / liquido que no moja

3706
Klinkenberg factor; slip .... , – a factor indicating the pressure dependence of measured permeability of a porous medium to gas. This permeability is larger than that to a liquid due to the 'slip', i.e. the flow rate of gas around the surface of the grains is not zero. / factor van Klinkenberg / facteur de Klinkenberg / Klinkenberg Faktor / factor de Klinkenberg

3707
tortuosity; detour factor, – factor indicating the average detour of the microscopic flow path around sand grains, compared with the length of a macroscopic path line. A detour factor is by nature larger than unity, but some authors define tortuosity as the reciprocal of the detour factor. / omwegfaktor / tortuosité / Tortuosität / tortuosidad

3708
path line, – line along which liquid or gaseous particles move through the subsurface under the influence of a difference in potential; under steady conditions equal to stream line. / stromingsweg; stroombaan / trajectoire d'une particule / – – / línea de flujo

stream line, – curve whose direction in all points and at the same moment is equal to the direction of the fluid velocity in those points. / stroomlijn / ligne de courant / – – / línea de corriente

3709
storage coefficient; storativity, – the volume of water that an aquifer releases from or takes into storage per unit surface area and per unit change of head. Symbol: S. / bergingscoëfficiënt / coefficient d'emmagasinement / – – / coeficiente de almacenaje

elastic storage, – the storativity in a confined aquifer, where change in pressure causes compression (or dilatation) of the aquifer and compression (or expansion) of the water and subsequent change in water content. / elastische berging / – – / – – / depósito elástico

3710
dispersion, – growth of a transition zone between different types of ground water (e.g. chemical composition, temperature) by the combined effects of interweaving flow through connected pores and molecular diffusion. / dispersie / dispersion / Zerstreuung; Dispersion / dispersión

3711
pumping test, – water is pumped out of a borehole and rate recorded; observations are made of water level in hole and in nearby holes during and after pumping. Data obtained are used to determine geohydrological parameters of the aquifer concerned around the hole. / pompproef / essai de pompage / Pumpversuch / ensayo de bombeo

falling head test, – a borehole is filled with water and the rate of fall of water level is observed to determine geohydrological parameters. Applied in poorly pervious formations, especially for engineering purposes. / proef met variabel waterniveau / essai à charge variable / Versuch mt abnehmendem Wasserdruck / ensayo de inyección a nivel variable

constant head test, – a borehole is filled with water and water is added to keep water level constant; the amount of water added per unit of time is used to compute geohydrological parameters. Applied in poorly pervious formations, especially for engineering purposes. / proef met constant waterniveau / essai à charge constante / Versuch mit konstantem Wasserdruck / ensayo de inyección a nivel constante

[2550] GEOLOGY OF PETROLEUM AND NATURAL GAS

[2551] *Types of Hydrocarbons; Origin and some Products*

3712
petroleum; oil, – a mixture of natural hydrocarbons in the liquid phase (locally occurring in the earth's crust). / aardolie; petroleum / pétrole; huile / Erdöl / petróleo

crude oil; crude, – petroleum as produced, i.e. before refining. / ruwe olie / pétrole brut; brut / Rohöl / crudo

3713
condensate, – a liquid (at standard conditions) hydrocarbon mixture – consisting mainly of very light components – that is produced together with gas by pressure decrease **(retrograde condensation)** from reservoirs in which both gas and condensate exist as a homogeneous phase. / condensaat / condensat / Kondensat / condensado

natural gas; gas, – a mixture of natural hydrocarbons mainly methane often with an admixture of other gases (e.g.nitrogen) that are in the gaseous phase at standard conditions. / aardgas / gaz; gaz naturel / Erdgas / gas natural

3714
wet oil, – oil containing a percentage of water, i.e. before dehydration. / natte olie / pétrole brut à forte teneur en eau / wasserhaltiges Rohöl / crudo húmedo

clean oil; dry oil, – oil without an appreciable content of water. / droge olie; schone olie / huile sèche / reines Öl / petróleo seco

3715
wet gas; rich gas , – natural gas with an appreciable content of liquid hydrocarbons, that may be suitable for extraction. / rijk gas / gaz riche; gaz humide / Nassgas / gas rico

lean gas, – natural gas without an appreciable content of liquid hydrocarbons. / arm gas / – – / Trockengas / gas pobre

3716
sweet, – said of a crude oil or natural gas, containing negligible amounts of hydrogen sulphide and mercaptans. / – – / non corrosif / – – / – –

sour, – said of a crude oil or natural gas, containing objectionable amounts of hydrogen sulphide and mercaptans. / – – / corrosif / sauer / – –

mercaptan, – a series of sulphur compounds, derived from hydrocarbons by the replacement of a hydrogen atom by a group containing one sulphur and one hydrogen atom; undesirable because of their bad odour even in small concentrations. / mercaptaan / mercaptan / Mercaptan / mercaptano

3717
alkanes; paraffin, – a class of saturated, i.e. carrying its full complement of hydrogen atoms, straight-chain (normal) or branched-chain (iso)hydrocarbons; general structure C_nH_{2n-2}. / alkanen; paraffine / alcanes; paraffine / Alkane; (Paraffine) / parafina

cyclanes; cycloalkanes; napthenes; cycloparaffin, – – a class of saturated hydrocarbons with a ring-type structure; general structure C_nH_{2n}. / cycloalkanen / cyclanes / Cycloalkane; (Naphtene) / naftenos

alkenes; olefins, – a class of unsaturated, i.e. the carbon atoms are linked by two or three bonds, straight or branched-chain hydrocarbons; rarely occurring in crude oils; general structure C_nH_{2n}. / alkenen / alcènes / Alkene; (Olefine) / olefinas

aromatics; cycloolefins, – a special class of unsaturated hydrocarbons with a ring-type structure, like benzene; general structure C_6H_6. / aromaten / aromatiques / Aromaten / aromáticos

3718
paraffin base; paraffinic, – crude oils consisting mainly of alkanes (paraffin).

naphthenic base; naphthenic, – crude oils consisting mainly of cycloalkanes (naphthenes).

asphaltic base; asphaltic, – crude oils consisting mainly of asphaltenes, and with densities below 35° API.

asphaltenes, – the portion of crude oils that is soluble in solvents such as benzene, chloroform, and carbon disulphide, but insoluble in low-boiling (C_3 to C_7) alkanes.

mixed base, – crude oils consisting of both alkanes and cycloalkanes with a certain proportion of armomatics.

NOTE:
– this is a very general classification; most crudes exhibit considerable overlap between types.

3719
bitumen, – a vague term, not recommended for specific purposes, with a wide range of varying definitions, e.g.
1. any naturally occurring hydrocarbon ranging from liquid (petroleum) to highly viscous (asphalt) varieties; hence excluding gaseous and solid (coal) members;
2. any naturally occurring hydrocarbon, including the gaseous and solid (e.g. asphaltite) members, but excluding coal;
3. hydrocarbon mixtures of artificial (pyrogenous) origin.

/ bitumen / bitume / Bitumen / bitumen

naphthabitumen, – the soluble (in carbon disulphide) organogenic matter: natural gas, petroleum, ozokerite, natural asphalt, asphaltite.

kerabitumen, – the insoluble, noncoaly organogenic matter of oil shales, oil-source rocks, recent sediments, etc.

NOTE:
– see table 1 – Caustobiolites

3720
kerogen, –
1. the increasingly accepted definition by Breyer: the organic constituents of sedimentary rocks that are neither soluble in aqueous alkaline nor in the common organic solvents;
2. a complex of diagenetically transformed vegetal and animal matter in solid state and of sapropelic origin. It yields

oil on destructive distillation (pyrolysis). / kerogeen / kérogène / Kerogen / kerogeno

kerogenite; kerogenous rock, — a deposit that contains kerogen in such amounts that oil is produced on distillation. / kerogeniet / kérogenite / Kerogenit / kerogenita

3721
sapropelite, — consolidated sapropel, yielding oil on destructive distillation. / sapropeliet / sapropélite / Sapropelit / sapropelita

oil shale, — a sapropelite containing a variable amount of transformed organic matter (from 10–67%), that on destructive distillation yields shale oil. A well-known misnomer, the fine-grained rock is neither a shale nor does it contain oil. The shaly appearance results from the foliated arrangement of the organic matter within the mineral matrix. It is considered to be a rich oil-source rock and is possibly suitable for the production of petroleum by special extraction processes. / olieklei / argile bitumineuse; schiste bitumineux / Ölschiefer / arcilla bituminosa

sapropelic coal, — an oil shale containing more than 67% of organic matter (or less than 33% of ash).

3722
oil-source rock; source rock, — a rock capable of generating and expelling petroleum in the past, the present or the future. An oil-source rock is always a gas-source rock also. / oliemoedergesteente / roche-mère à huile / Erdölmuttergestein / roca madre de petróleo

gas-source rock, — a rock capable of generating and expelling natural gas in the past, the present or the future. A naturally occurring gas originates from two sources:
- humic coal (predominantly methane) without the formation of petroleum,
- oil-source rocks together with or after the formation of petroleum.

/ gasmoedergesteente / roche-mère à gaz / Gasmuttergestein / roca madre de gas

3723
lipid, — a general name for plant and animal products typified by esters of higher fatty acids but including certain other oil-soluble, water-insoluble substances. Considered to be a precursor of petroleum. / lipide; vetzuur / lipide / Lipide / lipido

wax, — plant or animal wax, i.e. a mixture of oxygen-containing molecules (esters) derived from long-chain fatty acids and alcohols with an even number of carbon atoms. Considered to be a precursor of oil of vegetal origin. / was / cire / Wachs / cera

carbohydrate, — a class of neutral compounds composed of carbon, hydrogen, and oxygen in units of 5 or 6 carbon atoms, usually connected by oxygen atoms, and having the general formula $C_n(H_2O)_m$. Includes sugars, starch, and cellulose, and is formed by all green plants through photosynthesis. / koolhydraat / hydrate de carbone; glucide / — — / —
—

3724
extractable organic matter, — organic matter that can be extracted from sedimentary rocks by means of organic solvents such as benzene, methanol, ether and chloroform. / extraheerbaar organisch materiaal / matière organique extractible / extrahierbares organisches Material / materia orgánica extraible

3725
maturity, — degree of similarity of extractable rock hydrocarbons to petroleum in chemical composition. / rijpheid / maturité; maturation / Maturität; (Reifegrad) / madurez

Distinguished are:

immature, — the indigenous organic matter yields petroleum on destructive distillation. / onrijp / immature / unreif / inmaduro

mature, — the indigenous extractable organic matter is similar to petroleum in composition. / rijp / mûr / reif / maduro

post-mature, — most of the heavier hydrocarbons have been converted to light ones and solid carbon. / overrijp / — — / überreif / — —

3726
maturity index, — chemical parameter indicating the maturity of an oil-source rock, especially the normal alkane carbon-number and the naphthene ring-number distributions are used. / rijpheidsindex / indice de maturation / Reifeparameter / índice de madurez

3727
carbon number, — the number of carbon atoms in one molecule, expressed e.g. as C_{14}. / koolstofgetal / nombre de carbone / Kohlenstoffzahl / número de carbonos

3728
odd predominance, — refers to the phenomenon that normal alkanes with an odd number of carbon atoms predominate over those with an even number of carbon atoms. In general, normal alkanes in the range C_{25} to C_{35} are taken into consideration. / — — / règle de prédominance du nombre impair / — — / predominancia impar

3729
carbon preference index; CPI, — the mean of two ratios that are determined by dividing the sum of concentrations of odd-carbon-numbered normal alkanes by the sum of even-carbon-numbered normal alkanes over given concentration ranges. One ratio is obtained by dividing the sum of the percentages of the odd C_{25} to C_{33} by the sum of the even C_{25} to C_{34}. The other ratio is obtained by changing the denominator to the sum of the percentages C_{24} to C_{32}.

3730
pyrolysis, —
1. decomposition of organic matter by heating in the absence of oxygen. This process is the basis for several methods of source-rock analysis,
2. destructive distillation, nearly identical with 1, although generally not carried out in a completely oxygen-free atmosphere. Applied to coals and oil shales (e.g. Fischer assay), but also to sedimentary rocks.

/ pyrolyse / pyrolyse / Pyrolyse / pirólisis

3731
modified Fischer assay, — a method used to determine the yield of shale oil from an oil shale by the low-temperature pyrolysis or destructive distillation of a rock sample.

3732
expulsion, — release of hydrocarbons from organic matter within source rocks, their transport within these rocks and their transfer into other, more permeable formations or into fractures, presumably, together with water, through compaction of the argillaceous source rocks. / uitdrijving; expulsie / expulsion / Primärmigration / expulsión

3733
transformation of oil, — changes in the chemical composition of oil after expulsion from its source rock. These changes can occur by physical, thermal and/or bacterial processes. / olietransformatie / transformation de l'huile / — — / transformación del crudo

3734
native asphalt, — a class of naturally occurring hydrocarbons with low fusing points, densities of about 1.0 to 1.1 and fixed-carbon values of 4 to 20%. / natuurlijk asfalt / asphalte naturel / Naturasphalt / asfalto natural

asphalt, —
1. native asphalt;
2. pyrogenous asphalts, i.e. man-made asphalts from thermal treatment of residual oils.

/ asfalt / asphalte / Asphalt / asfalto

3735
ozokerite, — a mineral wax, yellow to dark-brown in colour, composed of the higher members (C_{22} to C_{29}) of the paraffin and mononaphthene series; believed to be a shallow alteration product of a paraffinic oil. / ozokeriet; aardwas / ozocérite; cire fossile / Ozokerit; Erdwachs / ozokerita

3736
montan wax, — a mineral wax extracted from certain lignites. / montaanwas / cire de montagne; cire de lignite / Montanwachs / cera montana

3737
asphaltite, — a group name for solid forms of hydrocarbons, harder and less fusible than native asphalts. They may be associated with mineral matter. Asphaltites are derived either from caustobiolites or sapropelites and more or less altered during or after migration. Usually found in veins and fissures. / asfaltiet / asphaltite / Asphaltit / asfaltita

pyrobitumen, — use not recommended; a confusing nongenetic name covering some insoluble asphaltites (asphaltic pyrobitumen) and humic coal (nonasphaltic pyrobitumen) as well as sometimes sapropelites. / pyrobitumen / pyrobitume / Pyrobitumen / pirobitumen

3738
gilsonite, — a soluble asphaltite; density 1.03-1.10, fixed carbon 10-20%, with black colour, brilliant lustre, brown streak and conchoidal fracture.

glance pitch, — a soluble asphaltite; density 1.10-1.15, fixed carbon 20-35%, with black streak and brilliant conchoidal fracture.

grahamite, — a soluble asphaltite; density 1.15-1.20, fixed carbon 35-55%, with jet-black lustre, conchoidal fracture, black streak; is fusible and brittle; often mixed with mineral up to 50%.

3739
wurtzilite, — a massive, black or light-brown elastic asphaltite resisting the usual organic solvents; density 1.05-1.07, fixed carbon 2-25%, with brilliant lustre, conchoidal fracture and light-brown streak; is infusible and decomposes before melting.

albertite, — an asphaltite; density 1.07-1.10, fixed carbon 25-50%, with brilliant lustre, conchoidal fracture and hardness 1-2; is practically insoluble in carbon disulphide but partly soluble in turpentine; decomposes prior to fusion. Brown to black streak.

impsonite, — a black variety of asphaltite; density 1.10-125, fixed carbon 50-85%; slightly soluble in carbon disulphide, almost insoluble in turpentine; black streak, infusible.

anthraxolite, — a coal-like, lustrous, probably highly coalified asphaltite, rich in carbon (85-95%); hardness 3-4, density near 2; insoluble in organic solvents, practically infusible.

3740
elaterite, — a rare, massive, amorphous, rubbery, dark-brown hydrocarbon or asphaltite ranging from soft and elastic when fresh to hard and brittle when exposed, and having a conchoidal fracture. Gives a brown streak and melts in candle flame without decrepitation.

coorongite, — a substance resembling rubber, composed of algae, bits of plant tissue and diatoms. It is often compared with elaterite and is perhaps an asphaltite but more probably a very immature sapropelite.

3741
API gravity, — an arbitrary scale, the standard method of the American Petroleum Institute used for reporting the density of a crude oil or a petroleum product. The degree API is related to the density of water by the formula:

$$\text{degree API} = \frac{141.5}{\text{density of water at } 60^\circ\text{F}} - 131.5$$

3742
shale oil, — oil obtained by the destructive distillation of oil shale.

3743
tar, — condensates from the thermal treatment (destructive distillation, pyrolysis, etc.) of organic materials; generally of liquid consistency; mainly soluble in carbon disulphide; comparatively volatile at high temperatures. Also a naturally occurring weathered oil. / teer / goudron; brai / Teer / alquitrán

pitch, — undistilled residue from the thermal treatment of organic matter (destructive distillation, pyrolysis, etc.); generally of viscous to solid consistency; comparatively nonvolatile; solubility in carbon disulphide depends considerably on severity of thermal treatment. Also a naturally occurring weathered oil. / pek / brai; poix / Pech / pez

3744
gasoline, — the portion of petroleum boiling in the temperature range of 30° to 200°C. / benzine / essence / Benzin / gasolina

naphtha, — the portion of petroleum boiling in the temperature range of 100° to 200°C. / nafta / naphte; essence lourde / – – / nafta

kerosine, – a petroleum distillate with a boiling range of about 180° to 230°C and with about 11 to 13 carbon atoms per molecule. / kerosine / kérosène; kérosine / Kerosin / keroseno

diesel fuel; .... oil; dieseline, – a refined petroleum distillate (a gas oil) containing mainly saturated hydrocarbons of 13 to 17 carbon atoms. / dieselolie / combustible pour moteur Diesel; gas-oil / Dieselöl / combustible diesel

gas oil, – a general name for a variety of refined petroleum distillates boiling mainly in the range of about 230° to 300°C and containing hydrocarbons with about 13 to 17 carbon atoms per molecule. The term is sometimes applied to fractions boiling even higher. / gasolie / gas-oil; gazole / Gasöl I / gasóleo

3745
heavy hydrocarbon, – a term applied to the total of saturated and aromatic hydrocarbons boiling above 325°C separated by fractionation from the extractable organic matter. / zware olie / pétrole lourd; huile lourde / – – / hidrocarburo pesado

3746
LNG: liquefied natural gas, – natural gas liquefied by compression for commercial purposes.

NGL; natural gas liquids, – liquid hydrocarbons extracted from produced natural gas.

3747
oil equivalent, – measure of the energy contained in oil; one million tonnes of oil equivalent, 1 MTOE = $41.9 \cdot 10^{15}$ J. / olie-equivalent / – – / – – / equivalente en petróleo

[2552] *Migration*

3748
migration, – any underground movement of hydrocarbons not caused by artificial means. / migratie / migration / Migration / migración

3749
primary migration, – migration of hydrocarbons, probably mostly in disseminated form, from source rock to reservoir rock. / primaire migratie / migration primaire / primäre Migration; Primärmigration / migración primaria

secondary migration; remigration, – migration of hydrocarbons (in bulk) after having been at rest in an accumulation, due to changes in physical conditions and/or tectonic mobilisation. / secundaire migratie; hermigratie / migration secondaire; remigration / sekundäre Migration; Sekundärmigration / migración secondaria

3750
vertical migration, – migration in a (near-) vertical direction, generally across bedding planes. / verticale migratie / migration verticale / vertikale Migration / migración vertical

lateral migration; horizontal ...., – migration generally (near-) parallel to the bedding planes. / laterale migratie; horizontale .... / migration latérale; .... horizontale / laterale Migration; horizontale .... / migración lateral; .... horizontal

3751
migration path; .... channel, – permeable parts of a rock complex through which hydrocarbons preferentially migrate(d): fault planes, fissures, permeable streaks, bed boundaries, etc. / migratieweg; ....kanaal / parcours de migration / Migrationsweg / vía de migración; camino de ....

3752
oil intrusion, – the movement of oil in bulk from a migration path into a permeable rock body. / olieintrusie / intrusion de pétrole; invasion de pétrole / Ölintrusion / intrusión de petróleo

3753
seepage, – a natural emanation of oil or gas. / sijpelplaats / suintement / Ausbiss / filtración

pitch lake; asphalt lake, – / asfaltmeer / lac d'asphalte; mare d'asphalte / Asphaltsee; Pech .... / lago de asfalto

[2553] *Accumulation*

3754
accumulation, – a more or less continuous, more or less extensive body of hydrocarbons contained in the pore space of a rock body; also, the process of formation of such a body of hydrocarbons. / accumulatie / accumulation; gisement / Lagerstätte; Akkumulation; Ansammlung; Speicherung / yacimiento; acumulación

3755
reservoir, – a body of rock containing an accumulation of hydrocarbons. / reservoir / réservoir; magasin / Speicher; Träger / almacén; yacimiento

3756
trap, – the configuration of rocks, impermeable to hydrocarbons, partly enveloping the reservoir, so as to prevent escape of accumulating, or accumulated, hydrocarbons. / fuik / piège / Falle / trampa

NOTE:
– the Dutch term 'fuik' is added by the editor, but is not generally accepted.

3757
structural trap, – a trap caused by the tectonics of the formations concerned. / structurele fuik / piège structural / strukturelle Falle / trampa estructural

anticlinal trap, – a trap caused by the anticlinal shape of the reservoir formation. / anticlinale fuik / piège anticlinal / Antiklinalfalle / trampa anticlinal

fault trap, – a trap in which a sealing fault provides (part of) the trapping configuration. / breukfuik / piège de faille / Bruchfalle; Störungs.... / trampa por falla

3758
stratigraphic trap, – a trap caused by the original, depositional configuration of the reservoir formation. / stratigrafische fuik / piège stratigraphique / stratigraphische Falle / trampa estratigráfica

wedge-out trap; pinch-out trap, – a trap caused by the reduction to zero of the reservoir rock thickness in an updip direction. / uitwiggingsfuik / piège de pincement / Auskeilungsfalle / trampa por cuña

permeability trap; shale-out trap, – a trap caused by the deterioration in an updip direction of the porosity and permeability of the reservoir rock, e.g. by the increase of the argillaceous content. / doorlatendheidsfuik / piège de perméabilité / Permeabilitätsfalle / trampa por permeabilidad

3759
unconformity trap, – a trap in which the reservoir rock is covered unconformably by the caprock. / discordantiefuik / piège par discordance / Diskordanzfalle / trampa por discordancia

3760
buried-hill trap, – a trap caused by fossil topography in which the reservoir rock formed a hill. / – – / piège par relief enterré / paläotopographische Falle / trampa por paleorelieve

3761
hydrodynamic trap, – a hypothetical trapping possibility, in which the oil/water interface is inclined as a consequence of flow of water. / hydrodynamische fuik / piège hydrodynamique / hydrodynamische Falle / trampa hidrodinámica

3762
closure; closure area, – the maximum horizontal area over which the trapping configuration can be effective; this area is outlined by the lowest closed contour line on top reservoir, i.e. at the depth at which the oil/water interface would stand if the trap were filled to capacity. / sluiting; sluitingsareaal / fermeture; aire de fermeture / Strukturschluss / cierre

vertical closure, – the vertical distance between the highest point of the reservoir and the depth of the spill point. / verticale sluiting / fermeture verticale / vertikaler Strukturschluss / cierre vertical

spill point, – the deepest point to which the trapping configuration is effective, i.e. the point at which hydrocarbons would escape (spill-over), if the trap were filled beyond capacity. / overlooppunt / point de fuite / unterster Strukturschluss; tiefster .... / punto de escape

3763
dip-closed, – is a reservoir if the trap is formed essentially by the dip of the sealing formations. / – – / fermeture par pendage / durch Einfallen geschlossen / cierre por pendiente

fault-closed, – is a reservoir if the trap is partly formed by a sealing fault. / – – / fermeture par faille / durch Störung geschlossen / cierre por falla

3764
cap rock, – rock, essentially impervious to hydrocarbons, covering the reservoir. / deklaag / roche couverture / Deckgestein / cobertura

3765
seal; sealing formation, – rocks impermeable to hydrocarbons and, in consequence, capable of forming a trap or a permeability barrier in a reservoir. / afsluitende formatie / roche couverture / Abdichtung / sello; formación sellada; .... impermeable

sealing fault, – a fault with sufficiently impermeable gouge to form a trap or a permeability barrier in a reservoir. / afsluitende breuk / scellement par recouvrement / abdichtende Störung / falla sellante

tar seal, – a deposit of tar (weathered oil) forming an effective trapping seal at the outcrop or subcrop of a reservoir formation. / – – / scellement par couche asphaltique / abdichtende Teerkappe / sello de alquitrán

3766
oil sand; oil sandstone; oil-bearing sand; oil-bearing sandstone, – sand (sandstone) containing a (near-)maximum saturation of petroleum. / oliezand; oliezandsteen; oliehoudend zand; oliehoudende zandsteen / sable pétrolifère; grès .... / Ölsand; Ölsandstein; ölführender Sand; .... Sandstein / arena petrolífera; arenisca ....

gas sand; gas sandstone; gas-bearing sand; gas-bearing sandstone, – sand (sandstone) with (near-)maximum saturation of gas. / gaszand; gaszandsteen; gashoudend zand; gashoudende zandsteen / sable gazifère; grès .... / Gassand; Gassandstein; gasführender Sand; .... Sandstein / arena gasífera; arenisca ....

water sand; water sandstone; water-bearing sand; water-bearing sandstone, – sand (sandstone) with (nearly) 100% water saturation. / waterzand; waterzandsteen; waterhoudend zand; waterhoudende zandsteen / sable aquifère; grès .... / Wassersand; Wassersandstein; wasserführender Sand; .... Sandstein / arena acuífera; arenisca ....

3767
wet sand; wet formation, – sand or formation with a high water saturation and therefore tending to produce water, either with or without hydrocarbons. / nat zand; natte formatie / sable humide; formation .... / verwässerter Sand; verwässerter Träger / arena húmeda; formación ....

3768
tar sand, – generally used but actually incorrect term for a sand or sandstone with a high saturation of heavy oil or asphalt, that cannot be produced by conventional, primary recovery techniques. / teerzand / sable asphaltique; .... bitumineux / Teersand / arena asfáltica

3769
tight rock, – a rock of too low permeability to produce fluid or gases or to allow fluid injection in a well. / dicht gesteente / roche peu-perméable; .... colmatée / dichtes Gestein / roca compacta

3770
homogeneous reservoir, – a reservoir in which rock composition and physical properties do not change appreciably. / homogeen reservoir / réservoir homogène / homogener Speicher / almacén homogéneo

reservoir isotropy, – the degree to which the physical properties of the reservoir rock remain constant in all directions. /

reservoirisotropie / isotropie du réservoir / Speicherisotropie / isotropía del almacén

3771
permeability barrier, – a zone of low permeability (parallel to the bedding or otherwise) in a reservoir, acting as an obstruction to fluid movements and pressure equalisation during production. / permeabiliteitsbarrière / barrière de perméabilité / Permeabilitätsbarriere / barrera de permeabilidad

3772
shale break; clay ...., – a (thin) layer of clay or claystone intercalated in a reservoir, tending to form a permeability barrier. / kleilaag / intercalation argileuse / Tonmittel / intercalación arcillosa

[2554] *Interrelation of Water, Oil and Gas*

3773
oil field; gas ...., – an area under which – or a part of the earth's crust in which – occur one or several, more or less contiguous, accumulations of oil/gas, capable of commercial production. / olieveld; gasveld / champ pétrolifère; .... pétrolier; .... de gaz / Ölfeld; (Erd)gasfeld / campo petrolífero; .... gasífero

oil pool, – a contiguous accumulation of oil, capable of commercial production; one or more adjacent pools can make up an oil field. / – – / gisement de pétrole / Ölansammlung / bolsa; (yacimiento)

3774
gas cap, – accumulation of gaseous hydrocarbons, overlying an oil accumulation in a reservoir. / gaskap / chapeau de gaz; calotte de gaz / Gaskappe / acumulación de gas

gas/oil contact; GOC; gas/oil interface, – the interface, generally horizontal, between gas cap and oil zone. / gas/oliecontact / niveau gaz/pétrole; (ligne de) contact gaz/pétrole / Gas/Öl Kontakt; Gas/Öl Spiegel / contacto; nivel gas/petróleo

3775
oil zone, – the part of the reservoir saturated with oil, and from which oil can be produced. / oliezone / zone pétrolifère / (Erd)ölzone / zona petrolífera

oil column; oil height, – the vertical distance between the gas/oil contact or, if no gas cap is present, the top of the reservoir and the oil/water contact; i.e. the vertical thickness of the oil zone. / oliekolom; hoogte van de .... / hauteur impregnée par l'huile / Ölsäule / columna de petróleo; altura de ....

hydrocarbon column; .... height, – the vertical distance between the highest point of the reservoir and the hydrocarbon/water contact. / koolwaterstofkolom; hoogte van de .... / hauteur de la colonne d'hydrocarbure / Kohlenwasserstoffsäule / – –

3776
oil ring; oil rim, – if in the gas cap the gas occupies the full thickness of the reservoir zone, oil may occur in a downdip ring or rim (partially) surrounding the gas cap. / oliering / anneau de pétrole / Ölring / anillo de petroleo

attic oil, – oil contained in the part of the reservoir situated above the highest drainage point, and that, in consequence, cannot normally be produced without additional drilling. / – – / huile d'amont pendage / Hangendöl / – –

3777
oil/water contact; OWC, – the, generally horizontal, interface between the oil-saturated part of the reservoir and the underlying water-bearing part; the term is a simplification, because in reality there is always a, more or less thick, transition zone. / olie/watercontact / contact pétrole/eau; niveau .... / Öl/Wasserkontakt; ....spiegel / nivel petróleo/agua; contacto ....

water table,– in oilfield usage, synonymous with oil/water contact. / waterspiegel / – – / Wasserspiegel / nivel de agua

transition zone, – vertical interval in the reservoir, whose length depends on porosity, in which the water saturation changes from 100% at the bottom to irreducible water saturation at the top. / overgangszone / zone de transition / Übergangszone / zona de transición

3778
free water level, – the depth to which water will stand in a well if the latter is left undisturbed for some time. / vrije waterspiegel / niveau hydrostatique / freier Wasserspiegel / nivel de agua libre

3779
bottom water, – water underlying the oil accumulation, i.e. in cases where the height of the oil column is less than the thickness of the reservoir. / bodemwater / eau de fond; eau de mur; eau sous-jacente / Liegendwasser / agua de fondo

edge water, – water underlying an oil accumulation along its edges, i.e. in cases where the height of the oil column exceeds the thickness of the reservoir. / randwater / eau de bordure; eau marginale / Randwasser / agua marginal

3780
intermediate water, – a zone with a relatively high water saturation (that, in consequence, will tend to produce water), intercalated in the reservoir between zones with (near-)irreducible water saturation and correspondingly high hydrocarbon saturation. / tussenwater / eau intermédiaire / Zwischenwasser / agua intermedia

3781
tarmat, – a somewhat ambiguous term, denoting a thin zone of asphalt at the oil/water contact and sometimes acting as a permeability barrier. / – – / – – / Asphaltimprägnierung / tapa de alquitrán

[2560] OIL WELLS AND FORMATION TESTS

NOTE:
– see also section
[2820] Drilling; Methods and some Technical Terms.

3782
exploration well, – a well drilled on a geologically or geophysically defined prospect before a discovery is made. / exploratieboring; ....put; verkenningsboring / sondage d'exploration; forage de recherche / Explorationsbohrung; Suchbohrung / sondeo de exploración

wildcat, –
1. an exploration well drilled in a geologically and geophysically poorly known area;
2. synonymous with exploration well in the more or less official classification of the American Association of Petroleum Geologists.

3783
stratigraphic test; stratigraphical well, – an exploration well drilled primarily for stratigraphical (geological) information; usually not expected to find hydrocarbons. / stratigrafische boring / sondage stratigraphique / stratigraphische Bohrung / sondeo estratigráfico

slim hole, – a hole drilled with unusually small diameter, mainly for stratigraphic information. / – – / filiforage / – – / sondeo delgado

3784
discovery well, – a well that encounters commercially producible hydrocarbons on a prospect. / ontdekkingsboring / puits de découverte / Fundbohrung; fündige Bohrung / pozo de descubrimiento

success ratio, – ratio of discovery wells to total number of exploration wells drilled in a particular area, region or geological province. / – – / taux de succès / Erfolgsquote; Erfolgsrate / relación exploración descubrimiento; coeficiente de éxito

3785
appraisal well, – a well drilled for the purpose of determining the reserves in an accumulation after the discovery has been made. / evaluatieboring / sondage d'évaluation / Bewertungsbohrung / pozo de evaluación

delineation well, – an appraisal well drilled specifically for early definition of the limits of a field or reservoir. / begrenzingsboring / sondage de délimitation / Begrenzungsbohrung / pozo de delimitación

outstep well; step-out, – an appraisal well drilled at some distance from existing productive wells for the purpose of investigating the continuity of the accumulation. / uitstapboring / puits d'extension / Erweiterungsbohrung / pozo de extrapolación

3786
development well; production well, – a well drilled for the purpose of production from ('development' of) a known reservoir or field. / productieput / puits de production / Förderbohrung; Produktionssonde / pozo de producción

3787
formation test, – operations to investigate what fluid can be produced from a certain interval in a well. /laagbeproeving / essai / Formationstest / ensayo; prueba; testificación

drillstem test; DST; Johnson test, – a formation test through drillpipe by means of special equipment (Johnson formation tester). / boorpijpproef / essai aux tiges / Gestängetest / ensayo por tubería de perforación

wireline test, – formation test by means of a fluid sampler run on cable. / kabelproef / test sur câble; essai avec appareil sur cable / Kabeltest / testificación por cable

3788
production test, – prolonged formation test, usually with the well already completed for sustained production. / productieproef / essai de production / Fördertest / prueba; ensayo de producción

[2570] PRODUCTION

3789
production well; productive well, –
1. a well drilled with the intention to exploit an oil or gas reservoir;
2. a well capable of producing oil and/or gas in commercial quantities.

/ productieput, productieve put / puits de production; .... productif / Fördersonde; ....bohrung / pozo productivo

3790
flowing well, – an oil well in which liquids and gas, solely driven by the pessure in the reservoir, flow from this reservoir to the surface. / spuitende put / puits éruptif / fliessende Sonde / pozo surgente

gusher (USA); – a highly productive flowing well.

3791
pumping well, – an oil well, in which the liquids are moved to the surface by means of a pump; this pump can be moved by rods and a pumping unit at the surface, or driven hydraulically or electrically. / pompput / puits pompé / Pumpsonde / pozo bombeado

gaslift well, – an oil well in which high-pressure gas is injected, generally through the annular space between casing and tubing; the gas leaves the well through the tubing, via gaslift valves, entraining the reservoir fluids. / gasliftput / puits produisant par allègement de gaz / Sonde mit Gaslift / operación gaslift

3792
offset well, – oil or gas well drilled near the boundary of a concession to prevent oil and gas from being withdrawn through

wells located in adjacent concessions. / — — / puits de limite / Grenzbohrung / pozo del límite de permiso

3793
injection well, — a well used for the injection of gas or water into the reservoir, usually for pressure maintenance, in other cases for temporary storage of produced gas or permanent storage of water produced with the hydrocarbons. / injectieput / puits d'injection / Einpress-sonde / pozo de inyección

3794
dry hole; — a well that did not encounter hydrocarbons in commercially producible quantities. / droge put / puits sec / Fehlbohrung / pozo seco

wet hole, — a well that produces water, either alone or in association with hydrocarbons. / natte put; verwaterde put / puits producteur d'eau / verwässerte Bohrung / pozo con producción de agua

tight hole, —
1. a well that does not produce any fluid and does not accept injection fluid due to lack of permeability in the formation;
2. a well on which no information is released for commercial reasons.

/ — — / puits en formation compacte; puits au secret / — — / pozo impermeable

3795
(well) completion, — installing the necessary equipment in a well and at the well head to make the well suitable for sustained production. / putafwerking / complétion / Komplettierung / puesta en producción

liner, —
1. a casing string that does not reach entirely up to the surface;
2. a perforated pipe installed in the well over the producing interval to prevent entry of rock material.

/ zeefpijp / tube crépine / Filterrohr / (tubo) filtro

barefoot hole, — a producing well without casing or liner over the producing interval. / open gat / nu-pied; puits tubé au toit / unverrohrtes Bohrloch / — —

3796
Christmas tree; production wellhead, — an assembly of valves and other fittings, installed at the wellhead to control flowing production. / spuitkruis / arbre de Noëll / Eruptionskreuz; Bohrlochkopf / árbol de navidad; (cruz de producción)

3797
separator, — a vessel in which gas or water is separated from the produced oil. / afscheider / séparateur / Separator / separador

dehydration, — the separation, at the surface, of oil and water produced from a well, either chemically or electrically. / ontwatering / déshydratation / Entwässerung / deshidratación

3798
(well) gauging, — measurement of the production of the oil, gas and water from a well over a certain period (usually a few hours to a day); this is repeated at regular intervals. / meten van de productie / jaugeage / Messung der Produktion / medir la producción

manifold, — an assembly of valves by which the production from a particular well can be separated from the mainstream of a cluster of wells to allow separate gauging. / verdeelstuk / claviature; collecteur; manifold / Verteiler / manifold; conducto; cañería

3799
gun perforation, — perforating casing installed in the well by means of a 'gun', run on cable, in order to allow fluids from the formation to enter the well, while keeping formation particles out. / — — / perforation à balles / Schussperforation / perforador de balas

3800
(reservoir) stimulation, — techniques aimed at improving the permeability of the reservoir rock in the immediate vicinity of the well in order to increase the flow from the reservoir to the well. / stimulatiebehandeling / stimulation / Stimulationsbehandlung / estimulación

acidising; acidisation; acid treatment, — stimulation by dissolving certain components of the reservoir rock by, usually hydrochloric, acid. / zuurbehandeling / acidification / Säurebehandlung / acidificación

hydraulic fracturing; fracture treatment; hydrofrac, — stimulation by injecting under high pressure a liquid to which granular material, usually sand, has been added; thus fractures are created in the reservoir, starting from the well bore, that are permanently filled with the sand and serve as permeable channels between the reservoir and the well. / hydraulisch breken / fracturation hydraulique / Hydrofrac; fracen; aufbrechen / fracturación

3801
production rate, — a well's production per time unit, usually 24 hours. / productiepeil / taux de production / Produktionsrate / marcha de la producción; valor medio de ....

3802
gas-oil ratio; GOR, — the quantity of gas produced from a well per unit quantity of oil produced from that well. / gas-olieverhouding / proportion gaz-huile; facteur de gaz / Gas-Öl Verhältnis / relación gas-petróleo

gas-liquid ratio; GLR, — the quantity of gas produced from a well per unit quantity of liquids (oil + water) produced. / gas-vloeistofverhouding / proportion gaz-liquide / Gas-Flüssigkeitsverhältnis / relación gas-líquido

water cut, — the amount of water in the produced liquid, expressed as a percentage (fraction) of the total liquid production. / watergehalte / teneur d'eau / Wassergehalt / procentaje de agua; contenido de agua

3803
disposal water, — water that is produced with oil or gas and that cannot be usefully utilized, e.g. for re-injection or recycling. / afvalwater / eau de gisement residuaire / Abwasser; Lagerstättenwasser / agua a disposición

3804
bottom sediment and water; bottom settling and water;

BS&W, – the percentage content of water and other substances that settle out of a sample of produced liquid when left standing. / – – / teneur en eau et sédiments / Bodensatz und Wasser / – –

3805
initial production, – the productive capacity of a well during the initial production period, usually taken from a few days to a few months, depending on the production characteristics. / beginproductie / production initiale / Anfangsproduktion / producción inicial

3806
productive capacity, – the amount of oil or gas a well or group of wells is capable of producing per unit time; the actual production rate may be less than the productive capacity for various reasons. / productiepotentieel; ....vermogen / capacité de production / Förderkapazität / capacidad de producción

well capacity; well potential, – the productive capacity of a single well. / putpotentieel / production potentielle du puits / Förderkapazität der Sonde / capacidad de producción del pozo

3807
peak production, – the maximum productive capacity of a well or group of wells; because of the natural decline this will in many cases be equal to the initial production. / topproductie; maximale productie / production culminante / Höchstförderung / producción máxima; máximo de producción

3808
productivity index, – the amount of oil or gas that moves radially from the reservoir into the well per unit time and per unit pressure difference between the reservoir and the bottom of the well. / productiviteitsindex / indice de productivité / Produktivitätsindex / índice de productividad

specific productivity index, – productivity index expressed per unit length of productive interval in the well (or per unit length of net sand in the productive interval). / specifieke productiviteitsindex / indice de productivité spécifique / spezifischer Produktivitätsindex / índice especifico de productividad

3809
injectivity index, – the amount of fluid or gas that can be injected into the reservoir from a well per unit time and per unit pressure difference between the bottom of the well and the reservoir. / injectiviteitsindex / indice d'injectivité / Injektivitätsindex / índice de inyección posible

3810
downtime, – a period during which a potentially productive well does not actually produce, for whatever reason. / – – / temps mort / Stillstandszeit; Stillegungszeit / tiempo perdido

3811
shut-in well; closed-in well, – a productive well in which production has been stopped by closing the wellhead valves. / ingesloten put / puits fermé / (ein)geschlossene Sonde / pozo cerrado

3812
dead well, – a formerly flowing well that does no longer produce by whatever cause, e.g. mechanical failure, plugging of tubing, lack of reservoir pressure. / dode put / puits stérile / nicht fordernde Bohrung / pozo muerto

3813
well repair; workover; clean-out job, – operations in a well to improve its mechanical condition in order to enhance the productive capacity. / putbehandeling; schoonmaken / nettoyage / Behandlung; Aufwältigung; Reinigung / limpieza

3814
sanding-up, – fine formation particles enter the well and hamper or prevent production, generally leading to a clean-out job. / verzanden / ensablement / versanden / enarenarse; ensuciarse de arena

mechanical properties log; MPL, – vertical profile of a ratio related to sonic interval travel time and interval density; used for indicating potential sand-trouble intervals in producing wells.

3815
formation damage, – through contact with the drilling fluid the permeability of the reservoir rock in the vicinity of the well may be impaired; in pressure build-up analysis this shows as **'skin effect'**. / – – / détérioration de la formation / Formationsschädigung / deteriozo de la formación

3816
decline, – the decrease of the production from a well, or group of wells, with time. / afname / déclin / Abfall / caída; bajada; disminución

decline curve, – a curve depicting the decrease of production. / afnamecurve / courbe de déclin / Abfallkurve / curva de caída

3817
cumulative production, – the total quantity of oil or gas produced from a well or group of wells up to a certain moment. / cumulatieve productie / production cumulée / kumulative Produktion; Kumulativproduktion / producción cumulativa

ultimate production, – the total quantity of hydrocarbons produced at the end of a well's, or field's, life, i.e. at the moment of abandonment. / totale productie / production totale; .... finale / Gesamtproduktion / producción final; .... total

3818
abandonment, –

1. the procedures applied to put a well permanently out of use, with the necessary safeguards to prevent leakages;
2. the moment when production from a well or field is finally stopped, usually for economic or reservoir technical reasons.

/ verlaten / abandon / Verlassen / abandono

[2580] RESERVOIR ENGINEERING

NOTE:
- see also section
  [2540] Rock Properties and Fluid Flow

[2581] *Reservoir Conditions*

3819
reservoir pressure; fluid ....; pore ...., – the pressure prevailing in the fluids occupying the pore volume in the reservoir rock. / reservoirdruk / pression de gisement / Lagerstättendruck / presión del yacimiento

grain pressure; matrix ...., – the pressure acting between the individual rock particles. / korreldruk / pression de réservoir; .... matricielle / petrostatischer Druck; Korndruck / presión granular

overburden pressure; rock ...., – the total pressure at any point in the earth's crust, resulting from the mass of the overlying rocks and fluids (equals grain pressure plus fluid pressure). / gronddruk / pression de surcharge; .... des terrains sus-jacents / Druck der Deckschichten; Gebirgsdruck / presión de sobrecarga

3820
hydrostatic pressure, – the pressure exerted by a column of water from the surface to a point in the subsurface. / hydrostatische druk / pression hydrostatique / hydrostatischer Druck / presión hidrostática

overpressure; geopressure; excess pressure, – fluid pressure in the reservoir rock in excess of hydrostatic pressure. / overdruk / surpression; pression excédentaire / Porenwasserüberdruck / sobrepresión

3821
pounds per square inch; psi, – antiquated unit of pressure; as a coincidence, on the average rock pressure increases with 1 psi per foot of depth. 1 psi = 6.89 kPa ($= 689 \cdot 10^{-4}$ bar $= 7.031 \cdot 10^{-2}$ kgf/cm^2).

[2582] *Reservoir Processes*

3822
primary recovery, – the recovery of hydrocarbons from a reservoir, utilising only the natural energy available in the reservoir. / primaire winning / récupération primaire / primäre Förderung / recuperación primaria

3823
gas-cap drive; gas ...., – recovery of oil in which the expansion of the gas cap is the main driving force. / gaskapstuwing / drainage par poussée de gaz; drainage par expansion du gaz libre / Gaskappentrieb / presión de gas libre; empuje por gas libre

gas-solution drive, – recovery of oil in which the expansion of the, initially, undersaturated oil and the dissolved gas provide the main driving force. / – – / drainage par expansion de gaz dissous / Gasentlösungstrieb / presión de gas disuelto; empuje por gas disuelto

3824
water drive, – recovery of hydrocarbons in which the main driving force is the expansion of the water in the water-bearing portion of the reservoir rock. / waterstuwing / drainage par poussée d'eau; .... hydraulique / Wassertrieb / presión hidráulica; empuje hidrostático

3825
edge-water drive, – water drive by advance of the edge water. / randwaterstuwing / drainage par poussée d'eau marginal / Randwassertrieb / empuje por agua marginal

bottom-water drive, – water drive by upward movement of the bottom water. / bodemwaterstuwing / drainage par poussée d'eau de fond / Liegendwassertrieb / empuje par agua del fondo

3826
edge-water encroachment, – advance of edge water from the aquifer into the oil zone during production (without actual edge-water drive). / opdringen van het randwater / envahissement de l'eau de bordure / Vordringen des Randwassers / avance del agua marginal

piston-like displacement, – a sharp interface between encroaching water and oil is maintained during water drive. / – – / déplacement piston / – – / – –

by-passing, – the advance of the edge water in water drive is irregular and bodies of oil are by-passed by 'tongues' (**tonguing**) or 'fingers' (**fingering**) of water.

coning, – around a producing well the oil/water contact is drawn up in the shape of a cone, or the gas/oil contact is similarly drawn down. / kegelvorming / formation de cone / Kegelbildung / en forma de cono

3827
compaction drive, – recovery of hydrocarbons in which the compaction of the reservoir rock under the pressure of the overburden is the main driving force. / winning door compactie / drainage par compaction / Kompaktionstrieb / empuje por compresión

3828
gravity drainage, – recovery of oil under the influence of gravity only. / drainering door zwaartekracht / drainage par gravité / Entölung durch Schwerkraft / drenaje por gravedad

3829
combination drive, – recovery of hydrocarbons by a process in which various mechanisms are active. / gecombineerde stuwing / drainage multiple; .... combiné / kombinierte Förderung / empuje múltiple

3830
depletion, – as hydrocarbons are withdrawn from the reser-

voir the pressure decreases as a result of insignificant influx of water from the aquifer. / depletie / déplétion / Depletion / agotamiento

3831
secondary recovery; supplemental ...., – methods to enhance the recovery of hydrocarbons by artifically supplying energy to the reservoir. / secundaire winning / récupération secondaire / sekundäre Förderung / recuperación secundaria

3832
pressure maintenance, – injection of fluids into the reservoir to maintain the pressure at a level favourable for production. / op druk houden / maintien de pression; recompression / Druckerhaltung / mantenimiento de presión

3833
water flooding; artificial water drive, – injection of water into a reservoir as a secondary recovery method. / waterinjectie / injection d'eau / Wasserfluten / inyección de agua

flank drive, – the reservoir is flooded with water injected downflank from the oil zone. / – – / injection périphérique / Flankenfluten / – – / – –

line drive, – the reservoir is flooded with water injected into a line of wells favourably located in the reservoir. / – – / injection en ligne / – – / – –

pattern drive, – the reservoir is flooded with water injected into wells arranged amongst the production wells in a regular grid pattern. / – – / injection par réseau incullé / – – / – –

3834
cross flow, – water injected into one reservoir zone passes (via fractures) into another reservoir zone, where its effects are usually undesirable.

thief sand, – an exceptionally permeable streak in the reservoir, into which flows an undesirably high proportion of the injected fluids. / – – / sable de capture / – – / – –

3835
gas injection, – injection of gas into the gas cap of a reservoir as a secondary recovery method. / gasinjectie / injection de gaz / Gasinjektion; Gaseinpressung / inyección de gas

recycling of gas, – re-injection of produced gas into the gas cap of a reservoir after liquid components have been removed from the gas. / hercirculeren van gas; herinjectie van gas / recyclage de gaz / Wiedereinpressung von Gas / recirculación de gas

3836
conservation of gas (pressure), – production regulated so as to minimize gas production in order to conserve energy in the reservoir. / gasconservering / conservation de gaz / Gasschonung / conservación de gas

3837
tertiary recovery, –
1. originally measures aimed at recovering additional oil from the reservoir after a secondary recovery project;
2. at present such methods may also be applied earlier in the development of the reservoir.

/ tertiaire winning / récupération tertiaire / tertiäre Förderung / recuperación terciaria

3838
in situ combustion; fire flooding; fire drive, – measure to improve ultimate recovery by causing combustion of reservoir oil around wells, into which air is injected, so as to increase reservoir temperature and decrease oil viscosity, i.e. to increase the mobility of the oil. / ondergrondse verbranding / combustion in situ / In-situ Verbrennung / combustión en 'situ'

3839
steam drive, – injecting steam into injection wells, in order to increase the mobility of the oil, which is then driven towards the production wells. / stoominjectie / injection de vapeur / Dampfeinpressung; ....injektion / inyección de vapor

hot water drive, – injection of hot water for the same purpose as in steam injection. / heet-waterinjectie / injection d'eau chaud / Heisswasserinjektion / inyección de agua caliente

steam soak; huff-and-puff method, – intermittent injection of steam into wells, in order to increase the mobility of the oil in the vicinity, that is then produced from the same wells.

3840
miscible flooding, – injection into the reservoir of a fluid, which is miscible with the oil, to increase its mobility and thus, as it were, to wash out the remaining oil saturation. / – – / déplacement miscible / chemisches Fluten / desplazamiento miscible

3841
retrograde condensation, – as a consequence of special phase-behaviour conditions, gases may condense in the reservoir as the pressure decreases during production. / retrograde condensatie / condensation rétrograde / retrograde Kondensation / condensación retrógrada

3842
well spacing, –
1. the (average) distance between producing wells in a field;
2. arrangement of producing wells in a regular grid pattern.

/ locatieafstand; putpatroon / espacement des puits / Bohrlochabstand; ....raster / espaciado

3843
infilling well, – an additional well drilled to increase the density of the well pattern. / tussenput / puits intermédiaire / Zwischensonde / pozo intermedio

interspacing, – drilling infillings wells in a regular spacing grid, e.g. in the centres of gravity of triangles of wells. / – – / – – / – – / interespaciado

3844
drainage area, – (circular) area around a production well, from which the well is considered to withdraw the bulk of its production. / drainagegebied / aire de drainage / Entölungsgebiet; Einzugs.... / área de drenaje

3845
offset area, – a strip on either side of a common concession

boundary to which agreements between the concessionaires, regarding well spacing and/or withdrawal rates, apply. / schutgebied / zone de limite / Schutzgebiet / área de enfrentación; área protegida

stripper well, – a well producing in the last stages of a strong water-drive reservoir, when only small amounts of oil are produced with large quantities of water. / – – / puits marginal / Grenzförderbohrung / pozo marginal

[2583] *Reservoir-Engineering Methods*

3846
reservoir performance, – a general term describing the behaviour of a reservoir during production, as regards productivity, pressure drop, etc. / reservoirgedrag / comportement du réservoir / Leistung des Reservoirs; Verhalten des .... / comportamiento del yacimiento

3847
bottom-hole pressure; BHP, – pressure in a well at the face of the producing formation. / bodemdruk / pression de fond / Sohlendruck; Boden.... / presión de fondo

static pressure, – BHP after a prolonged closed-in period. / statische druk / pression statique / statischer Druck / presión estatica

flowing pressure, – BHP while the well is flowing. / spuitende druk / pression d'écoulement / Fliessdruck / presión en erogación

3848
pressure bomb; bottom-hole pressure gauge, – apparatus designed to register the pressure in a well over a certain interval of time. / bodemdrukmeter / enrégistrateur de pression de fond / Bodendruckschreiber / registrador de presión de fondo; bomba de presión

3849
initial pressure, – pressure in the reservoir before first production commenced. / begindruk / pression initiale / Anfangsdruck / presión inicial

abandonment pressure, – pressure in the reservoir when production is finally terminated. / verlatingsdruk / pression d'abandon / Enddruck / presión de abandono

3850
pressure build-up (curve), – (the curve depicting) the increase of the bottom-hole pressure from the flowing pressure after the well has been closed-in; after a long enough time the pressure will build up to the static bottom-hole pressure. / drukopbouw / remontée de pression / Druckaufbau / registro de subida de presión

pressure drawdown, – the difference between the static and the flowing bottom-hole pressures; it is closely related to production rate. / drukval / décompression / Druckentlastung; Druckdifferenz / decompresión

3851
skin effect, – a reduction in permeability around the wellbore, caused by contact with the drilling fluid; it affects the shape of the build-up curve. / – – / effet pelliculaire / – – / efecto pelicular

3852
pressure gradient, – the change in pressure over a certain depth interval (in a well, in a reservoir, etc.). / drukgradiënt / gradient de pression / Druckgradient / gradiente de presión

3853
pressure drop, – the change in pressure over a certain time interval during production. / drukverval / chute de pression / Druckgefälle / caída de presión

3854
isobaric map; pressure map, – a map with lines of equal pressure depicting the variation in pressure over the area of the reservoir; the pressures are corrected to a datum level. / drukkaart / carte isobare / Druckkarte / mapa isobárico

datum plane, – an arbitrary level in the reservoir to which bottom-hole pressures measured in different wells at different depths in the reservoir are corrected for comparison purposes. / referentievlak / surface de référence / Bezugsfläche / nivel de referencia

3855
bottom-hole sample, – a sample of well fluid taken in a well at the face of the producing formation and brought to the surface in a sealed container. / bodemmonster / échantillon de fond / Bodenprobe / muestra de fondo

3856
PVT analysis, – laboratory measurement of the (phase) behaviour of a sample of reservoir fluid when pressure, volume and temperature are varied.

3857
undersaturated oil, – oil that releases no free gas at reservoir pressure and temperature conditions. / onderverzadigde olie / pétrole sous-saturé / untersättigtes Öl / petróleo subsaturado

bubble point; .... pressure, – the pressure at which the first free gas appears when a volume of undersaturated oil is allowed to expand. / verzadigingspunt / point de bulle / Sättigungsdruck; Gasentlösungsdruck / punto de burbujeo

3858
dew point; dew pressure, – the pressure at which the first fluid appears if a volume of gas is compressed (or expands in retrograde condensation). / dauwpunt / point de rosée / Taupunkt / punto de rocío

3859
tank oil; stock-tank oil; standard tank oil, – produced oil as stored in the field tank, i.e. after expansion to standard surface conditions, removal of components that are in the gaseous phase at those conditions and dehydration. / tankolie / huile de stockage / Tanköl / petróleo en tanque

oil originally in place; STOOIP, – the volume that all oil originally present in the reservoir would occupy if converted to stock-tank conditions.

gas initially in place; GIIP, – the volume that all gas originally present in the reservoir would occupy if converted to standard conditions.

3860
formation-volume factor (oil), – the volume occupied at reservoir conditions by a quantity of stock-tank oil together with its corresponding dissolved gas, i.e. after an imaginary transfer back to the reservoir. / volumefactor (voor olie) / facteur volumétrique (d'huile) / Formationsvolumenfaktor (für Öl) / factor volumétrico (de petróleo)

shrinkage factor, – the factor to be applied to a volume of reservoir fluid to derive the corresponding volume of stock-tank oil, i.e. the reciprocal of the oil formation-volume factor. / schrinkfactor / facteur de contraction / Schrumpfungsfaktor / factor de merma

3861
formation volume factor (gas), – the volume that one unit of volume of gas at standard conditions will occupy at reservoir conditions. / volumefactor (voor gas) / facteur volumétrique (de gaz) / Formationsvolumenfaktor (für Gas) / factor volumétrico (de gas)

3862
(gas) deviation factor, – a factor to be applied in gas calculations to allow for behaviour deviating from that of an ideal gas. / gasdeviatiefactor / facteur de déviation de gaz / Gasdeviationsfaktor / factor de deviación de gas

3863
flash expansion, – instantaneous expansion of an oil/gas sample from reservoir (or separator) conditions to standard conditions in PVT analysis. / – – / expansion éclair; .... instantanée / Blitz-Expansion / – –

3864
movable oil, – the part of the oil in a reservoir that can ideally be replaced by water, i.e. the volume in excess of the residual oil saturation. / – – / huile mobile / förderbares Öl / – –

3865
material balance; .... calculation, – a fundamental calculation in reservoir engineering, in which the volume of cumulative production at a certain date is equated to the difference between the initial volume of hydrocarbons in the reservoir and the volume present at the reference date. / materiaalbalans; ....berekening / bilan matière; calcul de .... / Materialbilanz; ....berechnung; Stoffhaushaltsrechnung / balance material

3866
reservoir model; .... simulation model, – a model (physical or numerical, mathematical) in which the geometry and physical properties of the reservoir and its production and pressure behaviour are simulated, with a view to enable predictions of future behaviour to be made. / reservoirmodel / modèle du réservoir / Lagerstättenmodell / modelo de yacimiento

numerical simulator; .... reservoir simulator, – a numerical, computerised reservoir model. / computer reservoirmodel / modèle numérique; .... d'ordinateur / numerisches Modell / modelo numérico

history matching, – adapting the characteristics of a numerical reservoir model so as to make its behaviour accurately match the known production and pressure history of the reservoir in order to improve the reliability of predictions of future behaviour by extrapolation.

3867
ultimate recovery, – the amount of oil/gas recovered from a reservoir at abandonment. / totale productie / récupération finale / Gesamtproduktion / recuperación total

recovery factor, – the ratio of ultimate recovery to the volume of oil/gas originally in place. / winningsfactor / facteur de récupération / Gewinnungsfaktor; Ausbautefaktor / factor de recuperación

[2584] *Reserves*

NOTES:
- in using the following section, the reader should bear in mind that there is no uniformity within the petroleum industry in the usage and definition of terms on reserves and their estimation and classification.
- see also sections
  [2750] Estimation of Ore Reserves,
  [2760] Mineral Resources,
- classifications in sections [2584] and [2760] are not comparable.

3868
reserves, – volume of hydrocarbons (oil, gas) that is known or believed to be present in the underground and suitable for eventual production. / reserves / réserves / Vorräte; Reserven / reservas

3869
estimated ultimate recovery, – the volume of hydrocarbons estimated, at any time during its production period, to be recovered from a reservoir (or field, etc.) when production is finally stopped (i.e. at abandonment); at any time equals cumulative production plus estimated reserves. / geschatte totale productie / récupération finale estimé; .... ultime estimée / geschätzte Gesamtproduktion / recuperación final estimada

3870
reserve estimate, – estimate of the volume of reserves present in a reservoir (or field, etc.); the estimate is affected by uncertainties regarding: existence of the accumulation, geological conditions, reservoir properties, productivity, economic conditions, etc. / reserveschatting / estimation des réserves / Vorratschätzung / estimación de reservas

3871
volumetric method, – calculation of reserves from estimates of volumetric parameters, such as: rock volume, net sand thickness, porosity, water saturation, formation volume factor, recovery factor. / volumetrische methode / méthode volumétrique / Volumen-Methode / método volumétrico

3872
gross rock volume, – the volume of oil-bearing rock in the reservoir, without allowance for the presence of intercalated non-productive (tight) rock. / bruto reservoirvolume / volume brut du réservoir / brutto Speichervolume / volumen bruto del yacimiento

area-depth graph, – a method of estimating gross rock volume by plotting the area enclosed by each contour line on top reservoir against the corresponding depth.

3873
net sand thickness, – the (vertical) thickness of reservoir rock (as encountered in a well) after subtraction of intercalations of non-productive (tight) rock. / netto zanddikte / épaisseur net des roches pétrolifères / netto Gesteinsmächtigkeit / espesor neto de la arena; espesor neto de la capa productiva

3874
decline method, – estimation of reserves by extrapolation of decline curves of production rate or reservoir pressure against cumulative production. / curve van afname-methode / méthode des courbes de déclin / Abfallkurven-Methode / método de las curvas de declinación

3875
cut-cumulative method, – estimation of reserves by extrapolation of a curve of percentage water in produced oil (=**'water-cut'**) against cumulative production.

3876
expectation-curve method, – a method of volumetric reserve estimation and classification in which the estimate is expressed as a cumulative probability curve showing for each of the total range of possible values the probability that at least this quantity can be recovered. / verwachtingscurvemethode / méthode des courbes d'expectation / Erwartungskurven-Methode / método de curvas de expectativa

expectation (of reserves), – the quantity indicated by the 'area under the expectation curve'.

3877
probabilistic combination, – a method by which arithmetic operations on a number of parameters can be carried out probabilistically by using the appropriate probability distribution for each parameter. (Used e.g. for the addition of reserves for a number of reservoirs determined by the expectation curve method.) / waarschijnlijkheidscombinatie / combinaison probabiliste / Wahrscheinlichkeitskombination / combinación probabilista

3878
reserve classification, – the subdivision of estimated reserves into classes in accordance with the degree of reliability with which the reserves in each class can be estimated. / classificatie van reserves / classification des réserves / Vorratsklassifizierung; ....kategorien / clasificación de reservas

3879
proven reserves; proved ...., –

1. reserves that are known, or can be assumed with reasonable certainty, to be commercially productive under present technological and economic conditions and for which sufficient information is available to allow a reasonably accurate estimate;
2. according to the American Petroleum Institute, API, (1977) '.... estimated quantities of .... crude oil which geological and engineering data demonstrate with reasonable certainty to be recoverable in future years from known reservoirs under existing economic and operating conditions'.

/ bewezen reserves / réserves prouvées / nachgewiesene Reserven; sichere Vorräte / reservas comprobadas

3880
probable reserves, – reserves for which the probability of existence (say, probability 50%) and the accuracy of estimation are less than for proven reserves. / waarschijnlijke reserves / réserves probables / wahrscheinliche Vorräte; .... Reserven / reservas probables

3881
possible reserves, – reserves for which the probability of existence (say, probability 25%) and the accuracy of estimation are less than for probable reserves. / mogelijke reserves / réserves possibles / mögliche Vorräte; .... Reserven / reservas posibles

3882
inferred reserves; speculative ...., – terms indicating reserves with low probabilities of existence (and probably low accuracy of estimation). / veronderstelde reserves; speculatieve .... / réserves spéculatives; .... estimées / vermutete Vorräte; spekulative Reserven / reservas especulativas; .... inferidas

3883
indicated additional reserves, – simplified API definition: reserves assumed with reasonable certainty to be recoverable from known reservoirs by the application of improved recovery techniques (e.g. secondary and tertiary recovery). / – – / réserves additionnelles / – – / reservas adicionales indicadas

3884
discovery, – API definition: '.... proved reserves credited to new fields and new reservoirs in old fields as the result of successful exploratory drilling and associated development drilling during the current year'. / ontdekking / découverte / Fund / descubrimiento

3885
extension, – API definition: '.... reserves credited to a reservoir because of enlargement of its proved area ....' (as the result of drilling additional wells). / uitbreiding / extension / Erweiterung / extensión

3886
revision, – API definition: 'changes in earlier estimates, either upward or downward, resulting from new information', e.g. obtained from development drilling and production history / herwaardering / révision / Revision / revisión

3887
prospective area, – area that is thought to be underlain by potentially productive accumulation(s) of hydrocarbons. / prospectief gebied / zone prometteuse / erdölhoffiges Gebiet / área probable

3888
proved-up area, – area shown to be overlying productive rocks by the drilling of wells. / – – / zone prouvée / – – / – –

3889
proven area; proved acreage, – area underlain by proven reserves. / bewezen gebied / zone prouvée / – – / área comprobada

3890
developed area; drilled-up area, – area of a field in which all wells required for proper production have already been drilled. / afgeboord gebied / zone développée / aufgebohrtes Gebiet; erschlossenes .... / área explotada

3891
associated gas, – gas occurring in a reservoir together with oil, either in solution or as free gas, and, therefore, usually produced together with the oil. / geassocieerd gas / gaz associé / assoziiertes Gas; Erdölgas / gas asociado

non-associated gas, – gas occurring by itself in a reservoir and that, therefore, can be produced without affecting oil production. / niet-geassocieerd gas / – – / freies Gas / gas no asociado; gas independiente

3892
barrel, – the American unit of volume, in which traditionally oil production and reserves are expressed, equivalent to 42 US gallons (= 159 litres). / vat / barril / Fasz / barril

barrel per acrefoot, – the American volumetric unit, in which initial oil in place is expressed, i.e. the amount of oil in barrels present in an area of one acre and nett sand thickness of one foot. The unit is equal to 159 litres of oil in a rock volume of 1223 m^3.

cubic foot, ... feet, – unit, in which gas production and reserves are expressed. It equals $283 \cdot 10^{-4}\,m^3$. / kubieke voet / pied cube / Kubikfuss / pié cúbico

[2590] OIL ECONOMICS

NOTE:
– see also section
[2770] Mine Valuation

3893
cash-out, – all outlay of cash for a project: capital, operating costs, taxes, etc., but not including depreciation. / uitgaven / dépenses / Ausgaben / desembolsos

cash-in, – all intake of cash for a project, mainly sales. / inkomsten / récettes / Einnahmen; Erlöse / entradas

3894
annual cash surplus; .... deficit, – the difference between the annual cash-in and cash-out. / jaarlijkse cash flow / cash flow annuel / jährlicher Cash-Flow / cash-flow anual; flujo de caja anual; flujo anual de tesorería

cumulative cash surplus; .... deficit, – the cumulative of the annual cash surpluses/deficits. / geaccumuleerde cash flow / cash flow cumulé / kumulativer Cash-Flow / flujo de caja cumulativo; flujo cumulativo de tesorería

ultimate cash surplus; .... deficit, – the cumulative cash surplus resp. deficit over the total life of the project. / uiteindelijke cash flow / cash flow total / Gesamtgewinn / flujo de caja final; flujo final de tesorería

3895
cash flow (curve), – a (curve depicting a) series of figures referring to successive years, such that the figure referring to any year represents the annual cash surplus or deficit during that year. / cash flow (grafiek) / cash flow (graphique); .... (courbe) / Cash-Flow (Kurve) / (curva de) flujo de caja; flujo de tesorería

3896
pay-out time, – the time elapsed from the beginning of a project until the time at which the cumulative cash flow is zero, i.e. the point at which the cumulative cash-flow curve crosses the zero line. / 'break-even' tijdstip; terugbetalings .... / délai de récupérationtemps de .... / Kapitalrückflusszeit / duración de amortización

3897
present value, – the present value of an amount in the future is equivalent to the amount that has to be put at a given time (reference date, usually the beginning of a venture) at a given compounded interest rate so that it will be equal to that future amount at the future date; amounts in the past are 'grossed up' in the same way. / contante waarde / valeur actuelle / Gegenwartswert / valor actual

discount rate; deferment rate, – the interest rate at which present values are calculated. / rentevoet / taux d'actualisation / Zinsfuss / tipo de interés

3898
discounted cash-flow method; DCF ...., – a method of assessing the potential profitability of a venture by calculating a cumulative discounted cash flow in which the discount rate is chosen such that the cumulative ultimate present value cash surplus is nil. / DCF methode / méthode des valeurs actuelles / Methode des inneren Zinsfusses / método DCF

earning power, – the discount rate calculated by the DCF method, used as a measure of the profitability of a venture. / rendement / taux de rentabilité interne; TRI;TIR / Verzinsung / rendimiento; tasa de rentabilidad

3899
money of the day terms; MOD, – a cash flow is expressed in MOD if the figures represent the actual amounts paid or received at the time the transaction takes place. / geïnfleerde waarde / monnaie courante / Wert zum Tageskurs / en moneda actual; en valor ....

constant value money terms; CVM; real terms, – a cash flow is expressed in constant value money terms, if each unit of cash has the same purchasing power for general goods and services throughout the life of the project. / constante koopkracht / monnaie constante / konstante Kaufkraft / en moneda constante; en valor ....

3900
lifting costs, – the costs directly concerned with the lifting of the oil from the reservoir to the surface. / opvoerkosten / coût d'extraction / Förderungskosten; technische Förderkosten / gastos de extracción

3901
operating agreement, – an agreement regulating the exploitation of a field, parts of which belong to different owners (concessionaires). / werkovereenkomst / accord d'opération / Arbeitsvereinbarung / convenio de operación

(operating) unit agreement, – an agreement for joint operation of a field, parts of which belong to different owners (concessionaires); usually one of the unit partners acts as unit operator. / overeenkomst tot gezamenlijke exploitatie / accord d'opérations unifiées / Gemeinschaftsvereinbarung / convenio de operación unificada

unitisation, – the process of arranging an operating unit agreement. / – – / unification / Unitisierung / unificación

3902
allowable, – maximum production rate imposed on a well or field for purposes of rationalisation of production (by agreement or by government regulation). / toegestane productiehoeveelheid / rythme de production plafonné / erlaubte Förderrate / admisible

proration, – the assigning of allowables to wells or fields on the basis of their production potential. / – – / fixation de plafonds / – – / prorrata

[2600] METALLIC MINERAL DEPOSITS

NOTE:

– in this chapter 'mineral' mostly stands for those minerals that, if present in sufficiently concentrated form, can be of economic value. 'Ore' is restricted, as far as historically possible, to the meaning as given below, in short a profitably extractable mineral association of metal compounds. All other usage is discouraged.

[2610] GENERAL TERMS

3903
metallogeny, – the study of the formation and evolution of metallic mineral deposits as an aspect of the geological evolution of the earth's crust. / metallogenese / gîtologie / metallogenese / metalogenía

metallogenic, metallogenetic,– pertaining to the formation of metallic mineral deposits. /metallogenetisch / métallogénique; gîtologique / metallogenetisch / metalogenético

3904
metallotect, – any geological feature influencing the concentration of elements to form a mineral deposit; partly synonymous with ore control. / – – / métallotecte / – – / metalotecto

strangulation, – local narrowing in the geochemical dispersion path of an element resulting in a local concentration of elements and a favourable area for the formation of mineral deposits. / vernauwing / étranglement / – – / estrangulación

3905
metallogenic map, – a map on which is shown the distribution of particular assemblages or provinces of mineral deposits and their relationship to such geologic features as tectonics and petrography. / metallogenetische kaart / carte métallogénique /metallogenetische Karte / mapa metalogenético

metallogenic epoch, – a clearly defined geological span of time, related to a geotectonic event, within which a number of mineral deposits have been formed. / ertsvormingstijdperk / époque métallogénique / Erzbildungsperiode; Erzausscheidungs.... / época metalogenética

3906
phase, – of mineral/ore formation, the long period of time of mineral accumulation or concentration of a single genetic process. / fase / phase / Phase / época

stage, – of mineral/ore formation, the shorter period of time, several of which may occur within the limits of a single phase, during which ore minerals of a definite composition accumulate in a more or less stable geological and physicochemical environment; successive stages are separated by a break in the mineralization.

mineral generation, – the mineral association characterizing a stage of mineral/ore formation. / – – / – – / Erzgeneration / generación mineral

NOTE:

– it should be noted that here the terms 'epoch' and 'stage' are used in a sense different from the formal stratigraphic ones.

3907
paragenesis,

1. as used in the USA, the distribution in time, or the sequence of formation of minerals;
2. as widely used in Europe, an association of minerals having a common origin, representing equilibrium conditions.

/ paragenese / paragenèse / Paragenese / paragénesis

zoning, – any regular pattern in the distribution of minerals or elements in the earth's crust; it may be shown in a single ore body, in a mineral district, or in a large region. Although zoning is related to the spacial distribution of elements and minerals, both time and space must be considered in the study of zonal phenomena. / zonale rangschikking / zonalité / zonale Verteilung / zonado; zonalidad

3908
metallogenic province, mineral ....; metallographic ...., minerogenic ....; an area in the order of n.10^5 km^2 and larger, characterized by the relatively abundant occurrence of geologically related mineral deposits. / mineraalprovincie; metaal....; erts.... / province métallogénique / Lagerstättenprovinz; Erz....; Mineral....; Metall.... / provincia metalogénica

belt, a province that is distinctly elongated in shape. / gordel / ceinture / Gürtel / cinturón

region, – a province between n.10^3 and n.10^5 km^2 in size. / regio / région / Unterprovinz; Subprovinz / región

metallogenic zone, a region, distinctly elongated in shape. / zone / – – / zone / – –

3909
mineral district; ore ...., – part of a metallogenic region in size n.10 km^2 and smaller. / ertsdistrict, mineraal...., metaal.... / district minier; district minéralisé / Lagerstättenbezirk / distrito minero

mineral field; ore ....,– part of a mineral district containing several mineral deposits. / ertsveld / champ minier; champ minéralisé / Lagerstättenfeld / campo minero

3910
mineral deposit, – a natural concentration of useful mineral substances, which under favourable circumstances can be profitably extracted. / delfstofafzetting / gîte minéral / Minerallagerstätte / yacimiento mineral

ore deposit,
1. in a wider sense: synonymous with mineral deposit;
2. in a stricter sense: a mineral deposit from which under the prevailing economic and technical situation a metal can be profitably extracted.

/ ertsafzetting / gisement métallifère / Erzlagerstätte / yacimiento mineral

ore body, – generally a solid and fairly continuous mass of ore, which may include low-grade and waste as well as pay ore, but is individualized by form or character from adjoining country rock. / ertslichaam / corps minéralisé / Erzkörper; Lagerstätte.... / masa mineralizada

3911
nonmetallic, – pertaining to a mineral substance without metallic lustre and having an economic importance not related to its metal content but to its technical usefulness in other ways. / niet-metallisch / non-métalique / nichtmetallisch / no metálico

metallic, – pertaining to a metal, a mineral with a metallic lustre, or a mineral from which one or more metals may be extracted. / metallisch / métallique / metallisch / metálico

polymetallic, – pertaining to a mineral or an ore from which more than one metal may be extracted. / polymetallisch / polymétallique / polymetallisch / polimetálico

monometallic, – pertaining to a mineral or an ore from which only one metal may be extracted. / monometallisch / monométallique / monometallisch / monometálico

3912
mineral; mineral substance; .... product, – in the practical sense, a natural raw material of the earth's crust, that under favourable circumstances may be affected with some kind of utility value to mankind. / minerale delfstof / minéral; substance minérale / Mineral; mineralische Substanz; mineralischer Rohstoff / mineral

ore mineral, – a homogeneous mineral species that may be used to extract one or more metals. / ertsmineraal / minerai / Erzmineral; Erzart / mena mineral

3913
ore, – a rock-building mineral association from which under the prevailing technical and economic conditions, one or more metals or compounds may be profitably extracted. / erts / minerai / Erz / mena

protore, – primary, low-grade metalliferous material that may be concentrated into ore by supergene processes. / proto-erts / protore; mineralisation primaire / Protore; Ausgangserz / mineral primario

3914
simple ore – ore that yields a single metal. / enkelvoudig erts / minerai simple; .... monominéral / einfaches Erz; monomineralisches Erz / mena monomineral

complex ore, – ore that yields several metals or that is composed of several ore minerals. / gemengd erts; samengesteld erts / minerai complexe / gemengtes Erz; zusammengesetztes Erz; Komplexerz / mena compleja

3915
high-grade ore, – /hoogwaardig erts / minerai à haute teneur; ... riche / hochwertiges Erz; Reicherz / mena rica

lean ore; low-grade ore, – poor ore or ore containing a lower proportion of metal than is usually worked. / arm erts; erts van laag gehalte / minerai à faible teneur; .... pauvre / geringwertiges Erz; Armerz / mena no explotable

3916
pay ore; workable ore, – those parts of an ore body that are both, sufficiently rich and large to contribute materially to the mine's profit. / winstgevend erts; ontginbaar erts / minerai exploitable / bauwürdiges Erz / mena explotable

3917
ore bearing, – ertshoudend / métallifère; minéralisé / erzhaltig; erzführend / – –

barren; sterile, – not containing minerals of value. / steriel / stérile / taub / esteril

3918
mother rock, – the crystallized magma from which an ore deposit originated. / moedergesteente / roche mère / Muttergestein / roca madre

parent rock; primary rock, – the unaltered rock from which the overlying residual weathering mantle originated. / oorsprongsgesteente / roche primaire / – – / – –

3919
host rock; invaded rock, – a rock body or formation serving as receptacle for a mineral deposit. / ontvangend gesteente;

gastheer.... / roche magasin / Wirtgestein; umgebendes ....; Neben.... / – –

3920
country rock, – the sterile rock located on both sides of an ore body or enveloping it. / nevengesteente / roche encaissante / Nebengestein / roca encajante

wall rock, – a term often used for country rock in connection with veins or dykes.

3921
gangue; veinstone,– the mineral aggregate dispersed throughout the ore body; it is composed of mostly non-metallic minerals without economic value to the mining operation. / ganggesteente (applicable to veins only) / gangue / taubes Mittel / ganga

gangue mineral, – a nonmetallic, or a non-economic metallic, mineral associated with ore minerals. / ganggesteentemineraal (applicable to veins only) / minéral de gangue / Gangmineral (applicable to ore bodies only) / – –

3922
Gangart, – in German mining terminology, the non-economic syngenetic minerals accompanying ore minerals in ore veins.

Lagerart, – in German mining terminology, the non-economic syngenetic minerals accompanying the ore substance in sedimentary ore deposits.

Ganggestein, – in German mining terminology, the detached pieces of more or less altered country rock enclosed in ore veins.

3923
Lagergestein, – in German mining terminology, the enclosed pieces of older rocks in a sedimentary ore deposit.

le stérile, – in French mining terminology, the combination of gangue and sterile country rock.

3924
waste, – general term for all non-economic material (gangue, barren rock, etc.) encountered during mining and milling an ore body. / afval / stérile / Abfall; Abraum / esteriles

overburden; cover; caprock; capping, – non-economic, surface material of any nature, consolidated or unconsolidated, that overlies a mineral or ore deposit; applies in particular to deposits mined from the surface by open cuts. / dekterrein; deklagen / morts-terrains; recouvrement / Deckschichten / montera

3925
ore microscopy; mineragraphy (obsolete), – the study of minerals in polished sections under the reflecting light microscope. / ertsmicroscopie / microscopie métallographique / Erzmikroskopie / mineragrafía; microscopía de menas

ore petrology, – the study of ores as rocks, hence the petrology of mineral deposits; roughly synonymous with metallogeny. / ertspetrologie / étude minéralogique du minerai / Erzpetrographie / petrología de menas

[2620] PROCESSES OF FORMATION

[2621] *Early Theories*

3926
descension theory, – (Werner and neptunists, 18th century), the theory that the material in veins was introduced in solution from above. / theorie der neptunische ertsvorming / théorie per descensum / Deszensionstheorie / neptunismo

3927
ascension theory, – (Elie de Beaumont, 1847), the theory that the matter filling fissure veins was introduced in solution from below. /theorie der plutonische ertsvorming / théorie per ascensum / Aszensionstheorie; Thermaltheorie / plutonismo

3928
theory of lateral secretion, – (Sandberger, 1877), the theory that the contents of a vein or lode are derived from the adjacent country rock by a leaching process, whereby either superficial water or thermal water is involved. / theorie van de laterale secretie / théorie de la sécrétion latérale / Lateralsekretionstheorie / secreción lateral

[2622] *Some Physico-Chemical Concepts*

NOTE:
– additional terms have been compiled in chapter [0400] Geochemistry and in section [1720] Soil-Forming Processes

3929
leaching, – the dissolution and removal of soluble constituents from a soil or rock by the natural action of percolating water. When the dissolved constituents are not transported over a considerable distance the term redistribution of matter is used. Hence leaching describes solution and transport. / logen; uitlogen / lessivage / Auslaugung / lixiviación

bacterial leaching, – leaching stimulated by the biochemical action of bacteria.

electrochemical leaching, – leaching stimulated by the existence of an electrochemical potential.

organometallic leaching, – leaching stimulated by the formation of soluble complex compounds.

3930
elutriation, – the washing away of the finer or smaller-mass particles in a soil by the natural action of rainwater run-off and percolation. / elutriatie / élutriation / – – / elutración

eluviation, – the downward movement of soluble and suspended material (particularly colloids) in a soil, from the A-horizon to the B-horizon, by percolating water. / inspoeling; eluviatie / éluviation; lessivage / Eluviation; Verlagerung / eluviación

illuviation, – the accumulation of material in a lower soil horizon supplied through the process of eluviation. / inspoeling; illuviatie / illuviation / Illuviation; Anreichung; Einwaschung; Einlagerung / illuviación

3931
transvaporisation, – incorporating of water from wall rock into magma. / transvaporisatie / transvaporisation / Transvaporisation / transevaporización

3932
coagulation, – the aggregation of particles in a colloidal solution to larger particles that may precipitate. / coagulatie / – – / (Koagulation) / coagulación

3933
filter pressing; filtration .....– the squeezing out of (residual magmatic) liquids from between crystal masses mashed together (by earth movements or other means); a process that, besides the formation of some types of silicate rocks, causes according to various authors the formation of some types of strictly magmatic ore deposits. / filterperseffect / effet de filtrepresse / Filterpressung; Abpressungsfiltration / filtración a presión

3934
sulphato-sulphidic proces, – this process is active in the cementation zones of sulphidic ore deposits. The sequence of formation of the secondary sulphides is controlled by the solubility products and the electrochemical properties of the sulphides, as indicated by the Schürmann series.

3935
Schürmann series, – a salt solution of a metal in this series is decomposed – and the metal is precipitated as sulphide – by the solid sulphide of all metals to the right in this series; i.e. Hg – Ag – Cu – Bi – Cd – Pb – Zn – Ni – Co – Fe – Mn. / reeks van Schürmann / – – / – – / series de Schürmann

[2623] *Petrological and Geological Concepts*

3936
mineralization, –
1. the process by which matter is introduced into a rock resulting in the formation of new minerals of potentially economic value or of a mineral deposit;
2. the state resulting from this process.

/ mineralisatie, verertsing / minéralisation / Vererzung / mineralización

mineralizing agent; mineralizer, – the gaseous or liquid agents of hypogene origin from which the ore minerals crystallize, or that help the crystallization of the ore minerals, to form mineral and ore deposits, especially those of pneumatolytic and hydrothermal origin. / mineralisator / agent minéralisateur; fluide ....; venue minéralisante / Mineralisator; Mineralbildner / agente mineralizador

metallization, – the processes by which metals are introduced into a rock or a certain district, resulting in the formation of a mineral deposit. Largely synonymous with mineralization.

3937
supergene, – pertaining to the formation of mineral concentration and enrichment at or near the surface resulting from weathering and downward movement of water. / supergeen / supergène / supergen; sekundäre Ausreicherung / supergénica

hypogene, – pertaining to the formation of mineral deposits through the action of ascending water, often tacitly assumed to be genetically connected with a cooling magma. / hypogeen / hypogène / hypogen / hipogénico

3938
magmatic differentiation, – a general term for the processes whereby different types of igneous rocks, including the strictly magmatic ore deposits, are produced from a common magmatic source; they are also held responsible for the formation of mineralizing solutions and vapours which create the mineral veins. / magmatische differentiatie / différenciation magmatique / magmatische Differentiation / diferenciación magmático

diagenetic differentation, – separation and migration of solid and fluid phases during diagenesis. / diagenetische differentiatie / différenciation diagenetique / diagenetische Differentiation / diferenciación diagenético

3939
liquation, – a process of differentiation in which two immiscible liquids separate from their common solution, e.g. from a magma; it is held responsible for the formation of some magmatic sulphide deposits. (The term has also been applied to the separation of residual liquid from crystals already formed). / liquide ontmenging / liquation; séparation de liquides non miscibles / liquide Entmischung; liquidmagmatische Entmischung / inmiscibilidad líquida

3940
rock opening; ground preparation, – forming of cavities that serve as conduits for mineralizing solutions (plumbing system) and as place of deposition of epigenetic mineral deposits. / gesteenteholte / cavité / – – / cavidad en las rocas

Bateman (1955) distinguished:

original cavities
- pore spaces
- crystal lattices
- vesicles or 'blow holes' (in lavas)
- lava drain channels
- cooling cracks
- igneous breccia cavities
- bedding planes

induced cavities
- fissures

- faulting
- shear-zone cavities
- cavities due to folding and warping
- volcanic pipes
- tectonic breccias
- collapse breccias
- solution caves
- rock alteration openings

cavity filling, – deposition of minerals from solutions in rock openings in contradistinction to replacement. / holte-opvulling / remplissage de cavité / – – / relleno de cavidades; relleno de huecos

3941
feeder; feeding channel,– the channel along which mineralizing solutions move up from depth. / voedingskanaal / fissure nourricière / Zufuhrkanal / vía de aporte

3942
accumulation, – in relation to mineral deposits, the amassing of certain minerals through naturally occurring processes and leading to the formation of a mineral deposit. /ophoping; concentratie / concentration / – – / concentración

relative accumulation; residual ...., – the increase in content of certain elements as the result of removal of most other elements originally present in the material or horizon under consideration.

absolute accumulation, – the increase in content of certain elements resulting from the supply from outside sources after transport over perceptible distances.

3943
alluvial mineral concentration, – concentration of weathering-resistant, mostly heavy, mineral grains through the elutriation or washing away of the surrounding finer, less resistant and lighter mineral particles by the action of running water. / alluviale mineraal concentratie / concentration (de minerai) alluvionnaire / alluviale Bildung von Minerallagerstätte / concentración aluvial

residual accumulation; .... ore formation, – the in situ concentration of certain elements at the surface through weathering and selective leaching of the local bedrock with simultaneous removal of other elements by surface and ground-water flow, e.g. laterite, bauxite. / residuaire aanrijking / concentration résiduaire / residuale Anreichung / concentración residual

3944
mechanical ore formation, – concentration through sorting of heavy, chemically resistent minerals.

3945
enrichment, – the additional deposition of economically valuable minerals in an existing rock with an initially low content in economically valuable minerals. / verrijking / enrichissement / Anreicherung / enriquecimiento

supergene enrichment; cementation, – in the description of mineral deposits indicating the filling of interstices in porous or shattered rocks with sulphides of supergene derivation below a zone of weathering and leaching.

3946
magmatic formation of a mineral deposit, – in a stricter sense the mineral deposits formed in intimate connection with the process of consolidation of a magma, including disseminated crystallization without concentration in the early stages, crystal segregations, crystallization of (injected) residual liquids and crystallization from immiscible liquid separations and accumulations. In a wider sense, all mineral deposits formed in connection with igneous activity. / magmatische delfstofafzetting / gisement magmatique / magmatische Lagerstätte / formación magmática de yacimientos

3947
dissemination, – a term applied to dispersed crystallizations of early formed crystals of ore minerals in deep-seated magmas, but also to dispersed formations of ore minerals in a more general sense. / verspreide uitkristallisatie; doorspikkeling / dissémination / disperse Auskristallisation; Einsprengung / diseminacio

3948
magmatic segregation, – the accumulation in place, before or after crystallization of a magmatic differentiate or immiscible liquid. / segregatie; uitscheiding / ségrégation / Segregation; Seigerung; Sonderung; Ausscheidung / segregación magmática

liquid gravitative accumulation, – the gravitative sinking of a heavy residual liquid to the bottom of a magma chamber; a process sometimes held responsible for the formation of some kinds of strictly magmatic ore deposits. / gravitatieve uitscheiding in vloeibare toestand / ségrégation magmatique par gravité / flüssige Abseigerung / concentración gravitativa

3949
injection, – the intrusion or penetration in veins and dykes of hot, sulphidic melts into the silicate host rock. / injectie / injection / Injektion / inyección

3950
pneumatolysis, – the process whereby minerals are produced from, and existing rocks are altered by, volatile components and gaseous emanations from solidifying magma. In European literature the term is used for processes under supercritical conditions, between about 600° and 400°C. According to R. Bunsen, also the fumarolic alteration of rocks. / pneumatolyse / pneumatolyse / Pneumatolyse / (p)neumatolisis

3951
emanation; exhalation,– the release of gaseous products (from magmatic fluids; also the products themselves. / emanatie; exhalatie / émanation; exhalation / Exhalation; (Emanation) / exhalación

3952
sublimation, – in mineralogy the transportation of elements and compounds as hot vapours and the subsequent deposition at reduced temperatures, of one or several mineral phase(s). / sublimatie / sublimation / Sublimation / sublimación

3953
hydrothermal process, – the formation of mineral deposits and rocks through the action of hot water; more specifically related to the cooling stage of a magma where the hypogene

aqueous fluids have a temperature below 500°C (American authors) or 400°C (European authors). / hydrothermale ontstaanswijze / processus hydrothermal / hydrothermale Entstehung / proceso hidrotermal

3954
hydrometasomatism; hydrometamorphosis, – alteration of rocks or minerals by mostly thermal-aqueous solutions, involving addition, removal or exchange of material. / hydrometasomatose / hydrométasomatose / Hydrometasomatose / alteración hidrotermal

3955
impregnation, – the introduction of minerals into an existing rock through metasomatic processes or by the filling of pore spaces and other cavities, mostly in a diffused or disseminated distribution. / impregnatie / impregnation/ Imprägnation / impregnación

infiltration, – the movement of aqueous solutions into a rock through its interstices and fractures. / infiltratie / infiltration / – – / infiltración

filtrational, – according to Smirnov, said of an ore-forming fluid or mineralizer composed of non-magmatic subsurface water.

3956
metasomatism; replacement, – a common phenomenon in the formation of mineral deposits. / metasomatose; vervanging; verdringing / métasomatose; substitution; remplacement / Metasomatose; Verdrängung; Ersetzung; Umwandlung; Umbildung; Austausch; Wechselreaktion; Substitution / metasomatismo

3957
autohydration, – the production of new minerals in an igneous rock, by the action of its own magmatic water, on already existent liquid magmatic minerals. / autohydratie / autohydratation / Autohydratation / autohidratación

autometasomatism; self-replacement,– the replacement of early formed minerals in an igneous rock by new minerals through the action of its own mineralizing agents. / autometasomatose / autométamorphisme / Autometasomatose / autometasomatismo

autopneumatolysis – the production of new minerals in an igneous rock by the action of its own gaseous mineralizing agents, e.g. the formation of pneumatolytic minerals in association with liquid magmatic ore minerals. / autopneumatolyse / autopneumatolyse / Autopneumatolyse / autoneumatolisis

3958
hydatogenesis; hydatomorphism, – the process whereby rocks and mineral deposits are formed from aqueous solutions, whether the water is magmatic or sedimentary. / hydatogenese / hydatogenèse / Hydatogenese / – –

hydatopneumatogenesis: ...pneumatolysis, – the process whereby rocks and mineral deposits are formed by both aqueous and gaseous agents. / hydatopneumatogenese / – – / – – / – –

hydatopyrogenesis, – the process whereby rocks and mineral deposits are formed by both aqueous and igneous agents. / hydatopyrogenese / hydatopyrogenèse / – – / – –

3959
rejuvenation, – the process by which the paragenesis of a mineral deposit is changed into a younger paragenesis characteristic for a higher temperature of formation. / verjonging; rejuvenatie / réjuvénation / Verjüngung; Rejuvenation / rejuvenecimiento

3960
hysterogenous, – pertaining to the formation of a mineral deposit on the earth's surface from the debris of other rocks.

3961
wall-rock alteration, – the alteration induced in the wall rock of mineralized veins as a result of increased temperature and the migration of substances, in part newly introduced. Besides the four main types listed below, it is customary, but not recommended, to name the process in its more advanced stage after the main mineral formed, e.g. alunitization, albitization. / omzetting; vervanging / altération / Umsetzung; Verdrängung / alteración

argillic alteration, – the development of clay minerals and related minerals at the expense of intermediate and calcic plagioclase;

potassium alteration, – new formation of K silicates, as feldspars and micas;

propylitic alteration, – the new formation of a mineral assembly containing albite, chlorite, epidote (incl. zoisite and clinozoisite) and carbonate and several other minerals including sulphides, particularly in andesitic and related rocks;

sericitic alteration, – the new formation of sericite at the expense of K-feldspar and chlorite, often accompanied by new quartz and pyrite.

3962
listvenitization, – a Russian term denoting an alteration with new formation of Mg-Fe carbonates, sericite and pyrite replacing dark silicates and feldspars and resulting in a quartz-carbonate-sericite assembly containing sulphides; particularly in basic and ultrabasic rocks.

beresite, – a Russian term denoting an alteration product, closely related to sericite alteration, composed of sericite and pyrite.

3963
synantetic, – pertaining to the formation of a new mineral between two other minerals by the interaction between the latter; as in reaction rims between e.g. silicates and oxidic minerals in a liquid magmatic mineral deposit. / synantetisch / synantétique / synantetisch / – –

3964
sulphurization, – the conceptual process of the formation of sulphides through the reaction between cations such as iron, nickel and copper in solid solution in common rock-forming minerals or in an igneous magma, and sulphur from an external source. / – – / sulfuration / – – / sulfurización

3965
sequential deposition, – the apparent chronologic succession in which the formation of the various minerals of a paragene-

sis takes place. / opeenvolgend / succession minéralogique / – – / deposición secuencial

telescoping, – overlapping of minerals formed in one zone by those formed in another, generally hotter zone. Restricted largely to deposits formed under shallow conditions, where changes in temperature and pressure are rapid. / – – / teléscopage / Teleskoping / telescoping; solape

[2630] CLASSIFICATION

[2631] *Related to Time and Depth of Formation*

3966
syngenetic; idiogenous, – formed contemporaneously with the enclosing rocks, hence very likely under similar conditions. / syngenetisch / syngénétique / syngenetisch / singenético

epigenetic, – of later origin than the enclosing rocks, hence very likely formed under different conditions. / epigenetisch / épigénétique / epigenetisch / epigenético

diplogenetic, – as used by Lovering, the composing elements are partly syngenetic and partly epigenetic.

3967
supergene; descendent; hypergene (Russian authors), – relating to processes, active at the surface in the zone of weathering and of descending waters. / supergeen / supergène / supergen / supergénico

hypogene; ascendent, – relating to processes, active below the zone of weathering, implying ascending waters. / hypogeen / hypogène / hypogen / hipogénico

[2632] *Related to the Source of the Material*

3968
authigenous; authigenic; indigenous (Routhier), – terms meaning taken form – at whatever time – in the place where it is now found and where the surrounding rock was also formed. / authigeen / authigène / authigen / autóctono

allogene; allothigene, – terms indicating that material has been introduced from outside the host rock of which it now constitutes a part. / allogeen / allogène / allothigen / alóctono

3969
exogene; allothigenic; allogene (Routhier), – in metallogeny, according to Amstutz, formed from material derived from outside the enclosing rock.

endogene; authigenic; indigenous (Routhier), – in metallogeny, according to Amstutz, formed from material derived from within the enclosing or host rock

3970
consanguineous, – in metalogeny, according to Routhier, a deposit whose material derives from – more or less – the same source as that of the material making up the host rock, independent of the time of its crystallization. / verwant / familier / verwandt / consanguíneo

foreign, – in metalogeny, according to Routhier, a deposit whose material derives from outside the host rock, at whatever time, and the source of which is much different from that of the material making up the host rock. / vreemd / étranger / – – / foráneo

3971
lithogene, – according to Lovering formed by concentration of elements derived from surrounding rock masses by the process of lateral secretion or regional metamorphism.

terrigene, – according to Routhier formed by concentration of elements derived from neighbouring continental areas as result of weathering and erosion.

inherited, – according to Roger, a term indicating that the metallic concentrations are derived from an old store of metal ions present locally and that has been reworked during one or more periods of geological history. / geërfd / hérité / – – / residual

[2633] *Various Bases of Classification*

3972
primary deposit, – an often misleading term for deposits unaffected by supergene enrichment. / primaire afzetting / gîte primaire / primäre Lagerstätte; protogene .... / depósito primario

secondary deposit, – an often misleading term for deposits resulting from supergene enrichment and alteration. / secundaire afzetting / gîte secondaire / sekundäre Lagerstätte; deuterogene .... / depósito secundario

NOTE:
– the term 'secondary' is also used in the sense of 'younger' when describing hypogene mineral associations.

3973
pyrogenic (obsolete), – applied to deposits related to the crystallization of magma. / pyrogeen; primair-magmatisch / magmatique / pyrogen; primärmagmatisch / – –

magmatogenic, – applied to mineral deposits that originated through magmatic processes in the widest sense. / magmatogeen / magmatique / – – / magmatogénico

3974
diagenetic, – applied to mineral deposits generated through processes during lithification and before metamorphism. / diagenetisch / diagenétiqué / – – / diagenético

metamorphogenic, – applied to mineral deposits generated through metamorphic processes that caused the concentration of certain elements. / – – / métamorphogène / – – / de origen metamórfico

3975
magmatic mineral deposit, – deposits resulting from the solidification of a magma; they belong mainly to two separate periods in the cooling history:

early magmatic deposit, – formed during the early stage of consolidation of generally basic magmas. / vroeg-magmatische afzetting / gisement magmatique de cristallisation précoce / frühmagmatische Lagerstätte / yacimiento magmático de cristalización precoz /

Distinction is made between:

disseminated deposit, – formed through scattered crystallization of valuable minerals without local concentration. / spikkelerts / gîte de dissémination / – – / – –

magmatic segregation deposit, – formed through crystal segregations. / magmatische segregatie-afzetting / gîte de ségrégation magmatique / magmatische Segregationslagerstätte / segregación magmático

magmatic injection deposit, – crystallization from immiscible liquid accumulations injected in the country rock. / geïnjecteerde magmatische afzetting / gîte d'injection magmatique / Abpressungslagerstätte; abgepresste Erzinjektion / depósito de inyección

late magmatic deposit, – formed during the late stage of consolidation from (injected) residual liquids after the crystallization of the main silicate minerals and with transitions to the pneumatolytic and pegmatitic deposits. / laat-magmatische afzetting / gîte magmatique de cristallization tardive / spät magmatische Lagerstätte / yacimiento magmático de cristalización tardía

3976
liquid-magmatic; orthotectic; orthomagmatic, – various terms for mineral deposits formed by concentration in molten magma during the early stages of its consolidation; mostly of basic affiliation. / liquid-magmatisch; orthotektisch / orthomagmatique / liquidmagmatisch; orthotektisch; orthomagmatisch / ortomagmático

3977
pneumatolytic; pneumatogenic, – said of a deposit formed by gaseous agents implying a relation with the latest or pneumatolytic phase of consolidation of a generally granitic magma. / pneumatolytisch / pneumatolitique / pneumatolitisch / (p)neumatolítico

3978
pegmatitic, – said of a deposit formed during the late or pegmatitic stage of consolidation of a generally granitic magma. / pegmatitisch / pegmatitique / pegmatitisch / pegmatítico

3979
emanation deposit, exhalation ...., – a mineral deposit of gaseous magmatic origin. These deposits are subdivided according to the circumstances under which the emanations occurred:
aeric, subaeric, – in the athmosphere;
subaquatic, – under water:
epicrustal, – at shallow depth in the earth crust;
hypabyssal, – at intermediate depth;
abyssal, – at great depth.
/ ematogene afzetting; exhalatieve .... / gisement exhalatif / Exhalationslagerstätte / depósito exhalativo

3980
contact-metamorphic deposit; pyrometasomatic ...., – mineral deposits formed at or near the contact of a magma with certain kinds of invaded rocks, very often limestones; characterized by an assemblage of distinctive high-temperature, pneumatolytic minerals. / contactmetamorfe afzetting; pyrometasomatische .... / gîte pyrometasomatique / Kontaktlagerstätte; kontaktpneumatolytische Verdrängungslagerstätte; pyrometasomatische Lagerstätte / depósito metamórfico de contacto

3981
pneumotectic deposit, – a term for transitional forms between straight magmatic and hydrothermal mineral deposits, in which the effects of mineralizers are very apparent. / pneumotektische afzetting; pneumatolytische ....; magmatisch-pneumatolytische .... / gîte pneumatolytique / pneumotektische Lagerstätte; liquidmagmatisch-pneumatolytische Übergangslagerstätte / depósito (p)neumatolítico

3982
hydrothermal deposit, – a mineral deposit produced by hot ascending emanations rich in water, according to the original meaning, of any derivation. / hydrothermale afzetting / gîte hydrothermal / hydrothermale Lagerstätte / depósito hidrotermal

hydrothermal s.s.; fugitive (H. Schmitt), – water is derived from magmatic sources.

hydatogene (Zirkel, Maucher), – water is derived from nonmagmatic sources.

3983
hypothermal (W. Lindgren); katathermal; high-hydrothermal, – said of a mineral deposit (excepting contact-metamorphic deposits) formed under the highest temperature conditions of the hydrothermal process (about 300° to 500°C) and at high pressure, generally in genetic connection with intrusive rocks that consolidated at no excessive depth. / hypothermaal; katathermaal / hypothermal; katathermal / hypothermal; katathermal; hochthermal / hipotermal

mesothermal (W. Lindgren); middle-thermal, – said of a mineral deposit formed under the intermediate temperature conditions of the hydrothermal process (about 175° to 300°C) and generally in genetic connection with intrusive rocks which consolidated at a rather considerable depth. / mesothermaal / mésothermal / mesothermal; mittelthermal / mesotermal

epithermal (W. Lindgren); low-hydrothermal, – said of a mineral deposit formed under the lowest temperature conditions of the hydrothermal process (below about 175°C), often near the surface and in Tertiary effusive rocks of mostly intermediate to acid composition. / epithermaal / épithermal / epithermal; niederthermal / epitermal

3984
leptothermal (L.C. Graton), – said of a mineral deposit formed under conditions about intermediate between those characteristic for the mesothermal and epithermal zones. / leptothermaal / leptothermal / leptothermal / leptotermal

xenothermal (A.F. Buddington), said of a mineral deposit formed under high temperature conditions at shallow to moderate depth. / xenothermaal / xénothermal / xenothermal / xenotermal

telethermal, (L.C. Graton); telemagmatic hydrothermal (P. Niggli), – said of a hydrothermal mineral deposit formed at relatively low temperature and near the surface, particularly occurring in limestones. / telethermaal / téléthermal / magmafern hydrothermal / teletermal

3985
intramagmatic, – said of a mineral deposit developed mainly inside its eruptive parent rock. / intramagmatisch / intramagmatique / intramagmatisch / intramagmático

extramagmatic, – said of a mineral deposit developed outside the direct sphere of Influence of an igneous rock mass i.e. including tele-, crypto-, apo-, and perimagmatic deposits. / extramagmatisch / – – / – – / extramagmático

perimagmatic, – said of a mineral deposit developed mainly beyond the rim of an igneous intrusion. / perimagmatisch / périmagmatique / perimagmatisch / perimagmático

3986
apomagmatic, – said of a mineral deposit of magmatic origin developed in surroundings that do not reveal its immediate relation to a body of igneous parent rock; the existence of the latter may, however, still be derived from the existence of dykes, phenomena of contact metamorphism, etc. / apomagmatisch / apomagmatique / apomagmatisch / apomagmático

cryptomagmatic, – said of a mineral deposit of magmatic origin developed in surroundings that do not reveal in any way the relationship to a body of igneous parent rock. / cryptomagmatisch / cryptomagmatique / kryptomagmatisch / criptomagmático

telemagmatic, – said of a mineral deposit of apparently magmatic origin but far removed from an igneous source. / telemagmatisch / télémagmatique / telemagmatisch / telemagmático

3987
marginal deposit, – a term applied to magmatic segregation deposits at the bottom and periphery of the intrusive parent rock, especially to Ni-Cu sulphide deposits of this kind at Sudbury (Canada). / bodemrandstandige afzetting / (gîte de ségrégation périphérique) / Bodensatzlagerstätte / depósito periférico

offset deposit, – a term applied to magmatic segregation deposits of a mixed magmatic and hydrothermal character that were injected into the country rock at a moderate or small distance from the parent rock, especially to Ni-Cu sulphide deposits of this kind at Sudbury (Canada). / – – / gîte de départ immédiat / – – / depósito 'offset'

3988
plutonic, – said of a mineral deposit of magmatic origin that has been formed under abyssal conditions. / plutonisch / plutonique / plutonisch / plutónico

3989
subvolcanic, – according to European terminology said of a mineral deposit of magmatic origin formed at moderate or shallow depth, i.e. under epicrustal conditions, very often in Kenozoic lavas; a term sometimes substituted by **volcanic**. / subvulkanisch / sub-volcanogénique / subvulkanisch / subvolcánico

volcanic, – said of a mineral deposit of magmatic origin, designated as 'young' by European students, and that has been formed under near-surface conditions and very often in Tertiary or younger volcanic rocks. In a strict sense, the deposit was formed in relation to surface and submarine eruptions. / vulkanisch / volcanogénique / vulkanisch / volcánico

3990
batholith-related deposit, – W. M. Emmons classified the lode deposits associated with batholiths on the basis of erosion, depth and the resulting presence of the barren core, the mineralized hood along the upper contacts and the roof of the batholith. His six classes are:

cryptobatholitic, – dykes, sills and mineral veins in the roof of the supposed but not exposed batholith;

acrobatholitic, – mineral deposits in and around the exposed cupolas composed of the mineralized hood; the barren core is not yet exposed;

epibatholitic, – mineral deposits in and around the outer rim (mineralized hood) of small exposures of the intrusive with barren core material in the centre;

embatholitic, – mineral deposits in and around the outer rim of large exposures of the batholith, the latter equalling the surface area occupied by roof pendants and hostrock fingers;

endobatholitic, – mineral deposits outside and in the narrow outer rim of large exposures of the batholith, which occupies a surface area much larger than that of the roof pendants;

hypobatholitic, – mineral deposits present in the deeply eroded parts of the batholith composed of barren core with only few small remnants of roof pendants.

3991
replacement deposit, – deposit mainly formed through metasomatic or replacement processes. / metasomatische afzetting; vervangings.... / gîte de remplacement; gîte métasomatique / Verdrängungslagerstätte / depósito de sustitución

3992
impregnation deposit, – deposit mainly formed through the process of impregnation. / – – / gîte d'impregnation / Imprägnationslagerstätte / depósito de impregnación

3993
injection deposit, – deposit mainly formed through the process of injection. / – – / gîte d'injection / Eindrängungslagerstätte; Spalten....; vererzte Ruschelzone / depósito de inyección

3994
secondary hydrothermal; (epirogenetically) regenerated, – said of a mineral deposit formed by metal displacements from deposits existing in the local basement or formed in an earlier phase of orogeny, e.g., the formation of an Alpine mineral deposit through the re-arrangement or displacement of Hercynian mineral concentrations. / secundair-hydrothermaal / régénéré / sekundär-hydrothermal; (epirogenetisch) regeneriert / hidrotermal secundario

3995
topomineral, – term indicating that a spatial, or topographic,

relationship between mineralization and a certain rock type or structure, is essential to the paragenesis and the shape of a mineral deposit. / topomineraal / topominéral / topomineralisch / topomineral

topomineralische Reaktionslagerstätte, – a German term used by H. Schneiderhöhn to denote mineral deposits whose shape and composition are largely determined by metasomatic reactions of the mineralizing agencies with susceptible rocks or mineral assemblages of the direct surrounding (limestones, basic rocks).

3996
dépot de couverture, –
1. According to Routhier a sulphide-containing mineral deposit occurring as part of sedimentary formations, in particular present a short distance above the abraded basement;
2. in a more general sense all mineral deposits present as part of the sedimentary formations laying on top of the local basement.

3997
metamorphic, – said of a mineral deposit that has been metamorphosed after the original mineral concentration came into being, with the resulting changes in texture, structure and mineralogy. / metamorf / métamorphisé / metamorph / metamórfico

detrital, – said of a mineral deposit derived from weathering detritus through the process of mechanical ore formation. / detritisch / détitique / – – / detrítico

placer, – a detrital deposit formed by the action of running water or wind, in particular used for alluvial gold, diamond and cassiterite deposits. / placer / placer / Seife / placer

3998
residual; .... placer, – said of a mineral deposit formed through residual accumulation. / residuair / résiduel / Residual; .... Trümmererz; Rückstandserz / depósito residual

infiltration deposit; metathetic ....; ground-water ...., – Russian terms for supergene weathering-related deposits, the economic minerals of which are derived from leaching and, after underground transport have been deposited elsewhere.

3999
eluvial deposit; .... placer, – a residual mineral deposit formed in situ. / eluviale ertsafzetting / gîte éluvionnaire / eluviale Lagerstätte; .... Seife / depósito eluvial

NOTE:
– the cassiterite-bearing eluvial deposits of Bangka, Belitung are called **kulit** (Malayan).

4000
alluvial deposit; stream placer, – a mineral deposit formed by alluvial mineral concentration. / alluviale ertsafzetting / gîte alluvionnaire / alluvial Lagerstätte; fluviatile Seife; allluviale .... / depósito aluvial

NOTE:
– the cassiterite-bearing alluvial sands of Bangka, Belitung are called **kaksa** (Chinese), in the Malay Peninsula **karang**. They mostly rest on a granitic or sedimentary bed rock or **kong** (Chinese), and they are mined in open pits called **kolong** (Malayan).

4001
Depending on the geomorphological position, mineral deposits transitional between eluvial and alluvial are recognized. Smirnov classifies these as follows:

deluvial placer, – formed in scree or talus;
proluvial placer, – formed in colluvial deposits;
terrace placer, – formed in river terrace deposits.

4002
evaporitic, – said of mineral deposits, originated through evaporation e.g. K and Mg salts in the last stages of sea-water evaporation or Na from seeping subsurface waters.

[2640] STRUCTURE AND SHAPE OF MINERAL DEPOSITS

[2641] *General Terms*

4003
massive ore, –
1. almost entirely composed of ore minerals, while the gangue minerals form a negligible portion;
2. also used to denote ore deposits with massive structure or of undefined massive shape.

/ massief erts / amas minéralisé / massiges Erz; Derberz; massige Erzlagerstätte / mena masiva

4004
disseminated ore, – the ore minerals form a small proportion of the ore and are scattered through the gangue as small specks, blebs or grains, generally accompanied by small veinlets. / spikkelerts / minerai disseminé / eingesprengtes Erz / mena diseminada

4005
ore shoot; ore chute, – richer portion of an ore deposit; generally forming bands or columns more or less steeply inclined in the plane of an ore vein or lode and of considerably larger size than ore pockets and ore bunches. / rijke ertszuil / colonne minéralisée / Erzfall; Erzmittel; Reicherzzone; Adelsvorschub; Adelszone / zona de enriquecimiento

4006
course of ore, –
1. a tabular, more or less horizontal mass of ore of unusually large extent and richness;
2. a tabular ore shoot;
3. a rich ore horizon or ore zone.

4007
bonanza, – an exceptionally rich ore shoot or bunch of ore, particularly with reference to gold and silver.

Edelfall; edle Geschicke, – old German terms for shoots of precious metal ore.

4008
ore horizon; ore zone, – a bed, horizon or zone of country rock that has been favourable to mineralization and in which many ore occurrences are located. / ertshorizon; ertszone / horizon minéralisé; niveau .... / Erzhorizont; Erzzone / horizonte mineralizado

4009
wall; cheek, – the portions of the country rock bounding the ore body. / wand / éponte; paroi / Wände; Nebengestein / hastiales

hanging wall, – the rock mass above an inclined ore vein or lode. / dak / toit / Hangende / hastial superior; techo

foot wall; footwall; lying wall, – the rock mass below an inclined ore vein or lode. / vloer / mur / Liegende / hastial inferior; muro

4010
roof, –
1. the layer directly above an ore bed;
2. the wall above a horizontal, stratiform ore body.

/ dak / toit / Dach / techo

floor, –
1. the layer directly beneath an ore bed;
2. the wall below a horizontal, stratiform ore body.

/ vloer / mur / Sohle / muro

4011
channel; channelway, – the channel along which mineralizing solutions or liquids have moved into the site of deposition of an ore body; the channels or feeders may be observed in the form of more or less mineralized fissures, veins or pipes. / voedingskanaal; toevoerkanaal / fissure nourricière / Zufuhrkanal / vía de aporte

4012
salband; selvage; .... clay; flucan; gouge; pug, –
1. a layer or sheet of argillaceous material extending along one or both of the walls of an ore vein between the vein itself and the enclosing wall rock;
2. the terms selvage and salband are sometimes used to denote the portion of the ore vein in contact with the wall rock.

/ kleiband / salbande (argileuse) / Salband; Besteg; Lettenbesteg; Ganglette / salbanda

4013
frozen wall; frozen to the wall; frozen to the country, – pertaining to an ore vein, whose material is in direct contact with, and closely adheres to the country rock or walls. / vergroeide wand; aangebakken wand / éponte soudée / fest aufgewachsenes Salband / hastial sin salbanda

free wall, – pertaining to an ore, whose material is separated from the walls by selvage or gouge; the vein filling may scale off cleanly from the gouge. / vrije wand / – – / freies Salband / hastial con salbanda

4014
vug; vugh; druse, – hollows or openings in mineral veins; the walls are usually encrusted with mineral layers of different composition. / gangholte / géode; cavité; druse / Druse / géoda; drusa

4015
zoned ore deposit, – an ore deposit characterized by two or more, mostly concentric or almost concentric zones, consisting of ores of different structure, composition, mineral or metal contents. / gezoneerde ertsafzetting / gîte zoné / zonare Erzlagerstätte; zonierte .... / yacimiento zonado

[2642] *Attitude of Ore Bodies*

4016
strike; course; direction of the strike, – the direction of a horizontal line in the plane of an ore vein or a tabular or lenticular ore body; measured with reference to a meridian. / strekking / direction / Streichen / rumbo

Stehende, – old German miner's term indicating an ore vein striking between N-S and NE-SW; used as suffix, e.g. Frisch-Glück-Stehende.

Morgen; Morgengang, – old German miner's terms indicating an ore vein striking between NE-SW and E-W; used as suffix, e.g., Gottlob-Morgengang.

Flache, – old German miner's term indicating an ore vein striking between E-W and NW-SE; used as suffix, e.g., Laura-Flache

Spaat; Spat, – old German miner's term indicating an ore vein striking between NW-SE and N-S; used as suffix, e.g., Halsbruchner-Spat.

4017
dip, – the vertical angle between a horizontal plane and the plane of an ore vein or tabular or lenticular ore body. / helling / pendage / Einfallen; Fallen / buzamiento

4018
hade; underlie, – the angle between the vertical and the plane of an ore vein or tabular or lenticular ore body, i.e. the complement of the dip. / hellingscomplement / complément du pendage / lotrechte Abweichung / complemento de buzamiento

plunge, – the vertical angle between the horizontal and the line of maximum elongation of an elongated ore body. / duiking / plongement / Abtauchen / hundimiento

4019
pitch; rake, – the inclination of an elongated ore shoot within a vein; the pitch is measured in the plane of the vein by the angle between the axis of the ore shoot and the strike of the vein. / duiking / plongement; pitch / Gefälle; Einschieben; Einschiessen / pitch

4020
heave, – old English miner's term indicating the horizontal distance between corresponding traces of a disrupted vein measured at right angles to the fault strike, i.e. the horizontal separation of the parts of a displaced vein.

slide; leap, – old English miner's terms to designate the throw of a displaced vein, i.e. the vertical separation of the parts of a displaced vein.

[2643] *Bedded Ore Deposits*

4021
bedded ore deposit; bedded ore; ore bed,– an ore deposit of sedimentary origin that is interbedded with other sediments, or is situated on the earth surface as an alluvial or other superficial accumulation. / ingeschakelde ertsafzetting; laagafzetting / gîte interstratifié; lit minéralisé; couche .... / eingelagerte Erzlagerstätte; Flöz / yacimiento interestratificado

NOTE:
– True bedded deposits are always sediments and are younger than the rocks below them and older than those above, if any such exist; they include placers and other alluvial deposits, as well as e.g. sedimentary iron, manganese and sulphide, besides a great variety of non-metallic deposits.

4022
concordant ore deposit, – an ore deposit, the contacts of which are parallel to the bedding of the adjacent rocks. / concordante ertsafzetting / gisement concordant / konkordantes Erzlager / depósito concordante

peneconcordant ore deposit, – an ore deposit that in general is concordant to the bedding of the adjacent rocks, but that in detail cuts across it, being nearly concordant to the bedding. / peneconcordante ertsafzetting / gisement peneconcordant / penekonkordantes Erzlager / depósito penecordante

discordant ore deposit, – an ore deposit, the contacts of which cut across the bedding of the adjacent rocks. / discordante ertsafzetting / gisement discordant / diskordantes Erzlager / déposito discordante

4023
stratified ore deposit; layered ...., – an ore deposit showing stratification, layering or bedding or forming part of a stratified, layered or bedded sequence of rocks; the strata, layers or beds may be of sedimentary or magmatic origin. / gelaagde ertsafzetting / gîte stratifié / geschichtete Erzlagerstätte / depósito estratificado

4024
stratiform ore deposit, – an ore deposit of any origin having the shape of a concordant stratum or layer in a sequence of rocks. / stratiforme ertsafzetting; laagvormige .... / gîte stratiforme / stratiforme Erzlagerstätte; flözartige .... / depósito estratiforme

stratabound ore deposit, – an ore deposit whose location in a certain stratum or in a certain stratigraphic position is implied by the origin of the deposits; they may include syngenetic as well as epigenetic deposits of diverse shape and form. / laaggebonden ertsafzetting / gîte stratoïde / schichtgebundene Erzlagerstätte / depósito estratoide

4025
tabular ore deposit, – an ore deposit of tabular or sheet-like form; in older classifications tabular deposits include all deposits now commonly referred to as stratified deposits. / – – / gisement tabulaire / – – / depósito tabular

4026
lenticular ore deposit; .... ore body; .... vein, – an ore deposit, ore body or vein showing lenticular shapes, occurring as a single lenticular mass or in a series of smaller lenticles, that follow one another in such a way as to constitute an interrupted vein. Lenticular veins usually occur concordantly intercalated in foliated and deformed rocks as phyllites and slates, where they form a series of lenticles, pinching out in short distances or showing pinch-and-swell structure. / lensvormige ertsafzetting; ertslens ; lensvormige ader / gisement lenticulaire; lentille / linsenförmiger Erzkörper; Erzlinse; Linsengang / depósito lenticular.

4027
blanket, – a miner's term for a horizontal, tabular ore body.

manto, – a miner's term derived from Spanish and denoting a flat-lying bedded ore deposit.

4028
ore roll; roll; roll-type deposit,– a curved layer of ore that is not concordant with the enclosing strata, but assumes discordant C-shaped or S-shaped cross sections. The dominant form of some uranium deposits in sandstones are referred to as roll-type deposits. / ertsrol / – – / – – / – –

4029
ore channel, –
1. an ore body elongated in channel-like shapes that pinches and swells, branches, changes dip or plunges, and erratically develops bulges above, below or to either side of the main trend; an often observed shape of uranium ore bodies believed to have formed in stream channels (**palaeostream channel, palaeochannel**) in fluviatile sandstones, e.g. on the Colorado Plateau, USA;
2. a little used term for a lode, referring to both the ore and gangue minerals; use in this sense is not recommended.

/ ertskanaal / chenal minéralisé / Erzkanal / paleocanal

4030
run, – very elongated irregular ore bodies occurring along certain channels or brecciated, fractured or other zones in the plane of stratification of certain beds; common form of ore bodies in limestones and dolomites in the lead-zinc districts of central USA.

4031
ore sill, – a liquid-magmatic ore body (i.e. concentrated in molten magma) emplaced in the form of a sill. / intrusieve ertsplaat / gisement en sill / intrusives Erzlager / sill mineralizado

ore dyke, – a liquid-magmatic ore body emplaced in the form of a dyke. / intrusieve ertsgang / gisement en dyke / intrusiver Erzgang / dique mineralizado

vein dyke, – a liquid-magmatic or pegmatitic ore body emplaced in the form of a vein or dyke.

[2644] *Ore Masses*

4032
(ore) mass, a general term used to describe ore bodies of widely differing size, lateral extension and form, but of no defined shape. / ertsmassa / amas de mineral; amas minéralisé / Stock / masa

ore stock; ore chamber; ore pocket; – terms in common use to describe larger ore masses. / – – / poche de minerai; poche minéralisé / Stock, Butz; Tasche / bolsada

(ore) bunch, ore blow; ore nest; ore kidney,– terms in common use to describe smaller ore masses. / nest / mouche / Butz; Nest / nido

4033
podiform ore deposit, – an ore deposit showing the form of pods; chromite deposits in Alpine-type ultramafics are predominantly of the podiform type, those in layered mafic intrusions predominantly of the stratiform type. / peulvormige ertsafzetting; podiforme ertsafzetting / gisement podiforme / schlauchförmige Erzlagerstätte / depósito podiforme

4034
(ore) pod, – a flat elongated ore body, usually of large dimensions. / ertslineaal / pod / Erzschlauch / – –

4035
sackform ore deposit, – an ore deposit in the form of rounded, but irregular masses. / zakvormige ertsafzetting / sacs / sackförmiger Erzlagerstätte / masa de mineral en sacos

4036
cuneiform; cuneate ore deposit; .... vein,– an ore deposit or vein characterized by a cross section showing the shape of an inverted wedge; usually believed to have been formed in tension fractures. / wigvormige ertsafzetting, ... ader / gîte cunéiforme; filon .... / – – / cuneiforme

4037
ore pipe; ore chimney,– a vertically elongated body of ore with a roughly circular or oval-shaped cross section. / ertspijp / cheminée minéralisée; pipe .... / Erzschlauch / pipa; chimenea

[2645] *Ore Veins*

4038
ore vein; metalliferous vein; lode; (lead), –

1. a sheet-like or tabular ore body, younger (epigenetic) than the enclosing wall rock. The terms vein and lode are often used almost interchangeably, but a vein usually has definite boundaries and a sharp distinction between vein filling and country rock is possible, whereas in a lode zones of brecciated material and replaced wall rock are often associated with the vein fillings in such a way that often a sharp boundary between lode and country rock cannot be recognized;
2. in the USA the term lode is occasionally used for composite veins;
3. the term lode is occasionally used to denote a mineral deposit in consolidated rock as opposed to a placer deposit.

/ ertsgang / filon; .... minéralisé; .... métallifère; gîte filonien / Gang; Erzgang; Trum(m) / filón metalífero; yacimiento filoniano

4039
veinlet, – a small or narrow vein. / ader / filonnet / veinule / Ader; Gänchen / filoncillo

(ore) stringer; ore tape; ore thread; ore ribbon, – small irregular ore veinlet or thin ore band having the shape of a stringer. / ertssnoer / filet minéralisé; stringer / Erzschnur; Erzfuge / venilla de mineral

4040
ore streak; ore schlieren, – irregular, more or less platy streaks of ore. / ertsslieren / traînées minéralisées / Erzschlieren / hiladas de mineral

4041
dilatation vein, – a vein occupying the space afforded by the opening or pulling apart of the vein walls, as contrasted with veins formed by wall-rock replacement. The term is sometimes also used to denote lenticular veins. / dilatatieader / filon de dilatation / – – / filón de dilatación

4042
replacement vein, – a vein formed by gradual transformation or metasomatic replacement of the wall rocks. / vervangingsgang / filon de substitution / Verdrängungsgang / filón de sustitución

4043
contact vein; .... deposit, – an ore vein or other ore body occurring at the contact of different rocks, often at the contact between a sedimentary and an igneous rock. / contactgang / – – / Kontaktgang / filón de contacto

4044
fissure vein; true vein; rake vein; right vein, – an ore vein filling a fissure of indefinite length and depth, and in general traversing the country rocks independent of their structure. / spleetgang / fissure / Spaltengang / filón en fisura

4045
composite vein; .... lode, – a series of more or less parallel ore veins, often discontinuous and forming lenticles in a shattered or shear zone. / samengestelde ertsgang / filon composé / zusammengesetzter Erzgang / filón compuesto

simple vein, – a vein occupying one fissure only. / enkelvoudige ertsgang / filon simple / einfache Gang / filón simple

4046
complex vein; .... lode, – a fissure vein showing two or more stages of opening and filling of the vein; complex veins include crustified, comby and other banded veins. / complexe ertsgang / filon complex / komplexer Erzgang / filón complejo

leader; .... vein; divider,–

1. a usually thin ore vein with the rock on either side of the vein impregnated with ore minerals to such extent as to render the whole worth working as a lode;
2. narrow crack in the middle of a lode, from which the mineralizing solutions spread out into the wall rocks.

4047
bunchy lode; .... vein, – pertaining to occurring in detached patches; each separate patch is called a bunch of ore.

4048
dradge lode; .... vein; brangled lode; .... vein, – a lode or vein in which the ore does not occur in courses, shoots or bunches, but is disseminated throughout the vein or gangue.

4049
banded vein; ribbon vein, – a vein showing bands parallel to the walls, consisting of layers of different mineral composition, grain size, structure or colour; the vein is said to show a banded or ribbon structure. / gebandeerde gang / filon rubané / gebänderter Gang / filón bandeado

crustified vein, – a banded vein composed of a succession of crusts of ore and gangue material, successively deposited as vein fillings; the crusts may show concentric, radial concretionary structures or comb structures consisting of a columnar development of minerals normal to the walls of the vein or crusts; crustified veins showing combs or comb structure are called **comby veins** or **comby lodes**. / overkorstingsader; .... met kamstructuur / filon concrétionné / Erzgang mit krustenförmige Struktur; .... mit Kammstruktur; krustenartiger Erzgang / – –

4050
brecciated vein, – a fissure filled with fragments of country rock and older vein material broken from the sides of the fissure and cemented by ore and gangue minerals. / brecciegang / filon brèchique / Brecciengang / filón brechoide

cockade ore; ring ore, – ore formed by deposition of successive crusts of minerals around breccia fragments in a vein. / kokardenerts / – – / Kokardenerz; Ringelerz / – –

chambered vein, – a vein or lode showing irregularities and brecciation, especially in and along the hanging wall. / kamervormig adernetwerk / poche filonienne / zusammengesetzter Gang im Sinne Cottas / bolsa filoniana

4051
Rasenläufer, – old German term for veins that pinch out at limited depth.

4052
fahlband, – sparse dissemination of sulphides following certain lines or band along the strike of schists and other metamorphic rocks. The German term 'fahl', meaning rusty brown, refers to the oxidized outcrops; fahlbands were first described from the silver district of Kongsberg, Norway, where they are several thousand metres long and between a tenth of a metre to over one hundred metres thick. / fahlband / fahlbande / Fahlband / fahlband

4053
saddle vein; .... reef, – saddle-shaped bedded ore vein localized in the crests of folds. The common form of gold veins at Bendigo, Australia. / zadelgang / gîte de charnière anticlinale / Sattelgang / filón en charnela anticlinal

trough vein, ... reef, – trough-shaped bedded ore vein localized in the troughs of folds; often occurring together with saddle veins. / troggang / gîte en gouttière; gîte de charnière synclinale / Muldengang / filon en charnela sinclinal

bed vein; bedded vein, – ore vein following the bedding planes in sedimentary rocks. / laaggang / filon-couche / Lagergang / filón capa

4054
pinch, – most veins exhibit irregularity in width; thin places in an ore vein or mineral zone. / insnoering; vernauwing / reserrement / Einschnürung; Verdrückung; Auskeilen / estrechamiento

swell, – thick places in an ore vein or mineral zone. / verdikking / gonflement; épaississement; élargissement / Anschwellung; Verdickung / ensanchamiento

pinch-and-swell structure, – the structure of a vein showing a succession of thin and thick places: / – – / filon en chapelets / Perlschnurartiger Gang / estructura arrosariada

4055
pitch-and-flat structure, – structure of a vein localized by bedding planes and steeply inclined fractures. The horizontal extensions of the vein along bedding planes are referred to as **flats** or **sheets;** the short-cutting branches along fractures connecting two flats are referred to as **pitches.**

step vein, – a vein alternately cutting through the strata of country rock and running parallel to them. / bajonetgang / – – / Hakengang; Stufengang / filón en bayoneta

4056
gash vein, – a term generally used to describe vein fillings of mineralized joints and solution openings, usually more or less vertical, in limestone; gash veins do not extend upward or downward beyond a given bed of the limestone.

4057
ladder vein, – more or less regularly spaced, short, transverse fissure fillings in dykes; the fissures extend roughly parallel to each other from wall to wall of the dyke, resembling the rungs of a ladder. / laddergang / filon en échelle / Leitergang / filón en escalera

4058
stockwork, –

1. an ore body formed by a mass of rock intersected by large numbers of small veins or stringers, sometimes more or less parallel, sometimes crossing and interlacing in all directions;
2. a mass of parallel or interlacing lodes, which occur in such a large number and close proximity that the intervening rock may be so highly mineralised that the whole can be worked as a single ore body;
3. the Carbonas of Cornwall, England, are essentially similar to stockworks but have also been classed as a disseminated deposit.

/ ertsadercomplex; (stockwerk) / stockwerk / Stockwerk; Trüm(m)erstock / stockwork

4059
Gangtrüm(m)erstock, – old German term for a vein-like ore stock sending out or terminating in a large number of small vein branches.

4060
sheeted vein deposit; .... zone deposit; shear zone ...., – a mineral deposit consisting of veins or lodes in a zone of shear faulting (shear zone); the deposit may consist of a system of closely spaced parallel veins separated by plates of country rock or of a number of wide tabular massive lodes or lenslike masses. / verertste schuifzone / gisement de shear zone; zone

de cisaillement minéralisé / vererzte Scherzone / depósito de zona de cizalla

shatter zone deposit, – a mineral deposit consisting of a network pattern of veins in a zone of randomly fissured or cracked rocks.

stringer lode; .... lead, – a shattered zone cemented by a network of small discontinuous ore stringers. / verertste vergruizingszone / – – / Trüm(m)erzone; Durchtrüm(m)erung / – –

[2646] *Vein Systems*

4061
field of veins, – an area or field intersected by veins. / veld van gangen / champ filonien / Gangfeld; Gangrevier / campo filionano

4062
main vein; master vein; major vein; trunk vein; mother lode; champion lode, –
1. the principal vein or lode passing trhough a particular ore district;
2. the term mother lode is also used for the original lode from which a placer deposit is derived.

/ hoofdgang; moedergang / filon principal; filon majeur / Hauptgang; Muttergang / veta madre; veta grande; filón principal

4063
(parallel) vein system, – a series of parallel or nearly parallel veins formed at the same time. / parallel gangenstelsel, ... gangsysteem / système filonien; filons parallèles / Gangsysteem; Gangzug; Parallelgänge / sistema de filones

radial vein system; fan-shaped ...., – a series of veins of the same age arranged in a radiating or fan-shaped pattern. / radiaal gangenstelsel / filons rayonnants / Radialgänge; Strahlengänge / filones radiales

4064
net-vein system, – a network of branching veins of essentially the same age, usually following a pre-existing network of fissures. / netvormig gangenstelsel / filons réticulés / Netzgänge / filones reticulades

4065
network of veins, –
1. in general a network of veins striking in different directions;
2. a network of veins formed by intersecting vein systems.

/ netwerk van gangen / réseau filonien / Gangnetz / haz de filones

4066
bunch of veins, – a series of parallel or almost parallel, closely spaced veins. / gangenbundel / faisceau de filons / Gangbüschel / haz de filones (paralelos)

cluster of veins; vein swarm, – an aggregate of irregularly striking veins and veinlets. / gangenzwerm / essaim filonien / Gangschwarm / – –

4067
linked vein system, – more or less parallel veins linked together by cross veinlets. / geschakeld gangenstelsel / filons interconnectés / Kettengang; Gangzug / filones interconectados

4068
Gefahrten, – old German term indicating a series of small veins parallel to a main vein.

4069
horsetail structure, –
1. a system of small, closely spaced, often curved fissure veins emanating from and concentrated mainly at one side of a major vein;
2. pattern of veins resembling a horsetail, especially when viewed in plan.

/ paardestaartvormige structuur / structure en queue de cheval / Pferdeschwanzstruktur / estructura en cola de caballo

4070
branching of veins, – the splitting up of veins in two or more branches. / vertakking van gangen / ramification de filons / Zertrüm(m)erung der Gänge / ramificación de filones

splitting up of a vein, – a vein may terminate by splitting up in a number of branches that pinch out after some distance. / uitsplitsing van een gang / – – / Zerschlagen des Ganges / – –

4071
branch; offshoot; spur; apophysis, – a short irregular branch of a vein. A **dropper** is a branch that after leaving the main vein disappears in the country rock; a **feeder** is a branch that joins the vein from the surrounding rocks. / aftakking; apofyse / veine; ramification; apophyse / Trum(m); Seiten trum(m); Nebentrum(m); Ast; Ausläufer; Ausreiser; Abkommende; Apophyse / ramificación

4072
Bogentrum(m), – German term for a vein branch that splits off and after some distance unites again with the main vein.

Diagonaltrum(m); Quertrum(m), – German terms for a vein branch connecting two parallel or nearly parallel veins.

horse; rider, – the comparatively large mass of country rock that is enclosed by a Bogentrum.

4073
junction, – of veins, two veins joining without intersection or displacement. / samentreffen / jonction / Zusammentreffen / cruce

simple junction, – a joining of two veins that converge at an acute angle, and run parallel for some distance. / scharing / jonction simple / Schaarung / – –

bifurcation, – two veins joined by a simple junction and running parallel may after some distance diverge again. / bifurcatie / bifurcation / Gabelung / bifurcación

4074
vein intersection; vein crossing, – the intersection of an older vein by a younger one. / gangdoorsnijding / croisement de filons / Kreuzung der Gänge; Gangkreuz / cruce de filones

4075
Winkelkreuz, – German term denoting the intersection of two veins striking perpendicular to each other. / – – / croisement orthogonal / Winkelkreuz / cruce perpendicular; intersección perpendicular

Schaarkreuz, – German term denoting the intersection of two veins whose strikes make an acute angle. / – – / croisement oblique / Schaarkreuz / cruce oblicuo; intersección oblicua

Durchfallungskreuz, – German term denoting the intersection of two veins with parallel strikes but different dips. / – – / croisement isogonal / Durchfallungskreuz / intersección isogonal

4076
hammock structure, – a structure resulting from the intersecting of two systems of veins at an acute angle.

4077
cross course; cross vein; counter vein, –

1. a vein intersecting an older one, which frequently shows displacement. Cross courses are sometimes metalliferous, in other cases they are composed of gangue or gouge or they are filled with drag ore,
2. the term cross vein is also used to denote veins crossing the bedding in stratified rocks as opposed to bed veins parallel to the bedding.

/ dwarsgang / filon transversal; croiseur /quergang; durchgreifender Gang / filón crucero

4078
displacement; dislocation, – of a vein, the separation or apparent relative displacement of the parts of a vein, intersected by a fault or a younger vein. / dislocatie / rejet / Verwerfung / desplazamento

4079
drag, – of a vein, the bending and displacement of veins in proximity to a fault. / sleuring / crochon déviation / Schleppen; Auslenkung; Ablenkung / desviación

drag ore, – fragments of ore torn from an ore body by a fault and scattered along the fault plane. / meegesleurd erts / minerai entraîné / durchbewegtes Erz; geschlepptes Erz / – –

[2650] TEXTURE AND STRUCTURE OF METALLIC MINERALS

NOTE:
– in this section a number of terms have been compiled that denote properties, but not exclusively so, of metallic minerals. See also sections:
[0940] The Constituents of Igneous Rocks
[1030] Texture and Structure (of metamorphic rocks)
[1950] Sedimentary Texture
[2340] Concretions

[2651] *Crystalline Structures*

4080
coniform, – / kegelvormig / cunéiforme / koniform / cónico

4081
filiform, – a texture of threadlike crystals of a mineral embedded in another mineral. / filigraantextuur / filiforme / Filigranstruktur / filiforme

4082
oleander-leaf texture, – in general, a mineral grain, crystal grain, or crystal aggregate having the shape of an oleander leaf. / oleanderbladtextuur / texture en feuille de laurier / Oleanderblattstruktur / textura en hoja de laurel

4083
chain structure, – a crystal aggregate structure (e.g. in chromites) in which a series of connected crystals resembles a chain. / kettingstruktuur / – – / Kettentextur / estructura en cadena

4084
comb texture; combstructure, – a term used for vein fillings in which individual crystals have grown perpendicular to the walls. / kamvormig / texture en peigne / Kammstruktur / textura en peine

4085
nodular, – structure composed of nodules or pertaining to the shape of nodules. / nodulair; knollig / noduleux / knollenförmig; knollig / nodular

nodule, – a small, hard, round to irregularly shaped body, usually with features (compositional, structural, physical), different from those of the host material. Due to these features it can be separated as a discrete body. / nodule; knol / nodule / Knoll / nódulo

4086
granule texture, – a texture showing oval or rounded mineral grains in a matrix; the grains are not of clastic origin and lack internal structure. A term used for the round and oval grains in iron formations. / korreltextuur / texture granulaire / Kornstruktur / textura granular.

4087
spherulitic; spheroidal – , a texture showing a ground mass embedding spherulites, i. e. globular masses built up by acicular (slender, needle-like) crystals that radiate from a central point or area. / sferulitisch / sphérolitique / sphärolitisch / esferulítica

4088
dendrite, – a mineral with a dentritic texture, i.e. crystallized in a branching, tree-like form, on or in another mineral (e.g. moss agate), on a bedding plain or in a fracture. Generally they are oxidic minerals (e.g. manganese oxide) or native elements (e.g. gold, silver). / dendriet / dendrite / Dendrit / dendrita

4089
cell texture; cell structure, – formed by segregation on exsolution or by mineral replacement of organic structures, e.g. cell walls, resulting in a network pattern around grain surfaces. / celtextuur / texture cellulaire / Zellstruktur; Zellen.... / textura celular

4090
slug, – a mining and mineral term denoting
1. a piece of alluvial gold up to about 500 grammes;
2. a lump of metal or valuable mineral, e.g. cassiterite.

[2652] *Metacolloidal Structures*

4091
colloform structure; colloidal ...., – rarely occurring minerals or mineral aggregates with a globular structure composed of concentric laminations or concentric bands formed through colloidal (or gel) precipitation. / colloforme structuur; colloidale .... / texture colloforme; .... colloidale / kolloforme Textur / estructura coloforme

metacolloidal structure, – a commonly occurring structure that developed through solidification and/or recrystallization. The structure manifests itself in concentrically built up concretions, globules, spherulites, framboids, botryoids and reniform or discontinuously festooned aggregates. / metacolloidale structuur / – – / metakolloidale Textur / estructura metacoloforme

4092
concretionary structure; concentric ....; scaly ....,– closely related to metacolloidal structures, and characterized by concentric shells of slightly varying properties due to variation during growth, e.g. concretionary iron stone. / concretionaire structuur; concentrische .... / texture concrétionnée / konkretionäre Textur / estructura concrecionada

4093
concretion, – hard, compact accumulation of mineral matter, usually a monomineralic aggregate of spherical, discoid or irregular shape, and generally formed by orderly and localized precipitation from aqueous solutions (often around a nucleus) e.g. iron oxide, pyrite. / concretie / concrétion / Konkretion / concreción

4094
botryoidal, – having the form of a bunch of grapes. Generally relates to mineral and crystal aggregates having a surface of spherical aggregates of radiating crystals. / druiventrosvormig; botryoidaal / botryoïdal / traubig / botroidal

framboidal, – a structure showing aggregates in spheroidal clusters resembling the fruits of a raspberry; more regular than botryoidal. / framboidaalstruktuur / – – / Himbeerstruktur / framboidal

mammillary, – a structure resembling breasts or portions of spheres. Mostly mineral aggregates that possess a knobby surface; similar to but on a larger scale than botryoidal. / knobbelig / mamelonné / – – / mamelar

4095
reniform, – / niervormig / réniforme / nierenförmig; nierig / arriñonado

4096
cockade structure; cocarde ...., – successive crusts of different minerals deposited upon fragments of rock or minerals, e.g. solution-breccia fragments. / cocardestruktuur / structure en cocarde / Kokardentextur; Ringel.... / – –

4097
crackled texture, – minute cracks developed by shrinkage during crystalization; mostly pertaining to metacolloidal mineral aggregates. / craquelétextuur / texture craquelée / rissige Struktur / – –

[2653] *Intergrowth Structures*

4098
intergrowth; locking; interlocking; clustering, – a state of contact between crystals due to simultaneous crystallization, unmixing from a host mineral or replacement. / vergroeiing / développement enchevêtré / Verwachsung / intercrecimiento

4099
simple intergrowth – rectilinear or gently curved boundaries of the components. / eenvoudige vergroeiing / – – / – – / intercrecimiento simple

mottled; spotty; amoeba,– a type of intergrowth with more distinctly curved to lobate boundaries of the components. / amoeboïde / – – / – – / – –

4100
myrmekitic intergrowth; eutectic ....; graphic ...., – interpenetrating or interfingering growth of two or more minerals present in comparable amounts. The grain boundaries are mutually rounded, and the minerals generally show uniform optical orientation. / grafische vergroeiing; myrmekitische .... / structure graphique; .... myrmékite / graphische Verwachsung; myrmekitische .... / intercrecimiento gráfico

4101
emulsion intergrowth; disseminated ....; buckshot ....; peppered ....,– denotes the disseminated occurrence within a host mineral of granules, drops and blebs of minerals without their crystal habit. / gedissemineerd; emulsie / – – / – – / – –

4102
atoll intergrowth; mantled ....; ring ....; shell ....; core ...., – a second mineral or invading material occurs as a core in

another mineral / atolvergroeiing; ringvormige .... / – – / – – / – –

concentric-spherulitic; multiple shell, – a type of intergrowth with alternating concentric shells of different composition. / concentrisch-spherulitisch / – – / – – / zonado esferulítico

4103
vein intergrowth; sandwich ...., – streaks or veins of mineral running in various directions through different (mineral) matter; more or less rectilinear component boundaries. / aderstructuur / – – / – – / – –

lamellae; layered; lattice; polysynthetic, – pertaining to a less common pattern of small-scale rhythmic alternations of different minerals; often an exsolution pattern or crystallographically orientated solid-solution pattern. / lamellair; traliestructuur / lamellair / Gittertextur; lamellar; lamellenartig; schichtig / – –

4104
boxwork structure; network ....; reticulate ....; (Widmanstätten intergrowth), – a regular or random pattern of intersecting mineral plates, blades or phases resulting from either a dissolution of host rock in which originally minerals were deposited along cleavage or fracture planes, or the formation of a new mineral along crystallographic planes of the parent solid solution. / netwerkstructuur / texture cloisonnée / – – / – –

4105
symplectic structure; symplectitic ....,– produced by the complex intergrowth of two different minerals. The component boundaries show alternating embayments or scallops. / symplektische vergroeiing / symplectite / symplektische Verwachsung / – –

4106
exsolution structure; unmixing ....; segregation ....,– any intergrowth type of minerals formed by exsolution of one or more guest minerals from a host mineral (e.g. by unmixing under certain conditions of cooling); structure types formed are crystallographic or orientated intergrowth structures; cell, cellular, net or mesh structures; oleander leaf structure; emulsion, pseudoeutectic or graphic structure. / ontmengingsstruktuur / texture d'exsolution / Entmischungsstruktur / estructura de exolución

[2654] *Stereometric Analysis*

4107
stereometric analysis, – in applied mineralogy, metallurgy, ore dressing, ceramic and refractory materials science and technology, stereometric analysis is concerned with the characterization of three-dimensional, (micro)structural features. Quantitative information can be obtained by suitable automatic measurements, using optical-electronic systems. Measurements are performed in plane surfaces. / stereometrische analyse / analyse stéréométrique / stereometrische Analyse / análisis tridimensional

4108
metric and topological (micro)structural characteristic, – some of the most important characteristics, as calculated from the parameters measured (point, linear, area and volume elements) and as obtained with linear scanners or television microscopes, are: mean grain size, grain-size distribution, surface density, volume fraction, specific surface, form factor, intergrowth or locking index. / (micro)structueel kenmerk / caractéristique structural; charactéristique textural / (mikro)texturelles Kennzeichen / – –

4109
volume fraction, – the percentage by volume of mineral phases present in an aggregate. / volumepercentage / fraction volumique / – – / porcentaje en volumen

specific surface, – the ratio of component surface to component volume. / specifiek oppervlak / surface spécifique / spezifische Oberfläche / superficie específica

4110
form factor, – expresses the ratio of the length (or greatest diameter) of a mineral or particle to its width (or smallest diameter). / vormfactor / – – / Form-Faktor / – –

4111
surface density, – in intergrowth structures the percentage of surface of a mineral in contact with one or more other minerals. It is a function of the measuring units and of the grid constant applied, the area and the number of intersection points measured in the contact surface between the minerals in question. / oppervlaktedichtheid / densité de surface de contact / Oberflächendichte / – –

4112
index of intergrowth; locking index, – the percentage of a mineral intergrown with other minerals. It is a function of the ratio of the surface density of that mineral to the total surface densities of all mineral phases in the aggregate. / vergroeiingsgraad / coefficient de mixité; indice de .... / Verwachsungsindex; ....grad / índice de intercrecimiento

liberation value, – in ore dressing, the degree in which a mineral is free (liberated) from other minerals. It is a function of the index of intergrowth, but in practice not simply inversely proportional to it. / vrijheidsgraad / coefficient de libération; indice de .... / Freiheitsgrad / grado de liberación

4113
coordination of phases, – the preference of a mineral to have a common boundary with one or more other minerals. It expresses the paragenetic relationships of a mineral. Grain sequences and common boundary characteristics can be used to indicate stages of diagenesis, metamorphism and metasomatism. / coördinatie van fasen / – – / Verwachsungsverhältnisse / – –

convex-concave statistics, – the statistical assessment of minerals in establishing their preference to curving of their surfaces or boundaries. Either concave or convex boundaries can be indicative of crystal-growth characteristics, replacement, paragenetic sequence etc. / convex-concaaf statistiek / – – / – – / – –

[2660] VARIOUS MINERAL DEPOSITS

4114
(banded) iron formation, – a sedimentary rock, typically thin-bedded and/or finely laminated, containing at least 15% iron of sedimentary origin, and commonly but not necessarily containing layers of chert; virtually restricted to the Precambrian. / ijzerformatie / jaspilite; quartzite ferrugineux / Eisenquarzite; Eisenglimmerschiefer; gebänderte Eisenjaspilite; gebänderte Eisenerze / taconita

synonyms:
itabirite (Brasil); banded hematite quartzite (India); banded ironstone (South-Africa); taconite (USA); banded jaspilite (Australia); quartz-banded ore (Scandinavia).

4115
ironstone, – any rock containing a substantial proportion of an iron compound, or any iron ore from which the metal may be smelted commercially; specifically an iron-rich sedimentary rock, either deposited directly as a ferruginous sediment or resulting from chemical replacement, with typically oolitic and pisolitic structures in the oxide and silicate ironstones, and generally of Phanerozoic age. / oölitisch ijzererts / roche ferrugineuse; Minette / Eisenstein; oolithische (Braun)eisenerze; Roteisenerze / hierro oolítico

4116
bog iron deposit, – sedimentary iron concentration, mainly composed of goethite, in swamps, lakes and sluggish creeks, in particular in recently glaciated regions, but also in volcanic lakes and streams, e.g. in Japan. / ijzeroerafzetting / fer des marais / Sumpferze; Raseneisenerze / concreciones férricas; hierro de pantano

4117
clay ironstone, – term applied to sheet-like deposits of concretionary masses consisting of argillaceous siderite, as occur with carbonaceous strata of the Coal Measures in Europe. / klei-ijzersteen / argile à passées ferrugineuses / Toneisenstein / arcilla ferruginosa

blackband ironstone, – a dark carbonaceous variety of clay ironstone. / koolrijke klei-ijzersteen / – – / Kohleneisenstein / arcilla ferruginosa negra

clay band, – an argillaceous, light-coloured variety of clay ironstone. / klei-ijzersteen / minerai de fer argileux / Toneisenstein / – –

4118
bean iron ore; pea iron ore, – a coarse pisolitic iron ore, the pisoliths and concretions consisting of goethite and limonites. / ijzerboonerts / fer pisolithique / Bohneisenerz / – –

4119
flaxseed ore, – iron ore composed of disk-shaped hematitic oolites that have been somewhat flattened parallel to the bedding plane.

4120
fossil ore, – iron ore in which shell fragments have been replaced and cemented together by hematite and/or siderite.

4121
hardpan, – an imprecise general term for a relatively hard, impervious, and often clayey layer of soil lying at or just below the surface, produced as a result of cementation of soil particles by precipitation of relatively insoluble materials such as silica, iron oxide, calcium carbonate, and organic matter, offering exceptionally great resistance to digging or drilling, and permanently hampering root penetration and downward movement of water.

4122
duricrust, – a product of terrestrial processes within the zone of weathering in which either iron and aluminium sesquioxides (in the case of ferricretes and alcretes) or silica (in the case of silcrete) or calcium carbonate (in the case of calcrete) or other compounds (in the case of magnesicrete) and the like have dominantly accumulated in and/or replace a preexisting soil, rock, or weathered material, to give a substance that may ultimately develop into an indurated mass and may constitute important mineral deposits.

horizon cuirassé (Maignien), – largely synonymous with duricrust. Within it are distinguished:

cuirasse, – the horizon cuirassé as visible in (natural) outcrop.

cuirasse sensu stricto, – the strongly indurated rock, breaks with difficulty but can be shaped with the hammer.

carapace, – less strongly indurated; easy to work with the pick.

4123
ferrite; ferricrete, – a duricrust of high iron content, the iron being present as goethite, hematite or 'limonite'.

silcrete, a siliceous natural material formed in a zone of silica accumulation produced by superficial (physico-chemical) and not by normal sedimentary, metamorphic, volcanic or plutonic processes. Largely synonymous with siliceous duricrust.

synomyms:
gibbers, surface quartzite, desert sandstone (Australia).

calcrete, – any material formed by the cementation and/or partial or complete replacement of a pre-existing soil (including all unconsolidated material above bed rock) by - predominantly - $CaCO_3$. The material may exhibit consistencies from very loose to hard and structures varying from powdery or gravelly to massive. The boundary between calcified soils and calcretes is set at 50% total carbonate (Ca + Mg) by mass and 50% K-fabric (i.e. fine-grained authigenic carbonate occurs as an essentially continuous medium). Largely synonymous with calcareous duricrust.

synonyms:
kankar (India); nari (Israel); tafezza; chebi-chebi; mbuga limestone; steppenkalk; giglin; vlei kalk (Africa); tosca (Spain); croûte calcaire (France); Kalkkruste (Germany); caliche (USA); travertine (Australia).

4124
laterite, – a weathering product rich in oxides of iron, aluminium or both. It is nearly devoid of bases and original silicates, but it may contain large amounts of quartz and kaolinite. It is either hard or capable of hardening on exposure to wetting and drying. / lateriet / latérite / Laterit / laterita

synonyms:

brickstone (India): ferricrete; murrum; ironstone (Africa); canga (Brasil); pisolite (Australia); plinthite; ironstone cap (USA); krikil (in part) (Indonesia); Krusteneisenstein, Lateriteisenerz (Germany.)

bauxite – a highly aluminous duricrust (or **alcrete**) and a potential ore of aluminium mainly composed of gibbsite and boehmite; named after Les Beaux near Arles, France. / bauxiet / bauxite / Bauxit / bauxita

4125
gossan; iron hat; ironstone (incorrect), – an iron-rich and mostly silica-bearing weathering product with caracteristic cellular and ruggy appearance, overlying a sulphide deposit. / ijzeren hoed / chapeau de fer / eiserner Hut / montera de oxidaciôn

4126
limonite, – a general field term for a mixture of reddish, yellowish, brownish or blackish-brown colour, of several iron minerals (hematite, Fe_2O_3; goethite, Fe_2O_3. H_2O); lepidocrocite, $Fe_2O_3.H_2O$), with or without presumably adsorbed additional water. A common secondary material formed through oxidation in the weathering environment; also precipitated from solutions in the same supergene environment. It often constitutes large volumina of gossans over sulphide deposits. / limoniet / limonite / Limonit / limonita

4127
jasper, – a dense, cryptocrystalline, opaque to slightly translucent variety of quartz (chert or chalcedony) associated with and/or containing impurities of iron oxide or iron hydroxide that give the rock various colours, characteristically red, yellow, brown, black, also green and grayish blue; it often constitutes large volumina of gossans over sulphide deposits. / jasper / jasper / Jaspis / jaspe

4128
saprolite, – a general name for thoroughly decomposed earthy, but non-transported bedrock, mostly with rock structures well preserved and often capped by a duricrust. (May be qualified by the original rock if known, e.g. granitic saprolite). / saproliet / saprolite / Saprolit / saprolita

synonyms:
kong (Bangka, Belitung, Indonesia, a Chinese word); karang (Malay Peninsula); piçarra (Brasil).

4129
red-bed deposit, – deposits of either various uranium and vanadium minerals or chalcosite and pyrite nodules occurring in specific thin layers accompanied by plant and wood remains. The deposit belongs to the terrestric environment, and is largely composed of sandstone, siltstone and shale; it is predominantly red in colour due to the presence of hematite usually coating individual grains. No traces of igneous or hydrothermal activity are present. / – – / gisement (de type) couche rouge / Uran-Kupfer-Konzentrationslagerstätte / areniscas rojas

4130
Kupferschiefer, – a thinly layered, sulphide-bearing marly shale, rich in bitumen and containing up to a few percent Cu, occurring above the reddish Zechstein conglomerate (or reddish sandstone, arkose and shale of the Rotliegend formation and below the Zechstein limestone in central Europe. / koperschalie / schistes cuprifères / Kupferschiefer / pizarras cupríferas

NOTE:
– the Copper Belt deposits of Zambia also are stratiform copper deposits not related to volcanic activity.

4131
kuroko (Japanese: black ore), – a stratabound, polymetallic, sulphide-sulphate deposit genetically related to Miocene felsic submarine volcanism.

4132
Mississippi valley type, –
1. zinc-lead deposit, often with barite or fluorite, occurring in a stratabound configuration within an unmetamorphosed carbonate host rock and with little or no obvious connections with igneous activity;
2. in a stricter sense only those deposits that have a **J-type lead** isotope composition, i.e. an excess of radiogenic lead isotopes indicating an apparent future age for the deposit;
3. the remaining deposits may then be divided in those with
   – **normal-type lead** isotopes, giving approximately correct ages, and those with
   – **B-type lead** isotopes, giving an apparent age older than the surrounding rocks.

4133
porphyry-type ore deposit; disseminated ore, – a type of large low-grade Cu, Mo or Sn deposits characterized by the disseminated occurrence, in grains and veinlets, of Cu-Fe, Cu and Mo sulphides or Sn oxides throughout a large volume of rock. They occur in quartz-bearing igneous rocks and also in schists, silicified limestone and volcanic rocks in close association with acid to intermediate igneous rocks. / porfier(koper)erts / gisement type porphyry; porphyre (cuprifère) / Porphyr(kupfer)erz; Imprägnations(kupfer)erz / – –

stockscheider, – a pegmatitic or granophyric quartz-rich marginal zone found between younger (tin) granite stocks and surrounding older granitic rocks; term originates from the Erzgebirge (Krusné Hory), Germany and Czecho-Slovakia.

4134
greisen, – the product of pneumatolitic metasomatism of granites, composed chiefly of quartz and micas with accessory topaz, tourmaline, cassiterite, wolframite, fluorite or sometimes sulphides; may be a tin ore.

skarn, – a Swedish mining term for the silicate gangue (amphibole, pyroxene, garnet, etc) of certain iron-ore and sulphide deposits of Archaean age, particularly those that have replaced limestone and dolomite (the term is also extended to cover analogous products of contact metamorphism in younger formations).

4135
banket, – the auriferous conglomerate of the Transvaal. More widely the term is also applied to other compact siliceous vein-quartz conglomerates.

reef; ledge – a miner's term for the metalliferous rock being mined, e.g. a gold-bearing quartz vein, the gold reefs of South Africa.

4136
paystreak, – in a placer-type deposit the rich metalliferous gravels on the bedrock. / – – / – – / Bodenseife / – –

largely synonymous with:

kaksa (Chinese word used in Bangka, Belitung, (Indonesia) in the valley deposits); kulit (Malayan word meaning skin, used in Indonesia for deposits on the divide areas); cascalha (Brasil, in the valley deposits).

[2670] ORE DRESSING

4137
ore dressing; mineral beneficiation, – the separation of ore minerals from the associated gangue. / ertsscheiding / traitement des minerais; minéralurgie / Aufbereitung / beneficio de minerales

4138
run-of-mine ore, – broken ore of current quality, as hoisted or otherwise removed from mine workings. / doorsnee van de exploitatie / (minerai) tout venant / Fördererz / mineral en bruto; mena en bruto

crude ore, – broken ore without any treatment by ore dressing. / ruw erts / minerai brut / Roherz / mineral crudo; mena cruda

4139
shipping ore, – ore prepared for transportation by vessel, rail, etc.; direct-shipping ore refers to ore leaving the mine without passing a concentration plant. / erts klaar voor verscheping / minerai marchand / Verschiffungserz; Verlade-Erz / mineral de embarque

4140
feed, – input of raw material, i.e. ores. / voeding / alimentation; charge / Aufgabe / alimentación; carga

mill heads, – input of concentration plant. / wasserijvoeding / alimentation / Aufgabe / cabezas de la concentradora

4141
concentrate, – the upgraded marketable product of mineral beneficiation. / concentraat / concentré / Konzentrat / concentrado

4142
middlings, – intermediate products of concentration plant; usually recirculated to a preceding stage of treatment. In general they consist of intergrown particles of ore and gangue. / tussenproduct / mixtes / Mittelgut / mixtos

4143
flow sheet, – scheme of operations for mineral beneficiation or any other industrial process. / stamboom / rhéogramme / Stammbaum; Fluszbild / flujograma

4144
screen analysis; sieve ...., – / zeefanalyse / analyse granulométrique / Siebanalyse / análisis granulométrico

grizzly, – / staafzeef / grille fixe / Stabrost / parrilla

on-stream analysis, – instantaneous chemical analysis of a continuous flow of ore. / continu-analyse / analyse continue; analyse en continu / Durchlaufanalyse / análisis continuo

4145
batch process, – a discontinuous process. / discontinu proces / procédé discontinu / diskontinuierliches Verfahren / proceso discontinuo

4146
grinding circuit, – sequence of operations in the grinding section of an ore-dressing plant. / maalcircuit / circuit de broyage / Mahlkreislauf / circuito de molienda

4147
autogenous grinding, – grinding in a short rotating mill with lumps of the same ore as the sole grinding medium; the diameter of such a mill is several metres. / autogeen malen / broyage autogène / Autogenmahlung / molienda autógena; autopulverización

4148
cone crusher, – / kegelbreker / concasseur à cône / Kegelbrecher / trituradora cónica

jaw crusher, – / kaakbreker / concasseur à mâchoires / Backenbrecher / machacadora de mandibulas

4149
mill, –
1. ertswasserij; kolen.... / lavoir; laverie / Aufbereitungsanlage / ingenio; lavadero; planta de beneficio
2. molen / broyeur / Mühle / molino

ball mill, – kogelmolen / broyeur à boulets / Kugelmühle / molino de bolas

rod mill, – / staafmolen / broyeur à barres / Stabmühle / molino de barras

4150
liberate, – to grind an ore sufficiently fine, freeing the valuable mineral from the associated gangue. / ontsluiten / libérer / aufschlieszen / liberar

pellet, – rounded agglomeration of ore minerals, about 1 cm in diameter. / knikker; pellet / boulette; pellet / Kügelchen; Pellet / bolilla; pellet

comminution, – grinding to powder size. / fijnmalen / broyage; comminution / Zerkleinerung / trituración fina

4151
calcining; roasting, – the expellation of the volatile content of an ore by burning or heating. / calcineren; roosten / calciner / kalzinieren / calcinar; tostar

4152
flotation, – mineral separation of fine-ground ore by means of a froth created in water; the addition of certain reagents

cause some minerals to float while others sink. / flotatie / flottation / Flotation / flotación

bulk flotation, – flotation of complex ores resulting in a combined concentrate of different ore minerals. / bulkflotatie / flottation collective / Sammelflotation / flotación colectiva

selective flotation, – flotation of one specific ore mineral. / selectieve flotatie / flottation sélective / selektive Flotation / flotación selectiva

4153
gravity concentration, – ore-dressing method based on the specific gravity of minerals. / zwaartekrachtscheiding / concentration gravimétrique / Schwerkraftscheidung; ....aufbereitung / concentración por gravedad

heavy-media separation; HMS; sink and float, – a gravity-concentration method by means of a heavy suspension. The carefully controlled density of the suspension causes heavier minerals to sink and lighter ones to float, gravity being the sole impelling force. / zware-suspensiescheiding / séparation en milieu dense / Schwimm-Sinkscheidung; Schwerflüssigkeits-Aufbereitung / separación en medio pesado

4154
magnetic separation, – ore-dressing method based on the magnetic properties of minerals. / magneetscheiding / séparation magnétique / Magnetscheidung / separación magnética

4155
jig, – an ore-dressing machine. Crushed ore is fed into a box with a perforated bottom and no tap; pulsating water currents cause heavier minerals to sink to the bottom where they are drawn off. / jig; (deintoestel) / jig / Setzmaschine / criba pulsante

4156
rake classifier, – an inclined sedimentation tank with a rake mechanism to move the coarse fraction away from the settling zone. / – – / classificateur à râteau / Rechenklassierer / clasificador de rastrillo

spiral classifier, – an inclined sedimentation tank with a spiral mechanism to move the coarse fraction from the settling zone. / spiraalscheider / classificateur à spirale / Spiralklassierer / espiral de separación

4157
shaking table; concentrating ...., – ore-dressing device based on the combined effect of size and density of minerals flowing in a pulp over a flat surface with parallel riffles and a horizontal shaking movement. / schudtafel / table à secousses / Schüttelherd / mesa concentradora

4158
sluice box, – / wasgoot / sluice / Waschrinne / lavadero

4159
thickener, – a settling tank for a dilute suspension; usually a large, circular vessel with a peripheral overflow and a central bottom discharge. The settled particles are removed by a rake mechanism. / indikker / épaississeur / Eindicker / espesador

4160
slime, – a suspension of particles smaller than 0.1 mm. / slib / boue; schlamm / Schlamm / lodo; lama

4161
deslime, – to remove the colloidal and semi-colloidal particles from a pulp in suspension. / ontslikken / déschlammer / entschlämmen / deslamar

4162
leaching, – / logen; uitlogen / lessivage; lixiviation / laugen / lixiviación

4163
byproduct, – a secondary product extracted from the same raw material as the main product. / bijproduct; nevenproduxt / sous-produit / Nebenprodukt / subproducto

4164
tailings, – reject of a mineral-beneficiation process. / afvalproduct / rejet / Berge / desecho

4165
effluent, – / afvalwater / effluent / Abwasser / efluente

4166
dump. – rejects of mine or plant. / steenberg / terril; haldes / Halde / desmonte; desechadero

[2700] MINERAL ECONOMICS

[2710] SAMPLING

NOTE:
– for further terms on samples and sampling, see sections
[1970] Grain-Size Determination
[2820] Drilling; Methods and some Technical Terms
[2931] Samples and Sampling Procedures

[2711] *Sampling of Ore in Place*

4167
drill sampling, – / boorbemonstering / échantillonnage de forage / Probenahme beim Bohren / muestreo de sondeo

face sampling, – sampling of ore in situ, restricted to ore exposed in mine workings, outcrops, benches, trenches, pits, etc. / (werk)frontbemonstering / échantillonnage des fronts d'attaque / Probenahme im Anstehenden / muestreo de frentes de trabajo

4168
channel sample; groove ...., — a sample taken from a chiseled (moiled) groove across a vein or formation. / sleufmonster / échantillon prélevé par saignée / Schlitzprobe / muestra por ranura

chip sample; pick ...., — a series of chips of ore or rock taken across an exposure at more or less uniformly distributed intervals. / scherfmonster / esquille / Hackprobe; Pick.... / muestra cincelada

4169
grab sample, — sample taken at random from broken ore. Car, chute, bench, stope, etc. samples, taken after blasting, can be distinguished. / grijpmonster / échantillon au hasard / Stichprobe; Hand.... / muestra tomada al azar

4170
stream sampling, — of mineral products moving on conveyors, hydraulic or other means of transport; is usually carried out intermittently at fixed time intervals. / stroombemonstering / échantillonnage au courant / — — / muestreo por captura de una corriente

[2712] *Sample Spacing*

4171
sample spacing, — refers to the spatial arrangement of points to be sampled with regard to grid or pattern and to distances of intervals. /spatiëring van monsterplaatsen / réseau d'échantillonnage / Probenraster / distribución de muestreo; red. de ....

4172
equidistant; linear spacing, — fixed distaces between sampled points in a single row. / (spatiëring) op gelijke afstanden / équidistant / gleiche Abstände / equidistante

checkerboard (spacing), — two-dimensional spacing according to a grid of squares. / dambord patroon /réseau à maille carrée / rechteckiges Verteilungsnetz; .... Verteilungsgitter / distribución escaqueada

4173
systematic spacing, — sampled spots follow a definite geometric pattern. / systematische spatiëring / distribution systématique des échantillons / systematische Beprobung / distribución sistemática

4174
random spacing, — locations to be sampled are selected haphazardly. / willekeurige spatiëring / distribution aléatoire / Probenahme mit zufälligen Abständen / distribución al azar

[2713] *Miscellaneous Terms*

4175
pilot sampling, — to test statistical parameters, like variability or grade, sometimes precedes a sampling campaign; it may supply valuable information as to sampling method and spacing to be adopted. / proefbemonstering / échantillonnage pilote / Versuchsprobenahme / muestreo piloto

4176
compound sample; composite ...., — a weighted mixture of two or more individual spot samples prepared as an aggregate for testing special properties or amenability to processing. / samengesteld monster / échantillon composite / Sammelprobe / mezcla de muestras

4177
check sampling, — various procedures to control precision and accuracy of the outcome of sampling; e.g. by retaking samples at spots sampled before. / controle-bemonstering / échantillonnage de controle / Kontrollprobenahme / muestreo de control

4178
paystreak, — portion of a mineral deposit (in particular a placer deposit) that carries profitable ore. / economisch winbare zone / — — / bauwürdiger Bereich / zona aprovechable

oreshoot, — sizable aggregation of ore of good grade; the term applies in particular to vein deposits. / winbare ertszone / minerai exploitable localisé / Erzfall; Veredlungszone / bolsada de mineral aprovechable

4179
salted sample, — the sample has been artificially enriched usually with fraudulent intent. / vervalst monster / échantillon salé / 'gesalzene' Probe / muestra salada; .... enriquecida

[2720] SAMPLE REDUCTION AND CHEMICAL ANALYSIS

4180
sample reduction, — practised to reduce weight and volume of original samples to portions adequate for chemical analysis or other methods of testing with minimum loss of representativeness. / monsterverkleining / réduction d'échantillon / Probenverjüngung / reducción de muestra

4181
coning and quartering, — procedure consists in piling the ore into a conical heap, spreading this out into a circular cake, dividing the cake radially into quarters, taking opposite quarters as sample and rejecting the other two. / kwarteren / quartage / Aufschütten zum Kegel und Kreuzteilen; Vierteln / partir una muestra; cuartear una ....

4182
sample splitter; riffler, — apparatus to divide a sample into two equal portions. / monsterverdeler / diviseur à riffles / Probenteiler; Riffel.... / partidor

core splitter, – apparatus to split drill cores lengthwise into halves, usually one half being analysed and the other stored for future reference or logging. /kernsplijtpers / diviseur de carotte /Kernspaltgerät / partidor de testigo

4183
assay, –
1. general term for an analysis of chemical composition. / analyse / analyse / Analyse / ensaye; ensayo
2. more specifically **fire assay,** testing of the content of precious metals by pyrometallurgical methods. / analyse / essai / – – / ensaye; ensayo

4184
scorification, – the step in fire assaying in which the assay portion is fused with granulated lead to produce a **'lead button'** that contains the precious metals. / verslakking / scorification / Verschlackung / escorificación

cupelling, – follows scorification in fire assaying and reduces the lead button to gold bead in a **'cupel'.** / cupellering; copellering / coupellation / Abtreiben; Treiben / copelación

[2730] STATISTICAL ANALYSIS OF SAMPLES

[2731] *Grade Computation*

4185
weighted average grade, – weighting refers to multiplying individual grades by factors that represent the mass or volume allotted to the corresponding spot sample; in tabular ore bodies this factor may be the length of a channel sample; the sum of weighted grades divided by the sum of weighting factors gives an estimate of mean tenor. / gewogen gemiddeld gehalte / moyenne pondérée de la teneur / gewogenes Mittel / promedio pesado de la ley

4186
extreme values; erratic ...., – frequently considered as 'outliers', as unreliable and not in line with the main body of sampling observations; statistical tests are available – based upon concepts of frequency distributions – to decide how to handle extreme values. / extreme monsterwaarden / valeurs extrêmes / Extremwerte; Ausreisser; stark abweichender Einzelwert / valores extremos

range, – the spread or difference between the largest and smallest observations, e.g. of ore grade. / spreiding / amplitude / Streubereich / extensión

[2732] *Statistics of Grade Distribution*

4187
frequency distribution, – frequencies of sample valves, grouped according to successive intervals, as related to the midpoints of the intervals; statistical formulae or laws can be fitted to such experimental distributions. / frequentieverdeling / distribution de fréquences / Häufigkeitsverteilung / distribución de frecuencias

4188
cumulative frequency curve, – shows for grouped data how many sample values are below and how many above a given grade. / cumulatieve frequentiekromme / courbe cumulative de fréquence / Summenlinie / curva cumulativa de frequencias

histogram, – a graph of frequency distribution obtained by drawing rectangles whose bases coincide with class intervals and whose areas are porportional to class frequencies; for equal class intervals, the heights of the rectangles are proportional to frequencies. / histogram; kolommendiagram / histogramme / Histrogramm / histograma

4189
normal distribution; standard ...., – first studied in connection with errors of measurement, and thus referred to as 'normal'; the main standard for all distributions of observations with two parameters, i.e. arithmetic mean and standard deviation. Areas under the bell-shaped Gaussian curve, extending from minus to plus infinity, represent probabilities. / normale verdeling; verdeling van Gauss / distribution normale; .... Laplace-Gauss / Normalverteilung / distribución normal

lognormal distribution; Galton ...., – if the logarithms of a variable have a normal distribution, the variable itself is said to have a lognormal distribution. On arithmetic scale lognormal distributions extend from zero to plus infinity and are positively skewed; they have two parameters. / log-normale verdeling / distribution lognormale; .... de Galton / Lognormalverteilung / distribución lognormal

4190
variance, – of a set of n observations is given by the sum of the squared deviations from the mean divided by n-1. / variantie / variance / Varianz / varianza

standard deviation, – is given as the positive square root of variance; it is by far the most used measure of **dispersion** (= variability). / standaard deviatie; .... afwijking / écart-type / Standardabweichung / deviación normal

coefficient of variability, – standard deviation divided by the corresponding mean as a more absolute measure of dispersion. / variatie-coëfficiënt / coefficient de dispersion / Variationskoeffizient / coeficiente de variación

4191
confidence interval; .... limits, – an interval for which one can assert with a specified degree of probability that it will contain the value of the parameter it is intended to estimate. This parameter is often the unknown, true mean grade, in which case the lower confidence limit is only of practical interest and a 'one-sided' estimate is given. / betrouwbaarheidsinterval; ....drempel / intervalle de confiance / Vertrauensbereich / intervalo de confianza

4192
biased estimate, –
1. an estimate, like for instance median grade, is said to be biased, if its value does not equal the parameter it is intended to estimate, like average grade;

2. in sampling, a bias is a systematic error introduced mostly if certain components of the population are more likely to be chosen than others; bias impairs the accuracy of an estimate, not the precision. / systematische fout; stelselmatige afwijking / estimation biaisée; échantillonnage .... / systematischer Fehler / error sistemático

[2740] GEOSTATISTICS

4193
geostatistics, –
1. in the broadest sense, applications of the theories of mathematical statistics in the fields of the geosciences;
2. as coined by G. Mathéron, the study of the spatial distribution of (ore) grade, based upon the theory of regionalised variables, which exhibit spatial (structural) and random behaviour. A probabilistic model (the intrinsic hypothesis) applies the formalism of random functions and leads to the use of the variogram as the fundamental function in computations of errors of estimation.

/ geostatistiek / géostatistique / Geostatistik / geo-estadística

4194
regionalisation, – the concept of Mathéron that (ore) grade and other localized geological observations are not stochastically distributed in geometric space, but display continuity and interdependence; analogous to the feature of trend, observable in most time series. / regionalisatie / régionalisation / Regionalisierung / regionalización

4195
autocorrelation; serial correlation, – the internal correlation (= relationship) among successive observations in spatial (or time) series; usually expressed as a function of the distance (or time) lag between observations. / kettingcorrelatie / autocorrélation / Autokorrelation / autocorrelacion

correlogram, – a graph of the autocorrelation coefficient plotted as a function of the number of lags (in distance or time) between observations. / correlogram / corrélogramme / Korrelogram / correlograma

4196
variogram, – according to Mathéron, a graph of a special measure of average dispersion plotted as a function of the distance (lag) between sampled spots. Here dispersion is not based on deviations with respect to average grade, but on deviations from spatially preceding sample values. / variogram / variogramme / Variogramm / variograma

covariogram, – a graph of the (estimated) covariance plotted as a function of the number of lags in distance or time between observations. / covariogram / covariogramme / Kovariogramm / covariograma

4197
kriging, – optium estimation of block grades: a geostatistical procedure, developed by the South-African Krige, to estimate the average tenor of a panel of ore (reserves) from the weighted average of samples inside as well as outside the panel thus avoiding any bias. The procedure makes it possible to find a set of weights that minimizes the estimation variance under the condition that the sum of the weights is equal to one. / kriging / krigeage / Kriging / krigeaje

4198
moving average; running ...., – an artificially constructed series in space (or time) in which each actual value is replaced by the mean of itself and selected equal numbers of the values immediately preceding and following it. It is a method to smooth fluctuations that increases the degree of autocorrelation. / voorschrijdend gemiddelde / lissage par moyenne mobile; .... par moyenne glissante / gleitendes Mittel / valor medio que mueve

[2750] ESTIMATION OF ORE RESERVES

4199
panel, – of ore reserve, in reserve estimates the aggregate volume of ore partitioned into blocks or panels, usually following bench plans, levels, etc. so as to correspond with planned units of future exploitation. / blok; veld / panneau / Abbaufeld; ....block; Flözteil / panel

4200
minimum stoping width; .... stoping height, – in reserve estimates of thin veins or other tabular deposits, overbreakage and resulting ore dilution can be taken into account on the basis of the minimum working room required for a given method of exploitation. / minimale stopebreedte / largeur minimale d'abat(t)age / Mindestmächtigkeit / ancho mínimo por explotar.

4201
minimum bench width, – limiting factor in the planning of mining in opencast; related to size of equipment and to slope stability. / minimale breedte voor trapfront / largeur minimale des terrasses / Mindeststrossenbreite / ancho mínimo delante de los bancos

4202
stripping ratio, – ratio in terms of thicknesses of overburden to be removed to ore to be exploited in opencast; mainly applied to stratiform deposits. / afdekverhouding / taux de recouvrement; coefficient de .... / Deckgebirgekoeffizient / coeficiente de recubrimiento

4203
cut-off grade, – the lowest average grade for exploitation units that qualify as ore in a given deposit; i.e. minimum grade for blocks of ore to be included in estimates of reserves. Stripping ratios and other limiting economical factors may likewise give rise to cut-off values. / (economisch) grensgehalte / teneur limite; .... liminaire; seuil d'exploitabilité / Grenzgehalt; Bauwürdigkeitsgrenze / tenor de cierre (económico)

4204
paylimit, –
1. synonymous with cut-off grade;
2. the break-even grade in terms of yield against direct operating costs; generally a lower gigure than cut-off as based upon optimized return on investments.

/ drempel voor lonend gehalte / teneur limite; .... payante / marginaler Erlös; ....Erzwert / tenor limite de aprovechabilidad

[2760] MINERAL RESOURCES

NOTES:
– see also section
[2584] Reserves (of hydrocarbons); although comparable in phrraseology, the terms are not identical;
– in the classification of mineral resources, as given here, the classification proposed by V.E. McKelvey and adopted by US Bureau of Mines and US Geological Survey (March, 1974) is followed. Older and now obsolete near-synonyms are put between brackets.

4205
mineral resources, – natural concentrations of mineral-bearing material in such a form that economic extraction of a mineral commodity is potentially feasible. The term includes reserves as well as hitherto undiscovered resources. / minerale hulpbronnen / ressources minérales / Mineralressourcen / recursos minerales

NOTE:
– see table 12 – Classification of Mineral Resources

4206
identified resources, – specific bodies of mineral-bearing material whose location, quality and quantity are known from geological evidence supported by engineering measurements with respect to the demonstrated category. / aangetoonde hulpbronnen / ressources identifiées / nachgewiesene Ressourcen / recursos identificados

4207
demonstrated reserves; reasonably assured ...., – a collective term for the sum of material in both measured and indicated reserves. / redelijk verzekerde reserve; aangetoond erts / réserves prouvés; .... démontrées / erkannte Vorräte / reserva razonable de mineral

4208
reserves; ore ...., – that portion of the identified resources from which a usable mineral (or energy) commodity can be economically and legally extracted at the time of determination. / (erts)reserves / réserves de minerai / (Erz-) Vorräte / reservas de mineral

4209
measured reserves; (proved ....), – material for which estimates of quality and quantity have been computed, within a margin of error of less than 20 per cent, from sample analyses and measurements from closely spaced and geologically well-known sample sites. / bewezen reserve / réserves reconnues; minerai démontré / sichere Vorräte; vorgerichtete .... / mineral reconocido; .... probado

indicated reserves; (probable ....), – material for which estimates of the quality and quantity have been computed partly from sample analyses and measurement and partly from reasonable geological projections. / waarschijnlijke reserve / réserves probables / angedeutete Vorräte / mineral probable

4210
inferred reserves; (possible ....), – material in unexplored extensions of demonstrated ore for which estimates of quality and size are based on geological evidence and projection. / mogelijke reserve / réserves possibles / vermutete Vorräte / mineral posible

4211
identified subeconomic resources, – materials that are not reserves, but may become so as a result of changes in economical and legal conditions. / aangetoonde mineralisatie van nog niet-winbaar gehalte / association minérale démontrée subéconomique / nicht bauwürdige Vorräte; Ausserbilanzvorräte / recursos de posible interés económico

4212
paramarginal, – the portion of subeconomic resources that borders on being capable of economical production or is not commercially available solely because of legal or political circumstances. / paramarginaal / paramarginal / paramarginal / paramarginal

submarginal, – the portion of subeconomic resources that would require a price more than 1.5 times the price at the time of determination, or a major cost-reducing advance in technology. / submarginaal / submarginal / submarginal / submarginal

4213
undiscovered resources, – unspecified bodies of mineralized material surmised to exist on the basis of broad geological knowledge and theory. / onontdekte hulpbronnen / ressources non-découvertes / unbekannte Ressourcen; unentdeckte .... / recursos sin identificación

4214
hypothetical resources, – undiscovered materials that may reasonably be expected to exist in a known mining district under known geological conditions. / hypothetische hulpbronnen / ressources hypothétiques / hypothetische Ressourcen; prognostische Vorräte / recursos hipotéticos

speculative resources, – undiscovered materials that may occur either in known types of deposits in a favourable geological setting where no discoveries have been made, or in as yet unknown types of deposits that remain to be recognized. / speculatieve hulpbronnen / ressources spéculatives / spekulative Ressourcen; prognostische Vorräte / recursos especulativos

4215
potential resources, – total of resources minus reserves. / potentiële hulpbronnen / ressources potentielles / potentielle Vorräte; .... Reserven / recursos potenciales

[2770] MINE VALUATION

NOTE:
– see also section
[2590] Oil Economics

4216
mine valuation; mine evaluation; mine appraisal, – to determine the financial value of a mining property. / waardebepaling van mijnbezit / évaluation; expertise / Bewertung; Lagerstättenbewertung; Bewertung von Bergwerkseigentum / valuación; peritaje; valoración

4217
Hoskold formula, – H. D. Hoskold, 1877, based the value of a mine – a wasting asset – on the premiss of a sinking fund in which part of the earnings accumulates at a safe rate of interest to the amount of the initial investment at the time the reserves are exhausted; the 'two rate' valuation formula contains next to the rate of compound interest for the sinking fund a risk rate for the investor. / formule van Hoskold / formule de Hoskold / Hoskold Formel / fórmula de Hoskold

Morkill formula, – a single rate variant of the Hoskold formula that coincides with present practice of valuation on the basis of discounted cash flow, / formule van Morkill / formule de Morkill / Morkill Formel / fórmula de Morkill

4218
annuity, – annual profits or payments. / annuïteit / annuité / Annuität / anualidad

4219
sinking fund, – profits set apart at interest to extinguish a debt. / amortizatiefonds / caisse d'amortissement / Tilgungsfonds / fondo de amortización

4220
deferred value, – value after a period of delay (e.g. on account of mine development). / uitgestelde waarde / valeur retardée / aufgeschobener Wert / valor a plazo

4221
cash flow, – balance of monies received and expended; different from profit or loss, because depreciation is not taken into account; hence cash flow = profit + depreciation.

DCF (method), – application of discounted cash flow to present values for valuation purposes.

4222
depletion record, – bookkeeping of ore reserves, i.e. balance of exploited ore against discovery and development. / balans van de ertsreserve / comptabilité des réserves / Reservenbilanz / balanza de agotamiento

depletion allowance, – reduction in taxes or other forms of gouvernment take to allow for depletion of reserves. / – – / – – / Steuermässigung für Substanzverzehr / – –

4223
wholesale price index, – index to compensate the effect of inflation on time series of commodity prices. / groothandelsprijsindex / indice des prix de gros / Index der Grosshandelspreis / índice de precios al por mayor

4224
royalty, – share of the production or profit reserved by the owner or by the government for permitting the exploitation of the property. / cijns; royalty / redevance / Abgabe; Förderabgabe; Förderzins; Bergregal (historical) / regalía; derechos

4225
stockpiling, – to accumulate products, e.g. when shipping facilities are suspended, but also intentionally for commercial, political or strategic reasons. / voorraadopslag / mise en stock; stockage / Bestandslagerung; Aufhaldung / amontonamiento; almacenamiento

bufferstock, – stockpile, specifically intended for regulatory reasons; to deaden sudden changes in price or availability. / buffervoorraad / stock tampon / Ausgleichsvorrat; Pufferstock / reserva reguladora de abastecimiento

[2800] EXPLORATION AND SURVEYING METHODS

[2810] GEOLOGICAL MAPPING AND RECORDING

4226
field geology, – geological investigation and mapping in the open field, i.e. at the surface of the earth. / veldgeologie / géologie de terrain / Feldgeologie / geología de campo

subsurface geology; underground ...., – geological investigation and mapping below the surface of the earth, i.e. on information obtained from drilling, mining, tunnels and by geophysical methods. / ondergrondse geologie / géologie souterraine / Untertage-Geologie / geología subterránea

marine geology; submarine ....; geological oceanography, – geological investigation and mapping of the sea floor and coasts. / mariene geologie / géologie marine / Meeresgeologie / geología marina

4227
geologist's compass, – compass, specially designed for use during geological field work. / geologisch kompas / boussole de géologue / Geologenkompass / brújula geológica

clinometer, – device, designed to measure angles in a vertical plane. / klinometer / clinomètre / Klinometer / clinómetro

4228
strike and dip readings, – / strekking- en hellingmetingen / observations des directions et pendages / Streich und Fallmessungen / medidas de dirección y de brizamiento

azimuth; bearing, – the horizontal direction of a line given as an angle measured clockwise from a reference direction, usually north. / azimut / azimut / Azimut / azimut

4229
hand specimen, – a rock sample of hand size. / handstuk / échantillon à main / Handstück / muestra de mano

sample bag; .... pouch, – small bag, used to store samples, taken in the field. / monsterzakje / sac à échantillons / Probenbeutel / bolsa para muestra

4230
peel technique, – method to preserve an (unconsolidated) rock or soil surface by covering it with e.g. cellulose acetate that after drying can be peeled off. / lakfilmmethode / méthode du laquefilm / Lackfilm-Methode / método de 'peel'; .... película

lacquer peel, – / lakfilm / laquefilm / Lackfilm / laca

4231
outcrop; exposure, – the place, strip of land or area, where a rock or rock body emerges at the surface, not concealed by soil, weathered material, buildings, etc. / ontsluiting; dagzoom / affleurement ; ligne d'.... / Aufschluss; Ausbiss; Ausstrich / afloramiento

to outcrop; outcropping; to crop out, – the emergence at the surface of a rock or rock body, not concealed by soil, weathered material, buildings, etc. / dagzomen; aan de dag komen / affleurer / ausbeissen; ausstreichen / aflorar

incrop, – a non-current term denoting a rock body that is extensively although discontinuously concealed.

4232
blanket, – a relatively thin cover of (unconsolidated) deposits concealing the solid rock. / dek; bedekking / manteau / Bedeckung; Überdeckung / recubrimiento

4233
outlier, – an erosional remnant separated from the main area of outcrop. / – – / lambeau; témoin / – – / vestigio

inlier, – a rock body emerging at the surface and surrounded by usually younger rocks. / – – / boutonnière / – – / testigo

4234
window, – an inlier below an unconformity plane or below an overthrust sheet. / venster / fenêtre / Fenster / ventana

subcrop, – the rock or rock body directly beneath an unconformity, usually related to subsurface conditions.

4235
test pit, – a shallow hole dug through the weathered zone to expose and investigate the unweathered rock underneath. / proefkuil / puits de recherche / Schürfloch; ....schacht / pozillo; calicata

trench, – elongated shallow excavation. / sleuf; greppel / tranchée; fouille / Graben / trinchera

4236
auger, – a short spiral-shaped tool run on a torque bar to drill through the weathered zone to collect soil and/or fresh rock samples. / spiraalboor; avegaar / tarière / Spiralbohrer; Drill.... / barrena; sonda de cinta; broca

4237
core; drillcore, – a cylindrical sample of rock taken from the bottom of the borehole by means of a core barrel. / kern; boorkern / carotte / Bohrkern / testigo

core box, – tray to collect drillcores. / kernkist; lade voor boorkernen / caisse de carottes / Kernkiste / caja de testigos

4238
sidewall sample, – sample taken from the wall of the drillhole by a tool run on wireline. / wandkern / échantillon (de carottier) latéral / Seitenkern; Schuszkern / testigo lateral

(drill) cuttings; ditch samples, – formation particles removed by the bit and the mud from the bottom of the borehole and transported to the surface by the mud flush. / boorgruis / débris de forage / Spülprobe / detritos de perforación

sludge sample, – a mineral-survey term for all or part of the drill cuttings collected, dried, and saved for assaying or chemical analysis.

4239
(geological) cartography, – the art of map construction and the science upon which it rests. The practice of making charts or maps. / kartografie / cartographie / Kartografie / cartografía

geological map, – map recording geological features in a region or area, particularly at the surface of the earth. / geologische kaart / carte géologique / geologische Karte / mapa geológico

4240
legend, – a concise description of the mapped units and of the symbols used on a map. / legenda / légende / Legende / leyenda

profile-type legend, – legend in use for the geological map of the Netherlands, showing in colours and symbols the column of lithologically different strata to a depth that can be reached by simple hand-auger equipment. / profieltype legenda / – – / – – / – –

scale, – the relation between map distance in the terrain expressed as a fraction or a ratio. / schaal / échelle / Maszstab / escala

4241
subcrop map, – map recording geological features directly below an unconformity, particularly in the subsurface.

worm's-eye map, – map recording geological features of the bottom surface above an unconformity.

4242
palaeogeological map, – map recording the reconstructed geology of a surface in the geological past. / palaeogeologische kaart / carte paléogéologique / paläogeologische Karte / mapa paleogeológico

NOTE:
– in the following, examples of maps are given; they may

pertain to the surface as well as to the subsurface. Many more types of specialized maps have been devised and are in use.

4243
isopach; isopachous line, – a line on a map drawn through points of equal true thickness of a designated stratigraphic unit. / isopach / isopaque / Mächtigkeitslinie; Isopache / isopaca

isopach map; isopachous map, – / isopachenkaart / carte isopaque / Isopachenkarte; Karte gleicher Mächtigkeiten / mapa de isopacas

true thickness, – the thickness of a stratum or group of strata measured perpendicular to the bedding planes. / ware dikte / épaisseur vraie; .... normale / wahre Mächtigkeit / espesor verdadero

4244
isochore, – a line drawn through points of equal vertical interval between two stratigraphic surfaces, regardless of the degree of dip. / isochoor / isochore / – – / isocora

isochore map, – / isochorenkaart / carte des isochores / – – / mapa de isocoras

4245
facies map, – a map showing the distribution of various facies types occurring within a designated geological unit. / faciëskaart / carte de faciès / Fazieskarte / mapa de facies

4246
isopleth map, – map showing the quantitative spatial distribution of some property or attribute. / isopletenkaart / carte des isoplèthes / – – / isopleta

4247
structural map; tectonic map, – map recording tectonic features. / structuurkaart; tektonische kaart / carte structurale / tektonische Karte / mapa estructural ; mapa tectónico

(structural) contour map, – map showing structure contours projected vertically onto a horizontal plane. / contourkaart / carte des courbes structurales / Strukturkarte / mapa de curvas estructurales

4248
contour (line); isohypse, –
1. an imaginary line connecting the points on any surface, e.g. geological, water table, or topographical, that have the same elevation;
2. the line on a map or chart representing points of equal elevation.

/ contourlijn; isohypse / courbe de niveau; isohypse / Niveaulinie; Isohypse / isoipsa

4249
structural datum; key horizon, – the stratigraphic surface on which contours are drawn. / structureel gidsniveau / niveau de référence / Leithorizont; Bezugsniveau / nivel de referencia

structural elevation, – the elevation of a specified stratigraphic surface above or below the datum plane. / structurele hoogte / élévation structurale / Höhe über dem Bezugsniveau; Höhe unter dem .... / elevación estructural

4250
geological section; .... profile, –
1. a natural rock-cut;
2. the representation of such in drawing;
3. a more or less interpretative vertical section of part of the earth's crust based on either surface or subsurface information or both, generally along a straight line.

/ geologisch profiel; geologische doorsnede / coupe géologique; profil .... / geologischer Profilschnitt / corte geológico

apparent dip, – of planar features in a geological section. / schijnbare helling / pendage apparent / scheinbares Einfallen / buzamiento aparente

4251
serial section, – a series of parallel sections illustrating the development of a certain stratigraphic or tectonic event. / in serie geplaatst profiel / coupe sériée / Profil-Serie / perfiles seriados; cortes ....

4252
fence diagram, – a number of sections arranged in a grid, so as to facilitate a three-dimensional view. / schuttingdiagram / coupes en réseau / Paneeldiagramm / diagrama compartimentado; .... en paneles

ribbon diagram, – a section not drawn along a straight line, but along a curved or broken one, so as to combine features of interest into one section. / – – / – – / gebogener Profilschnitt; geknickter .... / diagrama curvilíneo

4253
composite section, – a profile compiled with the aid of various more or less parallel sections. / gecombineerd profiel / coupe synthétique / Sammelprofil; Profilserie / corte compuesto; perfil ....

4254
generalized section, – a drawing showing characteristics of various distinct areas grouped together in one section. / algemeen profiel / coupe généralisée / generalisierter Profilschnitt / corte generalizado

ideal section, – a section showing, in addition to factual data, the hypothetic stratigraphical or structural conditions as interpreted by the author. / ideaal profiel / coupe idéalisée / idealisierter Profilschnitt / corte ideal

4255
columnar section; stratigraphical column, – a chart showing in a vertical column the sequence and the thicknesses of the strata for a given area, with their lithologies, fossil content and other relevant information. / stratigrafische kolom; normaal profiel / profil stratigraphique / Normalprofil; Säulen.... / columna estratigráfica

4256
well log, – a columnar section derived from information obtained in a borehole. / profiel van de boring / coupe de sondage / Bohrprofil / corte de sondeo

4257
standard section, – a profile showing as complete as possible a sequence of the strata in a certain area, thus affording a standard for correlation. / standaardprofiel / coupe standardisée / Standardprofil / corte tipo

4258
block diagram; stereogram, – a diagrammatic representation of a block of the earth's crust with an impression of relief and perspective; geological information may be shown on the visible sides. / blokdiagram; stereogram / bloc-diagramme; stéréogramme / Blockbild / bloque diagrama; estereograma

4259
rosette diagram, – diagram to represent the attitudes of any planar structural feature in an area, e.g. joints, faults, veins. Strikes and dips are plotted on two semicircles respectively that are divided in sectors, each representing a strike or dip interval; the number of features in each sector is indicated by some scale. / roosdiagram / diagramme en rose ; .... en rosette / Richtungsrose / diagrama en roseta; .... en rosa

point diagram; – diagram to represent the attitudes of any planar structural feature in an area. Their poles on a semisphere are projected on a circle, either by stereographic or equal-area projection. From such a diagram a **contour diagram** may be prepared. / puntendiagram / diagramme polaire; .... circulaire / Polpunktdiagramm; Dichteplan / diagrama estereográfico

4260
geological province, – an area in which the geological history has been essentially identical or that is characterized by particular structural or physiographic features. / geologische provincie / province géologique / – – / provincia geológica

[2820] DRILLING; METHODS AND SOME TECHNICAL TERMS

4261
well; borehole; drillhole; hole. – / boring; boorgat; put / sondage; puits; forage / Bohrung; Bohrloch; Sonde / pozo; perforación; sondeo

4262
cable-tool drilling, – drilling by means of tools suspended on a steel cable. / kabelboren / forage au cable / Seilbohren; pennsylvanisch Bohren / perforación con cable

percussion drilling, – drilling by means of reciprocating tools. / stotend boren; pulsboren / forage par percussion / Stossbohren / perforación a percusión

churn drilling, – drilling method in which the bit strikes the rock through free fall over a limited distance. / stotend boren / forage par battage / Freifallbohren / perforación por batido

4263
rotary drilling, – drilling by means of rotating drilling tools (drillpipe and bit). / draaiend boren / forage-rotary / Rotarybohren / perforación a rotación; .... rotativa

rotary air-blast drilling; RAB ...., – a rotary method where compressed air is used to flush the cuttings from the hole.

rotary-percussion drilling, – combination of rotary and percussion drilling techniques. / draaiend pulsboren / forage â rotary-percussion / Rotary-Stossbohren / perforación a rotación-percusión

airlift drilling, – rotary drilling with compressed air as circulant. / – – / forage à l'air / im Lufthebeverfahren bohren / – –

4264
turbodrilling; turbine drilling, – drilling by means of a turbine-driven drilling bit on drillpipe. / turbineboren / turboforage / Turbinenbohren / perforación con turbina

4265
diamond drill, – any one of a number of different sizes and kinds of machines designed to impart a rotary and longitudinal movement to drillpipe and a diamond-inset bit; generally used in mineral prospecting and development but also to drill blastholes and to do various types of soil and foundation-testing work. / diamantboor / appareil de forage au diamant; sondeuse au .... / Diamantbohrer / perforación con diamante

4266
drill carriage; jumbo, – movable platform, stage or frame, used for engineering purposes, incorporating several rock drills, and that usually travels on a tunnel track. / boorwagen / jumbo / Bohrwagen; Jumbo / – –

4267
rig, – a drilling machine, complete with auxiliary equipment such as drawworks, pumps and engines. / boorinstallatie / installation de forage / Bohranlage / equipo de perforación

derrick, – the steel tower placed over a borehole to support the drilling tools for hoisting. / toren / derrick; tour de forage / Turm / torre

mast; drilling mast, – a portable derrick, usually having two sections that can either be telescoped or 'jack-knifed' and lowered for transport by truck. / mast / mat de forage / Mast / mástil

4268
(water) swivel; gooseneck, – a device connecting the mud hose to the drill string and designed to permit the latter to be rotated in the borehole while drilling mud is pumped into it. / zwanenhals; spoelkop / tête d'injection / Spülkopf / cabeza de inyección

4269
kelly, – the square, upper length of drilling string. / meeneemstang / tige carrée d'entrainement / Kelly; Mitnehmerstange / barra cuadrada; .... de accionamiento

turn table; rotary ...., – a rotable disc set in the centre of the derrick floor; it contains a square aperture, in which the kelly is fastened with clamps, permitting vertical movement. / draaitafel / table de rotation / Drehtisch / mesa giratoria

4270
drillstring; drilling ....; drillstem, – the assemblage of drill-

pipe and collars, and bit. / boorserie / train de tiges / Bohrgarnitur / columna perforadora

drillpipe; drilling rod, – hollow, coupled rods connecting the bit and core barrel in a borehole to the swivel head of a rotary-drill rig on the surface. / boorstang; ....pijp / tige de forage / Bohrstange / varilla de perforación

drill collar, – heavy, thick-walled type of drillpipe, placed at the lower end of the drillstring above the bit. / zwaarstang / masse-tige / Schwerstange / lastrabarrena

4271
(rock) bit, – device that, attached to the drillpipe, acts as rock-cutting tool. Various types are in use, such as soft-formation drag bits, hard-formation roller bits or bits designed for coring; the cutting points or edges may be of hardened steel or for extremely hard rock of diamond. / beitel; kroon / trépan; couronne; outil de forage / Meiszel; Krone / trépano

core bit, – an annular-shaped bit designed to cut a core sample of rock in boreholes. The cutting points may be serrations, diamonds or other hard substances inset in the face of the bit. / kernkroon / trépan carottier / Kernmeissel / barrena cortatestigos

4272
core barrel, – a length of pipe, designed to form the coupling unit between core bit and drillstring; it carries or contains the core produced until raised to the surface. / kernbuis / tube carottier / Kernrohr / tubo sacatestigos; tubo portatestigos

core catcher; core lifter, – a split, fluted ring of spring steel used in a core-barrel assembly to hold and retain core while the core barrel is being hoisted from a borehole. / kernvanger / arrache-carotte / Kernfänger / altrapatestigos

NOTE:
– see table 13 – Bit Designation and Core Size, as in use for mineral surveys and engineering.

4273
core recovery, – the amount of core retained, generally expressed as a percentage of the total obtainable, i.e. the length of the core barrel. / kernopbrengst / récupération de carotte / Kerngewinn / recuperación (de testigo)

4274
conductor; standpipe, – a relatively short length of pipe driven into the soil as the first step of spudding in a borehole. / mantelbuis / tube conducteur; tube guide / Standrohr / tubo; conducción vertical

4275
casing, – steel tube, cemented in a drillhole as protection against, e.g., collapse of the hole, unwanted fluid movement. / verbuizing / tubage au cuvelage / Verrohrung; Casing / tubería de revestimiento

tubing, – steel tube hung in the well through which the production is brought to the surface or liquids are injected into a reservoir rock. / stijgbuis / colonne de production; tubage de .... / Steigrohr / tubería de producción

4276
cement-bond log, – vertical profile of quality of cement-to-casing bond around a cemented casing. It is based on the wave-train intensity of acoustic signals; unsupported free casing makes for strong first arrivals.

4277
open hole, –
1. the uncased part of a borehole. / open gat; onverbuisd gat; onbekleed gat / trou non tubé; trou ouvert / freies Loch; unverrohrtes Loch / pozo desnudo; pozo descubierto;
2. a borehole free of any obstructing object or material.

4278
bailer, – pipe section with valve at the bottom to clean boreholes. / – – / curette; tube à sédiment / Schlammbüchse / cuchara

4279
annular space, –
1. the space between drilling string and casing;
2. the space between casing and tubing.

/ annulaire ruimte / espace annulaire / Ringraum / espacio anular

packer, – an assemblage of expandable rubber rings sealing the annular space at its lower end. / pakker / packer; garniture d'étanchéité; obturateur / Packer / empaquetadura; obturador de empaque

4280
round-trip, – to pull and run drillstring or drillpipes, usually to change the bit or empty the core barrel. / toer / aller-retour / Marsch / maniobra

slip; safety clamp, – any of several types of rod clamps used at the head of a borehole to hold the drillstring while being pulled or lowered. / klembeugel; ....blok / colliers de serrage; coins de retenue / Klemme; Schelle / cuña

4281
(drilling) mud; mudflush; drilling fluid, – water with clay in suspension and with other substances (chemicals for stabilisation, weighting material) added, used for circulation in the borehole during drilling in order to cool the bit, counterbalance formation pressure, stabilize the borehole and bring up drill cuttings. / (boor)spoeling; dik.... / boue (de forage) / (Bohr)spülung / lodo (de perforación)

sludge, – term for drilling mud, used in mineral surveys. / – – / – – / Bohrschlamm / limo

4282
mudscreen, – a vibrating screen through which the mud returning from the hole is sieved retaining the cuttings. / schudzeef / armoire à boue / Schüttelsieb / tamiz de lodo

4283
straight circulation, – circulation takes place downwards inside the drilling string, upwards in the annular space; the usual condition in deep drilling. / normale circulatie / circulation directe / normale Zirkulation / circulación directa

reverse circulation; counterflush, – circulation takes place downwards in the annular space, upwards inside the drillstring. / omgekeerde circulatie / circulation inverse / Verkehrtspülung / circulación inversa

4284
lost circulation, – a condition that occurs when the drilling fluid escapes into crevices or (highly) permeable walls of a borehole and does not or only partly return to the head of the hole. / spoelingverlies / circulation perdue / Spülungsverlust / perdida de circulación

4285
counterflush drilling, – reverse-circulation drilling that, in combination with a hollow bit, allows continuous coring without round-trips. / – – / forage à injection inversée / Saugbohren / – –

4286
stuck drillpipe, – drillpipe sticks in the hole so that it cannot be rotated or moved up and down. / vastgelopen pijp; vastzittende pijp / tige bloqué; tige coincé / festgeklemmtes Gestänge / tubería agarrada

4287
fishing, – operations to remove equipment accidentally left in the drillhole, e.g. parted drillpipe, often a consequence of stuck pipe. / vissen; vangen / repêchage / Fangarbeit / pescar

fishing tool, – apparatus of various types used on the end of a drillstring to fish or remove from the hole lost pieces of drilling equipment or tramp iron. / vanggereedschap; vanggerei / outils de repêchage / Fanggerät / herramientas de pesca; .... de recuperación

lost hole, – a borehole in which the target could not be reached because of caving, squeezing, loose ground, or inability to recover lost tools or junk. / – – / trou perdu / Fehlbohrung / pozo perdido

4288
blowout, – unintentional and uncontrolled eruption of formation content (oil, gas, water) from the well at the surface. / eruptie; spuitende put; spuiter / éruption / Ausbruch / erupción

wild well, – well blowing out. / – – / – – / – – / pozo incontrolado

4289
blowout preventer; BOP, – apparatus installed at the wellhead to keep the well under control during drilling. / eruptieafsluiter / bloc obturateur de puits; (BOP) / Ausbruchschliesser; BOP; Präventer / dispositivo contra erupciones

4290
killing a well, – stopping a threatening or actual blowout, usually by pumping heavy mud until the uncontrolled flow is stopped. / put doodpompen / tuer un puits / Sonde totdrücken / controlar el pozo; 'matar' el pozo

capping a well, – stopping a blowout from the surface by installing special equipment at the wellhead. / – – / fermeture d'un puits; / – – / tapar el pozo

relief well, – a directional well deviated towards a wild well with the purpose of killing the well. / ontlastingsboring / puits de secours; .... d'intervention / Entlastungsbohrung / pozo de auxilio

4291
directional drilling, – drilling deviated wells so as to reach specified subsurface targets, e.g. drainage points located favourably in the reservoir. / gericht boren / forage dirigé / Richtbohren; abgelenkte Bohrung / pozo dirigido

sidetrack(ed) hole, – a deviated hole drilled from a point in an existing drillhole, usually to by-pass a fish or to correct the deviation angle. / – – / – – / abgelenkte Bohrung / pozo desviado

4292
to deflect, – to intentionally change the course of a borehole at a point some distance below the well head. / afbuigen / dévier / ablenken / desviar

4293
deviated hole, – a drillhole that, accidentally or on purpose, deviates appreciably from the vertical. / gedeviëerd gat; scheef gat / sondage dévié / abgelenktes Bohrloch / pozo inclinado; pozo desviado

crooked hole, – a drillhole with (undesirably) strong bends. / krom gat / – – / – – / agujero curvo

4294
borehole survey; deviation ....; directional ...., – the process of determining the course of, and the target point reached by a borehole, using one of several different azimuth and dip-recording devices small enough to be lowered into a borehole; also, the record of the information thereby obtained. / boorgatopmeting / mesure et contrôle du sondage / Bohrlochvermessung / reconocimiento del pozo; .... de la desviación

deflectometer, – a device, usually installed in a borehole, to measure displacements perpendicular or oblique to the direction of the borehole by precise measurement of relative changes in direction between tube segments. / deflectometer / déflectomètre / Deflektometer / deflectometro

inclinometer, – a sonde or torpedo to be lowered into a borehole for the determination of deviation of borehole inclination from the vertical. / inclinometer / inclinomètre / Inklinometer; (Neigungsmessgerät) / inclinometro

4295
acid-dip survey, – a method of determining the angular inclination of a borehole in which a glass test tube-like bottle partly filled with a dilute solution of hydrofluoric acid is inserted in a watertight metal case. When the assemblage is lowered into a borehole and left for 20 to 30 minutes, the acid etches the bottle at a level plane from which the inclination of the borehole can be measured. / etscylinder-hellingmeting / – – / – – / reconocimiento del buzamiento

Tro-Pari survey instrument, – trade name of a single-shot borehole surveying instrument combining a compass and inclinometer, which is locked in place by the action of a preset time clock.

Eastman survey instrument, – various models of a particular make of mechanical and photographic borehole-drift indicators; the single-shot models are small enough to be used in EX (smallest size: 2.15 cm) diamond-drill holes.

multiple-shot instrument, – a borehole-survey instrument capable of taking and recording a series of inclination and bearing readings on a single trip into the borehole.

4296
to ream, – to enlarge the diameter of an existing hole. / ruimen / aléser / erweitern; räumen / ensanchar; rectificar

to underream, – to enlarge or ream a borehole below the casing. / onderruimen / élargir / nachbohren; nachräumen / ensanchar al fondo

4297
pilot hole, –
1. a small hole drilled ahead of a full-sized, or larger, borehole. / loodsgat / avant-trou / vorgebohrtes Loch / pozo piloto;
2. a borehole drilled in advance of mine workings to locate water-bearing fissures or formations. / verkenningsgat / avant-trou / – – / – –

4298
rate of penetration, – velocity of penetration of a bit during drilling. / indringingssnelheid; boorsnelheid / vitesse de pénétration / Eindringungsgeschwindigkeit / velocidad de penetración

drillability, – index value of the resistance of a rock to drilling, used for engineering purposes. / boorbaarheid / forabilité / Bohrbarkeit / sondeabilidad

4299
boring record; .... log, – record of observations made during boring about such aspects as rate of penetration, core recovery, water losses, etc. / boorstaat; boorrapport / coupe de sondage; log de ....; relevé de .... / Bohrprotokoll; Bohrbericht / registro de sondeo

[2830] REMOTE SENSING

[2831] *General Terms*

4300
remote sensing, – acquisition of information on some property of an object or phenomenon by a device that is not in direct contact with the object investigated, and that makes use of the wavelength regions from ultraviolet to short radio waves. By some authors conventional geophysical methods are included, which use is not recommended. / teledetectie / télédétection / Fernerkundung / teledetección

4301
active (sensing) system, – a system possessing its own source of radiation, e.g. radar, and that records this radiation when reflected by the surface or object under investigation. / actief systeem / système actif / actives System / sistema activo

passive (sensing) system, – a system that records radiation from an outside source, that is reflected by the target, e.g. aerial photography, or emitted, e.g. spectroscopy. / passief systeem / système passif / passives System / sistema pasivo

4302
remote-sensing platform, – the platforms in use are:

artificial satellite, – man-made body that revolves in a more or less fixed orbit around the earth or other body of the solar system. / kunstmatige satelliet / satellite artificiel / künstlicher Satellit / satélite artificial

spacecraft, – manned or unmanned space vehicle that is maneuverable. / ruimtevaartuig; ....schip / engin spatial / Raumschiff / vehículo espacial

aircraft, – / vliegtuig / avion / Flugzeug / avión

4303
flight path, – the path with reference to the earth, made or followed in air or space by a man-made body. / vluchtweg / trajectoire de vol / Flugweg / trayectoria de vuelo

flight line, – a line drawn on a map or chart to represent a flight path. / vluchtlijn / ligne de vol / Kurslinie / línea de vuelo

4304
electromagnetic spectrum, – the array of electromagnetic radiation, expressed in wavelengths, frequencies or photon energy. / electromagnetisch spectrum / spectre électromagnétique / elektromagnetisches Spektrum / espectro electromagnético

NOTE:
– see table 14 – The Electromagnetic Spectrum

4305
continuous spectrum, – a spectrum, not structured in detail, gradually varying in intensity from one end of the spectrum to the other. / continu spectrum / spectre continu / Kontinuspektrum; kontinuierliches Spektrum / espectro continuo

absorption spectrum, – a spectrum, containing absorption bands and absorption lines, resulting from the passage of radiation through a selectively radiation-absorbing medium. / absorptiespectrum / spectre d'absorption / Absorptionsspektrum / espectro de absorción

emission spectrum, – a radiation spectrum with different relative intensities, emitted by a radiating body. / emissiespectrum / spectre d'émission / Emissionsspektrum / espectro de emisión

4306
spectral band, – an interval in the electromagnetic spectrum. / spectrale band / bande spectrale / Spektralband / banda espectral

4307
near infrared; solar ...., – the shorter infrared wavelengths extending between about $7 \cdot 10^{-7}$ and around $2 \cdot 10^{-6}$ to $3 \cdot 10^{-6}$ m. It includes the radiation reflected from living vegetal tissue (peaking around $85 \cdot 10^{-8}$ m); it is present during daylight hours only. / nabij infrarood / proche infrarouge / nahes Infrarot / infrarrojo cercano

photographic infrared, – the portion of the spectrum with wavelengths just beyond the red end of the visible spectrum, generally about $7 \cdot 10^{-7}$ to 10^{-6} m, or defined as the useful limits of film sensitivities. / fotografisch infrarood / infrarouge photographique / fotografisches Infrarot / infrarrojo fotografiable

4308
thermal infrared; middle ...., – the wavelengths between around $2 \cdot 10^{-6}$ to $3 \cdot 10^{-6}$ and $25 \cdot 10^{-6}$ m. / thermisch infrarood; midden.... / infrarouge thermique / thermales Infrarot / infrarrojo térmico; .... medio

thermal band, – a general term for wavelengths transmitted through the atmospheric window at $8 \cdot 10^{-6}$ to $13 \cdot 10^{-6}$ m; occasionally also used for the windows around $3 \cdot 10^{-6}$ to $6 \cdot 10^{-6}$ m. / thermische band / bande thermique / thermisches Band / banda térmica

4309
far infrared, – the wavelengths between about $25 \cdot 10^{-6}$ and 10^{-4}m. / ver infrarood / infrarouge lointain / fernes Infrarot / infrarrojo lejano

4310
radiometer, – an instrument for measuring the intensity of radiation in some band of wavelengths. / stralingsmeter / radiomètre / Radiometer; Strahlungsmesser / radiómetro

4311
specific illumination; luminous flux, – the intensity of light impinging on unit area. / specifieke verlichting / illumination spécifique / spezifische Beleuchtung; Lichtstrom / iluminación específica; flujo luminoso

total illumination, – the total amount of light impinging on a surface, measured per unit time. / verlichting / illumination totale / – – / iluminación total

4312
irradiance, – the measure, in power units (W), of radiant flux incident on a surface. / instraling / irradiance / Bestrahlungsstärke / – –

irradiancy, – the measure of radiant flux density incident on a surface (W cm^{-2}). / instralingsdichtheid / – – / – – / – –

4313
radiance, – emitted radiant flux in power units (W) of a body. / stralingsvermogen / radiance / Strahlungsvermögen / – –

radiancy, – emitted radiant flux density from a body (W cm^{-2}). / stralingsdichtheid / – – / Emissionsvermögen / – –

4314
absorptance; absorption factor, – the ratio of the radiant flux absorbed by a body to that incident upon it; either measured over all wavelengths or at a specified wavelength only. / absorptiefactor / absorptivité / Absorptionsgrad / absortancia; índice de absorción

4315
transmittance, – the ratio of the amount of radiant energy transmitted through a body to the amount of energy incident upon it. / stralingsdoorlatendheid / transmissivité / Transmissionsgrad; Durchlassgrad / índice de transmisión

attenuance, – the ratio of the amount of radiant energy decreased by a body to the amount of energy incident upon it. Attenuation of energy is due to reflection, absorption and diffusion; it is the complement of transmittance. / afzwakking / atténuation / – – / – –

4316
reflectance; albedo, – the ratio of the amount of radiant energy that a body reflects to the amount of radiant energy incident upon it. / albedo / réflectivité; albedo / Reflexionsgrad; Rückstrahlvermögen; Albedo / reflectancia; albedo

scatter, –
1. the process by which a rough surface reflects incident radiation;
2. the process of diffusion of a portion of the incident radiation in all directions by small particles suspended in a medium of different refraction indices.

/ verstrooiing / dispersion / Streuung / dispersión

4317
diffuse reflection, – scattering of radiation in many directions by reflection from a surface. / diffuse terugkaatsing; .... reflectie / réflexion diffuse / diffuse Reflexion / reflexión difusa

diffuse reflector, – a surface that reflects incident rays in a multiplicity of directions, due either to irregularities of the surface or because the material is optically inhomogeneous. / diffuse reflector / – – / – – / reflector difuso

4318
specular reflection, – reflection at which all or most angles of reflection are equal to the angle of incidence. / spiegelende terugkaatsing; .... reflectie / réflexion spéculaire / gerichtete Reflexion; Spiegelung / reflexión especular

specular reflector, – a surface that reflects the incident rays like a mirror. / spiegelende reflector / réflecteur spéculaire / – – / reflector especular

4319
resolution cell, – the ground element that is distinguishable on a remote-sensing image. / oplossingselement / tache de résolution / Auflösungselement / resolución espacial

scintillation, – interference of signals from different objects, situated within the same resolution cell, potentially causing loss of returns. / scintillatie; flonkering / scintillation / Szintillation / centelleo

4320
image; imagery, – a two-dimensional representation of a part of the earth's surface (or other body in the solar system) obtained by an artificial satellite or aircraft with a photographic, television, scanning, or radar system. Photographic films are returned to earth, while television, scanning and radar systems may transmit the obtained information by radio to a receiving station on earth, where the image is processed. / afbeelding / image / Abbildung; Bild / imagen

4321
ground information; .... data; .... truth, – information collected on the ground as an aid to the interpretation of remote-sensing surveys. / oppervlakte-gegevens / réalité de terrain / Geländedaten / información en el terreno

4322
boresight camera, – a camera mounted with its optical axis parallel to the axis of a sensor, e.g. radar, to photograph the area being sensed, thus providing location data.

4323
ground control, – relative positions, in horizontal as well as vertical sense of points on the earth's surface, as obtained by ground surveys. / topografische opname / contrôle au sol / (topographische) Geländeaufnahme / control de tierra

4324
photogrammetric control; minor .... – relative positions of points on the earth's surface, obtained by photogrammetric methods only. / fotogrammetrische opname / contrôle photogrammétrique / photogrammetrische Geländeaufnahme / control fotogramétrico

radial triangulation; principle point method; radial plot; minor control plot, – a method of triangulation using overlapping aerial photographs. / luchtfototriangulatie / triangulation par fente radiale / Radialtriangulation / triangulación radial

4325
control point, – any point in a control system that can be identified on a photograph. / controlepunt / point de contrôle / Paszpunkt; Kontroll.... / punto de control

4326
multiband system; multichannel ...., – a system for data collection, by which a target is observed simultaneously in different bands of the spectrum. / meervoudige-band systeem / système multispectral / Multispektral-System / sistema multibando

[2832] *Aerial Photography*

4327
photogeology, – the geological interpretation of, especially aerial, photographs, also of photographs taken from artificial satellites; the compilation of geological maps from photographs. / fotogeologie / photogéologie / Photogeologie / fotogeología

4328
photogrammetry, – method to make reliable measurements and maps of the earth's surface by means of photographs taken from an artificial satellite, an aircraft or the surface. / fotogrammetrie / photogrammétrie / Photogrammetrie; Bildmessung / fotogrametría

4329
aerial camera, – a camera designed for use in aircraft. / luchtfotocamera / chambre de prise de vue aérienne / Luftbildkamera / cámara aérea

mapping camera; surveying ...., – a camera designed for topographic surveying. / karteringscamera / chambre métrique / Messkamera / cámara cartográfica; .... agrimensora

4330
multilens camera, –

1. a camera, possessing two or more lenses pointing at the same target, that can produce multiband photographs, if different film/filter combinations are used;
2. a camera, possessing two or more lenses at an angle to one another, by which two or more overlapping photographs can be made simultaneously.

/ multilenscamera / chambre multiobjectifs / Mehrlinsenkamera / cámara multilente

multiband camera, – a camera, exposing different areas of one film, or more than one film, using two or more lenses equipped with different filters or one lens and a beam splitter, to provide two or more photographs in different spectral bands. / multiband-camera / chambre multibande / Multispektralkamera / cámara multibanda

4331
metric camera, – a specially constructed and calibrated camera used to obtain geometrically accurate photographs for use in photogrammetric instruments. / metrische camera / chambre métrique / Messkamera / cámara métrica

4332
continuous-strip camera, – a camera in which a continuous-strip exposure is made by moving the film continuously past a narrow slit at a speed proportional to the angular velocity of the object to be photographed in respect to the camera. / – – / chambre à défilement (continu) / Streifenkamera / – –

4333
horizon camera, – a camera used in combination with another aerial camera to photograph the horizon at the same moment that the other photographs are taken. From the horizon photographs, the tilts of the other photographs can be measured. / horizoncamera / – – / Horizontkamera / cámara de horizonte

4334
orthochromatic film, – film recording black to white tone values that correspond to those of the colours present in nature; in practice film, sensitive to blue and green only. / orthochromatische film / pellicule orthochromatique; film .... / orthochromatischer Film / película ortocromática

panchromatic film; broad-band film, – film sensitive to a large interval of the spectrum, e.g. to the entire visible part. / panchromatische film / pellicule panchromatique; film .... / panchromatischer Film / película pancromática

4335
colour-blind film, – film sensitive to violet, blue and blue-green. / – – / – – / – – / película alocromática

colour-sensitive film, – film sensitive to both, the higher and the lower part of the visible spectrum; it may be either orthochromatic or panchromatic. / – – / film couleur / – – / – –

4336
infrared film, – film, especially sensitive to near-infrared and blue light; blue light is cut out by a red filter. / infraroodfilm / film infrarouge / Infrarot-Film / película infrarroja

4337
false-colour film; colour-infrared film, – film, sensitive for the ultraviolet, visible and photographic infrared parts of the spectrum, recorded in a spectral colour of the frequency band one higher than the natural radiation, e.g. infrared in

red, red in green, green in blue. Blue and ultraviolet are cut out by a yellow filter. / onjuiste-kleurenfilm / film fausse-couleur; film infrarouge couleur / Falschfarbenfilm / película falso color; .... color-infrarroja

4338
filter, – any material which selectively modifies the radiation that is transmitted through an optical system. / filter / filtre / Filter / filtro

NOTE:
– many special-purpose filters are in use, of which a few examples are given.

4339
contrast filter, – / contrastfilter / filtre de contraste / Kontrastfilter / filtro de contraste

polarizing filter, – / polarisatiefilter / filtre de polarisation / Polarisationsfilter / filtro de polarización

vignetting filter, – / vignetteringsfilter / – – / Vignettierungsfilter / filtro fotográfico

4340
exposure, – the total quantity of light received per unit area of film. / belichting / exposition / Belichtung / exposición

exposure latitude, – the ratio of maximum to minimum exposure limits yielding satisfactory results. / belichtingsspeelruimte / – – / Belichtungsspielraum / margen de exposición

4341
sensitometry, – the measurement of the response of a photosensitive material under the influence of light. / sensitometrie / sensitométrie / Sensitometrie / sensitometría

4342
density, – a measure of the amount of silver deposited by exposure and development in a given area of a photograph, expressed as the percentage of light passing through that area. / dichtheid / densité / Dichte / densidad

4343
contrast, – in photography, the degree of differentiation between variations in shade between black and white. / contrast / contraste / Kontrast / contraste

subject contrast, – the difference in light intensity between the brightest light end the deepest shadow in a photographed subject. / onderwerpcontrast / contraste objet / Subjektkontrast / contraste del tema; .... del motivo

gradation, – the range of tones from the brightest light to the deepest shadow. / gradatie / gradation / Gradation / gradación

4344
granularity, – the appearance of small dark dots on a developed photographic image, particularly evident on enlargements. / korreligheid / granularité / Körnigkeit / granulado

4345
texture, – in a photograph, the frequency of change and arrangement of tones, e.g. fine, medium, coarse, stippled or mottled. / textuur / texture / Textur / textura

4346
colour composite, – a colour picture produced by assigning different colours to images taken in different spectral bands, and printing them together in overlap. / – – / composition colorée / Farbkomposit / falso color compuesto

4347
resolution, – the ability of a photographical system to render a sharply defined image. / oplossend vermogen / résolution / Auflösung(svermögen) / resolución

optical definition, –
1. the ability of a lens to record fine detail;
2. the distinctness or clarity of detail on a photograph.
/ optische scherpte / pouvoir de résolution / optische Schärfe / definición óptica

4348
camera station; exposure ...., – the point in space, air or on the ground, occupied by the camera at the moment of exposure. / – – / – – / Aufnahmeort / situación de la cámara; punto de toma

air station, – the point in the air, occupied by the camera at the moment of exposure. / – – / – – / Aufnahmeort / punto de toma aérea

4349
aerial photograph; .... photo, – / luchtfoto / photographie aérienne / Luftbild / fotografía aérea

survey photograph; .... photo, – an aerial photograph taken for topographic survey purposes. / – – / mission photographique / Messbild / – –

satellite photograph, – photograph taken from an artificial satellite, rendering a small-scale image of a large part of the earth's surface, usually as a colour composite. / satellietfoto / photographie de satellite / Satellitenbild / fotografía de satélite

4350
air base, – the distance between two aerial camera stations. / luchtbasis / base / Aufnahmebasis / base aérea

4351
tilt, – the angle between the optical axis of the camera and the vertical. / scheefstelling / déversement / Nadirdistanz / inclinación

relative tilt, – the tilt of a nearly vertical photograph with reference to an arbitrary, not necessarily horizontal plane, e.g. that of one of the adjoining photographs in a strip. / relatieve scheefstelling / déversement relatif / relative Nadirdistanz / inclinación relativa

4352
image-motion compensator, – a device installed with certain aerial cameras to compensate for the forward motion of an aircraft to avoid blurring of the photographs taken. / – – / – – / Bildwanderungs-Kompensator / cinederivómetro

4353
vertical (aerial) photograph, – an aerial photograph made with the optical axis of the camera as nearly vertical as practicable. / vertikale luchtfoto / photographie verticale / Senkrechtaufnahme / fotografía (aérea) vertical

oblique (aerial) photograph, – an aerial photograph taken with the camera axis in a direction between horizontal and vertical. / scheve luchtfoto / photographie oblique / Schrägaufnahme / fotografía (aérea) oblicua

lateral-oblique photograph, – an oblique aerial photograph taken with the camera axis as nearly as possible normal to the flight line. / zijwaarts scheve luchtfoto / photographie oblique latérale / Lateral-Schrägaufnahme / fotografía lateral-oblicua

4354
horizon photograph, – an aerial photograph of the horizon taken simultaneously with another one to obtain an indication of the orientation of the latter. / horizonfoto / – – / Horizontbild / fotografía de horizonte

4355
wing photograph, – a photograph taken by one of the side lenses of a multilens camera.

4356
split-vertical photography, – photography with two aerial cameras, operated simultaneously and mounted at opposite angles from the vertical, in order to obtain a small overlap on the produced photographs. / – – / – – / Konvergentbilder / – –

4357
composite photograph, – an assemblage, made of the separate photographs made by each lens of a multilens camera in a simultaneous exposure; it equals a photograph taken with a single wide-angle lens. / samengestelde foto / composition photographique / – – / fotografía compuesta

4358
photo strip, – any number of photographs taken along a flight line, usually at an approximately constant altitude. / fotostrook / bande / Bildstreifen / pasada fotográfica

run, – the line followed by an aircraft in making a photo strip. / vluchtlijn / ligne de vol / Flugweg / línea de vuelo

frame, – any individual member of a continuous sequence of photographs. / foto / cliché / Bild / fotograma

4359
overlap, – the amount by which a photograph includes the area covered by another one, customarily expressed as a percentage or a fraction. / overlapping / recouvrement / Überdeckung / solape

longitudinal overlap; forward lap (USA), – the overlap between aerial photographs in the same flight line. / vliegrichting-overlapping / recouvrement longitudinal / Längsüberdeckung / solape longitudinal

lateral overlap; side lap (USA), – the overlap between aerial photographs of adjacent parallel flight lines. / zijwaartse overlapping / recouvrement latéral / Querüberdeckung / solape lateral

4360
effective area, – for any aerial photograph, which is one of a series in a flight strip, the central part of the photograph bounded by the bisectors of the overlaps with the adjacent photographs. / effectief oppervlak / surface exploitable / überdeckter Bildteil / área útil

4361
stereoscopic coverage, – aerial photographs taken with sufficient overlap to permit complete stereoscopic examination. / stereoscopische dekking / couverture stéréoscopique / stereoskopische Überdeckung / recubrimiento estereoscópico

4362
stereoscopic pair; stereopair, – two photographs with sufficient overlap to enable stereoscopic examination. / stereopaar; stereoscopisch paar / couple stéréoscopique / Stereopaar / par estereoscópico

stereoscopic triplet; stereo...., – a series of three photographs, of which the end members overlap sufficiently on the central one to enable complete stereoscopical examination of the latter. / stereotriplet / – – / Stereo-Triplet / terna estereoscópica

4363
stereobase; stereoscopic base, – a line representing distance and direction between corresponding image points on a stereopair correctly oriented and adjusted for comfortable stereoscopic vision. / stereobasis / – – / Stereobasis / base estereoscópica

4364
parallax, – the apparent displacement of the position of an object with respect to its surroundings caused by a shift in the position of the observer. / parallax; verschilzicht / parallaxe / Parallaxe / paralaje

4365
stereocomparator; stereometer, – a stereoscope equipped with special aids for measuring parallax. / stereometer; stereocomparator / stéréocomparateur / Stereokomparator / estereómetro

parallax wedge, – a simplified stereometer for obtaining object heights at examination of stereoscopic pairs, consisting of two converging series of dots or graduated lines on a transparent templet, that can be stereoscopically fused into a single line or row. / parallaxwig / barre à parallaxe / Stereometer / cuña de paralajes

4366
stereoscope, – a binocular instrument for viewing a stereopair of photographs in order to obtain the impression of a three-dimensional model. / stereoscoop / stéréoscope / Stereoskop / estereoscopio

4367
orthostereoscopy, – stereoscopic viewing in which horizontal and vertical scales appear to be the same. / orthostereoscopie / orthostéréoscopie / Orthostereoskopie / ortoestereoscopio

hyperstereoscopy, – stereoscopic viewing in which the vertical scale is exaggerated in comparison to the horizontal one, being the usual condition in viewing aerial photographs in stereopairs. / hyperstereoscopie / hyperstéréoscopie / Hyperstereoskopie / hiperestereoscopio

inverted stereo; pseudoscopic ...., – stereoscopic viewing in which the impression of relief is the reverse of that actually existing, which takes place when the order of a stereo pair is inverted. / omgekeerd ruimtebeeld / stéréoscopie inversée / pseudostereoskopie / estereoscopio de inversión

4368
fiducial mark; collimating mark, – one of the four index marks, rigidly connected with the camera lens, of which each forms an image on the corresponding edge of the negative. / ijkmerk / repères de fond de chambre / Rahmenmarke / marcas fiduciales; .... de colimación

fiducial axis; photograph axis; geometric axis; plate axis, – the line joining opposite fiducial marks on a photograph. / ijkingsas / – – / – – / ejes fiduciales

4369
principal point, – on an aerial photograph the crossing point of the two fiducial axes; if truly not-tilted it lies vertically below the camera at the moment of exposure. / hoofdpunt / point principal / Hauptpunkt / punto principal

4370
calibration templet, – a calibrated templet, used for the rapid and accurate marking of principal points on a series of photographs; and, for a multilens camera, used in assembling the photographs into a composite one. / ijk-malplaat / – – / – – / patrón de calibre

4371
photo base, – the distance between the principal points of two adjoining vertical aerial photographs. / fotobasis / base / Bildbasis / base del fotograma

4372
transposed principal point, – the principal point, stereoscopically transferred to the adjoining photographs of a stereotriplet. / nevenpunt / – – / Hauptpunktübertragung / punto principal conjugado

centre line, – a line extending from the principal point through the transposed principal points. / middenlijn / ligne des centres / Fluglinie / línea de centros

4373
parallactic grid, – a uniform pattern of rectangular lines engraved on some transparent material and placed over the photographs of a stereoscopic pair or in the optical system of a stereoscope, to provide a continuous floating-mark system. / parallactisch rooster / grille parallactique / parallaktisches Netz / plantilla paraláctica

perspective grid, – a network of lines, drawn or superimposed on a photograph, representing the perspective of a systematic network of lines on the ground or datum plane. / perspectivisch rooster / grille perspective / perspektivisches Netz; (Möbius-Netz) / plantilla perspectiva

4374
grid method, – a method of plotting detail from an oblique photograph by superimposing a perspective or a map grid on the photograph and transferring the detail by eye. / roostermethode / – – / Möbius-Netz-Methode / método del reticulado

4375
gridded oblique, – an oblique air photograph printed with a superimposed grid to assist in the determination of positions of areas or points on the photograph. / – – / – – / Schrägbild mit überlagertem Netz / reticulado oblicuo

4376
rectification, – the process to project a tilted or oblique photograph in such a way, that aberrations in the positions of points on the photograph with relation to their true positions are corrected. / onthoeking / redressement / Entzerrung / rectificación

4377
transforming printer, – a specially designed projection printer for use with a particular multilens camera for rectification of the oblique negatives taken by that camera. / onthoeker / redresseur d'images / – – / impresor de transformación

transformed print, – a photographic print made by a transforming printer. / onthoekte afdruk / image redressée / – – / fotografía transformada

4378
ratio printing, – the printing of a photograph copy at enlarged or reduced scale with respect to the original, in order to adjust its scale to that of an adjoining photograph. / – – / – – / – – / relación de impresión

4379
relief displacement; height ...., – owing to the perspective of an aerial photograph, the difference in position of a photographed point with respect to the datum position of that point. / reliëfverplaatsing / – – / (radiale) Geländeversetzung / desplazamiento del relieve; .... de las alturas

tilt displacement, – displacement of photographed points on a tilted photograph, radially outward or inward with respect to the isocentre. / scheefstellingsverplaatsing / – – / – – / desplazamiento por basculamiento

4380
mosaic; laydown, – an assemblage of overlapping aerial photographs, forming a continuous representation of a part of the earth's surface. / mozaiek / mosaïque / Mosaik / mosaico

strip mosaic, – a mosaic consisting of one strip of aerial photographs taken on a single flight. / fotostrookmozaiek / – – / Streifenmosaik / mosaico de banda

4381
uncontrolled mosaic, – a mosaic composed of unrectified prints, details of which have been matched from print to print without ground control or other information. / niet gecorrigeerde mozaiek / mosaïque non contrôlée / Mosaik aus nicht entzerrten Bildern / mosaico no controlado

semi-controlled mosaic, – a mosaic of rectified or unrectified prints composed according to a common basis of orientation other than ground control. / ten dele gecorrigeerde mozaiek

/ mosaïque semi-contrôlée / Mosaik aus grob entzerrten Bildern / mosaico semicontrolado

controlled mosaic, – a mosaic composed according to ground control, and in which prints are used that have been ratioed and rectified as shown to be necessary by the control. / gecorrigeerde mozaiek / mosaïque contrôlée / Mosaik aus entzerrten Bildern / mosaico controlado

4382
photo index, – composite photograph mosaic map made by assembling individual photographs into their proper relative positions and copying the assembly by photographical means at a reduced scale. / – – / photo index / – – / fotografía índice

4383
photo map, – a map composed from photographs. / fotokaart / photocarte / Fotokarte / fotomapa; ....plano

[2833] Ra*dio* D*etection* a*nd* R*anging*

4384
radar, – a method and the equipment to obtain images of the earth's surface and/or locate and identify distant objects by radio techniques. The transmitted radio waves, usually in the 3 cm band, are scattered by material objects, and the returning waves are recorded. By emitting pulses of high-frequency waves accurate measurement of distance is possible. Its widest application is found in navigation; it is in use also to obtain geomorphological and hydrological data. / radar / radar / Radar / radar

4385
sidelooking radar; SLAR, – an all-weather, day and night airborne radar, particularly able to image large areas of terrain, and that produces photo-like pictures. / – – / radar latéral / seitwärts schauendes Radar / radar lateral

brute-force radar, – an airborne sidelooking radar system equipped with a long physical transmitting and receiving antenna, which narrows the beam width and increases azimuth resolution. The received signal is used directly to produce an image. / – – / – – / nicht kohärentes Radar / radar de aperture real

4386
synthetic-aperture radar; SAR, – radar which integrates multiple returns from the same ground cell, in this way constructing a synthetically long apparent or effective aperture, and that uses Doppler effects for the production of phase-history film or tape that can be digitally or optically processed to produce an image. / – – / radar à ouverture synthétique / Radar mit synthetischer Apertur / radar de apertura sintética

4387
coherent radar, – a radar system in which the phase relation between transmitted and received signals is measured and utilized (normally to obtain Doppler information). / – – / – – / kohärentes Radar / radar coherente

continuous-wave radar; CW radar, – a species of radar, that transmits modulated or unmodulated continuous waves, and of which the simplest forms detect only targets that move in respect to the radar system. Continuous-wave radar systems may be used in airborne mapping of ground features. / – – / – – / – – / radar de onda contínua

4388
data film; signal film. – the film on which phase history and intensity of the returned waves are recorded. / opslagfilm; registratiefilm / – – / Datenfilm / película registradora

phase history, – the development of the wave phase with time. / faseverloop / – – / Phasenverlauf / – –

optical correlator, – an instrument that makes use of the phase histories, as recorded on the data film, to make a radar image. / optische correlator / corrélateur optique / optischer Korrelator / correlacionador óptico

4389
radar beam, – the beam of energy produced by the radar transmitter. / radarstraal / faisceau radar / Radarkeule / – –

boresight, – the direction of maximum transmitted signal from the transmitting antenna.

aspect angle; look angle; look direction, – the angle at which the radar beam is directed towards an object. / invalshoek / – – / Depressionswinkel / – –

4390
return; echo, – radar wave reflected by an object and recorded. / echo / écho / Echo / eco

ground return, – radar echoes reflected from the terrain. / grondecho / – – / Geländeecho / eco del terreno

4391
no-return, – an area on the radar imagery that is devoid of returns. / echoloos / – – / ohne Echo / – –

radar shadow, – a no-return area extending beyond an elevated object that, cutting the radar beam, prevents illumination of the area behind it. / radarschaduw / – – / Radarschatten / sombra de radar

4392
radar cross section, – the ratio of power return in a radar echo to power received by the target reflecting the signal. / radardoorsnede / – – / – – / – –

radar reflectivity, – in general, a measure of the efficiency of a radar target in reflecting a radar signal. / radarreflectiviteit / – – / Radarreflektivität / reflectividad de radar

4393
range, – the distance from the radar antenna to the target; in sidelooking radar, the perpendicular distance from target to flight direction. / afstand; afstandsbereik / – – / Strecke / – –

slant range, – distance along a straight line from the radar antenna to the target. / ruimtelijke afstand / – – / Schrägstrecke / – –

ground range, – the distance of flight line to target. / grondafstand / – – / Grundstrecke / – –

4394
range direction, – the azimuth of the line from radar antenna to target. / radarazimuth / – – / Radarazimut / – –

4395
slant-range image, – a sidelooking radar image in which objects are located at distances corresponding to their slant-range distances. / – – / – – / Schrägstreckenbild / – –

ground-range image, – a sidelooking radar image in which objects at datum level are located at distances corresponding to their separation in the terrain. / – – / – – / Grundstreckenbild / – –

4396
range streaking, – extensions of normal target images in the range direction.

4397
extended range reflection, – a pulse reflected from an object beyond normal range and recorded while the subsequent pulse returns.

4398
range marks, – artificially produced marks on the sidelooking radar-image film, indicating increments of distance in either ground range or slant range. / radar-afstandsmarkering / – – / Abstandsmarken / – –

4399
layover, – displacement of the top of an elevated feature with respect to its base on the sidelooking radar image.

4400
radar map, – a map made through the application of radar techniques. / radarkaart / carte radar / Radarkarte / mapa de radar

[2834] *Scanning*

4401
scanning, – the sweep of an antenna, prism mirror or other element in an instrument, airborne or in a spacecraft or artificial satellite, across the flight direction, as it is applied in remote-sensing instruments. / aftasten / baloyage / Abtasten / barrido

4402
line scanning, – method of scanning, by which the scanning motion describes lines perpendicular to the direction of movement of the remote-sensing platform. / lijnaftasting / baloyage linéaire / Zeilenabtasten / barrido lineal

circular scanning, – method of scanning, by which a mirror revolves around an axis perpendicular to the surface that is scanned. / cirkelvormige aftasting / baloyage circulaire / konisches Abtasten / barrido circular

4403
scanner, – a radiometer that, by using the rotational or oscillating movement of a plane mirror or prism, scans a path normal to its movement. The incoming radiation is converted by an electro-optical system in such a way, that a photograph-like picture of the scanned object is produced. / aftaster / radiomètre à baloyage / Abtaster / barredor; equipo de barrido

multispectral scanner, – a scanner that can scan an object simultaneously in different wavelength bands, resulting in the production of two or more different images of the same object. / multispectrale aftaster / radiomètre multispectral à baloyage / multispektraler Abtaster / barredor multiespectral

4404
barrel distortion, – a type of distortion in scanning imagery, caused by the combination of the forward motion of the remote-sensing platform and the scanner-mirror rotation. If straight lines are cut obliquely by the flight line, they will be distorted in a sigmoidal way on the image. / – – / – – / S-förmige Verzeichnung / distorsión en barril

[2840] DEEP SEISMIC METHODS

NOTE:
– see also section
[0320] Seismology

[2841] *Field Procedures*

4405
refraction method, – seismic procedure utilizing seismic waves transmitted along the top of high-velocity layers. / refractiemethode / sismique réfraction / Refraktionsverfahren / método de refracción

reflection method, – seismic procedure utilizing body waves returned to the surface by reflection at interfaces. / reflectiemethode / sismique réflexion / Reflexionsverfahren / método de reflexión

4406
well-shooting survey, – method of determining the velocity distribution by generating waves at the surface and recording these in a deep well with the aid of a well seismometer. It can also provide check shots for adjusting CVL-derived time-depth relations. / putschieten / carottage sismique / Bohrlochversenkmessung / sondeo sísmico

continuous velocity log; CVL, – detailed record of the velocity-depth relation obtained by means of seismic transducers and receivers lowered into a well. / continue snelheidslog / carottage continu des vitesses / akustik Log; sonic Log / diagrafía de velocidades continuas

4407
coverage, – segment of an interface in the subsurface that is covered with reflection points. / bedekking / couverture / Überdeckung / cobertura

multiple coverage, – term used to indicate that the same part of the subsurface is covered by different shots for the purpose of stacking. / multipele bedekking; meervoudige .... / couverture multiple / Mehrfachüberdeckung / cobertura múltiple

4408
continuous profiling, – field procedure in which the reflecting horizons are covered along the shooting line with a set of closely and regularly spaced reflection points. / continu profielschieten / tirer en profil continu / kontinuierliches Profilschiessen / corte continuo

4409
weight dropping, – seismic reflection method in which waves are generated by dropping weights. / valmethode / sismique par chute de poids / Fallgewichtsseismik / caída de pesos

vibroseis, – seismic reflection method in which the signal is transmitted into the earth by means of a vibrator. The signal lasts several seconds and sweeps gradually through a certain frequency range. After cross-correlating the recording with the transmitted signal, an acceptable power of resolution can be reached.

4410
air gun, – device used in marine seismic exploration for generating waves by letting compressed air abruptly expand in the water.

sparker, – marine seismic source that generates waves by means of an electric discharge through the water. / sparker / sparker / Sparker / sparker

4411
shotpoint, – location of the shothole or shothole pattern. In the latter case, the centre of the pattern is often used to indicate the location of the whole pattern. / schotpunt / point de tir / Schusspunkt / punto de tiro

shothole pattern, – group of shotholes fired simultaneously. / schotgatpatroon / point de tir en nappe / Schiessanordnung / – –

4412
channel, – path along which data flow from the detector station to the recording device. / kanaal / canal / Kanal / canal

4413
detector, – tool used to transform seismic particle movement or pressure variations into electric signals. / detector / détecteur / Detektor / detector

4414
seismometer; geophone, – detector responding to the particle velocity caused by seismic waves. / seismometer; geofoon / géophone / Seismometer; Geophon / sismómetro; geófono

particle velocity, – speed at which particles move during passage of a seismic wave. / deeltjessnelheid / vitesse des particules / Partikelgeschwindigkeit / velocidad de las partículas

4415
pressure detector; hydrophone, – detector responding to the pressure variations caused by seismic waves. / drukdetector; hydrofoon / hydrophone / Druckaufnehmer; Hydrophon / detector de presión; hidrófono

4416
seismometer array; .... pattern; bunched seismometers, – group of seismometers in an ordered geometrical arrangement and connected to one recording channel. / seismometerpatroon / réseau de géophones / Geophonanordnung / red de sismómetros

seismometer spacing; detector ...., – distance between adjacent seismometers in an array. / seismometerafstand / intervalle entre géophones / Seismometerabstand / intervalo entre sismómetros

4417
detector station, – location of a seismic array; in general the centre of the array is used to indicate this location. / detectiestation / station de réception / Empfangstation / estación detectora .

seismometer spread; spread, – set of detector stations used to record the seismic waves generated by one shot. Generally the stations in a spread are located at regular intervals along a line. / seismometeropstelling / dispositif des géophones / Seismometeraufstellung; Seismometerauslage / tendido sismométrico

4418
station spacing, – distance between adjacent detector stations in a spread. / stationsinterval / intervalle entre les stations / Stationsabstand / intervalo de estaciones

4419
straddle spread; centre ....; split ...., – layout with the shotpoint at the centre of the seismometer spread. / centrale opstelling / tir central / Zentralaufstellung / tiro central

cross spread, – seismometer spread perpendicular to the line along which the shotholes are located. / dwarsopstelling / dispositif transversal / Queraufstellung / tendido perpendicular

[2842] *Seismic Recording and Processing*

4420
record; seismogram ,– set of traces pertaining to one shot. / record; seismogram / enregistrement; sismogramme / Aufnahmegeschuss / registro

trace, – registration of a directly observed or processed seismic vibration as a function of time. / snaar; spoor / trace / Spur / traza

4421
analog recording, – recording method in which data are represented by some physical quantity such as height of a curve, degree of blackening or degree of magnetisation. / analoge registratie / enregistrement analogique / analoge Registrierung / registro analógico

wiggle recording; squiggle ...., – analog optical recording method by means of a curve; its height indicates the signal level. / wriggel-opname / enregistrement galvanométrique / Linienschrift-Aufnahme / – –

variable-area recording; VAR, – analog optical recording method by means of a trace of constant width that is partly black, partly white; the width of the black part indicates the signal level. / veranderlijke-oppervlakteregistratie / enregistrement en aire variable / Flächenschrift-Aufzeichnung / registro de superficie variable

variable-density recording, – analog optical recording method by means of a trace of constant width and variable degree of blackening; the degree of blackening indicates the signal level. / veranderlijke-dichtheidsregistratie / enregistrement en densité variable / Dichteschrift-Aufzeichnung / registro de densidad variable

4422
digital recording, – recording procedure in which data are sampled and stored on a digital tape. / digitale opname / enregistrement digital / digitale Registrierung / registro digital

4423
sampling, – in digital recording, reading the signal at regular time intervals (= sampling interval). / bemonstering / échantillonage / Digitalisieren / muestreo

aliasing, – phenomenon that sinusoidal waves of different frequencies can produce the same sampled signal. / vouwen van frequenties / aliasage / Alias-Effekte / – –

4424
time multiplexing, – sampling procedure in which the first samples of all the traces are recorded first, followed by the second samples etc. / tijdmultiplexen / multiplexage / Multiplexen / – –

demultiplexing, – recording the samples of a time-multiplexed recording so that the samples of each trace are grouped together. / demultiplexen / démultiplexage / Demultiplexen / – –

4425
automatic gain control; automatic volume control; AGC; AVC, – automatic procedure of normalising average signal amplitude by slowly varying amplification. / automatische volumeregeling; automatische sterkteregeling; ASR / contrôle automatique de gain / automatische Amplitudenregelung / control automático de ganancia; control automático de volumen

4426
gain ranging, – instrumental recording procedure where the amplification of the signal is automatically stepped up or down bij a factor 2, 4, 8 or 16 in response to changes in signal level

4427
shot break; time ...., – registration of the moment at which the seismic wave is initiated. / – – / instant de tir / Abriss / – –

4428
first arrival; first kick; initial, – first arrival of seismic energy on a trace. / eerste inzet / première arrivée / Ersteinsatz / primera llegada

uphole time, – first-arrival time recorded at the surface near the shothole. / schietgatstijd / temps vertical / Aufzeit / tiempo vertical

4429
datum correction, – correction applied to travel times to allow for the fact that source and detector do not lie on the reference level (= datum). / correctie naar het reductieniveau / correction de plan de référence / Korrektion auf Bezugsniveau / corrección del nivel de referencia

4430
weathering correction, – correction to allow for the time delay caused by an irregularly shaped low-velocity zone immediately underlying the surface. / verweringscorrectie / correction de la zone altérée / Verwitterungskorrektur / corrección de la zona de alteración

4431
static correction, – correction of the reflection time depending on shot and geophone positions, but not on reflector depth. It is generally composed of datum and weathering corrections. / statische correctie / correction statique / statische Korrektur / corrección estática

4432
normal-moveout correction; NMO ....; dynamic ...., – correction applied to observed reflection times so as to eliminate the influence of the horizontal distance between shothole and detector station. / horizontale-afstandscorrectie / correction dynamique / dynamische Korrektur / corrección dinámica

4433
stacking, – summing of the traces of a common-midpoint gather after correction. / stapeling / sommation / Stapeln / – –

common-midpoint gather; common-depth-point ....; CDP ....; common-subsurface-point ...., – set of seismic recordings that have the same midpoint between shotpoint and detector station. / groep met gemeenschappelijk middenpunt / groupement en position miroir / Sortierung nach gemeinsamen Untergrundpunkten / – –

4434
stacking velocity, – velocity used for computing the normal-moveout correction applied before adding traces of a CDP gather in the stacking procedure. / stapelingssnelheid / vitesse de sommation / Stapelgeschwindigkeit / – –

4435
signal, – expression of a specific wave phenomenon on a seismogram. Often signal is used in the more limited sense of primary reflection. / signaal / signal / Signal / señal

4436
reflection, – phenomenon of seismic waves being partly returned at an interface. Also used to indicate the signals on seismic sections that correspond with these reflected waves. / reflectie / réflexion / Reflexion / reflexión

4437
primary reflection, – signal due to a body wave reflected at one interface only. / primaire reflectie / réflexion primaire / Primärreflexion / reflexión primaria

4438
seismic noise, – any part of the seismic recording that is considered undesirable and hampers the recognition of the desired signal. / seismische ruis / bruit sismique / seismischer Noise; seismisches Rauschen / ruido sísmico

ambient noise; background noise, – seismic disturbance not caused by the shot. / achtergrondsruis / bruit de fond / regelloses Noise / ruido de fondo

shot-generated noise, – seismic noise generated by the shot / schotruis / bruit du tir / Schuss-Noise / ruido del tiro

4439
ground roll; roller, – seismic surface wave generated by the shot. / oppervlaktegolf / onde de surface / störende Oberflächenwelle; Roller / onda superficial

4440
bubble pulse, – undesirable repetition of the outgoing signal, caused by oscillation of the gas bubble generated by certain marine seismic sources.

reverberation; ringing, – resonance phenomenon that causes repetitions of reflection signals at only slowly decreasing amplitude, often observed on marine seismograms. / reverberatie / réverbération / Reverberation / reverberación

4441
multiple reflection, – signal due to a body wave that has been reflected at more than one interface. / meervoudige reflectie; herhaalde .... / réflexion multiple / Mehrfach-Reflexion; multiple Reflexion / reflexión múltiple

4442
ghost, – undesired repetition of the signal caused by additional reflection at an interface above the level of the source and/ or the detector. / – – / fantôme / – – / fantasma

4443
signal-to-noise ratio, – ratio of the power of desired reflections to that of undesired noise. / signaal-ruis-verhouding / rapport signal-bruit / Nutz-zu-Störsignalverhältnis / relación señal-ruido

4444
diffraction, – special type of wave that is generated when an incident wave meets a sharp corner or edge. / diffractie / diffraction / Diffraktion / difracción

4445
filter, – a device that converts an incoming signal into a modified outgoing signal. For electrical filters the incoming and outgoing signals are varying voltages or currents. Digital filters operate on digitized signals by means of computer operations. / filter / filtre / Filter / filtro

input, – incoming signal of a filter. / invoer / signal d'entrée / Eingangsignal / señal de entrada

output. – outgoing signal of a filter. / uitvoer / signal de sortie / Ausgangsignal / señal de salida

4446
impulse response, – output of a filter when the input is a single pulse of short duration. / impulsresponsie / réponse impulsionelle / Impulsantwort / respuesta punctual

4447
operator, – impulse response of a digital filter. / operator / opérateur / Operator / operador

4448
frequency response; transfer function, – response of a filter to a sinusoidal input of unit amplitude as a function of the frequency. This response can be expressed in terms of the amplitude and the phase of the output with respect to the input. / frequentie-karakteristiek; overdrachtsfunctie / réponse en fréquence / Frequenzgang / respuesta de frecuencia

amplitude response; .... characteristic, – the amplitude part of a frequency response. / amplitudekarakteristiek / réponse en amplitude / Amplitudengang / respuesta de amplitud; característica de ....

phase response; phase characteristic, – the phase shift part of a frequency response. The phase shift is the time shift of a sinusoidal component of the output with respect to the corresponding component of the input, in general expressed in terms of an angle, a shift of a complete period amounts to 2π radians (=360°). / fasekarakteristiek / réponse en phase / Phasengang / respuesta de fase; característica de fase

4449
power, – average of the square of the trace over a certain time interval. / vermogen / puissance / Leistung / potencia

4450
cross correlation, – the integral (sum) of the product of two signals shifted in time with respect to each other, as a function of the shift. / kruiscorrelatie / hétérocorrélation / Kreuzkorrelation / correlación cruzada

autocorrelation; autocovariance, – cross correlation of a signal with itself. Strictly speaking, the autocorrelation should be normalised to unity for the time shift zero, and the autocovariance should be the integral divided by the length of the integration interval (resp. the sum divided by the number of terms). Often, however, neither of these two definitions is adhered to and the normalisation is not applied. / autocorrelatie; autocovariantie / autocorrélation; autocovariance / Autokorrelation; Autokovarianz / autocorrelación

4451
Fourier transform; frequency spectrum, – relations between amplitude and frequency, and between phase and frequency of the sinusoidal components into which a signal can be decomposed. / Fouriergetransformeerde; frequentiespectrum

/ transformée de Fourier; spectre des fréquences / Fouriertransformierte; Frequenzspektrum / transformación de Fourier

amplitude spectrum, – the amplitude part of a Fourier transform. / amplitudespectrum / spectre des amplitudes / Amplitudenspektrum / espectro de amplitudes

phase spectrum, – the phase part of a Fourier transform. The phase of a sinusoidal component is its time shift with respect to a sinusoidal signal of the same frequency having its peak at time zero. In general it is expressed in terms of an angle, a shift of a complete period amounts to 2π radians (=360°). / fasespectrum / spectre des phases / Phasenspektrum / espectro de fases

power (density) spectrum, – square of the amplitude spectrum. / vermogensdichtheidsspectrum / spectre des puissances / Leistungsdichtespektrum / espectro de potencias

4452
convolution, – cross correlation of two signals after reversing one of them in time. Often used for computing the output of a filter. In this case the two signals are the input and the impulse response. / convolutie / convolution / Konvolution; Faltung / convolución

4453
power of resolution, – a measure of the extent to which it is possible to distinguish two signals separated by a short time interval. / oplossend vermogen / pouvoir de résolution / Auflösungsvermögen / poder de resolución

4454
deconvolution, – filtering procedure to compensate for the effect of the earth's transmission properties (e.g. non-elastic absorption) and the filtering characteristics of the recording equipment. The purpose is to increase the power of resolution. / deconvolutie / déconvolution / Dekonvolution / deconvolución

dereverberation; deringing, – kind of deconvolution for removing reverberations from seismic traces. / dereverberatie; ontzingen / déréverbération / Dereverberation / desreverberación

4455
whitening, – filtering procedure that makes the amplitude spectrum of the signal constant over a certain frequency interval. Its purpose is to shorten the signal.

4456
velocity filter, – filter that discriminates between signals having different apparent velocities. / snelheidsfilter / filtre de vitesses / Geschwindigkeitsfilter / filtro de velocidades

4457
migration, – procedure applied in reflection seismology for the purpose of shifting reflections from positions vertically below the shotpoint-detector midpoint to their correct spatial positions. / migratie / migration / Migration / migración

[2843] *Interpretation*

4458
seismic section; record ...., – display of seismic traces in proper spatial sequence. / seismische sectie / coupe sismique; section .... / seismische Sektion / corte sísmico

4459
picking; correlation, – process indentifying peaks and troughs on a seismic section as equivalent phases of reflections from the same boundary. / correlatie; pikken / pointé; corrélation / Korrelation / correlación

4460
line-up, – alignment of peaks of troughs across several traces of a seismic section. Reflections are mostly recognised by their line-ups. / oplijning / alignement / – – / alineación

4461
interface; horizon, – surface separating two media having different seismic properties. / grensvlak; horizon / surface de séparation; horizon / Grenzfläche; Horizont / superficie de separación

reflection horizon, – set of reflection signals displayed on a seismic section and corresponding to one interface. / reflectiehorizon / horizon de réflexion / Reflexionshorizont / horizonte de reflexión

4462
phantom horizon, – horizon obtained by averaging the dips of nearby reflecting horizons, thus indicating the trend of the dip but generally not corresponding to an actual interface. / schijnhorizon / horizon fantôme / Phantom-Horizont / horizonte fantasma

4463
T-X graph, – plot of arrival time of a seismic wave phenomenon against the horizontal distance from the source. / T-X kromme / report T-X / T-X Diagramm; Laufzeitskurvendarstellung / gráfica T-X

intercept time, – time at which the extension of a segment of a refraction T-X graph meets the time axis. / snijdingstijd / ordonnée à l' origine / Interceptzeit / ordenada en el origen

4464
diffraction curve; maximum-convexity curve, – relation between diffraction time and horizontal distance from the diffraction point to either the shot or the detector position. / diffractiekromme / courbe de diffraction / Diffraktionskurve / curva de difracción

4465
average velocity, – the distance travelled by a seismic wave divided by the time required for covering this distance. / gemiddelde snelheid / vitesse moyenne / mittlere Geschwindigkeit / velocidad media

4466
root-mean-square velocity; RMS ...., – the velocity that determines the normal-move-out curve in the case of horizon-

tal layering. / middelbare snelheid / vitesse moyenne quadratique / RMS Geschwindigkeit / velocidad cuadrática media

4467
interval velocity, – average velocity for the interval between two interfaces. / intervalsnelheid / vitesse d'intervalle / Intervallgeschwindigkeit / velocidad de intervalo

4468
apparent velocity, – distance covered by a seismic wave per unit time in a certain specified direction, mostly taken along the surface. / schijnbare snelheid / vitesse apparente / scheinbare Geschwindigkeit / velocidad aparente

4469
velocity function, – relation between velocity of propagation and location. / snelheidsfunctie / loi des vitesses / Geschwindigkeitsfunktion / función de velocidad

4470
wavefront, – surface of equal travel time. / golffront / front d'onde / Wellenfront / frente de onda

4471
trajectory; travel path; ray path, – line along which the seismic energy is assumed to travel in conventional geometric wave-propagation theory. / stralenpad; trajectorie / trajectoire; rayon / Trajektorie; Strahlenweg / trayectoria

4472
normal incidence, – property of a reflection indicating that at the point of reflection its trajectory is perpendicular to the interface. Normal-incidence reflections are recorded when shot and detector are situated close together. / loodrechte inval / incidence perpendiculaire / senkrechtes Einfallen / incidencia normal

4473
reflection coefficient; reflectivity, – the ratio of the amplitude of the reflected wave to that of the incident wave. / reflectiecoëfficiënt; reflectiviteit / coefficient de réflexion; réflectivité / Reflexionskoeffizient; Reflektivität / coeficiente de reflexión

4474
acoustic impedance; .... resistance, – product of rock density and velocity of propagation of seismic waves; the contrast in acoustic impedance across an interface determines its reflectivity at normal incidence. / acoustische impedantie / impédance acoustique; résistance .... / akustische Impedanz; Schallwiderstand / impedancia acústica

4475
attenuation, – weakening of seismic energy as a consequence of absorption and spreading during wave propagation. / attenuatie / atténuation / Dämpfung / atenuación

4476
bright spot, – part of a reflection horizon having an anomalously high amplitude.

dim spot, – part of a reflection horizon having an anomalously low amplitude.

4477
flat spot, – reflection caused by a difference in the fluid content of the pores within the same formation (e.g. an oil-water contact). It is recognized by its approximately flat attitude on the seismic section contrasting with that of the reflections from layer boundaries.

4478
synthetic, – term applied to seismic responses computed on the basis of an assumed subsurface model, as in 'synthetic seismogram' and 'synthetic section'. / synthetisch / synthétique / synthetisch / sintético

modelling, – the computation of the synthetic seismic response for an assumed subsurface configuration, often with the intention of comparing the computed results with actually observed data. / berekening aan het model / établissement d'un modèle / Modellieren / modelado

[2850] SHALLOW SEISMIC AND ACOUSTIC METHODS

NOTES:

- Whereas the methods compiled in section [2840] Deep Seismic Methods, developed mainly by the petroleum industry to obtain structural and stratigraphic information at depth, shallow seismic and acoustic methods are being used to observe sea-bottom and top bed-rock configuration and reflections in between, mainly for metallic mineral exploration and for marine engineering purposes.
- some terms, due to differences in development and purpose, have different meaning, others may be interchangeable; duplication of terms has been avoided as far as possible.

[2851] *General Terms*

4479
shallow seismic profiling, – a marine, continuous profiling method in which a seismic sound source is moved near the surface of the sea transmitting short bursts of seismic waves with a wide frequency pattern in a rapid succession. Reflections at the sea floor and at interfaces below it are detected and recorded during the time intervals between bursts. The trajectories are nearly vertical (normal incidence). / ondiep seismisch profileren / sismique réflexion continu faible profondeur / Flachwasserseismik / perfil sísmico poco profundo

4480
acoustic profiling, – a marine, continuous profiling method in which an acoustic source is moved near the surface of the sea transmitting short pulses of a discrete frequency in a rapid succession. Reflections at the sea floor and at acoustic discontinuities below it are detected and recorded during the time intervals between pulses. The trajectories are vertical (normal incidence). / acoustisch profileren; sonische opnamemethode / – – / akustische Vermessung / perfil acústico

4481
seismic wave, – complex sound wave with a wide frequency spectrum showing a peak in the lower frequency range. / seismische golf / onde sismique / seismische Welle / onda sísmica

shock wave, – a seismic wave resulting from a sudden expansive burst like an explosion. / schokgolf / onde de choc / Stosswelle / onda de choque

4482
acoustic wave; sonic wave, – sound wave of a discrete frequency. / acoustische golf; sonische golf / onde acoustique; onde sonore / Schallwelle / onda acústica

pulse, – an electric, electromagnetic or acoustic wave train of short duration containing one major frequency. / puls / impulsion / Impuls / impulso

pulse length, – duration of the transmitted acoustic pulse. Sometimes also used to indicate the duration of the electric pulse, fed into the source. / pulslengte / durée de l'impulsion / Impulsdauer / – –

4483
sledge; sledging, – a method of acoustic profiling in which a transducer is mounted in a sledge that is towed by the survey vessel over the sea floor. Due to the acoustic coupling of transducer and sea bed the multiple reflections from the sea floor are eliminated. Especially important in shallow water. / slede / traîneau / (tiefgeschleppter) Geräteträger / – –

4484
deep-ocean profiling, – acoustic survey of the bed of the ocean. Although the travelling times may be as long as 30 sweep times and more, the reflections of ocean floor and lower interfaces line up in the same way as if the travel time is shorter than one sweep. During this process, many outgoing and ascending wave fronts traverse one another. / profileren bij oceaandiepte / profilage grande profondeur en mer / seismische Registrierungen in der Tiefsee / – –

4485
side-scan sonar, – a marine surveying method to observe the configuration of and objects on the sea floor by acoustical means. Usually two high-frequency transceivers are used, mounted over either side of a vessel or mounted in a fish towed at depth. They produce sound beams with wide vertical and very narrow horizontal angle in opposite directions, so as to obtain oblique recordings of the sea floor at both sides of the surveyed line. / – – / sonar latéral / – – / – –

[2852] *Design and Equipment*

4486
sound source, – a device for producing in water a short burst of seismic waves or a powerful acoustic pulse of short duration. / geluidsbron / source sismique / Schallquelle / fuente de sonido

4487
transmitter, driver, – generator of the high-energy electric signal to be fed into the sound source. This equipment may vary from a complicated digital or analog power amplifier (transducer sources) to a simple gate for passing a high-current discharge (sparker source and Boomer source). / zender / émetteur / Sender / emisor

4488
transducer source, – an electro-acoustic, ceramic or magnetostrictive transformer converting an alternating electric current into acoustic waves or reversely. / electro-acoustische omvormer / transducteur électro-acoustique / Schallgeber / – –

transceiver, – transformer used alternatively as a transducer source and as a detector of returned pulses. It may be mounted in the ship's hull or over the side (outrigger) or it can be towed. The frequency spectrum is not very wide. / – – / sondeur acoustique / Schallkopf / – –

4489
transmit-receive switch, – an automatic device to disconnect the receiver from the transceiver during the transmission of the pulse and to re-establish the connection during the time of detection. / zend-ontvangschakelaar / commutation; inverseur / Unterbrecherkontakt / – –

4490
source level, – the total force supplied by a transducer operating at maximum output. / geluidsniveau / puissance d'émission / maximale Senderstärke / – –

4491
directional characteristic; beamed transmission, – the sound beam radiated by a transducer operating at a given frequency is defined by the source level and the beam angle, i.e. the angle between the transducer axis of radiation and the directions in which the sound pressure is half the pressure measured in the axis at the same distance from the transducer face. The directional characteristics improve with increase of the area of the vibrating (sensing) face. Improvement of the directional characteristics improves the efficiency of the transmission. With respect to detection (transceiver) there is an improvement in the signal-to-noise ratio. / gerichtheidsdiagram / caractéristique directionnelle / Richtcharakteristik / – –

4492
frequency modulation; tjrp, – the transmitted acoustic pulse sweeps gradually through a certain frequency range. After cross correlation of the received reflections with the transmitted signal an acceptable vertical resolution can be obtained, notwithstanding the comparatively long duration of the acoustic pulse. / frequentiemodulatie / modulation de fréquence / Frequenzmodulation / modulación de frecuencia

4493
sparker source, – device generating short bursts of seismic waves by means of an electric discharge in water. It generates a wide frequency spectrum and has no useful directional characteristics. It is towed by the survey vessel. / sparker / étinceleur; sparker / Funkengenerator; Sparker / sparker

spark-array, – arrangement of several sparker sources in a frame or as a splice of a composite cable that are fired simultaneously. The purpose of the arrangement is to improve the efficiency of the transmission of seismic waves. / sparkerkabel / dispositif d'émission sparker / Bündelung von Funkenschallquellen / – –

4494
Boomer source, – a device generating short bursts of seismic waves in water by a forceful repulsion of an aluminium plate by eddy currents if a high electric current is allowed to pass through a coil perpendicular to that plate. It is towed by the survey vessel. It is a broad-band source without useful directional characteristics.

4495
deep-towed sparker; .... Boomer, – a sparker or a spark-array respectively a Boomer source towed in a special fish at a water depth of about 40 m. The fish contains a hydrophone system and must be provided with a design for heave compensation. The purpose is to eliminate bubble pulse and ghost reflection from the surface of the sea, thus improving the vertical resolution.

4496
hydrophone, – pressure detector used in water for detecting reflections. / hydrofoon / hydrophone / Hydrophon / hidrófono

hydrophone string, – a number of hydrophones regularly spaced in a short hose (up to about 20 m) of neutral buoyancy, feeding one common amplifier. The purpose is to improve the directional characteristics of the detecting system in the direction of the surveyed profile, thus improving the signal-to-noise ratio. The string is towed by the survey vessel or by a special fish. / hydrofoonkabel / câble d'hydrophone / Hydrophonkabel / – –

4497
(hydrophone) response, – the rate of conversion at a given frequency of sound pressure into amplitude of the electric signal. / gevoeligheid / réponse sismique / Hydrophoncharakteristik / respuesta

4498
operating frequency, –
1. the major frequency of the acoustic pulse;
2. the pass-band frequency range, used with broad-band (i.e. seismic) sources.

/ werkfrequentie / fréquence de travail / Arbeitsfrequenz; genutztes Frequenzband; Durchlassbereich / – –

4499
receiver, – amplifier of the electric signal produced by hydrophone or transceiver. Usually some analog electronic processing is carried out during the survey, and then the necessary circuitry is contained in the receiver. / ontvanger / récepteur / Verstärker / receptor

programmed amplifier, – a receiver, containing the circuitry to carry out some analog electronic processing during the survey. / geprogrammeerde ontvanger / amplificateur programmable / programmierter Verstärker / receptor programado

4500
recorder; graphic ...., – apparatus to make a visual presentation of the detected signals, usually in variable density recording. / schrijver / enregistreur / Aufnahmegerät; Schreiber / carta de registro

recording, – the display in a time-time graph on a paper film, as obtained through the recorder, and in which, by slowly moving the paper the successive traces are printed close together. The emission times of the successive outgoing acoustic or seismic signals are presented as a horizontal line at the upper margin of the paper, while the successive reflections from the sea bottom and from deeper interfaces line up as in a geological section. / registratie; sonogram / enregistrement; sonogramme / Aufnahme; Registrierung / registro

4501
multichannel recorder; dual ...., – a recorder with the facility to make a visual presentation of two adjoining or more than two channels. The functions of the channels must have the same dependence with respect to one of the two time axes, i.e. ship's speed or reflection-time axis. / twee-kanalenschrijver; meer.... / enregistreur multicanaux / Mehrkanal-Aufnahmegerät / registro multicanal

digital recorder, – a dual or multichannel recorder with the facility to vary by digital means the vertical scale of the display without changing the depth range. / digitale schrijver / enregistreur digital / digitales Aufnahmegerät / registro digital

glass-fibre recorder, – a recorder operating on basis of an oscilloscope tube together with fibre-glass light propagation. It is possible to change both the vertical and the horizontal scale without changing the depth range. / glasdraadschrijver / enregistreur-oscilloscope / Glasfaser-Aufnahmegerät / – –

4502
phasing, – the facility, when operating in greater water depths, to display the recording at a higher level in order to avoid on the recording paper a wide blank space above the sea-floor recording and to improve depth range below it. / fasering / décalage; déphasage / Registrierverzögerung / – –

4503
sweep, – one trace of the graphic recorder, corresponding to one burst of seismic waves or to one acoustic pulse. / lijn / balayage / einmaliger Durchlauf / – –

sweep time, – the time interval representing one sweep. / – – / temps de balayage / dargestellter Zeitabschnitt / – –

4504
repetition rate; pulse rate, – the number of transmitted signals, either acoustic pulses or short-burst seismic waves, produced per second, usually monitored by the rotation rate of the recorder. / herhalingsfrequentie; puls.... / fréquence d'émission / Pulsrate / – –

keying function, – of the recorder, its rotation rate (in revolutions per minute).

4505
depth range, – the depth or the corresponding time (in milliseconds) limited by the width of the recording paper and the keying function of the recorder. / dieptebereik / échelle de profondeur / Tiefenbereich / – –

4506
half-wave rectification, – application of rectifying circuitry to

an alternating current (electric signal) so as to eliminate either the positive or the negative portion of that current. The result is a pulsating direct current. The effect in the recording is that half of the signal is printed white giving rise to a characteristic phase alignment. The method is necessary if the transmitted signal is of long duration (sparker, Boomer). / enkele gelijkrichting / redressement par demi-alternance / Halbwellen-Gleichrichtung; Einweg-.... / – –

full-wave rectification, – application of rectifying circuitry to an alternating current (electric signal) so as to bring about a varying direct current with a frequency double the original alternating current. In the recording the effect is that the reflections are printed black. A lighter shade of grey or white indicates that no reflections are present. This procedure is advantageous if the transmitted pulse is of very short duration (transducer source). / dubbele gelijkrichting / redressement total / Zweiweg-Gleichrichtung / – –

4507
dynamic range, – of recorder, adjustable dynamic range to match the recorder display with the dynamic range of the receiver output. / dynamisch bereik / niveau dynamique / Dynamikbereich; maximal aufzeichenbares Amplitudenverhältnis / – –

[2853] *Processing*

4508
time-varied gain control; TVG, – analog or real-time digital processing of the signal with the purpose to match the large dynamic range of the hydrophone (transceiver) signal with the smaller dynamic range of the recorder and of human observational capability. This is done by a gradual increase of the gain of the receiver after transmission of the seismic or acoustic signal. In advanced TVG designs the gain increases from a preset level at a preset time to another preset level to be reached after another preset time. / tijd-afhankelijke gevoeligheidregeling / contrôle de gain / zeitabhängige Verstärkungsregelung / – –

automatic gain control; AGC; **automatic volume control;** AVC, – automatic analog procedures of normalizing signal amplitude by varying the amplification rapidly. / automatische volumeregeling; .... sterkteregeling / contrôle de gain automatique / automatische Amplitudenregelung; .... Schwundausgleich / control automático de volumen

4509
matched filter, – analog or real-time digital procedure to allow the signal to pass a filter designed for frequencies other than the operating frequency, but yet containing significant information. / – – / boîte de filtres / Formfilter / – –

4510
moving-window stack, – real-time digital procedure in which after rapid digital sampling and time-multiplexing of the signal the data are averaged with the corresponding samples of previous sweeps after a certain weighting factor, which dwindles gradually with time, has been assigned to the samples of the previous sweeps. The 'dwindling factor' can be varied by the operator. In this way, the device will be adding new information to the process of averaging at the same time progressively forgetting older information. / – – / faisceau à fenêtre réglable / gleitende Stapelung / – –

envelope stack, – the above averaging process carried out after full-wave rectification. / – – / faisceau total / einhüllende Stapelung / – –

phase stack, – the above averaging process carried out after half-wave rectification or without prior rectification. / – – / faisceau de déphasage / Phasenstapelung / – –

4511
smoothing; heave compensation, –
1. a real-time digital procedure eliminating by averaging the interferences (undulations) caused by the heave of hydrophone or transceiver;
2. a procedure in which a device is used, based on inertia, that monitors circuitry to delay or advance the instant of transmission of the outgoing signal in order to compensate heave of hydrophone or transceiver;
3. combination of 1 and 2.

/ vereffening / fonctionnement régulier; compensation des mouvements; lissage / Hubkompensation; Hubglättung / – –

heave, – the vertical component of the movement of a craft due to waves; heave is observed as an undulating variation of the recorded times to be converted to depth. / verticale opheffing / mouvement verticale / Hub / – –

[2854] *General Phenomena and Interpretation*

4512
cavitation, – if during the process of generating an acoustic pulse or seismic waves in water the pressure becomes lower than the local hydrostatic pressure a vacuum may be brought about. Not only considerable interference follows from the subsequent collapse of these vacuums, but, as many will be filled instantaneously by vapour, there is a considerable absorption of useful acoustic or seismic energy. / cavitatie / cavitation / Kavitation / cavitación

4513
penetration, – depth of propagation of wave energy in consolidated and in unconsolidated rock. / penetratie / pénétration / Eindringtiefe / penetración

recorded penetration, – penetration as far as can be observed in the recordings. / geregistreerde penetratie / pénétration enregistrée / registrierte Eindringtiefe / penetración registrada

4514
absorption, – the weakening of the intensity of sound by friction or by elasticity of rock. Absorption increases considerably with frequency. / absorptie / absorption / Absorption / absorción

spreading, – the weakening of the intensity of sound with increasing distance from the source due to dilatation. / spreiding / dispersion / geometrische Absorption / – –

4515
irradiated area, – the area of an acoustic discontinuity (inter-

face), hit by the transmitted wave front. / aangestraald oppervlak / interface / Bildbereich / – –

Fresnel area, – the portion of the irradiated area represented in the detected signal. / Fresnel-oppervlak / surface de Fresnel / Fresnel-Zone / – –

overrun factor, – the degree of overlap of successive Fresnel areas. / – – / coefficient de couverture / Überdeckungsgrad / – –

4516
vertical resolution, – the extent to which it is possible to discriminate between successive reflections, separated by a short time interval, in one trace. / vertikaal oplossend vermogen / pouvoir de résolution vertical / vertikales Auflösungsvermögen / poder de resolución vertical

horizontal resolution, – the extent to which it is possible to discriminate between small objects. The measure is the corresponding Fresnel area. / horizontaal oplossend vermogen / pouvoir de résolution horizontal / horizontales Auflösungsvermögen / poder de resolución horizontal

4517
signal, – the perceptible phenomena that succeed the transmission of a seismic thrust or an acoustic pulse. Usually, these phenomena are received as a complex alternating current composed of a large variety of frequencies and containing many phase shifts. / signaal / signal / Signal / señal

seismic signal, – the outgoing and returning seismic waves that correspond to the received electric signal. / seismisch signaal / signal sismique / seismisches Signal / señal sísmica

acoustic signal, – the outgoing and returning acoustic waves that correspond to the received electric signal. / acoustisch signaal / signal acoustique / akustisches Signal / señal acústica

4518
side reflection, – the reflection at an object (e.g. pipeline or wreck) located within the irradiated area but outside the Fresnel area of the sea floor. It arrives later than the reflection from the sea floor, and if such an object is passed the successive side reflections line up as a hyperbole pointing like an arrow to the point of passage of the object. / zijreflectie / réflexion latérale / Seitenreflexion / – –

4519
stratum reflection; interreflection, – multiple reflection that originates within the geological profile, in contrast to those that originate at the sea bed **(bottom reflection).** / laagreflectie / réflexion d'interface / Tiefenreflexion / – –

bottom reflection, – / bodemreflectie / réflexion du fond / Bodenreflexion / – –

4520
detection of gas, – due to the extreme reflectivity of trapped gas, usually dispersed in the interstitial water as tiny bubbles, the incoherent energy returning, even from outside the Fresnel area, is readily detected, and is shown by a marked lengthening of the reflected pulse. The margin of the area where gas occurs is often recorded as a side reflection. No recorded penetration is obtained below a zone containing gas. / gasdetectie / détection de gaz / Gasanomalie / – –

gas seepage, – if the continuous profiler passes over a location where gas escapes continuously from the sea bed, so that dispersed bubbles rise to the surface in a constant stream, no recording of reflections at lower interfaces is possible, and the recording shows a complete blank. / gasontsnapping / fuite de gaz / Gasaustritt / – –

[2860] MAGNETOMETRICAL, GRAVIMETRICAL AND RADIOMETRICAL METHODS

NOTE:
– in this section terms have been assembled that are of special interest for exploration; they are additional to sections
[0330] Geomagnetism and Rock Magnetism,
[0340] Gravity and Isostasy
[0500] Isotope Geology

[2861] *Magnetometrical Equipment*

4521
vertical magnetometer, – instrument to measure changes in the vertical component of the magnetic field intensity. / verticale magnetometer / magnétomètre à componente verticale / Vertikal-Magnetometer / magnetómetro vertical

horizontal magnetometer, – instrument to measure changes in the horizontal component of the magnetic field intensity. / horizontale magnetometer / magnétomètre à componente horizontale / Horizontal-Magnetometer / magnetómetro horizontal

4522
airborne magnetometer, – instrument trailed behind an airplane, by which the total intensity of the magnetic field is measured. / vliegtuigmagnetometer / magnétomètre aéroporté / Flugmagnetometer / magnetómetro aeroportado

stinger, – in airborne magnetic surveying an extension of the tail of the aircraft to house the magnetometer; in this mode of operation compensation for the magnetic effect of the aircraft is necessary.

bird, – a magnetometer suspended below an aircraft to minimize the effect of the aircraft's magnetic field. / – – / oiseau / Sonde / magnetómetro pájaro

4523
flux-gate magnetometer, – magnetometer with two permalloy cores (the flux gate) near-saturated by the weak earth's normal field; a sinusoidal magnetizing field completes the saturation. The magnetization of the permalloy cores is distorted, the amount of which is proportional to the ambient field.

4524
nuclear magnetometer; high-sensitivity magnetometer, — magnetometer based on nuclear resonance, used for high-accuracy airborne surveying. / nucleaire magnetometer / magnétomètre à resonance nucléaire / Kernresonanzmagnetometer / magnetómetro nuclear

proton magnetometer, precession magnetometer, — a magnetometer based on proton resonance, commonly used for total magnetic field measurements. / proton magnetometer / magnétomètre à proton / Protonenmagnetometer; Protonenpräzessionsmagnetometer / magnetómetro a protón

4525
magnetic gradiometer, — instrument for measuring gradient of magnetic field; in airborne work usually two high sensitivity magnetometers in birds at different altitudes below the aircraft, thus giving measurement of vertical gradient; at sea two proton magnetometers at different distances from the ship, thus giving the horizontal gradient. / gradiometer / gradiomètre magnétique / — — / gradiométro magnético

[2862] *Magnetometrical Measurement and Interpretation*

4526
ground station; base ...., — fixed station in or near to survey area for monitoring changes in the earth's magnetic field. / basis station / station de base / Basisstation / estación base

4527
fiducial, — point of measurent of magnetic field; can be based on either a time or a distance interval. / — — / — — / Zuordnungsmarken / estación magnética

4528
diurnal variation, — fluctuations in the intensity and direction of the magnetic field with an approximately daily period. / dagelijkse schommeling; .... gang / variation diurne / Tagesgang / variación diurna

4529
magnetic storm, — rapid irregular variations of the earth's magnetic field; surveying is usually impossible during such periods. / magnetische storm / orage magnétique / magnetischer Sturm / tormentas magnéticas

4530
heading effect, — effect of direction of traverse of ship or aircraft during survey operations. / richtingeffect / effet de cap / Richtungs-Effekt / — —

4531
(intensity of) magnetization, — strictly magnetic moment per unit volume; in exploration often taken as the product of susceptibility and inducing magnetic field. / magnetisatie / intensité (d'aimantation) / — — / intensidad de magnetización

4532
anomaly, — difference between the observed and the normal value of a component of the magnetic field. / anomalie / anomalie / Anomalie / anomalía

isogam, — line of equal magnetic anomaly. / isogam / isogamme / Isogamme / isogama

4533
downward continuation, — interpretation method in which the values of a component of the magnetic field at lower levels are computed from the values at the surface. / voortzetting naar beneden / prolongement vers le bas / (analytische) Fortsetzung nach unten / prolongación hacia abajo

4534
second derivative, — of the magnetic field intensity or of one of its components with respect to the vertical coordinate. / tweede afgeleide / dérivée seconde / zweite Ableitung / derivada segunda

4535
international geomagnetic reference field; IGRF, — a mathematical model of the earth's magnetic field, often used in exploration as the background or regional field for the observed data. / internationaal geomagnetisch referentieveld / champ géomagnétique de référence internationale / internationales geomagnetisches Referenzfeld / campo de referencia geomagnética internacional

[2863] *Gravimetrical Methods*

4536
observed gravity, — value of gravity measured at station and corrected only for drift and closure error. / waargenomen zwaartekracht / champ de pesanteur mesuré / gemessene Schwere / gravedad observada

4537
density, — mass per volume; lateral changes in density cause the gravity anomalies of interest to the explorer. Unit: kg.m^{-3}. / dichtheid / densité / Dichte / densidad

gravity unit, — a practical unit of gravitational acceleration, equal to 10^{-6} m.s^{-2} (= 10^{-1} milligal, use of which is discouraged). / eenheid van zwaartekracht / unité gravimétrique / Schwerkrafteinheit / unidad de gravedad

4538
free air gravity, — gravity data corrected for latitude and elevation but not for the Bouguer plate effect. / vrije lucht zwaartekracht / pesanteur à l'air libre / Freiluftschwere / gravedad aire libre

Eötvös correction, — correction to gravity measurements made from a moving base, e.g. ship or aircraft, that corrects for velocity and changes of velocity of the base relative to the earth's rotation. / Eötvös correctie / correction d'Eötvös / Eötvös-Korrektion / corrección de Eötvös

4539
stabilized platform, — platform on which a marine or airborne gravimeter is mounted; the platform is designed to minimize

the accelerations of the platform support. / gestabiliseerd platform / plateforme stabilisée / stabilisierte Plattform / plataforma establizadora

4540
drift, – gradual change in the reading of a gravimeter that is caused by changes in the properties of the gravimeter spring. / drift / dérive / Gang / deriva

4541
closure error, – observational error after closing a loop of stations; usually corrected for by linear adjustments to observed values in appropriate loop or by a least squares adjustment in loop network. / sluitfout / erreur de fermeture; écart de .... / Schleiffenschlussfehler / error de cierre

4542
residual gravity map, – gravity map from which the gradual changes of gravity have been removed. / residuaire zwaartekrachtskaart / carte de la résiduelle du champ de pesanteur / Karte des Restfeldes / mapa de gravedad residual

4543
downward continuation, – interpretation method in which the gravity values at lower levels are computed from the gravity values at the surface. / voortzetting naar beneden / prolongement vers le bas / (analytische) Fortsetzung nach unten / prolongación hacia abajo

4544
second derivative, – of the acceleration of gravity with respect to the vertical coordinate. / tweede afgeleide / dérivée seconde / zweite Ableitung / derivada segunda

[2864] *Radiometrical Methods*

4545
radioactivity surveying; gamma-ray ...., – measurements of variations in gamma radiation with the objective of mapping the distribution of radioactive elements (usually K, U, Th). / radioactiviteitskartering; gammastraling.... / prospection radiométrique; .... gamma; méthode .... / radiometrische Übersichtsvermessung / cartografía radioactiva

4546
Geiger counter; Geiger-Müller ...., – instrument to detect and measure gamma radiation, ionizing gas atoms within a thin-walled tube. The ionized rays trigger discharges that are counted. / Geiger teller / compteur Geiger / Geiger Zähler; Geiger-Müller .... / contador Geiger

scintillation counter, – a device sensitive to gamma radiation, especially used in boreholes. It consists of an activated crystal, mounted on top of a photomultiplier tube. The gamma pulses are transformed into light flashes (scintillations) inside the crystal, liberating electrons in a photocathode. The signals are amplified and continuously recorded as a function of depths. The more sensitive scintillation counters have replaced the Geiger counters in much modern work. They have been constructed to differentiate between radiations of different energy. / scintillatieteller / scintillomètre / Szintillationszähler / escintilómetro

4547
channel spectrometer, – instrument that counts individual radioactive nuclides of a number of separated energies. / meerkanalen-spectrometer / spectromètre multicanaux / Mehrkanal-Spektrometer / espectrómetro de canales

differential spectrometer, – counter that records radiation falling within predetermined upper and lower energy limits only. / enkelkanaal spectrometer / spectrométre différentiel / Einkanal-Spektrometer; (....-Diskriminator) / espectrómetro diferencial

integral spectrometer, – counter that records radiation above a predetermined energy level only. / drempelspectrometer / spectromètre à seuil / Integral-Diskriminator / espectrómetro integral

4548
spectral log, – a gamma-ray record showing both energy and intensity of gamma rays over a measured distance, usually in a drillhole. / gammastraalspectrumlog / diagraphie gamma / Gamma-Log / registro espectral de rayos gamma

4549
radioactive equilibrium, – after a certain time the decaying radioactive parent element reaches equilibrium with its radioactive products, measured by the proportion of beta and gamma rays, and compared with such known value for a radioactive element in equilibrium. / radioactief evenwicht / équilibre radioactif / radioaktives Gleichgewicht / equilibrio radioactivo

spectral drift, – unreliability of radioactive element estimates resulting from radioactive disequilibrium. / spectrale gang / – – / – – / deriva espectral

4550
curie, – unit of the rate of radioactive disintegration, defined, independent of the species of nuclide, as $3.7 \cdot 10^{10}$ disintegrations per second, i.e. the approximate specific activity of one gramme of radium. Symbol: Ci. / curie / curie / Curie / curie

roentgen, – unit of the quantity of gamma radiation as measured by the amount of ionization produced, i.e. the quantity that will give rise to $2.08 \cdot 10^9$ ion pairs per cm^3 of dry air at standard temperature and pressure (or $1.61 \cdot 10^{12}$ ion pairs per gramme of air), that at an exposure of 1 roentgen absorbs an amount of energy equal to $88 \cdot 10^{-7}$ J. / roentgen / roentgen / Roentgen / roentgen

4551
rad; radiation-absorbed-dose, – equivalent to 10^{-5} J of energy absorbed per gramme of soft tissue from any type of ionizing radiation. Symbol: D; 1 rad = 10^{-2} J.kg^{-1}. / geabsorbeerde dosis; rad / rad / Rad / dra

rem; roentgen-equivalent-man, – the absorption by the human body of any type of ionizing radiation that produces an effect equivalent to the absorption of 1 roentgen of gamma radiation. / eenheid van dosisequivalent; rem / rem / Rem / ehr

[2870] ELECTRICAL AND ELECTROMAGNETIC METHODS

4552
telluric current, – natural electrical currents in the earth, caused by variations in the terrestrial magnetic field and by the rotation of the earth through external (ionospheric) electrical fields; in particular during magnetic storms they may vary rapidly in direction and intensity. / tellurische stroom / courant tellurique / Erdstrom / corriente telúrica

4553
telluric method, – measurement of components of telluric currents at the surface of the earth for geological reconnaissance. / tellurische methode / méthode tellurique / tellurisches Verfahren / método telúrico

magnetotelluric method, – measuring components of electrical and magnetic fields, due to fluctuating telluric currents; enables to obtain deep structural information. / magnetotellurische methode / méthode magnéto-tellurique / Magnetotellurik; MT / método magneto-telúrico

4554
electric(al) method, – measurement of an electrical effect at or near the surface in order to elucidate characteristic qualities of the subsurface; use is made of either natural or artificial (dc or ac) electrical potential fields; main applications: mineral prospecting, structural geology, ground-water surveys. / electrische methode / méthode électrique / elektrisches Verfahren / método eléctrico

4555
direct current; dc, – electric current of constant strength and constant direction of flow. / gelijkstroom / courant continu / Gleichstrom / corriente continua

commutated current, – electric current of constant strength of which the direction of flow is reversed at constant intervals of time. / gecommuteerde gelijkstroom / courant pulsé / kommutierter Strom / corriente conmutada

alternating current; ac, – electric current of which the strength and direction of flow change periodically according to a sine function. Reversals are expressed in number per second; unit: hertz, Hz. / wisselstroom / courant alternatif / Wechselstrom / corriente alterna

4556
electric current strength, – the amount of electricity flowing through a medium; symbol: I, unit: ampère, A. / electrische stroomsterkte / intensité de courant électrique / elektrische Stromstärke / intensidad de corriente (eléctrica)

current density, – current strength per unit area; unit: ampère per square metre, $A.m^{-2}$. / stroomdichtheid / densité de courant électrique / Stromdichte / densidad de corriente (eléctrica)

potential difference; voltage; electromotive force, – the difference in potential e.g. between two electrodes, i.e. the power that is dissipated or generated in a circuit per unit of current strength; symbol: V, unit: volt, V, unit: watt per ampère, $V = WA^{-1}$. / spanning / tension / Spannung / voltaje

electric resistance, – the resistance to electric flow through a medium; symbol: R, unit: ohm, $\Omega = VA^{-1}$. / weerstand / résistance électrique / Widerstand / resistencia eléctrica

electric conductance, – the ease with which a current can flow through a medium; the reciprocal of resistance; symbol: G, unit: siemens, $S = \Omega^{-1}$. / geleiding / conductivité électrique / Leitfähigkeit / conductancia

resistivity, – the resistance that is offered by a cube of unit dimensions of a material to the flow of electric current; symbol: ϱ, unit: ohmmetre, $\Omega m = \Omega \frac{m^3}{m^2}$ / soortelijke weerstand / résistivité / spezifischer Widerstand / resistividad específica

4557
self-potential method; spontaneous polarization ....; SP ...., – measurement of natural electrical fields, due to electrochemical effects around sulphide-mineral bodies or to physical effects of flowing fluids; used for mineral prospecting and in borehole logging. / spontane-potentiaalmethode / méthode de potentiel spontané / Eigenpotentialverfahren / método de potencial espontáneo

polarization, – the production of a double layer of charge at interfaces as a result of an applied electric current or of a magnetic field. / polarisatie / polarisation / Polarisation / polarización

4558
equipotential-line method, – a resistivity method in which a current is caused to flow in the earth between two electrodes. The line of points constituting loci of a given voltage difference is mapped with a third electrode. / equipotentiaalkartering / méthode des équipotentielles / Potentialkartierung / método de líneas equipotenciales

4559
mise-à-la-masse method, – a resisivity method in which one current electrode is placed in the accessible part of a conducting body (for instance a mineral deposit at the surface or in a borehole). The second current electrode is at a great distance. The measuring (potential) electrodes are moved around with the purpose to map the potential differences existing around the conducting body in order to map its dimensions and outline. / mise à la masse methode / mise à la masse / Methode 'mise-à-la-masse' / método de toma de tierra

4560
potential-drop ratio method, – when an electrical current is supplied to the ground, comparison of potential differences along surface profiles may yield information about inclined to vertical resistivity anomalies. / potentiaalverval-methode / méthode de chute de potentiel / Potential-Gradientenverfahren / método de relación de caída de potencial

4561
resistivity method, – studying the electrical resistivity of the subsoil by means of an electrode array consisting of current electrodes (dc) and measuring electrodes. This method (particularly variant 1) has found wide application in ground-water surveys, in which it is used to gain information on lithology and structure of aquifers and on water quality towards depth. / (electrische) weerstandsmethode / méthode de résistivité (électrique) / Widerstandsverfahren / método de resistividad

Two variants can be distinguished:

depth sounding or **vertical profiling** with an expanding electrode array, in order to probe the stratification of the subsurface;

trenching or **horizontal profiling** by moving an electrode system with fixed spacing along a traverse for the detection of resistivity contrasts.

4562
electrode array; .... configuration; .... disposition, – in the resistivity method, various systems of two current electrodes (C_1, C_2) for dc supply to the ground and of two (or more) measuring (potential, P_1, P_2) electrodes for the observation of the resulting potential difference. / electrodenopstelling / disposition des électrodes / Elektrodenanordnung; ....konfiguration / disposición de electrodos

Some available possibilities are:

Wenner array, – with four equidistant electrodes in line: C_1 P_1 P_2 C_2 with distances $P_1P_2 = \frac{1}{3}(C_1C_2)$.

Schlumberger array, – with a similar in-line disposition, but distance $P_1P_2 \leq \frac{1}{3}(C_1C_2)$.

dipole-dipole array, – in which the measuring dipole P_1 P_2 may have any position with respect to the current dipole C_1C_2.

Lee array, – a 5 electrode configuration ($C_1P_1P_0P_2C_2$) with distances $C_1P_1 = P_1P_2 = P_2C_2$

4563
gradient survey, – a variation of the Schlumberger array, in which the potential electrodes are moved along a rectangular grid that is situated between two current electrodes at some distance. / – – / méthode du rectangle / Gradienten-Verfahren / registro de gradientes

line electrode, – the current electrode consists of a long wire on the ground; various configurations are possible also, e.g. a series of electrodes placed along a straight line. / lijnelectrode / électrode linéaire / Linienelektrode / línea electródica

4564
apparent resistivity, – in electrical surveys interpretation, a quantity calculated from the values of current strength and potential difference with the aid of a geometric factor that is dependent on both, electrode array and distances; if the subsurface were uniform and isotropic this quantity would be equal to the specific resistivity of this infinitely thick layer. / schijnbare weerstand / résistivité apparente / scheinbarer Widerstand / resistividad aparente

4565
field curve, – in electrical surveys a plot of the measured quantity(-ties) as a function of an independent variable; e.g. in a resistivity survey the apparent resistivity versus electrode spacing. / veldkromme / boucle expérimentale / Messkurve / curva de campo

standard curve, – graph of the theoretical electrical effect of a comparatively simple subsurface model; sets of standard graphs are obtained by a systematic variation of essential parameters (dimensions, electrical qualities). / standaardkromme / abaque / Standardkurve / curva estandard

4566
quantitative interpretation, – attempt to match a field curve with the theoretical effect of a certain subsurface configuration; if successful this configuration is thought to represent the geological information contained in the electrical measurement. The interpretational process may be either of the direct or the indirect type. / kwantitatieve interpretatie / interprétation quantitative / quantitative Auswertung / interpretación cuantitativa

In **direct interpretation** the geological information is inferred directly from the electrical data. / directe interpretatie / interprétation directe / direkte Auswertung / interpretación directa

In **indirect interpretation** a preliminary model is conceived and its electrical effect calculated; subsequently the model is adapted in order to obtain an acceptable fit with the field curve. In this operation sets of standard graphs may be of considerable use, but presently electronic computer techniques provide more sophisticated tools. / indirecte interpretatie / interprétation indirecte / indirekte Auswertung / interpretación indirecta

4567
principle of equivalence, – constitutes a drawback in the quantitative interpretation of electrical surveys; electrical surface phenomena are not unique and different geological subsurface configurations may cause similar electrical effects, in which respect they are equivalent. / gelijkwaardigheidsprincipe / principe de l'équivalence / Äquivalenzprinzip / principio de equivalencia

suppression, – is caused by the limited resolving power of an exploration method: when under certain conditions an electrical effect becomes marginal, it may easily escape detection. / onderdrukking / suppression / Schichtunterdrückung / supresión

4568
horizontal layer model, – an interpretational model that consists of homogeneous and horizontal layers and is used in resistivity surveys for geological and hydrogeological purposes. / horizontale lagen model / modèle stratiforme / Horizontalschichtfall / modelo de estratos horizontales

Stefanescu integral, – the mathematical expression of the conduction of the electrical current through a horizontal layer model. The integral can be developed either into an infinite series (image method) or by asymptotic expansion. In the form of a convolution integral it can be solved by means of processes currently available in the electrical filtering technique. The latter solution provides a versatile tool for the interpretation of resistivity field curves. / integraal van Stefanescu / intégrale de Stefanescu / Stefanescu-Integral / integral de Stefanescu

4569
induced polarization method; IP ...., – mainly used for mineral prospecting with some application in ground-water surveys. The induced polarization effect is measured by a similar outfit and the same electrode array that are used in resistivity surveys, observing either the slow decay of voltage in the ground after a short current pulse **(time-domain method)**, or low frequency variation of earth impedance **(frequency-domain method)**. / geïnduceerde polarisatiemethode / méthode de polarisation provoquée; PP / Induzierte Polarisation; I.P. / método de polarización inducida

Two types of induced polarization are distinguished:

electrode polarization, – a mineral grain, with electronic conduction, embedded in a fluid-filled porous rock (where conduction is mainly ionic), is surrounded by a host of ions, the passage of which is blocked.

membrane polarization, – a weaker effect is caused by negatively charged clay-mineral surfaces that attract positive ions from the surrounding electrolyte.

4570
time domain, – when in the induced polarization method, the measurements are performed on the decay curve of the residual voltage after turning off an applied direct current, the observations are said to have been carried out in the time domain. / tijddomein / domaine temporel / Zeitbereich / dominio del tiempo

chargeability, – a measure of the IP effect in the time domain, defined as the ratio of the difference between original and residual voltage and the original voltage: $m = (V_o - V_1)/V_o$.

normalized time integral, – unit M (in millivolt times seconds per volt, mVs/V) in time-domain-induced polarization. M is the integrated area under the induced polarization decay curve between times t_1 and t_2, normalized by the primary voltage V_p:

$$M = \frac{1}{V_p}\int_{t_1}^{t_2} V_t\,dt$$

4571
frequency domain, – when in the induced polarization method, a steady alternating current is applied, this results in a certain phase lag between current and residual voltage that is dependent on the frequency of the current that is applied. / frequentiedomein / domaine fréquentiel / Frequenzbereich / dominio de frecuencia

metal factor, – in the frequency domain observations are usually made at two (low) frequencies; the calculated apparent resistivities are expressed as the metal factor:

$$MF = \frac{R_1 - R_2}{R_1 \cdot R_2}\, 2\pi \cdot 10^5.$$

4572
electromagnetic method, – mainly used for mineral prospecting. In a transmitter unit an alternating electrical current generates a primary magnetic field. If a conductive body is present below the surface of the earth, an electromagnetic field is induced that is recorded by a receiving unit. / electromagnetische methode / méthode électro-magnétique / elektromagnetisches Verfahren / método electro-magnético

ELF, – an extremely low frequency (30-300 Hz) electromagnetic exploration method. The energy in this range originates mostly from lightning strikes.

VLF, – a very low frequency (3-30 kHz) electromagnetic exploration method. The energy originates from long-range radio positioning stations.

4573
audio-frequency magnetic method; AFMAG (trade mark), – mineral prospecting and mapping technique that uses as an energy source the naturally occurring components of the terrestrial magnetic field in the audio and sub-audio frequency range; these result mainly from electrical disturbances in the atmosphere. Azimuth and inclination of the two major axes of the integration ellipsoid (polarization ellipse) of both the primary and the secondary field are measured.

4574
Turam (trade mark), – electromagnetic survey method in which the gradient of the field around a fixed transmitter is measured with two coils at fixed separation that are moved in lines perpendicular to the transmitter axis. Measurements are made of the ratios of the amplitudes of and the phase (angle) difference between the two magnetic fields.

4575
horizontal-loop method; HEM; Slingram (trade mark), – an electromagnetic survey method in which transmitter and receiver coils are horizontal. / horizontale spoel-methode / méthode à boucle horizontale; Slingram / Slingram / método de bucles horizontales; Slingram; MEH

vertical-loop method; VEM, – the two coils are held vertically. / verticale spoel-methode / méthode à boucle verticale / VEM / método de bucles verticales; MEV

4576
Input System; Barringer Input (trade names), – a time-domain induced polarization electromagnetic airborne survey system in which measurements are made during off-periods between source pulses.

4577
cross-over point, – the point at which a reversal of electromagnetic dip direction occurs over the apex of a conductor. / overgangspunt / – – / Umkehrpunkt / punta de inflexión

zero level, – reading at points having assumed normal field. This level is usually selected and has a bearing on the interpretation of anomalies. / nulniveau / niveau zéro / Nullniveau / nivel cero

[2880] PHYSICAL BOREHOLE MEASUREMENTS

[2881] *Logging Devices*

4578
well logging, – recording, as a function of depth, the magnitude of a specific physical attribute (resistivity, natural gamma radiation, sonic velocity, etc.) of the formation surrounding the borehole. / (fysische) boorgatmeting / diagraphie / Bohrlochmessung / diagrafía

logging tool; .... device, – / (boorgat)sonde / sonde (de diagraphie / (Bohrloch)Messgerät; ....-Sonde / sonda (de diagrafía)

4579
spontaneous potential log; SP, – vertical profile of the difference between the potential of a movable electrode in the borehole and the fixed potential of a surface electrode. It depends on permeability and shaliness of formation, and on salinities present. Mainly used in petroleum and groundwater surveys. Symbol/subscript: SP, unit: millivolt, mV. / spontane potentiaal log / diagraphie de potentiel spontané / SP-Log; Eigenpotentialkurve / diagrafía de potencial espontáneo

4580
redox log, – the record of the continuously measured reduction-oxidation potential (Eh) of the formations adjacent to the borehole; it indicates the proportions of oxidized and reduced forms of various minerals and ions in the formation waters. Applied mainly in mineral surveys. Unit: millivolt, mV.

4581
induced polarisation log; IP log, – the record of the induced polarisation effect in a borehole. Applied mainly in mineral surveys.

Several electrode arrays are in use (C_1, C_2 current electrodes; P_1, P_2 measuring electrodes):

pole-pole array; normal ....; two ...., P_1C_1 downhole; P_2C_2 at surface, C_2 at 'infinity'.

pole-dipole array; lateral ....; three ...., – $P_2P_1C_1$ downhole at equal distances; C_2 at surface at 'infinity'.

directional array, – to find the strike of an electrically anomalous subsurface body. P_1 downhole; P_2C_1, C_2 at surface, C_1 and C_2 at 'infinity'.

downhole-radial array, – a five-electrode array moved along radii around borehole: P_1P_2 at surface; two current electrodes fixed respectively downhole and at top, third current electrode at 'infinity'.

4582
resistivity log, – vertical profile of electric resistivity measured in the borehole between two electrodes. Dependent on depth of investigation, actual formation parameters and bed thickness, it can measure true resistivity (R_t), flushed/invaded zone resistivity (R_{xo}), etc. Mainly used in petroleum and ground-water surveys. Unit Ωm – most formation resistivities fall between 0.2 and 1000 Ωm. Reciprocal: **conductivity** unit: Sm^{-1}. / weerstandslog / diagraphie de résistivité / Widerstandslog / diagrafía de resistividad

4583
normal (resistivity) logging tools, – measure the potential difference (between a surface electrode and a downhole electrode) resulting from a constant electric current, maintained between two other electrodes. Reference distances between the two electrode systems (indicating also increasing depths of penetration into the formation surrounding the borehole):

short normal; SN, – 16″ between downhole potential and current electrodes. In ground-water surveys distances of 20 cm (in 10-20 cm diameter holes) or of 40 cm (in 60-100 cm diameter holes) are often used.

long normal; LN, – 64″ between electrodes. In ground-water surveys: 100 cm and 200 cm in resp. 10-20 and 60-100 cm holes.

NOTE:
– the use of inches and feet in this and following definitions is due to factory specifications.

4584
induction log, – vertical profile of formation conductivity, measured according to the electromagnetic induction principle, in boreholes containing low-conductivity fluids (oil-base muds or air). Mainly used in petroleum surveys; to some extent in mineral surveys. / inductielog / diagraphie par induction / Induction Log / diagrafía de inducción

induction-logging tools, – a family of tools providing Induction-Electrical-Survey (IES) with varying depths of investigation. The Dual-Induction-Laterolog (DIL) and the 6FF40 systems find the widest applications.

4585
lateral resistivity log, – device measuring the potential difference between two downhole electrodes, indicating apparent formation resistivity. It is provided with an 18′8″ electrode spacing. / lateraallog / diagraphie latérale de résistivité; .... inverse de .... / Laterale; Inverse / sonda lateral

4586
focused-resistivity logging tools, – family of resistivity tools using devices to focus the measuring current, resulting in improved vertical resolving power for a given depth of investigation. Available systems cover deep investigation (determination of true formation resistivity, R_t: Laterog–7 and –3, Dual 'deep' Laterog LLd) to shallow depths of investigation (determination of flushed zone resistivity, R_{xo}: Laterog–8, Dual 'shallow' Laterog LLs, or microspherically-focused log MSFL). / gefocusseerde weerstandslog / sonde de diagraphie de résistivité focalisées / fokussierende Widerstandsmeszgeräte / diagrafía de electrodos de focalización

4587
microlog, – the record of contact-resistivity measurements of mud cake and formation immediately adjoining the borehole through a very short-spaced (~1″) in-line three-electrode system. Used to determine R_{xo} (flushed zone resistivity) and permeable-bed boundaries. Used mainly in petroleum surveys. / microlog / microlog / Mikrolog / – –

microlaterolog-logging tool; MLL, – short-spaced concentric electrode-contact device focuszing the current into the formation adjoining the borehole. Measures R_{xo} mainly, virtually undisturbed by mud-cake resistivity.

proximity-logging tool, – contact device measuring borehole resistivity on same principle as microlaterolog, with somewhat larger electrode-spacings. Also measures R_{xo} but is less sensitive to mud-cake thickness.

4588
electromagnetic (travel time) logging tool; EMP, – device measuring at very high frequencies the travel time of electromagnetic waves between two electrodes. It reflects the oil/water saturation near the borehole, and for heavy-oil or tarsands with little invasion measures initial hydrocarbon saturation. The tool is not yet widely available for application (1978).

4589
magnetic susceptibility log, – record of the magnetic susceptibility of the rocks surrounding a borehole using electromagnetic induction. Applied mainly in mineral surveys. Unit: dimensionless units × 10^{-6}.

magnetic vector logging tool, – an instrument that measures the three orthogonal components of the total magnetic field, using a fluxgate magnetometer. Applied mainly in mineral surveys. Unit: nanotesla, nT = 10^{-5} gamma).

4590
gamma-ray log, – vertical profile of total natural radioactivity of the borehole-wall formations. In most cases it is a measure for the formations' shaliness. It is used for the definition of clay- and shale-layer thicknesses and can be used in a cased hole. Application in petroleum- and ground-water surveys; to some extent in mineral surveys. / gamma log / diagraphie de radioactivité naturelle / Gamma Ray Log / diagrafía gamma-ray

gamma-ray logging tool, – equipment detecting on scintillation counters the (total) natural radiation of ^{40}K, ^{238}U, ^{232}Th of the surrounding formations. Units: counts per second, API units (the difference between the high and low radioactivity sections in the calibration pit of the American Petroleum Institute is defined as 200 API units). / gamma sonde / sonde de radioactivité naturelle / Gammasonde / sonda de radiación natural

4591
gamma-ray spectral log; spectrometrical gamma-ray log; spectral log, – equipment recording the natural gamma radiation attributable to a specific element in the borehole-wall formations by measuring radiative energy in selective windows in the energy spectrum. Useful in evaporite or ore-composition analysis, especially the evaluation of uranium deposits.

selective gamma-gamma log, – similar to gamma-ray spectral log except that a gamma-ray source is used instead of natural gamma rays. At energies below about 0.2 MeV the chemical composition of the rock has a large influence on the resulting gamma-ray distribution leading to the possibility of determining the heavy element concentration and chemical composition of the rock.

4592
X-ray fluorescence, – a method using an isotopic source of gamma rays to irradiate the borehole wall that in turn emits X-rays with energies characteristic of the elements in the rock; used in mineral surveys.

4593
(formation) density log, – vertical profile of changes in (electron) density of the formation around the borehole. The detected intensity of scattered gamma radiation, induced by irradiating the formation with medium-energy gamma rays, reflects the electron density of the formation, that is about proportional to true bulk density. Is mainly used as a 'porosity' indicator, and, in combination with e.g. neutron devices, for lithology determination and gas-oil discrimination. Available with different spacings (depths of penetration). / dichtheidslog / diagraphie de densité / Dichtelog / diagrafía de densidad

formation density compensated tool; FDC, – provides density measuring equipment (dual-spacing) in which the correction for mud-cake density and thickness is automatically made; included as a separate curve on the log.

4594
neutron log, – vertical profile of neutron-capture rate in the borehole-wall formations. As slowing down and capture of neutrons mainly depend on the amount of hydrogen present in the formation, in clean (non-clayey) formations the neutron log reflects the amount of liquid-filled porosity. By cross plotting neutron log with density logs or sonic logs, lithologies can be identified. Can be used in cased holes. / neutronlog / diagraphie de neutrons / Neutron Log / diagrafía neutron

neutron logging tools, – family of logging tools recording the intensity of scattered neutron or induced gamma radiation, resulting from fast-neutron irradiation of the borehole-wall formations. It includes the **GNT** (gamma-ray/neutron tools), the **SNP** (sidewall neutron porosity) and the **CNL** (compensated neutron log). Principally used in petroleum surveys.

4595
thermal decay-time logging tool; TDT; neutron-life time log, – equipment measuring the decay time of thermal neutrons (mainly due to neutron capture) in the borehole-wall formations. The formation capture-cross section (Σ) is constituted by proportional contributions from matrix, clay, formation water and hydrocarbon capture-cross sections respectively. In combination with other tools it can be used to indicate changes in fluid saturation for instance (remaining) hydrocarbon content or water saturation, also in cased (old) wells, or during secondary recovery projects. The newest version (TDT-K) provides a dual (far and near) detector set-up, for count-rate comparison.

4596
nuclear magnetic (resonance) logging tool; NML, – equipment responding to presence of movable fluid in the formations' pores. It measures proton relaxation times in interstitial water, while, through a free fluid index (FFI) it indicates the amount of movable fluid in the pores, and turns out to be correlatable to the formation's permeability. It is not yet fully operational (1978).

circumferential microsonic logging tool; CAD, – equipment recording sonic travel times and amplitudes between (four) sets of transmitter-receivers, spaced 90° apart along the tool's circumference. Its set-up aims at detecting formation fracture by comparing differential attenuation between sets. The service is not yet commercially available (1978).

4597
sonic log; acoustic velocity log; continuous log, – vertical profile of the interval transit time Δt needed by a compressional acoustic wave pulse in the kc/s range to travel along the borehole from transmitter to receiver. Logging service is offered with transmitter-receiver distances ranging from 3 to 15 feet (long-spaced sonic). Also complete wave trains, displaying wave amplitude in addition to arrival time, can be provided. Is generally used as a 'porosity' tool, or in conjunction with density and/or neutron tool for determination of lithology by cross plotting. / acoustische log / diagraphie acoustique / Sonic-Log; Akustiklog / diagrafía sónica

borehole-compensated sonic logging tool; BHC, – device using a configuration of sonic-wave transmitting and receiving devices that allows compensation for sonde-tilt and hole-size variations.

cycle skipping, – jump in sonic interval-transit time occurring when the first arrival at the far receiver is not strong enough to actuate it. This receiver is then triggered by a later arrival. / – – / présence de cycles manqués / – – / – –

4598
dipmeter log, – vertical profile of stratum inclination. It is derived from relative shifts between 3 or 4 resistivity logs taken simultaneously, and equally spaced around the borehole. It

shows calculated dip and azimuth of the layer plane involved, and it may, in specific dip patterns, sometimes indicate structural features. It is nowadays generally run in its high-resolution version (HDT), recording four high-resolution microresistivity curves; an additional electrode on one pad yields a fifth curve at displaced depth. Thus it accommodates various correlative options (interval length, step distance and search angle). / — — / pendagemétrie / — — / — —

HDT-dipmeter tool, — 4-arm high-resolution version of electric dipmeter.

4599
directional survey, — measurement, either continuously or at discrete levels, of the drift of a borehole, i.e. azimuth and inclination with the vertical. Often applied as part of a dipmeter survey.

4600
poteclinometer, — device used for continuous directional surveys; a pendulum moves a variable-resistant arm so that the resistance is a measure of the angle with the vertical; a compass needle moves another arm providing a measure of the azimuth.

photoclinometer, — device that records photographically azimuth and angle of inclination at discrete levels. Used in mineral surveys.

4601
temperature log, — vertical profile of observed temperatures in a well. Due to drilling disturbance of original formation temperature (unless taken months after drilling), in petroleum-exploration wells the profile is generally only a relative, qualitative indicator of subsurface temperatures. Often used for locating top cement, where setting heat causes a local temperature anomaly. In ground-water observation wells, where thermal equilibrium has long been restored, detailed measurement of temperature as function of depth, provides information on geohydrological conditions, e.g. infiltration and seepage, that have led to the temperature gradients observed. / temperatuurlog / diagraphie de température; (thermométrie) / Temperaturslog / diagrafía de termometría

4602
borehole gravimeter, — measures the gravity effects in boreholes to determine average rock densities.

4603
caliper log, — vertical profile of the diameters in a borehole. Caliper logging belongs to the standard combination of tools, employed in the open hole. It is indispensable for the interpretation of the various other logs, and it is used for a volumetric calculation of the amount of cement, needed for the setting of casing. / diameterlog / diagraphie de diamétreur / Kaliberlog / calibrador

[2882] *Formation Properties and Log Interpretation*

4604
quick-look interpretation, — well-site (mostly qualitative) log interpretation based on direct comparison (or overlay) of several logs taken. It aims at helping decide on future operations (completing or abandoning, testing, etc.). / — — / interprétation à la tête du puits; .... sur le champ / Vorauswertung / — —

4605
Archie's formulas, — empirical relationships that form the basis of electric log interpretation. They relate resistivity ratios with porosity, matrix texture, and oil/water saturation:

$$F = \phi^{-m} = \frac{R_o}{R_w} \text{ and } S_w^{-n} = \frac{R_t}{R_o}$$

/ formules van Archie / lois d'Archie / Archie's Formeln / ecuaciones de Archie

4606
cementation factor, — exponent in Archie's empirical formula relating formation resistivity factor (F) and porosity (∅). It reflects matrix properties, such as degree of cementing, tortuosity etc. Symbol: m. / cementatiefactor / facteur de cimentation / Zementationsfaktor / factor de cementación

saturation exponent, — exponent in Archie's empirical formula, relating water saturation (S_w) to the formation's resistivity when fully saturated with formation water (R_o) and the true formation resistivity (R_t). Symbol: n. / verzadigingsexponent / exposant de terme saturation / Sättigungsexponent / exponente de saturación en agua

4607
formation resistivity, — characteristic resistance to flow of electric current in a formation. Most rock grains are non-conducting, the electrical current flowing largely through the interstitial pore fluid. Resistivity depends on salinity of the formation water and on other factors such as shaliness and hydrocarbon saturation. / specifieke formatieweerstand / résistivité de la couche / spezifischer Widerstand / resistividad de la formación

true resistivity, — characteristic resistance of uninvaded zone, i.e. of formation with unchanged pore contents. Symbol: R_t. / originele formatieweerstand / résistivité vraie de la zone vierge; .... non envahie / wahrer Gesteinswiderstand / resistividad verdadera de la formación

4608
formation-water resistivity, — characteristic resistance of formation water, dependent e.g. on salinity and temperature. It can often be determined from the SP-log in clean (non-clayey) water-bearing intervals. Symbol: R_w. / specifieke weerstand van formatiewater / résistivité de l'eau de gisement / Porenwasserwiderstand / resistividad del agua de formación

4609
formation resistivity factor, — the ratio of the resistivity of a water-saturated rock sample (R_o) and the resistivity of the saturating water (R_w). It reflects porosity, pore structure (tortuosity) and pore-size distribution; in the fresh-water zone it is a function of grain size. Symbol: F. / formatiefactor / facteur de formation / Formationsfaktor / factor de formación

apparent resistivity, — the resistivity recorded in resistivity logging; it departs from the true formation resistivity due to the mud-filled borehole and the invasion of mud filtrate; it depends on the ratios electrode distance/hole diameter and electrode distance/bed thickness and on the resistivity contrasts between successive beds. / schijnbare weerstand / résistivité apparente / scheinbarer Widerstand / resistividad aparente

4610
hydrocarbon saturation, – fraction (or percentage) of pore volume, filled with hydrocarbons. Symbol: S_o. / olie-gas verzadiging / saturation en hydrocarbures; teneur en .... / Ölsättigung / coeficiente de saturación en hidrocarburos

residual oil saturation, – percentage of pore volume still containing hydrocarbons after flushing or flooding. Symbol: S_{or}. / residuele olieverzadiging / saturation résiduelle en pétrole / Restölsättigung / coeficiente de saturación en petróleo

4611
water saturation, – the fraction (or percentage) of the pore volume, filled with water.
fresh formation: symbol S_w.
flushed (invaded) zone: symbol S_{xo}.
/ waterverzadiging / saturation en eau; teneur en eau / Wassersättigung / saturación en agua de la formación

residual water saturation; irreducible ...., – the fraction or percentage of the pore volume that still contains (immovable) water, remaining after intense displacement/flushing of the original pore contents (by mercury, toluene, etc.). Symbol: S_{wr}. / residuele waterverzadiging / saturation résiduelle en eau / Haftwassersättigung / saturación residual en agua de la formación

4612
capillary pressure curve, – a plot relating for a rock sample the saturations of wetting and non-wetting fluids with the pressure that has to be applied on the non-wetting fluid to establish these saturations. It reflects the pore-size distribution of the rock. / capillaire druk curve / courbe de pression capillaire / Kapillardruckkurve / – –

4613
hydrogen index, – indicator of total hydrogen content; it is proportional to the quantity of hydrogen per unit volume, taking the hydrogen index of fresh water at surface conditions as unit of reference. Symbol: H. / waterstof-index / indice d'hydrogène / Wasserstoff-Index / índice de hidrógeno

4614
bulk density, – effective density of the formations around the borehole, generally porous and fluid filled. It is constituted from proportional contributions by matrix density (ϱ_m) and fluid density (ϱ_f): $\varrho_b = \varnothing\varrho_f + (1 - \varnothing)\,\varrho_m$ Symbol: ϱ_b. / formatiedichtheid / densité in situ / Gesamtdichte / densidad de formación

4615
liquid junction potential; diffusion ...., – net potential across the junction between solutions of different salinity, resulting from differences in ion mobility, especially between the drilling mud in the borehole and the interstitial water in the borehole-wall formations. Symbols: E_j; E_d. Unit: mV. / contactpotentiaal / potentiel de contact / Flüssigkeitskontaktpotential; Konzentrationspotential / potencial de contacto

4616
membrane potential, – electrochemical potential across clay formations, due to an electric current, resulting from a selective cation flow across the clay. Symbol: E_m, unit: mV. / membraanpotentiaal / potentiel de membrane / Membranpotential / potencial de membrana

4617
streaming potential; electro-kinetic potential, – electric potential difference across a mud cake created by the flow of mud filtrate through this cake. Symbol: E_k, unit: mV. / stromingspotentiaal / potentiel électro-cinétique; .... d'électrofiltration / Filtrationspotential; Strömungspotential; elektrokinetisches Potential / potencial electro-cinético

4618
mud cake, – thin cover deposited on the borehole wall by solid particles of the drilling mud, filtered off during mud(-filtrate) penetration into the formation. / spoelingkoek / cake de boue; gâteau de boue / Filterkuchen / costra de lodo

4619
invasion, – penetration of mud filtrate into the permeable borehole wall, creating a disturbed zone of different hydrocarbon saturation or water resistivity. / invasie / invasion / Infiltration / invasión

invaded zone, – zone surrounding the borehole, in which mud filtrate has replaced a (major) part of the original pore contents, thereby locally changing such reservoir parameters as hydrocarbon saturation and formation-water resistivity. / invasiezone; infiltratiezone / zone envahie / infiltrierte Zone / zona invadida

4620
flushed zone, – zone very closely surrounding the borehole, in which penetrating mud filtrate has flushed away all movable original formation water and hydrocarbons. / gespoelde zone / zone lavée / geflutete Zone / zona lavada por el filtrado

transition zone, – zone surrounding the borehole, beyond the flushed zone, in which the penetrating mud filtrate has partially replaced the movable original pore fluids. Such zones are characterised by a progressive change in resistivity. / overgangszone / zone de transition / Übergangszone; Transition zone / zona de transición

4621
base line, – lines drawn through the minimum (shale-base line) and the maximum deflections (sand-base line) characteristic of respectively impervious clays and permeable sands on an SP log. More specifically the fairly constant reading of SP or gamma-ray log opposite a sufficiently thick bed of clays **(shale line)** or clean sands **(sand line)**. / basislijn / ligne de base (des argiles, ... des sables) / Tonlinie; Sand.... / línea de base

4622
bed thickness, – log determined thickness of a zone of constant physical properties that differ from those of adjacent zones. / laagdikte / épaisseur de la couche / Schichtmächtigkeit / espesor de la capa

4623
shaliness; shale content, – the amount of clay (or 'shale' in

driller's jargon) in dispersed or laminar form, present in the formation. It is related to electrical conductivity through ion-exchange processes. / kleiïgheid / teneur en argile / Tongehalt / contenido en arcilla

cation-exchange capacity; CEC, – measure of the amount of counter-cations associated with the clay present in the formation sample; indicator of shaliness of formation. Dimensions: milli-equivalent per millilitre of pore volume. Symbol: Q. / kation uitwisselend vermogen / capacité d'échange de cations / Kationen-Austauschkapazität; (KAK) / capacidad de intercambio de cationes

4624
evaporite-log evaluation, – procedure in which the composition of evaporites is determined by cross plotting, e.g. of density, sonic, neutron and gamma-ray log data.

[2890] GEOCHEMICAL METHODS

4625
geochemical exploration, – any method of mineral exploration based on systematic measurement of one or more chemical properties of a naturally occurring material. / geochemische opsporing / prospection géochimique; exploration .... / geochemische Prospektion / exploración geoquímica

4626
background, – the range of abundances of an element in normal, not mineralized naturally occurring material; although a regional concept it varies considerably with type of bedrock, history of weathering and soil formation, environment, etc. Often quoted as the mean value of the range. / achtergrond-concentratie / fond / Spiegelwert / fondo

4627
threshold value, – the limit of normal background variation, separating background values from anomalous values. A local threshold value, at a higher or lower level than the regional value, may occur in restricted areas, e.g. resulting from pervasive incipient mineralization, hydrothermal alteration or leaching. / drempelwaarde / teneur significative; seuil d'anomalie / Schwellewert / valor umbral

4628
contrast, – the ratio of anomalous values to background values, expressed as: maximum value to threshold value, maximum value to mean background value, threshold value to mean background value. / contrast / contraste / Kontrast / contraste

4629
halo; aureole, – the nearly equidimensional dispersion pattern spreading symmetrically upwards or outwards from the source of material. / aureool; hof / halo; auréole / Aureole; Dispersionshof; Hof / halo; aureola

The following types of halo are distinguished:

4630
genetic, – halo resulting from primary dispersion. / genetisch / primaire / primär / genético

superimposed, – halo formed in the regolith by the movement of material in subsurface waters. / – – / superimposée / – – / sobreimpuesto

wall-rock halo; .... aureole, – formed in the igneous or metamorphic rocks adjoining a hypogene mineral deposit. / halo in nevengesteente / auréole dans l'encaissant / Dispersionshof im Nebengestein / halo en la roca encajante

alteration halo, – the zone of hypogene alteration surrounding particularly the porphyry-copper type of deposits but also many other deposits. / – – / auréole d'altération hypogène / Alterationshof / halo de alteración

4631
superjacent dispersion pattern, – a pattern developed more or less directly above the bedrock source. / bovenliggend / susjacent / – – / suprayacente

lateral dispersion pattern, – a pattern displaced to one side with respect to the bedrock source and partly or completely underlain by barren bedrock. / lateraal / latéral / – – / lateral

fan, – a dispersion pattern that spreads predominantly to one side of the source of material. / waaier- / cône / – – / abanico

train, – a linear dispersion pattern resulting from movement along well defined drainage channels. Also used for the linear distribution of boulders caused by glacial transport. / lineair / trainée / – – / lineal

4632
(geochemical) anomaly, – a departure from the geochemical patterns that are normal for a given area or geochemical province; a measurement or an abundance that deviates from the norm or background. / anomalie / anomalie / Anomalie / anomalía

4633
allogenic; additive, – anomaly that is the expression of lateral dispersion of elements from an outside source superimposed on a pre-existing local background. / allogeen / allogène; de surimposition / – – / superimpuesta

autigenic, – anomaly that is the result of a rearrangement of locally available elements, often combined with a negative anomaly as a result of leaching. / autigeen / autigène / autigen / autigénica

4634
hydrochemical, – anomaly occurring in natural waters. / hydrochemisch / hydrogéochimique / hydrochemisch / hidroquímica

hydromorphic, – anomaly formed by the movement of solutions followed by precipitation. / hydromorf / hydromorphe / hydromorph / hidromórfica

leakage, – anomaly formed by leakage of metal-ion-containing solutions through fractures and other channels from a deep-seated source into a near-surface environment. / – – / anomalie de fuite / – – / escape

4635
break-in-slope; seepage, – anomaly formed by the seepage of

ground water toward the earth's surface, occurring at the intersection of the local topography with the ground-water table. / – – / anomalie hydromorphe d'émergence / – – / percolación; filtración

seepage, – anomaly formed specifically where a ground-water flow invades a chemically different environment. / – – / anomalie hydromorphe de précipitation / – – / filtración

4636
soil, – anomaly formed in (residual) soil as the result of weathering of the underlying bedrock. / bodem- / en sol / im Boden / suelo

4637
displaced, – anomaly found at a certain distance from the primary source, mostly downslope or downstream. / verplaatst / déplacée / verlagert / desplazada

false; non-significant, – anomaly, spatially unrelated to a primary source and mostly the result of contamination, sampling or analytical errors, or scavenging action by iron or manganese oxides. / schijnbaar / fausse / Schein- / falsa

4638
geochemical isograd, – on a geochemical map, the contours of equal element content. / geochemische isograad / isanomal / Linie gleicher Gehalt / isograda geoquímica

4639
orientation survey; pilot ....; **feasibility study**; preliminary ...., – a limited survey or study, preliminary to a large or regional survey, to determine the optimum field-analytical and interpretative parameters to be used in the actual survey. / oriëntatie-onderzoek / étude d'orientation / Übersichtsprospektion / estudio de orientación; .... piloto; .... de factibilidad; .... preliminar

4640
hydrochemical exploration; geochemical water survey, – a survey in which the sampling is restricted to naturally occurring waters (ground water, surface water). / hydrochemische verkenning / prospection hydrogéochimique / hydro-geochemische Prospektion / exploración hidroquímica

4641
drainage(-basin) survey, – a regional survey in which samples of the water and/or sediments of the drainage channels are taken in a systematic way, with the understanding that each sample represents the best composite of materials from the catchment area upstream from the sampling site. / onderzoek in het afwateringsnet / prospection dans le réseau hydrographique / – – / prospección en redes de drenaje

stream-sediment survey, – a type of drainage survey, in which the active stream sediment, and/or the sediments from the adjacent flood plain, or the heavy minerals separated from these sediments, are sampled and analysed. / onderzoek van riviersediment / prospection de sédiments du ruisseau / Bachsedimentprospektion / prospección en sedimentos fluviales

4642
precipitation barrier, – an area of rapid precipitation of the metallic element content in natural water, caused by changing conditions of the water (mostly acidity or oxygen availability). / neerslagdrempel / barrière de précipitation / Fällungsbarriere / barrera de precipitación

4643
cut-off, – the point where the maximum metal value is found when proceeding in upstream direction. / – – / coupure / – – / – –

masking, – the effect of certain layers (glacial till, alluvium, bentonite and others) to cause the absence of an anomaly in soils or stream sediments overlying a distinct metal source. / – – / écran / Tarnung / enmascarar

4644
soil(-sampling) survey, – a survey using residual soils for sampling and analysis. / geochemisch bodemonderzoek / prospection géochimique en sol / geochemische Bodenprospektion / prospección en suelos

rock(-sampling) survey; lithogeochemical ...., – a survey using fresh bedrock for sampling and analysis. / geochemisch gesteente-onderzoek / prospection géochimique en roche; .... lithogéochimique / Festgesteinsprospektion / prospección en roca

4645
biogeochemical exploration; geochemical plant survey, – a survey using selected (parts of) species of animals or plants or their dead remains (skeletons, humus), for sampling and analysis. / biogeochemisch onderzoek / prospection biogéochimique / biogeochemische Prospektion / exploración biogeoquímica

geobotanical exploration; vegetation survey, – a visual survey using the distribution of specific plants, selected parts thereof, or plant associations or deviating growth patterns to study geochemical distribution patterns. / geobotanisch onderzoek / prospection géobotanique / geobotanische Prospektion / exploración geobotánica; estudio de la vegetación

4646
indicator plant; geobotanical indicator, – plant species or characteristic variations in growth habits of plants that are restricted in their distribution to rocks or soils with definite physical or chemical properties. / indicatorplant / plante indicatrice; indicateur géobotanique / Indikatorpflanze / indicador vegetal; .... geobotánico

specific, – the plant's development is dependent on the presence or a certain abundance of one or more metal elements.

symptomatic, – the plant's development is altered, but its distribution is unaffected, by the presence of certain metal elements.

4647
accumulator plant, – a plant that takes up a particular element in quantities considerably in excess of that in other plants. / – – / plante accumulatrice / – – / planta acumuladora

converter plant, – a plant that takes up a chemical compound from the soil, builds it into its living structure, and at death returns it to the soil in soluble form. / – – / – – / – – / planta convertidora

4648
vapour survey; gas ...., – a survey that samples and analyses the vapour phase present in, flowing through or hovering over the soil. / damponderzoek / prospection géochimique par le gaz / Gasprospektion / prospección por gases

4649
pathfinder; indicator element, – a mineral or element found in close association with the element being sought; it can be more readily detected or has more, obviously anomalous features than the element that is the main object of search. / gidselement / traceur / Pfadfinderelement; Indikatorelement / elemento indicador

4650
scavenger, – a chemical compound with the capability to adsorb or absorb other chemical elements in such quantities that the dispersion pattern of the element prospected for is seriously disturbed. / – – / piège chimique / – – / – –

4651
detection limit, – the minimum content of an element that can be measured by a specific analytical method; it is also dependent upon matrix (host rock material) and other elements present. / detectiegrens / limite de détection / Nachweisgrenze / limite de detección

4652
contamination, –
1. pollution of the natural geochemical environment arising from human activity;
2. the addition of unwanted extraneous material to a sample during its processing.

/ besmetting; vervuiling / contamination / Kontamination / contaminación

4653
heavy metals, – in geochemical context the elements Zn, Cu, Pb and additionally Ni, Co and Ag. / zware metalen / métaux lourds / Schwermetalle / metales pesados

4654
extractable metals, – metals that can be extracted from a sample by any given chemical treatment. / extraheerbare metalen / métaux extractibles / extrahierbare Metalle / metales extraibles

4655
cold extractable heavy metals; cxHM, – an analytical procedure at room temperature using a variety of dilute acids and buffers to maintain pH within specific limits, resulting in either selective or restrictive extractions. / koud-extraheerbare zware metalen / métaux lourds extractibles à froid / kalt extrahierbare Metalle / metales extraibles en frío

Partly synonymous:

cxME cold extractable metals
THM total heavy metals
ammonium citrate soluble heavy metals test
Bloom test

hot extractable metals; hx, – all metal ions, except those incorporated in the stable rock-forming silicates, and extracted through vigorous attack by acids at temperatures just below their boiling points. / warm-extraheerbare metalen / métaux extractibles à chaud / heiss extrahierbare Metalle / metales extraibles en caliente

4656
EDTA, – a much used organic extraction reagent composed of ethylene-diamine-tetra-acetic acid.

MIBK, – an organic extraction reagent composed of methyl-iso-butyl ketone.

4657
gridding, line cutting, – preparing an area for systematic (geochemical) prospecting by marking or cutting lines following a predetermined grid.

4658
costean, – a Cornish term for pitting and trenching as a means for producing artificial prospecting outcrops.

4659
pan, – a dish of approximately 50 cm diameter with a flat, conical or rounded bottom used to concentrate through 'washing' heavy minerals from sands and gravels. / pan; waspan / batée / Pfanne / lavadero

sluice (box), – a channel constructed of timber through which water flows to wash out heavy minerals from gravel and sand. / wasgoot / couloir de lavage / Waschrinne / lavadero

[2900] ENGINEERING GEOLOGY

[2910] GENERAL TERMS

4660
engineering geology, – study of geological factors affecting the planning, design, construction, operation and maintenance of engineering structures. / ingenieursgeologie / geologie de l'ingénieur / Ingenieurgeologie / geología del ingeniero

4661
applied geology, – study of geological factors affecting economic, engineering, water supply, or military problems. / toegepaste geologie / géologie appliquée / angewandte Geologie / geología aplicada

4662
geotechnics; geotechnique, – application of geological and engineering principles and methods to solve civil engineering problems. It includes soil mechanics, rock mechanics and the engineering aspects of applied geology. / geotechniek / géotechnique / Geotechnik / geotécnica

4663
engineering geophysics, – application of geophysical principles and techniques to the solution of civil engineering problems. / ingenieursgeofysica / géophysique de l'ingéniérie / Ingenieurgeofysik / geofísica del ingeniero

engineering geomorphology, – study of geomorphological factors affecting the planning, design, construction, operating and maintenance of engineering structures. / ingenieursgeomorfologie / géomorphologie de l'ingéniérie / Ingenieurgeomorphologie / geomorfología del ingeniero

soil mechanics, – the theoretical and applied science of the mechanical behaviour of soils, representing a branch of applied mechanics concerned with the response of soil to the force fields of its physical environment. / grondmechanica / mécanique des sols / Bodenmechanik / mecánica de suelos

4664
rock mechanics, – the theoretical and applied science of the mechanical behaviour of rocks, representing a branch of applied mechanics concerned with the response of rock to the force fields of its physical environment. / gesteentemechanica / mécanique des roches / Felsmechanik / mecánica de rocas

4665
environmental geology, – collection, study and application of geological data and principles for planning, development and management of the most efficient and beneficial use of the geological environment and its resources. / milieugeologie / géologie de l'environnement; (écogéologie) / Umweltgeologie / geología ambiental

[2920] BEHAVIOUR AND PROPERTIES OF SOIL AND ROCK

[2921] *General Aspects of Material Behaviour*

4666
deformation, – change of form or volume of a body due to the influence of an external force. / vervorming; deformatie / déformation / Deformation; Verformung / deformación

strain, – ratio between change of length, △1, (or volume) of a body and its original length. / relatieve vervorming / déformation relative / Dehnung / deformación relativa

4667
stress, – force per unit area. / spanning / contrainte / Spannung / esfuerzo

stress-strain diagram, – graph showing the relationship for a material between the stress applied to it and the resulting length (or volume) strain. / spanning-vervorming diagram / graphique contrainte-déformation / Spannungs-Dehnungs Diagramm / diagrama de esfuerzos-deformación relativa

4668
effective stress, – stress transmitted from particle to particle in a soil or rock mass. In water-saturated soil and rock the effective stress is the difference between the total stress and the pressure of the water in the pores, voids and cracks. / effectieve spanning / contrainte effective / wirksame Spannung / esfuerzo efectivo

4669
elasticity, – deformation of a material that disappears after removal of the deforming force(s); elastic deformation is independent of time; stress and strain are linearly related. / elasticiteit / élasticité / Elastizität / elasticidad

plasticity, – property of material to deform its shape or volume permanently without rupture, and that, once begun, is continuous without increase of stress. / plasticiteit / plasticité / Plastizität / plasticidad

4670
thixotropy, – the property of certain colloïdal substances (e.g. a thixotropic clay) to weaken or change from a gel to a sol when shaken but to increase in strength upon standing. / thixotropie / thixotropie / Thixotropie / thixotropía

4671
viscosity, – the property of a substance to offer internal resistance to flow. In viscous bodies the rate of strain is directly proportional to the level of shear stress. / viscositeit / viscosité / Viskosität; Zähigkeit / viscosidad

rheidity, – the time required for deformation by viscous flow to become more than 1000 times the elastic deformation; it depends on temperature, confining pressure and stress conditions.

rheology, – the study of flow phenomena in general. / rheologie / rhéologie / Rheologie / reología

4672
creep, – continuously increasing, usually slow, deformation (strain) of material, resulting from a small constant stress acting over a long period of time. / kruipen / fluage / Kriechen / – –

flow, – continuously increasing deformation of material resulting from a small constant stress. / vloeien / écoulement / Fliessen / – –

4673
ductile, – property of a material to be deformed considerably (5-10% strain) before rupture occurs. / kneedbaar; vervormbaar / ductile / duktil / dúctil

brittle, – property of a material to fracture at small amounts of deformation (less than 3-5% strain). / bros / fragile / spröde / frágil

4674
fatigue, – property of a material to fail after many repetitions of a deforming stress, which by itself is not high enough to cause failure. / (materiaal)moeheid / fatigue / Ermüdung / fatiga

4675
hysteresis, – property of material to recover to its original form after removal of the deforming force(s); it does not happen instantaneously but needs a certain amount of time. / hysterese / hysteresis; hystérèse / Hysterese / hystéresis

4676
failure; rupture, – breaking of material that has been stressed beyond its failure strength. / breuk / rupture / Bruch / fractura

progressive failure, – gradual development of a failure plane in a body or mass due to a changing stress distribution. / voortschrijdende breuk / rupture progressive / progressiver Bruch / fractura progresiva

4677
shear, – mode of failure of a body or mass, whereby the portion on one side of a plane or surface slides past the portion on the opposite side in a direction parallel to the plane. The term shear is often used to indicate 'shear strain'. / schuif / cisaillement / Schub / cizalla; cizalladura

4678
dilatancy, – increase of bulk volume deformation due to fracturing, relative movement or rotation of soil grains in a soil mass or rock blocks in a rock mass. Dilatancy is accompanied by an increase in pore volume. / dilatantie / dilatance / Dilatanz / dilatancia

4679
compaction, – any process (such as by weight of overburden or engineering structure, or by desiccation) leading to a loss of pore space and an increase of density. Compaction may be achieved by e.g. rolling, tamping, vibr(oflot)ation, dewatering. / verdichting; compactie / compactage; serrage / Verdichtung / compactación

[2922] *General Aspects of Material Properties*

4680
homogeneous, – a body or mass is homogeneous when samples from different locations but of the same size and orientation have the same properties. / homogeen / homogène / homogen / homogéneo

inhomogeneous, – a body or mass is inhomogeneous when samples from different locations but of the same size and orientation have different properties. / inhomogeen / hétérogène / inhomogen / inhomogéneo

4681
isotropic, – a body or mass is isotropic when its properties are the same in all directions. / isotroop / isotrope / isotrop / isótropo

anisotropic, – a body or mass is anisotropic when its properties are different in different directions. / anisotroop / anisotrope / anisotrop / anisótropo

4682
strength; resistance, – limit value of differential stress that can be withstood by a material without being broken or considerably deformed (in plastic deformation). Strength has the dimension of stress. / sterkte; vastheid / résistance / Festigkeit / resistencia

4683
compressive strength, – / druksterkte; drukvastheid / résistance à la compression / Druckfestigkeit / resistencia a la compresión

uniaxial compressive strength, – / eenassige druksterkte / résistance à la compression uniaxale / einaxiale Druckfestigkeit / resistencia a la compresión uniaxial

triaxial compressive strength, – / drieassige druksterkte / résistance à la compression triaxiale / dreiaxiale Druckfestigkeit / resistencia a la compresión triaxial

4684
tensile strength, – / treksterkte; trekvastheid / résistance à la traction / Zugfestigkeit / resistencia a la tracción

shear strength, – / schuifweerstand / résistance au cisaillement; résistance aux tensions tangentielles / Scherfestigkeit / resistencia a la cizalladura

4685
peak strength, – in cases where the strength of a material is dependent on the amount of strain, the peak strength is the maximum value of strength; it is observed usually at small amounts of strain. / – – / résistance au pic / Spitzenfestigkeit / resistencia punta

residual strength, – in cases where the strength of a material decreases with increasing deformation, the residual strength is the value to which the strength is lowered after a considerable amount of deformation. / – – / résistance résiduelle / Restscherfestigkeit / resistencia residual

4686
friction, – resistance against relative movement along a boundary plane between two bodies. / wrijving / frottement / Reibung / fricción

internal friction, – resistance against relative movement along a plane through a body. / inwendige wrijving / frottement interne / innere Reibung / fricción interna

4687
(natural) angle of repose, – the maximum angle of slope at which loose, cohesionless material comes to rest. The angle is dependent on the frictional characteristics of the material, the size and roughness of the material particles. / maximale hellingshoek; hoek van het natuurlijke talud / angle de talus naturel; angle de repos / natürlicher Böschungswinkel / ángulo de reposo

4688
Mohr (stress) diagram, – a diagram with shear stress (τ) and normal stress (σ) as axes, usually respectively ordinate and abscis. / (spannings)diagram van Mohr / diagramme de Mohr / Morsches Spannungsdiagramm / diagrama (de esfuerzos) de Mohr

shear stress, – the magnitude of stress along a shear plane. / schuifspanning / contrainte de cisaillement; .... tangentielle / Schubspannung / esfuerzo de cizalla

normal stress, – the magnitude of stress perpendicular to a shear plane. / normaalspanning / contrainte normale / Normalspannung / esfuerzo normal

4689
friction criterion; Coulomb ...., – enables the calculation of the shear stress, necessary for relative movement along a slip

plane. It is derived from the normal stress across this plane and the friction coefficient, f:

$$f = \frac{\tau}{\sigma}$$

/ wrijvingskriterium / critère de frottement; .... de Coulomb / Coulombsches (Reibungs)gesetz / criterio de fricción

angle of (internal) friction, – in a Mohr diagram the angle (φ) of the line $f = \frac{\tau}{\sigma}$ with the abscis. / wrijvingshoek / angle de frottement / Reibungswinkel / ángulo de fricción

friction coefficient; internal friction, – the tangent of the angle of friction. / wrijvingscoëfficiënt / coefficient de frottement / Reibungskoeffizient; ....beiwert / coeficiente de fricción

4690
Mohr circle, – a type of Mohr diagram, representing the two-dimensional state of stress in a particular point and considered in the plane of least (P_1) and greatest (P_3) principal stress. P_1 and P_3 are plotted on the normal-stress axis; sets of stresses in planes inclined to P_1 appear on the circumference of a circle with its centre in a zero point, midway between the plots of P_1 and P_3. / cirkel van Mohr / circle (des contraintes) de Mohr / Mohrscher Spannungskreis / círculo de Mohr

4691
Mohr envelope, – enveloping curve or straight line of a set of Mohr circles for rock types, e.g. under varying conditions of confining pressure. The envelope is tangent at the point representing rupture and thus will give also the orientation of the shear planes. / omhullende van Mohr / courbe intrinsique (de Mohr) / Mohrsche Hüllkurve / curva intrínsica de Mohr

4692
principal stress, – one of the three components of the stress field inside the earth's crust (or any other stress field) that act on the sides of a unit cube. They are vector quantities; distinguished are the axes of least, intermediate and greatest principal stress. / hoofdspanning / contrainte principale / Hauptspannung / esfuerzo principal

4693
cohesion, – the component of the shear stress that is independent of normal stress. / cohesie / cohésion / Kohäsion / cohesión

apparent cohesion, – may be observed along rock-separation planes due to interlocking of surface roughness or in moist sandy soils due to surface tension in capillary interstices. / schijnbare cohesie / cohésion apparente / scheinbare Kohäsion / cohesión aparente

4694
Mohr-Coulomb fracture criterion, – a criterion for rupture to occur, taking into acount the cohesion (c) which appears as the intercept on the ordinate in a Mohr diagram:
$\tau = c + \sigma \tan\varphi$.
/ breukkriterium van Mohr-Coulomb / critère de rupture Mohr-Coulomb / Mohr-Coulombsches Bruchkriterium / criterio de fractura Mohr-Coulomb

4695
modulus of elasticity; Young's modulus, – ratio of stress to corresponding strain in elastic deformation. The modulus of elasticity is a property of materials that show linear proportionality between applied stress and resulting strain untill rupture occurs. / elasticiteitsmodulus; elasticiteitscoëfficient / module d'élasticité; module de Young / Elastizitätsmodul; Youngscher Modul / módulo de elasticidad; coeficiente de elasticidad; módulo de Young

4696
modulus of deformation, – term used instead of modulus of elasticity for materials that do not show linear proportionality between applied stress and resulting strain. / vervormingsmodulus / module de déformation / Verformungsmodul / módulo de deformación; coeficiente de deformación

4697
yield stress; .... point; .... strength; elastic limit, – the threshold value of (differential) stress at which permanent (non-elastic) deformation first occurs in a material. / elasticiteitsgrens / limite élastique; limite d'élasticité / Elastizitätsgrenze; Fliessgrenze / límite elastico; límite de elasticidad

4698
bulk modulus; modulus of incompressibility; volume elasticity, – a modulus of volumetric deformation or elasticity for hydrostatic states of stress. Symbol K. / modulus van samendrukbaarheid / module de compression / Kompressionsmodul / módulo de elasticidad

(bulk) compressibility; coefficient of compressibility, – the reciprocal value of bulk modulus, K^{-1}. / coëfficiënt van samendrukbaarheid / compressibilité / Kompressibilität / coeficiente de compresibilidad

4699
shear modulus; modulus of rigidity, – modulus of elasticity for strain due to shear stress. / afschuivingsmodulus / module de (deformation par) cisaillement / Schubmodul / módulo de cizalla; coeficiente de ....; coeficiente de cizalladura; módulo de ....

4700
Poisson's ratio, – ratio of the lateral unit strain to the longitudal unit strain in a body that has been stressed longitudinally at stresses below its yield stress. / coëfficiënt van Poisson / coefficient de Poisson / Poisson-Zahl / coeficiente de Poisson

Poisson's number, – the reciprocal value of the Poisson's ratio. / getal van Poisson / – – / Querdehnzahl; Querverformungszahl / número de Poisson

4701
hardness, – property of minerals, rocks or other materials that may be described either as resistance to scratching or as the height of rebound of a dropped object. / hardheid / dureté / Härte / dureza

Mohs (hardness) scale, – the relative resistance to scratching. / (hardheids)schaal van Mohs / échelle (de dureté) de Mohs / Mohssche (Härte)skala / escala (de dureza) de Mohs

Shore (hardness) scale, – the height of rebound of a small standard object, dropped from a fixed height on the surface of a specimen. / (hardheids)schaal van Shore / échelle (de dureté) de Shore / Shore (Härte)skala / escala (de dureza) de Shore

[2923] *Engineering Properties and Behaviour of Soils*

4702
gradation, – the proportion of material of each particle size, or the frequency distribution of various sizes, constituting a soil or sediment. / korrelverdeling / gradation; granulometrie / Kornverteilung / gradación (de tamaños)

well graded, – an engineering term pertaining to soil or rock consisting of particles of many sizes, having a uniform or equable distribution of particles from coarse to fine. / continue korrelverdeling / granulométrie continue / gut abgestufte Kornverteilung / granulometría continua

poorly graded, – an engineering term pertaining to soil or rock consisting of particles all having approximately the same size. / discontinue korrelverdeling / granulométrie discontinue / schlecht abgestufte Kornverteilung / granulometría discontinua

NOTE:
– for terms pertaining to the texture of soils and rocks, that are not special engineering terms, one is referred to sections
[0943] [0944] Textures, Structures (of igneous rocks),
[1030] Texture and Structure (of metamorphic rocks,
[1740] [1750] Soil Texture, Structure,
[1950] Sedimentary Texture.

4703
consolidation, – of soils, gradual or slow reduction in volume and increase in density of a soil mass in response to increased load or compressive stress; e.g. the adjustment of a saturated soil involving the squeezing of water from pores. / consolidatie / consolidation / Konsolidierung / consolidación

preconsolidation, – the greatest effective stress to which a soil has been subjected. / voorconsolidatie / préconsolidation; consolidation préalable / Vorkonsolidierung / preconsolidación

overconsolidation, – consolidation greater than normal for the existing overburden, resulting from desiccation or overburden which since has been removed. / overconsolidatie / surconsolidation / Überkonsolidierung / supraconsolidación; sobreconsolidación

4704
settlement, –
1. gradual downward movement of an engineering structure, due to compression of the soil below the foundation;
2. gradual lowering of overlying strata due to extraction of mined material (coal, oil, gas, water);
3. gradual lowering of superficial material (such as coastal sediments) due to compaction of the underlying sediment.

/ klink; inklinking; zetting; zakking / tassement; affaissement; enfoncement / Setzung / asentamiento

differential settlement, – nonuniform settlement; the uneven downward movement of different parts of an engineering structure that may cause damage to the structure. / zettingsverschil / tassement differentiel / unterschiedliche Setzung / asentamiento diferencial; .... desigual; .... no homogéneo

subsidence, – synonym for meaning 2 and 3 of settlement. / verzakking; inzakking / affaissement / Bodensenkung / subsidencia

4705
earth pressure; soil pressure, – the lateral pressure acting between earth materials and a structure, such as a wall. / gronddruk / poussée des terres / Erddruck / presión lateral del terreno; .... del suelo

earth pressure at rest, – earth pressure when the soil is in its natural state, without having been permitted to yield or without having been compressed. / neutrale gronddruk / poussée des terres au repos / Ruhedruck / presión del terreno en reposo; .... natural del terreno

4706
active earth pressure, – the minimum value of earth pressure, when a soil mass is permitted to yield sufficiently to cause its internal shearing resistance along a potential failure surface to be completely mobilized. / actieve gronddruk; steundruk / poussée active des terres / aktiver Erddruck / presión activa del terreno

passive earth pressure, – the maximum value of earth pressure, when a soil mass is compressed sufficiently to cause its internal shearing resistance along a potential failure surface to be completely mobilized. / passieve gronddruk; grondweerstand; tegendruk van de grond / butée / passiver Erddruck / presión pasiva del terreno

4707
bearing capacity, – load per unit area that the ground can support safely without excessive yield. / draagvermogen / capacité portante / Tragfähigkeit; Tragvermögen / capacidad de carga

4708
plate-bearing test, – field test to determine the bearing capacity of the ground. Plates are usually 60 to 100 cm in diameter and loaded by jacking from a special heavy truck-trailer. When applied in rock tunnels for the determination of the modulus of deformation the type of test is called **plate-loading test**. / plaatbelastingsproef / essai de charge sur plaque; essai à la plaque / Plattendruckversuch / placa para ensayo de carga

California Bearing Ratio; CBR, – test to determine the shear strength in vertical direction of a soil by penetration of a steel rod of a small diameter. / CBR-waarde / indice portant Californien; (CBR) / CBR-Wert / índice de penetración California; (CBR)

4709
penetration resistance, –
1. unit load required to produce a specified penetration into a soil at a specified rate;
2. number of blows of a hammer of specified mass, falling from a specified height to produce a specified penetration of a body of specified form, usually a cone, into the soil.

/ penetratieweerstand / résistance à la pénétration / Eindringwiderstand / resistencia a la penetración

4710
sounding, – pushing, driving or hammering a rod down through the soil for the determination of the depth of boundaries between strata of different strengths; sounding is often used in soft materials. / sondering / sondage / Sondierung / sondeo

4711
dynamic penetration; .... sounding, – a rod with a cone at its end is driven into the ground by a hammer of specified mass falling from a specified height; the number of blows necessary for a specified amount of advance is counted. / dynamische sondering / pénétromètre dynamique / Rammsondierung / dinámica de penetración; sondeo dinámico

static penetration; .... sounding, – a rod with a cone at its end is pushed into the ground with constant speed; the force necessary for this penetration is recorded as **penetration resistance**. / statische sondering / pénétromètre statique / Drucksondierung / estática de penetración; sondeo estático

4712
standard penetration test, – determination of the penetration resistance of a standard sampling spoon when hammered into the undisturbed and cleaned bottom of a borehole; number of blows necessary for the penetration of 1 foot is recorded . / standaard-sondeerproef / essai de pénétration standard / – – / prueba de penetración standard

4713
consistency, – the relative ease or difficulty with which a soil is deformed. The consistency depends on the water content. Four different states of consistency are distinguished: solid state, semi-solid state, plastic state, liquid state. / consistentie / consistance / Konsistenz / consistencia

solid state, – / vaste toestand / état solide / fester Zustand / estado sólido

semi-solid state, – / halfvaste toestand / état semi-solide / halbfester Zustand / estado semisólido

plastic state, – / plastische toestand / état plastique / weicher Zustand / estado plástico

liquid state, – / vloeibare toestand / état liquide / breiiger Zustand / estado líquido

4714
Atterberg limit; consistency limit, – water content at the boundary between two states of consistency. / grens van Atterberg; consistentiegrens / limite de consistence; limite d'Atterberg / Atterbergsche Konsistenzgrenze / límite de consistencia; .... de Atterberg

4715
liquid limit, – water content at boundary between liquid state and plastic state. / vloeigrens / limite de liquidité / Fliessgrenze / límite de líquidez

plastic limit, – water content at boundary between plastic state and (semi-)solid state. / plasticiteitsgrens; uitrolgrens / limite de plasticité / Plastizitätsgrenze; Ausrollgrenze / límite de plasticidad

shrinkage limit, – water content at boundary between semi-solid and solid state. Below this limit a decrease in moisture content will not cause a decrease in volume. / krimpgrens / limite de retrait / Schwindgrenze; Schrumpfgrenze / límite de retracción

4716
plasticity index, – range of water content at which a soil behaves plastically. Numerical value is liquid limit minus plastic limit. / plasticiteitsgetal; plasticiteitsindex / indice de plasticité / Plastizitätszahl / índice de plasticidad

liquidity index, – expression for the consistency of a soil at its natural water content. Numerical value is the natural water content minus the plastic limit, and divided by the plasticity index. / vloeibaarheidsindex / indice de liquidité / Fliesszahl / índice de líquidez

consistency index, – ratio of the liquid limit minus the natural water content to the plasticity index. / consistentiecoëfficiënt / indice de consistance / Konsistenzzahl / índice de consistencia

4717
sensitivity, – the effect of remolding on the consistency of a cohesive soil. / gevoeligheid / sensibilité; (sensitivité) / Empfindlichkeit; Sensitivität / sensitividad

sensitive clay, – a clay of which the strength is decreased to a small fraction of its original value after remolding at constant moisture content. / gevoelige klei / argile sensible (au remaniement) / strukturempfindlicher Ton / arcilla sensitiva

4718
liquefaction, – the sudden large decrease of the shearing resistance of a cohesionless soil, caused by a collapse of the structure by shock or strain, and associated with a sudden but temporary increase of pore-fluid pressure. It causes the temporary transformation of the material into a fluid mass. / vloeibaar worden; liquefactie / liquéfaction / Verflüssigung / licuefacción

4719
quick clay, – a clay that loses all its strength after being disturbed, and behaves like a liquid afterwards. / vloeiklei / argile thixotropique; .... sensible / Quickton / arcilla fluida

quicksand; driftsand, – oversaturated sand forming a highly mobile mass that yields easily to pressure and tends to suck down heavy objects touching its surface. / drijfzand / sable boulant / Schwimmsand / arenas movedizas

4720
swelling, – increase of volume due to increase of water content or due to freezing. / zwelling; opzwelling / gonflement / Schwellen / hinchazón

4721
frost heaving, – upheaval of soil due to the freezing of soil water and subsequent formation of subsurface ice lenses. / opvriezing / soulèvement dû au gel; gonflement .... / Frosthebung / elevación del suelo debido a helada

[2924] *Engineering Properties and Behaviour of Rocks*

4722
rock; rock material; intact rock, – in the engineering sense, rock is a naturally formed material with a uniaxial compressive strength over a certain minimum value, usually taken as 1 $MN.m^{-2}$ ($\approx$ 10 $kg.cm^{-2}$), and composed of mineral grains. Rock is considered to be intact, if it does not contain discon-

tinuities of sedimentological, structural or other origin. / gesteente; gesteentemateriaal / matrice; (roche) / Festgestein / roca; material rocoso; roca sin discontinuidades

rock mass, – rock as it occurs in situ, including discontinuities. / gesteentemassa / massif rocheux / Fels / masa rocosa

4723
discontinuity; joint; separation plane, – in rock mechanics, usually any open joint or joint that potentially, without stress being applied, can become opened, whatever its origin is sedimentary, tectonic or otherwise. Discontinuities separate the blocks of intact rock in a rock mass. / discontinuïteit; diaklaas / discontinuité; diaclase / Diskontinuität; Trennfläche; Kluft / discontinuidad; diaclasa

fault, – in rock mechanics, any separation plane in a rock mass along which appreciable relative movement has taken place. / breuk / faille / Störung / falla

NOTE:
– for further terms on joints see section [1320] Fracturing

4724
joint system; .... set, – group of parallel or almost parallel joints in a rock mass. / groep diaklazen / famille de joints parallèles / Kluftschar / sistema de diaclasas

4725
(joint) spacing, – perpendicular distance between discontinuities. Normally refers to the mean or modal spacing of a set of joints. / diaklaasafstand / espacement (des diaclases) / Kluftabstand / espaciado de diaclasas

joint frequency; jointing intensity, – the inverse value of joint spacing, i.e. number of discontinuities per metre. / frequentie van diaklazen / fréquence des diaclases / Klüftigkeitsziffer; Kluftdichte / frecuencia de diaclasas

4726
persistence; continuity; extent, – measure, in percentage, of the areal extent or penetration length of a discontinuity. Termination of the discontinuity in solid rock or against other discontinuities reduces the persistence. / – – / continuité; degré de séparation / Klufterstreckung / persistencia; continuidad; extensión

4727
(joint) roughness, – small-scale surface irregularities on a joint plane. / ruwheid / rugosité / Rauhigkeit / rugosidad

(joint) waviness, – surface irregularities with longer wave lengths. / gegolfdheid / ondulation / Welligkeit / ondulación

4728
(joint) aperture; .... opening, – perpendicular distance between adjacent walls of a discontinuity, in which the intervening space is filled with air, water or loose material. / diaklaasopening / ouverture; épaisseur / Öffnungsweite; Kluftweite / apertura (de diaclasa)

4729
(joint) filling, – material that occupies the space between the adjacent rock walls of a discontinuity. The filling is usually weaker than the surrounding rock, but sometimes is present in form of hard mineral coatings of e.g. quartz, calcite. / diaklaasvulling / remplissage / (Kluft)füllung / relleno (de diaclasa)

gouge, – claylike material occurring between the walls of a fault as a result of wear during relative movement. / – – / (argile de faille) / Tonpaste / milonita; arcilla de falla

4730
rock block, – block of intact rock terminated by the surrounding joint planes. / rotsblok / bloc rocheux / Kluftkörper; Grundkörper / bloque rocoso

unit-rock block, – imaginary block of intact rock, of which form and size are determined by mean orientation and mean spacing of the joint sets in the rock mass. The unit-rock block is used as a simplification of complicated joint occurrences to allow for geomechanical or analytical model investigations. / – – / bloc unitaire / Einheitskluftkörper / unidad de bloque rocoso

4731
loosening, – decrease of tightness of a rock mass as a result of opening of joint planes due to a decrease of confining stress. Loosening is observed near all free rock-mass surfaces and may cause problems in surface or subsurface excavation works as the rock-mass strength is decreased. / losser worden / décompression; (ameublissement); foisonnement / Auflockerung / desintegración

loosening zone, – / zone van losser gesteente / zone décomprimée / Auflockerungszone / zona de desintegración

4732
velocity index, – ratio of wave-propagation velocity in a natural rock mass to the velocity in the rock material of which the mass is composed. The velocity index is related to the jointing intensity of a rock mass. / snelheidsindex / indice de qualité (sismique) / Geschwindigkeitsindex / índice de velocidad

4733
point-load strength; PLS, – the load, necessary to split rock samples by application of a concentrated load using a pair of conical plates. Samples may be of a cylindrical form or irregular lumps. Used as index number for rock-strength classification. / breuksterkte bij puntbelasting / résistance à la charge concentrée / Punktlastfestigkeit / resistencia al 'punto de carga'

(elastic) rebound number, – index value used for rock-quality classification based on the rebound of objects, dropped from a fixed height or driven with a fixed speed against the rock surface. Equipment used is **Shore scleroscope** or **Schmidt hammer**. / terugspringgetal / indice sclérométrique / Rückprall / – –

4734
slake durability, – resistance offered by a rock sample to weakening and disintegration when subjected to cycles of drying and wetting. / – – / résistance au délitage / Zerfallsdauer / durabilidad

soundness, – resistance to disintegration from repeated freezing and thawing. / vorstbestendigheid / résistance au gel / Frostbeständigkeit / – –

weatherability, – susceptibility to weathering. / verweerbaarheid / altérabilité / Verwitterbarkeit / – –

[2925] *Hydraulics*

NOTE:
– in this section a few terms concerning engineering practice have been compiled; see also
[2520] Geology of Subsurface Waters,
[2540] Rock Properties and Fluid Flow.

4735
hydraulic head, –
1. the height of the free surface of a body of water relative the earth's surface or a datum level. / waterdrukhoogte / charge (hydraulique) / hydraulische Druckhöhe / carga hydráulica
2. the difference in height between the free water surface above (upstream) and below (downstream) a hydraulic work, such as a dam, a sluice, a weir. / niveauverschil; verval / différence de charge; chute / hydraulischer Druckhöhenunterschied; Fall / diferencia de carga

4736
hydraulic grade (line), –
1. in free water, the water surface;
2. in subsurface water, the line connecting the points to which the water would rise in pipes open to the surface.

/ verhanglijn / surface piézométrique / – – / (línea de) nivel hidráulico;

hydraulic gradient, – slope of the hydraulic grade line. / verhang / gradient hydraulique / Druckhöhengefälle; hydraulisches Gefälle / gradiente hydráulico

4737
hydraulic thrust; water thrust, – pressure exerted on (parts of) a rock or soil mass or an engineering structure due to differences in hydraulic head at the different sides of the mass or structure. / hydraulische zijwaartse druk / poussée hydraulique / Wasserschub; hydraulischer Schub; Strömungsdruck / presión hidráulica

4738
seepage pressure, – the force that is transferred from water flowing through a porous medium to the medium itself by viscous friction. / stromingsdruk / pression de filtration / Sickerströmungsdruck / presión de corriente de percolación

4739
drainage; unwatering; dewatering, – removal of surface water or ground water from an area or mass of soil or rock by free flow or by pumping. / drainage; ontwatering / drainage; épuisement; assèchement / Drainung; Entwässerung / drenaje

4740
buoyancy; (hydrostatic) uplift, – upward pressure exerted by a fluid on a body wholly or partly immersed in it. The upward pressure equals the loss in weight of the volume of fluid displaced by the body. / drijvende kracht; opwaartse waterdruk / sous-pression; poussée d'Archimède / (hydrostatischer) Auftrieb / presión hidrostática ascendente; presión de Arquímedes

[2930] METHODS OF INVESTIGATION

[2931] *Samples and Sampling Procedures*

NOTE:
– for further terms on sampling see sections
[1970] Grain-Size Determination
[2710] Sampling of Ore in Place
[2720] Sample Reduction and Chemical Analysis
[2730] Statistical Analysis of Samples
[2820] Drilling; Methods and some Technical Terms

4741
sample; specimen, – part of soil or rock that is used for the determination of material or other properties. / monster / échantillon; éprouvette / Probe / muestra

sampling, – procedure of taking samples. / bemonstering / prélèvement d'échantillons; échantillonnage / Probeentnahme / desmuestre

4742
undisturbed sample, – a sample of soil or rock that has been subjected to so little disturbance during sampling that it is suitable for testing to determine the properties of the material in situ / ongestoord monster; ongeroerd .... / échantillon intact; .... non remanié / ungestörte Probe / muestra original; .... no contaminada

disturbed sample, – a sample that is disturbed during the sampling procedure. / gestoord monster; geroerd .... / échantillon remanié / gestörte Probe / muestra contaminada

4743
remolded sample, – sample whose internal structure has been modified by kneading. / verkneed monster / échantillon trituré / geknetete Probe / muestra triturada

4744
(irregular) lump; block, – sample of irregular form as used for some types of index testing like point-load testing. / onregelmatig gevormd monster; brok / échantillon de forme quelconque; .... non façonné; .... irrégulier / Gesteinsprobe beliebiger Form / muestra de forma irregular

4745
sampler; sampling tool, – equipment used for taking samples. / bemonsteringsapparaat / outil de prélèvement d'échantillons / Probeentnahmegerät / muestreador; testiguero

4746
drive sampler, – sampler for soil that is forced into the ground by pushing or jacking (static method) or hammering (dynamic method). / – – / carottier à foncer / – – / muestreador dirigido

piston sampler, – drive sampler in which the lower end is closed by a piston operated from the surface by a piston rod. Upon reaching the desired depth the piston is released from the sampling tube. For sampling the tube is then forced further into the ground while the piston is withdrawn into the tube. / monsternameapparaat met zuiger / carottier à piston / Kolbenentnahmegerät / muestreador con pistón

4747
drop sampler, – equipment for under water sampling of soil; the sampler is dropped to the bottom that is penetrated due to the weight of the sampler. / – – / carottier par battage; .... par gravité / Schwerelot / muestreador sumergible

4748
sampling tube, – usually a thin-walled tube that is fixed to the sampler head with setscrews. It is withdrawn with a sample of the material into which the tube was driven at the bottom of an already existing hole. / steekbuis / tube carottier / Entnahmestutzen / tubo de muestreo

sampling spoon, – piece of equipment that is fixed to the sampler head to sample low-cohesive material that would fall out of a sampling tube. / steekapparaat / cuillère pour prélèvement d'échantillon / Schappe / cuchara de muestreo

4749
continuous sample, – uninterrupted soil or rock sample of up to several metres length that is contained in an impervious sleeve and stands inside the driving tube. / ononderbroken kernmonster / échantillon continu; carotte continue / zusammenhängende Bohrprobe / testigo continuo

4750
integral sample, – intact core from a borehole in soft rock which is strengthened for this purpose with a steel rod and injected cement. / – – / carotte intégrale / – – / muestra integral; testigo integral

4751
washboring, – boring procedure applicable in material that can be washed without appreciable chopping activity. Water is pumped through a washpipe to the bottom of the hole and returns with the washed material between washpipe and casing. / spoelboring / sondage par lancage / Spülbohrung / sondeo por inyección

washpipe, – / spoelbuis / tube de lancage / Spülrohr / tubo de lavado

4752
rock-quality designation; RQD, – percentage of the total length of a borehole that is brought out as cores with a length of over 10 cm.

[2932] *Laboratory and Field Testing; Instrumentation*

4753
triaxial cell, – apparatus in which a cylindrical sample of rock or soil encased in an impervious membrane is subjected to a confining pressure and then loaded axially. / triaxiaal apparaat; celapparaat / cellule triaxiale; appareil triaxial / Triaxial-Apparat; triaxiale Zelle; dreiaxiale Zelle / celula triaxial

4754
shear box, – apparatus in the form of a split box in which a sample of soil or rock is placed; the top part of the box and sample are subjected to a horizontal (shear) load with respect to the lower part; the sample may be sheared under a vertical (normal) load. / schuifapparaat / boîte de cisaillement / Schergerät / caja corte

4755
permeameter, – apparatus for the determination of the permeability of soil or rock; water is passed through a sample of the material to be investigated under various conditions of head. / doorlatendheidsapparaat / perméamètre; appareil de perméabilité / Durchlässigkeitsgerät / permeámetro

4756
consolidometer; oedometer, – apparatus in the form of a ring, in which a sample of clay-rich soil is placed between porous stones and subjected to increments in vertical loads; the deformation of the soil sample is observed in relation to time. / consolidatiering; oedometer / anneau de consolidation; oedomètre / Ödometer; Kompressionsgerät / edómetro

4757
Proctor compaction test, – remolded soil is compacted in a standardized steel cylinder by a standard number of blows with a standard impact. / verdichtingsproef van Proctor / essai (de compactage de) Proctor / Proctor Verdichtungsversuch / ensayo Proctor de compactación

4758
penetrometer; sounding device, – instrument to determine the resistance of soil or soft rock against penetration of a needle, cone or body of other shape when pushed (static penetrometer) or hammered (dynamic penetrometer) into it. / sondeerapparaat / pénétromètre; sonde / Sonde; Rammsonde; Penetrometer / penetrómetro

cone penetrometer; – penetrometer of conical shape. / conussondeerapparaat / pénétromètre à cône / Kegeleindringungsgerät; Kegelgerät / penetrómetro estático

4759
vane-shear test, – an in situ shear test in which a rod with thin radial vanes at the end is forced into the soil and the resistance to rotation of the rod is determined. / vinproef; torsievinproef / essai au scissomètre / Flügelsondenversuch / ensayo de corte; .... de molinete

vane apparatus, – vane used for vane-shear test. / vinapparaat / scissomètre; appareil à palettes; moulinet / Flügelsonde / molinete

4760
pressure meter; dilatometer; – cylindrical probe or sonde that

is lowered into a borehole and exerts hydraulically a radial load on the borehole wall; the relation between exerted load and measured deformation of the borehole is recorded. / pressiometer; boorgatdrukmeter / pressiomètre; dilatomètre / Dilatometer; Pressiometer; Bohrlochverformungssonde / dilatómetro; presurómetro

4761
Brazilian test, – a disc of rock (diameter and thickness about equal) is loaded along a diameter until failure occurs. /Braziliaanse proef / essai Brésilien/ Brasilientest / ensayo Brasileño

pierced-disc test; ring test, – a disc or rock with a small central hole is loaded across a diameter until failure occurs. The central hole acts as a stress concentrator.

4762
pressure-chamber test, – a section of a tunnel is closed off and filled with water; by rising the water pressure the tunnel walls are loaded and the deformation of the rock mass around the tunnel is determined in relation to the load applied. / – – / essai de pression en caverne / Druckkammerversuch / ensayo de cámara de presión

4763
radial jack test, – in a section of a tunnel plate-load tests are executed simultaneously in many directions perpendicular to the tunnel axis; deformation in the tested directions is measured in relation to the applied load. / – – / essai de compression radiale / Radialpressenversuch / ensayo de gatos radiales

flat jack test, – in a slot in rock a thin, flat jack of large dimension is pumped up with oil under pressure; deformation of the surrounding rock is measured. / – – / essai au vérin plat / Druckkissenversuch / ensayo de gato plato

4764
anchor pull out test; pull test, – measurement of the short-term strength of a rock bolt or rock anchor installed under field conditions by increasing the pulling force up to the level that the bolt or anchor is pulled out or broken. / – – / essai de traction sur boulon; .... d'arrachage de boulon / Ankerzugversuch / ensayo de anclaje a tracción

4765
torsional shear test, – in a borehole, a core is sheared off the bottom of the hole by rotation; the torsional force necessary for shearing is recorded. / torsieschuifproef / essai de torsion / Torsionsscherversuch / ensayo de resistencia a la torsión

4766
borehole stress-relief test, – as applied in situ, the bottom of a borehole is carefully flattened and cleaned, and strain gauges are glued on it; after coring a small extra length of hole, the new core expands by stress relief; the amount of expansion is an indication for the natural state of stress. / boorgatontspanningsproef / essai de libération des contraintes naturelles en forage / Bohrlochentspannungsversuch; Überbohrversuch zur Spannungmessung / ensayo de liberación de las tensiones naturales en el fondo del taladro

4767
strain gauge; strain gage (USA), – a device with which strain can be measured by an electrical signal. / rekstrookje / extensomètre à résistance électrique / Dehnungsmeszstreifen; (DMS) / medidor de deformaciones (eléctrico)

strain meter, – a device with which strain can be measured by reading a mechanically operated scale. / rekmeter / déformètre; comparateur / Dehnungsmesser / medidor de deformaciones (mecánico)

4768
settlement gauge, – a device that is buried in material of which the settling has to be determined; its functioning can be based on the principle of uniform water levels in the settlement gauge and a connected, remote read-out unit. / zettingsmeter / appareil pour mesurer le tassement; tassomètre / Setzungsmessgerät / celula de medición de asientos

4769
pressure gauge, – a device for measuring the pressure of liquids or gases. / manometer / manomètre / Druckmesser; Druckgeber; Manometer / manómetro

4770
load cell; dynamometer, – a device for measuring load. / krachtmeter / dynamomètre / Kraftmessdose / célula de carga; dinamómetro

4771
(earth) pressure cell, – a device that is embedded in a soil fill, cemented in the wall of a tunnel or between rock and tunnel lining to measure the prevailing stresses. / gronddrukmeter / appareil pour mesurer la pression des terres; cellule de pression / Erddruckmessdose / célula de presión total

4772
anchor-load cell, – device to measure the force to which an anchor is tensioned. / ankerkrachtmeter / appareil pour mesurer la traction d'ancrage / Ankerkraftmessdose / célula de carga de anclaje

4773
extensometer, – a device, installed in a borehole, to measure changes in length between well head and fixed points at several locations in the borehole, by means of transmission rods or tensioned wires. / extensometer / extensomètre en sondage / Extensometer / extensómetro

4774
pendulum; – a device to determine differential horizontal movement between two points vertically above each other; it consists of a plumb hanging on a wire fixed at the upper pendulum end; movements of the plumb are recorded. / slinger / pendule / Pendel / péndulo

inverted pendulum; – a pendulum where the wire is fixed at its lower end and tensioned by a floating top part; movements of the top part are recorded. / omgekeerde slinger / pendule inversé / – – / péndulo invertido

[2933] *Model Testing and Geotechnical Analysis*

4775
model, – simplification of actual conditions and concentration on the essential aspects of an engineering project; models are used to make complex engineering problems accessible to mathematical and/or experimental investigations. / model / modèle / Modell / modelo

protoype, – the actual engineering work (if not unique, as e.g. a tunnel or bridge) or a component of it; model analysis or model testing is used for the prediction of prototype behaviour, but for certain types of engineering work prototype testing may be feasible as for instance for pile-loading tests, anchor pull tests, testing of tunnel sections. / prototype / prototype / Prototyp / prototipo

4776
mathematical model; numerical model, – complex engineering problems are translated to mathematical equations, which include effects resulting from material behaviour such as elasticity, plasticity and viscosity. Mathematical models can be based either on continuum mechanics or on discontinuum mechanics with consideration of separation planes. / rekenmodel / modèle mathématique / Rechenmodell / modelo matemático

4777
numerical analysis, – analysis of engineering behaviour based on a mathematical model. / rekenkundige analyse / analyse mathématique / numerische Analyse / análisis numérico

stability analysis, – analysis of stability of an engineering work with help of a mathematical model. / stabiliteitsanalyse / analyse de stabilité / Standsicherheitsanalyse / análisis de estabilidad

safety factor, – numerical expression of degree of safety, usually as the ratio of maximum allowable load to acting load. / veiligheidsfactor; veiligheidscoëfficiënt / coefficient de sécurité / Sicherheitsfaktor; Sicherheitskoeffizient / coeficiente de seguridad

4778
physical model, – materialized simplification of complex engineering conditions; physical models of equivalent and analogue types are used. / fysisch model / modèle physique / physikalisches Modell / modelo físico

model test, – investigation of behaviour of a physical model under certain conditions. / modelproef / essai sur modèle / Modellversuch / ensayo sobre modelo

equivalent model, – a type of physical model; reproduction of the conditions that characterize the engineering problems on a certain (usually reduced) scale using model materials. Physical properties and quantities of test arrangement and model materials must follow the laws of similitude in order to allow for the extrapolation of model-test results to the prototype behaviour. / equivalent model / modèle équivalent / Equivalentmodell / modelo equivalente

4779
geomechanical model, – equivalent physical model for which geological conditions are simplified and carefully reproduced. These models can be made three-dimensional and are particularly used in those cases where the mathematical formulation and solution of the problem becomes too difficult or where the relationships between certain parameters are not known. Special problems to be solved with this type of model test are arch dams on geologically complex foundations. / geomechanisch model / modèle géomécanique / geomechanisches Modell / modelo geomecanico

4780
base-friction model, – in this equivalent physical model, gravitational loading is simulated by frictional forces, exerted on the base of a two-dimensional model when the base on which the model rests is pulled away. This model test is mainly used for the purpose of demonstration of already known mechanical or geometrical effects. / – – / – – / – – / modelo por rozamiento en la base

4781
centrifuge model, – in this equivalent physical model the exaggerated gravity makes it possible to use model materials of relatively high strength. A disadvantage of centrifuge models lies in the severe restrictions of size. / centrifugemodel / modèle centrifuge / Zentrifugalmodell / modelo centrifugo

4782
analogue model, – this type of physical model can be used when a physical process is found available that is described by the same mathematical formulation as the mechanical behaviour of the material of the prototype to be investigated. Examples of analogue models are electrical resistance analogue and photo-elastic models. / analoogmodel / modèle analogique / Analogmodell / modelo analógico

4783
electrical (resistance) analogue model, – analogue physical model for study of water seepage and drainage problems by means of the flow of electricity through a conductor that can be described by the same differential equation as the flow of water. / elektrische weerstand analoog model / modèle analogique électrique / elektrisches Widerstands-Analogmodell / modelo analógico de resistencia eléctrica

photo-elastic model, – analogue physical model based on the principle that stress distribution in some transparent materials can be made visible in polarized light. / foto-elastisch model / modèle photo-élastique / photo-elastisches Modell / modelo fotoelástico

[2934] *Engineering-Geological Mapping*

4784
engineering geological map, – regional map showing engineering geological information on the area covered. / ingenieursgeologische kaart / carte géologique pour l'ingénieur / Ingenieurgeologische Karte / mapa geotécnico

engineering geological plan, – a plan is usually covering a smaller area than a map and thus can show more detailed information than a map; engineering geological plans are used for site investigations. / ingenieursgeologische detailkaart; .... plattegrond / – – / Ingenieurgeologischer Plan / plano geotécnico

4785
land classification; terrain ...., — subdivision of the land into units on the basis of aspects like landform or use; land classification on the basis of landform is used when very large areas have to be mapped as for instance for the selection of a road alignment. / landclassificatie / classification des terrains / Landklassifikation / clasificación de terrenos; mapa de ordenación del territorio

land unit; terrain ...., — unit of land that has uniform characteristics in the used classification system and is different in characteristics from the surrounding units. / landeenheid / — — / Landeinheit / unidades de terrenos

4786
land evaluation; terrain ...., — evaluation of the suitability of land or terrain for a certain purpose, such as agriculture, waste disposal, urban extension. / landevaluatie / évaluation des terrains / Landbewertung / evaluación del terreno; .... del sitio

4787
homogeneous zone; .... area, — expressions used in engineering geology to indicate zones or areas of the investigated terrain that possess uniform geotechnical characteristics. / homogene zone / zone homogène / Homogenbereich / zona homogénea; area ....

4788
ground, — engineering geological term for all natural materials found at the earth's surface; a general term for soil, rock or overburden. / grond / terrain / Grund / terreno; suelo

overburden, — loose, non-consolidated material on top of bedrock. / bovengrond; opliggende grond / sol superficiel; (terrain de) couverture / Überlagerung / recubrimiento

bedrock, — consolidated rock underlying the superficial materials (overburden). / vast gesteente; rotsbodem / terrain rocheux; assise rocheuse / Felsuntergrund / subsuelo consolidado; roca subyacente; substrato

rock head, — thickness of rock above a subsurface work like tunnel or cavern. / gesteente-overdekking / (hauteur de) couverture / Felsüberlagerung / recubrimiento rocoso

4789
site; building site, — location where a civil engineering work is under construction or planned to be constructed. / bouwplaats; werk / site (de l'ouvrage); emplacement (de l'ouvrage); emplacement du chantier / Baustelle; Standort / emplazamiento de la obra

4790
site investigation; — engineering geological investigation to obtain sufficient knowledge and understanding of ground conditions to ensure that the engineering work may be designed, constructed and subsequently operated with maximum economy and in complete safety. / terreinonderzoek; grondverkenning / reconnaissance du terrain; recherches sur place; reconnaissance géotechnique / Baugrunduntersuchung; Standorterkundung / reconocimiento del emplazamiento

4791
feasibility study, — stage in the planning of an engineering work in which the possibilities for realisation of the work are investigated. / uitvoerbaarheidsstudie / étude de faisabilité / Ausführbarkeitsuntersuchung / estudio de viabilidad

[2940] PROJECTS AND ACTIVITIES

[2941] *Soil and Rock Support*

4792
anchor, — a steel rod or cable installed in a borehole, either in soil or rock. One end is firmly anchored in the hole by means of a mechanical device and/or grout; the other (threaded) end projects out of the well head and is equipped with a nut and anchor plate which bears against the ground surface. Anchors can be pretensioned by tightening the nut. / anker / tirant (d'ancrage) / Anker / anclaje

anchoring; anchorage, — verankering / ancrage / Verankerung / anclaje

4793
bolt, — a short anchor with steel rod used exclusively in rock. / — — / boulon / Felsnagel / bulón

bolting, — / — — / boulonnage / Nagelung / bulonaje

4794
anchor cable, — cable consisting of a number of steel wires connecting anchoring part and head of an anchor. / ankerkabel / câble d'ancrage / Ankerkabel / cable de anclaje

anchor rod, — steel rod connecting anchoring part and head of an anchor or bolt. / ankerstang / tige d'ancrage / Ankerstange / barra de anclaje

anchor plate, — plate, pressed by the anchor head to the surface of soil or rock. / ankerplaat / plaque d'ancrage / Ankerplatte / placa de anclaje

anchor head, — end part of the anchor consisting of a threaded steel rod and a nut which can be screwed onto the rod to tension the anchor. / ankerkop / tête d'ancrage / Ankerkopf / cabeza del anclaje

4795
Perfo-Anchor, — trade name; anchor fastened in a borehole by a perforated tube filled with cement and placed at the bottom; the cement is squeezed out by the anchor rod when pushed into the opening of the tube.

prestressed anchor, — anchor which has been prestressed by tightening the nut in order to exert a certain load on the anchored mass. / voorgespannen anker / tirant précontraint / Vorspannanker / anclaje pretensado

4796
underpinning, — support in the form of concrete or brickwork to prevent (parts of) a rock mass (or artificial construction) to fall or slide downwards along a slope or excavation or into a tunnel. / ondersteuning; ondervanging / reprise en sous-oeuvre / Unterfangung / recalce

4797
retaining wall, – wall constructed in front of a mass of soil or rock to prevent it moving downward along a slope or into a cut. / steunmuur / mur de soutènement / Stützmauer / muro de sostenimiento; muro de contención

buttress, – column, wider at the bottom than at the top, designed both, to transfer loads vertically to a foundation and to oppose possible thrust at its top. Used in engineering geology for support of large slabs of rock that may slide along an inclined plane. / steunbeer / contrefort / Stützpfeiler / contrafuertes

4798
reinforced earth, – when a fill is made behind a retaining wall, after construction of the wall the segments of the wall can be individually anchored in the fill material using narrow bands of steel of several metres length. / – – / terre armée / bewehrte Erde / tierra armada

4799
sheet piling, – line of interlocking piles, usually of steel, each driven individually into the ground to form a continuous wall as an obstruction to water seepage or movement of ground material, as e.g. a cofferdam, a sea wall. / damwand / rideau de palplanches / Spundwand / tablestacado

[2942] *Soil and Rock Quality Improvement*

4800
to seal, – to cover the surface of soil or rock with a layer or coating. Embankments or bottoms of reservoirs are sealed to prevent water infiltration and leakage; cuts, natural slopes and tunnel walls are sealed to protect them against weathering, disintegration and loosening. Gunite, shotcrete and asphalt are sealing materials. / afdichten / recouvrir; protéger / abdichten; versiegeln / sellar

sealing, – / afdichting / (couche de) protection / Abdichtung; Versiegelung / sellado

4801
gunite, – a mixture of sand, cement and water, sprayed on the surface for water proofing; it is usually sprayed on wire mesh to give it more strength and prevent cracking. / – – / gunite / Gunit / gunita

shotcrete, – quick-setting mortar sprayed on the surface; shotcrete is used with a steel reinforcement to replace precast or formed concrete for tunnel lining and support of rock cuts. / – – / béton projeté / Spritzbeton / hormigón proyectado

4802
injection, – pumping of grout or other fluid substances under pressure through a borehole into cracks and voids of soil or rock; permeability of injected ground is strongly decreased and strength increased after hardening of injection fluid. / injectie / injection / Injektion; Verpressung / inyección

4803
grout, – injection fluid consisting of a mixture of cement and water (and in cases bentonite and other additives); sand may be added if the permeability of the injected mass is to be high. / injectiecement / coulis (de ciment) / Zementmilch; Zementschlämme / lechada

grouting, – pumping grout through a borehole into soil or rock. / cement injecteren / injecter / Zement injizieren; Zement einpressen / inyectar lechada

grouthole; borehole specially made for grout injection. / injectieboorgat / forage d'injection / Injektionsbohrloch / taladro de inyección

4804
grout curtain; grouted cut-off wall, – alignment of grout injections to prevent ground-water flow in a certain direction; often applied below dams and dam abutments. / injectiescherm / voile d'injection; rideau ....; écran injecté; parafouille .... / Abdichtungsschleier; Injektionsschleier / pantalla de inyecciones

drainage screen, – alignment of boreholes downstream of the grout curtain to collect water that has passed the grout curtain. / drainagescherm / écran de drainage / Drainageschirm / pantalla drenante

4805
blending, – process to improve the poor grading of a soil by adding the missing grain sizes. / bijmenging / mélange / Beimischung / mezcla de suelos

4806
soil stabilisation, – chemical or mechanical treatment designed to increase or maintain the stability of a mass of soil or otherwise to improve its engineering properties. Used methods are electroosmosis, freezing, mixing with lime, cement, tar or other chemical soil-stabilizing agents. / grondversteviging; grondstabilisatie / stabilisation du sol / Bodenverfestigung / estabilización de suelos

4807
cement stabilisation, – soil stabilisation by mixing cement through the soil. / cementstabilisatie / stabilisation au ciment / Verfestigung mit Zement; Vermörtelung / estabilización de suelos con cemento

soil liming; lime stabilisation, – soil stabilisation by mixing lime through the soil. / kalkstabilisatie / stabilisation à la chaux / Verfestigung mit Kalk / estabilización con cal

freezing process, – temporary soil-stabilisation method by freezing; it can also be used in soft, water bearing rock. / bevriezingsproces / congélation / Gefrierverfahren / congelación de suelos

4808
compaction, – process by which pore space of a soil mass is reduced and density increased, enhancing bearing capacity, reducing tendency to settle or deform under load and increasing general stability of the soil. / verdichting / compactage; serrage / Verdichtung / compactación

4809
rolling, – compaction of soil by repeated passing over it with a heavy roller. / walsen / cylindrage; roulage / walzen / compactación por rodillo

tamping, – compaction of soil by repeated loading with a heavy plate or a tamping roller. / aanstampen / damage / stampfen / compactación por impacto

vibration; vibrocompaction, – compaction of soil by means of a vibrating plate compactor. / trillen / vibration / Rüttelverdichtung / vibración; vibrocompactación

4810
watering, – compaction of soil by adding water to it. / inwateren / arrosage / wässern / compactación con ayuda de agua

vibroflotation, – compaction of soil by penetration with a vibrator using a high-pressure jet of water; specially applicable in sands, where a compact column of sand is left, capable of bearing a considerable load. / vibroflotatie / vibroflotation / Vibroflotation / vibroflotación

4811
dewatering; drainage, – general term applied to the natural or man-induced removal of surface or ground water from a given area, soil or rock mass either by gravity or pumping; dewatering increases the bearing capacity and general stability of the ground. / drainage; ontwatering / drainage; épuisement; assèchement / Dränung; Entwässerung / drenaje; achicar

drain, – borehole, conduit or channel by which a mass of soil or rock is drained. / drainageleiding / drain / Dränleitung / dren

drainpipe, – tubes of burned clay or concrete, laid in the ground to collect and carry off (excess) ground water. / draineerbuis / tuyau de drainage / Dränrohr / tubo drenante

weephole, – void created during construction in retaining walls, canal linings, foundations, etc. to permit drainage of water collecting behind and beneath structures in order to reduce hydrostatic head. / waterafvoerbuis / barbacane / Entwässerungsöffnung / mechinal

4812
filter, – a device to prevent sediment to enter into the drainage system. / filter / filtre / Filter / filtro

4813
sanddrain, – vertical column of sand in impervious soil to enable drainage of surface water into deeper, pervious soil layers. / drainagekolom; zandpaal / drain de sable; pieu ....; pieu drainant / Sanddrän / dren vertical de arena

[2943] *Blasting; Excavation*

4814
to blast, – to remove and/or disintegrate a ground mass with explosives. / doen springen; opblazen / faire sauter; tirer / sprengen / volar

4815
detonation, – extremely rapid and violent chemical reaction causing the production of a large volume of gas. / detonatie / détonation / Detonation / detonación

explosive, – chemical substance that can be detonated for blasting purposes. / springstof; explosief / explosif / Sprengstoff / explosivo

4816
charge, – the amount of explosive used in a blasthole. / lading / charge / Ladung / carga

blasting cap; detonator; initiator, – a small tube containing a flashing mixture for firing explosives. / ontsteking / détonateur / Sprengkapsel / detonador

stemming, –
1. the material used to fill a blasthole after the explosive charge has been inserted into it in order to prevent rapid escape of explosion gases;
2. act of pushing and tamping material into the blasthole.

/ – – / bourrage / Versatz / retacado

4817
bottom charge, – concentrated explosive charge at the bottom of a borehole. / bodemlading / charge de fond / Bodenladung / carga de fondo

chamber blasting; coyote-hole blasting, – method of quarry blasting in which large explosive charges are confined in small tunnel chambers inside the quarry face. / – – / abattage en chambre; fourneau de mine / Kammersprengung / – –

4818
burden, – distance between charge and free surface in direction of throw. / voorgift / ligne de moindre résistance; couverture / Vorgabe / línea de menor resistencia

throw, – the broken rock blown out during blasting. / afslag / projections / Ausblasung / proyección de rocas

4819
crater, – excavation, generally of conical shape, generated by an explosive charge. / krater / cratère / Sprengtrichter / cráter

cavity, – underground hole created by detonation of a fully contained explosive. / explosieholte / cavité / Sprenghöhle / cavidad

4820
blasting index; blastibility, – index value of the resistance of a rock formation to blasting. / springbaarheid / résistance à l'explosif / Sprengbarkeit / índice de voladura

explosive factor; powder factor, – the ratio of mass or volume of rock, which an explosive charge is expected to break, to the mass of the explosive charge. / springstoffactor / charge spécifique / Sprengstoff-faktor / – –

4821
fragmentation; shattering, – breaking of a rock mass into smaller blocks due to blasting. / fragmentatie / fragmentation / Zerkleinerung; Zertrümmerung / fragmentación

shattering zone, – zone influenced by the blasting in such a way that it is broken into smaller blocks. / gebroken zone / zone fracturée / Zertrümmerungszone / zona fracturada

4822
round, – a set of holes drilled and charged in a tunnel or quar-

ry that are fired instantaneously or with short-delay detonators. / keten; ronde / volée / Abschlag / voladura

spacing, – the distance between adjacent blastholes in a direction parallel to the face. / boorgatafstand / espacement / Bohrlochabstand / espaciamiento

delay, – time interval (fraction of a second) between detonation of explosive charges; frequently used in blasting of a round. / vertraging / retard / Sprengverzögerung / retardo

4823
controlled blasting, – forms of blasting designed to preserve the integrity of the remaining rocks such as in case of smooth blasting or presplit blasting. / gecontroleerd springen / tir controlé; découpage contrôlé / schonendes Sprengen / voladura controlada

smooth blasting, – method of accurate perimeter blasting which leaves the remaining rock almost undamaged using many blastholes and small charges per hole. / zacht springen / découpage fin / schonendes Sprengen / – –

presplitting, – a blasting technique, whereby a smooth fracture plane is created as final contour of the excavation; carried out before the rock mass to be excavated is drilled or blasted. / voorsplijting / prédécoupage / vorspalten / – –

4824
cut; excavation, – relatively large hollow in the earth's surface made by cutting or excavation of rock and/or soil. / ingraving / fouille; tranchée; déblai / Einschnitt / excavación

4825
fill; embankment, – soil or loose rock used to raise the surface of low-lying land. / grondophoging / remblai; terrassement / Auffüllung; Anschüttung / relleno

backfill, – earth or loose rock used to replace material removed during excavation or to be placed behind structures such as bridge abutments. / achtervulling / remblai (de remplissage) / Hinterfüllung / – –

hydraulic fill, – earth fill transported and deposited by means of water pumped through a pipeline. / opgespoten grond / remblai hydraulique / Aufspülung / – –

4826
cut and fill, – the excavation of soil or rock in one location and use of it as fill in a nearby location; often applied in construction of roads, canals and embankments. / – – / équilibre des déblais et des remblais / – – / excavación y relleno

4827
slope ratio; slope gradient, – inclination, steepness of the surface of a cut or fill; an important design criterion, usually expressed as the ratio of vertical to horizontal distance (1:4 for example). / taludhelling / inclinaison (du talus); pente (du talus); fruit / Bösschungswinkel / inclinación de talud

4828
to rip, – to loosen and/or disintegrate ground material with a ripper, consisting of steel teeth connected to a tractor or bulldozer. / losscheuren / ripper / reiszen / ripar

rippability, – susceptibility of the ground to ripping. / losscheurbaarheid / rippabilité / – – / ripabilidad

(to) scrape, – to cut ground material with a shovel connected to a tractor or bulldozer. / – – / décaper / schrappen / – –

4829
muck; mud, – mixture of fine particles and water resulting from excavation works in cuts and quarries. In tunnelling muck is used for all disintegrated rock formed by blasting or by free flow of incompetent material into the underground cavity. / modder / boue; déblai / Schlick; Schlamm / lodo

4830
heave, – upward movement of the bottom of an excavation or tunnel as a result of decompression. / oppersing / soulèvement / Hebung; Sohlhebung / levantamiento

[2944] *Urban Geology; Foundations*

4831
waste disposal, – removal of domestic and industrial waste products and their introduction into the natural environment. / afvalverwijdering / décharge / Abfallbeseitigung; Müllbeseitigung / basureras

waste tip; waste dump, – location where waste is deposited, respectively above and below surface. / vuilstortplaats / décharge en tas / Mülldeponie; Müllkippe / basureros

fill, – man-made deposit of waste materials and/or natural ground. / opvulling / décharge; remblai / Auffüllung / relleno

sanitary landfill, – waste dump where waste is buried with sediment every day, resulting in a deposit of alternate layers of compacted waste and sediment. / – – / décharge contrôlée / – – / – –

4832
underground disposal, – of waste
1. liquids and solutions through boreholes in deep, confined aquifers,
2. solids, e.g. radioactive wastes, either in a mine or solution cavity in rock salt or as a cement grout in artificial fractures in shales.

/ ondergrondse afvalverwijdering / rejet souterrain / unterirdische Abfallbeseitigung / – –

4833
settlement, – gradual downward movement of an engineering structure or superficial soil material due to compaction of the soil below it. / klink; zakking; zetting / tassement; enfoncement / Setzung / asentamiento

differential settlement, – non-uniform settlement leading to uneven lowering of different parts of an engineering structure, often resulting in damage to the structure. / zettingsverschil / tassement différentiel / unterschiedliche Setzung / asentamiento diferencial

tilt, – inclination of engineering structures due to differential settlement. / scheefzakken; overhellen / basculement; renversement / Neigung / basculamiento

4834
foundation engineering, – application of the earth sciences, applied mechanics and structural engineering to the solution of foundation problems. / funderingstechniek / technique des fondations / Gründungstechnik / ingeniería de cimientaciones

4835
foundation, – lower part of a structure, that transmits the load of the structure to the ground. / fundering; fundatie / fondation / Gründung; Fundierung / cimientación

footing, – lower portion of the foundation of a structure that transmits loads directly to the ground. / fundamentbodemplaat / semelle / Gründungssohle / solera de cimientación; zapata de ....

shallow foundation, – / oppervlaktefundering / fondation en surface / Flachgründung / cimientación superficial

shallow footing, – / ondiepe fundamentplaat / semelle superficielle / flaches Einzelfundament / zapata

4836
foundation course; .... stratum, – the ground stratum or layer on which the foundation is resting. / funderingslaag / couche de fondation / Gründungsschicht / nivel de cimientación; estrato de ....

4837
spread footing, – a type of foundation that takes the weight of part of the building and spreads it over a larger area in order to decrease the foundation stress. / verbrede fundering / fondation superficielle / Flachgründung / cimientación superficial

individual footing, – spread footing, supporting a single column. / kolomfundering / semelle isolée / Einzelfundament / zapata individual

strip footing, wall ....; continuous ...., – elongated spread footing supporting a wall or row of columns. / funderingsstrook / semelle filante / Streifenfundament / zapata corrida

raft foundation; mat ...., – continuous, reinforced concrete slab covering the entire base of the structure. / plaatfundering / radier; fondation sur radier / Plattengründung / losa de cimientación

4838
pile, – a long, slender stake or structural element of timber, concrete or steel, that is driven, jetted or cast in situ in the ground for the purpose of supporting a load or of compacting a soil. / paal; heipaal / pieu / Pfahl; Rammpfahl / pilote

pile driving, – emplacement of piles by hammering. / heien (van palen) / battage de pieux / Pfahlrammung; Rammarbeit / pilotaje

pile driver; pile driving rig,– equipment for driving piles. / heistelling / sonnette de battage / Pfahlramme / equipo para hinca de pilotes

4839
point-bearing pile; end-bearing pile; a pile that is driven so deep that its lower end reaches a firm non-plastic material; in design friction along the pile is usually neglected. / stuitpaal / pieu-colonne / Standpfahl / pilote columna

friction pile; floating pile, – pile which is supported by the frictional forces exerted on its sides by the soil in which it is driven. / kleefpaal / pieu flottant / Reibungspfahl; schwebender Pfahl / pilote de fricción; .... flotante

skin friction, – friction exerted on the sides of a pile. / kleef; mantelwrijving / frottement latéral / Mantelreibung / rozamiento lateral

negative skin friction, – downward dragging friction force exerted on a pile by settlement of the surrounding soil. / negatieve kleef / frottement négatif / negative Mantelreibung / rozamiento negativo

[2945] *Roads and Bridges*

4840
subgrade, – natural ground surface on top of which the different layers of the road pavement are emplaced. / ondergrond / terrain naturel / Planum / explanación

pavement, – all layers of different materials forming road foundation and road surfacing together constitute the pavement. / weglichaam / chaussée / Oberbau / pavimento

4841
road foundation, – lower part of the pavement overlying the subgrade or an embankment; the road foundation consists of coarse-grained selected soil or crushed rock; road foundation may be subdivided into a lower layer, subbase, and an upper layer, road base. / wegfundering / corps de la chaussée / Tragschicht / terraplén

subbase, – / – – / couche de fondation / zweite Tragschicht / subbase

road base, – / – – / couche de base / erste Tragschicht / base

4842
surfacing, – top layer of pavement consisting of base course and wearing course. / verharding / couche de surface / Decke / capa de superficie

base course, – layer of crushed stone, gravel or stabilized soil directly underlying the wearing course. / – – / couche de liaison / Binderschicht / base

wearing course, – top pavement layer of concrete, asphalt concrete, asphalt or mixtures of coarse soil with bituminous materials. / deklaag; slijtlaag / revêtement (d'usure) / Deckschicht / capa de rodadura

4843
macadam, – flexible pavement consisting of rolled and compacted crushed stone, using water only as a binder. / macadam / macadam / Makadam / macadam

4844
ballast, – natural or artificial material that supports a railroad track, usually crushed stone. / ballast / ballast / Schotter / balasto

4845
pier, – intermediate support of a bridge, transferring vertically part of the bridge load to the ground. / pijler / pile / Pfeiler / pila

abutment, – terminal support of a bridge usually of concrete, often taking an inclined load. / (brugge)hoofd / culée / Widerlager / estribo

4846
culvert, – pipe of larger diameter through an embankment to allow water to be drained across the embankment. / duiker / ponceau; rigole; conduite enterrée / Durchlass / drenaje enterrado

[2946] *Underground Excavation; Tunnelling*

4847
tunnel, – elongated, essentially linear, underground excavation with a length greatly exceeding its width or height, open at both ends. / tunnel / tunnel / Tunnel / túnel

adit; drift; gallery, – tunnel open to the surface at one end only. / steengang / tunnel d'accès / Stollen / galería

4848
shaft, – underground tunnel-shaped excavation with a vertical or near-vertical direction. / schacht / puits / Schacht / pozo

4849
cavern; chamber, – large underground excavation, e.g. for the emplacement of power stations, storage rooms. / caverne / chambre / Kaverne / caverna; cámara

4850
portal, – tunnel entrance, usually constructed of concrete to support unstable ground. / tunnelportaal / tête / Portal / boquilla

4851
arch, – curved upper part of an underground excavation. / gewelf / voûte / Gewölbe / bóveda

crown; roof, – highest part of an underground excavation. / dak / clé (de voûte) / Firste / clave

4852
face; heading, – working face at the end of a tunnel. / tunnelfront / front de taille; .... d'avancement / Ortsbrust / frente

wall, – side of a tunnel. / wand / mur; piédroit / Ulme / pareo

floor; invert, – the bottom part of a tunnel. / vloer / radier / Sohle / solera

4853
shield tunnelling, – tunnelling method used in soft ground; a shield is pressed into the tunnel face, cutting the tunnel outline ahead of the working face. / schildtunnelbouw / creusement au bouclier / Schildvortrieb / perforación con escudo

4854
mole, – tunnelling machine cutting tunnel with a circular cross section. / tunnelgraafmachine / tunnelier / Tunnelvortriebsmaschine / topo

4855
squeezing ground, – plastic material that can be squeezed into a tunnel under the influence of the stress field around a tunnel. / persbare grond / terrain plastifié / kriechende Boden / suelo extruible

running ground, – water-saturated ground that can flow freely into an underground excavation. / vloeigrond / terrain boulant / Fliesserde / – –

swelling ground, – ground that swells when it is exposed to humidity of the atmosphere in the excavation. / zwellende grond / terrain gonflant / schwellende Boden / suelo expansivo

4856
pay line, – limits of the tunnel cross section for which payment will be made for excavation. / – – / ligne payée / Sollprofil / límite de abono

overbreak, – material which has been excavated outside the pay line. / – – / hors profil / Mehrausbruch / sobre excavación; sobre límite de abono

4857
rock burst, – sudden explosive-like detachment of slabs of sound rock; it is due to relief of residual stress. / steenslag / écaillage brutal; coup de terrain / Gebirgsschlag / golpe de montaña

4858
support, – steel, wooden or concrete equipment used temporarily to prevent failure of ground around underground excavations. / ondersteuning / soutènement / Ausbau / entibación; sostenimiento

lining, – permant protection and support of underground excavations by means of masonry, concrete, steel, etc., cast in situ around the walls. / – – / revêtement / Auskleidung / revestimiento

[2947] *Embankments, Dams and Reservoirs*

4859
embankment dam, – dam consisting of an impermeable core packed in and supported by permeable shoulders of soil and/or rock. / grondophogingsdam / barrage en remblais / Schüttdamm / presa de tierra

earth(-fill) dam, – embankment dam with shoulders of soil. / aarden dam / barrage en terre; digue en terre / Erddamm / presa de tierra

rock(-fill) dam, – embankment dam with shoulders of rocky material. / dam van stortsteen / barrage en enrochements / Steinschüttdamm / presa de escollera

4860
slope, – the outer surface of an embankment dam. / talud / talus / Böschung / talud

berm, – a horizontal bench in the outer surface. / berm / berme / Berme / berma

4861
core, – impermeable, central part of an embankment dam, consisting of clay, clay-rich soil or concrete. / kern / noyau / Kern / núcleo

cut-off, – downward extension of impermeable core into the ground below the dam to control seepage below the dam; the cut-off may be extended into the dam abutments and can be made in the form of a clay-filled trench, a grout curtain or a concrete or sheet-piling wall. / afdichting / parafouille / Abdichtung / – –

cut-off trench, – / afdichtingssleuf / tranchée parafouille / Abdichtungsgraben / zanja de impermeabilización

cut-off wall, – / afdichtingsmuur; schermmuur / mur parafouille / Dichtungssporn; Abdichtungsmauer / pantalla de impermeabilización

4862
saturation line; phreatic line; line of seepage, – the free water table in an embankment dam as seen in cross section, causing seepage, if (ground-)water levels at either side are different. / kwellijn; verwekingslijn / ligne d'écoulement; ligne de saturation / Sickerlinie; Sättigungslinie / línea de saturación

filter (layer), – provision in an embankment dam for percolating water to escape without damage to the embankment; filters consist of coarse-grained material with high permeability. / filter(laag) / couche filtrante; filtre; tapis filtrant / Filter(schicht) / filtro

4863
water tightness, – property of a reservoir or a dam preventing water loss. / waterdichtheid / étanchéité / Wasserdichtigkeit / estanqueidad

leakage, – loss of water by percolation through a dam or the ground surrounding a reservoir. / lekkage / débit de fuite / Wasseraustritt; Sickerverlust / filtración

piping, – internal erosion of foundation or embankment caused by seepage; may attain dangerous proportions if hydraulic gradient is high. / welvorming / (formation de) renard / hydraulischer Grundbruch; Quellenbildung / sifonamiento

4864
gravity dam, – massive and impermeable, concrete monolith with an essentially triangular cross section. / massieve stuwdam / barrage poids / Gewichtsmauer; Schwergewichtsmauer / presa de gravedad

buttress dam, – less massive form of gravity dam, consisting of a heavy buttress or butresses supporting a thinner concrete wall. / – – / barrage à contreforts / Pfeilerstaumauer / presa de contrafuertes

4865
arch dam, – relatively thin concrete wall, horizontally curved and shaped as an arch or double arch. / boogdam / barrage voûte / Bogenmauer / presa de arco gravedad

cupola dam; thin concrete shell, curved horizontally and vertically. / koepeldam / voûte à double courbure / – – / presa de bóveda

4866
abutment, – valley side, supporting a dam. / – – / appui (latéral); culée / Widerlager / estribo

4867
penstock; pressure shaft, – steel pipes (of large diameter) leading from lake over the surface or through a shaft to powerhouse to transfer the water under high pressure to the hydro-electric generators. / drukleiding; drukschacht / conduite forcée / Druckschacht; Druckrohrleitung / tubería a presión

4868
spillway, – structure that allows water to pass over or around a dam when the water level in the reservoir rises above a maximum. / overlaat / déversoir de crue; évacuateur de crues / Überlauf / aliviadero

4869
cofferdam, – temporary dam constructed to protect an engineering work under construction against flooding. / kofferdam / batardeau / Fangedamm; Kofferdamm / ataguia

4870
pumped storage project, – hydro-electric power scheme, comprising of two reservoirs at different elevations; energy is stored by pumping water from the lower to the higher reservoir. / – – / station de transfert d'énergie par pompage; centrale de pompage / Pumpspeicherwerk / sistema de bombeo

[2948] *Construction Materials*

4871
(borrow) pit, – location where soil is excavated to be used for constructional or other industrial use. / winplaats; grondontgravingsplaats / lieu d'emprunt; puits d'emprunt / Schürfgrube / prestamo

borrow material, – / ontgraven materiaal / matériaux d'emprunt / Aushubmaterial / prestamo

4872
aggregate, – materials, which when bound together into a conglomerated mass by a matrix, form concrete, mastic, mortar, plaster, etc. / toeslagstof / granulats / Zuschlagstoff / aridos

4873
washing, – rejection of fine particles from a soil with the help of water. / uitspoeling / délavage / ausspülen; aûswaschen / lavado

screening; sieving, – separation of different grain-size classes to reach the proper texture for the construction material. / zeven / criblage; tamisage / Siebung / cribado

4874
quarry, – location where rock is extracted suitable for building stone, crushed rock or rip-rap. / steengroeve / carrière / Steinbruch / cantera

4875
stripping, – removal of soil and unsuitable rock from the surface of the rock to be quarried. / afgraven / découverte / Abtrag / desbroce

4876
rip-rap, – broken stone or boulders used as a protective layer on the upstream face of a dam or on sea walls to protect these against wave action and erosion by water. / steengroeve-afval / tapis de blocs / 'rip-rap' / escollera

dimension stone, – blocks of stone with even surfaces and of a specified shape and size. / natuurbouwsteen / blocs / Baustein / tamaño de bloques

crushed rock, – rock that is crushed to the desired size by crushing plants. / gebroken steen / pierre concassée / Splitt; Schotter / piedra machacada

Tables

1. Caustobiolites [0400] [2410] [2551] 369
2. Metamorphic Facies [1050] 369
3a. Grain-Size Classes of Igneous Rocks [0943] 370
3b. The overlapping Grain-Size Classes of Volcanic Ejecta [1224] 370
4. The Stratigraphical Hierarchies [1422] 370
5. Soil-textural Classes according to various Usages [1740] 371
6. Grades of Clastic Sediments [1970] 372
7. Particle Size and Composition of Limestones [2122] 372
8. Depostional Textures of Limestones [2123] 373
9. Coal-Rank Classification [2423] 374
10. Coal Macerals [2430] 375
11. Coal Structure [2430] 375
12. Classification of Mineral Resources [2760] 376
13. Bit Designation and Core Size as in use in Mineral Surveys and Engineering [2820] 376
14. The Electromagnetic Spectrum [2831] 376

Table 1
Caustobiolites

1. deposits containing syngenetic, largely autochthonous organic constituents:

kerogenous deposits (sapropelites) kerogenous clays, marls, limestones sapropelic coal	carbonaceous deposits (humolites) peat humic coal (brown coal and hard coal)

2. epigenetic, largely allochthonous natural derivatives of 1:

from kerogenous deposits natural gas petroleum asphalt mineral wax asphaltite	from carbonaceous deposits natural gas

Table 2
Metamorphic Facies (adapted from Barth, Hyndman and Mason by H. G. van Dorssen)

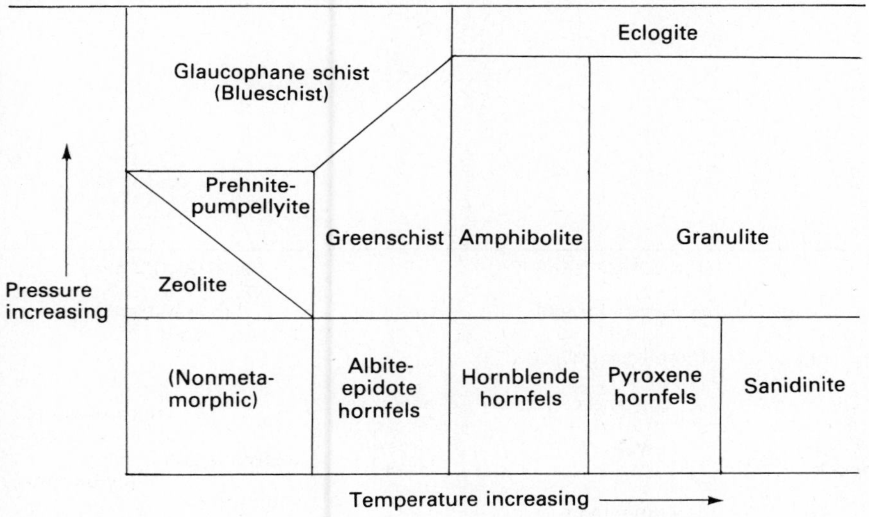

Table 3a
Grain-Size Classes of Igneous Rocks

	mm		
very coarse			
	30		
coarse			
	5	macrocrystalline	
medium			
	1		
	0.5		
fine		mesocrystalline	
	0.1		
very fine			
	0.01	microcrystalline	aphanitic
dense			
	0.001		
		cryptocrystalline	aphanitic

Table 3b
The overlapping Grain-Size Classes of Volcanic Ejecta

block	> 50	mm
gravel	2–75	mm
lapilli	1–64	mm
sand	0.05–2	mm
ash	0.04–4	mm
dust	< 0.05	mm

Table 4
The Stratigraphical Hierarchies

LITHOSTRATIGRAPHY	CHRONOSTRATIGRAPHY			BIOSTRATIGRAPHY	
lithostratigraphic units (rock units)	geochronologic units (time units)	chronostratigraphic units (time-rock units)	examples	biostratigraphic units	
	eon	eonothem	Phanerozoic		taxon-range zone
	era	erathem	Mesozoic		
	period	system	Jurassic	superbiozone	e.g.
group	epoch	series	Liassic		genus zone
formation				biozone	
member		superstage			species zone
bed	age	stage	Sinemurian		
	subage	substage		subbiozone	
(zone)	chron	chronozone		zonule	

Table 5
Soil-Textural Classes according to various Usages (grain sizes in mm)

stones	coarse fragments	grind	éléments grossiers	Boden-skelett	grava
2 ——	2 ——	2 ——	2 ——	2 ——	2 ——
					arena muy gruesa
	very coarse sand				
				Grobsand	1.5 ——
					arena gruesa
	1 ——				1 ——
	coarse sand	grof zand	sable grossier	0.63 ——	
0.6 ——					arena media
	0.5 ——				
medium sand	medium sand			Mittelsand	
	0.25 ——				0.25 ——
0.2 ——	fine sand	0.2 ——	0.2 ——	0.2 ——	arena fina
					0.125 ——
fine sand	0.1 ——				
		fijn zand	sable fin	Feinsand	arena muy fina
	very fine sand				
				0.063 ——	0.0625 ——
0.06 ——					
	0.05 ——	0.05 ——	0.05 ——		
coarse silt			limon grossier	Grobschluff	limo
0.02 ——			0.02 ——	0.02	
medium silt	silt	silt		Mittelschluff	
0.006 ——			limon fin	0.006 ——	
fine silt				Feinschluff	0.0039 ——
0.002 ——	0.002 ——	0.002 ——	0.002 ——	0.002 ——	
clay	clay	lutum, klei	argile	Ton	arcilla
(1)	(2)	(3)	(4)	(5)	(6)

(1) – UK – Soil Survey Field Handbook 1974
(2) – USA – Soil Taxonomy 1975
(3) – Neth. – SSI-Systeem van Bodemclassificatie 1966
(4) – France – Bases et Techniques d'une Cartographie des Sols 1967
(5) – BRD – Kartieranleitung 1971
(6) – Spain – Estratigrafía (Corales et al.) 1977

Table 6
Grades of Clastic Sediments (after W. H. Twenhofel)

phi scale	limiting dimensions mm	particle	aggregate	lithified product	
		boulder	boulder gravel	... conglomerate	rudites
8	256				
		cobble	cobble gravel	... conglomerate	
6	64				
		pebble	pebble gravel	... conglomerate	
2	4				
		granule	granule gravel	... conglomerate	
1	2				
		very coarse sand grain	very coarse sand	very coarse sandstone	arenites
0	1				
		coarse sand grain	coarse sand	coarse sandstone	
−1	$\frac{1}{2}$				
		medium sand grain	medium sand	medium sandstone	
−2	$\frac{1}{4}$				
		fine sand grain	fine sand	fine sandstone	
−3	$\frac{1}{8}$				
		very fine sand grain	very fine sand	very fine sandstone	
−4	$\frac{1}{16}$				
		silt particle	silt	siltstone	lutites
−8	$\frac{1}{256}$				
		clay particle	clay	claystone	

Table 7
Particle Size and Composition of Limestones (>10% allochems): (modified after: R. L. Folk (1962), AAPG Memoir 1, p. 62-84)

	sparite cement more abundant than micrite matrix	micrite matrix more abundant than sparite cement
>25% intraclasts	intrasparites	intramicrites
<25% intraclasts		
>25% oolites		
<25% oolites	oosparites	oomicrites
volume ratio bioclasts to pellets:		
>3:1	biosparites	biomicrites
<3:1 – >1:3	biopelsparites	biopelmicrites
<1:3	pelsparites	pelmicrites

ADJECTIVES and PREFIXES
dolomitized – the rock contains more than 10% replacement dolomite
dolomitic – the rock contains more than 10% dolomite of uncertain origin
primary dolomite; dolo- – the rock consists of primary (i.e. considered to be directly precipitated) dolomite
sandy, silty, clayey – the rock contains 10 to 50% of terriginous material; adjective depends on dominant grain size

Table 8
Depositional Textures of Limestones (modified after R. J. Dunham (1962), AAPG Memoir 1, p. 108-121)

<table>
<tr><td colspan="6">ALLOCHTHONOUS:
ORIGINAL COMPONENTS NOT ORGANICALLY BOUND DURING DEPOSITION</td><td colspan="3">AUTOCHTHONOUS:
ORIGINAL COMPONENTS ORGANICALLY BOUND DURING DEPOSITION</td></tr>
<tr><td colspan="4">LESS THAN 10% > 2 mm COMPONENTS</td><td colspan="2">MORE THAN 10% > 2 mm COMPONENTS</td><td rowspan="4">BY ORGANISMS WHICH ACT AS BAFFLES</td><td rowspan="4">BY ORGANISMS WHICH ENCRUST AND BIND</td><td rowspan="4">BY ORGANISMS WHICH BUILD A RIGID FRAMEWORK</td></tr>
<tr><td colspan="3">CONTAINS LIME MUD (< 0.02 mm)</td><td>NO LIME MUD</td><td rowspan="3">MUD SUPPORTED</td><td rowspan="3">GRAIN SUPPORTED</td></tr>
<tr><td colspan="2">MUD SUPPORTED</td><td colspan="2" rowspan="2">GRAIN SUPPORTED</td></tr>
<tr><td>LESS THAN 10% GRAINS (0.02 – 2 mm)</td><td>MORE THAN 10% GRAINS (0.02 – 2 mm)</td></tr>
<tr><td>MUD-STONE</td><td>WACKE-STONE</td><td>PACK-STONE</td><td>GRAIN-STONE</td><td>FLOAT-STONE</td><td>RUD-STONE</td><td>BAFFLE-STONE</td><td>BIND-STONE</td><td>FRAME-STONE</td></tr>
</table>

Table 9
Coal-Rank Classification

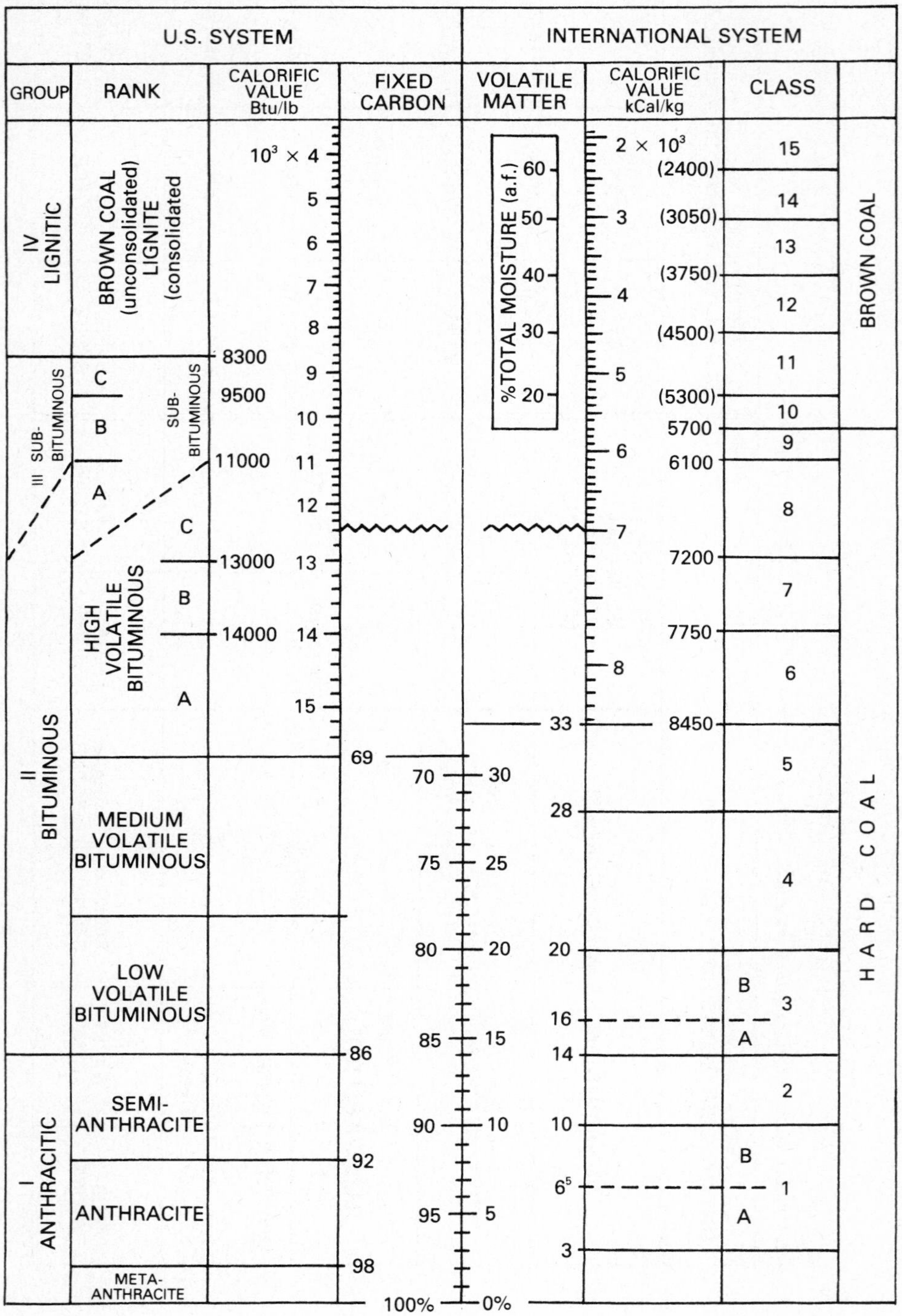
U.S. SYSTEM
INTERNATIONAL SYSTEM
GROUP
RANK
CALORIFIC VALUE Btu/lb
FIXED CARBON
VOLATILE MATTER
CALORIFIC VALUE kCal/kg
CLASS
IV LIGNITIC
BROWN COAL (unconsolidated) LIGNITE (consolidated)
III SUB-BITUMINOUS
C
B
A
SUB-BITUMINOUS
8300
9500
11000
HIGH VOLATILE BITUMINOUS
13000
14000
II BITUMINOUS
MEDIUM VOLATILE BITUMINOUS
LOW VOLATILE BITUMINOUS
I ANTHRACITIC
SEMI-ANTHRACITE
ANTHRACITE
META-ANTHRACITE
$10^3 \times 4$
5
6
7
8
9
10
11
12
13
14
15
69
70
75
80
85
86
90
92
95
98
100%
30
25
20
15
10
5
0%
%TOTAL MOISTURE (a.f.)
60
50
40
30
20
33
28
20
16
14
10
6^5
3
2×10^3
3
4
5
6
7
8
(2400)
(3050)
(3750)
(4500)
(5300)
5700
6100
7200
7750
8450
15
14
13
12
11
10
9
8
7
6
5
4
3
B
A
2
1
B
A
BROWN COAL
HARD COAL

Table 10
Coal Macerals

brown coal	hard coal	
maceral group		maceral
huminite	vitrinite	collinite telinite vitrodetrenite
liptinite	exinite	sporinite cutinite alginite resinite liptodetrinite
inertinite	inertinite	sclerotinite semifusinite fusinite micrinite macrinite

Table 11
Coal Structure

LITHOTYPE	MICROLITHO-TYPE	Principal Groups of constituent macerals		
vitrain	vitrite	vitrinite	V	monomacerals
fusain	fusite	inertinite	I	
–	liptite	exinite	E	
clarain	clarite	– { vitrinite + exinite		bimacerals
durain	durite	– { inertinite + exinite		
–	vitrinertite	– { vitrinite + inertinite		
transitions:	duro-clarite	V + E + I		trimacerals
	claro-durite	I + E + V		

Table 12
Classification of Mineral Resources (McKelvey diagram)

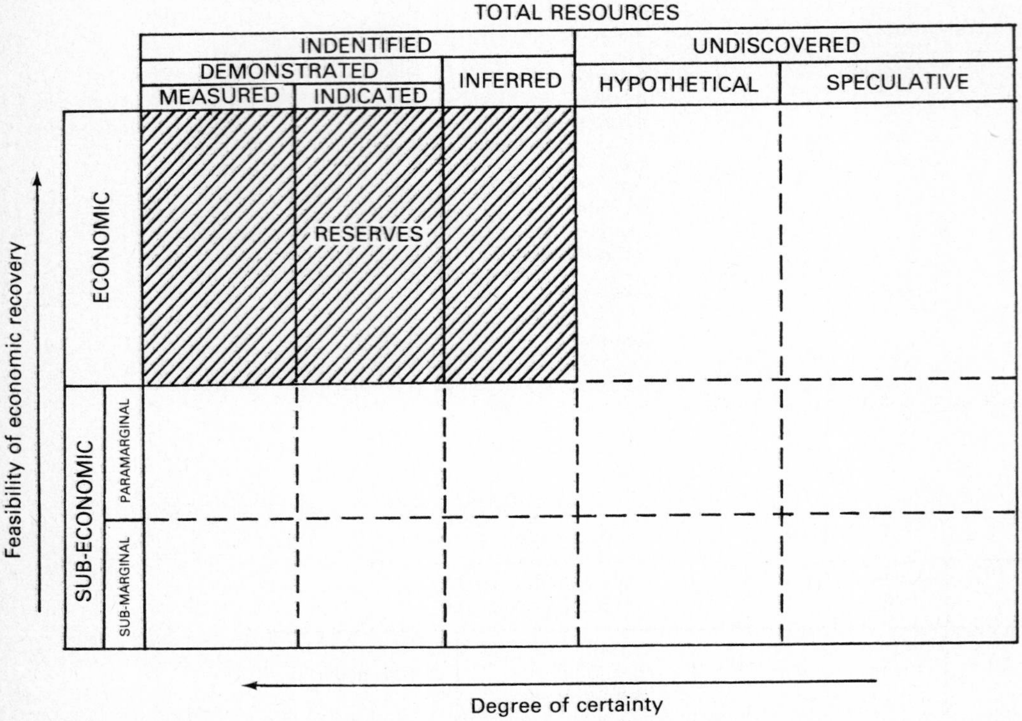

Table 13
Bit Designation and Core Size as in use in Mineral Surveys and Engineering

bit designation	diameter of core
EX	2.15 cm
AX	3.01 cm
BX	4.20 cm
NX	5.47 cm
HX	7.62 cm

Table 14
The Electromagnetic Spectrum (approximate wave lengths in m)

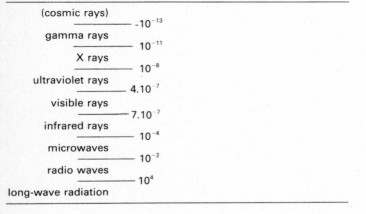

(cosmic rays)
—— -10^{-13}
gamma rays
—— 10^{-11}
X rays
—— 10^{-8}
ultraviolet rays
—— 4.10^{-7}
visible rays
—— 7.10^{-7}
infrared rays
—— 10^{-4}
microwaves
—— 10^{-2}
radio waves
—— 10^{4}
long-wave radiation

English

(including terms in non-nomenclature languages)

A

aa lava 1260
abandonment 3818
ablation 2013
ablation 2380
abrasion 2938
abrasion, eolian – 2015
abrasion, glacial – 2015
abrasion, marine – 2016
abrasion, wind – 2015
abrasion, wind – 2154
absorptance 4314
absorption 0406
absorption 0112
absorption 4514
absorption factor 4314
absorption spectrum 0491
abutment 4845
abutment 4866
abyss 2104
abyssal 0617
abyssal 2847
abyssal 3979
abyssal phase 1197
abyssal plain 2792
abyssopelagic 1910
abyssopelagic 2847
ac 4555
acceleration, peak – 0120
accelerogram 0120
accelerometer 0120
accessory 0594
accessory, characterizing – 0595
accessory, distinctive – 0595
accretion 1822
accretion 2988
accretion hypothesis 0022
accretion theory 0048
accretion, vertical – 2513
accumulation 2987
accumulation 3754
accumulation 3942
accumulation, absolute – 3942
accumulation area 2380
accumulation clock 0429
accumulation, crystal – 0674
accumulation, liquid gravity – 3948
accumulation, relative – 3942
accumulation, residual – 3942
accumulation, residual – 3943
accumulation, zone of – 2380
accumulative phase 0700
accumulative rock 0830
accumulator plant 4647
ACF diagram 0989
achondrite 0072
acicular 0522
acicular 0755
acicular 3185
acid 0601
acid-dip survey 4295
acid, humic – 3525
acidic 0601
acidification 2738
acidisation 3800
acidising 3800
acidite 0601
acidity 0422
acid treatment 3800
acme 1985
acme zone 1773
acrobatholitic 3990
acrozone 1781
action, volcanic – 1149
action, wind – 0043
activity, effusive – 1213
activity, element of eruptive – 1209
activity, ephemeral – 1154
activity, explosive – 1216
activity, fumarolic – 1375
activity, Mount Katmai – 1165
activity, period of – 1208
activity, pyroclastic – 1309
activity, volcanic – 1149
actualism 0006
actuogeology 0008
actuopalaeontology 1886
adamantine 0484
adaptation 1980
adaptive 1980
adcumulate 0832
adcumulus growth 0832
additive 4633
adinole 0933
adit 4847
adsorption 0406
adsorption complex 2755
adularization 0932
aeolian 2870
aeolianite 2870
aeolotropy 0110
aeon 1767
aeration, zone of – 3651
aeric 3979
aerobic 0424
aerobic 1906
aerobic decay 2932
aerolite 0070
affluent 2207
AFMAG 4573
AFM diagram 0698
AFM diagram 0989
aftereffect, elastic – 0114
after-phase 1167
aftershock 0147
afterworking, elastic – 0114
AGC 4425
AGC 4508
age 1762
age, absolute – 0429
age, concordant – 0430
age, discordant – 0430
agent, mineralizing – 3936
agglomerate 1344
agglomerate, lava of tuff – 1346
agglomerate, vent – 1248
agglutinate 1346
aggradation 2245
aggregate 2724
aggregate 4872
aggregate, crystalline – 0771
aggressive 0619
agmatite 1042
agonic line 0284
agpaitic 0693
air base 4350
aircraft 4302
air gun 4410
Airy's hypothesis 0354
Airy phase 0190
A'KF diagram 0989
aklé 2184
akyrosome 1039
alas 2646
albedo 4316
albertite 3739
albitization 0932
alcrete 4124
aleurite 3205
algae, calcareous – 3359
algal 3314
algal ball 3363
algal biscuit 3363
algal head 3363
algal mat 3364
algal ridge 3361
algal rim 3362
algal structure 3358
alginite 3572
aliasing 4423
alimentation, region of – 2380
alkalic 0694
alkali-calcic series 0692
alkali-calc index 0692
alkalic series 0692
alkali-lime index 0692
alkaline 0694
alkalinity 3677
alkanes 3717
alkenes 3717
allochem 3271
allochromatic 0483
allochronic 1818
allochronous 1818
allochthonous 0727
allochthonous 1685
allochthonous 1815
allochthonous 2945
allogene 0660
allogene 3968
allogene 3969
allogenic 0660
allogenic 2946
allogenic 4633
allopatric 1817
allothigene 0660
allothigene 3968
allothigenetic 2946
allothigenic 0660
allothigenic 2946
allothigenic 3969
allothigenous 0660
allothigenous 2946
allothimorphic 0983
allotriblast 0980
allotriomorphic 0752
allotriomorphic-granular 0808
allotype 1949
allowable 3902
alluviation 2886
alluvium 2886
alpine 0600
alpine-type 0600
alpine-type 1078
alpinotype 0600
alteration 3425
alteration, argillic – 3961
alteration, potassium – 3961
alteration, propylitic – 3961
alteration, sericitic – 3961
alteration, wall-rock – 3961
altiplanation 2590
alunitization 0932
amoeba 4099
amphibious 1858
amphibolite 0934
amphibolite, feather – 0934
amphibolization 0932
amplifier, programmed – 4499
amplitude 0164
amplitude 1634
amplitude characteristic 4448
amplitude response 4448
amplitude spectrum 4451
amygdale 0782
amygdaloid 0872
amygdaloidal 0872
amygdule 0782
anaclinal 2288
anadiagenesis 3427
anaerobic 0424
anaerobic 1906
anaerobic decay 2932
anagenesis 3427
analcimization 0932
analysis, combination – 3594
analysis, mechanical – 2711
analysis, mechanical – 3160
analysis, numerical – 4777
analysis, on-stream – 4144
analysis, particle-size – 2711
analysis, proximate – 3594
analysis, PVT – 3856
analysis, screen – 4144
analysis, sieve – 3157
analysis, sieve – 4144
analysis, spectral – 0221
analysis, stability – 4777
analysis, stereometric – 4107
analysis, ultimate – 3594
anamigmatization 1025
anapeirean rocks 0690
anaseism 0187
anastomosed 2218
anastomosing 2218
anastomosing 2501
anastomosis 2115
anatectite 1026
anatexis 1025
anatexis, differential – 1029
anatexis, partial – 1029
anatexite 1026
anchieutectic 0634
anchimetamorphism 0888
anchimetamorphism 3420
anchimonomineral 0589
anchimonomineralic 0589
anchor 4792
anchorage 4792
anchor cable 4794
anchor head 4794
anchoring 4792
anchor-load cell 4772
anchor plate 4794
anchor, prestressed – 4795
anchor pull out test 4764
anchor rod 4794
andesite line 0691
anelasticity 0111
angle, aspect – 4389
angle, critical – 0213
angle, look – 4389
angle of emergence 0215
angle of repose 3137
angle of repose 4687
angle of rest 3137
angle, optic – 0555
angle, optical – 0555
angular 3189
angular blocky 2729
angular distance 0155
anhedral 0523
anhedral 0752
anhedron 0752
anhydrite 3381
anhydrous 0609
anisotropic 4681
anisotropic, optically – 0561
anisotropy 4681
anisotropy 0110
anisotropy, magnetic– 0303
annuity 4218
anogene 0616
anogenic 0616
anomaly 4532
anomaly, Bouguer – 0368
anomaly, free-air – 0368
anomaly, geochemical – 4632
anomaly, geothermal – 0390
anomaly, gravity – 0367
anomaly, isostatic – 0369
anomaly, local isostatic – 0369
anomaly, magnetic– 0286
anomaly, modified Bouguer – 0370
anomaly, regional isostatic – 0369
anomaly, transient magnetic – 1192
anorogenic 0726
anorogenic period 1106
anoxic 0424
anoxic decay 2932
antecedent 2290
anteclise 1084

anthodite 2150
anthracite 3561
anthracography 3565
anthracology 3565
anthraxylon 3587
anticenter 0154
anticentre 0154
anticline 1618
anticline, doubly plunging – 1638
anticline, residual – 1697
anticline, synformal – 1621
anticlinorium 1671
antidune 3008
antiepicentre 0154
antiferromagnetism 0301
antiform 1618
aoritic 1035
apalhraun 1259
aperture 1440
aperture, joint, – 4728
apex 1635
aphanite 0798
aphanitic 0798
aphanophyre 0824
aphanophyric 0824
aphotic 1905
aphrolith 1260
aphyric 0798
API gravity 3741
aplitic 0811
apomagmatic 3986
apophysis 0725
apophysis 4071
Appalachian relief 2564
apron, frontal – 2424
apron of rubble 2007
aquafact 2445
aqueo-igneous 0656
aqueous 2880
aquiclude 3640
aquifer 3634
aquifer, artesian – 3637
aquifer, confined – 3636
aquifer, perched – 3635
aquifugue 3640
aquitard 3639
arabesquitic 0827
arborescent 1821
arc, flashing – 1371
arch 4851
archaeomagnetism 0294
archetype 1979
archibenthic 1908
Archie's formulas 4605
arch, natural – 2079
arch, pressure – 2113
arculite 0768
area, effective – 4360
area, irradiated – 4515
area, scorched – 1279
areic 2257
arenaceous 3225
arenarious 3225
arenite 3225
arenite, quartz – 3230
arête 2617
argillaceous 3211
argillite 3211
argon, atmospheric – 0450
argon, excess – 0450
argon, radiogenic – 0450
arkose 3233
armouring 0787
aromatics 3717
array, directional – 4581
array, down-hole-radial – 4581
array, electrode – 4562
array, lateral – 4581
array, normal – 4581
array, pole-dipole – 4581
array, pole-pole – 4581
array station 0119
array, three – 4581
array, two – 4581
arrival 0172
arrival, first – 4428
artenkreis 1935
arterite 1045
as 2420
ascendent 3967
aschistite 0668
aseismic region 0253
ash 3599
ash avalanche, hot – 1369
ash cloud 1363
ash cloud, electrified – 1365
ash cone 1427
ash, extraneous – 3600
ash fall 1363
ash flow cooling unit 1340
ash flow, hot – 1369
ash flow, incandescent – 1369
ash flow sheet 1340
ash, inherent – 3600
ash, intrinsic – 3600
ash, mantle of – 1320
ash number 3599
ash, pit – 3600
ash, potential – 3600
ash, rain of – 1363
ash, secondary – 3600
ash, sedimentary – 3600
ash shower 1363
ash slide, hot – 1369
ash, volcanic – 1317
asphalt 3734
asphaltenes 3718
asphaltic base 3718
asphaltite 3737
asphalt lake 3753
asphalt, native – 3734
assay 4183
assay, fire – 4183
assemblage 1805
assemblage 3179
assemblage, life – 1896
assemblage, metamorphic – 0885
assemblage zone 1774
assimilation, marginal – 0683
assmiliation 0681
association 1892
association 3179
association, rock – 0630
association, soil – 2673
astenosphere 0090
asterism 0489
asthenolith 1122
astrobleme 0075
ataxite 0069
ataxite 1243
atectite 1032
atexite 1032
Atlantic suite 0690
atmosphere 0084
atoll 3340
atrio 1461
attenuance 4315
attenuation 0219
attenuation 4475
Atterberg limit 4714
attitude 1480
attrition, wind – 2154
attritus 3586
attritus, opaque – 3586
attritus, translucent – 3586
audio-frequency magnetic method 4573
augen 0952
augengneiss 0936
augen structure 0952
auger 4236
aulacogen 1085
aureole 0899
aureole 4629
aureole, wall-rock – 4630
autallotriomorphic 0811
authigenesis 3434
authigenetic 3434
authigenic 2946
authigenic 3434
authigenic 3553
authigenic 3968
authigenic 3969
authigenous 3434
authigenous 3968
autigenic 4633
autochthonous 0727
autochthonous 1685
autochthonous 1813
autochthonous 2945
autocorrelation 4195
autocorrelation 4450
autocovariance 4450
autoecology 1888
autohydration 3957
autoinjection 0621
autointrusion 0621
autolith 0664
autometamorphism 0659
autometasomatism 0659
automorphic 0750
automorphic-granular 0808
autopneumatolysis 0659
autopneumatolysis 3957
avalanche 2358
avalanche chute 2362
avalanche cone 2362
avalanche, debris – 1368
avalanche, debris – 2047
avalanche, drift – 2359
avalanche, dry – 2359
avalanche, fire – 1367
avalanche, glowing – 1367
avalanche, ground – 2359
avalanche, hot – 1367
avalanche, ice – 2361
avalanche, internal – 1472
avalanche, mixed – 1368
avalanche, powdery – 2359
avalanche track 2362
avalanche trench 2362
avalanche, volcanic – 1367
avalanche, wet – 2359
avalanche, wind-slab – 2360
avalanching 2972
AVC 4425
AVC 4508
average, moving – 4198
average, running – 4198
avulsion 2231
axial angle, optic – 0555
axial plane 1631
axial plane, optic – 0554
axial surface 1631
axial surface trace 1631
axiolite 0767
axiolith 0767
axiolitic 0767
axis, a- 1016
axis, b- 1016
axis, beta– 1020
axis, c- 1016
axis, crystallographic – 0511
axis, f- 1017
axis, fabric – 1016
axis, fiducial – 4368
axis, fold – 1630
axis, geomagnetic – 0280
axis, geometric – 4368
axis, girdle – 1020
axis, optic – 0554
axis, photograph – 4368
axis, pi– 1020
axis, plate – 4368
axis pole 0280
axis, reference – 1016
azimuth 4228
azimuth, epicentral– 0155
azimuth, station – 0160
azoic 1738

B

bacilite 0762
back-arc basin 1136
backdeep 1110
backfill 4825
background 4626
backset bed 3141
backset lamina 3141
backshore 2487
backslope, cuesta – 2558
backswamp 2239
backtrough 1110
backwash 2813
backwash rush 2813
badland 2030
bafflestone 3299
bahada 2581
bahamite 3303
bailer 4278
bajada 2581
baking 0897
ball 2486
ballas 0495
ballast 4844
ball, offshore – 2459
band 3101
band 3122
band, dirt – 2379
band, dirt – 3506
band, dust – 2379
banded 3122
band, hard – 2201
banding 0860
banding 0945
banding 3122
banding, crystallization– 0862
banding, trough – 0870
band, spectral – 4306
band, thermal – 4308
bank 2297
bank 2469
bank 2802
bank, concave – 2235
bank, convex – 2235
bank, cut – 2236
banket 4135
bankfull stage 2251
bank, inner – 2235
bank, outer – 2235
bank, undercut – 2236
bar 2201
bar 2244
bar 2459
bar, bay-mouth – 2463
barchane 2177
bar, cuspate – 2465
barefoot hole 3795
bar finger 2897
bar, flying – 2466
bar, glacial rock – 2625
barkhan 2177
bar, longshore – 2459
bar, looped – 2465
bar, lunate – 2462
bar, marsh – 2516
bar, offshore – 2459
bar, point – 2225
barranca 1474
barranco 1474
barrel 3892
barrel per acrefoot 3892
barren 1738
barren 3917

barrier 1916
barrier 2201
barrier 2461
barrier, bay-mouth – 2463
barrier, cuspate – 2465
barrier, fixed coastal – 2462
barrier, geographical – 1916
barrier island 2468
barrier, precipitation – 4642
Barringer Input 4576
bar, scroll – 2233
Barth norm 0606
basalt, alkali – 0695
basalt, alkalic – 0695
basalt, alkali-olivine – 0695
basalt, cone – 1258
basalt, flood – 1257
basalt globe 1256
basalt, high-alumina – 0695
basalt, hyalo– 1262
basalt plain 1257
basalt, plateau – 1257
base 0626
base course 4842
base level of erosion 2025
base level of erosion, general – 2025
base level of erosion, local – 2025
base level of erosion, permanent – 2025
base level of erosion, temporary – 2025
base level of karst erosion 2066
base line 4621
base saturation 2756
base station 0353
base, stereoscopic – 4363
base, wave – 2442
basic 0603
basic behind 1066
basic front 1066
basification 0684
basin 0029
basin 1086
basin 1641
basin 2793
basin 3648
basin, back-arc – 1136
basin, catchment – 2380
basin, closed – 2579
basin, crater – 1412
basin, deflation – 2187
basin, drainage – 2192
basin, epicontinental – 1086
basin, fault – 1591
basin, flood – 2239
basin, geyser – 1404
basin, glacial – 2620
basin, glacial rock – 2620
basin, hydrographical – 2192
basin, intramontane – 1112
basin, marsh – 2517
basin, multi-ringed – 0030
basin, multi-ringed – 0053
basin, reception – 2276
basin, river – 2192
basin, terminal – 2629
basin, tidal – 2508
basin, tongue – 2629
basin, volcanic – 1463
basiophitic 0846
basioxyophitic 0846
basis 0626
basite 0603
batch process 4145
batholite 0729
batholith 0729
batholyte 0729
batholyth 0729
bathyal 2846
bathylimnion 2334
bathylite 0729
bathylith 0729
bathypelagic 1910
bathypelagic 2845
bathyseism 0158
baumpot 3503
bauxite 4124
B-axis 0235
B⊥B′-tectonite 1003
B∧-B′-tectonite 1003
bay 2451
bayou 2502
beach 2485
beach, back – 2840
beach, barrier – 2461
beach cusp 2493
beach face 2487
beach face 2840
beach ridge 2490
beach ridge, normal – 2490
beach scarp 2489
bearing 4228
bearing capacity 4707
beat 0178
Becke line 0559
bed 1750
bed 3098
bed, bottomset – 2896
bedded 3095
bedding 3095
bedding, angular cross – 3142
bedding, avalanche – 3136
bedding, centroclinal – 1241
bedding, chevron cross – 3146
bedding, concordant – 3115
bedding, contorted – 3051
bedding, crescent-type cross – 3144
bedding, cross – 3133
bedding, current – 3134
bedding, delta – 3139
bedding, false – 3135
bedding, festoon cross – 3145
bedding, flaser – 3148
bedding, flaser cross – 3148
bedding, foreset – 3140
bedding, graded – 3130
bedding, herringbone cross – 3146
bedding, inclined – 1482
bedding, inclined – 3136
bedding joint 3099
bedding, lenticular cross – 3150
bedding, low-angle cross – 3143
bedding, massive – 3108
bedding, parallel – 3115
bedding, periclinal – 1241
bedding, planar cross – 3142
bedding plane 1750
bedding plane 3096
bedding plane, stripped – 2557
bedding, quaquaversal – 1241
bedding, rhythmic – 3121
bedding, ripple cross – 3147
bedding, sorted – 3131
bedding, tangential cross – 3142
bedding, tidal – 3134
bedding, trough-cross – 3144
bedding, wavy – 3138
bedding, zigzag cross – 3146
bed, dirt – 3506
bed, foreset – 2896
bed form 2246
bed, high-water – 2242
bed, impervious – 3640
bed, key – 1751
bed load 2952
bed, marker – 1751
bed, mean-water – 2242
bed, plane – 3017
bedrock 2011
bedrock 4788
bed, rootlet – 3515
beds, dipping – 1482
bed, semipervious – 3639
bed, topset – 2896
bed, water-bearing – 3634
beef 3466
beekite 3457
bell-jar intrusion 0734
belonite 0761
belonite 1424
belteroporic 1009
belt, fold – 1674
belt, metallogenic – 3908
bench 2059
bench 2297
benchland, piedmont – 2604
bench lava 1285
bench, rock – 2307
bench, structural – 2555
bench, wave-cut – 2479
bench width, minimum – 4201
bend, boathook – 2209
Benioff zone 1134
benthic 1907
benthic 2844
benthonic 1907
benthonic 2844
benthos 1907
bentonite 3218
beresite 3962
bergschrund 2382
berm 2297
berm 2488
berm 4860
berm crest 2488
berm edge 2488
BHC 4597
BHP 3847
biaxial 0553
bibalve 1859
bifurcation 2210
bifurcation 4073
bight 2450
bind 3502
bindstone 3299
biocalcitutite 3274
biocenose 1896
biocenosis 1896
biochemical oxygen demand 3678
biochron 1768
bioclast 3168
bioclastic 3168
bioclastic 3273
biocoenosis 1896
biodololutite 3274
bioerosion structure 1883
biofacies 2861
biogenesis 1792
biogenetic 1792
biogenetic law 1968
biogenic 1792
biogeochemistry 0395
bioherm 3331
biohorizon 1771
bio-interval zone 1772
bio-interval zone 1776
biolith 3269
biolithite 3269
biome 1895
biomicrite 3288
biomicrite, packed – 3288
biomicrite, sparse – 3288
biomicrorudite 3289
biomicrosparite 3289
bionomics 1887
biopelmicrite 3291
biopelsparite 3290
biospararenite 3287
biosparite 3287
biosparrudite 3287
biosphere 1788
biostratification structure 1884
biostratigraphical unit 1770
biostratigraphical zone 1769
biostratigraphy 1709
biostratonomy 1791
biostrome 3331
biota 1804
biotic 1903
biotitization 0932
biotope 1902
bioturbation 1885
bioturbation 3067
bioturbation structure 1875
bioturbation structure 1885
biozone 1769
biozone, sub– 1782
biozone, super– 1782
bird 4522
bireflection 0575
birefringence 0560
bisectrix 0556
bisectrix, acute – 0556
bisectrix, obtuse – 0556
bit, core – 4271
bit, rock – 4271
bittern 3377
bitter salt 3383
bitumen 3719
bituminite 3575
bituminization 0426
blade 2103
bladed 0772
blanket 4027
blanket 4232
blanket deposit 3109
blanket, ejecta – 0055
blanket, polar – 0040
blast 0274
blast 4814
blastetrix 1014
blastibility 4820
blasting cap 4816
blasting, chamber – 4817
blasting, controlled – 4823
blasting, coyote-hole – 4817
blasting index 4819
blasting, sand – 2939
blasting, smooth – 4823
blasting, wet – 2939
blasto– 0971
blastomylonite 0916
blastomylonite 1541
bleached horizon 2700
bleached layer 2700
bleicherde 2700
blending 4805
block 1087
block 1327
block 3243
block 4744
block, downthrown – 1559
block, downthrown – 1590
block, erratic – 2949
block-faulted area 1586
block field 2057
block field 2656
blocking 2340
block, perched – 2163
block stream 2057
block, upthrown – 1559
block, upthrown – 1589
blocky 2729
blocky, angular – 2729
blocky, subangular – 2729
blood rain 2159
Bloom test 4655
blowhole 1274
blowhole 2109
blow, ore – 4032
blowout 2187
blowout 4288
blowout preventer 4289

blowpipe flame 1350
blowpiping, volcanic – 1178
blue-bands structure 2374
bluff 2516
bluff, river – 2236
boart 0495
bocca 1440
BOD 3678
body, eruptive – 1239
body, extrusive – 1239
bog 3532
bog burst 3541
bog, hanging – 3534
bog, highmoor – 3542
bog iron deposit 4116
bog, lowmoor – 3533
bog moat 3537
bog pool 3540
bog, quaking – 3530
bog, sphagnum – 3543
bog, valley – 3534
boiling, retrograde – 1174
bolson 2580
bolt 4793
bolting 4793
bomb 1328
bomb, almond-shaped – 1329
bomb, bread-crust – 1329
bomb, cored – 1329
bomb, cow-dung – 1329
bomb, exploding – 1329
bomb, explosive – 1329
bomb, flotation – 1296
bomb, globular – 1329
bom, bipolar fusiform – 1329
bomb, olivine – 1329
bomb, pancake – 1329
bomb, peridotite – 1329
bomb pit 1418
bomb, pumiceous – 1329
bomb, ribbon – 1329
bomb, rotational – 1329
bomb sag 1418
bomb, slag – 1329
bomb, spherical – 1329
bomb, spheroidal – 1329
bomb, spindle-shaped – 1329
bomb, tear-shaped – 1326
bomb, tear-shaped – 1329
bomb, twisted – 1329
bomb, unipolar – 1329
bonanza 4007
bone bed 3245
Boomer, deep-towed – 4495
Boomer source 4494
BOP 4289
border, basic- 0707
border, chilled – 0707
borderland, continental – 2789
border, mafic – 0707
bore 2823
borehole 4161
borehole stress-relief test 4766
borehole survey 4294
boresight 4389
boring 1882
boring record 4299
bornhardt 2608
borrow material 4871
borrow pit 4871
boss 0730
boss, ring – 0742
bostonitic 0839
botn 2610
botryoid 2136
botryoidal 0777
botryoidal 4094
bottom 3106
bottom charge 4817
bottom load 2952
bottom sediment and water 3804
bottomset bed 2896
bottom settling and water 3804
boudinage 1609
Bouger correction 0359
Bouguer plate 0359
Bouguer reduction 0359
boulder 1327
boulder 3243
boulder choke 2114
boulder clay 3250
boulder, erratic – 2422
boulder, exotic – 2949
boulder pavement 2160
boulder stream 2057
boule 0497
boundstone 3298
box, shear – 4754
box, sluice – 4659
boxwork 2118
brachyanticline 1640
brachyhaline 2832
brachysyncline 1640
bradyseism 0096
bradyseism 1194
bradytely 1977
braided 2218
braided terrain 0041
branch 1582
branch 1965
branch 4071
branching 2210
Brazilian test 4761
breakage 3431
break, clay – 3772
breakdown 2110
breakdown, block – 2110
breakdown, cavern – 2110
breaker 2811
breaker zone 2439
breaking down 1991
break-in-slope 4635
break, shale – 3772
break thrust 1577
breccia 3258
breccia, bone – 3261
breccia, clinker – 1268
breccia, collapse – 3263
breccia, contact – 0710
breccia, crumble – 1249
breccia, crush – 1539
breccia, dessication – 3048
breccia, edgewise – 3262
breccia, eruptive – 1345
breccia, explosion – 1346
breccia, fault – 1539
breccia, flow – 1268
breccia, friction – 1539
breccia, injection – 0711
breccia, intraformational – 3264
breccia, intrusion – 0710
breccia, lava – 1345
breccia, mud – 3260
breccia, palagonite – 1347
breccia, plutonic – 1042
breccia, reef – 3355
breccia, reef-rock – 3342
breccia, sedimentary – 3258
breccia, tectonic – 1539
brecciation 2933
breccia, vent – 1248
breccia, volcanic – 1345
breccia, volcanic friction – 1248
brickstone 4124
bridge 2102
bridge, natural – 2079
bridging 3441
brilliant 0493
brine 3676
brittle 0471
brittle 4673
brotocrystal 0756
brown-coal gel 3525
BS&W 3804
B-tectonite 1003
Btu 3605
bubble impression 3062
bubble mark 3062
bubble point 3857
bubble pulse 4440
bubble sand 3063
buckling 1665
buckling hypothesis 1118
buffer 0422
bufferstock 4225
building code 0256
bulgannyakh 2643
bulge, structural – 1639
bulk modulus 4698
Bullard's method 0363
bump 2112
bump 0115
bunch, ore – 4032
buoyancy 4740
burden 4818
burial diagenesis 3417
buried 2595
burrow 1882
burst 0115
burst, glacier – 1357
burst, rock – 2112
butte 2561
butte, volcanic – 1477
buttress 2606
buttress 4797
by-passing 3826
byproduct 4163
bysmalith 0734

C

cabochon 0500
CAD 4596
cadacryst 0844
caking power 3595
caking property 3595
cala 2452
calc-alkalic 0694
calc-alkalic series 0692
calcarenaceous 3277
calcarenite 3276
calcareous 2736
calcareous deposit 3267
calcareous, non– 2736
calcibreccia 3279
calcic series 0692
calcification 3458
calcigravel 3278
calcilutite 3274
calcimicrite 3282
calcining 4151
calcipelite 3274
calcirudite 3278
calcisiltite 3275
calcite bubble 2146
calcite ice 2144
calcite scale, floating – 2144
calcrete 4123
caldera 0036
caldera 1456
caldera, collapse – 1459
caldera complex 1457
caldera, conca – 1459
caldera, cryptovolcanic – 1183
caldera, erosion – 1473
caldera island 1417
caldera, phreatic – 1458
caldera ring 1461
caldera, simple – 1457
caldera, summit – 1460
caldera, sunken – 1459
caldron 1463
calibration templet 4370
caliche 4123
California Bearing Ratio 4708
caloric system, high– 0389
caloric system, low– 0389
calorific value – 3604
calorific value, gross – 3604
calorific value, net – 3604
camera, aerial – 4329
camera, boresight – 4322
camera, continuous-strip – 4332
camera, horizon – 4333
camera, mapping – 4329
camera, metric – 4331
camera, multiband – 4330
camera, multilens – 4330
camera, surveying – 4329
canal, collapse – 1290
canga 4124
canyon, lateral – 2384
canyon, submarine – 2795
cap 1696
capacity 2958
capacity, bearing – 4707
capacity, productive – 3806
cap, blasting – 4816
cape 2446
capillary 0774
capillary fringe 2767
capillary water 2767
capillary zone 2767
capillary zone 3657
capping 3924
capping, surface – 2758
cap rock 1696
cap rock 3764
cap rock 3924
cap, spherical – 0363
capture by intercision 2322
capture by lateral erosion 2322
capture, elbow of – 2326
capture, river – 2322
capture theory 0047
carapace 1273
carat weight 0482
carbargillite 3505
carbohydrate 3723
carbominerite 3591
carbonado 0496
carbonate bank, offshore – 3323
carbonate body, built-out – 3327
carbonate body, built-up – 3328
carbonate buildup 3324
carbonate compensation depth 0423
carbonate content, calcium – 2735
carbonate content, free – 2735
carbonate deposit 3266
carbonate mass 3324
carbonate platform 3322
carbonate ramp 3322
carbonate rock 3266
carbonate sediment 3266
carbonation 0932
carbonation 2000
carbonatization 0932
carbonatization 2000
carbon-14 dating 0451
carbon, fixed – 3601
carbonification 3519
carbonite 3564
carbonization 3607
carbon-nitrogen ratio 0425
carbon number 3727
carbon preference index 3729
carbon ratio 3601
cartography, geological – 4239
cascade 2203
cascalha 4136
cascalho 0496
cash deficit, annual – 3894
cash deficit, cumulative – 3894
cash deficit, ultimate – 3894
cash flow 4221

cash flow curve 3895
cash-flow method, discounted – 3898
cash-in 3893
cash-out 3893
cash surplus, annual – 3894
cash surplus, cumulative – 3894
cash surplus, ultimate – 3894
casing 4275
cast 1825
cast, external – 1825
cast, flute – 3026
cast, groove – 3033
cast, internal – 1825
cast, load – 3059
cast, mud-crack – 3047
cast, salt-crystal – 3042
cast, sole – 3019
cataclasis 0914
cataclasite 0914
cataclinal 2288
catagenesis 3419
cataract 2204
cataract, glacier – 2377
catastrophism 0005
catastrophist geology 0005
catazone 0993
catchment area 2191
catchment area 3670
catclay soil 2739
category 1927
catena 2674
cation-exchange capactiy 4623
cations, exchangeable – 2755
cat run 2098
cat's eye 0488
cauda 0175
cauldron 1463
cauldron 2793
cauldron bottom 3503
cauldron subsidence 0743
cauldron subsidence 1186
cauldron subsidence, subterranian – 1186
cauldron subsidence, underground – 1186
causeway, giant's – 1301
caustic 0216
caustobiolith 0427
cave 2095
cave, bedding – 2096
cave, blowing – 2108
cave, breathing – 2108
cave coral 2133
cave deposit 2121
cave, fault – 2096
cave, fissure – 2096
cave formation 2124
cave, ice – 2107
cave-in lake 2646
cave loam 2122
cave, master – 2101
caveology 2094
cave pearl 2149
cave pisolite 2149
cavern 2095
cavern 4849
cavern, lava – 1291
cavernous 1305
cavernous 3064
cavern, sea – 2472
cave, sea – 2472
cave, shaft – 2097
cavitation 2120
cavitation 4512
cavities, induced – 3940
cavities, original – 3940
cavity 4819
cavity, ceiling – 2106
cavity filling 3940
cavity, miarolitic – 0873
cay, sand – 2443
CBR 4708
CD 0454
CDP 4433
CEC 4623
ceiling cavity 2106
ceiling pocket 2106
ceiling tube 2106
cell, load – 4770
cell, rhombohedral – 0516
cell, triaxial – 4753
cellular 0874
cellular 1305
cellular 3064
cell, unit – 0510
celyphitic rim 0786
cement 3433
cementation 3433
cementation 3945
cementation factor 4606
cementation, granular – 3437
cementation, overgrowth – 3437
cementation, rim – 3437
cementation, syntaxial – 3437
cement, blocky – 3439
cement, druse – 3438
cement, drusy – 3438
cemented 2733
cement stabilisation 4807
cenote 2105
cenotypal 0636
cenotype 0636
cenozone 1774
centre line 4372
centroclinal bedding 1241
centrum 0156
cepstrum 0224
chadacryst 0844
chain, volcanic – 1448
chalazoidite 1321
chalk 3315
chalkification 3415
chamber 4849
chamber blasting 4817
chamber, dyke – 1204
chamber, fissure – 1204
chamber, ore – 4032
chamber, volcanic 1202
Chandler wobble 0200
channel 0042
channel 2191
channel 3022
channel 4011
channel 4412
channel, cross – 2486
channel, ebb – 2510
channel, eruption – 1221
channel, false – 2212
channel, feeding – 3941
channel fill 3022
channel, flood – 2510
channel, meander – 2100
channel, meandering – 2217
channel of ascent 1221
channel, palaeostream – 4029
channel, rip – 2486
channels, interlaced – 2243
channels, intertwined – 2243
channel, sinuous – 2217
channels, tangled – 2243
channel, straight – 2217
channel, supply – 1290
channel, tidal – 2509
channel, volcanic – 1221
channel, wash-out – 2026
channelway 4011
chaotic terrain 0038
characteristic, directional – 4491
characteristic, microstructural – 4108
charcoal 3607
charcoal, mineral – 3580
charge 4816
chargeability 4570
charge, bottom – 4817
charnockite, m– 0990
chasm 2104
chasma 0034
chatoyancy 0488
chattermark 2398
Chayes point counter 0612
chebi-chebi 4123
cheek 4009
cheironym 1956
chelation 0409
chemical oxygen demand 3678
chenier 2491
chenier plain 2491
chert 3366
chert, diatomaceous – 3371
chert ironstone 3373
chert, radiolarian – 3369
chilled zone 0707
chill zone 0707
chimney, volcanic – 1221
china clay 3217
chip 3186
chipping 2941
chloritization 0932
choke, boulder – 2114
chondrite 0072
chondrule 0072
chonolite 0732
chonolith 0732
chorismite 0794
chorismitic 0794
Christmas tree 3796
chroma 2677
chromocratic 0588
chron 1761
chronohorizon 1758
chronomere 1760
chronostratigraphical horizon 1758
chronostratigraphical unit 1757
chronostratigraphy 1712
chronotaxial 1732
chronozone 1758
chute cut-off 2230
chymogenetic 1040
chymogenic 1040
cinder 1294
cinder cone 1427
CIPW norm 0606
CIPW norm 0686
circalittoral 2839
circle, pi– 1020
circulation, lost – 4284
circulation, reverse – 4283
circulation, straight – 4283
circulation, thermohaline – 2829
circulus 1808
cirque 2610
cirque, compound – 2613
cirque excavation, level of – 2615
cirque-floor level 2615
cirque, nivation – 2612
cirque niveau 2615
cirque platform 2613
cirque stairway 2614
cirque steps 2614
cirque threshold 2610
cirque, valley-head – 2611
cistern rock 0734
cladogenesis 1962
clan 0687
clarain 3580
clarite 3583
Clarke number 0399
Clarke value 0399
class 0685
class 1929
classification 1921
classification, chronostratigraphical, – 1712
classification, land – 2660
classification, land – 4785
classification, lithotratigraphical, – 1710
classification, natural – 1921
classification, reserve – 3878
classification, soil – 2661
classification, textural – 2713
classificaiton, terrain – 4785
classifier, rake – 4156
classifier, spiral – 4156
class, textural – 2713
clast 3166
clastic deposit 3164
clastic ratio 3166
clastolithic sedimentation 1201
clathrate 0854
clavalite 0763
clay 2714
clay 3212
clay band 4117
clay, boulder – 2422
clay, boulder – 3250
clay break 3772
clay, brown – 2923
clay, fat – 3216
clay filling 2122
clay, fire – 3217
clay, fire – 3517
clay fraction 2714
clay gall 3246
clay ironstone 3488
clay, lean – 3216
clay, marly – 3223
clay mineral, mixed-layer – 3214
clay mineral, three-layers – 3214
clay mineral, two-layers – 3214
clay pan 2734
clay, quick– 4719
clay, red – 2923
clay, root – 3515
clay, selvage – 4012
clay, sensitive – 4717
clay shale 3220
claystone 3219
clay, varved – 3128
clay with flints 2931
clean-out job 3813
cleat 3508
cleat, butt – 3508
cleat, end – 3508
cleat, face – 3508
cleavability 1596
cleavable 1596
cleavage 0469
cleavage 1596
cleavage, crenulation – 1600
cleavage, false – 1600
cleavage fan 1602
cleavage fan, convergent – 1602
cleavage fan, divergent – 1602
cleavage, flow – 1599
cleavage, fracture – 1601
cleavage, herringbone – 1600
cleavage, pencil – 1610
cleavage plane 0540
cleavage plane 1598
cleavage, shaly – 1599
cleavage, shear – 1600
cleavage, slaty – 1599
cleavage, strain-slip – 1600
cleft 1471
cleft 2474
cliff 2470
cliff, abandoned – 2477
cliff, ancient – 2477
cliff fall 2473
cliff, plunging – 2475
cliff, river – 2236
cliffs, line of – 2478
climax community 1891

climax formation 1891
cline 1984
clinker breccia 1268
clinker field 1260
clinker, welded – 1268
clinoform 2916
clinograph 0124
clinolimnion 2334
clinometer 4227
clino-nonconformity 1721
clinothem 2916
clod 2725
close 1626
closure 3762
closure area 3762
closure error 4541
closure, vertical – 3762
cloud, ash – 1363
cloud, caulifower – 1362
cloud, eruption – 1360
cloud, explosion – 1360
cloud, glowing – 1366
cloud, pine-tree shaped – 1362
clunch 3502
cluster 0749
clustering 4098
clusterite 2136
cluster, volcanic – 1450
CNL 4594
coagulation 0413
coagulation 3932
coal 3496
coal, algal – 3563
coal ball 3507
coal, banded – 3497
coal bed 3499
coal, bituminous – 3559
coal, boghead – 3563
coal, bright brown – 3558
coal, brown – 3495
coal, cannel – 3563
coal, channel – 3563
coal, cinder – 3564
coal, coking – 3608
coal, component of – 3585
coal, crumble – 3556
coal, dry – 3560
coal, earth – 3555
coal equivalent, standard – 3606
coal, fat – 3559
coal, finger – 3564
coal, flame – 3559
coal, float – 3512
coal, gas – 3559
coal, gas-flame – 3559
coal grade 3552
coal gravel 3512
coal, hard – 3496
coal, humic – 3497
coalification 3519
coalification, degree of – 3521
coalification gradient 3522
coal, leaf – 3556
coal, lean – 3560
coal, lipid – 3497
coal liquefaction 3609
coal, long-flame – 3559
coal, lustrous brown – 3558
coal measures 3498
coal, mixed – 3497
coal, moor – 3556
coal, mother of – 3580
coal, needle – 3556
coal, non-banded – 3497
coal, paper – 3563
coal, pebble – 3514
coal petrography 3565
coal pipe 3503
coal, pitch – 3557
coal quality 3551
coal rank 3552
coal, sapropelic – 3497
coal, sapropelic – 3721
coal scare 3511
coal seam 3499
coal series 3498
coal, smut – 3555
coal, soft brown – 3555
coal split 3506
coal, steam – 3559
coal streak 3499
coal structure 3565
coal type 3552
coal type 3581
coarse fragment 2722
coarse-grained 0796
coarse-grained, very – 0796
coarse textured 2721
coast 2437
coast, abrasion – 2528
coast, accretion – 2528
coast, Atlantic-type – 2522
coast, composite – 2527
coast, compound – 2526
coast, conformable – 2521
coast, contraposed – 2527
coast, Dalmatian-type – 2521
coast, depressed – 2524
coast, diagonal – 2522
coast, drowned – 2525
coast, embayed – 2533
coast, emerged – 2525
coast, fault – 2520
coast, fault-line – 2520
coast, fault-scarp – 2520
coast, ingression – 2529
coast, lava-flow – 1280
coast line 2437
coast, lobate – 2533
coast, longitudinal – 2521
coast, monoclinal – 2519
coast, multicycle – 2530
coast, neutral – 2523
coast of advance 2528
coast of progradation 2528
coast of retrogradation 2528
coast of transverse deformation 2527
coast, Pacific-type – 2521
coast, prograded – 2528
coast, ragged – 2532
coast, raised – 2524
coast, stable – 2526
coast, stationary – 2526
coast, submerged – 2525
coast, transverse – 2522
coast, unconformable – 2522
coated 3402
coating 3402
cobble 3242
cobblestone 3242
coccolith 1840
cockpit karst 2080
COD 3678
coda 0175
coercive force 0307
coercivity 0307
coesite 0076
coeval 1730
cofferdam 4869
cohesion 4693
cohesion, apparent – 4693
coke 3608
coke, natural – 3564
coking capacity 3596
collapse, external – 1464
collapse, roof – 1188
collapse sink 2072
collinite 3569
collision, continental – 1131
colloform 3093
colluvium 2868
colony 1806
colour 0546
colour composite 4346
colour index 0688
column 2138
columnar 0755
columnar 0772
columnar 2730
column, eruption – 1361
column, magmatic – 1226
column, stratigraphical – 4255
comagmatic 0629
comagmatic region 0628
combination 0521
combustion, in situ – 3838
commensalism 1899
comminution 1319
comminution 4150
common-depth-point gather 4433
common-midpoint gather 4433
common-subsurface-point gather 4433
community 1890
community, climax – 1891
compact 0859
compaction 3428
compaction 4679
compaction 4808
compaction, chemical – 3432
compaction, differential – 3428
compaction, mechanical – 3429
compaction, vadose – 3416
compartment 0362
compass, geologist's – 4227
compensation depth 0345
compensation, heave – 4511
compensation, isostatic – 0339
compensation, isostatic mass – 0342
compensation, isostatic pressure – 0344
compensation level 0345
compensation, local – 0340
compensation mass 0341
compensation, regional – 0340
competence 2958
competent 1624
complementary 0669
complex 1756
complex 1928
complexing 0408
complexing, organo-metal – 0409
complexity 0275
complex, metamorphic – 0885
complex, soil – 2673
component, accessory – 0594
component, auxiliary – 0594
component, essential – 0594
component, gaseous – 1172
component, gaseous – 0650
component, principal – 0594
component, vertical – 1563
component, volatile – 0650
component, volatile – 1172
composition, mineralogical – 0593
composition plane 0536
composition, standard mineral – 0606
composition surface 0536
compressibility 4698
compressibility, bulk – 4698
compressibility, coefficient of – 4698
compression 0187
concentrate 4141
concentrate, beach – 2912
concentrate, lag – 2967
concentration, alluvial mineral – 3943
concentration, gravity – 4153
concentric-spherulitic 4102
concordant 0717
concordant age 0430
concordant bedding 3115
Concordia-Discordia diagram 0435
concretion 2732
concretion 3472
concretion 4093
concurrent-range zone 1777
condensate 3713
condensation 0412
condensation, retrograde – 3713
condensation, retrograde – 3841
condominant 1816
conductance, electric – 4556
conduction, heat – 0381
conductivity 4582
conductivity, hydraulic – 3700
conductivity, thermal – 0382
conductor 4274
conduit 1221
cone 0059
cone, adventive – 1437
cone, ash – 1427
cone, breached – 1471
cone, central – 1443
cone, cinder – 1427
cone, composite – 1429
cone, double – 1442
cone, driblet – 1298
cone karst 2081
cone, lateral – 1437
cone, lava – 1427
cone, littoral – 1300
cone, minor – 1437
cone, mud – 1396
cone, normal – 1435
cone, parasitic – 1437
cone, puff – 1396
cone, pyrometric – 1196
cones, cluster of – 1450
cone, scoria – 1427
cone sheet 0742
cone, slag – 1427
cone, spatter – 1298
cone, subsidiary – 1437
cone, tephra – 1428
cone, transition – 2423
cone, tuff – 1428
cone, volcanic – 1434
confidence interval 4191
confidence limits 4191
confluence 2208
confluence step 2624
conformable 0719
conformable 1719
conformity 1719
congelifract 2656
congelifraction 1996
congeneric 1943
conglomerate 3251
conglomerate, basal – 3114
conglomerate, boulder – 3254
conglomerate, edgewise – 3262
conglomerate, intraformational – 3257
conglomerate, mud-flake – 3252
conglomerate, volcanic – 1343
conglomeratic 3251
coniform 4080
coning 3826
coning and quartering 4181
connate fluid 0650
conodont 1860
conoplain 2582
Conrad discontinuity 0086
conretionary 3472
consanguineous 0629
consanguineous 3970
consanguinity 0629
consequent 2284
consequent, radial – 2291

consistence 2746
consistency 2746
consistency 4713
consistency index 4716
consistency limit 4714
consolidated 3422
consolidation 3422
consolidation 4703
consolidometer 4756
conspecific 1943
constituent, accessory – 0594
constituent, auxiliary – 0594
constituent, banded – 3579
constituent, essential – 0594
constituent, fugitive – 0650
constituent, fugitive – 1172
constituent, opaque – 0610
constituent, principal – 0594
contact aureole 0899
contact breccia 0710
contact, faced – 3449
contact, fitted grain – 3449
contact, sutured – 3449
contact zone 0899
contamination 0681
contamination 3672
contamination 4652
contemporaneous 1729
contemporaneous 1818
contemporaneous 3474
continental 2865
continental collision 1131
continental drift 1115
continental shelf 2788
continuation, downward – 4533
continuation, downward – 4543
continuity 4726
continuum 1810
contortion 3051
contortion, intraformational – 3051
contour line 4248
contraction hypothesis 1114
contrast 4343
contrast 4628
contrast, subject – 4343
control, ground – 4323
control, minor – 4324
control, photogrammetric – 4324
control point 4325
convection, heat – 0381
convergence 1725
convergence 1973
convergence 2787
convergence, adaptive – 1973
convergence, metamorphic – 0913
convergency 1725
convergency 1973
converter plant 4647
convexity curve, maximum – 4464
convolution 3052
convolution 4452
coordination number 0408
coorongite 3740
Copper Belt 4130
coprolith 1863
coquina 3308
coquinite 3308
coquinoid 3309
coral, ahermatypic – 3335
coral cap 3329
coral, cave – 2133
coral formation 2133
coral, hermatypic – 3335
coralline 1821
cordillera 1103
core 1070
core 1628
core 3155
core 4237
core 4861
core barrel 4272
core bit 4271
core box 4237
core catcher 4272
core, detached – 1628
core, earth's – 0093
core, inner – 0093
core lifter 4272
core, outer – 0093
corer 3155
core recovery 4273
corer, gravity – 3156
corer, piston – 3156
corer, vibro-piston – 3156
core stone 2609
cornice, snow – 2356
corona 0786
coronite 0786
corrasion 2014
corrasion 2938
corrasion, eolian – 2015
corrasion, eolian – 2154
corrasion, glacial – 2015
corrasion, stream – 2015
corrasion, wind – 2015
corrasion, wind – 2154
correction, datum – 4429
correction, dynamic – 4432
correction, elevation – 0357
correction, ellipticity – 0098
correction, free-air – 0358
correction, isostatic – 0361
correction, latitude – 0355
correction, normal-moveout – 4432
correction, static – 4431
correction, station – 0160
correction, terrain – 0360
correction, tidal – 0365
correction, topographic – 0356
correction, weathering – 4430
correlation 1716
correlation 4459
correlation, cross – 4450
correlation, serial – 4195
correlator, optical – 4388
correlogram 4195
corrie 2610
corrosion 0642
corrosion 2017
corrosion 2942
corrosion border 0786
corrosion, magmatic – 0642
corrosion rim 0786
corrosion zone 0786
corrugation 3051
corrugation, intraformational – 3051
cortex 3403
coseismal line 0210
coseismic line 0210
coset 3111
cosmogony 0003
cospecific 1943
costean 4658
cotype 1951
Coulomb criterion 4689
counterflush 4283
country, frozen to the – 4013
country rock 0709
coupling 0273
course 4016
course, base – 4842
course, lower – 2195
course, middle – 2195
course, upper – 2195
course, wearing – 4842
covariogram 4196
cove 2474
cover 3924
coverage 4407
coverage, multiple – 4407
coverage, stereoscopic – 4361
coyote-hole blasting 4817
CPI 3729
crack, cooling – 0713
crack, frost – 2640
crack, heat – 1993
cracking, endogenous – 3508
crack, mud – 3046
crack, radiating – 1468
crack, shrinkage – 1993
crack, sun – 1993
crag and tail form 2421
crater 0054
crater 1412
crater 4819
crater, active – 1211
crater, adventive – 1439
crater basin 1412
crater, breached – 1471
crater, cone-in-cone – 1445
crater, double – 1442
crater, horseshoe-shaped – 1471
crater, impact – 0027
crater, inbreak – 1455
crater island 1417
crater lake 1454
crater, lateral – 1439
crater, meteor – 0075
crater, parasitic – 1439
crater, pit – 1455
crater, primary – 0027
crater ring 1451
crater, secundary – 0027
craters, row of – 1449
crater, subterminal – 1438
crater, summit – 1436
crater, twin – 1442
craton 1080
crawl 2098
crawl passage 2098
crawl way 2098
creek, marsh – 2517
creep 0044
creep 0114
creep 4672
creep, continuous – 2039
creep, frost – 2041
creep recovery 0114
creep, rock – 2040
creep, seasonal – 2039
creep, soil – 2038
creep, soil – 2780
creep, talus – 2040
crenulation 1643
crescumulate 0834
crestal surface 1632
crest, dividing – 2316
crest line 1632
crest-line apex 1635
crest-line culmination 1635
crest point 1633
crest, ripple – 3001
crest, sharp – 2617
crest, smoking – 2158
crevasse 2240
crevasse 2375
crevasse, longitudinal – 2376
crevasse, marginal – 2376
crevasse splay 2241
crevasse, transverse – 2376
crevassing 2240
crinoidal 3313
crocydite 1047
crop out, to – 4231
cross 2289
cross bedding 3133
cross course 4077
cross groove 3034
cross lamination 3133
cross-over point 4577
cross-ripple mark, rectangular – 3013
cross section, wetted – 2261
cross stratification 3133
crotivina 2745
crotovine 2745
crown 0705
crown 4851
crude 3712
crumb 2727
crumbling 1991
crusher, cone – 4148
crusher, jaw – 4148
crush rock, flinty – 0916
crust 2759
crust, earth's – 0086
crust, fusion – 0073
crust, rigid – 0348
crust stone 2145
crust, thickness of the M – 0347
crust, weathering – 2010
cryoconite 2387
cryopedology 2634
cryoplanation 2650
cryoturbation 2647
cryptobatholitic 3990
cryptoclastic 3166
cryptocrystalline 0800
cryptoexplosion structure 1184
cryptogenic 1976
cryptographic 0848
cryptomaceral 3567
cryptomagmatic 3986
cryptovolcanic structure 1184
cryptovolcanism 1181
cryptozoon 1844
crystal 0506
crystal accumulation 0674
crystal, corroded – 0756
crystal defect 0531
crystal, embayed – 0756
crystal flotation 0673
crystal form 0520
crystal form 0753
crystal lattice 0507
crystall class 0512
crystalline rock 0583
crystalline texture 0816
crystallinity 0791
crystallite 0760
crystallitic 0760
crystallization 0632
crystallization banding 0862
crystallization differentiation 0670
crystallization, fractional – 0670
crystallization, heterad – 0833
crystallization index 0702
crystallization, mimetic – 1008
crystallization, rhythmic– 0864
crystallizing force 0909
crystalloblast 0908
crystalloblastesis 0908
crystalloblastic order 0909
crystalloblastic series 0909
crystalloblastic strength 0909
crystallography 0505
crystallography, optical – 0544
crystallothrausmatic 0867
crystal mush 0640
crystal, primary precipitate – 0830
crystal, regenerated – 0910
crystal sedimentation 0673
crystal settling 0673
crystal structure 0509
crystal system 0514
CTB treaty 0266
cubic 0515
cubichnia 1880
cuesta 2558
culmination 1635
culvert 4846

cumulate 0830
cumulite 0765
cumulophyre 0821
cumulophyric 0821
cumulo volcano 1422
cumulus 0830
cumulus crystal 0830
cuneiform 4036
cupel 4184
cupelling 4184
cupola 0724
cupola 1205
cupola, water – 1359
curie 4550
Curie point 0308
currant, tidal – 2821
current, alternating – 4555
current bedding 3134
current, commutated – 4555
current density 4556
current, density – 2979
current, direct – 4555
current, ebb – 2822
current, flood – 2822
current, longshore – 2825
current mark 2086
current mark 3020
current, ocean – 2785
current, rip – 2826
current ripple mark 3004
current, sea – 2785
current strength, electric – 4556
current, surface – 2785
current, suspended – 2981
current, suspension – 2980
current, telluric – 4552
current, turbidity – 2981
current velocity 2260
curtain 2135
curvature value 0336
curve, field – 4565
curve, standard – 4565
cusp 2304
cusp, beach – 2493
cut 2020
cut 4824
cutan 2691
cut and fill 4826
cut-cumulative method 3875
cutinite 3572
cut-off 4643
cut-off 4861
cut-off, chute – 2230
cut-off, meander – 2229
cut-off, neck – 2230
cutting out of beds 1557
cutt-off wall, grouted – 4804
CVL 4406
CVM 3899
C-wave 0193
cwm 2610
cxHM 4655
cxME 4655
cyclanes 3717
cycle, arid – 2567
cycle, desert – 2567
cycle, fluvial – 2567
cycle, geochemical – 0396
cycle, geographical – 2567
cycle, glacial – 2567
cycle, hydrological – 3612
cycle, karst – 2567
cycle, marine – 2567
cycle, normal – 2567
cycle of erosion 2566
cycle of sedimentation 3119
cycle, orogenetic – 1105
cycle, orogenic – 1105
cycle, periglacial – 2567
cycle, petrogenetic– 0582
cycle, shore-line – 2567
cycle skipping 4597
cycle, volcanic – 1156
cycloalkanes 3717
cyclographic projection 1021
cycloolefins 3717
cycloparaffin 3717
cyclopean 0960
cyclothem 3119
cyst 1841

D

dactylite 0849
dactylitic 0849
dactylotype 0849
dagala 1289
damage, formation – 3815
dam, arch – 4865
dam, buttress – 4864
dam, cupola – 4865
dam, earth-fill – 4859
dam, embankment – 4859
dam, gravity – 4864
damming 2340
damp, black – 3554
damp, choke – 3554
damp, fire – 3554
damping 0127
damping, aperiodical – 0129
damping, critical – 0129
damping, dynamic – 0127
damping factor 0128
damping ratio 0128
damping, viscous – 0127
damp, white – 3554
dam, rock-fill – 4859
darcy 3697
Darcy's law 3696
dark-coloured 0588
Darwinism 1966
dating, absolute – 0429
dating, isotopic – 0429
dating, radioactive – 0429
dating, radiometric – 0429
datum correction 4229
datum plane 0357
datum plane 3854
datum, structural – 4249
dc 4555
DCF method 3898
DCF method 4221
debitumenization 3520
debouchure 2196
debris 2005
debris plain 2006
debris slide 2046
decalcification 2679
decay 1998
decay, aerobic – 2932
decay, anaerobic – 2932
decay, anoxic – 2932
decay clock 0429
decay constant 0431
decay, oxic – 2932
decementation 3444
declination, magnetic – 0284
decline 3816
decline curve 3816
decline method 3874
décollement 1663
décollement structure 3053
decomposition 1998
decomposition 2741
decomposition 2929
decomposition, humic – 1999
deconvolution 4454
decoupling 0273
decussate 0949
dedolomitization 0930
dedolomitization 3458
deep 2247
deep 2794
deep-seated 0617
deep-sea trench 1138
defect, line – 0531
defect, plane – 0531
defect, point – 0531
deferment rate 3897
definition, optical – 4347
deflation 2155
deflation basin 2187
deflation hole 2187
deflation residue 2160
deflectometer 4294
deflect, to – 4292
deformation 1075
deformation 4666
deformation, deep-burial plastic – 3447
deformation fabric 1007
deformation lamella 0539
deformation lamella 0999
deformation, modulus of – 4696
deformation, penecontemporaneous – 3050
deformation plane 0238
deformation, shallow-burial plastic – 3447
degeneration 1982
deglaciation 2367
deglacieration 2367
degradation 2012
degradation 2589
degradation, faceted surface of – 2601
degradation, intersecting surfaces of – 2601
degranitization 1065
dehydration 3797
delay 4822
dell 2281
delta 2494
delta 2895
delta, arcuate – 2485
delta, bay – 2498
delta, bird-foot – 2485
delta, cuspate – 2495
delta, digitate – 2495
deltaic 2895
delta, inner tidal – 2499
delta, lobate – 2495
delta, outer tidal – 2499
delta, protruding – 2497
delta, storm – 2500
delta, sub– 2496
delta-value 0454
delta, wave – 2500
demagnetization, AC – 0321
demagnetization, thermal – 0321
demoiselle 2029
demultiplexing 4424
dendrite 0525
dendrite 3400
dendrite 4088
dendritic 2292
dendrochronology 1737
dendrogram 1960
dendroid 1821
dense 0799
density 4342
density 4537
density, apparent – 2751
density, apparent – 3602
density, bulk – 2751
density, bulk – 4614
density current 2979
density, helium – 3602
density, liquid apparent – 3602
density, packing – 3082
density stratification 2882
density, surface – 4111
density, true – 3602
densofacies 0987
denudation 2012
denudation 2013
denudation, inset surfaces of – 2601
denudation, step-like surfaces of – 2601
denudation, stepped surfaces of – 2601
denudation, stepping surfaces of – 2601
denudation, surface of – 2591
denudation, top level of – 2598
depletion 3830
depletion allowance 4222
depletion record 4222
depocentre 2986
deposit 2985
deposit, alluvial – 2886
deposit, alluvial – 4000
deposit, aquoglacial – 2402
deposit, batholith-related – 3990
deposit, bedded ore – 4021
deposit, blanket – 3109
deposit, bog iron – 4116
deposit, calcareous – 3267
deposit, carbonaceous – 3493
deposit, carbonate – 3266
deposit, cave – 2121
deposit, clastic – 3164
deposit, colluvial – 2868
deposit, concordant ore – 4022
deposit, contact – 4043
deposit, contact-metamorphic – 3980
deposit, cuneate ore – 4035
deposit, deep-sea – 2922
deposit, discordant ore – 4022
deposit, disseminated – 3975
deposit, early magmatic – 3975
deposit, eluvial – 3999
deposit, emanation – 3979
deposit, exhalation – 3979
deposit, flash-flood – 2884
deposit, flood-plain – 2885
deposit, fluvioglacial – 2402
deposit, fragmental – 3164
deposit, glacial – 2402
deposit, glacioaqueous – 2402
deposit, glaciolacustrine – 2402
deposit, hydrothermal – 3982
deposit, impregnation – 3992
deposit, infiltration – 3998
deposit, injection – 3993
deposit, intertidal – 2907
deposition 2985
deposition, sequential – 3965
deposit, lag – 2160
deposit, lahar – 1356
deposit, late magmatic – 3975
deposit, lenticular ore – 4026
deposit, magmatic injection – 3975
deposit, magmatic mineral – 3975
deposit, magmatic segregation – 3975
deposit, marginal – 3987
deposit, mineral – 3910
deposit, mudflow – 1356
deposit, offset – 3987
deposit, ore – 3910
deposit, peneconcordant ore – 4022
deposit, pneumotectic – 3981
deposit, podiform ore – 4033
deposit, porphyry type ore – 4133
deposit, primary – 3972
deposit, pyroclastic – 3170
deposit, pyroclastic – 1310
deposit, pyrometasomatic – 3980
deposit, regressive – 3112
deposit, replacement – 3991

deposit, residual – 2930
deposit, roll-type – 4028
deposit, sackform ore – 4035
deposit, secondary – 2993
deposit, secondary – 3972
deposit, sedentary – 2855
deposit, sedimentary – 2850
deposit, shear zone – 4060
deposit, sheeted zone – 4060
deposit, stratabound ore – 4024
deposit, stratified ore – 4023
deposit, stratiform ore – 4024
deposit, tabular ore – 4025
deposit, tidal – 2907
deposit, transgressive – 3113
deposit, volcanoclastic – 1310
deposit, wash-over – 2911
depression, cone of – 3667
depression, intermediate – 1137
depression, ring – 1185
depression, thaw – 2646
depression, volcano-tectonic – 1182
depth 1526
depth, focal – 0157
depth hoar 2351
depth, mean hydraulic – 2262
depth zone 0993
dereverberation 4454
deringing 4454
derivative, second – 4534
derivative, second – 4544
derived 2947
dermolith 1261
deroofing 0624
deroofing, extrusion by – 1232
derrick 4267
descendent 3967
desert 2577
desert armor 2160
desert armor 3405
desert armour 2160
desert armour 3405
desert lacquer 3405
desert patina 3405
desert pavement 2160
desert pavement 3405
desert polish 3405
desert, rocky – 2578
desert rose 3406
desert sandstone 4123
desert, sandy – 2578
desert, stony – 2578
desert varnish 3405
desilication 0929
deslime 4161
desmosite 0933
desorption 0407
desquamation 1994
dessication breccia 3048
dessication crack 3044
destruction by insolation 1992
desulphurization 0415
detachment 1663
detectability, threshold of – 0270
detection capability 0270
detection limit 4651
detection threshold 0270
detector 4413
detector, pressure – 4415
detector spacing 4416
detector station 4417
detention time 3627
detonation 1217
detonation 4815
detonator 4816
detour factor 3707
detrital 3165
detrital 3997
detrition 2005
detritus 2005
detritus 3165
detritus, incandescent – 1315
detritus, nonluminous – 1315
detritus, reef – 3355
deuteric 0658
developed area 3890
development, accelerated – 2574
development, ascending – 2574
development, declining – 2574
development, descending – 2574
development, uniform – 2574
development, waning – 2574
development, waxing – 2574
deviation factor, gas – 3862
deviation survey 4294
devitrification 0919
devolatilization 3520
dewatering 4739
dewatering 4811
dew point 3858
diabasic 0845
diablastic 0950
diabrochite 1068
diachronism 1736
diachronous 1736
diachyte 1028
diadochy 0477
diadysite 1046
diagenesis 3388
diagenesis, advanced – 3418
diagenesis, burial – 3417
diagenesis, early – 3418
diagenesis, environmental – 3398
diagenesis, field of – 3390
diagenesis, late – 3418
diagenesis, organic – 3521
diagenesis, phreatic – 3399
diagenesis, vadose – 3399
diagenetic 3974
diagenetic environment, shallow-burial – 3393
diagenetic environment 3391
diagenetic environment, deep-burial – 3393
diagenetic environment, near-surface – 3392
diagram, beta– 1021
diagram, block – 4258
diagram, contour – 4259
diagram, fence – 4252
diagram, pi– 1018
diagram, point – 4259
diagram, ribbon – 4252
diagram, rosette – 4259
diamagnetism 0298
diameter, effective – 3196
diameter, equivalent – 3195
diameter, mean – 3199
diameter, median – 3199
diameter, modal – 3199
diameter, nominal – 3195
diameter, sedimentation – 3195
diamict 3124
diamictite 3124
diaphthoresis 0904
diaphthorite 0904
diapir 1693
diapirism 1693
diaplectic 0895
diaschistite 0668
diastem 1724
diastem 3125
diastrophism 1074
diatectite 1027
diatexis 1027
diatexite 1027
diatomaceous chert 3371
diatomaceous earth 3370
diatomaceous shale 3371
diatom earth 3370
diatomite 3370
diatreme 1221
dictyonite 1043
die out downward, to – 1526
die out upward, to – 1526
diesel fuel 3744
dieseline 3744
diesel oil 3744
differentation, magmatic – 3938
differentiate 0667
differentiated 0667
differentiation 0398
differentiation 0667
differentiation by liquid immiscibility 0678
differentiation, crystallization – 0670
differentiation, diagenetic – 3475
differentiation, diagenetic – 3938
differentiation, filtration – 0675
differentiation, flowage– 0677
differentiation, gaseous transfer – 0679
differentiation, geochemical – 0398
differentiation, gravitational – 0672
differentiation, gravitative – 0672
differentiation index 0701
differentiation, liquation – 0678
differentiation, metamorphic – 0911
differentiation, pneumatolitic – 0679
differentiation, volatile transfer– 0679
diffluence pass, glacial – 2621
diffluence step 2624
diffraction 0217
diffraction 4444
diffraction curve 4464
diffusion, metamorphic – 0912
diffusion potential – 4615
digestion 0681
digitation 1687
dike 0741
dike 1516
dike 3073
diktyonite 1043
diktytaxitic 0842
dilatancy 0109
dilatancy 4678
dilatation 0187
dilation 1050
dilatometer 4760
dilatometer test 3596
dimension stone 4876
dimorphism 1811
dip 1484
dip 4017
dip angle 1485
dip, angle of – 1485
dip, apparent – 1484
dip, apparent – 4250
dip-closed 3763
dip, direction of – 1490
dip flattens out 1487
diplogenetic 3966
dip, magnetic – 0283
dipole-dipole array 4562
dipole field, axial – 0279
dipole field, eccentric – 0279
dip, original – 1486
dip-pole 0282
dip, primary – 1486
dip separation 1550
dip, slackening of the – 1487
dip slip 1564
dip slope 2559
dip, true – 1484
dipyrization 0932
direction, a- 1016
direction, b- 1016
direction, c- 1016
direction, look – 4389
direction of the strike 4016
dirt, nitrous – 2123
disaggregation 1989
discharge 1217
discharge 2249
discharge efficiency 2250
discharge, specific – 3703
discoidal 3184
disconformity 1721
discontinuity 0099
discontinuity 4723
discontinuity layer 2333
discontinuity, seismic – 0099
discordant 0718
discordant age 0430
discount rate 3897
discovery 3884
disc test, pierced – 4761
disequilibrium clock 0429
disequilibrium dating, uranium-series – 0443
disintegration 1989
disintegration 2929
disintegration 3526
disintegration, block – 1990
disintegration, granular – 1990
disintegration, mineral – 1990
disk-shaped 3184
dislocation 0531
dislocation 1522
dislocation 4078
dislocation, edge – 0531
dislocation, mixed – 0531
dislocation, screw – 0531
disomatic crystal 0663
dispersal fan 2948
dispersal shadow 2948
dispersion 0197
dispersion 0403
dispersion 0572
dispersion 1812
dispersion 3710
dispersion 4190
dispersion, axial – 0572
dispersion, crossed – 0572
dispersion curve 0226
dispersion, horizontal – 0573
dispersion, inclined – 0573
dispersion, inverse – 0197
dispersion, normal – 0197
dispersion pattern 0403
dispersion pattern, lateral – 4631
dispersion pattern, linear – 4631
dispersion pattern, primary – 0404
dispersion pattern, secondary – 0405
dispersion pattern, superjacent – 4631
dispersion, secondary – 0405
disphotic 1904
displaced 4637
displacement 4078
displacement, height – 4379
displacement meter 0121
displacement, piston-like – 3826
displacement plane 0238
displacement, relief – 4379
displacement, tilt – 4379
disposal, underground – 4832
disposal, waste – 4831
disruption theory 0046
dissection 2019
dissemination 3947
dissipating area 2258

dissipation 0219
dissolution 0681
dissolved load 2956
distal 2997
distance, angular – 0155
distance, epicentral – 0155
distortion, barrel – 4404
distributary 2211
distribution, lognormal – 4189
distribution, normal – 4189
distribution, polymodal – 3197
distribution, standard – - 4189
district, mineral – 3909
district, ore – 3909
divarication 2210
divergence 1969
divergent 0776
diverter 2325
divide 2312
divide 2314
divide, consequent – 2317
divide, creeping – 2319
divide, leaping – 2320
divide, longitudinal – 2318
divide, migrating – 2319
divider 4046
divide, shifting – 2319
divide, subsequent – 2318
divide, tidal – 2511
divide, valley-floor – 2315
divide, water – 2316
doab 2312
docrystalline 0804
dohyaline 0804
doleritic 0845
dolina 2072
doline 2072
doline, collapse – 2072
doloclast 3167
doloclast 3270
dololutite 3274
dolomicrite 3282
dolomite 3318
dolomite, calcareous – 3319
dolomite, calcitic – 3319
dolomite mottling 3461
dolomite, primary – 3320
dolomite, primary – 3380
dolomitite 3318
dolomitization 3460
dolomitization, over– 3460
dolosparite 3280
dolostone 3318
DOM 3521
domain 1006
dome 0035
dome 0059
dome 0728
dome 1641
dome, air – 3065
dome, basaltic – 1420
dome, cumulo – 1422
dome, laccolithic – 1206
dome, piercement – 1695
dome, plug – 1422
dome, pressure – 2113
dome, salt – 1695
dome, sand – 3065
dome, twin – 1442
dome, volcanic – 1422
dome, water – 1359
domichnia 1880
dominant 1816
dopatic 0818
dopplerite 3525
dormancy 1169
dosemic 0818
doublet 0480
down-dip 1489
downfall, intercrateral – 1472
down-faulted 1559
downstream side 2964
downthrow 1559
downtime 3810
draa 2180
drag 1537
drag 4079
dragging 1537
drain 4811
drainage 2253
drainage 4739
drainage 4811
drainage area 3648
drainage area 3844
drainage basin 2192
drainage-basin survey 4641
drainage class 2766
drainage density 2255
drainage, endoreic – 2256
drainage, ephemeral – 2253
drainage, episodical – 2253
drainage, exoreic – 2256
drainage, external – 2256
drainage, gravity – 3828
drainage, intermittent – 2253
drainage, internal – 2256
drainage net 2254
drainage, out-flowing – 2256
drainage pattern 2254
drainage pattern, annular – 2254
drainage pattern, bird-foot – 2254
drainage pattern, craw-foot – 2254
drainage pattern, dendritic – 2254
drainage pattern, digitate – 2254
drainage pattern, pennate – 2254
drainage pattern, radial – 2254
drainage pattern, rectangular – 2254
drainage pattern, tree-like – 2254
drainage pattern, trellised – 2254
drainage, perennial – 2253
drainage, periodical – 2253
drainage ratio 2250
drainage screen 4804
drainage, seasonal – 2253
drainage, soil – 2768
drainage texture 2255
drainage, through-flowing – 2256
drainpipe 2098
drainpipe 4811
draping 1664
drawdown 3667
dredge 3154
dreikanter 2162
drewite 3379
driblet cone 1298
driblet spire 1298
drift 2961
drift 4540
drift 4847
drift, ablation – 2412
drift, beach – 2962
drift, englacial – 2408
drift, glacial – 2402
drift, glacial – 2403
drift, glaciofluvial – 2402
drifting, beach – 2962
drifting, littoral – 2962
drift line 2962
drift, littoral – 2962
drift plain, fluvioglacial – 2424
driftsand 4719
drift, sand – 2156
drift, spectral – 4549
drift, superglacial – 2406
driftwood 2970
drillability 4298
drill carriage 4266
drill collar 4270
drillcore 4237
drillcuttings 4238
drill, diamond – 4265
drilled-up area 3890
drillhole 4261
drilling, airlift – 4263
drilling, cable-tool – 4262
drilling, churn – 4262
drilling, counterflush – 4285
drilling, directional – 4291
drilling fluid 4281
drilling, percussion – 4262
drilling, RAB – 4263
drilling, rotary – 4263
drilling, rotary air-blast – 4263
drilling, rotary-percussion – 4263
drilling, turbine – 4264
drillpipe 4270
drillpipe, stuck – 4286
drillstem 4270
drillstem test 3787
drillstring 4270
dripstone 2127
drive, artificial water – 3833
drive, bottom-water – 3825
drive, combination – 3829
drive, compaction – 3827
drive, edge-water – 3825
drive, fire – 3838
drive, flank – 3833
drive, gas – 3823
drive, gas-cap – 3823
drive, gas-solution – 3823
drive, hot water – 3839
drive, line – 3833
drive, pattern – 3833
driver 4487
driver, pile – 4838
drive, steam – 3839
drive, water – 3824
dropper 4071
drum fire 1237
drumlin 2420
drumlin, rock – 2420
drumlin, true – 2420
druse 0873
druse 4014
drusy 0873
drying up 2342
DST 3787
dualistic hypothesis 0019
dubiocrystalline 0801
ductile 4673
ductolith 0746
dullam 0499
dump 4166
dumpling, volcanic – 1296
dune 2165
dune 2872
dune 3007
dune, anchored – 2175
dune, beach – 2168
dune, blow-out – 2187
dune, clay – 2872
dune, cliff – 2169
dune, coastal – 2166
dune, continental – 2170
dune, crescentic – 2177
dune, embryonic – 2176
dune, exterior – 2167
dune, fixed – 2175
dune, horseshoe – 2177
dune, hummocky – 2190
dune, inland – 2170
dune, inner – 2167
dune, interior – 2167
dune, lee – 2174
dune, linear – 2171
dune, littoral – 2166
dune, longitudinal – 2171
dune low 2188
dune, migrating – 2175
dune, mobile – 2175
dune, moving – 2175
dune, outer – 2167
dune, parabolic – 2177
dune, phytogenic – 2172
dune, psammogenic – 2172
dune, sand – 2872
dune, seif – 2171
dune, shifting – 2175
dune slack 2188
dune, stabilized – 2175
dune, stationary – 2175
dune, transverse – 2171
dune, wandering – 2175
dune, wind-shadow – 2174
durability, slake – 4734
durain 3580
duration 0240
duricrust 4122
duripan 2734
durite 3583
dust 3206
dust band 2379
dust cloud 1363
dust, cosmic – 0065
dust fall 2159
dust, meteoritic – 0065
dust shower 2159
dust, volcanic – 1316
dust well 2387
dy 3527
dyke 0741
dyke 1516
dyke chamber – 1204
dyke, clastic – 1516
dyke, clastic – 3073
dyke complex 0749
dyke, headed – 0746
dyke, hollow – 1245
dyke, neptunean – 3073
dyke, ore – 4031
dyke, radial – 1227
dyke, radiating – 1227
dyke, replacement – 0745
dyke ridge 0741
dyke, ring – 0742
dyke rock 0741
dyke, roofed – 0747
dyke, secundine – 0744
dyke set 0749
dyke swarm 0749
dyke system 0749
dyke, vein – 4031
dyke wall 0741
dyke, welded– 0744
dyke with expanded summit 0734
dykite 0741
dynamofluidal 0973
dynamometamorphism 0892
dynamometer 4770
dynamo theory 0278
dyngja 1420
dyscrystalline 0797
dysodil 3563
dysphotic 1904

E

eagre 2823
earth 3086
earth, fine – 2719
earth flow 2051
earth pressure 4705
earth pressure at rest 4705
earth pyramide 2029
earthquake 0138
earthquake, artificial – 0265

earthquake, collapse – 0141
earthquake control 0262
earthquake, cryptovolcanic – 0140
earthquake engineering 0254
earthquake, impact – 0141
earthquake, inland – 0148
earthquake magnitude 0220
earthquake mechanism 0232
earthquake, micro– 0146
earthquake, multiple – 0143
earthquake prediction 0261
earthquake prevention 0262
earthquake, principal – 0142
earthquake-proof construction 0260
earthquake record 0116
earthquake regulation 0262
earthquake, relay – 0145
earthquake-resistant design 0260
earthquake rift 2552
earthquake risk 0255
earthquake scarplet 2552
earthquake, silent – 0114
earthquake sound 0203
earthquakes, swarm – 0144
earthquake, submarine – 0148
earthquake, tectonic – 0139
earthquake, volcanic – 0140
earthquake, volcanic – 1193
earth, radiolarian – 3368
earth, reinforced – 4798
earth sciences 0002
earth, seat – 3515
earth's heat 0374
earth, siliceous – 3367
earthy 3086
Eastman survey instrument 4295
ebb channel 2510
ebb current 2822
ebullitim 1358
ebullition 1358
échelon faults, en – 1585
échelon folds, en – 1676
echo 4390
ecology 1887
ecostratigraphy 1711
ecosystem 1893
ecotype 1942
ectectite 1034
ectexis 1034
ectexite 1034
ectinite 1055
eddy 2269
edge 0518
edge 1686
edge, calcite – 2148
edged, sharp– 3190
edge-water encroachment 3826
edifice, volcanic – 1413
EDTA 4656
effect, chilling – 1271
effect, heading – 4530
effect, indirect – 0366
efficiency, seismic – 0205
effluent 4165
effusion 1213
effusion, fissure – 1233
effusive rock 1240
Eh 0421
einkanter 2162
ejecta, accessory – 1314
ejecta, accidental – 1314
ejecta, allothigenous – 1314
ejecta, authigenous – 1314
ejecta blanket 0055
ejecta, cognate – 1313
ejecta, essential – 1314
ejecta, juvenile – 1314
ejectamenta 1313
ejectamenta, clastic – 1313
ejecta, noncognate – 1314
ejecta, resurgent – 1314
elastic 0472
elasticity 4669
elasticity, modulus of – 4695
elastic limit 4697
elastic rebound number 4733
elastic-rebound theory 0108
elaterite 3740
electrode array 4562
electrode configuration 4562
electrode disposition 4562
electrode, line – 4563
electrode system, permanent – 3673
electromotive force 4556
eleutheromorphic 0983
elevation, structural – 4249
ELF 4572
elimination of beds 1557
ellipsoid, international – 0331
ellipticity correction 0098
elongation 0567
elutriation 3159
elutriation 3930
eluvial horizon 2701
eluviation 2678
eluviation 3930
eluvium 2930
emanation 3951
emanation, volcanic – 1215
emanation, volcanic – 1376
embankment 4825
embatholitic 3990
embayment 0756
embayment 2451
embrechite 1055
emergence 1728
emergence 2857
emergence angle 0215
emergent 0174
emersio 0174
emission 1210
emission, explosive – 1216
emission of gas 1215
emission of vapour 1215
EMP 4588
emplacement 0618
emulsion intergrowth 4101
emulsion stage 0652
enclosure 0662
encrinal 3313
encrinite 3313
encrinitic 3313
endemic 1814
endobatholitic 3990
endoblastesis 0657
endobiont 1854
endoblastic 0657
endocast 1825
endogene 3969
endogene effect 0704
endogenetic 0081
endometamorphism 0898
endomorph 0789
endomorphism 0898
endoscope 0503
endoskeleton 1861
endosphere 0089
end with depth, to – 1526
energy, form – 0909
energy, geothermal – 0388
energy, kinetic eruptive – 1171
energy, potential eruptive – 1171
energy release 0204
energy, volcanic – 1171
engineering, earthquake – 0254
engineering, foundation – 4834
engineering seismology 0254
enrichment 3945
enrichment, supergene – 3945
entectite 1036
entexis 1036
entexite 1036
enthalpy 0389
enthalpy system, high– 0389
enthalpy system, low– 0389
entrenched part 2390
envelope 1070
environment 1900
environment 2859
environmental 2859
environment, diagenetic – 3391
environment, high-energy – 2881
environment, low-energy – 2881
environment, primary – 0402
environment, secondary – 0402
environment, sedimentary – 2859
eolian 2870
eolianite 2870
eolotropy 0110
eometamorphism 0902
eon 1767
eonothem 1767
Eötvös 0337
Eötvös correction 4538
epeirogeny 1090
epibatholitic 3990
epibenthos 1908
epibiont 1855
epibole 1773
epibolite 1049
epicenter 0154
epicentral azimuth 0155
epicentral distance 0155
epicentre 0154
epicontinental 2915
epicontinental basin 1086
epifauna 1853
epigenesis 3419
epigenetic 3474
epigenetic 3553
epigenetic 3966
epigenic 2290
epilimnion 2332
epimagma 1201
epimagmatic 0658
epineritic 2841
epinorm 0606
epipedon 2697
epipelagic 1910
epipelagic 2845
epiphyte 1856
epirogenesis 1090
epirogeny 1090
epitactic 0542
epitaxy 0542
epitaxy relationship 0542
epithermal 3983
epixenolith 0663
epizone 0993
epoch 1764
epoch, metallogenic – 3905
epoch, polarity – 0292
equal-area net 1015
equant 0754
equant 3181
equator, geomagnetic – 0281
equator, magnetic– 0283
equidimensional 0522
equidimensional 0754
equidimensional 3181
equidistant 4172
equiform 0754
equilibrium, isostatic – 0346
equilibrium, radioactive – 4549
equipotential-line method 4558
equivalence, principle of – 4567
equivalent value 3675
era 1766
erathem 1766
erg 2183
erg 2578
erodibility 2936
erosion 2012
erosion 2936
erosion, accelerated – 2777
erosion, accelerated – 2036
erosion, base level of – 2025
erosion, chemical – 2017
erosion, cosmic – 0028
erosion, cycle of – 2566
erosion, differential – 2024
erosion, downward – 2022
erosion, eolian – 2015
erosion, glacial – 2015
erosion, grooving – 2022
erosion, gully – 2018
erosion, gully – 2778
erosion, headward – 2021
erosion, headward migration – 2021
erosion, headwater – 2021
erosion, interrupted cycle of – 2571
erosion, lateral – 2022
erosion, marine – 2016
erosion, plain of marine – 2536
erosion, platform of level of – 2591
erosion residual 2607
erosion, retrogressive – 2021
erosion, rill – 2033
erosion, rill – 2778
erosion, sheet – 2032
erosion, sheet – 2779
erosion, sheet-flood – 2032
erosion, soil – 2036
erosion, soil – 2776
erosion, stream – 2015
erosion system 2573
erosion, vertical – 2022
erosion, wave – 2016
erosion, wind – 2015
erratic 2422
erratic 2949
erratic block 2949
error, closure – 4541
eruption 1211
eruption, areal – 1232
eruption, central – 1228
eruption centre 1220
eruption cloud 1360
eruption column 1361
eruption cylinder 1224
eruption, detritus – 1214
eruption, eccentric – 1229
eruption, embryonic – 1219
eruption, explosive – 1216
eruption, fissure – 1233
eruption, flank – 1230
eruption, Hawaiian-type – 1158
eruption, indirect – 1218
eruption, intermittent – 1212
eruption, lava – 1214
eruption, magmatic – 1219
eruption, mechanism of the – 1170
eruption, mixed – 1219
eruption, mud – 1395
eruption, paroxysmal – 1164
eruption, Peléan-type – 1163
eruption, phreatic – 1351
eruption, Plinian-type – 1161
eruption point 1220
eruption, pseudovolcanic 1351
eruption, secondary – 1351
eruption, semivolcanic – 1351
eruption, Strombolian-type 1159
eruption, subaerial – 1236
eruption, subaquatic – 1236
eruption, subglacial – 1236
eruption, submarine – 1236
eruption, summit – 1229

eruption, Vesuvian-type – 1160
eruption, Vulcanian-type – 1160
eruptive 0616
eruptive body 1239
eruptive capacity, kinetic – 1171
eruptive energy, kinetic – 1171
eruptive energy, potential – 1171
eruptive phase 1208
escarpment 2799
escarpment, fault 2548
esker 2420
estavelle 2092
estimate, biased – 4192
estuarine 2904
estuary 2456
estuary 2784
etched terrain 0040
etch figure 0530
etch figure 3409
etch pit 0530
etchplain 2592
ethmolith 0737
eucrystalline 0795
eugeosyncline 1097
eugranitic 0810
euhaline 2834
euhedral 0523
euhedral 0750
euhedron 0750
eulittoral 2838
euphotic 1904
eurybathic 1911
eurybenthic 1909
euryhaline 1912
eurythermal 1913
eurytopic 1914
eustatic change of level 2858
eutaxite 0861
eutaxite 1243
eutaxitic 0861
eutrophic 3549
euxinic 2925
evaluation, land – 4786
evaluation, mine – 4216
evaluation, terrain – 4786
evaporation 0412
evaporation 2342
evaporative unit 3605
evaporite 3375
evaporite-log evaluation 4624
evaporitic 4002
evapotranspiration 3621
evasion 0273
events, premonitory – 1191
evolution 1964
evolutionary zone 1779
evolution, convergent – 1973
evolution, iterative – 1971
evolution, line of – 1965
evolution, magmatic – 0666
evolution, parallel – 1971
evolution, quantum – 1971
evolution, rectilinear – 1975
exaration 2396
excavation 4824
excavation, gaseous – 1178
excavation of valleys 2270
exfoliation 1994
exhalation 1215
exhalation 1376
exhalation 3951
exhumed 2595
exinite 3571
exinoid 3570
exocast 1825
exogene 3969
exogene effect 0704
exogenetic 0081
exogeology 0024
exometamorphism 0898
exomorphic zone 0899
exomorphism 0898
exoskeleton 1861
exosphere 0083
exotic block 2949
exotic boulder 2949
exotic constituent 2949
expansion, flash – 3863
expansion hypothesis 1123
expectation curve method 3876
expextation of reserves 3876
exploration, biogeochemical – 4645
exploration, geobotanical – 4645
exploration, geochemical – 4625
exploration, hydrochemical – 4640
explorationist 0017
exploration well 3782
explosion 1217
explosion caldera 1458
explosion crater 1431
explosion focus 1177
explosion funnel 1432
explosion graben 1235
explosion level 1177
explosion, level of – 1177
explosion, nucleair – 0268
explosion, phreatomagmatic – 1219
explosion point 1176
explosion, underground nuclear – 0268
explosion vent 1431
explosive 4815
explosive factor 4820
explosive phase 1208
explosive phenomena 1175
explosivity index 1175
exposure 4231
exposure 4340
exposure latitude 4340
expulsion 3732
exsolution, 0476
exsudatinite 3576
exsurgence 2090
extension 3885
extensometer 0122
extensometer 4773
extent 4726
extent, downward – 1526
extent in lateral direction 1524
extent, lateral – 1524
extent, vertical – 1526
extinction 0564
extinction 1986
extinction angle 0566
extinction, oscillatory – 0968
extinction, undulatory – 0565
extinction, undulatory – 0968
extinction, wavy – 0968
extraclast 2945
extraclast 3272
extramagmatic 3985
extraneous 3553
extrusion 1207
extrusion by deroofing 1232
extrusive body 1239
extrusive phase 1197
extrusive rock 1240
eyed structure 0952

F

F 2266
fabric 0793
fabric 1004
fabric 1605
fabric 3077
fabric analysis 0995
fabric axis 1016
fabric, deformtion – 1007
fabric, depositional – 1011
fabric, depositional – 3077
fabric diagram 1015
fabric domain 1006
fabric element 1012
fabric, growth – 1010
fabric, growth-zone – 1010
fabric, megascopic – 1005
fabric, mesoscopic – 1005
fabric, microscopic – 1005
fabric, primary – 3077
fabric, rock – 1004
fabric, se– 1004
fabric, si– 1004
fabric, structural – 1004
fabric, sub– 1005
fabric, tectonic – 1007
face 0517
face 1529
face 1622
face 4852
facet 2086
facet, fault – 2553
facet, spur-end – 2553
facies 1715
facies 2860
facies, bio- 2861
facies, diagenetic – 3435
facies fauna 1850
facies fossil 1807
facies, litho- 2861
facies, metamorphic – 0987
facies, mineral – 0585
facies, sedimentary – 2860
facies series, metamorphic – 0991
facies, shelf – 2915
facies, tecto– 2861
facies, tectonic – 0996
facies, volcanic – 1238
facing 1622
faecal pellet 1863
fahlband 4052
failure 4676
failure, progressive – 4676
fall 2203
fall, wall – 2111
false 4637
false alarm 0269
false form 0982
family 0687
family 1930
family, soil – 2664
fan 4631
fan, alluvial – 2007
fanglomerate 2885
fan of lava 1287
fan, rock – 2585
faro 3339
fascicular 0773
fatigue 4674
fault 1521
fault 4723
fault, active – 1527
fault, antithetic – 1588
fault, arcuate – 1535
fault, associated – 1581
fault, auxiliary – 1581
fault, bedding – 1547
fault block 1531
fault, border – 1593
fault, boundary – 1593
fault, branching – 1582
fault, closed – 1534
fault-closed 3763
fault, cross – 1545
fault, curved – 1535
fault, curvilinear – 1567
fault, dead – 1527
fault, deep-reaching – 1525
fault deflection 1536
fault deviation 1536
fault, dextral – 1561
fault, diagonal – 1545
fault, dip – 1545
fault, dominant – 1580
faulted down 1559
fault escarpment 2548
fault facet 2553
fault facet, triangular – 2553
fault gouge 1540
fault, gravitational – 1707
fault, growth – 1707
fault, high-angle – 1549
fault, hinge – 1567
faulting, splay – 1583
faulting, zone of – 1532
fault ledge 2548
fault, left lateral – 1561
fault, levelled – 2549
fault, longitudinal – 1546
fault, low-angle – 1549
fault, low-dipping – 1549
fault, main – 1580
fault, major – 1580
fault, marginal – 1593
fault, minor – 1580
fault, near-surface – 1525
fault, normal – 1558
fault, oblique – 1545
fault, open – 1534
fault, peripheral – 1592
fault plane 0238
fault plane 1528
fault, plane – 1535
fault, profound – 1525
fault propagation 0239
fault, ramifying – 1582
fault, rejuvenated – 1527
fault, renewed – 1527
fault, reverse – 1558
fault, revived – 1527
fault, right lateral – 1561
fault, ring– 0743
fault, rotational – 1567
fault scarp 2548
fault scarp, exhumed – 2548
fault scarp, rejuvenated – 2548
fault scarp, resurrected – 2548
fault scarp, revived – 2548
fault, sealing – 3765
fault, secondary – 1580
faults, en échelon – 1585
fault set 1584
faults, framework of – 1578
faults, group of parallel – 1584
fault, shallow – 1525
fault, sinistral – 1561
faults, intersecting – 1579
faults, network of – 1578
faults, radial – 1594
faults, radiating – 1594
faults, step – 1587
fault, steep-dipping – 1549
fault, strike – 1544
fault, strike-slip – 1560
fault, subsidiary – 1581
fault surface 1528
fault, synthetic – 1588
fault system 1578
fault system, conjugate – 1584
fault, tear – 1560
fault tectonics 1520
fault, trailed – 1536
fault, transcurrent – 1560
fault, transform – 1141
fault, transform – 1560
fault, translational – 1566
fault, transverse – 1546
fault valley 2547
fault, wrench – 1560
fault zone 1532
fauna 1795
fauna, depauperate – 1851

fauna, dwarf – 1851
fauna, facies – 1850
faunal 1795
faunula 1796
FDC 4593
feasibility study 4639
feasibility study 4791
features, planar – 0894
feed 4140
feedback 0133
feeder 0723
feeder 2207
feeder 3941
feeder 4071
feidsh 2186
feldsparphyric 0818
feldspathization 1064
fels 0935
felsenmeer 2057
felsic 0597
felsite 0597
felsite 0798
felsitic 0800
felsitoid 0798
felsophyre 0824
felsophyric 0824
felty 0825
femic 0607
fenite 0925
fenitization 0925
ferricrete 4123
ferricrete 4124
ferrimagnetism 0302
ferrite 4123
ferro-femic index 0698
ferromagnesian 0599
ferromagnetism 0299
ferruginous 3464
festoon 2649
fiamme 1342
fiard 2454
fiber 2705
fibroblastic 0963
fibrocrystalline 0815
fibrous 0522
fibrous 0773
fiducial 4527
field curve 4565
field, geomagnetic – 0276
field, mineral – 3909
field moisture capacity 2774
field, near – 0274
field, ore – 3909
field strength, magnetic – 0317
filiform 4081
filiform 0774
fill 3105
fill 4825
fill 4831
fill, cut and – 4826
fill, hydraulic – 4825
filling by plant growth 2341
filling, joint – 4729
filling up 2341
film 3100
film, broad-band – 4334
film, colour-blind – 4335
film, colour-infrared – 4337
film, colour-sensitive – 4335
film, data – 4388
film, false colour – 4337
film, infrared – 4336
film, intergranular – 2371
film, orthochromatic – 4334
film, panchromatic – 4334
film, signal – 4388
filter 4338
filter 4445
filter 4812
filter, band-pass – 0137
filter, band-reject – 0137
filter, contrast – 4339
filter, cut-off – 0137
filtering, frequency – 0136
filtering, matched – 0227
filtering, velocity – 0228
filter layer 4862
filter, matched – 4509
filter, notch – 0137
filter, polarizing – 4339
filter pressing 0675
filter pressing 3933
filter, velocity – 4456
filter, vignetting – 4339
filtration 0414
filtrational 3955
filtration differentiation 0675
filtration pressing 3933
fine earth 2719
fine-grained 0796
fine-grained, very – 0799
fines 3207
fine-textured 2721
fingering 3826
fining-upward sequence 3129
fiord 2453
fire 0485
fire clay 3217
fire clay 3517
fire flooding 3838
fire fountain 1283
fire globe 1372
fire pit 1284
fire spitting 1237
fire, spouts of – 1283
firn 2363
firnification 2363
firn line 2364
firth 2453
Fischer assay, modified – 3731
fishing 4287
fishing tool 4287
fissile 1597
fissility 1597
fission-track dating 0442
fissure 1512
fissure 2375
fissure, concentric – 1469
fissure, contraction – 0713
fissure, cooling – 1517
fissure, desiccation – 1518
fissure filling 1515
fissure, flank – 1468
fissure, frost – 2640
fissure, gaping volcanic – 1234
fissure, infilled – 1514
fissure, joint – 1513
fissure, radial – 1468
fissure, tension – 1519
fissure vent 1225
fissure, volcanic – 1465
fixism 1113
fjord 2453
flag 3132
flaggy 3132
flagstone 3132
flake 2111
flake 3186
flake fall 2111
flame, blowpipe – 1350
flaming orifice 1386
flank outflow 1266
flaser 0952
flaser 3103
flaser bedding 3148
flaser structure 0952
flat 2506
flat 4055
flat, alkali – 2580
flatiron 2562
flat, mud – 2910
flatness 3182
flat, piedmont – 2604
flat, sand – 2910
flattening 0330
flattening 1613
flaw 0487
flexible 0472
flexing 1537
flexure 1616
flight line 4303
flight of terraces 2296
flight path 4303
flint 3491
flint curtain 3491
flints, clay with – 2931
floatstone 3296
flocculation 2992
floccule 2992
flood basalt 1257
flood basin 2239
flood channel 2510
flood current 2822
flood, flash – 2252
flooding, miscible – 3840
flooding, water – 3833
flood, lava – 1276
flood plain 2239
flood-plain lobe 2225
flood-plain scroll 2234
flood, sheet – 2032
flood zone 1773
floor 0705
floor 3501
floor 4010
floor 4852
flora 1797
floral 1797
flotation 0673
flotation 4152
flotation, bulk – 4152
flotation, selective – 4152
flow 2050
flow 2249
flow 4672
flowage 1701
flowage, primary– 0631
flowage structure 0868
flow banding 0860
flow, base – 3628
flow breccia 1268
flow breccia, welded – 1268
flow, cross – 3834
flow, flexural – 1665
flow, helical – 2264
flow joint 0714
flow, laminar – 2264
flow layering 0868
flow line 0868
flow mark 3020
flow marking 2086
flow meter 3669
flow, overland – 3617
flow regime 2960
flow sheet 4143
flow, spiral – 2264
flowstone 2128
flow structure 0868
flow structure, intrastratal – 3052
flow structure, linear – 0869
flow structure, planar – 0869
flow structure, platy – -869
flow, subsurface – 3623
flow surface 1002
flow, tectonic – 0998
flow texture 0868
flow, turbulent – 2264
flucan 4012
fluid, connate – 0650
fluidization 1370
fluid, non-wetting – 3705
fluid, wetting – 3705
fluolite 1262
fluorescence 0467
fluorescence 0491
fluorescence, X-ray – 4592
flute 2086
flute 3025
flute cast, corkscrew – 3026
flute, solution – 2087
fluting 3025
fluvial 2883
fluviatile 2883
fluvioglacial 2883
fluviomarine 2902
flux density, magnetic – 0317
fluxing, gas – 1178
fluxion banding 0860
fluxion structure 0868
flux, luminous – 4311
flux, magnetic – 0317
fluxoturbidite 2984
flux, volatile – 0650
flysch 2926
foam 1308
foam, volcanic – 1308
foamy 0875
focal depth 0157
focal mechanism 0232
focal parameter 0151
focal sphere 0233
focal time 0159
focus 0156
focus, depth of – 0157
focus, eruptive – 1177
focus, high – 0157
focus, volcanic – 1177
fodinichnia 1881
foId, neutral – 1656
fold 1616
fold, anticlinal – 1618
fold, asymmetric – 1644
fold axis 1630
fold, back– 1689
fold, backward – 1689
fold belt 1674
fold, box – 1649
fold, buckle – 1665
fold bundle 1674
fold, chevron – 1651
fold, concentric – 1652
fold, concertina – 1651
fold, conical – 1655
fold, cross – 1678
fold, cylindrical – 1654
fold, diapiric – 1698
fold, drag – 1570
fold, fan-shaped – 1650
fold, inclined – 1646
folding, back – 1689
folding, backward – 1689
folding, disharmonic – 1663
folding, double – 1668
folding, enterolithic – 3076
folding, flow – 1701
folding, incipient – 1666
folding, multiple – 1668
folding, penecontemporaneous – 1667
folding phase 1107
folding phase 1669
folding, precursory – 1667
folding, superimposed – 1668
fold, intrafolial – 1660
fold, knee – 1648
fold, main – 1642
fold, major – 1642
fold, minor – 1642
fold nappe 1682
fold, non-cylindrical – 1654
fold overthrust 1576
fold, overturned – 1647
fold, parallel – 1652
fold, parasitic – 1677
fold, plunging – 1637
fold, polycIinal – 1657
fold, ptygmatic – 1051

fold, ptygmatic – 1658
fold, reclined – 1656
fold, recumbent – 1680
fold, refolded – 1668
fold, rootless – 1659
fold scarp, monoclinal – 2551
folds, cascading – 1706
folds, en échelon – 1676
fold, shear – 1662
fold, similar – 1653
fold, slump – 3055
fold, small – 1642
folds, packet of recumbent – 1681
folds, piled-up recumbent – 1681
fold, supratenuous – 1664
fold, symmetrical – 1644
fold, synclinal – 1619
fold system 1670
fold tectonics 1615
fold, unsymmetric – 1644
fold, upright – 1645
fold, zigzag – 1651
foliaceous 0775
foliate 0948
foliated 0775
foliation 0948
foliation 1604
foliation, closed – 0948
foliation, open – 0948
fondoform 2917
fondothem 2917
foot, cubic – 3892
footing 4835
footing, continuous – 4837
footing, individual – 4837
footing, shallow – 4835
footing, spread – 4837
footing, strip – 4837
footing, wall – 4837
footprint 1877
footwall 1529
footwall 4009
force of crystallization 0909
ford 2247
förde 2454
foredeep 1109
foredune 2168
foreign 3970
foreland 1109
foreland 2447
foreland 2492
foreland, cuspate – 2447
fore-phase 1157
foreset bed 2896
foreset bed 3140
foreset bedding 3140
foreset bedding, overturned – 3054
foreset lamina 3140
foreset lamination 3140
foreshock 0147
foreshore 2487
foretrough 1109
forking 2110
formation 1754
formation, banded iron – 4114
formation, climax – 1891
formation resistivity factor 4609
formation, sealing – 3765
formation, wet – 3767
form energy 0909
formenkreis 1936
form factor 4110
form genus 1932
form, negative volcanic – 1410
form, positive volcanic – 1409
fossa 0034
fosse 1453
fossil 1789
fossil 2595
fossil, facies – 1807
fossil, guide – 1785
fossiliferous 1789
fossil, index – 1785
fossil, key – 1785
fossil, zonal guide – 1785
foundation 4835
foundation coefficient 0258
foundation course 4836
foundation engineering 4834
foundation, mat – 4837
foundation, raft – 4837
foundation, road – 4841
foundation, shallow – 4835
foundation stratum 4836
foundering, roof – 1188
fountain, water – 1359
Fourier transform 4451
Fr 2266
fractionalization 0670
fractionation 0414
fractionation 0670
fractionation, crystal – 0670
fractionation, liquid – 0680
fracture 0469
fracture 0541
fracture 1494
fracture, conchoidal – 0541
fracture criterion, Mohr-Coulomb – 4694
fractured 1494
fracture, extension – 1503
fracture porosity 3694
fracture, ring– 0743
fracture treatment 3800
fracture zone 1595
fracturing 1494
fracturing, hydraulic – 3800
fragipan 2734
fragmentation 4821
framboid 3484
framboidal 3484
framboidal 4094
frame 4358
framestone 3299
framework 3079
framework 3680
framework stability 3424
frayed end 3414
free air gravity 4538
free period 0130
freezing process 4807
frequency, beat – 0178
frequency, corner – 0225
frequency curve, cumulative – 4188
frequency distribution 4187
frequency domain 0223
frequency domain 4571
frequency modulation 4492
frequency, operating – 4498
frequency response 4448
frequency spectrum 4451
freshet 2252
Fresnel area 4515
fretted terrain 0039
friable 3087
friction 4686
friction, angle of – 4689
friction coefficient 4689
friction criterion 4689
friction, internal – 4686
friction, internal – 4689
friction, skin – 4839
fringe, capillary – 3657
fritting 0897
front 1686
front, basic – 1066
front normal 0550
front range 2605
frost crack 2640
frost crack, seasonal – 2641
frosted 3408
frost fissure 2640
frost heaving 2642
frost heaving 4721
frosting 3408
frost mound 2643
frost, mush – 2639
frost splitting 1996
frost wedging 1996
frost work 1996
frost work 2394
frothy 0875
Froude number 2266
frustule 1842
frustule 3370
fuel ratio 3601
fugitive constituent 0650
full 2486
Fuller's earth 3218
fumarole 1379
fumarole, acid – 1388
fumarole, alkalic – 1388
fumarole area 1380
fumarole, chlorine – 1387
fumarole, cold – 1389
fumarole, crater – 1383
fumarole, croicolitic – 1384
fumarole, dry – 1387
fumarole, eruptive – 1385
fumarole field 1380
fumarole, fissure – 1383
fumarole, leucolitic – 1384
fumarole mound 1390
fumarole, permanent – 1381
fumarole pimple 1390
fumarole, primary – 1382
fumarole, roaring – 1385
fumarole, rootless – 1382
fumarole, secondary – 1382
fumarole, steam – 1389
fumarole, temporary – 1381
fumarole, tunnel – 1383
fumarolic activity 1375
fumarolic stage 1168
fume, volcanic – 1349
fume, volcanic – 1377
function, keying – 4504
funnel sink 2072
furrow 1474
furrow 1538
furrow and ridge, longitudinal – 3028
fusain 3580
fusinite 3574
fusion crust 0073
fusion, differential – 1029
fusion, selective – 1029

G

gabbroid 0812
gage, strain – 4767
Gaia theory 0395
gain control, automatic – 4425
gain control, automatic – 4508
gain control, time-varied – 4508
gain ranging 4426
gal 0324
gallery 4847
galliard 3518
Galton distribution 4189
gamma 0318
gangue 3921
gangue mineral 3921
ganister 3518
gap 1556
gap 2796
gap, air – 2327
gap, stratigraphic – 1552
gap, water – 2327
gap, wind – 2327
gas 3713
gas, associated – 3891
gas-barren ground 1378
gas cap 3774
gas content 0650
gas, detection of – 4520
gas, dissolved– 0641
gas, dissolved – 1172
gas-emission phase 1162
gas field 3773
gas film 1370
gas fluxing 1178
gash 1512
gash, tension – 1519
gasification, coal – 3609
gas initially in place 3859
gas injection 3835
gas in solution 1172
gas, juvenile – 1348
gas, lean – 3715
gas, liquefied natural – 3746
gas-liquid ratio 3802
gas liquids, natural – 3746
gas, magmatic – 1348
gas, marsh – 3529
gas, natural – 3713
gas, non-associated – 3891
gas oil 3744
gas/oil contact 3774
gas/oil interface 3774
gas-oil ratio 3802
gasoline 3744
gas phase 1162
gas phase, intermediate – 1162
gas, phreatic - 1351
gas pit 3066
gas pressure – 1173
gas pressure, conservation of – 3836
gas, recycling of – 3835
gas, resurgent magmatic – 1172
gas, rich – 3715
gas sand 3766
gas sandstone 3766
gas seepage 1401
gassi 2186
gas skin 1370
gas-source rock 3722
gas, streaming – 0676
gas, substitute natural – 3609
gas survey 4648
gastrolith 1864
gas, volcanic – 1349
gas, wet – 3715
gauge, pressure – 4769
gauge, strain – 4767
gauging, well – 3798
gauss 0318
geanticline 1099
Geiger counter 4546
Geiger-Müller counter 4546
gelifluction 2042
gelisol 2635
gelivity 1996
gemmology 0478
gem stone 0478
generation 0591
generation 2573
generation, mineral – 3906
generitype 1948
genetic 4630
genotype 1979
gens 1934
gentle 1626
genus 1933
genus, form – 1932
genus, type – 1947
genus zone 1778
geobarometry 0994
geobios 1802
geocentric latitude 0098
geochemical thermometer 0379

geochemistry 0395
geochemistry, stable isotope – 0453
geochronology 1713
geochronology, isotope – 0428
geode 3481
geodepression 1120
geodesy 0013
geodynamics 0080
geognosy 0001
geogony 0003
geography, physical – 0012
geohydrology 3611
geoid 0328
geoisotherm 0377
geoisothermic plane 0377
geologist 0001
geology 0001
geology, applied – 0016
geology, applied – 4661
geology, catastrophist – 0005
geology, dynamical – 0010
geology, economic – 0016
geology, engineering – 4660
geology, environmental – 4665
geology, experimental – 0014
geology, field – 4226
geology, historical – 0011
geology, marine – 4226
geology, mathematical – 0015
geology, physical – 0010
geology, stable-isotope – 0453
geology, structural – 1478
geology, submarine – 4226
geology, subsurface – 4226
geology, underground – 4226
geomagnetic field 0276
geomagnetic latitude 0281
geomagnetic reference field, international – 4535
geomagnetism 0276
geometry, hydraulic – 2263
geomorphology 1987
geomorphology, engineering – 4663
geonomy 0079
geopetal 3107
geopetal structure 2996
geophagous 1857
geophone 0118
geophone 4414
geophysics 0078
geophysics, engineering – 4663
geopiezometry 0994
geopressure 3820
geostatistics 4193
geosyncline 1094
geotechnics 4662
geotechnique 4662
geotectonics 1073
geotherm 0378
geothermal 0376
geothermal energy 0388
geothermal field 0392
geothermal gradient 0380
geothermal step 0380
geothermics 0374
geothermometer 0379
geothermometer 0994
geothermometry 0379
geothermometry 0994
geotumor 1120
germanotype 1078
gerontic 1866
geyser 1404
geyser basin 1404
geyser crater 1405
geyserite 1406
geyser jet 1405
geyser, mud – 1398
geyser pipe 1405
geyser shaft 1405
geyser, siliceous – 1406
ghost 0802
ghost 4442
ghourd 2179
ghyll 2104
giant's causeway 1301
gibbers 4123
gigantism 1867
GIIP 3859
gilgai 2763
gilsonite 3738
girdle 0501
girdle 1019
girdle axis 1020
gizzard stone 1864
gjá 1234
glacial 2873
glacial erratic 2949
glaciation 2367
glacier 2366
glacier, alpine-type – 2428
glacier burst 1357
glacier, carrying – 2385
glacier cave 2391
glacier, cirque – 2430
glacier, cold – 2368
glacier, continental – 2425
glacier, dendritic – 2432
glacier, foot – 2435
glacier, hanging – 2433
glacier, Himalayan-type – 2432
glacierit 2433
glacier, Malaspina-type – 2435
glacier meal 2875
glacier milk 2392
glacier mill 2389
glacier, mountain – 2426
glacier, Mustag-type – 2431
glacier, piedmont – 2435
glacier, plateau – 2426
glacier, Pyrenean-type – 2430
glacier, reconstructed – 2434
glacier, remanié – 2434
glacier, riding – 2385
glacier, salt – 1699
glacier snout 2390
glacier stairway 2622
glacier table 2386
glacier, temperate – 2368
glacier terminus 2390
glacier, tidal – 2436
glacier, tide – 2436
glacier, tidewater – 2436
glacier tongue 2390
glacier, trunk – 2432
glacier, Turkestan-type – 2429
glacier, valley – 2427
glacier, wall-sided – 2433
glacier, warm – 2368
glaciofluvial 2883
glaciology 2345
glaciomarine 2903
glacis 2582
glare 0577
glare, primary – 0577
glare, secondary – 0577
glass 0627
glass, basalt – 1262
glass, volcanic – 1262
glassy 0484
glassy 0805
glauconitic 3177
glauconitization 3459
gleysation 2684
glide plane 0539
gliding, gravitational – 1703
gliding plane 1705
gliding tectonics 1702
global tectonics new, – 1125
globe lightning 1372
globosphaerite 0765
globulite 0765
globulith 0731
glomerocryst 0771
glomerophyric 0821
glomeroporphyritic 0821
gloss 3407
GLR 3802
gneiss 0936
gneiss, composite – 1044
gneiss dome, mantled – 0890
gneissic structure 0951
gneiss, mixed – 1046
gneissoid 0871
gneissose structure 0951
gneissosity 0951
gneiss, pencil – 0936
gneiss, permeation – 1059
gneiss, primary– 0631
gneiss, veined – 1044
GNT 4594
GOC 3774
gooseneck 4268
GOR 3802
gossan 4125
gouge 4012
gouge 4729
gouge, clay – 1540
gouge, fault – 1540
gour 2147
graben 1590
graben 2546
graben, explosion – 1235
graben, volcanic sector – 1464
gradation 4343
gradation 4702
grade, cut-off – 4203
graded bedding 3130
graded, poorly – 4702
graded, well– 4702
grade line, hydraulic – 4736
grade scale 3161
grade, weighted average – 4185
gradient 4827
gradient, geothermal – 0380
gradient, hydraulic – 4736
gradient of the bed 1488
gradient, reversed – 2626
grading 0494
grading 3130
grading, reversed – 3130
gradiometer, magnetic – 4525
grahamite 3738
grain 0504
grain 3078
grain boundary 0532
grained, coarse– 3194
grained, even– 0807
grained, fine– 3194
grained, medium– 3194
grain growth 3454
grain interpenetration 3448
grain size 0792
grain size 2710
grain size 3193
grain-size distribution 2709
grain-size frequency distribution 3197
grain-size grade 3162
grain size, median – 3199
grainstone 3297
grain supported 3294
granide 0637
granite, central – 0729
granite, graphic – 0847
granite tectonics 0712
granitic 0810
granitification 1062
granitisation 1062
granitization 1062
granitoid 0810
granofels 0935
granolite 0810
granophyric 0848
granoschistose 0953
granospherite 0779
granular 0806
granular 2727
granularity 0792
granularity 3081
granularity 4344
granulation 1990
granulation 2934
granule 3240
granulite 0937
granulitic 0954
granuloblastic 0960
granulometry 3160
granulose 0960
granulyte 0937
grape formation 2136
grapestone 3302
graph, area-depth – 3872
graphic 0847
graphiphyre 0848
graphophyre 0848
gravel 3244
gravel bar 2244
gravel, boulder – 3249
gravel, granule – 3249
gravel, lag – 2160
gravel, lag – 2656
gravel, lag – 2967
gravelly 3244
gravel, volcanic – 1324
gravimeter 0352
gravimeter, borehole – 4602
gravimeter, null-reading type of – 0352
gravimetry 0349
gravitation 0322
gravitational constant 0325
gravitational tectonics – 1702
gravity 0324
gravity, acceleration of – 0324
gravity determination 0349
gravity field 0326
gravity, force of – 0323
gravity, free air – 4538
gravity, gradient of – 0335
gravity, international formula– 0333
gravity, map, regional – 0371
gravity map, residual – 4542
gravity measurement 0349
gravity meter 0352
gravity, normal value of – 0332
gravity, observed – 4536
gravity potential 0327
gravity station 0353
gravity unit 4537
Gray-King assay 3596
graywacke 3234
great group 2665
greensand 3231
greenschist 0938
greenstone 0938
greenstone belt 1082
gregaritic 0822
greisen 0924
greisen 4134
greisening 0924
greisenization 0924
Grenzhorizont 3544
greywacke 3234
gridding 4657
grid method 4374
grid, parallactic – 4373
grid, perspective – 4373
grike 2087
grinding 2940
grinding, autogenous – 4147
grinding circuit 4146
grit 3229
grizzly 4144
groove 1538

groove, glacial – 2398
groove mark, ruffled – 3035
grooving marks 1270
grotto 2095
ground 4788
ground control 4323
ground data 4321
ground factor 0259
ground, frost – 2635
ground, frozen – 2635
ground, gas-barren – 1378
ground, hard – 3126
ground information 4321
groundmass 0626
groundmass 3080
ground, patterned – 2652
ground, polygonal – 2653
ground preparation 3940
ground range 4393
ground-range image 4395
ground return 4390
ground, running – 4855
ground, squeezing – 4855
ground, striated – 2653
ground, striped – 2653
ground, swelling – 4855
ground truth 4321
ground vein 2641
ground water 3658
ground-water artery 3638
ground-water balance 3633
ground water, confined – 3659
ground-water decrement 3633
ground-water deposit 3998
ground-water divide 3647
ground-water encroachment, velocity of – 3703
ground-water increment 3633
ground water, phreatic – 2765
ground-water storeys 3641
ground-water table 2765
group 1755
group 1928
group-velocity method 0229
grout 4803
grout curtain 4804
grouthole 4803
grouting 4803
growth hillock 0528
growth, horizontal – 2513
growth, incipient stage of – 1666
growth line, curved – 0479
growth, outward – 2513
growth pyramid 0526
growth sector 0526
growth spiral 0529
growth, upward – 2513
guano 2878
guest 0844
guide fossil 1785
gulch 2026
gully 0042
gully 2026
gully 2474
gully 2509
gullying 2018
gunite 4801
gusher 3790
Gutenberg discontinuity 0092
Gutenberg-Oldham discontinuity 0092
guyot 2798
gypsification 3459
gypsite 3381
gypsum 3381
gypsum flower 2151
gyrogonite 1843
gyrolith 1843
gyttja 3528

H

habit 0521
habit 0592
habitat 1901
hadal 2848
hadal 2924
hade 1548
hade 4018
hailstone impression 3041
hailstone imprint 3041
half-life 0431
haline 0805
hälleflinta 0939
halmyrolysis 3476
halo 4629
halo, alteration – 4630
halocline 2830
halokinesis 1694
halolimnic 1915
halophyte 1846
halo, wall-rock – 4630
hamada 2578
hardening, case – 3411
hardness 0473
hardness 4701
hardness, total – 3677
hardpan 2733
hardpan 4121
Harker variation diagram 0697
harmonic 0167
hartschiefer 1543
hat, iron – 4125
Hawaiian-type eruption 1158
Hayford zone 0362
haystack hill 2074
HDT-dipmeter tool 4598
head 1686
head 2446
head, hydraulic – 4735
heading 4852
heading effect 4530
headland 2446
headland, winged – 2446
head test, constant – 3711
head test, falling – 3711
headwater basin 2276
headwater region 2193
hearth, central – 1202
heat capacity 0386
heat capacity, specific – 0386
heat content 0389
heat-current density 0384
heat-flow rate 0385
heat flow, terrestrial – 0383
heat-flow unit 0385
heat, mass transfer of – 0381
heat, radioactive – 0375
heat, radiogenic – 0375
heave 1551
heave 4020
heave 4830
heave 4511
heave, apparent – 1553
heave, stratigraphic – 1552
heaving, frost – 2642
heaving, frost – 4721
helicitic 0955
helictite 2132
helluhraun 1252
HEM 4575
hemicrystalline 0804
hemicyclothem 3120
hemidiatreme 1206
hemipelagic mud 2920
hervidero 1395
heteradcumulate 0833
heteroblastic 0956
heterochronous 1734
heterokinetic 1050
heteromorphic 0635
heteromorphism 0635
heterospory 1834
heterotactic 1007
heterothrausmatic 0867
heterotopical 1731
hexagonal 0515
hexahedrite 0068
HFU 0385
hiatal 0820
hiatus 1723
hierarchy 1925
high-energy environment 2881
Hilt's law 3522
hinge line 1629
hinge point 1629
hinge zone 1629
hinterland 1110
histogram 4188
history matching 3866
HMS 4153
hoar, depth – 2351
hoarfrost formation 2143
hodochrone 0171
hodograph 0171
hogback 0741
hogback 2562
hole 2247
hole 4261
hole, barefoot – 3795
hole, blow – 1274
hole, churn – 2248
hole, crooked – 4293
hole, cryoconite – 2387
hole, deviated – 4293
hole, dry – 3794
hole, eddy – 2248
hole, kettle – 2632
hole, light – 2076
hole, lost – 4287
hole, open – 4277
hole, pilot 4297
hole, sidetracked – 4291
hole, slim – 3783
hole, swallow – 2075
hole, wet – 3794
hollow 1410
hollow 2095
hollow 3539
holoblast 0984
holocrystalline 0803
holohyaline 0805
hololeucocratic 0587
holomarine 2900
holosiderite 0066
holostratotype 1740
holotype 1949
homeoblastic 0960
homeocrystalline 0807
homeomorphism 1820
homeomorphy 1820
homeothrausmatic 0867
homogeneous 0716
homogeneous 4680
homogeneous area 4787
homogeneous zone 4787
homoiohaline 2831
homologous 1972
homomelanocratic 0588
homomorphy 1820
homonym 1955
homoplastic 1980
homoseism 0210
homoseismal line 0210
homospore 1835
homospory 1834
homotactic 1007
homotaxial 1733
homotype 1954
honeycomb 2001
hook 2464
hopper crystal 3382
horizon 1749
horizon 4461
horizon, A– 2698
horizon, A1– 2698
horizon, A2– 2700
horizon, Ap– 2699
horizon, B– 2702
horizon, bleached – 2700
horizon, buried – 2693
horizon, C– 2706
horizon, Cg– 2707
horizon, chronostratigraphical – 1758
horizon, colour-B– 2703
horizon, Cr– 2707
horizon, E– 2700
horizon, eluvial – 2701
horizon, illuvial – 2702
horizon, key – 4249
horizon, lithostratigraphical – 1749
horizon, recurrence – 3544
horizon, soil – 2687
horizontal-loop method 4575
horizon, textural-B– 2703
hornfels 0940
horn, glacial – 2618
hornito 1298
horse 4072
horseback 3503
horst 1589
Hoskold formula 4217
host 0844
hot-dry-rock system 0394
hot spot 1144
hot-water system 0393
hourglass structure 0534
hue 2677
huff-and-puff method 3839
hum 2074
humification 2683
humification 3525
huminite 3568
hummock 3539
hummock, earth – 2643
hummock, peat – 2643
hummock, tundra – 2643
humocoll 3529
humocollinite 3568
humodetrinite 3568
humolite 3497
humotelinite 3568
hump 2400
humus 2743
humus 3525
humus, mild – 2744
humus, raw – 2744
hver 1407
hx 4655
hyalinocrystalline 0823
hyaloclastite 1347
hyalocrystalline 0804
hyalomylonite 0916
hyalo-ophitic 0840
hyalopilitic 0837
hybrid 0681
hybridism 0681
hybridization 0681
hydatogenesis 3958
hydatomorphism 3958
hydatopneumatogenesis 3958
hydatopyrogenesis 3958
hydatopyrogenic 0656
hydration 0419
hydraulic head 4735
hydraulic regime 2267
hydraulic theory 2263
hydrite 3590
hydrobios 1802
hydrocarbon column 3775
hydrocarbon column height 3775
hydrocarbon, heavy – 3745
hydrocarbon saturation 4610
hydrochemical 4634

hydroexplosion 1351
hydrofrac 3800
hydrogen index 4613
hydrogeochemistry 3611
hydrogeology 3611
hydrology 3610
hydrolysis 0418
hydrometamorphosis 3954
hydrometasomatism 3954
hydromorphic 4634
hydromorphic characteristic 2690
hydrophone 4415
hydrophone 4496
hydrophone response 4497
hydrophone string 4496
hydroplutonic 0656
hydrosphere 0085
hydrostatic uplift 4740
hydrothermal, low – 3983
hydrothermal process 3953
hydrothermal, secondary – 3994
hydrothermal stage 0654
hypabyssal 0617
hypabyssal phase 1197
hypautomorphic 0751
hypautomorphic-granular 0808
hyperbole method 0152
hypercyclothem 3120
hyperite texture 0853
hypermelanic 0588
hypermelanic 0599
hypersaline 2835
hypersaline 3377
hyperstereoscopy 4367
hyperthermal region 0391
hypidioblast 0980
hypidiomorphic 0751
hypidiomorphic-granular 0808
hypidiomorphic-granular 0954
hypidiotopic 3471
hypoabyssal 3979
hypobatholitic 3990
hypocenter 0156
hypocentre 0156
hypocrystalline 0804
hypogene 0617
hypogene 3937
hypogene 3967
hypohyaline 0804
hypolimnion 2334
hypomagma 1199
hypostratotype 1746
hypothermal 3983
hysteresis 4675
hysteresis, magnetic – 0305
hysterogenous 3960

I

ice, anchor – 2639
ice cap 2425
ice carapace 2425
ice crowding 2401
ice crust 2352
ice-crystal imprint 3043
ice, dead – 2369
ice, efflorescent – 2639
icefall 2377
ice, firn – 2363
ice, fossil – 2369
ice, ground – 2639
ice, inland – 2425
ice mantle 2425
ice, needle – 2639
ice push 2401
ice-pushed ridge 2419
ice pyramid 2377
ice-rafted 2969
ice rafting 2969
ice-scour notch 2395
ice shed 2381
ice sheet 2425
ice sheet, continental – 2425
ice shove 2401
ice, stagnant – 2369
ice thrust 2401
ice wedge 2640
ichnocenose 1874
ichnocenosis 1874
ichnocoenose 1874
ichnofossil 1873
ichnofossil 3067
ichnology 1871
ichor 1061
identification capability 0271
identification threshold 0271
idioblast 0980
idioblastic series 0909
idiochromatic 0483
idiomorph 0750
idiomorphic 0523
idiomorphic 0750
idiomorphic-granular 0808
idiotopic 3471
idogenous 3966
igneous 0616
igneous province 0628
ignimbrite 1340
IGRF 4535
illam 0499
illite 3214
illumination, conoscopic – 0547
illumination, orthoscopic – 0547
illumination, specific – 4311
illumination, total – 4311
illuvial horizon 2702
illuviation 2678
illuviation 3930
image 4320
image-motion compensator 4352
imagery 4320
imatra stone 3486
imbibition 1060
imbrication 1571
imbrication 3040
imitation stone 0481
immature 3176
immature 3725
impact 0026
impact crater 0027
impactite 0076
impact theory 0049
impedance, acoustic – 4474
impetus 0173
impounding 2340
impregnation 3396
impregnation 3955
imprint 0970
imprint 1826
imprint, ice-crystal – 3043
impsonite 3739
impulse response 4446
impulsive 0173
incidence, angle of – 0211
incidence, grazing – 0212
incidence, normal – 4472
incidence, plane of – 0211
incision 2020
inclination, angle of – 1485
inclination, magnetic – 0283
inclinometer 4294
inclusion 0662
inclusion 0789
inclusion, accidental– 0663
inclusion, cognate– 0664
inclusion, endogenous– 0664
inclusion, exogenous– 0663
inclusion, noncognate– 0663
inclusion, surmicaceous – 1072
incompetent 1624
incompressibility, modulus of – 4698
incrop 4231
incrustation 3387
indentation 2450
index fossil 1785
indicator element 4649
indicator, geobotanical – 4646
indicator plant 4646
indicatrix 0552
indigenous 1813
indigenous 3968
indigenous 3969
indirect effect 0366
induction, magnetic – 0317
induration 3423
inequigranular 0807
inertinite 3573
inertite 3582
infauna 1853
infiltration 2768
infiltration 3396
infiltration 3619
infiltration 3955
infiltration capacity 3620
infiltration rate 3620
inflection line 1627
inflection point 1627
inflexion line 1627
inflexion point 1627
infracrustal 0584
infralittoral 2839
infraneritic 2841
infrared, far – 4309
infrared, middle – 4308
infrared, near – 4307
infrared, photographic – 4307
infrared, solar – 4307
infrared, thermal – 4308
infraspecific 1944
infrastructure 1077
ingredient, banded – 3579
inherent 3553
inherited 3971
inhomogeneous 4680
initial 1108
initial 4428
initiator 4816
injected body 0715
injected mass 0715
injection 0618
injection 3949
injection 4802
injection breccia 0711
injection complex 0620
injection, satellitic – 1203
injection, sedimentary – 3070
injection, sole – 0738
injectivity index 3809
inlet 2338
inlet, tidal – 2508
inlier 4233
inner arc, volcanic – 1137
input 4445
Input System 4576
inselberg 2608
insequent 2287
inset 0758
inshore 2840
insolation, destruction by – 1992
inspection, on-site – 0267
instant 1759
insular slope 2791
integration 0910
intensity scale 0243
intensity, seismic – 0243
interbedded 3116
interbiohorizon zone 1772
interbiohorizon zone 1776
intercalated 3117
intercalation 3117
interception 3615
intercept time 4463
interclast 2945
intercumulus 0831
intercumulus liquid 0831
intercumulus material 0831
interface 3096
interface 4461
interference colour 0568
interference colour, anomalous – 0568
interference figure 0569
interfingering 3118
interflow 3623
interfluve 2312
intergranular 0841
intergranular 0954
intergranular 0998
intergranular film 2371
intergrowth 0783
intergrowth 4098
intergrowth, atoll – 4102
intergrowth, buckshot – 4101
intergrowth, core – 4102
intergrowth, disseminated – 4101
intergrowth, eutectic – 4100
intergrowth, graphic – 0847
intergrowth, graphic – 4100
intergrowth, index of – 4112
intergrowth, mantled – 4102
intergrowth, myrmekitic – 4100
intergrowth, peppered – 4101
intergrowth, ring – 4102
intergrowth, sandwich – 4103
intergrowth, shell – 4102
intergrowth, simple – 4099
intergrowth, vein – 4103
interjection 0717
interkinematic 1079
interlayering 3116
inter-limb angle 1626
interlocking 4098
interlocking texture 0816
intermediate 0602
interpenetration, grain – 3448
interprecipitate liquid 0831
interprecipitate material 0831
interpretation, direct – 4566
interpretation, indirect – 4566
interpretation, quantitative – 4566
interpretation, quick-look – 4604
interreflection 4519
intersecting 2596
intersection, vein – 4074
intersertal 0840
interspacing 3843
interstice 3083
interstice 3683
interstice, primary – 3684
interstice, secondary – 3685
interstitial 3083
interstratification 3116
interstratified 3116
interstream area 2312
intertidal zone 2836
intertodetrinite 3574
intertonguing 3118
interval 1718
interval zone 1772
interval zone, biostratigraphical – 1776
interzone, barren – 1775
intraclast 3272
intragranular 0998
intramagmatic 3985
intramicarenite 3284
intramicrite 3284
intramicrudite 3284
intramontane basin 1112
intraspararenite 3283
intrasparite 3283
intrasparrudite 3283

intrazone, barren – 1775
intrusion 0618
intrusion 0715
intrusion, bell-jar – 0734
intrusion breccia 0710
intrusion, composite – 0721
intrusion, multiple– 0721
intrusion, oil – 3752
intrusion, sedimentary – 3070
intrusive 0619
intrusive 0715
intrusive body 0715
intrusive, composite – 0721
intrusive, compound – 0722
intrusive mountain 0729
intrusive, multiple – 0721
intrusive phase 1197
intrusive rock 0715
intrusive, simple – 0720
intrusive vein 0748
intumescence of lava 1272
invasion 0618
invasion 4619
invasive 0619
inversion 1088
inversion, relief – 2563
invert 4852
inverted 1491
involution 1690
involution 2648
ion exchange 0420
ion-exchange capacity 0420
ionium-thorium dating 0444
ion potential 0419
IP log 4581
IP method 4569
iridescence 0486
iron hat 4125
iron meteorite 0066
iron pan 2734
irons 0066
ironstone 3464
ironstone 4115
ironstone 4124
ironstone 4125
ironstone, banded – 4114
ironstone, blackband – 3504
ironstone, blackband – 4117
ironstone cap 4124
ironstone, cherty – 3373
ironstone, clay – 3488
ironstone, clay – 3504
ironstone, clay– 4117
irradiance 4312
irradiancy 4312
irradiation 0492
irruption 0618
irruptive 0619
irruptive 0715
isanomalic map 0372
isbrae 2366
island 2443
island arc 1135
island, attached – 2534
island, continental – 2444
island, detached – 2534
island formed by aggradation 2535
island, land-tied – 2534
island, marsh – 2507
island, oceanic – 2444
island, rock – 2227
island shelf 2791
island, upheaved – 2535
island, uplifted – 2535
island, volcanic – 1416
isocal 3523
isocarb 3523
isochore 4244
isochromatic curve 0571
isochron 0210
isochron 0433
isoclinal 1626
isoclinic line 0283
isodynamic line 0285
isofacial 0992
isogal map 0372
isogam 4532
isogam map 0372
isogonic line 0284
isograd 0992
isogradal 0992
isograde 0992
isograd, geochemical – 4638
isogyre 0570
isohypse 4248
isohyps, water-table – 3644
isolation 1981
isomesical 1735
isometric 0515
isometric 0522
isomorph 1974
isomorphism 0464
isopach 4243
isophysical 0992
isopic 2863
isoporic line 0287
isoseism 0245
isoseismal 0245
isoseismal, innermost – 0246
isoseismal line 0245
isoseismic line 0245
isoseist 0245
isospore 1835
isospory 1834
isostasy 0338
isostatic isocorrection line map 0373
isotherm 0377
isothrausmatic 0867
isotime curve 0210
isotope dilution analysis 0432
isotope fractionation curve 0455
isotope fractionation factor 0455
isotope fractionation, kinetic – 0457
isotopic dating 0429
isotropic 4681
isotropic, optically – 0561
isotropy 4681
isotypism 0463
isovol 3523
itabirite 4114

J

jack test, flat – 4763
jack test, radial – 4763
jade 0498
jama 2076
jasper 3492
jasper 4127
jaspilite 3492
jaspilite, banded – 4114
jet 3557
jig 4155
Johnson test 3787
joint 1495
joint 4723
jointage 1495
joint aperture 4728
joint, bedding – 3099
joint, cross – 1506
joint, diagonal – 1507
joint, dip – 1506
jointed 1495
joint, extension – 1503
joint face 1496
joint, feather – 1505
joint filling 4729
joint, flow – 0714
joint frequency 4725
jointing 1495
jointing, columnar – 1501
jointing, cylindrical – 1302
jointing intensity 4725
jointing, prismatic – 1501
jointing, rhomboidal – 1502
jointing, sheet – 1500
jointing, slab – 1500
jointing, spheroidal – 1304
jointing, topographic – 1995
joint, longitudinal – 1507
joint, main – 1509
joint, major – 1509
joint, master – 1509
joint, minor – 1509
joint, non-systematic – 1510
joint, oblique – 1507
joint, open – 1513
joint opening 4728
joint pattern 1508
joint, pinnate shear – 1505
joint, release – 1506
joint roughness 4727
joint, secondary – 1509
joint set 1498
joint set 4724
joint, shear – 1504
joint spacing 4725
joints, system of conjugate – 1499
joint, strike – 1506
joint surface 1496
joint system 1498
joint system 4724
joint, systematic – 1510
joint system, conjugate – 1499
joint, tension – 1503
joint, transverse – 1507
joint waviness 4727
jökulhlaup 1357
jökull 2366
jökull 2425
jumbo 4266
junction 2208
junction 4073
junction, accordant – 2208
junction, barbed – 2209
junction, deferred – 2209
junction, discordant – 2208
junction, shifted – 2209
junction, simple – 4073
Jurassian relief 2565

K

kainotype 0636
kaksa 4000
kaksa 4136
Kalb line 0559
kame 2420
kame terrace 2420
kankar 4123
kaolin 3217
kaolin-coal tonstein 3505
kaolinite 3214
kaolinization 0926
kaolization 0926
karang 4000
karang 4128
karren 2086
Karrenfeld 2088
karst 2064
karst bridge 2079
karst, cockpit – 2080
karst, cone – 2081
karst, covered – 2069
karst, deep – 2067
karst depression 2071
karst erosion, base level of – 2066
karst forming 2063
karst funnel 2072
karstification 2063
karst landscape 2064
karst, naked – 2068
karst phenomena 2065
karst pit 2076
karst region 2064
karst, shallow – 2067
karst spring 2089
karst, subterranean – 2070
karst, tower – 2082
karst window 2078
katagenesis 1977
katagenesis 3419
kataseism 0187
katathermal 3983
katazone 0993
kelly 4269
kelyphite 0786
kelyphitic border 0786
kelyphitic rim 0786
kerabitumen 3719
kerogen 3720
kerogenite 3720
kerosine 3744
kettle 2632
kettle, giant's – 2389
key bed 1751
key fossil 1785
kick, first – 4428
kidney, ore – 4032
kink 1661
kink band 1661
kipuka 1289
Klinkenberg factor 3706
klippe 1692
knick 2199
knitted 0964
knob and kettle topography 2630
knoll 2800
knotted 0965
Koenigsberger ratio 0310
kolong 4000
kong 4000
kong 4128
koum 2578
kriging 4197
krikil 4124
krotovina 2745
kulit 3999
kulit 4136
Kupferschiefer 4130
kuroko 4131
kurtosis 3198
kyriosome 1039

L

labradorescence 0486
laccolite 0734
laccolith 0734
laccolithic dome 1206
lacquer peel 4230
lacustrine 2888
lag concentrate 2967
lagg 3537
lag gravel 2656
lag gravel 2967
lagoon 2458
lagoon 3345
lagoonal 2905
lagoon channel 3348
lagoon flat 3346
lagoon floor 3345
lagoon shelf 3346
lagoon, tidal – 2458
laguna 3345
lagune 3345
lahar 1352
lahar, cold – 1355
lahar deposit 1356

lahar, hot – 1353
lahar, outbreak – 1354
lahar, rain – 1355
lake 2330
lake ball 2891
lake, cave-in – 2646
lake, crater – 1454
lake, explosion – 1433
lake, ice-marginal – 2393
lake, lava – 1281
lake, mort – 2232
lake, oxbow – 2232
lake, paternoster – 2620
lake platform 2337
lake, playa – 2343
lake, proglacial – 2393
lake, relic – 2344
lake, residual – 2344
lake rim 2335
lake, rock-basin – 2620
lake, salt – 2893
lake, thaw – 2646
lake, thermokarst – 2646
lake, tidal – 2506
lake, volcanic barrier – 1419
lake without outlet 2338
Lamarckianism 1967
Lamarckism 1967
lamella 2705
lamella, deformation – 0999
lamellae 4103
lamellae, shock – 0894
lamellar 0775
lamellate 0775
lamella, textural subsoil – 2705
lamina 3101
lamina, backset – 3141
lamina, foreset – 3140
laminated 3101
laminated terrain 0040
lamination 3101
lamination 2374
lamination 1277
lamination, climbing ripple – 3149
lamination, convolute – 3052
lamination, cross – 3133
lamination, foreset – 3140
lamination, igneous – 0868
lamination, ripple – 3147
lamination, ripple-drift cross – 3149
lamination, wavy – 3138
lamprophyric 0809
landfill, sanitary – 4831
landform, structural – 2554
landform, tectonic – 2537
landscape, morvan – 2596
landscape, storeyed – 2599
landslide 2044
landslide, cliff – 2473
landslide scar 2052
landslide slip 2044
landslide topography 2056
landslide track 2054
landslip, cliff – 2473
land unit 4785
lap, forward – 4359
lapidary 0502
lapilli 1324
lapilli, accretionary – 1321
lapilli, crystal – 1325
lapilli, filiform – 1326
lapilli, porous – 1324
lapilli, scoriaceous – 1324
lap, side – 4359
Larsen variation diagram 0697
late kinematic 1079
late orogenic 1108
laterite 4124
lateritization 2681
late-volcanic phase 1197
lath-like 0522
latitude correction 0355
latitude, geocentric – 0098
latitude, geomagnetic– 0281
lattice 0964
lattice 4103
lattice, crystal – 0507
lattice orientation 1013
lattice plane 0508
lattice point 0508
lattice row 0508
laug 1407
lava 1242
lava, aa – 1260
lava, agglomerate – 1269
lava, aphrolithic – 1260
lava, autobrecciated – 1248
lava ball 1296
lava ball, accretionary – 1296
lava bed 1277
lava, block – 1259
lava, breccia – 1345
lava, capillary – 1326
lava cascade 1287
lava cauldron 1284
lava cave 1291
lava cavern 1291
lava clot 1297
lava colonnade 1301
lava column 1246
lava cone 1427
lava, congealed – 1242
lava, corded – 1252
lava cupula 1422
lava delta 1280
lava desert 1278
lava dome 1422
lava dome with tongue-like offshoot 1423
lava eruption 1214
lava, farinaceous – 1263
lava field 1278
lava flood 1276
lava flow 1263
lava flow, clastogenetic – 1282
lava flow, effluent – 1266
lava flow, interfluent – 1267
lava flow, parasitic – 1265
lava flow, rootless – 1282
lava flow, superfluent – 1266
lava, fluent – 1261
lava, foamy – 1307
lava fountain 1282
lava, grooved – 1270
lava lake 1281
lava levee 1295
lava, lobe of – 1264
lava, mud – 1355
lava, mud – 1397
lava nodule, basaltic – 1256
lava of tuff agglomerate 1346
lava, pahoehoe – 1250
lava pellet 1324
lava, pillow – 1253
lava pit 1284
lava plain 1278
lava plateau 1278
lava plug 1477
lava, pumiceous – 1307
lava rag 1297
lava ring 1284
lava river 1263
lava, ropy – 1252
lava scarp 1286
lava scratches 1270
lava sheet 1275
lava sheeth 1292
lava, solidified – 1242
lava stalactite 1293
lava stalagmite 1293
lava stream 1263
lava, thread-lace – 1307
lava toadstool, squeezed-up – 1288
lava toes 1251
lava tongue 1264
lava tower 1299
lava tree-cast 1292
lava tree-mold 1292
lava tube 1290
lava, tuff – 1341
lava tunnel 1290
lava upheaval 1272
laydown 4380
layer 3097
layer, active – 2647
layer, annual – 3127
layer, C– 2706
layer, D– 2708
layered 3095
layered 4103
layer, gravelly – 2651
layering 3095
layering, annual – 3127
layering, cryptic – 0865
layering, differentiated – 1606
layering, flow – 0868
layering, harrisitic – 0863
layering, rhytmic – 0863
layering, Willow-Lake type – 0863
layering, zebra – 0863
layer, mineral-graded – 0866
layer, ploughed – 2699
layer, plowed – 2699
layer, soil – 2688
layer, suprapermafrost – 2647
layer, tilled – 2699
layover 4399
leaching 2678
leaching 3939
leaching 4162
leaching, bacterial – 3929
leaching, electrochemical – 3929
leaching, organometallic – 3929
lead 4038
lead, anomalous – 0441
lead, B-type – 4132
lead, B-type – 0441
lead button 4184
lead, common – 0437
leader 4046
leading edge 1687
lead-isotope growth curves 0438
lead, J-type – 0441
lead, J-type – 4132
lead-lead dating 0434
lead-model age 0440
lead model, multi-stage – 0439
lead model, single-stage – 0439
lead, normal type – 4132
lead, primeval – 0437
lead, primordial – 0437
lead, radiogenic – 0437
leaf 3509
leakage 4634
leakage 4863
leap 4020
lebensspur 1875
lebensspur, surface – 1875
lectostratotype 1745
lectotype 1951
ledge 0527
ledge 4135
ledge, calcite – 2148
Lee array 4562
lee slope 2965
leeward 2964
legend 4240
legend, profile-type – 4240
length 1524
length of the equivalent pendulum 0126
lens 1752
lens 3102
lenticle 1752
lenticle 3102
lenticular 3102
lentil 1752
lentil 3102
lepidoblastic 0961
leptite 0937
leptothermal 3984
leptynolite 0940
leucocratic 0587
leucosome 1038
levee, lava – 1295
levee, natural – 2238
level, crowning – 2597
level, hydrostatic – 3646
leveling 2589
level, phreatic – 3642
level, piezometric – 3646
level, standing – 3666
level, summit – 2597
liberate 4150
liberation value 4112
lifting costs 3900
light-coloured 0587
light hole 2076
lightning, globe – 1372
light ray 0548
lignite 3495
lily pad 2141
liman 2457
limb 1625
limb, drawn-out middle – 1680
limb, floor – 1576
limb, inverted – 1647
limb, inverted – 1680
limb, lower – 1576
limb, middle – 1576
limb, normal – 1647
limb, normal – 1680
limb, oversteepened – 1647
limb, reversed – 1647
limb, reversed – 1680
limb, stretched-out middle – 1680
limb, upper – 1576
limeclast 3167
limeclast 3270
lime ice 2144
lime mud 3295
lime-mud mound 3325
lime stabilisation 4807
limestone 3305
limestone, argillaceous – 3221
limestone, calcarenitic – 3277
limestone, coral – 3310
limestone, dolimitic – 3316
limestone, lithographic – 3306
limestone, pelleted – 3307
limestone, reef – 3310
limestone, sandy – 3317
liming, soil – 4807
limnetic 2890
limnic 2888
limnobios 1802
limnology 2331
limnoplankton 1801
limonite 4126
lineage 1965
lineage zone 1779
linear parallel 0972
lineation 0973
lineation 1607
lineation 2999
lineation, crenulation – 1612
lineation, current – 3027
lineation, intersection – 1611
lineation, parting – 3027
lineation, secondary – 0973
lineation, stretching – 1613
line cutting 4657
line, isopachous – 4243

liner 3795
line-up 4460
lining 4858
lining, plastic – 1226
linophyre 0829
linophyric 0829
lipid 3723
lipotectite 1031
lipotexite 1031
liptinite 3570
liptite 3582
liptodetrinite 3572
liquation 0678
liquation 3939
liquefaction 3084
liquefaction 4718
liquefaction, coal – 3609
liquid-dominated system 0393
liquidity index 4716
liquid limit 4715
liquid limit, upper – 2749
liquid, magmatic – 3976
liquid, residual – 0647
liquor, residual – 0646
listric surface 1573
listvenitization 3962
lithic 3178
lithification 3421
lithocalcilutite 3274
lithodlolutite 3274
lithofacies 2861
lithogene 3971
lithogenesis 2853
lithohorizon 1749
lithoidal 0835
lithology 2853
lithophagous 1857
lithophysa 0780
lithophysal 0876
lithosphere 0088
lithostratigraphical horizon 1749
lithostratigraphical zone 1747
lithostratigraphic unit 1748
lithostratigraphy 1709
lithothamnion ridge 3361
lithotope 2862
lithotype 3579
lithozone 1747
lit-par-lit 1049
litter 2742
littoral 2836
littoral 2906
LN 4583
LNG 3746
load 2950
load, bed – 2952
load, bottom – 2952
load cast 3059
load cell 4770
load cell, anchor– 4772
load, dissolved – 2956
load, saltation – 2954
load, sediment – 2951
load, solution – 2956
load, stream – 2951
load structure 3059
load, suspended – 2955
load, suspension – 2955
load, traction – 2952
loam 2720
loam 3215
loam, cave – 2122
loam, residual – 2122
lobate scarp 0032
lobe, flood-plain – 2225
lobe, frontal – 1687
locality, type – 1946
local-range zone 1784
loch, sea – 2453
locking 4098
locking index 4112
lode 4038
lode, brangled – 4048
lode, bunchy – 4047
lode, champion – 4062
lode, comby – 4049
lode, complex – 4046
lode, composite – 4045
lode, dradge – 4048
lode, mother – 4062
lode, stringer – 4060
loess 2871
loess doll 3489
loess kindchen 3489
loess nodule 3489
loess puppet 3489
log, acoustic velocity – 4597
log, boring – 4299
log, caliper – 4603
log, cement-bond – 4276
log, continuous – 4597
log, continuous velocity – 4406
log, dipmeter – 4598
log, formation density – 4593
log, gamma-ray – 4590
log, gamma-ray spectral – 4591
logging device 4578
logging tool 4578
logging tool, borehole compensated 4597
logging tool, circumferential microsonic – 4596
logging tool, electromagnetic travel time – 4588
logging tool, gamma-ray – 4590
logging tool, magnetic vector – 4589
logging tool, microlaterolog – 4587
logging tool, nuclear magnetic resonance – 4596
logging tool, proximity – 4587
logging tools, focused resistivity – 4586
logging tools, induction – 4584
logging tools, neutron – 4594
logging tools, normal restistivity – 4583
logging tool, thermal decay-time – 4595
log, induced polarisation – 4581
log, induction – 4584
log, lateral resistivity – 4585
log, magnetic susceptibility – 4589
log, mechanical properties – 3814
log, neutron – 4594
log, neutron-life time – 4595
log, redox – 4580
log, resistivity – 4582
log, selective gamma-gamma – 4591
log, sonic – 4597
log, spectral – 4548
log, spectral – 4591
log, spectrometrical gamma-ray – 4591
log, spontaneous potential – 4579
log, temperature – 4601
longitudinal 2286
longitudinal 2289
long normal 4583
longshore current 2825
longulite 0762
loop 2465
loosening 4731
loosening zone 4731
lopolith 0735
Love wave 0189
low 2460
low 2486
low-energy environment 2881
lowland, inner – 2560
low-velocity channel 0100
low-velocity layer 0100
low-velocity zone 0100
LQ 0189
lubinite 2153
luminescence 0467
lump 3301
lump, irregular – 4744
lump, mud – 3072
lump, slag – 1297
luster 3407
lustre 0484
lutaceous 3203
lutite 3203
lutum 3203
L-wave 0188
lydian stone 3369
lydite 3369
lynchet 2062

M

maar 1433
macadam 4843
macaluba 1395
maceral 3566
maceral, active – 3578
maceral association 3581
maceral group 3566
maceral, perhydrous – 3577
maceral, subhydrous – 3577
maceral suite 3566
maceral variety 3567
maceration 3594
macrinite 3574
macrocrystalline 0797
macroevolution 1970
macrofabric 1005
macromeritic 0797
macropolyschematic 0794
macroscopic 0797
macroseism 0138
macroseismic region 0248
macrospheric 1868
macrospore 1837
macula 0649
maculose 0965
maculose rock 0941
mafelsic 0598
mafic 0599
mafic front 1066
mafite 0599
magma 0639
magma 1198
magma blister 0648
magma chamber 1202
magma column 1226
magma, crateral – 1244
magma, igneous – 1198
magma, internal molten – 1198
magma, latent – 0643
magma, parental – 0645
magma, pocket of – 1202
magma, primary – 0644
magma province 0628
magma, reborn – 0634
magma, rest – 0646
magma system 0394
magmatic 0616
magmatic differentiation 0667
magmatic evolution 0666
magmatic gas, resurgent – 1172
magmatic phase 1108
magmatic stoping 0623
magmatite 0616
magmatogenic 3973
magnesium front 1066
magnetic moment 0296
magnetism, rock – 0293
magnetization 0295
magnetization, anhysteretic remanent – 0314
magnetization, chemical – 0312
magnetization, depositional remanent – 0312
magnetization, intensity of – 4531
magnetization, remanent– 0306
magnetization, residual – 0306
magnetization, reversed – 0315
magnetization, saturation – 0297
magnetization, thermo-remanent – 0309
magnetization, viscous – 0313
magnetizing force 0317
magnetohydrodynamics 0278
magnetometer 0316
magnetometer, airborne – 4522
magnetometer, astatic – 0319
magnetometer, flux-gate – 4523
magnetometer, high-sensitivity – 4524
magnetometer, horizontal – 4521
magnetometer, nuclear – 4524
magnetometer, precession – 4524
magnetometer, proton – 4524
magnetometer, spinner – 0320
magnetometer, vertical – 4521
magnetopause 0277
magnetosheath 0277
magnetosphere 0277
magnetotelluric method 4553
magnification curve 0133
magnification, dynamic – 0131
magnification, static – 0131
magnitude 0220
magnitude scale 0220
magnitude, unified – 0220
malleable 0471
malloseismic region 0250
malpais 1279
mammillary 4094
manganese nodule 3477
mangrove marsh 2515
manifold 3798
mantle convection 1124
mantle, earth's – 0091
mantle, lower – 0091
mantle of ash 1320
mantle of scoriae – 1330
mantle plume 1145
mantle rock 2008
mantle, upper – 0091
manto 4027
map, detailed soil – 2671
map, engineering geological 4784
map, facies – 4245
map, geological – 4239
map, isanomalic – 0372
map, isobaric – 3854
map, isochore – 4244
map, isogal – 0372
map, isogam – 0372
map, isopach – 4243
map, isopachous – 4243
map, isopleth – 4246
map, metallogenic – 3905
map, palaeogeological – 4242
map, photo – 4383
mapping unit 2672
map, pressure – 3854
map, radar – 4400
map, reconnaissance soil – 2671
map, residual gravity – 4542
map, soil – 2671
map, structural – 4247
map, structural contour – 4247
map, subcrop – 4241
map, tectonic – 4247

map unit 2672
map, worm's-eye – 4241
marble 0920
mare 0051
margarite 0766
marginal deep 1111
marginal trough 1111
margin, chilled – 0707
marine 2900
marine, deep – 2918
marine, glacial – 2903
marine, normal – 2901
marine, restricted – 2901
marine, shallow – 2914
mark 3018
mark, bounce – 3038
mark, brush – 3039
mark, chevron – 3036
mark, collimating – 4368
mark, crinkle – 3056
mark, current – 3020
mark, drag – 3032
marker bed 1751
mark, fiducial – 4368
mark, flow – 3020
mark, flute – 3025
mark, frondescent – 3057
mark, groove – 3033
mark, impact – 3037
marking 3018
marking, cabbage-leaf – 3057
marking, rill – 3029
marking, plumose – 1511
mark, prod – 3037
mark, rill – 3029
mark, ripple – 3000
mark, roll – 3038
mark, saltation – 3039
mark scour – 3021
marks, grooving – 1270
mark, skip – 3039
mark, slide – 3058
mark, sole – 3019
mark, thumb – 0074
mark, swash – 3030
mark, swirl – 0481
mark, tool – 3032
mark, water-level – 3031
marl 3221
marl, clayey – 3223
marl, limy – 3224
marl mud 3222
marlstone 3221
marmarization 0920
marmarosis 0920
marmorization 0920
marmorosis 0920
marsh 3532
marsh, algal – 2515
Marshall line 0691
marsh basin 2517
marshes, belt of – 2512
marsh gas 3529
marsh, high – 2514
marsh, marine – 2514
marsh, salt – 2514
marsh, salt – 2909
mascon 0052
masking 4643
mass, deficiency of – 0343
mass, excess of – 0343
massive 0859
mass movement 2037
mass, stationary – 0125
mass, steady – 0125
mass transfer of heat 0381
mass wasting 2037
mast 4267
master event 0153
material balance 3865
material balance calculation 3865
material, detrital – 2005
matrix 0626
matrix 3080
matrix, opaque – 3586
matrix porosity 3687
matter, amorphous opaque – 3589
matter, granular opaque – 3589
Matterhorn peak 2618
matter, massive opaque – 3589
matter, opaque – 3588
mature 3176
mature 3725
mature, post– 3725
maturity 3725
maturity index 3726
maximum 1020
mbuga limestone 4123
meal, glacial – 2875
meander 2219
meander belt 2224
meander channel 2100
meander core 2227
meander cut-off 2229
meander, cut-off – 2232
meander, enclosed – 2221
meander, entrenched – 2221
meander, free – 2220
meander, free-swinging – 2220
meander, incised – 2221
meander, inclosed – 2221
meander, ingrown – 2222
meander, inherited – 2221
meander, intrenched – 2221
meander loop 2232
meander neck 2226
meander scar 2228
meander spur 2227
meander spur, cut-off – 2227
meander, terrace – 2223
meander, valley – 2221
measure 3123
mechanical analysis 2711
mechanics, rock – 4664
mechanics, soil – 4664
mediosilicic 0602
Mediterranean suite 0690
medium-grained 0796
medium, porous – 3680
medium-textured 2721
megacryst 0608
megacrystalline 0797
megacyclothem 3120
megafabric 1005
megagametophyte 1838
megalospheric 1868
megaripple 3007
megascopic 0797
megaseismic region 0249
megasporangium 1837
megaspore 1837
meiofauna 1853
meizoseismal area 0246
melanes 0599
melange 2927
melanic 0588
melanocratic 0588
melanosome 1038
melatope 0570
melikaria 3487
melt 1198
melting, differential – 1029
melting, partial – 1029
melting, zone– 0625
member 1753
membrane, semi-permeable – 3397
Mercalli scale, modified – 0243
mercaptan 3716
merismite 1048
merocrystalline 0804
meroplankton 1801
merzlota 2637
mesa 2556
mesh 3157
mesh number 3157
mesh texture 0856
mesitis 0913
mesocratic 0587
mesocrystalline 0797
mesocumulate 0832
mesofabric 1005
mesohaline 2832
mesolimnion 2333
mesonorm 0606
mesopelagic 1910
mesopelagic 2845
mesostasis 0626
mesothermal 3983
mesotrophic 3549
mesotype 0587
mesozone 0993
meta- 0883
meta-anthracite 3562
metabasite 0883
metabentonite 3218
metablastesis 0884
metablastic 0884
metabolism 1803
metacryst 0976
metacrystal 0976
metal factor 4571
metalimnion 2333
metallic 0484
metallic 3911
metallization 3936
metallogenetic 3903
metallogenic 3903
metallogeny 3903
metallo-organic compound 0410
metallotect 3904
metals, cold extractable heavy – 4655
metals, extractable – 4654
metals, heavy – 4653
metals, hot extractable – 4655
metaluminous 0693
metamorphic 3997
metamorphic assemblage 0885
metamorphic aureole 0899
metamorphic belt, paired – 1139
metamorphic complex 0885
metamorphic convergence 0913
metamorphic differentiation 0911
metamorphic diffusion 0912
metamorphic facies 0987
metamorphic grade 0992
metamorphic overprint 0970
metamorphic rank 0992
metamorphic rock 0880
metamorphic zone 0899
metamorphism 0879
metamorphism, allochemical – 0881
metamorphism, burial – 0887
metamorphism, cataclastic – 0892
metamorphism, caustic – 0897
metamorphism, constructive – 0887
metamorphism, contact – 0896
metamorphism, degree of organic – 3521
metamorphism, dislocation – 0892
metamorphism, dynamic – 0892
metamorphism, dynamo – 0892
metamorphism, dynamothermal – 0887
metamorphism, endomorphic – 0898
metamorphism, eo – 0902
metamorphism, exomorphic – 0898
metamorphism, geothermal – 0887
metamorphism, igneous – 0896
metamorphism, impact – 0893
metamorphism, injection – 0889
metamorphism, isochemical – 0881
metamorphism, kinetic – 0892
metamorphism, load – 0887
metamorphism, local – 0891
metamorphism, mechanical – 0892
metamorphism, optalic – 0897
metamorphism, organic – 3521
metamorphism, pluri– 0901
metamorphism, plutonic – 0993
metamorphism, prograde – 0903
metamorphism, progressive – 0903
metamorphism, regional – 0887
metamorphism, retrograde – 0904
metamorphism, retrogressive – 0904
metamorphism, shock – 0893
metamorphism, static – 0887
metamorphism, superimposed – 0901
metamorphism, thermal – 0896
metamorphism, thermo– 0896
metamorphite 0880
metamorphogenic 3974
metamorphosis 1794
metaripple 3014
metasom 1067
metasomatism 3956
metasomatism, contact – 0900
metasomatism, regional – 1056
metasomatose 3453
metasome 1067
metaster 1041
metatect 1040
metatectite 1031
metatexis 1030
metatexite 1030
metathetic deposit 3998
metatype 1954
meteor 0063
meteoric water line 0458
meteorite 0063
meteorite, iron – 0066
meteorite, stony – 0071
meteorite, stony-iron – 0070
meteoritics 0062
meteoroid 0063
meter, strain – 4767
method, audio-frequency magnetic – 4573
method, cut-cumulative – 3875
method, decline – 3874
method, electrical – 4554
method, electromagnetic – 4572
method, equipotential-line – 4558
method, expectation curve – 3876
method, frequency-domain – 4569
method, horizontal-loop – 4575
method, induced polarization – 4569
method, magnetotelluric – 4553
method, mise-à-la-masse – 4559
method, potential-drop ratio – 4560
method, principle point – 4324
method, resistivity – 4561
method, self-potential – 4557
method, spontaneous polarization – 4557
method, telluric – 4553

method, time-domain – 4569
method, vertical-loop – 4575
method, volumetric – 3871
Mevdedev/Sponheuer/Karnik scale 0243
mianthite 1033
miarolitic 0873
MIBK 4656
micaceous 0775
micaceous 3177
micrinite 3574
micrite 3281
microaphanitic 0800
micrococquina 3308
microcryptocrystalline 0800
microcrystalline 0797
microcrystalline 3091
microearthquake 0146
microevolution 1970
microfabric 1005
microfelsitic 0800
microfold 1643
microfossil 1790
microgametophyte 1838
microlite 0759
microlith 0759
microlithotype 3581
microlitic 0836
microlog 4587
micromeritic 0797
micrometeorite 0064
micropalaeontology 1786
microplankton 1830
microscope, petrographic – 0611
microscope, polarizing – 0611
microseism 0097
microseismic region 0251
microspheric 1868
microsporangium 1836
microspore 1836
microtectonics 0995
microtectonics 1479
microtexture 0790
mictite 1023
middlings 4142
mid-oceanic ridge system 1129
migma 1022
migmatite 1022
migmatitization 1022
migmatization 1022
migration 3748
migration 4457
migration, capillary – 3624
migration channel 3751
migration, horizontal – 3750
migration, lateral – 3750
migration path 3751
migration, primary – 3749
migration, secondary – 3749
migration, vertical – 3750
mill 4149
mill, ball – 4149
mill heads 4140
milligal 0367
mill, rod – 4149
mimetic 1008
mine appraisal 4216
mineragraphy 3925
mineral 0460
mineral 3912
mineral, accessory – 0594
mineral, accessory – 3174
mineral, antistress – 0979
mineral, artificial – 0461
mineral assemblage 0586
mineral association 0586
mineral association, critical – 0988
mineral, auxiliary – 0594
mineral beneficiation 4137
mineral, clay – 3213
mineral, contact – 0978
mineral, critical – 0988
mineral deposit 3910
mineral deposit, formation of a magmatic – 3946
mineral, diagnostic – 0689
mineral, essential – 0594
mineral, heavy – 3173
mineral inclusion 3553
mineral, index – 0988
mineral, industrial – 0462
mineralization 2638
mineralization 3936
mineral, light – 3173
mineral matter 3553
mineral, normative – 0607
mineral, occult – 0757
mineralogy 0459
mineraloid 0460
mineral, ore – 0462
mineral, ore – 3912
mineral, primary – 0596
mineral, principal – 0594
mineral product 3912
mineral, released – 0757
mineral, rock-forming – 0462
mineral, secondary – 0596
mineral, specific – 0594
mineral, stable – 3175
mineral standard 0607
mineral, stress – 0979
mineral substance 3912
mineral, symptomatic – 0689
mineral, synthetic – 0461
mineral, typomorphic – 0988
mineral, varietal – 0595
minor element 0400
miocrystalline 0804
miogeosyncline 1096
miospore 1833
mise-à-la-masse method 4559
Mississippi valley type 4132
mixed base 3718
mixoeuhaline 2834
mixohaline 2833
mixtite 3124
MLL 4587
MM-scale 0243
moat 1453
móberg 1347
mobile belt 1093
mobile component 1058
mobilisat 1057
mobilism 1113
mobility 3699
mobility ratio 3699
mobilizate 1057
mobilization 1057
MOD 3899
modal 0593
mode 0166
mode 0593
mode, fundamental – 0167
mode, higher – 0167
model 4775
model, analogue – 4782
model, base-friction – 4780
model, centrifuge – 4781
mode, leaking – 0195
model, electrical resistance analogue – 4783
model, equivalent – 4778
model, geomechanical – 4779
model, horizontal layer – 4568
modelling 4478
model, mathematical – 4776
model, numerical – 4776
model, photo-elastic – 4783
model, physical – 4778
model test 4778
moder 2744
mofette 1402
mogote 2074
Mohorovičić discontinuity 0087
Mohr circle 4690
Mohr-Coulomb fracture criterion 4694
Mohr envelope 4691
Mohr stress diagram 4688
Mohs hardness scale 4701
Mohs, scale of – 0473
moisture, available – 2773
moisture, bed – 3598
moisture content 2769
moisture, free – 3598
moisture, inherent – 3598
moisture pression 2771
moisture pressure 2770
moisture-release curve 2772
moisture-retention curve 2772
moisture, soil – 2764
moisture, soil – 3652
moisture suction 2770
moisture tension 2771
moisture, total – 3598
molasse 2926
mold 1824
mold, external – 1824
molding, pressure – 3446
mold, internal – 1824
mole 3675
mole 4854
moment 1759
monadnock 2608
money of the day terms 3899
money terms, constant value – 3899
monistic hypothesis 0020
monoclinic 0515
monocline 1617
monogene 0589
monogenic 0589
monolith 1424
monometallic 3911
monomict 3265
monomictic 3265
monomineralic 0589
monophyletic 1960
monoschematic 0794
monotypic 1945
monotypical 1945
mons 0037
montan wax 3736
montmorillonite 3214
monzonitic 0813
moon milk 2153
moonquake 0061
moor 3532
moor pound 3540
moor, transition – 3536
mor 2744
moraine 2403
moraine, ablation – 2412
moraine amphitheatre 2418
moraine, border – 2416
moraine, carried – 2405
moraine, deposited – 2410
moraine, end – 2413
moraine, englacial – 2408
moraine, flank – 2416
moraine, fluted – 2409
moraine, frontal – 2413
moraine, frontal – 2416
moraine, ground – 2409
moraine, interlobate – 2407
moraine in transit 2405
moraine, lateral – 2411
moraine, longitudinal – 2415
moraine, medial – 1295
moraine, medial – 2407
moraine, median – 2407
moraine, moving – 2405
moraine, push – 2419
moraine rampart 2414
moraine, recessional – 2417
moraine, scoria – 1295
moraine, stadial – 2417
moraine, stranded – 2410
moraine, subglacial – 2409
moraine, surface – 2406
moraine, terminal – 2413
moraine, thrust – 2419
moraine, upsetted – 2419
morainic-belt topography 2630
Morkill formula 4217
morphogenetic zone 1779
morphology 1793
morphotectonics 1076
mortar 0966
morvan landscape 2596
mosaic 0960
mosaic 4380
mosaic, controlled – 4381
mosaic, semi-controlled – 4381
mosaic, strip – 4380
mosaic, uncontrolled – 4381
mosor 2608
mother liquid 3377
mother liquor 3377
mother-of-pearl 0503
motion, line of – 2372
mottle 2692
mottled 2692
mottled 3462
mottled 4099
mould 1824
mouldering 3526
mould, external – 1824
mould, internal – 1824
moulin 2389
mound, frost – 2643
mountain, anticlinal – 2541
mountain, block – 2539
mountain building 1100
mountain, central – 0054
mountain chain 1102
mountain chain, bordering – 2542
mountain, cirque – 2618
mountain, dome – 2538
mountain, fault-block – 2539
mountain, fault-wedge – 2539
mountain, fold – 2538
mountain, folded – 2538
mountain, horst – 2540
mountain knot 2605
mountain milk 2153
mountain, monoclinal – 2541
mountain, remainder – 2607
mountain, residual – 2607
mountains, high – 2575
mountain spur 2605
mountain, synclinal – 2541
mountain system 1102
mountain, table – 2556
Mount Katmai activity 1165
mouth 2196
movement, componental – 0999
movement, eustatic – 2858
movement, negative – 2858
movement, positive – 2858
MPL 3814
MSK-scale 0243
muck 3527
muck 4829
mud 3210
mud 4829
mud accumulation, linear – 3325
mud ball 3247
mud ball, armoured – 3247
mud ball, sulphur – 1321
mud ball, volcanic – 1321
mud, blue – 2921
mud cake 4618
mud cone 1396
mud, coral – 3311
mud-crack cast, crumpled – 3047

mud crack, incomplete – 3046
mud-crack polygon 3046
mud, drilling – 4281
muddy 3210
mud eruption 1395
mud field 1390
mud flake 3248
mud flat 2910
mud flow 2051
mud flow 2974
mudflow deposit 1356
mudflow, hot volcanic – 1353
mud flow, sulphur – 1399
mudflow, volcanic – 1352
mudflow, volcanic – 1355
mudflush 4281
mud geyser 1398
mud, green – 2921
mud, hemipelagic – 2920
mud lava 1397
mud, lime – 3295
mud lump 3072
mud, marl – 3222
mud, marly – 3222
mud pot 1398
mudscreen 4282
mudstone 3219
mudstone, carbonate – 3282
mudstone, conglomeratic – 3256
mudstone, dolomite – 3282
mudstone, lime – 3296
mudstone, pebbly – 3219
mudstone, pebbly – 3256
mud, sulphur – 1399
mud-supported 3294
mud volcano 1395
mud volcano 1700
mud volcano 3071
mulching, self- 2762
mull 2744
mullion 1608
multiband system 4326
multichambered 1870
multichannel system 4326
multilocular 1870
multiple-shot instrument 4295
multiplexing, time – 4424
Munsell colour system 2676
Munsell soil-colour charts 2676
murbruk 0966
murrum 4124
mutation 1982
mylonite 0915
mylonite 1541
mylonitic 0969
mylonitization 0915
mylonization 0915
myrmekite 0785
myrmekitic 0852

N

nacre 0503
nannofossil 1790
nannoplankton 1799
naphtenic base 3718
naphtha 3744
naphthenes 3717
nappe 1683
nappe, fold – 1682
nappe, overthrust – 1575
nappe, plunging – 1688
nappes, overlapping – 1691
nappes, packet of superim-posed – 1691
nappes, pile of – 1691
nappe tectonics 1679
nari 4123
nealogic 1866
neanic 1866
near field 0274
nearshore 2889
nebular hypothesis 0021
nebula, solar – 0021
nebulite 1054
neck 1477
neck cut-off 2230
needle-like 0522
nekton 1798
nektonic 1798
nematoblastic 0962
neo-Darwinism 1966
neo-Lamarckism 1967
neomagma 0644
neomineralization 0905
neomorph 1983
neontology 1787
neosome 1037
neostratotype 1745
neotype 1951
neovolcanism 1150
nepionic 1866
neptunism 0004
neritic 2841
neritic 2915
neritic, inner – 2841
neritic, outer – 2841
nesophitic 0846
nest 3105
nested 1444
nest, ore – 4032
Neuman lines 0068
neutral 0602
névé 2363
névé 2364
névé basin 2364
névé field 2364
névé line 2364
NGL 3746
niche 1902
nick 2199
nick, cyclic – 2199
nickpoint 2199
nickpoint, cyclic – 2199
Niggli norm 0606
nip 2470
nip 3509
nivation 2357
nivation cirque 2612
niveo-aeolian 2870
niveolian 2870
NML 4596
NMO 4432
nodal line 0234
nodal plane 0234
nodular 0867
nodular 3473
nodular 4085
nodule 0771
nodule 3473
nodule 4085
nodule, loess – 3489
nodule, manganese – 3477
nodule, septarian – 3487
noise, ambient – 4438
noise, background – 4438
noise level 0097
noise, seismic – 0097
noise, seismic – 4438
noise, shot-generated – 4438
noise, signal-generated – 0177
noise, subterranean – 1237
nomenclature, binary – 1922
nomenclature, binomial – 1922
nomenclature, binominal – 1922
nomenclature, trinominal – 1922
nomenclature, uninominal – 1922
nomen conservandum 1958
nomen nudum 1958
nomen rejectum 1958
nonconformity, clino– 1721
nonmetallic 3911
norm 0606
nose, structural – 1639
notch 1471
notch 2020
notch 2472
novaculite 3366
nuciform 2729
nucleation 1360
nuée ardente 1366
null vector 0235
nunatak 2618
nunatak, volcanic – 1289
nut 2729

O

obduction 1133
obelisk 1424
oblique, gridded – 4375
obsequent 2285
obsequent 2550
observations, density of – 2675
observations, volcano-logical – 1190
obsidian 1262
obsidian, basalt – 1262
ocean 2782
oceanic 2842
oceanic 2918
oceanization 1142
oceanography 2781
oceanography, geological – 4226
oceanology 2781
ocellus 0855
octahedrite 0067
oedometer 4756
offlap 1727
offlap 3112
offset 1555
offset area 3845
offsetting of strata 1523
offshoot 1582
offshoot 2605
offshoot 4071
offshore 2840
ogive 2378
ogive, false – 2378
oikocryst 0844
oil 3712
oil, attic – 3776
oil-bearing sand 3766
oil, clean – 3714
oil column 3775
oil column height 3775
oil, crude – 3712
oil, dry – 3714
oil equivalent 3747
oil field 3773
oil intrusion 3752
oil, movable – 3864
oil originally in place 3859
oil pool 3773
oil rim 3776
oil ring 3776
oil sand 3766
oil sandstone 3766
oil saturation, residual – 4610
oil shale 3721
oil, shale – 3742
oil-source rock 3722
oil, standard tank – 3859
oil, stock-tank – 3859
oil, transformation of – 3733
oil, undersaturated – 3857
oil/water contact 3777
oil, wet – 3714
oil zone 3775
oldland 2560
olefins 3717
oligohaline 2832
oligomict 3265
oligomictic 3265
oligotrophic 3549
olistolith 2977
olistostrome 2977
ombrogenous 3550
omission of beds 1557
oncolite 3300
onion-skin weathering 1994
onlap 1727
onlap 3113
onset 0172
on-site inspection 0267
ontogeny 1794
ontology, comparative – 0007
ooid 3479
oolite 3479
oolith 3479
oolitic 3479
oomicrite 3286
oomicrudite 3286
oospararenite 3285
oosparite 3285
oosparrudite 3285
ooze 3210
ooze, calcareous – 3307
ooze, diatom – 3370
ooze, pelagic – 2920
ooze, radiolarian – 3368
ooze, siliceous – 3367
opacite 0610
opalescence 0486
opaque 0545
open 1626
opening, joint – 4728
operating agreement 3901
operator 4447
ophicalcite 0930
ophiolite 0696
ophiolite 1098
ophiolite, non-sequence type – 0696
ophiolite, sequence type – 0696
ophiolitic suite 0696
ophitic 0845
ophthalmite 1052
Oppel zone 1780
opthalmite 1052
optical sign 0557
orbicular 0867
orbicule 0867
order 0685
order 1929
order, soil – 2666
ore 3913
ore, bean iron – 4118
ore bearing 3917
ore bed 4021
ore, bedded – 4021
ore body 3910
ore body, lenticular – 4026
ore bunch 4032
ore, bunch of – 4047
ore channel 4029
ore chimney 4037
ore chute 4005
ore, cockade – 4050
ore, complex – 3914
ore control 3904
ore, course of – 4006
ore, crude – 4138
ore deposit, zoned – 4015
ore, disseminated – 4004
ore, disseminated– 4133
ore, drag – 4079
ore dressing 4137
ore dyke 4031
ore, flaxseed – 4119
ore formation, mechanical – 3944
ore formation, residual – 3943
ore, fossil – 4120
ore, high-grade – 3915

ore horizon 4008
ore, lean – 3915
ore, low-grade – 3815
ore mass 4032
ore, massive – 4003
ore microscopy 3925
ore, pay – 3916
ore, pea iron – 4118
ore pipe 4037
ore pod 4034
ore, quartz-banded – 4114
ore reserves 4208
ore, ring – 4050
ore roll 4028
ore, run-of-mine – 4138
ore schlieren 4040
ore, shipping – 4139
oreshoot 4005
oreshoot 4178
ore sill 4031
ore, simple – 3914
ore stock 4032
ore streak 4040
ore stringer 4039
ore vein 4038
ore, workable – 3916
ore zone 4008
organic matter, extractable – 3724
organo-metal complexing 0409
organometallic compound 0410
orientation diagram 1015
orientation, dimensional – 1013
orientation, lattice – 1013
orientation, preferred – 1013
orient of pearl 0486
orifice, flaming – 1386
orifice, volcanic – 1221
origin, centre of – 1969
origin time 0159
orocline 1140
orogen 1101
orogenesis 1100
orogenic belt 1093
orogenic belt 1101
orogenic period 1106
orogenic phase 1107
orogeny 1100
orterde 2704
ortho 0986
orthochem 3271
orthochemical 3268
orthoconglomerate 3255
orthocumulate 0832
orthodolomite 3320
orthodolomite 3380
orthogenesis 1975
orthogeosyncline 1095
orthogneiss 0936
orthogonal 0206
orthomagmatic 3976
orthomagmatic stage 0651
orthophyric 0826
orthoquartzite 3230
orthorhombic 0515
orthoselection 1975
orthostereoscopy 4367
orthotectic 0616
orthotectic 3976
orthotectic stage 0651
ortstein 2704
oscillation, free – 0198
oscillation hypothesis 1119
oscillation, spheroidal – 0198
oscillation, toroidal – 0198
oscillation, torsional – 0198
otolith 1865
oulopholite 2151
outburst 0115
outburst 1211
outcrop 4231
outcrop bending 2058
outcrop curvature 2058
outcropping 4231
outcrop, to – 4231
outer arc, non-volcanic – 1137
outflow, flank – 1266
outlet 2338
outlier 2561
outlier 4233
output 4445
outwash, glacial – 2402
outwash plain 2424
overbreak 4856
overburden 3501
overburden 3924
overburden 4788
overconsolidation 4703
overflow 2251
overflow level 2251
overflow, summit – 1266
overfold 1647
overgrowth 3455
overgrowth, epitaxial – 3456
overgrowth, syntaxial – 3456
overhang 2472
overlap 1556
overlap 4359
overlap, lateral – 4359
overlap, longitudinal – 4359
overlap, regressive – 1727
overlap, stratigraphic – 1552
overlap, transgressive – 1727
overlap zone 1777
overpressure 3820
overpressure, tectonic – 0886
overprint 0970
overrun factor 4515
oversaturated 0605
overthrust 1568
overthrust block 1569
overthrust fold 1576
overthrust mass 1569
overthrust plane 1568
overthrust sheet 1575
overtone 0167
overturned 1491
OWC 3777
oxbow 2232
oxic 0424
oxic decay 2932
oxidation 0416
oxidation potential 0421
oxidation-reduction-potential 0421
oxidation-reduction-reaction 0416
oxygen demand 3678
oxymesostasis 0846
oxyophitic 0846
ozokerite 3735

P

Pacific suite 0690
packer 4279
packing 3082
packing density 3082
packstone, lime – 3297
pahoehoe lava 1250
pahoehoe, slab – 1250
pair, stereoscopic – 4362
palaeobotany 1827
palaeochannel 4029
palaeoclimatology 0011
palaeocurrent 2864
palaeoecology 1889
palaeogeography 0011
palaeomagnetism 0294
palaeontology 1786
palaeosol 3126
palaeosome 1037
palaeotemperature 2864
palaeotemperature determination 0456
palaeotypal 0636
palaeovolcanism 1150
palaeozoology 1847
palaetiology 0007
palagonite 1254
palagonite breccia 1347
palagonite tuff 1347
palagonitization 1254
paleontology 1786
palette 2118
palichnology 1872
palichthyology 1848
palimpsest 0971
palingenesis 1024
palingenesis 1968
palingenetic 1024
palingenetic magma 0634
palinspastic reconstruction 0009
palsa 2644
paludal 2887
paludal 3532
paludification 3524
palustral 2887
palynology 1828
palynomorph 1829
pan 2517
pan 4659
panautomorphic 0808
panautomorphic-granular 0808
pandemic 1814
panel 4199
panfan 2584
Pangaea 1116
panidiomorphic 0808
panidiomorphic-granular 0808
panidiomorphic-granular 0954
pan, induced – 2761
pan, marsh – 2517
panplane 2604
pan, pressure – 2761
pan, salt – 2517
pan, salt – 2909
pan, traffic – 2761
para 0986
paracme 1985
paraconglomerate 3256
paracrystalline 0997
paraffin 3717
paraffin base 3718
paragenesis 0586
paragenesis 3907
paragenesis, diagenetic – 3394
paragenesis, salt – 3378
paragenetic 0586
paralic 2898
parallax 4364
parallax wedge 4365
parallel 0972
paramagnetism 0298
paramarginal 4212
paramorphism 0465
parastratotype 1743
paratectonic 0906
paratype 1950
parautochthonous 0727
paravolcanic phenomena 1180
parent material 2657
parent rock 0883
parent rock 2657
particle 3078
particle diameter 3193
particle size 2710
particle size 3193
particle-size analysis 2711
particle-size distribution 2709
particle-size fraction 2712
parting 0469
parting 0541
parting 3506
parting, dirt – 3506
parting, polyhedric – 1303
parting, shaly – 1599
partition 2102
pascichnia 1881
pass 2503
passage, interdune – 2186
passive 0619
pasture, salt – 2514
patera 0037
path 0206
path, brachistochronic – 0209
path difference, optical – 0563
pathfinder 4649
path, least-time – 0209
path line 3708
path, minimum time – 0209
path, sheep – 2060
patina 3403
paulopost 0658
pavement 4840
pavement, stone – 2656
paylimit 4204
pay line 4856
pay-out time 3896
paystreak 4136
paystreak 4178
PDB 0454
peak and trough 0164
peak, central – 0054
peak strength 4685
peak zone 1773
pearl, cave – 2149
pearl, cultured – 0503
pearly 0484
peat 3494
peat, banded – 3548
peat, calluna – 3545
peat dolomite 3507
peat, eutrophic – 3549
peat formation 3524
peat gel 3525
peat, heath – 3545
peat, highmoor – 3542
peatification 3524
peat, limnic – 3531
peat, lowmoor – 3535
peat, marsh – 3548
peat, meadow – 3536
peat, mesotrophic – 3549
peat, moor – 3542
peat, moorland – 3542
peat, moss – 3543
peat, oligotrophic – 3549
peat, pitch – 3525
peat, sphagnum – 3543
peat, terrestrial – 3546
peat, tundra – 3547
peat, wood – 3536
pebble 3241
pebble bar 2244
pebble, pitted – 3410
ped 2724
ped face 2724
pediment 2582
pediment, desert – 2582
pediment pass 2586
pediment, rock – 2582
pediplain 2584
pediplane 2584
pedogenesis 2659
pedology 2858
pedon 2689
ped surface 2724
peel technique 4230
pegmatitic 0848
pegmatitic 3978
pegmatitic stage 0653
pegmatitization 1063
pegmatoid 0814
pegmatophyric 0848
pegostylite 2152
pelagic 1910

pelagic 2844
pelagic 2919
pelagic ooze 2920
pelagochthonous 3512
Pelean-type eruption 1163
Pele's hair 1326
Pele's tear 1326
pelite 3208
pelitic 3208
pellet 3094
pellet 4150
pellet, faecal – 1863
pellet, faecal – 3094
pellet, fecal – 3094
pellet, lava – 1324
pellet, volcanic – 1321
pelmicrite 3293
pelsparite 3292
pendant 0706
pendent formation 2130
pendulum 0351
pendulum 4774
pendulum, inverted – 4774
pendulum, length of the equivalent – 0126
pendulum, reversible – 0351
penecontemporaneous 3474
peneplain 2589
peneplain, incipient – 2593
peneplain, local – 2593
peneplain, partial – 2593
peneplain, primary – 2593
peneplain torso 2594
peneplanation 2589
peneplane 2591
peneseismic region 0252
penetration 4513
penetration, dynamic – 4711
penetration, rate of – 4298
penetration, recorded – 4513
penetration resistance 4709
penetration resistance 4711
penetration, static – 4711
penetration test, standard – 4712
penetrometer 4758
penetrometer, cone – 4758
penstock 4867
pepino-hill 2074
peralkaline 0693
peraluminous 0693
percentage, 15-atmosphere – 2773
percentage, 15-bar – 2773
percolation 3622
percrystalline 0804
percussion figure 3410
percussion mark 3410
Perfo-Anchor 4795
perforation, gun – 3799
pergelisol 2637
perhyaline 0804
periclinal bedding 1241
periclinal structure 1641
periglacial 2633
perilith 1329
perimagmatic 3985
perimeter, wetted – 2261
period 0163
period 1765
period, free – 0130
perlitic 0878
permafrost 2637
permeability 3697
permeability barrier 3771
permeability, diagenetic – 3701
permeability, effective – 3698
permeability, fracture – 3701
permeability, hydraulic – 3700
permeability, intrinsic – 3697
permeability, magnetic – 0304
permeability, primary – 3701
permeability, relative – 3698
permeability, secondary – 3701
permeability, soil – 2753
permeability, solution – 3701
permeameter 4755
permeation 1059
permissive 0619
perpatic 0818
persemic 0818
persilicic 0601
persistence 4726
perthite 0785
perthitic 0851
petre dirt 2123
petrifaction 3457
petroblastesis 1059
petrochemistry 0580
petrofabric 1004
petrofabric analysis 0995
petrofabric diagram 1015
petrofabrics 0995
petrogenesis 0581
petrogenetic grid 0991
petrogeny 0581
petrogeometry 0995
petrography 0579
petrography, coal – 3565
petrography, sedimentary – 2851
petroleum 3712
petrology 0578
petrology, ore – 3925
petrology, sedimentary – 2851
petrology, structural – 0995
petromorph 2126
petromorphology 0995
petrotectonics 0996
pF 2771
pH 0422
phacoid 3055
phacolite 0733
phacolith 0733
phaenotype 1979
phanerite 0795
phaneritic 0795
phanerocryst 0758
phanerocrystalline 0795
phanerocrystalline-adiagnostic 0800
phantom´ 0802
phantom horizon 4462
phase 0165
phase 3906
phase, abyssal – 1197
phase, biochemical – 3519
phase, biogenetic – 3519
phase characteristic 4448
phase, eruptive – 1208
phase, explosive – 1208
phase, extrusive – 1197
phase, geochemical – 3519
phase history 4388
phase, hypabyssal – 1197
phase, intrusive – 1197
phase lag 0134
phase, late-volcanic – 1197
phase lead 0134
phase, long wave – 0188
phase, preliminary – 0181
phase, principal wave – 0188
phase, repose – 1169
phase response 0135
phase response 4448
phases, coordination of – 4113
phase shift 0134
phase spectrum 4451
phase-velocity method 0230
phase, volcanic – 1197
phasing 4502
phenoblast 0981
phenoclast 3239
phenocryst 0758
phenocryst, pseudo– 0981
phenotype 1979
phi-grade scale 3163
phi scale 3163
phlebite 1044
phosphatization 3459
phosphorescence 0467
phosphorite 3480
photic 1904
photo, aerial – 4349
photo base 4371
photoclinometer 4600
photogeology 4327
photogrammetry 4328
photograph, aerial – 4349
photograph, composite – 4357
photograph, horizon – 4354
photograph, lateral-oblique – 4353
photograph, oblique aerial – 4353
photograph, satellite – 4349
photograph, survey – 4349
photograph, vertical aerial – 4353
photograph, wing – 4355
photography, split-vertical – 4356
photo index 4382
photo map 4383
photo strip 4358
photo, survey – 4349
photosynthesis 1803
phreatic diagenesis 3399
phreatic ground water 2765
phreatic level 2765
phreatic level 3642
phreatic line 4862
phreatomagmatic explosion 1219
phyletic 1960
phyllite 0943
phyllite-mylonite 0918
phyllonite 0918
phyllonite 1541
phyllonitization 0918
phylogenesis 1959
phylogenetic tree 1960
phylogenetic zone 1779
phylogeny 1959
phylozone 1779
phylum 1929
phyteral 3592
phytoclast 3168
phytopalaeontology 1827
phytoplankton 1800
piçarra 4128
picking 4459
picurite 3557
piedmont bench 2604
piedmont flat 2604
piedmont slope 2583
piedmont stairway 2604
pier 4845
piezocrystallization 0633
piezogene 0977
piezoglypt 0074
piezometer 3646
piezometric head 3646
pile 4838
pile driver 4838
pile driving 4838
pile driving rig 4838
pile, end-bearing – 4839
pile, floating – 4839
pile, friction – 4839
pile, point-bearing – 4839
piling, sheet – 4799
pillar 2140
pillar, earth – 2029
pillow lava 1253
pilotaxitic 0825
pilotaxitic 0838
pinch 4054
pinch out 3104
pingo 2643
pinnacle 2805
pipe 1221
pipe amygdule 0782
pipe, clastic – 3074
pipe, gravel – 2077
piperno 1342
pipernoid 1342
pipe, sand – 2077
pipe, solution – 2077
piping 4863
pipkrake 2639
pisolite 3478
pisolite 4124
pisolite, cave – 2149
pisolite, volcanic – 1321
pisolith 3478
pisolitic 3478
pit 2104
pit 3409
pit, bomb – 1418
pit, borrow – 4871
pitch 1614
pitch 3743
pitch 4019
pitch 4055
pitch, glance – 3738
pitch lake 3753
pitchstone 1262
pit, fault – 2547
pit, open – 2105
pit, test – 4235
pit water 3598
pivotability 3191
placer 3236
placer 3997
placer, beach – 2912
placer, deluvial – 4001
placer, proluvial – 4001
placer, residual – 3998
placer, stream – 4000
placer, terrace – 4001
plain 0029
plain, coastal – 2440
plain, desert – 2584
plain of lateral planation 2604
plain of marine erosion 2536
plain, outwash – 2424
plain, salt – 2580
plain, sand – 2157
planation 2395
planation, zone of – 2604
plane, auxiliary– 0238
plane, axial – 1631
plane, datum – 0357
plane, datum – 3854
plan, engineering geological – 4784
plane of action 0238
plane of bedding 3096
plane parallel 0972
planetesimal hypothesis 0022
planetology 0023
planet, terrestrial – 0025
plankton 1799
plankton bloom 2843
planktonic 1799
planophyre 0829
planophyric 0829
plant, accumulator – 4647
plant bullion 3507
plant, converter – 4647
plant, indicator – 4646
plant survey, geochemical – 4645
plastic 2747
plasticity 2747
plasticity 4669
plasticity constants 2748
plasticity index 4716
plastic limit 4715

plastic limit, lower – 2749
plateau 2797
plateau, basalt – 1257
plateau, basaltic – 1257
plateau, elevated – 0033
plateau, pressure – 1278
plateau, stripped structural – 2555
plate bearing test 4708
plate-loading test 4708
plate tectonics 1125
platform 1083
platform 2141
platform, abrasion – 2479
platform, built – 2480
platform, marine-built – 2480
platform, marine-cut – 2479
platform, marine-erosion – 2479
platform of solid lava 1285
platform, stabilized – 4539
platform, wave-beveled – 2479
platform, wave-built – 2480
platform, wave-cut – 2479
platform, wave-eroded – 2479
platy 0522
platy 0755
platy 2728
platy 3183
playa 2580
playa 2892
playa, coastal – 2892
pleistoseismal line 0246
pleochroism 0562
pleocrystalline 0803
pleomorphism 1811
plesiotype 1952
plexus 1961
plication 1643
Plinian-type eruption 1161
plinthite 4124
plot, minor control – 4324
plot, radial – 4324
ploughsole 2761
plowpan 2761
plowsole 2761
PLS 4733
plucking 2396
plug 3105
plug dome 1422
plug, lava – 1477
plug, plutonic – 0734
plug, salt – 1695
plug, volcanic – 1477
plumbing system 3940
plumbline 0329
plumbline, deflection of the – 0329
plunge 1614
plunge 4018
plunge pool 2028
plunging 2811
plurimetamorphism 0901
pluton 0715
plutonian 0617
plutonic 0617
plutonic 3989
plutonic breccia 1042
plutonism 0004
plutonism 0615
plutonite 0617
pluton, partial – 0721
pneumatogenic 0655
pneumatogenic 3977
pneumatolitic differentiation 0679
pneumatolysis 3950
pneumatolytic 3977
pneumatolytic stage 0653
pocket 2649
pocketing 2086
pocket, ore – 4032
podzolisation 2680
poecilitic 0844
poeciloblastic 0974
poikilitic 0844
poikiloblast 0981
poikiloblastic 0974
poikilohaline 2831
poikilophitic 0846
point 0494
point 2449
point bar 2225
point diagram 1018
point group 0512
point-load strength 4733
Poisson's number 4700
Poisson's ratio 4700
polar blanket 0040
polarisation 0236
polarity epoch 0292
polarity event 0292
polarity, reversed – 0290
polarization 4557
polarization, electrode – 4569
polarization, magnetic – 0 7
polarization, membrane – 4569
pole 1018
pole diagram 1018
pole, geomagnetic– 0280
pole, magnetic – 0282
pole, pi– 1020
pole strength 0295
polish 1538
polish 3404
polishing 2399
polish, limit of glacial – 2631
polje 2074
pollen 1839
polygene 0590
polygenetic 0590
polygenic 0590
polygon, ice-wedge – 2655
polygon, stone – 2653
polygon, Tamyr – 2655
polygon, tundra – 2655
polyhaline 2832
polymetallic 3911
polymetamorphic diaphthoresis 0904
polymetamorphism 0901
polymict 3265
polymictic 3265
polymineralic 0589
polymorphism 0464
polymorphism 1811
polyphyletic 1960
polyschematic 0794
polysynthetic 4103
polytopic 1819
polytypic 1945
polytypism 0463
pond 2330
ponding 2340
ponding 2971
pond, tidal – 2506
ponor 2074
pool 2247
pool 2330
population 1941
porcellanite 3374
pore 3083
pore 3683
pore filling 3440
pore space 2752
pore throat 3695
pore velocity 3703
pore volume 3681
porosity 3681
porosity, diagenetic – 3685
porosity, effective – 3704
porosity, fabric-selective – 3693
porosity, fracture – 3685
porosity, fracture – 3694
porosity, functional – 3695
porosity, grainsupported – 3686
porosity, interconnected – 3695
porosity, intergranular – 3688
porosity, interparticle – 3688
porosity, intragranular – 3688
porosity, intraskeletal – 3688
porosity, matrix – 3687
porosity, minus-cement – 3442
porosity, moldic – 3691
porosity, mudsupported – 3686
porosity, primary – 3684
porosity, secondary – 3685
porosity, sedimentary – 3684
porosity, shelter – 3689
porosity, soil – 2752
porosity, solution – 3685
porosity, texture-related – 3693
porosity, texture-selective – 3693
porosity, total – 3681
porosity, vug – 3692
porphyritic 0817
porphyro-aphanitic 0817
porphyroblast 0981
porphyroblastic 0956
porphyroclast 0981
porphyrocryst 0758
porphyrogranulitic 0846
porphyroid 0957
porphyry 0817
porphyry, glass – 0823
porphyry type ore deposit 4133
portal 4850
position, structural – 1480
posoder factor 4820
postcrystalline 0997
post-eruption phenomena 1374
postkinematic 0726
postkinematic 0906
postkinematic 1079
postkinetic 0726
postorogenic 0726
postorogenic 1108
posttectonic 0726
posttectonic 0906
postvolcanic phenomena 1374
pot 2104
pot 2105
pot 2248
potassium-argon dating 0449
potassium-calcium dating 0449
poteclinometer 4600
potential 3645
potential difference 4556
potential-drop ratio method 4560
potential, electro-kinetic – 4617
potential, liquid junction – 4615
potential, membrane – 4616
potential, streaming – 4617
pothole 2027
pothole 2104
pothole 2105
pothole 2248
pothole 3503
pothole, glacial – 2389
pothole, moulin – 2389
pot, mud – 1398
Potsdam system 0334
Potsdam value 0334
pounds per square inch 3821
powder factor 4820
power 4449
power density spectrum 4451
power, earning – 3898
power, reflecting – 0574
pozzuolana 1338
pradolina 2392
Pratt's hypothesis 0345
preadaptation 1980
precipitate, primary – 0830
precipitation 0411
precipitation 3614
precipitation barrier 4642
preconsolidation 4703
precrystalline 0997
precursor 0150
predominance, odd– 3728
prekinematic 0906
prekinematic 1079
presplitting 4823
pression, moisture – 2771
pressure, abandonment – 3849
pressure, active earth – 4706
pressure at rest, earth – 4705
pressure bomb 3848
pressure, bottom-hole – 3847
pressure, bubble point – 3857
pressure build-up curve 3850
pressure burst 0115
pressure cell, earth – 4771
pressure chamber test 4762
pressure curve, capillary – 4612
pressure, dew point – 3858
pressure drawdown 3850
pressure drop 3853
pressure, earth – 4705
pressure, excess – 3820
pressure, flowing – 3847
pressure, fluid – 3819
pressure fringe 0958
pressure, gas – 1173
pressure gauge 4769
pressure gauge, bottom-hole – 3848
pressure gradient 3852
pressure, grain – 3819
pressure, hydrostatic – 3820
pressure, initial – 3849
pressure, magmatic – 0622
pressure maintenance 3832
pressure, matrix – 3819
pressure meter 4760
pressure, moisture – 2770
pressure, overburden – 3819
pressure, passive earth – 4706
pressure plateau 1278
pressure, pore – 3819
pressure release 1173
pressure, reservoir – 3819
pressure, rock – 3819
pressure, seepage – 4738
pressure shadow 0958
pressure shaft 4867
pressure, soil – 4705
pressure, soil-water – 2770
pressure solution 3445
pressure, static – 3847
pressure, volcanic – 0622
pretectonic 0906
price index, wholesale – 4223
primary environment 0402
principal point 4369
principal point, transposed – 4372
printer, transforming – 4377
printing, ratio – 4378
print, transformed – 4377
priority, law of – 1924
prism 2730
prismatic 0522
prismatic 0755
prismatic 2730
prism, sedimentary – 3105
probabilistic combination 3877
process 2568
process, sulphato-sulphidic – 3934
production, cumulative – 3817
production, initial – 3805
production, peak – 3807
production rate 3801
production, ultimate – 3817
productivity index 3808

productivity index, specific – 3808
profile, buried – 2693
profile, geological – 4250
profile, graded – 2198
profile, graded shore – 2441
profile, interrupted – 2198
profile, irregular – 2198
profile of equilibrium 2198
profile, shore – 2441
profile, soil – 2686
profile, terminal – 2198
profile, ungraded – 2198
profiling, acoustic – 4480
profiling, continuous – 4408
profiling, deep-ocean – 4484
profiling, horizontal – 4561
profiling, shallow seismic – 4479
profiling, vertical – 4561
projection, cyclographic – 1021
promontory 2446
promontory, structural – 1639
propagation, fault – 0239
prophyroclastic 0966
propylite 0921
propylitization 0921
proration 3902
prospective area, 3887
protactinium-thorium dating 0445
Protector compaction test 4757
protoclastic 0857
protodolomite 3321
protolith 0883
protomylonite 0916
protore 3913
prototectite 0644
prototype 1979
prototype 4775
protrusion 1247
protuberance 1409
proved acreage 3889
proved-up area 3888
provenance 2943
provenance, area of – 2943
proven area 3889
province, circumpacific – 0690
province, faunal – 1849
province, geochemical – 0401
province, geological – 4260
province, geothermal – 0390
province, heat-flow – 0390
province, igneous 0628
province, intrapacific – 0690
province, metallogenic – 3908
province, metallographic – 3908
province, mineral – 3908
province, minerogenic – 3908
province, petrographic – 0628
province, sediment-petrological – 2852
proximal 2997
psammite 3226
psammitic 3225
psammitic 3226
psephite 3238
psephitic 3238
pseudoanticline 1620
pseudoaspite 1430
pseudobomb 1296
pseudofossil 1789
pseudomoraine 1295
pseudomorph 0982
pseudomorph 1823
pseudomorphism 0465
pseudomorphism 0982
pseudomorphism 3453
pseudomorphous after 0982
pseudonodule 3060
pseudopalagonite 1255
pseudoporphyritic 0819
pseudoporphyritic 0956
pseudoporphyroblastic 0959
pseudostratification 0869
pseudosyncline 1621
pseudotachylite 0917
pseudotachylite 1542
pseudovolcano 1189
psi 3821
ptygma 1051
ptygmatic 1051
pucker, frontal – 1687
puckering 1643
puddingstone 3253
pug 4012
pull test 4764
pulse 4482
pulse length 4482
pulse repetition rate 4504
pulverization 1319
pumice 1308
pumice flow, incandescent – 1369
pumiceous 0875
pumiceous 1308
pumicites 1308
pumping test 3711
push, lava – 1288
putrefaction 3526
puy 1425
puzzolana 1338
PVT analysis 3856
P-wave 0182
pycnocline 2830
pyramidal 0522
pyramid, earth – 2029
pyramid pebble 2162
pyritization 3458
pyrobitumen 3737
pyroclast 1309
pyroclast 3170
pyroclastic deposit 1310
pyroclastic rock 1311
pyroclastics 1309
pyrogenesis 0616
pyrogenic 0616
pyrogenic 0655
pyrogenic 3973
pyrogenous 0616
pyrolysis 3730
pyromagma 1200
pyromeride 0876
pyrometamorphism 0896
pyrometasomatism 0900
pyrometer 1195
pyrometer, optical – 1195
pyromorphism 0896

Q

quad 3605
quagmire 3530
quantum evolution 1971
quaquaversal bedding 1241
quarry 4874
quarrying 2396
quartering, coning and – 4181
quartz-banded ore 4114
quartzite 3230
quartzite, banded hematite – 4114
quartzite, surface – 4123
quickclay 4719
quicksand 3085
quicksand 4719
quicksand, volcanic – 1318
Q-wave 0189

R

rabbit run 2098
race 1939
rad 4551
radar 4384
radar beam 4389
radar, brute-force – 4385
radar, coherent – 4387
radar, continuous-wave – 4387
radar cross section 4392
radar, CW – 4387
radar map 4400
radar reflectivity 4392
radar shadow 4391
radar, sidelooking – 4385
radar, synthetic-aperture – 4386
radiance 4313
radiancy 4313
radiated 0776
radiation-absorbed-dose 4551
radiation, heat – 0381
radioactive dating 0429
radioactivity, hypothesis of – 1117
radiocarbon dating 0451
radiolarian chert 3369
radiolarian earth 3368
radiolarian ooze 3368
radiolarite 3368
radiolite 0778
radiolitic 0843
radiometer 4310
radiometric dating 0429
radius, hydraulic – 2262
raft 0661
rafting 2968
rafting, ice – 2969
rain, blood – 2159
rain drop impression 3041
rain drop imprint 3041
rain, eruption – 1373
rain print 3041
rain splash 2035
rain, volcanic – 1373
rake 4019
ramification 2210
rampart 1452
rampart, spatter – 1452
rang 0685
range 1781
range 4186
range 4393
range, depth – 4505
range direction 4394
range, dynamic – 4507
range, dynamic – 0132
range marks 4398
range-overlap zone 1777
range reflection, extended – 4397
range streaking 4396
range zone 1781
rapakivi 0828
rapid 2202
rarefaction 0187
rassenkreis 1938
rattlestone 3485
ravine 2026
ravinement 2018
ray 0056
Rayleigh wave 0189
ray parameter 0207
ray path 0206
ray path 4471
ray system 0056
ray-velocity surface 0549
Re 2265
reaction rim 0786
real terms 3899
ream, to – 4296
rebound number, elastic – 4733
recapitulation theory 1968
receiver 4499
recharge, artificial – 3671
recharge, induced – 3671
record 4420
record, earthquake – 0116
recorder 4500
recorder, digital – 4501
recorder, dual – 4501
recorder, glass-fibre – 4501
recorder, graphic – 4500
recorder, multichannel – 4501
recording 4500
recording, analog – 4421
recording, digital – 4422
recording, squiggle – 4421
recording, variable-area – 4421
recording, variable-density – 4421
recording, wiggle – 4421
record, seismic – 0116
recovery, creep – 0114
recovery, estimated ultimate – 3869
recovery factor 3867
recovery, primary – 3822
recovery, secondary – 3831
recovery, supplemental – 3831
recovery, tertiary – 3837
recovery, ultimate – 3867
recrystallization 0905
recrystallization 3452
recrystallization, strain – 0967
rectangular 2292
rectification 4376
rectification, full-wave – 4506
rectification, half-wave – 4506
redbed 2879
redbed deposit 4129
redeposition 2993
redox potential 0421
reduction 0416
reduction 1991
reduction, bacterial – 0417
reduction, free-air – 0358
reduction, isostatic – 0361
reduction sphere 3463
reduction, topo-isostatic – 0364
reef 2443
reef 2804
reef 3333
reef 4135
reefal 3334
reef, algal – 3360
reef, back – 3344
reef, barrier – 3338
reef cap 3329
reef complex 3337
reef core 3341
reef edge 3350
reef flank 3353
reef flat 3343
reef, fore – 3349
reef, fringing – 3338
reef front 3351
reef, inter- 3357
reef, knoll – 3329
reef, off– 3356
reefoid 3334
reef, patch – 3347
reef, pinnacle – 3347
reef, regressive – 3327
reef, ring – 3340
reef rock 3342
reef, saddle – 4053
reef slope 3354
reef talus 3353
reef tract 3326
reef, transgressive – 3328
reef, trough – 4053
reef wall 3336
reefy 3334
reen 2062
re-entrant 2450
reference, level of – 0354

refining, zone– 0625
reflectance 0574
reflectance 4316
reflection 4436
reflection, bottom – 4519
reflection coefficient 4473
reflection, diffuse – 4317
reflection, external – 0576
reflection horizon 4461
reflection, internal – 0576
reflection method 4405
reflection, multiple – 4441
reflection, primary – 4437
reflection, side – 4518
reflection, specular – 4318
reflection, stratum – 4519
reflection, total – 0213
reflectivity 0574
reflectivity 4473
reflector, diffuse – 4317
reflector, specular – 4318
refolding 1668
refraction, index of – 0551
refraction method 4405
refraction, wave – 2810
refractive index 0551
refractory 0788
reg 2578
regenerated, epirogenetically – 3994
regeneration sequence 3538
regime, hydraulic – 2267
regionality, degree of – 0340
regionalization 0255
regonalization 4194
region, frontal – 1686
region, metallogenic – 3908
regmaglypt 0074
regolith 2008
regression 1726
regular 0515
rejuvenation 2572
rejuvenation 3959
rejuvenescence 2572
relaxation 1179
relic 0985
relic 1985
relic, armoured– 0787
relic, metastable – 0985
relic sediment 2856
relic, stable – 0985
relic structure 0971
relic, unstable – 0985
relief 0558
relief, alpine – 2576
relief, Appalachian – 2564
relief, bold – 2576
relief, geochemical – 0401
relief, initial – 2569
relief intensity 2575
relief inversion 2563
relief, Jurassian – 2565
relief, multiconvex – 2600
relief, steep – 2576
rem 4551
remigration 3749
remote sensing 4300
remote-sensing platform 4302
reniform 0776
reniform 4095
rent, volcanic – 1466
repacking 3430
repetition of beds 1557
repetititon rate 4504
repichnia 1879
replacement 0931
replacement 3453
replacement 3956
repose, angle of – 4687
repose period 1169
resequent 2550
resedimentation 2993
resequent 2284
reserve estimate 3870
reserves 3868
reserves 4208
reserves, demonstrated – 4207
reserves, indicated – 4209
reserves, indicated additional – 3883
reserves, inferred – 3882
reserves, inferred – 4210
reserves, measured – 4209
reserves, possible – 3881
reserves, possible – 4210
reserves, probable – 3880
reserves, probable – 4209
reserves, proved – 3879
reserves, proved – 4209
reserves, proven – 3879
reserves, reasonably assured – 4207
reserves, speculative – 3882
reservoir 3755
reservoir formation 3679
reservoir, homogeneous – 3770
reservoir isotropy 3770
reservoir model 3866
reservoir performance 3846
reservoir rock 3679
reservoir simulation model 3866
reservoir simulator, numerical – 3866
reservoir stimulation 3800
residence time 0397
residence time 3627
residual 0241
residual 3998
residual error 0241
residual liquid 0647
residue, insoluble – 2930
resinite 3572
resinoid 3570
resinous 0484
resistance 4682
resistance, acoustic – 4474
resistance, electric – 4556
resistance, penetration – 4709
resistance, penetration – 4711
resister 3175
resistivity 4556
resistivity, apparent – 4564
resistivity, apparent – 4609
resistivity factor, formation – 4609
resistivity, formation – 4607
resistivity, formation-water – 4608
resistivity method 4561
resistivity, true – 4607
resolution 4347
resolution cell 4319
resolution, horizontal – 4516
resolution, power of – 4453
resolution, vertical – 4516
resorption 0642
resorption border 0786
resources, hypothetical – 4214
resources, identified – 4206
resources, identified subeconomic – 4211
resources, mineral – 4205
resources, potential – 4215
resources, speculative – 4214
resources, undiscovered 4213
response, amplitude – 4448
response curve 0133
response, impulse – 4446
response, phase – 4448
restite 1032
resurgence 2089
resurrected 2595
retardation 0563
retention, specific – 3704
retention, surface – 3618
reticular texture 0856
reticulate texture 0856
reticulite 1331
return 4390
return, no- 4391
returns 0191
revealed 2595
reverberation 4440
reversal, geomagnetic – 0291
reversal, self– 0311
revision 3886
reworked 2947
Reynolds number 2265
rheidity 4671
rheid mechenism 0108
rhenium-osmium dating 0446
rheology 0102
rheology 4671
rheomorphism 1057
rhizoconcretion 3490
rhombochasm 1143
rhythmite 3121
ria 2452
ribbon 0945
ribbon, ore – 4039
Richter scale 0220
rider 4072
ridge 2486
ridge 2796
ridge, anticlinal – 2541
ridge, beach – 2490
ridge, dividing – 2316
ridge, fault – 1589
ridge, glacial pressure – 2419
ridge, homoclinal – 2541
ridge, ice-contorted – 2419
ridge, serrate – 2617
ridge, spur – 2605
ridge, synclinal – 2541
ridge system, mid oceanic – 1129
ridge, wrinkle – 0031
ridge, wrinkle – 0058
riegel 2625
riffle 2247
riffler 4182
rift 1092
rift 1590
rift, concentric – 1469
rift, earthquake – 2552
rift, radial – 1468
rift, transverse – 1470
rift valley 1590
rift valley 2546
rift zone 1467
rig 4267
rigidity, modulus of – 4699
rill 2026
rill 2194
rille 0057
rill marking 3029
rill wash 2033
rill, wet-weather – 2026
rimstone bar 2147
rimstone pool 2147
ring complex 0742
ring depression 1185
ring dyke 0742
ring-fracture intrusion 0742
ring-fracture stoping 1187
ringing 4440
rip 4828
rip current 2826
ripening 2685
rippability 4828
ripple 2814
ripple 3000
ripple, adhesion – 3016
ripple amplitude 3002
ripple crest 3001
ripple cross bedding 3147
ripple, giant – 3007
ripple height 3002
ripple, incomplete – 3015
ripple index 3002
ripple length 3002
ripple mark 3000
ripple mark, crescentic – 3005
ripple mark, current – 3004
ripple marked 3000
ripple mark, flat-topped – 3015
ripple mark, interference – 3013
ripple mark, linguoid – 3006
ripple mark, longitudinal – 3011
ripple mark, lunate – 3005
ripple mark, oscillation – 3012
ripple mark, rectangular cross – 3013
ripple mark, rhomboid – 3010
ripple mark, transverse – 3003
ripple mark, wave – 3012
ripple, starved – 3015
ripple through 3001
ripplet, wind anti– 3016
ripple, wave-current – 3011
ripple, wavelength 3002
ripple, wind – 3016
ripple, wind granule – 3016
rip-rap 4876
rise 2089
rise 2796
rise in level 2513
Rittmann norm 0606
river 2191
river abstraction 2322
river bed 2242
river beheading 2323
river capture 2322
river, competent – 2329
river deflection 2213
river deviation 2213
river diversion 2213
river, false – 2212
river head 2193
river inversion 2214
riverless 2257
river, misfit – 2329
river piracy 2322
river, pirate – 2325
river system 2205
river, tidal – 2504
river, underfit – 2329
RMS velocity 4466
road base 4841
roasting 4151
roche moutonnée 2400
rock 3389
rock 4722
rock, balanced – 2163
rock, beach – 3412
rock bench, wave-cut – 2479
rock block 4730
rock block, unit– 4730
rock breaking 1989
rock burst 0115
rock burst 2112
rock burst 4857
rock, carbonate – 3266
rock, country – 1530
rock, country – 3501
rock, country – 3920
rock, crushed – 4876
rock, eruptive – 1238
rock fall 2049
rock flour 3209
rock, foliated – 1604
rock, fused – 1198
rock head 4788
rock, host – 3919
rock, intact – 4722
rock, invaded – 3919
rock, irruptive– 0617
rock island 2227
rock, kerogenous – 3720

rockkindred 0630
rock mass 4722
rock material 4722
rock milk 2153
rock mill 2248
rock, molten – 1198
rock, mother – 2944
rock, mother – 3918
rock, mushroom – 2163
rock opening 3940
rock, parent – 0883
rock, parent – 2657
rock, parent – 2944
rock, parent – 3918
rock, pedestal – 2163
rock, primary– 0631
rock, primary – 3918
rock, pyroclastic – 1311
rock-quality designation 4752
rock roll 3510
rock salient 2606
rock salt 3382
rock step 2200
rock stream 2057
rock survey 4644
rock, tight – 3769
rock, trap – 0638
rock, tuffogenic – 1332
rock unit 1748
rock, volcanic – 1238
rock, wall – 3501
rock waste 2005
rodding structure 1608
rod, drilling – 4270
rod-shaped 3185
roentgen 4550
roentgen-equivalent-man 4551
Roga test 3595
roll 4028
rollability 3191
roller 4439
roll, ground – 4439
rolling 4809
roof 0705
roof 3501
roof 4010
roof 4851
roof fall 2110
roof foundering 1188
roof pendant 0706
root 0343
root 1104
root zone 1684
rose 3092
rosette 3092
Rosiwall method 0613
rotary table 4269
rotation, angle of – 1126
rotation, external – 1000
rotation, internal – 1000
rotation, pole of – 1126
rotation, tectonic – 1000
rotting 3526
rough 3190
roughness, joint – 4727
round 4822
rounded 3188
rounded, well– 3188
roundness 3187
roundness class 3187
roundness index 3187
roundstone 3242
round-trip 4280
royalty 4224
RQD 4752
R-tectonite 1003
rubble 2005
rubble 3259
rubidium-strontium dating 0447
rudaceous 3237
rudimentary 1978
rudite 3237
rudstone 3297
rugged 3190
rumbling, subterranean – 1237
run 0723
run 4030
run 4358
runic 0847
runnel 2194
runnel 2486
runnel, solution – 2087
runoff 3616
runoff coefficient 2250
runoff, ground-water – 3628
runoff, surface – 3617
runup 2813
runway 2191
rupture 4676
rupture velocity 0239
rust 2692
rust mottle 2692
R-wave 0189

S

saccharoidal 0811
saccharoidal 3090
saddle 1635
safety clamp 4280
safety factor 4777
sag 2649
sag and swell surface 2630
sag, bomb – 1418
sagger 3517
sagre 3517
sag structure 3059
salar 2894
salband 0707
salband 4012
salic 0607
salient 2304
salient, rock – 2606
saliferous 3376
salina 2580
salina 2894
salinity 2827
salinity 3376
salinity 3675
saltation 1982
saltation 2954
saltation load 2954
salt-crystal cast 3042
salt dome 1695
salt glacier 1699
salting 2514
salt marsh 2909
salt pan 2909
saltpetre, chile – 3384
saltpetre dirt 2123
salt pillow 1695
salt plug 1695
salt polygon 3049
salt stock 1695
salt tectonics 1694
salt wall 1695
salt wedging 1997
sammelkristallization 0910
sample 4741
sample bag 4229
sample, bottom-hole – 3855
sample, channel – 3593
sample, channel – 4168
sample, chip – 4168
sample, column – 3593
sample, composite – 4176
sample, compound – 4176
sample, continuous – 4749
sample, disturbed – 4742
sample, grab – 3153
sample, grab – 4169
sample, integral – 4750
sample, pick – 4168
sample, pillar – 3593
sample pouch 4229
sampler 4745
sampler, drive – 4746
sampler, drop – 4747
sample reduction 4180
sample, remolded – 4743
sampler, piston – 4746
sample, salted – 4179
samples, ditch – 4238
sample, sidewall – 4238
sample, sludge – 4238
sample spacing 4171
sample, undisturbed – 4742
sampling 4423
sampling 4741
sampling, check – 4177
sampling, drill – 4167
sampling, face – 4167
sampling, pilot – 4175
sampling spoon – 4748
sampling, stream – 4170
sampling tool 4745
sampling tube 4748
sand 3227
sand 2718
sand, aeolian – 2871
sand avalanche 2973
sand bar 2244
sand blasting 2939
sand, blowing – 2156
sand, blown – 2871
sand, cavernous – 3063
sand cay 2443
sand, coral – 3311
sand, cover – 2871
sand crystal 3443
sand dome 3065
sanddrain 4813
sand drift 2961
sand, drift – 2156
sand, drift – 4719
sand, drifting – 2156
sand fall 2973
sandfall face 2185
sand flat 2910
sand flow 2973
sand flow, hot – 1369
sand, flying – 2156
sand fraction, median of the – 2718
sand, gas-bearing – 3766
sand, glass – 3227
sanding up 2990
sanding-up 3814
sand key 2443
sand levee 2182
sand line 4621
sand, loamy – 3227
sand, musical – 3088
sand plain 2157
sand, quick– 4719
sand, roaring – 3088
sand, running – 3085
sand, scoriaceous – 1323
sandsea 1462
sandsea 2183
sand, shoestring – 2899
sand, singing – 3088
sand size 3227
sand, sounding – 3088
sand, squeaking – 3088
sandstone 3228
sandstone, clean – 3228
sandstone, crystal – 3465
sandstone, feldspathic – 3232
sandstone, gas-bearing – 3766
sandstone, oil-bearing – 3766
sandstone, orthoquartzitic – 3230
sandstone, quartzitic – 3230
sandstone, sheet – 3109
sandstone, water-bearing – 3766
sand, thief – 3834
sand trap 3152
sandur 2424
sand, very coarse – 3249
sand, vesicular – 3063
sand, volcanic – 1322
sand volcano 3071
sand, water-bearing – 3766
sand wave 3007
sand wedge 2641
sand, wet – 3767
sand, whistling – 3088
sand, wind-blown – 2871
sand, wind-driven – 2156
saprolite 2009
saprolite 4128
sapropel 3528
sapropelite 3497
sapropelite 3721
SAR 4386
satellite, artificial – 4302
saturated 0605
saturation 3650
saturation, base – 2756
saturation exponent 4606
saturation, hydrocarbon – 4610
saturation, irreducible water – 4611
saturation line 0697
saturation line 4862
saturation magnetization 0297
saturation percentage, base – 2756
saturation, residual oil – 4610
saturation, residual water – 4611
saturation, water – 4611
saturation, zone of – 3656
saussurite 0927
saussuritization 0927
scale 1538
scale 4240
scaling 1994
scallop 2086
scallop 2087
scallop 3025
scanner 4403
scanner, multispectral – 4403
scanning 4401
scanning, circular – 4402
scanning, line – 4402
scapolitization 0932
scarp 2053
scarp, beach – 2489
scarp, cuesta – 2558
scarp, downwarping – 2551
scarp, fault – 2548
scarp, fault-line – 2550
scarp, lava – 1286
scarplet, earthquake – 2552
scarp, lobate – 0032
scarp, monoclinal – 2551
scatter 4316
scatter diagram 1018
scattering 0218
scavenger 4650
SCE 3606
scheiteltiefe 0214
schist 0944
schistic 0975
schist, knotted – 0941
schistoid 0975
schistose 0975
schistose 1603
schistosity 0975
schistosity 1603
schistosity, linear – 1603
schist, spotted – 0941
schliere 0665
schlieren 0665
schlieren, ore – 4040
schlieric 0665

Schlumberger array 4562
Schmidt hammer 4733
Schmidt net 1015
Schürmann series 3935
Schürmann's law 3522
scintillation 4319
scintillation counter 4546
sclerotinite 3574
scopulite 0769
scoria 1294
scoriaceous 1294
scoriaceous cover 1330
scoria cone 1427
scoriae, mantle of – 1330
scoria moraine 1295
scoria, thread-lace – 1331
scorification 4184
scouring 2936
scouring, glacial – 2395
scour mark, transverse – 3021
scour, wind – 2015
scrape 4828
scratch 1538
scratching 2397
scree 2007
scree 2867
screening 4873
scroll bar 2233
scroll, flood-plain – 2234
se 1004
sea 2782
sea 2807
sea, adjacent – 2783
sea cliff 2470
sea-floor spreading 1128
seaknoll 2800
seal 3765
seal 4800
sea level, mean – 2817
sealing 4800
sealing, surface – 2760
seal, tar – 3765
seam 3123
sea, marginal – 2783
seam, guide – 3500
seam, marker – 3500
seam nodule 3507
seamount 2798
seapage, line of – 4862
seapeak 2798
seaquake 0148
seascarp 2799
seat earth 3515
seat rock 3516
seat stone 3516
sebkha 2580
sebkha 2892
sebkha, coastal – 2892
secondary environment 0402
secretion, theory of lateral – 3928
sectile 0471
section 0685
section, columnar – 4255
section, composite – 4253
section, generalized – 4254
section, geological – 4250
section, ideal – 4254
section, record – 4458
section, seismic – 4458
section, serial – 4251
section, standard – 4257
secular variation 0289
sedentary 1919
sedentary deposit 2855
sediment 2850
sedimentary deposit 2850
sedimentary rock 2850
sedimentary structure 2994
sedimentation 2989
sedimentation balance 3158
sedimentation, clastolithic – 1201
sedimentation, crystal – 0673
sedimentation, cycle of – 3119
sedimentation, cyclic – 3121
sedimentation, internal – 3436
sedimentation rate 2989
sedimentation, rhythmic – 3121
sediment, carbonate – 3266
sediment charge 2957
sediment discharge 2957
sediment, internal – 3436
sediment load 2951
sedimentology 2849
sediment-petrological province 2852
sediment, relic – 2856
sediment transport rate 2957
seepage 3629
seepage 3753
seepage 4635
seepage, gas – 1401
seepage, gas – 4520
seepage pressure 4738
seepage velocity 3703
Seger cone 1196
seggar 3517
segregation 3395
segregation banding 0946
segregation, magmatic – 3948
seiche 2816
seism 0138
seismic activity 0242
seismic constant 0257
seismic-electric effect 0113
seismic event 0095
seismic focus 0156
seismic intensity 0243
seismicity 0242
seismicity, specific – 0242
seismic record 0116
seismic region 0247
seismic risk 0255
seismic wave 0161
seismic zoning map 0244
seismogram 0116
seismogram 4420
seismograph 0116
seismograph, broad-band – 0116
seismograph vault 0116
seismology 0094
seismology, engineering – 0254
seismology, forensic – 0264
seismology, nucleair – 0263
seismology, planetary – 0094
seismometer 0118
seismometer 4414
seismometer array 4416
seismometer pattern 4416
seismometers, bunched – 4416
seismometer spacing 4416
seismometer spread 4417
seismometer, strong motion – 0120
seismoscope 0117
selection, natural – 1966
selenochemistry 0045
selenography 0045
selenology 0045
selenophysics 0045
self-potential method 4557
self-replacement 3957
selvage 0707
selvage 4012
selvedge 0707
semi-anthracite 3560
semicrystalline 0804
semifusinite 3574
semipegmatitic 0844
semi-thermal region 0391
sempatic 0818
sensing system, active – 4301
sensing system, passive – 4301
sensitivity 4717
sensitometry 4341
separation, heavy-media – 4153
separation, magnetic – 4154
separation, normal – 1551
separation, normal horizontal – 1555
separation plane 4723
separation, stratigraphic – 1552
separation, vertical – 1551
separator 3797
septaria 3487
sequence 2573
sequence 3110
sequence, fining-upward – 3129
sequence, volcanic – 1238
sefac 2377
sere 1894
seriate 0820
sericitization 0932
series 1764
series, charnockite – 0990
series, igneous rock – 0630
series, rock – 0630
series, soil – 2663
serir 2578
serpentinite 0923
serpentinization 0923
serrate texture 0816
sessile 1919
set 3111
setting edge 0501
settlement 4704
settlement 4833
settlement, differential – 4704
settlement, differential – 4833
settlement gauge 4768
settling 2991
settling, crystal – 0673
settling velocity 2991
shadow zone 0216
shaft 2104
shaft 4848
shake 2095
shaking table 4157
shale 3220
shale break 3772
shale, carbonaceous – 3505
shale content 4623
shale, diatomaceous – 3371
shale line 4621
shale, oil – 3721
shale, paper – 3220
shale, siliceous – 3367
shaliness 4623
shaly 3220
shark-tooth projection 1251
shatter cone 0076
shattering 4821
shattering, frost – 1996
shattering zone 4821
shatter zone deposit 4060
shear 0105
shear 4677
shear box 4754
shearing off 1663
shear modulus 4699
shear plane 2055
shear strain 0105
shear strength 4684
shear stress 4688
shear test, torsional – 4765
shear test, vane– 4759
shear zone 1533
sheen 0490
sheet 0739
sheet 2129
sheet 4055
sheet erosion 2779
sheet, extruded – 1275
sheeting 1995
sheet piling 4799
sheet sandstone 3109
sheet wash 2032
shelf 2788
shelf edge 2788
shelf, insular – 2791
shelf, off– 2918
shelf, stable – 1089
shelfstone 2148
shell, multiple – 4102
shield 1081
shield 2118
shield cluster, volcanic – 1450
shield volcano 0035
shield volcano 1420
shield volcano, pseudo – 1430
shingle 1571
shingle 2912
shingle bar 2244
shistovite 0076
shoal 2247
shoal 2469
shoal 2803
shock 0138
shock, distant – 0150
shock, local – 0149
shock, main – 0142
shock wave 0274
shock zone 0894
shoestring sand 2899
shore 2438
shore, back – 2840
shore, exposed – 2336
shore face 2485
shore face 2840
shore, fore – 2840
Shore hardness scale 4701
shore, lee – 2336
shore line 2437
shore line, crenulate – 2532
shore line, graded – 2531
shore line, indented – 2532
shore line, inner – 2461
shore line, outer – 2461
shore line, smooth – 2531
shore line, straight – 2531
shore line, ungraded – 2531
shore profile 2441
Shore scleroscope 4733
shore, sheltered – 2336
shore, weather – 2336
short normal 4583
shot break 4427
shotcrete 4801
shothole pattern 4411
shotpoint 4411
shoulders – 4859
shrinkage crack 3044
shrinkage factor 3860
shrinkage hypothesis 1114
shrinkage limit 4715
SI 0703
si 1004
side, downthrown – 1559
siderite 0066
siderolite 0070
side, upthrown – 1559
sieve 0974
sieve analysis 3157
sieving 4873
sif 2178
sigma-t 2827
signal 4435
signal 4517
signal, acoustic – 4517
signal, seismic – 4517
signal-to-noise ratio 4443
signal, transient – 0176
significant, non– 4637
sign, optic – 0557
silcrete 4123
siliceous 3365
silicic 0605
silicification 0929

silicification	3458
silification	0929
silk	0490
silk	2178
silky	0484
sill	0740
sill	2796
sillar	1340
sill, ore –	4031
silt	2715
silt	3204
silting up	2341
silting up	2990
siltstone	3204
silty	3204
simulator, numerical –	3866
simultaneous	1729
sink	1455
sink	2072
sink	2075
sink and float	4153
sink, collapse –	2072
sink, funnel –	2072
sinkhole	2072
sinking fund	4219
sink, peripheral –	1697
sink, solution –	2072
sink, thaw –	2646
sink, volcanic –	1456
sinter	3385
sinter, siliceous –	1406
sinuosity	2216
sinus	0051
site	4789
site, building –	4789
site investigation	4790
size grade	3162
skär	2455
skarn	4144
skedophyre	0829
skedophyric	0829
skeletal	3169
skeleton	1861
skeleton, volcanic –	1476
skerries	2455
skewness	3198
skialith	1071
skin	2691
skin	3100
skin, calcite –	2142
skin effect	3815
skin effect	3851
skin friction	4839
skin friction, negative –	4839
slack	3513
slag	1294
slaggy	1294
slag lump	1297
slake durability	4734
slaking	2758
slant range	4393
slant-range image	4395
SLAR	4385
slate	0942
slate, knotted –	0941
slate, pencil –	0942
slate, spotted –	0941
sledge	4483
sledging	4483
sleet	2349
slicken	2731
slickenside	1538
slickenside	2731
slide	2043
slide	2976
slide	4020
slide, earth –	2046
slide, rock –	2048
sliding	2043
sliding	2976
slime	4160
Slingram	4575
slip	4280
slip along shear planes	2373
slip band	0539
slip, bedding-plane –	1623
slip, dip –	1564
slip, direction of the net –	1562
slip face	2185
slip face	2966
slip factor	3706
slip, flexural –	1665
slip, net –	1562
slip plane	0539
slip plane	2055
slip, strike –	1565
slip surface	1002
slip surface	2055
slip, vertical –	1563
slope	4827
slope	4860
slope, basal –	2603
slope, constant –	2602
slope, continental –	2790
slope crack	2277
slope, cut-bank –	2236
slope, front –	2558
slope, frontal –	2480
slope, island –	2791
slope, offshore –	2480
slope ratio	4827
slope recession	2602
slope retreat	2602
slope, slip-off –	2237
slope, undercut –	2236
slope, waning –	2602
slope, wash –	2603
slope, waxing –	2602
slot, solution –	2087
sludge	3210
sludge	4281
slug	4090
sluice	4659
sluice box	4158
slump	2045
slump	3055
slump ball	3055
slump fold	3055
slumping	2045
slumping	2975
slump mass	3055
slurry	3210
slush, snow –	2348
smoke ring	1364
smoothing	4511
SMOW	0454
smuth	3555
SN	4583
SNG	3609
snowbreak	2355
snow, buckshot –	2350
snow, corn –	2350
snow cornice	2356
snowdrift	2353
snow, drifted –	2347
snow, dry –	2349
snow, dust –	2346
snow, firn –	2363
snow, granular –	2350
snow limit	2354
snow line	2254
snow, overhanging –	2356
snow, penitent –	2365
snow, powder –	2346
snow recrystallization	2363
snow, sand –	2346
snowslide	2358
snowslip	2358
snow slush	2348
snow, spring –	2350
snow, sugar –	2350
snow, wet –	2349
snow, wild –	2346
snow, wind-packed –	2347
snow wreath	2353
SNP	4594
sod	2698
soda straw	2131
soil	2657
soil	4636
soil, ABC–	2695
soil, AC–	2695
soil, acid –	2737
soil, acid sulphate –	2739
soil, alkaline –	2737
soil association	2673
soil, azonal –	2669
soil, buried –	2693
soil classification	2661
soil complex	2673
soil creep	2780
soil erosion	2776
soil family	2664
soil-forming factor	2659
soil fraction	2712
soil genesis	2659
soil horizon	2687
soil, intrazonal –	2669
soil layer	2688
soil liming	4807
soil, lunar –	0060
soil mapping	2670
soil moisture	2764
soil order	2666
soil-organic matter	2740
soil pressure	4705
soil profile	2686
soil, residual –	2009
soil sampling survey	4644
soil science	2658
soil separate	2712
soil series	2663
soil slip	2046
soil stabilisation	4806
soil stripe	2654
soil structure	2723
soil suitability	2660
soil survey	4644
soil taxonomy	2662
soil texture	2709
soil ulmin	3525
soil-water pressure	2770
soil wedge	2641
soil, zonal –	2667
solar nebula	0021
solar system	0018
sole	1572
sole	3106
sole cast	3019
sole mark	3019
solfatara	1391
solfatara, acid –	1393
solfatara field	1392
solfataric area	1392
solfataric stage	1168
solidification, dermolithic –	1200
solidification index	0703
solifluction	2041
solifluction	2780
soligenous	3550
solum	2694
solution	0411
solution, gas in–	0641
solution, intrastratal –	3413
solution load	2956
solution opening	3690
solution, pressure –	3445
solution ripple	2086
solution sink	2072
somma	1461
somma volcano	1445
sonar, side-scan –	4485
sonic boom	0177
sorption	0407
sorted	3202
sorted, well–	3202
sorting	3200
sorting coefficient	3201
sorting, crystal –	0671
sounding	4710
sounding, deep seismic –	0094
sounding, depth –	4561
sounding device	4758
sounding, dynamic –	4711
sounding, static	4711
soundness	4734
sound source	4486
sounds, volcanic –	1237
sour	3716
source	0156
source area	2943
source level	4490
source parameter	0151
source rock	2944
source rock	3722
source window	0231
space, annular –	4279
spacecraft	4302
space group	0513
spacing	1497
spacing	4822
spacing, checkerboard –	4172
spacing, joint –	4725
spacing, linear –	4172
spacing, random –	4174
spacing, systematic –	4173
spar	3280
sparite	3280
spark-array	4493
sparker	4410
sparker, deep-towed –	4495
sparker source	4493
spate	2252
spatter cone	1298
spatter rampart	1452
speciation	1969
species	1937
species, climax –	1891
species, cosmopolitan –	1809
species, type –	1948
specimen	4741
specimen, hand –	4229
spectral log	4548
spectrometer, channel –	4547
spectrometer, differential –	4547
spectrometer, integral –	4547
spectrum, absorption –	4305
spectrum, amplitude –	4451
spectrum, continuous –	4305
spectrum, electromagnetic –	4304
spectrum, emission –	4305
spectrum, phase –	4451
spectrum, power –	0222
spectrum, power density –	4451
spelean	2877
speleogen	2125
speleolite	2124
speleology	2094
speleothem	2124
speleothem	2877
sphaeroid	3180
sphaerolite	0778
sphaerolite	3482
sphaerolitic	0876
sphaerosiderite	3483
sphenochasm	1143
sphenolith	0736
spherical cap	0363
sphericity	3180
spheroid	0330
spheroid	3180
spheroidal	0867
spheroidal	3180
spheroidal	4087
spherosiderite	3483
spherulite	3482
spherulite	4087

spherulite	0778
spherulite, compound–	0779
spherulitic	0876
spherulitic	3482
spherulitic	4087
spicularite	3372
spicule	1862
spiculite	0764
spiculite	3372
spilite	0922
spilitization	0922
spilling	2811
spill point	3762
spillway	2239
spillway	4868
spillway, glacial –	2393
spilosite	0933
spine	1424
spinifex texture	0850
spire	1298
spire	1424
spit	2464
spit, flying –	2466
spit, recurved –	2464
S-plane	1002
splash cup	2139
splash, rain –	2035
splash zone	2439
splay	1583
splay, crevasse –	2241
splitter, core –	4182
splitter, sample –	4182
splitting	1995
splitting	2210
splitting	2935
splitting	3159
SP log	4579
SP method	4557
spongework	2115
spongework, bedding-plane –	2116
spongework, joint-plane –	2116
spongework, roof –	2117
spongolite	3372
spongy	0875
spoor	1876
sporangium	1832
spore	1832
sporinite	3572
sporomorph	1831
sporophitic	0846
spot, bright –	4476
spot, dim –	4476
spot, flat –	4477
spotted	0965
spotty	4099
spray zone	2439
spread	4417
spread, centre –	4419
spread, cross –	4419
spreading	4514
spread, lateral –	2513
spread, split –	4419
spread, straddle –	4419
spring	3630
spring, artesian –	3631
spring, boiling –	1407
spring, hot –	3632
spring, hypogene –	1408
spring, karst –	2089
spring, mineral –	3632
spring, perched karst –	2091
spring, pulsating –	1404
spring, thermal –	1407
spring, thermal –	3632
spring, vaucluse –	3631
spring, vauclusian –	2093
spring, warm –	3632
sprouts	1260
sprouts, spinose –	1260
spumous	0875
spur	2806
spur	4071
spur-end facet	2553
spur, little-trimmed –	2228
spur, rocky –	2606
spur, truncated –	2628
spur, valley –	2228
squeeze	2098
squeeze-up	1288
sserir	2578
S-surface	1002
stabilisation, cement –	4807
stabilisation, lime –	4807
stabilisation, soil –	4806
stability analysis	4777
stack	2443
stack, envelope –	4510
stacking	4433
stack, moving-window –	4510
stack, phase –	4510
stage	1762
stage	2568
stage	3906
stage, concluding –	1167
stage, fumarolic –	1168
stage of maturity	2570
stage of old age	2570
stage of senility	2570
stage of youth	2570
stage, preliminary –	1157
stage, solfataric –	1168
staining	0492
stairway, giant –	2622
stairway, glacial –	2614
stairway, glacial –	2622
stalactite	2131
stalactite, botryoidal –	2136
stalactite, lava –	1293
stalactite, sheet –	2134
stalactite, straw –	2131
stalactite, tubular –	2131
stalacto-stalagmite	2138
stalagmite	2137
stalagmite, lava –	1293
stalagmite, mushroom –	2141
stalagmite, splash –	2141
stalagmite, stool –	2141
stalagmite, terraced –	2141
standard	0454
standard curve	4565
standard deviation	4190
standpipe	4274
state, liquid –	4713
state, plastic –	4713
state, semi-solid –	4713
state, solid –	4713
state, steady –	0397
station, air –	4348
stationary block	1569
station, base –	4526
station, camera –	4348
station correction	0160
station, exposure –	4348
station, ground –	4526
station spacing	4418
statistics, convex-concave –	4113
steady state	0397
steam eruption, secondary –	1403
steam soak	3839
steam vent	1389
S-tectonite	1002
steephead	2083
Stefanescu integral	4568
Steinmann's trinity	0696
Steinmann trinity	1098
stemming	4816
stenobathic	1911
stenobenthic	1909
stenohaline	1912
stenothermal	1913
stenotopic	1914
step	0527
step	1538
step, geothermal –	0380
step, glacial –	2623
step-out	3785
steptoe	1289
stereo base	4363
stereocomparator	4365
stereogenetic	1041
stereogenic	1041
stereogram	4258
stereo, inverted –	4367
stereometer	4365
stereopair	4362
stereo, pseudoscopic –	4367
stereoscope	4366
stereosome	1041
stereotriplet	4362
sterile	3917
stickiness	2750
sticky	2750
sticky point	2750
stictolite	1053
stictolith	1053
stinger	4522
stock	0730
stock, ore –	4032
stockpiling	4225
stockscheider	4133
stockwork	4058
stomach stone	1864
stone	0071
stone	2722
stone	3242
stone circle	2653
stone, composite –	0480
stone, core –	2609
stone, dimension –	4876
stone field	2057
Stoneley wave	0195
stone pavement	2656
stone polygon	2653
stone, pumice –	1308
stone ring	2653
stone river	2057
stone, rocking –	2163
stone stripe	2654
stone, synthetic –	0479
stone, water-faceted –	2445
stony	0835
stony-iron meteorite	0070
stony-irons	0070
STOOIP	3859
stoopway	2098
stoped block	0661
stoping	0623
stoping, arrested –	0623
stoping height, minimum –	4200
stoping, magmatic –	0623
stoping, overhand –	0623
stoping, overhead –	0623
stoping, ring-fracture –	1187
stoping, underhand –	0623
stoping width, minimum –	4200
storage capacity	3625
storage coefficient	3709
storage, elastic –	3709
storage project, pumped –	4870
storativity	3709
storm, magnetic –	4529
storm surge	2824
stoss side	2963
strain	0104
strain	4666
strain gauge	4767
strain meter	0122
strain meter	4767
strain release	0107
strain shadow	0958
strain, transient –	0114
strand	2485
strand plain	2492
strangulation	3904
strata, displaced –	1523
strata, inclined – -	1482
strata, tilted –	1482
strata, upturned –	1481
stratification	3095
stratification, cross –	3133
stratification, density –	2882
stratification plane	3096
stratified	3095
stratotype	1740
stratotype, boundary –	1744
stratotype, component –	1742
stratotype, composite –	1742
stratotype, unit –	1741
stratovolcano	1429
stratum	1750
stratum	3097
stratum, foundation –	4836
stratum, water-bearing –	3634
streak	0466
streak	3101
streaking, mineral –	0973
stream	2191
stream, allochthonous –	2259
stream, allogenous –	2259
stream, captor –	2325
stream, capturing –	2325
stream development	2197
stream, englacial –	2388
streamer	2173
stream, glacial –	2392
stream line	2372
stream line	3708
stream load	2951
stream, main –	2206
stream, master –	2206
stream, robber –	2325
stream-sediment survey	4641
stream, side –	2207
streamsink	2075
stream, stem –	2206
stream, subglacial –	2388
stream, supraglacial –	2388
stream system	2205
stream, trunk –	2206
strength	0243
strength	4682
strength, compressive –	4683
strength, peak –	4685
strength, point-load –	4733
strength, residual –	4685
strength, shear –	4684
strength, tensile –	4684
strength, triaxial compressive –	4683
strength, uniaxial compressive –	4683
strength, yield –	4697
stress	0103
stress	4667
stress, deviatory –	0103
stress, differential –	0103
stress, effective –	4668
stress, normal –	4688
stress, principal –	0103
stress, principal –	4692
stress-relief test, borehole –	4766
stress shadow	0958
stress, shear –	4688
stress-strain curve	0106
stress-strain diagram	4667
stress, yield –	4697
stria	1538
striae, glacial –	2397
striation	0468
striation	1538
striation	2397
striation	2999
striation, limit of glacial –	2631
strike	1483
strike	2286

strike 4016
strike and dip readings 4228
strike, azimuth of the – 1483
strike, bearing of the – 1483
strike, direction of the – 1483
strike line 1483
strike separation 1554
strike slip 1565
stringbog 2636
string, drilling – 4270
stringer lead 4060
stringer, ore – 4039
stripe 0945
stripe, soil – 2654
stripe, sorted – 2654
stripe, stone – 2654
stripped 2595
stripping 4875
stripping ratio 4202
stromatactis 3437
stromatholith 1049
stromatite 1049
stromatolite 3332
stromatolite, proto– 3332
Strombolian-type eruption 1159
strontium, common – 0448
strontium, initial – 0448
strontium, radiogenic – 0448
structural analysis 0995
structural petrology 0995
structure 0858
structure 2568
structure, air-heave 3065
structure, banded– 0860
structure, banded– 0945
structure, banded – 4049
structure, bioerosion – 1883
structure, biostratification – 1884
structure, bioturbation – 1875
structure, bioturbation – 1885
structure, bird's eye – 3304
structure, blue-bands – 2374
structure, boxwork – 4104
structure, brecciated – 2933
structure, cell – 4089
structure, chain – 4083
structure, chickenwire – 3468
structure, chute and pool – 3009
structure, cocarde – 4096
structure, cockade – 4096
structure, colloform – 4091
structure, colloidal – 4091
structure, columnar – 1501
structure, comb – 4084
structure, concentric – 4092
structure, concretionary – 4092
structure, cone-in-cone – 1445
structure, cone-in-cone – 3469
structure, cryptovolcanic – 1184
structure, cut and fill – 3024
structure, deformation – 3050
structure, diapiric – 1693
structure, directional – 2998
structure, dish – 3069
structure, dwelling – 1880
structure, exsolution – 4106
structure, feeding – 1881
structure, flame – 3061
structure, fluidal – 0868
structure, fluid-escape – 3068
structure, geopetal – 2996
structure, grade of – 2726
structure, granular – 3081
structure, gravity-collapse – 1704
structure, halokinetic – 1694
structure, halotectonic – 1694
structure, hammock – 4076
structure, horsetail – 3401
structure, horsetail – 4069
structure, imbricate – 1571
structure, imbricated – 3040
structure, intrastratal flow – 3052
structure, linear – 1607
structure, load – 3059
structure, metacolloidal – 4091
structure, mosaic – 3089
structure, network – 4104
structure, piercement – 1693
structure, pinch-and-swell – 1609
structure, pinch-and-swell – 4054
structure, pit and mound – 3066
structure, pitch-and-flat – 4055
structure, platy – 0868
structure, plumose – 1511
structure, porphyroblastic – 3470
structure, primary sedimentary – 2995
structure, prismatic – 1501
structure, reticulate – 4104
structure, rhyotaxitic – 0868
structure, rib and furrow – 3151
structure, ribbon – 4049
structure, rodding – 1608
structure, ruin marble – 3467
structure, sag – 3059
structure, scaly – 4092
structure, schistose – 1603
structure, scour and fill – 3024
structure, secondary sedimentary – 2995
structure, sedimentary – 2994
structure, segregation – 4106
structure, shingle-block – 1571
structure, soil – 2723
structure, spur-and-groove – 3351
structure, symplectic – 4105
structure, symplectitic – 4105
structure, tepee – 3075
structure type 2726
structure, unmixing – 4106
structure, vermicular – 34667
structure, volcanic – 1413
structure, volcano-tectonic – 1182
structure, water-escape – 3068
study, preliminary – 4639
stylolite 3451
subaerial 2866
subaeric 3979
subage 1763
subalkalic 0694
subaluminous 0694
subangular 2729
subangular 3189
subaquatic 3979
subaqueous 2880
subarkose 3233
subbase 4841
subbiozone 1782
subcrater 1439
subcrop 4234
subduction zone 1134
subfacies, metamorphic – 0987
subfamily 1930
subgenus 1933
subgrade 4840
subgraywacke 3234
subgreywacke 3234
subhedral 0523
subhedral 0751
subhedron 0751
subidioblast 0980
subidiomorphic 0751
subjacent body 0728
subjacent mass 0728
subglacial 2873
sublimate 1394
sublimate product 1394
sublimation 0412
sublimation 3952
sublittoral 2839
sublittoral, inner – 2839
sublittoral, outer – 2839
submaceral 3567
submarginal 4212
submarine 2900
submature 3176
submergence 1728
submergence 2857
subophitic 0845
subrounded 3188
subsequent 2286
subsequent 3474
subsidence 2857
subsidence 4704
subsidence, cauldron – 0743
subsidence, cauldron – 1186
subsidence, external – 1464
subsilicic 0603
subsoil 2696
subspecies 1939
substage 1763
substitution, atomic – 0477
substitution, coupled – 0477
substrate 3107
substratum 0348
substratum 2696
substratum 3107
substratum of a volcano 1414
subterranean 2876
subtidal 2908
subtuberant mountain 0729
subvolcanic 0617
subvolcanic 3989
subvolcano 1206
subzone 1782
success ratio 3784
sucrosic 0811
suction, moisture – 2770
suctive 0619
sugary 0811
sugary 3090
suite 0630
sulphurization 3964
sulphur mud 1399
sulphur-mud column 1400
sulphur-mud flow 1399
sulphur-mud pool 1398
sulphur-mud tree 1400
summit concordance 2598
summit level 2597
summit levels, accordance of – 2598
summit levels, stairway of – 2599
summit level, storeyed – 2599
summit overflow 1266
sun crack 3045
superfamily 1930
supergene 3937
supergene 3967
superimposed 2290
superimposed 4630
superindividual 1012
supermature 3176
superparamagnetism 0300
superposed 2290
superposition 1714
superposition, law of – 1714
superprint 0970
superspecies 1935
superstage 1763
superstructure 1414
superzone 1782
support 4858
suppression 4567
supracrustal 0584
supralittoral 2098
supralittoral 2837
suprastructure 1077
supratidal 2908
supravolcano 1413
surf 2812
surface, crestal – 1632
surface density 4111
surface, enveloping – 1672
surface, initial – 2569
surface, listric – 1573
surface, median – 1673
surface, polished – 1538
surface, sag and swell – 2630
surface, specific – 2754
surface, specific – 4109
surface texture 3192
surface wash 2031
surfacing 4842
surf zone 2439
surf zone 2812
surging 2811
surreitic 1050
survey, directional – 4294
survey, directional – 4599
survey, drainage – 4641
survey, gas – 4648
survey, geochemical water – 4640
survey, gradient – 4563
surveying, gamma-ray – 4545
surveying, radioactivity – 4545
survey, lithogeochemical – 4644
survey, orientation – 4639
survey, pilot – 4639
survey, rock sampling – 4644
survey, soil – 2670
survey, soil – 4644
survey, vapour – 4648
survey, vegetation – 4645
susceptibility, magnetic – 0298
suspended load 2955
suspension 0411
suspension current 2980
suspension load 2955
sutured 0960
suture zone 1132
suturic 3450
swale 2234
swale 2486
swallet 2075
swallow hole 2075
swallow hole, coastal – 2075
swamp, coastal – 2514
swamp, mangrove – 2515
swamp, paralic – 2514
sward 2698
swarm earthquakes 0144
swash 2813
S-wave 0184
sweep 4503
sweep time 4503
sweet 3716
swell 2807
swell 4054
swelling 4720
swelling index, free – 3595
swelling number, crucible – 3595
swirl 2269
swirl mark 0481
swirls 0665
swivel, water – 4268
symbiosis 1898
symmict 3124
sympatric 1817
symplectic 0849
symplectitic 0849
symplektic 0849
symplektitic 0849
symptoms, premonitory – 1191
synantetic 0784
synantetic 3963
synchronic 1818
synchronous 1729

synchronous 1818
syncline 1619
syncline, antiformal – 1620
syncline, doubly plunging – 1638
syncline, rim – 1697
synclinorium 1671
syndepositional 2854
syndiagenesis 3398
syneclise 1084
synecology 1888
syneresis crack 3044
synform 1619
syngenesis 3398
syngenetic 2854
syngenetic 3474
syngenetic 3553
syngenetic 3966
synkinematic 0726
synkinematic 0906
synkinematic 1079
synneusis 0821
synonym 1955
synonymy 1955
synorogenic 0726
synorogenic 1108
synsedimentary 2854
syntectite 0682
syntectonic 0726
syntexis 0682
synthetic 4478
syntype 1951
system 1765
system, active – 4301
system, passive – 4301
system, passive – 4301
system, solar – 0018

T

table 0501
table, concentrating – 4157
table land, stripped structural – 2555
tablemount 2798
table, turn – 4269
tabular 0522
tabular 0755
tachylite 1262
taconite 4114
tafezza 4123
tafone 2002
tailings 4164
talik 2638
talik, intrapermafrost – 2638
talik, open – 2638
talik, subpermafrost – 2638
talik, suprapermafrost – 2638
talus 2867
talus cone 2007
talus, continental – 2790
talus fan 2007
talus, insular – 2791
talus slope 2007
tamping 4809
Tamyr polygon 2655
tank oil 3859
tape, ore – 4039
taphocoenosis 1896
taphonomy 1897
taphrogenesis 1091
taphrogeny 1091
tar 3743
tarmat 3781
tarrace, ice-marginal – 2420
tar sand 3768
tar seal 3765
tautonym 1957
taxite 1243
taxogenesis 1963
taxon 1926
taxon 2662
taxonomy 1920
taxonomy, soil – 2662
taxon-range zone 1783
T.D. curve 0171
TDS 3675
TDT 4595
tectofacies 2861
tectogen 1101
tectogenesis 1100
tectonic analysis 0996
tectonic fabric 1007
tectonic overpressure 0886
tectonic rotation 1000
tectonics 1073
tectonics, gliding – 1702
tectonics, granite – 0712
tectonite 1001
tectonite 1605
tectonite, fusion – 1001
tectonite, primary – 1001
tectonite, secondary – 1001
tectonite, slip – 1002
tectonosphere 0080
tectotope 2862
teilzone 1784
tektite 0077
telemagmatic 3986
telescoping 3965
teleseism 0150
telethermal 3984
telinite 3569
telluric 0002
telluric method 4553
temblor 0138
temperature system, low – 0393
tenacity 0470
tension, moisture – 2771
tephra 1310
tephra 3170
tephra cone 1428
tephrochronology 1312
terminate at depth, to – 1526
termination, hacksaw – 3414
termination, saw-toothed – 3414
termination, serrate – 3414
terra 0050
terrace 2801
terrace, alluvial – 2298
terrace, alternate – 2304
terrace bluff 2295
terrace, built – 2298
terrace, built-up – 2298
terrace, climatic – 2308
terrace, coastal – 2484
terrace, congelifractate – 2309
terrace, constructional – 2298
terrace, continental – 2790
terrace, continuous – 2299
terrace, cut – 2298
terrace, cyclic – 2299
terrace, defended-cusps – 2305
terrace, depositional – 2337
terrace, diastrophic – 2307
terrace, dipping – 2311
terrace, drift – 2298
terrace edge 2295
terrace, emerging – 2311
terrace, erosion – 2298
terrace, erosional – 2337
terrace, eustatic – 2308
terrace face 2295
terrace, field – 2061
terrace, fill-in fill – 2301
terrace, fill – 2298
terrace flight 2296
terrace, fluvioglacial – 2309
terrace front 2295
terrace, ice-marginal – 2420
terrace intersection – 2311
terrace, kame – 2420
terrace level 2294
terrace, local – 2299
terrace, main – 2299
terrace, marine – 2484
terrace, meander-scar – 2304
terrace, meander-spur – 2303
terrace, noncyclic – 2299
terrace, plunging – 2311
terrace, polygenic – 2302
terrace, reef-front – 3352
terrace, river – 2293
terrace, rock – 2298
terrace, rock-defined – 2305
terrace, rock-perched – 2305
terrace scarp 2295
terrace, shore – 2337
terrace, shore – 2484
terraces, inset – 2300
terrace, slip-off slope – 2302
terrace slope 2295
terraces, matched – 2300
terraces, matching – 2300
terraces, paired – 2300
terraces, stepped – 2296
terraces, stepping – 2296
terrace, stream – 2293
terrace, stream-built – 2298
terrace, stream-cut – 2298
terrace, structural – 2307
terrace, structural – 2555
terracette 2059
terrace, valley-plain – 2299
terrace, wave-built – 2337
terrace, wave-built 2479
terrace, wave-built – 2480
terrace, wave-cut – 2337
terrain, braided – 0041
terrain, chaotic – 0038
terrain, etched – 0040
terrain, fretted – 0039
terrain, laminated – 0040
terrain unit 4785
terrestrial 2865
terrigene 3971
terrigenous 2902
tesselon 2641
test-ban treaty 0266
test, formation – 3787
test, plate bearing – 4708
test, production – 3788
test-site 0267
test, stratigraphic – 3783
Tethys 1127
tetragonal 0515
textural class 2713
textural classification 2713
texture 0790
texture 3077
texture 4345
texture, cell – 4089
texture, comb – 4084
texture, crackled – 4097
texture, granular – 3081
texture, granule – 4086
texture, hornfelsic – 0940
texture, kneaded – 2649
texture, ocellar – 0855
texture, oleander-leaf – 4082
texture, pegmatitic – 3470
texture, soil – 2709
texture, surface – 3192
thalweg 2279
thanatocoenosis 1896
thaw depression 2646
thaw lake 2646
thaw sink 2646
theory, ascension – 3927
theory, descension – 3926
theory, hydraulic – 2263
thermal 0376
thermal aureole 0899
thermal diffusivity 0387
thermally normal region 0391
thermal, middle – 3983
thermistor 0379
thermocline 2333
thermocline 2830
thermocouple 0379
thermogene 0977
thermohaline 2829
thermokarst 2645
thermokarst lake 2646
thermometer, geochemical – 0379
thermometer, geological – 0994
thermometry, geological – 0994
thermometry, isotope – 0456
thick-bedded 3108
thickener 4159
thickness, bed – 4622
thickness, net sand – 3873
thickness, true – 4243
thin-bedded 3108
thin out 3104
thixotropic 4670
thixotropy 4670
THM 4655
tholeiite 0695
tholoid 1422
tholus 0037
thorium-lead dating 0436
thread, ore – 4039
threshold, glacial – 2625
threshold treaty 0266
threshold value 4627
thrombolite 3330
through, ripple – 3001
throw 1551
throw 4818
throw, stratigraphic – 1552
thrust, break – 1577
thrust, hydraulic – 4737
thrust, distance of – 1568
thrust plane 1568
thrust plane, basal – 1572
thufur 2643
thumb marks 0074
thunderstorm, volcanic – 1373
tidal current 2821
tidal flat 2505
tidal-flat area 2505
tidalite 2907
tidal range 2819
tidal section 2504
tide 2818
tide, earth – 0199
tide, high – 2818
tide, low – 2818
tide, mean high – 2820
tide, mean low – 2820
tide, range of the – 2819
tide, red – 2843
tight 1626
tight hole 3794
tightness, water – 4863
tight rock 3769
till 2404
till 2874
till 3250
till, ablation – 2404
tillite 2874
till, lodgment – 2404
till, marine – 2903
tilloid 2978
till-plains topography 2630
tilt 0124
tilt 4351
tilt 4833
tilth 2757
tiltmeter 0124
tilt, relative – 4351
time break 4427
time-distance curve 0171
time-distance graph 0171
time domain 0223
time domain 4570

time integral, normalized – 4570
time-stratigraphical unit 1757
time, travel – 0171
time, uphole – 4428
tind 2618
tjäle 2635
tjäle, perenne – 2637
tjrp 4492
tombolo 2467
tongue 0725
tongue 1752
tongue 2448
tonguing 3826
tonstein 3505
tool, formation density compensated – 4593
tool, HDT-dipmeter – 4598
tool, sampling – 4745
top 0705
top 3106
top eruption 1229
topogenous 3550
topography, knob and kettle – 2630
topography, morainic-belt – 2630
topography, sag and swell – 2630
topography, till-plains – 2630
topomineral 3995
topotactic 0543
topotaxy 0543
topotype 1953
topozone 1784
topset bed 2896
tor 2609
torbanite 3563
torrential 2884
torsion balance 0350
tortuosity 3707
touchstone 3369
tourmalinization 0932
tower karst 2082
T-plane 1017
trace 1492
trace 1876
trace 4420
trace, crawling – 1879
trace element 0400
trace fossil 1873
trace fossil 3067
trace, grazing – 1881
trace, resting – 1880
trachyophitic 0846
trachytic 0838
trachytoid 0839
track 1877
track, pasture – 2060
track, sheep – 2060
trackway 1877
traction 2953
traction load 2952
trail 1878
trailway 1878
train 4631
trajectory 4471
transceiver 4488
transducer source 4488
transfer function 4448
transfluence 2383
transformation 1062
transformation 3426
transfusion 1069
transgression 1726
transgressive 0718
transition cone 2423
transition zone 0091
transition zone 3777
transition zone 4620
transit time 0171
translation 2370
translational movement 1566
translation glide line 1017
translation glide plane 1017
translucent 0545
transmissibility 3702
transmission, beamed – 4491
transmissivity 3702
transmit-receive switch 4489
transmittance 4315
transmitter 4487
transparent 0545
transport, eolian – 2154
transport, mass – 2037
transport, tectonic – 0998
transport, wind-borne – 2154
transvaporisation 3931
transverse 2289
trap 0638
trap 3756
trap, anticlinal – 3757
trap, buried-hill – 3760
trap, fault – 3757
trap, hydrodynamic – 3761
trapp 0638
trap, permeability – 3758
trappide 0638
trap, pinch-out – 3758
trap, sand – 3152
trap, shale-out – 3758
trap, stratigraphic – 3758
trap, structural – 3757
trap, unconformity – 3759
trap, wedge-out – 3758
travel path 4471
travel time 0171
travel-time curve 0171
travel-time table 0171
travertine 4123
treatment, artificial – 0492
tree-mold, lava – 1292
trellis 2292
tremblor 0138
tremometer 0118
tremor 0146
tremor, earth – 0146
tremor, earthquake – 0146
tremor, volcanic – 1193
trench 2793
trench 4235
trench, cut-off – 4861
trench, fault – 1590
trenching 4561
trend 1493
treptomorphism 0881
triangulation, radial – 4324
triaxial cell 4753
tribe 1931
tribus 1931
tributary 2207
trichite 0770
triclinic 0515
trigonal 0515
trimacerite 3584
trimorphism 1869
tripartite method 0152
triple junction 1130
triplet 0480
triplet, stereoscopic – 4362
tripoli 3371
tritium dating 0452
trituration 1319
tromometer 0118
trona 3386
Tro-Pari survey instrument 4295
trough 2793
trough banding 0870
trough bench 2619
trough end 2619
trough, fault – 1590
trough, fault – 2546
trough headwall 2619
trough line 1636
trough, offshore – 2460
trough shoulder 2619
trough surface 1636
truncation, surface of – 2591
tsunami 2816
tube 2099
tube, ceiling – 2106
tube, feeding – 1290
tubing 4275
tufa, reef – 3312
tuff 1332
tuff 3171
tuffaceous 1332
tuffaceous 3171
tuff, ash – 1333
tuff, ash flow – 1340
tuff, chaotic – 1333
tuff cone 1428
tuff, crystal – 1333
tuff, crystal-lithic – 1333
tuff, crystal-vitric – 1333
tuff, dust – 1333
tuff, explosion – 1335
tuff, flood – 1340
tuff flow, hot – 1369
tuff flow, incandescent – 1369
tuffisite 1339
tuffite 1332
tuffite 3171
tuff, lapilli – 1333
tuff lava 1341
tuff, lithic – 1333
tuff, lithic-crystal – 1333
tuff, palagonite – 1347
tuff, pisolitic – 1333
tuff, pumice – 1333
tuff, reworked – 1337
tuff, scoria – 1333
tuff, sedimentary – 1334
tuff, stratified dry – 1336
tuff, subaeric – 1335
tuff, subaqueous – 1335
tuff, vitric – 1333
tuff, welded – 1341
tumefaction of lava 1272
tumescence 1194
tumulus 1272
tundra 2636
tundra hummock 2643
tundra pond 2646
tunnel 4847
tunnelling, shield – 4853
Turam 4574
turbidite 2983
turbidity current 2981
turbidity current, nose of a – 2982
turbidity current, steady – 2981
turbidity current, tail of a – 2982
turbodrilling 4264
turf 2698
TVG 4508
T-wave, 0201
twin 0535
twin, contact – 0537
twin, interpenetration – 0537
twin law 0535
twinning 1000
twinning, polysynthetic – 0538
twin, penetration – 0537
T-X graph 0171
T-X graph 4463
type area 1739
type locality 1739
type locality 1946
type region 1739
type section 1740

U

ubehebe 1426
U-figure 2754
ulmification 3524
ultrabasic 0604
ultrabasite 0604
ultramafic 0600
ultramafite 0600
ultrametagranite 1026
ultrametamorphism 0882
ultramylonite 0916
ultramylonite 1541
ultravolcanian 1166
unconformable 1720
unconformity 1720
unconformity, angular – 1721
unconformity, erosional – 1722
unconformity, nonangular – 1721
unconformity, parallel – 1721
unda 2913
undaform zone 1913
undathem 2913
undation 1122
undation hypothesis 1121
underclay 3515
undercliff 2476
undercutting 2471
undercutting action 2023
underflow 3626
underlayer 3107
underlie 4018
undermining 2471
undermining action 2023
underpinning 4796
underream, to – 4296
undersaturated 0605
undersaturated, critically – 0605
underthrust 1574
undertow 2826
uniaxial 0553
uniformitarianism 0006
unit agreement, operating – 3901
unit, biostratigraphical – 1770
unit, British thermal – 3605
unit, chrono-stratigraphical – 1757
unit, evaporative – 3605
unitisation 3901
unit, lithostratigraphic – 1748
unit, time-stratigraphical – 1757
universal stage 0614
unloading 1995
unmixing, tectonic – 0947
unsaturated 0605
unsaturated zone 3651
unwatering 4739
up-dip 1489
upheaval of plug 1247
upland 2518
upland, fretted – 2616
uplands 0029
upland, truncated – 2588
uplift, hydrostatic – 4740
uprush 2813
upstream side 2963
upthrow 1559
upwelling 1358
upwelling 2786
uralitization 0928
uranium-lead dating 0434
uranium-series disequilibrium dating 0443
U-stage 0614
uvala 2073

V

vacuole 0874

vadose compaction 3416
vadose diagenesis 3399
vagility 1917
vagrant 1918
vale, inner – 2560
validity 1923
valle 0034
valley, anticlinal – 2544
valley, asymmetric – 2274
valley, beheaded – 2324
valley, betrunked – 2481
valley, blind – 2083
valley bottom 2278
valley, collapse – 2085
valley, cul-de-sac – 2083
valley, dead – 2328
valley, death – 1402
valley development 2270
valley, drowned – 2282
valley, dry – 2084
valley, dry – 2328
valley, dune – 2188
valley, extended – 2282
valley, fault – 2547
valley, fault-line – 2547
valley, fault-zone – 2547
valley fill 2278
valley, flood-plain – 2278
valley floor 2278
valley-floor side strip 2603
valley, hanging – 2483
valley, hanging glacial – 2627
valley head 2276
valley, homoclinal – 2544
valley, ice-marginal – 2392
valley line 2279
valley, monoclinal – 2544
valley mouth 2280
valley, multiple-cycle – 2275
valley, open – 2276
valley, perched – 2627
valley plain 2278
valley, poison – 1402
valley profile, longitudinal – 2271
valley profile, transverse – 2271
valley, proglacial – 2393
valley, rift – 2546
valleys, formation of – 2270
valley side 2277
valley-side slope 2277
valley spur 2228
valley, structural – 2543
valley, submarine – 2795
valley, synclinal – 2544
valley system, dismembered – 2482
valley system, engrafted – 2283
valley, tectonic – 2543
valley, trough-shaped – 2273
valley, two-cycle – 2275
valley, two-storey – 2275
valley, U-shaped – 2273
valley, V-shaped – 2272
valley wall 2277
valley, warped – 2543
valuation, mine – 4216
value 2677
value, deferred – 4220
value, present – 3897
values, erratic – 4186
values, extreme – 4186
vane apparatus 4759
vane-shear test 4759
vapour-dominated system 0393
vapour survey 4648
vapour tension, retrograde increase of – 1174
VAR 4421
variability, coefficient of – 4190
variance 4190
variation diagram 0697
variation, diurnal – 4528
variation, secular – 0289
variegated 3462
variety 1940
variogram 4196
variole 0781
variolite 0781
variolitic 0877
variometer 0288
varve 3128
varvite 3128
vault 0116
vegetation survey 4645
vein, banded – 4049
vein, bed – 4053
vein, bed – 4077
vein, bedded – 4053
vein, brangled – 4048
vein, brecciated – 4050
vein, bunchy – 4047
vein, chambered – 4050
vein, comby – 4049
vein, complex – 4046
vein, composite – 4045
vein, contact – 4043
vein, counter – 4077
vein, cross – 4077
vein crossing 4074
vein, crustified – 4049
vein, cuneate – 4035
vein deposit, sheeted – 4060
vein, dilatation – 4041
vein, dradge – 4048
vein dyke 4031
vein, eruptive – 0748
vein filling 1515
vein, fissure – 1514
vein, fissure – 4044
vein, gash – 4056
vein, ground – 2641
vein, intrusive – 0748
vein, ladder – 4057
vein, leader – 4046
vein, lenticular – 4026
veinlet 4039
vein, main – 4062
vein, major – 4062
vein, master – 4062
vein, metalliferous ore – 4038
vein, rake – 4044
vein, replacement – 4042
vein, ribbon – 4049
vein, right – 4044
vein, saddle – 4053
veins, branching of – 4070
veins, bunch of – 4066
veins, cluster of – 4066
veins, field of – 4061
vein, simple – 4045
veins, network of – 4065
vein, splitting up of a – 4070
vein, step – 4055
veinstone 3921
vein swarm 4066
vein system, fan-shaped – 4063
vein system, linked – 4067
vein system, net– 4064
vein system, parallel – 4063
vein sytem, radial – 4063
vein, trough – 4053
vein, true – 4044
vein, trunk – 4062
velocity, apparent – 0208
velocity, apparent – 4468
velocity, average – 4465
velocity, first critical – 2265
velocity function 4469
velocity, group – 0170
velocity index 4732
velocity, interval – 4467
velocity, particle – 4414
velocity, phase – 0169
velocity, root-mean-square – 4466
velocity, rupture – 0239
velocity, second critical – 2266
velocity, stacking – 4434
velocity, thread of maximum – 2268
velocity, threshold – 2959
velocity, wave – 0168
VEM 4575
veneer 3126
Vening Meinesz, axis of – 1137
venite 1045
vent agglomerate 1248
vent breccia 1248
vent, cylindrical – 1224
vent, feeding – 1221
vent, fissure – 1225
ventifact 2161
vent, rootless – 1223
vent, shifting of the volcanic – 1231
vent, steam – 1389
vent, volcanic – 1221
vermiform 0774
vertical 0329
vertical-loop method 4575
vesicle cylinder 1306
vesicular 0874
vesicular 1305
vesicule 0874
vestigial 1978
Vesuvian-type eruption 1160
vibration 4809
vibration, free – 0198
vibration measurer 0123
vibrocompaction 4809
vibroflotation 4810
vibrograph 0123
vibroseis 4409
virgation 1675
viscosity 4671
vitrain 3580
vitreous 0805
vitrifaction 0708
vitrinertite 3583
vitrinite 3569
vitriphyric 0823
vitrite 3582
vitrodetrinite 3569
vitrophyre 0823
vitroporphyric 0823
vitrous 0484
VLF 4572
void 3064
void ratio 3682
voids 3683
volatile 0650
volatile component 1172
volatile matter 3598
volcanic 1148
volcanic 3989
volcanic centre 1220
volcanic chamber 1202
volcanic cycle 1156
volcanic energy 1171
volcanic mass 1413
volcanic phase 1197
volcanic pile 1413
volcanic tremor 1193
volcanic wreck 1476
volcanism 1147
volcanism, epigenetic – 1152
volcanism, external – 1207
volcanism, orogenic – 1151
volcanism, subrecent – 1150
volcanite 1238
volcano 1411
volcano, abortive – 1222
volcano, active – 1153
volcano, bedded – 1429
volcano belt 1447
volcano, central – 1434
volcanoclastic 3172
volcanoclastic deposit 1310
volcano, cleaved – 1475
volcano, composite – 1441
volcano, compound – 1441
volcano, cumulo – 1422
volcano, dissected – 1475
volcano, dormant – 1155
volcano, embryonic – 1222
volcanoes, chain of – 1448
volcanoes, cluster of – 1450
volcanoes, row of – 1448
volcano, extinct – 1155
volcano, furrowed – 1474
volcanogenic 1148
volcano girdle 1447
volcano, imbricated – 1421
volcano, island – 1416
volcanology 1146
volcano, mud – 1395
volcano, mud – 1700
volcano, mud – 3071
volcano, sand – 3071
volcano, shield – 0035
volcano, shield – 1420
volcano, simple - 1441
volcano, solitary – 1446
volcano, somma – 1445
volcano, stratified – 1429
volcano, subactive – 1153
volcano, subaerial – 1415
volcano, subaquatic – 1415
volcano, submarine – 1416
volcano, substratum of a – 1414
volcano-tectonic structure – 1182
volcano theory 0049
volcano, twin – 1442
volcano-watching 1190
volcano zone 1447
voltage 4556
volume 2249
volume control, automatic – 4425
volume control, automatic – 4508
volume factor gas, formation – 3861
volume factor oil, formation – 3860
volume fraction 4109
volume, gross rock – 3872
volumetric method 3871
vortex 2269
vug 4014
vugh 4014
Vulcanian-type eruption 1160
vulcanic 1148
vulcanism 1147
vulcanology 1146

W

wacke 3235
wackestone, lime – 3296
wadi 2587
wall 0705
wall 1453
wall 1529
wall 4009
wall 4852
wall, cut-off – 4861
wall fall 2111
wall, foot – 4009
wall, free – 4013
wall, frozen – 4013
wall, frozen to the – 4013
wall, hanging – 1529
wall, hanging – 4009
wall, lying – 4009

wall, retaining – 4797
wall rock 0709
wall rock 1530
wall rock 3501
wall rock 3920
want 3510
warping 2513
wash 2869
washboring 4751
washing 2937
washing 4873
wash-out 3023
wash-out 3510
wash-out channel 2026
washover 2500
washover fan 2500
washpipe 4751
wash, rill – 2033
wash, sheet – 2032
wash, subsurface – 2034
wash, surface – 2031
wash zone 2439
wastage, surface – 2013
waste 2005
waste 3924
waste disposal 4831
waste dump 4831
waste mantle 2009
waste tip 4831
water, atmospheric – 3613
water, available – 2773
waterbloom 2843
water, bottom – 3779
water, brackish – 3676
water, connate – 3661
water content 2769
water control 2775
watercut 3802
watercut 3875
water, deep – 2809
water, disposal – 3803
water, edge – 3779
waterfall 2203
water, flood – 2251
water, formation – 3662
water fountain 1359
water, fresh – 3676
water, funicular – 3654
water, high – 2818
watering 4810
water, intermediate – 3780
water, juvenile – 1348
water, juvenile – 3664
water level, free – 3778
water line, meteoric – 0458
water, low – 2818
water, magmatic – 1348
water mass 2828
water, mean high – 2820
water, mean low – 2820
water, metamorphic – 0976
water, meteoric – 3614
water, native – 3662
water of compaction 3660
waterparting 2314
water, pendular – 3653
water, phreatic – 3659
water, pit – 3598
water quality 3674
water, regenerated – 0976
water, regenerated – 3663
water, saline – 3676
water sand 3766
water sandstone 3766
water saturation 4611
water, shallow – 2809
watershed 2276
watershed 2314
watershed 3648
watershed, zig-zag – 2321
water, soil – 3652
water, subsurface – 3649
water, surface – 3649
water table 3643
water table 3777
water table, apparent – 3643
water-table class 2766
water-table isohyps 3644
water thrust 4737
water tightness 4863
water type 2828
water, vadose – 3655
wave, acoustic – 4482
wave, air – 0203
wave base 2442
wave, bodily seismic – 0179
wave, body – 0179
wave, boundary – 0196
wave, channel – 0195
wave, circumferential – 0188
wave complexity 0275
wave, compressional – 0183
wave, conical – 0186
wave, converted – 0185
wave, core – 0180
wave, coupled – 0193
wave current ripple 3011
wave, dilatational – 0183
wave, distortional – 0184
wave, equivoluminal – 0184
wave front 0549
wave front 4470
wave, gravitational – 0202
wave guide 0101
wave, guided – 0195
wave, head – 0186
wave height 2808
wave in deep water 2809
wave in shallow water 2809
wave, interface – 0196
wave, irrotational – 0182
wave, leaky – 0195
wave length 0162
wave length 2808
wave, long – 0188
wave, longitudinal – 0182
wave, mantle – 0194
wave number 0162
wave of translation 2815
wave path 0206
wave period 2808
wave, pressure – 0182
wave, primary – 0182
wave, push – 0182
wave, push-pull – 0128
wave, querwelle – 0189
wave ray 0206
wave, rotational – 0184
wave, secondary – 0184
wave, seismic – 0161
wave, seismic – 4481
wave, shake – 0184
wave, shear – 0184
wave, shock – 0274
wave, shock – 4481
wave slowness 0168
wave, solitary – 2815
wave, sonic – 4482
wave surface 0549
wave, surface – 0188
wave, tangential – 0184
wave, tidal – 2816
wave, transformed – 0185
wave, transverse – 0184
wave velocity 0168
wave, visible – 0192
wave, wind – 2807
waviness, joint – 4727
wax 3723
wax, montan – 3736
waxy 0484
WDC 0082
weatherability 4734
weathering 1988
weathering 2928
weathering, chemical – 1998
weathering, chemical – 2928
weathering, cosmic – 0028
weathering, differential – 2004
weathering, frost – 1996
weathering, mechanical – 1989
weathering, mechanical – 2928
weathering, onion-skin – 1994
weathering, physical – 1989
weathering, selective – 2004
weathering, spheroidal – 2003
weathering, submarine – 3476
weathering waste 2005
wedge 1571
wedge, ice – 2640
wedge out 3104
wedge out 3509
wedge, sand – 2641
wedge, soil – 2641
wedge, tidal – 2509
weephole 4811
weight dropping 4409
welding, pressure – 3446
well 2105
well 3665
well 4261
well, appraisal – 3785
well capacity 3806
well, capping a – 4290
well, closed-in – 3811
well completion 3795
well, dead – 3812
well, delineation – 3785
well, development – 3786
well, discovery – 3784
well, exploration – 3782
well, flowing – 3790
well, gaslift – 3791
wellhead, production – 3796
well, infilling – 3843
well, injection – 3793
well, killing a – 4290
well log 4256
well logging 4578
well, offset – 3792
well, outstep – 3785
well potential 3806
well, production – 3786
well productive – 3789
well, pumping – 3791
well, relief – 4290
well repair 3813
well-shooting survey 4406
well, shut-in – 3811
well spacing 3842
well, stratigraphical – 3783
well, stripper – 3845
well, wild – 4288
Wenner array 4562
Wentworth scale 3163
whale back 2182
whirl 2269
whirlpool 2269
whisker 0524
whitening 4455
whiting 3379
Widmanstätten figures 0067
Widmanstätten intergrowth 4104
Wiechert-Gutenberg discontinuity 0092
wildcat 3782
wildflysch 2926
wilting percentage, permanent – 2773
wilting point 2773
wind action 0043
wind-faceted stone 2162
wind grinding 2154
windkanter 2162
window 4234
window, karst – 2078
window, source – 0231
wind ripple 3016
wind scour 2154
windward 2963
wind wave 2807
wind wear 2154
winnowing 2937
wireline test 3787
workover 3813
World Data Centre 0082
world quake 0138
wreck, volcanic – 1476
wrinkle 0058
wrinkle ridge 0031
Wulff net 1015
wurtzilite 3739
W-wave 0191

X

xenoblast 0980
xenoblastic 0980
xenocryst 0663
xenolith 0663
xenomorphic 0752
xenomorphic-granular 0808
xenomorphic-granular 0954
xenothermal 3984
xenotopic 3471
xeolite 3205
xylanthrax 3607
xylite 3495
xylith 3587
xylophagous 1857

Y

yardang 2164
yield 0272
yield point 4697
yield, safe – 3668
yield, specific – 3704
yield strength 4697
yield stress 4697
Young's modulus 4695

Z

Zavaritsky diagram 0699
zeolitization 0932
zero level 4577
zibar 2181
zone 0362
zone 0519
zone 1717
zone, acme – 1773
zone, assemblage – 1774
zone, bio-interval – 1772
zone, biostratigraphical – 1769
zone, concurrent-range – 1777
zone, evolutionary – 1779
zone, flood – 1773
zone, flushed – 4620
zone, folded – 1674
zone, front – 1686
zone, genus – 1778
zone, interbiohorizon – 1772
zone, interval – 1772
zone, invaded – 4619
zone, lineage – 1779
zone, lithostratigraphical – 1747
zone, local-range – 1784
zone, metallogenic – 3908
zone, morphogenetic – 1779
zone, overlap – 1777
zone, peak – 1773
zone, phylogenetic – 1779
zone, range – 1781
zone, range-overlap – 1777

zone, shattering – 4821
zone, sub– 1782
zone, subbio– 1782
zone, taxon-range – 1783
zone, teil– 1784
zone, topo – 1784
zoning 0474
zoning 0533
zoning 3907
zoning, inverse – 0475
zoning, normal – 0475
zonule 1782
zoobenthos 1852
zoochlorellae 1845
zooplankton 1800
zooxanthellae 1845
Z-phenomenon 0237

Nederlands

A

aangepast 0719
aangroei 2513
aangroei 2988
aanpassing 1980
aanrijking, residuaire – 3943
aanslibbing 2513
aanslibbing 2886
aanspoelsel, lijn van – 2962
aanstampen 4809
aantappingskloof 2326
aanwas 2988
aardachtig 3086
aardbeving 0138
aardbeving, crypto-vulkanische – 0140
aardbeving, diepe wereld– 0158
aardbeving, instortings– 0141
aardbeving, kunstmatige – 0265
aardbeving, meervoudige – 0143
aardbeving, onderzeese – 0148
aardbevingpreventie 0262
aardbevingsbestendig bouwplan 0260
aardbevingsgeluid 0203
aardbevingsleer 0094
aardbevingsmechanisme 0232
aardbevingsregulatie 0262
aardbevingsrisico 0255
aardbevingsspleet 2552
aardbevingstrede 2552
aardbevingsvoorspelling 0261
aardbevingsvrij bouwwerk 0260
aardbevingszwerm 0144
aardbeving, tektonische – 0139
aarde 3086
aardgas 3713
aardkern 0093
aardkorst 0086
aardmagneetveld 0276
aardmagnetisme 0276
aardmantel 0091
aardolie 3712
aardpijp 2077
aardpyramide 2029
aardrijkskunde, natuurkundige – 0012
aardschok 0138
aardstoot 0138
aardstorting 2046
aardstroom 2051
aardtrilling 0146
aardverschuiving 2044
aardverschuiving 2045
aardwarmte 0374
aardwarmte, nuttige – 0388
aardwas 3735
aardwetenschappen 0002
ablatie 2013
ablatie 2380
ablatiedal 2384
ablatie, eolische – 2155
ablatiekloof 2384
ablatiemorene 2412
abrasie 2016
abrasie 2938
abrasiekust 2528
abrasieplat 2479
abrasieterminante 2442
abrasievlak 2479
abrasievlakte 2536
absorptie 0112
absorptie 0406
absorptie 4514
absorptiefactor 4314
absorptiespectrum 0491
absorptiespectrum 4305
abyssaal 0617
abyssaal 2847
abyssale vlakte 2792
abyssopelagisch 1910
abyssopelagisch 2847
accelerometer 0120
accessorisch bestanddeel 0594
accretie 1822
accumulatie 2987
accumulatie 3754
accumulatieterras 2298
accumulatieterras, ineengeschakeld – 2301
accumulatieterras, marien – 2480
accumulatievlak, litoraal – 2480
ACF diagram 0989
achondriet 0072
achterdiep 1110
achtergrond-concentratie 4626
achterland 1110
achterland 2518
achtervulling 4825
acme 1985
acme-zone 1773
actief 0619
actualisme 0006
actuogeologie 0008
adamant 0484
adaptatie 1980
adaptief 1980
ader 4039
ader, intrusieve – 0748
ader, lensvormige – 4026
adernetwerk, kamervormig – 4050
aderstructuur 4103
ader, wigvormige – 4036
adhaesieribbel 3016
adinole 0933
adsorptie 0406
adsorptiecapaciteit 0420
adsorptiecomplex 2755
adularisatie 0932
adventiefkegel 1437
adventiefkrater 1439
aeolianiet 2870
aeolisch 2870
aeolotropie 0110
aeon 1767
aeonotheem 1767
aërobe ontbinding 2932
aeroliet 0071
aëroob 0424
aëroob 1906
afaniet 0798
afanitisch 0798
afbeelding 4320
afbladdering 1994
afbraak 2741
afbuigen 4292
afdamming 2340
afdekverhouding 4202
afdichten 4800
afdichting 4800
afdichting 4861
afdichtingsmuur 4861
afdichtingssleuf 4861
afdruk 1824
afdruk 1826
afdruk, onthoekte – 4377
afgeboord gebied 3890
afgeleide, tweede – 4534
afgeleide, tweede – 4544
afgeplat 3183
afgerond 3188
afgerond-blokkig 2729
afgerond, matig – 3188
afgerond, sterk – 3188
afgerond, zwak – 3189
afgeschoven 1559
afgietsel 1825
afgietsel op basisvlak 3019
afgietsel, uitwendig – 1825
afgietsel van stroom-uitschulping 3026
afglijden 2976
afglijding 2043
afglijding 2972
afglijding 2976
afglijding door de zwaartekracht 1703
afglijdingsbal 3055
afglijdingsbreuk 1707
afglijdingshelling 2237
afglijdingsmassa 3055
afglijdingsmeander 2222
afglijdingsplooi 3055
afglijdingsspoor 3058
afglijdingsstructuur 1704
afglijdingsterras 2302
afglijdingsvlak 2055
afgraven 4875
afgrond 2104
afirisch 0798
afkoelingsbast 0713
afkoelingseffect 1271
afkoelingsrand 0707
afkoelingsspleet 1517
afleidingskloof 2326
AFM diagram 0698
AFM diagram 0989
afname 3816
afnamecurve 3816
afname–methode, curve van – 3874
afotisch 1905
afplatting 3182
afplatting 0330
afronding 3187
afrondingsindex 3187
afschaving 2395
afscheider 3797
afschilfering 1994
afschuiving 1558
afschuiving 2045
afschuiving 2975
afschuivingsmodulus 4699
afschuiving van het klif 2473
afschuring 2936
afschuring, glaciale – 2395
afslag 4818
afslibbaar 2716
afslibbare delen 2716
afslibbing 3159
afslijping 2938
afsluitingsrichel 2335
afspoeling 2031
afspoeling, vlakte– 2032
afspoeling, vlakte– 2779
afstamming 1964
afstammingsleer 1959
afstammingslijn 1965
afstammingsreeks 1965
afstand 4393
afstand, ruimtelijke – 4393
afstandsbereik 4393
afstandsbereik 4393
afstandscorrectie, horizontale – 4432
afstandscorrectie, horizontale – 4432
afstorting 2473
afstorting in de krater 1472
afstorting in de krater 1472
afstortingslawine, gemengde – 1368
afstortingslawine, gemengde – 1368
afstroming, oppervlakkige – 3623
aftakking 1582
aftakking 2211
aftakking 4071
aftasten 4401
aftaster 4403
aftaster, multispectrale – 4403
aftasting, cirkelvormige – 4402
afval 3924
afvalproduct 4164
afvalverwijdering 4831
afvalverwijdering, ondergrondse – 4832
afvalwater 3803
afvalwater 4165
afvloeiingscoëfficient 2250
afvloeiingsloos 2257
afvoer 2249
afvoer 3616
afvoer, basis– 3628
afvoer, grondwater– 3628
afvoer, oppervlakte– 3617
afvoer, specifieke – 2250
afwatering 2768
afwateringsgebied 2192
afwateringsnet, onderzoek in het – 4641
afwatering, zonder bovengrondse – 2257
afwijking, stelselmatige – 4192
afwijking uit het lood 1548
afzetting 2850
afzetting 2985
afzetting, alluviale – 2886
afzetting, biogene – 3269
afzetting, blokken– 3249
afzetting, bodemrandstandige – 3987
afzetting, colluviale – 2868
afzetting, contactmetamorfe – 3980
afzetting, diepzee– 2922
afzetting, ematogene – 3979
afzetting, exhalatieve – 3979
afzetting, fluvioglaciale – 2402
afzetting, getijde– 2907
afzetting, glaciale – 2402
afzetting, hernieuwde – 2993
afzetting, hydrothermale – 3982
afzetting, klastische – 3164
afzetting, klastisch-vulkanische – 1310
afzetting, kort na – 3474
afzetting, laat-magmatische – 3975
afzetting, magmatische geïnjecteerde – 3975
afzetting, magmatische segregatie – 3975
afzetting, magmatisch-pneumatolytische – 3981
afzetting, metasomatische – 3991
afzetting, organogene – 3269
afzetting, pneumatolytische – 3981
afzetting, pneumotektische – 3981
afzetting, primaire – 3972
afzetting, pyroklastische – 3170
afzetting, pyrometasomatische – 3980
afzetting, regressieve – 3112
afzetting, riviervlakte– 2885
afzetting, secundaire – 2993
afzetting, secundaire – 3972
afzettingsgesteente 2850
afzetting, stortvloed– 2884
afzetting, transgressieve – 3113
afzetting, vaste-koolwaterstofhoudende – 3493
afzetting, vroeg-magmatische – 3975
afzetting, vulkanoklastische – 1310
afzondering 1981
afzwakking 4315
agglomeraat 1344
agglomeraatlava 1269

aggregaat, kristallijn – 0771
agmatiet 1042
agoon 0284
agpaitisch 0693
agressief 0619
Airy 0345
Airy fase 0190
A'KF diagram 0989
aktuopaleontologie 1886
albedo 4316
albitisatie 0932
aleuriet 3205
aleuroliet 3205
algen– 3314
algenbal 3363
algendijk 3361
algendom 3363
algenknobbel 3363
algenkool 3563
algenmat 3364
algenmoeras 2515
algenrand 3362
algenstruktuur 3358
alkali 0694
alkalibazalt 0695
alkalikalk index 0692
alkalikalkreeks 0692
alkaliniteit 3677
alkalische reeks 0692
alkanen 3717
alkenen 3717
allochemisch deeltje 3271
allochromatisch 0483
allochroon 1818
allochthoon 0727
allochthoon 1685
allochthoon 1815
allochthoon 2945
allochthoon deeltje 3272
allogeen 0660
allogeen 2946
allogeen 3968
allogeen 4633
allopatrisch 1817
allothigeen 2946
allotriomorf 0752
allotype 1949
alluviatie 2886
alluvium 2886
alpine gebergtevorming 1078
alpinotype 0600
alteratie 3425
altiplanatie 2590
alunitisatie 0932
amandel 0782
amandelsteen 0872
amfibisch 1858
amfiboliet 0934
amfibolisatie 0932
amoeboïde 4099
amortizatiefonds 4219
amplitude 1634
amplitudekarakteristiek 4448
amplitudespectrum 4451
amplitudo 0164
amygdaloïdaal 0872
anaclinaal 2288
anadiagenese 3427
anaërobe ontbinding 2932
anaëroob 0424
anaëroob 1906
anagenese 3427
analcimisatie 0932
analoogmodel 4782
analyse 4183
analyse, continu– 4144
analyse, elementaire – 3594
analyse, mechanische – 2711
analyse, rekenkundige – 4777
analyse, stereometrische – 4107
anaseisme 0187
anastomose 2115
anastomose, breukvlak– 2116
anastomose, dak– 2117
anastomose, laagvoeg– 2116
anastomosering 2501
anatexiet 1026
anatexis 1025
anatexis, differentiële – 1029
anatexis, selectieve – 1029
anchieutektisch 0634
anchimetamorfose 3420
anchimetamorfose 0888
andesietlijn 0691
anelasticiteit 0111
anhedrisch 0523
anhydriet 3381
anisotroop 4681
anisotroop, optisch – 0561
anisotropie 0110
anisotropie 4681
anisotropie, magnetische – 0303
anker 4792
ankerkabel 4794
ankerkop 4794
ankerkrachtmeter 4772
ankerplaat 4794
ankerstang 4794
anker, voorgespannen – 4795
annuïteit 4218
anomalie 4532
anomalie 4632
anomalie, Bouguer– 0368
anomalie, gemodificeerde Bouguer– 0370
anomalie, geothermische – 0390
anomalie, isotatische – 0369
anomalie, lokale isostatische – 0369
anomalie, magnetische – 0286
anomalie, regionale isostatische – 0369
anomalie, vrijelucht– 0368
anomalie, zwaartekrachts– 0367
anorogeen 0726
anorogene tijd 1106
anthraciet 3561
anticentrum 0154
anticlinaal 1618
anticlinaal, pseudo– 1620
anticlinaal, tweezijdig duikende – 1638
anticline 1618
anticlinorium 1671
antiduin 3008
antiferromagnetisme 0301
antiform 1618
apex 1635
aplitisch 0811
apofyse 0725
apofyse 4071
apomagmatisch 3986
arabeskachtig 0827
aragonietnaaldenslib 3379
archaeomagnetisme 0294
archetype 1979
archibenthisch 1908
Archie, formules van – 4605
arculiet 0768
areaaleruptie 1232
areniet 3225
argilliet 3211
argon, atmosferisch – 0450
argon, overmaat – 0450
argon, radiogeen – 0450
arkose 3233
arm, dode – 2212
aromaten 3717
array station 0119
arteriet 1045
as 3599
as, a– 1016
as, b– 1016
as, beta– 1020
as, c– 1016
aschistiet 0668
asdek 1320
aseismisch gebied 0253
as, f– 1017
asfalt 3734
asfaltiet 3737
asfaltmeer 3753
asfalt, natuurlijk – 3734
as, geomagnetische – 0280
asgetal 3599
as, gordel– 1020
askegel 1427
as, kristallografische – 0511
aslawine, hete – 1369
as, maaksel– 1016
as, optische – 0554
ASR 4425
asregen 1363
assemblagezone 1774
assenbeeld 0569
assendispersie 0572
assenhoek, optische – 0555
assenvlak 1631
assenvlak, spoor van het – 1631
assen, vlak van de optische – 0554
assimilatie 0681
assimilatie, rand– 0683
associatie 1892
associatie 586
associatie, metamorfe – 0885
associatie, mineralen– 3179
asstroom, hete – 1369
asthenoliet 1122
asthenosfeer 0090
astuf 1333
as, vulkanische – 1317
aswolk 1363
aswolk, electrisch geladen – 1365
ataxiet 0069
ataxiet 1243
atexiet 1032
atlantische serie 0690
atmosfeer 0084
atol 3340
atolvergroeiing 4102
atrio 1461
attenuatie 0219
attenuatie 4475
Atterberg, grens van – 4714
Atterberg, waarden van – 2748
aulacogeen 1085
aureool 4629
authigeen 2946
authigeen 3434
authigeen 3968
authigeen 4633
authigenese 3434
autochthoon 0727
autochthoon 1685
autochthoon 1813
autochthoon 2945
autochthoon deeltje 3272
autocorrelatie 4450
autocovariantie 4450
autoecologie 1888
autohydratie 3957
autointrusie 0621
autoklaas 0115
autoklaas 2112
autoliet 0664
autolieten 1313
autometamorfose 0659
autometasomatose 0659
autometasomatose 3957
automorf 0750
autopneumatolyse 0659
autopneumatolyse 3957
avegaar 4236
axioliet 0767
axiolithisch 0767
azimut 4228
azoïsch 1738

B

baai 2451
baaienkust 2533
bacilliet 0762
backset laag 3141
bahada 2581
bahamiet 3303
bajonetgang 4055
ballast 4844
band 3122
bandenstructuur 0860
bandenstructuur 0945
bandering 3122
banding 0860
banding 0945
band, spectrale – 4306
band, thermische – 4308
band, vuile – 2379
bank 2244
bank 2297
bank 2469
bank 2506
bank 2733
bank 2802
barchaan 2177
barranco 1474
barrière 1916
barrière, geografische – 1916
barstvorming 3431
B-as 0235
basen, omwisselbare – 2755
basen, uitwisselbare – 2755
basenverzadiging 2756
basiet 0603
basificatie 0684
basis 0626
basisch 0603
basisch front 1066
basisconglomeraat 3114
basislijn 4621
basisstation 0353
basisvlak-sculptuur 3019
bastaard 0681
batholiet 0729
bathyaal 2846
bathylimnion 2334
bathypelagisch 1910
bathypelagisch 2845
bauxiet 4124
bazaltbol 1256
bazalt, kegel– 1258
bazaltplateau 1257
bazaltplaveisel 1301
B ⊥ B'-tektoniet 1003
B ∧ B'-tektoniet 1003
Becke, lijn van – 0559
beddingmateriaal 2952
bedekking 4232
bedekking 4407
bedekking, horizontale – 1551
bedekking, meervoudige – 4407
bedekking met water 2857
bedekking, multipele – 4407
bedekking, schijnbare – 1553
bedekking, stratigrafische – 1552
bedekking, terugwijkende – 3112
bedekking, terugwijkende – 1727
bedekt 2595
beekiet 3457
beek, inglaciale – 2388
beek, subglaciale – 2388
beek, supraglaciale – 2388
beenderenlaag 3245
begin– 1108

begindruk 3849
beginproductie 3805
begraven 2595
begrenzingsboring 3785
behandeling, kunstmatige – 0492
beitel 4271
beitsen 0492
bekken 0029
bekken 1086
bekken 2793
bekken 3648
bekken, epicontinentaal – 1086
bekken, gesloten – 2579
bekken, glaciaal – 2620
bekken, intramontaan – 1112
bekken, tong– 2629
bekrassing 2397
belichting 4340
belichting, conoscopische – 0547
belichting, orthoscopische – 0547
belichtingsspeelruimte 4340
beloniet 0761
bemonstering 4423
bemonstering 4741
bemonstering, boor– 4167
bemonstering, controle– 4177
bemonstering, proef– 4175
bemonsteringsapparaat 4745
bemonstering, stroom– 4170
bemonstering, werkfront– 4167
benedenloop 2195
Benioff, zone van – 1134
benthonisch 1907
benthos 1907
bentoniet 3218
benzine 3744
bepleistering 1226
bereik, dynamisch – 0132
bereik, dynamisch – 4507
bergafglijding 2048
berg, centrale – 0054
berging, elastische – 3709
bergingscoëfficiënt 3709
bergkam, scherpe – 2617
berg, onderzeese – 2798
bergstorting 2049
berg, vuurspuwende – 1411
berm 2297
berm 4860
besmetting 4652
bestanddeel, exotisch – 2949
bestanddeel, gasvormig – 0650
bestanddeel, gasvormig – 1172
bestanddeel, mobiel – 1058
bestanddeel, opaak – 0610
bestanddeel, vluchtig – 0650
bestanddeel, vluchtig – 1172
bestraling 0492
bethonisch 2844
betrouwbaarheidsdrempel 4191
betrouwbaarheidsinterval 4191
beving, plaatselijke – 0149
beving, verre – 0150
beving, vukanische – 0140
beving, vulkanische – 1193
bevriezingsproces 4807
beweging, eustatische – 2858
beweging, helicoïdale – 2264
beweging, laminaire – 2373
beweging langs glijvlakken 2373
beweging, negatieve – 2858
beweging, positieve – 2858
beweging, schroefdraadvormige – 2264
bewegingslijn 2372
bewezen gebied 3889
bezinking 2991
bezinkingssnelheid 2991
bifurcatie 2210
bifurcatie 4073
bijmenging 4805
bijproduct 4163
binnenbocht 2235
binnenboog, vulkanische – 1137
binnendelta 2499
binnenduin 2167
binnenkern 0093
binnenmantel 0091
binnenmorene 2408
binnenrif 3344
biochron 1768
biocoenose 1896
biofacies 2861
biogeen gesteente 3269
biogenese 1792
biogenetisch 1792
biogenetische grondwet 1968
biogeochemie 0395
bioherm 3331
biohorizon 1771
bioklast 3168
bioklastisch 3168
bioklastisch 3273
bioliet 3269
biomicriet 3288
biomicriet, bioklast-arme – 3288
biomicriet, bioklast-rijke – 3288
biomicrorudiet 3289
biomicrospariet 3289
bionomie 1887
bioom 1895
biopelmicriet 3291
biopelspariet 3290
biosfeer 1788
biospariet 3287
biosparrudiet 3287
biosparzandsteen 3287
biostratigrafie 1710
biostratigrafische eenheid 1770
biostratonomie 1791
biostroom 3331
biota 1804
biotietnest 1072
biotisch 1903
biotitisatie 0932
biotoop 1902
bioturbatie 1885
bioturbatie 3067
biozone 1769
biozone, super– 1782
bireflectie 0575
bisectrice 0556
bisectrice, scherpe – 0556
bisectrice, stompe – 0556
bitterzout 3383
bitumen 3719
bituminisatie 0426
blaasjes, cylindervormige – 1306
blaasjes-zand 3063
blaasschacht 1224
bladerig 0775
bladerig 0948
bladerig gesteente 1604
bladerkool 3556
blasto– 0971
blastomyloniet 0916
blastomyloniet 1541
blauwe-banden-structuur 2374
blauwe-bladen-structuur 2374
blazer 1385
bloedregen 2159
bloedverwantschap 0629
bloeitijd 1985
blok 1087
blok 1327
blok 3243
blok 4199
blok, afgeschoven – 1559
blokcement 3439
blokdiagram 4258
blokenveld 2057
blokkengrind 3249
blokkig 2729
blok, opgeschoven – 1559
blok, opgeslokt– 0661
blootgelegd 2595
blubber 3210
bocca 1440
bocht 2450
bodem 0705
bodem 2657
bodem- 4636
bodemassociatie 2673
bodembewoners 1907
bodemcomplex 2673
bodemdelen, minerale – 2719
bodemdruk 3847
bodemdrukmeter 3848
bodemerosie 2036
bodemerosie 2776
bodem, geribde – 2653
bodemgeschiktheid 2660
bodemhorizont 2687
bodem, intrazonale 2668
bodemkartering 2670
bodemklei 3515
bodemkunde 2658
bodemlaag 2688
bodemlaag 2896
bodemlaag 3515
bodemlading 4817
bodemlast 2952
bodemmonster 3855
bodemonderzoek, geochemisch – 4644
bodemprofiel 2686
bodemreflectie 4519
bodemschommelingen 1194
bodemstructuur 2246
bodemstructuur 2723
bodemtractie 2953
bodem, vlakke – 3017
bodemvloeiing 2041
bodemvocht 2764
bodemvocht 3652
bodemvormende factor 2659
bodemvorming 2659
bodemwater 3779
bodemwaterstuwing 3825
bodem, zonale – 2667
boetelingensneeuw 2365
bolbliksem 1372
bolkap 0363
bolletje 3094
bolschaalverwering 1994
bolson 2580
bolvormigheid 3180
bom 1328
bom, bandvormige – 1329
bom, bipolaire – 1329
bom, bolvormige – 1329
bom, druppelvormige – 1329
bom, exploderende – 1329
bom, gedraaide – 1329
bom, lintvormige – 1329
bom, peervormige – 1329
bom, rotatie– 1329
bom, samengestelde – 1329
bom, spindelvormige – 1329
bom, unipolaire – 1329
bont 3462
boogdam 4865
boogdelta 2495
boomvormig 1821
boorbaarheid 4298
boordkorstbom 1329
boorgang 1882
boorgat 3665
boorgat 4261
boorgatafstand 4822
boorgatdrukmeter 4760
boorgatmeting, fysische – 4578
boorgatopmeting 4294
boorgatsonde 4578
boorgruis 4238
boorinstallatie 4267
boorkern 3155
boorkern 4237
boorkernen, lade voor – 4237
boorpijp 4270
boorpijpproef 3787
boorrapport 4299
boorserie 4270
boorsnelheid 4298
boorspoeling 4281
boorstaat 4299
boorstang 4270
boorwagen 4266
boren, draaiend – 4263
boren, draaiend puls– 4263
boren, gericht – 4291
boren, stotend – 4262
boring 3665
boring 4261
boringsdichtheid 2675
boring, stratigrafische – 3783
bornhardt 2608
bostonitisch 0839
bosveen 3536
botryoïdaal 0777
botryoïdaal 4094
botsfiguur 3410
bottenbreccie 3261
bottenlaag 3245
boudinage 1609
Bouguercorrectie 0359
Bouguerplaat 0359
Bouguerreductie 0359
bouwcode 0256
bouwplaats 4789
bouwselstabiliteit 3424
bouwvoor 2699
bovenbouw 1414
bovengrond 4788
bovenkant 3106
bovenliggend 4631
bovenloop 2195
bovenstructuur 1077
bovenvleugel 1576
bovenvleugel 1680
bovenzijde 3106
brachyanticlinaal 1640
brachysynclinaal 1640
bradiseisme 1194
bradyseisme 0096
bradytelie 1977
branding 2812
brandingsgrot 2472
brandingskloof 2474
brandingsluifel 2472
brandingsnevels, zone van – 2439
brandingsnis 2472
brandingsstrook 2812
brandingsvlakte 2479
brandingszone 2439
Braziliaanse proef 4761
break-even tijdstip 3896
brecciatie 2933
breccie 3258
breccie, afbrokkelings– 1249
brecciegang 4050
breccie, instortings– 3263
breccie, intraformationele – 3264
breccie, intrusie– 0710
breccie, rif– 3355
breccie, rifsteen– 3342
breccie, samenzakkings– 3263
breccie, sedimentaire – 3258
breccie, vulkanische – 1345
breedte, geocentrische– 0098
breedte, geomagnetische – 0281
breedte voor trapfront, minimale – 4201
breker 2811
brekingsindex 0551

bres 1471
breuk 0469
breuk 0541
breuk 1466
breuk 1494
breuk 1521
breuk 4676
breuk 4723
breuk, afbuiging van een – 1536
breuk, afsluitende – 3765
breuk, antithetische – 1588
breukbekken 1591
breukbekken 2547
breukbreccie 1539
breukdal 2547
breuk, diagonale – 1545
breuk, diep-doordringende – 1525
breuken, doorsneden met – 1494
breuken, en échelon – 1585
breuken, groep van evenwijdige – 1584
breuken, netwerk van – 1578
breuken, radiaire – 1594
breuken, radiale – 1594
breuken, snijdende – 1579
breukfuik 3757
breuk, gebogen – 1535
breuk, gesloten – 1534
breukgrot 2096
breuk, herleefde – 1527
breuk in het laagvlak 1547
breukklei 1540
breukkuil 0743
breuklijndal 2547
breuklijnkust 2520
breuklijntrede 2550
breuk, longitudinale – 1546
breuk, meegesleepte – 1536
breuk met kromlijnige bewegingsrichting 1567
breuk, open – 1534
breuk, oppervlakkige – 1525
breuk, perifere – 1592
breuk, schelpachtige – 0541
breukschol 1531
breuk, secundaire – 1580
breuksplijting 1601
breuk, steilstaande – 1549
breuksterkte bij puntbelasting 4733
breuk, synthetische – 1588
breuksysteem 1578
breuksysteem, conjugaat – 1584
breuktektoniek 1520
breuk, transformatie– 1141
breuk, transformatie– 1560
breuk, translatie– 1566
breuk, transversale – 1546
breuktrede 2548
breuktrede, blootgelegde – 2548
breuktrede, verjongde – 2548
breuktrog 2546
breuk, uitgewerkte – 1527
breuk, vereffende – 2549
breuk, vertakte – 1582
breukvlak 0238
breukvlak 1528
breukvlakfacet 2553
breuk, vlakke – 1535
breukvoortplanting 0239
breuk, voortschrijdende – 4676
breukvorming 1494
breukwand 1529
breukwand, benedenliggende – 1529
breukwand, benedenliggende – 1529
breukwand, bovenliggende – 1529
breukwand, onderliggende – 1529
breuk, werkende – 1527
breukzone 1467
breukzone 1532
breukzone 1595
breuk, zwakhellende – 1549
briklaag 2703
briklaag 2734
brillant 0493
broek 3532
brok 4744
bron 2193
bron 3630
bron, artesische – 3631
brongebied 2193
brongebied 2276
brongegeven 0151
bron, hypogene – 1408
bron, intermitterend spuitende – 1404
bron, kokende – 1407
bron, minerale – 3632
bron, pulserende – 1404
bron, thermale – 1407
bron, thermale – 3632
bron, vaucluse– 3631
bronvenster 0231
bros 0471
bros 4673
brug 2102
bruggehoofd 4845
brug, natuurlijke – 2079
bruinkool 3495
B-tektoniet 1003
buffer 0422
buffervoorraad 4225
buis 2099
buitenbocht 2235
buitenboog, niet-vulkanische – 1137
buitenboog, niet-vulkanische – 1137
buitendelta 2499
buitenduin 2167
buitengaats 2840
buitenkern 0093
buitenmantel 0091
bukgang 2098
Bullard, methode van – 0363
bult 3539
bultrots 2400
bultstalagmiet 2141
bysmaliet 0734

C

cabochon 0500
cala 2452
calcietblaasje 2146
calcietrijp 2143
calcietschub, drijvende – 2144
calcietvlies 2142
calcificatie 3458
calcilutiet 3274
calcimicriet 3282
calcineren 4151
calcirudiet 3278
caldeira 1456
caldera 1456
calderacomplex 1457
caldera-eiland 1417
caldera, enkelvoudige – 1457
caldera gevormd door een freatische uitbarsting 1458
calderawal 1461
calorisch veld, hoog– 0389
calorisch veld, laag– 0389
camera, metrische – 4331
canyon 0034
canyongeul 2100
canyon, onderzeese – 2795
capaciteit 2249
capaciteit 2958
capaciteit, warmte– 0386
capillaire zone 2767
capillaire zone 3657
carbonaatbank in open zee 3323
carbonaatcompensatiediepte 0423
carbonaathelling 3322
carbonaatlichaam, opgebouwd – 3328
carbonaatlichaam, uitgebouwd – 3327
carbonaatmassa 3324
carbonaatopbouwsel 3324
carbonaatvlakte 3322
carbonatisatie 0932
carbonatisatie 2000
cascade 2203
cascadeplooien 1706
cash flow, geaccumuleerde – 3894
cash flow grafiek 3895
cash flow, jaarlijkse – 3894
cash flow, uiteindelijke – 3894
cataclinaal 2288
cataract 2204
catastrofisme 0005
catena 2674
caustic 0216
caverne 4849
caverneus 0874
caverneus 1305
caverneus 3064
cavitatie 2120
cavitatie 4512
CBR-waarde 4708
CD 0454
celapparaat 4753
cel, elementair– 0510
cellen- 3064
cellig 0874
cellig 1305
cel, rhomboëdrische – 0516
celtextuur 4089
cement 3433
cementatie 3433
cementatiefactor 4606
cementatie, korrelige – 3437
cementstabilisatie 4807
cement, syntaxiaal – 3437
cenozone 1774
centrifugemodel 4781
chadakrist 0844
Chandler schommeling 0200
charnockiet, m– 0990
chasma 0034
chatoyantie 0488
cheironiem 1956
chelaat, metalloörganisch – 0410
chelaat, organometallisch – 0410
chelatie 0409
chelatie, organometallische – 0409
chenier 2491
cheniervlakte 2491
chevronspoor 3036
chilisalpeter 3384
chloritisatie 0932
chondriet 0072
chonoliet 0732
chorismiet 0794
chroma 2677
chron 1761
chronomeer 1760
chronostratigrafie 1712
chronostratigrafische eenheid 1757
chronostratigrafische horizon 1758
chronotax 1732
chronozone 1758
cijns 4224
circalitoraal 2839
circulatie, normale – 4283
circulatie, omgekeerde – 4283
circulatie, thermohaliene – 2829
circulus 1808
CJPW norm 0686
cladogenese 1962
Clarke, getal van – 0399
classificatie, bodem– 2661
classificatie, land– 4785
classificatie van reserves 3878
clavaliet 0763
clinolimnion 2334
coagulatie 0413
coagulatie 3932
cocardestruktuur 4096
coccoliet 1840
cockpit-landschap 2080
coda 0175
coërcitiefkracht 0307
cohesie 4693
cohesie, schijnbare – 4693
cokes 3608
cokeskool 3608
cokes, natuurlijke – 3564
colloform 3093
colluvium 2868
comagmatisch 0629
comagmatisch gebied 0628
commensalisme 1899
compactie 3428
compactie 4679
compactie, chemische – 3432
compactie, differentiële – 3428
compactie, mechanische – 3429
compactiewater 3660
compartiment 0362
compensatiediepte 0345
compensatie, isostatische – 0339
compensatie, lokale – 0340
compensatiemassa 0341
compensatieniveau 0345
compensatie, regionale – 0340
competent 1624
competentie 2958
complementair 0669
complex 1756
complex 1928
complexe verbindingen, vorming van – 0408
complexiteit 0275
component, verticale – 1563
compressie 0187
conca 1459
concentraat 4141
concentratie 3942
concentratie, alluviale mineraal – 3943
concentratie, residuaire – 2967
concentrisch-spherulitisch 4102
concordant 0717
concordant 1719
concordantie 1719
Concordia-Discordia diagram 0435
concretie 2732
concretie 3472
concretie 4093
concretie-houdend 3472
concretionnair 3472
condensaat 3713
condensatie 0412
condensatie, retrograde – 3841
condominant 1816
confluentietrap 2624
conform 0719
conform 1719
conformiteit 1719
congenerisch 1943
conglomeraat 3251
conglomeraat, blokken– 3254
conglomeraat,

intraformationeel– 3257
conglomeraat, slikplakjes– 3252
conglomeraat, vulkanisch – 1343
conglomeratisch 3251
conodont 1860
Conrad discontinuïteit 0086
consequent 2284
consequent, radiaal – 2291
consistentie 2746
consistentie 4713
consistentiecoëfficiënt 4716
consistentiegrens 4714
consolidatie 4703
consolidatiering 4756
contactaureool 0899
contactgang 4043
contacthof 0899
contactpotentiaal 4615
contact, sutuur– 3449
contact, vlak – 3449
contactzone 0899
contaminatie 0681
continentaal 2865
continentale rand 2788
continentale verschuiving 1115
continentbotsing 1131
contourkaart 4247
contourlijn 4248
contractiehypothese 1114
contrast 4343
contrast 4628
contrastfilter 4339
controlepunt 4325
conussondeerapparaat 4758
convectie, warmte– 0381
convergentie 1725
convergentie 1973
convergentie 2787
convergentie, metamorfe – 0913
convolutie 3052
convolutie 4452
coördinatiegetal 0408
copellering 4184
coproliet 1863
coroniet 0786
corrasie 2014
corrasie 2938
corrasie, eolische – 2154
corrasie, fluviatiele – 2015
correctie, breedte– 0355
correctie, getij– 0365
correctie, hoogte– 0457
correctie naar het reductieniveau 4429
correctie, stations– 0160
correctie, statische – 4431
correctie, terrein– 0360
correctie, topografische – 0356
correctie, verwerings– 4430
correctie, vrijelucht– 0358
correlatie 1716
correlatie 4459
correlatie, kruis– 4450
correlator, optische – 4388
correlogram 4195
corrosie 0642
corrosie 2017
corrosie 2942
corrosiebasis van de karst 2066
corrosiedoline 2072
cosmogonie 0003
cotype 1951
covariogram 4196
craqueletéxtuur 4097
crevassecomplex 2241
crevassegeul 2240
crinoiden– 3313
crinoidenarm 3313
crotivina 2745
cryptogeen 1976
cryptografisch 0848
cryptokristallijn 0800
cryptomaceraal 3567
cryptomagmatisch 3986
cryptovulkanisme 1181
cryptozoön 1844
cuesta 2558
cuestafront 2558
cuestavlakte 2558
culminatie 1635
cumuliet 0765
cumulofier 0821
cumulofirisch 0821
cumulovulkaan 1422
cupellering 4184
curie 4550
Curiepunt 0308
cutan 2691
cycloalkanen 3717
cyclotheem 3119
cyclus, aride – 2567
cyclus, erosie – 2566
cyclus, fluviatiele – 2567
cyclus, glaciale – 2567
cyclus, karst– 2567
cyclus, mariene erosie– 2567
cyclus, periglaciale – 2567
cyclus, sedimentatie– 3119
cyclus, vulkanische – 1156
cyst 1841

D

dactiliet 0849
dactylitisch 0849
dag komen, aan de – 4231
dagzomen 4231
dagzoom 4231
dagzoomafstand, kortste – 1555
dagzoomhiaat 1556
dagzoomverdubbeling 1556
dak 0705
dak 3501
dak 4009
dak 4010
dak 4851
dakfranje 0706
dak, instorten van het – 1188
dak, loslaten van het – 2110
dakpansgewijze ligging 1727
dakpansgewijze ligging 3040
dakpansgewijze ligging 3113
dal, afgeknot – 2481
dal, anticlinaal – 2544
dal, asymmetrisch – 2274
dalbegin 2276
dal, blind – 2083
dalbodem 2278
dal, droog – 2084
dal, droog – 2328
daleinde 2276
dalgletsjer 2427
daling 2857
dalingskust 2524
dalkaar 2611
dalmeander 2221
dal, meercyclisch – 2275
dalmonding 2280
dal, monoclinaal – 2544
dalnet, uiteengevallen – 2482
dalnet, verbrokkeld – 2482
dal, onthoofd – 2324
dal, ontworteld – 2481
dalopvulling 2278
dal, polycyclisch – 2275
dalspoor 2228
dal, synclinaal – 2544
dalsysteem, geënt – 2283
dalsysteem, samengetrokken – 2283
dal, tektonisch – 2543
dal, tektonisch bepaald – 2543
daltorso 2327
daltrap 2199
daltrede 2199
daltrede, glaciale – 2623
dal, tweecyclisch – 2275
dal, U-vormig – 2273
dalveen 3534
dal, verdronken – 2282
dal, verlaten – 2328
dal, verlengd – 2282
dalvorming 2270
dal, V-vormig – 2272
dalwand 2277
dalwaterscheiding 2315
dalwaterscheidingshoogte 2315
dal, zwevend – 2483
dal, zwevend glaciaal – 2627
dam, aarden – 4859
dam, grondophoging– 4859
dampdrukverhoging, retrograde – 1174
dampkring 0084
dampkringwater 3613
damponderzoek 4648
dam van stortsteen 4859
damwand 4799
darcy 3697
Darcy, wet van – 3696
Darwinisme 1966
datering, absolute – 0429
datering, isotopische – 0429
datering, radiometrische – 0429
dauwpunt 3858
DCF methode 3898
debiet 2249
decementatie 3444
declinatie, magnetische – 0284
décollement structuur 3053
deconvolutie 4454
dedolomitisatie 0930
deelbaar 1597
deelbaarheid 1597
deeltje 3078
deeltjesdoorsnede 3193
deeltjesgrootte 3193
deeltjessnelheid 4414
deflatie 2155
deflectometer 4294
deformatie 1075
deformatie 4666
deformatie direct na afzetting 3050
deformatielamel 0999
deformatielamelle 0539
deformatiemaaksel 1007
deformatievlak 0238
degeneratie 1982
deglaciatie 2367
deinend 2811
deining 2807
deintoestel 4155
dek 4232
dekblad 1683
dekblad, duikend – 1688
dekbladen, overgrijpende – 1691
dekbladenstapeling 1691
dekbladentektoniek 1679
dekblad, overschuivings– 1575
dekking, stereoscopische – 4361
deklaag 3126
deklaag 3764
deklaag 4842
deklagen 3924
dekterrein 3501
dekterrein 3924
dekzand 2871
delfstofafzetting 3910
delfstofafzetting, magmatische – 3946
delfstoflaag 3123
delfstof, minerale – 3912
delle 2281
delta 2494
delta 2895
delta– 2895
deltaïsch 2895
delta, sub– 2496
delta-waarde 0454
demagnetisatie, thermische – 0321
demagnetisatie, wisselveld– 0321
demping 0127
demping 0219
demping, aperiodische – 0129
demping, dynamische – 0127
demping, kritische – 0129
dempingsfactor 0128
dempingsverhouding 0128
demping, viskeuze – 0127
demultiplexen 4424
dendriet 0525
dendriet 3400
dendriet 4088
dendritisch 1821
dendrochronologie 1737
denudatie 2012
denudatie 2013
denudatieniveau, bovenste – 2598
denudatiesysteem 2573
denudatieterras 2307
denudatieterras 2555
denudatievlakte 2591
denudatievlakten, in-elkaar-geschakelde – 2601
denudatievlakten, trapvormige – 2601
denudatievlakte, polygenetische – 2601
depletie 3830
depocentrum 2986
depressie 1635
depressie, cryptovulkanische – 1183
depressie, intermediaire – 1137
depressie, tektono-vulkanische – 1182
depressie, vulkano-tektonische – 1182
dereverberatie 4454
desilicificatie 0929
desintegratie 2929
desmosiet 0933
desorptie 0407
detailkaart, ingenieursgeologische – 4784
detectiedrempel 0270
detectiegrens 4651
detectiestation 4417
detectievermogen 0270
detector 4413
detonatie 1217
detonatie 4815
detritisch 3165
detritisch 3997
detritus 2005
detritus 3165
detritus, niet-lichtende – 1315
detritus, rif– 3355
deuterisch 0658
devritificatie 0919
diablastisch 0950
diachronisme 1736
diachroon 1736
diaftorese 0904
diaftoriet 0904
diagenese 3388
diagenese, begravingsdiepte-gerelateerde – 3417
diagenese, freatische – 3399
diagenese, gevorderde – 3418
diagenese, late – 3418
diagenese, milieu-gerelateerde – 3398
diagenese, vadose – 3399

diagenese, vroege – 3418
diagenetisch 3974
diagenetische omgeving, diepe – 3393
diagenetische omgeving, ondiepe – 3393
diagenetische omgeving, oppervlakte nabije – 3392
diagenetisch gezelschap 3394
diagenetisch veld 3390
diagonaalkust 2522
diagram, beta– 1021
diaklaas 1495
diaklaas 4723
diaklaasafstand 4735
diaklaas, diagonale – 1507
diaklaas, longitudinale – 1507
diaklaasnetwerk 1498
diaklaas, ongevulde – 1513
diaklaas, open – 1513
diaklaasopening 4728
diaklaas, overlangse – 1507
diaklaaspatroon 1508
diaklaas, secundaire – 1509
diaklaasspleet 1513
diaklaassysteem 1498
diaklaas, transversale – 1507
diaklaasvlak 1496
diaklaasvulling 4729
diaklazen, frequentie van – 4725
diaklazen, groep – 1498
diaklazen, groep – 4724
diaklazen, groepering van – 1498
diaklazen, systeem van – conjugate – 1499
diamagnetisme 0298
diamantboor 4265
diameter, effectieve – 3196
diameter, equivalente – 3195
diameter, gemiddelde – 3199
diameterlog 4603
diameter, mediaan– 3199
diameter, modale – 3199
diapier 1693
diapierplooi 1698
diapirisme 1693
diaplectisch 0895
diaschistiet 0668
diasteem 1724
diasteem 3125
diastrofisme 1074
diatexiet 1027
diatexis 1027
diatomeeënaarde 3370
diatomeeën-hoornsteen 3371
diatomeeënschalie 3371
diatomeeën-skeletje 3370
diatomeeënslik 3370
diatomiet 3370
diatrema 1221
dicht 0799
dicht 1626
dichtgroeiing 2341
dichtheid 4342
dichtheid 4537
dichtheid, schijnbare – 2751
dichtheid, schijnbare – 3602
dichtheidsgelaagdheid 2882
dichtheidslog 4593
dichtheidsstroming 2979
dichtheid, werkelijke – 3602
dichtslibbing 2341
dichtslibbing 2990
dicht, zeer – 1626
diep 2794
diepte 2247
dieptebereik 4505
dieptegesteente 0617
diepte, grootst bereikte – 0214
dieptemagma 1199
diepzeeklei, bruine – 2923
diepzeeklei, rode – 2923
diepzeetrog 1138
dieselolie 3744
differentiaat 0667
differentiatie 0398
differentiatie 0667
differentiatie, diagenetische – 3475
differentiatie, diagenetische – 3938
differentiatie, diffusieve – 0680
differentiatie door gefractioneerde kristallisatie 0670
differentiatie, geochemische – 0398
differentiatie, gravitatieve – 0672
differentiatieindex 0701
differentiatie, magmatische – 3938
differentiatie, metamorfe – 0911
differentiatie, pneumatolytische – 0679
differentiatie, vloei– 0677
diffluentiepas, glaciale – 2621
diffluentietrede 2624
diffractie 0217
diffractie 4444
diffractiekromme 4464
diffusie, metamorfe – 0912
dik-gelaagd 3108
dikspoeling 4281
dikte, ware – 4243
diktyoniet 1043
dilatantie 0109
dilatantie 4678
dilatatie 0187
dilatatieader 4041
dimorphisme 1811
dipoolveld, axiaal – 0279
dipoolveld, excentrisch – 0279
disconformiteit 1721
discontinuïteit 4723
discontinuïteitsvlak 0099
discontinuïteitsvlak 2333
discordant 0718
discordant 1720
discordantie 1720
discordantiefuik 3759
dislocatie 0531
dislocatie 1522
dislocatie 4078
dislocatiebeving – 0139
dislocatie, rand– 0531
dislocatie, schroef– 0531
dispersie 0197
dispersie 0403
dispersie 0572
dispersie 1812
dispersie 3710
dispersiecurve 0226
dispersie, gekruiste – 0572
dispersie, hellende – 0573
dispersie, horizontale – 0573
dispersie, inverse – 0197
dispersie, normale – 0197
dispersie, omgekeerde – 0197
dispersiepatroon 0403
dispersie, primaire – 0404
dispersie, secundaire – 0405
distaal 2997
divergentie 1969
dodendal 1402
doline 2072
dolomicriet 3282
dolomiet 3318
dolomietfragment 3167
dolomietfragment 3270
dolomiet, kalkige – 3319
dolomiet, orthochemische – 3320
dolomiet, primaire – 3380
dolomiet, primaire – 3320
dolomietslibsteen 3282
dolomietspariet 3280
dolomiet-vlekkig 3461
dolomitisatie 3460
dolomitisatie, voortgaande – 3460
dom 0059
dominant 1816
dooizakking 2646
doorbraak– 2289
doorbraak– 2290
doorbraakdal, kort – 2327
doorbraakdal, verlaten – 2327
doorbraakgeul 2240
doorbraaklahar 1354
doorkliefd 1495
doorlaatvermogen 3702
doorlatendheid 2753
doorlatendheid 3697
doorlatendheid, effectieve – 3698
doorlatendheid, primaire – 3701
doorlatendheid, relatieve – 3698
doorlatendheidsapparaat 4755
doorlatendheid, water– 3700
doorlatenheidsfuik 3758
doorloper 3175
doorschijnend 0545
doorsmelting van het dak 0624
doorsnede, geologische – 4250
doorsnede, natte – 2261
doorsnijden met dalen 2019
doorspikkeling 3947
doorzichtig 0545
doppleriet 3525
dosisequivalent, eenheid van – 4551
dosis, geabsorbeerde – 4551
doublet 0480
draa 2180
draagvermogen 4707
draaibreuk 1567
draaikolk 2269
draaitafel 4269
draaitafel, universele– 0614
drainage 2768
drainage 4739
drainage 4811
drainagegebied 3844
drainagekolom 4813
drainageleiding 4811
drainagepatroon 2254
drainagescherm 4804
draineerbuis 4811
drainering door zwaartekracht 3828
dreg 3154
drempel 2201
drempel 2335
drempel 2796
drempelberg 2625
drempel, glaciale – 2625
drempelverdrag 0266
drempelwaarde 4627
driehoeksdelta 2495
driekanter 2162
driepuntsmethode 0152
drift 4540
drijfhout 2970
drijfzand 3085
drijfzand 4719
drijfzand, vulkanisch – 1318
droobraakdal, droog – 2327
droog 0609
drooglopen 2857
droogvalling 2857
druipsteen 2127
druiventrosstalactiet 2136
druiventrosvormig 4094
drukafname 1173
drukcompensatie, isostatische – 0344
druk curve, capillaire – 4612
drukdetector 4415
drukgewelf 2113
drukgradiënt 3852
druk houden, op – 3832
druk, hydraulische zijwaartse – 4737
druk, hydrostatische – 3820
drukkaart 3854
drukkoepel 2113
drukleiding 4867
druk, magmatische – 0622
drukontlasting 1173
drukopbouw 3850
drukschacht 4867
drukschaduw 0958
drukspliјting 1596
druk, spuitende – 3847
druk, statische – 3847
druksterkte 4683
druksterkte, drieassige – 4683
druksterkte, eenassige – 4683
drukval 3850
drukvastheid 4683
drukverval 3853
drumlin 2420
dualistische hypothese 0019
dubbelbreking 0560
dubbelvulkaan 1442
dubbelvulkaan 1445
dubiokristallijn 0801
duiker 4846
duiking 4018
duiking 4019
duin 2165
duin 2872
duindel 2188
duin, klei– 2872
duinkruin, rokende – 2158
duin, organogeen – 2172
duinpan 2189
duinrest 2190
duinvallei 2188
duin, vastgelegd – 2175
duin, zand– 2872
dun-gelaagd 3108
duur 0240
dwars– 2289
dwarsbreuk 1545
dwarsdiaklaas 1506
dwarsduin 2171
dwarsgang 4077
dwarskust 2522
dwarsopstelling 4419
dwarsplooi 1678
dwarsprofiel van een dal 2271
dwarsspleet 2376
dy 3527
dynamotheorie 0278
dysfotisch 1904

E

ebgeul 2510
ebschaar 2510
ebstroom 2822
echo 4390
echoloos 4391
ecologie 1887
economisch winbare zone 4178
ecostratigrafie 1711
ecosysteem 1893
ecotype 1942
ectexiet 1034
ectexis 1034
ectiniet 1055
edelsteen 0478
edelsteenkunde 0478
edelsteen, synthetische – 0479
eenassig 0553
eersteling 0759

effect, indirect – 0366
efflata 1313
efflata, autogene – 1314
effusie 1213
effusiegesteente 1240
Eh 0421
eigen periode 0130
eigentrilling 0198
eiland 2443
eiland, aangehecht – 2534
eiland, afgesnoerd – 2534
eiland, continentaal – 2444
eilandenboog 1135
eiland, oceanisch – 2444
eilandsglooiing 2791
eilandsplat 2791
eiland, vastgeslibd – 2534
eindfase 0175
eindmorene 2413
eindmorenereliëf 2630
eindverhang 2198
ejectamenta 1313
elasticiteit 4669
elasticiteitscoëfficiënt 4695
elasticiteitsgrens 4697
elasticiteitsmodulus 4695
elastisch 0472
elastische vereffening, theorie der – 0108
electrodenopstelling 4562
electrodenopstelling, vaste – 3673
ellepticiteitscorrectie 0098
ellipsoïde, internationale – 0331
elongatie 0567
elutriatie 3159
elutriatie 3930
eluviale horizon 2701
eluviatie 2678
eluviatie 3930
eluvium 2930
emanatie 3951
emanatie, vulkanische – 1215
emanatie, vulkanische – 1376
embrechiet 1055
emergent 0174
emergentiehoek 0215
emersie 1728
emissiespectrum 4305
emulsie 4101
emulsief stadium 0652
endemisch 1814
endobiont 1854
endoblastese 0657
endoblastisch 0657
endogeen 0081
endomorfose 0898
endoscoop 0503
endosfeer 0089
endoskelet 1861
energie, geothermische – 0388
energie, milieu van hoge – 2881
energie, milieu van lage – 2881
energie, vrijkomende – 0204
energie, vulkanische – 1171
entexiet 1036
entexis 1036
enthalpie 0389
eometamorfose 0902
eon 1767
eonotheem 1767
Eötvös 0337
Eötvös correctie 4538
epeirogenese 1090
epibenthos 1908
epibiont 1855
epicentraal azimuth 0155
epicentrale afstand 0155
epicentrum 0154
epicontinentaal 2915
epicontinentaal bekken 1086
epifauna 1853
epifiet 1856
epigenese 3419
epigenetisch 2290
epigenetisch 3474
epigenetisch 3553
epigenetisch 3966
epilimnion 2332
epimagma 1201
epineritisch 2841
epinorm 0606
epipedon 2697
epipelagisch 1910
epipelagisch 2845
epitactisch 0542
epitaxiale rand 3456
epitaxie 0542
epithermaal 3983
epizone 0993
equator, geomagnetische – 0281
equidimensionaal 0754
equigranulair 3081
equipotentiaal-kartering 4558
equivalent 1733
era 1766
eratheem 1766
erodeerbaarheid 2936
erosie 2012
erosie 2936
erosie, aantapping door zijdelingse – 2322
erosiebasis 2025
erosiebasis, algemene – 2025
erosiebasis, locale – 2025
erosiebasis, plaatselijke – 2025
erosiebasis, tijdelijke – 2025
erosie, bodem– 2776
erosiecaldera 1473
erosie, chemische – 2017
erosiecyclus 2566
erosiecyclus, onderbroken – 2571
erosie, diepte– 2022
erosie, fluviatiele – 2015
erosie, geul– 2778
erosie, glaciale – 2015
erosie, kosmische – 0028
erosie, mariene – 2016
erosie-nonconformiteit 1722
erosie, priel– 2778
erosiesculptuur 3021
erosiesculptuur, transversale – 3021
erosie, selectieve – 2024
erosieterminante 2198
erosieterras 2298
erosieterras 2337
erosieterras, marien – 2479
erosie, terugschrijdende – 2021
erosie, versnelde – 2777
erosie, verticale – 2022
erosie, vlakte– 2032
erosie, vlakte– 2779
erosie, wind- 0043
erosie, zijwaartse – 2022
erraticum 2949
erts 3913
erts, aangetoond – 4207
ertsadercomplex 4058
ertsafzetting 3910
ertsafzetting, alluviale – 4000
ertsafzetting, concordante – 4022
ertsafzetting, discordante – 4022
ertsafzetting, eluviale – 3999
ertsafzetting, gelaagde – 4023
ertsafzetting, gezoneerde – 4015
ertsafzetting, ingeschakelde – 4021
ertsafzetting, laaggebonden – 4024
ertsafzetting, laagvormige – 4024
ertsafzetting, lensvormige – 4026
ertsafzetting, peneconcordante – 4022
ertsafzetting, peulvormige – 4033
ertsafzetting, podiforme – 4033
ertsafzetting, stratiforme – 4024
ertsafzetting, wigvormige – 4036
ertsafzetting, zakvormige – 4035
erts, arm – 3915
ertsdistrict 3909
erts, enkelvoudig – 3914
ertsgang 4038
ertsgang, complexe – 4046
ertsgang, enkelvoudige – 4045
ertsgang, intrusieve – 4031
ertsgang, samengestelde – 4045
erts, gemengd – 3914
erts, hoogwaardig – 3915
ertshorizon 4008
ertshoudend 3917
ertskanaal 4029
erts klaar voor verscheping 4139
ertslens 4026
ertslichaam 3910
ertslineaal 4034
ertsmassa 4032
erts, massief – 4003
erts, meegesleurd – 4079
ertsmicroscopie 3925
ertsmineraal 0462
ertsmineraal 3912
erts, ontginbaar – 3916
ertspetrologie 3925
ertspijp 4037
ertsplaat, intrusieve – 4031
erts, proto– 3913
ertsprovincie 3908
ertsreserve, balans van de – 4222
ertsrol 4028
erts, ruw – 4138
erts, samengesteld – 3914
ertsscheiding 4137
ertsslieren 4040
ertssnoer 4039
erts van laag gehalte 3915
ertsveld 3908
ertsvormingstijdperk 3905
ertsvorming, theorie der neptunische – 3926
ertsvorming, theorie der plutonische – 3927
ertswasserij 4149
erts, winstgevend – 3916
ertszone 4008
ertszone, winbare – 4178
ertszuil, rijke – 4005
eruptie 1211
eruptie 4288
eruptie-afsluiter 4289
eruptie, centrale – 1228
eruptie, embryonale – 1219
eruptie, excentrische – 1229
eruptie, freatische – 1351
eruptiefzuil 1221
eruptiefzuil 1477
eruptiegesteente 1238
eruptie, indirecte – 1218
eruptie, intermitterende – 1212
eruptiekanaal 1221
eruptiekanaal, spleetvormig – 1225
eruptiekolom 1361
eruptielichaam 1239
eruptie, lineaire – 1233
eruptie, magmatische – 1219
eruptie, modder– 1395
eruptie, paroxysmale – 1164
eruptiepunt 1220
eruptieregen 1373
eruptie, semivulkanische – 1351
eruptiespleet 1466
eruptie, subaerische – 1236
eruptiewolk 1360
esskool 3559
estavelle 2092
estuarien 2904
estuarium 2456
estuarium 2784
etage 1762
ethmoliet 0737
etsfiguur 0530
etsfiguur 3409
etsput 0530
eufotisch 1904
eugeosynclinaal 1097
euhalien 2834
euhedrisch 0523
eulitoraal 2838
eurybathisch 1911
eurybenthisch 1909
euryhalien 1912
eurytherm 1913
eurytoop 1914
eutaxiet 1243
eutaxitisch 0861
eutaxtiet 0861
eutroop 3549
euxinisch 2925
evaluatieboring 3785
evaporiet 3375
evapotranspiratie 3621
evenwicht, dynamisch – 0397
evenwicht, isostatisch – 0346
evenwicht, radioactief – 4549
evenwichtsverhang 2198
evolutie 1964
evolutie, convergente – 1973
evolutie, gerichte – 1975
evolutie, iteratieve – 1971
evolutie, magmatische – 0666
evolutie, parallele – 1971
exhalatie 1215
exhalatie 1376
exhalatie 3951
exogeen 0081
exogeologie 0024
exomorfose 0898
exosfeer 0083
exoskelet 1861
expansiehypothese 1123
exploitatie, doorsnee van de – 4138
exploitatie, overeenkomst tot gezamenlijke – 3901
exploratieboring 3782
exploratieput 3782
explosie 1217
explosiecaldera 1458
explosief 4815
explosie, freatomagmatische – 1219
explosiegat 1440
explosiehaard 1177
explosieholte 4819
explosiekrater 1431
explosieniveau 1177
explosiepijp 1221
explosiepunt 1176
explosieslenk 1235
explosietrechter 1432
explosietuf 1335
explosieve verschijnselen 1175
explosiviteit, indexcijfer voor de – 1175
expulsie 3732
extensometer 0122
extensometer 4773
extern effect 0704
extramagmatisch 3985
extrusie 1207
extrusiefase 1197
extrusiegesteente 1240

extrusielichaam 1239

F

facies 1715
facies 2860
facies, diagenetische – 3435
faciesfauna 1850
faciesfossiel 1807
facieskaart 4245
facies, metamorfe – 0987
facies, mineraal– 0585
faciesreeks, metamorfe – 0991
facies, shelf– 2915
facies, vulkanische – 1238
fahlband 4052
fakoliet 0733
familie 0687
familie 1930
familie, bodem– 2664
faneriet 0795
fanerokristallijn 0795
fantoom 0802
faro 3339
fase 0165
fase 3906
fase, abyssische – 1197
fase, biochemische – 3519
fasecurve 0135
fase, eruptieve – 1208
fase, explosieve – 1208
fase, geochemische – 3519
fase, hoofd– 0188
fase, hoofd– 0190
fase, hypoabyssische – 1197
fase, inleidende – 1157
fasekarakteristiek 4448
fase, magmatische – 1108
fasen, coördinatie van – 4113
fase, postvulkanische – 1197
faseresponsie 0135
fasering 4502
fasesnelheid 0169
fasesnelheidsmethode 0230
fasespectrum 4451
faseverloop 4388
faseverschuiving 0134
faseversnelling 0134
fasevertraging 0134
fase, vulkanische – 1197
fauna 1795
fauna- 1795
fauna, dwerg– 1851
faunula 1796
Fedorov, tafel van – 0614
felsisch 0597
felsisch gesteente 0597
felsofier 0824
felsofirisch 0824
femisch 0607
feniet 0925
fenitisatie 0925
fenoklast 3239
fenokrist 0758
fenotype 1979
ferrimagnetisme 0302
ferromagnetisme 0299
fiber 2705
fibroblastisch 0963
fibrokristallijn 0815
fijnkorrelig 0796
fijnkorrelig 3194
fijnkorrelig materiaal 3207
fijnkorrelig, zeer – 0799
metamorfose, isochemische – 0881
filigraantextuur 4081
film 3100
film 3402
film, onjuiste kleuren– 4337
film, orthochromatische – 4334
film, panchromatische – 4334
filter 4338
filter 4445
filter 4812
filter, afsnij– 0137
filter, doorlaat– 0137
filterlaag 4862
filterperseffect 0675
filterperseffect 3933
filtersnelheid 3703
filtratie 0414
firn 2363
firnbekken 2364
firngebied 2364
firngrens 2364
firnkom 2364
firnlijn 2364
firnsneeuw 2363
firnveld 2364
firnvorming 2363
fixisme 1113
fjord 2453
flank 0707
flank 1625
flank, abnormale – 1647
flankeruptie 1230
flank, normale – 1647
flard 2649
flebiet 1044
flexibel 0472
flexuur 1616
flexuurkust 2519
flexuurtrede 2551
flikkerboog 1371
flocculatie 2992
flonkering 4319
flora 1797
flotatie 0673
flotatie 4152
flotatie, bulk– 4152
flotatie, selectieve – 4152
fluïdisatie 1370
fluorescentie 0467
fluorescentie 0491
fluviatiel 2883
fluvioglaciaal 2883
fluviomarien 2902
flux, magnetische – 0317
fluxoturbidiet 2984
flysch 2926
focale bol 0233
foliatie 0948
foliatie 1604
foliatie, aaneengesloten – 0948
foliatie, onderbroken – 0948
förde 2454
foreset gelaagdheid 3140
foreset gelaagdheid, overkiepte – 3054
foreset laag 3140
formatie 1754
formatie, afsluitende – 3765
formatiedichtheid 4614
formatiefactor 4609
formatie, natte – 3767
formatie, steenkoolhoudende – 3498
formatiewater 3662
formatieweerstand, originele – 4607
formatieweerstand, specifieke – 4607
fosfatisering 3459
fosforescentie 0467
fosforiet 3480
fossa 0034
fossiel 1789
fossiel 2595
fossielhoudend 1789
fossiel, zonaal gids– 1785
fotisch 1904
foto 4358
fotobasis 4371
fotogeologie 4327
fotogrammetrie 4328
foto, samengestelde – 4357
fotostrook 4358
fotostrookmozaiek 4380
fotosynthese 1803
Fourier-getransformeerde 4451
fout, systematische – 4192
fractie 2712
fractionatie 0414
fragipan 2734
fragmentatie 4821
framboïdaal 3484
framboïdaalstruktuur 4094
framboïde 3484
freatisch vlak 2765
frequentiedomein 0223
frequentiedomein 4571
frequentie, hoek– 0225
frequentie-karakteristiek 4448
frequentiekromme, cumulatieve – 4188
frequentiemodulatie 4492
frequentiescheiding 0136
frequentiespectrum 4451
frequenties, vouwen van – 4423
frequentieverdeling 4187
Fresnel-oppervlak 4515
front 1686
frontlaag 2896
frontmorene 2413
frontmorene 2416
frontnormaal 0550
frontscharnier 1687
frontzone 1686
Froude 2266
frustula 1842
fuik 3756
fuik, anticlinale – 3757
fuik, hydrodynamische – 3761
fuik, stratigrafische – 3758
fuik, structurele – 3757
fumarole 1379
fumarole, alkalische salmiak– 1388
fumarole, bestendige – 1381
fumarole, chloride – 1387
fumarole, eruptieve – 1385
fumarole, koude – 1389
fumarole, kroikolitische – 1384
fumarole, leucolitische – 1384
fumarolenheuvel 1390
fumarolenpuist 1390
fumarolenstadium 1168
fumarolentoestand 1168
fumarolenveld 1380
fumarolenwerking 1375
fumarole, primaire – 1382
fumarole, secundaire – 1382
fumarole, tijdelijke – 1381
fumarole, vlammende – 1386
fumarole, wortelloze – 1382
fumarole, zure – 1388
fundamentbodemplaat 4835
fundamentplaat, ondiepe – 4835
fundatie 4835
fundering 4835
funderingslaag 4836
funderingsstrook 4837
funderingstechniek 4834
fundering, verbrede – 4837
fyletisch 1960
fylliet 0943
fylloniet 0918
fyllonitisatie 0918
fylogenese 1959
fylogenetische stamboom 1960
fylogenie 1959
fylum 1929
fytoklast 3168
fytopalaeontologie 1827
fytoplankton 1800

G

gaffelverdeling 2210
gal 0324
gamma 0318
gammastraalspectrumlog 4548
gammastralingkartering 4545
gang 0741
gang 1514
gang, dagelijkse – 4528
gangdoorsnijding 4074
gangenbundel 4066
gangencomplex 0749
gangen, netwerk van – 4065
gangenstelsel, geschakeld – 4067
gangenstelsel, netvormig – 4064
gangenstelsel, parallel – 4063
gangenstelsel, radiaal – 4063
gangen, veld van – 4061
gangen, vertakking van – 4070
gangenzwerm 4066
gang, gebandeerde – 4049
ganggesteente 3921
ganggsteente 0741
gang, holle – 1245
gangholte 4014
gang, klastische – 1516
gang, klastische – 3073
gang, radiaal uitstralende – 1227
gang, radiale – 1227
gang, spectrale – 4549
gangsysteem 0749
gang, uitsplitsing van een – 4070
gangverschil 0563
gang, vervangings– 0745
gangvulling 1515
gaping, horizontale – 1551
gaping, schijnbare – 1553
gaping, stratigrafische – 1552
garvenschist 0934
gas, arm – 3715
gasbelafdruk 3062
gasconservering 3836
gasdetectie 4520
gasdeviatiefactor 3862
gas, doorsijpeling van – 1401
gasdruk 1173
gasfase, intermediaire – 1162
gas, freatisch – 1351
gas, geassocieerd – 3891
gasgehalte 0650
gas, hercirculeren van – 3835
gas, herinjectie van – 3835
gasinjectie 3835
gas, juveniel – 1348
gaskap 3774
gaskapstuwing 3823
gaskool 3559
gasliftput 3791
gas, magmatisch – 1348
gasmoedergesteente 3722
gas, niet-geassocieerd – 3891
gasolie 3744
gas/oliecontact 3774
gas-olieverhouding 3802
gasomhulling 1370
gasontsnapping 4520
gas, opgelost – 0641
gas, opgelost – 1172
gas, resurgent magmatisch – 1172
gas, rijk – 3715
gassen, uitholling door – 1178
gast 0844
gastheer 0844
gastheergesteente 3919
gastroliet 1864
gasveld 3773
gasvlamkool 3559

gas-vloeistofverhouding 3802
gasvormige bestanddelen 3597
gas, vulkanisch – 1349
gaszand 3766
gaszandsteen 3766
gat, gedeviëerd – 4293
gat, krom – 4293
gat, onbekleed – 4277
gat, onverbuisd – 4277
gat, open – 4277
gat, scheef – 4293
gauss 0318
Gauss, verdeling van – 4189
geanticlinaal 1099
gebandeerd 3122
gebergteketen 1102
gebergteknoop 2605
gebergte, rand– 2542
gebergte-uitloper 2605
gebergtevorming 1100
gedaanteverwisseling 1794
gedifferentieerd 0667
gedissemineerd 4101
geërfd 3971
gegolfdheid 4727
gehalte, aangetoonde mineralisatie van nog nietwinbaar – 4211
gehalte, drempel voor lonend – 4204
gehalte, economisch grens– 4203
gehalte, gewogen gemiddeld – 4185
Geiger teller 4546
gekliefd 1495
gekorreldheid 2934
gelaagd 3095
gelaagdheid 3095
gelaagdheid, afstortings– 3136
gelaagdheid, centroclinale – 1241
gelaagdheid, concordante – 3115
gelaagdheid, cryptische – 0865
gelaagdheid, delta– 3139
gelaagdheid, diagonale – 3133
gelaagdheid, diagonale – 3135
gelaagdheid, dooreengeplooide – 3051
gelaagdheid, evenwijdige – 3115
gelaagdheid, foreset – 3140
gelaagdheid, gegradeerde – 3130
gelaagdheid, gesorteerde – 3131
gelaagdheid, getijden– 3134
gelaagdheid, geultjes-kriskras– 3144
gelaagdheid, golvende – 3138
gelaagdheid, guirlandekriskras– 3145
gelaagdheid, harrisitische – 0863
gelaagdheid, hellende – 1482
gelaagdheid, hellende – 3136
gelaagdheid, hoekige kriskras– 3142
gelaagdheid, jaar– 3127
gelaagdheid, kriskras – 3133
gelaagdheid, periclinale – 1241
gelaagdheid, plutonische – 0869
gelaagdheid, rhythmische – 3121
gelaagdheid, ribbel-kriskras– 3147
gelaagdheid, ritmische – 0863
gelaagdheid, stromings– 3134
gelaagdheid, tangentiale kriskras– 3142
gelaagdheid, zigzag-kriskras– 3146
gelamineerd 3101
geldigheid 1923
geleiding 4556
geleiding, warmte– 0381
gelijkkorrelig 0807
gelijkrichting, dubbele – 4506
gelijkrichting, enkele – 4506
gelijksoortig 1943
gelijkstroom 4555
gelijkstroom, gecommuteerde – 4555
gelijktijdig 1729
gelijktijdig 1730
gelijktijdig 1818
gelijkvormig 0754
gelijkwaardigheidsprincipe 4567
gelivatie 1996
geluiden, vulkanische – 1237
geluidsbarriereknal 0177
geluidsbron 4486
geluidsniveau 4490
gematteerd 3408
gemeenschap 1805
gemeenschap 1892
gemiddelde, voortschrijdend – 4198
generatie 0591
generitype 1948
genetisch 4630
genotype 1979
gens 1934
genus 1933
genus, vorm– 1932
genus-zone 1778
geobarometrie 0994
geobios 1802
geochemie 0395
geochemie 0580
geochemie, stabiele isotopen– 0453
geochronologie 1713
geochronologie, isotopen– 0428
geode 3481
geodepressie 1120
geodesie 0013
geofoon 4414
geofysica 0078
geofysica, ingenieurs– 4663
geognosie 0001
geogonie 0003
geohydrologie 3611
geoïde 0329
geoisotherm 0377
geoisothermvlak 0377
geologie 0001
geologie, algemene – 0010
geologie, catastrofen – 0005
geologie, economische – 0016
geologie, experimentele – 0014
geologie, historische – 0011
geologie, ingenieurs– 4660
geologie, mariene – 4226
geologie, milieu– 4665
geologie, ondergrondse – 4226
geologie, stabiele isotopen– 0453
geologie, structurele – 1478
geologie, toegepaste – 0016
geologie, toegepaste – 4661
geologie, veld– 4226
geologie, wiskundige – 0015
geoloog 0001
geomorfologie 1987
geomorfologie, ingenieurs– 4663
geonomie 0079
geopetaal 3107
geophaag 1857
geostatistiek 4193
geosynclinaal 1094
geotechniek 4662
geotektoniek 1073
geotherm 0378
geothermie 0374
geothermisch 0376
geothermisch bedrag 0380
geothermische gradiënt 0380
geothermische metingen 0379
geothermisch veld 0392
geothermometrie 0994
geotumor 1120
geraamte 1861
geremaniëerd 2947
geribbeld 3000
gerichtheidsdiagram 4491
gerommel, onderaards – 1237
gerontisch 1866
geschakeerd 3462
geslacht 1933
gesorteerd 3202
gesorteerd, goed – 3202
gesorteerdheid 3200
gespleten 1495
gesteente 3389
gesteente 4722
gesteenteassociatie 0630
gesteentebrokjes, met veel – 3178
gesteente, dicht – 3769
gesteentegang 1516
gesteentegruis 2005
gesteenteholte 3940
gesteentemassa 4722
gesteentemateriaal 4722
gesteentemeel 3209
gesteentemeel, glaciaal – 2875
gesteente-onderzoek, geochemisch – 4644
gesteente, ontvangend – 3919
gesteente, primair – 0631
gesteente, pyroklastisch – 1311
gesteentereeks 0630
gesteente, sedimentair – 2850
gesteenteslag 0115
gesteenteslijpsel 3209
gesteente, vast – 2011
gesteente, vast – 4788
gesteente, vulkanisch – 1238
gesteente, zone van losser – 4731
gestrekt 0972
getij 2818
getij, aard– 0199
getijdeafzetting 2907
getijdengletsjer 2436
getijdenmeer 2506
getijdentraject 2504
getijdenzone 2836
getijgeul 2509
getijschaar 2509
getijstroom 2504
getijstroom 2821
getuigeberg 2561
geul 0042
geul 2192
geul 3022
geulen, vlechtwerk van – 2243
geulerosie 2018
geulerosie 2778
geulopvulling 3022
geultje 2026
geultjeskarren 2087
gevlekt 2692
gevlekt 3462
gevoeligheid 4497
gevoeligheid 4717
gevoeligheidregeling, tijd-afhankelijke – 4508
gewelf 4851
geyser 1404
geyserbekken 1404
geyseriet 1406
geyser, ketel van de – 1405
geyser, kiezelzuurhoudende –. 1406
geyser, waterstraal van een – 1405
gidselement 4649
gidsfossiel 1785
gidslaag 1751
gidslaag 3500
gidsniveau, structureel – 4249
gigantisme 1867
gilgai reliëf 2763
gips 3381
gipsbloem 2151
gipsiet 3381
gipsvorming 3459
git 3557
glaciaal 2873
glaciatie 2367
glaciologie 2345
glaciomarien 2903
glacis 2582
glans 0484
glans 0490
glans 3407
glanskool 3580
glanskool, gestreepte – 3580
glas 0484
glas 0627
glasachtig 0805
glasdraadschrijver 4501
glasdraden 1326
glasdruppel 1326
glasslier 0481
glastuf 1333
glas, vulkanisch – 1262
glaszand 3227
glauconietisch 3177
glauconitisatie 3459
glazig 0805
gletscher 2366
gletsjer 2366
gletsjerbeek 2392
gletsjer, dragende – 2385
gletsjerfront 2390
gletsjer, gebergte– 2426
gletsjer, gematigde – 2368
gletsjer, geregenereerde – 2434
gletsjergroef 2398
gletsjer, herstelde – 2434
gletsjerijsval 2377
gletsjer, koude – 2368
gletsjerkrassen 2397
gletsjerkunde 2345
gletsjerlahar 1357
gletsjermelk 2392
gletsjermolen 2389
gletsjer, plateau– 2426
gletsjerpoort 2391
gletsjer, rijdende – 2385
gletsjerspleet 2375
gletsjertafel 2386
gletsjertong 2390
gletsjer van het alpine type 2428
gletsjer van het dendritische type 2432
gletsjer van het firnbekken-type 2428
gletsjer van het Himalaya type 2432
gletsjer van het lawine type 2429
gletsjer van het Mustag type 2431
gletsjer van het Turkestan type 2429
gletsjer, warme – 2368
glijdvlak 0539
glijhelling 2966
glijvlak 1705
glijvlak 2055
glimmerachtig 0775
glimmerhoudend 3177
glis 0487
globosferiet 0765
globuliet 0765
gloedlawine 1367
gloedwolk 1366
glomerokrist 0771
glomeroporfirisch 0821
gneis 0936

gneisachtig 0871
gneiskoepel, omhulde – 0890
gneis, primaire – 0631
gneis, stengel – 0936
gneisstructuur 0951
gneistextuur 0951
golf, aardbevings– 0161
golf, aardkern– 0180
golf, acoustische – 4482
golfbasis 2442
golf, compressie– 0183
golf, condensatie– 0183
golf, dilatatie– 0183
golf, eenling– 2815
golffront 0549
golffront 4470
golf, geleide– 0195
golfgeleider 0101
golf, gertransformeerde – 0185
golfgetal 0162
golf, gravitatie– 0202
golf, grens– 0196
golfhoogte 2808
golf, kanaal– 0195
golf, kop– 0186
golf, koppelings– 0193
golf, lange – 0188
golf, lek– 0195
golflengte 0162
golflengte 2808
golf, lichaams– 0179
golf, longitudinale – 0182
golf, mantel– 0194
golfoploop 2813
golfoploopricheltje 3030
golfoppervlak 0549
golf, oppervlakte– 0188
golf, oppervlakte– 4439
golf, P– 0182
golfperiode 2808
golfribbel 3012
golf, ruimte– 0179
golf, S– 0184
golf, seismische – 4481
golfsnelheid 0168
golf, sonische – 4482
golfterugloop 2813
golf, translatie– 2815
golf, transversale – 0184
golfweg 0206
golf, zichtbare – 0192
golf, zwaartekracht– 0202
gordel 1020
gordel 3908
gordelas 1020
gordijn 2135
gors 2514
gour 2147
graafgang 1882
graasspoor 1881
graat 2617
gradatie 0494
gradatie 4343
gradering 3130
gradiënt van de laag 1488
gradiometer 4525
graft 2062
granitisatie 1062
granitisch 0810
granoblastisch 0960
granofirisch 0848
granosferiet 0779
granulair 2727
granulaire analyse 2711
granulaire samenstelling 2709
granuliet 0937
granulitisch 0954
granulometrie 3160
grasplag 2698
grauwacke 3234
gravimeter 0352
gravimetrie 0349
gravitatie 0322
gravitatieconstante 0325
gravitatietektoniek 1702
grein 0504
greisen 0924
greisenvorming 0924
grenslaag 3544
grensvlak 4461
greppel 4235
griffellei 0942
griffelsplijting 1610
grijpmonster 3153
grind 2722
grind 3244
grindbank 2244
grind, grof strand– 2912
grindhoudend 3244
grind, residuair – 2967
groef 2796
groef 3033
groefafgietsel 3033
groef, kruisende – 3034
groeilijn, gebogen – 0479
groeisector 0526
groeispiraal 0529
groeiterras 0528
groenschist 0938
groenzand 3231
groep 1755
groep 1928
groep 3111
groep, meervoudige – 3111
groepssnelheidmethode 0229
groeve 1538
grofkorrelig 0796
grofkorrelig 3194
grofkorrelig, zeer – 0796
grond 2657
grond 4788
grond, ABC– 2695
grond, AC– 2695
grondafstand 4393
grond, azonale – 2669
grond, basische – 2737
grond, begraven – 2693
gronddruk 3819
gronddruk 4705
gronddruk, actieve – 4706
gronddrukmeter 4771
gronddruk, neutrale – 4705
gronddruk, passieve – 4706
grondecho 4390
grondfactor 0259
grondijs 2639
grond, intrazonale – 2668
grondlaag 2688
grondlawine 2359
grondmassa 0626
grondmassa 3080
grondmorene 2409
grondmorenereliëf, onregelmatig – 2630
grondontgravingsplaats 4871
grond, opgespoten – 4825
grondophoging 4825
grond, opliggende – 4788
grond, persbare – 4855
grondstabilisatie 4806
grondtoon 0167
grondverkenning 4790
grondversteviging 4806
grondwater 3658
grondwaterader 3638
grondwater, afgesloten – 3659
grondwaterbalans 3633
grondwater, freatisch – 2765
grondwater met schijnspiegel 3635
grondwaterscheiding 3647
grondwatersnelheid, werkelijke – 3703
grondwaterspiegel 2765
grondwaterspiegel 3643
grondwaterspiegel, hoogtelijn van de – 3644
grondwaterstandsverlaging 3667
grondwatertafel 2765
grondwatertrap 2766
grondwaterverdiepingen 3641
grondweerstand 4706
grond, zonale – 2667
grond, zure – 2737
grot 2095
grotklei 2122
grotklei, opvulling met – 2122
grotparel 2149
grotpisoliet 2149
grotsediment 2121
grotten– 2877
grottenkunde 2094
grotverval 2110
guano 2878
Gutenberg discontinuïteit 0092
guyot 2798
gyrogoniet 1843
gyttja 3528

H

haak, V– 2465
haakwal 2464
haard 0156
haard, centrale – 1202
haarddiepte 0157
haardmechanisme 0232
haardparameter 0151
haard, perifere – 1203
haardtijd 0159
haarvormig 0774
habitus 0521
habitus 0592
hadaal 2848
hadaal 2924
haf 2458
hagelsteenindruk 3041
hal 2635
halfglazig 0804
haling 2816
halmyrolyse 3476
halocline 2830
halofiet 1846
halo in nevengesteente 4630
halokinese 1694
halolimnisch 1915
halveringstijd 0431
handstuk 4229
hanger 2130
hangwater 2767
hardheid 0473
hardheid 3677
hardheid 4701
hardheid, tijdelijke – 3677
hardkop 2608
Harker variatiediagram 0697
harmonische 0167
harmonische, hoger – 0167
hars 0484
harst 2352
hartschiefer 1543
Hawaii-werking 1158
Hayford, zone van – 0362
hechtwater 3661
heideplag 2698
heideveen 3545
heien van palen 4838
heipaal 4838
heistelling 4838
helderheid 2677
helicitisch 0955
helictiet 2132
helling 1484
helling 4017
helling, afnemen van de – 1487
hellingafwaarts 1489
helling, buitenste – 2480
helling, continentale – 2790
helling, geringer worden van de – 1489
helling, in de richting tegengesteld aan de – 1489
helling, in de richting van de – 1489
hellingknik 2277
hellingmeting, etscylinder– 4295
helling, oorspronkelijke – 1486
hellingopwaarts 1489
helling, primaire – 1486
hellingpuin 2867
helling, schijnbare – 1484
helling, schijnbare – 4250
hellingscomplement 4018
hellingshoek 1485
hellingshoek, maximale – 3137
hellingshoek, maximale – 4687
hellingsmeter 0124
hellingsrichting 1490
hellingterugtrekking 2602
hellingterugwijking 2602
hellingveen 3534
hellingverzakking 1464
hemicyclotheem 3120
hemidiatrema 1206
hemipelagisch slik 2920
herhalingsfrequentie 4504
herhaling van lagen 1557
herkomstgebied 2943
hermigratie 3749
herplooiing 1668
hersedimentatie 2993
herwaardering 3886
heteroblastisch 0956
heterochroon 1734
heteromorf 0635
heteromorfie 0635
heterosporie 1834
heterotaktisch 1007
heterotoop 1731
hexaëdriet 0068
hexagonaal 0515
hiaat 1723
hierarchie 1925
Hilt, wet van – 3522
hindernis 1916
hindernisduin 2174
histogram 4188
hittebarst 1993
hittescheur 1993
hoed 1696
hoed, ijzeren – 4125
hoefijzermeer 2232
hoekdiscordantie 1721
hoekig 3189
hoek, kritische – 0213
hof 4629
hol 2095
holoblast 0984
holohyalien 0805
holokristallijn 0803
hololeukokraat 0587
holomarien 2900
holomelanokraat 0588
holostratotype 1740
holotype 1949
holte 3064
holte, blaasvormige – 0874
holte, miarolitische – 0873
holte-opvulling 3940
homeoblastisch 0960
homeomorfie 1820
homogeen 4680
homogeen 0716
homogene zone 4787
homoiohalien 2831
homoloog 1972
homoniem 1955

homoplastisch 1980
homoseiste 0210
homospore 1835
homosporie 1834
homotaktisch 1007
homotax 1733
homotype 1954
hongerrivier 2329
honingraat 2001
hoofdbestanddeel 0594
hoofdbeving 0142
hoofdbreuk 1580
hoofddiaklaas 1509
hoofdgang 4062
hoofdplooi 1642
hoofdpunt 4369
hoofdrivier 2206
hoofdspanning 0103
hoofdspanning 4692
hoofdzaal 2101
hooggebergte 2575
hoogplateau 0033
hoogte, structurele – 4249
hoogveen 3546
hoogwater 2818
hoogwater, gemiddeld – 2820
hoornrots 0940
horizon 1749
horizon 4461
horizon, A– 2698
horizon, A1– 2698
horizon, A2– 2700
horizon, Ap– 2699
horizon, B– 2702
horizon, banden-B– 2705
horizon, begraven – 2693
horizon, C– 2706
horizoncamera 4333
horizon, chronostratigrafische – 1758
horizon, D– 2708
horizon, eluviale – 2701
horizonfoto 4354
horizon, G– 2707
horizon, inspoelings– 2702
horizon, kleur-B– 2703
horizont 2688
horizontaalverschuiving 1560
horizon, textuur-B– 2703
hornito 1298
horst 1589
horstgebergte 2540
Hoskold, formule van – 4217
houtskool 3607
houtskool, fossiele – 3580
huidje 2129
huidje 2691
huidje 3100
huidje 3402
hulpbronnen, aangetoonde – 4206
hulpbronnen, hypothetische – 4214
hulpbronnen, minerale – 4205
hulpbronnen, onontdekte – 4213
hulpbronnen, potentiële – 4215
hulpbronnen, speculatieve – 4214
hulpvlak 0238
hum 2074
humificatie 2683
humificatie 3525
humus 2743
humus 3525
humus, ruwe – 2744
humuszuur 3525
hyalo-ofitisch 0840
hyalopilitisch 0837
hybride 0681
hybridisatie 0681
hydatogenese 3958
hydatopneumatogenese 3958
hydatopyrogeen 0656
hydatopyrogenese 3958
hydratie 0419
hydrobios 1802
hydrochemisch 4634
hydroexplosie 1351
hydrofoon 4415
hydrofoon 4496
hydrofoonkabel 4496
hydrogeochemie 3611
hydrogeologie 3611
hydrolaccoliet 2643
hydrologie 3610
hydrolyse 0418
hydrometasomatose 3954
hydromorf 4634
hydromorf kenmerk 2690
hydrosfeer 0085
hydrothermaal 0654
hydrothermaal, secundair– 3994
hydrothermale ontstaanswijze 3953
hypabyssaal 0617
hypautomorf 0751
hypautomorf 0808
hyperboolmethode 0152
hypercyclotheem 3120
hypersalien 2835
hypersalien 3377
hyperstereoscopie 4367
hypidioblast 0980
hypidiomorf 0751
hypidiomorf 0808
hypidiotoop 3471
hypocentrum 0156
hypogeen 3937
hypogeen 3967
hypohyalien 0804
hypokristallijn 0804
hypolimnion 2334
hypomagma 1199
hypostratotype 1746
hypothermaal 3983
hysterese 4675
hysterese, magnetische – 0305

I

ichnocenose 1874
ichnocoenose 1874
ichnofossiel 1873
ichnofossiel 3067
ichnologie 1871
ichor 1061
identieficatiedrempel 0271
identieficatievermogen 0271
idioblast 0980
idiochromatisch 0483
idiomorf 0523
idiomorf 0750
idiotoop 3471
ignimbriet 1340
ijkingsas 4368
ijk-malplaat 4370
ijkmerk 4368
ijsbodem 2635
ijsdek 2425
ijs, dood – 2369
ijsdruk 2401
ijs, fossiel – 2369
ijsgat, dood– 2632
ijsgrot 2107
ijskap 2425
ijskelder 2107
ijskorst 2352
ijskristalafdruk 3043
ijslawine 2361
ijsnis 2610
ijspyramide 2377
ijsscheiding 2381
ijsstuwing 2401
ijsval 2377
ijswig 2640
ijswig, opgevulde – 2641
ijzerband 2734
ijzerbooneerts 4118
ijzererts, oölitisch – 4115
ijzerfiber 2734
ijzerformatie 4114
ijzerhoudend 3464
ijzerkiezel 3492
ijzermeteoriet 0066
ijzeroerafzetting 4116
ijzersteen 3464
ijzersteen, klei– 4117
illuviatie 3930
imbricatie 3040
imitatie 0481
impactiet 0076
impactspoor 3037
impedantie, acoustische – 4474
impregnatie 3396
impregnatie 3955
impulsresponsie 4446
inbraakdelta 2500
inclinatie, magnetische – 0283
inclinometer 4294
incompetent 1624
incrustatie 3387
indamping 0412
indamping 2342
indicatrix 0552
indikker 4159
indringingssnelheid 4298
inductie, magnetische – 0317
infauna 1853
infiltratie 2768
infiltratie 3396
infiltratie 3619
infiltratie 3955
infiltratiecapaciteit 3620
infiltratie, kunstmatige – 3671
infiltratiesnelheid 3620
infiltratiezone 4619
inflectielijn 1627
inflectiepunt 1627
infracrustaal 0584
infralitoraal 2839
infraneritisch 2841
infraroodfilm 4336
infrarood, fotografisch – 4307
infrarood, midden – 4308
infrarood, nabij – 4307
infrarood, thermisch – 4308
infrarood, ver – 4309
infraspecifisch 1944
ingenieursseismologie 0254
ingeschakeld 3117
ingraving 4824
ingressiekust 2529
inham 2450
inham 2474
inheems 1813
inhomogeen 4680
injecteren, cement – 4803
injectie 3949
injectie 4802
injectieboorgat 4803
injectiecement 4803
injectieput 3793
injectiescherm 4804
injectie, sedimentaire – 3070
injectiviteitsindex 3809
inkapseling 1690
inkeping 1471
inklinking 4704
inknikkingshypothese 1118
inkoling 3519
inkolingsgraad 3552
inkolingsgradiënt 3522
inkomsten 3893
inmonding 2208
inpersingsfase 1197
inschakeling 3117
insequent 2287
inslag 0026
inslagaardbeving 0141
inslagkrater 0027
inslagtheorie 0049
inslagtrechter 1418
insluitsel 0662
insluitsel 0789
insluitsel, verwant – 0664
insluitsel, vreemd – 0663
insnijding 2020
insnijdingsmeander 2221
insnoering 3510
insnoering 4054
inspectie, plaatselijke – 0267
inspoeling 2678
inspoeling 3930
inspoelingshorizon 2702
inspoelingslaag 2702
instorting 2473
instortingsaardbeving 0141
instortingsbekken, ondergronds – 1186
instortingscaldera 1459
instortingsdal 2085
instortingsdoline 2072
instortingskanaal 1290
instortingskrater 1455
instraling 4312
instralingsdichtheid 4312
instulping 2649
intensiteit 0243
intensiteitsschaal 0243
intercalatie 3117
interceptie 3615
interferentiefiguur 0569
interferentiekleur 0568
interferentiekleur, anomale – 0568
interferentieribbel 3013
intergranulair 0841
intergranulair 0998
interkinematisch 1079
intermediair 0602
intern effect 0704
interpretatie, directe – 4566
interpretatie, indirecte – 4566
interpretatie, kwantitatieve – 4566
intersertaal 0840
interstitiëel 3083
interval 1718
intervalzone 1772
intervalzone, biostratigrafische – 1776
interzone, steriele – 1775
intragranulair 0998
intramagmatisch 3985
intramicareniet 3284
intramicriet 3284
intramicrudiet 3284
intraspariet 3283
intrasparrudiet 3283
intrasparzandsteen 3283
intrazone, steriele – 1775
intrusie 0618
intrusief 0619
intrusief 0715
intrusief complex 0620
intrusief, deel– 0721
intrusief, enkelvoudig – 0720
intrusief, gemengd – 0721
intrusief, meervoudig – 0721
intrusief, samengesteld – 0722
intrusie, gemengde – 0721
intrusiegesteente 0715
intrusie, meervoudige – 0721
intrusieplaat 0740
intrusieplug 0734
intrusie, sedimentaire – 3070
intrusietektoniek 0712

intrusiva 0715
inval, loodrechte – 4472
inval, scherende – 0212
invalshoek 0211
invalshoek 4389
invalshoek, kritische – 9229
invalsvlak 0211
invangtheorie 0047
invasie 4619
invasiezone 4619
inversie 1088
invoer 4445
involutie 2648
inwateren 4810
inzakking 4704
inzakking, ketelvormige – 1186
inzakkingsstructuur 3059
inzet 0172
inzet, eerste – 4428
inzet, geleidelijk beginnende – 0174
inzet, scherpe – 0173
inzijging 3629
inzinking 1463
inzinking 2793
inzinking, vulkanische – 1410
ionenpotentiaal 0419
ionenuitwisseling 0420
ionenuitwisselingscapaciteit 0420
ionium-thorium datering 0444
iridescentie 0486
isanomalieënkaart 0372
isochoor 4244
isochorenkaart 4244
isochromaat 0571
isochroon 0210
isochroon 0433
isoclinaal 1626
isocline 0283
isodyname 0285
isofasische lijn 0571
isogallenkaart 0372
isogam 4532
isogammenkaart 0372
isogoon 0284
isograad 0992
isograad, geochemische – 4638
isogyre 0570
isohypse 4248
isohypse van de grondwaterspiegel 3644
isolement 1981
isomesisch 1735
isometrisch 0522
isometrisch 3181
isomorf 1974
isomorfie 0464
isopach 4243
isopachenkaart 4243
isopisch 2863
isopletenkaart 4246
isopore 0287
isoseismische lijn 0245
isoseiste 0245
isospore 1835
isosporie 1834
isostasie 0338
isostatische isocorrectielijnenkaart 0373
isotopen fractionering, kinetische– 0457
isotopen-fractioneringscurve 0455
isotopen-fractioneringsfactor 0455
isotopen-verdunningsanalyse 0432
isotroop 4681
isotroop, optisch – 0561
isotropie 4681
isotypie 0463
isovole 3523

J

jade 0498
jasper 4127
jaspiliet 3492
jaspis 3492
jeugdstadium 2570
jig 4155

K

kaakbreker 4148
kaap 2446
kaap, hoge – 2446
kaar 2610
kaarbodemniveau 2615
kaardrempel 2610
kaargebergte 2616
kaargletsjer 2430
kaarling 2618
kaarniveau 2615
kaar, onecht – 2611
kaars 2140
kaar, samengestelde – 2613
kaart, bodem– 2671
kaart, bodemkundige detail– 2671
kaart, bodemkundige overzichts– 2671
kaarteenheid 2672
kaarterras 2613
kaart, foto– 4383
kaart, geologische – 4239
kaart, ingenieursgeologische – 4784
kaart, metallogenetische – 3905
kaart, palaeogeologische – 4242
kaart, radar– 4400
kaartrap 2614
kaart, tektonische – 4247
kabelboren 4262
kabelproef 3787
Kalb, lijn van – 0559
kaldera 0036
kalium-argon datering 0449
kalium-calcium datering 0449
kalkafzetting 3267
kalkalgen 3359
kalkalkali– 0694
kalkalkalisch 0694
kalkalkalische reeks 0692
kalkarm 2736
kalkbreccie 3279
kalkgehalte, koolzure – 2735
kalkhoudend 2736
kalk, kalkzandsteenachtige – 3277
kalkkiezel 3278
kalkloos 2736
kalkreeks 0692
kalk, rif– 3310
kalkrijk 2736
kalkslib 3295
kalkslib 3307
kalkslibsteen 3274
kalkslibsteen 3275
kalkslibsteen 3296
kalkstabilisatie 4807
kalksteen 3305
kalksteen, crinoidenrijke – 3313
kalksteen, dolomitische – 3316
kalksteenfragment 3167
kalksteenfragment 3270
kalksteen, kleiige – 3221
kalksteen, lithografische – 3306
kalksteen, pellet– 3307
kalkzandig 3277
kalk, zandige – 3317
kalkzandsteen 3276
kame 2420
kamers bevattende, veel – 1870
kameterras 2420
kamstructuur, overkorstingsader met – 4049
kamvormig 4084
kamwaterscheiding 2316
kanaal 4412
kanalenkust 2521
kandelkool 3563
kanteling 0124
kaolien 3217
kaolinisatie 0926
kap 1696
karaat 0482
karbonaatgesteente 3266
karbonaatgesteente, sedimentair – 3266
karren 2086
karrenveld 2088
karst 2064
karst, bedekte – 2069
karstbron 2089
karst, corrosiebasis van de – 2066
karst, diepe – 2067
karstgebied 2064
karstlaagte 2071
karstlandschap 2064
karst, onbedekte – 2068
karst, onderaardse – 2070
karst, ondiepe – 2067
karstpijp 2076
karstrestberg 2974
karstvenster 2078
karstverschijnselen 2065
karstvorming 2063
karteringscamera 4329
kartografie 4239
katagenese 1977
katagenese 3419
kataklase 0914
kataklasiet 0914
kataseisme 0187
katathermaal 3983
katazone 0993
kategorie 1927
kationen, omwisselbare – 2755
kationen, uitwisselbare – 2755
kation uitwisselend vermogen 4623
katteklei 2739
katteoog 0488
kaustobioliet 0427
kegel 0059
kegel, aan de kust gevormde – 1300
kegelberg 1435
kegelbreker 4148
kegel, centrale – 1443
kegelkarst 2081
kegel, parasitische – 1437
kegelvormig 4080
kegelvorming 3826
kei 3242
kei 3243
keienleem 2874
keienleem 3250
keienvloer 2651
keienvloer in een woestijn 3405
keileem 2404
keileem 2422
keileem 2874
keileem 3250
kelder 0116
kelyfitische rand 0786
kenmerkend bestanddeel 0594
kenmerk, microstructureel – 4108
kern 0730
kern 1628
kern 3155
kern 4237
kern 4861
kern, afgeknepen – 1628
kernapparaat 3155
kernblok 2609
kernbuis 4272
kernexplosie 0268
kernexplosie, ondergrondse – 0268
kernkist 4237
kernkroon 4271
kernmonster, ononderbroken – 4749
kernopbrengst 4273
kernsplijtpers 4182
kernvanger 4272
kernvorming 1360
kerogeen 3720
kerogeniet 3720
kerosine 3744
ketel 2793
keteldal 1463
keten 4822
keten, anticlinale – 2541
ketengebergte 1103
keten, monoclinale – 2541
keten, synclinale – 2541
kettingcorrelatie 4195
kettingstruktuur 4083
kiezel 3241
kiezelaarde 3367
kiezelig 3365
kiezel-ijzersteen 3373
kiezellei 3367
kiezelslib 3367
kiezelwoestijn 2578
kiste 1841
klappersteen 3485
klasse 0685
klasse 1929
klassificatie *) 1921
klassificatie, biostratigrafische – 1710
klassificatie, chronostratigrafische – 1712
klassificatie, natuurlijke – 1921
klast 3166
klastische korrel 3166
klastisch fragment 3166
kleef 4839
kleefgrens 2750
kleef, negatieve – 4839
kleefpaal 4839
kleefribbel 3016
klei 2714
klei 3212
kleibal 3246
kleibal, gepantserde – 3247
kleiband 2734
kleiband 4012
klei, gevoelige – 4717
kleiig 2721
kleiig 3211
kleiigheid 4623
klei-ijzersteen 3488
klei-ijzersteen 3504
klei-ijzersteen, koolrijke – 3504
klei-ijzersteen, koolrijke – 4117
klei, keramische – 3517
kleilaag 3772
klei, magere – 3216
klei, mergelige – 3223
kleimineraal 3213
kleischalie 3220
kleisteen 3219
kleisteen, koolhoudende – 3505
klei, vette – 3216
klei, vuurvaste – 3217
klembeugel 4280
klemblok 4280
kleur 0546
kleurindex 0688

*) Zie ook: classificatie

kleurtoon 2677
kleurverzadiging 2677
kleverig 2750
kleverigheid 2750
klieving 1495
klieving, cylindrische – 1302
klieving, kogelvormige – 1304
klieving, plaatvormige – 1500
klieving, polyëdrische – 1303
klieving, romboëdervormige – 1502
klieving, sferoïdale – 1304
klieving, zuilvormige – 1501
klif 2470
klif, deels verdronken – 2475
klif, dood – 2477
klifduin 2169
klifje 2470
klifkust, vereffende – 2478
klif, onderzees – 2799
klif, secundair – 2476
klif, verlaten – 2477
kling 2103
klink 4704
klink 4833
Klinkenberg, factor van – 3706
klinograaf 0124
klinometer 4227
klino-nonconformiteit 1721
klip 2443
klip 2805
klippe 1692
klont 3301
kluit 2725
kneedbaar 4673
knieplooi 1648
knikker 4150
knikplooi 1661
knikzone 1661
knobbelig 4094
knol 0771
knol 3473
knol 4085
knollig 3473
knollig 4085
knooplijn 0234
knoopvlak 0234
Koenigsberger verhouding 0310
koepel 0724
koepel 1205
koepel 1641
koepeldam 4865
koepelgebergte 2538
koepelvulkaan 0035
kofferdam 4869
kofferplooi 1649
kogel 0867
kogelmolen 4149
kokardenerts 4050
koken, retrograad – 1174
kolenequivalent, standaard – 3606
kolengruis 3513
kolenliquefactie 3609
kolenvergassing 3609
kolenwasserij 4149
kolkgat 2027
kolkgat 2028
kolkgat 2248
kolom 2730
kolomfundering 4837
kolommendiagram 4188
kolom, stratigrafische – 4255
kolonie 1806
kom 2239
kom 2247
kom 2330
kompas, geologisch – 4227
kool 3496
kool, aardachtige – 3555
kool, bitumineuze – 3559
kool, bladerige – 3563
kool, doffe – 3580
kool, humeuze – 3497
koolhydraat 3723
koollaag 3499
koollaagje 3499
koollens 3499
koolpetrografie 3565
koolstof-14 datering 0451
koolstof, gebonden – 3601
koolstofgetal 3727
koolstof-stikstofverhouding 0425
koolstructuur 3565
kool, veredelde – 3564
koolwaterstofkolom 3775
koolwaterstofkolom, hoogte van de – 3775
koolzuurinwerking 2000
koopkracht, constante – 3899
kop 1686
kopererts, porfier– 4133
koperschalie 4130
kopje 2190
koppeling 0273
koraalachtig 1821
koraal, ahermatypisch – 3335
koraal, hermatypisch – 3335
koralenkalk 3310
korrel 3078
korrel 3240
korrelcontact, in elkaar passend – 3449
korreldeformatie 3446
korreldruk 3819
korrelgrens 0532
korrelgrootte 2710
korrelgrootte 3162
korrelgrootte 3193
korrelgrootte 0792
korrelgrootte analyse 2711
korrelgrootte-analyse, mechanische – 3160
korrelgrootte-indeling 2713
korrelgrootte-klasse 2712
korrelgrootte-klasse 3162
korrelgrootte, mediaan– 3199
korrelgrootte, modale – 3199
korrelgrootten-indeling 3161
korrelgrootteverdeling 2709
korrelgrootteverdeling 3197
korrelgrootteverdeling, polymodale – 3197
korrelig 0806
korrelig 3081
korreligheid 3081
korreligheid 4344
korrelineengroeiing 3448
korreling 2934
korrelklasse 2712
korrelondersteuning 3294
korrelpakking 3082
korrelskelet 3079
korrelsneeuw 2350
korreltextuur 4086
korrelverdeling 4702
korrelverdeling, continue – 4702
korrelverdeling, discontinue – 4702
korrelvergroting 3454
korrelverspreidingsverandering 3430
korst 2759
korstcement 3438
korst, dikte van de – 0347
korst, starre – 0348
korstvorming 2760
kracht, drijvende – 4740
krachtmeter 4770
kras 1538
krater 0054
krater 1405
krater 1412
krater 4819
kraterbekken 1412
kratereiland 1417
kraterfumarole 1383
krater, hoefijzervormige – 1471
kratermeer 1454
krater met centrale boccas 1445
krater, parasitaire – 1439
kraterpijp 1221
kraterpijpagglomeraat 1248
kraterpijpbreccie 1248
kraterpijpverplaatsing 1231
kraterpijpwrijvingsbreccie 1248
krater, primaire – 0027
kraterrij 1449
krater, secundaire – 0027
krater, subterminale – 1438
kraterwal 1451
kraterwand 1453
krater, werkende – 1211
kraton 1080
kriging 4197
krijtkalk 3315
krimpgrens 4715
krimpscheur 1993
krimpscheur 3044
krimpscheur 3046
krimpscheur, beginnende – 3046
krimpscheur-opvulling 3047
krimpscheur-polygoon 3046
kringloop, geochemische – 0396
kringloop, hydrologische – 3612
kringloop, orogenetische – 1105
kringloop, petrogenetische– 0582
kringloop van het water 3612
kriskrasgelaagdheid 3133
kriskraslaminatie 3133
kristal 0506
kristalafscheiding 0671
kristalbrei 0640
kristalfout 0531
kristal, gecorrodeerd – 0756
kristal, geregenereerd – 0910
kristalklasse 0512
kristallapilli 1325
kristallentuf 1333
kristalliet 0760
kristallijn gesteente 0583
kristallijn, zichtbaar – 0795
kristalliniteit 0791
kristallisatie 0632
kristallisatie, collectieve – 0910
kristallisatie, gefractioneerde – 0670
kristallisatieïndex 0702
kristallisatiekracht 0909
kristallisatie, mimetische – 1008
kristallisatie, ritmische – 0864
kristallitisch 0760
kristalloblast 0908
kristalloblastese 0908
kristalloblastische reeks 0909
kristallografie 0505
kristallografie, optische – 0544
kristalroos 3092
kristalrooster 0507
kristalrozet 3092
kristalsedimentatie 0673
kristalstelsel 0514
kristalstructuur 0509
kristalvorm 0520
kristalvorm 0753
kristalzandsteen 3465
krommingswaarde 0336
kronkelberg 2227
kronkelhals 2226
kronkeling, dooreen– 3051
kronkelwaard 2225
kronkelwaardgeul 2234
kroon 4271
krotovina 2745
kruimel 2727
kruimelig 3087
kruin 1633
kruinlijn 1632
kruinvlak 1632
kruip 0044
kruip 0114
kruip 2038
kruipen 2038
kruipen 4672
kruipen van puin 2040
kruipgang 2098
kruipherstel 0114
kruipspoor 1878
kruipspoor 1879
kryokoniet 2387
kryokonietgat 2387
kryopedologie 2634
kryoturbatie 2647
kryptoklastisch 3166
kubisch 0515
kust 2437
kust– 2906
kust, aangroeiende – 2528
kust behorend, tot de – 2836
kust, breuk– 2520
kust, concordante – 2521
kustdelta, open – 2497
kust, discordante – 2522
kustduin 2166
kust, epigenetische – 2527
kust gelegen, uit de – 2840
kust, gelobde – 2533
kust, gerafelde – 2532
kust, getande – 2532
kust, gladde – 2531
kustklif 2470
kustlijn 2437
kustlijn, binnen– 2461
kustlijn, buiten– 2461
kustlijn, vereffende – 2531
kust, meercyclische – 2530
kust met lavastromen 1280
kustmoeras 2514
kust-nabij 2889
kust, onbewegelijke – 2526
kust, opgeheven – 2524
kust-playa 2892
kust, polycyclische – 2530
kustprofiel 2441
kustprofiel, vereffend – 2441
kust, rechtlijnige – 2531
kust, samengestelde – 2527
kust-sebkha 2892
kust, stabiele – 2526
kuststroming 2825
kustterras 2484
kusttype, Dalmatisch – 2521
kusttype, gemengd – 2526
kust van het Atlantische type 2522
kust van het Pacifische type 2521
kustvlakte 2440
kwarteren 4181
kwartsareniet 3230
kwartsiet 3230
kwel 3629
kwelder 2514
kwelder 2909
kweldereiland 2507
kwelderklif 2516
kwelderkom 2517
kwelderkreek 2517
kwelderpan 2517
kwelderwal 2516
kwelderzoom 2512
kwellijn 4862

L

laag 1750

laag 3097
laag 3098
laag 3123
laag, actieve – 2647
laag, afgesloten watervoerende – 3636
laagafzetting 4021
laag, artesische watervoerende – 3637
laag, backset – 3141
laagbeproeving 3787
laag, D– 2708
laagdikte 4622
laag, foreset – 3140
laaggang 4053
laaggrot 2096
laag, jaar– 3127
laag, ondoorlatende – 3640
laagopeenvolging 3110
laagpakket 1753
laagreflectie 4519
laag, slecht doorlatende – 3639
laagte, perifere – 2560
laagveen 3546
laagvlak 3096
laagvlak, blootgelegd – 2557
laagvoeg 3099
laagwater 2818
laagwater, gemiddeld – 2820
laag, watervoerende – 3634
laag, zeer dunne – 3122
laat-kinematisch 1079
laatorogenetisch 1108
labradorescentie 0486
lacustrien 2888
lacustrisch 2888
laddergang 4057
lading 4816
ladu 1367
lagen, gekantelde – 1482
lagen, hellende – 1482
lagen, opgerichte – 1481
lagen, verplaatste – 1523
lage-snelheidskanaal 0100
lagg 3537
lagunair 2905
lagune 2458
lagune 3345
lagunebodem 3345
lagune, getijden– 2458
lagunekanaal 3348
laguneplat 3346
lagunevlakte 3346
laharafzetting 1356
lahareruptie 1403
lahar, gewone – 1355
lahar, koude – 1355
lahar, warme – 1353
lakfilm 4230
lakfilmmethode 4230
lakkoliet 0734
Lamarckisme 1967
lamellair 4103
lamellen, schok– 0894
lamina 3101
lamina, backset – 3141
laminatie 1277
laminatie 3101
laminatie, convolute – 3952
laminatie, foreset – 3140
laminatie, golvende – 3138
laminatie, klimmende ribbel– 3149
laminatie, kriskras– 3133
laminatie, ribbel– 3147
lamprofierisch 0809
landbeving 0148
landclassificatie 2660
landduin 2170
landeenheid 4785
landevaluatie 4786
landijs 2425
landschap, chaotisch – 0038
landschap, trapsgewijs gelegen – 2599
landtong 2448
landveen 3542
langsdiaklaas 1506
lapies 2086
lapilli 1324
lapillituf 1333
Larsen variatiediagram 0697
last 2950
lateraal 4631
lateraallog 4585
lateriet 4124
lateritisatie 2681
latvormig 0522
lava 1242
lava, aa- 1260
lavabal 1296
lavabal, aangegroeide – 1296
lavabank 1277
lava, blok– 1259
lavabrok 1327
lavadek 1275
lavadelta 1280
lavadom 1422
lavadom met tongvormige uitloper 1423
lavafontein 1282
lava, gekraste – 1270
lava, gestolde – 1242
lavagrot 1291
lavahelling 1286
lavahol 1291
lavaholte met stamafdruk 1292
lava, huid– 1261
lavakegel 1427
lavaklodder 1297
lava, koek– 1253
lavakoker 1292
lavakolom 1246
lavakorst 1273
lavakrassen 1270
lavakurk 1288
lavalawine, gloeiende – 1367
lavameer 1281
lavameerkrater 1284
lavamorene 1295
lavamuur 0741
lavaorgel 1301
lava, pahoehoe – 1250
lavaplaat 1275
lava, plaat van gestolde – 1285
lavaring 1284
lava, schol– 1259
lavaschoorsteen 1298
lavaspetters, kegel van – 1298
lavastalagmiet 1293
lavastalaktiet 1293
lavastroom 1263
lavastroom, wortelloze – 1282
lavatong 1264
lavatoren 1299
lava, touw– 1252
lavatunnel 1290
lava-uitlopers 1251
lava-uitvloeiing, parasitische – 1265
lavaval 1287
lavaveld 1278
lavavenster 1289
lavavlakte 1278
lavavloed 1276
lavawaaier 1287
lawinebaan 2362
lawine, droge – 2359
lawinegeul 2362
lawinekegel 2362
lawine, natte – 2359
lawinevergletsjeringtype 2429
lectotype 1951
leem 2717
leem 2720
leem 3215
legenda 4240
legenda, profieltype – 4240
lei 0942
leisplijting 1599
leisteen 0942
lekkage 4863
lektostratotype 1745
lemig 2721
lengte– 2289
lengteduin 2171
lengtekust 2521
lengtemorene 2415
lengteprofiel van een dal 2271
lengtespleet 2376
lens 1752
lens 3102
lensje 3102
lensvormig 3102
lensvormig lichaam 1752
lenticulaire structuur 0952
lenzen 0952
lepidoblastisch 0961
leptiet 0937
leptinoliet 0940
leptothermaal 3984
leukokraat 0587
leukosoom 1038
levensgemeenschap 1890
levensgemeenschap 1896
levensgemeenschap, climax– 1891
levensspoor 1875
levensspoor, oppervlakte– 1875
licht 2721
lichtschacht 2076
lichtstraal 0548
lido 2461
ligging, tektonische – 1480
ligniet 3495
lij 2964
lijhelling 2965
lijn 4503
lijnaftasting 4402
lijnelectrode 4563
lij-oever 2336
lijzijde 2964
liman 2457
limnetisch 2890
limnisch 2888
limnobios 1802
limnologie 2331
limnoplankton 1801
limoniet 4126
lineatie 0973
lineatie 1607
lineatie 2999
lipide 3723
liquatie 0678
liquefactie 3084
liquefactie 4718
liquid-magmatisch 3976
lithificatie 3421
lithofacies 2861
lithofyse 0780
lithogenese 2853
lithoidaal 0835
lithologie 2853
lithophaag 1857
lithosfeer 0088
lithostratigrafie 1709
lithostratigrafische eenheid 1748
lithostratigrafische horizon 1749
lithostratigrafische zone 1747
lithothamniumdijk 3361
lithotoop 2862
lithotype 3579
lithozone 1747
litoraal 2836
lit-par-lit 1049
littoraal 2906
locatieafstand 3842
loefzijde 2963
loess 2871
loesspoppetje 3489
log, acoustische – 4597
logen 3939
logen 4162
log, gamma – 4590
log, gammastraal-spectrum – 4548
log, inductie – 4584
log, neutron – 4594
log, spontane potentiaal – 4579
log, temperatuur – 4601
longuliet 0762
lood, anomaal – 0441
lood-evolutiemodel, enkelvoudig – 0439
lood-evolutiemodel, meervoudig – 0439
lood, gewoon – 0437
lood-lood datering 0434
lood-modelouderdom 0440
lood-ontwikkelingscurven 0438
lood, radiogeen – 0437
loodsgat 4297
loodzand 2700
loopsneeuw 2351
looptijd 0171
looptijd 1781
looptijdkromme 0171
looptijdtabel 0171
looptijdzone 1781
looptijdzone, overlappende – 1777
looptijdzone, taxon – 1783
loopzand 3085
loper 2175
lopoliet 0735
losbarsting 1217
losbreken 2396
losrukken 2396
losscheurbaarheid 4828
losscheuren 4828
losser worden 4731
Love golf 0189
lozing 1210
lubliniet 2153
luchtbasis 4350
luchtfoto 4349
luchtfotocamera 4329
luchtfoto, scheve – 4353
luchtfototriangulatie 4324
luchtfoto, vertikale – 4353
luchtfoto, zijwaarts scheve – 4353
luminescentie 0467
lushaakwal 2465
lutiet 3203
lutietisch 3203
lutumfractie 2714
lydiet 3369

M

maagsteen 1864
maaksel 0793
maaksel 1004
maaksel 1605
maaksel 3077
maakselanalyse 0995
maakselas 1016
maaksel, deformatie– 1007
maakseldiagram 1015
maakseldomein 1006
maakselelement 1012
maaksel, extern – 1004
maaksel, groei– 1010
maaksel, intern – 1004
maaksel, mega– 1005
maaksel, meso– 1005

maaksel, micro– 1005
maaksel, sedimentatie– 1011
maalcircuit 4146
maalstroom 2269
maanbaai 0051
maanbeving 0061
maanbodem 0060
maancontinent 0050
maanzee 0051
maar 1433
maastextuur 0856
macadam 4843
maceraal 3566
maceraal, actief – 3578
maceraal-groep 3566
maceratie 3594
macroëvolutie 1970
macrokirstallijn 0797
macropolyschematisch 0794
macroseisme 0138
macroseismisch gebied 0248
macrosferisch 1868
macrospore 1837
mafelsisch 0598
mafisch 0599
mafisch gesteente 0599
magerkool 3560
magma 0639
magma 1198
magmahaard 1202
magma, herboren – 0644
magmakamer 1202
magmakamer, gangvormige – 1204
magmakolom 1226
magma, latent – 0643
magma, moeder – 0645
magma, oer– 0644
magma, oppervlakte– 1244
magma, palingeen – 0644
magma, rest– 0646
magmatisch 0616
magmatische opslokking 0623
magmatisch, primair– 3973
magmatogeen 3973
magmazuil 1226
magneetscheiding 4154
magnetisatie 0295
magnetisatie 4531
magnetisatie, afzettings– 0312
magnetisatie, anhysteretische – 0314
magnetisatie, chemische – 0312
magnetisatie, omgekeerde – 0315
magnetisatie, verzadigings– 0297
magnetisatie, viskeuze – 0313
magnetisch moment 0296
magnetiserende kracht 0317
magnetisme, gesteente – 0293
magnetisme, remanent – 0306
magnetisme, thermoremanent – 0309
magnetohydrodynamica 0278
magnetolaag 0277
magnetometer 0316
magnetometer, astatische – 0319
magnetometer, horizontale – 4521
magnetometer, nucleaire – 4524
magnetometer, proton– 4524
magnetometer, verticale – 4521
magnetometer, vliegtuig– 4522
magnetopause 0277
magnetosfeer 0277
magnitude 0220
magnitudeschaal 0220
malen, autogeen – 4147
malen, fijn– 4150
malloseismisch gebied 0250
mangaanknol 3477
mangrovemoeras 2515
manometer 4769
mantelbuis 4274
mantelconvectie 1124
mantelpluim 1145
mantel, polaire – 0040
mantelwrijving 4839
margariet 0766
marien 2900
marien, beperkt – 2901
marien, diep – 2918
marien, normaal – 2901
marien, ondiep – 2914
marmer 0920
marmorisatie 0920
mascon 0052
massabeweging 2037
massacompensatie, isostatische – 0342
massaoverschot 0343
massa, stationaire – 0125
massatekort 0343
massa, trage – 0125
massatransport 2037
massief 0859
massief 1087
massief, onderliggend – 0728
mast 4267
mat 3408
materiaalbalans 3865
materiaalbalansberekening 3865
materiaal, detritisch – 2005
materiaal, ontgraven – 4871
matetexis 1030
matig 1626
matrix 0626
matrix 3080
mattering 3408
maximum 1019
meander 2219
meander, afgesneden – 2232
meanderafsnijding 2229
meanderbocht 2232
meanderdoorbraak 2229
meandergordel 2224
meanderhals 2226
meander, ingesneden – 2221
meanderklif 2228
meander, overerfde – 2221
meanderspoor 2227
meanderspoorterras 2303
meanderterras 2304
meanderterras, beschermd – 2305
meander, vrije – 2220
mechanica, gesteente– 4664
mechanica, grond– 4664
mediterrane serie 0690
medium, poreus – 3680
meeneemstang 4269
meer 2330
meer– 2888
meer, afvoerloos – 2338
meerafzetting, glaciale – 2402
meerbal 2891
meerkanalenschrijver 4501
meer, proglaciaal – 2393
megacyclotheem 3120
megagametofiet 1838
megakrist 0608
megalosferisch 1868
megaribbel 3007
megaseismisch gebied 0249
megasporangium 1837
megaspore 1837
meiofauna 1853
mélange 2927
melanokraat 0588
melanosoom 1038
melatoop 0570
melikaria 3487
membraanpotentiaal 4616
membraan, semipermeabel – 3397
mercaptaan 3716
mergel 3221
mergel, kalkige – 3224
mergel, kleiige – 3223
mergelslik 3222
merismiet 1048
merk 3018
meroplankton 1801
mesa 2556
mesohalien 2832
mesokraat 0587
mesokristallijn 0797
mesonorm 0606
mesopelagisch 1910
mesopelagisch 2845
mesostasis 0626
mesothermaal 3983
mesotroof 3549
mesozone 0993
meta– 0883
metaal 0484
metaaldistrict 3908
metaalprovincie 3908
meta-anthraciet 3562
metabasiet 0883
metabentoniet 3218
metablastese 0884
metablastisch 0884
metabolisme 1803
metakrist 0976
metalen, extraheerbare – 4654
metalen, koud-extraheerbare zware – 4655
metalen, warm-extraheerbare – 4655
metalen, zware – 4653
metallisch 3911
metallisch, niet– 3911
metallogenese 3903
metallogenetisch 3903
metalumineus 0693
metamorf 3997
metamorfe zone, dubbele – 1139
metamorf gesteente 0880
metamorfiet 0880
metamorfose 0879
metamorfose 1794
metamorfose, allochemische – 0881
metamorfose, begravings– 0887
metamorfose, belastings– 0887
metamorfose, contact– 0896
metamorfose, dislocatie– 0892
metamorfose, dynamo– 0892
metamorfose, gesuperponeerde – 0901
metamorfose, graad van – 0992
metamorfose, injectie – 0889
metamorfose, inslag– 0893
metamorfose, kaustische – 0897
metamorfose, lokale – 0891
metamorfose, organische – 3521
metamorfose, progressieve – 0903
metamorfose, regionale – 0887
metamorfose, retrograde – 0904
metamorfose, schok– 0893
metamorfose, statische – 0887
metamorfose, thermische – 0896
metaribbel 3014
metasomatose 3453
metasomatose 3956
metasomatose, contact– 0900
metasomatose, regionale – 1056
metasoom 1067
metatekt 1040
metatype 1954
meteoor 0063
meteoorkrater 0075
meteoorsteen 0063
meteorenstof 0065
meteoriet 0063
meteorietenkunde 0062
methode, electrische – 4554
methode, electromagnetische – 4572
methode, horizontale spoel– 4575
methode, magnetotellurische – 4553
methode, potentiaalverval– 4560
methode, tellurische – 4553
methode, verticale spoel– 4575
miarolitisch 0873
micriet 3281
micro-aardbeving 0146
microcoproliet 1863
microcoquina 3308
microëvolutie 1970
microfossiel 1790
microgametofiet 1838
microkristallijn 0797
microkristallijn 3091
microliet 0759
microlithotype 3581
microlitisch 0836
microlog 4587
micrometeoriet 0064
micropalaeontologie 1786
microplankton 1830
microplooi 1643
microscoop, polarisatie– 0611
microseisme 0097
microseismische beweging 0097
microseismisch gebied 0251
microsferisch 1868
microsporangium 1836
microspore 1836
microstructuur 1479
microtektoniek 1479
microtextuur 0790
middelgebergte 2575
middelkorrelig 0796
middelkorrelig 3194
middenlijn 4372
middenloop 2195
middenmorene 1295
middenmorene 2407
middenoceanische rug 1129
middenpunt, groep met gemeenschappelijk – 4433
middenvleugel 1576
middenvleugel, uitgerekte – 1680
migamatiet 1022
migma 1022
migmatisatie 1022
migratie 3748
migratie 4457
migratie, horizontale – 3750
migratiekanaal 3751
migratie, laterale – 3750
migratie, primaire – 3749
migratie, secundaire – 3749
migratie, verticale – 3750
migratieweg 3751
mijngas 3554
milieu 1900
milieu 2859
milieu, primair – 0402
milieu, secundair – 0402
milligal 0367
mimetisch 1008
mineraal 0460
mineraal, accessorisch – 3174
mineraalassociatie, kritische – 0988
mineraal, contact– 0978
mineraaldistrict 3908
mineraal, ganggesteente– 3921
mineraal, gesteentevormend – 0462
mineraal, industriëel – 0462

mineraal, karakteristiek – 0689
mineraal, kenmerkend – 0594
mineraalkrans 0786
mineraal, kritisch – 0988
mineraal, kunstmatig – 0461
mineraal,licht – 3173
mineraal, occult – 0757
mineraal, primair – 0596
mineraalprovincie 3908
mineraal, secundair – 0596
mineraal, specifiek – 0595
mineraal, typomorf – 0988
mineraal, zwaar – 3173
minerale bestanddelen 3553
mineralisatie 2682
mineralisatie 3936
mineralisator 3936
mineralogie 0459
mineraloide 0460
miogeosynclinaal 1096
miospore 1833
mise à la masse methode 4559
mixoeuhalien 2834
mixohalien 2833
mobilisaat 1057
mobilisatie 1057
mobilisme 1113
mobiliteit 3699
mobiliteitsverhouding 3699
modaal 0593
modder 4829
modderfontein 1398
modderig 3210
modderkegel 1396
modderkogel 1321
modderlava 1397
modder, slappe – 3210
modderstroom 2051
modderstroom 2974
modderstroomconglomeraat 1356
modderstroom, vulkanische – 1352
moddervulkaan 1395
moddervulkaan 1700
modderwel 1398
mode 0166
mode, fundamentele – 0167
mode, hogere – 0167
model 4775
model, berekening aan het – 4478
model, elektrische weerstand analoog – 4783
model, equivalent – 4778
model, foto-elastisch – 4783
model, fysisch – 4778
model, geomechanisch – 4779
model, horizontale lagen – 4568
modelproef 4778
moder 2744
modus 0593
moedergang 4062
moedergesteente 0883
moedergesteente 2944
moedergesteente 3918
moederloog 3377
moedermateriaal 2657
moeheid, materiaal– 4674
moeras– 2887
moeras 3532
moerasgas 3529
moerassig 3532
mofette 1402
Mohorovičić discontinuïteit 0087
Mohr, cirkel van – 4690
Mohr-Coulomb breukkriterium 4694
Mohr, omhullende van – 4691
Mohr, spanningsdiagram van – 4688
Mohs, hardheidsschaal van – 0473
Mohs, hardheidsschaal van – 4701
mol 3675
molasse 2926
moment 1759
monadnock 2608
mond 2196
monding 2196
monding, gelijkvloerse – 2208
mondingsgeul 2503
mondingsspoor, afgeknotte – 2628
monding, zwevende – 2208
monistische hypothese 0020
monoclinaal 1617
monocline 1617
monofyletisch 1960
monoklien 0515
monometallisch 3911
monomict 3265
monomineraal 0589
monoschematisch 0794
monotypisch 1945
mons 0037
monster 4741
monster, geroerd – 4742
monster, gestoord – 4742
monster, grijp– 4169
monsternameapparaat met zuiger 4746
monster, ongeroerd – 4742
monster, ongestoord – 4742
monster, onregelmatig gevormd – 4744
monsterplaatsen, spatiëring van – 4171
monster, samengesteld – 4176
monster, scherf– 4168
monster, sleuf– 4168
monsterverdeler 4182
monsterverkleining 4180
monster, verkneed – 4743
monster, vervalst – 4179
monsterwaarden, extreme – 4186
monsterzakje 4229
montaanwas 3736
monzonitisch 0813
morene 2403
morene, afgezette – 2410
moreneamfitheater 2418
morene, inwendige – 2408
morene, verlaten – 2416
morene, voortbewogen – 2405
morenewal 2414
morfologie 1793
morfotektoniek 1076
Morkill, formule van – 4217
morphogenetische zone 1779
mortel 0966
mosoor 2608
mosveen 3542
Mount Katmai werking 1165
mozaiek 4380
mozaiek, gecorrigeerde – 4381
mozaiek, niet gecorrigeerde – 4381
mozaiek, ten dele gecorrigeerde – 4381
mozaïektextuur 0960
mui 2486
mull 2744
multiband-camera 4330
multilenscamera 4330
multiplexen, tijd– 4424
Munsell kleurenkaart 2676
Munsell kleursysteem 2676
mutatie 1982
myloniet 0915
myloniet 1541
mylonitisatie 0915
mylonitisch 0969
myrmekiet 0785
myrmektitisch 0852

N

naaldijs 2639
naald, rots– 1424
naaldvormig 0522
naaldvormig 0755
naaldvormig 3185
nabeving 0147
nafase 1108
nafta 3744
nannofossiel 1790
nannoplankton 1799
naschok 0147
natuurbouwsteen 4876
nawerking 1167
nawerking, elastische – 0114
nawerking, vulkanische – 1374
neanisch 1866
nebuliet 1054
neck 1477
neer 2269
neerslaan 0411
neerslag 3614
neerslagdrempel 4642
nekton 1798
nektonisch 1798
nematoblastisch 0962
neo-Darwinisme 1966
neo-Lamarckisme 1967
neomineralisatie 0905
neomorf 1983
neontologie 1787
neosoom 1037
neostratotype 1745
neotype 1951
neovulkanisch 0636
neovulkanisme 1150
nepionisch 1866
neptunisme 0004
nerietisch 2915
neritisch 2841
nest 3105
netwerkstructuur 4104
Neumann, lijnen van – 0068
neus, anticlinale – 1639
nevenbestanddeel 0594
nevenbreuk 1581
nevengesteente 0709
nevengesteente 1530
nevengesteente 3501
nevengesteente 3920
nevenkrater 1439
nevenproduct 4163
nevenpunt 4372
nevenrivier 2207
niche 1902
niervormig 0776
niervormig 4095
nissenberg 2618
nivatie 2357
nivatiebekken 2612
nivatienis 2612
niveau, freatisch – 3642
niveau in rusttoestand 3666
niveauverschil 4735
niveo-aeolisch 2870
nodulair 4085
nodule 4085
nok 0705
nomenclatuur, binomische – 1922
nomenclatuur, trinomische – 1922
nomenclatuur, uninomische – 1922
nonconform 1720
nonconformiteit 1720
nonconformiteit, erosie– 1722
nonconformiteit, klino– 1721
nonconformiteit, parallele – 1721
nopjeslei 0941
nopjesschist 0941
nopjestextuur 0956
normaalspanning 4688
normatief mineraal 0607
normatieve samenstelling 0606
novaculiet 3366
nulmeter 0352
nulniveau 4577
nulvector 0235
nunatak 2168
nuttig effect 0205

O

obsequent 2285
obsidiaan 1262
oceaan 2782
oceanisch 2842
oceanisch 2918
oceanizatie 1142
oceanografie 2781
oceanologie 2781
octaëdriet 0067
oecologie 1887
oedometer 4756
oeralitisatie 0928
oerbank 2704
oerbank 2733
oerduintje 2176
oerlaag 2704
oerlood 0437
oeroppervlak 2569
oerreliëf 2569
oerstroomdal 2392
oertype 1979
oever 2438
oever– 2906
oeverbank 2337
oever, concave – 2235
oever, convexe – 2235
oeverfront 2485
oevergrondwater, infiltratie in – 3671
oever, holle – 2236
oevermorene 2416
oever-nabij 2889
oever, steile – 2236
oevers treden, buiten de – 2251
oeverterras 2337
oeverwal 2238
oeverwal van lava 1295
ofioliet 0696
ofioliet 1098
ofiolietserie 0696
ofitisch 0845
ofthalmiet 1052
ogen 0952
ogenstructuur 0952
ogive 2378
ogive, onechte – 2378
oikokrist 0844
oleanderbladtextuur 4082
olie, droge – 3714
olie-equivalent 3747
olieintrusie 3752
olieklei 3721
oliekolom 3775
oliekolom, hoogte van de – 3775
oliemoedergesteente 3722
olie, natte – 3714
olie, onderverzadigde – 3857
oliering 3776
olie, ruwe – 3712
olie, schone – 3714
olietransformatie 3733
olieveld 3773

olieverzadiging, residuele – 4610
olie/watercontact 3777
oliezand 3766
oliezandsteen 3766
oliezone 3775
olie, zware – 3745
oligohalien 2832
oligomict 3265
oligotroof 3549
olistoliet 2977
olistostroom 2977
olivijnbom 1329
ombrogeen 3550
ombuiging 1629
omgeving 1900
omgeving 2859
omgeving, diagenetische – 3391
omhulling 3402
omkering 1088
omkering van het reliëf 2563
omkering, zelf– 0311
omloopberg 2227
ompoling, geomagnetische – 0291
omtrek, natte – 2261
omvormer, electro-acoustische – 4488
omwegfaktor 3707
omzetting 3961
oncoliet 3300
onderbouw 1414
onderdompeling 1728
onderdompeling 2857
onderdrukking 4567
onderduiking 1614
onder-etage 1763
ondergeslacht 1933
ondergravende werking 2023
ondergrond 4840
ondergrondcoëfficent 0258
ondergronds 2876
onderkant 3106
onderlopen 2857
ondermijning 2471
onderruimen 4296
onderschuiving 1574
ondersoort 1939
ondersteuning 4796
ondersteuning 4858
onderstroom 2826
onderstroom 3626
onderstructuur 1077
ondertijd 1763
ondervanging 4796
onderverzadigd, kritisch – 0605
ondervleugel 1576
onderwatereruptie 1236
onderwerpcontrast 4343
onderzijde 3106
onderzoek, biogeochemisch – 4645
onderzoek, geobotanisch – 4645
ondiepte 2247
ondiepte 2803
oneffen 3190
ongelijkkorrelig 0807
ongericht 0949
onrijp 3176
onrijp 3725
ontdekking 3884
ontdekkingsboring 3784
ontduiking 0273
ontgassing 1215
ontgassing 3520
ontgassingskuil 3066
ontglazing 0919
onthoeker 4377
onthoeking 4376
onthoofdingsknie 2326
ontkalking 2679
ontkiezeling 0929
ontkoppeling 0273
ontlastingsboring 4290
ontleding 2929
ontmenging 0476
ontmenging, liquide – 3939
ontmengingsstructuur 4106
ontmenging, tektonische – 0947
ontmenging, vloeibare – 0678
ontogenese 1794
ontogenie 1794
ontologie, vergelijkende – 0007
ontploffing 1217
ontslikken 4161
ontsluiten 4150
ontsluiting 4231
ontspanning 1179
ontspanningsdiaklaas 1506
ontspanningsproef, boorgat– 4766
ontstaansgebied 1969
ontsteking 4816
onttrekkingskegel 3667
ontvanger 4499
ontvanger, geprogrammeerde – 4499
ontwatering 2253
ontwatering 2768
ontwatering 3797
ontwatering 4739
ontwatering 4811
ontwatering, externe – 2256
ontwatering, interne – 2256
ontwateringspatroon 2254
ontwikkeling, neergaande – 2574
ontwikkeling, opgaande – 2574
ontwikkeling, vormconstante – 2574
ontzingen 4454
onverzadigd 0605
onverzadigde zone 3651
onweer, vulkanisch – 1373
oöiede 3479
oöliet 3479
oölietbolletje 3479
oölietisch 3479
oömicriet 3286
oömicrudiet 3286
oorsprong 2193
oorsprongsgebied 2943
oorsprongsgesteente 2944
oorsprongsgesteente 3918
oorsprongstijd 0159
oöspareniet 3285
oöspariet 3285
oösparrudiet 3285
opaak 0545
opalescentie 0486
opblazen 4814
opborreling 1358
opbrengst 0272
opdooilaag 2647
opdroging 2342
opdruk 0970
opduiking 2857
opduikingskust 2525
opeenhoping 2987
opeenvolging 1714
opeenvolging, vulkanische – 1238
open gat 3795
operator 4447
opgelost materiaal 2956
opheffing, verticale – 4511
ophicalciet 0930
ophoping 3942
oplijning 4460
oplossing 0411
oplossing in de laag 3413
oplossing onder druk 3445
oplossingsdoline 2072
oplossingselement 4319
oplossingsgroef 2087
oplossingsresidu 2930
oplossingsruimte 3690
opname, digitale – 4422
opname, fotogrammetrische – 4324
opnamemethode, sonische – 4480
opname, topografische – 4323
opname, wriggel– 4421
Oppel-zone 1780
oppersing 4830
oppervlak, aangestraald – 4515
oppervlak, effectief – 4360
oppervlak, mediaan – 1673
oppervlak, omhullend – 1672
oppervlak, oorspronkelijk – 2569
oppervlaksmorene 2406
oppervlak, soortelijk – 2754
oppervlak, specifiek – 2754
oppervlak, specifiek – 4109
oppervlaktedichtheid 4111
oppervlaktefundering 4835
oppervlakte-gegevens 4321
oppervlaktegolf 4439
oppervlaktekarst 2067
oppervlaktemateriaal, los – 2008
oppervlaktewater 3649
opschuiving 1558
opslagfilm 4388
opslibbing 2513
opslibbing 2990
opslokking 1133
opslokking, gestuite – 0623
opslokkingszone 1134
opslokking van de vloer 0623
opslokking van het dak 0623
opsporing, geochemische – 4625
opstelling, centrale – 4419
opstuwing 2971
opstuwingsterras 2310
opvoerkosten 3900
opvriezing 4721
opvullen 2245
opvulling 2341
opvulling 3105
opvulling 4831
opvullingsdelta 2498
opwaseiland 2535
opwelling 1358
opwelling 2786
opwelving 1194
opwelving door vorst 2642
opzwelling 0109
opzwelling 4720
orbiculair 0867
orbicule 0867
orde 0685
orde 1929
orde 2666
organische bestanddelen 2740
organische stof 2740
organisch materiaal, extraheerbaar – 3724
organogeen gesteente 3269
orgelpijp, geologische – 2077
oriëntatie-onderzoek 4639
oriëntatie, voorkeurs– 1013
oriëntparel 0486
orocline 1140
orogeen 1101
orogenese 1100
orogenetische beweging 1100
orogenetische fase 1107
orogenetische periode 1106
ortho- 0986
orthochemisch 3268
orthochemisch deeltje 3271
orthoconglomeraat 3255
orthofirisch 0826
orthogenese 1975
orthogeosynclinaal 1095
orthogonaal 0206
orthokwartsiet 3230
orthomagmatisch stadium 0651
orthorhombisch 0515
orthoselectie 1975
orthostereoscopie 4367
orthotektisch 3976
oscillatiehypothese 1119
oscillatieribbel 3012
otoliet 1865
ouderdom, adsolute – 0429
ouderdom, concordante – 0430
ouderdom, discordante – 0430
ouderdomsstadium 2570
overbruggend 3441
overconsolidatie 4703
overdekking, gesteente– 4788
overdrachtsfunctie 4448
overdruk 3820
overdruk, tektonische – 0886
overgangskegel 2423
overgangspunt 4577
overgangsveen 3536
overgangszone 3777
overgangszone 4620
overgroei 3455
overgroeiingscement 3437
overheersend 1816
overhellen 4833
overhellend 1491
overkipt 1491
overkorstingsader 4049
overlaat 2339
overlaat 4868
overlapping 4359
overlapping, vliegrichting– 4359
overlapping, zijwaartse – 4359
overloopniveau 2251
overlooppunt 3762
overrijp 3725
overschoven blok 1569
overschuimend 2811
overschuivende schol 1569
overschuiving 1568
overschuivingsafstand 1568
overschuivingsvlak 1568
overschuivingsvlak, basaal – 1572
overslagdelta 2500
overstortend 2811
overstromingsgebied 2239
oververzadigd 0605
oxydatie 0416
oxydatie-reductie reactie 0416
ozokeriet 3735

P

paal 4838
paardestaart, oplossings– 3401
paarlmoer 0484
paarlmoer 0503
paar, stereoscopisch – 4362
pacifische serie 0690
pad 0206
paddenstoelplooi 1650
paddestoelrots 2163
pakker 4279
pakking 3082
pakkingsdichtheid 3082
palaeobotanie 1827
palaeoecologie 1889
palaeogeografie 0011
palaeoklimatologie 0011
palaeomagnetisme 0294
palaeontologie 1786
palaeosol 3126
palaeosoom 1037
palaeotemperatuur 2864
palaeotemperatuurbepaling 0456
palaeovulkanisch 0636

palaeovulkanisme 1150
palaeozoölogie 1847
palaetiologie 0007
palagoniet 1254
palagonietbreccie 1347
palagoniettuf 1347
palagonitisatie 1254
palichnologie 1872
palichthyologie 1848
palimpsest 0971
palingeen 0124
palingenese 1024
palingenese 1968
palinspastische reconstructie 0009
palynologie 1828
palynomorf 1829
pan 4659
panallotriomorf 0808
panautomorf 0808
pandemisch 1814
Pangaea 1116
panidiomorf 0808
pannekoekbom 1329
panxenomorf 0808
papsneeuw 2348
para- 0986
paraboolduin 2177
paracme 1985
paraconglomeraat 3256
paraffine 3717
paragenese 0586
paragenese 3907
paragenese, diagenetische – 3394
paragenese, zout– 3378
paragenetisch 0586
parakristallijn 0997
paralisch 2898
parallax 4364
parallaxwig 4365
paramagnetisme 0298
paramarginaal 4212
paramorfose 0465
parastratotype 1743
paratektonisch 0906
paratype 1950
parautochtoon 0727
parel, gecultiveerde – 0503
passief 0619
patera 0037
patina 3403
patroon 0793
patroon, dambord– 4172
PDB 0454
pediment 2582
pedimentpas 2586
pediplain 2584
pedogenese 2659
pedon 2689
pegmatitisatie 1063
pegmatitisch 3978
pegmatitisch stadium 0653
pegmatoïde 0814
pegostyliet 2152
peiling, diepe seismische – 0094
pek 3743
pekel 3676
peksteen 1262
pelagisch 1910
pelagisch 2844
pelagisch 2919
Pelea's haar 1326
Pelea's traan 1326
Pelée-type eruptie 1163
peliet 3208
pelietisch 3208
pellet 4150
pelmicriet 3293
pelspariet 3292
peneplain 2589
peneplainisatie 2589
peneseismisch gebied 0252
penetratie 4513
penetratie, geregistreerde – 4513
penetratieweerstand 4709
peralkalisch 0693
peralumineus 0693
percolatie 3622
pergelisol 2637
peridotietbom 1329
periglaciaal 2633
perimagmatisch 3985
periode 0163
periode 1765
periode, eigen– 0130
perlitisch 0878
permafrost 2637
permeabiliteit 3697
permeabiliteit, effectieve – 3698
permeabiliteit, magnetische – 0304
permeabiliteit, primaire – 3701
permeabiliteit, relatieve – 3698
permeabiliteitsbarrière 3771
permeabiliteit, secundaire – 3701
permeatie 1059
perthiet 0785
perthititisch 0851
petroblastese 1059
petrogenese 0581
petrogenetisch raster 0991
petrografie 0579
petrografie, sediment– 2851
petroleum 3712
petrologie 0578
petrologie, sediment– 2851
petrologie, structurele – 0995
petrotektoniek 0996
pF 2771
pF-kromme 2772
P-golf 0182
pH 0422
phacoïd 3055
phi-schaal 3163
phylloniet 1541
phylogenetische reeks 1934
phylogenetische zone 1779
phylozone 1779
phyteraal 3592
piedmontgletsjer 2435
piedmonthelling 2583
piedmont-trap 2604
piedmont-vlakte 2604
pieken 1260
piekversnelling 0120
piëzogeen 0977
piëzokristallatie 0633
piëzometer 3646
pijler 4845
pijp, klastische – 3074
pijp, vastgelopen – 4286
pijp, vastzittende – 4286
pijp, wortelloze – 1223
pikken 4459
pilaar 2138
pilotaxitisch 0825
pingo 2643
piperno 1342
piramidaal 0522
pisoliet 3478
pisolietbolletje 3478
pisoliettuf 1333
pisoliet, vulkanisch – 1321
pisolithisch 3478
plaat 0739
plaat 2506
plaatbelastingsproef 4708
plaatfundering 4837
plaatsname 0618
plaatvormig 0522
plaatvormig 0755
placer 3236
placer 3997
planair 0972
planeet, aardse – 0025
planetenstelsel 0018
planetesimaalhypothese 0022
planetisimaaltheorie 0048
planetologie 0023
plankton 1799
planktonisch 1799
plant, indicator– 4646
plas 2330
plasticiteit 2747
plasticiteit 4669
plasticiteitsconstanten 2748
plasticiteitsgetal 4716
plasticiteitsgrens 4715
plasticiteitsindex 4716
plastisch 2747
plat 0755
plat 3183
plat, continentaal – 2788
plateau 2797
plateaubazalt 1257
platentektoniek 1125
platform 1083
platform, gestabiliseerd – 4539
platheid 3182
platig 2728
platig 3132
plat-stengelig 0772
plattegrond, ingenieurs–geologische – 4784
plaveisel 0960
plaveisel van stenen 2656
playa 2580
playa 2892
playameer 2343
pleistoseismisch gebied 0246
pleistoseiste 0246
pleochroïsme 0562
plesiotype 1952
pletbaar 0471
plexus 1961
Plinisch eruptietype 1161
ploegzool 2761
plooi 1616
plooi, afglijdings– 3055
plooi, afstandsgetrouwe – 1652
plooi-as 1630
plooi, asymmetrische – 1644
plooibundel 1674
plooi, concentrische – 1652
plooi, congruente – 1653
plooi, cylindrische – 1654
plooidekblad 1682
plooi, duikende – 1637
plooi, eerste aanleg van een – 1666
plooien en échelon 1676
plooien, pakket van liggende – 1681
plooien, stapeling van liggende – 1681
plooi, evenwijdige – 1652
plooi, geplooide – 1668
plooi, hellende – 1646
plooiing, beginnende – 1666
plooiing, disharmonische – 1663
plooiing, enterolietische – 3076
plooiing fijne – 3051
plooiing, intra-formationele dooreen– 3051
plooiing in wording 1666
plooiingsfase 1107
plooiingsfase 1669
plooiingsgebergte 2538
plooiingshoek 1626
plooiingstektoniek 1615
plooiing, sterke dooreen– 3051
plooiingszone 1674
plooiing tijdens de sedimentatieperiode 1667
plooi, kegelvormige – 1655
plooi, liggende – 1680
plooi, meervoudige – 1668
plooi, niet-cylindrische – 1654
plooi, overhellende – 1647
plooioverschuiving 1576
plooi, parallelle – 1652
plooi, parasitaire – 1677
plooi, ptygmatische – 1658
plooi, rechte – 1645
plooi, secundaire – 1642
plooi, symmetrische – 1644
plooisysteem 1670
plooi, vormgetrouwe – 1653
plurimetamorfose 0901
plutonisch 0617
plutonisch 3988
plutonisme 0004
plutonisme 0615
plutoon 0715
pneumatogeen 0655
pneumatolyse 3950
pneumatolytisch 3977
pneumatolytisch stadium 0653
podzolering 2680
poedersneeuw 2346
poikilitisch 0844
poikiloblastisch 0974
poikiloblast 0981
poikilofitisch 0846
poikilohalien 2831
Poisson, coëfficiënt van – 4700
Poisson, getal van – 4700
polarisatie 0236
polarisatie 4557
polarisatiefilter 4339
polarisatiekleur 0568
polarisatie, magnetische – 0297
polarisatiemethode, geïnduceerde – 4569
polariteit, omgekeerde – 0290
polariteitsinterval 0292
polariteitstijdperk 0292
polijsting 2399
polijsting 3404
polje 2074
pollen 1839
pollenduin 2190
polyconide 1441
polyfyletisch 1960
polygeen 0590
polygoonbodem 2653
polyhalien 2832
polymetallisch 3911
polymetamorfose 0901
polymict 3265
polymineraal 0589
polymorfie 0464
polymorfie 1811
polyschematisch 0794
polytoop 1819
polytypie 0463
polytypisch 1945
pompproef 3711
pompput 3791
pool 1018
pooldiagram 1018
pool, geomagnetische – 0280
poolgordel 1020
pool, magnetische – 0282
populatie 1941
porcellaniet 3374
porfier 0817
porfierachtig 0819
porfieroblastisch 3470
porfieroïd 0957
porfiriet 0817
porfirisch 0817
porfiritisch 0817
porfiroblast 0981
porfiroblastisch 0956
porfiroklast 0981

porie 3083
porie 3683
poriën- 3083
poriëngehalte 3681
poriëngetal 3682
poriënruimte, beschikbare – 3695
poriënvolume 2752
poriënvolume 3681
poriënvullend 3440
porositeit 3681
porositeit, afdruk– 3691
porositeit, breuk– 3694
porositeit, effectieve – 3704
porositeit, gevormde – 3685
porositeit, grondmassa– 3687
porositeit, honingraat – 3689
porositeit, inkorrelige – 3688
porositeit, maaksel– 3693
porositeit, minus cement – 3442
porositeit, oorspronkelijke – 3684
porositeit, primaire – 3684
porositeit, secundaire – 3685
porositeit, totale – 3681
porositeit, tussen-korrelige – 3688
Postdam-waarde 0334
postkinematisch 0726
postkinematisch 0906
postkinematisch 1079
postkristallijn 0997
postorogeen 0726
posttectonisch 0726
posttektonisch 0906
postvulkanische verschijnselen 1374
potentiaal 3645
potentiaalmethode, spontane – 4557
Potsdam-systeem 0334
pozzolaanaarde 1338
Pratt 0345
pre-adaptatie 1980
prekinematisch 0906
prekinematisch 1079
prekristallijn 0997
pressiometer 4760
pretektonisch 0906
priel 2509
prielerosie 2778
prijsindex, groothandels– 4223
prikspoor 3037
prioriteitswet 1924
prisma 2730
prismatisch 0522
prismatisch 0755
prismatisch 2730
proces 2568
proces, discontinu – 4145
Proctor, verdichtingsproef van – 4757
productie, cumulatieve – 3817
productie, geschatte totale – 3869
productiehoeveelheid, toegestane – 3902
productie, maximale – 3807
productie, meten van de – 3798
productiepeil 3801
productiepotentieel 3806
productieproef 3788
productieput 3786
productieput 3789
productie, totale – 3817
productie, totale – 3867
productievermogen 3806
productiviteitsindex 3808
productiviteitsindex, specifieke – 3808
proefkuil 4235
profiel, ABC– 2695
profiel, AC– 2695
profiel, algemeen – 4254
profiel, begraven – 2693
profiel, evenwichts– 2198
profiel, gecombineerd – 4253
profiel, geologisch – 4250
profiel, ideaal – 4254
profiel, in serie geplaatst – 4251
profielknik 2199
profielknik, cyclische – 2199
profielknik, verjongings– 2199
profiel, normaal – 4255
profiel, onvereffend– 2198
profielschieten, continu – 4408
profiel van de boring 4256
profileren, acoustisch – 4480
profileren bij oceaandiepte 4484
profileren, ondiep seismisch – 4479
projectie, cyclografische – 1021
prop 3105
propje 3094
propje, faeces– 3094
propyliet 0921
propylitisatie 0921
prospectief gebied 3887
protactinium-thorium datering 0445
protodolomiet 3321
protoklastisch 057
protomyloniet 0916
prototype 1979
prototype 4775
provincie, erts- 3908
provincie, fauna– 1849
provincie, geochemische – 0401
provincie, geologische – 4260
provincie, geothermische – 0390
provincie, petrografische – 0628
provincie, sediment-petrologische – 2852
proximaal 2997
psammiet 3226
psammietisch 3226
psefiet 3238
psefietisch 3237
psefietisch 3238
pseudoaspite 1430
pseudofossiel 1789
pseudogelaagdheid 0869
pseudomorf 0982
pseudomorf 1823
pseudomorf naar 0982
pseudomorfose 0465
pseudomorfose 0982
pseudonodule 3060
pseudopalagoniet 1255
pseudoporfirisch 0819
pseudoporfiroblastisch 0959
pseudotachyliet 0917
pseudotachyliet 1542
pseudovulkaan 1189
ptygmatisch 1051
puddingsteen 3253
puimsteen 1308
puimsteenachtig 0875
puimsteenachtig 1308
puimsteenbom 1329
puimsteentuf 1333
puinafglijding 2046
puinafglijdingsbaan 2054
puinafglijdingsnis 2052
puinafglijdingsreliëf 2056
puinband 2379
puinhelling 2007
puinhelling, onderzeese – 2480
puinkegel 2007
puinlawine 2047
puinstorting 2046
puinveld 2006
puinverstopping 2114
puinvlakte 2006
puinvorming 2005
puinwaaier 2007
puinwaaier-conglomeraat 2985
puls 4482
pulsboren 4262
pulsherhalingsfrequentie 4504
pulslengte 4482
punt 2449
puntbelasting, breuksterkte bij – 4733
puntdelta 2495
puntendiagram 4259
puntenteller 0612
puntgroep 0512
puntje 0494
put 2104
put 2105
put 3665
put 4261
putafwerking 3795
putbehandeling 3813
put, dode – 3812
put doodpompen 4290
put, droge – 3794
put, ingesloten – 3811
putkrater 1455
put, natte – 3794
putpatroon 3842
putpotentieel 3806
put, productieve – 3789
putschieten 4406
put, spuitende – 3790
put, spuitende – 4288
puy 1425
puzzolaanaarde 1338
pycnocline 2830
pyritisatie 3458
pyritisering 3458
pyrobitumen 3737
pyrogeen 0655
pyrogeen 3973
pyrogenese 0616
pyroklast 1309
pyroklast 3170
pyroklastica 1309
pyroklastica, accessorische – 1314
pyroklastica, juveniele – 1314
pyrolyse 3730
pyromagma 1200
pyromeride 0876
pyrometamorfose 0896
pyrometasmatose 0900
pyrometer 1195
pyrometer, optische – 1195

R

rad 4551
radar 4384
radar-afstandsmarkering 4398
radarazimuth 4394
radardoorsnede 4392
radarreflectiviteit 4392
radarschaduw 4391
radarstraal 4389
radiaalstralig 0776
radiair 0843
radioactiviteitshypothese 1117
radioactiviteitskartering 4545
radiokoolstof datering 0451
radiolariënaarde 3368
radiolariën-hoornsteen 3369
radiolariënslik 3368
radiolariet 3368
radiolariet 3369
randbreuk 1593
randcement 3437
randdiep 1111
randgebied, continentaal – 2789
rand, geschulpte – 0032
randinzinking 1697
randkloof 2382
randlaagte 2560
randmorene 2416
randspleet 2376
randsynclinaal 1697
randwater 3779
randwater, opdringen van het – 3826
randwaterstuwing 3825
randzee 2783
randzone, basische – 0707
rang 0685
rangschikking, zonale – 3907
rapakivi 0828
ras 1939
ravijn 0042
ravijn 2026
Rayleigh golf 0189
reactierand 0786
reactieschaal 0786
reactiewaarde 3675
reagerend, niet– 0788
rechthoekig 2292
record 4420
redox potentiaal 0421
reductie 0416
reductie, bacteriële – 0417
reductie, isostatische – 0361
reductie, topo-isostatische – 0364
reductie, vrijelucht– 0358
reductiezone, bolvormige – 3463
reeks, charnockitische – 0990
reeks, phylogenetische – 1934
referentieniveau 0354
referentieniveau 0357
referentieveld, internationaal geomagnetisch – 4535
referentievlak 3854
reflectie 4436
reflectiecoëfficiënt 4473
reflectie, diffuse – 4317
reflectie, externe – 0576
reflectie, herhaalde – 4441
reflectiehorizon 4461
reflectie, interne – 0576
reflectie, meervoudige – 4441
reflectiemethode 4405
reflectie, primaire – 4437
reflectie, spiegelende – 4318
reflectie, totale – 0213
reflectiviteit 0574
reflectiviteit 4473
reflector, diffuse – 4317
reflector, spiegelende – 4318
refractie, golf– 2810
refractiemethode 4405
regendruppelindruk 3041
regeneratiecomplex 3538
regengeul 2026
regenlahar 1355
regio 3908
regionalisatie 0255
regionalisatie 4194
regionaliteit, graad van – 0340
registratie 4500
registratie, analoge – 4421
registratiefilm 4388
registratie, veranderlijke-dichtheids– 4421
registratie, veranderlijke-oppervlakte– 4421
regoliet 2008
regressie 1726
rejuvenatie 3959
rekdiaklaas 1503
rekenmodel 4776
rekgolf 0183
rekmeter 4767
rekristallisatie 0905
rekristallisatie 3452

rekristallisatie, dynamische – 0967
rekspleet 1519
rekstrookje 4797
relaisbeving 0145
relict 0971
relict 0985
relict 1985
relicthaak 2466
relict, instabiel – 0985
relictsediment 2856
relict, stabiel – 0985
reliëf 0558
reliëf, alpien – 2576
reliëf, apalachisch – 2564
reliëf-generatie 2573
reliëf, hooggebergte– 2576
reliëfintensiteit 2575
reliëf, jurassisch – 2565
reliëf, multiconvex – 2600
reliëf, oorspronkelijk – 2569
reliëfsequentie 2573
reliëfsuccessie 2573
reliëfverplaatsing 4379
relikt, gepantserd – 0787
reliktmeer 2344
rem 4551
rendement 0272
rendement 3898
rendement, seismisch – 0205
rentevoet 3897
resequent 2284
reserve, bewezen – 4209
reserve, mogelijke – 4210
reserve, redelijk verzekerde – 4207
reserves 3868
reserves, bewezen – 3879
reserveschatting 3870
reserves, erts– 4208
reserves, mogelijke – 3881
reserves, speculatieve – 3882
reserves, veronderstelde – 3882
reserves, waarschijnlijke – 3880
reserve, waarschijnlijke – 4209
reservoir 3755
reservoirdruk 3819
reservoirformatie 3679
reservoirgedrag 3846
reservoirgesteente 3679
reservoir, homogeen – 3770
reservoirisotropie 3770
reservoirmodel 3866
reservoirmodel, computer – 3866
reservoir volume, bruto – 3872
residu 0241
residuair 3998
residuaire concentratie 2967
residu, onoplosbaar – 2930
residu, oplossings– 2930
resonantietheorie 0046
resorptie 0642
responsiecurve 0133
responsie, impuls– 4446
restberg 2607
restiet 1032
restiet 1066
restmeer 2344
restvloeistof 0647
retentie, oppervlakte– 3618
reverberatie 4440
Reynolds 2265
rhenium-osmium datering 0446
rheologie 0102
rheologie 4671
rheomorfose 1057
rhizoconcretie 3490
rhombochasme 1143
ria 2452
ribbe 0518
ribbel 3000
ribbelgolflengte 3002
ribbelhoogte 3002
ribbelindex 3002
ribbelkam 3001
ribbellengte 3002
ribbel, linguoïdale – 3006
ribbel, longitudinale – 3011
ribbel met afgeplatte kam 3015
ribbel, rhomboïdale – 3010
ribbel, sikkel – 3005
ribbel, stroom-golf – 3011
ribbel, transversale – 3003
ribbeltrog 3001
Richter, schaal van – 0220
richting 1493
richtingeffect 4530
richting, t– 1017
rif 2443
rif 2804
rif 3333
rifachtig 3334
rif, algen– 3360
rif, barrière– 3338
rif, buiten het – 3356
rifcomplex 3337
riffel 3499
rifflank 3353
rifflankpuin 3353
rif, franje– 3338
riffront 3351
rifgesteente 3342
rifhelling 3349
rifhelling 3354
rifkap 3329
rifkern 3341
rifknobbel 3329
rif, knollen– 3329
rifmuur 3336
rifplat 3343
rifpuin 3355
rifrand 3350
rif, regressief – 3327
rift 1092
rift 1590
rif, transgressief – 3328
riftuf 3312
rif, tussen – 3357
rif, van een – 3334
rifzone 3326
rijp 3176
rijp 3725
rijpheid 3725
rijpheidsindex 3726
rijping 2685
rijzingseiland 2535
rijzingskust 2524
ril 0057
rimpel 1643
rimpel 2814
rimpelgroef 3035
rimpeling 3051
rimpelrug 0031
rimpelrug 0058
ringdal 1453
ringenkiezel 3457
ringintrusie 0742
ringvlakte 0030
ringvlakte 0053
ringwal 1452
ringwalberg 0054
rits 2491
rivier 2191
rivier– 2883
rivieraantapping 2322
rivier, aanvallende – 2325
rivierafleiding 2213
rivier, allochtone – 2259
rivierbedding 2242
rivierbekken 2192
rivierdichtheid 2255
rivieren, gebied tussen twee – 2312
rivierkronkel 2219
rivieromkering 2214
rivieromlegging 2213
rivieromlegging, spontane – 2213
rivieronthoofding 2323
rivier, passende – 2329
rivierroof 2322
riviersediment, onderzoek van – 4641
rivierstelsel 2205
rivierterras 2293
rivier, verarmde – 2329
rivier, verkommerde – 2329
rivierzwelling 2252
roentgen 4550
roestvlek 2692
rolbaarheid 3191
rolspoor 3038
rolsteen 3241
rolsteen 3242
rompgebergte 2588
ronde 4822
rondist 0501
roofrivier 2325
rookkolom 1361
rookring 1364
roosdiagram 4259
roosten 4151
rooster, kristal– 0507
roosterlijn 0508
roostermethode 4374
roosteroriëntatie 1013
rooster, parallactisch – 4373
rooster, perspectivisch – 4373
roosterpunt 0508
roostervlak 0508
Rosiwall, methode van – 0613
rotatie, externe – 1000
rotatiehoek 1126
rotatie, interne – 1000
rotatiepool 1126
rotatie, tektonische – 1000
rotsblok 4730
rotsbodem 4788
rotsmassa, vooruitspringende – 2606
rotsterras 2298
rotswoestijn 2578
rotting 3526
rottingsslik 3527
rottingsslik 3528
royalty 4224
R-tektoniet 1003
rubidium-strontium datering 0447
rudiet 3237
rudietisch 3237
rudimentair 1978
rug 0705
rug 2796
ruimen 4296
ruimte, annulaire – 4279
ruimtebeeld, omgekeerd – 4367
ruimtegroep 0513
ruimteschip 4302
ruimtevaartuig 4302
ruis 0097
ruis, achtergronds- 4438
ruisniveau 0097
ruis, schot- 4438
ruis, seismische – 4438
ruis, signaal opgewekte – 0177
ruitribbel 3010
rustperiode 1169
rustspoor 1880
ruw 3190
ruwheid 4727

S

saliniteit 3376
saliniteit 3675
salisch 0607
saltatie 1982
saltatie 2954
saltatielast 2954
saltatiespoor 3039
samendrukbaarheid, coëfficiënt van – 4698
samendrukbaarheid, modulus van – 4698
samendrukking, vadose – 3416
samenspoelsel 2869
samenstelling, kwantitatieve mineralogische – 0593
samenstelling, modale – 0593
samentreffen 4073
samenvloeiing 2208
samenvloeiing, gelijkvloers – 2208
samenvloeiing, verplaatste – 2209
samenvloeiing, verschoven – 2209
samenvloeiing, weerhaakvormige – 2209
saproliet 2009
saproliet 4128
sapropeel 3528
sapropeliet 3497
sapropeliet 3721
satellietfoto 4349
satelliet, kunstmatige – 4302
saussuriet 0927
saussuritisatie 0927
saxonische orogenese 1078
schaal 4240
schacht 2104
schacht 4848
schachtgrot 2097
schaduwzone, seismische – 0216
schalenkogel 1321
schalie 3220
schalie-achtig 3220
schalie-houdend 3220
schalie, papier– 3220
schaliesplijting 1599
scharing 4073
scharnierlijn 1629
scharnierpunt 1629
scheefstelling 4351
scheefstelling, relatieve – 4351
scheefstellingsverplaatsing 4379
scheefzakken 4833
scheiding, snelheids– 0228
scheidingsvlak 3096
scheiding, zware-suspensie– 4153
schelpenkalk 3308
schelpenkalksteen 3308
scheren 2455
schermmuur 4861
scherpkantig 3190
scherpte, optische – 4347
schervenlava 1250
scheur 1466
scheur, open – 1512
schiervlakte 2589
schiervlakte, oorspronkelijke – 2593
schiervlakterest 2594
schiervlaktevorming 2589
schietgatstijd 4428
schietlood 0329
schietloodafwijking 0329
schijfvormig 3184
schijnbaar 4637
schijnfossiel 1789
schijnhorizon 4462
schijnspiegel 3643

schild 1081
schildtunnelbouw 4853
schildvulkaan 0035
schildvulkaan 1420
schildvulkaangroep 1450
schilfer 2111
schilfer 3186
schilferig 0775
schist 0944
schistachtig 0975
schisteus 0975
schisteus 1603
schistositeit 0975
schistositeit 1603
schistositeit, stengelige – 1603
schiststructuur 1603
schizoliet 0668
Schmidt, net van – 1015
schokgolf 0274
schokgolf 4481
schokzone 0894
schollengebergte 2539
schollengebergte, scheefgesteld – 2539
schollengebied 1586
schommeling, dagelijkse – 4528
schommelsteen 2163
schoonmaken 3813
schoorwal 2463
schoorwal 2464
schor 2514
schor 2909
schorkreek 2517
schorre 2514
schot 2102
schoteltrog 1641
schotgatpatroon 4411
schotpunt 4411
schriftgraniet 0847
schriftgranitisch 0847
schrijver 4500
schrijver, digitale – 4501
schrinkfactor 3860
schub 1538
schub 1571
schub 3186
schubbig 0775
schubstructuur 1571
schubvorming 1571
schudtafel 4157
schudzeef 4282
schuif 0105
schuif 4677
schuifapparaat 4754
schuifdiaklaas 1504
schuifplooi 1662
schuifspanning 4688
schuifweerstand 4684
schuifzone 1533
schuifzone, verertste – 4060
schuimig 0875
schuimlava 1307
Schürmann, reeks van – 3935
Schürmann, wet van – 3522
schutgebied 3845
schuttingdiagram 4252
scintillatie 4319
scintillatieteller 4546
scopuliet 0769
sculptuur 3018
sebkha 2580
sebkha 2892
sectie 0685
sectie, seismische – 4458
sedentaat 2855
sedentair 1919
sediment 2850
sedimentafvoersnelheid 2957
sedimentatie 2989
sedimentatiebalans 3158
sedimentatie, interne – 3436
sedimentatie, klastolithische – 1201
sedimentatiekust 2528
sedimentatie, rhythmische – 3121
sedimentatiesnelheid 2989
sedimentatieterras 2337
sedimentgesteente 2850
sediment, intern – 3436
sedimentlast 2951
sedimentlast 2957
sedimentologie 2849
sediment, relict– 2856
Seger, kegel van – 1196
segregatie 3395
segregatie 3948
segregatiebanding 0946
seiche 2816
seismiciteit 0242
seismiciteit, specifieke – 0242
seismische constante 0257
seismisch-electrisch effect 0113
seismisch evenement 0095
seismische zoneringskaart 0244
seismisch gebeuren 0095
seismisch gebied 0247
seismograaf 0116
seismograafkelder 0116
seismogram 0116
seismogram 4420
seismologie 0094
seismologie, nucleaire – 0263
seismologie, planetaire – 0094
seismometer 0118
seismometer 4414
seismometerafstand 4416
seismometeropstelling 4417
seismometerpatroon 4416
seismoscoop 0117
sensitometrie 4341
septarie 3487
sequentie 3110
serac 2377
sere 1894
sericitisatie 0932
serie 1764
serie, bodem– 2663
serpentiniet 0923
serpentinisatie 0923
sessiel 1919
sfenochasme 1143
sfenoliet 0736
sfericiteit 3180
sferoïdaal 3180
sferoïde 0330
sferoïde 3180
sferoliet 0778
sferoliet 1321
sferoliet 3482
sferolietisch 3482
sferoliet, samengestelde – 0779
sferolitisch 0876
sferosideriet 3483
sferulitisch 4087
S-golf 0184
Shore, hardheidsschaal van – 4701
sideriet 0066
sideroliet 0070
sigma-t 2827
signaal 4435
signaal 4517
signaal, acoustisch – 4517
signaal, kortstondig – 0176
signaal-ruis-verhouding 4443
signaal, seismisch – 4517
signaal, voorbijgaand – 0176
sijpelplaats 3753
sikkelduin 2177
silex 3366
silicificatie 0929
silicificatie 3458
silt 2715
silt 3204
siltachtig 3204
siltig 3204
siltsteen 3204
sinter 3385
sinterbord 2141
sinterdam 2147
sinter, kiezelige – 1406
sinterzoom 2148
skapolitisatie 0932
skelet 1861
skelet 3680
skelet– 3169
skelet, inwendig – 1861
skelet, uitwendig – 1861
slak 1294
slakken, aaneengebakken – 1346
slakkenbom 1329
slakken, gekoekte – 1346
slakkenkegel 1427
slakkenkoek 1329
slakkenschede 1330
slakkenschoorsteen 1298
slakkentop van aa-lava 1260
slakkentuf 1333
slakkenwal 1452
slakkenzand 1323
slakkig 1294
slede 4483
sleepplooi 1570
sleepspoor 1878
sleepspoor 3032
slemp 2758
slempigheid 2758
slemp, oppervlakkige – 2760
slemp, oppervlakte– 2760
slenk 1590
slenk 3539
slenkdal 2546
slenk, explosie– 1235
slenklaagte 2546
slenk, vulkaansector – 1464
sleuf 4235
sleufmonster 3593
sleuring 1537
sleuring 4079
slib 2716
slib 2920
slib 4160
slib, kalk– 3295
slib, koraalhoudend – 3311
slibondersteuning 3294
slieren 0665
slijkband 2379
slijkpoel 1398
slijkveld 1390
slijpgrens 2631
slijping 2940
slijtlaag 4842
slik 3210
slikachtig 3210
slikbal 3247
slik, blauw – 2921
slikbrok 3248
slikbron 1398
slik, groen – 2921
slik, hemipelagisch – 2920
slikplaat 2910
slikplak 3248
slikvulkaan 3071
slikwad 2910
slinger 0350
slinger 4774
slingerlengte, equivalente – 0126
slinger, omgekeerde – 4774
slinger, reversie– 0351
sloef 2716
slotfase 1167
sluitfout 4541
sluiting 3762
sluitingsareaal 3762
sluiting, verticale – 3762
smelt 1198
smeltingskorst 0073
smeltpeer 0497
smeltschaal 2387
smeltschacht 1455
smelttuf 1340
smeltwaterafzetting 2402
smeltwaterbeek 2388
smeltwaterdal 2393
smeltwatergat 2389
smeltwaterkanaal 2393
smeltwaterpijp 2389
smeltwaterrug 2420
SMOW 0454
snaar 4420
sneeuwafglijding 2358
sneeuw, afsmelten van de – 2355
sneeuw, door wind samengepakte – 2347
sneeuw, droge – 2349
sneeuwgrens 2354
sneeuwlawine 2358
sneeuwluifel 2356
sneeuw, natte – 2349
sneeuwopwaaiing 2353
sneeuwschollenlawine 2360
snelheid, eerste kritische – 2265
snelheid, fase– 0169
snelheid, filter- 3703
snelheid, gemiddelde – 4465
snelheid, golf- 0168
snelheid, groeps– 0170
snelheid, interval – 4467
snelheid, middelbare – 4466
snelheid, schijnbare – 0208
snelheid, schijnbare – 4468
snelheidsfilter 4456
snelheidsfunctie 4469
snelheidsindex 4732
snelheidslaag, lage- 0100
snelheidslog, continue – 4406
snelheid, stapelings– 4434
snelheid, tweede kritische – 2266
snijdbaar 0471
snijdend, elkaar – 2596
snijdingstijd 4463
snoer 3122
solfatare 1391
solfatarenstadium 1168
solfatarentoestand 1168
solfatarenveld 1392
solfatare, zure – 1393
solifluctie 2041
solifluctie 2780
soligeen 3550
soll 2632
solum 2694
somma 1461
sommavulkaan 1445
sondeerapparaat 4758
sondeerproef, standaard– 4712
sonde, gamma – 4590
sondering 4710
sondering, dynamische – 4711
sondering, statische – 4711
sonogram 4500
soort 1937
soort, climax– 1891
soort, cosmopolitische 1809
sortering 3200
sorteringscoëfficiënt 3201
spaat 3280
spanning 0103
spanning 4556
spanning 4667
spanning, deviatorische – 0103
spanning, differentiële – 0103
spanning, effectieve – 4668
spanning, verschil– 0103
spanning-vervorming

diagram 4667
spanning-vervormingskromme 0106
spariet 3280
sparker 4410
sparker 4493
sparkerkabel 4493
spaterosie 2035
spatiëring 1497
spatiëring op gelijke afstanden 4172
spatiëring, systematische – 4173
spatiëring, willekeurige – 4174
spatkom 2139
spatstalagmiet 2141
spatzone 2439
speciatie 1969
species 1937
spectraal analyse 0221
spectrometer, drempel– 4547
spectrometer, enkelkanaal– 4547
spectrometer, meerkanalen– 4547
spectrum, continu – 4305
spectrum, electromagnetisch – 4304
spectrum, vermogensdichtheid– 0222
speleoliet 2124
speleologie 2094
speleotheem 2877
spelonk 2095
spicula 1862
spiculiet 0764
spiegel 1538
spikkelerts 3975
spikkelerts 4004
spiliet 0922
spilitisatie 0922
spilosiet 0933
spinifex textuur 0850
spiraalboor 4236
spiraalscheider 4156
spleet 1465
spleet 1512
spleet, concentrische – 1469
spleet, dwarslopende – 1470
spleeteffusie 1233
spleeteruptie 1233
spleetfumarole 1383
spleetgang 4044
spleet, gevulde – 1514
spleetgrot 2096
spleetkarren 2087
spleet, radiale – 1468
spleetvulling 1515
splijtbaar 1596
splijtbaarheid 1596
splijting 0469
splijting 1495
splijting 1995
splijting 2935
splijtingssporen-datering 0442
splijtingsvlak 1496
splijtvlak 0540
splijtvlak 1598
splijtvlak 3508
splijtvlakwaaier 1602
splitsen 3159
splitsing 0667
splitsing 2210
spoelboring 4751
spoelbuis 4751
spoelingkoek 4618
spoelingverlies 4284
spoelkop 4268
spoelzandwaaier 2424
spoelzone 2439
spongoliet 3372
sponzig 0875
spoor 1492
spoor 1876
spoor 4420
spoor, aangevreten – 2228
sporangium 1832
spore 1832
sporenelement 0400
sporenfossiel 1873
sporenfossiel 3067
sporomorf 1831
spreiding 1812
spreiding 4186
spreiding 4514
spreiding van de zeebodem 1128
springbaarheid 4820
springen, doen – 4814
springen, gecontroleerd – 4823
springen, zacht – 4823
springstof 4815
springstoffactor 4820
sproeizone 2439
spronghoogte, stratigrafische – 1552
spronghoogte, verticale – 1551
spronglaag 2333
sprongribbel, hydraulische – 3009
spuiter 4288
spuitgat 1274
spuitkruis 3796
staafmolen 4149
staafvormig 3185
staafzeef 4144
stabiliteitsanalyse 4777
stadium 2568
stadium, jeugdig – 2570
stadium, jong – 2570
stadium, oud – 2570
stadium, rijp – 2570
stadium van rijpheid 2570
stalactiet 2131
stalagmiet 2137
stam 0687
stam 1929
stamboom 4143
stand 1480
standaard 0454
standaard afwijking 4190
standaard deviatie 4190
standaardkromme 4565
standaardmineraal 0607
standaardprofiel 4257
standplaats 1901
stapeling 4433
station, basis– 4526
stationsazimuth 0160
stationsinterval 4418
statistiek, convex-concaaf - 4113
steekapparaat 4748
steekbuis 4748
steekkern 3155
steekvlam 1350
steen 3241
steen 3242
steenberg 4166
steengang 4847
steen, gebroken – 4876
steengroeve 4874
steengroeve-afval 4876
steenkern 1824
steenkool 3496
steenkoolhoudende formatie 3498
steenkrans 2653
steenlawine 1368
steenlawine 2047
steenmeteoriet 0071
steenmiddel 3506
steenpolygoon 2653
steenrichel 2654
steen, samengestelde – 0480
steenschaal 0088
steenslag 4857
steensnijder 0502
steenstroom 2057
steenwoestijn 2578
steenwording 3421
steenzout 3382
Stefanescu, integraal van – 4568
Steinmann, triniteit van – 0696
Steinmann, triniteit van – 1098
stekels 1260
S-tektoniet 1002
stengelig 0755
stengelig 0772
stengelig 0972
stenobaat 1911
stenobenthisch 1909
stenohalien 1912
stenotherm 1913
stenotoop 1914
steppenkalk 4123
sterduin 2179
stereffect 0489
stereobasis 4363
stereocomparator 4365
stereogram 4258
stereometer 4365
stereopaar 4362
stereoscoop 4366
stereotriplet 4362
sterkte 0243
sterkte 4682
sterkteregeling, automatische – 4425
sterkteregeling, automatische – 4508
sterkteschaal 0243
steunbeer 4797
steundruk 4706
steunmuur 4797
stigmaria-vloer 3515
stijgbuis 4275
stijghoogte 3646
stiktoliet 1053
stikvallei 1402
stimulatiebehandeling 3800
stockwerk 4058
stof 3206
stofband 2379
stof, kosmisch – 0065
stoflawine 2359
stofregen 2159
stofsneeuw 2346
stoftuf 1333
stof, vulkanisch – 1316
stofwisseling 1803
stollingsgesteente 0616
stollingsindex 0703
Stoneley golf 0195
stoomeruptie, secundaire – 1403
stoomfumarole 1389
stoominjectie 3839
stoomveld, natuurlijk – 0393
stootoever 2236
stootoeverhelling 2236
stootspoor 3037
stoottijd 0159
stopebreedte, minimale – 4200
storing, magnetische – 1192
storm, magnetische – 4529
stormstrand 2487
stormtij– 2908
stormvloed 2824
stortbeek– 2884
stortbeekkloof 2026
storthelling 2185
stortvloed 2252
stortvloedafzetting 2884
straal 0056
straalafbuiging 0217
straal, gemiddelde – 2262
straalgewijs 0776
straal, hydraulische – 2262
straalparameter 0206
stralenkrans 0056
stralenpad 4471
stralingsdichtheid 4313
stralingsdoorlatendheid 4315
stralingsmeter 4310
stralingsvermogen 4313
straling, warmte– 0381
strand 2485
strandbank 2459
strand, droog – 2487
strandduin 2168
strandhoorn 2493
strandloper 2168
strand, nat – 2487
strandplacer 2912
strandrug 2486
strandrug 2490
strandrug, normale – 2490
strandvlakte 2492
strandwal 2459
strandwal 2463
strandwaleiland 2461
strandwal, vaste – 2462
strandwal, vrije – 2461
stratigrafie 1708
stratotype 1740
stratotype, component – 1742
stratotype, grens – 1744
stratotype, samengesteld – 1742
stratovulkaan 1429
streep 1538
streepkleur 0466
strekking 1483
strekking 4016
strekking, azimuth van de – 1483
strekking- en hellingmetingen 4228
strekking, lengte in de – 1524
strekkingsbreuk 1544
strekkingsrichting 1483
strengenveen 2636
streperig 0665
streping 0468
striatie 2999
strohalmstalactiet 2131
stromataktis 3437
stromatiet 1049
stromatoliet 3332
stromatoliet, proto– 3332
Stromboli-type eruptie 1159
stroming, capillaire – 3624
stroming, dichtheids– 2979
stroming, laminaire – 2264
stromingsdruk 4738
stromingsmeter 3669
stromingspotentiaal
stroming, turbulente – 2264
stroming, zee– 2785
strontium, gewoon – 0448
strontium, initiëel – 0448
strontium, radiogeen – 0448
strooisel 2742
strooiwaaier 2948
stroom 2191
stroomafwaartse zijde 2964
stroombaan 2372
stroombaan 3708
stroombreccie 1268
stroombreccie, aaneengelaste – 1268
stroomdichtheid 4556
stroomdraad 2268
stroomgebied 2192
stroomgebied 3648
stroomlast 2951
stroomlijn 3708
stroomlineatie 3027
stroomontwikkeling 2197
stroom, oppervlakte– 2785
stroompje 2194
stroomregime 2960
stroomribbel 3004

stroomsnelheid 2260
stroomsnelheid, drempelwaarde van de – 2959
stroomspoor 3020
stroomspoor 3032
stroomsterkte, electrische – 4556
stroom, tellurische – 4552
stroom-uitschulping 3025
stroomversnelling 2202
stroom, vroegere – 2864
structuur 0858
structuur 2568
structuur 2723
structuur, afzettings– 3077
structuur, bio-erosie– 1883
structuur, biostratificatie– 1884
structuur, bioturbatie– 1885
structuurbodem 2652
structuur, breccie– 2933
structuur, colloforme – 4091
structuur, colloidale – 4091
structuur, concentrische – 4092
structuur, cryptoëxplosie – 1184
structuur, cryptovulkanische – 1184
structuur, deformatie – 3050
structuur, diapiere – 1693
structuurelement 2724
structuurelement, wand van – 2724
structuur, fluïdale – 0868
structuur, geopetale – 2996
structuurgraad 2726
structuur, halokinetische – 1694
structuur, halotektonische – 1694
structuur, hyperitische – 0853
structuurkaart 4247
structuur, kam– 3351
structuur, kippengaas– 3468
structuur, kneed– 2649
structuur, koolblad– 3057
structuur, lineaire – 1607
structuur, metacolloidale – 4091
structuur met inzakkingsballen 3059
structuur, mozaïek– 3089
structuur, paardestaartvormige – 4069
structuur, pluim– 1511
structuur, primaire – 3077
structuur, primaire sedimentaire – 2995
structuur, ruïnen-marmer – 3467
structuur, schotel– 3069
structuur, secundaire sedimentaire – 2995
structuur, sedimentaire – 2994
structuurterras 2307
structuurtrede 2200
structuurtype 2726
structuur, uitschuring- en opvullings– 3024
structuurvorm 2554
structuurvorm 2726
structuur, vulkano-tektonische – 1182
structuur, wormvormige – 3467
structuur, zuil– 1501
stuiflawine 2359
stuifmeel 1839
stuifzand 2156
stuifzand 2871
stuifzand 2961
stuitpaal 4839
stuitspoor 3038
stuwdam, massieve – 4864
stuwing 2340
stuwing, gecombineerde – 3829
stuwmeer, vulkanisch – 1419
stuwmorene 2419
stuwvloed 2823
stuwwal 2419
styloliet 3451
subaërisch 2866
subalkalisch 0694
subalumineus 0694
subaquatisch 2880
subaquatisch gevormd 2880
subarkose 3233
subfacies, metamorfe – 0987
subfamilie 1930
subgenus 1933
subglaciaal 2873
subgrauwacke 3234
sublimatie 0412
sublimatie 3952
sublimatieproduct 1394
sublitoraal 2839
submaceraal 3567
submarginaal 4212
submarien 2900
subofitisch 0845
subsequent 2286
subspecies 1939
substraat 2696
substraat 3107
substratum 0348
subvulkaan 1206
subvulkanisch 3989
suikerachtig 3090
sulfaatreductie 0415
super-etage 1763
superfamilie 1930
supergeen 3937
supergeen 3967
superindividu 1012
superparamagnetisme 0300
superpositie 1714
superpositie, principe van – 1714
superspecies 1935
superzone 1782
supracrustaal 0584
supralitoraal 2837
supravulkaan 1413
susceptibiliteit, magnetische – 0298
suspensie – 0411
suspensielast 2955
suspensiemateriaal 2955
suspensiestroom 2980
suturisch 3450
sutuur 1132
S-vlak 1002
symbiose 1898
symmict 3124
sympatrisch 1817
symplektitisch 0849
synaerese-scheur 3044
synantetisch 0784
synantetisch 3963
synchrone 1818
synchroon 1729
synchroon 1818
synclinaal 1619
synclinaal, pseudo– 1620
synclinaal, tweezijdig duikende – 1638
syncline 1619
synclinorium 1671
synecologie 1888
synform 1619
syngenese 3398
syngenetisch 2854
syngenetisch 3474
syngenetisch 3553
syngenetisch 3966
synkinematisch 0726
synkinematisch 0906
synkinematisch 1079
synoniem 1955
synonymie 1955
synorogeen 0726
synorogenetisch 1108
synsedimentair 2854
syntaxiale rand 3456
syntectiet 0682
syntectonisch 0726
syntexis 0682
synthetisch 4478
syntype 1951
systeem 1765
systeem, actief – 4301
systeem, meervoudigeband – 4326
systeem, passief – 4301

T

taaiheid 0470
tachyliet 1262
tafel 0501
tafelberg 2556
tafelberg 2798
tafrogenese 1091
tak 1965
talud 4860
taludhelling 4827
talud, hoek van het natuurlijke – 4687
tankolie 3859
taphocenose 1896
taphocoenose 1896
taphonomie 1897
tautoniem 1957
taxiet 1243
taxogenese 1963
taxon 1926
taxon 2662
taxonomie 1920
taxonomie, bodem– 2662
tectofacies 2861
tectonosfeer 0080
teeltkeus, natuurlijke – 1966
teer 3743
teerzand 3768
tefra 1310
tefrochronologie 1312
tegendruk van de grond 4706
teken 3018
teken, optisch – 0557
tekoniek, intrusie– 0712
tektiet 0077
tektoniek 1073
tektoniet 1001
tektoniet 1605
tektoniet, primaire – 1001
tektoniet, secundaire – 1001
tektoniet, smelt– 1001
tektotoop 2862
teledetectie 4300
telemagmatisch 3986
teleseisme 0150
telethermaal 3984
tellurisch 0002
temperatuurvereffeningscoëfficient 0387
terras 2801
terras, continentaal – 2790
terras, cyclisch – 2299
terras, doorlopend – 2299
terras, fluviatiel – 2293
terras, fluvioglaciaal – 2309
terras, marien – 2484
terrasmeander 2223
terras, niet-cyclisch – 2299
terrasniveau 2294
terras, onderduikend – 2311
terras, opduikend – 2311
terras, plaatselijk – 2299
terraspunt tussen twee meanderkliffen 2304
terrasrand 2295
terras, riffront– 3352
terrassen, ineengeschakelde – 2300
terrassen, in hoogte overeenkomende – 2300
terrassenkruising 2311
terrassenlava 1285
terrassentrap 2296
terras, structureel – 2555
terras, tektonisch – 2307
terraswand 2295
terreinonderzoek 4790
terrestrisch 2865
terrigeen 2902
terugbetalingstijdstip 3896
terugkaatsing, diffuse – 4317
terugkaatsing, spiegelende – 4318
terugkoppeling 0133
terugplooiing 1689
terugspringgetal 4733
terugtrekkingsmorene 2417
Tethys 1127
tetragonaal 0515
textuur 0790
textuur 2709
textuur 3077
textuur 4345
textuurindeling 2713
textuurklasse 2713
textuur, netvormige – 0856
textuur, oppervlakte– 3192
textuur, pegmatietische – 3470
textuur, vertande – 0816
T-golf 0201
thanatocoenose 1896
theorie, laterale secretie – 3928
therme 1407
thermisch gebied, hoog – 0391
thermisch gebied, midden – 0391
thermisch normaal gebied 0391
thermocline 2830
thermogeen 0977
thermokarst 2645
thermokarstkom 2646
thermokarstmeer 2646
thermokoppel 0379
thermometer, geochemische– 0379
thermometer, geologische – 0994
thermometrie, isotopen– 0456
tholeïet 0695
tholoïde 1422
tholus 0037
thorium-lood datering 0436
tijd 1762
tijddomein 0223
tijddomein 4570
tijdrekening, geologische – 1713
tijdstratigrafie 1712
tijdstratigrafische eenheid 1757
tijdvak 1764
tij, hoog – 2818
tij,laag – 2818
tijverschil 2819
tilliet 2874
till, mariene – 2903
tilloïed 2978
tiltmeter 0124
toelaat 2338
toendra 2636
toendrapolygoon 2655
toer 4280
toeslagstof 4872
toestand, halfvaste – 4713
toestand, plastische – 4713
toestand, vaste – 4713
toestand, vloeibare – 4713
toetssteen 3369
toevoerbuis 1405
toevoerkanaal 0723
toevoerkanaal 4011

toevoerpijp 0723
tombolo 2467
tong 0725
tong 1752
tongribbel 3006
top 0705
topcaldera 1460
top en dal
toperuptie 1229
tophoogte, gelijkheid in – 2598
topkrater 1436
toplaag 2896
topniveau 2597
topniveaus, trapsgewijs gelegen – 2599
topogeen 3550
topomineraal 3995
topotactisch 0543
topotaxie 0543
topotype 1953
topozone 1784
topproductie 3807
topvlakte 2597
tor 2609
toren 4267
torenkarst 2082
torsiebalans 0350
torsieschuifproef 4765
torsie-vinproef 4759
tourmalinisatie 0932
trachitisch 0838
trachitoïde 0839
tractie 2953
traject 0206
trajectorie 4471
tralie 0964
traliestruktuur 4103
transfluentie 2383
transformatie 3426
transgressie 1726
translatie 1566
translatie 2370
transport, tektonisch – 0998
transvaporisatie 3931
trap 0638
trapbreuken 1587
trapdal, glaciaal – 2622
trapmonding 2208
trechtergang 0742
trechtermonding 2456
trede 0527
trede 1538
trekduin 2175
treksterkte 4684
trekvastheid 4684
tremometer 0118
triaxiaal apparaat 4753
tribus 1931
trichiet 0770
trigonaal 0515
triklien 0515
trillen 4809
trilling 0146
trilling, sferoïdale – 0198
trillingswijze 0166
trilling, toroïdale – 0198
trilveen 3530
trimorfisme 1869
triplet 0480
tripoli 3371
tritium datering 0452
troebelingsstroom 2981
troebelingsstroom, achtereinde van een – 2982
troebelingsstroom, duurzame – 2981
troebelingsstroom, vooreinde van een – 2982
trog 2793
trogbekkenmeertje 2620
trogdal 2273
trogeinde 2619
troggang 4053
troglijn 1636
trogschouder 2619
trogvlak 1636
tromboliet 3330
trommelvuur 1237
tromometer 0118
trosvormig 0777
tsoenami 2816
tuf 3171
tufachtig 3171
tufagglomeraatlava 1346
tuf, chaotische – 1333
tuffeus 1332
tuffiet 1332
tuffiet 3171
tuffisiet 1339
tuf, gelaagde droge – 1336
tuf, gewelde – 1341
tuf, in water afgezette – 1335
tufkegel 1428
tuf, lithische – 1333
tuf, sedimentaire – 1334
tufstroom, hete – 1369
tuf, verplaatste – 1337
tuf, vulkanische – 1332
tumulus 1272
tunnel 4847
tunnelfront 4852
tunnelfumarole 1383
tunnelgraafmachine 4854
tunnelportaal 4850
turbidiet 2983
turbineboren 4264
tussenlaag 3117
tussenlaagje, steriel – 3506
tussenlagen, aanwezigheid van – 3116
tussenlagen, met – 3116
tussenlagen, vorming van – 3116
tussenliggend 3117
tussenproduct 4142
tussenput 3843
tussenruimte 3683
tussenruimte, primaire – 3684
tussenruimte, secondaire – 3685
tussenschakeling 3117
tussenwater 3780
T-vlak 1017
tweeassig 0553
twee-kanalenschrijver 4501
tweekanter 2162
tweekleppig 1859
tweeling 0535
tweeling, contact– 0537
tweeling, doordringings– 0537
tweelingkrater 1442
tweeling, penetratie– 0537
tweelingswet 0535
tweelingvulkaan 1442
tweeslachtig 1858
T-X kromme 4463
typegebied 1739
typegeslacht 1947
typelokaliteit 1946
typesectie 1740
typesoort 1948

U

uitbarsting 1211
uitbarsting, explosieve – 1216
uitbarsting, gemengde – 1219
uitbarsting, lava – 1214
uitbarsting, mechanisme van de – 1170
uitbarsting, onderzeese – 1236
uitbarstingsbestanddeel 1209
uitbarstingsbreccie 1346
uitbarstingsenergie, kinetische – 1171
uitbarsting, subglaciale – 1236
uitbarsting van los materiaal 1214
uitblazing 2155
uitblazingscylinder 1224
uitblazingsprodukten, mantel van – 0055
uitbreiding 3885
uitdoving 0564
uitdoving, onduleuze – 0565
uitdoving, onduleuze – 0968
uitdovingshoek 0566
uitdrijving 3732
uitdrogingsbreccie 3048
uitdrogingsscheur 3044
uitdrogingsscheur 3045
uitdrogingsspleet 1518
uiteenvallen in blokken 1990
uiteinde, gekorven – 3414
uiteinde, gerafeld – 3414
uiteinde, getand – 3414
uitgaven 3893
uitgestrektheid, laterale – 1524
uitgestrektheid naar de diepte 1526
uitkristallisatie, verspreide – 3947
uitlogen 3929
uitlogen 4162
uitloop 2338
uitloper 2806
uitrolgrens 2749
uitrolgrens 4715
uitscheiding 3948
uitscheiding in vloeibare toestand, gravitatieve – 3948
uitschulpen 3025
uitschulping 2086
uitschuring 2936
uitschuring, glaciale – 2395
uitslag 0164
uitsnijden van lagen 1557
uitspoeling 2678
uitspoeling 3023
uitspoeling 4873
uitspoelingshorizon 2701
uitstapboring 3785
uitsterven 1986
uitstoting 1247
uitstroming van gas 1215
uitvloeiingsgesteente 1240
uitvloeiing uit de glooiing 1266
uitvloeiing van lava, onderaardse – 1267
uitvloeiing van lava, terminale – 1266
uitvlokking 2992
uitvoer 4445
uitvoerbaarheidsstudie 4791
uitwaaiering 1675
uitwassen, stekelige – 1260
uitwerpen 1210
uitwerpselen, allogene – 1314
uitwerpselen, gloeiende – 1315
uitwerpselen, gruisvormige – 1313
uitwerpselen, klastische – 1313
uitwerpselen, niet-juveniele – 1314
uitwiggen 3104
uitwiggen 3509
uitwiggingsfuik 3758
ultrabasiet 0604
ultrabasisch 0604
ultramafisch 0600
ultramafisch gesteente 0600
ultrametamorfose 0882
ultramyloniet 0916
ultramyloniet 1541
ultravulkanisch 1166
undatie 1122
undatiehypothese 1121
uniformitarianisme 0006
uniformitarisme 0006
uranium-lood datering 0434
uvala 2073

V

vagiliteit 1917
vagrant 1918
validiteit 1923
vallei, onderzeese – 2795
valmethode 4409
vals alarm 0269
vangen 4287
vanggereedschap 4287
vanggerei 4287
variabel, continu – 0820
variabel, discontinu – 0820
variantie 4190
variatie-coëfficiënt 4190
variatiediagram 0697
variatie, seculaire – 0389
variëteit 1940
variogram 4196
variole 0781
varioliet 0781
variolitisch 0877
variometer 0288
vastelandsglooiing 2790
vastelandsplat 2788
vastelandsterras 2790
vastheid 4682
vat 3892
vauclusebron 2093
veen 3494
veen 3532
veendoorbraak 3541
veen, limnisch – 3531
veenmeer 3540
veenmoeras, hoogliggend – 3542
veenmoeras, laagliggend – 3533
veenrandmoeras 3537
veen, terrestrisch – 3546
veen, toendra– 3547
veenvorming 3524
veiligheidscoëfficiënt 4777
veiligheidsfactor 4777
veld 4199
veldkromme 4565
veldspatisatie 1064
veldsterkte, magnetische – 0317
veniet 1045
Vening Meinesz, as van – 1137
venster 4234
veraarding 3526
verankering 4792
verblijfplaats 1901
verblijftijd 0397
verblijftijd 3627
verbranding, ondergrondse – 3838
verbrandingswaarde 3604
verbrandingswaarde, bruto – 3604
verbrandingswaarde, netto – 3604
verbrokkelend, gemakkelijk – 3087
verbrokkeling 1991
verbuizing 4275
verdamping 0412
verdeelstuk 3798
verdeling, log-normale – 4189
verdeling, normale – 4189
verdichting 4679
verdichting 4808
verdikking 4054
verdringing 3956
verdrinking 2857
verdrinkingskust 2525
verdubbeling, horizontale – 1551

verdubbeling, schijnbare – 1553
verdubbeling, stratigrafische – 1552
verdwijnen naar boven 1526
verdwijnen naar de diepte 1526
verdwijngat 2075
vereffening 2589
vereffening 4511
vereffeningsvlakte 2591
vereffeningsvlakte, embryonale – 2593
vereffeningsvlakte in wording 2593
vereffening, zone met zijdelingse – 2604
verertsing 3936
vergipsing 3459
verglazing 0708
verglazing 0897
vergletsjering 2367
vergleying 2684
vergroeiing 0783
vergroeiing 4098
vergroeiing, eenvoudige – 4099
vergroeiing, grafische – 0847
vergroeiing, grafische – 4100
vergroeiing, myrmekitische – 4100
vergroeiing, ringvormige – 4102
vergroeiingsgraad 4112
vergroeiingsvlak 0536
vergroeiing, symplektische – 4105
vergrotingscurve 0133
vergrotingsfactor, dynamische – 0131
vergroting, statische – 0131
vergruizing 1991
vergruizingsbreccie 1539
vergruizingszone, verertste – 4060
verhang 4736
verhangknik 2199
verhanglijn 4736
verhang, tegengesteld – 2626
verhard 3422
verharding 3422
verharding 3423
verharding 4842
verharding van de buitenkant 3411
verheffing 2796
verheffing, vulkanische – 1409
verjonging 2572
verjonging 3959
verkalking 3458
verkarsting 2063
verkenning, hydrochemische – 4640
verkenningsboring 3782
verkenningsgat 4297
verkiezeling 0929
verkiezeling 3458
verkit 2733
verkitting 3433
verkoling 3607
verkorreling 1990
verkorreling 2934
verkrijting 3415
verlanding 2341
verlanding 2513
verlaten 3818
verlatingsdruk 3849
verlichting 4311
verlichting, specifieke – 4311
vermaling 2940
vermoddering 3526
vermogen 2249
vermogen 4449
vermogen, horizontaal oplossend – 4516
vermogen, oplossend – 4347
vermogen, oplossend – 4453
vermogensdichtheidsspectrum 4451
vermogen, vertikaal oplossend – 4516
vermolming 3526
vernauwing 3904
vernauwing 4054
verontreiniging 3672
verplaatsing, hellende – 1564
verplaatsing, horizontale – 1565
verplaatsing, richting van de ware – 1562
verplaatsingsmeter 0120
verplaatsing, verticale – 1563
verplaatsing, ware – 1562
verplaatst 4637
verplooiing 2648
verpulvering 1319
verrijking 3945
verschilzicht 4364
verschuiving, genivelleerde – 2549
verschuiving, ringvormige – 0743
verschuivingsklif 2548
verslakking 4184
verslemping, oppervlakkige – 2760
versnellingsregistratie 0120
versperring 2340
verspoeld 2947
verspreidingszone 1781
verspringen van lagen 1523
verstarring, oppervlakkige – 1200
verstening 3457
verstrooiing 0218
verstrooiing 0219
verstrooiing 4316
vertakking 1583
vertakking 1687
vertakking 2210
vertakt 2292
vertand 0960
verticaal 0329
vertraging 4822
vertweelinging, polysynthetische – 0538
verval 4735
vervalsconstante 0431
vervanging 0931
vervanging 3453
vervanging 3956
vervanging 3961
vervanging, atomaire – 0477
vervanging, gekoppelde – 0477
vervangingsafzetting 3991
vervangingsgang 4042
vervlakking 2589
vervloeiing 3084
vervormbaar 4673
vervorming 0104
vervorming 1075
vervorming 4666
vervorming, diepe plastische – 3447
vervorming, ondiepe plastische – 3447
vervorming, relatieve – 4666
vervormingsenergie, vrijgekomen – 0107
vervormingsmeter 0122
vervormingsmodulus 4696
vervuiling 4652
verwachtingscurvemethode 3876
verwant 3970
verweerbaarheid 4734
verwekingslijn 4862
verwelkingspunt 2773
verwering 1988
verwering 2928
verwering, biologische – 1999
verwering, chemische – 1998
verwering, chemische – 2928
verwering door temperatuurwisseling 1992
verwering, droge – 1992
verwering, fysische – 1989
verwering, kosmische – 0028
verwering, mechanische – 2928
verwering, mechanische – 1989
verweringsbodem 2009
verwering, selectieve – 2004
verweringskorst 2010
verweringslaag 2009
verweringspuin 2005
verweringsresidu 2930
verweven 0964
verwilderd 2218
verzadigd 0605
verzadigde zone 3656
verzadiging, olie-gas – 4610
verzadigingsexponent 4606
verzadigingsgraad 3650
verzadigingslijn 0697
verzadigingspunt 3857
verzakking 4704
verzakking, ringvormige – 1185
verzakking, vulkanotektonische – 1182
verzamelbekken 2380
verzamelgebied 2276
verzanden 3814
verzanding 2990
verzet 1555
verzuring 2738
vesiculair 1305
vetkool 3559
vetzuur 3723
vezelig 0522
vezelig 0773
vibroflotatie 4810
vibrograaf 0123
vignetteringsfilter 4339
vijver 2330
viltachtig 0825
vinapparaat 4759
vingervormig in elkaar grijpen 3118
vinproef 4759
virgatie 1675
viscositeit 4671
vissen 4287
vitrofier 0823
vitrofirisch 0823
vlak 0517
vlak, listrisch – 1573
vlakte 0029
vlakte, fluvioglaciale – 2424
vlaktemeander 2220
vlak van actie 0238
vlamkool 3559
vlechtend 2218
vlechting 2501
vleikalk 4123
vlek 2692
vleklei 0941
vleugel, afgeschoven – 1559
vleugel, omgekeerde – 1647
vleugel, omgekeerde – 1680
vleugel, opgeschoven – 1559
vliegtuig 4302
vlietsteen 2128
vloedbranding 2823
vloedgeul 2510
vloedgolf 2816
vloedklif 2489
vloedschaar 2510
vloedstroom 2822
vloeibaarheidsindex 4716
vloeibaar worden 4718
vloeien 4672
vloeigrens 2749
vloeigrens 4715
vloeigrond 4855
vloeiklei 4719
vloeilaminatie 0868
vloeilineatie 0868
vloeiplooiing 1701
vloei, primaire– 0631
vloeistof, bevochtigende – 3705
vloeistof, niet-bevochtigende – 3705
vloeistructuur 0868
vloer 3501
vloer 4009
vloer 4010
vloer 4852
vlok 2992
vluchtlijn 4303
vluchtlijn 4358
vluchtweg 4303
vocht, beschikbaar – 2773
vocht, gebonden – 3598
vochtgehalte 2769
vochtgehalte bij veldcapaciteit 2774
vochtkarakteristiek 2772
vochtspanning 2770
vochtspanning 2771
vocht, totaal – 3598
vocht, vrij – 3598
voeding 4140
voedingsgang 1881
voedingsgebied 2380
voedingsgebied 3670
voedingskanaal 3941
voedingskanaal 4011
voetindruk 1877
voet, kubieke – 3892
voetspoor 1877
vogelpootdelta 2495
voldersaarde 3218
volumefactor voor gas 3861
volumefactor voor olie 3860
volumemassa 2751
volumepercentage 4109
volumeregeling, automatische – 4425
volumeregeling, automatische – 4508
volumetrische methode 3871
voorberg 2561
voorboden 1191
voorconsolidatie 4703
voorde 2247
voordiep 1109
voorduin 2168
voorfase 1157
voorgebergte 2605
voorgift 4818
voorland 1109
voorland 2447
voorlandgletsjer 2435
voorloper 0147
voorloper 0150
voorloper 0181
voorplooiing 1667
voorraadopslag 4225
voorschok 0147
voorsplijting 4823
voortzetting naar beneden 4533
voortzetting naar beneden 4543
vormenkombinatie 0521
vormfactor 4110
vormoriëntatie 1013
vorm, tektonische – 2537
vorstbestendigheid 4734
vorstbulten, begroeide toendra – 2643
vorstscheur 2640
vorstspleet 2640
vorst, vereffening door – 2650
vorstverwering 1996
vorstverwering, gevoelig-

heid voor – 1996
vorstverweringspuin 2656
vreemd 3970
vrijheidsgraad 4112
vuilstortplaats 4831
Vulcano-type eruptie 1160
vulkaan 1411
vulkaan, alleenstaande – 1446
vulkaanbewaking 1190
vulkaan, dakpansgewijs gelaagde – 1421
vulkaandamp 1349
vulkaandamp 1377
vulkaan, eenassige – 1441
vulkaaneiland 1416
vulkaanembryo 1222
vulkaan, enkelvoudige – 1441
vulkaan, gegroefde – 1474
vulkaan, gekerfde – 1475
vulkaan, gekloofde – 1475
vulkaan, gemengde – 1429
vulkaangordel 1447
vulkaangrind 1324
vulkaan in ruststadium 1155
vulkaankegel 1434
vulkaanlichaam 1413
vulkaanmantelspleet 1468
vulkaanmassief 1413
vulkaan, meerassige – 1441
vulkaan met centrale krater 1434
vulkaan, modder– 1395
vulkaan, monoconische – 1441
vulkaan, omwalde – 1445
vulkaan, onderwater– 1415
vulkaan, onderzeese – 1416
vulkaanreeks 1448
vulkaanrij 1448
vulkaanruïne 1476
vulkaan, samengestelde – 1441
vulkaan, slijk– 1395
vulkaan, slik– 1395
vulkaanspleet 1466
vulkaanspleet, gapende – 1234
vulkaanspleet, open – 1234
vulkaan, subactieve – 1153
vulkaan, subaerische – 1415
vulkaantheorie 0049
vulkaan, uitgedoofde – 1155
vulkaan, uitgestorven – 1155
vulkaanvorm, negatieve – 1410
vulkaanvorm, positieve – 1409
vulkaan, vrijstaande – 1446
vulkaan, werkende – 1153
vulkaan, zand– 3071
vulkaanzone 1447
vulkanen, opeenhoping van – 1450
vulkaniet 1238
vulkanisch 1148
vulkanisch 3989
vulkanische bijverschijnselen 1180
vulkanisme 1147
vulkanisme, epigenetisch – 1152
vulkanisme, orogeen – 1151
vulkanisme, subrecent – 1150
vulkanogenetisch 1148
vulkanoklastisch 3172
vulkanologie 1146
vuur 0485
vuurspuwend 1237
vuursteen 3491
vuursteen-eluvium 2931
vuursteengordijn 3491
vuurwerk 1283

W

waaiduin 2187
waaier-lineair 4631
waaigat 2187
waardebepaling van mijnbezit 4216
waarde, contante – 3897
waarde, geïnfleerde – 3899
waarde, uitgestelde – 4220
waarnemingsdichtheid 2675
waarschijnlijkheidscombinatie 3877
wad 2505
waddeneiland 2507
waddengebied 2505
wadi 2587
wadkom 2508
walmorene 2414
walsen 4809
wand 0705
wand 4009
wand 4852
wand, aangebakken – 4013
wandgesteente 1530
wandkern 4238
wand, vergroeide – 4013
wand, vrije – 4013
wantij 2511
warmtecapaciteit, soortelijke – 0386
warmtegebied 1144
warmtegeleidingsvermogen 0382
warmte-inhoud 0389
warmte, radiogene – 0375
warmtestroomdichtheid 0384
warmtestroom, snelheid van – 0385
warmtestroom, terrestrische – 0383
warmtetransport, convectief – 0381
warmwaterveld 0393
warve 3128
warvengesteente 3128
warvenklei 3128
was 0484
was 3723
wasgoot 4158
wasgoot 4659
wasserijvoeding 4140
waterafvoerbuis 4811
water, atmosferisch – 3613
waterbeheersing 2775
waterbergend vermogen 3625
water, beschikbaar – 2773
waterbloei 2843
water, brak– 3676
water, connaat – 3661
waterdampfumarole 1389
waterdichtheid 4863
water, diep – 2809
waterdom 1359
waterdrukhoogte 4735
waterdruk, opwaartse – 4740
waterfontein 1359
water, freatisch – 3659
water, funiculair – 3654
watergang 2191
watergehalte 2769
watergehalte 3802
water, geregenereerd – 3663
waterinjectie 3833
waterinjectie, heet – 3839
water, juveniel – 1348
water, juveniel – 3664
waterkoepel 1359
waterkwaliteit 3674
waterlijn, meteorisch– 0458
water, magmatisch – 1348
watermassa 2828
water, metamorf – 0976
water, meteorisch – 3614
waterniveau, proef met constant – 3711
waterniveau, proef met variabel – 3711
water, ondergronds – 3649
water, ondiep – 2809
water, opkomend – 2251
water, pendulair – 3653
waterscheiding 2314
waterscheiding, consequente – 2317
waterscheidingskam 2316
waterscheidingslijn 2314
waterscheidingsrug 2312
waterscheiding, subsequente – 2318
waterscheiding, verschuivende – 2319
waterscheiding, verspringende – 2320
waterscheiding, zich verplaatsende – 2319
waterscheiding, zig-zaggende – 2321
waterspiegel 3777
waterspiegel, vrije – 3778
waterstandsmerk 3031
waterstof-index 4613
waterstuwing 3824
watertype 2828
water, vadoos – 3655
waterval 2203
waterveen 3535
waterverharding 2957
waterverzadiging 4611
waterverzadiging, residuele – 4611
water, wassend – 2251
waterzand 3766
waterzandsteen 3766
water, zoet– 3676
water, zout– 3676
WDC 0082
weefsel 0793
weerhaakwal 2464
weerstand 4556
weerstand, schijnbare – 4564
weerstand, schijnbare – 4609
weerstandslog 4582
weerstandslog, gefocusseerde – 4586
weerstandsmethode, electrische – 4561
weerstand, soortelijke – 4556
weerstand van formatiewater, specifieke – 4608
wegfundering 4841
weglichaam 4840
weg, minimum-tijd– 0209
wegschuiving 1663
weide, buitendijkse – 2514
welvorming 4863
Wentworth-schaal 3163
wereldbeving 0138
werk 4789
werkfrequentie 4498
werking, effusieve – 1213
werking, explosieve – 1216
werking, kortstondige – 1154
werking, periode van – 1208
werking, pyroklastische – 1309
werking, vulkanische – 1149
werkovereenkomst 3901
Widmansätten, figuren van – 0067
wildflysch 2926
windcorrasie 2015
windcorrasie 2154
winderosie 2015
windkei 2162
windkuil 2187
windkuilduin 2187
windoever 2336
windribbel 3016
windschaduwduin 2174
windtransport 2154
windvaan 2173
winning door compactie 3827
winning, primaire – 3822
winning, secundaire – 3831
winningsfactor 3867
winning, tertiaire – 3837
winning, veilige – 3668
winplaats 4871
winterbed 2242
wisselstroom 4555
woestijn 2577
woestijnlak 3405
woestijn, lavasteen – 1279
woestijnpantser 2160
woestijnroos 3406
wolk, bloemkoolvormige – 1362
wolk, pijnboomvormige – 1362
wolzakverwering 2003
woongang 1880
woonplaats 1901
wormvormig 0774
wortel 0343
wortel 1104
wortelgebied 1684
wortelvloer 3515
wrijfspiegel 1538
wrijfspiegel 2731
wrijving 4686
wrijving, inwendige – 4686
wrijvingsbreccie 1539
wrijvingscoëfficiënt 4689
wrijvingshoek 4689
wrijvingskriterium 4689
Wulff, net van – 1015

X

xenoblast 0980
xenoblastisch 0980
xenokrist 0663
xenoliet 0663
xenomorf 0752
xenothermaal 3984
xenotoop 3471
xyliet 3495
xylophaag 1857

Y

yardang 2164

Z

zadel 1635
zadelgang 4053
zak 2649
zakking 4704
zakking 4833
zand 2716
zand 2718
zand 3227
zand, aeolisch – 2871
zandafglijding 2973
zandbank 2244
zandbank 2459
zand, caverneus – 3063
zand, dek– 2871
zanddikte, netto – 3873
zandeilandje 2443
zandfractie, mediaan van de – 2718
zand, gashoudend – 3766
zandgolf 3007
zandig 2721
zandig 3225
zand, koraal– 3311
zandkorrelgrootte 3227
zandkristal 3443

zand, lemig – 3227
zandloperstruktuur 0534
zand, nat – 3767
zand, oliehoudend – 3766
zand, opvulling met – 2990
zand-opwelving 3065
zandpaal 4813
zandplaat 2910
zandsteen 3228
zandsteen, gashoudende – 3766
zandsteen, kwartsietische – 3230
zandsteen, oliehoudende – 3766
zandsteen, schone – 3228
zandsteen, veldspaathoudende – 3232
zandsteen, veldspaatrijke – 3232
zandsteen, waterhoudende – 3766
zandstraling 2939
zandstroom 2973
zand, stuif– 2871
zand, stuivend – 2156
zand, stuivend – 2871
zandstuiving 2156
zandtransport, eolisch – 2961
zandval 2973
zandval 3152
zandvlakte, stuif– 2157
zandvulkaan 3071
zand, vulkanisch – 1322
zandwad 2910
zand, waterhoudend – 3766
zandwind bewerkte steen, door – 2161
zandwoestijn 2578
zandzee 1462
zand, zingend – 3088
Zavaritsky, diagram van – 0699
zee 2782
zee– 2900
zeebeving 0148
zeeduin 2166
zeef- 0974
zeefanalyse 3157
zeefanalyse 4144
zeefopening 3157
zeefpijp 3795
zeegang 2807
zeegat 2508
zeeloper 2168
zeeniveau, gemiddeld – 2817
zeeponoor 2075
zeereep 2168
zender 4487
zend-ontvangschakelaar 4489
zeolitisatie 0932
zetting 4704
zetting 4833
zettingsmeter 4768
zettingsverschil 4704
zettingsverschil 4833
zeven 4873
ziften 2937
zigzagplooi 1651
zijde 0484
zijdeglans 0490
zijmorene 2411
zijreflectie 4518
zijrivier 2207
zomerbed 2242
zone 0362
zone 0519
zone 1717
zone 3908
zone, gebroken – 4821
zone, gespoelde – 4620
zonering 0474
zonering 0533
zonering, inverse – 0475
zonering, normale – 0475
zone, sub– 1782
zone, vol-capillaire 2767
zonnestelsel 0018
zonula 1782
zoöbenthos 1852
zoöchlorellae 1845
zoolmerk 3019
zoöplankton 1800
zoöxanthellae 1845
zoutgehalte 2827
zoutgehalte 3376
zoutgletsjer 1699
zouthoudend 3376
zoutkristal-pseudomorfose 3042
zoutkussen 1695
zoutmeer 2893
zoutmuur 1695
zoutparagenese 3778
zoutpijler 1695
zoutpolygoon 3049
zouttektoniek 1694
zoutvlakte 2580
zoutwigwerking 1997
zuigspanning 2770
zuil 2138
zuur 0601
zuurbehandeling 3800
zuur gesteente 0601
zuurgraad 0422
zuurstofbehoefte 3678
zwaar 2721
zwaarstang 4270
zwaartekracht 0323
zwaartekracht, eenheid van – 4537
zwaartekracht, gradiënt van de – 0335
zwaartekracht, internationale formule 0333
zwaartekracht, normaalwaarde van de – 0332
zwaartekrachtsbepaling 0349
zwaartekrachtscheiding 4153
zwaartekrachtskaart, regionale – 0371
zwaartekrachtskaart, residuaire – 4542
zwaartekrachtsmeting 0349
zwaartekrachtspotentiaal 0327
zwaartekrachtstation 0353
zwaartekrachtsveld 0326
zwaartekracht, versnelling van de – 0324
zwaartekracht, vrije lucht – 4538
zwaartekracht, waargenomen – 4536
zwak 1626
zwanenhals 4268
zwavelmodder 1399
zwavelmodderboom 1400
zwavelmodderkogeltje 1321
zwavelmodderschoorsteen 1400
zwavelmodderstroom 1399
zwavelmodderzuil 1400
zwelkoepel 1272
zwelling 4720
zwerfduin 2175
zwerfkei 2422
zwerfsteen 2422
zwerfsteen 2949
zwervend 1918
zwin 2486

Français

A

abaissement 2012
abaissement axial 1635
abandon 3818
abannet 2077
abaque 4565
abattage en chambre 4817
abîme 2076
abîme 2104
ablation 2013
ablation 2380
ablation éolienne 2155
abondance 2249
abrasion 2016
abrasion 2938
abrasion, niveau limite d'– 2442
abrasion, plaine d'– 2536
absorption 0112
absorption 0406
absorption 4514
absorption, spectre d'– 0491
absorptivité 4314
abyssal 0617
abyssal 2847
abyssale, plaine – 2792
abyssopélagique 2847
accélération de pointe 0120
accélérogramme 0120
accéléromètre 0120
accolement, plan d'– 0536
accrétion 1822
accrétion 2988
accrétion, hypothèse de l'– 0022
accrétion, théorie de l'– 0048
accroissement, ligne courbé d'– 0479
accumulation 2987
accumulation 3754
accumulation cristalline 0674
accumulation, domaine d'– 0700
accumulation éolienne 2961
ACF, diagramme – 0989
achondrite 0072
aciculaire 0522
aciculaire 0755
aciculaire 3185
acide 0601
acide humique 3525
acide, roche – 0601
acide ulmique 3525
acidification 3800
acidité 0422
acme 1985
activité de type Mont Katmai 1165
activité effusive 1213
activité éphémère 1154
activité éruptive, élément d'– 1209
activité explosive 1216
activité explosive, indice d'– 1175
activité fumerollienne 1375
activité hawaïenne 1158
activité péléenne 1163
activité plinienne 1161
activité pyroclastique 1309
activité strombolienne 1159
activité volcanique 1149
activité vulcanienne 1160
actualisation, taux d'– 3897
actualisme 0006
actuopaléontologie 1886
adamantin 0484
adaptation 1980
adaptive 1980
adcumulat 0832
adhérence 2750
adhésivité, point d'– 2750
adinole 0933
adsorbant, complex – 2755
adsorption 0406
adularisation 0932
aeration, zone d'– 3651
aérobie 0424
aérobie 1906
aérolite 0071
aétite 3485
afaissement 4704
affaissement volcanique 1410
affleurement 4231
affleurement, ligne d'– 4231
affleurer 4231
affluent 2207
affouillement, trace d'– 3021
affouillement transversale, trace d'– 3021
affrontement, zone d'– 1132
AFM diagramme 0698
AFM, diagramme – 0989
âge 1762
âge absolu 0429
âge concordant 0430
âge discordant 0430
agent minéralisateur 3936
agglomérat 1344
agglomérat de cratère 1248
agglomération monominéral 0771
agglutinant, pouvoir – 3595
aggradation 2245
aggrégat cristallin 0771
agmatite 1042
agone 0284
agpaitique 0693
agrégat 2724
agrégat 3301
agrégation, degré d'– 2726
aiguigeois 2075
aiguille 1424
aiguille 2805
aiguille glaciaire 2618
aiguilles, en forme d'– 0522
aile 1625
aimantation 0295
aimantation anhysterétique 0314
aimantation chimique 0312
aimantation des roches 0293
aimantation, intensité d'– 4531
aimantation inversée 0315
aimantation rémanente 0306
aimantation rémanente de dépot 0312
aimantation thermorémanente 0309
aimantation visqueuze 0313
aire d'alimentation 2192
aire de drainage 2192
air grisouteux 3554
Airy 0345
Airy phase 0190
A'KF, diagramme – 0989
albedo 4316
albitisation 0932
alcalin 0694
alcaline, série – 0692
alcalinité 3677
alcalino-calcique, série – 0692
alcalis-chaux, indice – 0692
alcanes 3717
alcènes 3717
aléser 4296
aleurite 3205
aleurolite 3205
aleurolite calcaire 3275
algaire, marge – 3361
algaire, matte – 3364
algaire, pelote – 3363
algues, à – 3314
algues calcaires 3359
algues, trottoir à – 3361
algues, trottoirs à – 2515
aliasage 4423
alignement 4460
alimentation 4140
alimentation, aire d'– 2380
alimentation, aire d'– 3670
alimentation artificielle 3671
alimentation, cheminée d'– 0723
alimentation induite 3671
alios 2704
alios ferrugineux 2734
aller-retour 4280
allochème 3271
allochromatique 0483
allochrone 1818
allochronique 1818
allochtone 0727
allochtone 1685
allochtone 1815
allochtone 2945
allochtone, constituant – 2949
allogène 0660
allogène 2946
allogène 3968
allogène 4633
allogène, constituant – 2949
allongement optique 0567
allopatrique 1817
allotype 1949
allure 1480
alluvionnement 2245
alluvionnement 2886
alluvions 2278
alluvions 2886
alluvions, cône d'– 2007
alpin, orogenèse de type – 1078
alpin, type – 0600
altérabilité 4734
altération 2928
altération 3962
altération chimique 1998
altération chimique 2928
altération cosmique 0028
altération mécanique 2928
altération météorique 1988
altération sous-marine 3476
altération superficielle, croûte d'– 2010
altération superficielle, manteau d'– 2010
alternance répétée 3119
altiplanation 2590
alunitisation 0932
alvéole 2001
amas 0730
amas boudiné 1288
amas minéralisé 4003
amas minéralisé 4032
ameublissement 4731
ameublissement, état d'– 2757
amortir en profondeur, s'– 1526
amortir vers la surface, s'– 1526
amortissement 0127
amortissement, caisse d'– 4219
amortissement critique 0129
amortissement dynamique 0127
amortissement, facteur d'– 0128
amortissement, rapport d'– 0128
amortissement visqueuse 0127
amphibie 1858
amphibolite 0934
amphibolitisation 0932
amphithéâtre morainique 2418
amplificateur programmable 4499
amplification dynamique 0131
amplification statique 0131
amplitude 0164
amplitude 1634
amplitude 4186
amygdalaire 0872
amygdalaire, roche – 0872
amygdale 0782
amygdale 3055
amygdaloïde, roche – 0872
anaclinale 2288
anaérobie 0424
anaérobie 1906
analcimisation 0932
analyse 4183
analyse continue 4144
analyse élémentaire 3594
analyse granulométrique 4144
analyse immédiate 3594
analyse mathématique 4777
analyse mécanique 2711
analyse mécanique 3160
analyse spectrale 0221
analyse stéréométrique 4107
anaséisme 0187
anastomose 2501
anastomosé 2218
anastomose par captures 2501
anatexie 1025
anatexie différentielle 1029
anatexite 1026
anchimétamorphisme 0888
anchimétamorphisme 3420
ancrage 4792
ancrage, câble d'– 4794
ancrage, plaque d'– 4794
ancrage, tête d'– 4794
ancrage, tige d'– 4794
andésitique, ligne – 0691
anélasticité 0111
angle critique 0213
angle d'équilibre 3137
angle de repos 4687
angulaire 2729
angulaire 3189
anhédrique 0523
anhydre 0609
anhydrite 3381
anisotrope 4681
anisotrope, optiquement – 0561
anisotropie 0110
anisotropie 4681
anisotropie magnétique 0303
anneau de lave 1284
annuité 4218
anomalie 4532
anomalie 4632
anomalie à l'air libre 0368
anomalie de Bouguer 0368
anomalie de Bouguer modifiée 0370
anomalie de fuite 4634
anomalie de gravité 0367
anomalie géothermique 0390
anomalie hydromorphe d'émergence 4635
anomalie hydromorphe de précipitation 4635
anomalie isostatique 0369
anomalie isostatique locale 0369
anomalie isostatique régionale 0369
anomalie magnétique 0286
anorogénique 0726
anorogénique, période – 1106
anoxique 0424
anse 2232
anse 2450
antécédent 2290
antéclise 1084
antétectonique 0906
anthracite 3561
anticentre 0154
anticlinal 1618
anticlinal à double plongement 1638
anticlinal, faux – 1620
anticlinal, pseudo– 1697
anticlinorium 1671
antidune 3008
antiferromagnétisme 0301

antiripplet 3016
aphanitique 0798
aphanitique, roche – 0798
aphotique 1905
aphyrique 0798
aplanissement 2589
aplanissement latéral, zone d'– 2604
aplanissement partiel, niveau d'– 2603
aplanissement, surface d'– 2591
aplati 0522
aplati 0755
aplatissable 0471
aplatissement 0330
aplatissement 1613
aplatissement 3182
aplitique 0811
apomagmatique 3986
apophyse 0725
apophyse 4071
appareil de perméabilité 4755
appareil triaxial 4753
appui latéral 4866
aptitude des sols 2660
aquifère 3634
aquifère multicouche, système – 3641
arabesque, en – 0827
arbre de Noëll 3796
arc externe non volcanique 1137
arche naturelle 2079
archéomagnétisme 0294
archétype 1979
Archie, lois d'– 4605
Archimède, poussée d'– 4740
arc insulaire 1135
arc interne volcanique 1137
arc lumineux 1371
arculite 0768
ardoise 0942
aréique 2257
arénite 3225
arénite quartzique 3230
arête 0518
arête 2617
arête enchassée 0501
arêtes synclinales, enveloppe des – 1636
arêtes vives, à – 3190
arête synclinale 1636
argile 2714
argile 3212
argile 3213
argile à blocaux 2422
argile à blocaux 2874
argile à blocaux 3250
argile à passées ferrugineuses 3488
argile à passées ferrugineuses 3504
argile à passées ferrugineuses 4117
argile à silex 2931
argile à varves 3128
argile bitumineuse 3721
argile de décalcification 2122
argile de faille 4729
argile de friction 1540
argile de frottement 1540
argile de trituration 1540
argile grasse 3216
argile maigre 3216
argile marneuse 3223
argile plastique 3216
argile réfractaire 3217
argile réfractaire 3517
argile résiduelle 2122
argile schisteuse 3220
argile sensible 4719
argile sensible au remaniement 4717
argile, teneur en – 4623
argile thixotropique 4719
argileuse, fraction – 2714
argileux 3211
argilite 3211
argilite 3219
argilite 3220
argon, atmosphérique 0450
argon, excès d'– 0450
argon radiogénique 0450
arkose 3233
armoire à boue 4282
aromatiques 3717
arrachage de boulon, essai d'– 4764
arrachement 2396
arrachement, surface d'– 2059
arrangement des particules 3082
arrière-fosse 1110
arrière-pays 1110
arrivée 0172
arrivée, première – 4428
arrondi 3188
arrondi, bien – 3188
arrondissement 3187
arrondissement, indice d'– 3187
arrosage 4810
artérite 1045
ascension du bouchon 1247
asismique, région – 0253
asphalte 3734
asphalte, lac d'– 3753
asphalte, mare d'– 3753
asphalte naturel 3734
asphaltite 3737
assèchement 4739
assèchement 4811
assemblage 1805
assemblage métamorphique 0885
assimilation 0681
assimilation de bordure 0683
assimilation marginale 0683
assise 3097
assise rocheuse 4788
association 1892
association 3179
association des sols 2673
association magmatique 0630
association minérale démontrée subéconomique 4211
association minéralogique 0586
association typomorphe 0988
astérisme 0489
asthénolite 1122
asthénosphère 0090
astroblème 0075
ataxite 0069
ataxite 1243
atectonique 0726
atexite 1032
atlantique, série – 0690
atmosphère 0084
atoll 3340
atrio 1461
atténuation 0219
atténuation 4315
atténuation 4475
Atterberg, limite d'– 4714
atterrissement de vagues, marque d'– 3030
attrition 2940
attrition éolienne 2154
auge 2793
auge, épaulement de l'– 2619
auge, fin d'– 2619
aulacogène 1085
auréole 4629
auréole d'altération hypogène 4630
auréole dans l'encaissant 4630
auréole de contact 0899
auréole de réaction 0786
auréole kélyphitique 0786
authigène 2946
authigène 3968
authigenèse 3434
authigénétique 3434
autigène 4633
autochtone 0727
autochtone 1685
autochtone 1813
autochtone 2945
autoclase 0115
autocorrélation 4195
autocorrélation 4450
auto-écologie 1888
autohydratation 3957
autoinjection 0621
autolite 0664
autométamorphisme 0659
autometamorphisme 3957
autométasomatose 0659
automorphe 0750
automulchage 2762
autopneumatolyse 0659
autopneumatolyse 3957
avalaison 2252
avalanche 2358
avalanche 2972
avalanche chaudie 2359
avalanche, cône d'– 2362
avalanche, couloir d'– 2362
avalanche de fond 2359
avalanche de glace 2361
avalanche de pierres 2047
avalanche de pierres 1368
avalanche de plaques de neige 2360
avalanche froidie 2359
avalanche, glissoir d'– 2362
avalanche humide 2359
avalanche incandescente 1367
avalanche intercratérique 1472
avalanche mixte 1368
avalanche poudreuse 2359
avalanche sèche 1369
avalanche sèche 2359
avalanche, sillon d'– 2362
avalanche terrière 2359
avalanche volante 2359
avalasse 2252
avancée de roche 2606
avant-dune 2168
avant-fosse 1109
avant-pays 1109
avant-trou 4297
aven 2076
aven 2104
avion 4302
axe a 1016
axe b 1016
axe c 1016
axe cristallographique 0511
axe de la déformation 1016
axe fluvial 2206
axe géomagnétique 0208
axe optique 0554
axe pi 1020
axes optiques, angle des – 0555
axes optiques, plan des – 0554
axial, plan – 1631
axial, trace du plan – 1631
axiolite 0767
axiolithique 0767
azimut 4228
azimut de l'épicentre 0155
azimut d'une station 0160
azoïque 1738

B

bâche 2486
bacillite 0762
badland 2030
bahada 2581
bahamite 3303
baie 2451
baignoir 2387
baïne 2486
bain magmatique 1198
balance de matériel, calcul de – 3865
balayage 4503
balayage, temps de – 4503
ballas 0495
ballast 4844
baloyage 4401
baloyage circulaire 4402
baloyage linéaire 4402
banc 2244
banc 2469
banc 2486
banc 2506
banc 2802
banc 2803
banc 3097
bande 2705
bande 4358
bande boueuse 2379
bande pierreuse 2654
bande sale 2379
bande spectrale 4306
bande terreuse 2654
bande thermique 4308
banquette 2059
banquette 2297
banquette rocheuse 2479
barbacane 4811
barchane 2177
bar digitée 2897
bariolage 2692
bariolé 3462
barkhane 2177
barrage 2340
barrage à contreforts 4864
barrage en enrochements 4859
barrage en remblais 4850
barrage en terre 4859
barrage poids 4864
barrage voûte 4865
barranca 1474
barranco 1474
barre 2201
barre 2459
barre 3506
barre en croissant 2462
barre glaciaire 2625
barre sous-marine littoral 2459
barrière 1916
barrière géographique 1916
barril 3892
basalte alcalin 0695
basalte des cônes 1258
basalte des plateaux 1257
basalte, globe de – 1256
basculement 0124
basculement 4833
base 3106
base 4350
base 4371
base des vagues, niveau de – 2442
base général, niveau de – 2025
base graveleuse 2651
base karstique, niveau de – 2066
base partiel, niveau de – 2025
bases, saturation en – 2756
basification 0684
basique 0603
basique, front – 1066
basite 0603
bas, ligne de – 4621
basse température,

système à – 0393
bassin 0029
bassin 1086
bassin 2793
bassin annulaire 0030
bassin annulaire 0053
bassin d'alimentation 2380
bassin d'effondrement 1591
bassin de geysers 1404
bassin de réception 2276
bassin de surcreusement 2620
bassin de vallée glaciaire 2620
bassin du cratère 1412
bassin d'un système de chenaux de marée 2508
bassin épicontinental 1086
bassin fermé 2579
bassin fluvial 2192
bassin hydrographique 2192
bassin intramontagneux 1112
bassin lacustre glaciaire 2620
bassin marginal 1136
bassin terminal 2629
bassin versant 2192
bassin versant 3648
bastion 2606
batardeau 4869
batée 4659
batholite 0729
bathyal 2846
bathylimnion 2334
bathypélagique 1910
bathypélagique 2845
bathyséisme 0158
battage de pieux 4838
battage sonnette de – 4838
battance 2758
battement 0178
bauxite 4124
Becke, frange de – 0559
Becke, ligeré de – 0559
bedrock 4788
beekite 3457
beine 2337
beine lacustre 2337
bélonite 0761
Benioff, zone de – 1134
benthique 1907
benthonique 1907
benthonique 2844
benthos 1907
bentonite 3218
berme 2297
berme 4860
bétoire 2076
béton projeté 4801
biaxe 0553
bifurcation 2210
bifurcation 4073
bilan d'eau d'une nappe 3633
bilan matière 3865
biocénose 1896
biochrone 1768
bioclaste 3168
bioclastique 3168
bioclastique 3273
biocoenose 1896
biofaciès 2861
biogenèse 1792
biogénétique 1792
biogénétique, loi – 1968
biogéochimie 0395
bioherme 3331
biolite 3269
biolithite 3269
biome 1895
biomicrite richement fossilifère 3288
bionomie 1887
bios 1804
biosphère 1788
biostratigraphie 1710
biostratigraphique, unité – 1770
biostratinomie 1791
biostrome 3331
biotique 1903
biotitisation 0932
biotope 1902
bioturbation 1885
bioturbation 3067
biozone 1769
biozone, super– 1782
biréflectance 0575
biréfringence 0560
bissectrice 0556
bissectrice aigue 0556
bissectrice obtuse 0556
bitume 3719
bituminisation 0426
bivalve 1859
blasto 0971
blastomylonite 1541
bloc 1087
blocage 2340
bloc anguleux 3242
bloc arrondi 3243
bloc charrié 1569
bloc-diagramme 4258
bloc digéré 0661
bloc erratique 2422
bloc faillé 1531
bloc rocheux 4730
blocs 4876
bloc soulevé 1589
blocs, région compartimentée en – 1586
blocs, région désintégrée en – 1586
bloc surélevé 1589
bloc unitaire 4730
bloc volcanique 1327
boart 0495
boghead 3563
bois flottant 2970
bolson 2580
bombe 1328
bombe allogène 1329
bombe ayant subi une rotation 1329
bombe de péridotite 1329
bombe de ponce 1329
bombe d'olivine 1329
bombe en bouse de vache 1329
bombe en croûte de pain 1329
bombe en forme de galette 1329
bombe en forme de goutte 1329
bombe en fuseau 1329
bombe explosive 1329
bombe fusiforme 1329
bombe rubanée 1329
bombe scoriacée 1329
bombe sphéroïdale 1329
bombe tordue 1329
bombe unipolaire 1329
bone bed 3245
BOP 4289
bordure basique 0707
bordure continentale 2789
bordure figée 0707
bosse 2642
bosse 3539
bosse glaciaire 2625
bosse ruiniforme 2609
bostonitique 0839
botryoïdal 0777
botryoïdal 4094
bouche 1274
bouche 1440
bouche 2196
bouche active 1211
boucle 2232
boucle expérimentale 4565
boucle volcanique 1412
bouclier 1081
boudinage 1609
boue 3210
boue 4160
boue 4829
boue bleue 2921
boue brune 2923
boue calcaire 3295
boue corallienne 3311
boue, coulée de – 2051
boue craquelée 3046
boue hémipélagique 2920
boue marneuse 3222
boue rouge 2923
boue silicieuse 3367
boue sulfureuse 1399
boue sulfureuse, colonne de – 1400
boue sulfureuse, coulée de – 1399
boueuse, coulée – 2974
boueux 3210
boue verte 2921
boue, volcan de – 1395
Bouguer correction de – 0359
Bouguer, couche plane de – 0359
boule 2609
boule d'algues 2891
boule lacustze 2891
boule synthétique 0497
boulette 3094
boulette 4150
boulon 4793
boulonnage 4793
bourrage 4816
bourrelet arqué 2233
bourrelet de poussée 2419
bourrelet de rive 2238
boursoufline 0044
boursouflure 2642
boussole de géologue 4227
boutonnière 4233
boxwork 2118
boyau 2099
brachyanticlinal 1640
brachysynclinal 1640
bradyséisme 1194
bradytélie 1977
brai 3743
branche 1582
branche 1965
branloire 3530
bras, faux – 2212
bras mort 2212
brèçhe 1471
brèche 3252
brèche 3258
brèche calcaire 3279
brèche de broyage 1539
brèche de cratère 1248
brèche d'écroulement 1249
brèche de dessication 3048
brèche de faille 1539
brèche d'effondrement 3263
brèche de friction 1539
brèche de friction 1248
brèche d'épanchement 1268
brèche d'épanchement unie 1268
brèche de projection 1346
brèche d'injection 0711
brèche intraformationnelle 3264
brèche intrusive 0710
brèche ossifère 3261
brèche palagonitique 1347
brèche récifale 3342
brèche récifale 3355
brèche sédimentaire 3258
brèche volcanique 1345
bréchification 2933
Brésilien, essai – 4761
brillance 2677
brillance 3407
brillant 0493
brisant 2811
brisure 2199
brossage, marque de – 3039
broyage 4150
broyage autogène 4147
broyage, circuit de – 4146
broyage, zone de – 1532
broyeur 4149
broyeur à barres 4149
broyeur à boulets 4149
bruit 0097
bruit de fond 4438
bruit du tir 4438
bruit sismique 4438
bruits souterrains 1237
bruits volcaniques 1237
brut 3712
Bullard, méthode de – 0363
bulle gazeuse, marque de – 3062
burinage 2397
butée 4706
butte à lentille de glace 2643
butte, avant – 2561
butte de schorre 2507
buttes gazonnées 2643
butte témoin 2561
butte-témoin karstique 2074

C

cabochon 0500
caillou 3241
caillou à facettes 2162
caillou façonnée par l'eau 2445
cailloutis 2867
cake de boue 4618
calanque 2452
calcaire 2736
calcaire 3305
calcaire à pelotes 3307
calcaire corallien 3310
calcaire dolomitique 3316
calcaire, fragment – 3270
calcaire gréseux 3317
calcaire lithographique 3306
calcaire micritique 3281
calcaire microcristallin 3281
calcaire, non – 2736
calcaire récifal 3310
calcaire, teneur en – 2735
calcarénite 3276
calcarénite fine 3275
calcification 3458
calcilutite 3274
calciner 4151
calcique, série – 0692
calcirudite 3278
calcisiltite 3274
calcite flottante 2144
calco-alcalin 0694
calco-alcaline, série – 0692
caldeira 1456
caldeira d'effondrement 1459
caldeira d'érosion 1473
caldeira d'explosion 1458
caldeira formée par une éruption phréatique 1458
caldeiras, ensemble de – 1457
caldeira simple 1457
caldeira sommitale 1460
caldère 0036
Californien, indice portant – 4708
calorifique, pouvoir – 3604
calotte 1273
calotte continentale 2425
calotte glaciaire 2425
calotte glaciaire locale 2425
calotte sphérique 0363
canal 4412

cannelure 1538
cannelure 3027
cannelure 3033
cannelure crusée par les vagues 2472
cannelure d'avalanches 2362
cannelure glaciaire 2398
cannelures divergentes 3034
canyon sous-marin 2795
caoudeyre 2187
cap 2446
cap 2447
capacité 2958
capacité calorifique 0386
capacité calorifique spécifique 0386
capacité portante 4707
capillaire 0774
capillaire, eau – 2767
capillaire, frange – 2767
capillaire, potentiel – 2770
capture, coude de – 2326
capture fluviale 2322
capture, gorge de – 2326
capture par osculation 2322
capture par tangence 2322
capture, théorie de la – 0047
capuchon 1696
caractéristique directionelle 4491
carapace 1273
carapace 4122
carat 0482
carat, quart de – 0504
carbonatation 2000
carbonaté 2736
carbonates, profondeur de compensation des – 0423
carbonatisation 0932
carbone-azote, rapport – 0425
carbone-14, datation – 0451
carbone fixe 3601
carbone fixe, pourcentage de – 3601
carbone, nombre de – 3727
carbonification 3519
carbonification biochimique 3519
carbonification géochimique 3519
carbonification, gradient de – 3522
carbonisation 3607
carcasse 1861
carcasse 2588
carottage continu des vitesses 4406
carottage sismique 4406
carotte 3155
carotte 4237
carotte, arrache– 4272
carotte continu 4749
carotte, diviseur de – 4182
carotte intégrale 4750
carotte, récupération de – 4273
carottes, caisse de – 4237
carottier 3155
carottier à foncer 4746
carottier à piston 3156
carottier à piston 4746
carottier à vibropiston 3156
carottier par battage 4747
carottier par gravité 3156
carottier par gravité 4747
carottier, tube – 4748
carrière 4874
carte de faciès 4245
carte de la résiduelle du champ de pesanteur 4542
carte des courbes structurales 4247
carte des isochores 4244
carte des isoplèthes 4246
carte des sols 2671
carte des sols à grande échelle 2671
carte des sols à petite échelle 2671
carte des sols détaillée 2671
carte d'isocorrection isostatique 0373
carte géologique 4239
carte géoloqique pour l'ingénieur 4784
carte gravimétrique régionale 0371
carte isanomale 0372
carte isobare 3854
carte isogale 0372
carte isogamme 0372
carte isopaque 4243
carte métallogénique 3905
carte paléogéologique 4242
carte pédologique 2671
carte pédologique de synthèse ·2671
carte, photo– 4383
carte radar 4400
carte structurale 4247
cartographie 4239
cartographie des sols 2670
cascade 2203
cascade de lave 1287
cash flow annael 3894
cash flow courbe 3895
cash flow cumuli 3894
cash flow graphique 3895
cash flow total 3894
cassant 0471
cassure 0469
cassure 0541
cataclase 0914
cataclasite 0914
cataclinale 2288
cataclysme 0005
catagenèse 1977
catagenèse 3419
cataracte 2204
cataracte de feu 1287
cataracte de glace 2377
cataséisme 0187
catastrophisme 0005
catégorie 1927
catena 2674
cations, capacité d'échange de – 4623
cations échangeables 2755
caustobiolité 0427
cave d'observation sismologique 0116
caverne 2095
caverne suivant une diaclase 2096
caverneux 1305
caverneux 3064
cavitation 2120
cavitation 4512
cavité 4819
cavité 3940
cavité 4014
cavité anastomosée 2115
cavité miarolitique 0873
cavité, remplissage de – 3940
cavité suivant une faille 2096
caye de sable 2443
CBR 4708
CD 0454
ceinture 3908
ceinture mobile 1093
ceinture plissée 1674
cellulaire 3064
cellule triaxiale 4753
celluleux 0874
cendre de salissure 3600
cendre extrinsèque 3600
cendre intrinsèque 3600
cendres 3599
cendres de constitution 3553
cendre sédimentaire 3600
cendres epigénétiques 3553
cendres, pluie de – 1363
cendres secondaires 3553
cendre volcanique 1317
cénotypique 0636
cénozone 1774
centre d'émission, déplacement du – 1231
centre émissif cylindrique 1224
centre émissif sans racine 1223
centripète 3107
cercle de pierres 2653
chaîne de sols 2674
chaîne de volcans 1448
chaîne monoclinale 2541
chaleur de combustion 3604
chaleur de la terre 0374
chaleur due à la radioactivité 0375
chaleur terrestre 0374
chaleur, transfert massique de – 0381
chambre 4849
chambre à défilement continu 4332
chambre de prise de vue aérienne 4329
chambre magmatique 1202
chambre magmatique en forme de dyke 1204
chambre métrique 4329
chambre métrique 4331
chambre multibande 4330
chambre multiobjectifs 4330
champ de boue 1390
champ de fumerolles 1380
champ de gaz 3773
champ de lave 1278
champ de pierres 2057
champ de solfatares 1392
champ filonien 4061
champ magnétique, intensité de – 0317
champ minéralisé 3909
champ minier 3909
champ pétrolifère 3773
chandelle 2140
Chandler, agitation – 0200
Chandler, oscillation – 0200
chantoir 2075
chapeau 1696
chapeau de fer 4125
charactéristique textural 4108
charbon 3496
charbon à gaz 3559
charbon bitumineuse 3559
charbon brillant 3580
charbon cokéfiable 3608
charbon, couche de – 3499
charbon, couches à – 3498
charbon de bois 3607
charbon demi-gras 3559
charbon de terre 3496
charbon en galets 3514
charbon, équivalent – 3606
charbon feuilleté 3556
charbon feuilleté 3563
charbon fibreux 3580
charbon flambant 3559
charbon flambant sec 3559
charbon, galets de – 3512
charbon gras 3559
charbon humique 3497
charbon maigre 3560
charbon mat 3580
charbon menu 3513
charbon minéral 3496
charbon mixte 3497
charbon, pétrographie du – 3565
charbon piciforme 3557
charbon, qualité du – 3551
charbon, rang du – 3552
charbon remanié 3512
charbon sapropélique 3497
charbon semi-brillant 3580
charbon, texture du – 3565
charbon, veine de – 3499
charge 2950
charge 2954
charge 4816
charge 4140
charge constante, essai à – 3711
charge de fond 2952
charge de fond 4817
charge, différence de – 4735
charge dissoute 2956
charge en suspension 2955
charge, figures de – 3059
charge hydraulique 4735
charge solide 2951
charge solide 2951
charge variable, essai à – 3711
charnière frontale 1687
charnière, ligne – 1629
charnière, point – 1629
charnière, zone – 1629
charnockitique, série– 0990
charpente 3079
charriage, distance de – 1568
charriage du premier genre 1576
charriage, surface de – 1568
chatière 2098
chatoyant 0488
chaudière 2028
chaudron 1463
chaudron, subsidence en – 0743
chaudron, subsidence en – 1186
chaussée 4840
chaussée, corps de la – 4841
chaussée de géants 1301
cheire scoriacée 1260
chélate métallo-organique 0410
chélate organométallique 0410
chélation 0409
chélation organo-métallique 0409
cheminée 1221
cheminée 2076
cheminée de fée 2029
cheminée minéralisée 4037
chenal 0042
chenal 2191
chenal 3022
chenal à méandres 2217
chenal d'eau de fonte 2393
chenal de flot 2510
chenal de jusant 2510
chenal de marais 2517
chenal de marée 2509
chenal minéralisé 4029
chenal, remblaiement de – 3022
chenal sinueux 2217
chenal transversal de bas de plage 2486
chenière 2491
chevauchante, masse – 1569
chevauchée, masse – 1569
chevauchement 1568
chevauchements, plan basal des – 1572
chevauchement, surface de – 1568
chevron 2378
chevron 2562
chevron, marque en – 3036
chevron, trace en – 3036
chiest 2609
chironyme 1956
chloritisation 0932
choc, zone de – 0894

chondrite 0072
chondrule 0072
chonolite 0732
chorismite 0794
chorismité 0791
chorismitique 0794
chroma 2677
chrone 1761
chronologie géologique 1713
chronomère 1760
chronostratigraphie 1712
chronostratigraphique, horizon – 1758
chronostratigraphique, unité – 1757
chronotaxe 1732
chronozone 1758
chute 2203
chute 4735
chute d'eau 2203
chute de bancs 2110
cierge 2140
ciment 3433
cimentation 3433
cimentation, facteur de – 4606
cimentation intergranulaire 3437
cimentation par nourrissage des grains 3437
ciment, coulis de – 4803
ciment de type géode 3438
cimenté 2733
ciment grenu 3439
cinérite 1333
CIPW norm 0686
circalittoral 2839
circulation directe 4283
circulation inverse 4283
circulation perdue 4284
circulation thermohaline 2829
cire 3723
cire de lignite 3736
cire de montagne 3736
cire fossile 3735
cirque composé 2613
cirque d'érosion 1473
cirque de tête de vallée 2611
cirque en escalier 2614
cirque en gradins 2614
cirque étagé 2614
cirque glaciaire 2610
cirque, grand– 2613
cirque, seuil de – 2610
cirques, niveau de – 2615
cirques, niveau de fonds de – 2615
cirques, plate-forme de – 2613
cisaillement 0105
cisaillement 4677
cisaillement, boîte de – 4754
cisaillement minéralisé, zone de – 4060
cisaillement, module de – 4699
cisaillement, plan de – 0539
cisaillement, zone de – 1533
cladogenèse 1962
clan 0687
clarain 3580
Clarke, valeur de – 0399
classe 0685
classe 1929
classé 3202
classé, bien – 3202
classe cristalline 0512
classe de sols 2666
classement 3200
classement, coefficient de – 3201
classificateur à râteau 4156
classificateur à spirale 4156
classification 1921
classification biostratigraphique 1710
classification chronostratigraphique 1712
classification des réserves 3878
classification des sols 2661
classification des terrains 4785
classification granulométrique 2713
classification lithostratigraphique 1709
classification naturelle 1921
clavalite 0763
claviature 3798
clayat 3504
clé de voûte 4851
cliché 4358
clinolimnion 2334
clinomètre 0124
clinomètre 4227
clivabilité 1596
clivable 1596
clivage 0469
clivage 1596
clivage ardoisier 1599
clivage ardoisier 1610
clivage cassant 1601
clivage, plan de – 0540
clivage, plan de – 1598
cloche 2113
cloche de dissolution 2106
cloison 1451
cloison 2102
cluse 2545
cluse morte 2327
cluse sèche 2327
cluse vive 2327
coagulation 0413
coal ball 3507
coccolithe 1840
coenozone 1774
coercivité 0307
coeur 1628
cohésion 4693
cohésion apparente 4693
coin comblée, fente en – 2641
coin de glace 2640
coin, marque en – 3037
coins de retenue 4280
coke 3608
cokéfiant, pouvoir – 3596
coke naturel 3564
col 2226
col de diffluence glaciaire 2621
collant 2750
collecteur 3798
colliers de serrage 4280
colline sous-marine 2800
collision de plaques continentales 1131
colloforme 3093
colluvions 2868
colmatage 2341
colmatage 2513
colmatage argileux 2122
colonie 1806
colonnade volcanique 1301
colonne 2138
colonne de fumée 1361
colonne de production 4275
colonne de soufre 1400
colonne d'hydrocarbure, hauteur de la – 3775
colonne lavique 1246
colonne magmatique 1226
colonne minéralisée 4005
colonnes, structure en – 2730
coloration, indice de – 0688
coloré artificiel 0492
comagmatique 0629
combe 2544
comblement 2245
comblement 2341
comblement par végétation 2341
comblement sableux 2990
comblement silteux 2990
combustion in situ 3838
commensalisme 1899
comminution 4150
communauté 1890
commutation 4489
compacité 3082
compactage 4679
compactage 4808
compaction 3428
compaction chimique 3432
compaction différentielle 3428
compaction mécanique 3429
compaction par percolation 3416
compaction, soudure par – 3446
comparateur 4767
compartiment 0362
compartiment effondré 1559
compensation des mouvements 4511
compensation isostatique 0339
compensation isostatique des masses 0342
compensation isostatique des pressions 0344
compensation locale 0340
compensation, niveau de – 0345
compensation, profondeur de – 0345
compensation régionale 0340
compensatrice, masse – 0341
compétence 2958
compétent 1624
complétion 3795
complexation 0408
complexe 1756
complexe 1928
complexe annulaire 0742
complexe carbonaté 3324
complexe des sols 2673
complex intrusif 0620
complexité 0275
composant volatil 0650
composant volatil 1172
composition colorée 4346
composition minéralogique virtuelle 0606
composition modale 0593
composition photographique 4357
compressibilité 4698
compression 0187
compression, module de – 4698
compression radiale, essai de – 4763
compteur de points 0612
conca 1459
concasseur à cône 4148
concasseur à mâchoires 4148
concentration 3942
concentration de minerai alluvionnaire 3943
concentration gravimétrique 4153
concentration résiduaire 3943
concentration résiduaire par déflation 2967
concentré 4141
concordance, en – 1719
concordance stratigraphique 1719
concordant 0717
concordant 1719
Concordia-Discordia, courbe – 0435
concrétion 2124
concrétion 2732
concrétion 3472
concrétion 4093
concrétion marneuse 3486
concrétionnaire 3472
condensat 3713
condensation 0412
condensation nucléaire 1360
condensation rétrograde 3841
condominant 1816
conduction de chaleur 0381
conductivité électrique 4556
conductivité hydraulique 3700
conductivité thermique 0382
conduite enterrée 4846
conduite forcée 4867
conduit naturel 3638
cône 0059
cône 1435
cône 4631
cône adventif 1437
cône central 1443
cône d'agglutinat 1298
cône de boue 1396
cône de cendre 1427
cône de débris 1427
cône de lave 1420
cône de lave 1427
cône de scories 1427
cône de transition 2423
cône de tufs 1428
cône double emboité 1445
cone, formation de – 3826
cône intérieur 1443
cône littoral 1300
cône rocheux 2585
cône sheet 0742
cône simple 1441
cône volcanique 1434
confiance, intervalle de – 4191
confluence à gradin 2208
confluence, gradin de – 2624
confluent 2208
confluent en crochet 2209
confluent en hamecon 2209
confluent entraîné 2209
conforme 0719
congélation 4807
congénérique 1943
congère 2353
conglomérat 3251
conglomérat à galets d'argile 3235
conglomérat de base 3114
conglomérat grossier 3254
conglomérat intraformationel 3257
conglomératique 3251
conglomérat volcanique 1343
conodonte 1860
Conrad, discontinuité de – 0086
consanguin 0629
consanguinité 0629
conséquent 2284
consistance 2746
consistance 4713
consistance, indice de – 4716
consistence, limite de – 4714
consolidation 3422
consolidation 4703
consolidation, anneau de – 4756
consolidation préalable 4703
consolidé 3422
conspécifique 1943
construction anti-sismique 0260
construction, code de – 0256
contact linéaire grain à grain 3449
contact rectiligne euhédral 3449
contact suturé 3449
contact, zone de – 0899
contamination 0681
contamination 0682
contamination 3672
contamination 4652
contemporain 1730
contemporain 1818
continental 2865

continent lunaire 0050
continuité 4726
continuité, en – 1719
continuité stratigraphique 1719
contortion 3051
contraction, facteur de – 3860
contraction, hypothèse de la – 1114
contrainte 0103
contrainte 4667
contrainte de cisaillement 4688
contrainte-déformation, graphique – 4667
contrainte differentielle 0103
contrainte effective 4668
contrainte normale 4688
contrainte principale 0103
contrainte principale 4692
contraintes naturelles en forage, essai de libération des – 4766
contraintes, relâchement des – 0107
contrainte tangentielle 4688
contraste 4343
contraste 4628
contraste objet 4343
contrefort 2505
contrefort 2606
contrefort 4797
contrepente 2626
contrôle au sol 4323
contrôle photogrammétrique 4324
contrôle, point de – 4325
convection de chaleur 0381
convection du manteau 1124
convergence 1725
convergence 1973
convergence 2787
convergence adaptive 1973
convergence métamorphique 0913
convolution 3052
convolution 4452
convolution, intraformationelle 3051
co-occurence, zone de – 1777
coordinance 0408
coprolite 3094
coprolithe 1863
coral ahermatypique 3335
coral hermatypique 3335
cordillère 1103
cordon appuyé 2462
cordon en boucle 2465
cordon en V 2465
cordon libre, île de – 2468
cordon littoral 2461
cordon littoral 2463
corne 0935
corne 2649
cornéenne 0940
corniche circulaire 1415
corniche de neige 2356
coronite 0786
corps minéralisé 3910
corrasion 2014
corrasion 2015
corrasion 2938
corrasion éolienne 2154
correction à l'air libre 0358
correction d'altitude 0357
correction de latitude 0355
correction de la zone altérée 4430
correction de l'influence luni-solaire 0365
correction d'ellipticité 0098
correction de plan de référence 4429
correction de plateau 0356
correction de station 0160
correction dynamique 4432
correction statique 4431
correction topographique 0360
correction topo-isostatique 0364
corrélateur optique 4388
corrélation 1716
corrélation 4459
corrélation, hétéro – 4450
corrélogramme 4195
corrosif 3716
corrosif, non – 3716
corrosion 0642
corrosion 2942
corrosion 2017
corrosion, figure de – 0530
corrugation 3051
cosmogonie 0003
co-spécifique 1943
côte 0518
côté 1625
côte 2437
côte 2558
côte à baies 2533
côte abritée 2336
côté affaissé 1559
côte affaissée 2524
côte à mouvement négatif 2525
côte à mouvement positif 2525
côte à structure longitudinale 2521
côte à structure transversale 2522
côte à type dalmate 2521
côte, au large de la – 2840
côte, avant– 2485
côte, avant– 2840
côte composée 2527
côte concordante 2521
côte contraposée 2527
côte crénelee 2532
côte, front de – 2558
côte d'abrasion 2528
côte d'accumulation 2528
côte d'affaissement 2524
côte déchiquetée 2532
côte de faille 2520
côte de flexure 2519
côte de lave 1280
côte de ligne de faille 2520
côte d'émersion 2525
côte d'érosion marine 2528
côte de soulèvement 2524
côte de submersion 2525
côte d'immersion 2525
côte d'ingression 2529
côte discordante 2522
côte du type atlantique 2522
côte du type mixte 2526
côte du type pacifique 2521
côte échancrée 2533
côte épigénique 2527
côte exposée 2336
côte indentée 2532
côte, ligne de – 2437
côte lobée 2533
côte longitudinale 2521
côte oblique 2522
côte polycyclique 2530
côte rectiligne 2531
côté relevé 1559
côte soulevée 2524
côte stable 2526
côte, trait de – 2437
côte transversale 2522
cotype 1951
couche 1750
couche 2688
couche 3097
couche active 2647
couche annuelle 3127
couche arable 2699
couche basale 2896
couche caractéristique 3500
couche de base 4841
couche de charbon 3123
couche de fondation 4836
couche de fondation 4841
couche de lave 1277
couche de liaison 4842
couche de surface 4842
couche d'ossements 3245
couche durcie 2734
couche dure 2734
couche en deux passées 3506
couche filtrante 4862
couche frontale 2896
couche granoclassée 3130
couche guide 1751
couche imperméable 3640
couche minéralisé 4021
couche minéralisée 3123
couche repère 1751
couche repère 3500
couche rouge 2879
couche rouge, gisement – 4129
couches basculées 1482
couches disloquées 1523
couche semi-perméable 3639
couches inclinées 1482
couche sommitale 2896
couches redressées 1481
coulée 2050
coulée boueuse 2051
coulée boueuse 2974
coulée de blocaille 2057
coulée de blocs 2057
coulée de boue 2884
coulée de boue 1352
coulée de boue 1397
coulée de cendres incandescentes 1369
coulée de pierres 2057
coulée pierreuse 2057
coulée volcanique 1263
couleur 0546
couleur de la trace 0466
couloir 2474
couloir 2796
couloir de lavage 4659
couloir interdunaire 2186
couloir interdunaire 2188
Coulomb, critère de – 4689
coup de gouge 2398
coup de terrain 4857
coupe de sondage 4256
coupe généralisée 4254
coupe géologique 4250
coupe idéalisée 4254
coupellation 4184
coupes en réseau 4252
coupe sériée 4251
coupe sismique 4458
coupe standardisée 4257
coupe synthétique 4253
couplage 0273
couplage réactif 0133
couple stéréoscopique 4362
coupole 0724
coupole 1205
coupole 1422
coupole 1641
coupole, eau soulevée en – 1359
coupole ellipsoïdale 1640
coupole salifère 1695
coupure 4643
courant alternatif 4555
courant boueux chaud 1353
courant continu 4555
courant continu de turbidité 2981
courant de flot 2822
courant de jusant 2822
courant de marée 2821
courant dense 2979
courant de retour 2826
courant de surface 2785
courant de turbidité 2981
courant de turbidité, partie frontale d'un – 2982
courant de turbidité, queue d'un – 2982
courant électrique, densité de – 4556
courant électrique, intensité de – 4556
courant hélicoidal 2264
courant laminaire 2264
courant, ligne de – 2268
courant, ligne de – 3708
courant littoral de houle 2825
courant marin 2785
courant pulsé 4555
courant tellurique 4552
courant torrentueux 2252
courant, trace de – 3020
courant turbide stratifié 2981
courant turbulent 2264
courant, vitesse du – 2260
courbe de niveau 4248
courbe pi 1020
couronne 0362
couronne 4271
couronne réactionnelle 0786
cours d'eau capteur 2325
cours d'eau de fonte 2388
cours d'eau, développement du – 2197
cours d'eau intérieur 2388
cours d'eau secondaire 2207
cours d'eau sous-glaciaire 2388
cours inférieur 2195
cours moyen 2195
cours supérieur 2195
couverture 4407
couverture 4818
couverture, coefficient de – 4515
couverture de grès 3109
couverture, hauteur de – 4788
couverture multiple 4407
couverture polaire 0040
couverture, roche – 3764
couverture, roche – 3765
couverture stéréoscopique 4361
couverture, terrain de – 4788
covariogramme 4196
craie 3315
cran 2483
craquelure thermique 1993
cratère 0054
cratère 1412
cratère 4819
cratère à bouches multiples 1445
cratère actif 1211
cratère adventif 1439
cratère citerne 1455
cratère d'effondrement 1455
cratère de météorite 0075
cratère d'explosion 1431
cratère d'impact 0027
cratère d'impact primaire 0027
cratère d'impact secondaire 0027
cratère double 1442
cratère égueulé 1471
cratère, enceinte du – 1451
cratère endormi 1155
cratère en phase de sommeil 1155
cratère, micro– 3066
cratères, alignement de – 1449
cratères, chaîne de – 1449
cratères jumelés 1442
cratère sommital 1436
cratère subterminal 1438
cratère terminal 1436
craton 1080

crayification 3415
creek 2517
crénelure 1643
crescumulat 0834
crête 1633
crête 2486
crête 2490
crête anticlinale 1632
crête anticlinale 2541
crête-crête 0164
crête de partage 2316
crête de poussée 2419
crête fumante 2158
crête monoclinale 2541
crête, plan de – 1632
crêtes, uniformité du niveau des – 2598
crête synclinale 2541
crêt monoclinal 2562
creusement 2020
creusement 2022
creusement au bouclier 4853
creusement des vallées 2270
creux 2247
creux 2808
creux de déflation 2187
crevasse 1465
crevasse 2375
crevasse de gel 2640
crevasse de tremblement de terre 2552
crevasse latérale 2376
crevasse longitudinale 2376
crevasse marginale 2376
crevasse sismique 2552
crevasse transversale 2376
criblage 4873
crique 2474
cristal 0506
cristal corrodé 0756
cristal creux en forme de tube 2131
cristal cumulus 0830
cristal de sable 3443
cristallin, défaut – 0531
cristalline, forme – 0753
cristalline, forme – 0520
cristalline, roche – 0583
cristalline, structure – 0509
cristallinité 0791
cristallin, réseau – 0507
cristallin, système – 0514
cristallisation 0632
cristallisation, force de – 0909
cristallisation fractionnée 0670
cristallisation hétéradcumulaire 0833
cristallisation, indice de – 0702
cristallisation mimétique 1008
cristallisation rhytmique 0864
cristallisé, mal – 0797
cristallite 0760
cristallitique 0760
cristalloblastèse 0908
cristalloblastique, série – 0909
cristallographie 0505
cristal regénéré 0910
croc 2190
crochon 4079
croisement isogonal 4075
croisement oblique 4075
croisement orthogonal 4075
croiseur 4077
croissance 3455
croissance 2513
croissance épitaxique 3456
croissance, figure de – 0528
croissance, pression de – 0909
croissance, secteur de – 0526
croissance, spirale de – 0529
croissant 2233
croissant de lit majeur 2234
croissant de plage 2493
crotovina 2745
crotovine 2745
croûte 2759
croûte calcaire 4123
croûte de fusion 0073
croûte plissée 0058
croûte rigide 0348
croûte terrestre 0086
cryoconite 2387
cryonivellement 2650
cryopédologie 2634
cryoturbation 2647
cryptoclastique 3166
cryptocristallin 0800
cryptogène 1976
cryptographique 0848
cryptomacéral 3567
cryptomagmatique 3986
cryptovolcanisme 1181
cryptozoon 1844
crystallophyllienne, roche – 0880
cubique 0515
cuesta 2558
cuesta, front de la – 2558
cuesta, revers de la – 2558
cuirasse 4122
cuisson 0897
culée 4845
culée 4866
cumulat 0830
cumulite 0765
cumulus 0830
cunéiforme 4080
cupule 2086
curette 4278
curie 4550
Curie, point de – 0308
cutane 2691
cuve 2248
cuve d'effondrement 1455
cuvette 1641
cuvette 2517
cuvette à sel 2517
cuvette de déflation 2187
cuvette de failles 2547
cuvette d'effondrement 1591
cuvette de surcreusement 2620
cuvette terminale 2629
cuvette thermokarstique 2646
cyclanes 3717
cycle aride 2567
cycle d'érosion 2566
cycle d'érosion glaciaire 2567
cycle d'érosion interrompu 2571
cycle d'érosion littorale 2567
cycle d'érosion périglacial 2567
cycle de sédimentation 3119
cycle fluviatile 2567
cycle géochimique 0396
cycle hydrologique 3612
cycle karstique 2567
cycle littoral 2567
cycle marin 2567
cycle naturel de l'eau 3612
cycle orogénique 1105
cycle pétrogénétique 0582
cycles manqués, présence de – 4597
cycle volcanique 1156
cyclothème 3119
cylindrage 4809
cyste 1841

D

dalle 3132
dalles, en – 3132
damage 4809
darcy 3697
Darcy, loi de – 3696
Darwinisme 1966
datation absolu 0429
datation isotopique 0429
datation radiométrique 0428
DBO 3678
DCO 3678
débimètre 3669
débit 2249
débit columnaire 1501
débit de sécurité 3668
débit d'esquilles 2941
débit en bâtons de craie 1610
débit en dalles 1500
débit en plaques 1500
débit prismatique 1501
débit rhomboédrique 1502
débit solide 2951
débit solide 2957
déblai 4824
déblai 4829
débordement 2251
débordement, niveau de – 2251
débordement transgressif 1727
débordement transgressif 3113
débris 2005
débris 3166
débris calcaire 3167
débris calcaire 3270
débris, champ de – 2006
débris de forage 4238
débris de plantes 3168
débris dolomitique 3167
débris récifaux 3355
débris superficiel, manteau de – 2008
débris volcanique 3170
début brusque 0173
début peu marqué 0174
décalage 4502
décalage de couches 1523
décalcification 2679
décantation 2991
décantation clastolitique 1201
décantation cristalline 0673
décantation, vitesse de – 2991
décapage 2936
décaper 4828
décapitation fluviale 2323
décarbonatation 2679
décharge 4831
décharge contrôlée 4831
décharge en tas 4831
déclenchement 1663
déclin 3816
déclinaison magnétique 0284
déclin, courbe de – 3816
déclin, méthode des courbes de – 3874
décollement 1663
décollement 2110
décollement 2473
décollement, structure de – 3053
décomposition 2741
décomposition 2929
décomposition 3526
décomposition aérobié 2932
décomposition anaérobié 2932
décomposition arénacée 1990
décomposition chimique 1998
décomposition, désagrégation et – 1988
décomposition en boules 2003
décomposition humique 1999
décomposition sélective 2004
décompression 3850
décompression 4731
décomprimée, zone – 4731
déconvolution 4454
découpage contrôlé 4823
découpage fin 4823
découplage 0273
découverte 3884
découverte 4875
dédolomitisation 0930
déferlement 2812
déferlement frontal 2811
déferlement glissant 2811
déferlement plongeant 2811
déferlement, zone de – 2439
déferlement, zone de – 2812
déflagration sonique 0177
déflation 2155
déflation, cuvette de – 2187
déflation, résidu de – 2160
déflectomètre 4294
défluent 2211
déformation 0104
déformation 1075
déformation 4666
déformation, axe de la – 1016
déformation, bande de – 0539
déformation-contrainte, courbe – 0106
déformation de compaction 3446
déformation en lamelles 0999
déformation, module de – 4696
déformation par glissement 1704
déformation pénécontemporaine 3050
déformation, plan de – 0238
déformation plastique d'enfouissement profond 3447
déformation plastique précoce 3447
déformation relative 4666
déformètre 4767
dégagement gazeux explosif 1217
dégénérescence 1982
déglaciation 2367
délavage 4873
délitement 2110
délit schisteux 1599
delta 2494
delta 2895
delta arqué 2495
delta construit par les vagues débordantes 2500
delta de fond de baie 2498
delta de lave 1280
delta de marée externe 2499
delta de marée interne 2499
delta de tempête 2500
delta digité 2495
delta en patte d'oie 2495
delta en pointe 2495
deltaïque 2895
delta lobé 2495
delta par rupture de cordon 2500
delta saillant 2497
delta, sous– 2496
delta, valeur de – 0454
déluge de lave 1276
demi-anthracite 3560
demi-période 0431
demoiselle 2029
démultiplexage 4424
dendrite 0525
dendrite 3400
dendrite 4088
dendritique 1821
dendrochronologie 1737
dendroïde 1821
dense 0799
densité 4342
densité 4537
densité apparente 2751
densité apparente 3602
densité, courant de – 2979
densité de surface de contact 4111
densité in situ 4614
densité vraie 3602

dénudation 2012
dénudation 2013
dénudation, niveau supérieur de – 2598
dépenses 3893
déphasage 0134
déphasage 4502
déplacée 4637
déplacé en bas 1559
déplacement miscible 3840
déplacement piston 3826
déplacement subit du lit 2231
déplacement véritable 1562
déplétion 3830
dépocentre 2986
dépôt 2985
dépôt calcaire 3267
dépôt carbonaté 3266
dépôt clastique 3164
dépôt d'eau de fonte 2402
dépôt de couverture 3996
dépôt de lahar 1356
dépôt de mer profonde 2922
dépôt de plaine d'inondation 2885
dépôt en nappe 3109
dépôt fluvioglaciaire 2402
dépôt glaciaire 2402
dépôt glacio-lacustre 2402
dépôt houiller 3493
dépôt intertidal 2907
dépôt pelliculaire 2142
dépôt, processus de – 2985
dépôt pyroclastique 3170
dépôt regressif 3112
dépôt résiduel 2930
dépôt secondaire 2993
dépôt souterrain 2121
dépôt tidal 2907
dépôt transgressif 3113
dépôt volcanique clastique 1310
dépôt volcanoclastique 1310
dépression 1635
dépression 2189
dépression annulaire 1185
dépression, cône de – 3667
dépression cryptovolcanique 1183
dépression entre arcs 1137
dépression karstique 2071
dépression latérale 2239
dépression périphérique 1697
dépression périphérique 2560
dépression volcano-tectonique 1182
déréverbération 4454
dérive 2961
dérive 4540
dérivé 2947
dérive côtière 2962
dérive des continents 1115
dérivée seconde 4534
dérivée seconde 4544
dérive littorale 2825
dérive littorale 2962
derrick 4267
désagrégation 2929
désagrégation en blocs 1990
désagrégation en boules 1304
désagrégation en grains 2934
désagrégation et décomposition 1988
désagrégation granulaire 1990
désagrégation mécanique 1989
désagrégation physico-chimique 1988
désagrégation physique 1989
désagrégation sélective 2004
désagrégation thermique 1992
désaimantation par champ alternatif 0321
desaimantation thermique 0321
descendance 1964
déschlammer 4161
déséquilibre radioactif, datation par – 0443
désert 2577
désert de lave 1279
désert de pierres 2578
désert de sable 2578
désert rocheux 2578
déservoir de crue 4868
déshydratation 3797
désilicification 0929
désintégration, constante de – 0431
desmosite 0933
désorption 0407
désoxydation 0416
desquamation 1994
desquamation 2935
dessication, amorce de polygone de – 3046
dessication, craquelure de – 3044
dessication, polygone de – 3046
désulfuration 0415
détecteur 4413
détection, capacité de – 0270
détection, limite de – 4651
détection, seuil de – 0270
détente 1179
détonateur 4816
détonation 1217
détonation 4815
détournement fluvial 2213
détritique 3165
détritique 3997
détritique, matériel – 3165
deutérique 0658
développée, zone – 3890
développement enchevêtré 4098
développement parallèle 1971
déversement 4351
déversement fluvial 2213
déversement relatif 4351
déversoir 2339
deviation 4079
déviation de gaz, facteur de – 3862
dévier 4292
dévitrification 0919
diablastique 0950
diachrone 1736
diachronisme 1736
diaclase 1495
diaclase 4723
diaclasé 1495
diaclase croisée 1506
diaclase de cisaillement 1504
diaclase de décompression 1506
diaclase de tension rempli 1519
diaclase d'extension 1503
diaclase diagonale 1507
diaclase directionnelle 1506
diaclase longitudinale 1507
diaclase oblique 1507
diaclase, plan de – 1496
diaclase principale 1509
diaclase satellite de cisaillement 1505
diaclases conjuguées, système de – 1499
diaclases, dessin des – 1508
diaclase secondaire 1509
diaclases, fréquence des – 4725
diaclases parallèles, groupe de – 1498
diaclases, réseau de – 1498
diaclases, système de – 1498
diaclase transversale 1507
diadochie 0477
diadysite 1046
diagenèse 3388
diagenèse avancée 3418
diagenèse de compaction 3427
diagenèse de nappe phréatique 3399
diagenèse d'enfouissement 3417
diagenèse de percolation 3399
diagenèse et métamorphisme de la matière organique 3521
diagenèse liée au milieu 3398
diagenèse précoce 3418
diagenèse très avancée 3418
diagenétique 3974
diagénétique, domaine – 3390
diagénétique, milieu – 3391
diagénétique, milieu précoce – 3393
diagénétique, milieu profond – 3393
diagénétique, milieu superficiel – 3392
diagramme beta 1021
diagramme circulaire 4259
diagramme en rose 4259
diagramme en rosette 4259
diagramme polaire 4259
diagraphic inverse de résistivité 4585
diagraphie 4578
diagraphie acoustique 4597
diagraphie de densité 4593
diagraphie de diamétreur 4603
diagraphie de neutrons 4594
diagraphie de potentiel spontané 4579
diagraphie de radioactivité naturelle 4590
diagraphie de résistivité 4582
diagraphie de résistivité focalisées, sonde de – 4586
diagraphie de température 4601
diagraphie gamma 4548
diagraphie latérale de résistivité 4585
diagraphie par induction 4584
diagraphie, sonde de – 4578
diamagnétisme 0298
diamètre efficace 3196
diamètre équivalent 3195
diamètre modal 3199
diamètre moyen 3199
diamictite 3124
diaphthorèse 0904
diapirisme 1693
diapir, pli – 1693
diaplectique 0895
diastème 1724
diastème 3125
diastrophisme 1074
diatomée, frustule de – 3370
diatomées, boue à – 3370
diatomées, schiste à – 3371
diatomées, silex à – 3371
diatomées, terre à – 3370
diatomite 3370
diatrème 1221
diatrème 1477
Diesel, combustible pour moteur – 3744
différenciat 0667
différenciation 0398
différenciation 0667
différenciation diagénétique 3475
différenciation diagénétique 3938
différenciation géochimique 0398
différenciation, indice de – 0701
différenciation magmatique 3938
différenciation métamorphique 0911
différenciation par cristallization fractionée 0670
différenciation par densité 0672
différenciation par diffusion 0680
différenciation par gravité 0672
différenciation par immiscibilité 0678
différenciation par transfert gazeux 0679
différenciation pneumatolitique 0679
différencié 0667
diffluence, gradin de – 2624
diffraction 0217
diffraction 4444
diffraction, courbe de – 4464
diffusion capillaire 3624
diffusion métamorphique 0912
diffusivité thermique 0387
digitation 1687
digue en terre 4859
dilatance 0109
dilatance 4678
dilatation 0187
dilatation, essai de – 3596
dilatomètre 4760
dilution isotopique, analyse par – 0432
dimorphisme 1811
dipôle axial, champ d'un – 0279
dipôle excentrique, champ d'un – 0279
diramation 2210
direction 1483
direction 4016
direction, azimut de la – 1483
direction générale 1493
direction, sens de la – 1483
directions et pendages, observations des – 4228
discoidal 3184
disconformité 1722
discontinuité 4723
discontinuité stratigraphique 1720
discontinuité, surface de – 0099
discordance angulaire 1721
discordance de ravinement 1722
discordance, en – 1720
discordance érosionelle 1722
discordance plate 1721
discordance stratigraphique 1720
discordant 0718
discordant 1720
dislocation 1522
dislocation-coin 0531
dislocation, théorie de la – 0046
dislocation-vis 0531
disparaître en haut 1526
disparaître en profondeur 1526
disparition sous 1614
dispersion 0197
dispersion 0218
dispersion 0219
dispersion 0403
dispersion 0572
dispersion 3710
dispersion 4316
dispersion 4514
dispersion, aire de – 2948
dispersion, à l'envers – 0197
dispersion, auréole de – 0403
dispersion, coefficient de – 4190
dispersion, courbe de – 0226
dispersion croisée 0572
dispersion des axes optiques 0572
dispersion diagonale 0573
dispersion droite 0572
dispersion, figure de – 0403
dispersion horizontale 0573

dispersion normal 0197
dispersion normale 0572
dispersion oblique 0573
dispersion primaire 0404
dispersion renversé 0197
dispersion secondaire 0405
dispositif transversal 4419
disposition structurale 1480
dissection 2019
dissémination 3947
dissipation 0219
dissolution, coupole de – 2106
dissolution du ciment 3444
dissolution intraformationnelle 3413
dissolution par pression 3445
dissolution, vide de – 3690
distal 2997
distribution aléatoire 4174
distribution lognormale 4189
distribution normale 4189
distribution polymodale 3197
district minéralisé 3909
district minier 3909
divarication 2210
divergence 1969
diversion fluviale 2213
diviain 3580
diviseur à riffles 4182
division en polygones 1303
division polyédrique 1303
doléritique 0845
doline 2072
doline de dissolution 2072
doline d'effondrement 2072
doline en puits 2105
dolomicrite 3282
dolomie 3318
dolomie calcaire 3319
dolomie de précipitation primaire 3320
dolomie orthochimique 3320
dolomie primaire 3380
dolomite 3318
dolomitique, fragment – 3270
dolomitique, intraclaste – 3270
dolomitisation 3460
dolomitisation, holo– 3460
domaine fréquentiel 4571
domaine temporel 4570
dôme 0035
dôme 0059
dôme 1641
dôme de sable 3065
dôme de sel 1695
dôme éruptif 1422
dôme éruptif avec lobe de lave 1423
dôme gneissique 0890
dôme volcanique 1422
dominant 1816
dopplérite 3525
dorsale 2796
dorsales médio-océaniques, système de – 1129
doublet 0480
draa 2180
dragée 2149
drague 3154
drain 4811
drainage 2253
drainage 2768
drainage 4739
drainage 4811
drainage, aire de – 3844
drainage, arrangement de – 2254
drainage combiné 3829
drainage, écran de – 4804
drainage multiple 3829
drainage par compaction 3827
drainage par expansion de gaz dissous 3823
drainage par expansion du gaz libre 3823
drainage par gravité 3828
drainage par poussé d'eau marginal 3825
drainage par poussée d'eau 3824
drainage par poussée d'eau de fond 3825
drainage par poussée de gaz 3823
drainage par poussée hydraulique 3824
drainage, tuyau de – 4811
drain de sable 4813
drumlin 2420
druse 4014
dualistique, hypothèse – 0019
ductile 4673
dune 2165
dune 2872
dune 3007
dune, chaîne de – 2180
dune chevauchant une falaise 2169
dune confuse 2190
dune continentale 2170
dune côtière 2166
dune d'argile 2872
dune de sable 2872
dune d'estran 2168
dune d'obstacle 2174
dune, embryon de – 2176
dune embryonnaire 2176
dune en croissant 2177
dune en forme arquée 2177
dune en forme de croissant 2177
dune extérieure 2167
dune externe 2167
dune fixée 2175
dune intérieure 2170
dune interne 2167
dune littorale 2166
dune longitudinale 2171
dune migrante 2175
dune mobile 2175
dune morcelée 2190
dune mouvante 2175
dune parabolique 2177
dune phytogène 2172
dune psammogène 2172
dune résiduelle 2190
dune stabilisée 2175
dune stationnaire 2175
dune transversale 2171
durcissement externe 3411
durée 0240
durée de l'impulsion 4482
durée de parcours 0171
durée de permanence 0397
dureté 0473
dureté 3677
dureté 4701
dy 3527
dyke 1516
dyke de remplacement 0745
dyke rayonnant 1227
dyke, ring – 0742
dykes, essaim de – 0749
dykes, faisceau de – 0749
dykes, réseau de – 0749
dynamomagnétique, théorie – 0178
dynamométamorphisme – 0892
dynamomètre 4770
dysphotique 1904

E

eau atmosphérique 3613
eau capillaire 2767
eau chaude, système d' – 0393
eau connée 3661
eau cunéiforme 3653
eau de bordure 3779
eau de bordure, envahissement de l'– 3826
eau de compression 3660
eau de fond 3779
eau de formation 3662
eau de gisement residuaire 3803
eau de surface 3649
eau disponible 2773
eau douce 3676
eau du mur 3779
eau d'une nappe captive 3659
eau d'une nappe libre 3659
eau du sol 2764
eau du sol 3652
eau en crue 2251
eau funiculaire 3654
eau intermédiaire 3780
eau juvénile 1348
eau juvénile 3664
eau marginale 3779
eau, masse d'– 2828
eau métamorphique 0976
eau météorique 3614
eau peu profonde 2809
eau, pression de succion de l'– 2770
eau profonde 2809
eau, qualité de l'– 3674
eau salée 3676
eau saumâtre 3676
eau sous-jacente 3779
eau souterraine 3649
eau souterraine 3658
eau suspendue 3655
eau, teneur d'– 3802
eau, teneur en – 2769
eau, teneur en – 4611
eau type 2828
eau utile 2773
eau vadose 3655
eaux météoriques, ligne des – 0458
eaux rouges 2843
eaux souterraines, ligne de partage des – 3647
éboulement 2048
éboulement 2049
éboulement 2473
éboulement, chemin d'– 2054
éboulement, relief d'– 2056
éboulement, trajectoire d'– 2054
éboulis 2005
éboulis, cône d'– 2007
éboulis de gélivation 2656
éboulis, formation d'– 2005
éboulis récifaux 3355
ébullition 1358
écaillage 1571
écaillage brutal 4857
écaille 1538
écaille 1571
écaille 3186
écart avec la verticale 1548
écart-type 4190
échancrure 1471
échancrures, pointe entre deux – 2304
échange ionique 0420
échange ionique, capacite d'– 0420
échantillon 3153
échantillon 4741
échantillonage 4423
échantillonage par la méthode des carrés 3159
échantillon à main 4229
échantillon au hasard 4169
échantillon composite 4176
échantillon continu 4749
échantillon, cuillère pour prélèvement d'– 4748
échantillon de carottier latéral 4238
échantillon de fond 3855
échantillon de forme quelconque 4744
échantillon de saignée 3593
échantillon intact 4742
échantillon irrégulier 4744
échantillonnage 4741
échantillonnage au courant 4170
échantillonnage biaisée 4192
échantillonnage de controle 4177
échantillonnage de forage 4167
échantillonnage des fronts d'attaque 4167
échantillonnage pilote 4175
échantillonnage, réseau d'– 4171
échantillon non façonné 4744
échantillon non remanié 4742
échantillon prélevé par saignée 4168
échantillon, réduction d'– 4180
échantillon remanié 4742
échantillon salé 4179
échantillons, distribution systématique des – 4173
échantillons, outil de prélèvement d'– 4745
échantillons prélèvement d'– 4741
échantillons, sac à – 4229
échantillon trituré 4743
échelle 4240
échelle de profondeur 4505
échelle des classes granulométriques, – 3161
échelon 1538
écho 4390
éclair en boule 1372
éclat 0115
éclat 0484
éclat 0490
écogéologie 4665
écologie 1887
écorce, épaisseur de l'– 0347
écorce terrestre 0086
écostratigraphie 1711
écosystème 1893
écotype 1942
écoulement 3616
écoulement 4672
écoulement, coefficient d'– 2250
écoulement de base 3628
écoulement de surface 3617
écoulement endoréique 2256
écoulement exoréique 2256
écoulement, fraction d'– 2250
écoulement hélicoidal 2264
écoulement hypodermique 3623
écoulement laminaire 2264
écoulement, ligne d'– 0868
écoulement, ligne d'– 4862
écoulement par gravité 1703
écoulement, plan d'– 1705
écoulement, régime d'– 2960
écoulement sous-fluviale 3626
écoulement souterrain 3628
écoulement, tectonique d'– 1702
écoulement turbulent 2264
écran 4643
écran injecté 4804
écroulement 2049
ectinite 1055
écueil 2443
écueils 2455
édifice externe 1414
édifice volcanique 1413
effet de cap 4530

effet indirect 0366
effet pelliculaire 3851
effluent 4165
effondrement 2110
effondrement de dégel 2646
effondrement du toit 2110
effondrement latéral 1464
effritement 1991
effusion 1213
effusion de flanc 1266
effusion fissurale 1233
effusion terminale 1266
Eh 0421
éjecta authigènes 1314
éjecta clastiques 1313
éjecta consanguins 1313
éjecta consanguins remaniés 1314
éjecta énallogènes 1314
élargir 4296
élargissement 3455
élargissement 4054
élasticité 4669
élasticité, limite d'– 4697
élasticité, module d'– 4695
élastique 0472
électrode linéaire 4563
électrodes, disposition des – 4562
élément en trace 0400
élément gazeux 0650
élément gazeux 1172
élément mineur 0400
élévation 2796
élévation structurale 4249
ellipsoïde international 0331
élutriation 3159
élutriation 3930
éluvial, horizon – 2701
éluviation 2678
éluviation 3930
éluvions 2855
éluvions 2930
émanation 3951
émanation gazeuse 1215
émanation gazeuse 1376
émanation volcanique 1215
émanation volcanique 1376
embouchure 2196
embouchure à gradin 2208
embouchure de vallée 2280
embouchure en raccord 2208
embranchement 1929
embréchite 1055
embruns, zone des – 2439
embut 2076
émergence 0174
émergence 2857
émergence, angle d'– 0215
émergent 0174
émersion 1728
émersion 2857
émetteur 4487
émiettement 1991
émission 1210
émission gazeuse 1215
émission, puissance d'– 4490
emmagasinement, capacité d'– 3625
emmagasinement, coefficient d'– 3709
émoulage éolienne 2154
emplacement de l'ouvrage 4789
emplacement du chantier 4789
empreinte 1824
empreinte 1826
empreinte 1877
empreinte de cristal de glace 3043
emprunt, lieu d'– 4871
emprunt, matériaux d'– 4871
emprunt, puits d'– 4871
émulsif, stade – 0652
encapuchonnement 1690
enclave 0662
enclave enallogène 0663
enclave homéogène 0664
enclave surmicacée 1072
enclos 1461
encorbellement 2472
encrines, à – 3313
encroûtement 3387
endémique 1814
endobionte 1854
endofaune 1853
endogène 0081
endogène, effet – 0704
endométamorphisme 0898
endoscope 0503
endosphère 0089
endosquelette 1861
enduit 2691
énergie, environnement à basse – 2881
énergie, environnement à haute – 2881
énergie éruptive cinétique 1171
énergie géothermique 0388
energie relâchée 0204
énergie, système à basse– 0389
énergie, système à haute– 0389
énergie volcanique 1171
enfoncement 4704
enfoncement 4833
enfouie 2595
engin spatial 4302
engrené 0960
enherbement 2341
ennoiement 2857
ennoyage 1728
ennoyage sous 1614
ennoyage, zone d'– 1635
enregistrement 4420
enregistrement 4500
enregistrement analogique 4421
enregistrement digital 4422
enregistrement en aire variable 4421
enregistrement en densité variable 4421
enregistrement galvanométrique 4421
enregistreur 4500
enregistreur digital 4501
enregistreur multicanaux 4501
enregistreur-oscilloscope 4501
enrichissement 3945
ensablement 2990
ensablement 3814
ensellement 1635
ensemble 1805
ensevelie 2595
entectique, matériel – 1036
entexie 1036
enthalpie 0389
enthalpie, système à basse– 0389
enthalpie, système à haute– 0389
entonnoir 1432
entonnoir 2075
entrainement gazeux 0676
entrée 2338
entrelacée, texture – 0816
entroques, à – 3313
enveloppe 1672
environment 1900
environnement 2859
environnement, d'– 2859
éolianite 2870
éolien 2870
éon 1767
éonothème 1767
Eötvös 0337
Eötvös, correction d'– 4538
épaisseur 4728
épaisseur de la couche 4622
épaisseur net des roches pétrolifères 3873
épaisseur normale 4243
épaisseur, vraie – 4243
épaississement 4054
épaississeur 4159
épandage, champ d'– 2258
épanouissement 1985
éperon 2227
éperon 2228
éperon 2606
éperon 2806
éperon façonné 2228
éperon tronqué 2628
épibionte 1855
épibole 1773
épibolite 1049
épicentrale, distance – 0155
épicentre 0154
épicontinental 2915
épifaune 1853
epigenèse 3419
épigénétique 2290
épigénétique 3474
épigénétique 3966
épigénie 3425
épigénie 3453
épigénique 2290
épilimnion 2332
épimagma 1201
épinéritique 2841
épines 1260
épinorme 0606
épipedon 2697
épipélagique 2845
épiphyte 1856
épirogenèse 1090
épirogénie 1090
épitactique 0542
épitaxie 0542
épithermal 3983
épizone 0993
éponte 0705
éponte 1529
éponte 4009
éponte soudée 4013
époque 1764
époque métallogénique 3905
éprouvette 4741
epsomite 3383
épuisement 4739
épuisement 4811
équateur geómagnëtique 0281
équateur magnétique 0283
équidimensional 3181
équidimensionnel 0754
équidistant 4172
équiforme 0754
équigranulaire 0806
équigranulaire 0807
équigranulaire 3081
équilibre des déblais et des remblais 4826
équilibre dynamique 0397
équilibre isostatique 0346
équilibre radioactif 4549
équipotential, courbe – 3644
équivalence, principe de l'– 4567
équivalent 1733
équivalent 3675
érathème 1766
ère 1766
érodibilité 2936
érosion 2012
érosion 2776
érosion 2936
érosion accélérée 2777
érosion, capacité à l'– 2936
érosion, chenal d'– 3510
érosion chimique 2017
érosion cosmique 0028
érosion, cycle d'– 2566
érosion différentielle 2024
érosion du sol 2036
érosion en nappe 2779
érosion en nappe 2032
érosion en ravelines 2778
érosion en ravins 2018
érosion en ravins 2778
érosion en rigoles 2033
érosion en rigoles 2778
érosion éolienne 2015
érosion éolienne 2154
érosion fluviatile 2015
érosion glaciaire 2015
érosion latérale 2022
érosion linéaire 2012
érosion marine 2106
érosion, niveau de base de l'– 2025
érosion par impact de la pluie 2035
érosion, plaine d'– 2591
érosion régressive 2021
érosion remontante 2021
érosion, surface d'– 2591
érosion, système d'– 2573
érosion verticale 2022
erratique 2949
erreur de fermeture 4541
éruptif 0616
éruptif, appareil – 1239
éruption 1211
éruption 4288
éruption aréale 1232
éruption boueuse 1395
éruption centrale 1228
éruption, centre d'– 1220
éruption de flanc 1230
éruption de fragments pyroclastiques 1214
éruption de lave 1214
éruption embryonnaire 1219
éruption excentrique 1229
éruption explosive 1216
éruption fissurale 1233
éruption hawaïenne 1158
éruption indirecte 1218
éruption latérale 1230
éruption linéaire 1233
éruption magmatique 1219
éruption, mécanisme de l'– 1170
éruption mixte 1219
éruption normale 1228
éruption paroxysmale 1164
éruption péléenne 1163
éruption phréatique 1351
éruption plinienne 1161
éruption secondaire 1403
éruption sommitale 1229
éruption sous-marine 1236
éruption strombolienne 1159
éruption subaérienne 1236
éruption subaquatique 1236
éruption subglaciale 1236
éruption, trace d'– 3065
éruption vulcanienne 1160
escalier en cirques 2614
escarpement 2053
escarpement de faille 2548
escarpement de faille exhumé 2548
escarpement de faille rajeuni 2548
escarpement de faille rajeuni par l'érosion 2548
escarpement de lave 1286
escarpement de ligne de faille 2550
escarpement lobé 0032
escarpement sous-marin 2799
esker 2420

espace annulaire 4279
espace, groupe d'– 0513
espacement 1497
espacement 4822
espacement des diaclases 4725
espacement des puits 3842
espèce 1937
espèce climax 1891
espèce cosmopolite 1809
espèce type 1948
esquille 3186
esquille 4168
essai 3787
essai 4183
essai à la plaque 4708
essai aux tiges 3787
essai avec appareil sur cable 3787
essai de production 3788
essaim filonien 4066
essence 3744
essence lourde 3744
estavelle 2092
estimation biaisée 4192
estran 2485
estuaire 2456
estuaire 2784
estuaire 2904
estuairien 2904
étachéité, garniture d'– 4279
étage 1762
étanchéité 4863
étang 2330
état initial 1157
état liquide 4713
état plastique 4713
état semi-solide 4713
état solide 4713
étendue dynamique 0132
étier 2509
étier 2517
étinceleur 4493
étranger 3970
étranglement 3509
étranglement 3904
étreinte 3510
eucristallin 0795
eugéosynclinal 1097
euhalin 2834
eulittoral 2838
euphotique 1904
eurybathe 1911
euryhalin 1912
eurytherme 1913
eurytope 1914
eutaxite 0861
eutaxite 1243
eutaxitique 0861
eutrophe 3549
euxinique 2925
évacuateur de crues 4868
évaluation 4216
évaluation des terrains 4786
évaporation 0412
évaporation, assèchement par – 2342
évaporite 3375
évapotranspiration 3621
évasion 0273
événement majeur 0153
évent 1221
éventail 1287
évolué 3176
évolué, peu – 3176
évolution 1964
évolution ascendante 2574
évolution convergente 1973
évolution descendante 2574
évolution des sols 2659
évolution magmatique 0666
évolution rectiligne 1975
évolution répétitive 1971
évolution uniforme 2574
exaration 2396
excavation 2471
excentrique 2132
excrément fossile 1863
exfoliation 1994
exfoliation 1995
exhalaison volcanique 1215
exhalaison volcanique 1376
exhalation 3951
exhaussement 2513
exhaussement 2857
exhumée 2595
exogène 0081
exogène, effet – 0704
exogéologie 0024
exométamorphisme 0898
exosphère 0083
exosquelette 1861
expansion du fond océanique 1128
expansion éclair 3863
expansion, hypothèse de l'– 1123
expansion instantanée 3863
expectation, méthode des courbes d'– 3876
expertise 4216
exploitabilité, seuil d'– 4203
exploration géochimique 4625
explosif 4815
explosifs, phénomènes – 1175
explosion 1217
explosion, centre d'– 1177
explosion mixte 1219
explosion, niveau d'– 1177
explosion nucléaire 0268
explosion nucléaire souterraine 0268
explosion phréatomagmatique 1219
explosion, point d'– 1176
exposition 4340
expulsion 3732
exsolution 0476
exsurgence, source d'– 2090
extension 1781
extension 3885
extension en profondeur 1526
extension latérale 1524
extension latérale 2513
extensomètre 0121
extensomètre 0122
extensomètre à résistance électrique 4767
extensomètre en sondage 4773
extinction 0564
extinction 1986
extinction, angle d'– 0566
extinction ondulée 0968
extinction onduleuse 0565
extinction roulante 0968
extraclaste 3272
extraction, coût d'– 3900
extrusion 1207

F

fabrique 0793
fabrique 1004
face 0517
face 2724
face amont 2963
face au vent 2963
face aval 2964
face sous le vent 2964
facette 2562
facette de faille 2553
facette indicatrice d'une faille 2553
facettes, à – 2596
faciès 0521
faciès 1715
faciès 2860
faciès de plateforme 2915
faciès diagénétique 3435
faciès, faune de – 1850
faciès, fossile de – 1807
facies métamorphique 0987
facies métamorphique, de même – 0992
facies métamorphique, série de – 0991
faciès minéralogique 0585
faciès volcanique 1238
facteur volumétrique de gaz 3861
facteur volumétrique de l'huile 3860
fagne 3542
fahlbande 4052
faille 1521
faille 4723
faille accessoire 1581
faille à déplacement curviligne 1567
faille à déplacement rectiligne 1566
faille anormale 1558
faille antithétique 1588
faille à pente raide 1549
faille à rejet en direction 1560
faille à rejet horizontal 1560
faille ayant rejoué 1527
faille bordière 1593
faille courbée 1535
faille d'accompagnement 1581
faille dans le plan des couches 1547
faille décalée 1536
faille de décrochement 1560
faille de raccord 1560
faille de raccordement 1141
faille d'étirement 1576
faille de transformation 1141
faille de transformation 1560
faille, déviation d'une – 1536
faille dextre 1561
faille diagonale 1545
faille directe 1558
faille directionnelle 1544
faille disjonctive 1534
faille en retour 1588
faille faiblement inclinée 1549
faille fermée 1534
faille gravitationnelle 1707
faille inverse 1558
faille limite 1593
faille longitudinale 1546
faille maîtresse 1580
faille majeure 1580
faille marginale 1593
faille, miroir de – 1538
faille morte 1527
faille nivelée 2549
faille normale 1558
faille oblique 1545
faille orthogonale 1545
faille ouverte 1534
faille périphérique 1592
faille perpendiculaire 1545
faille, plan de – 0238
faille, plan de – 1528
faille plane 1535
faille plate 1535
faille principale 1580
faille profonde 1525
faille, propagation de – 0239
faille rajeunie 1527
faille ramifiée 1582
faille rasée 2549
faille rotatoire 1567
failles à gradins 1587
faille sans relief 2549
faille satellite 1581
failles conjuguées, système de – 1584
failles croisées 1579
faille secondaire 1580
failles en échelons 1585
failles en escalier 1587
faille sénestre 1561
failles en réseau 1578
failles étagées 1587
failles, faisceau de – 1584
failles parallèles, système de – 1584
failles radiales 1594
failles rayonnantes 1594
failles, relais de – 1585
failles, réseau de – 1578
failles, style tectonique à – 1520
failles, système de – 1578
failles, tectonique de – 1520
faille superficielle 1525
faille synthétique 1588
failles, zone de – 1467
faille transformante 1141
faille transformante 1560
faille transversale 1546
faille verticale de décrochement 1560
faille vivante 1527
faisabilité, étude de – 4791
faisceau à fenêtre réglable 4510
faisceau de – 3111
faisceau de déphasage 4510
faisceau total 4510
falaise 2470
falaise d'éboulement 2476
falaise, fausse – 2476
falaise littorale 2470
falaise morte 2477
falaise plongeante 2475
falaises alignées 2478
familier 3970
famille 0687
famille 1930
famille de sols 2664
fanconglomérat 2885
fanglomérat 2885
fantôme 0802
fantôme 4442
fantôme, horizon – 4462
faro 3339
fatigue 4674
fauchage, phénomène de – 2058
faune 1795
faune naine 1851
faunique 1795
faunistique 1795
faunule 1796
fausse 4637
fausse alerte 0269
feldspathisation 1064
felsique 0597
felsique, roche – 0597
felsophyre 0824
felsophyrique 0824
fémique 0607
fenestré 0964
fenêtre 1289
fenêtre 4234
fenêtre de la source 0231
fenêtre karstique 2078
fénite 0925
fénitisation 0925
fente 1465
fente 1466
fente 1512
fente 2935
fente concentrique 1469
fente de contraction 1993
fente de dessiccation 1518
fente de gel 2640

fente de retrait 3044
fente de retrait, remplissage de – 3047
fente filonienne 1514
fente radiale 1468
fente transversale 1470
fente volcanique béante 1234
fermé 1626
fermeture 3762
fermeture, aire de – 3762
fermeture, écart de – 4541
fermeture par faille 3763
fermeture par pendage 3763
fermeture verticale 3762
fer pisolithique 4118
ferrimagnétisme 0302
ferromagnésien 0599
ferromagnétisme 0299
ferrugineux 3464
feston 2649
feu, crachant du – 1237
feu, de – 0485
feu, gerbes de – 1283
feuillet 0739
feuillet 3101
feuilleté 0775
feuilletée 2728
feuillet frontal 3140
feuillet inverse 3141
feuillets bleus, structure en – 2374
feuillets, en – 3101
feu roulant 1237
fibreux 0522
fibreux 0773
fibroblastique 0963
fibrocristallin 0815
figure 3018
figure basale 3019
figure de corrosion 3409
filaments vitreux 1326
filet 2194
filet charbonneux 3499
filet minéralisée 4039
filiforage 3783
filiforme 4081
film couleur 4335
film fausse-couleur 4337
film infrarouge 4336
film infrarouge couleur 4337
film orthochromatique 4334
film panchromatique 4334
filon 0741
filon 1514
filon 4038
filon 4050
filon clastique 1516
filon clastique 3073
filon complexe 4046
filon composé 4045
filon concrétionné 4049
filon couche 0740
filon-couche 4053
filon creux 1245
filon cunéiforme 4036
filon de dilatation 4041
filon de substitution 4042
filon en chapelets 4054
filon en échelle 4057
filonienne, roche – 0741
filon intrusif 0748
filon majeur 4062
filon metallifère 4038
filon minéralisé 4038
filonnet 4039
filon principal 4062
filon rubané 4049
filons, croisement de – 4074
filons, faisceau de – 4066
filon simple 4045
filons interconnectés 4067
filons parallèles 4063
filons, ramification de – 4070
filons rayonnants 4063
filons réticulés 4064
filon transversal 4077
fils étirés 1326
filtrage des fréquences 0136
filtration 0414
filtration, pression de – 4738
filtre 4338
filtre 4445
filtre 4812
filtre 4862
filtre de contraste 4339
filtre de polarisation 4339
filtre de vitesses 4456
filtre éliminateur de fréquence 0137
filtre passe-bande 0137
filtre passe-bas 0137
filtre passe-haut 0137
filtre ponctuel 0137
filtre-presse, effet de – 0675
filtre-presse, effet de – 3933
filtres, boîte de – 4509
fines, les – 3207
fissile 1597
fissilité 1597
fission, datation par des traces de – 0442
fissure 1465
fissure 1512
fissure 1513
fissure 4044
fissure concentrique 1469
fissure de contraction 0713
fissure de flanc 1468
fissure de refroidissement 1517
fissure de retrait 0713
fissure de retrait 1993
fissure d'éruption 1225
fissure d'extension 1519
fissure endogène 3508
fissure excentrique 1470
fissure filonienne 1514
fissure fumerollienne 1383
fissure nourricière- 3941
fissure nourricière 4011
fissure radiale 1468
fissure, remplissage de – 1515
fissure vide 1513
fissure volcanique béante 1234
fixation de plafonds 3902
fixisme 1113
fjord 2453
flambage 1665
flambage, hypothèse du – 1118
flambement 1665
flambement, hypothèse du – 1118
flamme d'émission volcanique 1350
flanc 1625
flanc inférieur 1576
flanc inverse 1647
flanc inverse 1680
flanc médian 1576
flanc médian étiré 1680
flanc normal 1647
flanc normal 1680
flanc renversé 1647
flanc renversé 1680
flanc supérieur 1576
flanc supérieur 1680
flèche 2464
flèche à crochet 2464
flèche à pointe libre 2464
flèche de jonction 2467
flèche-isthme 2467
flèche littorale 2463
flèche relicte 2466
flèche témoin 2466
flétrissement, point de – 2773
fleur de gypse 2151
fleuve 2191
flexible 0472
flexure 1616
floculat 2992
floculation 2992
floral 1797
flore 1797
flot de fond 2826
flottation 0673
flottation 4152
flottation collective 4152
flottation sélective 4152
fluage 4672
fluidalité magmatique 0868
fluidalité primaire 0631
fluide minéralisateur 3936
fluide mouillant 3705
fluide non-mouillant 3705
fluidisation 1370
fluolite 1262
fluorescence 0467
fluorescence 0491
flute 3025
flutes, formation de – 3025
fluvial 2883
fluviatile 2883
fluvioglaciaire 2883
fluvio-marin 2902
flux magnetique 0317
fluxoturbidite 2984
flux thermique 0385
flux thermique de la terre 0383
flux thermique terrestre 0383
flysch 2926
focale, sphère – 0233
foisonnement 4731
foliation 0948
foliation 1604
foliation continue 0948
foliation discontinue 0948
folié 0775
folié 0948
fonctionnement régulier 4511
fond 4626
fondation 4835
fondation en surface 4835
fondations, techniques des – 4834
fondation superficielle 4837
fondation sur radier 4837
fond durci 3126
fond induré 3126
fondrière 3532
fontaine 1359
fontaine de lave 1282
fontaines de feu 1283
fonte des neiges 2355
forabilité 4298
forage 3665
forage 4261
forage à injection inversée 4285
forage à l'air 4263
forage à rotary-percussion 4263
forage au cable 4262
forage au diamant, appareil de – 4265
forage, boue de – 4281
forage de recherche 3782
forage dirigé 4291
forage, installation de – 4267
forage, mat de – 4267
forage, outil de – 4271
forage par battage 4262
forage par percussion 4262
forage-rotary 4263
forage, tour de – 4267
förde 2454
formation 1754
formation carbonifère 3498
formation climax 1891
formation, détérioration de la – 3815
formation, facteur de – 4609
formation humide 3767
formes, cercle de – 1936
forme structurale 2554
forme tectonique 2537
fosse 0034
fosse 2247
fosse 2793
fosse 2794
fossé 1453
fossé 2546
fossé 2793
fossé d'effondrement 1590
fossé d'effondrement 2546
fossé d'explosion 1235
fossé latéral 2384
fosse marginale 1111
fosse océanique 1138
fossile 1789
fossile 2595
fossile caractéristique 1785
fossile caractéristique de zone 1785
fossile de trace 1873
fossile indicateur 1785
fossile repère 1785
fossilifère 1789
fouille 4235
fouille 4824
foulon, terre à – 3218
fourier, transformée de – 4451
fourneau de mine 4817
foyer 0156
foyer central 1202
foyer, mécanisme au – 0232
foyer négatif 0157
foyer périphérique 1203
fractionnement 0414
fractionnement cinétique des isotopes 0457
fractionnement isotopique, courbe de – 0455
fractionnement isotopique, facteur de– 0455
fraction volumique 4109
fracturation 1494
fracturation cylindrique 1302
fracturation granulaire 3431
fracturation hydraulique 3800
fracture 0469
fracture 0541
fracture 1494
fracturé 1494
fracture annulaire 0743
fracture conchoïdale 0541
fracture, zone de – 1595
fragile 4673
fragipan 2734
fragmentation 4821
framboid 3484
framboidal 3484
frange capillaire 3657
fréquence, courbe cumulative de – 4188
fréquence d'émission 4504
fréquence des battements 0178
fréquence de travail 4498
fréquence, domaine de – 0223
fréquences, distribution de – 4187
fréquences, spectre des – 4451
Fresnel, surface de – 4515
friable 3087
frittage 0897
front 1686
front basique 1066
front d'avancement 4852
front de taille 4851
frottement 4686
frottement, angle de – 4689
frottement, coefficient de – 4689

frottement, critère de – 4689
frottement interne 4686
frottement latéral 4839
frottement négatif 4839
Froude 2266
fruit 4827
frustule 1842
fuite, débit de – 4863
fuite, point de – 3762
fumée en forme de pin-parasol 1362
fumée volcanique 1349
fumée volcanique 1377
fumerolle 1379
fumerolle acide 1388
fumerolle alcaline 1388
fumerolle ammoniacale 1388
fumerolle aqueuse 1389
fumerolle boueuse 1398
fumerolle croicolitique 1384
fumerolle d'acide chlorhydrique 1387
fumerolle de cratère 1383
fumerolle de fissure 1383
fumerolle d'origine secondaire 1382
fumerolle en flammes 1386
fumerolle éruptive 1385
fumerolle froide 1389
fumerolle leucolitique 1384
fumerolle permanente 1381
fumerolle primaire 1382
fumerolles, alignement de – 1383
fumerolle sans racine 1382
fumerolles, phase des – 1168
fumerolle sulfurique 1388
fumerolle temporaire 1381
fumerollien, stade – 1168
fusain 3580
fusion des neiges 2355
fusion du toit 0624
fusion partielle 1029
fusion par zone 0625

G

gabbroïque 0812
gain automatique, contrôle de – 4508
gain, contrôle automatique de – 4425
gain, contrôle de – 4508
gaine de lave 1292
gaine de scories 1330
gal 0324
galerie 1877
galerie de reptation 1878
galet 3242
galet de boue 3247
galet erratique 2422
galet mou 3246
galet plat argileux 3248
galet roulé 3242
galets, banc de – 2244
galets de plage 2912
Galton, distribution de – 4189
gamma 0318
gamma, méthode – 4545
gangue 3921
garde 1427
gas-oil 3744
gastrolithe 1864
gâteau de boue 4618
gauss 0318
gayet 3497
gayet 3563
gaz 3713
gaz associé 3891
gaz, calotte de – 3774
gaz, chapeau de – 3774
gaz, conservation de – 3836
gaz, dégagement de – 1215
gaz des marais 3529
gaz, détection de – 4520
gaz diffuse, source de – 1401
gaz dissous 0641
gaz dissous 1172
gazéification du charbon 3609
gaz, facteur de – 3802
gaz, fuite de – 4520
gaz-huile, proportion – 3802
gaz humide 3715
gaz juvénile 1348
gaz-liquide, proportion – 3802
gaz naturel 3713
gazole 3744
gaz/pétrole, contact – 3774
gaz/pétrole, niveau – 3774
gaz phréatique 1351
gaz, pression de – 1173
gaz, recyclage de – 3835
gaz résurgent 1172
gaz riche 3715
gaz sous pression, ouverture causée par des – 1178
gaz, teneur en – 0650
gaz, trace de – 3065
gaz volcanique 1349
géanticlinal 1099
Geiger, compteur – 4545
gélivation 1996
gélivation 2394
gélivité 1996
gemmologie 0478
génération 0591
génération de formes de relief 2573
généro type 1948
génotype 1979
genre 1933
genre de forme 1932
gens 1934
géobarométrie 0994
géobios 1802
géochimie 0395
géochimie des isotopes stables 0453
géochronologie 1713
géochronologie isotopique 0428
géode 3481
géode 4014
géodépression 1120
géodésie 0013
géodynamique interne 0080
géognosie 0001
géogonie 0003
géographie physique 0012
géohydrologie 3611
géoide 0328
géoisotherme 0377
géologie 0001
géologie appliquée 0016
géologie appliquée 4661
géologie de l'environnement 4665
géologie de l'ingénieur 4660
géologie des catastrophes 0005
géologie des isotopes stables 0453
géologie des phénomènes actuels 0008
géologie de terrain 4226
géologie économique 0016
géologie expérimentale 0014
géologie générale 0010
géologie historique 0011
géologie marine 4226
géologie mathématique 0015
géologie souterraine 4226
géologie structurale 1478
géologue 0001
géomagnétique 0281
géomagnétisme 0276
géomorphologie 1987
géomorphologie de l'ingéniérie 4663
géonomie 0079
géophage 1857
géophone 4414
géophones, dispositif des – 4417
géophones, intervalle entre – 4416
géophones, réseau de – 4416
géophysique 0078
géophysique de l'ingéniérie 4663
géostatistique 4193
géosynclinal 1094
géotechnique 4662
géotectonique 1073
géothermal 0376
géothermie 0374
géothermique 0376
géothermique, champ – 0392
géothermique, degré – 0380
géothermique, gradient – 0380
géothermomètre 0379
géothermomètre 0994
géothermométrie 0379
géothermométrie 0994
géothermometrie isotopique 0456
géotumeur 1120
gerbe en panache 1362
gerbe éruptive 1361
gerbes enflammées 1283
gérontique 1866
geyser 1404
geyser, cheminée de – 1405
geyser, cratère de – 1405
geysérite 1406
geyser, jet de – 1405
geyser siliceux 1406
gigantisme 1867
gilgai 2763
gisement 3754
gisement concordant 4022
gisement de shear zone 4060
gisement discordant 4022
gisement en dyke 4031
gisement en sill 4031
gisement exhalatif 3979
gisement lenticulaire 4025
gisement magmatique 3946
gisement métallifère 3910
gisement pénéconcordant 4022
gisement tabulaire 4025
gisement type porphyry 4133
gîte alluvionnaire 4000
gîte cunéiforme 4036
gîte de charnière anticlinale 4053
gîte de charnière synclinale 4053
gîte de départ immédiat 3987
gîte de remplacement 3991
gîte de ségrégation périphérique 3987
gîte d'impregnation 3992
gîte d'inclusion 3975
gîte d'injection 3993
gîte d'injection magmatique 3975
gîte éluvionnaire 3999
gîte en gouttière 4053
gîte filonien 4038
gîte hydrothermal 3982
gîte interstratifié 4021
gîte magmatique de cristallisation précoce 3975
gîte magmatique de cristallisation tardive 3975
gîte métasomatique 3991
gîte minéral 3910
gîte pneumatolytique 3981
gîte primaire 3972
gîte pyrométasomatique 3980
gîte secondaire 3972
gîte stratifié 4023
gîte stratiforme 4024
gîte stratoïde 4024
gîte zoné 4015
gîtologie 3903
gîtologique 3903
givre 0487
givre calcaire 2143
glace, aiguilles de – 2639
glace fossile 2369
glace morte 2369
glace, poussée de la – 2401
glace profonde 2639
glaces, poussée des – 2401
glaciaire 2873
glaciation 2367
glaciation régionale 2425
glacier 2366
glacier, arche du – 2391
glacier chevauchant 2385
glacier débouchant en mer 2436
glacier de cirque 2430
glacier de marée 2436
glacier de montagne 2426
glacier de piedmont 2435
glacier de plateau 2426
glacier de type Mustag 2431
glacier de type Turkestan 2429
glacier de vallée 2427
glacier de vallée simple 2428
glacier du type alpin 2428
glacière 2107
glacier, front du – 2390
glacier, lait de – 2392
glacier, lobe du – 2390
glacier polaire 2368
glacier polysynthétique 2432
glacier porteur 2385
glacier régénéré 2434
glacier suspendu 2433
glacier, table de – 2386
glacier tempéré 2368
glaciologie 2345
glacio-marin 2903
glacis 2582
glacis rocheux désertique 2582
glauconieux 3177
glauconitisation 3459
gleyification 2684
glissement 2043
glissement 2473
glissement 2976
glissement 3055
glissement de débris 2046
glissement de loupes 2045
glissement de montagne 2048
glissement de terrain 2044
glissement en forme d'écailles 2045
glissement en masse 2048
glissement entre les feuillets 2373
glissement, pli de – 3055
glissement, structure de – 3058
glissement, surface de – 1002
glissement, surface de – 1705
glissement, surface de – 2055
glissement, surface de – 2731
glissement, tectonique de – 1702
glissoir 2054
globe 1372
globosphérite 0765
globulite 0765
gloméroporphyrique 0821
glucide 3723
gneiss 0936
gneiss en crayon 0936
gneissique 0871
gneissique, texture – 0951
gonflement 4054

gonflement 4720
gonflement dû au gel 4721
gonflement, indice de – 3595
gore blanc 3505
gorge de torrent 2026
goudron 3743
gouffre 2097
gouffre 2104
gouffre absorbant 2075
gouffre de la mort 1402
goulet 3348
gour 2147
goutte 1326
gouttelette 1326
gouttière 2234
gradation 4343
gradation 4702
gradient hydraulique 4736
gradin de faille 2548
gradin de flexure 2551
gradin glaciaire 2623
gradin marginal 2604
gradin sismique 2552
gradin structural 2200
gradiomètre magnétique 4525
graduation de la qualité 0494
grain 0792
grain 0792
grain 3078
grain 3081
graine 0093
grain fin, à – 0796
grain fin, à – 3194
grain grossier, à – 0796
grain grossier, à – 3194
grain hétérogène à – 0820
grain, joint de – 0532
grain moyen, à – 0796
grain moyen, à – 3194
grain, taille du – 3193
grain très fin, à – 0799
grain très grossier, à – 0796
granitisation 1062
granitoïde 0810
granoblastigue 0960
granoclassement 3130
granophyrique 0848
granosphérite 0779
grantie graphique 0847
granulaire 2727
granularité 4344
granulats 4872
granule 2727
granule 3240
granulométrie 3160
granulométrie 4702
granulométrie continue 4702
granulométrie discontinue 4702
granulométrie par tamisage 3157
granulométrique, analyse – 2711
granulométrique, classe – 2713
granulométrique, classe – 3162
granulométrique, composition – 2709
granulométrique, distribution statistique – 3197
granulométrique, fraction – 2710
granulométrique, fraction – 2712
granulométrique, médiane– 3199
graphique 0847
gras 0484
grauwacke 3234
graveleux 3244
gravier 3241
gravier 3244
gravier, banc de – 2244
gravier fin 3249
gravier grossier 3249
gravier volcanique 1324
gravimètre 0352
gravimètre basé sur la méthode zéro 0352
gravimétrie 0349
gravimétrique, station – 0353
gravitation 0322
gravitation, constante de la – 0325
gravité 0324
gravité, formule internationale 1930 de – 0330
gravité, gradient de la – 0335
gravité, potentiel de la – 0327
gravité, valeur normale de – 0332
greisen 0924
greisenification 0924
grêlons, empreinte de – 3041
grenu 0806
grenue 2727
grenue 3081
grès 3228
grès aquifère 3766
grès de plage 3412
grès feldspathique 3232
grès gazifère 3766
grès oolitique 3285
grès pétrolifère 3766
grès propre 3228
grès quartzitique 3230
grès subarkosique 3233
grès très fin 3204
grève 2485
griffe 3499
grille fixe 4144
grille parallactique 4373
grille perspective 4373
grisou 3554
grondements souterrains 1237
grossier, élément – 2722
grotte de lave 1291
grotte formée par les brisants 2472
grotte glacée 2107
groupe 1755
groupe 1928
groupement 1890
groupe ponctuel 0512
grumeau 2727
grumuleuse 2727
guano 2878
gué 2247
guirlande 1020
gunite 4801
Gutenberg, discontinuité de – 0092
guyot 2798
gypse 3381
gypse, transformation en – 3459
gypsite 3381
gyrogonite 1843

H

habitat 1901
habitus 0521
habitus 0592
hadal 2848
hadal 2924
haldes 4166
halmyrolyse 3476
halo 4629
halocinèse 1694
halocline 2830
halophyte 1846
Harker, diagramme de variation selon – 0697
harmonique 0167
hartschiefer 1543
hauteur 2808
haut-fond 2247
haut-fond 2469
haut-fond 2803
Hayford, zone de – 0362
hélicitique 0955
hemicyclothème 3120
hémidiatrème 1206
herbu 2514
hérisson de boue 3247
hérité 3971
hétéradcumulat 0833
hétéroblastique 0956
hétérochrone 1734
hétérogène 4680
hétérogranulaire 0807
hétéromorphe 0635
hétéromorphisme 0635
hétérosporie 1834
hétérotactique 1007
hétérotope 1731
heure initiale 0159
hexaédrite 0068
hexagonal 0515
hiatus 1723
hiérarchie 1925
Hilt, loi de – 3522
hiperthermal, région – 0391
histogramme 4188
hodochrone 0171
hogback 2562
holoblaste 0984
holocristallin 0803
holohyalin 0805
hololeucocrate 0587
holomarin 2900
holomélanocrate 0588
holosidérite 0066
holostratotype 1740
holotype 1949
homéoblastique 0960
homéomorphe 1980
homéomorphie 1820
homogène 0716
homogène 4680
homogène, zone – 4787
homohalin 2831
homologue 1972
homonyme 1955
homoséiste 0210
homoséiste, ligne – 0210
homospore 1835
homosporie 1834
homotactique 1007
homotaxe 1733
homotype 1954
horizon 1749
horizon 2687
horizon A 2698
horizon A1 2698
horizon A2 2700
horizon Ap 2699
horizon argileux 2734
horizon argil-illuvial 2703
horizon argillique 2703
horizon B 2702
horizon-B de couleur 2703
horizon B2 de texture 2703
horizon B en bandes 2705
horizon C 2706
horizon C1 2707
horizon cendreux 2700
horizon chronostratigraphique 1758
horizon cuirassé 4122
horizon D 2708
horizon éluvial 2701
horizon enfoui 2693
horizon illuvial 2702
horizon induré 2733
horizon limite 3544
horizon lithostratigraphique 1749
horizon minéralisé 4008
hornito 1298
horst 1589
Hoskold, formule de – 4217
hôte 0844
houille bitumineuse 3559
houille claire 3580
houille foliaire 3580
houille gras 3559
houille grenue 3580
houille mate 3580
houille mate fibreuse 3580
houille moyenne 3580
houille papyracée 3563
houille sèche à longue flamme 3559
houille semi-brillante 3580
houille terne 3580
houillification 3519
houle 2807
huile 3712
huile d'amont pendage 3776
huile de stockage 3859
huile, hauteur impregnée par l'– 3775
huile lourde 3745
huile mobile 3864
huile sèche 3714
huile, transformation de l'– 3733
hum 2074
humidité de constitution 3598
humidité libre 3598
humidité totale 3598
humification 2683
humification 3525
humus 2743
humus 3525
humus brut 2744
humus doux 2744
hyalin 0805
hyalo-ophitique 0840
hyalopilitique 0837
hyaloporphyrique 0823
hybridation 0681
hybride 0681
hydatogenèse 3958
hydatopyrogenèse 3958
hydratation 0419
hydrate de carbone 3723
hydrobios 1802
hydrocarbures, teneur en – 4610
hydrogène, indice d'– 4613
hydrogéochimie 3611
hydrogéochimique 4634
hydrogéologie 3611
hydroisohypse 3644
hydrolaccolite 2643
hydrologie 3610
hydrologie souterraine 3611
hydrolyse 0418
hydromagnétique, théorie – 0278
hydrométasomatose 3954
hydromorphe 4634
hydromorphie, caractéristique d'– 2690
hydrophone 4415
hydrophone 4496
hydrophone, câble d'– 4496
hydrosphère 0085
hydrothermal, processus – 3953
hydrothermal, stade – 0654
hypabyssique 0617
hyperalcalin 0693
hyperalumineux 0693
hyperboles, méthode des – 0152
hypercyclothème 3120
hypéritique, structure – 0853
hypersalé 3377
hypersalin 2835
hyperstéréoscopie 4367
hypidioblaste 0980
hypidiomorphe 0751
hypidiomorphe 0808
hypoabyssal 0617
hypocentre 0156

hypocristallin 0804
hypogène 3937
hypogène 3967
hypolimnion 2334
hypomagma 1199
hypothermal 3983
hypothermale, région – 0391
hystérèse 4675
hystérèse magnétique 0305
hysteresis 4675

I

ichnocénose 1874
ichnofossile 1873
ichnofossile 3067
ichnologie 1871
ichor 1061
identification, capacité d'– 0271
identification, seuil d'– 0271
idioblaste 0980
idiochromatique 0483
idiomorphe 0523
idiomorphe 0750
idiomorphe 3471
igné 0616
ignée, roche – 0616
ignimbrite 1340
igue 2076
île 2443
île continentale 2444
île d'accumulation 2535
île d'émersion 2535
île détachée 2534
île de type frison 2507
île dunaire 2507
île océanique 2444
île rattachée 2534
île volcanique 1416
illumination conoscopique 0547
illumination orthoscopique 0547
illumination spécifique 4311
illumination totale 4311
illuvial, horizon – 2702
illuviation 2678
illuviation 3930
image 4320
image redressée 4377
imatra 3486
imbrication 3040
imbrication 3118
imitation de pierre 0481
immature 3725
immergé 2880
impact 0026
impactite 0076
impact, marque d'– 3037
impacts, théorie des – 0049
impédance acoustique 4474
impetus 0173
impluvium 2192
impregnation 1060
impregnation 3396
impregnation 3955
impulsion 0173
impulsion 4482
incidence, angle d'– 0211
incidence perpendiculaire 4472
incidence, plan d'– 0211
incidence rasante 0212
incision 2020
inclinaison 1484
inclinaison, angle d'– 1485
inclinaison devient faible, l'– 1487
inclinaison magnétique 0283
inclinaison originelle 1486
inclinaison s'adoucit, l'– 1487
inclinaison s'atténue, l'– 1487
inclinomètre 4294
inclusion 0789
incompétent 1624
incurvation 2649
indentation 2450
indicateur géobotanique 4646
indicatrice 0552
indices, ellipsoïde des – 0552
induction magnetique 0317
induration 3423
infiltration 2768
infiltration 3396
infiltration 3619
infiltration 3955
infiltration, capacité d'– 3620
infiltration, taux d'– 3620
inflexion, ligne d'– 1627
inflexion, point d'– 1627
infracrustal 0584
infralittoral 2839
infranéritique 2841
infrarouge lointain 4309
infrarouge photographique 4307
infrarouge, proche – 4307
infrarouge thermique 4308
infraspécifique 1944
infra-structure 1077
iniection, voile d' – 4904
initiale 1108
injecter 4803
injection 3949
injection 4802
injection d'eau 3833
injection d'eau chaud 3839
injection de gaz 3835
injection de vapeur 3839
injection en ligne 3833
injection, forage d' – 1919
injection par réseau incullé 3833
injection periphérique 3833
injection, rideau d'– 4804
injection, tête d'– 4268
injectivité, indice d'– 3809
inlandsis 2425
inondation 2857
inselberg 2608
inséquent 2287
insolation, craquelure d'– 3045
inspection in situ 0267
instant de tir 4427
intensité 0243
intensité 2677
intensité, échelle d'– 0243
intercalaire, événement – 0292
intercalation 3117
intercalation argileuse 3772
intercalé 3117
interception 3615
intercinématique 1079
intercrête 3001
intercumulus 0831
interdigitation 3118
interface 3096
interface 4515
interférence, image d'– 0569
interfluve 2312
intergranulaire 0841
intermédiaire 0602
interpénétration 2648
interpénétration granulaire 3448
interprétation à la tête du puits 4604
interprétation directe 4566
interprétation indirecte 4566
interprétation quantitative 4566
interprétation sur le champ 4604
intersertal 0840
interstice 3082
interstice 3683
interstitiel 3083
interstratification 3116
interstratifié 3116
interstratifié 3117
intervalle 1718
intervalle stérile interzonal 1775
intraclaste 3272
intramagmatique 3985
intrasparite 3283
intrusif 0619
intrusif 0715
intrusion 0618
intrusion composée 0721
intrusion composée 0721
intrusion de pétrole 3752
intrusion sédimentaire 3070
intrusions, groupe d'– 0722
intrusion simple 0720
intrusive, roche – 0715
intumescence 1194
intumescence de lave 1272
intumescence profonde 1120
invasion 4619
invasion de pétrole 3752
inverseur 4489
inversion 1088
inversion, auto– 0311
inversion d'une rivière 2214
inversion du relief 2563
inversion géomagnétique 0291
involution 2648
ionium-thorium, datation – 0444
iridescence 0486
irradiance 4312
irradiation 0492
isanomal 4638
isochore 4244
isochromatique, courbe – 0571
isochrone 0210
isochrone 0433
isoclinal 1626
isocline 0283
isodyname 0285
isofaciès 2863
isogamme 4532
isogone 0284
isograde 0992
isohypse 4248
isolement 1981
isomésique 1735
isométrique 0522
isomorphe 1974
isomorphisme 0464
isopaque 4243
isopique 2863
isopore 0287
isosiste 0245
isosiste, ligne – 0245
isospore 1835
isosporie 1834
isostasie 0338
isotherme, surface – 0377
isotrope 4681
isotrope, optiquement – 0561
isotropie 4681
isotypisme 0463
issue 2338
isthme 2226

J

jade 0498
jaillissement 1358
jais 3557
jama 2076
jardang 2164
jaspe 3492
jaspe 4127
jaspilite 3492
jaspilite 4114
jaugeage 3798
jayet 3557
jet d'eau 1359
jet de boue 1398
jet de rive 2813
jet de rive, zone lavée par le – 2439
jeune 3176
jig 4155
joint 3506
joint de stratification 3099
joints de retrait cylindriques 1302
joints parallèles, famille de – 4724
jonction 4073
jonction simple 4073
jonction triple 1130
jumbo 4266

K

Kalb, frange de – 0559
kame 2420
kaolin 3217
kaolinisation 0926
karst 2064
karst à dolines 2080
karst à tourelles 2082
karst complet 2067
karst couvert 2069
karstification 2063
karst imparfait 2067
karstique, paysage – 2064
karstique, région – 2064
karstiques, phénomènes – 2065
karst nu 2068
karst parfait 2067
karst partiel 2067
karst profond 2067
karst sous-jacent 2070
karst superficiel 2067
katagenèse 3419
katathermal 3983
katazone 0993
kélyphitique, couronne – 0786
kérogène 3720
kérogenite 3720
kérosène 3744
kérosine 3744
Klinkenberg, facteur de – 3706
Koenigsberger, rapport de – 0310
kreek 2517
krigeage 4197

L

labradorescence 0486
lac 2330
laccolite 0734
lac-croissant 2232
lac de barrage volcanique 1419
lac de cratère 1454
lac de lave 1281
lac de playa 2343
lac en forme de croissant 2232
lac proglaciaire 2393
lac relique 2344
lac résiduel 2344
lac salé 2893
lac sans écoulement 2338
lac sans issue 2338
lac thermokarstique 2646
lacune 1723
lacustre 2888
lagg 3537
lagon 3345
lagon, chenal du – 3348
lagon, plancher du – 3345
lague 2486
lagunaire 2905
lagune 2458
lagune à marée 2458
lahar causé par des pluies 1355

lahar chaud 1353
lahar de fonte glaciaire 1357
lahar de rupture 1354
lahar froid 1355
lait de lune 2153
Lamarckisme 1967
lambeau 4233
lambeau de toit 0706
lambeau témoin 0706
lamellaire 0775
lamellaire 2728
lamellaire 4103
lamelles de choc 0894
lames claires 3580
lame tranchante 2103
laminaire 0972
lamination 1277
lamination 3101
lamination à feuillets frontaire 3140
lamination contournée 3052
lamination de courant 3134
lamination de rides migrantes 3149
lamination inclinée 3140
lamination oblique 3133
lamination vers l'amont 3149
lamine 3101
lamines, en – 3101
lamitute 3128
lamprophyrique 0809
langue de terre 2448
langue glaciaire 2390
lanze 3132
lapidaire 0502
lapié à rigoles 2087
lapié de diaclases 2087
lapiés 2086
lapiés, champ de – 2088
lapiez 2086
lapilli 1324
lapillis cristallins 1325
Laplace-Gauss, distribution – 4189
laquefilm 4230
laquefilm, méthode du – 4230
largeur minimale d'abattage 4200
largeur minimale des terrasses 4201
Larsen, diagramme de variation selon – 0697
latéral 4631
latérite 4124
latéritisation 2681
latitude géocentrique 0098
latitude géomagnetique 0281
lattes, en – 0522
lave 1242
lave aa 1260
lave à blocaux 1259
lave à blocs scoriacés 1260
lave accrûe, boule de – 1296
lave à l'état solide 1242
lave boueuse 1397
lave, boule de – 1296
lave, canal de – 1290
lave, colonne de – 1246
lave cordée 1252
lave, coulants de – 1251
lave, coulée de - 1263
lave d'agglomérat 1269
lave de tufs agglomérés 1346
lave en blocs 1259
lave en boyaux 1252
lave en courrelets 1252
lave en coussins 1253
lave, enduit de – 1226
lave en échaudé 1307
lave en galettes 1253
lave en gratons 1260
lave en oreillers 1253
lave en tessons 1250
lave farineuse 1263
lave figée 1242
lave, fleuve de – 1263
lave fondue, jets de – 1298
lave, langue de – 1264
lave, larme de – 1326
lave liquide, jets de – 1282
lave lisse 1261
lave, lobe de – 1264
lave morte 1285
lave pahoehoe 1250
lave parasite, coulée de – 1265
lave, paroi tapissée de – 1226
lave, placage de – 1226
lave, rayures sur – 1270
laverie 4149
lave sans racine, coulée de – 1282
lave scoriacée 1260
lave sousjacente, coulée de – 1267
lave souterraine, coulée de – 1267
lave striée 1270
lave, torrent de – 1263
lavoir 4149
lèche de vagues 3030
lectostratotype 1745
lectotype 1951
légende 4240
lenteur 0168
lenticulaire 3102
lenticulaire, amas – 1752
lentille 1752
lentille 3055
lentille 3102
lentille 4026
lépidoblastique 0961
leptite 0937
leptothermal 3984
leptynique 0954
leptynite 0937
leptynolite 0940
lessivage 2678
lessivage 3929
lessivage 3930
lessivage 4162
lessivage latéral 2034
leucocrate 0587
leucosome 1038
levée de lave 1295
levée externe de marais 2516
levée naturelle 2238
lèvre abaissée 1559
lèvre affaissée 1559
lèvre relevée 1559
lèvre soulevée 1559
libération, coefficient de – 4112
libérer 4150
lido 2461
ligne de faîte 2316
ligne des centres 4372
lignée 1965
ligne neutre 0570
ligne payée 4856
lignite 3495
lignite brun 3555
lignite brun terreux 3555
lignite clivé 3556
lignite friable 3556
lignite noir brillant 3558
lignite tendre 3555
lignite xyloide 3495
liman 2457
limite élastique 4697
limnétique 2890
limnique 2888
limnique 2890
limnologie 2331
limnoplancton 1801
limon 2715
limon 2720
limon 3215
limoneuse, texture – 2721
limoneux 3203
limoneux 3204
limonite 4126
linéaire 0972
linéation 0973
linéation 1607
linéation 2999
linéation de crénelure 1612
linéation d'étirement 1613
linéation d'intersection 1611
lipide 3723
liquation 0678
liquéfaction 3084
liquéfaction 4718
liquéfaction du charbon 3609
liqueur mère 3377
liquidation 3939
liquide dominante, système à – 0393
liquide intercumulus 0831
liquide résidual 0647
liquidité, indice de – 4716
liquidité, limite de – 2749
liquidité, limite de – 4715
lisière 0527
lissage 4511
lit 3098
lit 3122
litage 3095
litage annuel 3127
litage en alternance répétée 3121
litage entrecroisée 3133
litage incliné 3140
litage massif 3108
litage oblique 3133
litage par différenciation 1606
litage périodique 3127
litage rhytmique 0863
lit contourné 3051
lit d'avalanche 3136
lit de fond 2952
lit de hautes-eaux 2242
lité 3095
lit frontal 3140
lithification 3421
lithique 3178
lithofaciès 2861
lithogenèse 2853
lithologie 2853
lithophage 1857
lithophyse 0780
lithosphère 0088
lithostratigraphie 1709
lithostratigraphique, horizon – 1749
lithostratigraphique, niveau – 1749
lithostratigraphique, unité – 1748
lithostratigraphique, zone – 1747
lithothamniées, trottoir à – 3361
lithotope 2862
litière 2742
lit inverse 3141
lit majeur 2242
lit minéralisé 4021
lit mineur 2242
lit ordinaire 2242
lit par lit 1049
lit plan 3017
lits épais, en – 3108
lits minces, en – 3108
littoral 2836
littoral 2906
lixiviation 4162
loess 2871
loess, coin de – 2641
loess, poupée du – 3489
longitudinale 2289
longulite 0762
lopolite 0735
Love, onde de – 0189
lublinite 2153
lumachelle 3308
lumachelle consolidée 3308
lumachellique 3309
luminescence 0467
luminosité 2677
lutite 3203

M

maar 1433
macadam 4843
macaroni 2131
macéral 3566
macéral actif 3578
macéral, variété de – 3567
maclage polysynthétique 0538
macle 0535
macle de contact 0537
macle de pénétration 0537
macle, loi de – 0535
macrocristallin 0797
macroévolution 1970
macroséisme 0138
macrosismique, région – 0248
macrospore 1837
mafique 0599
mafique, front – 1066
mafique, roche– 0599
magasin 3755
magasin, roche – 3679
magma 0639
magma 1198
magma neuf 1200
magma palingénétique 0644
magma parental 0645
magma primaire 0644
magma résiduel 0646
magma superficiel 1244
magmatique 0616
magmatique 3973
magnétique, susceptibilité – 0298
magnétique terrestre, champ – 0276
magnétisante, force – 0317
magnétisme terrestre 0276
magnétomètre 0316
magnétomètre à componente horizontale 4521
magnétomètre à componente verticale 4521
magnétomètre aéroporté 4522
magnétomètre à proton 4524
magnétomètre à résonance nucléaire 4524
magnétomètre astatique 0319
magnétopause 0277
magnétosphère 0277
magnitude 0220
magnitude, échelle de – 0220
maillé 0964
maille 3157
maille élémentaire 0510
maillée, texture – 0856
maille rhomboédrique 0516
malléable 0471
mallosismique, région – 0250
mamelon 1422
mamelon central 2227
mamelonné 4094
manganèse, nodule de – 3477
manifestations prévolcaniques 1191
manifestations ultimes 1167
manifold 3798
manomètre 4769

manteau 4232
manteau de cendre 1320
manteau inférieur 0091
manteau supérieur 0091
manteau terrestre 0091
marais 3532
marais à mangrove 2515
marais à sphaignes 3543
marais, bordure de – 2512
marais côtier 2514
marais, de – 2887
marais élevé 3542
marais, fer des – 4116
marais littoral 2909
marais salé 2514
marais salé 2909
marais, terre ferme au bord du – 2518
marbre 0920
marche 0527
marche, différence de – 0563
mardelle 2076
mardelle 2632
mare 2330
mare 2632
marécage 3532
marécage bordier 3537
marécageux 3532
marée 2818
marée, amplitude de – 2819
marée basse 2818
marée haute 2818
marée terrestre 0199
margarite 0766
marias plat 3533
marigot 2509
marigot 2517
marin 2900
marmite 2105
marmite de géant 2028
marmite de géant 2248
marmite de géant 2389
marmite torrentielle 2027
marmorisation 0920
marnage 2819
marne 3221
marne argileuse 3223
marne calcaire 3224
marque 3018
marque de niveau 3031
marque d'objet 3032
marque glaciaire 2395
mascaret 2823
mascon 0052
masse charriée 1569
masse, déficit de – 0343
masse, excès de – 0343
masse extrusive 1239
masse stationaire 0125
masse-tige 4270
masse volumique 2751
massif 0859
massif ancien 2560
massif arasé 2588
massif éruptif 1239
massif rocheux 4722
massif surélevé 2540
mat 3408
matérial éjecté 0055
matériel cryoclastique 2656
matériel de départ 2657
matériel détritique 2005
matériel-originel 2657
matériel-parental 2657
matériel pyroclastique 1309
matière minérale 3553
matière organique extractible 3724
matrice 3080
matrice 4722
maturation 2685
maturation 3725
maturation, indice de – 3726
maturité 3725
maximum 1019
maximum et minimum 0164
méandre 2219
méandre abandonné 2232
méandre de terrasse 2223
méandre de vallée 2221
méandre divagant 2220
méandre, échancrure de – 2228
méandre encaissé 2221
méandre encaissé 2222
méandre hérité 2221
méandre imprimé 2221
méandre libre 2220
méandre, lobe de – 2225
méandre recoupé 2232
méandre, recoupement d'un – 2229
méandre sculpté 2222
méandres, lit des – 2224
méandres, zone des – 2224
méat 2099
mécanique des roches 4664
mécanique des sols 4664
méditerranéenne, série – 0690
mégacristal 0608
mégacyclothème 3120
mégagamétophyte 1838
mégalosphérique 1868
mégaride 3007
mégasismique, région – 0249
mégasphérique 1868
mégasporange 1837
mégaspore 1837
méiofaune 1853
mélange 2927
mélange 4805
mélanocrate 0588
mélanosome 1038
mélatope 0570
melikaria 3487
membrane semi-permeable 3397
meneau 1608
mer 2782
mer adjacente 2783
mer, basse – 2818
mer bordière 2783
mercaptan 3716
mer de rochers 2057
mer du vent 2807
mer, haute – 2818
meringuée 1288
mer lunaire 0051
mer moyenne, basse – 2820
mer moyenne, haute – 2820
mer, niveau moyen de la – 2817
meroplancton 1801
mer, pleine – 2818
mésocrate 0587
mésocristallin 0797
mésocumulat 0832
mésohalin 2832
mésolimnion 2333
mésonorme 0606
mésopélagique 2845
mésostase 0626
mesothermal 3983
mesothermale, région – 0391
mésotrophe 3549
mésotype 0587
mésozone 0993
méta- 0883
méta-alumineux 0693
méta-anthracite 3562
métabasite 0883
métabentonite 3218
métablastèse 0884
métablastique 0884
métabolisme 1803
métalimnion 2333
métallifère 3917
métallique 0484
métallique 3911
métallique, non– 3911
métallogénique 3903
métallotecte 3904
métamorphique, degré – 0992
métamorphique, roche – 0880
métamorphiques, couple de zones – 1139
métamorphiques couplées, zones – 1139
métamorphisé 3997
métamorphisme 0879
métamorphisme allochimique 0881
métamorphisme de choc 0893
métamorphisme de contact 0896
metamorphisme de la matière organique, diagenèse et – 3521
métamorphisme d'enfouissement 0887
métamorphisme d'impacte 0893
métamorphisme général 0887
métamorphisme isochimique 0881
métamorphisme optalique 0897
métamorphisme prograde 0903
métamorphisme régional 0887
métamorphisme régressif 0904
métamorphisme thermique 0896
métamorphisme topochimique 0881
métamorphogène 3974
métamorphose 1794
métaride 3014
métasomatose 3956
métasomatose régionale 1056
métatectique, matériel – 1040
métatype 1954
métaux, extractibles – 4654
métaux extractibles à chaud 4655
métaux, lourds – 4653
métaux lourds extractibles à froid 4655
météore 0063
météorite 0063
météorite pierreuse 0071
météorites, science des – 0062
météoroide 0063
méthode à boucle horizontale 4575
méthode à boucle verticale 4575
méthode de chute de potentiel 4560
méthode de polarisation provoquée 4569
méthode de potential spontané 4557
méthode de résistivité électrique 4561
méthode des équipotentielles 4558
méthode du rectangle 4563
méthode électrique 4554
méthode électro-magnétique 4572
méthode magnéto-tellurique 4553
méthode tellurique 4553
miarolitique 0873
micacé 0775
micacé 3177
micrite 3281
micrite 3296
micrite à bioclastes abondants 3288
micrite à bioclastes épars 3288
micrite dolomitique 3282
micrite peu fossilifère 3288
microcoprolithe 1863
microcristallin 0797
microcristallin 3091
microévolution 1970
microfalaise 2470
microfalaise 2516
microfossile 1790
microgamétophyte 1838
microlite 0759
microlithotype 3581
microlitique 0836
microlog 4587
microlumachelle 3308
micrométéorite 0064
micropaléontologie 1786
micropegmatitique 0848
micropegmatoïde 0848
microplancton 1830
microplankton 1830
micropli 1643
microscope polarisant 0611
microscopie métallographique 3925
microséisme 0097
microséisme 0146
microsismique, agitation – 0097
microsismique, région – 0251
microsphérique 1868
microsporange 1836
microspore 1836
microstructure 1479
microtectonique 1479
microtexture 0790
mictite 1023
migma 1022
migmatisation 1022
migmatite 1022
migmatitisation 1022
migration 3748
migration 4457
migration horizontale 3750
migration latérale 3750
migration, parcours de – 3751
migration primaire 3749
migration secondaire 3749
migration verticale 3750
milieu 1900
milieu poreux 3680
milieu primaire 0402
milieu secondaire 0402
mimétique 1008
minerai 0462
minerai 3912
minerai 3913
minerai à faible teneur 3915
minerai à haute teneur 3915
minerai, amas de – 4032
minerai brut 4138
minerai complexe 3914
minerai de fer argileux 4117
minerai démontré 4209
minerai disseminé 4004
minerai entraîné 4079
minerai, étude minéralogique du – 3925
minerai exploitable 3916
minerai exploitable localisé 4178
minerai marchand 4139
minerai métallique 0462
minerai monominéral 3914
minerai pauvre 3915
minerai, poche de – 4032
minerai riche 3915
minerai simple 3914
minerai tout venant 4138
minéral 0460
minéral 3912
minéral accessoire 0594
minéral accessoire 3174
minéral argileux 3213
minéral artificiel 0461
minéral auxiliaire 0594
minéral caractéristique 0595

minéral caractéristique 0689
minéral constituant 0462
minéral de gangue 3921
minéral de métamorphisme de contact 0978
minéral de synthèse 0461
minérale industrielle, substance – 0462
minéral essentiel 0594
minéral industriel 0462
minéralisation 2682
minéralisation 3936
minéralisation littorale 2912
mineralisation primaire 3913
minéralisé 3917
minéral léger 3173
minéral lourd 3173
minéral métallique 0462
minéral normatif 0607
mineral occulte 0757
minéralogie 0459
minéral opaque 0462
minéral opaque 0610
minéral primaire 0596
mineral principal 0594
minéral résistant 3175
minéral secondaire 0596
minéral stable 3175
minéral synthétique 0461
minéral typomorphe 0988
minéralurgie 4137
minéral virtuel 0607
minette 4115
miogéosynclinal 1096
miospore 1833
miroir 1538
miroir, groupement en position – 4433
mise à la masse 4559
mission photographique 4349
mixité, coefficient de – 4112
mixohalin 2833
mixohalin 2834
mixtes 4142
mobile, composant – 1058
mobile, zone – 1093
mobilisat 1057
mobilisation 1057
mobilisme 1113
mobilité 3699
mobilité, taux de – 3699
modal 0593
mode 0166
mode 0593
mode à perte 0195
mode fundamental 0167
mode harmonique 0167
modèle 4775
modèle analogique 4782
modèle analogique électrique 4783
modèle centrifuge 4781
modèle d'ordinateur 3866
modèle équivalent 4778
modèle, essai sur – 4778
modèle, établissement d'un – 4478
modèle géomécanique 4779
modèle mathématique 4776
modèle numérique 3866
modèle photo-élastique 4783
modèle physique 4778
modèle stratiforme 4568
moder 2744
modéré 1626
modulation de fréquence 4492
mofette 1402
mofette 3554
Mohorovičić, discontinuité de – 0087
Mohr, cercle des contraintes de – 4690
Mohr-Coulomb, critère de rupture – 4694
Mohr, courbe intrisique de – 4691
Mohr, diagramme de – 4688
Mohs, échelle de dureté de – 0473
Mohs, échelle de dureté de – 4701
mol 3675
molasse 2926
môle tectonique 1589
mollisol 2647
moment 1759
moment magnétique 0296
moment, valeur du – 0336
monadnock de position 2608
monadnock de résistance 2608
monistique, hypothèse – 0020
monnaie constante 3899
monnaie courante 3899
monoclinal 1617
monoclinique 0515
monogénique 3265
monolithe 1424
monométallique 3911
monominéral 0589
monophylétique 1960
monotypique 1945
mons 0037
montagne à cirques 2618
montagne à structure de voûtement 2538
montagne à structure en dôme 2538
montagne à structure faillée 2539
montagne à structure plissée 2538
montagne, avant– 2605
montagne-blocs 2539
montagne centrale 0054
montagne déchiquetée par des cirques 2616
montagne de plissement 2538
montagne, éperon de – 2605
montagne fracturée 2539
montagne fracturée penchée 2539
montagne, haute – 2575
montagne marginale 2542
montagne moyenne – 2575
montagne plissée-faillée 2539
montagne résiduelle 2607
montagne rongée par des cirques 2616
montagnes, chaîne de – 1102
montagnes, noeud de – 2605
montagne sous-marine 2798
montagne tabulaire 2556
mont insule 2608
mont sous-marin 2798
monzonitique 0813
moraine 2403
moraine cannelée 2409
moraine d'ablation 2404
moraine d'ablation 2412
moraine de fond 2404
moraine de fond 2409
moraine de fond 2874
moraine de lave 1295
moraine déposée 2410
moraine de poussée 2419
moraine de retrait 2417
moraine frontale 2413
moraine frontale 2416
moraine inférieure 2409
moraine intérieure 2408
moraine interne 2408
moraine latérale 2411
moraine longitudinale 2415
moraine marginale 2415
moraine médiane 1295
moraine médiane 2407
moraine mouvante 2405
moraine profonde 2409
moraine rempart 2414
moraine riveraine 2416
moraine superficielle 2406
Morkill, formule de – 4217
morne 2608
morphologie 1793
morphotectonique 1076
mortier, en – 0966
morts terrains 3501
morts-terrains 3924
mosaïque 4380
mosaïque contrôlée 4381
mosaïque, en – 0960
mosaïque non contrôlée 4381
mosaïque semi-contrôlée 4381
motte 2725
motte de lave 1297
motte de magma 1297
mouche 4032
mouchetage dolomitique 3461
moucheté 3462
mouille 2247
moulage 1825
moulage basal 3019
moulage de flute 3026
moulage externe 1825
moule de tronc d'arbre 1292
moule externe 1824
moule interne 1824
moulin 2389
moulinet 4759
moulinet hydrométrique 3669
mouvement en masse 2037
mouvement eustatique 2858
mouvement, ligne de – 2372
mouvement négatif 2858
mouvement positif 2858
mouvement verticale 4511
moyenne glissante, lissage par – 4198
moyenne mobile, lisage par – 4198
mull 2744
multiloculaire 1870
multiplexage 4424
Munsell, code – 2676
mur 0705
mur 1451
mur 3501
mur 3515
mur 3516
mur 4009
mur 4010
mur 4852
mûr 3725
mur à radicelles 3515
mur argileux 3515
mur de soutènement 4797
mûre de caverne 2149
mur parafouille 4861
mutation 1982
mylonite 0915
mylonite 1541
mylonitique 0969
mylonitisation 0915
myrmékite 0785
myrmékite, structure – 4100
myrmékitique 0852

N

nacre 0503
nannofossile 1790
nannoplancton 1799
naphte 3744
nappe 1683
nappe 3109
nappe artésienne 3637
nappe captive 3636
nappe de charriage 1575
nappe de chevauchement 1575
nappe de lave 1275
nappe de recouvrement 1682
nappe libre, niveau d'une – 3642
nappe perchée 3635
nappe phréatique 2765
nappe plongeante 1688
nappes chevauchantes 1691
nappes, pile de – 1691
nappes, tectonique en – 1679
néanique 1866
nébuleuse, hypothèse de la– 0021
nébuleuse solaire 0021
nébulite 1054
neck 1477
necton 1798
nectonique 1798
neige à pénitents 2365
neige coulante 2351
neige entassée par le vent 2347
neige flasque 2348
neige folle 2346
neige humide 2349
neige poudreuse 2346
neige pourrie 2350
neige sèche 2349
neiges éternelles, limite des – 2354
neige soufflée 2353
nématoblastique 0962
Néo-darwinisme 1966
néoformation de minéraux 0905
Néo-lamarckisme 1967
néomorphe 1983
néontologie 1787
néosome 1037
néostratotype 1745
néotype 1951
néovolcanisme 1150
népionique 1866
neptunisme 0004
nerf 3506
nerf stérile 3506
néritique 2841
néritique 2915
nettoyage 3813
Neumann, lignes de – 0068
névé 2363
névé 2364
névé, limite du – 2364
niche 1902
niche d'arrachement 2052
niche de départ 2052
niche de rongement glaciaire 2395
nivation 2357
nivation, niche de – 2612
niveau de bruit 0097
niveau de référence 4249
niveau des crêtes 2597
niveau dynamique 4507
niveau hydrostatique 3778
niveau local 2025
niveau minéralisé 4008
niveau piézometrique 3646
niveau statique 3666
niveau temporaire 2025
niveau zéro 4577
nivelage 2589
niveo-éolien 2870
nodale, surface – 0234
nodal, ligne – 0234
nodal, plan – 0234
nodulaire 3473
nodule 0771
nodule 3473
nodule de manganèse 3477
nodule de sidérite 3483

noduleux 4085
nomenclature binominale 1922
nomenclature trinominale 1922
nomenclature uninominale 1922
non-orienté 0949
norme 0606
novaculite 3366
noyau 1628
noyau 4861
noyau étranglé 1628
noyau externe 0093
noyau interne 0093
noyau terrestre 0093
nuage de cendres 1363
nuage de poussières 1363
nuée ardente 1366
nuée de cendres électrisées 1365
nuée en forme de chou-fleur 1362
nuée volcanique 1360
nunatak 2618
nu-pied 3795

O

obduction 1133
obséquent 2285
observations, généralité des – 2675
obsidienne 1262
obturateur 4279
obturateur de puits, bloc – 4289
océan 2782
océanique 2842
océanique 2918
océanisation 1142
océanographie 2781
océanologie 2781
ocelle 3055
octaédrite 0067
oedomètre 4756
oeuil de chat 0488
ogive 2378
oiseau 4522
oligogénique 3265
oligohalin 2832
oligotrophe 3549
olistolite 2977
olistostrome 2977
ombilic 2793
ombre, zône d'– 0216
ombrogène 3550
oncolithe 3300
ondation 1122
ondation, hypothèse de l'– 1121
onde acoustique 4482
onde couplée 0193
onde de choc 0274
onde de choc 4481
onde de compression 0183
onde de dilatation 0183
onde de gravitation 0202
onde de manteau 0194
onde de surface 0188
onde de surface 4439
onde de volume 0179
onde d'interface 0196
onde du noyeau 0180
onde, front d'– 4470
onde, guide d'– 0101
onde guidée 0195
onde longitudinale 0182
onde, longueur d'– 0162
onde, longueur d'– 2808
onde, nombre d'– 0162
onde, normale au plan d'– 0550
onde P 0182
onde, période d'– 2808
onde, plan d'– 0549
onde S 0184
onde sismique 0161
onde sismique 4481
onde solitaire 2815
onde sonore 4482
ondes R 0191
onde, surface d'– 0549
onde T 0201
onde transformée 0185
onde transversale 0184
onde visible 0192
ondulation 2642
ondulation 4727
ondulation, indice d'– 3002
ontogénèse 1794
ontogénie 1794
ontologie comparative 0007
oogone 1843
oolithe 3479
oolithique 3479
oolithique, roche – 3479
opalescence 0486
opaque 0545
opérateur 4447
opération, accord d'– 3901
opérations unifiées, accord d'– 3901
ophicalcite 0930
ophiolite 0696
ophiolite 1098
ophiolitique, série – 0696
Oppel-zone 1780
optique cristalline 0544
orage magnétique 4529
orage volcanique 1373
orbiculaire 0867
orbicule 0867
ordonnée á l'origine 4463
ordre 0685
ordre 1929
ordre 2666
organique, matière – 2740
orgue de lave 1301
orient 0486
orientation crystallographique 1013
orientation, étude d'– 4639
orientation préférentielle 1013
origine 2193
origine, centre d'– 1699
origine, région d'– 2943
orocline 1140
orogène 1101
orogenèse 1100
orogenèse de type saxon 1078
orogénie 1101
orogénique, période – 1106
orogénique, phase – 1107
ortho- 0986
orthochème 3271
orthoconglomérat 3255
orthocumulat 0832
orthogenèse 1975
orthogéosynclinal 1095
orthogonal 0206
orthomagmatique 3975
orthomagmatique, stade – 0651
orthophyrique 0826
orthoquartzite 3230
orthorhombique 0515
orthostéréoscopie 4367
oscillation, hypothèse de l'– 1119
oscillation libre 0198
oscillation libre, période d'– 0130
oscillation toroidale 0198
otolithe 1865
oued 2587
oule 2610
oulle 2248
ouralitisation 0928
ouvala 2073
ouvert 1626
ouverture 4728
oxique 0424
oxydation 0416
oxydo-réduction, potentiel d'– 0421
oxydo-réduction, réaction d'– 0416
ozocérite 3735

P

pacifique, série – 0690
packer 4279
paille 0487
palagonite 1254
palagonitisation 1254
palatre 2887
paléobotanique 1827
paléoclimatologie 0011
paléocourant 2864
paléoécologie 1889
paléogéographie 0011
paléomagnétisme 0294
paléontologie 1786
paléontologie végétale 1827
paléosol 3126
paléosome 1037
paléotempérature 2864
paléothermometrie 0456
paléovolcanique 0636
paléovolcanisme 1150
paléozoologie 1847
palétiologie 0007
palettes, appareil à – 4759
palichnologie 1872
palichthyologie 1848
palingenèse 1024
palingenèse 1968
palingénétique 1024
palinspastique, reconstruction – 0009
paludéen 3532
palus 3532
palustre 3532
palynologie 1828
palynomorphe 1829
panaché 2692
panache mantellique 1145
pandémique 1814
pan effondré 1559
panidiomorphe 0808
panneau 4199
pan salé 2909
paquets de mer, zone des – 2439
para- 0986
paracme 1985
paraconglomérat 3256
paraffine 3717
parafouille 4861
parafouille injecté 4804
paragenèse 0586
paragenèse 3179
paragenèse 3907
paragenèse diagénétique 3394
paragenétique 0586
paralique 2898
parallaxe 4364
parallaxe, barre à – 4365
paramagnétisme 0298
paramarginal 4212
paramètre de la source 0151
paramètre hypocentral 0151
paramorphisme 0465
parastratotype 1743
paratype 1950
parautochtone 0727
paravolcanique, manifestations – 1180
parcours 0206
paroi 1453
paroi 1529
paroi 4009
paroi inférieure 1529
parois 0705
paroi supérieure 1529
partage, col de – 2315
partage conséquente, ligne de – 2317
partage des eaux en vallée, ligne de – 2315
partage des eaux, ligne de – 2314
partage des eaux, ligne de – 2511
partage des glaces, ligne de – 2381
partage en zigzag, ligne de – 2321
partage migrante, ligne de – 2319
partage se déplacant brusquement, ligne de – 2320
partage se déplacant lentement, ligne de – 2319
partage subséquente, ligne de – 2318
particule 3078
particules, diamètre des – 3193
particules, dimension des – 3193
particules élémentaires minérales 2719
particules fines 3207
partie encaissée 2390
passage surbaissé 2098
passe 2503
passe 2508
passée 3122
passée stérile 3506
pâte 0626
pâté de corall 3347
patera 0037
patine 3403
patine désertique 3405
pavage de blocs 2656
pavage de pierres 2656
pavage désertique 2160
pavage désertique 3405
pavé désertique 2160
paysage de moraine de fond 2630
paysage de moraines terminales 2630
paysage étagé 2599
PDB 0454
pédiment 2582
pédiment, col de – 2586
pédiments coalescents 2584
pédiments soudés 2584
pédiplaine 2584
pédogenèse 2659
pédogenèse, facteur de – 2659
pédologie 2658
pedon 2689
pédoncule 2226
pegmatitique 0847
pegmatitique 3978
pegmatitique, stade – 0653
pegmatitisation 1063
pegmatoïdique 0814
pégostylite 2152
pélagique 1910
pélagique 2844
pélagique 2919
Pélé, cheveux de – 1326
Pélé, larme de – 1326
pelite 3208
pelitique 3208
pellet 4150
pellicule 3100
pellicule 3126
pellicule gazeuse 1370
pellicule orthochromatique 4334
pellicule panchromatique 4334

pelote fécale 1863
pendage 1484
pendage 4017
pendage, angle de – 1485
pendage apparent 1484
pendage apparent 4250
pendage, azimut du – 1490
pendage, complément du – 4018
pendage, en amont – 1489
pendage, en aval – 1489
pendagemétrie 4598
pendant 2130
pendeloque 2130
pendule 0351
pendule 4774
pendule inversé 4774
pendule, longueur équivalent du – 0126
pendule réversible 0351
pénécontemporain 3474
pénéplaine 2589
pénéplaine d'emblée 2593
pénéplaine embryonnaire 2593
pénéplaine inachevée 2593
pénéplaine locale 2593
pénéplaine naissante 2593
pénéplaine partielle 2593
pénéplaine-reliquat 2594
pénéplanation 2589
pénésismique, région – 0252
pénétration 4513
pénétration enregistrée 4513
pénétration, résistance à la – 4709
pénétration standard, essai de – 4712
pénétration vitesse de – 4298
pénétromètre 4758
pénétromètre à cône 4758
pénétromètre dynamique 4711
pénétromètre statique 4711
pente à sédimentation carbonatée 3322
pente continentale 2790
pente de glissement 2237
pente d'équilibre 2198
pente d'équilibre 3137
pente de rebondissement 2236
pente du talus 4827
pente glaciaire, rupture de – 2623
pente insulaire 2791
pente limite 2198
pente récifale 3354
pente, rupture de – 2199
pente sous le vent 2965
pente structurale 2559
percé 2290
percé 2290
percée 2327
percée de levée naturelle 2240
percolation 3622
percussion, figure de – 3410
percussion, marque de – 3410
perforation 1882
perforation à balles 3799
pergélisol 2637
périglaciaire 2633
périmagmatique 3985
périmètre mouillé 2261
périmorphose d'un tronc 1292
période 0163
période 1765
perlé 0484
perle de caverne 2149
perle de culture 0503
perle de soufre 1321
perlitique 0878
perméabilité 2753
perméabilité 3697
perméabilité, appareil de – 4755
perméabilité, barrière de – 3771
perméabilité effective 3698
permeabilité magnétique 0304
perméabilité primaire 3701
perméabilité rélative 3698
perméabilité secondaire 3701
perméamètre 4755
perte marine 2075
perthite 0785
perthitique 0851
perturbation magnétique 1192
pesanteur 0323
pesanteur, accélération de la – 0324
pesanteur à l'air libre 4538
pesanteur, champ de – 0326
pesanteur, intensité de la – 0324
pesanteur mesuré, champ de – 4536
pesanteur, mesure de la – 0349
pétrification 3457
pétrochimie 0580
pétrofabrique 0995
pétrofabrique 1605
pétrogenèse 0581
pétrographie 0579
pétrographie sédimentaire 2851
pétrole 3712
pétrole, anneau de – 3776
pétrole brut 3712
pétrole brut à forte teneur en eau 3714
pétrole/eau, contact – 3777
pétrole/eau, niveau – 3777
pétrole, gisement de – 3773
pétrole lourd 3745
pétrole sous-saturé 3857
petrolier, champ – 3773
pétrolifère, zone – 3775
pétrologie 0578
pétrologie sédimentaire 2851
peuplements turficales, série de – 3538
pF 2771
pH 0422
phacolite 0733
phanérogène 0795
phase 0165
phase 1758
phase 3906
phase abyssale 1197
phase d'activité 1208
phase de paroxysmes 1208
phase de repos 1169
phase des ondes longues 0188
phase de tranquillité 1169
phase d'extrusion 1197
phase d'intrusion 1197
phase éruptive 1208
phase explosive 1208
phase finale 0175
phase finale 1167
phase fumerollienne 1168
phase gazeuse intermédiaire 1162
phase hypabyssale 1197
phase magmatique 1108
phase post-volcanique 1197
phase préliminaire 0181
phase principale 0190
phase solfatarienne 1168
phase terminale 1167
phase volcanique 1197
phénoblaste 0981
phénoclaste 3239
phénocristal 0758
phénomènes annonciateurs 1191
phénomènes prémonitoires 1191
phénotype 1979
Phi, échelle en – 3163
phlébite 1044
phosphatisation 3459
phosphorescence 0467
phosphorite 3480
photique 1904
photogéologie 4327
photogrammétrie 4328
photographie aérienne 4349
photographie de satellite 4349
photographie oblique 4353
photographie oblique latérale 4353
photographie verticale 4353
photo index 4382
photosynthèse 1803
phréatique, surface de la nappe – 2765
phtanite 3369
phylétique 1960
phyllade 0943
phyllade en crayon 0942
phyllonite 0918
phyllonite 1541
phyllonitisation 0918
phylogenèse 1959
phylogénie 1959
phylogénique, arbre – 1960
phylum 1929
physique du globe 0078
phytoclaste 3168
phytoplancton 1800
pied cube 3892
piedmont, gradin de – 2604
piedmont, plate-forme de – 2604
piédroit 4852
piège 3756
piège anticlinal 3757
piège chimique 4650
piège de faille 3757
piège de perméabilité 3758
piège de pincement 3758
piège hydrodynamique 3761
piège par discordance 3759
piège par relief enterré 3760
piège stratigraphique 3758
piège structural 3757
piémont, glacis de – 2583
pierre 3242
pierre agglomérié 0480
pierre concassée 4876
pierre précieuse 0478
pierre précieuse synthétique 0479
pieu 4838
pieu-colonne 4839
pieu de sable 4813
pieu drainant 4813
pieu flottant 4839
piézocristallisation 0633
piézomètre 3646
pile 4845
pilier 2138
pilier de terre 2029
pilier prismatique complet 3593
pilotaxitique 0825
pinacle 2805
pinacle corallien 3347
pincement 3509
pingo 2643
pipe minéralisée 4037
piperno 1342
pipkrakes 2639
pisolite 2149
pisolite volcanique 1321
pisolithe 3478
pisolithique, roche – 3478
piste 1877
piste 1878
pitch 4019
placer 3236
placer 3997
placer marin 2912
plage 2485
plage, arrière– 2487
plage, arrière– 2840
plage, avant– 2840
plage, bas de – 2487
plage, bas de – 2840
plage, basse – 2487
plage, crête de – 2490
plage, gradin de – 2488
plage, haute – 2487
plage, levée de – 2490
plage, microfalaise de – 2489
plage progradée 2492
plaine 0029
plaine à chenières 2491
plaine alluviale fluvio-glaciaire 2424
plaine alluviale pro-glaciaire 2424
plaine côtière 2440
plaine de cendres 1462
plaine de troncature 2591
plaine d'inondation 2239
plaine littorale 2440
planaire 0972
plan auxiliaire 0238
plancher stalagmitique 2128
plancton 1799
planctonique 1799
plan de construction anti-sismique 0260
plan de stratification découvert 2557
planète terrestre 0025
planétologie 0023
planèze 1278
plankton 1799
plante accumulatrice 4647
plante indicatrice 4646
plaque 3132
plaque de bruyère 2698
plaque de gazon 2698
plaque de lave figée 1285
plaque, essai de charge sur – 4708
plaques, en – 3132
plaques, tectonique de – 1125
plasticité 2747
plasticité 4669
plasticité, indice de – 4716
plasticité, limite de – 4715
plasticité, limite inférieure de – 2749
plasticité, limites de – 2748
plasticité, limite supérieure de – 2749
plastique 2747
plat 3183
plate 2728
plateau 2141
plateau 2797
plateau basaltique 1257
plateau continental 2788
plateau insulaire 2791
plateau surélevé 0033
plate-forme 1083
plate-forme carbonatée 3322
plate-forme carbonatée isolée 3323
plate-forme continentale 1788
plate-forme d'abrasion 2479
plate-forme d'accumulation 2480
plate-forme d'aplanissement 2479
plate-forme d'érosion marine 2479
plate-forme insulaire 2791
plate-forme stabilisée 4539
plate-forme stable 1089
plate-forme structurale 2555
platine universelle 0614
playa 2580
playa 2892
pleins-bords, niveau de – 2251
pleistosiste 0246

pleistosiste, ligne – 0246
pleistosiste, région – 0246
pléochroisme 0562
plésiotype 1952
pli 1616
pli, amorce du – 1666
pli, angle du – 1626
pli annexe 1642
pli à noyau de percement 1693
pli à noyau de percement 1698
pli à noyau percant 1698
pli anticlinal 1618
pli à rebours 1689
pli asymétrique 1644
pli, axe de – 1630
pli coffré 1649
pli concentrique 1652
pli conique 1655
pli couché 1680
pli cylindrique 1654
pli de cisaillement 1662
pli de fond 1120
pli déjeté 1646
pli d'entraînement 1570
pli déversé 1647
pli diapir 1698
pli dissymétrique 1644
pli droit 1645
pli écaille 1576
pli, embryon – 1666
pli en chevron 1651
pli en éventail 1650
pli en genou 1648
pli en retour 1689
pli-faille 1577
pli-faille couché 1576
pli majeur 1642
pli mineur 1642
pli-nappe 1682
pli non-cylindrique 1654
pli parallèle 1652
pli parasite 1677
pli plongeant 1637
pli polyclinal 1657
pli principal 1642
pli ptygmatique 1658
pli renversé 1647
pli replié 1668
pli replissé 1668
pli reployé 1668
plis à relais 1676
plis chouchés empilés 1681
pli secondaire 1642
pli semblable 1653
plis en cascades 1706
plis en chaise 1706
plis en coulisses 1676
plis en échelons 1676
plis, faisceau de – 1674
plis frontaux, région des – 1686
plis, groupement de – 1670
pli similaire 1653
plissement à rebours 1689
plissement contemporain de la sédimentation 1667
plissement disharmonique 1663
plissement dysharmonique 1663
plissement embryonnaire 1666
plissement en retour 1689
plissement entérolithique 3076
plissement par fluage 1701
plissement, phase de – 1107
plissement, phase de – 1669
plissement précurseur 1667
plissement, tectonique de – 1615
plissement, zone de – 1674
plissotement 1643
plis superposés, paquet de – 1681
plis, système de – 1670
pli symétrique 1644
pli synclinal 1619
pli transverse 1678
plomb, âge conventionnel du – 0440
plomb anormal 0441
plomb commun 0437
plomb, fil à – 0329
plomb-plomb, datation – 0434
plomb primitif 0437
plomb radiogénique 0437
plombs à plusieurs stades, modèle d'évolution des – 0439
plombs à un seul stade, modèle d'evolution des – 0439
plombs, courbes d'évolution des – 0438
plongement 4018
plongement 4019
pluie de poussière 2159
pluie d'éruption 1373
pluie de sang 2159
pluie, empreinte de gouttes de – 3041
pluriloculaire 1870
plutonique 0617
plutonique 3988
plutonique, roche – 0617
plutonisme 0004
plutonisme 0615
pluviosité 3614
pneumatogène 0655
pneumatolitique 3977
pneumatolyse 3950
pneumatolytique, stade – 0653
poche 2649
poche 3105
poche filonienne 4050
pod 4034
podzolisation 2680
poecilitique 0844
poecilitique, crystal – 0981
poeciloblastique 0974
poecilophitique 0846
poikilohalin 2831
point 0494
point brillant 0216
point chaud 1144
point de bulle 3857
pointe 2449
pointé 4459
point noir 0570
point principal 4369
Poisson, coefficient de – 4700
poix 3743
polarisation 4557
polarisation anomale, teinte de – 0568
polarisation, figure de – 0569
polarisation magnétique 0297
polarisation, teinte de – 0568
polarité, époque de – 0292
polarité inversée 0290
polarization 0236
pôle 1018
pôle de la guirlande 1020
pôle géomagnétique 0280
pôle magnétique 0282
pôles, diagramme de – 1018
poli 3404
polissage 2399
polissage glaciaire, limite du – 2631
polje 2074
pollen 1839
polyédrique 2729
polygénétique 0590
polygénique 3265
polygone de pierres 2653
polygone de sel 3049
polyhalin 2832
polymétallique 3911
polymétamorphisme 0901
polyminéral 0589
polymorphisme 0464
polymorphisme 1811
polyphylétique 1960
polytope 1819
polytopique 1819
polytypique 1945
polytypisme 0463
pompage, centrale de – 4870
pompage, essai de – 3711
ponce 1308
ponceau 4846
ponceux 0875
ponceux 1308
pont 2102
pont argileux 3441
pont naturel 2079
population 1941
porcelanite 3374
pore 3083
pore, remplissage de – 3440
pores, volume des – 3681
porosité 2752
porosité 3681
porosité effective 3704
porosité efficace 3695
porosité fissurale 3694
porosité initiale 3442
porosité intergranulaire 3688
porosité intragranulaire 3688
porosité matricelle 3687
porosité primaire 3684
porosité secondaire 3685
porosité, taux de – 3682
porosité texturale 3693
porosité totale 3681
porosité vacuolaire 3692
porphyre 0817
porphyre cuprifère 4133
porphyrique 0817
porphyrite 0817
porphyroblaste 0981
porphyroblastique 0956
porphyroclaste 0981
porphyroclastique 0966
porphyroïde 0819
porphyroïde 0957
portée 2249
position débordée 3113
position en retrait 1727
position en retrait 3112
postcinématique 1079
postorogénique 0726
post-orogénique 1108
posttectonique 0726
post-tectonique 0906
post-volcaniques, phénomènes – 1374
potassium-argon, datation – 0449
potassium-calcium, datation – 0449
potentiel 3645
potentiel de contact 4615
potentiel d'électrofiltration 4617
potentiel de membrane 4616
potentiel électro-cinétique 4617
potentiel ionique 0419
Potsdam, système de – 0334
poulier 2464
pourissement 3526
pourrière 2187
pourriture 3526
poussée active des terres 4706
poussée des terres 4705
poussée des terres au repos 4705
poussée hydraulique 4737
poussière 3206
poussière cosmique 0065
poussière d'abrasion 3209
poussière glaciaire 3209
poussière météorique 0065
poussière volcanique 1316
pouvoir de résolution 4347
pouvoir réflecteur 0574
pouzzolane 1338
PP 4569
Pratt 0345
préadaptation 1980
précinématique 1079
précipitation 0411
précipitation atmosphérique 3614
précipitation, barrière de – 4642
précipitation chimique, de – 3268
préconsolidation 4703
précurseur 0150
précurseurs, signes – 1191
prédécoupage 4823
prédominance du nombre impair, règle de – 3728
pré-littoral 2839
prépondérant 1816
pré-salé 2514
pressiomètre 4760
pression, baisse de – 1173
pression capillaire, courbe de – 4612
pression, cellule de – 4771
pression, chute de – 3853
pression d'abandon 3849
pression d'écoulement 3847
pression de fond 3847
pression de fond, enrégistrateur de – 3848
pression de gisement 3819
pression de reservoir 3819
pression des terrains susjacents 3819
pression des terres, appareil pour mesurer la – 4771
pression de succion de l'eau 2770
pression de surcharge 3819
pression en caverne, essai de – 4762
pression excédentaire 3820
pression gazeuse 1173
pression gazeuse, augmentation rétrograde de – 1174
pression, gradient de – 3852
pression hydrostatique 3820
pression initiale 3849
pression magmatique 0622
pression, maintien de – 3832
pression matricielle 3819
pression, remontée de – 3850
pression statique 3847
primaire 4630
priorité, loi de – 1924
prismatique 0522
prismatique 0755
prismatique 0772
prismatique 2730
prismatique 3185
prismatique aplati 0772
prix de gros, indice des – 4223
probabiliste, combinaison – 3877
procédé discontinu 4145
processus 2568
Proctor, essai – 4757
production assuré, débit de – 3668
production, capacité de – 3806
production culminante 3807
production cumulée 3817
production finale 3817
production initiale 3805
production plafonné, rythme de – 3902
production potentielle du puits 3806
production, puits de – 3789
production, taux de – 3801

production totale 3817
productivité, indice de – 3808
productivité spécifique, indice de – 3808
produit, sous– 4163
produits pyroclastiques 1309
profil 2686
profilage grande profondeur en mer 4484
profil continu, tirer en – 4408
profil définitif 2198
profil d'équilibre 2198
profil d'équilibre définitif 2198
profil géologique 4250
profil, hors – 4856
profil littoral 2441
profil littoral à forme d'équilibre 2441
profil longitudinal d'une vallée 2271
profil longitudinal, ressaut du – 2199
profil non régularisé 2198
profil régularisé 2198
profil stratigraphique 4255
profil terminal d'équilibre 2198
profondeur du foyer 0157
profondeur focale 0157
profondeur hypocentrale 0157
profondeur maximal 0214
profondeur, zone de – 0993
projection hawaïenne 1158
projection intermittente 1212
projection planosphérique 1021
projections 1313
projections 4818
projections incandescentes 1315
projections refroidies 1315
prolongement vers le bas 4533
prolongement vers le bas 4543
prometteuse, zone – 3887
promontoire 2446
promontoire ailé 2446
propagation, courbe de – 0171
propagation minimum, temps de – 0209
propagation, table de temps de – 0171
propagation, temps de – 0171
propagation, vitesse de – 0168
propylite 0921
propylitisation 0921
prospection biogéochimique 4645
prospection dans le réseau hydrographique 4641
prospection de sédiments du ruisseau 4641
prospection gamma 4545
prospection géobotanique 4645
prospection géochimique 4625
prospection géochimique en roche 4644
prospection géochimique en sol 4644
prospection géochimique par le gaz 4648
prospection hydrogéochimique 4640
prospection lithogéochimique 4644
prospection radiométrique 4545
protactinium-thorium, datation – 0445
protection, couche de – 4800
protéger 4800
protoclastique 0857
protodolomite 3321
protomylonite 0916
protore 3913
prototype 4775
protubérance 1288
prouvée, zone – 3888
prouvée, zone – 3889
provenance 2934
provenance, région de – 2943
province faunistique 1849
province géochimique 0401
province géologique 4260
province géothermique 0390
province magmatique 0628
province métallogénique 3908
province pétrographique 0628
province pétrographique sédimentaire 2852
proximal 2997
psammite 3226
psammitique 3226
pséphite 3238
pséphitique 3237
pséphitique 3238
pseudoaspite 1430
pseudofossil 1789
pseudomorphe 0982
pseudomorphe 1823
pseudomorphisme 0465
pseudomorphose d'apres 0982
pseudomorphose de cristal de sel 3042
pseudonodule 3060
pseudopalagonite 1255
pseudoporphyroblastique 0959
pseudotachylite 0917
pseudotachylite 1542
pseudo-volcan 1189
ptygmatique 0151
puissance 0272
puissance 4449
puissance calorifique 3604
puits 1274
puits 2104
puits 3665
puits 4261
puits 4848
puits au secret 3794
puits de découverte 3784
puits de lave 1284
puits de limite 3792
puits de production 3786
puits de recherche 4235
puits de secours 4290
puits d'extension 3785
puits d'injection 3793
puits d'intervention 4290
puits en formation compacte 3794
puits éruptif 3790
puits fermé 3811
puits, fermeture d'un – 4290
puits intermédiaire 3843
puits marginal 3845
puits naturel 2077
puits pompé 3791
puits producteur d'eau 3794
puits productif 3789
puits produisant par allègement de gaz 3791
puits sec 3794
puits stérile 3812
puits tubé au toit 3795
puits, tuer un – 4290
pulvérisation 1319
pureté 2677
pustule fumerollienne 1390
putréfaction 3526
puy 1425
pycnocline 2830
pyramidal 0522
pyramide de terre 2029
pyritisation 3458
pyrobitume 3737
pyroclaste 1309
pyroclaste 3170
pyrogène 0655
pyrolyse 3730
pyromagma 1200
pyroméride 0876
pyrométamorphisme 0896
pyrométasomatose 0900
pyromètre 1195
pyromètre à couple 0379
pyromètre optique 1195

Q

quadratique 0515
qualité sismique, indice de – 4732
quartage 4181
quartzite 3230
quartzite ferrugineux 4114

R

rabattement de nappe 3667
rabotage 2395
raccordement normal 2208
raccordement parfait 2208
race 1939
racine 0343
racine 1104
racine 2226
racines, région des – 1684
racines, zone de – 1684
rad 4551
radar 4384
radar à ouverture synthétique 4386
radar, faisceau – 4389
radar latéral 4385
radiance 4313
radié 0776
radié 0843
radier 4837
radier 4852
Radiolaires, boue à – 3368
Radiolaires, silex à – 3369
Radiolaires, terre à – 3368
radiolarite 3368
radiomètre 4310
radiomètre à baloyage 4403
radiomètre multispectral à baloyage 4403
rai 0056
rainure 0057
rainure 2472
rainure 3033
rajeunissement 2572
rameau 1965
ramification 0723
ramification 2210
ramification 4071
rang 0685
rapakivi 0828
rapide 2202
rapine fluviale 2322
ravin 0042
ravinement 2018
ravinement 2778
ravinement 3023
Rayleigh, onde de – 0189
rayon 4471
rayon hydraulique 2262
rayon lumineux 0548
rayon moyen 2262
rayonnement de chaleur 0381
rayon sismique 0206
rayon sismique, paramètre de – 0207
rayure 1538
rayure de polissage rapide 0481
raz de marée 2816
raz de marée de tempête 2824
réactionnelle, bordure – 0786
re-arrangement mécanique 3430
rebondissement élastique, théorie du – 0108
rebord continental 2788
rebroussement 1537
récepteur 4499
réception, station de – 4417
récettes 3893
réchauffement radioactif, hypothèse du – 1117
recherches sur place 4790
récif 2443
récif 2804
récif 3333
récifal 3334
récifal, chapeau – 3329
récifal, complexe – 3337
récifale, d'origine – 3334
récifale, paroi – 3336
récifale, roche – 3342
récifale, terrasse – 3352
récifale, zone – 3326
récifal, flanc – 3353
récifal, front – 3351
récif algaire 3360
récifal, inter– 3357
récifal, intumescence – 3329
récifal, noyau – 3341
récifal, platier – 3343
récif, arrière – 3344
récif, avant – 3356
récif-barrière 3338
récif externe 3349
récif frangeant 3338
récif interne 3344
récif régressif 3327
récif transgressif 3328
recompression 3832
reconnaissance du terrain 4790
recoupement par court-circuit 2230
recoupement par tangence 2230
recouvrement 3501
recouvrement 4359
recouvrement 3924
recouvrement, coefficient de – 4202
recouvrement horizontal 1551
recouvrement horizontal apparent 1553
recouvrement latéral 4359
recouvrement longitudinal 1556
recouvrement longitudinal 4359
recouvrement stratigraphique 1552
recouvrement, taux de – 4202
recouvrir 4800
récristallisation 0905
récristallisation 3452
recul des versants 2602
récupération, délai de – 3896
récupération, facteur de – 3867
récupération finale 3867
récupération finale estimée 3869
récupération primaire 3822
récupération secondaire 3831
récupération, temps de – 3896
récupération tertiaire 3837
récuperation ultime estimée 3869
redéposition 2993
redevance 4224
redox, potentiel – 0421
redressement 4376
redressement par demi-alternance 4506
redressement total 4506
redresseur d'images 4377
réduction 0416
réduction à l'air libre 0358
réduction bactérienne 0417
réduction chimique, zone

sphérique de – 3463
réduction isostatique 0361
référence internationale, champ géomagnétique de – 4535
référence, niveau de – 0357
référence, surface de – 0354
référence, surface de – 3854
réflectance 0574
réflectence 0574
réflecteur spéculaire 4318
réflection totale 0213
réflectivité 0574
réflectivité 4316
réflectivité 4473
reflet 0577
réflexion 4436
réflexion, coefficient de – 4473
réflexion diffuse 4317
réflexion d'interface 4519
réflexion du fond 4519
réflexion externe 0576
réflexion, horizon de – 4461
réflexion interne 0576
réflexion latérale 4518
réflexion multiple 4441
réflexion primaire 4437
réflexion, sismique – 4405
réflexion spéculaire 4318
refractaire 0788
réfraction 2810
réfraction, double – 0560
réfraction, indice de – 0551
réfraction, sismique – 4405
refroidissement, effet de – 1271
régénéré 3994
régime hydraulique 2267
régime hydrique, contrôle du – 2775
régime imbriqué 1571
région 3908
régionalisation 0255
régionalisation 4194
régionalité, degré de – 0340
régolite 2008
régression 1726
rejet 4078
rejet 4164
rejet, composante verticale du – 1563
rejeté en profondeur 1559
rejet en direction 1554
rejet horizontal 1551
rejet horizontal apparent 1553
rejet horizontal normal 1555
rejet incliné 1550
rejet incliné 1564
rejet longitudinal 1556
rejet net 1562
rejet net, sens de – 1562
rejet parallèle aux couches 1552
rejet perpendiculaire aux couches 1552
rejet souterrain 4832
rejet stratigraphique 1552
rejet transversal 1555
rejet vertical 1551
réjuvénation 3959
relaxation 1179
relief 0558
relief appalachien 2564
relief escarpé 2576
relief fort 2576
relief jurassien 2565
relief multiconvexe 2600
relief primitif 2569
relief, vigueur du – 2575
relique 0971
relique 0985
relique 1985
relique cuirassé 0787
relique instable 0985
relique sédimentaire 2856
relique stable 0985
rem 4551
remanié 2947
remblai 4825
remblai 4831
remblai de remplissage 4825
remblaiement 3105
remblaiement de vallée 2278
remblaiement fluviatile 2885
remblai hydraulique 4825
remblayage 3022
remigration 3749
rempart 1452
rempart 1461
rempart morainique 2414
remplacement 0931
remplacement 3453
remplacement 3956
remplissage 3105
remplissage 4729
remplissage filonien 1515
renard, formation de – 4863
rendement 0272
rendement sismique 0205
réniforme 0776
réniforme 4095
rentabilité interne, taux de – 3898
rentrant 2450
renversé 1491
renversement 4833
repêchage 4287
repêchage, outils de – 4287
repères de fond de chambre 4368
répétition de couches 1557
replat 3346
replat boueuse 2910
replat sableuse 2910
replat sous-marin 2801
replat structural 2307
replat structural 2555
réplique 0147
replissement 1668
reploiement 1668
réponse, courbe de – 0133
réponse en amplitude 4448
réponse en fréquence 4448
réponse en phase 4448
réponse impulsionelle 4446
réponse sismique 4497
repos, angle de – 4687
reprise en sous-oeuvre 4796
reptation 0114
reptation 2038
reptation 2780
reptation de débris 2040
reptation des éboulis 2040
reptation permanente 2039
reptation saisonnière 2039
réseau à maille carrée 4172
réseau, configuration du – 2254
réseau cristallin 0507
réseau dendritique, système en – 2292
réseau, densité de – 2255
réseau endoréique 2256
réseau en treillis, système en – 2292
réseau exoréique 2256
réseau filonien 4065
réseau fluvial 2205
réseau, noeud du – 0508
réseau orthogonal, système en – 2292
réseau radial, système en – 2291
réseau, rangée du – 0508
resédimentation 2993
réséquent 2284
reserrement 4054
réserves 3868
réserves additionnelles 3883
réserves, comptabilité des – 4222
réserves de minerai 4208
réserves démontrées 4207
réserves, estimation des – 3870
réserves estimées 3882
réserves possibles 3881
réserves possibles 4210
réserves probables 3880
réserves probables 4209
réserves prouvées 3879
réserves prouvées 4207
réserves reconnues 4209
réserves spéculatives 3882
réservoir 3755
réservoir, comportement du – 3846
réservoir homogène 3770
réservoir, isotropie du – 3770
réservoir magmatique 1202
réservoir, modèle du – 3866
réservoir, roche – 3679
réservoir, volume brut du – 3872
résidu 0241
résidu argileux 2122
résidu de déflation 2160
résidu de déflation 2967
résiduel 3998
résiduel, formation – 2930
résidu insoluble 2930
résidu non-soluble 2930
résineux 0484
résistance 4681
résistance acoustique 4474
résistance à la charge concentrée 4733
résistance à la compression 4683
résistance à la compression triaxiale 4683
résistance à la compression uniaxiale 4683
résistance à la traction 4684
résistance à l'explosif 4820
résistance au cisaillement 4684
résistance au délitage 4734
résistance au gel 4734
résistance au pic 4685
résistance électrique 4556
résistance, ligne de moindre – 4818
résistance résiduelle 4685
résistivité 4556
résistivité apparente 4564
résistivité apparente 4609
résistivité de la couche 4607
résistivité de l'eau de gisement 4608
résistivité non envahie 4607
résistivité vraie de la zone vierge 4607
résolution 4347
résolution horizontal, pouvoir de – 4516
résolution, pouvoir de – 4453
résolution, tache de – 4319
résolution vertical, pouvoir de – 4516
résorption 0642
ressant 1538
ressaut 2470
ressaut cyclique 2199
ressaut de faille 2548
ressaut glaciaire 2623
ressources hypothétiques 4214
ressources identifiées 4206
ressources minérales 4205
ressources non-découvertes 4213
ressources potentielles 4215
ressources spéculatives 4214
reste problématigue 1789
restite 1032
restite 1066
résurgence 2089
rétablissement élastique 0114
retard 0563
retard 4822
rétention au champ, capacité de – 2774
rétention d'eau, courbe de – 2772
rétention, durée de – 3627
rétention initiale 3618
rétention superficielle 3618
rétention utile, capacité de – 2773
rétention volumique 3704
retenue 2340
réticulaire, plan – 0508
réticulite 1331
retourné 1491
retrait 2813
retrait, forme de – 1993
retrait, limite de – 4715
rétrométamorphisme 0904
retroussement 1537
réverbération 4440
revêtement 3402
revêtement 4858
revêtement d'usure 4842
revêtu 3402
révision 3886
Reynolds 2265
rhenium-osmium, datation – 0446
rhéogramme 4143
rhéologie 0102
rhéologie 4671
rhéomorphisme 1057
rhizoconcrétion 3490
rhombochasme 1143
rhomboédrique 0515
ria 2452
Richter, échelle de – 0220
ricochet, marque de – 3038
ride 2796
ride 2814
ride 3000
ridé 3000
ride aplanie 3015
ride aplatie 3015
rideau 2059
rideau 2062
rideau 2135
rideau de palplanches 4799
ride, crête de – 3001
ride, creux de – 3001
ride de courant 3004
ride de vagues 3012
ride de vagues et courants 3011
ride d'interférence 3013
ride d'oscillation 3012
ride en croissant 3005
ride éolienne 3016
ride, hauteur de – 3002
ride incomplète 3015
ride linguoïde 3006
ride longitudinale 3011
ride, longueur d'onde de – 3002
ride losangique 3010
ride transversale 3003
rift 1092
rigole 4846
rigoles, lacis de – 2243
rigoles, réseau de – 2243
rigoles, tresse de – 2243
rimaye 2382
rip current 2826
rippabilité 4828
ripper 4828
rivage 2438
rivage extérieur 2461
rivage intérieur 2461

rivage, ligne de – 2437
rivage, près du – 2889
rivage régularisé 2531
rive concave 2235
rive concave 2236
rive convexe 2235
rive extérieure 2235
rive intérieure 2235
rive plate 2235
rive raide 2236
rivière 2191
rivière allogène 2259
rivière à marées 2504
rivière compétente 2329
rivière conquérante 2325
rivière dépérie 2329
rivière incompétente 2329
rivière, lit de la – 2242
rivière maîtresse 2206
rivière principale 2206
rivière sous-adaptée 2329
roche 3389
roche 4722
roche bordière 1530
roche carbonatée 3266
roche colmatée 3769
roche d'épanchement 1240
roche effusive 1240
roche encaissante 0709
roche encaissante 1530
roche encaissante 3920
roche en dalles 3132
roche en plaques 3132
roche éolisée 2161
roche éolisée à deux facettes 2162
roche éolisée en pyramide 2162
roche extrusive 1240
roche ferrugineuse 3464
roche ferrugineuse 4115
roche ferrugineuse silicifiée 3373
roche magasin 3919
roche mère 2944
roche mère 3918
roche-mère à gaz 3722
roche-mère à huile 3722
roche moutonnée 2400
roche originelle 0883
roche peu-perméable 3769
roche primaire 3918
roche pyroclastique 1311
rocher branlant 2163
rocher-champignon 2163
rocher en forme de champignon 2163
rocher perché 2163
roche solide 2011
roches verdes, zone de – 1082
roche volcanique 1238
rocs, chaos de – 2057
roentgen 4550
rose 3092
rose des sables 3406
rosée, point de – 3858
Rosiwall, méthode de – 0613
rotation, angle de – 1126
rotation, aptitude à la – 3191
rotation, externe – 1000
rotation, interne – 1000
rotation, pôle de – 1126
rotation, tectonique – 1000
rouille 2692
roulage 4809
roulage, marque de – 3038
roulante, extinction – 0565
rubanement 0860
rubanement de ségrégation 0946
rubanement par cristallisation 0862
rubanné 3095
rubannement 3095
rubannement 3122
rubidium-strontium, datation – 0447
rudimentaire 1978
rudite 3237
rugosité 4727
rugueux 3190
ruine volcanique 1476
ruissellement, transport par – 2031
ruissellement de surface 3617
ruissellement en filets 2033
ruissellement en nappe 2032
ruissellement en surface 2032
ruissellement, marque de – 3029
ruissellement, rigole de – 2026
rupture 4676
rupture progressive 4676
rupture spontanée 0115
rupture spontanée 2112
rupture, vitesse de – 0239
rythme élémentaire 3121

S

sable 2718
sable 3227
sable aquifère 3766
sable asphaltique 3768
sable à structure spongieuse 3063
sable, avalanche de – 2973
sable, banc de – 2244
sable bitumineux 3768
sable boulant 3085
sable boulant 4719
sable, chute de – 2973
sable, coin de – 2641
sable corallien 3311
sable, coulée de – 2973
sable de capture 3834
sable de couverture 2871
sable éolien 2156
sable éolien 2871
sable fluent 3085
sable gazifère 3766
sable humide 3767
sable limoneux 3227
sable, médiane du – 2718
sable mouvant 3085
sable mouvant volcanique 1318
sable musical 3088
sable pétrolifère 3766
sable, plaine de – 1462
sable, plaine de – 2157
sable pour verrerie 3227
sable, remplissage avec du – 2990
sable scoriacé 1323
sableux 3225
sable, vague de – 3007
sable vert 3231
sable vésiculaire 3063
sable volcanique 1322
saccaroïde 0811
saccharoïde 3090
sacs 4035
saillant 2304
saillant 2447
saillant anticlinal 1639
saillant de roche 2606
saillant triangulaire 2447
saillie de roche 2606
salar 2580
salbande 0707
salbande argileuse 4012
salbeux 3227
salifère 3376
salifère, tectonique – 1694
salina 2580
salinité 2827
salinité 3376
salinité 3675
salique 0607
salle 2101
salpètre naturel du Chili 3384
saltation 1982
saltation 2954
saltation, marque de – 3039
sapement 2023
sapement 2471
saprolite 4128
sapropèle 3528
sapropélite 3721
satellite artificiel 4302
saturation 2677
saturation 3650
saturation en eau 4611
saturation en hydrocarbures 4610
saturation, exposant de – 4606
saturation, ligne de – 0697
saturation, ligne de – 4862
saturation magnétique 0296
saturation résiduelle en eau 4611
saturation résiduelle en pétrole 4610
saturation, zone de – 3656
saturé 0605
saturée, zone non– 3651
saumure 3676
saussurite 0927
saussuritisation 0927
sauter, faire – 4814
scapolitisation 0932
scellement par couche asphaltique 3765
scellement par recouvrement 3765
scellement superficiel 2760
schiste 0944
schiste bitumineux 3721
schiste carton 3220
schiste charbonneux 3505
schiste noduleux 0941
schistes cuprifères 4130
schistes siliceux 3367
schistes silicifiés 3367
schiste tacheté 0941
schisteux 0975
schisteux 1603
schisteux 3220
schiste vert 0938
schistification 3509
schistosité 0975
schistosité cristallophyllienne 1603
schistosité de crénelure 1600
schistosité de fracture 1601
schistosité linéaire 1603
schlamm 4160
schlieren 0665
Schmidt, canevas de – 1015
schorre 2514
Schürmann, loi de – 3522
scialet 2104
sciences de la terre 0002
scintillation 4319
scintillomètre 4546
scissomètre 4759
scissomètre, essai au – 4759
sclérométrique, indice – 4733
scopulite 0769
scoriacé 1294
scorie 1294
scories agglomérées 1346
scories agglutinées 1346
scories, rempart de – 1452
scorification 4184
sebkha 2580
sebkha 2892
sebkha côtière 2892
secousse prémonitoire 0147
secousses, série de – 0144
secousse volcanique 1193
sécrétion latérale, théorie de la – 3928
sectile 0471
section 0685
section mouillée 2261
section sismique 4458
section transversale d'une vallée 2271
sécurité, coefficient de – 4777
sédentaire 1919
sédiment 2850
sédimentaire, roche – 2850
sédimentation 2989
sédimentation, balance de – 3158
sédimentation en alternance répétée 3121
sédimentation interne 3436
sédimentation rhythmique 3121
sédimentation, taux de – 2989
sédiment carbonaté 3266
sédiment interne 3436
sédimentologie 2849
sédiment résiduel 2855
Seger, cône de – 1196
ségrégation 3395
ségrégation 3948
ségrégation cristalline 0671
ségrégation magmatique par gravité 3948
ségrégation tectonique 0947
seiche 2816
séisme 0138
séisme continental 0148
séisme contrôlé 0262
séisme d'impact 0141
séisme local 0149
séisme lointain 0150
séisme lunaire 0061
séisme, micro– 0146
séisme mineur 0146
séisme mondial 0138
séisme, prévention de – 0262
séisme, prévision de – 0261
séisme sous-marin 0148
séisme volcanique 0140
séisme volcanique 1193
séismicité 0242
séismicité spécifique 0242
séismogramme 0116
séismographe 0116
séismographe à large bande 0116
séismologie 0094
séismologie de l'ingénieur 0254
séismologie nucléaire 0263
séismologie planétaire 0094
séismomètre 0118
séismoscope 0117
sélection naturelle 1966
sel gemme 3382
sel, glacier de – 1699
sel, mur de – 1695
sel, noyau de – 1695
sels grimpants 2142
sel sublimé 1394
semelle 4835
semelle de labour 2761
semelle filante 4837
semelle isolée 4837
semelle superficielle 4835
semi-anthracite 3560
semicristallin 0804
semi-profondeur, de – 0617
sensibilité 4717
sensitivité 4717
sensitométrie 4341
sentier de moutons 2060
sentier de vaches 2060
séparateur 3797

séparateur de sable 3152
séparation, degré de – 4726
séparation de liquides non-miscibles 3939
séparation en milieu dense 4153
séparation magnétique 4154
séparation, surface de – 3096
séparation, surface de – 4461
septaria 3487
séquence 2674
séquence 3110
séquence morphogénique 2573
sérac 2377
séricitisation 0932
série 1764
série 1781
série 1894
série chevauchante 1576
série chevauchée 1576
série de roches ignées 0630
série de sols 2663
série turficale 3538
serpentinisation 0923
serpentinite 0923
serrage 3509
serrage 4679
serrage 4808
serré 1626
serrée 3509
serrée 3510
sessile 1919
seuil 2201
seuil 2247
seuil 2335
seuil 2796
seuil d'anomalie 4627
seuil rocheux glaciare 2625
Shore, échelle de dureté de – 4701
sidérite 0066
sidérite, nodule de – 3483
sidérolite 0070
sigma-t 2827
signal 4435
signal 4517
signal acoustique 4517
signal-bruit, rapport – 4443
signal d'entrée 4445
signal de sortie 4445
signal sismique 4517
signal transitoire 0176
signe optique 0556
silex 3366
silex 3491
silex, argile à – 2931
siliceux 3365
silicification 3458
silification 0929
sill 0740
sillon 2796
sillon d'avant-côte 2460
silt 3204
siltstone 3204
sinuosité 2216
sinus 0051
sismique, constante – 0257
sismique, événement – 0095
sismique par chute de poids 4409
sismique réflexion continu faible profondeur 4479
sismique, région – 0247
sismiques, carte des zones – 0244
sismo-électrique, effet – 0113
sismogramme 0116
sismogramme 4420
sismographe 0116
sismographe à large bande 0116
sismologie 0094
sismologie planétaire 0094
sismomètre 0118
sismoscope 0117
site de l'ouvrage 4789
slikke 2505
slikke 2910
Slingram 4575
sluice 4158
slumping 2975
slush 3210
SMOW 0454
société 1890
socle 2791
soffine 1385
soie, de – 0490
sol 2657
sol ABC 2695
sol AC 2695
sol acide 2737
solaire, système – 0018
sol alcalin 2737
sol à polygones de toundra 2655
sol azonal 2669
sol de végétation 3515
sol, eau du – 2764
sol, en – 4636
sol enfoui 2693
solfatare 1391
solfatare acide 1393
solfatares, phase des – 1168
sol figuré 2652
sol gelé 2635
solidification, indice de – 0703
solidification superficielle 1200
soliflucion 2041
solifluxion 2041
solifluxion 2780
soligène 3550
sol intrazonal 2668
sol lunaire 0060
sol organique 3527
sol polygonal 2653
sol rayé 2653
sol résiduel 2009
sol résiduel 2855
sol, science du – 2658
sol strié 2653
sol structuré 2652
sol sulphaté acide 2739
sol superficiel 4788
sol, texture du – 2709
solum 2694
solution 0411
sol zonal 2667
somma 1461
sommation 4433
sommavolcan 1445
sommeil, période de – 1169
sommet 0705
sommet 1635
sommet 3106
sommets, escalier de – 2599
sonar latéral 4485
sondage 4261
sondage 4710
sondage, coupe de – 4299
sondage de délimitation 3785
sondage d'évaluation 3785
sondage dévié 4293
sondage d'exploration 3782
sondage, log de – 4299
sondage, mesure et contrôle du – 4294
sondage par lancage 4751
sondage, relevé de – 4299
sondage sismique profond 0094
sondage stratigraphique 3783
sonde 4758
sonde de radioactivité naturelle 4590
sondeur acoustique 4488
sondeuse au diamant 4265
sonogramme 4500
soufflard 1389
soufflard 2108
soufflard 2109
soufrière 1391
soulèvement 4830
soulèvement dû au gel 4721
soulèvement, niveau limite de – 2598
soulèvement volcanique 1409
source 3630
source artésienne 3631
source bouillante 1407
source d'affleurement 2091
source de boue sulfureuse 1398
source hypogène 1408
source jaillissante intermittente 1404
source karstique 2089
source minérale 3632
source sismique 4486
source thermale 1407
source thermale 3632
source vauclusienne 2093
source vauclusienne 3631
sous-charriage 1574
sous-espèce 1939
sous-étage 1763
sous-facies métamorphique 0987
sous-famille 1930
sous-genre 1933
sousjacent, corps – 0728
sousjacent, massif – 0728
sous-marin 2840
sous-marin 2900
sous-pression 4740
sous-saturqe 0605
sous-sol, coefficient de – 0258
sousterrain 2876
soutènement 4858
soyeux 0484
sparker 4410
sparker 4493
sparker, dispositif d'émission – 4493
spatial, groupe – 0513
spatite 3280
spéciation 1969
spectre continu 4305
spectre d'absorption 4305
spectre d'émission 4305
spectre des amplitudes 4451
spectre des phases 4451
spectre des puissances 0222
spectre des puissances 4451
spectre électromagnétique 4304
spectromètre à seuil 4547
spectromètre différentiel 4547
spectromètre multicanaux 4547
spéléo– 2877
spéléolite 2124
spéléologie 2094
spéléothème 2877
spéléothème 2127
sphénochasme 1143
sphéricité 3180
sphéroïdal 3180
sphéroïde 0330
sphéroïde 3180
sphérolite 0778
sphérolite 1321
sphérolite 3482
sphérolite composée 0779
sphérolitique 0876
sphérolitique 3482
sphérolitique 4087
sphérule de boue de soufre 1321
spicule 1862
spiculite 0764
spilite 0922
spilitisation 0922
spilosite 0933
spongolite 3372
sporange 1832
sporangium 1832
spore 1832
sporomorphe 1831
squelette 1861
squelette 3680
squelettique 3169
S, surface – 1002
stabilisation à la chaux 4807
stabilisation au ciment 4807
stabilisation du sol 4806
stabilité, analyse de – 4777
stabilité structurale 3424
stade 2568
stade de jeunesse 2570
stade de maturité 2570
stade de post-maturité 2570
stade de sénilité 2570
stade de veillesse 2570
stade explosif 1176
stade final 2570
stade initial 2570
stade pénultième 2570
stade solfatarien 1168
stalactite 2131
stalactite à plateau 2141
stalactite coralliforme 2133
stalactite de lave 1293
stalactite en tuyau de plume 2131
stalactite fistuleuse 2131
stalagmite 2137
stalagmite coralliforme 2133
stalagmite d'éclaboussement 2141
stalagmite d'éclaboussure 2141
stalagmite de lave 1293
stalagmite en pile d'assiettes 2141
stampes 3501
standard 0454
station composite 0119
station de base 0353
station de base 4526
station de transfert d'énergie par pompage 4870
stations, intervalle entre les – 4418
Stefanescu, intégrale de – 4568
Steinmann, triade de – 1098
Steinmann, trilogie de – 0696
sténobathe 1911
sténobenthique 1909
sténohalin 1912
sténotherme 1913
sténotope 1914
stéréocomparateur 4365
stéréogramme 4258
stéréoscope 4366
stéréoscopie inversée 4367
stérile 3917
stérile 3924
stérile, le – 3923
stériles 3501
stimulation 3800
stock 0730
stockage 4225
stock, mise en – 4225
stock tampon 4225
stockwerk 4058
Stonely, onde de – 0195
stoping 0623
strate 1750
strate 3097
stratification 3095
stratification à élément frontal déversé 3054
stratification à feuillet frontal déversé 3054
stratification à lits frontaire 3140
stratification de courant 3134
stratification deltaïque 3139
stratification de marée 3134

stratification de rhythme annuel 3127
stratification en alternance répétée 3121
stratification en gouttière 3144
stratification entrecroisée 3144
stratification, fausse – 3135
stratification inclinée 1482
stratification lenticulaire 3150
stratification massive 3108
stratification oblique 3133
stratification oblique 3136
stratification oblique angulaire 3142
stratification oblique de rides 3147
stratification oblique en arêtes de hareng 3146
stratification oblique en chevrons 3146
stratification oblique en feston 3145
stratification oblique plane 3142
stratification oblique tabulaire 3140
stratification oblique tangentielle 3142
stratification ondulée 3138
stratification parallèle 3115
stratification par densité 2882
stratification périclinale 1241
stratification, plan de – 3096
stratifications obliques combés, petites – 3148
stratification tidale 3134
stratifié 3095
stratigraphie 1708
stratotype 1740
stratovolcan 1429
striage 2397
striage de la lave 1270
striation 0468
striation 2999
striation glaciaire, limite de la – 2631
strie 1538
strie irrégulière 0481
stries 1270
stries glaciaires 2397
stringer 4039
stromatactis 3437
stromatolite 3332
stromatolite, proto– 3332
strontium commun 0448
strontium primaire 0448
strontium radiogénique 0448
structural, élément – 2724
structure 0858
structure 2568
structure 2723
structure 3077
structure algaire 3358
structure à renflements et étranglements 1609
structure basale 3019
structure brêchique 2933
structure centripète 2996
structure chiffonnée 2649
structure collonnaire 2730
structure cryptoexplosive 1184
structure cryptovolcanique 1184
structure de bioérosion 1883
structure de biostratification 1884
structure de bioturbation 1885
structure de charge 3059
structure de creusement et de remblaiement 3024
structure de déformation 3050
structure de nutrition 1881
structure d'étirement 1607
structure d'involution 2649
structure directionnelle 2998
structure empâtée, à – 3294
structure en arêtes et creux 3151
structure en bandes bleues 2374
structure en cage à poulets 3468
structure en cocarde 4096
structure en coupelle 3069
structure en écailles 1571
structure en feuillets 3101
structure en flamme 3061
structure en lamines 3101
structure en queue de cheval 4069
structure en sablier 0534
structure entérolithique 3076
structure foliacée, roche à – 1604
structure fibreuse de remplissage 3401
structure fluidale 0868
structure foliacée, roche à – 1604
structure frondescente 3057
structure graphique 4100
structure halocinétique 1694
structure halotectonique 1694
structure imbriquée 3040
structure jointive, à – 3294
structure litée 0945
structure oeillée 0952
structure porphyroblastique 3470
structure primaire 3077
structure prismée 1501
structure protogneissique 0960
structure rubanée 0860
structure rubanée 2374
structure ruiniforme 3467
structure sédimentaire 2994
structure sédimentaire primaire 2995
structure sédimentaire secondaire 2995
structure tourmentée 2649
structure type 3579
structure, type de – 2726
structure vermiculaire 3467
structure volcano-tectonique 1182
structurologique, diagramme – 1015
stylolite 3451
subaérien 2866
subâge 1763
subalcalin 0694
subalumineux 0693
subangulaire 2729
subangulaire 3189
subaquatique 2880
subarrondi 3188
subduction, zone de – 1134
subglaciaire 2873
subgrauwacke 3234
sub-idiomorphe 3471
sublimation 0412
sublimation 3952
sublimation, produit de – 1394
sublimé de sel 1394
sublittoral 2839
submacéral 3567
submarginal 4212
submergence 2857
submersion 1728
subophitique 0845
subséquent 2286
subsidence 2857
subsidence en chaudron souterraine 1186
subsidence volcanique, secteur de – 1464
substance minérale 3912
substitution 0931
substitution 3956
substitution couplée 0477
substitution ionique 0477
substratum 0348
substratum 2696
substratum 3107
substratum d'un volcan 1414
subtidal 2908
subvolcan 1206
sub-volcanogénique 3989
suc 1477
succession minéralogique 3965
succession volcanique 1238
suintement 3629
suintement 3753
sulfuration 3964
super-étage 1763
superfamille 1930
supergène 3937
supergène 3967
superimposée 4630
superparamagnétisme 0300
superposition 1714
superposition, critère de – 1714
superposition, loi de – 1714
superstructure 1077
superzone 1782
suppression 4567
suppression des couches 1557
supracrustal 0584
supralittoral 2837
suprastructure 1414
supratidal 2908
surconsolidation 4703
surface axiale 1631
surface dénudative à facettes 2601
surface dénudative à méplats 2601
surface enveloppe 1672
surface, état de – 3192
surface exploitable 4360
surface libre 3643
surface listrique 1573
surface médiane plis 1673
surface piézométrique 4736
surface primitive 2569
surfaces dénudatives étagées 2601
surfaces emboîtées 2601
surface spécifique 4109
surface spécifique du sol 2754
surimposé 2290
surimposition, de – 4633
surimpression 0970
surpression 3820
surpression tectonique 0886
sursaturé 0605
susjacent 4631
suspension 0411
suture 1132
suturé 3450
symbiose 1898
sympatrique 1817
symplectique 0849
symplectite 4105
synantétique 0784
synantétique 3963
synchrone 1729
synchronique 1818
syncinématique 0726
syncinématique 0906
syncinématique 1079
synclinal 1619
synclinal à double plongement 1638
synclinal bordier 1697
synclinal, faux – 1621
synclinorium 1671
synéclise 1084
synécologie 1888
synérèse, fissure de – 3044
syngenèse 3398
syngénétique 2854
syngénétique 3474
syngénétique 3966
synneusis 0821
synonyme 1955
synonymie 1955
synorogénique 0726
synorogénique 1108
synsédimentaire 2854
syntectite 0682
syntectonique 0726
syntectonique 0906
synthétique 4478
syntype 1951
système 1765
système actif 4301
système filonien 4063
système fluvial 2205
système, multispectral – 4326
système passif 4301

T

table 0501
table à secousses 4157
table de rotation 4269
tabulaire 0522
tabulaire 0755
tabulaire 3183
tache 2692
tacheté 0965
tacheté 2692
tacheté 3462
tachylite 1262
tafone 2002
taille des sables, de la – 3227
tallard 2507
talus 2867
talus 4860
talus croulant 2185
talus croulant 2966
talus d'éboulis 2007
talus frontal 2480
talus, inclinaison du – 4827
talus naturel, angle de – 4687
talus récifal 3353
tamisage 4873
tampon 0422
taphocénose 1896
taphocoenose 1896
taphonomie 1897
taphrogenèse 1091
tapis de blocs 4876
tapis filtrant 4862
tapon 1422
tardicinématique 1079
tardi-orogénique 1108
tarière 4236
tassement 4704
tassement 4833
tassement, appareil pour mesurer le – 4768
tassement différentiel 4704
tassement différentiel 4833
tassomètre 4768
tautonyme 1957
taux de succès 3784
taxinomie 1920
taxite 1243
taxon 1926
taxon 2662
taxonomie 1920
taxonomie des sols 2662
taxon, zone d'extension d'un – 1783
tectite 0077
tectofaciès 2861
tectonique 1073
tectonique, cassante – 1520
tectonique globale 1125

tectonite 1001
tectonite 1605
tectonosphère 0080
teinte 2677
télédétection 4300
télémagmatique 3986
teléscopage 3965
téléséisme 0150
téléthermal 3984
tellurique 0002
témoin 4233
temps, domaine de – 0223
temps mort 3810
temps vertical 4428
ténacité 0470
teneur en eau et sédiments 3804
teneur liminaire 4203
teneur limite 4203
teneur limite 4204
teneur, moyenne pondérée de la – 4185
teneur payante 4204
teneur significative 4627
tension 4556
tension de l'eau 2771
téphra 1310
téphrochronologie 1312
terminaison 3414
terminaison en biseau 3104
terminaison en biseau 3509
terminaison en dent de scie 3414
terni 3408
ternissement 3408
terrain 4788
terrain anastomosé 0041
terrain boulant 4855
terrain buriné 0040
terrain chaotique 0038
terrain fritté 0039
terrain gonflant 4855
terrain houiller 3498
terrain laminé 0040
terrain naturel 4840
terrain plastifié 4855
terrain, réalité de – 4321
terrain rocheux 4788
terrains, morts – 3501
terrasse 2801
terrasse, abrupt de – 2295
terrasse alluviale 2298
terrasse armée 2305
terrasse climatique 2308
terrasse construite 2298
terrasse construite 2337
terrasse construite 2480
terrasse continentale 2790
terrasse continue 2299
terrasse cyclique 2299
terrasse d'accumulation 2298
terrasse d'accumulation 2337
terrasse d'accumulation 2480
terrasse de barrage 2310
terrasse de culture 2061
terrasse de dépôt 2298
terrasse de glissement 2302
terrasse de glissement latéral 2302
terrasse de kames 2420
terrasse de lave figée 1285
terrasse de lobe 2303
terrasse de matériel congélifracté 2309
terrasse de méandre à armature rocheuse 2305
terrasse de méandres 2304
terrasse d'éperon 2303
terrasse de profil d'équilibre 2298
terrasse de remblaiement 2298
terrasse de remblaiement emboîtée 2301
terrasse d'érosion 2298
terrasse d'érosion 2337
terrasse de stabilité 2298
terrasse diastrophique 2307
terrasse émergeante 2311
terrasse, escarpement de – 2295
terrasse eustatique 2308
terrasse, fausse – 2302
terrasse fluviale 2293
terrasse fluvioglaciaire 2309
terrasse lacustre 2337
terrasse littorale 2484
terrasse locale 2299
terrassement 4825
terrasse, niveau de – 2294
terrasse non-cyclique 2299
terrasse plongeante 2311
terrasse polygénique 2302
terrasse principale 2299
terrasse, rebord de – 2295
terrasse régionale 2299
terrasse rocheuse 2298
terrasses appariées 2300
terrasses couplées 2300
terrasses emboîtées 2300
terrasses en gradins 2296
terrasses, entrecroisement de – 2311
terrasses étagées 2296
terrasses étagées 2300
terrasses polycycliques 2300
terrasse structurale 2555
terrasse, talus de – 2295
terre à – 3086
terre armée 4798
terreau 3525
terre fine 2719
terre réfractaire 3517
terre silicieuse 3367
terrestre 2865
terreux 3086
terre végétale 3525
terrier 1882
terrier d'habitation 1880
terrigène 2902
terril 4166
territoire raviné 2030
teste de substitution 1825
test sur câble 3787
tête 1686
tête 4850
téthys 1127
tétragonal 0515
texture 0790
texture 3077
texture 4345
texture argileuse 2721
texture cellulaire 4089
texture cloisonnée 4104
texture colloforme 4091
texture colloidale 4091
texture concrétionnée 4092
texture craquelée 4097
texture d'exsolution 4106
texture en feuille de laurier 4082
texture en mosaïque 3089
texture en peigne 4084
texture granulaire 4086
texture graphique 0847
texture ocellaire 0855
texture pegmatitique 3470
texture réticulée 0856
texture sableuse 2721
texture spinifex 0850
texture symplectique 0783
thalweg 2279
thanatocénose 1896
thanatocoenose 1896
théorie per ascensum 3927
théorie per descensum 3926
thermal 0376
thermique 0376
thermique, densité de flux – 0384
thermocline 2830
thermohalin 2829
thermokarst 2645
thermométrie 4601
tholéite 0695
tholoide 1422
tholus 0037
thorium-plomb, datation – 0436
thrombolite 3330
tige bloqué 4286
tige carrée d'entrainement 4269
tige coincé 4286
tige de forage 4270
tillite 2874
tillite 2903
tillites 2969
tilloïde 2978
tine 2248
TIR 3898
tirant d'ancrage 4792
tirant précontraint 4795
tir central 4419
tir contrôlé 4823
tir en nappe, point de – 4411
tirer 4814
tir, point de – 4411
toit 0705
toit 1633
toit 3501
toit 4009
toit 4010
toit, écroulement du – 1188
tombolo 2467
tonalité 2677
tongue 1752
topogène 3550
topominéral 3995
topotactique 0543
topotaxie 0543
topotype 1953
topozone 1784
torrent boueux froid 1355
torrent de feu 1263
torrent glaciaire 2392
torrentiel 2884
torsion, balance de – 0350
torsion, essai de – 4765
tortuosité 3707
toundra 2636
tourbe 3494
tourbe à bois 3536
tourbe à sphagnum 3543
tourbe à sphaignes 3543
tourbe de bruyère 3545
tourbe de haute tourbière 3542
tourbe de prairie 3536
tourbe de toundra 3547
tourbe de tourbière basse 3535
tourbe humosapropélique 3548
tourbe lacustre 3531
tourbe, marais à – 3532
tourbe terrestre 3546
tourbière 3532
tourbière basse 3533
tourbière branlante 3530
tourbière cordée 2636
tourbière, debâcle de 3541
tourbière de pente 3534
tourbière de vallée 3534
tourbière haute 3542
tourbière, mare de – 3540
tourbière réticulée 2636
tourbification 3524
tourbillon 1364
tourbillon 2269
tourbillon figée de lave 1292
tourbillon mobile 2269
tour de lave 1299
tourmalinisation 0932
trace 1492
trace 1876
trace 4420
trace d'activité 1875
trace de locomotion 1879
trace de pacage 1881
trace de station 1880
trace de vie 1875
trace de vie superficielle 1875
trace fossile 1873
traceur 4649
trachytique 0838
traction 2953
traction d'ancrage, appareil pour mesurer la – 4772
traction sur boulon, essai de – 4764
traînage, trace de – 3032
train de tiges 4270
traîne 2173
traîneau 4483
trainée 4631
traînees minéralisées 4040
traité limité 0266
traitement artificiel 0492
traitement des minerais 4137
trajectoire 2372
trajectoire 4471
trajectoire de vol 4303
trajectoire d'une particule 3708
trajet 0206
tranchée 4235
tranchée 4824
tranchée parafouille 4861
transducteur électro-acoustique 4488
transfluence 2383
transformation 3426
transgression 1726
transgressivité parallèle 1721
transition, zone de – 3777
transition, zone de – 4620
translation 2370
translation continentale 1115
translucide 0545
transmissivité 3702
transmissivité 4315
transparent 0545
transport en masse 2037
transport éolien 2154
transport éolien 2961
transport par l'eau interstitielle 2034
transport, taux de – 2957
transvaporisation 3931
transversale 2289
trapp 0638
travertin, barre de – 2147
tremblement crypto-volcanique 0140
tremblement de relé 0145
tremblement de terre 0138
tremblement de terre artificiel 0265
tremblement de terre, bruit de – 0203
tremblement de terre, crevasse de – 2552
tremblement de terre d'effondrement 0141
tremblement de terre, gradin de – 2552
tremblement de terre multiple 0143
tremblement de terre principal 0142
tremblement de terre, risque de – 0255
tremblement de terre tectonique 0139
tremblement de terre volcanique 1193
trémie 2114
trépan 4271

trépan carottier 4271
TRI 3898
triangulation par fente radiale 4324
tribu 1931
tributaire 2207
trichite 0770
triclinique 0515
trigonal 0515
trimorphisme 1869
tripartie, méthode – 0152
triplet 0480
tripoli 3371
tritium, datation – 0452
trituration 1319
triturée, roche finement – 2875
troncon de vallée 2327
trottoir 2479
trou 2793
trou d'eau 2247
trou d'eau 2632
trou de cryoconite 2387
trou de midi 2387
trou d'impact 1418
trou non tubé 4277
trou ouvert 4277
trou perdu 4287
trou souffleur 2108
trou souffleur 2109
tsunami 2816
tubage au cuvelage 4275
tubage de production 4275
tube à sédiment 4278
tube carottier 4272
tube conducteur 4274
tube crépine 3795
tube de lancage 4751
tube guide 4274
tuf 3385
tufacé 1332
tuf à cristaux 1333
tuf à lapilli 1333
tuf à ponces 1333
tuf à scories 1333
tuf chaotique 1333
tuf d'explosion 1335
tuf en poussière 1333
tuffacé 3171
tuffisite 1339
tuffite 1332
tuffite 3171
tuf intrusif 1339
tuf lithique 1333
tuf palagonitique 1347
tuf pisolitique 1333
tuf remanié 1337
tuf sédimentaire 1334
tuf soudé 1340
tuf soudé 1341
tuf stratifié 1336
tuf subaquatique 1335
tuf transporté 1337
tuf volcanique 1332
tuf volcanique 3171
tuméfaction 1272
tumule 1272
tunnel 4847
tunnel d'accès 4847
tunnel de lave 1290
tunnel fumerollien 1383
tunnelier 4854
turbidite 2983
turbidité, courant de – 2981
turboforage 4264
T-X, report – 4463
type, genre – 1947
type, localité – 1739
type, localité – 1946
type, région – 1739

U

ultrabasique 0604
ultrabasite 0604
ultramafique 0600
ultramafique, roche – 0600
ultramétamorphisme 0882
ultramylonite 0916
ultramylonite 1541
ultravulcanien 1166
uniaxe 0553
unification 3901
uniformitarisme 0006
unité cartographique 2672
unité gravimétrique 4537
unité sédimentaire composite 3111
upwelling 2786
uranium-plomb, datation – 0434

V

vacuolaire 0874
vacuolaire 1305
vacuole 0874
vacuoles cylindriques 1306
vagile 1918
val 2544
valeur actuelle 3897
valeur retardée 4220
valeurs actuelles, méthode des – 3898
valeurs extrêmes 4186
validité 1923
vallée 0034
vallée à fond 2278
vallée à gradins glaciaires 2622
vallée anticlinale 2544
vallée à profil en U 2273
vallée à profil en V 2272
vallée asymétrique 2274
vallée aveugle 2083
vallée décapitée 2324
vallée de dépression axiale 2545
vallée de faille 2547
vallée d'effondrement 2085
vallée d'effondrement 2546
vallée de la mort 1402
vallée de ligne de faille 2547
vallée de zone de faille 2547
vallée dissymétrique 2274
vallée emboîtée 2275
vallée en abaissement d'axe des plis 2545
vallée en auge 2273
vallée en fosse d'effondrement 2546
vallée entaillée 2272
vallée en vallée 2275
vallée, fin de – 2276
vallée, fond de – 2278
vallée glaciaire perchée 2627
vallée glaciaire suspendue 2627
vallée, gradin de – 2199
vallée marginale proglaciaire 2392
vallée monoclinale 2544
vallée morte 2084
vallée morte 2328
vallée ouverte 2276
vallée perchée 2483
vallée polycyclique 2275
vallée, profil transversal d'une – 2271
vallée prolongée 2282
vallée reculée 2083
vallée, ressaut de – 2199
vallées démembré, réseau de – 2482
vallée sèche 2084
vallée sèche 2328
vallées, formation des – 2270
vallées intégrées, réseau de – 2283
vallée sous-marine 2795
vallée structurale 2543
vallée submergée 2282
vallée suspendue 2483
vallée synclinale 2544
vallée tectonique 2543
vallée tronconnée 2324
vallée tronquée 2481
vallée, versant d'une – 2277
valleuse 2483
vallon 2281
vallon 2544
vallon, tête de – 2276
vallum morainique 2416
van 2610
vannage 2937
vapeur dominante, système à – 0393
vapeurs volcaniques 1349
vapeurs volcaniques 1377
variance 4190
variation, diagramme de – 0697
variation diurne 4528
variation séculaire 0289
variété 1940
variogramme 4196
variole 0781
variolite 0781
variolitique 0877
variomètre 0288
varve 3128
varvite 3128
vase 2920
vase 3307
vase micritique 3281
vase putride 3528
vase sableuse 2910
vecteur nul 0235
veine 4071
veine d'eau 3638
veine de charbon 3123
veine siliceuse 3491
veinule 4039
vélocité critique primaire 2265
vélocité critique secundaire 2266
Vening Meinesz, axe de – 1137
vénite 1045
vent, action du – 0043
vent de sable 2156
venue de surface 1239
venue éruptive externe 1239
venue intrusive profonde 0715
venue minéralisante 3936
vergence 1622
verglas 2352
vérin plat, essai au – 4763
vermiculaire 0774
vernis calcaire 2142
verre 0627
verre volcanique 1262
verrou 2335
verrou 2625
verrue fumerollienne 1390
versant colluvial 2603
versant concave 2236
versant convexe 2237
versant convexe 2602
versant, inflexion du – 2602
versant, partie concave du – 2602
versant, partie convexe du – 2602
versant, rupture de – 2277
verticale 0329
verticale, déviation de la – 0329
vésiculaire 0874
vésiculaire 1305
vésicule 0874
vésiculé 1305
vey 2505
vey 2910
vibration 4809
vibration sphéroïdale 0198
vibroflotation 4810
vibrographe 0123
vide 3064
vide 3683
vides, volume des – 2752
vire 2297
virgation 1675
viscosité 4671
vitesse apparente 0208
vitesse apparente 4468
vitesse, couche à faible – 0100
vitesse d'écoulement souterrain 3703
vitesse de filtration 3703
vitesse de groupe 0170
vitesse d'entraînement 2959
vitesse de phase 0169
vitesse de phase, méthode de – 0230
vitesse de propagation 0168
vitesse de sommation 4434
vitesse des particules 4414
vitesse d'intervalle 4467
vitesse moyenne 4465
vitesse moyenne quadratique 4466
vitesses de groupe, méthode des – 0229
vitesses, loi des – 4469
vitesses, séparation des – 0228
vitraon 3580
vitreux 0484
vitreux 0805
vitrification 0708
vitrophyre 0823
volatiles, matières – 3597
volatiles, perte des matières – 3520
volcan 1411
volcan à cratère central 1434
volcan actif 1153
volcan assoupi 1155
volcan à structure imbriquée 1421
volcan avorté 1222
volcan bouclier 0035
volcan complexe 1441
volcan de boue 1395
volcan de boue 1700
volcan de boue 3071
volcan démantelé 1475
volcan de sable 3071
volcan disséqué 1475
volcan dormant 1155
volcan, embryon de – 1222
volcan embryonnaire 1222
volcan en bouclier 1420
volcan endormi 1155
volcan en repos 1155
volcan éteint 1155
volcanique 1148
volcanique, théorie – 0049
volcanisme 1147
volcanisme d'orogène 1151
volcanisme épigène 1152
volcanisme extrusive 1207
volcanisme orogénique 1151
volcanisme, séquelles du – 1374
volcanisme souterrain 1181
volcanisme subactuel 1150
volcan isolé 1446
volcan mixte 1429
volcanoclastique 3172
volcanogène 1148
volcanogénique 3989
volcanologie 1146
volcan raviné 1474
volcans alignés 1448

volcans en bouclier, groupe de – 1450
volcans, groupe de – 1450
volcans jumelés 1442
volcan sommeillant 1155
volcan sous-marin 1416
volcans, surveillance des – 1190
volcan subactif 1153
volcan subaérien 1415
volcan subaquatique 1415
volcan superficiel 1413
volcan, vestige de – 1476
volée 4822
vol, ligne de – 4303
vol, ligne de – 4358
volumétrique, méthode – 3871
voûte 4851
voûte à double courbure 4865
voûte de lave avec soupirail 1292

W

wadi 2587
WDC 0082
Wentworth, échelle de – 3163
whisker 0524
Widmanstätten, figures de – 0067
wildflysch 2926
Wulff, canevas de – 1015

X

xénoblaste 0980
xénoblastique 0980
xénocrystal 0663
xénolithe 0663
xénomorphe 0752
xénomorphe 0808
xénomorphe 3471
xénothermal 3984
xylophage 1857

Y

yardang 2164
yeux 0952
Young, module de – 4695

Z

Zavaritsky, diagramme de – 6690
zéolitisation 0932
zonage 0474
zonage direct 0475
zonage inverse 0475
zonage normal 0475
zonalité 3907
zone 0362
zone 0519
zone 1717
zoné 3122
zone 3908
zone broyée 1532
zone de limite 3845
zone des brisants 2439
zone d'extension 1781
zone envahie 4619
zonée, structure – 0533
zone failleuse 1532
zone fracturée 4821
zone intertidale 2836
zone lavée 4620
zone volcanique 1447
zonule 1782
zoobenthos 1852
zoochlorelles 1845
zooplancton 1800
zooxanthelles 1845

Deutsch

A

Abart 1940
Abbau 2741
Abbaublock 4199
Abbaufeld 4199
Abbildung 4320
Abblasung 2155
Abblätterung 1994
Abbruch 2799
Abbruchkante 2516
Abbruchwand 2516
Abdachungs– 2284
Abdämmung 2340
abdichten 4800
Abdichtung 3765
Abdichtung 4800
Abdichtung 4861
Abdichtungsgraben 4861
Abdichtungsmauer 4861
Abdichtungsschleier 4804
Abdruck 1824
Abdruck 1826
Abfall 3816
Abfall 3924
Abfallbeseitigung 4831
Abfallbeseitigung, unterirdische – 4832
Abfallkurve 3816
Abfallkurven-Methode 3874
Abfangen 3615
Abfluss 3616
Abflussfaktor 2250
Abflusskoeffizient 2250
Abfluss, oberirdischer – 3617
abfluss, Trockenwetter– 3628
abfluss, Zwischen– 3623
Abfolge 3110
Abgabe 4224
abgedeckt 2595
abgeschoben 1559
abgleiten 2976
Abgleiten 2972
Abgleitung durch die Schwerkraft bedingt 1703
Abgliederungsinsel 2534
Abguss 1825
Abhobelung 2395
Abkommende 4071
Abkühlungskluft 1517
ablagern 2985
Ablagerung 2850
Ablagerung 2886
Ablagerung 2985
Ablagerung 3170
Ablagerung, alluviale – 2886
Ablagerung, erneute – 2993
Ablagerung, Gezeiten– 2907
Ablagerung, glaziale – 2402
Ablagerung, glazifluviatile – 2402
Ablagerung, glaziomarine – 2903
Ablagerung, kolluviale – 2868
Ablagerung, kurz nach – 3474
Ablagerung, pyroklastische – 1310
Ablagerung, regressive – 3112
ablagerung, Schichtflut– 2884
Ablagerung, sedentäre – 2855
Ablagerung, sekundäre – 2993
Ablagerungsmilieu 2859
Ablagerung, Tiefsee– 2922
Ablagerung, transgressive – 3113
Ablagerung, vulkanoklastische – 1310
Ablation 2013
Ablation 2380
Ablation, äolische – 2155
Ablationsmoräne 2412
Ablationstal 2384
Ableitung, zweite – 4534
Ableitung, zweite – 4544
ablenken 4292
Ablenker 2325
Ablenkung 4078
Ablenkungsknie 2326
Ablenkungsschlucht 2326
Abplattung 0330
Abplattung 3182
Abpressungsfiltration 0675
Abpressungsfiltration 3933
Abpressungslagerstätte 3975
Abrasion 2938
Abrasion, marine – 2016
Abrasionsebene 2536
Abrasionsfläche 2536
Abrasionsküste 2528
Abrasionsplatte 2479
Abrasionsterminante 2442
Abraum 3924
Abrieb 2938
Abriss 4427
Abrisskante 2053
Abrissnarbe 2052
Abrissnische 2052
Abrutsch 3055
Abrutschen 2972
Abrutschen 2975
Abrutschmasse 3055
Abrutschung, schuppenförmige – 2045
Absatz 2470
Absatz, glazilimnischer – 2402
Abscherung 1663
Abscherungsstruktur 3053
Abschiebung 1558
Abschlag 4822
Abschlämmbares 2716
Abschleifung, glättende – 2399
Abschleifungsstrom 2826
Abschuppung, schalenförmige – 1994
Abseigerung, flüssige – 3948
Absenkung 2857
Absetzen von Schichten 1523
absinken 2991
Absonderung 0469
Absonderung 0713
Absonderung 1981
Absonderung in plattigen Kluftkörpern 1500
Absonderung in rhomboidalen Kluftkörpern 1502
Absonderung, kugelige – 1304
Absonderung, polyedrische – 1303
Absonderung, prismatische – 1501
Absonderung, säulenförmige – 1501
Absonderung, zylindrische – 1302
Absorption 0112
Absorption 0406
Absorption 4514
Absorption, geometrische – 4514
Absorptionsgrad 4314
Absorptionsspektrum 0491
Absorptionsspektrum 4305
Absperrung 2340
Absplittern 2941
Abspülung 2031
Abspülung, flächenhafte – 2032
Abstammung 1964
Abstände, gleiche – 4172
Abstandsmarken 4398
Absturz im Krater 1472
Abtasten 4401
Abtasten, konisches – 4402
Abtaster 4403
Abtaster, multispektraler – 4403
Abtauchen 1614
Abtauchen 4018
Abtrag 4875
Abtragung 2012
Abtragung 2013
Abtragung 2936
Abtragung, flächenhafte – 2032
Abtragungsfläche 2591
Abtragungsflächen, ineinandergeschachtelte – 2601
Abtragungsfläche, polygenetische – 2601
Abtragungsniveau, oberes – 2598
Abtragungsstufe 2558
Abtreiben 4184
Abwasser 3803
Abwasser 4165
Abweichung, lotrechte – 4018
abyssal 2847
abyssisch 0617
abyssopelagisch 1910
abyssopelagisch 2847
Abzweigung 1582
Abzweigung 2211
ACF-Diagramm 0989
Achondrit 0072
Achse, a– 1016
Achse, b– 1016
Achse, beta– 1020
Achse, c– 1016
Achse, f– 1017
achse, Gefüge– 1016
Achse, geomagnetische– 0280
achse, Gürtel– 1020
Achse, kristallographische – 0511
Achsenbild 0569
Achsendepression 1635
Achsenebene 1631
Achsenebene, optische – 0554
Achsenebene, Spur der – 1631
Achsenwinkel, optischer – 0555
Achse, optische – 0554
Ackerkrume 2699
Acme 1985
Acme-Zone 1773
Adaptation 1980
adaptiv 1980
Adelsvorschub 4005
Adelszone 4005
Ader 4039
ader, Eruptiv– 0748
Adhäsionsrippel 3016
Adinol 0933
Adkumulat 0832
Adsorption 0406
Adularisation 0932
Adularisierung 0932
Adventivkegel 1437
Adventivkrater 1439
Aeolotropie 0110
aerob 0424
aerob 1906
Aerolit 0071
AFC-Diagramm 0698
AFM-Diagramm 0989
Agglomerat 1344
Agglomeration 0771
Agglomerationstheorie 0048
Agglomeratlava 1269
Aggregat 2724
Aggregat, kristallinisches – 0771
Aggregatoberfläche 2724
Aggregats, Fläche eines – 2724
Aggregierungsstufe 2726
aggressiv 0619
Agmatit 1042
Agon 0284
agpaitisch 0693
Ahnenreihe 1960
Airy 0345
Airy Phase 0190
A'KF-Diagramm 0989
Akkreszenz 1822
Akkumulation 2987
Akkumulation 3754
Akkumulation, Kristall– 0674
Akkumulationsterrasse 2298
Akkumulationsterrasse 2480
aktiv 0619
Aktualismus 0006
Aktuogeologie 0008
Aktuopaläontologie 1886
Akustiklog 4597
Akyrosom 1039
Akzelerogram 0120
Akzelerometer 0120
Albedo 4316
Albitisation 0932
Albitisierung 0932
Aleurit 3205
Aleurolit 3205
Algen– 3314
Algenbarre 3361
Algendom 3363
Algenkuppel 3363
Algenmarsch 2515
Algenmatte 3364
Algenmatte, unverfestigte – 3332
Algenrand 3362
Algenrücken 3361
Algentextur– 3358
Alias-Effekte 4423
Alkali– 0694
Alkalikalk 0694
Alkali-Kalkindex 0692
Alkalikalkreihe 0692
Alkalinität 3677
Alkalireihe 0692
Alkane 3717
Alkene 3717
alllothigen 2946
Allochem 3271
allochron 1818
allochthon 0727
allochthon 1685
allochthon 1815
allochthon 2945
allopatrisch 1817
allothigen 0660
allothigen 3968
allotriomorph 0523
allotriomorph 0752
allotriomorphkörnig 0808
Allotypus 1949
Alluvion 2886
Alluvionen 2278
alpinotyp 0600
alpinotyp 1078
Alter 1762
Alter, absolutes – 0429
Alterationshof 4630
Alter, diskordantes – 0430
Alter, konkordantes – 0430
Altgestein 0709
Altrumpf 2594
altvulkanisch 0636
Altwasser 2232
Alunitisation 0932
Alunitisierung 0932
amphibisch 1858
Amphibolisation 0932
Amphibolisierung 0932
Amphibolit 0934
Amplitude 0164
Amplitude 1634
Amplitudengang 4448
Amplitudenregelung, automatische – 4425
Amplitudenregelung, automatische – 4508

Amplitudenspektrum 4451
Amplitudenverhältnis, maximal aufzeichenbares – 4507
amygdaloïdisch 0872
Anadiagenese 3427
anaerob 0424
anaerob 1906
Anaklinal– 2288
Analcimisation 0932
Analogmodell 4782
Analogmodell, elektrisches Widerstands– 4783
Analyse 4183
analyse, Elementar– 3594
analyse, Immediat– 3594
analyse, Kurz– 3594
Analyse, mechanische – 2711
Analyse, numerische – 4777
Analyse, stereometrische – 4107
Anasëism 0187
Anatexis 1025
Anatexis, differenzielle – 1029
Anatexis, partielle – 1029
Anatexis, selektive – 1029
Anatexit 1026
Anbauterrasse 2061
anchieutektisch 0634
Anchimetamorphose 0888
Anchimetamorphose 3420
Andesit-Linie 0691
Anelastizität 0111
Anfangsdruck 3849
Anfangsproduktion 3805
angerundet 3188
Angliederungsinsel 2534
Angriffsküste 2528
Anhäufung, klastolitische – 1201
Anhydrit 3381
anisotrop 4681
Anisotropie 0110
Anisotropie 4681
Anisotropie, magnetische – 0303
anisotrop, optisch – 0561
Anker 4792
Ankerkabel 4794
Ankerkopf 4794
Ankerkraftmessdose 4772
Ankerplatte 4794
Ankerstange 4794
Ankerzugversuch 4764
Anlandung 2513
Anmoor 3527
Annuität 4218
Anomalie 4532
Anomalie 4632
anomalie, Bouguer– 0368
anomalie, Freiluft– 0368
Anomalie, geothermische – 0390
Anomalie, isostatische – 0369
Anomalie, magnetische – 0286
anomalie, modifizierte Bouguer– 0370
Anomalie, regionale isostatische – 0369
Anordnung 1480
anorogen 0726
anorogene Periode 1106
anorogene Zeit 1106
Anpassung 1980
Anreicherung 2678
Anreicherung 3945
Anreicherungshorizont 2702
Anreichung 3930
Anreichung, residuale – 3943
Ansammlung 3754
Anschmelzung, oberflächliche – 0708
Anschüttung 4825
Anschwellung 4054
Anschwemmungsküste 2528
Ansiedlung 1806
Anteklise 1084
antezedent 2290
Anthrazit 3561
Antiferromagnetismus 0301
Antiform 1618
Antiform-Mulde 1620
Antiklinale 1618
Antiklinale, Pseudo– 1620
Antiklinalfalle 3757
Antiklinalkamm 2541
Antiklinaltal 2544
Antikline 1618
Antiklinorium 1671
Antizentrum 0154
An- und Abschwellen 1609
Anwachs 2988
Anwachsküste 2528
Anwachslinie, gebogene – 0479
Anwuchs 2513
Anzapfung 2322
Anzapfungsknie 2326
Anziehung 0322
Äolianit 2870
äolisch 2870
Äon 1767
Äonothem 1767
Apamoor 2636
Apex 1635
aphanide 0798
Aphanit 0798
aphanitisch 0798
aphotisch 1905
aphyrisch 0798
apomagmatisch 3986
Apophyse 0725
Apophyse 4071
Äquator, geomagnetischer – 0281
Äquator, magnetischer – 0283
Aquifer 3634
äquivalent 1733
Äquivalent, chemisches – 3675
Äquivalentdurchmesser 3195
Äquivalenzprinzip 4567
Ära 1766
Aragonitnadel–Schlamm 3379
Ärathem 1766
Arbeitsfrequenz 4498
Arbeitsvereinbarung 3901
Archäomagnetismus 0294
archibenthisch 1908
Archie's Formeln 4605
areïsch 2257
Arenit 3225
Argillit 3211
Argon, atmosphärisches – 0450
Argon, radiogenes – 0450
Argon, überschuss – 0450
Arkose 3233
Arkulit 0768
Arm, blinder – 2212
Armerz 3915
Aromaten 3717
Array-Station 0119
Art 1937
Artenkreis 1935
Artentod 1986
Arterit 1045
Art, kosmopolitische – 1809
Asche 3599
Asche, epigenetische – 3600
Aschendecke 1320
Aschenkegel 1427
Aschenlawine, heisse – 1369
Aschenregen 1363
Aschenstrom, heisser – 1369
Aschentuff 1333
Aschenwolke 1363
Aschenwolke, elektrisch geladene – 1365
Asche, syngenetische – 3600
Asche, vulkanische – 1317
aseismisches Gebiet 0253
Asphalt 3734
Asphaltimprägnierung 3781
Asphaltit 3737
Asphaltsee 3753
Aspite 1420
Assimilation 0681
assimilation, Rand– 0683
Assoziation 1892
Assoziation 3179
Ast 1965
Ast 4071
Asterismus 0489
Asthenolith 1122
Asthenosphäre 0090
Ästuar 2456
Ästuar 2784
Ästuar– 2904
Aszensionstheorie 3927
Ataxit 0069
Ataxit 1243
Atectit 1032
Atektit 1032
Atexit 1032
atlantische Reihe 0690
atlantische Sippe 0690
Atmosphäre 0084
Atoll 3340
Atrio 1461
Atterbergsche Konsistenzgrenze 4714
Ätzfigur 0530
Ätzfigur 3409
Aufbereitung 1319
Aufbereitung 4137
Aufbereitungsanlage 4149
Aufblätterung 1994
aufbrechen 3800
Aufeis 2639
Aufeishügel 2643
Auffüllung 2341
Auffüllung 4825
Auffüllung 4831
Aufgabe 4140
aufgearbeitet 2947
aufgebohrtes Gebiet 3890
Aufhaldung 4225
aufhören, nach oben – 1526
aufhören, nach unten – 1526
Aufkochen 1358
Auflagerung, übergreifende – 1727
Auflaufen der Wellen 2813
Auflockerung 4731
Auflockerungssone 4731
Auflösung 2929
Auflösung, diagenetische – 3413
Auflösungselement 4319
Auflösungsvermögen 4347
Auflösungsvermögen 4453
Auflösungsvermögen, horizontales – 4516
Auflösungsvermögen, vertikales – 4516
Aufnahme 4500
Aufnahmebasis 4350
Aufnahmegerät 4500
Aufnahmegerät, digitales – 4501
Aufnahmegeschuss 4420
Aufnahmeort 4348
Aufragung 2800
Aufschiebung 1558
Aufschlickung 2513
aufschliessen 4150
Aufschluss 4231
Aufschmelzen, zonares – 0625
Aufschmelzung 0623
Aufschmelzung des Liegenden 0623
Aufschotterungsterrasse 2298
Aufschütten zum Kegel und Kreuzteilen 4181
Aufschüttungsebene, glaziale – 2424
Aufschüttungsinsel 2535
Aufschüttungsmäander 2220
Aufschüttungsterrasse 2298
Aufschüttungsterrasse, eingeschaltete – 2301
Aufschüttungsterrasse, marine – 2480
Aufsprudeln 1358
Aufspülung 4825
Auftauchen 1728
Auftauchen 2857
Auftauchküste 2525
Auftreten, allmähliges – 0174
Auftrieb 2786
Auftrieb, hydrostatischer – 4740
Aufwallung 1358
Aufwältigung 3813
Aufwölbung 1194
Aufwölbung 2796
Aufwölbung, lakkolithische – 1206
Aufwuchs 2513
Aufzeichnung, Dichteschrift– 4421
Aufzeichnung, Flächenschrift– 4421
Aufzeichnung, Linienschrift– 4421
Aufzeit 4428
Augen 0952
Augenstruktur 0952
Aulakogen 1085
Aureole 4629
Ausbau 4858
Ausbautefaktor 3867
ausbeissen 4231
Ausbiss 3753
Ausbiss 4231
Ausblasen 2937
Ausblasung 2155
Ausblasung 4818
Ausbruch 1211
Ausbruch 4288
Ausbruch, exzentrischer – 1229
Ausbruch, gemischter – 1219
ausbruch, Lava– 1214
ausbruch, Lockermassen– 1214
Ausbruch, magmatischer – 1219
Ausbruch, phreatischer – 1351
Ausbruch, phreatomagmatischer – 1219
Ausbruchs, Ausklingen des – 1167
Ausbruchsbeben 1193
ausbruch, Schlamm– 1395
Ausbruchschliesser 4289
Ausbruchsenergie, kinetische – 1171
Ausbruchsenergie, potentielle – 1171
Ausbruchskapazität zu einem bestimmten Zeitpunkt 1171
Ausbruchsmechanismus 1170
Ausbruchsöffnung 1440
Ausbruchstätigkeit, einmalige – 1154
Ausbruchstätigkeit, gemischte – 1219
Ausbruch, subaerischer– 1236
Ausbruch, subglazialer – 1236
Ausbruchswolke 1360
Ausbruch, untermeerischer – 1236
ausdünnen 3104
Ausfall von Schichten 1557
ausflocken 2992
Ausfluss 2338
Ausführbarkeitsuntersuchung 4791

Ausfüllungsdelta 2498
Ausgaben 3893
Ausgangserz 3913
Ausgangsgestein 2657
Ausgangssignal 4445
ausgefranst 3414
ausgegraben 2595
Ausgleichsfläche 0345
Ausgleichsgefälle 2198
Ausgleichsküste 2531
Ausgleichstiefe 0345
Ausgleichsvorrat 4225
Ausguss 1824
Ausguss 1825
Ausguss einer Kolkmarke 3026
Ausguss einer Schleifmarke 3033
Ausguss einer Sohlmarke 3019
Ausguss, innerer – 1824
Aushauchung 1376
Aushubmaterial 4871
auskeilen 3104
auskeilen 3509
Auskeilen 4054
Auskeilungsfalle 3758
Auskleidung 4858
auskolken 3025
Auskristallisation, disperse – 3947
Ausläufer 2806
Ausläufer 4071
Auslaugungshorizont 2701
Auslenkung 4079
Auslesedecke 2160
Auslesse, natürliche – 1966
Auslieger 2561
Auslogung 3929
Auslöschung 0564
Auslöschungswinkel 0566
Auslöschung, undulierende – 0968
Auslöschung, undulöse – 0565
Auslöschung, undulöse – 0968
Auspressungsdifferentiation 0675
Ausreicherung, sekundäre – 3937
Ausreiser 4071
Ausreisser 4186
Ausrollgrenze 4715
Ausscheidung 3948
Ausschürfen 2396
Aussenbogen, nichtvulkanischer – 1137
aussenbürtig 0081
Aussendüne 2167
Aussengroden 2514
Aussenrand des Grodens, erhöhter – 2516
Aussenskelett 1861
ausspülen 4873
Ausspülen 3159
Aussterben 1986
ausstreichen 4231
Ausstrich 4231
Austausch 3956
Austrocknung 2342
Austrocknungsbreccie 3048
auswaschen 4873
Auswaschen 3159
Auswaschung 3510
Auswaschungshorizont 2701
Auswaschungsrinne 3510
Auswehung 2155
Auswertung, direkte – 4566
Auswertung, indirekte – 4566
Auswertung, quantitative – 4566
Auswurfdecke 0055
Auswürflinge 1313
Auswürflinge, abgekühlte – 1315
Auswürflinge, allothigene – 1314
Auswürflinge, authigene – 1313
Auswürflinge, glühende – 1315
Auswürflinge, juvenil-authigene – 1314
Auswürflinge, magmaeigene – 1313
autallotriomorph 0811
authigen 2946
authigen 3434
authigen 3968
authigen 4633
Authigenese 3434
autochthon 0727
autochthon 1685
autochthon 1813
autochthon 2945
Autogenmahlung 4147
Autohydratation 3975
Autointrusion 0621
Autokorrelation 4195
Autokorrelation 4450
Autokovarianz 4450
Autolith 0664
Autometamorphose 0659
Autometasomatose 0659
Autometasomatose 3957
automorph 0523
automorph 0750
Autopneumatolyse 0659
Autopneumatolyse 3975
Axiolit 0767
axiolithisch 0767
Azidität 0422
Azimut 4228
azoisch 1738

B

Bach, inglazialer – 2388
B-Achse 0235
Bachsedimentprospektion 4641
Bach, subglazialer – 2388
Backenbrecher 4148
Backfähigkeit 3595
Backriss 2020
Bahada 2581
Bahamit 3303
Balge 2509
Balje 2509
Ballas 0495
Bändchen 2705
Bänderton 3128
Bänderung 2374
Bänderung 3101
Bänderung 3122
Bänderung, harrisitische – 0863
bänderung, Kristallisations– 0862
Bänderung, kryptische – 0865
Bänderung, primäre – 0860
Bänderung, rhythmische – 0863
Band, thermisches – 4308
Bank 1750
Bank 2244
Bank 2297
Bank 2469
Bank 2506
Bank 2802
Bank 3097
Bank 3098
Bankung 3108
Barchan 2177
Barranco 1474
Barre 2201
Barre 2297
Barre 2459
Barre, sickelförmige – 2462
Barriere 1916
Barriere, geographische – 1916
Basalkonglomerat 3114
basalt, Alkali– 0695
Basaltkugel 1256
Basaltplateau 1257
Basaltsäulen-gepflasterte Verebnung 1301
Basalt, Schildvulkan– 1258
Basalttafelland 1257
Basensättigung 2756
Basifizierung 0684
Basis 0626
basisch 0603
basische Front 1066
Basisstation 0353
Basisstation 4526
Basit 0603
Bastion 2606
Batholith 0729
bathyal 2846
bathypelagisch 1910
bathypelagisch 2845
Bathyseism 0158
Baugrunduntersuchung 4790
Baumstammnegativ 1292
Baunormen 0256
Baustein 4876
Baustelle 4789
Bauwürdigkeitsgrenze 4203
Bauxit 4124
Bazillit 0762
B⊥B' Tektonit 1003
B∧B' Tektonit 1003
Beben, kontinentales – 0148
Beben, lokales – 0149
Beben, submarines – 0148
Beben, vulkanische – 1193
Becken 0029
Becken 1086
Becken 1641
Becken 2793
Becken 3648
Becken, geschlossenes – 2579
Becken im Rücken des Inselbogens 1136
Becken, intramontanes – 1112
Becken, tektonisches – 1591
Beckesche Linie 0559
Bedeckung 4232
Beekit 3457
Begleitverwerfung 1581
begraben 2595
Begrenzungsbohrung 3785
Behandlung 3813
Behandlung, künstliche – 0492
Beimischung 4805
Belastungsmarke 3059
Beleuchtung, konoskopische – 0547
Beleuchtung, orthoskopische – 0547
Beleuchtung, spezifische – 4311
Belichtung 4340
Belichtungsspielraum 4340
Belonit 0761
Belonit 1424
Belonosphärit, homogener – 0778
belteropor 1009
Benioff-Zone 1134
benthal 1907
benthonisch 1907
Benthos 1907
Bentonit 3218
Benzin 3744
Beprobung, systematische – 4173
Bereich, bauwürdiger – 4178
Bereich, dynamischer – 0132
Berge 3501
Berge 3506
Berge 4164
Bergemittel 3506
Berg, feuerspeiender – 1411
Bergfussebene 2583
Bergfussfläche 2583
Bergfusstreppe 2604
Bergmehl 2153
Bergmilch 2153
Bergregal 4224
Bergrutsch 2044
Bergschlag 0115
Bergschlag 2112
Bergschrund 2382
Bergsturz 2044
Bergsturz 2048
Bergsturz 2049
Bergsturzes, Bahn des – 2054
Bergsturzrelief 2056
Berg , unterseeischer – 2798
Berme 2488
Berme 4860
Bermenkante 2488
Beschleunigung, maximale – 0120
Beschleunigungsmesser 0120
Beschleunigungsregistrierung 0120
Bestandslagerung 4225
Bestandteile, flüchtige – 3597
Bestandteil, exotischer – 2949
Bestandteil, flüchtiges – 0650
Bestandteil, flüchtiges – 1172
Bestandteil, gasförmiges – 1172
Bestandteil, opaker – 0610
Besteg 4012
Bestrahlung 0492
Bestrahlungsstärke 4312
bethonisch 2844
Beule 1664
Bewegung an den Scherflächen 2373
Bewegung, negative – 2858
Bewegung, positive – 2858
Bewegungslinie 2372
Bewegungsrichtung, wahre – 1562
Bewertung 4216
Bewertungsbohrung 3785
Bewertung von Bergwerkseigentum 4216
Bezugsfläche 0354
Bezugsfläche 3854
Bezugsniveau 0354
Bezugsniveau 0357
Bezugsniveau 4249
Biegegleitfaltung 1665
biegsam 0472
Bifurkation 2210
Bild 4320
Bild 4358
Bildbasis 4371
Bildbereich 4515
Bildmessung 4328
Bildstreifen 4358
Bildteil, mehrfach überdeckter – 4360
Bildwanderungs-Kompensator 4352
Bimsstein 1308
bimssteinartig 0875
bimssteinartig 1308
Bimssteinbombe 1329
Bimsstein, schwammiger – 1331
Bimssteintuff 1333
Binderschicht 4842
Binnendüne 2170
Binneneis 2425
Binnenentwässerung 2256
Biochron 1768
Biofazies 2861
Biogenesis 1792
biogenetisch 1792
biogenetisches Grundgesetz 1968
Biogeochemie 0395
Bioherm 3331

Bioherm, Kalkschlamm– 3325
bioklastisch 3273
biolith 3269
Biolithit 3269
Biom 1895
Biomikrit 3288
Biomikrit, dicht gepackter – 3288
Biomikrit, locker gepackter – 3288
Biomikrosparit 3289
Biomikrudit 3289
Bionomie 1887
Biopelmikrit 3291
Biopelsparit 3290
Bios 1804
Biospararenit 3287
Biosparit 3287
Biosparrudit 3287
Biosphäre 1788
Biostratigraphie 1710
biostratigraphische Einheit 1770
Biostratinomie 1791
Biostratinomie 1897
Biostrom 3331
biotisch 1903
Biotitisation 0932
Biotitisierung 0932
Biotitputzen 1072
Biotop 1902
Bioturbation 1885
Bioturbation 3067
Biozone 1769
biozone, Sub– 1782
biozone, Super– 1782
Biozönose 1896
Bireflexion 0575
Bisektrix 0556
Bisektrix, spitze – 0556
Bisektrix, stumpfe – 0556
Bittersalz 3383
Bitumen 3719
Bituminisierung 0426
Bivalve 1859
Blähgrad 3595
Blähungsgrad 3595
Blänke 3540
Blasenabdruck 3062
Blasenraum 0874
Blasensand 3063
Blasen, zylindrische – 1306
Blaser 1385
blasig 0874
blasig 1305
Blasloch 1274
Blastetrix 1014
blasto– 0971
Blastomylonit 0916
Blastomylonit 1541
Blatt 0739
blätterig 0775
blätterig 0948
Blätterkohle 3556
Blätterkohle 3563
Blätterung 2374
Blätterung 3101
Blattverschiebung 1560
Blattverschiebung, linkshändige – 1561
Blattverschiebung, rechtshändige – 1561
Blattverschiebung, steile – 1560
Blaublätterstruktur 2374
Blauschlick 2921
Blei, anomales – 0441
Blei-Blei Datierung 0434
Bleicherde 2700
Blei-Entwicklungslinien 0438
Blei, gewöhnliches – 0437
Blei-Modellalter 0440
Bleimodell, einstufiges – 0439
Bleimodell, mehrstufiges – 0439
Blei, radiogenes – 0437
Blindtal 2083
Blitz-Expansion 3863
Block 1087
Block 3243
Block, ausgeworfener – 1327
Blockbild 4258
Block, gerundeter – 3243
Blocklehm 2422
Blockmeer 2057
Blockstrom 2057
Blockwerk 3249
Blockzement 3439
Blütenstaub 1839
Blütezeit 1985
Blutregen 2159
Bocca 1440
Bodden 2454
Boden 0705
Boden 2657
Boden, A-B-C– 2695
Boden, A-C– 2695
Bodenaggregat 2724
Boden, alkalischer – 2737
Bodenart 2713
Bodenarteneinteilung 2713
Boden, azonaler – 2669
Boden, begrabend – 2693
Bodenbildung 2659
Bodenbildung, Faktor de– 2659
Bodencatena 2674
Bodendruck 3847
Bodendruckschreiber 3848
Bodeneignung 2660
Bodeneinheit 2672
Bodeneinheit 2673
Bodeneis 2639
Bodenentwicklung 2659
Bodenerosion 2036
Bodenerosion 2776
Bodenfeuchte 2764
Bodenfeuchte 3652
Bodenfliessen 2041
Boden, fossiler – 3126
Bodenfracht 2952
Bodengare 2757
Bodengefüge 2723
Bodengesellschaft 2673
Bodenhorizont 2687
Boden, im – 4636
Boden, intrazonaler – 2668
Bodenkartierung 2670
Bodenkriechen 2038
Boden, kriechende – 4855
Bodenkriechen, periodisches – 2039
Bodenkriechen, permanentes – 2039
Bodenkunde 2658
Bodenladung 4817
Boden mit A-B-C-profil 2695
Boden mit A-C-Profil 2695
Bodenprobe 3855
Bodenprofil 2686
Bodenprospektion, geochemische – 4644
Bodenreflexion 4519
Bodensatzlagerstätte 3987
Bodensatz und Wasser 3804
Boden, saurer – 2737
Bodenschicht 2688
Boden, schwellende – 4855
Boden, schwerer – 2721
Bodenseife 4136
Bodensenkung 4704
Bodenskelett 2722
Boden, sulphatsaurer – 2739
Bodensystematik 2661
Bodentyp 2666
Bodentypenreliefsequenz 2674
Bodenunruhe, mikroseismische – 0097
Bodenverfestigung 4806
Bodenwasser 2764
Bodenwasser 3652
Boden, zonaler – 2667
Bogendelta 2495
Bogendüne 2177
Bogenmauer 4865
Bogentrum 4072
Bogheadkohle 3563
Bohneisenerz 4118
Bohrabstand 2675
Bohranlage 4267
Bohrbarkeit 4298
Bohrbericht 4299
Bohren, pennsylvanisch – 4262
Bohrgang 1882
Bohrgarnitur 4270
Bohrkern 4237
Bohrloch 3665
Bohrloch 4261
Bohrloch, abgelenktes – 4293
Bohrlochabstand 3842
Bohrlochabstand 4822
Bohrlochkopf 3796
Bohrlochmessgerät 4578
Bohrlochmessung 4578
Bohrloch, unverrohrtes – 3795
Bohrlochverformungssonde 4760
Bohrlochvermessung 4294
Bohrlochversenkmessung 4406
Bohrprobe, zusammenhängende – 4749
Bohrprofil 4256
Bohrprotokoll 4299
Bohrraster 3842
Bohrschlamm 4281
Bohrspülung 4281
Bohrstange 4270
Bohrung 3665
Bohrung 4261
Bohrung, abgelenkte – 4291
Bohrung, fündige – 3784
Bohrung, nicht fordernde – 3812
Bohrungsdichte 2675
Bohrung, stratigraphische – 3783
Bohrung, verwässerte – 3794
Bohrwagen 4266
Bolson 2580
Bombe, ausgeschwänzte – 1329
Bombe, bandförmige – 1329
Bombe, bipolare – 1329
Bombe, explodierende – 1329
Bombe, gedrehte – 1329
Bombe, kugelige – 1329
Bombe mit Kern 1329
Bombe, perilithische – 1329
Bombe, spindelförmige – 1329
bombe, Tränen– 1326
Bombe, tropfenförmige – 1329
Bombe, unipolare – 1329
Bombe, vulkanische – 1328
BOP 4289
Bort 0495
Böschung 4860
Böschung, meeresseitige – 2480
Böschungswinkel, natürlicher – 4687
Bösschungswinkel 4827
bostonitisch 0839
Bottomset Lage 2896
Boudinage 1609
Bouguer-Korrektur 0359
Bouguerplatte 0359
Brachyantikline 1640
Brachysynkline 1640
Bradyseismus 0096
Bradyseismus 1194
Brandschiefer 3505
Brandung 2812
Brandungserosion 2016
Brandungsfläche 2479
Brandungsgasse 2474
Brandungshöhle 2472
Brandungskehle 2472
Brandungsplatform 2479
Brandungsplatte 2479
Brandungsplatte, schmale – 2479
Brandungsplattform 2536
Brandungsufer 2336
Brandungszone 2439
Brandungszone 2812
Brasilientest 4761
Brauneisenerze, oolitische – 4115
braunkohle, erdige Weich– 3555
braunkohle, Glanz– 3557
braunkohle, Glanz– 3558
braunkohle, Hart– 3495
braunkohle, Weich– 3495
braunkohle, Weich– 3555
Breccie 3258
breccie, Abbröckelungs– 1249
breccie, Einsturz– 3263
breccie, Explosivausbruchs– 1346
Brecciegang 4050
breccie, Injektions– 0711
Breccie, intraformationelle – 3264
breccie, Intrusiv– 0710
Breccienschlot 1222
breccie, Riff– 3355
Breccie, Riffgesteins– 3342
Breccierung 2933
Breccie, sedimentäre – 3258
Breccie, tektonische – 1539
Breccie, vulkanische – 1345
Brecher 2811
Brecherzone 2439
Brechung 2810
Brechungsindex 0551
Breite, geomagnetische – 0281
Breite, geozentrische – 0098
Breitenerosion 2022
Brekzie 3258
Brennwert 3604
Bresche 1471
Bresche– 2289
Brilliantschliff 0493
Bröckel 2725
Brodel 2648
Brodeltextur 2649
Brotkrustenbombe 1329
Bruch 0469
Bruch 0541
Bruch 1494
Bruch 1521
Bruch 4676
Bruchbildung 1494
Bruchdom 2113
Brüche, zerlegt durch – 1494
Bruchfalle 3757
Bruchfaltengebirge 2539
Bruchfaltungs- 1078
Bruchfläche 0238
Bruchfortpflanzungsgeschwindigkeit 0239
Bruchgewölbe 2113
bruch, Kessel– 0743
bruch, Kreis– 0743
Bruchküste 2520
Bruchlinienküste 2520
Bruchlinienstufe 2550
Bruchlinietal 2547
Bruch, muscheliger – 0541
Bruch, progressiver – 4676
Bruchschieferung 1601
Bruchscholle 1531
Bruchschollengebiet 1586
Bruchschollengebirge 2539
Bruchsenke 2546
Bruchstück 3166

Bruchstufe 2548
Bruchstufe, aufgedeckte – 2548
Bruchstufenfacette 2553
Bruchstufe, wiederbelebte – 2548
Bruchtal 2547
Bruchtektonik 1520
Bruchwaldtorf 3536
Bruch, wiederaufgelebter – 1527
Bruchzone 1467
Bruchzone 1532
Bruchzone 1595
Bruchzonental 2547
Brunnen 2105
Brunnen 3665
B-Tektonit 1003
Bucht 2450
Bucht 2451
Buchtenküste 2533
Buckelsinter 2141
Bühne 2297
Bullard, Methode von – 0363
Bult 3539
Büsserschnee 2365
Butz 4032
Bv 2703
Bysmalith 0734

C

Cabochon 0500
Cala 2452
Caldera 1456
Canyon 2795
Carbargillit 3505
carbonatarm 2736
carbonatfrei 2736
Carbonatgehalt 2735
carbonathaltig 2736
carbonatreich 2736
Carbon-Ratio 3601
Cash-Flow, jährlicher – 3894
Cash-Flow, kumulativer – 3894
Cash-Flow Kurve 3895
Casing 4275
CBR-Wert 4708
CD 0454
cenotyp 0636
Chandler Schwankung 0200
Charakteristik 0133
Charakter, optischer – 0557
Charnockit, M– 0990
Chasma 0034
Chatoyieren 0488
Chedakristall 0844
Chelatbildung 0409
Chenier 2491
Chenierebene 2491
Chile Salzpeter 3384
Chloridfumarole 1387
Chloritisation 0932
Chloritisierung 0932
Chondrit 0072
Chondrule 0071
Chonolith 0732
chorismatisch 0794
Chorismit 0794
Chroma 2677
Chron 1761
Chronomere 1760
Chronostratigraphie 1712
chronostratigraphische Einheit 1757
chronostratigraphischer Horizont 1758
chronotax 1732
Chronozone 1758
chymogen 1040
CIPW norm 0686
circalitoral 2839
Clarain 3580
Clarke-Wert 0399
Clarke-Zahl 0399
Clavalit 0763
Coccolith 1840
Coda 0175
Conca 1459
Concordia-Discordia Diagramm 0435
Conodont 1860
Conrad-Diskontinuität 0086
Cotypus 1951
Coulombesches Reibungsgesetz 4689
Crinoiden– 3313
crinoidenführend 3313
Crinoidenkalkstein 3313
Crocydit 1047
Cryptozoon 1844
Cuesta 2558
Cumulit 0765
Curie 4550
Curiepunkt 0308
Cutan 2691

D

Dach 0705
Dach 4010
Dachaufschmelzung 0623
Dachfläche 3106
Dachfläche 3501
Dachschieferung 1599
Dachzeigellagerung 3040
Daktylit 0849
daktylitisch 0849
Damm, natürlicher – 2238
Dampfdrucksteigerung mit fortschreitender Kristallisation 1174
Dampfdrucksteigerung, thermisch rückläufige – 1174
Dampfeinpressung 3839
Dampferuption, sekundäre – 1403
Dampfförderung 1215
Dampfinjektion 3839
Dampfinjektion, intermittierende 3839
Dampfkraftwerk, geothermisches- 0393
Dämpfung 0127
Dämpfung 0219
Dämpfung 4475
Dämpfung, aperiodische – -129
Dämpfung, dynamische – 0127
Dämpfung, kritische – 0129
Dämpfungsfaktor 0128
Dämpfungsverhältnis 0128
Dämpfung, visköse – 0127
Dampf, vulkanischer – 1349
Dampf, vulkanischer – 1377
Darcy 3697
Darcysches Gesetz 3696
Darwinismus 1966
Datenfilm 4388
Datierung, absolute – 0429
Datierung, radiometrische – 0429
Dauer 0240
Dauerfrostboden 2635
Dauerfrostboden 2637
Decke 1683
Decke 3126
Decke 4842
Deckenaufspaltung 1687
Deckenbasalt 1257
Deckenbau 1679
Deckenbruch 2110
Deckeneinsturz 2110
Deckenerguss 1278
Deckengebirgs– 1078
Deckengruppe 1691
Deckenpaket 1691
Deckensandstein 3109
Deckenstapel 1691
Deckenstirn 1686
Deckensystem 1691
Deckentektonik 1679
Decken, übergreifende – 1691
Decken, überschiebende – 1691
Deckenwurzel 1684
Deckfalte 1682
Deckgebirgekoeffizient 4202
Deckgestein 3764
Deckgletscher 2425
Deckschale 1683
Deckschicht 4842
Deckschichten 3501
Deckschichten 3924
Deckung, horizontale – 1551
Deckung, scheinbare horizontale – 1553
Deckung, stratigraphische – 1552
Dedolomitisation 0930
Dedolomitisierung 0930
Deflation 2155
Deflationskessel 2187
Deflationsrückstand 2160
Deflationswanne 2187
Deflektometer 4294
Deformation 1075
Deformation 4666
Deformation kurz nach Ablagerung 3050
Deformationsfläche 0238
Deformationsgefüge 3050
Deformationskristalloblastese 0907
Deformationslamelle 0999
Deformationsmesser 0122
Deformationsstruktur 3050
Deformation unter geringmächtiger Sedimentbedeckung, plastische – 3447
Deformation unter mächtiger Sedimentbedeckung, plastische – 3447
Degeneration 1982
Deglaziation 2367
dehnbar 0471
Dehnung 4666
Dehnungsfuge 1519
Dehnungsmesser 4767
Dehnungsmessstreifen 4767
Dehnungsriss 1519
Deklination, magnetische – 0284
Dekonvolution 4454
dekussat 0949
Delle 2281
Delta 2494
Delta 2895
deltaisch 2895
Deltaschichtung 3139
delta, Sub– 2496
Delta, vorgeschobenes – 2497
delta-Wert 0454
Demultiplexen 4424
Dendrit 0525
Dendrit 3400
Dendrit 4088
dendritisch 1821
Dendrochronologie 1737
Denudation 2012
Denudation 2103
Denudation, äolische – 2155
Denudationsniveau, oberes – 2598
Denudationsterrasse 2307
Denudationsterrasse 2555
Depletion 3830
Depozentrum 2986
Depression, intermediäre – 1137
Depression, kryptovulkanische – 1183
Depressionswinkel 4389
Depression, vulkanische – 1410
Depression, vulkano-tektonische – 1182
Derberz 4003
Dereverberation 4454
Desilifizierung 0929
Desintegration 2929
Desmosit 0932
Desorption 0406
Destruktionsstufe 2558
Desulfurisierung 0415
Deszensionstheorie 3926
Detektionskapazität 0270
Detektionsschwelle 0270
Detektor 4413
Detersion 2399
Detonation 1217
Detonation 4815
Detraktion 2396
detritisch 3165
Detritus 3165
Detrituslawine, glühende – 1367
detritus, Riff– 3355
deuterisch 0658
diablastisch 0950
diachron 1736
Diachronismus 1736
Diadisit 1046
Diadochie 0477
Diadochie, gekoppelte – 0477
Diagenese 3388
Diagenesebereich 3390
Diagenese, fortgeschrittene – 3418
Diagenese, frühe – 3418
Diagenese, milieubezogene – 3398
Diagenese, phreatische – 3399
Diagenese, späte – 3418
Diagenese, vadose – 3399
diagenetischer Milieu – 3391
diagenetischer Milieu mit geringmächtiger Sedimentbedeckung – 3393
diagenetischer Milieu mit mächtiger Sedimentbedeckung – 3393
diagenetischer Milieu nahe der Oberfläche – 3392
Diagonalkluft 1507
Diagonalküste 2522
Diagonaltrum 4072
Diagonalverwerfung 1545
Diagramm, beta– 1021
Diamagnetismus 0298
Diamantbohrer 4265
Diamantglanz 0484
diamiktisch 3124
Diamiktit 3124
Diaphthorese 0904
Diaphthorit 0904
Diapir 1693
Diapir 1695
Diapirfalte 1698
Diapirismus 1693
diaplektisch 0895
Diastem 3125
Diastrophismus 1074
Diatexis 1027
Diatexit 1027
Diatomeenerde 3370
Diatomeenschale 3370
Diatomeenschiefer 3371
Diatomeenschlamm 3370
Diatomit 3370
Diatrema 1221
dicht 0799
Dichte 4342

Dichte 4537
Dichtelog 4593
Dichteplan 4259
Dichte, scheinbare – 2751
Dichte, scheinbare – 3602
Dichteschichtung 2882
Dichtestrom 2979
Dichte, wahre – 3602
Dichtlagerung 2760
Dichtschlämmung 2760
Dichtungssporn 4861
Dieselöl 3744
Differentation, diagenetische – 3938
Differentation, magmatische – 3938
Differentiat 0667
Differentiation 0398
Differentiation 0667
Differentiation, diagenetische – 3475
differentiation, Diffusions– 0680
Differentiation durch die Schwere 0673
Differentiation, geochemische – 0398
Differentiation, gravitative – 0672
differentiation, Kristallisations– 0670
Differentiation, metamorphe – 0911
Differentiation, pneumatolytische – 0679
Differentiationsindex 0701
Differentiationstrennung 0671
differentiiert 0667
Diffluenzpass, glazialer – 2621
Diffluenzstufe 2624
Diffraktion 0217
Diffraktion 4444
Diffraktionskurve 4464
Diffusion, metamorphe – 0912
digitalisieren 4423
Digitation 1687
Diktyonit 1043
Dilatanz 0109
Dilatanz 4678
Dilatation 0187
Dilatometer 4760
Dilatometer-Test 3596
Dimorphismus 1811
Dipolfeld, axiales– 0279
Dipolfeld, excentrisches – 0279
Dipol-Moment, magnetisches – 0296
diskoidal 3184
Diskontinuität 4723
Diskontinuitätsfläche 0099
diskordant 0718
diskordant 1720
Diskordanzfalle 3759
Diskordanzküste 2522
Diskordanz, stratigraphische – 1720
Diskriminator, Einkanal– 4547
Diskriminator, Integral– 4547
Dislokation 1522
Dislokationsbeben 0139
Dispersion 0197
Dispersion 0403
Dispersion 0572
Dispersion 3710
Dispersion der optischen Achsen 0572
Dispersion, gekreuzte – 0572
Dispersion, geneigte – 0573
Dispersion, horizontale – 0573
Dispersion, inverse – 0197
Dispersion, normale – 0197
dispersion, Primär– 0404
dispersion, Sekundär– 0405
Dispersionshof 4629
Dispersionshof im Nebengestein 4630
Dispersionskurve 0226
Dispersion, umgekehrte – 0197
Dissipation 0219
distal 2997
Divergenz 1969
DMS 4767
Doline 2072
Dolinenkarst 2080
Dolomikrit 3282
Dolomit 3318
Dolomitbruchstück 3270
Dolomit, Ca– 3321
Dolomitfleckigkeit 3461
Dolomitfragment 3270
Dolomit, frühdiagenetischer – 3320
Dolomitisierung 3460
Dolomit, primärer – 3320
Dolomit, primärer – 3380
Dolomitstein 3318
Dolomitstein, kalkiger – 3319
Dolomitstein, kalzitischer – 3319
Dom 0059
Dom 1641
Domgebirge 2538
Domvulkan 0035
Doppelbrechung 0560
Doppelkrater 1442
Doppelvulkan 1442
Dopplerit 3525
Drainage 2768
Drainageschirm 4804
Drainung 4739
Dränleitung 4811
Dränrohr 4811
Dränung 4811
Drapierfalte 1664
Drapierstruktur 1664
Dredge 3154
Dredsche 3154
Drehtisch 4269
Drehwaage 0350
Dreikanter 2162
Dreipunktmethode 0152
Drewit 3379
drift, Küsten– 2962
drift, Strand– 2962
Drillbohrer 4236
Druckaufbau 3850
Druckaufnehmer 4415
Druckausgleich, isostatischer – 0344
Druck der Deckschichten 3819
Druckdifferenz 3850
Druckentlastung 1173
Druckentlastung 3850
Druckerhaltung 3832
Druckfestigkeit, dreiaxiale – 4683
Druckfestigkeit, einaxiale – 4683
Druckgeber 4769
Druckgefälle 3853
Druckgradient 3852
Druckhöhe, hydraulische – 4735
Druckhöhengefälle 4736
Druckhöhenunterschied, hydraulischer – 4735
Druck, hydrostatischer – 3820
Druckkammerversuch 4762
Druckkarte 3854
Druckkissenversuch 4763
Drucklösung 3445
Druck, magmatischer – 0622
Druckmesser 4769
Druck, petrostatischer – 3819
Druckrohrleitung 4867
Druckschacht 4867
Druckschatten 0958
Drucksondierung 4711
Druck, statischer – 3847
Druck, tektonischer – 0886
Drumlin 2420
Druse 4014
Drusenraum, miarolitischer – 0873
Drusenzement 3438
dualistische Hypothese 0019
dubiokristallin 0801
Dublette 0480
duktil 4673
Düne 2165
Düne 2872
Düne aus Tonscherben 2872
Düne, befestigte – 2175
Düne, festliegende – 2175
Düneninsel 2468
Dünenkamm, rauchender – 2158
Dünental 2186
Dünental 2188
Dünenwanne 2189
Dünenwüste 2578
Düne, organogene – 2172
Düne, ruhende – 2175
düne, Sand– 2872
Düne, stillstehende – 2175
Dünung 2807
Durain 3580
Durchbruch 2240
Durchbruchfächer 2241
Durchbruchs– 2289
Durchbruchs– 2290
Durchbruchsberg 2215
Durchbruchstal, Kurzes – 2327
Durchfallungskreuz 4075
Durchfliessmessgerät 3669
Durchfluss 3622
Durchlass 4846
Durchlassbereich 4498
Durchlassgrad 4315
Durchlässigkeit 2753
Durchlässigkeit 3697
Durchlässigkeit, effektive – 3698
Durchlässigkeit, primäre – 3701
Durchlässigkeit, relative – 3698
Durchlässigkeit, sekundäre – 3701
Durchlässigkeitsgerät 4755
Durchlaufanalyse 4144
Durchlauf, einmaliger – 4503
Durchmesser, effektiver – 3196
durchmesser, Median– 3199
Durchmesser, mittlerer – 3199
Durchmesser, modaler – 3199
durchscheinend 0545
Durchschlagloch 1274
Durchschmelzung des Daches 0624
durchsichtig 0545
Durchtrümerung 4060
Durchwachsung 0783
Dy 3527
Dynamikbereich 4507
Dynamotheorie 0278
dysphotisch 1904

E

Ebbestrom 2822
Ebbrinne 2510
Echo 4390
Echo, ohne – 4391
Ecke 2306
Eckflur 2306
eckig 3189
Ectinit 1055
Eddy 2269
Edelfall 4007
Edelstein 0478
Edelsteinkunde 0478
Edelsteinschneider 0502
Effekt, indirekter – 0366
Efforation 2119
Effusion 1213
Effusivgestein 1240
Eh 0421
eigenfarbig 0483
Eigenperiode 0130
Eigenpotentialkurve 4579
Eigenpotentialverfahren 4557
Eigenschaft, Kohlen– 3551
Eigenschwingung – 0198
einachsig 0553
Einbruch am Aussenhang 1464
Einbruchskaldera 1459
Einbruchskessel 1183
Einbruchskrater 1455
Einbruchsstruktur 1704
Einbuchtung 2450
Eindampfung 0412
Eindellung 1418
Eindicker 4159
Eindrängungslagerstätte 3993
Eindringtiefe 4513
Eindringtiefe, registrierte – 4513
Eindringungsgeschwindigkeit 4298
Eindringwiderstand 4709
Eindunstung 2342
Einebnung 2589
Einebnungsfläche 2591
Einfallen 1484
Einfallen 1614
Einfallen 4017
Einfallen, entgegen dem – 1489
Einfallen geschlossen, durch – 3763
Einfallen, scheinbares – 1484
Einfallen, scheinbares – 4250
Einfallen, senkrechtes – 4472
Einfallens, in der Richtung des – 1489
Einfallens, Richtung des – 1490
Einfallsebene 0211
Einfall, streichendes – 0212
Einfallswinkel 0211
Einfallswinkel 1485
Einfangtheorie 0047
Eingangsignal 4445
eingebettet 3117
eingeengter Teil 2390
eingeschaltet 3117
Einheitsergiebigkeit 3702
Einheitskluftkörper 4730
Einkanter 2162
Einlagerung 2678
Einlagerung 3117
Einlagerung 3930
Einlagerungshorizont 2702
Einnahmen 3893
einpressen, Zement– 4803
Einpress-Sonde 3793
Einregelung 1013
Einrumpfung 2589
Einsatz 0172
Einsatz, deutlicher – 0173
Einschaltung 3117
Einschieben 4019
Einschiessen 4019
Einschlagerdbeben 0141
Einschlagtrichter 1418
Einschluss 0662
Einschluss 0789
Einschluss, enallogener – 0663
Einschluss, endogener – 0664
Einschluss, exogener – 0663
Einschluss, fremder – 0663
Einschluss, homogener – 0664
Einschluss, mixogener – 1071
Einschneiden 2936
Einschneidung 2020
Einschnitt 1471
Einschnitt 2020

Einschnitt 4824
Einschnürung 4054
Einsenkung 1183
Einsenkungshypothese 1118
Einsickerfähigkeit 3620
Einsickerung 3619
Einsprengling 0758
Einsprengung 3947
Einstülpung 2649
Einsturzbeben 0141
Einsturzdoline 2072
Einsturzkaldera 1459
Einsturzkrater 1455
Einsturztal 2085
Einteilung 1921
Einwaschung 2678
Einwaschung 3930
Einwaschungshorizont 2702
Einwicklung 1690
Einzelfundament 4837
Einzelfundament, flaches – 4835
Einzelwert, stark abweichender – 4186
Einzugsgebiet 2192
Einzugsgebiet 3670
Einzugsgebiet 3844
Eisdecke 2425
Eisdrift 2969
Eisdruck 2401
Eisenerze, gebänderte – 4114
Eisenglimmerschiefer 4114
Eisenjaspilite, gebänderte – 4114
Eisenmeteorit 0066
Eisenquartzite 4114
eisenschüssig 3464
Eisenstein 3464
Eis, fossiles – 2369
Eishöhle 2107
Eiskappe 2425
Eiskeil 2640
Eiskeilspalte, aufgefüllte – 2641
Eiskeilspaltenboden 2655
Eiskeller 2107
Eislawine 2361
Eisnadelabdruck 3043
Eispfeiler 2377
Eispressung 2401
Eisscheide 2381
Eisscheitel 2381
Eisschiebung 2401
Eisschub 2401
Eisstromnetz 2432
Eis verdriftet, vom – 2969
Ektekt 1034
Ektexis 1034
elastisch 0472
elastischen Zurückschnellens, Theorie des – 0108
Elastizität 4669
Elastizitätsgrenze 4697
Elastizitätsmodul 4695
Elektrodenanordnung 4562
Elektrodenkonfiguration 4562
Elementverteilung 0401
Ellipsoid, internationales – 0331
Elliptizitätskorrektur 0098
Elongation 0567
Elutriation 3159
Eluvialboden 2930
Eluvialhorizont 2701
Eluviation 2678
Eluviation 3930
Eluvium 2930
Emanation 3951
Emanation, vulkanische – 1215
Emanation, vulkanische – 1376
Embrechit 1055
Embryonaldüne 2176
Embryonalfaltung 1666
emergent 0174
Emergenzwinkel 0215
Emissionsspektrum 4305
Emissionsvermögen 4313
Empfangstation 4417
Empfindlichkeit 4717
emulsives Stadium 0652
Enddruck 3849
endemisch 1814
enden, gegen die Teufe hin – 1526
enden, nach oben – 1526
enden, nach unten – 1526
enden, zum Hangenden hin – 1526
Endmoräne 2413
Endmoränenlandschaft 2630
Endobiont 1854
Endoblastese 0657
endoblastisch 0657
endogen 0081
endogener Effekt 0704
Endomorphose 0898
Endoskop 0503
Endosphäre 0089
Endphase 1167
Endprofil 2198
Energieabstrahlung 0204
Energie, geothermale – 0388
eng 1626
eng, sehr – 1626
Entbasung 2738
entblösst 2595
Entekt 1036
Entexis 1036
Entgasung 1215
Entgasung 3520
Entglasung 0919
Entkalkung 2679
Entkopplung 0273
Entlastungsbohrung 4290
Entlastungskluft 1506
Entmagnetisierung durch Wechselfield 0321
Entmagnetisierung, thermische – 0321
Entmischung 0476
Entmischung, liquide – 0678
Entmischung, liquide – 3939
Entmischung, liquidmagmatische – 3939
Entmischungsstruktur 4105
Entnahmemenge, zulässige – 3668
Entnahmestutzen 4748
Entölung durch Schwerkraft 3828
Entölungsgebiet 3844
entschlämmen 4161
Entsilifizierung 0929
Entspannung 1179
entspannungsversuch, Bohrloch– 4766
Entstehungsgebiet 1969
Entwässerung 2253
Entwässerung 2768
Entwässerung 3797
Entwässerung 4739
Entwässerung 4811
Entwässerung, endoreische – 2256
Entwässerung, exoreische – 2256
Entwässerungsanordnung 2254
Entwässerungsgebiet 2192
Entwässerungsmuster 2254
Entwässerungsnetzes, Muster des – 2254
Entwässerungsöffnung 4811
Entwicklung, absteigende – 2574
Entwicklung, aufsteigende – 2574
Entwicklung, gleichförmige – 2574
Entwicklung, konvergente – 1973
Entwicklung, parallele – 1971
Entzerrung 4376
Eolinait 2870
Eometamorphose 0902
Eötvös 0337
Eötvös-Korrektion 4538
Epeirogenese 1090
epibenthos 1908
Epibiont 1855
Epibolit 1049
Epifauna 1853
Epigenese 3419
epigenetisch 2290
epigenetisch 3474
epigenetisch 3553
epigenetisch 3966
epikontinental 2915
Epikontinentalbecken 1086
Epilimnion 2332
Epimagma 1201
epineritisch 2841
Epinorm 0606
Epipedon 2697
epipelagisch 1910
epipelagisch 2845
Epiphyt 1856
Epirogenese 1090
epitaktisch 0542
Epitaxie 0542
epithermal 3983
Epizenterazimut 0155
Epizentraldistanz 0155
Epizentralentfernung 0155
Epizentrum 0154
Epizone 0993
Epoche 1764
Equivalentmodell 4778
Erdbeben 0138
Erdbebengeräusch 0203
Erdbebenkontrolle 0262
Erdbeben, Krypto-vulkanisches – 0140
Erdbebenkunde 0094
Erdbeben, künstliches – 0265
Erdbebenmechanismus 0232
Erdbeben, mehrfacher – 0143
Erdbebenrisiko 0255
Erdbebenschwarm 0144
Erdbebensichere Bauwerk 0260
Erdbebensichere Konstruktion 0260
Erdbebenspalte 2552
Erdbebenstufe 2552
Erdbeben, tektonisches – 0139
Erdbebenverhütung 0262
Erdbeben-Vorhersage 0261
Erdbülte 2643
Erddam 4859
Erddruck 4705
Erddruck, activer – 4706
Erddruckmessdose 4771
Erddruck, passiver – 4706
Erde 3086
Erde, bewehrte – 4798
Erdfall 2072
Erdgas 3713
Erdgasfeld 3773
Erdgezeit 0199
erdig 3086
Erdkern 0093
Erdkruste 0086
Erdloch 2076
Erdmagnetfeld 0276
Erdmagnetismus 0276
Erdmantel 0091
Erdöl 3712
Erdölgas 3891
erdölhoffiges Gebiet 3887
Erdölmuttergestein 3722
Erdpfeife 2077
Erdpfeiler 2029
Erdpyramide 2029
Erdrutsch 2046
Erdrutsch 2051
Erdrutsches, Bahn des – 2054
Erdschlipf 2051
Erdstoss 0138
Erdstreif 2654
Erdstrom 4552
Erdwachs 3735
Erdwärme 0374
Erdwissenschaften 0002
Erfolgsquote 3784
Erfolgsrate 3784
Ergussgestein 1240
Erhebung 2796
Erhebungsgrenze, obere – 2598
Erlöse 3893
Erlös, marginaler – 4204
Ermüdung 4674
Ernährungsgebiet 2192
Erodierbarkeit 2936
Erosion 2012
Erosion 2936
Erosion, äolische – 0043
Erosion, äolische – 2154
Erosion, beschleunigte – 2777
erosion, Boden– 2776
Erosion, Endkurve der – 2198
erosion, Flächen– 2779
Erosion, flächenhafte – 2032
Erosion, fluviatile – 2015
erosion, Graben– 2778
Erosion, kosmische – 0028
Erosion, lineäre – 2012
Erosion, linienhafte – 2012
Erosion, marine – 2016
erosion, Rinnen– 2778
Erosion, rückläufige – 2021
Erosion, rückschreitende – 2021
Erosionsbasis 2025
Erosionsbasis, allgemeine – 2025
Erosionsbasis, lokale – 2025
Erosionsbasis, örtliche – 2025
Erosionsdiskordanz 1721
Erosionsdiskordanz 1722
Erosionsdoline 2072
erosion, Seiten– 2022
Erosion, selektive – 2024
Erosionshaltstufe 2199
Erosionskaldera 1473
Erosionskessel 1473
Erosionskessel 2027
Erosionskessel, grosser – 2028
Erosionsmäander 2221
Erosionsmarke 3021
Erosionsmarke, transversale – 3021
Erosionsrinne 2026
Erosionssyklus 2566
Erosionssyklus, unterbrochener – 2571
Erosionsterminante 2198
Erosionsterrasse 2298
Erosionsterrasse, marine – 2479
Erosionstrichter 2276
erosion, Tiefen– 2022
erratischer Block 2949
erschlossenes Gebiet 3890
Ersetzung 3956
Erstarrung, oberflächliche – 1200
Erstarrungsgestein 0616
Erstarrungsindex 0703
Ersteinsatz 4428
Erstreckung, seitliche – 1524
Erstreckung, streichende – 1524
Eruption 1211
Eruption, indirekte – 1218
Eruption, intermittierende – 1212
Eruption, nichtextrusive – 1219
Eruption, paroxysmale – 1164

Eruption, semi-vulkanische – 1351
Eruptionskreuz 3796
Eruptionsprodukte, lockere – 1313
Eruptionspunkt 1220
Eruptionsregen 1373
Eruptionssäule 1361
Eruptionsspalte 1225
Eruptionsspalte 1466
Eruptions-Vorboten 1191
Eruptions-Vorzeichen 1191
Eruptionswolke 1360
Eruption, Tätigkeitselement der – 1209
eruption, Zentralschlot– 1228
Eruptivgang 1516
Eruptivgestein 0616
Eruptivgestein 1238
Eruptivgesteinsdecke 1275
Eruptivkörper 1239
Erwartungskurven-Methode– 3876
erweitern 4296
Erweiterung 3885
Erweiterungsbohrung 3785
Erz 3913
Erzart 0462
Erzart 3912
Erzausscheidungsperiode 3905
Erz, bauwürtiges – 3916
Erzbildungsperiode 3905
Erz, durchbewegtes – 4079
Erz, einfaches – 3914
Erz, eingesprengtes – 4004
Erzfall 4005
Erzfall 4178
Erzfuge 4039
erzführend 3917
Erzgang 4038
Erzgang, intrusiver – 4031
Erzgang, komplexer – 4046
Erzgang, krustenartiger – 4049
Erzgang mit krustenförmige Struktur 4049
Erzgang, zusammengesetzter – 4045
Erz, gemengtes – 3914
Erzgeneration 3906
Erz, geringwertiges – 3915
erzhaltig 3917
Erz, hochwertiges – 3915
Erzhorizon 4008
Erzinjektion, abgepresste – 3975
Erzkanal 4029
Erzkörper 3910
Erzkörper, linsenförmiger – 4026
Erzlager, diskordante – 4022
Erzlager, intrusives – 4031
Erzlager, konkordante – 4022
Erzlager, penekonkordante – 4022
Erzlagerstätte 3910
Erzlagerstätte, eingelagerte – 4021
Erzlagerstätte, flözartige – 4024
Erzlagerstätte, geschichtete – 4023
Erzlagerstätte, massige – 4003
Erzlagerstätte, sackförmige – 4035
Erzlagerstätte, schichtgebundene – 4024
Erzlagerstätte, schlauchförmige – 4033
Erzlagerstätte, stratiforme – 4024
Erzlagerstätte, zonare – 4015
Erzlagerstätte, zonierte – 4015
Erzlinse 4026
Erz, massiges – 4003
Erzmikroskopie 3925
Erzmineral 0462
Erzmineral 3912
Erzmittel 4005
Erz, monomineralisches – 3914
Erzpetrographie 3925
Erzprovinz 3908
Erzschlauch 4034
Erzschlauch 4037
Erzschlieren 4040
Erzschnur 4039
Erzwert, marginaler – 4204
Erzzone 4008
Erz, zusammengesetztes – 3914
Esskohle 3559
Esskohle 3560
Estavelle 2092
Ethmolith 0737
Eugeosynklinale 1097
eugranitisch 0810
euphotisch 1904
eurybatisch 1911
eurybenthisch 1909
euryhalin 1912
eurytherm 1913
eurytop 1914
eustatische Bewegung 2858
Eutaxit 0861
Eutaxit 1243
eutaxitisch 0861
eutroph 3549
euxinisch 2925
Evaporit 3375
Evapo-Tranpiration 3621
Evolution 1964
Evolution, gerichtete – 1975
Evolution, iterative – 1971
Evolution, magmatische– 0666
Exaration 2396
Exhalation 1376
Exhalation 3951
Exhalationslagerstätte 3979
exhumiert 2595
exogen 0081
exogener Effekt 0704
Exogeologie 0024
Exomorphose 0898
Exosphäre 0083
Expansionshypothese, thermische – 1123
Explorationsbohrung 3782
Explosion 1217
Explosionsgraben 1235
Explosionsherd 1177
Explosionskaldera 1458
Explosionskrater 1431
Explosionslichtbogen 1371
Explosionsniveau 1177
Explosionspunkt 1176
Explosionstuff 1335
Explosivausbruch 1216
explosive Erscheinungen 1175
Explosivitätsgrad 1175
Extensometer 4773
Extraklast 3272
Extrusion 1207
Extrusionsphase 1197
Extrusivgestein 1240
Extrusivkörper 1239
Extrusivmasse 1239

F

Fächerfalte 1650
Facies, diagenetische – 3435
Faciesserie, metamorphe – 0991
Fadenlapilli 1326
Fahlband 4052
Fährte 1877
Fall 4735
Falle 3756
Falle, hydrodynamische – 3761
Fallen 1484
Fallen 1614
Fallen 4017
Falle, paläotopographische – 3760
Falle, stratigraphische – 3758
Falle, strukturelle – 3757
Fallgewichtsseismik 4409
Fallrichtung 1490
Fallrichtung, entgegengesetzt der – 1489
Fallrichtung, in der – 1489
Fallsturz 2049
Fällung 0411
Fällungsbarriere 4642
Fallwinkel 1485
Fallwinkel einer Nerwerfung – 1549
Fallwinkel einer Verwerfung 1548
Falschalarm 0269
Falschfarbenfilm 4337
Falte 1616
Falte, abtauchende – 1637
Falte, Adventiv– 1677
Falte, allochthone – 1659
Falte, asymmetrische – 1644
Falte, aufrechte – 1645
Falte, geneigte – 1646
Falte, kegelförmige – 1655
Falte, kongruente – 1653
Falte, konzentrische – 1652
Falte, liegende – 1680
Fältelung 1643
Fältelung 3051
Fältelung, intraformationelle – 3051
Faltenachse 1630
Faltenbündel 1674
Faltendecke 1682
Faltendurchkreuzung 1678
Faltengebirge 2538
Falten, gestaffelte – 1676
Faltengürtel 1674
Falte, nicht-zylindrische – 1654
Falten, Paket liegender – 1681
Faltenschar 1674
Faltenspiegel 1632
Faltenspiegel 1672
Falten, Stapel liegender – 1681
Faltenstrang 1674
Faltensystem 1670
Faltentektonik 1615
Falten, versetzte – 1676
Falte, parallele – 1652
Falte, parasitäre – 1677
Falte, ptygmatische – 1658
Falte, schiefe – 1646
Falte, stehende – 1645
Falte, symmetrische – 1644
Falte, trikline – 1657
Falte, überkippte – 1647
Falte, überschobene – 1576
Falte, unsymmetrische – 1644
Falte, untergeordnete – 1642
Falte, vergente – 1646
Falte, wiedergefaltete – 1668
Falte, zylindrische – 1654
Faltung 4452
Faltung, Anfangsstadium der – 1666
Faltung, disharmonische – 1663
Faltung, enterolithische – 3076
Faltungsphase 1107
Faltungsphase 1669
Faltungszone 1674
Faltungstektonik 1615
Faltung, synsedimentäre – 1667
Faltung, überprägende – 1668
Faltung, wiederholte – 1668
Familie 0687
Familie 1930
Fangarbeit 4287
Fangedamm 4860
Fanggerät 4287
Fanglomerat 2885
Farbe 0546
Farbe 2677
Farbintensität 2677
Farbkomposite 4346
Farbtiefe 2677
Färbung 0492
Farbzahl 0688
faro 3339
faserig 0522
faserig 0755
faserig 0773
Faserkalzit 2143
Faserkohle 3580
Fass 3892
Fastebene 2589
Fastebene, Bildung einer – 2589
Fäulnis 3526
Faulschlamm 3528
Faulschnee 2350
Fauna 1795
fauna, Zwerg– 1851
Faunen– 1795
Faunenzone 1774
Faunula 1796
Fazies 1715
Fazies 2860
Faziesfauna 1850
Faziesfossil 1807
Fazieskarte 4245
Fazies, metamorphe– 0987
Fazies, Metamorphose– 0997
fazies, Mineral– 0585
Fazies, vulkanische – 1238
Federkielstalaktit 2131
Fehlbohrung 3794
Fehlbohrung 4287
Fehler 0487
Fehler, systematischer – 4192
Fehlordnung 0531
Feinboden 2719
Feinerde 2719
feingeschichtet 3108
feinkörnig 0796
feinkörnig 3194
feinkörniges Material 3207
feinkörnig, sehr – 0799
Feldintensität 0317
Feldkapazität 2774
Feldkapazität, nutzbare – 2773
Feldspatisierung 1064
Feldstärke 0317
Fels 0935
Fels 4722
Felsausläufer 2606
Felsband 2201
Felsebene, aride – 2582
Felsebenen, verwachsene aride – 2584
Felsen, äolisch bearbeiteter – 2161
Felsenmeer 2057
Felsenmeer 2656
Felsen, windpolierter – 2161
Felsenwüste 2578
Felsfussfläche, aride – 2582
felsisch 0597
felsisches Gestein 0597
Felsit 0597
Felskegel 2585
Felsnagel 4793
Felsophyr 0824
felsophyrisch 0824
Felsosphärit 0778
Felsriegel 2625
Felsrutsch 2048
Felsrutschung 2048
Felssporn 2606
Felsstufe 2200

Felssturz 2049
Felsterrasse 2298
Felsüberlagerung 4788
Felsuntergrund 4788
Felsvorsprung 2606
femisch 0607
Fenit 0925
Fenitisation 0925
Fenster 4234
Fenster, teleseismischer – 0231
Fernbeben 0150
Fernberg 2608
Ferner 2366
Fernerkundung 4300
Fernling 2608
Ferrimagnetismus 0302
Ferromagnetismus 0299
festgeheftet 1919
Festgestein 4722
Festgesteinsprospektion 4644
Festigkeit 4682
Festlandsvulkan 1415
Fettkohle 3559
Feuchte 2769
Feuchtigkeit, grobe – 3598
Feuchtschnee 2349
Feuchtschneelawine 2359
Feuer 0475
Feuerfontäne 1283
Feuersäule 1283
feuerspeiend 1237
fibroblastisch 0963
fibrokristallinisch 0815
Fiederkluft 1505
Fiedermarke 3036
Fiederstörung 1581
Fiederstörung 1583
Filigranstruktur 4081
Film 3100
Film 3402
Film, Infrarot– 4336
Film, orthochromatischer – 4334
Film, panchromatischer – 4334
Filter 4338
Filter 4445
Filter 4812
Filter, Abschneide– 0137
Filter, Durchlass– 0137
Filtergeschwindigkeit 3703
filter, Kerb– 0137
Filterkuchen 4618
Filterpressung 3933
Filterpressung 0675
Filterrohr 3795
Filterschicht 4862
filter, Sperr– 0137
filterung, Geschwindigkeits– 0228
Filtration 0414
Filtrationspotential 4617
filzartig 0825
filzig 0825
final 1108
Findling 2422
Fireclay 3517
Firneisbildung 2363
Firnfeld 2364
Firnfeldgletscher 2428
Firngebiet 2364
Firngrenze 2364
Firnkesselgletscher 2429
Firnlinie 2364
Firnmulde 2364
Firnmuldegletscher 2428
Firnschnee 2363
Firnstromgletscher 2431
Firste 3501
Firste 4851
Fixismus 1113
Fjord 2453
Flache 4016
Fläche 0517
Fläche, listrische – 1573
Flächenerguss 1276
Flächenerosion 2032
Flächeneruption 1232
Flächenkombination 0521
Flächenspülung 2032
Flächentreppe 2601
Flachgründung 4835
Flachgründung 4837
Flachmündung 2208
Flachufer 2235
Flachwasserseismik 4479
Flammkohle 3559
Flanke 1625
Flankenerguss 1266
Flankeneruption 1230
Flankenfluten 3833
Flaser 0952
Flaser 3103
Flaserschichtung 3148
Flasertextur 0952
Flaumschnee 2346
Fleck 2692
Fleckenschiefer 0941
Flecktextur 0965
Flexur 1616
Flexurküste 2519
Flexurstufe 2551
Fliessbreccie, verschweisste vulkanische – 1268
Fliessbreccie, vulkanische – 1268
Fliessdifferentiation 0677
Fliessdruck 3847
Fliessen 4672
Fliesserde 4855
Fliessfaltung 1701
Fliessfazette 2086
Fliessgefüge 0868
Fliessgeschwindigkeit des Grundwassers 3703
Fliessgrenze 4697
Fliessgrenze 4715
Fliessrichtung 0868
Fliessrinnen 0042
Fliesssand 3085
Fliessschicht 2647
Fliesszahl 4716
Flintstein 3491
Flocke 2992
Flora 1797
floral 1797
Flotation 0673
Flotation 4152
flotation, Sammel– 4152
Flotation, selektive – 4152
Flöz 3123
Flöz 3499
Flöz 4021
Flözteil 4199
Flözverdrückung 3510
Flügel 1625
Flügelsonde 4759
Flügelsondenversuch 4759
Fluglinie 4372
Flugsand 2156
Flugsand 2871
Flugsand 2961
Flugweg 4303
Flugweg 4358
Flugzeug 4302
Fluidaltextur 0868
Fluidisation 1370
Fluktuationstextur 0868
Fluoressenz 0467
Fluoressenz 0491
Fluss 2191
Fluss– 2883
Flussablenkung 2213
Flussbecken 2192
Flussbett 2242
Flussbild 4143
Flussdichte 2255
Flusserosion 2015
Fluss, fähiger – 2329
Flussfracht 2951
Flusshaupt 2193
Flüssigkeitskontaktpotential 4615
Fluss, landfremder – 2259
flusslos 2257
Flussmäander 2220
Fluss, magnetischer – 0317
Fluss-Mündungsgebiet 2784
Fluss, primärer – 0631
Flusssystem 2205
Flussterrasse 2293
Flussumkehr 2214
Fluss, unterfähiger – 2329
Flussverlegung 2213
Flutbrandung 2823
Fluten, chemisches – 3840
Flutrinne 2510
Flutstrom 2822
Flutwelle 2816
fluviatil 2883
fluvioglazial 2883
Fluvioglazialterrasse 2309
fluviomarin 2902
Fluxoturbidit 2984
Flysch 2926
Folge- 2284
Foramtionsschädigung 3815
Förde 2454
Förderabgabe 4224
Förderbohrung 3786
Förderbohrung 3789
Fördererz 4138
Förderkanal 1221
Förderkanal, wurzelloser – 1223
Förderkapazität 3806
Förderkapazität der Sonde 3806
Förderkosten, technische – 3900
Förderrate, erlaubte – 3902
Fördersonde 3789
Fördertest 3788
Förderung 1210
Förderung, kombinierte – 3829
Förderung, primäre – 3822
Förderung, sekundäre – 3831
Förderungskosten 3900
Förderung, tertiäre – 3837
Förderweite 1568
Förderzins 4224
Foreset Lage 2896
Foreset-Schichtung 3140
Formation 1754
Formationsfaktor 4609
Formationsglied 1753
Formationstest 3787
Formationswasser 3662
Formenergie 0909
Formenkreis 1936
Form Faktor 4110
Formfilter 4509
Formgattung 1932
Formregel 1013
Formregelung 1013
Form, tektonische – 2537
Formungsebene 0238
Förna 2742
Fortsetzung nach unten 4533
Fortsetzung nach unten 4543
fossil 2595
Fossil 1789
Fossilfestkalk 3269
fossilführend 1789
Fossilkalk, mässig fester – 3308
Fossilkalk, stark verfestigter – 3308
Fossil-Kammern 3690
Fossil, vermeintliches – 1789
fossil, Zonen-Leit – 1785
Fotosynthese 1803
Fouriertransformierte 4451
Fracen 3800
Fracht 2950
Fracht 2953
Fracht, bodennahe – 2952
Fraktion 2712
Fraktionierung 0414
fraktionierung, Kristallisations– 0670
Frakturieren 3800
framboid 3484
framboidal 3484
Freifallbohren 4262
Freiheitsgrad 4112
Freiluftschwere 4538
Fremdasche 3600
fremdfarbig 0483
Fremdlingsfluss 2259
Fremdmineral 0663
Frequenzband, genutztes – 4498
Frequenzbereich 0223
Frequenzbereich 4571
frequenz, Eck– 0225
Frequenzfilterung 0136
Frequenzgang 4448
Frequenzmodulation 4492
Frequenzspektrum 4451
frequenz, Träger– 0178
Fresnel-Zone 4515
Fressspur 1881
Fricktionsstreife 1538
Frittung 0708
Frittung 0897
Front, basische – 1066
Frostauffaltung 2642
Frostauftreibung 2642
Frostbeständigkeit 4734
Frostbeule 2642
Frostblähung 2642
Frostbodem 2635
Frostgefügeboden 2652
Frostglättung 2650
Frosthebung 2642
Frosthebung 4721
Frostmusterboden 2652
Frostriss 2640
Frostspalte 2640
Frostsprengung 1996
Frostsprengung, Empfindlichkeit für – 1996
Froststauchung 2647
Frostverbrodelung 2647
Frostverwitterung 1996
Frostverwitterungsschutt 2656
Frostwirkung 2394
Frostzerrung 2547
Froude 2266
Fruchtschiefer 0941
Frühstadium 2570
Frustel 1842
Fullererde 3218
Füllung 3105
Füllung 4729
Fumarole 1379
Fumarole, brausende – 1385
Fumarole, eruptive – 1385
Fumarole, flammende – 1386
Fumarole, kroikolitische – 1384
Fumarole, leucolitische – 1384
Fumarolenfeld 1380
Fumarolenhügel 1390
Fumarolenstadium 1168
Fumarolentätigkeit 1375
Fumarole, primäre – 1382
Fumarole, saure – 1388
Fumarole, ständige – 1381
Fumarole, vergängliche – 1381
Fumarole, wurzellose – 1382
Fund 3884
Fundbohrung 3784
Fundierung 4835
Funkengenerator 4493
Funkenschallquellen,

Bündelung von – 4493
Furche 1538
Furt 2247
Fusain 3580
Fussspur 1877

G

Gabbro 0812
gabbroid 0812
Gabelung 2210
Gabelung 4073
Gagat 3557
gal 0324
Gammasonde 4590
Gang 0741
Gang 1514
Gang 4540
Gangart 3922
Gangbüschel 4066
Gängchen 4039
Gang, durchgreifender – 4077
Gang, einfache – 4045
Ganges, Zerschlagen des – 4070
Gänge, Zertrümerung der – 4070
Gangfeld 4061
Gangfüllung 1515
Gang, gebänderte – 4049
Ganggestein 0741
Ganggestein 3922
Ganggestein, aschistes – 0668
Gang, hohler – 1245
Gang im Sinne Cottas, Zusammengesetzter – 4050
gang, Intrusiv– 0748
Gang, klastischer – 3073
Gangkomplex 0749
Gangkreuz 4074
Ganglette 4012
Gangmauer 0741
Gangmineral 3921
Gangnetz 4065
Gang, perlschnurartiger – 4054
Gangrevier 4061
gang, Ring– 0742
Gangschwarm 0749
Gangschwarm 4066
Gang, sedimentärer – 1516
Gangspalte 1514
Gangsystem 0749
Gangsystem 4063
Gangtrümerstock 4059
Gangunterschied 0563
gang, Verdrängungs– 0745
Gangzug 4063
Gangzug 4067
Ganister 3518
Garbenschiefer 0934
Gasanomalie 4520
Gas, assoziiertes – 3891
Gasaustritt 4520
Gasaustritt, brennender – 1386
Gasbeulen 3065
Gasdeviationsfaktor 3862
Gasdruck 1173
Gaseinpressung 3835
Gasentlösungsdruck 3857
Gasentlösungstrieb 3823
Gasflammkohle 3559
Gas-Flüssigkeitsverhältnis 3802
Gasförderung 1215
Gasförderung, diffuse – 1401
Gas, freies 3891
Gasgehalt 0650
Gas, gelöstes – 0641
Gas, gelöstes – 1172
Gashülle 1370
Gasinjektion 3835
Gas, juveniles – 1348
Gaskappe 3774
Gaskappentrieb 3823
Gaskohle 3559
Gaskrater 3066
Gas, magmatisches – 1348
Gasmuttergestein 3722
Gasöl I 3744
Gas/Öl Kontakt 3774
Gas/Öl Spiegel 3774
Gas-Öl Verhältnis 3802
Gasphase, intermediäre – 1162
Gas, phreatisches – 1351
Gasprospektion 4648
Gas, resurgentes magmatisches – 1172
Gassand 3766
Gassandstein 3766
Gasschonung 3836
Gasströme, Aushöhlung durch heisse – 1178
Gastkristall 0844
Gastrichter 3066
Gastrolith 1864
Gas, vulkanisches – 1349
Gas, Wiedereinpressung von – 3835
Gatt 2508
Gattung 1933
Gattungssone 1778
Geantiklinale 1099
gebändert 3122
gebirge, Rand– 2542
Gebirge, schlagendes – 2112
Gebirgsausläufer 2605
Gebirgsbildung 1100
Gebirgsdruck 3819
Gebirgskette 1102
Gebirgsknoten 2605
Gebirgsschlag 0115
Gebirgsschlag 4857
Gebirgstreppe 2604
Gebirgswurzel 1104
geblättert 0948
Gefahrten 4068
Gefälle 4019
Gefälle, ausgeglichenes – 2198
Gefälle, hydraulisches – 4736
Gefälle, normales – 2198
Gefälle, rückläufiges – 2626
Gefälle, widersinniges – 2626
Gefällsbruch 2199
Gefällsbruch, zyklischer – 2199
Gefällsknick 2199
Gefällskurve, unausgeglichene – 2198
Gefällsstufe 2199
Gefällsstufe, zyklische – 2199
gefleckt 2692
gefleckt 3462
Gefrierverfahren 4807
Gefüge 0793
Gefüge 1004
Gefüge 1605
Gefüge 2723
gefüge, Ablagerungs– 3077
gefüge, Absatz– 1011
Gefügeachse 1016
gefüge, Bioturbations– 1885
Gefüge, brecciöses – 2933
gefüge, Deformations– 1007
Gefügediagramm 1015
Gefügeelement 1012
gefügeelement höherer Ordnung 1012
Gefügeform 2726
gefüge, Geopetal– 1011
Gefüge, geopetales – 2996
gefüge, Intern– 1004
Gefügekunde 0995
Gefüge, lineares – 1607
gefüge, Mega– 1005
gefüge, Meso– 1005
gefüge, Mikro– 1005
gefüge, Primär– 3077
Gefügestandteil 3566
gefüge, Teil– 1005
Gefügetyp 2726
gefüge, Wachstums– 1010
Gegenrippel 3008
Gegenwartswert 3897
gehalt, Grenz– 4203
Gehalt, Linie gleicher 4638
Gehänge 2130
Gehängemoor 3534
Geiger-Müller Zähler 4546
Geiger Zähler 4545
geklüftet 1495
Gekriech 2038
Gekrösefaltung 3076
Gekröseschichtung 3051
Geländeaufnahme, photogrammetrische – 4324
Geländeaufnahme, topographische – 4323
Geländedaten 4321
Geländeecho 4390
Geländeversetzung, radiale – 4379
Gelegeufer 2336
Gemeinschaft 1805
Gemeinschaft 1892
Gemeinschaftsvereinbarung 3901
Gemengteil 0594
Gemengteil 3566
Gemengteil, mitbestimmender– 0594
Gemengteil, stellsvertretender – 0595
Generation 0591
Genotypus 1948
Genotypus 1979
Gens 1934
Genus 1933
Geobarometrie 0994
Geobios 1802
Geochemie 0395
Geochemie 0580
geochemie, stabile Isotopen– 0453
Geochronologie 1713
geochronologie, Isotopen– 0428
Geodäsie 0013
Geode 3481
Geodepression 1120
Geodynamik 0080
Geognosie 0001
Geogonie 0003
Geographie, physische – 0012
Geohydrologie 3611
Geoid 0328
Geo-Isotherm 0377
Geologe 0001
Geologenkompass 4227
Geologie 0001
Geologie, allgemeine – 0010
Geologie, angewandte – 4661
Geologie, angewandte – 0016
Geologie, ekonomische – 0016
Geologie, experimentelle – 0014
geologie, Feld– 4226
Geologie, historische – 0011
geologie, Ingenieurs– 4660
geologie, Katastrophen– 0005
Geologie, mathematische – 0015
geologie, Meeres– 4226
geologie, stabile Isotopen– 0453
Geologie, strukturelle – 1478
geologie, Umwelt– 4665
Geologie, Untertage– 4226
Geomagnetismus 0276
geomorfologie, Ingenieur– 4663
Geomorphologie 1987
Geonomie 0079
geopetal 3107
geophag 1857
Geophon 4414
Geophonanordnung 4416
Geophysik 0078
geophysik, Ingenieur– 4663
Geostatistik 4193
Geosynklinale 1094
Geotechnik 4662
Geotektonik 1073
geothermal 0376
Geothermik 0374
geothermischer Gradient 0389
geothermische Tiefenstufe 0380
Geothermometrie 0379
Geothermometrie 0994
Geotumor 1120
Geowissenschaften 0002
Geräteträger, tiefgeschleppter – 4483
Geräuscherscheinungen, vulkanische – 1237
Geräusche, unterirdische – 1237
Gerinne 2191
Gerippe 1861
gerippelt 3000
germanotyp 1078
Geröll 2912
Geröll 3241
Geröll 3242
Geröllbank 2244
Geröllhalde 2667
Geröllwüste 2578
gerontisch 1866
gerundet 3188
gerundet, gut – 3188
gerundet, Kanten – 3189
Gerüst 1861
Gerüst 3079
Gerüst 3680
Gerüststabilität 3424
Gesamtdichte 4614
Gesamtporenvolumen 2752
Gesamtproduktion 3817
Gesamtproduktion 3867
gesättigt 0605
gescheckt 3462
geschichtet 3095
Geschicke, edle – 4007
Geschiebe, erratisches – 2422
Geschiebelehm 2422
Geschiebelehm 2874
Geschiebelehm 3250
Geschiebemergel 2422
geschlossen 1626
Geschwindigkeit, erste kritische – 2265
geschwindigkeit, Gruppen– 0170
geschwindigkeit, Intervall– 4467
Geschwindigkeit, mittlere – 4465
geschwindigkeit, Partikel– 4414
geschwindigkeit, Phasen– 0169
Geschwindigkeit, scheinbare – 0208
Geschwindigkeit, scheinbare – 4468
Geschwindigkeit, Schicht erniedrigter – 0100
Geschwindigkeitsfilter 4456
Geschwindigkeitsfunktion 4469
Geschwindigkeitsindex 4732
geschwindigkeit, Stapel– 4434
Geschwindigkeit, zweite kritische – 2266
Geschwistervulkan 1442
Gesellschaft 1805
Gesims 2297
Gestänge, festgeklemmtes – 4286
Gestängetest 3787
Gestein 3389
Gestein, anstehendes – 2011

Gestein, dichtes – 3769
Gesteinglas 0627
gestein, Primär– 0631
Gestein, pyroklastisches – 1311
Gesteinsgang 1516
Gesteinsglas 1262
Gesteinsgruss 2005
Gesteinskunde 0579
Gesteinsmächtigkeit, netto – 3873
Gesteinsmehl 3209
Gesteinsmehl, glaziales – 2875
Gesteinsprobe beliebiger Form 4744
Gesteinsriegel 2625
Gesteinsschutt 2005
Gesteinsschwelle 2625
Gesteinsserfall 2005
Gesteinswiderstand, wahrer – 4607
Gestein, umgebendes – 3919
Gestein, vulkanisches – 1238
gestrickt 0964
Gewichtsmauer 4864
Gewinningsfaktor 3867
Gewitter, vulkanisches – 1373
Gewölbe 4851
Geyserit 1406
Geysir 1404
Geysirbecken 1404
Geysir, kieselsäurehaltiger – 1406
Geysirkrater 1405
Geysirschlot 1405
Geysirs, Wasserstrahl eines – 1405
gezähnt 3414
Gezeiten 2818
Gezeitenablagerung 2907
Gezeitenbecken 2508
Gezeitendelta, äusseres – 2499
Gezeitendelta, inneres – 2499
Gezeitengletscher 2436
Gezeitenrinne 2509
Gezeitenschichtung 3134
Gezeitenstrom 2821
Gezeitenströmung 2821
Gezeitentrichter 2509
Gezeitentümpel 2506
Gigantismus 1867
Gilgai 2763
Gipfelausbruch 1229
Gipfelflur 2597
Gipfelflurtreppe 2599
Gipfelhöhenkonstanz 2598
Gipfelhöhen, Konstanz der – 2598
Gipfelkaldera 1460
Gipfelkrater 1436
Gipfelniveau 2598
Gipfelstockwerk 2599
Gips 3381
Gips, Bildung von – 3459
Gipsgestein 3381
Gipsrosette 2150
Gitter 0964
Gitterebene 0508
Gittergerade 0508
gitter, Kristall– 0507
Gitterpunkt 0508
Gitterregel 1013
Gitterregelung 1013
Gittertextur 4103
Gjaspalte 1234
Glanz 0484
Glanz 0490
Glanz 3407
Glanz der Perlen 0486
Glanzkohle 3580
Glas 0627
glasartig 0805
Glasfäden 1326
Glasfaser-Aufnahmegerät 4501
Glasglanz 0484
Glassand 3227
Glastuff 1333
Glas, vulkanisches – 1262
Glattfläche 2731
Glättung 2399
glaukonitisch 3177
Glaukonitisierung 3459
glazial 2873
Glazialtal, hangendes – 2627
Glaziologie 2345
glaziomarin 2903
gleichalt 1729
gleichalt 1730
gleichalt 1818
gleichförmig 0719
gleichförmig 0754
Gleichgewicht, dynamisches – 0397
Gleichgewicht, isostatisches – 0346
Gleichgewicht, radioaktives – 4549
Gleichgewichtsgefälle 2198
Gleichgewichtskurve 2198
Gleichgewichtsprofil 2198
gleichkörnig 3081
gleichkörnig, mittel– 0820
Gleichrichtung, Einweg– 4506
Gleichrichtung, Halbwellen– 4506
Gleichrichtung, Zweiweg– 4506
Gleichstrom 4555
Gleitbrett 1683
Gleitebene 0539
Gleitfläche 1002
Gleitfläche 1538
Gleitfläche 1705
Gleitfläche 2055
Gleithang 2237
Gleithangterrasse 2302
Gleitmäander 2222
Gleittektonik 1702
Gletscher 2366
Gletscher, alpiner – 2428
Gletscherbach 2392
Gletscherbruch 2377
Gletschererosion 2015
Gletscherfall 2377
Gletscherfurche 2398
gletscher, Gebirgs– 2426
Gletscher, Hochpolar– 2368
Gletscherkratzen 2397
Gletscherkritzen 2397
Gletscherkunde 2345
Gletscherlawine 2361
Gletschermilch 2392
Gletschermühle 2389
gletscher, Plateau– 2426
Gletscher, regenerierter – 2434
Gletscherschrammen 2397
Gletscher, sich überschiebender – 2385
Gletscherspalte 2375
Gletscherstirn 2390
Gletschersturz 2377
Gletscher, temperierte – 2368
Gletschertisch 2386
Gletschertopf 2389
Gletschertor 2391
Gletscher, tragender – 2385
Gletschertrichter 2389
Gletschertypus, turkestanischer – 2429
Gletscherzunge 2390
glimmerartig 0775
glimmerig 3177
Globaltektonik 1125
Globosphärit 0765
Globulit 0765
glomeroporphyritisch 0821
Glutlawine 1367
Glutsäule 1283
Gluttuff 1340
Gluttuffstrom 1369
Glutwolke 1366
Gneis 0936
gneis, Ader– 1044
gneisähnlich 0871
gneis, Misch– 1046
gneis, Primär– 0631
Gneisskuppe, ummantelte – 0890
gneis, Stengel – 0936
Gneistextur 0951
Graben 0034
Graben 1092
Graben 1590
Graben 2793
Graben 4235
Grabenerosion 2778
graben, Explosions– 1235
graben im Vulkanmantel, Sektor – 1464
Grabensenke 1590
Grabensenke 2546
Grabental 2546
Grabgang 1882
Gradation 4343
Gradient der Schicht 1488
Gradienten-Verfahren 4563
Gradierung 3130
Graduieren 0494
Grain 0504
Granitisation 1062
Granitisierung 1062
Granittektonik 0712
granoblastisch 0960
granophyrisch 0848
Granosphärit 0779
Granulation 2934
Granulierung 2934
Granulit 0937
granulitisch 0954
Granulometrie 3160
grapestone 3302
graphische Implikations– 0847
Grasnarbe 2698
Grasplagge 2698
Grat 2617
Grauwacke 3234
Gravimeter 0352
Gravimeter mit Nullmethode 0352
Gravimetrie 0349
Gravitation 0322
Gravitationskonstante 0325
greenstone belt 1082
Greiferprobe 3153
Greisen 0924
Greisenalters, Stadium des – 2570
Greisenbildung 0924
Grenzbohrung 3792
Grenzfläche 3096
Grenzfläche 4461
Grenzförderbohrung 3845
Grenzgipfelflur 2597
Grenzverwerfung 1593
Griesschnee 2350
Griffelschiefer 0942
Griffelschieferung 1603
Griffelschieferung 1610
grit 3229
grobgeschichtet 3108
grobkörnig 0796
grobkörnig 3194
grobkörnig, sehr – 0796
Grobsand 3249
Groden 2514
Grodenpriel 2517
Grodensteilkante 2516
Grosshandelspreis, Index der – 4223
Grosszyklus 3120
Grotte 2095
Grubengas 3554
Grubenwetter 3554
Grund 4788
Grundbruch, hydraulischer – 4863
Grundfaktor 0259
Grundfalte 1120
Grundkörper 4730
Grundlawine 2359
Grundmasse 0626
Grundmasse 3080
Grundmoräne 2409
Grundmoräne 2874
Grundmoränenlandschaft 2630
Grundstrecke 4393
Grundstreckenbild 4395
Gründung 4835
Gründungsschicht 4836
Gründungssohle 4835
Gründungstechnik 4834
Grundwasser 3658
Grundwasserabfluss 3628
Grundwasserabfluss, unechter – 3623
Grundwasserabsenkung 3667
Grundwasserader 3638
Grundwasseranreicherung, künstliche – 3671
Grundwasser, freatisches – 2765
Grundwasser, freies – 3659
Grundwasser, gespanntes – 3659
Grundwasserhaushalt 3633
Grundwasserleiter 3634
Grundwasserleiter mit artesischem Wasser 3637
Grundwasserleiter mit gespanntem Wasser 3636
Grundwasserleiter mit schwebendem Wasser 3635
Grundwasseroberfläche 3643
Grundwasserscheide 3647
Grundwasserspiegel 2765
Grundwasserspiegel, schwebender – 3643
Grundwasserspiegels, Isohyps des – 3644
Grundwasserstand, freier – 3642
Grundwasserstauer 3640
Grundwasserstockwerke 3641
Grundwasserstufe 2766
grundwasser, Tal– 3626
Grünsand 3231
Grünschiefer 0938
Grünschlick 2921
Gruppe 1755
Gruppe 1928
Gruppengeschwindigkeitmethode 0229
Guano 2878
Gufferlinie 2407
Gültigkeit 1923
Gunit 4801
Gürtel 0501
Gürtel 1020
Gürtel 3908
Gürtelachse 1020
Gutenberg-Diskontinuität 0092
guyot 2798
Gyrogonites 1843
Gyttja 3528

H

Haareis 2639
haarförmig 0774
Haarkristall 0524
Habitus 0521
Habitus 0592

Hackprobe 4168
hadal 2848
hadal 2924
Haff 2458
Haftfestigkeit des Bodens 2750
Haftrippel 3016
Haftwasser 2767
Haftwasser 3653
Haftwassersättigung 4611
Hagelkorneindruck 3041
Hagelkornmarke 3041
Haken 2464
Hakengang 4055
Haken, gekrümmter – 2464
Hakenrelikt 2466
Hakenschlagen 2058
Hakenschlinge 2465
Haken, V-förmiger – 2465
Halbanthrazit 3560
Halbglanzkohle 3580
Halbkarst 2067
Halbmonddüne 2177
Halbmondkrater 1471
Halbwertsseit 0431
Halde 4166
Haldendüne 2187
Haldenhang 2603
Hallig 2507
Halmyrolyse 3476
Halokinese 1694
halokristallinisch 0804
Halophyt 1846
Halotektonik 1694
Halskuppe 1477
Handprobe 4169
Handstück 4229
Hängegletscher 2433
Hängemündung 2208
Hangende 4009
Hangendes 1529
Hangendes 3501
Hangendöl 3776
Hangendschenkel 1576
Hängetal 2483
Hängetal, glaziales – 2627
Hängetal, litorales – 2481
Hangkrater 1438
Hangschutt 2667
Hangversenkung 1464
Harker Variationsdiagramm 0697
Harmonikafalte 1651
Harmonikafalte 1661
harmonische 0167
Harnisch 1538
Harnisch 2731
Harnisch 3058
Harnischfläche 1538
Harsch 2352
Harscht 2352
Harst 2352
Härte 0473
Härte 3677
Härte 4701
Härtestufe 2200
Hartgrund 3126
Härtling 2608
Hartschiefer 1543
Härtung 3423
Harz 0484
Häufigkeitsverteilung 4187
Hauptabteilung 1929
Hauptbeben 0142
Hauptfalte 1642
Hauptfluss 2206
Hauptgang 4062
Hauptgemengteil 0594
Hauptkluft 1509
Hauptpunkt 4369
Hauptpunktübertragung 4372
Hauptspannung 0103
Hauptspannung 4692
Hauptstromrinne 2509
Hauptverwerfung 1580
Häutchen 2691
Häutchen 3100
Hawaiitätigkeit 1158
Hayfordzone 0362
Hebung 4830
Hebungsinsel 2535
Hebungsküste 2524
Heideplagge 2698
Heidetorf 3545
heimisch 1813
Heisswasserinjektion 3839
Heisswasserkraftwerk, geothermisches – 0393
Heisswasserkraftwerk niedriger Temperatur 0393
Heizwert 3604
Heizwert, oberer – 3604
Heizwert, unterer – 3604
helizitisch 0955
Heller 2514
Helligkeit 2677
Hemizyklothem 3120
Herausheben 2396
Herausreissen 2396
Herd 0156
Herdkugel 0233
Herdmechanismus 0232
Herdparameter 0151
Herdtiefe 0157
Herdtiefe, negative – 0157
Herdzeit 0159
Herkunft 2943
Herkunftsgebiet 2943
heteroblastisch 0956
heterochron 1734
Heteromorph 0635
Heteromorphie 0635
Heteromorphismus 0635
Heterosporie 1834
heterotaktisch 1007
heterotopisch 1731
Hexaedrit 0068
hexagonal 0515
Hiatus 1723
Hierarchie 1925
Hilfsebene 0238
Himbeer– 3484
Himbeerstruktur 4094
Hindernisdüne 2174
Hinterfüllung 4825
Hinterwasser 2239
Histogramm 4188
Hochebenenbildung 2590
Hochfluttbett 2242
Hochgebirge 2575
Hochland 2518
Hochlandeis 2426
Hochland, lunares – 0050
Hochmoor 3542
Hochmoortorf 3542
Hochplateau 0033
Hochscholle 1589
Hochscholle 2540
Höchstförderung 3807
hochthermal 3983
Hochwasser 2818
Hochwasser, mittleres – 2820
Hodographe 0171
Hof 4629
Höhe über dem Bezugsniveau 4249
Höhe unter dem Bezugsniveau 4249
Höhle 2095
Höhlen– 2877
Höhlenkunde 2094
Höhlenlehm 2122
Höhlenlehm, Ausfüllung mit – 2122
Höhlensediment 2121
Höhlenverfall 2110
Hohlkehle 2472
Hohlkerbe 2472
Hohlraum 3683
Hohlraum, blasenförmiger – 0874
Hohlraum, miarolitischer – 0873
Hohlraum, primärer – 3684
Hohlraum, sekundärer – 3685
Holoblast 0984
holohyalin 0805
Holokarst 2067
holokristallin 0803
hololeukokrat 0587
holomarin 2900
holomelanokrat 0588
holophanerokristallin 0795
Holostratotypus 1740
Holotypus 1949
Holzkohle 3607
Holzkohle, fossile – 3580
Homakonide 1435
Homate 1435
homogen 0716
homogen 4680
Homogenbereich 4787
homolog 1972
Homonym 1955
homöoblastisch 0960
Homöomorphie 1820
Homoseiste 0210
Homospore 1835
Homosporie 1834
homotaktisch 1007
homotax 1733
Homotypus 1954
Horizont 1749
Horizont 3122
Horizont 4461
Horizont, A– 2698
Horizont, Ae– 2700
Horizont, Ah– 2698
Horizont, Ai– 2698
Horizont, Al– 2700
Horizontaldistanz 1555
Horizontalschichtfall 4568
Horizontalversatz 1551
Horizontalversatz, stratigraphischer – 1552
Horizontalverschiebung 1560
Horizont, Ap– 2699
Horizont, B– 2702
Horizont, (B)– 2703
Horizont, begrabener – 2693
Horizontbild 4354
horizont, Bleich– 2700
Horizont, Bt– 2703
horizont, Bt-Illuvial– 2703
Horizont, C– 2706
Horizont, chronostratigraphischer – 1758
Horizont, D– 2708
horizont, Eluvial– 2701
Horizont, gebänderte B– 2705
Horizont, Gr– 2707
horizont, Illuvial– 2702
Horizontkamera 4333
Horizont, lithostratigraphischer – 1749
horizont, Tonanreicherungs– 2703
horizont, Tonanreicherungs– 2734
Horizont, verbraunter – 2703
Horizont, verlehmter – 2703
Hornfels 0940
Hornfelstextur 0960
Hornito 1298
Hornstein 3366
Hornstein, Diatomeen– 3371
Horst 1589
Horstgebirge 2540
Horstscholle 2540
Hoskold Formel 4217
hot spot 1144
Hub 4511
Hubglättung 4511
Hubkompensation 4511
Hufeisendüne 2177
Hülle, migmatische – 1070
Hum 2074
Humifizierung 2683
Humifizierung 3525
Huminsäure 3525
Humus 2743
Humus 3525
Humuskohle 3497
Hungerfluss 2329
Hüpfmarke 3039
Hut 1696
Hut, eiserner – 4125
Hutgestein 1696
hyalin 0805
Hyaloklastit 1347
hyalo-ophitisch 0840
hyalopilitisch 0837
hybrid 0681
Hybridisierung 0681
Hydatogenese 3958
hydatopyrogen 0656
Hydration 0419
Hydrierung 3609
Hydrobios 1802
hydrochemisch 4634
Hydroexplosion 1351
Hydrofrac 3800
Hydrogeochemie 3611
Hydrogeologie 3611
Hydrolakkolith 2643
Hydrologie 3610
Hydrologie des Grundwassers 3611
Hydrolyse 0418
Hydrometasomatose 3954
hydromorph 4634
hydromorphes Bodenmerkmal 2690
Hydrophon 4415
Hydrophon 4496
Hydrophoncharakteristik 4497
Hydrophonkabel 4496
hydroplutonisch 0656
Hydrosphäre 0085
hydrothermale Entstehung 3953
hydrothermales Stadium 0654
hydrothermal, magmafern– 3984
hydrothermal, sekundär– 3994
hypabyssal 0617
hypabyssisch 0617
hypautomorph 0751
hypautomorph 0808
Hyperbolmethode 0152
hyperitische Struktur 0853
hypersalin 3377
Hyperstereoskopie 4367
hyperthermale Region 0391
Hyperzyklothem 3120
Hypidioblast 0980
hypidiomorph 0751
hypidiomorphkörnig 0808
hypogen 3937
hypogen 3967
hypokristallin 0804
Hypolimnion 2334
Hypomagma 1199
Hypostratotypus 1746
hypothermal 3983
Hypozentrum 0156
Hysterese 4675
Hysterese, magnetische – 0305

I

Ichnofossil 1873
Ichnofossil 3067
Ichnologie 1871
Ichnozönose 1874
Ichor 1061
Identificationsschwelle 0271
Identifikationskapazität 0271
Idioblast 0980
idioblastische Reihe 0909
idiomorph 0523
idiomorph 0750
idiotopisch 3471
Ignimbrit 1340
Illuvialhorizont 2702
Illuviation 2678
Illuviation 3930
Imatra-Stein 3486
Imbibition 1060
Impaktit 0076
Impaktkrater 0054
Impakttheorie 0049
Impedanz, akustische – 4474
Impetus 0173
Imprägnation 3396
Imprägnation 3955
Imprägnationskupfererz 4133
Imprägnationslagerstätte 3992
Impuls 4482
Impulsantwort 4446
Impulsdauer 4482
Indikatorelement 4649
Indikatorpflanze 4646
Indikatorvergrösserung 0131
Indikatrix 0552
Induktion, magnetische – 0317
Industriemineral 0462
Infiltration 3396
Infiltration 4619
infrakrustal 0584
infralitoral 2839
infraneritisch 2841
Infrarot, fernes – 4309
Infrarot, fotografisches – 4307
Infrarot, nahes – 4307
Infrarot, thermales – 4308
infraspezifisch 1944
Ingenieurseismologie 0254
Ingressionsküste 2529
inhomogen 4680
initial– 1108
Injektion 3949
Injektion 4802
Injektionsbohrloch 4803
Injektion, sedimentäre – 3070
Injektionsfältelung 1051
Injektionsschleier 4804
Injektivfalte 1698
Injektivitätsindex 3809
injizieren, Zement – 4803
Inklinometer 4294
Inkohlung 3519
Inkohlungsgrad 3552
Inkohlungsgradient 3522
inkompetent 1624
Inkrustierung 3387
Inlandeis 2425
Innenbogen, vulkanischer – 1137
innenbürtig 0081
Innendüne 2167
Innenmoräne 2408
Innenreflexion 0576
Innenskelett 1861
Insel 2443
Inselberg 2608
Inselberg, glazialer – 2625
Inselbogen 1135
Inselhang 2791
insel, Kontinental– 2444
Insellagune 3340
Insel, landfest gewordene – 2534
Inselnehrung 2467
Insel, ozeanische – 2444
Inselschelf 2791
insequent 2287
Intensität 0243
Intensitätsskala 0243
Interceptzeit 4463
Interferenzfarbe 0568
Interferenzfarbe, anomale – 0568
Interferenzfigur 0569
Interferenz-Rippel 3013
intergranular 0841
intergranular 0998
interkinematisch 1079
Interkumulus 0831
intermediär 0602
intersertal 0840
intersitial 3083
Intertidenbereich 2836
Interval 1718
Interzeption 3615
intragranular 0998
Intraklast 3272
intramagmatisch 3985
Intramikarenit 3284
Intramikrit 3284
Intramikrudit 3284
Intraspararenit 3283
Intrasparit 3283
Intrasparrudit 3283
Intrusion 0618
Intrusion, mehrfache – 0721
Intrusion, sedimentäre – 3070
Intrusionsphase 1197
Intrusion, zusammengesetzte – 0721
intrusiv 0619
Intrusiv, einfaches – 0720
Intrusivgang 0740
Intrusivgestein 0715
Intrusiv-Komplex 0620
Intrusivkörper 0715
Intrusivlager 0740
Intrusiv, mehrfaches – 0721
intrusiv, Teil– 0721
Intrusiv, zusammengesetztes – 0721
Intrusiv, zusammengesetztes – 0722
Inundationsgebiet 2239
Inverse 4585
Inversion 1088
Inversion des Reliefs 2563
Ionenaustausch 0420
Ionenaustausch-Kapazität 0420
Ionenpotential 0419
Ionium-Thorium Datierung 0444
IP 4569
Isanomalenkarte 0372
isochromatische Kurve 0571
Isochrone 0210
Isochrone 0433
Isodyname 0285
isofaziell 0992
Isogamme 4532
Isogammenkarte 0372
Isogon 0284
Isograd 0992
Isogyre 0570
Isohypse 4248
Isokalore 3523
Isokarbe 3523
isoklinal 1626
Isoklinaltal 2544
Isokline 0283
Isolation 1981
isomesisch 1735
isometrisch 0522
isometrisch 0754
isometrisch 3181
isomorph 1974
Isomorphie 0464
Isonomalenkarte, gravimetrische – 0372
Isopache 4243
Isopachenkarte 4243
isopisch 2863
Isopor 0287
Isoseiste 0245
Isospore 1835
Isosporie 1834
Isostasie 0338
Isotherm 0377
Isotopendatierung 0429
Isotopenfraktionierung, kinetische – 0457
Isotopenfraktionierungsfaktor 0455
Isotopenfraktionierungskurve 0455
Isotopenverdünnungsanalyse 0432
isotopisch 2863
isotrop 4681
Isotropie 4681
isotrop, optisch – 0561
Isotypie 0463
Isovole 3523

J

Jade 0498
Jahreslage 3127
Jahresschichtung 3127
Jama 2076
Jaspilit 3492
Jaspis 3492
Jaspis 4127
Jett 3557
Jugendstadium 2570
Jumbo 4266
jungvulkanisch 0636

K

Kabeltest 3787
KAK 4623
Kalbsche Linie 0559
Kaldera 0036
Kaldera 1456
Kaldera, einfache – 1457
Kalderainsel 1417
Kalderakomplex 1457
Kaldera, phreatisch gebildete – 1458
Kalderawall 1461
Kaliberlog 4603
Kalium-Argon Datierung 0449
Kalium-Kalzium Datierung 0449
Kalkablagerung 3267
Kalkalgen 3359
Kalkalkali– 0694
Kalkalkalireihe 0692
Kalkarenit 3276
kalkarm 2736
Kalkbreccie 3279
Kalkbruchstück 3168
Kalkbruchstück 3270
kalk, Crinoiden– 3313
Kalkfragment 3270
kalkfrei 2736
Kalkgehalt 2735
kalkhaltig 2736
Kalkkies 3278
kalk, Korallen– 3310
Kalkkruste 4123
Kalklutit 3274
Kalkpelit 3274
kalkreich 2736
Kalkreihe 0692
kalk, Riff– 3310
Kalkrudit 3278
kalksandig 3277
Kalkschlamm 3295
Kalksiltit 3275
Kalkstein 3305
Kalkstein, dolomitischer – 3316
Kalkstein, kalkarenitischer – 3277
Kalksteinkies 3278
Kalkstein, kotpillenführender – 3307
Kalkstein, lithographischer – 3306
Kalkstein, sandiger – 3317
Kalzimikrit 3282
kalzinieren 4151
Kalziterbse 2149
Kalzitrosette 2150
Kalzittapete 2142
Kame 2420
Kamesterrasse 2420
Kammeis 2639
Kammersprengung 4817
Kammersystem 2292
Kammstruktur 4084
Kammstruktur, Erzgang mit – 4049
Kammwasserscheide 2316
Kanal 4412
Kandelit 3563
Kante 0518
Kaolin 3217
Kaolinisation 0926
Kaolinisierung 0926
Kap 2446
Kapazität 2958
Kapillardruckkurve 4612
Kapillarpotential 2770
Kapillarsaum 3657
Kapillar-Saum 2767
Kapillarwasser 2767
Kapillarwasser, geschlossenes – 3657
Kapillarwasser, offenes – 3654
Kapitalrückflusszeit 3896
Kar 2610
Karat 0482
Karbonatablagerung 3266
Karbonatbank im Schelfbereich 3323
Karbonatgestein 3266
Karbonat-Hügelstruktur 3324
Karbonatisation 0932
Karbonatisierung 0932
Karbonatisierung 2000
Karbonat-Kompensationstiefe 0423
Karbonatkörper 3324
Karbonatkörper, Hügelstruktur – 3328
Karbonatkörper, hügelstrukturierter – 3328
Karbonatmasse 3324
Karbonatplattform 3322
Karbonatplattform, flach einfallende – 3322
Karbonatsediment 3266
Kargebirge 2616
Kargletscher 2430
kar, Gross– 2613
Karling 2618
Karniveau 2615
Karplatte 2613
karren 2086
Karrenfeld 2088
Karriegel 2610
Karschwelle 2610
Karsohlenniveau 2615
Karst 2064
Karst, bedeckter – 2069
Karstbildung 2063

Karstbrunnen 2076
Karsterscheinungen 2065
Karstfenster 2078
Karst, flacher – 2067
Karstgebiet 2064
Karstinselberg 2074
Karstlandschaft 2064
Karst, nackter – 2068
Karstphänomene 2065
Karstquelle 2089
Karstquelle 3631
Karstquelle, starke – 2093
Karstrestberg 2074
Karst, Schichtquelle im – 2091
Karstschlot 2076
Karst, seichter – 2067
Karst, tiefer – 2067
Karsttrichter 2072
Karst, unterirdischer – 2070
Karstwanne 2071
Karstwanne 2074
karte, Boden– 2671
karte, Bodenspezial– 2671
karte, Bodenübersichts– 2671
Karte der regionalen Schwere 0371
Karte des Restfeldes 4542
karte, Foto– 4383
Karte, geologische – 4239
Karte gleicher Mächtigkeiten 4243
Karte, ingenieurgeologische – 4784
Karte, metallogenetische – 3905
Karte, paläogeologische – 4242
karte, Radar– 4400
Karterrasse 2613
Karte, tektonische – 4247
Kartiereinheit 2672
Kartografie 4239
Kartreppe 2614
Kar, unechtes – 2611
Kaskade 2203
Kaskadenfalten 1706
Kastental 2273
Katagenese 1977
Katagenese 3419
Kataklase 0914
Kataklasit 0914
Kataklinal– 2288
Katarakt 2204
Kataseisme 0187
Katastrofismus 0005
katathermal 3983
Katazone 0993
Kategorie 1927
Kationen, austauschbare – 2755
Kationen-Austauschkapazität 4623
Katzenauge 0488
Kaufkraft, konstante – 3899
Kaustik 0216
Kaustobiolit 0427
Kaverne 4849
kavernös 3064
Kavitation 2120
Kavitation 4512
Kees 2366
Kegel 0059
Kegel, an der Küste gebildeter – 1300
Kegelberg 1435
Kegelbildung 3826
Kegelbrecher 4148
Kegeleindringungsgerät 4758
Kegelgang 0742
Kegelgerät 4758
Kegelkarst 2081
Kegel, parasitärer – 1437
Kegel, zentraler – 1443
Keilschollengebirge 2539
Kelly 4269
Kelyphit 0786
Kelyphitrinde 0786
Kennelkohle 3563
Kennzeichen, mikrotexturell – 4108
Kerbtal 2272
Kern 1628
Kern 3155
Kern 4861
Kern, abgequetschter – 1628
Kern, äusserer – 0093
Kernexplosion 0268
Kernexplosion, unterirdische – 0268
Kernfänger 4272
Kerngerät 3155
Kerngewinn 4273
Kern, granitischer – 1070
Kern, innerer – 0093
Kernkiste 4237
Kernmeissel 4271
Kernrohr 4272
Kernspaltgerät 4182
Kernsprung 1993
Kernsprung 1995
Kernwaffen-Teststopvertrag 0266
Kerogen 3720
Kerogenit 3720
Kerosin 3744
Kerze 2140
Kessel 1463
Kessel 2793
Kesselbruch 1591
Kesselbruchfeld 2547
Kesseleinbruch 0743
Kesseleinbruch 1186
Kesseleinbruch, unterirdischer – 1186
Kettengang 4067
Kettengebirge 1103
Kettentextur 4083
Kies 3244
Kiesbank 2244
Kiesel 3491
Kieselerde 3367
kieselig 3365
Kieselschiefer 3367
Kieselschlamm 3367
kiesig 3244
Kieswüste 2578
Kippschalenverwerfung 1567
Kippung 0124
Kladogenese 1962
Klapperstein 3485
Klasse 0685
Klasse 1929
Klassifikation 1921
Klassifikation, biostratigraphische – 1710
klassifikation, Boden– 2661
Klassifikation, chronostratigraphische – 1712
klassifikation, Land– 4785
Klassifikation, lithostratigraphische – 1709
Klassifikation, natürliche – 1921
klassifizierung, Vorrats– 3878
Klast 3166
Klebepunkt 2750
kleindrusig 0873
Kleinfalte 1643
Kleinfältelungsachsen 1612
Kleintektonik 1479
Klemme 4280
Kliff 2470
Kliffdüne 2169
Kliffhöhle 2472
Kliff, kleines – 2470
Klifflinie 2478
Kliffrutschung 2473
Kliff, totes – 2477
Kliff, untertauchendes – 2475
Klinkenberg Faktor 3706
Klinograph 0124
Klinometer 4227
Klippe 1692
Klippe 2443
Kluft 1495
Kluft 4723
Kluftabstand 1497
Kluftabstand 4725
Kluft, ac– 1507
Kluft, bc– 1507
kluftbildung 1495
Kluft, D– 1507
Kluftdichte 4725
Klufterstreckung 4726
Kluftfläche 1496
Kluftfugenhöhle 2096
Kluftfüllung 4729
Kluft, hko– 1507
Klüftigkeit 1495
Klüftigkeitssiffer 4725
Kluftkarren 2087
Kluftkörper 4730
Kluft, L– 1507
Kluftmuster 1508
Kluftnetz 1498
Kluftnetz 1508
Kluft, Q– 1507
Kluftschaar, regelmässige – 1510
Kluftschar 1498
Kluftschar 4724
Kluftscharen, System sich kreuzender – 1499
Kluftspalte 1513
Kluft, streichende – 1506
Kluftsystem 1498
Klüftung 1495
Kluft, unregelmässige – 1510
Kluft, untergeordnete – 1509
Kluftweite 4728
Klumpen 2725
Knickband 1661
Knickpunt 2199
Knickzone 1661
Kniefalte 1648
Knittermarke 3056
Knochenbreccie 3261
Knoll 4085
Knolle 0771
Knolle 3473
knollenförmig 4085
knollig 3473
knollig 4085
Knotenfläche 0234
Knotenlinie 0234
Knotenschiefer 0941
Koagulation 0413
Koagulation 3932
Kochen 1358
Koenigsberger Q-Faktor 0310
Koerzitivkraft 0307
Kofferdam 4869
Kofferfalte 1649
Kohäsion 4693
Kohäsion, scheinbare – 4693
Kohle 3497
kohle, Faulschlamm– 3497
Kohleformation 3498
Kohleneisenstein 3504
Kohleneisenstein 4117
Kohlenflöz 3499
Kohlenformation 3498
Kohlengebirge 3498
Kohlengeröll 3512
Kohlengrus 3513
Kohlenhydrat 3723
Kohlenklein 3513
Kohlenpetrographie 3565
Kohlenschmitz 3499
Kohlenserie 3498
Kohlenstoff-14 Datierung 0451
Kohlenstoff, fixer – 3601
Kohlenstoff-Stickstoff-Verhältnis 0425
Kohlenstoffzahl 3727
Kohlentyp 3579
Kohlenwasserstoffsäule 3775
Kohlestruktur 3565
Kohleverflüssigung 3609
Kohlevergasung 3609
Kokardenerz 4050
Kokardenstruktur 4096
Koks 3608
Kokskohle 3608
koks, Natur– 3564
Kokungsgradbestimmung 3596
Kokungsvermoögen 3596
Kolbenentnahmegerät 4746
Kolbenlot 3156
Kolk 2027
Kolk 2247
Kolk 2330
Kolk 3540
Kolkloch 2027
Kolkmarke 3025
Kollapsstruktur 1704
kolloform 3093
Kolluvium 2868
Kolonie 1806
komagmatisch 0629
komagmatische Region 0628
Kommensalismus 1899
Kompaktion 3428
Kompaktion, chemische – 3432
Kompaktion differentielle – 3428
Kompaktion, mechanische – 3429
Kompaktionsbeule 1664
Kompaktionstrieb 3827
Kompaktion, vadose – 3416
Kompartiment 0362
Kompensation, isostatische – 0339
Kompensation, lokale – 0340
Kompensation, regionale – 0340
Kompensationsmasse 0341
kompetent 1624
Kompetenz 2958
komplementär 0669
Komplettierung 3795
Komplex 1756
Komplex 1928
Komplex 2673
Komplexbildung 0408
Komplexbildung, organometallische – 0409
Komplexerz 3914
Komplexität 0275
komponentengestützt 3294
Komponente, vertikale – 1563
Kompressibilität 4698
Kompression 0187
Kompressionsgerät 4756
Kompressionsmodul 4698
Kondensat 3713
Kondensation 0412
Kondensation, retrograde – 3841
Kondensationskernen, Bildung von – 1360
Konfluenz 2208
Konfluenzstufe 2624
Konfluenz, verlegte – 2209
Konfluenz, verschleppte – 2209
kongenerisch 1943
Konglomerat 3251
Konglomerat, Block– 3254
Konglomerat, intraformationelles – 3257
konglomeratisch 3251
konglomerat, Tongallen– 3252
Konglomerat, vulkani-

sches – 1343
Konide 1435
koniform 4080
Konihomate 1435
konkordant 0717
konkordant 1719
Konkordanzküste 2521
Konkordanz, stratigraphische – 1719
Konkretion 2732
Konkretion 3472
Konkretion 4093
konkretionär 3472
konsequent 2284
Konsistenz 2746
Konsistenz– 4713
Konsistenzzahl 4716
konsolidiert 3422
Konsolidierung 3422
Konsolidierung 4703
konspezifisch 1943
Kontaktaureole 0899
Kontaktgang 4043
Kontakthof 0707
Kontakthof 0899
Kontaktlagerstätte 3980
Kontaktmetamorphose, kaustische – 0897
Kontaktmetasomatose 0900
Kontakt, suturierter – 3449
Kontaktwirkung, endomorphe – 0898
Kontaktwirkung, exomorphe – 0898
Kontamination 4652
kontinental 2865
Kontinentalabfall, randlicher – 2789
Kontinentaldüne 2170
Kontinentalhang 2790
Kontinentalschelf 2788
Kontinentalterrasse 2790
Kontinentalverschiebung 1115
Kontinentprall 1131
Kontinent-Zusammenstoss 1131
Kontinuspektrum 4305
Kontraktionshypothese 1114
Kontraktionsspalte 0713
Kontrast 4343
Kontrast 4628
Kontrastfilter 4339
Kontrollprobenahme 4177
Kontrollpunkt 4325
Konvergentbilder 4356
Konvergenz 1725
Konvergenz 1973
Konvergenz 2787
Konvergenz, metamorphe – 0913
Konvolution 3052
Konvolution 4452
Konzentrat 4141
Konzentrationspotential 4615
Koordinationssahl 0408
Köpfen eines Flusses 2323
Kopplung 0273
Koprolith 1863
Koralle, hermatypische – 3335
korallenähnlich 1821
Korallenkalkstein 3310
Korallenkappe 3329
Kordillere 1103
korkenzieherkolk 3026
Korn 3078
Korn 3240
Korndruck 3819
Korndurchdringung 3448
Kornfraktion 2712
Korngefüge 2727
korngestützt 3294
Korngrenze 0532
Korngrösse 0792
Korngrösse 2710
Korngrösse 3193
Korngrössenanalyse 2711
Korngrössenanalyse, mechanische – 3160
Korngrössen, Einteilung der – 3161
Korngrössen-häufigkeitsverteilung 3197
Korngrössenklasse 3162
Korngrössenklasse 2713
Korngrössenverteilung 2709
Korngrössen-verteilung, polymodale – 3197
Korngrössenzusammensetzung 2709
körnig 0806
körnig 3081
körnig, gleichmässig – 0807
Körnigkeit 3081
Körnigkeit 4344
körnig, gleichmässig– 0807
Kornschnee 2350
Kornstruktur 4086
Körnung 2709
Körnung, bestimmung der – 2711
Kornverteilung 4702
Kornverteilung, gut abgestufte – 4702
Kornverteilung, schlecht abgestufte – 4702
Kornwachstum 3454
Korona 0786
Koronit 0786
Korrasion 2014
Korrasion 2015
Korrasion 2938
Korrektion auf Bezugsniveau 4429
korrektion, Stations– 0160
Korrektur, Bouguer– 0359
korrektur, Breiten– 0355
Korrektur, dynamische – 4432
korrektur, Gezeiten– 0365
Korrektur, statische – 4431
Korrelation 1716
Korrelation 4459
korrelation, Kreuz– 4450
Korrelator, optischer – 4388
Korrelogramm 4195
Korrosion 0642
Korrosion 2017
Korrosion 2942
Korrosionsbasis 2066
Korrosionsdoline 2072
Kosmogonie 0003
kosmopolitisch 1814
Kotpille 1863
Kotpille 3094
Kovariogramm 4196
Kraft, magnetisierende – 0317
Kraftmessdose 4770
Kraftwerk, geothermisches – 0392
Krater 1412
Kraterbecken 1412
Kraterfumarole 1383
Kraterinsel 1417
Krater mit zentralen Bochen 1445
Krater, parasitärer – 1439
krater, Primär– 0027
Kraterreihe 1449
Kraterring 1451
Kratersee 1454
krater, Sekundär– 0027
Krater, tätiger – 1211
Kratertrichter 1432
Kraterwall 1451
Kraterwall, hufeisenförmiger – 1471
Kraterwand 1453
Krater-Zwischental, hufeisenförmiges – 1461
Kraton 1080
Kreide 3315
Kreislauf, geochemischer – 0396
Kreislauf, hydrologischer – 3612
Kreislauf, petrogenetischer – 0582
kreuzend, sich – 2596
Kreuzschichtung 3146
Kreuzung der Gänge 4074
Kriechen 0044
Kriechen 0114
Kriechen 4672
kriechen, Boden– 2780
Kriechgang 2098
Kriechspur 1878
Kriechspur 1879
Kriging 4197
Kristalisationskraft 0909
Kristall 0506
Kristallapilli 1325
Kristallbrei 0640
Kristallform 0520
Kristallform 0753
Kristallgitter 0507
kristallines Gestein 0583
Kristallinität 0791
Kristallisation 0632
Kristallisation, Abbildungs– 1008
Kristallisation, fraktionierte – 0670
Kristallisation, mimetische – 1008
Kristallisation, rhythmische – 0864
Kristallisationsdifferentiation 0670
Kristallisationsindex 0702
Kristallisationsschieferung 0948
Kristallisationsschieferung 1604
kristallisiert, schlecht – 0797
Kristallit 0760
kristallitisch 0760
Kristallklasse 0512
Kristall, korrodierter – 0756
Kristallkumulus 0830
Kristalloblast 0908
Kristalloblastese 0908
Kristallographie 0505
Kristallographie, optische – 0544
Kristall, regeneriertes – 0910
Kristallsandstein 3465
Kristallstruktur 0509
Kristallsystem 0514
Kristalltuff 1333
Kristal, teilweise resorbierter – 0756
Kritzung 2397
Krokydit 1047
Krone 4271
Krotowine 2745
Krume 2699
Krümelgefüge 2727
Krümmelig 3087
Krümmungsgrösse 0336
Kruste 2759
Kruste, Mächtigkeit der – 0347
Krusteneisenstein 4124
Kruste, Neubildung ozeanischer – 1128
Kruste, starre – 0348
Kryokonit 2387
Kryokonitloch 2387
Kryokonitschal 2387
Kryopedologie 2634
Kryoturbation 2647
Kryptoexplosionsstruktur 1184
kryptogen 1976
kryptoklastisch 3166
kryptokristallin 0800
Kryptomaceral 3567
kryptomagmatisch 3986
kryptopegmatitisch 0848
Kryptovulkanismus 1181
Kubikfuss 3892
kubisch 0515
Kugel 0867
Kugelblitz 1372
Kügelchen 4150
kugelig 0867
kugelig 3180
Kugeligkeit 3180
Kugelkalotte 0363
Kugelkarst 2080
Kugelkohle 3514
Kugelmühle 4149
Kuhtritt 2060
Kulmination 1635
Kulturterrasse 2061
Kümmerfluss 2329
Kumulat 0830
Kumulativproduktion 3817
Kumulus 0830
Kupferschiefer 4130
Kuppe 0724
Kuppel 1205
Kuppel 1641
Kuppelgebirge 2538
Kupste 2190
Kupstendüne 2190
Kurslinie 4303
Küste 2437
Küste, aufgetauchte – 2525
Küste, diskordante – 2522
Küste, epigenetische – 2527
Küste, gebuchtete – 2533
Küste, gehobene – 2524
Küste, gelappte – 2533
Küste, geradlinige – 2531
Küste, gesunkene – 2524
Küste, gezahnte – 2532
Küste, indifferente – 2523
Küste, konkordante – 2521
Küste, mehrzyklische – 2530
Küsten– 2906
Küstendüne 2166
Küstenebene 2440
Küste, neutrale – 2523
küstenfern 2840
Küstenkliff 2470
Küstenlängstransport 2962
Küstenlinie 2437
küstenlinie, Aussen– 2461
Küstenlinie, glatte – 2531
küstenlinie, Innen– 2461
Küstennah 2889
Küstenplaya 2892
Küstenprofil 2441
Küstenprofil, ausgeglichenes – 2441
Küstensebkha 2892
Küstenströmung 2825
Küstensumpf 2514
Küstenterrasse 2484
Küstentypus, atlantischer – 2522
Küstentypus, dalmatischer – 2521
Küstentypus, gemischter – 2526
Küstentypus, pazifischer – 2521
Küste, stabile – 2526
Küste, transversale – 2522
Küste, untergetauchte – 2525
Küste, zerrissene – 2532
Küste, zusammengesetzte – 2527
Kyriosom 1039

L

Labradorisieren 0486
Lackfilm 4230

Lackfilm-Methode 4230
Ladung 4816
Ladungsstärke 0272
Lage 1750
Lage 3097
Lage 3098
Lagentextur 0945
Lagerart 3922
Lagergang 0740
Lagergang 1514
Lagergang 4053
Lagergestein 3923
Lagerstätte 3754
Lagerstätte, alluviale – 4000
Lagerstätte, deuterogene – 3972
Lagerstätte, eluviale – 3999
Lagerstätte, frühmagmatische – 3975
Lagerstätte, hydrothermale – 3982
lagerstätte, Kohlen– 3493
Lagerstättekörper 3910
Lagerstätte, magmatische – 3946
Lagerstättenbewertung 4216
Lagerstättenbezirk 3909
Lagerstättendruck 3819
Lagerstättenfeld 3909
Lagerstättenmodell 3866
Lagerstättenprovinz 3908
Lagerstättenwasser 3803
Lagerstätte, pneumotektische – 3981
Lagerstätte, primäre – 3972
Lagerstätte, protogene – 3972
Lagerstätte, pyrometasomatische – 3980
Lagerstätte, sekundäre – 3972
Lagerstätte, spätmagmatische – 3975
Lagerung, konkordante – 3115
Lagerung, regressive – 1727
Lagerung, regressive – 3112
Lagerungsgesetz 1714
Lage, taube – 3506
Lage, tektonische – 1480
Lagg 3537
lagunär 2905
Lagune 2458
Lagune 3345
lagune, Gezeiten– 2458
Lagunenboden 3345
Lagunen-Gezeitenfläche 3346
Lagunenkanal 3348
Lagunen-Schelf 3346
Laharablagerung 1356
lahar, Durchbruch– 1354
Lahar, heisser – 1353
Lahar, kalter – 1355
Lahar, Schmelzwasser– 1357
Lakkolith 0734
lakustrisch 2888
Lamarckismus 1967
lamellar 0775
lamellar 0972
Lamellartextur 4103
lamellenartig 4103
lamellen, Schock– 0894
Lamina 3101
Laminarbewegung 2373
Lamination 1277
Lamination, magmatische – 0868
laminiert 3101
lamprophyrisch 0809
Landbewertung 4786
Landeinheit 4785
Landschaft, äolisch überformte – 0040
Landschaft mit Stockwerkbau 2599
Landspitze 2449
Landzunge 2448
Längs– 2289
Längsdüne 2171
Längskluft 1507
Längsküste 2521
Längsmoräne 2415
Längsprofil eines Tales 2271
Längsspalte 2376
Längsstufe, Destruktions– 2555
Längsüberdeckung 4359
Lapilli 1324
Lapillituff 1333
Lappen 2649
Larsen Variationsdiagramm 0697
laterale 4585
Lateralerguss 1266
Lateralerosion 2022
Lateralkegel 1437
Lateralkrater 1439
Lateralsekretionstheorie 3928
Laterit 4124
Lateriteisenerz 4124
Lateritisierung 2681
Laubkohle 3556
Laufzeit 0171
Laufzeitkurve 0171
Laufzeitskurvendarstellung 4463
Laufzeittabelle 0171
laugen 4162
Lava 1242
Lava, Aa– 1260
Lavaausfluss, terminaler – 1266
Lavaausläufer 1251
Lavabank 1277
Lavabelag 1226
lava, Block– 1259
Lavadecke 1275
Lavadelta 1280
Lavadom 1272
Lavadom 1422
Lavadomkruste 1273
Lavadom mit zungenförmigem Auswuchs 1423
Lavadorn 1424
Lavaerguss, parasitärer – 1265
Lavaerguss, unterirdischer – 1267
Lava, erstarrte – 1242
Lavafächer 1287
Lavafäden 1326
Lavafall 1287
Lavafeld 1278
Lavafeld-Aufpressung 1278
Lavafeld-Aufwölbung 1278
Lavafenster 1289
Lavafladen 1329
Lavaflatschen 1329
Lavaflut 1276
Lavafontäne 1282
Lavagang 0741
lava, Gekröse– 1252
Lava, geschrammte – 1270
Lavagrotte 1291
Lavahang 1286
Lavahöhle 1291
Lavahülle 1292
Lavakaskade 1287
Lavakegel 1298
Lavakegel 1427
lava, Kissen– 1253
Lavakugel 1296
Lavakugel, angewachsene – 1296
Lavakuppel 1422
Lavamauer 0741
Lava mit geglätteter Oberfläche 1261
Lavamoräne 1295
Lavamulde 1290
Lavanadel 1424
Lavaorgel 1298
Lavaorgel 1301
Lavapfropfen 1288
Lavapilz 1288
Lavaring 1284
Lavarücken 0031
Lavasäule 1246
lava, Schollen– 1259
Lavaschornstein 1298
Lavaschrammen 1270
Lavasee 1281
Lavaseekrater 1284
lava, Seil– 1252
Lavastalagmit 1293
Lavastalaktit 1293
lava, Strick– 1252
Lavastrom 1263
Lavastrom, klastogener – 1282
Lavastromküste 1280
Lavastrom, wurzelloser – 1282
Lavaträne 1326
Lavatunnel 1290
Lavatunnel, eingestürzter – 1290
Lavaturm 1298
Lavaturm 1299
Lavauferterrasse 1285
Lava, weissglühende mehlige – 1263
lava, Wulst– 1252
Lavawüste 1279
lava, Zacken– 1260
Lavazehen 1251
Lavazunge 1264
Lawine, gemischte – 1368
Lawine, kalte – 2359
Lawine, nasse – 2359
Lawinenbahn 2362
Lawinengang 2362
Lawinengraben 2362
Lawinenkegel 2362
Lawinentrog 2362
Lawinenzug 2362
lawine, Schnee – 2358
Lawine, trockene – 2359
Lawine, warme – 2359
Lebensgemeinschaft 1890
Lebensgemeinschaft 1896
Lebensspur 1875
Lebewelt 1804
Lectostratotypus 1745
Lectotypus 1951
Leeblattschichtung 3136
Leehang 2965
Leeseite 2964
Leeufer 2336
leewärts 2964
Legende 4240
Lehm 2720
Lehm 3215
Leiste 2297
leistenförmig 0522
Leistung 4449
Leistungsdichtespektrum 4451
Leitbank 1751
Leitergang 4057
Leitfähigkeit 4556
Leitfähigkeit, hydraulische – 3700
Leitflöz 3500
Leitfossil 1785
Leithorizont 4249
leitung, Wärme– 0381
lenticulär 3102
lentikulare Textur 0952
lepidoblastisch 0961
Leptinolith 0940
Leptit 0937
leptothermal 3984
Lesedecke 2160
Lesedecke 2967
Lettenbesteg 4012
Leuchtspur 0063
leukokrat 0587
Leukosom 1038
Levée 2238
Lichtschacht 2076
Lichtstrahl 0548
Lichtstrom 4311
Lido 2461
Liefergebiet 2943
Liegende 4009
Liegendes 1529
Liegendes 3501
Liegendkörper 0728
Liegendschenkel 1576
Liegendwasser 3779
Liegendwassertrieb 3825
Liman 2457
limnetisch 2890
limnisch 2888
Limnobios 1802
Limnologie 2331
Limnoplankton 1801
Limonit 4126
linear 0972
Linearstreckung 0973
Lineartextur 1607
Lineation 1607
Lineation 2999
Linienelektrode 4563
Linse 1752
Linse 3102
Linse, kleine – 3102
linsenförmiger Körper 1752
Linsengang 4026
Linsenschichtung 3150
Linsig 3102
Lipide 3723
Lipotexit 1031
Liquation 0678
Liquefaktion 3084
liquidmagmatisch 3976
liquidmagmatisches Stadium 0651
Lithifikation 3421
lithisch 3178
Lithofazies 2861
Lithogenese 2853
Lithohamnien-Rücken 3361
lithoidisch 0835
Lithologie 2853
lithophag 1857
Lithophyse 0780
Lithosphäre 0088
Lithostratigraphie 1709
lithostratigraphische Einheit 1748
lithostratigraphischer Horizont 1749
lithostratigraphische Zone 1747
Lithotop 2862
Lithotyp 3579
litoral 2836
litoral 2906
lit-par-lit 1049
Loch 2247
Loch 3064
Loch, freies – 4277
Loch, unverrohrtes – 4277
Loch, vorgebohrtes – 4297
Lockerboden 2008
Log, Akustik– 4406
Log, Gamma– 4548
Log, Gamma Ray – 4590
Log, Induction– 4584
Log, Neutron– 4594
Log, Sonic – 4406
Log, Sonic– 4597
Longulit 0762
Lopolith 0735
Löss 2871
Lösskindl 3489
Lösung 0411
Lösungsfracht 2956
Lösungs-Hohlraum 3690
Lot 0329

Lot 3156
Lotabweichung 0329
Lotlinie 0329
Lotrichtung 0329
Love welle 0189
Lublinit 2153
Luftargon 0450
Luftbild 4349
Luftbildkamera 4329
Lufthebeverfahren bohren, im – 4263
Lumineszenz 0467
lump 3301
Lutit 3203
lutitisch 3203
Luvseite 2963
Lydit 3369

M

Mäander 2219
Mäander, abgeschnittener – 2232
Mäanderabschneidung 2229
Mäander, bewegliche – 2220
Mäanderdurchbruch 2229
Mäander, eingeschnittener – 2221
Mäander, eingesenkter – 2221
Mäander, ererbter – 2221
Mäander, freie – 2220
Mäandergürtel 2224
Mäanderhals 2226
Mäanderkerbe 2228
Mäanderkerben, Spitze zwischen zwei – 2304
Mäanderkerbeterrasse 2304
Mäanderlobus 2225
Mäanderschlinge 2232
Mäandersporntterrasse 2303
Mäanderstreifen 2224
Mäanderterrasse 2304
Mäanderterrasse mit geschüzten Spitzen 2305
Mäander, vererbter – 2221
Mäanderzunge 2225
Maar 1433
Maccaluba 1395
Maceral 3566
Maceralgruppe 3566
Maceral-Varietät 3567
Mächtigkeitslinie 4243
Mächtigkeit, wahre – 4243
mafelsisch 0598
mafisch 0599
mafisches Gestein 0599
Mafitit 0599
Magerkohle 3560
Magma 0639
Magma 1198
Magmagestein 0616
Magmaherd 1202
Magmaherd, sekundärer – 1203
Magmakammer 1202
Magmakammer, gangförmige – 1204
Magma, latentes – 0643
magma, Mutter– 0645
magma, Oberflächen– 1244
Magma, palingenes – 0644
magma, Primär– 0644
magma, Rest– 0646
Magmasäule 1226
magma, Stam– 0645
magmatisch 0616
magmatische Korrosion 0642
magmatische Phase 1108
magmatisch, primär– 3973
Magmatit 0616
magma, Ur– 0644
Magnetisierung 0295
Magnetisierung, chemische – 0312
Magnetisierung, umgekehrte – 0315
Magnetisierung, viskose – 0313
magnetismus, Gesteins– 0293
Magnetismus, remanenter – 0306
Magnetismus, thermoremanenter – 0309
Magnetohydrodynamik 0278
Magnetometer 0316
Magnetometer, astatisches – 0319
magnetometer, Flug– 4522
Magnetometer, Horizontal– 4521
magnetometer, Kernresonanz– 4524
magnetometer, Protonen– 4524
magnetometer, Protonenpräzessions– 4524
Magnetometer, Vertikal– 4521
Magnetopause 0277
Magnetosphäre 0277
Magnetotellurik 4553
Magnetscheidung 4154
Magnitude 0220
Magnitudenskala 0220
magnitude, Raumwellen– 0220
Mahlen 2940
Mahlkreislauf 4146
Mahlstrom 2269
Maibolt 2739
Makadam 4843
Makkalube 1395
Makroevolution 1970
makrokristallin 0797
makropolyschematisch 0794
Makroseisme 0138
Makroseismik 0138
makroseismisches Gebiet 0248
Makrospore 1837
Malaspinatypusgletscher 2435
malloseismische Region 0250
Mandel 0782
Mandelstein 0872
Manganknolle 3477
Mangrovesumpf 2515
Manometer 4769
Mantelaufwölbung 1145
Mantelkonvektion 1124
Mantel, oberer – 0091
Mantelreibung 4839
Mantelreibung, negative – 4839
Mantel, unterer– 0091
Mare 0051
Margarit 0766
marin 2900
Marke 3018
Marmor 0920
Marmorisierung 0920
Marsch 4280
Marschengürtel 2512
Marschensaum 2512
Marschinsel 2507
Maschendrahttextur 3468
Maschentextur 0856
Maschenweite 1497
Maschenweite 3157
Mascon 0052
Masse 1087
Massenausgleich, isostatischer – 0342
Massengestein 0616
Massenkonzentration 0052
Massenüberschuss 0343
Masse, stationäre – 0125
massig 0859
Massiv 1087
Massstab 4240
Mast 4267
Materialbilanz 3865
Materialbilanzberechnung 3865
Matrix 0626
Matrix 3080
mattiert 3408
Mattierung 3408
Mattkohle 3580
Maturität 3725
Maximum 1019
Maximum und Minimum 0164
Mazeration 3594
mechanik, Boden– 4664
mechanik, Fels– 4664
mediterrane Sippe 0690
Medium, poröses – 3680
Meer 2782
Meerball 2891
Meeresbucht 2451
Meeresbusen 2451
Meeresforschung 2781
Meereskunde 2781
Meeresniveau, mittleres – 2817
Meeresschwinde 2075
Meeresströmung, küstenparallele – 2825
Megagametophyt 1838
Megakristall 0608
megalosphärisch 1868
Megarippel 3007
megaseismisches Gebiet 0249
Megasporangium 1837
Megaspore 1837
Megazyklothem 3120
Mehlschnee 2350
Mehrausbruch 4856
Mehrfachüberdeckung 4407
mehrkammerig 1870
Mehrkanal-Aufnahmegerät 4501
Mehrlinsenkamera 4330
Meiofauna 1853
Meissel 4271
Mélange 2927
melanokrat 0588
Melanosom 1038
Melatop 0570
Melikaria 3487
Membranpotential 4616
Membran, semipermeabele 3397
Mercaptan 3716
Mergel 3221
Mergel, kalkiger – 3224
Mergelstein 3221
Mergel, toniger – 3223
Merismit 1048
Meroplankton 1801
Mesa 2556
Mesitis 0913
mesohalin 2832
Mesokarst 2067
mesokrat 0587
mesokristallin 0797
Mesokumulat 0832
Mesonorm 0606
mesopelagisch 1910
mesopelagisch 2845
Mesostatis 0626
mesothermal 3983
mesotroph 3549
mesotyp 0587
Mesozone 0993
Messbild 4349
Messkamera 4329
Messkamera 4331
Messkurve 4565
Meta- 0883
Meta-Anthrazit 3563
Metabasit 0883
Metabentonit 3218
Metablastese 0884
metablastisch 0884
Metabolismus 1803
Metakristall 0976
Metalle, extrahierbare – 4654
Metalle, heiss extrahierbare – 4655
Metalle, kalt extrahierbare – 4655
Metallglanz 0484
metallisch 3911
Metallogenese 3903
metallogenetisch 3903
Metallprovinz 3908
metaluminisch 0693
metamorph 3997
metamorpher Gürtel, paariger – 1139
metamorphes Gestein 0880
Metamorphite 0880
Metamorphose 0879
Metamorphose 1794
Metamorphose, allochemische – 0881
metamorphose, Belastungs– 0887
metamorphose, Dislokations– 0892
metamorphose, Druck– 0892
metamorphose, Dynamo– 0892
Metamorphosegrad 0992
metamorphose, Impact– 0893
metamorphose, Injektions– 0889
Metamorphose, isochemische – 0881
Metamorphose, kinetische – 0892
metamorphose, Kontakt– 0896
Metamorphose, lokale – 0891
Metamorphose, mechanische – 0892
metamorphose, metasomatische Kontakt– 0900
Metamorphose, organische – 3521
Metamorphose, progressive – 0903
metamorphose, Regional– 0887
Metamorphose, retrograde – 0904
Metamorphose, rückschreitende – 0904
metamorphose, Schock– 0893
Metamorphose, statische – 0887
Metamorphose, Stauungs– 0892
Metamorphose, thermische – 0896
metamorphose, Versenkungs– 0887
Metarippel 3014
Metasom 1067
Metasomatose 3453
Metasomatose 3956
metasomatose, Regional– 1056
Metaster 1041
Metatekt 1040
Metatektit 1031
Metatexis 1030
Metatexit 1030
Metatypus 1954
Meteoraufschlag 0026
Meteorit 0063
Meteoritenkunde 0062
Meteorit, Stein-Eisen – 0070
Meteorkrater 0027
Meteorkrater 0075
miarolitisch 0873
Microlithotyp 3552
Mictit 1023
Migma 1022
Migmatisierung 1022
Migmatit 1022
Migmatitisierung 1022
Migration 3748
Migration 4457
Migration, horizontale – 3750

Migration, laterale – 3750
Migration, primäre – 3749
Migration, sekundäre – 3749
Migrationsweg 3751
Migration, vertikale – 3750
Mikrit 3281
Mikrobeben 0146
Mikroerdbeben 0146
Mikroevolution 1970
Mikrofossil 1790
Mikrogametophyt 1838
mikrokristallin 0797
mikrokristallin 3091
Mikrolith 0759
Mikrolithotyp 3581
mikrolitisch 0836
Mikrolog 4587
Mikrometeorit 0064
Mikropaläontologie 1786
Mikroplankton 1830
Mikroschillkalk 3308
Mikroseismische Bodenunruhe 0097
mikroseismisches Gebiet 0251
Mikroskop, Polarisations – 0611
mikrosphärisch 1868
Mikrosporangium 1836
Mikrospore 1836
Mikrotektonik 1479
Mikrotextur 0790
Miktit 1023
Milieu 1900
Milieu 2859
milligal 0367
mimetisch 1008
Mindestmächtigkeit 4200
Mindeststrossenbreite 4201
Mineral 0460
Mineral 3912
Mineral, akzessorisches – 0594
Mineral, akzessorisches – 3174
Mineral-Assoziation 0586
Mineralassoziation, kritische – 0988
Mineralbestandteile 3553
Mineralbildner 3936
Mineralgehalt 3553
Mineralgemengteil, kritisches – 0988
Mineral, gesteinsbildendes – 0462
Mineralisierung 2682
Mineralisierung 3526
mineral, Kontakt– 0978
Mineral, künstliches – 0461
Minerallagerstätte 3910
Minerallagerstätte, alluviale Bildung von – 3943
mineral, Leicht– 3173
Mineralneubildung 0905
Mineral, normatives – 0607
Mineralogie 0459
Mineral, okkultes – 0757
Mineral, primäres – 0596
Mineralprovinz 3908
Mineralquelle 3632
Mineral, resistentes – 3175
mineral, Schwer– 3173
Mineral, sekundäres – 0596
Mineral, symptomatisches – 0689
Mineral, typomorphes – 0988
Mineral-Vergesellschaftung 0586
Mineral, wesentliches 0594
Miogeosynklinale 1096
Miospore 1833
mise-à-la masse, Methode – 4559
Mitnehmerstange 4269
Mittagsloch 2387
Mittel 3506
Mittelebene 1673
Mittelgebirge 2575
Mittel, gewogenes – 4185
Mittel, gleitendes – 4198
Mittelgut 4142
mittelkörnig 0796
mittelkörnig 3194
Mittellauf 2195
Mittelmoräne 1295
Mittelmoräne 2407
Mittelschenkel 1576
Mittelschenkel, ausgedünnter – 1680
Mittelschenkel, inverser – 1647
Mittelschenkel, inverser – 1680
Mittelschenkel, überkippter – 1647
Mittel, taubes – 3921
mittelthermal 3983
mixohalin 2833
mobiler Gürtel 1093
Mobilisat 1057
Mobilisierung 1057
Mobilismus 1113
Mobilität 3699
Mobilitätsverhältnis 3699
Möbius-Netz 4373
Möbius-Netz-Methode 4374
modal 0593
Modder 3210
Mode 0166
mode, Fundamental– 0167
Modell 4775
Modell, geomechanisches – 4779
Modellieren 4478
Modell, numerisches – 3866
Modell, photo-elastisches – 4783
Modell, physikalisches – 4778
Modellversuch 4778
Moder 2744
Modus 0593
Mofette 1402
Mohorovičić discontinuität 0087
Mohr-Coulombsches Bruchkriterium 4694
Mohrsche Hüllkurve 4691
Mohrscher Spannungskreis 4690
Mohrsches Spannungsdiagramm 4688
Mohssche Härteskala 4701
Mol 3675
Molasse 2926
Mollisol 2647
Moment 1759
Mondbeben 0061
Mondmilch 2153
monistische Hypothese 0020
monoklin 0515
Monoklinale 1617
Monoklinalkamm 2541
Monoklinaltal 2544
Monokline 1617
monometallisch 3910
monomikt 3265
monomineralisch 0589
monophyletische 1960
monoschematisch 0794
monotypisch 1945
mons 0037
Montanwachs 3736
monzonitisch 0813
Moor 3532
Moorausbruch 3541
moorig 3532
Moorkohle 3556
Moorsee 3540
Moor, Sphagnum– 3543
Moostorf 3543
Moräne 2403
Moräne, abgelagerte – 2410
Moräne, bewegte – 2405
Moränenamphitheater 2418
Moränewall 2414
Morgen 4016
Morgengang 4016
Morkill Formel 4217
Morphologie 1793
Morphotektonik 1076
Mörtel– 0966
Mosaik 4380
Mosaik aus entzerrten Bildern 4381
Mosaik aus grob entzerrten Bildern 4381
Mosaik aus nicht entzerrten Bildern 4381
Mosaikstruktur 3089
Mosaiktextur 0960
Mosor 2074
Mosor 2608
Mount Katmai-Tätigkeit 1165
MT 4553
Mudde 3548
mudstone, Kalk – 3296
Mugelkohle 3514
Mühle 4149
Mulde 1619
Mulde 2793
Mulde, Antiform– 1620
Mulde, konvergent eintauchende – 1638
Muldengang 4053
Muldenlinie 1636
Muldenspiegel 1636
Muldental 2273
Muldental, tektonisches – 2544
Mull 2744
Müllbeseitigung 4831
Mülldeponie 4831
Mullionsstruktur 1608
Müllkippe 4831
Multiplexen 4424
Multispektralkamera 4330
Multispektral-System 4326
Mündung 2196
Mündung, gleichsohlige – 2208
Mündungsrinne 2503
Mündung, ungleichsohlige – 2208
Munsell Farbtafeln 2676
Murbruk– 0966
Mure 2051
Murstrom, vulkanischer – 1352
Muschel 1858
Mustaggletschertypus 2431
Mutation 1982
Mutter der Perle 0503
Muttergang 4062
Muttergestein 0883
Muttergestein 2657
Muttergestein 2944
Muttergestein 3918
Mutterlauge 3377
Mylonit 0915
Mylonit 1541
mylonitisch 0969
Mylonitisierung 0915
Myrmekit 0785
myrmektitisch 0852

N

Nachahmung 0481
Nachbargestein 0709
Nachbeben 0147
nachbohren 4296
Nachbrechen 2473
Nachfolge- 2286
Nachhallerscheinungen 1374
Nachklangerscheinungen 1374
Nachläufer 0175
nachräumen 4296
Nachsackungsdoline 2072
Nachsackungserscheinung 2072
Nachstürzen 2473
Nachweisgrenze 4651
Nachwirkung, elastische – 0114
Nadel 1861
Nadeleis 2639
nadelförmig 0755
nadelförmig 3185
nadelig 0522
nadelig 0755
Nadelkohle 3556
Nadirdistanz 4351
Nadirdistanz, relative – 4351
Nagelkalk 3469
Nagelung 4793
Nährgebiet 2380
Nannofossil 1790
Nannoplankton 1799
Naphtene 3717
Narbe 1132
Narbenzone 1684
Nase, antiklinale – 1639
Nassgas 3715
Naturasphalt 3734
Naturbrücke 2079
Naturschacht 2076
neanisch 1866
Nebenausbruchstelle 1439
Nebenelement 0400
Nebenfalte 1642
Nebenfluss 2207
Nebengang 0725
Nebengemengteil 0594
Nebengestein 0709
Nebengestein 1529
Nebengestein 1530
Nebengestein 3501
Nebengestein 3919
Nebengestein 3920
Nebengestein 4009
Nebengrundwasserspiegel 3643
Nebenkrater 1439
Nebenprodukt 4163
Nebentrum 4071
Nebenverwerfung 1581
Nebulit 1054
Neck, Reztypus– 1206
Nehrung 2463
Nehrung 2464
Neigung 4833
Neigung, primäre – 1486
Neigungsmessgerät 4294
Neigung, ursprüngliche – 1486
Neigung verflacht sich 1487
Neigung wird schwächer 1487
Nekton 1798
nektonisch 1798
nematoblastisch 0962
Neo-Darwinismus 1966
Neo-Lamarckismus 1967
neomorph 1983
Neontologie 1787
Neosom 1037
Neostratotypus 1745
Neotypus 1951
Neovulkanismus 1150
nepiónico 1866
Neptunismus 0004
neritisch 2841
neritisch 2915
Nest 3105
Nest 4032
Netzgänge 4064
Netzgestein 1043
Netz, gitterförmiges – 2292
Netzleiste 2118
Netzleisten 3047
Netz, parallaktisches – 4373
Netz, perspektivisches – 4373
Netztextur 0856
Neukristallisation 0905
Neumann-Linien 0068
neutral 0602

Niedermoor 3533
Niedermoortorf 3535
Niederschlag 3614
niederthermal 3983
Niederwasser 2818
Niedrigwasser 2818
Niedrigwasser, mittleres – 2820
nierenförmig 0776
nierenförmig 4095
nierig 4095
Nische 2474
Nivation 2357
Nivationsnische 2612
Niveaulinie 4248
Niveau, piëzometrisches – 3646
Noise, regelloses – 4438
Noise, Schuss– 4438
Noise, seismischer – 4438
Nomenklatur, binominale – 1922
Nomenklatur, uninominale – 1922
Norm 0606
Normalgefälle 2198
Normalgefällskurve 2198
Normalbett 2242
Normalprofil 4255
Normalschwere, internationale Formel – 0333
Normalspannung 4688
Novaculit 3366
Nukleation 1360
Nullniveau 4577
Nunatak 2618
Nutz-zu-Störsignalverhältnis 4443

O

Obduktion 1133
Oberbau 1077
Oberbau 1414
Oberbau 4840
Oberfläche 3106
Oberflächenabfluss 3617
Oberflächendichte 4111
Oberflächenrückhaltung 3618
Oberflächenvulkanismus 1207
Oberflächenwasser 3649
Oberflächenwelle, störende – 4439
Oberflächerelief 3192
Oberfläche, spezifische – 2754
Oberfläche, spezifische – 4109
Oberkante 3106
Oberlauf 2195
Obermoräne 2406
Oberwelle 0167
obsequent 2285
Obsidian 1262
Ödland, gasverätztes – 1378
Ödland, stark zertaltes – 2030
Ödometer 4756
offen 1626
Öffnungsweite 4728
Öffnungswinkel 1626
Ogive 2378
Ökologie 1887
Ökostratigraphie 1711
Ökosystem 1893
Ökotyp 1942
Oktaëdrit 0067
Ölansammlung 3773
Oleanderblattstruktur 4082
Olefine 3717
Ölfeld 3773
Öl, förderbares – 3864
oligohalin 2832
oligomikt 3265
oligotroph 3549
Ölintrusion 3752
Olistholith 2977
Olistostrom 2977
Olivinbombe 1329
Öl, reines – 3714
Ölring 3776
Ölsand 3766
Ölsandstein 3766
Ölsättigung 4610
Ölsäule 3775
Ölschiefer 3721
Öl, untersättigtes – 3857
Öl/Wasserkontakt 3777
Öl/Wasserspiegel 3777
Ölzone 3775
ombrogen 3550
Omissionsfläche 3126
Onkoid 3363
Onkolith 3300
Ontogenese 1794
Ontogenie 1794
Ontologie, vergleichende – 0007
Ooid 3479
ooidisch 3479
Oolith 3479
oolithisch 3479
Oomikrit 3286
Oomikrudit 3286
Oospararenit 3285
Oosparit 3285
Oosparrudit 3285
opak 0545
Opalisieren 0486
Operator 4447
Ophicalcit 0930
Ophiolit 0696
Ophiolit 1098
Ophiolitserie 0696
ophitisch 0845
Ophthalmit 1052
Oppel-Zone 1780
Ordnung 1929
Ordnung 0685
organisches Material, extrahierbares – 3724
organische Substanz 2740
Orogen 1101
Orogenbogen 1140
orogene Periode 1106
orogene Phase 1107
orogener Gürtel 1093
Orogenese 1100
Orogenschlinge 1140
Orterde 2704
Ortho – 0986
Orthochem 3271
orthochemisch 3268
Orthodolomit 3320
Orthogenese 1975
Orthogeosynklinale 1095
Orthokonglomerat 3255
Orthokumulat 0832
orthomagmatisch 3976
orthophyrisch 0826
orthorhombisch 0515
Orthostereoskopie 4367
orthotektisch 3976
Ortsbrust 4852
Ortsinspektion 0267
Ortstein 2704
Ortstein 2733
Os 2420
Ossillationshypothese 1119
Ossillationsrippel 3012
Otolith 1865
Oxidation 0416
Oxidations-Reduktions-Reaktion 0416
Ozean 2782
Ozeanisation 1142
ozeanisch 2843
ozeanisch 2918
Ozeanographie 2781
Ozellarstruktur 0855
Ozokerit 3735

P

Packer 4279
Packschnee 2347
Packung 3082
Packungsdichte 3082
Pahoehoelava 1250
Palagonit 1254
Palagonitbildung 1254
Palagonitbreccie 1347
Palagonitformation 1347
Palagonittuff 1347
Paläoboden 3126
Paläobotanik 1827
Paläogeographie 0011
Paläoklimatologie 0011
Paläomagnetismus 0294
Paläontologie 1786
Paläoökologie 1889
Paläosol 3126
Paläosom 1037
Paläoströmung 2864
Paläotemperatur 2864
Paläotemperaturbestimmung 0456
paläotyp 0636
Paläovulkanismus 1150
Paläozoologie 1847
Palasom 1037
Palätiologie 0007
Palichnologie 1872
Palichthyologie 1848
Palingenese 1024
Palingenese 1968
palingenetisch 1024
palinspastische Rekonstruktion 0009
palustrisch 2887
Palynologie 1828
Palynomorph 1829
panallotriomorph 0808
pandemisch 1814
Paneeldiagramm 4252
Pangaea 1116
panidiomorph 0808
panidiomorphkörnig 0808
panxenomorph 0808
Panzerbildung 3411
Papierkohle 3563
Pappschnee 2348
Para– 0986
Parabeldüne 2177
Paracme 1985
Paraffine 3717
Paragenese 0586
Paragenese 3907
Paragenese, diagenetische – 3394
paragenetisch 0586
paragnese, Salz– 3378
parakinematisch 0906
Parakonglomerat 3256
parakristallin 0997
paralisch 2898
Parallaxe 4364
Parallelgänge 4063
Paramagnetismus 0298
paramarginal 4212
Paramorphose 0465
Parastratotypus 1743
paratektonisch 0906
Paratypus 1950
parautochthon 0727
Partikel 3078
Partikeldurchmesser 3193
Partikelgrösse 3193
passiv 0619
Passpunkt 4325
patera 0037
Patina 3403
pazifische Reihe 0690
pazifische Sippe 0690
PDB 0454
Pech 3743
Pechkohle 3557
Pechsee 3753
Pechstein 1262
Pediment 2582
Pedimentpass 2586
Pedionit 1275
Pedon 2689
pegmatitähnlich 0814
pegmatitisch 3978
pegmatitisches Stadium 0653
Pegmatitisierung 1063
Pegmatitstruktur 3470
pegmatoid 0814
pegmatophyrisch 0848
Pegostylit 2152
pelagisch 1910
pelagisch 2844
pelagisch 2919
Peleantätigkeit 1163
Peles Haar 1326
Peleträne 1326
Pelit 3208
pelitisch 3208
Pellet 4150
Pelmikrit 3293
Pelsparit 3292
Pendel 0351
Pendel 4774
Pendellänge, äquivalente – 0126
pendel, Reversions– 0351
Pendulärwasser 3653
Peneplain 2589
peneseismische Region 0252
Penetrometer 4758
peralkalisch 0693
peraluminisch 0693
Pergelisol 2637
Peridotitbombe 1329
Periglazial 2633
perimagmatisch 3985
Perimeter, benetztes – 2261
Periode 0163
Periode 1765
Perlenmutter 0503
perle, Zucht– 0503
perlitisch 0878
Perlmutter 0503
Perlmutterglanz 0484
Permeabilität 3697
Permeabilität, effektive – 3698
Permeabilität, magnetische – 0304
Permeabilität, relative – 3698
Permeabilitätsbarriere 3771
Permeabilitätsfalle 3758
Permeation 1059
perthititsch 0851
Pertit 0785
Petroblastese 1059
Petrogenese 0581
petrogenetischer Raster 0991
Petrographie 0579
Petrologie 0578
Petrotektonik 0996
pF 2771
Pfadfinderelement 4649
Pfahl 4838
Pfahlramme 4838
Pfahlrammung 4838
Pfahl, schwebender – 4839
Pfanne 4659
Pfannkuchenbombe 1329
Pfeiler 4845
Pfeilerstaumauer 4864
Pferdeschwanzstruktur 3401
Pflanzenasche 3600

Pflanzenpaläontologie 1827
Pflanzen-Sandkuppe 2190
Pflasterboden 2656
Pflasterstruktur 3089
Pflastertextur 0960
Pflugsohle 2761
pF-WG-Kurve 2772
pH 0422
Phakoid 3055
Phakolith 0733
phaneride 0795
Phanerit 0795
phanerokristallin 0795
Phänokristall 0758
Phänotypus 1979
Phantom 0802
Phantom-Horizont 4462
Phase 0165
Phase 3906
Phase, abyssische – 1197
Phase, biochemische – 3519
Phase, eruptive – 1208
Phase, explosive – 1208
Phase, geochemische – 3519
phase, Haupt– 0188
Phase, hypoabyssische – 1197
Phasengang 4448
Phasengeschwindigkeitsmethode 0230
Phasenkarakteristik 0135
Phasenspektrum 4451
Phasenstapelung 4510
Phasenverlauf 4388
Phasenverschiebung 0134
Phasenverspätung 0134
Phasenvorauseilung 0134
Phase, postvulkanische – 1197
phase, Vor– 0181
Phase, vulkanische – 1197
Phenoklast 3239
Phi 3163
Phi-Skala 3163
Phlebit 1044
Phosphatisierung 3459
Phosphoressenz 0467
Phosphorit 3480
photisch 1904
Photogeologie 4327
Photogrammetrie 4328
phyletisch 1960
Phyllit 0943
Phyllitmylonit 0918
Phyllonit 0918
Phylogenese 1959
phylogenetische Reihe 1960
phylogenetische Zone 1779
Phylogenie 1959
Phylonit 1541
Phylum 1929
Phyteral 3592
Phytoklast 3168
Phytopaläontologie 1827
Phytoplankton 1800
Pickprobe 4168
Piedmontfläche 2664
Piedmonttreppe 2604
piezogen 0977
Piezokristallisation 0633
Piezometer 3646
Pille 3094
pilotaxitic 0825
Pilzfels 2163
Pinge 2072
Pinienwolke 1362
Piperno 1342
Pipkrake 2639
pisolitisch 3478
Pisolit 2149
Pisolit 3478
Pisolith 3478
Pisolithtuff 1333
Pisolith, vulkanisch – 1321
Pitkrater 1455
Planetesimaltheorie 0022
Planetologie 0023
Planet, terrestrische – 0025
Plan, ingenieurgeologischer – 4784
Plankton 1799
planktonisch 1799
Planum 4840
plastisch 2747
Plastizität 2747
Plastizität 4669
Plastizitätsgrenze 4715
Plastizitätsgrenzen 2748
Plastizitätsgrenze, obere – 2749
Plastizitätsgrenze, untere – 2749
Plastizitätszahl 4716
Plate 2506
Plateau 2797
Plateaubasalt 1257
Platte 2506
Platte 3132
Plattendruckversuch 4708
Plattengefüge 2728
Plattengründung 4837
Plattensandstein 3132
Plattentektonik 1125
Plattform 1083
Plattform einer Schichtstufe, obere – 2558
Plattform, stabilisierte – 4539
plattig 0755
plattig 3132
plattig 3183
Plattigkeit 3182
plattstengelig 0772
Plättungslineation 1613
Playa 2580
Playa 2892
Playasee 2343
pleistoseismische Region 0246
Pleistoseiste 0246
Pleochroismus 0562
Plesiotypus 1952
Plete 1808
Plexus 1961
Plock, vulkanischer – 0734
Plurimetamorphose 0901
Pluton 0715
plutonisch 0617
plutonisch 3988
Plutonismus 0004
Plutonismus 0615
pneumatogen 0655
pneumatolitisch 3977
Pneumatolyse 3950
pneumatolytisches Stadium 0653
Podsolierung 2680
poikilitisch 0844
Poikiloblast 0981
poikiloblastisch 0974
poikilophitisch 0846
Poisson-Zahl 4700
Pol 1018
Polarablagerungen 0040
Polarisation 0236
Polarisation 4557
Polarisation, induzierte – 4569
Polarisationsfilter 4339
Polarisierung, magnetische – 0297
Polarität, reverse – 0290
Polaritätsepoche 0292
Poldiagramm 1018
Pole, geomagnetische – 0280
Polgürtel 1020
Polierschiefer 3371
Polimetamorfismo 0901
Politur 3404
Pollen 1839
Polpunktdiagramm 4259
Polumkehr-Ereignis 0292
Polyedergefüge 2729
polygen 0590
polygenetisch 0590
Polygonboden 2653
polyhalin 2832
polymetallisch 3911
Polymetamorphose 0901
polymikt 3265
polymineralisch 0589
Polymorphie 0464
Polymorphismus 1811
polyphyletisch 1960
polyschematisch 0794
polytopisch 1819
Polytypie 0463
polytypisch 1945
Population 1941
Pore 3083
Poren- 3083
Porenfüllung 3440
Porengehalt 3681
Porengehalt, nutzbarer – 3704
Porenvolumen 2752
Porenvolumen 3681
Porenwasserüberdruck 3820
Porenwasser, ursprüngliches – 3661
Porenwasserwiderstand 4608
Porenwinkelwasser 3653
porig 1305
Porosität 3681
Porosität, Gesteinslösungs– 3692
porosität, Hohlform– 3688
Porosität, Interpartikel– 3688
Porosität, Intrapartikel– 3688
Porosität, Klüft– 3694
Porosität, Lösungs– 3690
Porosität, Matrix– 3687
Porosität, minus-zement – 3442
Porosität, nutzbarer – 3695
Porosität, Partikel– 3688
Porosität, Partikellösungs– 3691
Porosität, primärer – 3684
porosität, Sack– 3689
porosität, Sackgassen– 3689
Porosität, sekundäre – 3685
Porosität vor der Zementbildung 3442
Porphyr 0817
porphyrartig 0819
porphyrisch 0817
Porphyrit 0817
porphyritisch 0817
porphyrkörnig 0820
Porphyrkristall 0758
Porphyrkupfererz 4133
Porphyroblast 0981
porphyroblastisch 3470
Porphyroid 0957
Porphyroklast 0981
Portal 4850
Porzellanit 3374
postkinematisch 0726
postkinematisch 0906
postkinematisch 1079
postkristallin 0997
postorogen 0726
postorogen 1108
posttektonisch 0906
postvulkanische Erscheinungen 1374
Potential 3645
Potential, elektrokinetisches – 4617
Potential-Gradientenverfahren 4560
Potentialkartierung 4558
Potsdamer Schweresystem 0334
Potsdamer Schwerewert 0334
Präadaptation 1980
präkinematisch 0906
präkinematisch 1079
präkristallin 0997
Prallhang 2236
Prallmarke 3038
Prallufer 2236
prätektonisch 0906
Pratt 0345
Präventer 4289
Pressiometer 4760
Pressschnee 2347
Priel 2509
Priel 2517
primär 4630
Primärdüne 2176
Primärmigration 3732
Primärmigration 3749
Primärreflexion 4437
Primärrumpf 2593
Prioritätsgesetz 1924
prismatisch 0522
prismatisch 0755
Prismengefüge 2730
Probe 4741
Probeentnahme 4741
Probeentnahmegerät 4745
Probe, geknetete – 4743
Probe, ‚gesalzene' – 4179
Probe, gestörte – 4742
Probenahme beim Bohren 4167
Probenahme im Anstehenden 4167
Probenahme mit zufälligen Abständen 4174
Probenbeutel 4229
Probenraster 4171
Proben teilen 3159
Probenteiler 4182
Probenverjüngung 4180
Probe, ungestörte – 4742
Problematikum 1789
Proctor Verdichtungsversuch 4757
Produkte, pyroklastische – 1309
Produkte, resurgent-authigene – 1314
produktion, geschätzte Gesamt– 3869
Produktion, kumulative – 3817
Produktion, Messung der – 3798
Produktionsrate 3801
Produktionssonde 3786
Produktivitätsindex, spezifischer – 3808
Produktivitätsindex 3808
Profilradius, mittlerer – 2262
Profilsäule 3593
Profilschiessen, kontinuierliches – 4408
Profilschnitt, gebogener – 4252
Profilschnitt, geknickter – 4252
Profilschnitt, generalisierter – 4254
Profilschnitt, geologischer – 4250
Profilschnitt, idealisierter – 4254
Profilserie 4253
Profil-Serie 4251
Projektion, zyklographische – 1021
Propfen 3105
prophyroblastisch 0956
Propylit 0921
Propylitbildung 0921
Propylitisation 0921
Propylitisierung 0921
Prospektion, biogeochemische – 4645
Prospektion, geobotanische – 4645
Prospektion, geochemische – 4625
Prospektion, hydro-geochemische – 4640

Protactinium-Thorium Datierung 0445
Protodolomit 3321
protogene Bildung 0664
Protoklase 0857
protoklastisch 0857
Protomylonit 0916
Protore 3913
Prototyp 1979
Prototyp 4775
provinz, Erz- 3908
provinz, Faunen– 1849
Provinz, geochemische – 0401
Provinz, geothermische – 0390
Provinz, petrographische – 0628
Provinz, sedimentpetrologische – 2852
proximal 2997
Psammit 3226
psammitisch 3225
psammitisch 3226
Psephit 3238
psephitisch 3237
psephitisch 3238
Pseudoaspite 1430
Pseudofossil 1789
pseudomorph 0982
pseudomorph 1823
pseudomorph nach 0982
Pseudomorphose 0465
Pseudomorphose 0982
Pseudomulde 1621
Pseudonodule 3060
Pseudopalagonit 1255
pseudoporphyritisch 0819
pseudoporphyroblastisch 0959
Pseudosattel 1620
pseudostereoskopisch 4367
Pseudotachylit 0917
Pseudotachylit 1542
Pseudovulkan 1189
ptygmatisch 1051
Puffer 0422
Pufferstock 4225
Pulsrate 4504
Pulverisierung 1319
Pulverschnee 2346
Pulverschneelawine 2359
Pumpsonde 3791
Pumpspeicherwerk 4870
Pumpversuch 3711
Punktdiamant 0494
Punktgruppe 0512
Punktlastfestigkeit 4733
Punktzähler 0612
Puy 1425
Puzzolan 1338
Puzzolanerde 1338
P-Welle 0182
pyramidal 0522
Pyritisierung 3458
Pyrobitumen 3737
pyrogen 3973
pyrogen 0655
Pyrogenese 0616
Pyroklast 3170
Pyroklast 1309
Pyroklastite 1309
Pyrolyse 3730
Pyromagma 1200
Pyromeride 0876
Pyrometamorphose 0896
Pyrometer 1195
Pyrometer, optisches – 1195

Q

Quarzarenit 3230
Quarzit 3230
Quarzsandstein 3230
Quastenmarke 3039
Quellbecken 1405
Quelle 3630
Quelle, artesische – 3631
Quelle, hypogene – 1408
Quelle, kochende – 1407
Quellenbildung 4863
Quellgebiet 2193
Quellkuppe 1422
Quellmulde 2276
Quellrohr 1405
Quellstalagmit 2152
Quelltrichter 2276
Quer– 2289
Queraufstellung 4419
Querdehnzahl 4700
Querdüne 2171
Quereinsattelung 1635
Querfalte 1678
Quergang 4077
Querkluft 1506
Querküste 2522
Querprofil eines Tales 2271
Querschnitt, benetzter – 2261
Querspalte 1470
Querspalte 2376
Quertrum 4072
Querüberdeckung 4359
Querverformingssahl 4700
Querverwerfung 1545
Quickton 4719

R

Rachel 2026
Rad 4551
Radar 4384
Radarazimut 4394
Radarkeule 4389
Radar, kohärentes – 4387
Radar mit synthetischer Apertur 4386
Radar, nicht kohärentes – 4385
Radarreflektivität 4392
Radarschatten 4391
Radar, seitwärts schauendes – 4385
Radialgang 1227
Radialgänge 4063
Radialpressenversuch 4763
Radialschlucht 1474
Radialschluchtenberg 1475
radialstrahlig 0776
radialstrahlig 0843
Radialtriangulation 4324
Radialverwerfungen 1594
Radioaktivitätshypothese 1117
Radiokohlenstoff-datierung 0451
Radiolarienerde 3368
Radiolarienschlamm 3368
Radiolarit 3369
Radiolarita 3368
Radiometer 4310
Rahmenmarke 4368
Rammarbeit 4838
Rammpfahl 4838
Rammsonde 4758
Rammsondierung 4711
Randkette 2542
Randmeer 2783
Randmoräne 2416
Randmulde 1697
Randschlucht 2384
Randschutt, vulkanischer – 1295
Randsenke 1111
Randsenke 1697
Randspalte 2376
Randstörung 1592
Randstufe 2604
Randsumpf 3537
Randverwerfung 1593
Randwasser 3779
Randwassers, Vordringen des – 3826
Randwassertrieb 3825
Rang 0685
Rang 3552
Rapakivi-Gefüge 0828
Raseneisenerze 4116
Raseneisenstein 2733
Raseneisenstein 2734
Rasenhügel 2643
Rasenläufer 4051
Rasse 1939
Rassenkreis 1938
Räuberfluss 2325
Rauchkringel 1364
rauh 3190
Rauhigkeit 4727
räumen 4296
Raumgewicht 2751
Raumgruppe 0513
Raum, interkolliner – 1461
Raumschiff 4302
Rauschen 0097
Rauschen, seismisches – 4438
Rauschen, signal-erzeugtes – 0177
Rayleigh Welle 0189
Reaktionshof 0786
Reaktionslagerstätte, topomineralische – 3995
Reaktionsrand 0786
Reaktionsrinde 0786
Rechenklassierer 4156
Rechenmodell 4776
Redox-Potential 0421
Redox-Reaktion 0416
Reduktion 0416
Reduktion, bakterielle – 0417
reduktion, Bouguer– 0359
reduktion, Freiluft- 0358
reduktion, Geländе– 0360
reduktion, Höhen– 0357
Reduktion, isostatische – 0361
Reduktionshof 3463
Reduktion, topographische – 0356
Reduktion, topographische-isostatische – 0364
Referenzbeben 0153
Referenzfeld, internationales geomagnetisches – 4535
Reflektion, Total– 0213
Reflektivität 4473
Reflexion 0574
Reflexion 0576
Reflexion 4436
Reflexion, diffuse – 4317
Reflexion, gerichtete – 4318
Reflexion, Mehrfach– 4441
Reflexion, multiple – 4441
Reflexionsbrecher 2811
Reflexionsgrad 4316
Reflexionshorizont 4461
Reflexionskoeffizient 4473
Reflexionsverfahren 4405
Refraktion 2810
Refraktionsverfahren 4405
Regel 1013
Regel nach Kornbau 1013
Regel nach Korngestalt 1013
Regelung 1013
Regelung nach Kornbau 1013
Regelung nach Korngestalt 1013
Regenerationskomplex 3538
regeneriert , epirogenetisch – 3994
Regenfurche 2026
Regenlahar 1355
Regenrille 2026
Regenriss 2026
Regentropfeneindruck 3041
Regentropfenmarke 3041
Regionalfeldes, Grad des – 0340
Regionalisierung 0255
Regionalisierung 4194
Registrierung 4500
Registrierung, analoge – 4421
Registrierung, digitale – 4422
Registrierverzögerung 4502
Regolithdecke 0060
Regolithschicht 0060
Regression 1726
Reibung 4686
Reibung, innere – 4686
Reibungsbeiwert 4689
Reibungsbreccie 1539
Reibungsbreccie, vulkanische – 1248
Reibungsgesetz 4689
Reibungskoeffizient 4689
Reibungspfahl 4839
Reibungswinkel 4689
Reicherz 3915
Reicherzone 4005
Reichweite 1781
reif 3176
reif 3725
Reifegrad 3725
Reifeparameter 3726
Reifestadium 2570
Reihe, charnockitische – 0990
Reihendüne 2171
Reinigung 3813
reissen 4828
Rejuvenation 3959
Rekristallisation 0905
Rekristallisation 3452
Rekristallisation, Strain– 0967
Relaisbeben 0145
Relief 0558
Relief, alpines – 2575
Relief, appalachisches – 2564
Reliefenergie 2575
Reliefgeneration 2573
Relief, geochemisches – 0401
relief, Hochgebirgs– 2576
Reliefinversion 2563
Relief, jurassisches – 2565
Relief, negatives vulkanisches – 1410
Relief, positives vulkanisches – 1409
Reliefsequenz 2573
relief, Steil– 2576
Reliefumkehr 2563
Reliefumkehrung 2563
Relikt 0971
Relikt 0985
Relikt 1985
Reliktendüne 2190
Relikt, gepanzertes – 0787
Relikt, instabiles – 0985
Reliktsediment 2856
Reliktsee 2344
Relikt, sekundäres – 0985
Relikt, stabiles – 0985
Rem 4551
Remanenz, anhysteretische – 0314
remanenz, Sedimentations– 0312
Resdiualkies 2967
Resedimentation 2993
resequent 2284
Reserven 3868
Reservenbilanz 4222
Reserven, mögliche – 3881
Reserven, nachgewiesene – 3879
Reserven, potentielle – 4215
Reserven, spekulative – 3882
Reserven, wahrscheinliche – 3880
Reservoirs, Leistung des – 3846
Reservoirs, Verhalten des – 3846

Residual 3998
Residualkonzentrat 2967
Residuallagerstätte 2930
Residuum 0241
Resorption 0642
Responzkurve 0133
Ressourcen, hypothetische – 4214
ressourcen, Mineral– 4205
Ressourcen, nachgewiesene – 4206
Ressourcen, spekulative – 4214
Ressourcen, unbekannte – 4213
Ressourcen, unentdeckte – 4213
Restberg 2607
Restit 1032
Restit 1066
Restling 2607
Restölsättigung 4610
Restsalzkissen 1697
Restscherfestigkeit 4685
Restschmelze 0647
Restsee 2344
Rest, unlöslicher – 2930
Retikulit 1331
Retrometamorphose 0904
Retromorphose 0904
Reverberation 4440
Revision 3886
Reynolds 2265
Rhenium-Osmium Datierung 0446
Rheologie 0102
Rheologie 4671
Rheomorphose 1057
Rhombochasmus 1143
Rhomboidrippel 3010
Rhythmit 3121
Ria 2452
Richtbohren 4291
Richtcharakteristik 4491
Richter Skala 0220
Richtungs-Effekt 4530
richtungslos 0859
Richtungsrose 4259
Riedel 2312
Riefung 2999
Riegel 2201
Riegel 2335
Riegel 2625
Riegelberg 2625
Rieselmarke 3029
Riesenquelle 2093
Riesenrippel 3007
Riesentopf 2389
Riesenwuchs 1867
Riff 2443
Riff 2469
Riff 2506
Riff 2804
Riff 3333
riffähnlich 3334
riff, Algen– 3360
riffartig 3334
Riff, Barriere– 3338
Riffbereich 3326
Riffel 2814
Riffelteiler 4182
riffenthaltend 3334
Riff-Flanke 3353
riff, Flecken– 3347
Riff-Front 3351
Riff gehörend, zum – 3334
Riffgerüst 3336
Riffgestein 3342
Riffhang 3354
Riffkalkstein 3310
Riffkern 3341
riff, Knollen– 3329
Riffkomplex 3337
riff, Krusten– 3347
riff, Kuppen– 3347
Riffmauer 3336
Riffplatte 3343
Riffrand 3350
Riff, regressives – 3327
riff, Ring– 3340
Riffront-Terrasse 3352
riff, Sand– 2459
riff, Säulen– 3347
riff, Saum– 3338
Riffseite, seewärtige – 3349
Rifftal 2460
Riff, transgressives – 3328
riff, Turm– 3347
riff, Unterwasser– 2459
riff, Wall– 3338
Rifkappe 3329
Rift 1092
Rift 1590
Rille 0057
Rillenabspülung 2033
Rillenkarren 2087
Rillenmarke 3033
Rinde, kelyphitische – 0786
Ringelerz 4050
Ringeltextur 4096
Ringgebirge 1461
Ringintrusion – 0742
Ringraum 4279
Ringspalte 1469
Ringtal 1453
Ringwall 1452
Rinne 2191
Rinne 2793
Rinne 3022
Rinnenerosion 2033
Rinnenerosion 2778
Rinnenfüllung 3022
Rinnenkarren 2087
Rinnen, verflochtene – 2243
Rinnsal 2191
Rinnsal 2194
Rippel 2814
Rippel 3000
Rippelhöhe 3002
Rippelindex 3002
Rippelkamm 3001
Rippel, komplexe – 3013
Rippel, longitudinale – 3011
Rippel mit gekappter Kamme 3015
Rippelschrägschichtung 3147
Rippel, sichelförmige – 3005
Rippeltal 3001
Rippel, transversale – 3003
Rippel, unvollständige – 3015
Rippstromrinne 2486
rip-rap 4876
Riss 1466
RMS, Geschwindigkeit – 4466
Roentgen 4550
Roherz 4138
Rohhumus 2744
Rohöl 3712
Rohöl, wasserhaltiges – 3714
Rohstoff, mineralischer – 3912
Roller 4439
Rollfähigkeit 3191
Rollmarke 3038
Rosette 3092
Rosiwall-Methode 0613
Rostfleck 2692
Rotarybohren 4263
Rotary-Stossbohren 4263
rotation, Extern– 1000
rotation, Intern– 1000
Rotationsbombe 1329
Rotationspol 1126
Rotationswinkel 1126
Rotation, tektonische – 1000
Roteisenerze 4115
Rotschicht 2879
R-Tektonit 1003
Rubidium-Strontium Datierung 0447
Rücken 2796
Rückfalte 1689
Rückfaltung 1689
Rückkopplung 0133
Rückland 1110
Rückprall 4733
Rücksenke 1110
Rückstand 2930
Rückstandserz 3998
Rückstrahlvermögen 4316
Rücktiefe 1110
Rücküberkippung 1689
Rückzugsmoräne 2417
rudimentär 1978
Rudit 3237
ruditisch 3237
Ruhedruck 4705
Ruheperiode 1169
Ruhespur 1880
Rumpfebene 2591
Rumpffläche 2591
Rumpfflächentreppe 2604
Rumpfgebirge 2588
Rumpftreppe 2601
Rumpftreppe 2604
Rundhöcker 2400
Rundung 3187
Rundungsindex 3187
runitische Textur 0816
Runse 2026
Runzelschieferung 1600
Runzelung 1643
Runzelung 3051
Runzelung, intraformationelle – 3051
Runzelungsachsen 1612
Ruschel 1538
Ruschelzone, vererzte – 3993
Rutsch 2043
Rutschen 2976
Rutschfalte 3055
Rutschfläche 1538
Rutschschramme 1538
Rutschstreife 1538
Rutschstruktur 1704
Rutschung 2037
Rutschung im Verband 2043
Rutschung mit gestörtem Verband 2050
Rutschungsmorphologie 2056
Rüttelverdichtung 4809

S

Sacktal 2083
Sackungsgefüge 3059
Salar 2580
Salband 0707
Salband 1530
Salband 4012
Salband, fest aufgewachsenes – 4013
Salband, freies – 4013
Salina 2580
Salinität 3376
salisch 0607
Salmiakfumarole, alkalische – 1388
Salpeter 2123
Salse 1395
Saltation 1982
Saltation 2954
Saltationsfracht 2954
Salzbeule 1695
salzführend 3376
Salzfumarole 1387
Salzgehalt 2827
Salzgehalt 3675
Salzgletscher 1699
Salzkissen 1695
Salzkristall-Pseudomorphose 3042
Salzlauge 3676
Salzmarsch 2514
Salzmarsch 2909
Salzmauer 1695
Salzpfanne 2892
Salzpfanne 2909
Salzpolygon 3049
Salzschnee 2350
Salzsee 2893
Salzsprengung 1997
Salzstock 1695
Salzsumpf 2892
Salzsumpf 2909
Salztektonik 1694
Salztonebene 2580
Sammelgebiet 2380
Sammelkristallisation 0910
Sammelprobe 4176
Sammelprofil 4253
Sand 2718
Sand 3227
Sand, Abgleiten von – 2973
Sand, äolischer – 2871
Sandbank 2244
Sandbank 2459
Sanddrän 4813
Sanderebene 2424
Sandfall 2973
Sandfalle 3152
Sandfeld 2157
Sandfraktion, Median der – 2718
Sand, gasführender – 3766
sandig 3225
Sandinselchen 2443
sand, Korallen– 3311
Sandkorngrösse 3227
Sandkristall 3443
Sand, lehmiger – 3227
Sandlinie 4621
Sand, ölführender – 3766
Sandschliff 2015
Sandschliff 2154
Sandsee 1462
Sandstein 3228
Sandstein, Feldspatführender – 3232
Sandstein, gasführender – 3766
Sandstein, ölführender – 3766
Sandstein, quarzitischer – 3230
Sandstein, reiner – 3228
Sandstein, wasserführender – 3766
Sandstrahlen 2939
Sandstrom 2973
Sandstrom, heisser – 1369
Sand, tönender – 3088
Sandtransport, äolischer – 2961
Sandtreiben 2156
Sanduhraufbau 0534
Sand, verwässerter – 3767
Sandvulkan 3071
Sand, vulkanischer – 1322
Sand, wasserführender – 3766
Sandwatt 2910
Sandwehen 2156
Sandwelle 3007
Sandwüste 2578
Saprolit 4128
Sapropel 3528
Sapropelit 3721
Sapropelkohle 3721
Satellitenbild 4349
Satellit, künstlicher – 4302
Sattel 1618
Sattel, beidseitig abtauchender – 1638
Sattelgang 4053
Sattelspiegel 1632

Sattel, Synform– 1621
Satteltal 2544
Sättigung 3650
Sättigungsbereich 3656
Sättigungsdruck 3857
Sättigungsexponent 4606
Sättigungsgrad 3650
Sättigungsintensität 0297
Sättigungslinie 0697
Sättigungslinie 4862
sauer 0601
sauer 3716
Sauerstoffbedarf 3678
Saugbohren 4285
Saugloch 2075
Säule 2138
Säulenprobe 3593
Säulenprofil 4255
säulig 0755
säulig 0772
Saumtiefe 1109
Saumtiefe 1111
Säurebehandlung 3800
saures Gestein 0601
Saussurit 0927
Saussuritbildung 0927
Saussuritisation 0927
Saussuritisierung 0927
saxonisch 1078
Schaarkreuz 4075
Schaarung 4073
Schacht 2104
Schacht 4848
Schachtbrettboden 2655
Schachtelrelief 2601
Schachthöhle 2097
Schafstieg 2060
Schale 1842
Schallgeber 4488
Schallkopf 4488
Schallwelle 4482
Schallwelle, sichtbare – 1371
Schallwiderstand 4474
Schalquelle 4486
Schappe 4748
Schar 2337
Schären 2455
scharf 0173
Schärfe, optische – 4347
Scharfkantig 3190
Scharnier 1629
Scharnierzone 1629
Scharte 1471
Schattenzone, seismische – 0216
Schaufelfläche 1573
schaumig 0875
Schaumlava 1307
scheibenförmig 3184
Scheinanomalie 4637
Scheitel 0705
Scheitel 1629
Scheitel 1635
Scheitelbereich 1629
Scheitelflur 2597
Scheitellinie 1629
Scheitellinie 1632
Scheitelpunkt 1633
Scheiteltal 2544
Scheiteltiefe 0214
Scheitelüberschiebung 1577
Schelffazies 2915
Schelfrand 2788
Schelf, stabiler – 1089
Schelle 4280
Schenkel 1625
Schenkel, Hangend– 1680
Schenkel, normaler – 1647
Scherbenlava 1250
Scherfalte 1662
Scherfestigkeit 4684
Schergerät 4754
Scherkluft 1504
Scherung 0105
Scherzone 1533
Scherzone, vererzte – 4060
Schicht 3097
Schicht 3098
Schicht- 2286
Schichtabspülung 2032
Schicht, Backset– 3141
Schicht, D– 2708
Schichtebene 3096
Schichten, aufgerichtete – 1481
Schichtenbau, mantelförmiger – 1241
Schichtenbau, periklinaler – 1241
Schichtenbau, trichterförmiger – 1241
Schichtenbau, zentroklinaler – 1241
Schichten, gekippte – 1482
Schichten, geneigte – 1482
Schichten, Kohleführende – 3498
Schichten, schräggestellte – 1482
Schichtenstellung 1480
Schichtenunterdrückung 1557
Schichten, versetzte – 1523
Schichten, verworfene – 1523
Schichtenwiederholung 1557
Schichtfläche 3096
Schichtfläche, blossgelegte – 2557
Schichtfluterosion 2032
Schichtfolge 3110
Schichtfolge, vulkanische – 1238
Schicht, Foreset- 3140
Schichtfuge 3099
Schichtfugenhöhle 2096
Schichtgefüge, biogenes – 1884
Schichtgestein 2850
Schichtgrenzhöhle 2096
schichtig 4103
Schichtkamm 2541
Schichtlücke 1723
Schichtmächtigkeit 4622
Schichtrippe 2562
Schicht, schwerdurchlässige – 3639
Schichtstufe 2558
Schichtstufenlandschaft 0040
Schichttafelland 2555
Schichtterrasse 2307
Schichtterrasse 2555
Schicht, undurchlässige – 3640
Schichtung 3095
Schichtung, geneigte – 1482
Schichtung, gradierte – 3130
Schichtung, Horizontal– 3017
Schichtung, Parallel– 3017
Schichtung, rhythmische – 3121
Schichtungsfläche 3096
Schichtung, Vorsetz– 3140
Schichtunterdrückung 4567
Schichtvulkan 1429
Schiefer 0942
Schiefer 0944
Schiefer 3220
schiefer, Blätter– 3220
schieferig 0972
schieferig 0975
schieferig 1603
Schieferigkeit 0975
Schieferigkeit, eigentliche – 0975
Schieferigkeit, gefältelte – 0975
Schieferigkeit, gewundene – 0975
Schieferigkeit, schuppige – 0975
Schieferigkeit, Zickzack– 0975
Schiefer, kristalliner – 1604
Schiefermylonit 1543
Schieferung 0948
Schieferung 0975
Schieferung 1596
Schieferung 1603
Schieferung 1604
Schieferung in Ton 1599
Schieferungsebene 1598
Schieferungsfächer 1602
Schieferungsfläche 1598
schiefrig 3220
Schiessanordnung 4411
Schild 1081
Schildvortrieb 4853
Schildvulkan 0035
Schildvulkan 1420
Schildvulkangebirge 1450
Schildvulkangruppe 1450
Schiller 0486
Schillkalk, mässig fester – 3308
Schillkalk, stark verfestigter – 3308
Schizolith 0668
Schlacke 1294
Schlackenagglomerat 1346
Schlackenbombe 1329
Schlackenfontäne 1282
Schlackenkegel 1427
Schlackenklumpen 1297
Schlackenmantel 1330
Schlackensand 1323
Schlackenschornstein 1298
Schlackentuff 1333
Schlackenwall 1452
schlackig 1294
Schlaglawine 2359
Schlagspur 3410
Schlamm 2920
Schlamm 3210
Schlamm 3307
Schlamm 4160
Schlamm 4829
Schlammbüchse 4278
Schlammfeld 1390
schlammgestützt 3294
schlammig 3210
Schlamm, Kalk– 3295
Schlammkegel 1396
Schlamm, korallenführender – 3311
Schlammkügelchen 1321
Schlammlava 1397
Schlammpfuhl 1398
Schlammsprudel 1398
Schlammstein 3219
Schlammstein, konglomeratischer – 3256
Schlammstrom 2051
Schlammstrom 2974
Schlammstrom, heisser – 1353
Schlammstrom, kalter – 1355
Schlammstrom, Schmelzwasser– 1357
Schlammstrom, vulkanischer – 1352
Schlammtuff 1356
Schlammvulkan 1700
Schlammvulkan 1395
Schlammvulkan 3071
Schlechte 3508
Schleifen 2940
Schleiffenschlussfehler 4541
Schleifmarke 3032
Schleifmarke 3033
Schleifmarken, sich kreuzende – 3034
Schleifung 2399
Schlenke 3539
Schleppen 4079
Schleppfalte 1570
Schleppung 1537
Schlick 3210
Schlick 4829
Schlickgeröll 3247
Schlickgeröll, gepanzertes – 3247
Schlick, hemipelgischer – 2920
schlickig 3210
Schlick, mergeliger – 3222
Schlickwatt 2910
Schlieren 0481
Schlieren 0665
schlierig 0665
Schliffgrenze 2631
Schliffkehle 2395
Schlitzprobe 3593
Schlitzprobe 4168
Schlotagglomerat 1248
Schlotbreccie 1248
Schlotpfropfen 1477
Schlotverlagerung 1231
Schlot, zylinderförmiger – 1224
Schlucht 2020
Schlucht 2026
Schluchtgang 2100
Schluckloch 2075
Schluff 2715
Schluff 3204
schluffig 3204
Schlundloch 2075
Schlussphase 1167
Schmelzbirne 0497
Schmelze 1198
Schmelze, Interkumulus– 0831
Schmelzherd 1202
Schmelzloch 2387
Schmelztuff 1340
Schmelzwasserablagerung 2402
Schmelzwasserabsatz 2402
Schmelzwasserbach 2388
Schmelzwassertal 2393
Schmidt'sches Netz 1015
Schmierkohle 3555
Schmutzband 2379
Schneebrettlawine 2360
Schnee-Erosion 2357
Schneegrenze 2354
Schneeschmelze 2355
Schneeverwehung 2353
Schneewächte 2356
Schneewehe 2353
Schnee, wilder – 2346
Schnee, windgepackter – 2347
schneidbar 0471
Schnitthang 2236
Schockzone 0894
Scholle, abgesunkene – 1559
Scholle, gehobene – 1559
Schollendom 1272
Schollengebirge 2539
Scholle, überschiebende – 1569
Scholle, überschobene – 1569
Scholle, verschluckte – 0661
Schotter 3244
Schotter 4844
Schotter 4876
Schotterterrasse 2298
Schotterwüste 2578
Schrägaufnahme 4353
Schrägaufnahme, Lateral – 4353
Schrägbild mit überlagertem Netz 4375
Schrägküste 2522
Schrägschichtung 3133
Schrägschichtung 3134
Schrägschichtung 3135
Schrägschichtung 3136
Schrägschichtung 3146
Schrägschichtung, ansteigende – 3149
Schrägschichtung, bogige – 4142
Schrägschichtung, ebene – 3142
Schrägschichtungsbogen 3151
Schrägschichtung, trogförmige – 3144

Schrägschichtung, winkelige – 3142
Schrägstrecke 4393
Schrägstreckenbild 4395
Schramme 1538
schrappen 4828
Schratten 2086
Schreiber 4500
Schreibkreide 3315
Schrift– 0847
Schriftgranit 0847
Schrumpfgrenze 4715
Schrumpfriss 3044
Schrumpfriss 3046
Schrumpfrisspolygon 3046
Schrumpfriss, unvollständige – 3046
Schrumpfungsriss 1993
Schrumpungsfaktor 3860
Schub 4677
Schubbahn 1705
Schubbrett 1683
Schubfläche 1568
Schubfläche, basale – 1572
Schub, hydraulischer – 4737
Schubmasse 1683
Schubmodul 4699
Schubspannung 4688
Schubweite 1568
Schubweite, horizontale – 1551
Schubweite, scheinbare horizontale – 1553
Schubweite, stratigraphische – 1552
Schuppe 1538
Schuppe 1571
Schuppe 3186
Schuppenstruktur 1571
schuppig 0775
schuppig 0948
Schürfgrube 4871
Schürfloch 4235
Schürfschacht 4235
Schürmannsche Regel 3522
Schürre 2536
Schüsseldoline 2073
Schusskanal 1221
Schusskern 4238
Schussperforation 3799
Schusspunkt 4411
Schuttbreccie 1249
Schüttdamm 4859
Schuttdecke 2008
Schüttelherd 4157
Schüttelsieb 4282
Schuttfächer 2007
Schuttfläche 2006
Schutthalde 2007
Schutthang 2007
Schuttkegel 2007
Schuttkriechen 2040
Schuttlawine 1368
Schuttlawine 2047
Schuttlawine, glühende – 1367
Schuttrutschung 2046
Schüttungswinkel 3137
Schuttwüste 2578
Schutzgebiet 3845
Schwalbbrecher 2811
schwammig 0875
Schwappmarke 3030
Schwarz-Weisstorfkontakt 3544
Schwebgut 2955
Schwefelschlamm 1399
Schwefelschlammkrater 1398
Schwefelschlammkügelchen 1321
Schwefelschlammsäule 1400
Schwefelschlammstrom 1399
Schwefelschornstein 1400
Schweisschlacken 1346
Schweisschlackenkegel 1298
Schweisstuff 1341
Schwelle 2201
Schwelle 2247
Schwelle 2796
Schwellen 4720
Schwellenvertrag 0266
Schwellewert 4627
Schwellungshügel 2643
Schwemm– 2869
Schwemmfächer 2948
Schwemmkegel 2948
Schwere 0324
Schwereanomalie 0367
Schwerebeschleunigung 0324
Schwerefeld 0326
Schwere, gemessene – 4536
Schweregleitung 1702
Schweregleitung 1703
Schweregradient 0335
Schwere, Karte der regionalen – 0371
Schwerelot 3156
Schwerelot 4747
Schweremessung 0349
schwere, Normal– 0332
Schwerepotential 0327
Schwerestation 0353
Schwerestörung 0367
Schwerestörung, lokale isostatische – 0369
Schwerflüssigkeits-Aufbereitung 4153
Schwergewichtsmauer 4864
Schwerkraft 0323
Schwerkraft 0324
Schwerkraftaufbereitung 4153
Schwerkrafteinheit 4537
Schwerkraftscheidung 4153
Schwermetalle 4653
Schwerstange 4270
Schwimmsand 3085
Schwimmsand 4719
Schwimmschnee 2351
Schwimm-Sinkscheidung 4153
Schwinde 2075
Schwindgrenze 4715
Schwingmoor 3530
Schwingrasen 3530
Schwingung, Sphäroidische – 0198
Schwingung, toroidale – 0198
Schwundausgleich, automatische – 4508
Schwundgebiet 2258
Schwundriss 3508
se 1004
Sebkha 2580
Sebkha 2892
Sediment 2850
Sedimentation 2989
Sedimentation, interne – 3436
Sedimentation, rhytmische – 3121
Sedimentationslücke, kleine – 1724
Sedimentationsrate 2989
Sedimentationsunterbrechung 1724
Sedimentationswaage 3158
Sedimentfazies 2860
Sedimentfracht 2951
Sedimentgefüge 2994
Sedimentgefüge, primäres – 2995
Sedimentgefüge, sekundäres – 2995
Sedimentgestein 2850
Sediment, internes – 3436
Sediment, Intertidal– 2907
Sediment, klastisches – 3164
Sedimentologie 2849
Sedimentpetrographie 2851
Sedimentpetrologie 2851
Sedimenttransports, Rate des – 2957
Sedimenttuff 1334
See 2330
See– 2888
See, abflussloser – 2338
Seeabsatz, glazialer – 2402
Seeball 2891
Seebeben 0148
Seegangsrippel 3012
Seegat 2508
See, vulkanisch aufgestauter – 1419
Segerkegel 1196
Segregation 3395
Segregation 3948
Segregationsbänderung 0946
Segregationslagerstätte, magmatische – 3975
Segregation, tektonische – 0947
Sehnenberg 2215
Seiche 2816
Seide 0490
Seidenglanz 0484
Seife 3236
Seife 3997
Seife, alluviale – 4000
Seife, eluviale – 3999
Seife, fluviatile – 4000
Seigerung 3948
Seilbohren 4262
seismische Konstante 0257
seismische Region 0247
seismische Registrierungen in der Tiefsee 4484
seismisches Ereignis 0095
Seismizität 0242
Seismizität, spezifische – 0242
seismo-elektrischer Effekt 0113
Seismogramm 0116
seismograph 0116
Seismograph, Breitband– 0116
Seismographenbunker 0116
Seismologie 0094
Seismologie, forensische – 0264
seismologie, Nuklear– 0263
Seismologie, Planetär– 0094
Seismometer 0118
Seismometer 4414
Seismometerabstand 4416
Seismometeraufstellung 4417
Seismometerauslage 4417
Seismoskop 0117
Seitenerosion, Anzapfung durch – 2322
Seitenkern 4238
Seitenkrater 1439
Seitenmoräne 2411
Seitenreflexion 4518
Seitentrum 4071
Sektion 0685
Sektion, seismische – 4458
Sekundärdüne 2168
Sekundärkliff 2476
Sekundärmigration 3749
Selbstmulcheffekt 2762
Semianthrazit 3560
Sender 4487
Senderstärke, maximale – 4490
Senke, kreisförmige – 1185
Senke, peripherische – 2560
Senke, vulkano-tektonische – 1182
Senkrechtaufnahme 4353
Senkung 2857
Senkungsküste 2524
Senkungsküste 2525
Senkungstrichter 3667
Sensitivität 4717
Sensitometrie 4341
Separator 3797
Septarie 3487
Sequenz 3110
Serac 2377
serial-porphyrisch 0820
serial-porphyrkörnig 0820
Sericitisation 0932
Sericitisierung 0932
Serie 1764
serie, Gesteins– 0630
Serpentine 2219
Serpentinisation 0923
Serpentinisierung 0923
Serpentinit 0923
sesshaft 1919
sessil 1919
Setzmaschine 4155
Setzung 4704
Setzung 4833
Setzungsmessgerät 4768
Setzung, unterschiedliche – 4704
Setzung, unterschiedliche – 4833
S-Fläche 1002
Shelf, stabiler – 1089
Shore Härteskala 4701
si 1004
Sicheldüne 2177
Sichelwanne 2398
Sicherheitsfaktor 4777
Sicherheitskoeffizient 4777
Sickerlinie 4862
Sickerströmungsdruck 4738
Sickerung 3629
Sickerverlust 4863
Siderit 0066
Siderolit 0070
Sieb– 0974
Siebanalyse 3157
Siebanalyse 4144
Siebung 4873
Sieden, retrogrades – 1174
sigma-t 2827
Signal 4435
Signal 4517
Signal, akustisches – 4517
Signal, seismisches – 4517
Signal, transientes – 0176
Silifizierung 0929
Silifizierung 3458
Silizifikation 0929
Sill 0740
Silt 3204
Siltstein 3204
Sink– 2991
Sinkgeschwindigkeit 2991
Sinter 3385
Sinterbecken 2147
Sinterdamm 2147
Sintergebilde 2124
Sinter, kieselsäurehaltiger – 1406
Sinterschale 2147
Sintertapete 2128
Sinterwanne 2147
Sinus 0051
sitzend 1919
Skapolitisation 0932
Skapolitisierung 0932
Skelett 1861
Skelett– 3169
Skopulit 0769
Slingram 4575
SMOW 0454
Sockel 1077
Sode 2698
Sohle 3501
Sohle 4010
Sohle 4852
Sohlendruck 3847
Sohlenfläche 2278
Sohlental 2278
Sohlfläche 3106
Sohlform 2246
Sohlhebung 4830

Sohlmarke 3019
Sole 3676
Solfatare 1391
Solfatarenfeld 1392
Solfatarenstadium 1168
Solfatare, saure – 1393
Solifluktion 2041
Solifluktion 2780
soligen 3550
Soll 2632
Sollprofil 4856
Solum 2694
Somma 1461
Sommavulkan 1445
Sonde 4261
Sonde 4522
Sonde 4758
Sonde, Bohrloch– 4578
Sonde, eingeschlossene – 3811
Sonde, fliessende – 3790
Sonde mit Gaslift 3791
Sonderung 3948
Sonde totdrücken 4290
Sondierung 4710
Sonnensystem 0018
Sorptionskomplex 2755
sortiert 3202
sortiert, gut – 3202
Sortierung 3200
Sortierungskoeffizient 3201
Spaat 4016
spaltbar 1596
Spaltbarkeit 0469
Spaltbarkeit 1596
Spalte 1465
Spalte 1512
Spalte 1513
Spalte, klaffende – 1513
Spalte, konzentrische – 1469
Spaltenerguss 1233
Spalteneruption 1233
Spaltenfrost 1996
Spaltenfrost-Verwitterung 1996
Spaltenfüllung 1515
Spaltenfumarole 1383
Spaltengang 1514
Spaltengang 4044
Spaltenlagerstätte 3993
Spaltental 2547
Spalte, offene – 1513
Spalte, querverlaufende – 1470
Spalte, radiale – 1468
Spaltfläche 0540
Spaltspurendatierung 0442
Spaltung 0667
Spaltung 2210
Spaltung 2935
Spannung 0103
Spannung 4556
Spannung 4667
spannung, Deviations– 0103
Spannungmessing, Überbohrversuch zur – 4766
Spannungsabfall 0107
Spannungs-Deformationskurve 0106
Spannungs-Dehnungs Diagramm 4667
Spannungsriss 1519
Spannung, wirksame – 4668
Sparit 3280
sparker 4410
Sparker 4493
Sparre 2378
Spat 3280
Spat 4016
spätkinematisch 1079
spätorogen 1108
Spätstadium 2570
Speicher 3755
Speicherfähigkeit 3625
Speicherformation 3679
Speichergestein 3679
Speicher, homogener – 3770
Speicherisotropie 3770
Speicherung 3754
Speichervolume, brutto – 3872
Speiloch 2092
Speisegebiet 2380
Spektral analyse 0221
Spektralband 4306
Spektrometer, Einkanal– 4547
Spektrometer, Mehrkanal– 4547
Spektrum, elektromagnetisches – 4304
Spektrum, kontinuierliches – 4305
spektrum, Leistungs– 0222
Speläologie 2094
Speleolit 2124
Speleothem 2877
Spelunke 2095
Spezies 1937
sphärisch 0867
Sphäroid 0330
Sphäroid 0867
Sphäroid 3180
sphäroidisch 0867
sphäroidisch 3180
Sphärokristall 0778
Sphärolit, Aschen– 1321
Sphärolith 0778
Sphärolith 3482
Sphärolith, zusammengesetzter – 0779
sphärolitisch 0876
sphärolitisch 3482
sphärolitisch 4087
Sphärosiderit 3483
Sphenochasmus 1143
Sphenolith 0736
Spiculit 0764
Spiculit 3372
Spiegel 1538
Spiegelung 4318
Spiegelwert 4626
Spielart 1940
Spilit 0922
Spilitisierung 0922
Spilosit 0933
Spiralbohrer 4236
Spiralklassierer 4156
Spitze 2449
Spitze 2805
Spitzenfestigkeit 4685
Splitt 4876
splitten 3159
Splitter 3186
Sporangium 1832
Spore 1832
Sporn 2227
Sporn, abgestumpfter – 2628
Sporn, abgestutzter – 2628
Sporn, angenagter – 2228
Sporomorph 1831
Spratzkegel, littoraler – 1300
Sprengbarkeit 4820
sprengen 4814
Sprengen, schonendes – 4823
Sprenghöhle 4819
Sprengkapsel 4816
Sprengstoff 4815
Sprengstoff-Faktor 4820
Sprengtrichter 1412
Sprengtrichter 4819
Sprengverzögerung 4822
Springquelle, intermittierende – 1404
Spritzbeton 4801
Spritzerzone 2439
Spritzzone 2439
spröde 0471
spröde 4673
Sprudel, heisser – 1407
Sprühzone 2439
Sprung 1521
Sprunghöhe, stratigraphische – 1552
Sprunghöhe, vertikale – 1551
Sprunghöhe, wahre – 1562
Sprungschicht 2333
Sprungschicht, Salzgehalt– 2830
sprungschicht, Temperatur– 2830
Sprungschwall 2252
Sprungweite, scheinbare horizontale – 1553
Spülbohrung 4751
Spülkopf 4268
Spülprobe 4238
Spülrinne 2026
Spülrohr 4751
Spülsaum 2962
Spülungsverlust 4284
Spülzone 2439
Spundwand 4799
Spur 1492
Spur 1876
Spur 4420
Spur, biogene – 1875
Spurenelement 0400
Spurenfossil 1873
Spurenfossil 3067
Spurenkunde 1871
Spur, organogene – 1875
Spurrichtung 1493
stabförmig 3185
Stabmühle 4149
Stabrost 4144
Stacheln 1260
Stadium 2568
Stadium der Unreife 2570
Stadium, greisenhaftes – 2570
Staffelstörungen 1587
Staffelverwerfungen 1587
Stalagmit 2137
Stalaktit 2131
Stamm 0687
Stamm 1929
Stammbaum 4143
Stammesgeschichte 1959
stampfen 4809
Standard 0454
Standardabweichung 4190
Standardkurve 4565
Standardprofil 4257
Standort 1901
Standort 4789
Standorterkundung 4790
Standpfahl 4839
Standrohr 4274
Standseife 2912
Standsicherheitsanalyse 4777
Stand, ufervoller – 2251
Stapelmoräne 2410
Stapeln 4433
Stapelung, einhüllende – 4510
Stapelung, gleitende – 4510
Stationsabstand 4418
Stationsazimut 0160
Staub 3206
Staubband 2379
Staubfall 2159
Staub, kosmischer – 0065
Staublawine 2359
Staubregen 2159
Staubsand 2156
Staubtuff 1333
Staub, vulkanischer – 1316
Stauchmoräne 2419
Stauchwall 2419
Staukuppe 1422
Staukuppenstrom 1423
Stausee, glazialer – 2393
Stauterrasse 2310
Stauung 2340
Stauwall 2419
Stechmarke 3037
Stefanescu-Integral 4568
Stehende 4016
Steigrohr 4275
Steilabfall einer Stufe 2558
Steilrand, sigmoidale – 0032
Steilufer 2236
Stein 3242
Steinband 2654
Steinbruch 4874
Steineis 2369
Stein, gerundeter – 3242
steinig 0835
Steinkern 1824
Steinkern 1825
steinkern, Skulptur– 1824
steinkern, Skulptur– 1825
Steinkohle 3496
Steinkohleneinheit 3606
Steinkranz 2653
Steinlawine 2047
Steinmann-Trilogie 0696
Steinmann Trinität – 1098
Steinmeteorit 0071
Steinpanzer 2160
Steinpflaster 2160
Steinplattenboden 2656
Steinpolygon 2653
Steinring 2653
Steinsalz 3382
Steinschale 0089
Steinschleifer 0502
Steinschüttdamm 4859
Steinsohle 2651
Steinstreif 2654
Stein, synthetischer – 0479
Steinwüste 2578
Stein, zusammengesetzter – 0480
S-Tektonit 1002
stengelig 0755
stengelig 0772
stengelig 0972
stenobathisch 1911
stenobenthisch 1909
stenohalin 1912
stenotherm 1913
stenotop 1914
Stereobasis 4363
stereogen 1041
Stereokomparator 4365
Stereometer 4365
Stereopaar 4362
Stereoskop 4366
Stereosom 1041
Steuermässigung für Substanzverzehr 4222
Stiche 2649
Stichflamme 1350
Stichprobe 4169
Stictolith 1053
Stillegungsseit 3810
Stillstandsküste 2526
Stillstandsseit 3810
Stimulationsbehandlung 3800
Stirneinrollung 1687
Stirnhang 2558
Stirnlappe 1687
Stirnmoräne 2413
Stirnmoräne 2416
Stirnregion 1686
Stirnscharnier 1687
Stirnschuppe 1687
Stirnzone 1686
Stock 0730
Stock 4032
Stockscheider 4133
Stockwerk 4058
Stockwerklandschaft 2599
Stockwerk, oberes – 1077
Stockwerk, tiefes – 1077

Stoffhaushaltsrechnung 3865
Stoffwechsel 1803
Stollen 4847
Stoneley Welle 0195
Störpegel 0097
Störung 1522
Störung 4723
Störung, abdichtende – 3765
Störung, abgelenkte – 1536
Störung, Abrutsch– 1707
Störung, divergierende – 1583
Störung, erdmagnetische – 1192
Störung geschlossen, durch – 3763
Störung, schichtenparallele – 1547
Störungsfalle 3757
Störungsfläche 1528
Störungshöhle 2096
Störungsletten 1540
Störungssone 1532
Störung, steile – 1549
Störung, synsedimentäre – 1707
Stossbohren 4262
Stosskuppe 1247
Stossmarke 3037
Stossufer 2236
Stosswelle 0274
Stosswelle 4481
Strahlengänge 4063
Strahlenkranz 0056
Strahlenweg 4471
Strahlparameter 0207
Strahl, seismischer – 0206
Strahlungsmesser 4310
Strahlungsvermögen 4313
strahlung, Wärme – 0381
Strand 2485
Strandböschung 2489
Stranddüne 2168
Strandebene 2492
Strandfläche 2492
Strandhorn 2493
Strandmulde 2486
Strand, nasser – 2487
Strandpriel 2486
Strandriff 2486
Strandrinne 2486
Strand, trockener – 2487
Strandwall 2459
Strandwall 2490
Strandwall, angelehnter – 2462
Strandwall, freier – 2461
Strandwallinsel 2468
Strandzone 2438
Strangmoor 2636
Stratigraphie 1708
Stratotypus 1740
Stratotypus, Grenz– 1744
Stratotypus, Teil– 1742
Stratotypus, zusammengesetzter – 1742
Stratovulkan 1429
Stratovulkan, schildförmiger – 1430
Strebepfeiler 2606
Strecke 4393
Streckungslineation 1613
Streichen 1483
Streichen 4016
Streichrichtung 1483
Streich und Fallmessungen 4228
Streifenart 3581
Streifenboden 2653
Streifenfundament 4837
Streifenkamera 4332
Streifenmosaik 4380
Streifung 0468
Streu 2742
Streubereich 4186
Streuung 0218
Streuung 4316
Strichdüne 2171
Strichfarbe 0466
Strieme 1538
Striemung 2999
Striemung 3058
Strom 2191
stromabgewandte Seite 2964
Stromatactis 3437
Stromatit 1049
Stromatolith 3332
Strombecken 2192
Strombolitätigkeit 1159
Strombreccie 1268
Strombreccie, verschweisste – 1268
Stromdichte 4556
Stromentwicklung 2197
Stromfeld 2960
Strom, kommutierter – 4555
Stromlinie 2372
Stromnetz 2205
Strom, Oberflächen – 2785
Stromschnelle 2202
Stromstärke, elektrische – 4556
Stromstrich 2268
Strömung, kapillare – 3624
Strömung, laminäre – 2264
strömung, Meeres– 2785
Strömungsdruck 4737
Strömungsgeschwindigkeit 2260
Strömungsmarke 3020
Strömungsmesser 3669
Strömungspotential 4617
Strömungsregime 2960
Strömungsrippel 3004
Strömungsstreifung 3027
Strömung, turbulente – 2264
strontium, Anfangs– 0448
Strontium, gewöhnliches– 0448
Strontium, radiogenes – 0448
Struckburschluss, tiefster – 3762
Strudel 2269
Strudelkessel 2027
Strudelkessel 2248
Strudelloch 2027
Strudelloch 2248
Strudeltopf 2027
Strudeltopf 2248
Struktur 0858
Struktur 2568
Struktur 3077
struktur, Bioerosions– 1883
Strukturboden 2652
Struktur, federförmige – 1511
Strukturform 2554
Struktur, gebänderte – 0860
Struktur, halokinetische – 1694
Struktur, halotektonische – 1694
Strukturkarte 4247
struktur, Kryptoexplosions– 1184
Struktur, kryptovulkanische– 1184
struktur, Pferdeschwanz– 4069
Struktur, rissige – 4097
Strukturschluss 3762
Strukturschluss, unterster – 3762
Strukturschluss, vertikaler – 3762
Struktur, Sporn-und-Rinnen– 3351
Struktur, vulkantektonische – 1182
Strunkpass 2327
Stufe 0527
Stufe 1538
Stufe 1762
Stufenabfall 2558
Stufenfläche 2558
Stufengang 4055
Stufenhang 2558
Stufenkar 2614
Stufenlehne 2558
Stufenmündung 2208
Stufenstirn 2558
Stufental, glaziales – 2622
Sturmdelta 2500
Sturmflut 2824
Sturm, magnetischer – 4529
Sturzbach 2884
Sturzbrecher 2811
Sturzflut 2252
Sturzseite 2185
Stützmauer 4797
Stützpfeiler 4797
Stylolith 3451
subaerisch 2866
Subalkali– 0694
Subalter 1763
subaluminisch 0694
subangular 3189
subaquatisch 2880
Subarkose 3233
Subduktionssone 1134
Subfazies, metamorphe – 0987
Subgenus 1933
subglazial 2873
subglaziär 2873
Subgrauwacke 3234
Subhaldenhang 2603
sub-idiotopisch 3471
Subjektkontrast 4343
Sublimation 0412
Sublimation 3952
Sublimationsprodukt 1394
sublitoral 2839
Submaceral 3567
submarginal 4212
submarin 2900
subophitisch 0845
Subpolyedergefüge 2729
Subprovinz 3908
subsequent 1108
subsequent 2286
Subsidenz 2857
Subspezies 1939
Substanz, mineralische – 3912
Substitution 3956
Substrat 0348
Substrat 2696
Substrat 3107
Subterminalkrater 1438
subtidal 2908
Subvulkan 1206
subvulkanisch 3989
Suchbohrung 3782
Sulzschnee 2350
Summenlinie 4188
Sumpferze 4116
Sumpfgas 3529
sumpfig 3532
Superfamilie 1930
supergen 3937
supergen 3967
Superparamagnetismus 0300
Superstufe 1763
suprakrustal 0584
supralitoral 2837
supratidal 2908
Suspension 0411
Suspensionsfracht 2955
Suspensionsstrom 2980
Suszeptibilität, magnetische – 0298
Sutur– 0960
Sutur 1132
suturiert 3450
suturisch 3450
S-Welle 0184
SWK 3544
Symbiose 1898
symmiktisch 3124
sympatrisch 1817
symplektisch 0849
synantetisch 0784
synantetisch 3963
Synäreseriss 3044
synchron 1729
synchron 1818
Syneklise 1084
Synform 1619
Synform-Sattel 1621
Syngenese 3398
syngenetisch 2854
syngenetisch 3474
syngenetisch 3553
syngenetisch 3966
synkinematisch 0726
synkinematisch 1079
Synklinale 1619
Synklinale, Pseudo– 1621
Synklinalkamm 2541
Synklinaltal 2544
Synkline 1619
Synklinorium 1671
Synonym 1955
Synonymie-Liste 1955
synorogen 0726
synorogen 1108
synsedimentär 2854
Syntektit 0682
Syntexis 0682
synthetisch 4478
Syntypus 1951
System 1765
System, actives – 4301
System, passives 4301
System, verzweigtes – 2292
System, zentrifugales – 2291
Szintillation 4319
Szintillationszähler 4546

T

Tachylyt 1262
Tafel 0501
Tafel 1083
Tafelberg 2556
Tafelberg 2798
tafelförmig 0755
tafelig 0522
tafelig 0755
Tagesgang 4528
Tal, abgeschnittenes – 2481
Tal, asymmetrisches – 2274
Talaue 2278
Talaufschüttung 2278
Talausgang 2280
Talbecken, glaziales – 2620
Talbeckensee 2620
Talbeginn 2276
Talbildung 2270
Tal, blindes – 2083
Talboden 2278
Tal, ertrunkenes – 2282
Talgehänge 2277
Tal, gekapptes – 2324
Tal, geköpftes – 2324
Talgletscher 2427
Talgrund 2278
Talhang 2277
Tal, hängendes – 2483
Tal-in-Tal 2275
Talkante 2277
Talkar 2611
Tallängsschnitt 2271
Tallehne 2277
Talleiste 2293
Tallinie 2279
Talmäander 2221
Tal, mehrzyklisches – 2275
Talmoor 3534
Talmündung 2280
Tal, offenes – 2276
Talquerschnitt 2271

Talschluss 2276
Talschnörkel 2234
Talsohle 2278
Talsporn 2227
Talsporn 2228
Talstufe 2199
Talstufe, glaziale – 2623
Talsystem, aufgelöstes – 2482
Talsystem, aufgepropftes – 2283
Tal, tektonisch angelegtes – 2543
Tal, tektonisches – 2543
Talterrasse 2293
Taltorso 2327
Tal, U-förmiges – 2273
Tal, verlängertes – 2282
Tal, V-förmiges – 2272
Talvorsprung 2228
Talwand 2277
Talwasserscheide 2315
Talweg 2279
Tal, zweizyklisches – 2275
Talzwischenscheide 2313
Tanköl 3859
Taphonomie 1897
Taphozönose 1896
Taphrogenese 1091
Tarnung 4643
Tasche 2649
Tasche 4032
Tätigkeit, effusive – 1213
Tätigkeit, explosive – 1216
Tätigkeit, hawaianische – 1158
Tätigkeit, kurzfristige – 1154
Tätigkeit, Mount Katmai– 1165
Tätigkeit, peleanische – 1163
Tätigkeit, plinianische – 1161
Tätigkeit, pyroklastische – 1309
Tätigkeitsperiode 1208
Tätigkeit, strombolianische – 1159
Tätigkeit, vulcanianische – 1160
Tätigkeit, vulkanische – 1149
taub 3917
Tauchdecke 1688
Tauniederung 2646
Taupunkt 3858
Tausenke 2646
Tautonym 1957
Taxit 1243
Taxogenese 1963
Taxon 1926
Taxon 2662
Taxonomie 1920
taxonomie, Boden– 2662
Teer 3743
Teerkappe, abdichtende – 3765
Teersand 3768
Teich 2330
teilbar 1597
Teilbarkeit 1597
Teilbewegung 0999
Teilchen 3078
Teilung 0541
Teilung 2210
Teilzone 1784
Tektit 0077
Tektofazies 2861
Tektonik 1073
tektonik, Granit– 0712
Tektonit 1001
Tektonit 1605
tektonit, Primär– 1001
tektonit, Schmelz– 1001
tektonit, Sekundär– 1001
Tektonosphäre 0080
Tektotop 2862
telemagmatisch 3986
Teleseisme 0150
Teleskoping 3965
tellurisch 0002
Temperaturleitfähigkeit 0387
Temperaturslog 4601
Temperatursprengung 1992
Temperaturtiefenkurve 0378
Tephra 1310
Tephrochronologie 1312
Terrain, chaotisches – 0038
Terrasse 2801
Terrasse, auftauchende – 2311
Terrasse, diastrophische – 2307
Terrasse, durchlaufende – 2299
Terrasse, eustatisch bedingte – 2308
Terrasse, klimatisch bedingte – 2308
terrasse, Land– 2558
Terrasse, lokale – 2299
Terrasse, marine – 2484
Terrassenböschung 2295
Terrassenfeld 2061
Terrassen, ineinandergeschachtelte – 2300
Terrassenkante 2295
Terrassenkreuzung 2311
Terrassenmäander 2223
Terrassen, nach Höhe übereinstimmende – 2300
Terrassensteilhang 2295
Terrassensteilrand 2295
Terrassenstockwerk 2296
Terrassentreppe 2296
Terrassenzug 2294
Terrasse, örtliche – 2299
Terrasse, untertauchende – 2311
terrestrisch 2865
terrigenisch 2902
Tethys 1127
tetragonal 0515
Tetragonalboden 2655
Teufenerstreckung 1526
Textur 0790
Textur 2709
Textur 3077
Textur 4345
Textur, flaserige – 0952
Textur, gerichtete – 2998
Textur, kolloforme – 4091
Textur, konkretionäre – 4092
Textur, metakolloidale – 4091
Textur, primäres Sediment– 2995
textur, Säulen– 1501
Textur, Sediment– 2994
Textur, sekundäre Sediment– 2995
Textur, wurmförmige – 3467
T-Fläche 1017
Thanatozönose 1896
thermal 0376
Thermalquelle 1407
Thermalquelle 3632
Thermaltheorie 3927
Therme 1407
Thermoelement 0379
thermogen 0977
Thermohalin 2829
Thermokarst 2645
Thermokarstniederung 2646
Thermokarstsee 2646
Thermokline 2333
Thermometer, geologisches – 0994
Thermometrie, geologische – 0994
thermometrie, Isotopen– 0456
Thixotropie 4670
thixotropisch 4670
Tholeit 0695
Tholoide 1422
tholus 0037
Thorium-Blei Datierung 0436
Thrombolith 3330
Tidefluss 2504
Tidegebiet 2504
Tidenhub 2819
Tidezone 2564
Tief 2508
Tief 2509
Tief 2794
Tiefenbereich 4505
Tiefenerosion 2778
Tiefenerstreckung 1526
Tiefengestein 0617
Tiefenreflexion 4519
Tiefensondierung, seismische – 0094
Tiefenstufe 0993
Tiefenwulst 1104
Tiefenzone 0993
Tiefherdbeben 0158
Tiefkarst 2067
Tiefländer 0028
Tiefscholle 1590
Tiefsee-Ebene 2792
Tiefseegraben 1138
Tiefseeton, brauner – 2923
Tiefseeton, roter – 2923
Tilgungsfonds 4219
Tillit 2874
Tilloid 2978
Tiltmeter 0124
Tischfels 2163
Tobel 2026
Tombolo 2467
Ton 2714
Ton 3212
Toneisenstein 3488
Toneisenstein 3504
Toneisenstein 4117
Toneisensteingeoden 3504
Ton, fetter – 3216
ton, Feuer– 3217
Ton, feuerfester – 3517
Tonfraktion 2714
Tongalle 3246
Tongalle 3248
Tongehalt 4623
Tongeröll 3246
Tongeröll 3248
Tongestein 3211
Tonglimmerschiefer 0943
tonig 3211
Tonlinie 4621
Ton, magerer – 3216
Ton, mergeliger – 3223
Tonmineral 3213
Tonmittel 3772
Tonpaste 4729
tonreich 2721
Tonschiefer 0942
Tonschiefer 3220
Tonstein 3219
Tonstein 3505
Ton, strukturempfindlicher – 4717
Tonverlagerung 2034
Topkante 3106
topogen 3550
topomineralisch – 3995
topotaktisch 0543
Topotaxie 0543
Topotypus 1953
Topset Lage 2896
Torf 3494
Torfbildung 3524
Torf, Calluna– 3545
Torfdolomit 3507
Torf, Sphagnum– 3543
Torf, terrestrischer – 3546
Torf, Tundra – 3547
Torsionsscherversuch 4765
Torsionstextur 1051
Tortuosität 3707
Toteis 2369
Toteisloch 2632
Toteispinge 2632
Totental 1402
Tracht 0521
trachytisch 0838
trachytoid 0839
Träger 3755
Träger, verwässerter – 3767
Tragfähigkeit 4707
Trägheitsmasse 0125
Tragschicht 4841
Tragschicht, erste – 4841
Tragschicht, zweite – 4841
Tragvermögen 4707
Trajektorie 4471
Transfluenz 2383
Transformation 3426
Transform-Störung 1141
Transform-Störung 1560
Transgression 1726
Transgressionskonglomerat 3114
Translation 1566
Translation 2370
Translationsstörung 1566
Transmissionsgrad 4315
Transport, tektonischer – 0998
Transvaporisation 3931
Transversalkluft 1507
Transversalspalte 1470
Transversalspalte 2376
Transversalstörung 1546
Transversalstörung, linkshändige– 1561
Transversalstörung, rechtshändige– 1561
Transversalverwerfung 1546
Trapp 0638
traubig 0777
traubig 4094
Trauf 2558
Treiben 4184
Treibholz 2970
Treibsand 2156
Treibsand, vulkanischer – 1318
Tremometer 0118
Tremor 0146
Trennfläche 4723
Triaxial-Apparat 4753
Tribus 1931
Trichit 0770
Trichterdoline 2072
Trichtergang 0742
Trichtermündung 2456
T-Richtung 1017
Triebsand 2156
trigonal 0515
triklin 0515
Trimorphismus 1869
tripel 3371
Triplet, Stereo– 4362
Triplette 0480
Tripoli 3371
Tritium Datierung 0452
Trittsiegel 1877
trocken 0609
Trockenfallen 2857
Trockengas 3715
Trockenriss 1518
Trockenriss 1993
Trockenriss 3044
Trockenriss 3045
Trockenschnee 2349
Trockental 2084
Trockental 2328
Trog 1641
Trogschluss 2619
Trogschulter 2619
Trogtal 2273
Trommelfeuer 1237
Tromometer 0118
Tropfstein 2127
Tropfsteinbildungen 2124
Trübestrom 2981

Trübestroms, Anfang eines – 2982
Trübestroms, Ende eines – 2982
Trübestrom, stetiger – 2981
Trugrumpf 2593
Trum 4038
Trum 4071
Trümerstock 4058
Trümerzone 4060
Trümmerbildung 2005
Trümmererz 3998
Trümmerfeld 2006
Trümmerkegel, abschliessender – 2114
Tsunami 2816
Tuff 3171
Tuffagglomeratlava 1346
Tuff, atmosphärisch extrudierter – 1335
Tuff, aufgearbeiteter – 1337
Tuff, chaotischer – 1333
tuff, geschichteter Trocken– 1336
Tuffisit 1339
Tuffit 1332
Tuffit 3171
tuffitisch 1332
Tuffkegel 1428
Tuff, lithischer – 1333
Tuff, subaquatisch extrudierter – 1335
Tuff, umgelagerter – 1337
Tuffvulkan 1428
Tuff, vulkanischer – 1332
Tümpel 2517
Tumulus 1272
Tundra 2636
Tunnel 4847
Tunnelfumarole 1383
Tunnelvortriebsmaschine 4854
Turbidit 2983
Turbiditstrom 2981
Turbinenbohren 4264
Turm 4267
Turmalinisation 0932
Turmalinisierung 0932
Turmkarst 2082
Tütenmergel 3469
T-Welle 0201
T-X Diagramm 4463
Typlokalität 1946
Typus-Art 1948
Typus-Gattung 1947
Typusgebiet 1739
Typuslokalität 1739

U

Überdeckung 4232
Überdeckung 4359
Überdeckung 4407
Überdeckungsgrad 4515
Überdeckung, stereoskopische – 4361
Übereindanderfolge, gleichförmige – 1719
Überfaltungsdecke 1682
Überflutung 1728
Überflutung 2857
Überflutungssediment 2885
Übergangskegel 2423
Übergangslagerstätte, liquidmagmatisch-pneumatolytische – 3981
Übergangsmoor 3536
Übergangssone 3777
Übergangssone 4620
Übergemengteil, charakteristischer– 0595
Übergemengteil, vikariierender – 0595
übergreifen 3113
Überhang 2472
Überindividuum 1012
überkippt 1491
Überkonsolidierung 4703
Überlagerung 1714
Überlagerung 4788
Überlagerung, ungleichförmige – 1721
überlappen 3113
Überlappungssone 1777
Überlauf 2339
Überlauf 4868
Überlaufen 2251
Überlaufniveau 2251
Überprägung 0970
überreif 3725
übersättigt 0605
Überschall-Knall 0177
Überschiebung 1568
Überschiebungsdecke 1575
Überschiebungsdecke 1683
Überschiebungsfläche 1568
Überschiebungsmasse 1683
Überschwemmungsbett 2242
Überschwemmungsgebiet 2239
Übersichtsprospektion 4639
Übersichtsvermessung, radioactive – 4545
Übertreten 2251
Überwachung vulkanischer Vorgänge 1190
überwiegend 1816
überzogen 3402
Überzug 3402
Ufer, äussere – 2235
Uferbank 2337
Uferbank, angeschwemmte – 2337
Uferbank, ausgewaschene – 2337
Ufer, exponiertes – 2336
Uferfiltration 3671
Ufer, innere – 2235
Ufer, konkave – 2235
Ufer, konvexe – 2235
Ufermoräne 2416
Uferterrasse 2337
Uferterrassenlava 1285
Uferwall 2490
Uferzone 2438
Ulme 4852
ultrabasisch 0604
Ultrabasit 0604
ultramafisch 0600
ultramafisches Gestein 0600
Ultramafit 0600
Ultrametamorphismus 0882
Ultrametamorphose 0882
Ultramylonit 0916
Ultramylonit 1541
ultravulkanisch 1166
Umbildung 3956
Umfang, benetzter – 2261
Umgebung 1900
Umgebung, primäre – 0402
Umgebung, sekundäre – 0402
Umgehung 0273
Umgestaltung 1794
Umhüllung, migmatische – 1070
Umkehrpunkt 4577
umkehr, Selbst– 0311
Umkehrung, geomagnetische – 0291
Umkristallisation 0905
Umlaufberg 2227
Umsetzung 3961
Umwandlung 3425
Umwandlung 3956
Undation 1122
Undationshypothese 1121
uneben 3190
ungesättigt 0605
ungesättigte Zone 3651
Ungleichgewichtsdatierung, radioaktive – 0443
Uniformitarismus 0006
Unitisierung 3901
Universaldrehtisch 0614
unreif 3176
unreif 3725
Unterart 1939
Unterbau 1077
Unterbrecherkontakt 4489
Unterfamilie 1930
Unterfangung 4796
Unterfläche 3106
Untergatuung 1933
Untergrundpunkten, Sortierung nach gemeinsamen – 4433
Untergrundskoeffizient 0258
Unterhöhlung 2471
unterirdisch 2876
Unterkante 3106
Unterlauf 2195
Untermoräne 2409
Unterprovinz 3908
untersättigt, kritisch – 0605
Unterschiebung 1574
Unterschneidung 2023
Unterschneidung 2471
Unterschneidungshang 2236
Unterspülung 2471
Unterströmung 2826
Unterstufe 1763
Untertauchen 1728
Untertauchen 2857
Untertauchküste 2525
Unterwaschung 2023
Unterwasserausbruch 1236
Unterwasservulkan 1415
Untiefe 2247
Untiefe 2469
Untiefe 2803
Uralitisation 0928
Uralitisierung 0928
Uran-Blei Datierung 0434
Uran-Kupfer-Konzentrationslagerstätte 4129
Urblei 0437
Urdüne 2176
Urnebel 0021
Urnebel-Hypothese 0021
Uroberfläche 2569
Urrelief 2569
ursprünglich 2284
Ursprungsseit 0159
Urstromtal 2392
Uvala 2073

V

Vagilität 1917
Varianz 4190
variation, Säkular– 0289
Variationsdiagramm 0697
Variationskoeffizient 4190
Varietät 1940
Variogramm 4196
Variol 0781
Variolit 0781
variolitisch 0877
Variometer 0288
Vauclusequelle 2093
VEM 4575
Vening Meinesz, Achse von – 1137
Venit 1045
Verankerung 4792
Verästelung 2501
Verbindung, metallorganische – 0410
Verbindung, organometallische – 0410
Verbreitung, geografische – 1781
Verbreitungszone 1781
Verbrennung, In-situ – 3838
Verbrennungswärme 3604
Verdichtung 4679
Verdichtung 4808
Verdickung 4054
Verdrängung 0931
Verdrängung 3453
Verdrängung 3956
Verdrängung 3961
Verdrängungsgang 4042
Verdrängungslagerstätte 3991
Verdrängungslagerstätte, kontaktpneumatolytische – 3980
Verdrängungsmetamorphose eines Baumstammes 1292
Verdriften 2968
Verdrückung 4054
Verebnungsfläche 2591
Verebnung, Zone lateraler – 2604
Veredelungszone 4178
Vereisung 2367
Vererzung 3936
Verfahren, diskontinuierliches – 4145
Verfahren, elektrisches – 4554
Verfahren, elektromagnetisches – 4572
Verfahren, tellurisches – 4553
Verfaulen 3526
Verfaulung 3526
verfestigt 2733
verfestigt 3422
Verfestigung 3422
Verfestigung 3423
Verfestigung mit Kalk 4807
Verfestigung mit Zement 4807
Verfirnung 2363
Verflüssigung 3084
Verflüssigung 4718
Verformung 1075
Verformung 4666
Verformungsmodul 4696
Verfüllung 2245
Vergesellschaftung 1892
vergesellschaftung, Gesteins– 0630
Vergesellschaftung, metamorphe – 0885
Verglasung 0708
Vergletscherung 2367
Vergleyung 2684
Vergrösserung, statische – 0131
Vergrusung 1991
Verjüngung 2572
Verjüngung 3959
Verkalkung 3458
Verkarstung 2063
Verkehrtspülung 4283
Verkieselung 0929
Verkieselung 3458
Verkittung 3433
Verknetung 2648
Verkohlung 3607
Verkreidung 3415
verkümmert 1978
Verlade-Erz 4139
verlagert 4637
Verlagerung 2678
Verlagerung 3930
Verlandung 2341
Verlandung 2513
Verlassen 3818
Vermessung, akustische – 4480
Vermodern 3526

Vermoderung 3526
Vermörtelung 4807
Verpressung 4802
Verrohrung 4275
Versanden 2990
versanden 3814
Versatz 4816
Versatz an einer Verwerfung 1550
Versatz, streichender – 1554
Versauerung 2738
Verschiebung 1521
Verschiebungsebene 0238
Verschiebungsrichtung, wahre – 1562
Verschiffungserz 4139
Verschlackung 4184
Verschlämmung 2758
Verschlicken 2990
Verschuppung 1571
Versenkungsdiagenese 3417
Versetzung 0531
versetzung, Schrauben– 0531
versetzung, Stufen– 0531
Versickerung 2768
Versickerung 3619
Versickerungsgeschwindigkeit 3620
Versickerungskapazität 3620
versiegeln 4800
Versiegelung 4800
Verstärker 4499
Verstärker, programmierter – 4499
Verstärkung, dynamische – 0131
Verstärkungsregelung, zeitabhängige – 4508
Versteinerung 3457
Verstümmelung 2648
Versturzkegel 2114
Versuchsprobenahme 4175
vertauben 3509
Verteiler 3798
verteilung, Lognormal– 4189
verteilung, Normal– 4189
Verteilungsgitter, rechteckiges – 4172
Verteilungsmuster 0403
Verteilungsnetz, rechteckiges – 4172
Verteilung, zonale – 3907
Vertikalerosion 2022
Vertorfung 3524
Vertrauensbereich 4191
Verunreinigung 3672
Verwachsung 0783
Verwachsung 4098
Verwachsung, graphische – 0847
Verwachsung, graphische – 4100
Verwachsung, myrmekitische – 4100
Verwachsung, runitische – 0847
Verwachsung, Schriftgranitische – 0847
Verwachsungsfläche 0536
Verwachsungsgrad 4112
Verwachsungsindex 4112
Verwachsungsverhältnisse 4113
Verwachsung, symplektische – 4105
Verwandlung 1794
Verwandschaft 0629
verwandt 0629
verwandt 3970
Verweildauer 0397
Verweildauer 3627
Verweilzeit 0397
Verwerfung 1521
Verwerfung 1522
Verwerfung 4078
Verwerfung, abgehobelte – 2549
Verwerfung, aktive – 1527
Verwerfung, antithetische – 1588
Verwerfung, diagonale – 1545
Verwerfung, ebene – 1535
Verwerfungen, gestaffelte – 1585
Verwerfungen, radiale – 1594
Verwerfungen, sich kreuzende – 1579
Verwerfungen, System sich kreuzender – 1584
Verwerfung, flache – 1549
Verwerfung, flachgründige – 1525
Verwerfung, gebogene – 1535
Verwerfung, geebnete – 2549
Verwerfung, geschlossene – 1534
Verwerfung, glatte – 1535
Verwerfung, inaktive – 1527
Verwerfung, klaffende – 1534
Verwerfung, kreisförmige – 0743
Verwerfung, lebende – 1527
Verwerfung mit Rotation 1567
Verwerfung, oberflächennahe – 1525
Verwerfung, rezente – 2552
Verwerfungsablenkung 1536
Verwerfungsabsturz 2548
Verwerfungsbreccie 1539
Verwerfung, schichtenparallele – 1547
Verwerfungsebene 1528
Verwerfung, sekundäre – 1580
Verwerfungsfläche 1528
Verwerfungsküste 2520
Verwerfungsnetz 1578
Verwerfungsschar 1584
Verwerfungssone 1532
Verwerfungsstufe 2548
Verwerfungssystem 1578
Verwerfungstal 2547
Verwerfung, steile – 1549
Verwerfungstektonik 1520
Verwerfung, streichende – 1544
Verwerfung, streichende – 1546
Verwerfung, synthetische – 1588
Verwerfung, tiefgreifende – 1525
Verwerfung, verzweigte – 1582
Verwerfung, wiederbelebte – 1527
Verwesung 2741
verwildert 2218
Verwitterbarkeit 4734
Verwitterung 1988
Verwitterung 2928
Verwitterung, biologische – 1999
Verwitterung, chemische – 1998
Verwitterung, chemische – 2017
Verwitterung, chemische – 2928
Verwitterung, kosmische – 0028
Verwitterung, mechanische – 1989
Verwitterung, mechanische – 2928
Verwitterung, physikalische – 1989
Verwitterungsauslese 2004
Verwitterungsboden 2009
Verwitterung, selektive – 2004
Verwitterungskorrektur 4430
Verwitterungskrume 2009
Verwitterungskruste 2010
Verwitterungsrinde 2010
Verwitterungsschutt 2005
Verwitterungsterrasse 2307
Verwitterungstrümmer 2005
Verwühlung 1885
Verzahnung, wechselseitige – 3118
Verzeichnung, S-förmige – 4404
Verzinsung 3898
Verzweigung 2210
Verzwillingung, polysynthetische – 0538
vesikular 1305
Vesuvtypusvulkan 1429
Vibrationskolbenlot 3156
Vibrationsmesser 0123
Vibroflotation 4810
Viehgangel 2060
vielkammerig 1870
Vierteln 4181
Vignettierungsfilter 4339
Virgation 1675
Viskosität 4671
Vitrian 3580
Vitrifizierung 0708
Vitrophyr 0823
vitrophyrisch 0823
vitroporphyrisch 0823
Vogelfussdelta 2495
volumenfaktor für Gas, Formations– 3861
volumenfaktor für Öl, Formations– 3860
Volumengewicht 2751
Volumen-Methode 3871
Vorauswertung 4604
Vorbeben 0147
Vorberg 2561
Vorbote 0150
Vordüne 2168
Vorfaltung 1667
Vorgabe 4818
Vorgang 2568
Vorgebirge 2446
Vorgebirge 2605
Vorgebirge mit zwei Nehrungsspitzen 2446
Vorhang 2135
vorherrschend 1816
Vorkonsolidierung 4703
Vorland 1109
Vorland 2447
Vorlandgletscher 2435
Vorlandvergletscherungstypus 2435
Vorläufer 0150
Vorläufer 0181
Vorphase 1157
Vorräte 3868
Vorräte, angedeutete – 4209
Vorräte, Ausserbilanz– 4211
Vorräte, erkannte – 4207
vorräte, Erz– 4208
Vorräte, mögliche – 3881
Vorräte, nicht bauwürdige – 4211
Vorräte, potentielle – 4215
Vorräte, prognostische – 4214
Vorräte, sichere – 3879
Vorräte, vermutete – 3882
Vorräte, vermutete – 4210
Vorräte, vorgerichtete – 4209
Vorratschätzung 3870
Vorratskategorien 3878
Vorrumpf 2593
Vorsenke 1109
Vorsetzschicht 3140
vorspalten 4823
Vorspannanker 4795
Vorstrand 2485
Vorstrand-Abhang 2480
Vortiefe 1109
V-Tal 2272
Vulcanotätigkeit 1160
Vulkan 1411
Vulkan, aktiver – 1153
Vulkan, alleinstehender – 1446
Vulkan, aufgesetzter – 1413
Vulkanbau mit Schuppenstruktur 1421
Vulkanbaute 1409
Vulkanbeben 0140
Vulkanberg 1409
Vulkanberge, radial gefurchter – 1474
Vulkan, derzeit ruhender – 1155
Vulkan, einachsiger – 1441
Vulkan, einfacher – 1441
Vulkanembryon 1222
Vulkan, erloschener – 1155
Vulkanform, negative – 1410
Vulkangebäude 1413
Vulkan, gemischter – 1429
Vulkangruppe 1450
Vulkangürtel 1447
Vulkan in derzeitigem Ruhezustand 1155
Vulkaninsel 1416
vulkanisch 1148
vulkanisch 3989
vulkanische Begleiterscheinungen 1180
Vulkanismus 1147
Vulkanismus, epigenetischer – 1152
Vulkanismus, orogener – 1151
Vulkanismus, subrezenter – 1150
Vulkanit 1238
Vulkankegel 1434
Vulkankette 1448
Vulkanmantelspalte 1468
Vulkanmassiv 1413
Vulkan, mehrachsiger – 1441
Vulkan mit Ringwall 1445
Vulkan mit zentralem Krater 1434
vulkanogenetisch 1148
vulkanoklastisch 3172
Vulkanologie 1146
Vulkan, polyzentrischer – 1441
Vulkanreihe 1448
Vulkan, rheuklastischer – 1429
Vulkanruine 1476
vulkan, Schlamm– 1395
Vulkanschlot 1221
Vulkan, schlummernder – 1155
Vulkanschotter 1324
Vulkanspalte, klaffende – 1234
Vulkans, Sockel des – 1414
Vulkan, subaktiver – 1153
Vulkans, Untergrund des – 1414
Vulkan, tätiger – 1153
Vulkantheorie 0049
Vulkan, untätiger – 1155
Vulkan, untermeerischer – 1416
Vulkanzone 1447

W

Waben 2001
Wachs 3723
Wachsglanz 0484
Wachstumshügel 0528
Wachstumssektor 0526
Wachstumsspirale 0529
Wackelstein 2163
Wadi 2587
Wahrscheinlichkeitskombination 3877
Walkerde 3218
Wallberg 2420
Walldüne 2171
Wallmoräne 2414
Walmtal 2545
walzen 4809
Walztextur 2649
Wand 0705
Wand 1529

Wände 4009
Wanderdüne 2175
Wandermoräne 2405
Wandersand 2156
Wandes, zurückfliehen eines – 2602
Wandes, zurückweichen eines – 2602
Wandes, zurückziehen eines – 2602
Wand, überhängende – 2472
Wanne 2579
Wärmefluss 0385
Wärmeflussdichte 0384
Wärmefluss, irdische – 0383
Wärmeinhalt 0389
Wärmeinhalt, System mit hohem – 0389
Wärmeinhalt, System mit niedrigem – 0389
Wärmekapazität 0386
Wärmekapazität, spezifische – 0386
Wärmeleitfähigkeit 0382
Wärmeproduktion, radioaktive – 0375
Wärmestrom 0385
Wärmestromdichte 0384
Wärmestrom, irdische – 0383
Wärmestroms, Region durchschnittlichen – 0391
Wärmestroms, Region erhöhten – 0391
Wärmetransport, konvektiver – 0381
Warve 3128
Warvengestein 3128
Waschberge 3600
Waschrinne 4158
Waschrinne 4659
Wasser, atmosphärisches – 3613
Wasseraufnahmefähigkeit 3625
Wasser, ausgepresstes – 3660
Wasseraustritt 4863
Wasserblüte 2843
wasser, Brack- 3676
Wasserdampffumarole, gasführende – 1389
Wasserdampffumarole, reine – 1389
Wasserdichtigkeit 4863
Wasserdom 1359
Wasserdruck, Versuch mit abnehmendem – 3711
Wasserdruck, Versuch mit konstantem – 3711
Wasserertrag 2249
Wasserfall 2203
Wasserfluten 3833
Wasserfontäne 1359
Wasser, fossiles – 3661
wasserfrei 0609
Wasserführung 2249
Wassergehalt 2769
Wassergehalt 3802
wassergehalt, Gesamt- 3598
Wassergehalt, hygroskopischer – 3598
wassergehalt, Rohkolhen– 3598
Wasser, juveniles – 1348
Wasser, juveniles – 3664
Wasserkörper 2828
Wasserkreislauf 3612
Wasserkuppel 1359
Wasser-Linie, meteorische – 0458
Wasser, magmatisches – 1348
Wassermasse 2828
Wassermenge 2249
Wasser, metamorphes – 0976
Wasser, meteorisches – 3614
Wässern 4810
Wasser, nutzbares – 2773
Wasserqualität 3674
Wasserriss 2026
wasser, Salz- 3676
Wassersand 3766
Wassersandstein 3766
Wassersättigung 4611
Wasserscheide 2312
Wasserscheide 2314
Wasserscheide 2511
Wasserscheide, durchgreifende – 2321
Wasserscheidekamm 2316
Wasserscheide, konsequente – 2317
Wasserscheide, sich langsam verschiebende – 2319
Wasserscheide, sich schnell verlegende – 2320
Wasserscheide, subsequente – 2318
Wasserscheide, verlegende – 2319
Wasserscheide, wandernde – 2319
Wasserscheide, zickzackförmige – 2321
Wasserschub 4737
Wasser, seichtes – 2809
wasserspannung, Boden– 2771
Wasserspannungskurve 2772
Wasserspende 2249
Wasserspiegel 3777
Wasserspiegel, freier – 3778
wasserspiegel, Ruhe– 3666
Wasserstandsmarke 3031
Wasser, steigendes – 2251
Wasserstoff-Index 4613
wasser, Süss- 3676
wasser, Tief– 2809
Wassertrieb 3824
Wasser, unterirdischer – 3649
Wasser, ursprüngliches – 3661
Wasser, vadoses – 3655
Wasserversorgung 2775
Wät 2505
Wattengebiet 2505
Wattrinne 2509
Wattsediment 2907
WDC 0082
Wechsel 1558
wechselkörnig 0820
Wechsellagerung 3116
Wechselreaktion 3956
Wechselschichtung, wellige – 3138
Wechselschlund 2092
Wechselstrom 4555
Wegaufnehmer 0121
weg, Extremal– 0209
Weidespur 1881
Weiher 2330
weit 1626
Welkepunkt 2773
welle, Bilatations– 0183
welle, Einzel– 2815
welle, Erdbeben– 0161
Welle, geführte – 0195
Welle, gekoppelte – 0193
welle, Gravitations– 0202
welle, Grenz– 0196
welle, Kanal– 0195
welle, Kern– 0180
welle, Kompressions– 0183
Welle, konvertierte – 0185
welle, Kopf– 0186
welle, Leck– 0195
welle, Leit– 0195
welle, Longitudinal– 0182
welle, Mantel– 0194
Wellenbasis 2442
Wellenfront 0549
Wellenfront 4470
Wellenfront, Normale zur – 0550
Wellengeschwindigkeit 0168
Wellenhöhe 2808
Wellenlänge 0162
Wellenlänge 2808
Wellenlänge 3002
Wellen-Langsamkeit 0168
Wellenleiter 0101
Wellenoberfläche 0549
Wellenperiode 2808
Wellenrippel 3012
Wellen, Wiederkehr– 0191
Wellenzahl 0162
welle, Oberflächen– 0188
welle, Oberflächenscherungs– 0189
Welle, P– 0182
welle, Quer– 0189
welle, Raum– 0179
Welle, S– 0184
Welle, seismische – 4481
Welle, sichtbare – 0192
welle, Streuungs– 0195
welle, Träger– 0178
welle, Transversal– 0184
welle, Verdichtungs– 0183
welle, Verdünnungs– 0183
Welle, zurückziehende – 2813
Welligkeit 4727
Wellungsebene 2591
Weltbeben 0138
Wendeachse 1627
Wendepunkt 1627
Wentworth-Skala 3163
Wert, aufgeschobener – 4220
werte, Extrem– 4186
Wert zum Tageskurs 3899
Wetter, matte – 3554
Wetter, schlagende – 3554
Wickelstruktur 3052
Widerlager 4845
Widerlager 4866
Widerstand 4556
Widerstand, scheinbarer – 4564
Widerstand, scheinbarer – 4609
Widerstandslog 4582
Widerstandsmessgeräte, fokussierende – 4586
Widerstand, spezifischer – 4556
Widerstand, spezifischer – 4607
Widerstandsverfahren 4561
Widmanstättische Figuren 0067
wiederaufgedeckt 2595
Wiesenmäander 2220
Wildfluss-ähnliche Struktur 0041
Wildflysch 2926
Wildschnee 2346
Windablation 2155
Winderosion 2154
Windgasse 2186
Windkanter 2162
Windkorrasion 2154
Windkuhle 2187
Windmulde 2187
Windrippel 3016
Windschattenufer 2336
Windschliff 2154
Windsee 2807
Windsichtung 2937
Windufer 2336
Windverfrachtung 2154
Winkeldiskordanz 1721
Winkelkreuz 4075
Winkel, kritischer – 0213
Wirbel 2269
Wirkung, abkühlende – 1271
Wirkungsgrad, seismischer – 0205
Wirtgestein 3919
Wirtkristall 0844
Wohnbaute 1880
Wohnort 1901
Wölbungsgebirge 2538
Wolke, blumenkohlförmige – 1362
Wollsackverwitterung 2003
Wühlgefüge 1885
Wulff'sches Netz 1015
Wurfschlacke 1297
Würgetextur 2649
wurmförmig 0774
Wurzel 0343
Wurzelboden 3515
Wurzelzone 1684
Wüste 2577
Wüstenlack 3405
Wüstenpflaster 2160
Wüstenpflaster 3405
Wüstenrose 3406

X

Xenoblast 0980
xenoblastisch 0980
Xenokristall 0663
Xenolith 0663
xenomorph 0752
xenothermal 3984
xenotopisch 3471
Xylit 3495
xylophag 1857

Y

Yardang 2164
Youngscher Modul 4695

Z

Zackenfirn 2365
Zackenoberfläche einer Aa-Lava 1260
Zähigkeit 4671
Zavaritsky-Diagramm 0699
Zeilenabtasten 4402
zeilling, Durchdringungs– 0537
Zeitabschnitt, dargestellter – 4503
Zeitbereich 0223
Zeitbereich 4570
Zeitrechnung, geologische – 1713
Zeitspanne 1781
Zelle, dreiaxiale – 4753
zelle, Elementar – 0510
Zellenstruktur 4089
Zelle, rhomboedrische – 0516
Zelle, triaxiale – 4753
zellig 0874
zellig 1305
Zellstruktur 4089
zellulär 3064
Zement 3433
Zementation 3433
Zementation, granulare – 3437
Zementationsfaktor 4606
Zementmilch 4803
Zementschlämme 4803
Zements, Lösung des – 3444
Zement, syntaxial– 3437
Zentralaufstellung 4419
Zentralberg 0054
Zentraleruption 1228
Zentralgranit 0729
Zentralherd 1202
zentralozeanischer Rücken, System – 1129
Zentrifugalmodell 4781
Zeolithisation 0932

Zeolithisierung 0932
zerbrochen 1494
Zerbröckelung 1991
Zerfall 1988
Zerfall 1990
Zerfall in Mineralkörner 1990
Zerfallsblei 0437
Zerfallsdauer 4734
Zerfallskonstante 0431
Zerfliessgrenze 2749
Zerkleinerung 4150
Zerkleinerung 4821
zerklüftet 1495
Zerklüftung 1495
Zerlegung 2928
Zerlegung, chemische – 2928
Zerlegung, mechanische – 2928
Zermalmung 1319
Zerrkluft 1503
Zerrspalte 1519
Zerschneidung 2019
Zerschneidung durch Erosionsrinnen, dichte – 2018
Zersetzung 1988
Zersetzung 2741
Zersetzung 2929
Zersetzung, aerobe – 2932
Zersetzung, anaerobe – 2932
Zerstörung 2929
Zerstreuung 3710
Zerstückelung 1990
Zertalung 2019
Zertrümmerung 4821
Zertrümmerungszone 4821
Zeugenberg 2561
Zickzackfalte 1651
Zickzackfalte 1661
Zinsfuss 3897
Zinsfusses, Methode des inneren – 3898
Zirkulation, normale – 4283
Zonarbau 0474
Zonarbau 0533
Zonarbau, invertierter – 0475
Zonarbau, normaler – 0475
Zone 0362
Zone 0519
Zone 1717
Zone, geflutete – 4620
Zone, infiltrierte – 4619
Zonenkarte, seismische– 0244
zone, Super– 1782
Zoobenthos 1852
Zoochlorellen 1845
Zooplankton 1800
Zooxanthellen 1845
Zubringerfluss 2207
Zuchtwahl, natürliche – 1966
Zucker– 3090
Zufluss 2207
Zufluss 2338
Zufuhrkanal 0723
Zufuhrkanal 3941
Zufuhrkanal 4011
Zufuhrspalte 0723
Zug 0104
zugedeckt 2595
Zugfestigkeit 4684
Zungenbecken 2629
zungenförmiger Körper 1752
Zungenrippel 3006
Zuordnungsmarken 4527
Zusammenballungstheorie 0022
Zusammenbruch des Herddaches 1188
Zusammenfluss 2208
Zusammenhalt der Aggregate 2726
Zusammensetzung, quantitative mineralogische – 0593
Zusammentreffen 4073
Zuschlagstoff 4872
Zuschüttung 2245
Zuschüttung 2341
Zustand, breiiger – 4713
Zustand, fester – 4713
Zustand, halbfester – 4713
Zustand, weicher – 4713
Zuwachs 2988
Zuwachsen 2341
Zwangsmäander 2221
zweiachsig 0553
Zweig 1965
Zweigestaltigkeit 1811
Zweigintrusion 1203
Zwickel-Poren 3688
Zwilling 0535
zwilling, Durchkreuzungs– 0537
zwilling, Ebenen– 0537
Zwillingsebene 0536
Zwillingsgesetz 0535
Zwillingskrater 1442
Zwillingsvulkan 1442
Zwischelage 3117
zwischengelagert 3117
Zwischenmittel 3506
Zwischenmoräne 2407
Zwischenraum 3683
Zwischensonde 3843
Zwischenwasser 3780
Zwischenzone, fossilleere – 1775
Zyklothem 3119
Zyklus, arider – 2567
zyklus, Erosions – 2566
Zyklus, fluviatiler – 2567
Zyklus, glazialer – 2567
zyklus, Karst– 2567
Zyklus, litoraler – 2567
Zyklus, mariner – 2567
Zyklus, orogener – 1105
Zyklus, periglazialer – 2567
zyklus, Sedimentations– 3119
Zyklus, vulkanischer – 1156
Zyst 1841

Español

A

abandono 3818
abanico 4631
abanico de lava 1287
abarrancamiento 2018
abertura 2240
abierto 1626
abigarrado 3462
abisal 0617
abisal 2847
abismo 0034
ablación 2013
ablación 2380
ablación 2395
abolladura 1418
abono, límite de – 4856
abono, sobre limíte de – 4856
abrasión 2154
abrasión 2938
abrasión eólica 2015
abrasión glacial 2015
abrasión glaciar 2395
abrasión marina 2016
abrasión marina, llanura de – 2536
absorción 0112
absorción 0406
absorción 4514
absorción, espectro de – 0491
absorción, índice de – 4314
absortancia 4314
acanaladura 1538
acantilado 2470
acantilado abandonado 2477
acantilado buzante 2475
acantilado, falso – 2476
acantilados, línea de – 2478
acarreado 2969
acarreo 2968
acarreo 2969
aceleración máxima 0120
acelerograma 0120
acelerómetro 0120
ACF, diagrama– 0989
achatamiento 0330
achicar 4811
acicular 0522
acicular 0755
acicular 3185
ácida 0601
acidificación 2738
acidificación 3800
ácido 3554
ácido humico 3525
acolchamiento 2762
acondrita 0072
acoplamiento 0273
acreción 1822
acreción 2988
acreción, hipótesis de la – 0022
acreción, teoría de la – 0048
acreción vertical 2513
acrozona 1781
actividad de tipo estromboliano 1159
actividad de tipo hawaiano 1158
actividad de tipo Monte Katmai 1165
actividad de tipo peleeano 1163
actividad de tipo pliniano 1161
actividad de tipo vulcaniano 1160
actividad efímera 1154
actividad efusiva 1213
actividad eruptiva, elemento de – 1209
actividad explosiva 1216
actividad explosiva, índice de – 1175
actividad fumarolica 1375
actividad piroclástica 1309
actividad sísmica 0242
actividad volcánica 1149
activo 0619
actualismo 0006
actualismo geológico 0008
actualismo paleontológico 1886
acuífero 3634
acuífero artesiano 3637
acuífero colgado 3635
acuífero confinado 3636
acumulación 0674
acumulación 2987
acumulación 3754
acumulación clasticolítica 1201
acumulación, zona de – 2380
acúmulado 0830
acumulativa, diagrama fase – 0700
acuñado 3104
acuñamiento progresivo 3509
acuñamiento tectónico 3509
adamantino 0484
adaptable 1980
adaptación 1980
adcumulado 0832
adherencia 2750
adherencia, punto de – 2750
adinola 0933
admisible 3902
adsorbente, complejo – 2755
adsorción 040
adularización 0932
aerobio 0424
aerobio 1906
aerolito 0071
afanita 0798
afanítica 0798
afanítica 0824
afídica 0798
afírica 0798
afloramiento 4231
aflorar 4231
afluente 2207
AFM, diagrama – 0698
AFM, diagrama– 0989
afótico 1905
agente mineralizador 3936
aglomerado 1344
aglomerado de chimenea 1248
aglomerado de lava 1269
aglutinante, poder – 3595
agmatita 1042
agónica 0284
agotamiento 3830
agotamiento, balanza de – 4222
agpaítico 0693
agradación 2245
agregado 2724
agregado 2725
agregado, cara de – 2724
agregado cristalino 0771
agua a disposición 3803
agua atmosférica 3613
agua, calidad del – 3674
agua caliente, sistema de – 0393
agua capilar 2767
agua colgada, superficie del – 3643
agua, contenido de – 3802
agua de compactación 3660
agua de fondo 3779
agua de formación 3662
agua del suelo 3652
agua dulce 3676
agua en crecida 2251
aguafacto 2445
agua fósil 3661
agua freática 2765
agua freática 3659
agua freática, superficie del – 3643
agua innata 3661
agua intermedia 3780
agua intersticial irreducible 3661
agua juvenil 1348
agua juvenil 3664
agua libre, nivel de – 3778
agua madre 3377
agua magmática 1348
agua marginal 3779
agua marginal, avance del – 3826
agua, masa de – 2828
agua metamórfica 0976
agua meteórica 3614
agua-nieve 2348
agua plutónica 3664
agua, porcentaje de – 3802
agua regenerada 3663
agua salada 3676
agua salobre 3676
aguas meteóricas, línea de – 0458
aguas profundas 2809
aguas someras 2809
agua subterránea 3649
agua subterránea 3658
agua subterránea confinada 3659
agua superficial 3649
agua tipo 2828
agua útil 2773
agua vadosa 3655
aguja 1424
aguja 2805
aguja glaciar 2618
agujero curvo 4293
aireación, zona de – 3651
aire viciado 3554
Airy 0345
Airy, fase – 0190
aislamiento 1981
A'KF, diagrama– 0989
albardón de lava 1295
albedo 4316
albitización 0932
albufera, de – 2905
alcalina, serie – 0692
alcalinidad 3677
alcalino 0694
alcalino-cálcica, serie – 0692
alcalis-calcio, índice – 0692
aleurita 3205
aleurolita 3205
algas calcareas 3359
algas, de – 3314
algas, estructura de 3358
alimentación 4140
alimentación, área de – 3670
alimentación, conducto de – 0723
alimentación inducida 3671
alimentador 0723
alineación 4460
alineación de crenulación 1612
alineación de intersección 1611
alios 2704
aliviadero 4868
almacén 3755
almacenaje, coeficiente de – 3709
almacenamiento 4225
almacenamiento, capacidad de – 3625
almacén homogéneo 3770
almacén, isotropía del – 3770
almacén, roca – 3679
alocromático 0483
alocrónico 1818
alóctono 0727
alóctono 1685
alóctono 1815
alóctono 2945
alóctono 3968
alógeno 2946
aloquímico 3271
alotigénico 0660
alotígeno 0660
alotipo 1949
alotriomórfico 0752
alpino, tipo – 0600
alpino, tipo – 1078
alquitrán 3743
alquitrán, capa de – 3781
alta mar 2840
alteración 3425
altéración 3962
alteración cósmica 0028
alteración magnética 1192
alteración superficial, costra de – 2010
altiplanización 2590
alto fondo 2247
altrapatestigos 4272
alud ardiente 1367
alud, camino del – 2362
alud, cono de – 2362
alud de ceniza ardiente 1369
alud de derrubios 1368
alud de fondo 2359
alud de hielo 2361
alud de nieve 2358
alud de placas de nieve 2360
alud en polvo de nieve 2359
alud mixto 1368
alunitización 0932
aluvión 2886
aluviones 2278
aluviones, cono de – 2007
alveola 2001
ambiental 2859
ambiente 1900
ambiente 2859
amfibolitización 0932
amígdala 0782
amigdalar 0872
amigdaloide 0872
amolar 2940
amontonamiento 4225
amortiguación 0127
amortiguación, cociente de – 0128
amortiguación dinámica 0127
amortiguación, factor de – 0128
amortiguación viscosa 0127
amortización, duración de – 3896
amortización, fondo de – 4219
amplificación, curva de – 0133
amplificación dinámica 0131
amplificación estática 0131
amplitud 0164
amplitud 1634
ampollada 2811
anaclinal 2288
anadiagénesis 3427
anaerobio 0424
anaerobio 1906
analcimización 0932
análisis continuo 4144
análisis elemental 3594
análisis espectral 0221
análisis granulométrico 4144
análisis immediato 3594
análisis mecánico 3160
análisis numérico 4777
análisis tridimensional 4107
anasismo 0187
anastomosado 2218
anastomosamiento 2501
anastomosamiento por capturas 2501
anatexia 1025
anatexia diferencial 1029
anatexia parcial 1029
anatexita 1026
anclaje 4792
anclaje a tracción, ensayo de – 4764
anclaje, barra de – 4794
anclaje, cabeza del – 4794
anclaje, cable de – 4794
anclaje, placa de – 4794

anclaje pretensado 4795
ancón 2450
anchieutectico 0634
anchimetamorfismo 0888
anchimetamorfismo 3420
ancho mínimo delante de los bancos 4201
ancho mínimo por explotar 4200
andesítica, línea – 0691
anelasticidad 0111
anfibío 1858
anfibolita 0934
anfiteatro morrénico 2418
angular 3189
ángulo crítico 0213
ángulo de buzamiento 1485
ángulo de reposo 3137
ángulo de reposo 4687
anhedral 0523
anhedral 0752
anhidrita 3381
anhidro 0609
anillo de lava 1284
anillo de petróleo 3776
anisotropía 0110
anisotropía 4681
anisotropía magnética 0303
anisótropo 4681
anisótropo, opticamente – 0561
ankerita carbonosa 3507
anomalía 4532
anomalía 4632
anomalía de aire libre 0368
anomalía de Bouguer 0368
anomalía de Bouguer modificada 0370
anomalía de la gravedad 0367
anomalía geotérmica 0390
anomalía isostática 0369
anomalía isostática local 0369
anomalía isostática regional 0369
anomalía magnética 0286
anorogénico 0726
anorogénico, período – 1106
antecedente 2290
anteclise 1084
antepaís 1109
anthodita 2150
anticlinal 1618
anticlinal doblemente buzante 1638
anticlinal residual 1697
anticlinal sinforme 1621
anticlinorio 1671
antiduna 3008
anti-epicentro 0154
antiferromagnetismo 0301
antiforma 1618
antracita 3561
anualidad 4218
aparición gradual 0174
aplanamiento 3182
aplanamiento lateral, zona de – 2604
aplanamiento parcial, nivel de – 2603
aplastamiento 0330
aplastamiento 1613
aplítica 0811
apófisis 0725
apogéo 1985
apogeo, zona de – 1773
apomagmático 3986
aporte, vía de – 3941
aporte, vía de – 4011
aprovechabilidad, tenor límite de – 4204
aprovechable, zona – 4178
aptitud del suelo 2660
aquirosoma 1039
arabesquítica 0827
arado, piso de – 2761
árbol de navidad 3796
arcilla 2714
arcilla 3212
arcilla bituminosa 3721
arcilla, capa de – 2734
arcilla con pedernal 2931
arcilla con suelos 3515
arcilla, contenido en – 4623
arcilla de decalcificación 2122
arcilla de falla 4729
arcilla de muro 3515
arcilla en bloques 2422
arcilla ferruginosa 3488
arcilla ferruginosa 3504
arcilla ferruginosa 4117
arcilla ferruginosa negra 4117
arcilla fluida 4719
arcilla, fracción – 2714
arcilla margosa 3223
arcilla oscura abisal 2923
arcilla plástica 3216
arcilla refractaria 3217
arcilla refractaria 3517
arcilla roja abisal 2923
arcilla sensitiva 4717
arcilla varvada 3128
arcilloso 3211
arco de islas 1135
arco de presión 2113
arco exterior no volcánico 1137
arco interior volcánico 1137
arco luminoso 1371
arcosa 3233
arculito 0768
archibentónico 1908
Archie, ecuaciones de – 4605
área probable 3887
área protegida 3845
área útil 4360
arena 2718
arena 3227
arena acuífera 3766
arena asfáltica 3768
arena, banco de – 2244
arena de coral 3311
arena de vidrio 3227
arena, ensuciarse de – 3814
arena eólica 2871
arena escoriácea 1323
arena, flujo de – 2973
arena gasífera 3766
arena húmeda 3767
arena, mar de – 1462
arena, mar de – 2183
arena margosa 3227
arena, mediano de la fracción – 2718
arena movediza 3085
arena movediza volcánica 1318
arena móvil 2156
arena musical 2088
arena, ola de – 3007
arena, pared de – 2973
arena petrolífera 3766
arenas movedizas 4719
arena, tempestad de – 2156
arena verde 3231
arena vesicular 3063
arena voladora 3085
arena volcánica 1322
arenisca 3228
arenisca acuífera 3766
arenisca cuarcítica 3230
arenisca feldespática 3232
arenisca gasífera 3766
arenisca petrolífera 3766
areniscas rojas 4129
arenita 3225
arenoso 3225
argilita 3211
argilita 3219
argón atmosférico 0450
argón, exceso de – 0450
argón radiogénico 0450
aridos 4872
arista 0518
arista 2617
aristas angudas, de – 3190
arkosa 3233
armazón 1861
armónico 0167
aromáticos 3717
arqueomagnetismo 0294
arquetipo 1979
Arquímedes, presión de – 4740
arrancamiento 2396
arranque 3431
arranque interrumpido 0623
arranque por realce 0623
arranque por rebaje 0623
arrastre 1537
arrecifal 3334
arrecifal, barrera – 3338
arrecifal, complejo – 3337
arrecifal, corteza – 3341
arrecifal, flanco – 3353
arrecifal, frente – 3351
arrecifal, inter– 3357
arrecifal, llanura – 3343
arrecifal, mancha – 3347
arrecifal, orla – 3338
arrecifal, pared – 3336
arrecife 2443
arrecife 2804
arrecife 3333
arrecife de algas 3360
arrecife regresivo 3327
arrecife transgresivo 3328
arriñonado 4095
arroyo 0057
arroyo 2194
arterita 1045
artesa 2273
artesa, extremo de la – 2619
ascenso del tapón 1247
asenderado 2060
asentamiento 2991
asentamiento 4704
asentamiento 4833
asentamiento desigual 4704
asentamiento diferencial 4704
asentamiento diferencial 4833
asentamiento no homogéneo 4704
asentamiento, velocidad de – 2991
asfaltita 3737
asfalto 3734
asfalto, lago de – 3753
asfalto natural 3734
asientos, celula de medición de – 4768
asimilación 0681
asimilación marginal 0683
asísmica, región – 0253
asociación 1805
asociación 1892
asociación 3179
asociación de rocas 0630
asociación de suelos 2673
asociación mineralógica 0586
astenolito 1122
astenosfera 0090
asterismo 0489
ataguía 4869
ataxita 1243
ataxito 0069
atectita 1032
atenuación 0219
atenuación 4475
atexita 1032
atlántica, serie – 0690
atmósfera 0084
atolón 3340
atrio 1461
Atterberg, límite de – 4714
augen, textura en – 0952
aulacogen 1085
aureola 4629
aureola de contacto 0707
aureola metamórfica – 0899
autigénesis 3434
autigénica 4633
autigénico 3434
autígeno 2946
autocorrelación 4195
autocorrelación 4450
autóctono 0727
autóctono 1685
autóctono 1813
autóctono 2945
autóctono 3968
autoecología 1888
autohidratación 3957
autointrusión 0621
autoinyección 0621
autometamorfismo 0659
autometamorfosis 0659
autometasomatismo 0659
autometasomatismo 3957
automórfica granular 0808
autoneumatolisis 0659
autopulverización 4147
avalancha 2252
avalancha 2972
avalancha de arena 2973
avalancha de nieve 2358
avalancha de piedras 2047
avión 4302
avulsión 2231
axial, superficie – 1631
axial, traza de la superficie – 1631
axiolítico 0767
axiolito 0767
azabache 3557
azimut 4228
azimut de la estación 0160
azimut epicentral 0155
azóico 1738

B

bacilito 0762
bahamita 3303
bahía 2451
bajada 2581
bajada 3816
bajamar 2818
bajamar media 2820
bajío 2506
bajío 2803
bajo 2803
bajocorrimiento 1574
bajo fondo 2506
balance de agua subterránea 3633
balance material 3865
balasto 4844
bancal 2516
bancal de lava 1285
banco 2244
banco 2297
banco 2802
banco carbonatado 3323
banda 2705
banda 3122
banda 3506
banda espectral 4306
banda estéril 3506
banda pedregosa 2654
banda polvorienta 2379
banda térmica 4308
banda terrosa 2654
bandeado 3122
bandeado críptico 0865
bandeado fluidal 0868

bandeado rítmico 0863
bandera 2135
bañado 2457
barbuja, marca de – 3062
barlovento 2963
barra 2201
barra 2459
barra 2823
barra arqueada 2233
barra cuadrada 4269
barra de accionamiento 4269
barranca 1474
barranco 2026
barras digitadas 2897
barredor 4403
barredor multiespectral 4403
barrena 4236
barrena cortatestigos 4271
barrera 1916
barrera geográfica 1916
barrido circular 4402
barrido lineal 4402
barridor, equipo de – 4403
barril 3892
barrizal 1398
barro 2920
barro 3210
barro 3307
barro carbonatado 3295
barro, colada de – 2051
barro, cráter de – 1398
barro de azufre 1399
barro, flujo de – 2974
barro siliceo 3367
barro sulforoso, columna de – 1400
barro sulfuroso, colada de – 1399
basalto acalino 0695
basalto de cono 1258
basalto de meseta 1257
basalto, globo de – 1256
basculamiento 0124
basculamiento 4833
base 4841
base 4842
base aérea 4350
base de las olas, nivel de – 2442
base del fotograma 4371
base estereoscópica 4363
base general, nivel de – 2025
base gravelosa 2651
base kárstico, nivel de – 2966
base, línea de – 4621
base parcial, nivel de – 2025
bases, saturación de – 2756
base temporal, nivel de – 2025
básica 0603
básico, frente – 1066
basificación 0684
basita 0603
basureras 4831
basureros 4831
batanero, tierra de – 3218
batial 2846
batidero 2812
batido 0178
batiente 2812
batilimnion 2334
batipelágico 1910
batipelágico 2845
batisismo 0158
batolito 0729
bauxita 4124
B ⊥ B', tectonita– 1003
B ∧ B', tectonita- 1003
Becke, línea de – 0559
belonito 0761
Benioff, zona de – 1134
bentónico 1907
bentónico 2844
bentonita 3218
bentos 1907
bentos errante 1918
berma 2488
berma 4860
biaxial 0553
bifurcación 2210
bifurcación 4073
biocenosis 1896
bioclástica 3168
bioclástica 3273
bioclasto 3168
bioesfera 1788
bioespararenita 3287
bioesparita 3287
bioesparrudita 3287
bioestratigrafía 1710
bioestratigráfica, unidad – 1770
bioestratonomía 1791
biofacies 2861
biogénesis 1792
biogenético 1792
biogeoquímica 0395
bioherma 3331
biohorizonte 1771
biolitito 3269
biolito 3269
bioma 1895
biomicrita 3288
biomicroesparita 3289
biomicrudita 3289
biopelmicrita 3291
biopelsparita 3290
biostroma 3331
biótico 1903
biotitización 0932
biotopo 1902
bioturbación 1885
bioturbación 3067
biozona 1769
biozona, super– 1782
bireflexión 0575
birefringente 0560
bisectriz 0556
bisectriz aguda 0556
bisectriz obtusa 0556
bismalito 0734
bitumen 3719
bituminización 0426
bivalvo 1859
blasto– 0971
blastomilonita 1541
bloque 1087
bloque 3243
bloque caido 2110
bloque de sal 1695
bloque diagrama 4258
bloque errático 2422
bloque fallado 1531
bloqueo 2340
bloque rocoso 4730
bloque rocoso, unidad de – 4730
bloques angulares, estructura en – 2729
bloques, área fallada en – 1586
bloques subangulares, estructura en – 2729
bloque volcánico 1327
boca 2503
bolilla 4150
bolita de fango de azufre 1321
bolsa 3773
bolsada 4032
bolsada de crioturbación 2649
bolsa filoniana 4050
bomba 1328
bomba de boñiga de vaca 1329
bomba de explosión 1329
bomba de olivina 1329
bomba de peridotita 1329
bomba de pumita 1329
bomba de rotación 1329
bomba en forma de cinta 1329
bomba en forma de corteza de pan 1329
bomba en forma de gota 1329
bomba en forma de tortilla 1329
bomba escoriácea 1329
bomba esférica 1329
bomba fusiforme 1329
bomba non núcleo 1329
bomba piriforme 1329
bomba torneada 1329
bomba unipolar 1329
bombeo, sistema de – 4870
boquilla 4850
borboteo 1358
borde básico 0707
borde dentado 3414
borde, zona de – 1070
borroso 0802
bostonítica 0839
botroidal 0777
botroidal 4094
Bouguer, anomalía de – 0368
Bouguer, corrección de – 0359
Bouguer, lámina de – 0359
bóveda 4851
bóveda 0116
bóveda de presión 2113
boxwork 2118
bradisismo 0096
bradisismo 1194
braditelia 1977
braquianticlinal 1640
braquisinclinal 1640
Brasileño, ensayo – 4761
brecha 1471
brecha 3258
brecha arrecifal 3355
brecha calcárea 3279
brecha de colapsamiento 3263
brecha de contacto 0710
brecha de desecación 3048
brecha de desmoronamiento 1249
brecha de explosión 1346
brecha de falla 1539
brecha de flujo 1268
brecha de flujo compactada 1268
brecha de fricción 1539
brecha de inyección 0711
brecha de roca arrecifal 3342
brecha de trituración 1539
brecha intraformacional 3264
brecha intrusiva 0710
brecha sedimentaria 3258
brecha volcánica 1345
brecha volcánica de fricción 1248
brechificacion 2933
brillante 0493
brillo 0490
broca 4236
brújula geológica 4227
B, tectonita– 1003
bucles, hipótesis de los – 1118
budinaje 1609
budinaje imperfecto 1609
bufadero 2109
bufador 2109
Bullard, método de – 0363
bulón 4793
bulonaje 4793
burbuja de calcita 2146
buzamiento 1484
buzamiento 4017
buzamiento aparente 4250
buzamiento, atenuación del – 1487
buzamiento, azimut del – 1490
buzamiento, complemento del – 1548
buzamiento, reconocimiento del – 4295

C

cabalgada, masa – 1569
cabalgamiento 1568
cabalgamiento, distancia de – 1568
cabalgamiento, manto de – 1575
cabalgamiento, plano de – 1568
cabeceo 2058
cabezo 2608
cabo 2446
cabrilla 2378
cabujón 0500
cadenas oceánicas, sistema axial de – 1129
cadena volcánica 1448
caída 3816
caída, curva de – 3816
caída de agua 2203
caída de pesos 4409
caja corte 4754
cala 2452
calcarenita 3276
calcarenítica 3277
calcáreo 2736
calcáreo, no – 2736
calcibrecha 3279
cálcica, serie – 0692
calcificación 3458
calcilutita 3274
calcimicrita 3282
calcinar 4151
calcirudita 3278
calcisiltita 3275
calcita flotante 2144
calco-alcalina, serie – 0692
calco-alcalino 0694
caldera 0036
caldera 1456
caldera, anillo de – 1461
caldera criptovolcánica 1183
caldera de desplome 1459
caldera de erosión 1473
caldera de explosión 1458
caldera en cima 1460
caldera freática 1458
caldera, hundimiento en – 1186
calderas, complejo de – 1457
caldera sencilla 1457
caldera subterránea, hundimiento en – 1186
caldero 1463
calibrador 4603
calicata 4235
California, índice de penetración – 4708
caliza 3305
caliza arcillosa 3221
caliza arenosa 3317
caliza arrecifal 3310
caliza calcarenítica 3277
caliza de corales 3310
caliza dolomítica 3316
caliza encrinítica 3313
caliza litográfica 3306
caliza mudstone 3296
caliza peletífera 3307
calor, contenido de – 0389
calor de la tierra 0374
calórico alto, sistema – 0389
calórico bajo, sistema – 0389
calorífico alto, sistema – 0389
calorífico bajo, sistema – 0389
calorífico, poder – 3604
calor, masa transferente de – 0381
calor radioactivo 0375
calor radiogénico 0375
calor terrestre 0374
calve 4851
cámara 4849
cámara aérea 4329

cámara agrimensora	4329
cámara cartográfica	4329
cámara de horizonte	4333
cámara de presión, ensayo de –	4762
cámara magmática	1202
cámara magmática en forma de dique	1204
cámara métrica	4331
cámara multibanda	4330
cámara multilente	4330
cámara, situación de la –	4348
cambio del curso	2213
cambio iónico, capacidad de –	0420
campo de barro	1390
campo de fumarolas	1380
campo de lava	1278
campo de piedras	2057
campo de piedras	2656
campo de solfataras	1392
campo filionano	4061
campo gasífero	3773
campo minero	3909
campo petrolífero	3773
canal	0042
canal	2191
canal	3022
canal	4412
canal abandonado	2212
canal de alud	2362
canal de creciente	2510
canal de desague anteglaciar	2393
canal de hundimiento	1290
canal de marea	2509
canal de vaciante	2510
canal erosivo	3510
canales de solución	2087
canales entrelazados	2243
canal meandriforme	2217
canal, relleno de –	3022
canal sinuoso	2217
canal transversal	2486
candal térmico –	0385
canería	3798
cantera	4874
canto	3242
canto armado	3247
canto blando	3246
canto de tres aristas	2162
canto de una arista	2162
canto errático	2422
caño	2517
cañón submarino	2795
caolín	3217
caolinización	0926
capa	1750
capa	3097
capa	3123
capa activa	2647
capa cementada	2734
capacidad	2958
capacidad calorífica	0386
capacidad de campo	2774
capacidad de carga	4707
capa de laboreo	2699
capa de lava	1277
capa de rodadura	4842
capa de suelo	2688
capa de superficie	4842
capa digitada	3506
capa enduricida	2733
capa esférica	0363
capa freática	2765
capa freática, clase de –	2766
capa gravelosa	2651
capa guía	1751
capa guía	3500
capa intrusiva	0740
capa roja	2879
capas acuíferas subterráneas superpuestas	3641
capas basculadas	1482
capas dislocadas	1523
capa semipermeable	3639
capas inclinadas	1482
capas levantadas	1481
capicidad calorífica específico	0386
capilar	0774
capilar, agua –	2767
capilar, franja –	2767
capilar, franja –	3657
capilar, zona –	3657
captura, codo de –	2326
captura, desfiladero de –	2326
captura fluvial	2322
captura por erosión lateral	2322
captura, teoría de la –	0047
capuchón	1696
cara	0517
característica de amplitud	4448
característica de fase	4448
cara de sotavento	2185
carbagilita	3505
carbohidrato	3723
carbón	3496
carbón arrastrado, fragmentos de –	3512
carbonatización	0932
carbonatización	2000
carbonato de calcio, contenido en –	2735
carbonatos, profundidad de precipitación de –	0423
carbón azabacheado	3557
carbón bituminoso	3559
carbón boghead	3563
carbón, calidad del –	3551
carbón, capa de –	3499
carbón, capa de –	3499
carbón, clase de –	3552
carbón de coque	3608
carbón de madera	3607
carbón de vapor	3559
carbón digitado	3506
cárbon, equivalente standard del –	3606
carbón, estructura del –	3565
carbón, finos de –	3513
carbón flotado	3512
carbón hojoso	3556
carbón húmico	3497
carbonificación	3519
carbonificación, gradiente de –	3522
carbonización	3607
carbonización natural	3519
carbón, lentejón de –	3499
carbón magro	3560
carbón, manto de –	3499
carbón mineral	3496
carbón, muro del –	3515
carbono-14, datación –	0451
carbono fijo	3601
carbono-nitrógeno, relación –	0425
carbono, proporción del –	3601
carbono, relación del –	3601
carbonos, nombre de	3727
carbono y volátiles, proporción de –	3601
carbón, petrografía de –	3565
carbón piciforme	3557
carbón piedra	3496
carbón, rango del –	3552
carbón sapropélico	3497
carbón, tipo de –	3552
cárcava	2026
carga	2950
carga	4140
carga	4816
carga, celula de –	4770
carga de anclaje, celula de –	4772
carga de corriente	2951
carga de fondo	4817
carga de saltación	2954
carga de sedimento	2951
carga de tracción	2952
carga, diferencia de –	4735
carga en solución	2956
carga en suspensión	2955
carga hidráulica	4735
carga sedimentaria	2957
carst	2064
carst, formación de –	2063
cárstico, paisaje –	2064
carstificación	2063
cartografía	4239
cartografía de suelos	2670
cartografía radioactiva	4545
cascada	2203
cascada de lava	1287
casquete esférico	0363
casquete polar	0040
cataclasis	0914
cataclasita	0914
cataclástica, roca –	0914
cataclinal	2288
catagénesis	1977
catarata	2204
catarata de hielo	2377
catasismo	0187
catastrofismo	0005
catazona	-0993
categoría	0685
categoría	1927
catena	2674
cationes cambiables	2755
cationes, capacidad de intercambio de –	4623
caudal	2249
caudal térmico	0385
cáustica	0216
cáustobiolito	0427
caverna	2095
caverna	4849
caverna de lava	1291
caverna marina	2472
caverna respirante	2108
caverna según fisura	2096
caverna según fractura	2096
caverna según la estratificación	2096
cavernosa	1305
cavernoso	3064
cavidad	2095
cavidad	4819
cavidad ascendente	2106
cavidades, relleno de –	3940
cavidad miarolítica	0873
cavitación	2120
cavitación	4512
cayo	2443
CBR	4708
CD	0454
cebollón	2609
celdilla cristalina	0510
célula cristalina	0510
celular	3064
celula triaxial	4753
cementación	3433
cementación, factor de –	4606
cementado	2733
cemento	3433
cemento de tipo drusa	3438
cemento en bloque	3439
cemento granular	3437
cemento syntaxial	3437
ceniza	3599
cenizas brutas	3600
cenizas, contenido en –	3599
cenizas del lavado	3600
cenizas en bocamina	3600
cenizas intrínsecas	3600
cenizas, lluvia de –	1363
cenizas secundarias	3600
ceniza volcánica	1317
cenote	2105
cenotipalo	0636
cenotipo	0636
cenozona	1774
ceolitización	0932
cera	3723
cera montaña	3736
céreo	0484
cerrado	1626
cerro testigo	2561
cesped	2698
césped almohadillado	2643
ciclo árido	2567
ciclo de erosión	2566
ciclo de erosión glacial	2567
ciclo de erosión litoral	2567
ciclo de erosión periglacial	2567
ciclo de sedimentación	3119
ciclo fluvial	2567
ciclo geoquímico	0396
ciclo hidrológico	3612
ciclo kárstico	2567
ciclo orogénico	1105
ciclo petrogenético	0582
ciclotema	3119
ciclo volcánico	1156
ciencias de la tierra	0002
cieno	3527
cierre	3762
cierre por falla	3763
cierre por pendiente	3763
cierre vertical	3762
cima	0705
cimentación, coeficiente de –	0258
cimientación	4835
cimientaciones, ingeniería de –	4834
cimientación, estrato de –	4836
cimientación, losa de	4837
cimientación, nivel de –	4836
cimientación, solera de –	4835
cimientación superficial	4835
cimientación superficial	4837
cimientación, zapata de –	4835
cinederivómetro	4352
cinemático tardío	1079
cintura	1020
cinturón	3908
cinturón orogénico	1093
cinturón volcánico	1447
CIPW norm	0686
circalitoral	2839
circo compuesto	2613
circo de cabecera de valle	2611
circo en escalera	2614
circo escalonado	2614
circo glaciar	2610
circo glaciárico	2430
circo, nivel de –	2610
circo, nivel de –	2615
circos, nivel de fondos de –	2610
circos, nivel de fondos de –	2615
circos, plataforma de –	2613
circulación directa	4283
circulación inversa	4283
circulación, perdida de –	4284
circulación termohalina	2829
círculo de piedras	2653
círculo-pi	1020
cirras recortadas	2616
cizalla	4677
cizalla, coeficiente de –	4699
cizalla, depósito de zona –	4060
cizalladura	4677
cizalladura, coeficiente de –	4699
cizalladura, módulo de –	4699
cizallamiento	0105
cizallamiento, zona de –	1533
cizalla, módulo de –	4699

cladogénesis 1962
clareno 3580
claridad 2677
Clarke, número de – 0399
clase 0685
clase 1929
clase cristalina 0512
clasificación 0494
clasificación 1921
clasificación 3200
clase textural 2713
clasificación bioestratigráfica 1710
clasificación, coeficiente de – 3201
clasificación cronoestratigráfica 1712
clasificación de reservas 3878
clasificación de suelos 2661
clasificación de terrenos 4785
clasificación de tierra 2660
clasificación litoestratigráfica 1709
clasificación natural 1921
clasificado 3202
clasificado, bien – 3202
clasificador de rastrillo 4156
clasto 3166
clasto calcáreo 3167
clasto calcáreo 3270
clasto dolomítico 3167
clasto dolomítico 3270
clavalito 0763
clinómetro 0124
clinómetro 4227
clivaje 1596
clivaje de crenulación 1600
cloritización 0932
cluse muerta 2327
cluse viva 2327
coagulación 0413
coagulación 3932
cobertura 3764
cobertura 4407
cobertura múltiple 4407
cobijadura 1568
cocción retrógrada 1174
coccolito 1840
cockpits, paisaje de – 2080
cocolito 1840
coda 0175
coercitividad 0307
coespecífico 1943
coetáneo 1730
coevo 1730
cohesión 4693
cohesión aparente 4693
colada 1275
colada 1277
colada de bloques 2057
colada de lodo 2051
colada de piedras 2057
colada estalagmitica 2128
colada pedregosa 2057
cola de presión 0958
colisión continental 1131
colmatación 2341
colmatación 2513
colmatación por arcilla 2122
colmatación por vegetación 2341
coloforme 3093
colonia 1806
color 0546
color compuesto, falso – 4346
colores, tabla de – 2676
color, índice de – 0688
columna 2138
columna eruptiva 1361
columna estratigráfica 4255
columna magmática 1226
columna perforadora 4270
columnar 0772
columnar 0755
columnar 2730
columnata volcánica 1301
coluvión 2868
comagmático 0629
combinación 0521
combustible, proporción – 3601
combustión en situ 3838
comensalismo 1899
compactación 3428
compactación 4679
compactación 4808
compactación con ayuda de agua 4810
compactación diferencial 3428
compactación mecánica 3429
compactación por impacto 4809
compactación por rodillo 4809
compactación química 3432
compactación vadosa 3416
compartimiento 0362
compensación isostática 0339
compensación isostática de masas 0342
compensación isostática de presiones 0344
compensación local 0340
compensación, nivel de – 0345
compensación, profundidad de – 0345
compensación regional 0340
compensadora, masa – 0341
competencia 2958
competente 1624
complejamiento 0408
complejidad 0275
complejo 1756
complejo 1928
complejo de suelos 2673
complejo intrusivo 0620
complejo metaloorgánico 0410
complejo metamórfico 0885
complejo organometálico 0410
complementario 0669
complemento de buzamiento 4018
componente accesorio 0594
componente auxiliare 0594
componente opaco 0610
componente principal 0594
componente subordinado 0594
componente volátil 0650
componente volátil 1172
compresibilidad, coeficiente de – 4698
compresión 0187
comprobada, área – 3889
comunidad 1890
comunidad, climax de – 1891
conca 1459
concentración 3941
concentración aluvial 3943
concentración gravitativa 3948
concentración por gravedad 4153
concentración residual 3943
concentrado 4141
concentradora, cabezas de la – 4140
concentrado residual 2967
concordancia estratigráfica 1719
concordante 0717
concordante 0719
concordante 1719
Concordia-Discordia, diagrama – 0435
concreción 2124
concreción 2732
concreción 3472
concreción 4093
concrecionario 3472
concreción dolomítica 3507
concreciones férricas 4116
condensación 0412
condensación de núcleos 1360
condensación retrógrada 3841
condensado 3713
condominante 1816
condrita 0072
condrula 0072
conducción 0381
conducción calorífica 0381
conducción de calor 0381
conducción vertical 4274
conductancia 4556
conductividad hidráulica 3700
conductividad térmica 0382
conducto 1882
conducto 3798
conducto volcánico 1221
conducto volcánico en forma de cilindro 1224
conducto volcánico en forma de fisura 1225
conducto volcánico sin raíz 1223
confianza, intervalo de – 4191
confluencia 2208
confluencia a nivel 2208
confluencia con desnivel 2208
confluencia diferida 2209
confluencia en horqueta 2209
confluencia, escalón de – 2624
conforme 0719
congelación de suelos 4807
congelación, dilatación por – 2401
congelación, presión de – 2401
congenérico 1943
congesta 2364
conglomerado 3251
conglomerado basal 3114
conglomerado de bloques 3254
conglomerado intraformacional 3257
conglomerado volcánico 1343
conglomerático 3251
cónico 4080
conjunto 1805
cono 0059
cono 1435
cono adventicio 1437
cono aluvial 2007
cono central 1443
cono compuesto 1429
cono de barro 1396
cono de ceniza 1427
cono de deyección 2007
cono de escorias 1427
cono de lava 1427
cono de salpicaduras de lava 1298
cono de transición 2423
conodonto 1860
cono, en forma de – 3826
cono lateral 1437
conolito 0732
cono litoral 1300
cono parasítico 1437
cono rocoso 2585
cono tobáceo 1428
cono volcánico 1434
Conrad, discontinudad de – 0086
consanguíneo 0629
consanguíneo 3970
consanguinidad 0629
consecuente 2284
consecuente, no – 2287
consistencia 2746
consistencia 4713
consistencia, índice de – 4716
consistencia, límite de – 4714
consolidación 3422
consolidación 4703
consolidado 3422
construcción anti-sísmica 0260
construcción, código de – 0256
contacto ajustado 3449
contacto por sutura 3449
contacto recto 3449
contacto, zona de – 0707
contacto, zona de – 0899
contador de puntos 0612
contaminación 3672
contaminación 4652
contemporáneo 1729
contemporáneo 1818
contemporáneo, no – 1818
continental 2865
continente lunar 0050
continuidad 4726
contorno freático 3644
contorsión 3051
contorsión intraformacional 3051
contracción, hipótesis de la – 1114
contrafuerte 2606
contrafuertes 4797
contrapendiente 2626
contraste 4343
contraste 4628
contraste del motivo 4343
contraste del tema 4343
control de tierra 4323
control fotogramétrico 4324
control, punto de – 4325
convección calorífica 0381
convección de calor 0381
convección del manto 1124
convergencia 1725
convergencia 1973
convergencia 2787
convergencia metamórfica 0913
convolución 3052
convolución 4452
coordinación, número de – 0408
copelación 4184
coprolito 1863
coque 3608
coque natural 3564
coquina 3308
coquinita 3308
coquización, poder de – 3596
coquizante, poder – 3596
coral ahermatípico 3335
coral hermatípico 3335
coraliforme 1821
corazón 1628
cordillera 1103
cordón en V 2465
cordón libre, isla de – 2468
cordón litoral 2459
cordón litoral 2461
corneana 0940
cornisa 2472
cornisa de nieve 2356
cornubianita 0940
corona 0786
corona 1020
corona, eje de la – 1020
corona, polo de la – 1020
coronita 0786
corrasión 2014
corrasión 2938
corrección de aire libre 0358
Corrección de Bouguer 0359
corrección de la elipticidad 0098
corrección de latitud 0355
corrección de la zona de alteración 4430
corrección del nivel de referencia 4429
corrección de terreno 0360
corrección dinámica 4432
corrección estática 4431
corrección isostática 0361
corrección luni-solar 0365
corrección por altitud 0357

corrección topográfica 0356
correción de estación 0160
corredor 2486
correlación 1716
correlación 4459
correlacionador óptico 4388
correlación cruzada 4450
correlograma 4195
corrido, bloque – 1569
corriente abajo 2964
corriente alóctono 2259
corriente alterna 4555
corriente arribo 2963
corriente conmutada 4555
corriente continua 4555
corriente costera 2825
corriente de barro volcánico 1352
corriente de barro volcánico 1355
corriente de ceniza incandescente 1369
corriente de densidad 2979
corriente de marea 2821
corriente de turbidez 2981
corriente de turbidez constante 2981
corriente de turbidez, extremo de una – 2982
corriente de turbidez, frente de – 2982
corriente eléctrica, densidad de – 4556
corriente eléctrica, intensidad de – 4556
corriente, línea de – 2268
corriente, línea de – 2372
corriente, línea de – 3708
corriente marina 2785
corriente principal 2206
corriente superficial 2785
corriente suspendida 2981
corriente telúrica 4552
corriente torrencial 2252
corriente turbulenta 2264
corriente, velocidad de – 2260
corrimiento 1568
corrimiento arrastre 1575
corrimiento, distancia de – 1568
corrimiento, plano de – 1568
corrimientos, plano basal de los – 1572
corrosión 0642
corrosión 2017
corrosión 2942
corrosión, figura de – 0530
corrosión, zona de – 0786
corrugación 3051
corrugación intraformacional 3051
corso alto 2195
corso bajo 2195
corso medio 2195
corso superior 2195
corte compuesto 4253
corte continuo 4408
corte de sondeo 4256
corte, ensayo de – 4759
corte generalizado 4254
corte geológico 4250
corte ideal 4254
corte sísmico 4458
cortes seriados 4251
corte tipo 4257
corteza 3402
corteza de fusión 0073
corteza, espesor de la – 0347
corteza rígida 0348
corteza terrestre 0086
cosmogonía 0003
costa 2437
costa abierta 2531
costa alzada 2524
costa compuesta 2527
costa contrapuesta 2527
costa de abrasión 2528
costa de acumulación 2528
costa de alzamiento 2524
costa de bahías 2533
costa de coladas de lava 1280
costa de elevación 2524
costa de emersión 2525
costa de erosión marina 2528
costa de escarpe de falla 2520
costa de falla 2520
costa de flexura 2519
costa de hunimiento 2524
costa de inmersión 2525
costa de levantamiento 2524
costa de línea de falla 2520
costa de tipo atlántico 2522
costa de tipo dálmata 2521
costa de tipo mixto 2526
costa de tipo pacífico 2521
costa emergida 2525
costa epigénica 2527
costa estable 2526
costa hundida 2524
costa indentada 2532
costa levantada 2524
costa, línea de – 2437
costa lobulada 2533
costa longitudinal 2521
costa oblicua 2522
costa partida 2532
costa polycíclica 2530
costa sumergida 2525
costa transversal 2522
costra 1273
costra 2145
costra 2759
costra 3126
costra de hielo 2352
costra de lodo 4618
cotipo 1951
covariograma 4196
covertura residual superficial 2009
cráter 0054
cráter 1412
cráter 4819
cráter adventicio 1439
cráter, borde de – 1451
cráter de barro 1398
cráter de bocas múltiples 1445
cráter de cumbre 1436
cráter de explosión 1431
cráter en actividad 1211
cráter en herradura 1471
cráter en hoya 1455
cráteres, hilera de – 1449
cráter fracturado 1471
cráter lateral 1439
cráter maclado 1442
cráter meteorítico 0027
cráter meteorítico 0075
cráter meteorítico secundario 0027
cráter parasítico 1439
cráter, pared de – 1453
cráter subterminal 1438
cratón 1080
crecimiento 3455
crecimiento, curva de – 0479
crecimiento, espiral de – 0529
crecimiento, figura de – 0528
crecimiento, sector de – 0526
crenulación 1643
cresacumulado 0834
cresta 2486
cresta anticlinal 2541
cresta cortante 2617
cresta de empuje 2419
cresta divisoria 2316
cresta fumadora 2158
cresta, línea de – 1632
cresta monoclinal 2541
cresta, punto de – 1633
cresta sinclinal 2541
crestas, uniformidad del nivel de – 2598
cresta, superficie de – 1632
crestón 2562
crestón aserrado 2562
criba 3152
cribado 4873
criba pulsante 4155
crinoidea 3313
crinoides 3313
criokarst 2646
crionivelamiento 2650
criopedologia 2634
crioplanización 2650
crioturbación 2647
criptoclástica 3166
criptocristalina 0800
criptógeno 1976
criptográfica 0848
criptomagmático 3986
criptovulcanismo 1181
cristal 0506
cristal corroido 0756
cristal cúmulo 0830
cristalina, forma – 0520
cristalina, forma – 0753
cristalinidad 0791
cristalinidad dudosa, de – 0801
cristalino, defecto – 0531
cristalino, sistema – 0514
cristalitico 0760
cristalito 0760
cristalización 0632
cristalización bandeada 0862
cristalización fraccionada 0670
cristalización, fuerza de – 0909
cristalización heteracumular 0833
cristalización, índice de – 0702
cristalización integral 0910
cristalización mimética 1008
cristalización, piezo– 0633
cristalización rítmica 0864
cristalizada, mal – 0797
cristaloblástica, fuerza – 0909
cristaloblástica, serie – 0909
cristaloblástico, orden – 0909
cristaloblasto 0908
cristalografía 0505
cristalografía óptica 0544
cristal regenerado 0910
crono 1761
cronoestratigráfica 1712
cronoestratigráfica, unidad – 1757
cronoestratigráfico, horizonte – 1758
cronotaxial 1732
cronozona 1758
crotovina 2745
cruce 4073
cruce de filones 4074
cruce oblicuo 4075
cruce perpendicular 4075
crucero 0469
crucero 1596
crucero, plano de – 0540
crudo 3712
crudo húmedo 3714
crudo, transformación del – 3733
cryptozoon 1844
cuarcita 3230
cuartear una muestra 4181
cuarteo 3159
cuarzarenita 3230
cubeta 2517
cubeta de deflación 2187
cubeta de denudación 2187
cubeta de fallas 2547
cubeta de valle glaciar 2620
cubeta lacustre glaciar 2620
cubeta terminal 2629
cubeta termokárstica 2646
cúbico 0515
cuchara 4278
cuchara de muestreo 4748
cuello 1477
cuenca 1086
cuenca 1641
cuenca 2793
cuenca 3648
cuenca cerrada 2579
cuenca de alimentación 2380
cuenca de géiseres 1404
cuenca de hundimiento tectónico 1591
cuenca del cráter 1412
cuenca de recepción 2193
cuenca de recepción 2276
cuenca de un sistema de canales de marea 2508
cuenca epicontinental 1086
cuenca hidrográfica 2192
cuenca intramontañosa 1112
cuenca mareal 2508
cuerpo carbonatado 3324
cuesta 2558
cuesta, espalda de la – 2558
cuesta, frente de – 2558
cueva de acantilado 2474
cueva glaciar 2391
cueva sopladora 2108
culminación 1635
cumbre 0705
cumbres escalonado, nivel de – 2599
cumulito 0765
cúmulo 0830
cumulofídico 0821
cuneiforme 4036
cuña 4280
cuña de arena 2640
cuña de hielo 2640
cuña de tierra 2641
cúpula 0705
cúpula 0724
cúpula 1205
cúpula 1641
cúpula de agua 1359
cúpula de lava 1422
curie 4550
Curie, punto de – 0308
curva de campo 4565
curva estandard 4565
curvatura, valor de – 0336
cúspide 2304
cuspilito 2493
cután 2691

Ch

chaflán 1583
chalazoidita 1321
Chandler, oscilación de – 0200
charca de pantano 3540
charnela frontal 1687
charnela, línea de – 1629
charnela, punto de – 1629
charnela, zona de – 1629
charnockita, m– 0990
chimenea 1221
chimenea 4037
chinenea con brechas 1248
chemenea volcánica, desplazamiento de la – 1231
chinarro 2364
chinarro, límite del – 2364
choque, marca de – 3037
choque, zona de – 0894

D

dactilita 0849
dactilítica 0849
darcy 3697
Darcy, ley de – 3696
darwinismo 1966
datación absolute 0429
datación isotópica 0429
datación radiométrica 0429
DCF, método – 3898
decalcificación 2679
decantación 0673
decantación 3159
decapitación fluvial 2323
decementación 3444
declinación magnética 0284
declinación, método de las curvas de – 3874
decompresión 3850
deconvolución 4454
defasaje 0134
definición óptica 4347
deflación 2155
deflación, abertura de – 1274
deflación, residuo de – 2160
deflectometro 4294
defluente 2211
deformación 0104
deformación 1075
deformación 4666
deformación, banda de – 0539
deformación, coeficiente de – 4696
deformación, eje de – 1016
deformaciones electrico, medidor de – 4767
deformaciones mecanico, medidor de – 4767
deformación, lamela de – 0539
deformación lamelar 0999
deformación, liberación de – 0107
deformación, módulo de – 4696
deformación penecontemporanea 3050
déformación, plano de – 0238
deformación plástica de poca profundidad 3447
deformación plástica profunda 3447
deformación relativa, diagrama de esfuerzos– 4667
deformación relativa 4666
deformación-tensión, curva de – 0106
degeneración 1982
degradación 2012
delta 2494
delta 2895
delta de fondo de bahía 2498
delta de lava 1280
delta de marea externa 2499
delta de marea interna 2499
delta de oleaje 2500
delta de tormento 2500
delta en pata de pájaro 2495
delta en punta 2495
deltáico 2895
delta saliente 2497
delta, sub– 2496
delta, valor de – 0454
demanda de oxígeno 3678
dendrita 0525
dendrita 3400
dendrita 4088
dendrítico 1821
dendrocronología 1737
dendroide 1821
densidad 4342
densidad 4537
densidad aparente 2751
densidad aparente 3602
densidad, corriente de – 2979
densidad de empaquetamiento 3082
densidad de formación 4614
densidad real 3602
denso 0799
denudación 2012
denudación 2013
denudación 2155
denudación con facetas, superficie de – 2601
denudación encajonadas, superficies de – 2661
denudación escalonadas, superficies de – 2601
denudación, nivel superior de – 2598
denudación, superficie de – 2591
denudación, techo de – 2598
depocentro 2986
deposición 2985
depósito 2985
depósito aluvial 2886
depósito aluvial 4000
depósito calcareo 3267
depósito carbonáceo 3493
depósito carbonatado 3266
depósito clástico 3164
depósito concordante 4022
deposito de cueva 2121
depósito de gravedad 2867
depósito de impregnación 3992
depósito de inyección 3975
depósito de inyección 3993
depósito de lahar 1356
depósito de llanura de inundación 2885
depósito de marea 2907
depósito de mar profundo 2922
depósito de morrena 2410
depósito de sustitución 3991
depósito de talud 2967
depósito de talud arrecifal 3353
depósito discordante 4022
depósito elástico 3709
depósito eluvial 3999
depósito estratificado 4023
depósito estratiforme 4024
depósito estratoide 4024
depósito exhalativo 3979
depósito fluvioglaciar 2402
depósito glaciar 2402
depósito glaciolacustre 2402
depósito hidrotermal 3982
depósito intermareal 2907
depósito lenticular 4026
depósito metamórfico de contacto 3980
depósito neumatolítico 3981
depósito offset 3987
depósito penecordante 4022
depósito periférico 3987
depósito piroclástico 3170
depósito pneumatolítico 3981
depósito podiforme 4033
depósito primario 3972
depósito regresivo 3112
depósito residual 2930
depósito residual 3998
depósito secundario 2993
depósito secundario 3972
depósito tabular 4025
depósito torrencial 2884
depósito transgresivo 3113
depósito volcánico clástico 1310
depósito volcanoclástico 1310
depresión 1635
depresión 2189
depresión 2793
depresión anular 1185
depresión, cono de – 3667
depresión intermedia 1137
depresión kárstica 2071
depresión lateral 2239
depresión periférica 2560
depresión volcáno-tectónico 1182
derechos 4224
deriva 2961
deriva 4540
deriva, bosque a la – 2970
deriva continental 1115
derivada segunda 4534
derivada segunda 4544
deriva de playa 2962
derivado 2947
deriva espectral 4549
deriva litoral 2962
dermolítica 1261
derrubios, cono de – 2007
derrubios, manto superficial de – 2008
derrumbe 2114
derrumbe intercrateral 1472
desacoplamiento 0273
desagregación 1991
desagregación por helada 1996
desagregación térmica 1992
desague 2338
desaparecer en profundidad 1526
desaparecer hacia arriba 1526
desarrollo, esbozo de – 1666
desarrollo, estadio incipiente de – 1666
desbordamiento 2251
desbordamiento, nivel de – 2251
desborde transgresivo 1727
desborde transgresivo 3113
desbroce 4875
descamación 1994
descarga 1210
descarga sedimentaria 2957
descomposición 2741
descomposición 2929
descomposición aerobia 2932
descomposición anaerobia 2932
descomposición húmica 1999
descubrimiento 3884
desdolomitización 0930
desecación, grieta de – 3044
desecación, grieta de – 3045
desecación, grieta de – 3046
desechadero 4166
desecho 4164
desembocadora 2196
desembocadora de un valle 2280
desembolsos 3893
desequilibrio radioactivo, datación por el – 0443
deshidratación 3797
deshielo 2367
desierto 2577
desierto arenoso 2578
desierto, baruiz de – 3405
desierto, cantal de – 3405
desierto de arena 2578
desierto de lava 1279
desierto de piedras 2578
desierto pedregoso 2578
desierto rocoso 2578
desierto, rosa del – 3406
desilificación 0929
desimanación por C.A. 0321
desimanación térmica 0321
desintegración 2929
desintegración 3526
desintegración 4731
desintegración, constante de – 0431
desintegración en bloques 1990
desintegración mineral 1990
desintegración por calor 1993
desintegración, zona de – 4731
deslamar 4161
desligamiento, cara de – 2731
deslizamiento 2043
deslizamiento 2976
deslizamiento, cicatriz del – 2052
deslizamiento de acantilado 2473
deslizamiento de derrubios 2046
deslizamiento del suelo 2780
deslizamiento de masas rocosas 2049
deslizamiento de montaña 2048
deslizamiento de suelos 2046
deslizamiento de tierras 2044
deslizamiento en forma de escamas 2045
deslizamiento en masa 2048
deslizamiento entre planos de cizalla 2373
deslizamiento, marca de – 3058
deslizamiento, plano de – 0539
deslizamiento por gravedad 1703
deslizamientos, relieve de – 2056
deslizamientos, topografía de – 2056
deslizamiento, superficie de – 1002
deslizamiento, superficie de – 1705
deslizamiento, superficie de – 2055
deslizamiento, tectónica de – 1702
deslizamiento, trayectoria del – 2054
deslumbramiento 0577
desmezcla 0476
desmonte 4166
desmosita 0933
desmuestre 4741
desnivel sísmico 2552
desorción 0406
despegadura 1663
despegue, estructura de – 3053
desplazada 4637
desplazado en profundidad 1559
desplazamento 4078
desplazamiento de inclinación 1564
desplazamiento de las alturas 4379
desplazamiento del relieve – 4379
desplazamiento de rumbo 1565
desplazamiento de talud 2040
desplazamiento lento de restos superficiales 2038
desplazamiento, línea de – 2962
desplazamiento miscible 3840
desplazamiento neto 1562
desplazamiento neto, componente vertical del – 1563
desplazamiento neto, dirección del – 1562
desplazamiento por basculamiento 4379
desplome de acantilado 2473
desplome de techo 1188
desplome lateral 1464
desprendimiento 3055
desprendimiento por deslizamiento 2051
desreverberación 4454
desulfurización 0415
desviación 4079
desviación de la plomada 0329
desviación fluvial 2213
desviación ortogonal 1523
desviación, reconocimiento de la – 4294
desviar 4292
desvitrificación 0919
desvolatización 3520
detección, capacidad de – 0270

detección, limite de – 4651
detección, umbral de – 0270
detector 4413
detectora, estación – 4417
detector de presión 4415
detonación 1217
detonación 4815
detonador 4816
detrítico 3165
detrítico 3997
detrítico arrecifal, material – 3355
detrito 3165
detritos de perforación 4238
detritus 2005
deutérico 0658
deviación de gas, factor de – 3862
deviación normal 4190
diablástica 0950
diaclasa 1495
diaclasa 4723
diaclasa, apertura de – 4728
diaclasa cruzada 1506
diaclasa de cizallamiento 1504
diaclasa de distensión 1506
diaclasa de extensión 1503
diaclasa de tensión 1503
diaclasa diagonal 1507
diaclasa direccional 1506
diaclasado 1495
diaclasa longitudinal 1507
diaclasamiento 1495
diaclasamiento cilíndrico 1302
diaclasamiento esferoidal 1304
diaclasa no sistemática 1510
diaclasa oblicua 1507
diaclasa, plano de – 1496
diaclasa plumosa de cizallamiento 1505
diaclasa principal 1509
diaclasa, relleno de – 4729
diaclasas conjugadas, sistema de – 1499
diaclasa secundaria 1509
diaclasa según la dirección 1506
diaclasas, frecuencia de – 4725
diaclasas, grupo de – 1498
diaclasa sistematica 1510
diaclasas, patrón de – 1508
diaclasas, sistema de – 1498
diaclasas, sistema de – 4724
diaclasa transversal 1507
diacronismo 1736
diácrono 1736
diadisita 1046
diaftoresis 0904
diaftorita 0904
diagénesis 3388
diagénesis ambiental 3398
diagénesis avanzada 3418
diagénesis, campo de – 3390
diagénesis freática 3399
diagénesis inicial 3418
diagénesis orgánica 3521
diagénesis por soterramiento 3417
diagénesis terminal 3418
diagénesis vadosa 3399
diagenético 3974
diagenético, ambiente – 3391
diagenético cerca de la superficie, ambiente – 3392
diagenético de poca profundidad, ambiente – 3393
diagnético profundo, ambiente – 3393
diagrafía 4578
diagrafía de densidad 4593
diagrafía de electrodos de focalización 4586
diagrafía de inducción 4584
diagrafía de potencial espontáneo 4579
diagrafía de resistividad 4582
diagrafía de termometría 4601
diagrafía de velocidades continuas 4406
diagrafía gamma-ray 4590
diagrafía neutron 4594
diagrafía, sonda de – 4578
diagrafía sónica 4597
diagrama-beta 1021
diagrama compartimentado 4252
diagrama curvilíneo 4252
diagrama en paneles 4252
diagrama en rosa 4259
diagrama en roseta 4259
diagrama estereográfico 4259
diagrama estructural 1015
diagrama-pi 1018
diamagnetismo 0298
diámetro efectivo 3196
diámetro equivalente 3195
diámetro mediano 3199
diámetro medio 3199
diámetro modal 3199
diamíctica 3124
diamíctico 3124
diapirismo 1693
diapiro 1693
diapléctico 0895
diastema 1724
diastema 3125
diastrofismo 1074
diatexia 1027
diatexita 1027
diatomeas, barro de – 3370
diatomeas, pedernal de – 3371
diatomeas, pizarra de – 3371
diatomeas, tierra de – 3370
diatomita 3370
diatrema 1221
dictionita 1043
diesel, combustible – 3744
diferenciación 0398
diferenciación diagenética 3475
diferenciación diagenética 3938
diferenciación geoquímica 0398
diferenciación gravitatoria 0672
diferenciación, índice de – 0701
diferenciación magmática 0667
diferenciación magmática 3938
diferenciación metamórfica 0911
diferenciación metamórfica estratificada 1606
diferenciación neumatolítica 0679
diferenciación por difusión 0680
diferenciación por filtración 0675
diferenciación por flujo 0677
diferenciado 0667
difluencia, escalón de – 2624
difracción 0217
difracción 4444
difracción, curva de – 4464
difusión metamórfica 0912
difusividad térmica 0387
digitación 1687
digitación 3506
dilatación 0187
dilatancia 0109
dilatancia 4678
dilatómetro 4760
dilatómetro, prueba del – 3596
dilución isotópica, análisis por – 0432
dimorfismo 1811
dinamomagnética, teoría – 0278
dinamómetro 4770
dipolar axial, campo – 0279
dipolar excéntrico, campo – 0279
dique 0741
dique 1516
dique anular 0742
dique circular 0742
dique clástico 1516
dique clástico 3073
dique cresta 0741
dique de substitución 0745
dique hueco 1245
dique mineralizado 4031
dique natural 2238
dique radial 1227
dique salino 1695
diques, grupo de – 0749
diques, sistema de – 0749
dirección 1483
dirección 1493
dirección y de brizamiento, medidas de – 4228
discoidal 3184
disconformidad 1720
discontinuidad 4723
discontinuidad sísmica 0099
discordancia angular 1721
discordancia estratigráfica 1720
discordancia no angular 1721
discordancia por erosión 1722
discordante 0718
discordante 1720
discristalino 0797
disección 2019
diseminación 3947
disfótico 1904
disgregación 2758
disipación 0219
disipación, area de – 2258
dislocación 0531
dislocación 1522
dislocación anular 0743
disminución 3816
dispersión 0197
dispersion 0218
dispersión 0219
dispersión 0403
dispersión 0572
dispersión 1812
dispersión 3710
dispersión 4316
dispersión, abanico de – 2948
dispersión axial 0572
dispersión cruzada 0572
dispersión, curva de – 0226
dispersión horizontal 0573
dispersión inclinada 0573
dispersión inversa – 0197
dispersíon normal 0197
dispersión primaria 0404
dispersión secundaria 0405
disposición 1480
distal 2997
distorsión en barril 4404
distribución 1781
distribución al azar 4174
distribución concomitante, zona de – 1777
distribución de la dispersión 0403
distribución escaqueada 4172
distribución lognormal 4189
distribución normal 4189
distribución polimodal 3197
distribución sistemática 4173
distrito minero 3909
disyunción columnar 1501
disyunción en lápices 1610
disyunción poligonal 1303
disyunción romboédrica 1502
divergencia 1969
divergente 0949
división en placas 1500
divisoria brusca 2320
divisoria consecuente 2317
divisoria de aguas 2314
divisoria de aguas 2316
divisoria de aguas subterráneas 3617
divisoria de cuencas hidrográficas, línea 2314
divisoria de hielos 2381
divisoria del valle 2315
divisoria de marea 2511
divisoria emigrante 2319
divisoria en zig-zag 2321
divisoria lenta 2319
divisoria subsecuente 2318
dolina 2071
dolina 2072
dolina de disolución 2072
dolina de hundimiento 2072
dolina en pozo 2105
dolina por colapso 2072
dolomía 3318
dolomía calcarea 3319
dolomía primaria 3320
dolomicrita 3282
dolomita 3318
dolomita primaria 3380
dolomitización 3460
dolomitización, supra– 3460
dominante 1816
dominio de frecuencia 4571
dominio del tiempo 4570
domo 0035
domo 0059
domo 1641
domo de arena 3065
domo de lava 1422
domo de sal 1695
domo eruptivo con lóbulo de lava 1423
domo gnéisico 0890
dopplerita 3525
dorsal 2796
dra 4551
draga 3154
dren 4811
drenaje 2253
drenaje 2768
drenaje 4739
drenaje 4811
drenaje, área de – 3844
drenaje endorreico 2256
drenaje enterrado 4846
drenaje exorreico 2256
drenaje por gravedad 3828
drenaje, red de – 2254
dren vertical de arena 4813
drumlin 2420
drusa 4014
dualística, hipótesis – 0019
dúctil 0472
dúctil 4673
duna 2165
duna 2872
duna continental 2170
duna creciente en forma arqueada 2177
duna de arcilla 2872
duna de arena 2872
duna de deflación 2187
duna de denudación 2187
duna de obstáculo 2174
duna de playa 2168
duna embrionaria 2176
duna en forma de mogote 2190
duna estacionaria 2175
duna exterior 2167
duna externa 2167
duna fija 2175
duna fitogenética 2172
duna interior 2167
duna interna 2167
duna litoral 2166
duna longitudinal 2171
duna móvil 2175
duna parabólica 2177
duna psamogenética 2172
dunar 2180

duna sobre el borde de un escarpe 2169
duna transversal 2171
duna viva 2175
durabilidad 4734
duración 0240
dureno 3580
dureza 0473
dureza 3677
dureza 4701

E

eco 4390
eco del terreno 4390
ecoestratigrafía 1711
ecología 1887
ecosistema 1893
ecotipo 1942
ectexia 1034
ectexita 1034
ectinita 1055
ecuador geomagnético 0281
ecuador magnético 0283
eczema salino 1695
edad 1762
edad absolute 0429
edad concordante 0430
edad discordante 0430
edafogénesis 2659
edafología 2658
edificio externo 1414
edificio volcánico 1413
edómetro 4756
efecto indirecto 0366
efecto pelicular 3851
eficiencia sísmica 0205
eflorescencia 2928
efluente 4165
efusión 1213
efusión de flanco 1266
efusión fisural 1233
efusión terminal 1266
Eh 0421
ehr 4551
eje-a 1016
eje-B 0235
eje-b 1016
eje-beta 1020
eje-c 1016
eje cristalográfico 0511
eje de deformación 1016
eje de fábrica 1016
eje de la corona 1020
eje-f 1017
eje fluvial 2206
eje geomagnético 0280
eje óptico 0554
ejes fiduciales 4368
ejes ópticos, ángulo de los – 0555
ejes ópticos, plano de los – 0554
elasticidad 4669
elasticidad, coeficiente de – 4695
elasticidad, límite de – 4697
elasticidad, módulo de – 4695
elasticidad, módulo de – 4698
elástico 0472
elástico, efecto – 0114
electródica, línea – 4563
electrodos, disposición de – 4562
electrodos permanentes, sistema de – 3673
elemento gaseoso 0650
elemento indicador 4649
elemento menor 0400
elemento volátil 0650
elevación 2796
elevación del suelo debido a helada 4721
elevación estructural 4249
elevación por helada 2642
elipsoide internacional 0331
elongación 0567
elutración 3930
eluviación 2678
eluviado, horizonte – 2701
eluvión 2930
emanación de gas 1215
emanación volcánica 1215
emanación volcánica 1376
embalsamiento 2340
embalse volcánico 1419
embocadero 2509
embrechita 1055
embudo de explosión 1432
embudo de impacto 1418
emergencia, ángulo de – 0215
emergente 0174
emersión 0174
emersión 1728
emersión 2857
emisión 1210
emisiones incandescentes 1315
emisiones no luminosas 1315
emisor 4487
empacado, sobre– 3430
empaque, obturador de – 4279
empaquetadura 4279
empaquetamiento 3082
empaquetamiento, densidad de – 3082
emplazamiento 0618
emplazamiento de la obra 4789
empuje hidrostático 3824
empuje múltiple 3829
empuje par agua del fondo 3825
empuje por agua marginal 3825
empuje por compresión 3827
empuje por gas disuelto 3823
empuje por gas libre 3823
emulsión, estadio de – 0652
enarenarse 3814
encarado 1622
encararse 1622
endémico 1814
endobionte 1854
endoblastesis 0657
endoblástico 0657
endoesqueleto 1861
endogenético 0081
endogénico 0081
endógeno 0081
endógeno, efecto – 0704
endomorfismo 0898
endoscopio 0503
endosfera 0089
endurecimiento 3423
energía cinética eruptiva 1171
energía geotérmica 0388
energía, liberación de – 0204
energía, medio de alta – 2881
energía, medio de baja – 2881
energía volcánica 1171
energio inyectado 0715
enfrentación, área de – 3845
enfriamiento, efecto de – 1271
enmascarar 4643
enriquecimiento 3945
enriquecimiento, zona de – 4005
ensanchamiento 4054
ensanchar 4296
ensanchar al fondo 4296
ensaye 4183
ensayo 3787
ensayo 4183
ensayo de bombeo 3711
ensayo de producción 3788
ensayo por tubería de perforación 3787
entalpía 0389
entalpía, system de alta – 0389
entexia 1036
entexita 1036
entibación 4858
entrada 2338
entrada de un valle 2280
entradas 3893
entrelazada, textura – 0816
eolianita 2870
eólico 2870
eón 1767
eonotema 1767
Eötvös 0337
Eötvös, corrección de – 4538
epibentos 1908
epibionte 1855
epibolita 1049
epicentral, distancia – 0155
epicentro 0154
epicontinental 2915
epifauna 1853
epífito 1856
epigénesis 3419
epigenética 3553
epigenético 3474
epigenético 3966
epilimnion 2332
epimagma 1201
epinerítico 2841
epinorma 0606
epipedon 2697
epipelágico 1910
epipelágico 2845
epirogénesis 1090
epitáctico 0542
epitaxia 0542
epitermal 3983
epizona 0993
época 1764
época 3906
época metalogenética 3905
epsomita 3383
equidimensional 0754
equidimensional 3181
equidistante 4172
equiformo 0754
equigranular 0806
equigranular 0807
equigranular 3081
equilibrio dinámico 0397
equilibrio isostático 0346
equilibrio radioactivo 4549
equivalencia, principio de – 4567
equivalente en petróleo 3747
era 1766
eratema 1766
erg 2183
erizo 2150
erosión 2012
erosión 2776
erosión 2936
erosión acelerada 2777
erosión, canal de – 3510
erosión, capacidad de – 2936
erosión, ciclo de – 2566
erosión cósmica 0028
erosión del suelo 2036
erosión diferencial 2024
erosión en barrancos 2018
erosión en cárcava 2778
erosión en surcos 2778
erosión eólica 2015
erosión eólica 2154
erosión fluvial 2015
erosión glacial 2015
erosión interrumpido, ciclo de – 2571
erosión laminar 2779
erosión lateral 2022
erosión marina 2016
erosión marina, llanura de – 2536
erosión, nivel de base de – 2025
erosión química 2017
erosión remonante 2021
erosión, sistema de – 2573
erosión superficial por numerosos abarrancamientos 2018
erosión, superficie de – 2591
erosión vertical 2022
errático 2949
errático, canto – 2949
error de cierre 4541
error sistemático 4192
erupción 0618
erupción 1211
erupción 4288
erupción apical 1229
erupción areal 1232
erupción, canal de – 1221
erupción central 1228
erupción, centro de – 1220
erupción de barro 1395
erupción de lava 1214
erupción de piroclastos 1214
erupción embrionaria 1219
erupciones, dispositivo contra – 4289
erupción excéntrica 1229
erupción explosiva 1216
erupción fisural 1233
erupción freática 1351
erupción indirecta 1218
erupción intermitente 1212
erupción lateral 1230
erupción magmática 1219
erupción, mecanismo de – 1170
erupción mixta 1219
erupción paroxísmica 1164
erupción secundaria de vapor 1403
erupción subacuática 1236
erupción subaérea 1236
erupción subglaciaria 1236
erupción submarina 1236
eruptiva 0616
eruptivo 0715
eruptivo, cuerpo – 1239
escala 4240
escalón 1538
escalón borde 0527
escalón de crecimiento 0527
escalón de playa 2488
escalón de terremoto 2552
escalón glaciar 2623
escalón orilla 0527
escalón sísmico 2552
escama 1538
escama 1571
escama de pared 2111
escape 4634
escapolitización 0932
escarcha calcárea 2143
escarpado lobulado 0032
escarpe 2470
escarpe 2799
escarpe de falla 2548
escarpe de falla exhumado 2548
escarpe de falla rejuvenecido 2548
escarpe de falle rejuvenecido por erosión 2548
escarpe de lava 1286
escarpe de línea de falla 2550
escarpe de terraza 2295
escarpe monoclinal 2551
escintilómetro 4546
escollera 4876
escollo 2805
escopulito 0769
escoria 1294
escoriáceo 1294
escorias aglutinadas 1346
escorias, terreno con – 1260
escorificación 4184
escorrentía 3616

escorrentía, coeficiente de – 2250
escorrentía de aguas subterráneas 3628
escorrentía subsuperficial 3623
escorrentía superficial 3617
escudo 1081
escudo 2118
escudo volcánico 0035
esfenolito 0736
esfericidad 3180
esferoidal 3180
esferoide 0330
esferoide 3180
esferolito 1321
esferosiderita 3483
esferulítica 0876
esferulítica 4087
esferulítico 3482
esferulito 0778
esferulito 3482
esferulito compuesto 0779
esfuerzo 4667
esfuerzo de cizalla 4688
esfuerzo efectivo 4668
esfuerzo normal 4688
esfuerzo principal 4692
esker 2420
espaciado 3842
espaciado de diaclasas 1497
espaciado de diaclasas 4725
espacial, grupo – 0513
espaciamiento 4822
espacio anular 4279
especiación 1969
especie 1937
especie cosmopolita 1809
especies, climax de – 1891
especie tipo 1948
espectro continuo 4305
espectro de absorción 4305
espectro de amplitudes 4451
espectro de emisión 4305
espectro de fases 4451
espectro de potencia 0222
espectro de potencias 4451
espectro electromagnético 4304
espectrómetro de canales 4547
espectrómetro diferencial 4547
espectrómetro integral 4547
espejo de falla 1538
espeleogeno 2125
espeleología 2094
espeleotema 2124
espeleotema 2877
espesador 4159
espesor de la capa 4622
espesor neto de la arena 3873
espesor neto de la capa productiva 3873
espesor verdadero 4243
espícula 1862
espiculito 0764
espilita 0922
espilitización 0922
espilosita 0933
espinas 1260
espolón 2806
espolón truncado 2628
espongolita 3372
esponjosa 0875
espora 1832
esporangio 1832
esporomorfo 1831
espumosa 0875
esquelético 3169
esqueleto 1861
esquirla 3186
esquisto 0944
esquisto duro 1543
esquistoida 0975
esquistosa 0975
esquistosidad 0975
esquistosidad 1603
esquistosidad exterior 1004
esquistosidad interior 1004
esquistosidad lineal 1603
esquistoso 1603
esquistos verdes, facies – 1082
esquisto verdre 0938
estabilidad, análisis de – 4777
estabilidad de la estructura 3424
estabilización con cal 4807
estabilización de suelos 4806
estabilización de suelos con cemento 4807
estación base 4526
estación de base 0353
estaciones, intervalo de – 4418
estación magnética 4527
estadio 2568
estadio de juventud 2570
estadio de madurez 2570
estadio de senilidad 2570
estado explosivo 1176
estado liquido 4713
estado plastico 4713
estado semisólido 4713
estado sólido 4713
estalactita 2131
estalactita botroidal 2136
estalactita coraliforme 2133
estalactita de lava 1293
estalactita de roca 2103
estalactita hojosa 2134
estalactita tubular 2131
estalagmita 2137
estalagmita coraliforme 2133
estalagmita de lava 1293
estalagmita de roca 2103
estalagmita en pila de platos 2141
estalagmita en seta 2141
estalagmita en terraza 2141
estallido de roca 2112
estancamiento 2340
estanque 2330
estanqueidad 4863
estaucamiento 2971
estenobatial 1911
estenobático 1911
estenohalino 1912
estenotermo 1913
estenotópico 1914
estereógeno 1041
estereograma 4258
estereómetro 4365
estereoscopio 4366
estereoscopio de inversión 4367
estereosoma 1041
esteriles 3924
esterilidad 3506
estero 2505
estero 2836
estictolita 1053
estilolito 3451
estimulación 3800
estración 0468
estrangulación 3904
estratificación 3095
estratificación anual 3127
estratificación centroclinal 1241
estratificación concordante 3115
estratificación contorsionada 3051
estratificación cruzada 3133
estratificación cruzada angular 3142
estratificación cruzada en artesa 3144
estratificación cruzada en zigzag 3146
estratificación cruzada festoneada 3145
estratificación cruzada planar 3142
estratificación cruzada tangencial 3142
estratificación deltáica 3139
estratificación entrecruzada 3134
estratificación entrecruzada de rizadura 3147
estratificación, falsa – 3135
estratificación fina 3108
estratificación foreset 3140
estratificación gradada 3130
estratificación gruesa 3108
estratificación inclinada 1482
estratificación inclinada 3136
estratificación lenticular 3150
estratificación masiva 3108
estratificación ondulada 3138
estratificación periclinal 1241
estratificación, plano de – 3096
estratificación producida por la acción de las mareas 3134
estratificación rítmica 3121
estratificado 3095
estratigrafía 1708
estrato 1750
estrato 3097
estrato guía 1751
estratotipo 1740
estratotipo, componente del – 1742
estratotipo compuesto 1742
estratotipo del límite 1744
estratovolcán 1429
estrechamiento 4054
estrechamiento progresivo 3509
estrechamiento tectónico 3509
estrecho 1626
estrella, efecto de – 0489
estría 1538
estriación 2397
estriación 2999
estría de solución 2087
estría glaciar 2398
estriaje 2397
estrías 1270
estrías glaciares 2397
estribo 4845
estribo 4866
estromalito, proto– 3332
estromatita 1049
estromatolito 3332
estroncio común 0448
estroncio inicial 0448
estroncio radiogénico 0448
estructura 0858
estructura 2568
estructura arrosariada 4054
estructura bandeada 0860
estructura bandeada 0945
estructura bioestratigráfica 1884
estructura chickenwire 3468
estructura coloforme 4091
estructura columnar 1501
estructura concrecionada 4092
estructura criptoexplosiva 1184
estructura criptovolcánica 1184
estructura cristalina 0509
estructura de bioerosión 1883
estructura de bioturbación 1885
estructura de carga 3059
estructura de colapso 1704
estructura de exolución 4106
estructura deformacional 3050
estructura de habitación 1880
estructura del suelo 2723
estructura de nutrición 1881
estructura directional 2998
estructura embricada 3040
estructura en bandas 2374
estructura en cadena 4083
estructura en cola de caballo 3401
estructura en cola de caballo 4069
estructura en llama 3061
estructura en mosaico 3089
estructura en reloj de arena 0534
estructura geopetal 2996
estructura gnéisica 0951
estructura, grado de – 2726
estructura halocinética 1694
estructura halotectónica 1694
estructura imbricada 1571
estructura laminada 2374
estructura linear 1607
estructura metacoloforme 4091
estructura néisica 0951
estructura ocellar 0855
estructura plumosa 1511
estructura ruinosa 3467
estructura sedimentaria 2994
estructura sedimentaria primaria 2995
estructura sedimentaria secundaria 2995
estructura subvolcánica 1414
estructura, tipo de – 2726
estructura volcano-tectónica 1182
estructura vermicular 3467
estructura zonada 0533
estuario 2456
estuario 2784
estuario, de – 2904
estudio piloto 4639
estudio preliminar 4639
etapa preliminar 1157
ethmolito 0737
eufótico 1904
eugeosinclinal 1097
euhalino 2834
euhedral 0523
euhedral 0750
eulitoral 2838
euribático 1911
euribentónico 1909
eurihalino 1912
euritermo 1913
euritópico 1914
eurybatial 1911
eutaxita 0861
eutaxita 1243
eutaxítica 0861
eutrófica 3549
euxínico 2925
evaluación del sitio 4786
evaluación del terreno 4786
evaporación 0412
evaporación, desecamiento por – 2342
evaporita 3375
evapotranspiración 3621
evasión 0273
evolución 1964
evolución ascendente 2574
evolución convergente 1973
evolución descendente 2574
evolución iterativa 1971
evolución, línea de – 1965
evolución magmatica 0666
evolución paralela 1971
evolución rectilínea 1975
evolución uniforme 2574
evolutiva, línea – 1965
excavación 4824
excavación magmática 0623
excavación, sobre – 4856
excavación y relleno 4826
excéntrica 2132
exfoliabilidad 1596
exfoliable 1596
exfoliación 0469
exfoliación 1596
exfoliación 1994
exfoliación, abanico de planos de – 1602

exfoliación de crenulación 1600
exfoliación paralela 1599
exfoliación pizarrosa 1599
exfoliación, plano de – 0540
exfoliación, plano de – 1598
exfoliación por fractura 1601
exhalación 1215
exhalación 1376
exhalación 3951
exhumada 2595
éxito, coeficiente de – 3784
exoesqueleto 1861
exogenético 0081
exogénico 0081
exógeno 0081
exógeno, efecto – 0704
exogeología 0024
exomorfismo 0898
exosfera 0083
exosqueleto 1861
exótico, constituyente – 2949
expansión, área de – 2258
expansión, hipótesis de la – 1123
expectativa, método de curvas de – 3876
explanación 4840
exploración biogeoquímica 4645
exploración descubrimiento, relación – 3784
exploración geobotánica 4645
exploración geoquímica 4625
exploración hidroquímica 4640
explosión 1217
explosiones en cadena 1237
explosión, foco de – 1177
explosión, fosa de – 1235
explosión freático-magmática 1219
explosión, nivel de – 1177
explosión nuclear 0268
explosión nuclear subterránea 0268
explosivo 4815
explosivos, fenómenos – 1175
explotación de seguridad 3668
explotada, área – 3890
exposición 4340
exposición, margen de – 4340
expulsión 3732
extensión 3885
extensión 4186
extensión 4726
extensión en profundidad 1526
extensión lateral 1524
extensión lateral 2513
extensómetro 0122
extensómetro 4773
extinción 0564
extinción 1986
extinción, ángulo de – 0566
extinción ondulante 0565
extinción ondulante 0968
extinción ondulosa 0968
extracción, gastos de – 3900
extraclasto 3272
extramagmático 3985
extraño 0727
extrusión 1207
extumescencia 1288
eyecciones 1313
eyecciones accesorias 1314
eyecciones accidentales 1314
eyecciones consanguíneas 1313
eyecciones esenciales 1314

F

fábrica 0793
fábrica 1004
fábrica 1605
fábrica 3077
fábrica de crecimiento 1010
fábrica deformada 1007
fábrica deposicional 1011
fábrica, diagrama de – 1015
fábrica, dominio de – 1006
fábrica, eje de – 1016
fábrica, elemento de – 1012
fábrica, mega– 1005
fábrica, meso– 1005
fábrica, micro– 1005
fábrica primaria 3077
fábrica, sub– 1005
faceta de falla 2553
faceta triangular 2553
facies 1715
facies 2860
facies de plataforma 2915
facies diagenético 3435
facies, fauna de – 1850
facies, fósil de – 1807
facies metamórfica 0987
facies metamórfica, serie de – 0991
facies mineralógica 0585
facies volcánica 1238
facoide 3055
facolito 0733
factibilidad, estudio de – 4639
factor volumétrico de gas 3861
factor volumétrico de petróleo 3860
fahlband 0452
faja orogénico 1093
faja plegada 1674
falda 2602
falsa 4637
falsa alarma 0269
falla 1521
falla 4723
falla abierta 1534
falla activa 1527
falla antitética 1588
falla arqueada 1535
falla auxiliar 1581
falla cerrada 1534
falla curvilínea 1567
falla deflectada 1536
falla, desviación de la – 1536
falla desviada 1536
falla de traslación 1566
falla dextral 1561
falla direccional 1544
falla en lisel 1583
falla enrasada 2549
falla, espejo de – 1538
falla gravitacional 1707
falla homotética 1588
falla horizontal de desgarze 1560
falla inactiva 1527
falla inversa 1558
falla longitudinal 1546
falla marginal 1593
falla muerta 1527
falla muy inclinada 1549
falla nivelada 2549
falla normal 1558
falla oblicua 1545
falla ortogonal 1545
falla paralela al plano de estratificación 1547
falla periférica 1592
falla perpendicular 1545
falla plana 1535
falla, plano de – 0238
falla plegada 1535
falla poco inclinada 1549
falla principal 1580
falla profunda 1525
falla, pulverización por – 1540
falla ramificada 1582
falla rejuvenecida 1527
falla rotacional 1567
fallas conjugadas, sistema de – 1584
fallas cruzadas 1579
fallas, divergencia de – 1583
falla secundaria 1580
falla sellante 3765
fallas en escalonamiento 1585
fallas en gradería 1587
fallas en relevo 1585
falla sinistral 1561
falla sintética 1588
fallas paralelas, grupo de – 1584
fallas radiales 1594
fallas, red de – 1578
fallas, sistema de – 1578
fallas, tectónica de – 1520
falla superficial 1525
falla, superficie de – 1528
falla transformante 1141
falla transformante 1560
falla transversal 1546
falla vertical de desgarze 1560
falla, zona de – 1532
familia 0687
familia 1930
familia de suelos 2664
fanerita 0795
fanerítica 0795
fanerocristalina 0795
fanglomerado 2885
fango 3210
fango, acumulación lineal de – 3325
fango azul 2921
fango carbonatado 3295
fango, corriente de – 2974
fango de coral 3311
fango hemipelágico 2920
fangolita 3219
fangolita conglomerática 3256
fango margoso 3222
fangoso 3210
fango verde 2921
fantasma 4442
fantasma, horizonte – 4462
faro 3339
fase 0165
fase abisal 1197
fase bioquímica 3519
fase de fumarola 1168
fase de gas 1162
fase de solfatara 1168
fase eruptiva 1208
fase explosiva 1208
fase extrusiva 1197
fase final 0175
fase final 1167
fase geoquímica 3519
fase hipabisal 1197
fase intrusiva 1197
fase magmática 1108
fase orogénica 1107
fase postvolcánica 1197
fase preliminar 0181
fase principal 0188
fase volcánica 1197
fatiga 4674
fauna 1795
fauna enana 1851
faunístico 1795
fáunula 1796
Fedorow, platina de – 0614
feldespatización 1064
felsita 0597
felsofídica 0824
felsofídico 0824
felsófiro 0824
fémico 0607
fenita 0925
fenitización 0925
fenoblasto 0981
fenoclasto 0981
fenoclasto 3239
fenocristal 0758
fenotipo 1979
ferrimagnetismo 0302
ferromagnesiano 0599
ferromagnetismo 0299
ferruginoso 3464
festón 2649
festón de calcita 2148
fibroblástica 0963
fibrocristalina 0815
fibroso 0522
fibroso 0773
figura grabada 3409
filamentos vítreos 1326
filetín 0501
filiforme 4081
filita 0943
filogénesis 1959
filogenético, árbol – 1960
filético 1960
filogenía 1959
filón bandeado 4049
filón brechoide 4050
filón capa 4053
filoncillo 4039
filón complejo 4046
filón compuesto 4045
filón crucero 4077
filón de contacto 4043
filón de dilatación 4041
filón de sustitución 4042
filón en bayoneta 4055
filón en charnela anticlinal 4053
filón en charnela sinclinal 4053
filón en escalera 4057
filón en fisura 4044
filones interconectados 4067
filones paralelos, haz de – 4066
filones radiales 4063
filones reticulades 4064
filón fisural 1514
filonita 0918
filonita 1541
filonitización 0918
filón metalífero 4038
filón principal 4062
filón simple 4045
filozona 1779
filtración 3629
filtración 3753
filtración 4635
filtración 4863
filtración a presión 3933
filtrado 0414
filtrado de velocidades 0228
filtrado en frecuencia 0136
filtro 4338
filtro 4445
filtro 4812
filtro 4862
filtro de contraste 4339
filtro de corte 0137
filtro de paso-banda 0137
filtro de polarización 4339
filtro de velocidades 4456
filtro fotográfico 4339
filum 1929
finocristalina 0796
finogranular 0796
finos 3207
fiordo 2453
fiordo 2454
firn 2363
fisibilidad 1597
físil 1597
fisión, datación por trazas de – 0442
fisura 1465
fisura 1512
fisura 1513
fisura 2376

fisura abierta 1513
fisura concéntrica 1469
fisura de contracción 0713
fisura de desecación 1518
fisura de enfriamiento 1517
fisura de retracción 1993
fisura de tensión 1519
fisura endógena 3508
fisura lateral 1468
fisura radial 1468
fisura, relleno de – 1515
fisura transversal 1470
fisura volcánica entreabierta 1234
fiteral 3592
fitoclasto 3168
fitoplancton 1800
fixismo 1113
fixista, doctrina – 1113
flanco 1625
flanco inferior 1576
flanco invertido 1647
flanco invertido 1680
flanco normal 1647
flanco normal 1680
flanco superior 1576
flanco superior 1680
flebita 1044
flecha 2464
flecha testigo 2466
flexura 1616
floculación 2992
floculo 2992
flora 1797
floral 1797
flor de yeso 2151
flotación 0673
flotación 4152
flotación colectiva 4152
flotación selectiva 4152
fluidización 1370
flujo 2050
flujo 2822
flujo anual de tesorería 3894
flujo cumulativo de tesorería 3894
flujo de caja anual 3894
flujo de caja, curva de – 3895
flujo de caja final 3894
flujo de tesorería 3895
flujo final de tesorería 3894
flujograma 4143
flujo, línea de – 3708
flujo luminoso 4311
flujo magnetico 0317
flujo primario 0631
flujo, régimen de – 2960
flujo subterráneo 3626
flujo tectónico 0998
flujo térmico de la tierra 0383
flujo térmico terrestre 0383
fluorescencia 0467
fluorescencia 0491
fluvial 2883
fluvioglacial 2883
fluviomarino 2902
fluxoturbidita 2984
focal, esfera – 0233
foco 0156
foliáceo 0775
foliación 0948
foliación 1604
foliación continua 0948
foliación discontinua 0948
foliado 0775
foliado 0948
fondo 0705
fondo 4626
foráneo 3970
foreset, estratificación – 3140
foreset, laminación – 3140
foreset sobreplegada, estratification – 3054
formación 1754
formación carbonífera 3498
formación, deteriozo de la – 3815
formación, factor de – 4609
formación húmeda 3767
formación impermeable 3765
forma del género 1932
forma estructural 2554
forma tectónica 2537
forma volcánica negativa 1410
forma volcánica positiva 1409
forro de lava 1292
fosa 0034
fosa 1453
fosa 2247
fosa 2793
fosa 2794
fosa de hundimiento 1590
fosa lateral 2384
fosa marginal 1111
fosa oceánica 1138
fosa tectónica 1590
fosa tectónica 2546
fosa tectónica circular 0743
fosfatización 3459
fosforescencia 0467
fosforita 3480
fósil 1789
fósil 2595
fósil característico 1785
fósil guía 1785
fósil-guía zonal 1785
fosilífero 1789
fótico 1904
fotogeología 4327
fotografía aérea 4349
fotografía aérea oblícua 4353
fotografía aérea vertical 4353
fotografía compuesta 4357
fotografía de horizonte 4354
fotografía de satélite 4349
fotografía índice 4382
fotografía lateral-oblícua 4353
fotografía transformada 4377
fotograma 4358
fotogrametría 4328
fotoplano 4383
fotosíntesis 1803
Fourier, transformación de – 4451
fraccionamiento 0414
fraccionamiento cinético de los isótopos 0457
fraccionamiento isotópico, curva de – 0455
fraccionamiento isotópico, factor de – 0455
fractura 0541
fractura 0469
fractura 0487
fractura 1092
fractura 1494
fractura 3431
fractura 4676
fracturación 1494
fracturación 3800
fractura concoidea 0541
fractura de contracción 0713
fracturado 1494
fractura progresiva 4676
fractura, zona de – 1595
frágil 4673
fragipan 2734
fragmentación 4821
framboidal 3484
framboidal 4094
framboide 3484
franja volcánica 1447
freática, capa – 2765
frecuencia de batido 0178
frecuencia de codo 0225
frecuencia, dominio de – 0223
frecuencias, curva cumulativa de – 4188
frecuencias, distribución de – 4187
frente 4852
frente básico 1066
frente dunar 2168
friable 3087
fricción 4686
fricción, ángulo de – 4689
fricción, coeficiente de – 4689
fricción, criterio de – 4689
fricción interna 4686
frontal, región – 1686
fronte 1686
Froude 2266
frústula 1842
frústula 3370
fuego 0485
fuente 3630
fuente, área – 2943
fuente artesiana 3631
fuente de fuego 1283
fuente kárstica colgada 2091
fuente mineral 3632
fuente termal 1407
fuente termal 3632
fuente vauclusiana 2093
fuente vauclusiana 3631
fumarola 1379
fumarola anhidra 1387
fumarola bufante 1385
fumarola croicolítica 1384
fumarola de cráter 1383
fumarola de tunél 1383
fumarola de vapor 1389
fumarola eruptiva 1385
fumarola fisural 1383
fumarola fría 1389
fumarola leucolítica 1384
fumarola llameante 1386
fumarola, montículo de – 1390
fumarola perenne 1381
fumarola primaria 1382
fumarola secundaria 1382
fumarola sin raíz 1382
fumarola temporal 1381
fuseno 3580
fusión 1198
fusión de las nieves 2355
fusión selectiva 1029
fusión, zona de – 0625

G

gabroidea 0812
gal 0324
galería 4847
galería en gatera 2098
galería meandriforme 2100
gamma 0318
ganancia, control automático de – 4425
ganga 3921
gap 1556
gas 3713
gas, acumulación de – 3774
gas asociado 3891
gas, conservación de – 3836
gas, contenido en – 0650
gas de pantano 3529
gas disuelto 0641
gas disuelto 1172
gases ardientes, excavación por – 1178
gas, escapes de – 1401
gas freático 1351
gasificación del carbón 3609
gas independiente 3891
gas juvenil 1348
gaslift, operación – 3791
gas-liquido, relación – 3802
gas natural 3713
gas no asociado 3891
gasóleo 3744
gasolina 3744
gas/petróleo, contacto – 3774
gas/petróleo, nivel – 3774
gas-petróleo, relación – 3802
gas pobre 3715
gas, presión de – 1173
gas resurgente 1172
gas rico 3715
gastrolito 1864
gas volcánico 1349
gatera 2098
gato plato, ensayo de – 4763
gatos radiales, ensayo de – 4763
gauss 0318
geanticlinal 1099
Geiger, contador – 4546
géiser 1404
géiser, cráter de – 1405
géiser, chimenea de – 1405
geíser de barro 1398
geiserita 1406
géiser silíceo 1406
géiser, surtidor de – 1405
gelividad 1996
gemología 0478
generación 0591
generación mineral 3906
género 1933
género, forma del – 1932
género tipo 1947
género-zona 1778
genético 4630
genotipo 1979
gens 1934
geobarometriía 0994
geóbios 1802
geocronología 1713
geocronología isotópica 0428
géoda 3481
géoda 4014
geodepresión 1120
geodesía 0013
geodinámica 0080
geo-estadística 4193
geófago 1857
geofísica 0078
geofísica del ingeniero 4663
geófono 4414
geognosía 0001
geogonía 0003
geografía física 0012
geohidrología 3611
geoide 0328
geoisoterma 0377
geoisotérmica, superficie – 0377
geoisotérmico, plano – 0377
geología 0001
geología ambiental 4665
geología aplicada 0016
geología aplicada 4661
geología catastrófica 0005
geología de campo 4226
geología del ingeniero 4660
geología de los isótopos estables 0453
geología económica 0016
geología estructural 1478
geología experimental 0014
geología general 0010
geología histórica 0011
geología marina 4226
geología matemática 0015
geología subterránea 4226
geólogo 0001
geomagnetismo 0276
geomorfología 1987
geomorfología del ingeniero 4663
geonomía 0079

geopetal 3107
geoquímica 0395
geoquímica de los isótopos estables 0453
geosinclinal 1094
geotectónica 1073
geoteécnica 4662
geotermal 0376
geotermal, función – 0378
geotérmia 0374
geotérmico, campo – 0392
geotérmico, desnivel – 0380
geotérmico, gradiente – 0380
geotérmico, salto – 0380
geotermismo 0374
geotermometría 0379
geotermometría 0994
geotermómetro 0994
germánico, tipo – 1078
geróntico 1866
gigantismo 1867
gilgai, relieve – 2763
girogonio 1843
glaciación 2367
glacial 2873
glaciar 2366
glaciar, arco del – 2391
glaciar cabalgado 2385
glaciar cabalgante 2385
glaciar de marea 2436
glaciar de meseta 2426
glaciar de montaña 2426
glaciar dendrítico 2432
glaciar de pie de monte 2435
glaciar de sal 1699
glaciar de tipo alpino 2428
glaciar de tipo Mustag 2431
glaciar de tipo Turkestan 2429
glaciar de valle 2427
glaciar fósil 2369
glaciar, frente del – 2390
glaciar, leche de – 2392
glaciar, mesa de – 2386
glaciar muerto 2369
glaciar polar 2368
glaciar regenerado 2434
glaciar suspendido 2433
glaciar templado 2368
glaciología 2345
glaciomarino 2903
glacis 2007
glacis 2582
glacis rocoso desértico 2582
glauconítico 3177
glauconitización 3459
gleyificación 2684
globosferito 0765
globulito 0765
glomerocristal 0771
glomerofídica 0821
gneis 0936
gneis apizarrado 0936
gneis con disyunción prismática 0936
gneisoide 0871
gneis primario 0631
golpe de gobia 2086
golpe de montaña 4857
gour 2147
gour, barra de – 2147
graben 1590
grada 1538
grada 2059
gradación 4343
gradación de tamaños 4702
gradiente de la capa 1488
gradiente estructural 2200
gradiente hidráulico 4736
gradiométro magnético 4525
grado, escala de – 3161
graduación 3130
gráfico 0847
granítica 0810
granitización 1062
granito gráfico 0847
granito, tectónica del – 0712
granizo, huella de – 3041
granizo, impresión de – 3041
grano 0504
grano 3078
granoblástica 0960
grano, contorno del – 0532
granoesquistosa 0953
granofídica 0848
granofídico 0847
grano fino, de – 0796
grano fino, de – 3194
grano grueso, de – 0796
grano grueso, de – 3194
grano heterométrico, con – 0820
grano, límite del – 0532
granolita 0810
grano medio, de – 0796
grano medio, de – 3194
grano muy fino, de – 0799
grano muy grueso, de – 0796
granosferito 0779
grano, tamaño de – 3193
granuda 0806
granulación 0967
granulación 2934
granulado 4344
granular 0806
granular 2727
granular 3081
granularidad 0792
granularidad 3081
granulita 0937
granulítica 0954
granulometría 3160
granulometría continua 4702
granulometría discontinua 4702
granulométrico, análisis – 2711
granulosidad 0792
grauvaca 3234
grava 2722
grava 3241
grava, banco de – 2244
grava, semejante a – 3244
grava volcánica 1324
gravedad 0324
gravedad, aceleración de la – 0324
gravedad aire libre 4538
gravedad, campo de la – 0326
gravedad, determinación de la – 0349
gravedad, formula internacional de – 0333
gravedad, fuerza de – 0323
gravedad, gradiente de la – 0335
gravedad, medición de – 0349
gravedad observada 4536
gravedad, potencial de la – 0327
gravedad, unidad de – 4537
gravedad, valor normal de la – 0332
gravera 2967
gravera 3244
gravera calcárea 3278
gravera de bloques 3249
gravera de gránulos 3249
gravilla 3240
gravimetría 0349
gravimétrica, estación – 0353
gravímetro 0352
gravímetro con indicator cero 0352
gravitación 0322
gravitación, constante de – 0325
greda 3215
greisen 0924
greisenización 0924
grieta 0487
grieta 1092
grieta 1465
grieta 1466
grieta 1518
grieta 2375
grieta de contracción 0713
grieta de contracción 3044
grieta de enframiento 2640
grieta lateral 2376
grieta longitudinal 2376
grieta marginal 2376
grieta sísmica 2552
grieta transversal 2376
grisú 3554
gruescocristalina 0796
gruescogranular 0796
grumo 2727
grupo 0687
grupo 1755
grupo 1928
grupo volcánico 1450
gruta 2095
gruta de acantilado 2472
gruta glaciar 2107
gruta helada 2107
guadi 2587
guano 2878
guía de onda 0101
gunita 4801
Gutenberg, discontinudad de – 0092
guyot 2798

H

habitat 1901
hábito 0521
hábito 0592
hadal 2848
hadal 2924
halmirolisis 3476
halo 4629
halocinesis 1694
haloclina 2830
halo de alteración 4630
halo en la roca encajante 4630
halófita 1846
halolímnico 1915
Harker, diagrama de variación de – 0697
hastial con salbanda 4013
hastiales 3501
hastiales 4009
hastial inferior 4009
hastial sin salbanda 4013
hastial superior 4009
Hayford, zona de – 0362
haz de filones 4065
helada 2367
helicítica 0955
hemiciclotema 3120
hemidiatrema 1206
hendidura 2474
hervidero 1395
heteracumulado 0833
heteroblástica 0956
heterócrono 1734
heterogranular 0807
heteromórfico 0635
heteromorfismo 0635
heteromorfo 0635
heterospórea 1834
heterotáctica 1007
heterotópico 1731
hexaedrito 0068
hexagonal 0515
hialino 0805
hialofítica 0840
hialopilítica 0837
hiato 1723
hibradación 0681
hibridisación 0681
híbrido 0681
hidatopirogénico 0656
hidatopyrógeno 0656
hidratación 0419
hidratación de sales 1997
hidrato de carbono 3723
hidróbios 1802
hidrocarburo pesado 3745
hidroexplosión 1351
hidrófono 4415
hidrófono 4496
hidrógeno, índice de – 4613
hidrogeología 3611
hidrogeoquímica 3611
hidrolacolito 2643
hidrólisis 0418
hidrología 3610
hidromorfa, característica – 2690
hidromórfica 4634
hidroplutonismo 0656
hidroquímica 4634
hidrosfera 0085
hidrotermal, altéración – 3954
hidrotermal, estadio – 0654
hidrotermal, proceso – 3953
hidrotermal secundario 3994
hielo, agujas de – 2639
hielo, huella de cristal de – 3043
hielo, manto de – 2425
hielo, masa de – 2425
hielo, pirámide de – 2377
hielo profundo 2639
hierro oolítico 4115
hiladas de mineral 4040
Hilt, ley de – 3522
hinchamiento al crisol, índice de – 3595
hinchazón 4720
hinterland 1110
hipabisal 0617
hipautomórfica 0808
hipérboles, método de las – 0152
hiperciclotema 3120
hiperestereoscopio 4367
hiperítica, estructura – 0853
hipersalina 3377
hipersalino 2835
hipertermal, región – 0391
hipidioblasto 0980
hipidiomorfa 0808
hipidiomórfica granular 0808
hipidiomórfico 0751
hipidiotópico 3471
hipocentro 0156
hipocristalino 0804
hipoestratotipo 1746
hipogénico 3937
hipogénico 3967
hipolimnion 2334
hipomagma 1199
hipotermal 3983
hipotermal, región – 0391
histéresis magnética 0305
histograma 4188
hodócrona 0171
hogar central 1202
hoja 0739
hojarasca 2742
hojosidad 1604
hojosidad por fracturas 1995
hojoso 0772
holoblasto 0984
holocristalino 0803
holoestratotipo 1740
holohialino 0805
hololeucocrático 0587
hololeucocrato 0587
holomarino 2900
holomelanocrático 0588
holomelanocrato 0588
holotipo 1949

hombrera 2619
homeomorfismo 1820
homoblástica 0960
homogénea, área – 4787
homogénea, zona – 4787
homogéneo 0716
homogéneo 4680
homohalino 2831
homólogo 1972
homónimo 1955
homoplástico 1980
homosísmica, línea – 0210
homotáctica 1007
homotaxial 1733
homotipo 1954
hondonada 2793
horizonte 1749
horizonte 2687
horizonte A1 2698
horizonte A2 2700
horizonte a capa superficial 2697
horizonte Ap 2699
horizonte B 2702
horizonte B coloreado 2703
horizonte blanqueado 2700
horizonte B textural 2703
horizonte C 2706
horizonte Cg 2707
horizonte cronoestratigráfico 1758
horizonte D 2708
horizonte de recurrencia 3544
horizonte eluviado 2701
horizonte eluvial 2701
horizonte en bandas 2705
horizonte fantasma 4462
horizonte iluviado 2702
horizonte litoestratigráfico 1749
horizonte mineralizado 4008
hormigón proyectado 4801
hornito 1298
Hoskold, fórmula de – 4217
hoya 2793
hoya de lava 1284
hoyo de explosión 1440
hoyo de kryokonita 2387
hueco 3064
huecos cilíndricos 1306
huecos, con – 1305
huecos, relleno de – 3940
huella animal 1875
huella superficial 1875
hulla de coque 3559
hulla de forja 3559
hulla de gas 3559
hulla, finos de – 3513
hulla flamante 3559
hulla grasa 3559
hulla grasa de llama corta 3559
hulla grasa de llama larga 3559
hulla lignitosa 3559
hulla magra antracitosa 3559
hulla seca de llama corta 3559
hulla seca de llama larga 3559
hum 2074
humedad, contenido de – 2769
humedad del suelo 2764
humedad intrínseca 3598
humedad libre 3598
humedad total 3598
humificación 2683
humificación 3525
humo volcánico 1349
humo volcánico 1377
humus 2743
humus 3525
humus bruto 2744
humus dulce 2744
hundimiento 4018
hundimiento de caverna 2110
hundimiento, hipótesis del – 1118
hundimiento por colapso 2072
hundimiento por disolución 2072
hundimiento volcánico 1410
hundimiento volcánico, sector de – 1464
hystéresis 4675

I

ibón de cráter 1454
ichor 1061
icnocenosis 1874
icnofósil 1873
icnofósil 3067
icnología 1871
identificación, capacidad de – 0271
identificación, umbral de – 0271
idioblasto 0980
idiocromático 0483
idiomórfica granular 0808
idiomórfico 0750
idiotópico 3471
ígnea 0616
ignimbrita 1340
ignívomo 1237
iluminación conoscopica 0547
iluminación específica 4311
iluminación ortoscópica 0547
iluminación total 4311
iluviación 2678
illuviado, horizonte – 2702
imagen 4320
imanación 0295
imanación anhisterética 0314
imanación de saturación 0297
imanación invertida 0315
imanación química 0312
imanación remanente 0306
imanación termoremanente 0309
imanación viscosa 0313
imantanda, fuerza – 0317
imatra, piedra de – 3486
imbibición 1060
imbricación 1571
imbricación 3040
imitación, piedra de – 0481
impactita 0076
impacto 0026
impactos, teoría de los – 0049
impedancia acústica 4474
impermeabilización superficial 2760
impermeable 3640
impetu 0173
impregnación 3396
impregnación 3955
impresión, relación de – 4378
impresor de transformación 4377
impronta 1826
impulsiva 0173
impulso 4482
incidencia, ángulo de – 0211
incidencia bajo ángulo crítico 0212
incidencia normal 4472
incidencia, plano de – 0211
incisión 2020
inclinación 1484
inclinación 4351
inclinación aparente 1484
inclinación, en el sentido de la – 1489
inclinación, en sentido contrario a la – 1489
inclinación magnética 0283
inclinación primitiva 1486
inclinación relativa 4351
inclusión 0789
incompetente 1624
incrustación 3387
indentación 2450
indicador geobotánico 4646
indicador vegetal 4646
indicatriz 0552
inducción magnética 0317
inequigranular 0807
infiltración 2768
infiltración 3396
infiltración 3619
infiltración 3955
infiltración, capacidad de – 3620
infiltración, grado de – 3620
inflexión, línea de – 1627
inflexión, punto de – 1627
infracortical 0584
infraespecífico 1944
infraestructura 1077
infralitoral 2839
infranerítico 2841
infrarrojo cercano 4307
infrarrojo fotografiable 4307
infrarrojo lejano 4309
infrarrojo medio 4308
infrarrojo térmico 4308
infrasaturada 0605
ingeniería sismológica 0254
ingenio 4149
inhomogéneo 4680
inicial 1108
inlandsis 2425
inmaduro 3176
inmaduro 3725
inmersión 1614
inmiscibilidad líquida 3939
insolación, destrucción por – 1992
inspección in situ 0267
intensidad 0243
intensidad, escala de – 0243
intercalación 3117
intercalación arcillosa 3772
intercalado 3117
intercambio iónico 0420
intercara 3096
intercepción 3615
intercinemático 1079
intercrecimiento 0783
intercrecimiento 4098
intercrecimiento gráfico 0847
intercrecimiento gráfico 4100
intercrecimiento, índice de – 4112
intercrecimiento simple 4099
intercúmulo 0831
interdigitada 3118
interespaciado 3843
interés, tipo de – 3897
interferencia anómalo, color de – 0568
interferencia, color de – 0568
interferencia, figura de – 0569
interfluvio 2312
intergranular 0841
intergranular 0998
interlazada, textura – 0816
intermareal 2836
intermareal, zona – 2505
intermedia 0602
interpenetración 2648
interpenetración de los granos 3448
interpretación cuantitativa 4566
interpretación directa 4566
interpretación indirecta 4566
intersección isogonal 4075
intersección oblicua 4075
intersección perpendicular 4075
intersertal 0840
intersticial 3083
intersticio 3083
intersticio 3683
intersticio por disolución 3690
intersticio primario 3684
intersticio secundario 3685
interstratificación 3116
interstratificado 3116
intervalo 1718
intervalo bioestratigráfico, zona de – 1776
intervalo, zona de – 1772
interzona esteril 1775
intraclasto 3272
intraespararenita 3283
intraesparita 3283
intraesparrudita 3283
intragranular 0998
intramagmático 3985
intramicarenita 3284
intramicrita 3284
intramicrudita 3284
intrusión 0618
intrusión 0715
intrusión compuesta 0721
intrusión de petróleo 3752
intrusión múltiple 0721
intrusión sedimentaria 3070
intrusivo 0619
intrusivo 0715
intrusivo compuesto 0721
intrusivo mezclo 0722
intrusivo múltiple 0721
intrusivo parcial 0721
intrusivo sencillo 0720
intrusivo simple 0720
intumescencia 1194
intumescencia de lava 1272
intumescencia profunda 1120
invasión 4619
inversión 1088
inversión, auto– 0311
inversión del relieve 2563
inversión de una corriente 2214
inversión geomagnética 0291
inversión polar, suceso de – 0292
invertido 1491
involución 1690
involución 2648
inyección 0618
inyección 3949
inyección 4802
inyección, cabeza de – 4268
inyección de agua 3833
inyección de agua caliente 3839
inyección de gas 3835
inyección de vapor 3839
inyecciones, pantalla de – 4804
inyección posible, índice de – 3809
inyección satélite 1203
inyección sedimentaria 3070
inyección, taladro de – 4803
inyectar lechada 4803
ionio-thorio, datación – 0444
iridiscencia 0486
irradiación 0492
isla 2443
isla continental 2444
isla de acumulación 2535
isla de caldera 1417
isla de cráter 1417
isla de emersión 2535
isla desunida 2534
isla dunar 2507
isla ligada 2534
isla oceánica 2444
isla separada 2534
isla unida 2534
isla volcánica 1416
isocalorífica 3523
isocarbonificada 3523
isoclina 0283
isoclinal 1626
isocora 4244
isocromática, curva – 0571
isócrona 0210
isócrona 0433

isodinámica 0285
isogama 4532
isogónica 0284
isograda 0992
isograda geoquímica 4638
isográdica 0992
isoipsa 4248
isoipsa freática 3644
isomésico 1735
isométrico 0522
isomorfismo 0464
isomorfo 1974
isopaca 4243
isópico 2863
isopleta 4246
isópora 0289
isosísmica, línea – 0245
isosista 0245
isospora 1835
isospórea 1834
isostasía 0338
isotérmico 0377
isotipismo 0463
isotropía 4681
isótropo 4681
isótropo, opticamente – 0561
isovolatil 3523

J

jade 0498
jaspe 3492
jaspe 4127
jaspilita 3492
jerarquía 1925
junción triple 1130

K

Kalb, línea de – 0559
kame 2420
karst 2064
karst con dolinas irregulares 2080
karst con torretas 2082
karst cubierto 2069
karst descubierto 2068
karst desnudo 2068
karst, formación de – 2063
karst fósil 2069
kárstica, región – 2064
kárstico, paisaje – 2064
kársticos, fenómenos – 2065
karstificación 2063
karst profundo 2067
karst somero 2067
karst subterráneo 2070
karst superficial 2067
katagénesis 1977
katagénesis 3419
kelifítica, aureola – 0786
kerogenita 3720
kerogeno 3720
keroseno 3744
Klinkenberg, factor de – 3706
Koenigsberger, cociente de – 0310
Koenigsberger, relación – 0310
krigeaje 4197
kryokonita 2387

L

laberinto 2115
laberinto según plano de diaclase 2116
laberinto según plano de estratificación 2116
labio hundido 1529
labio hundido 1559
labio levantado 1529
labio levantado 1559
labio muro 1529
labio techo 1529
labio yacente 1529
laboreo 2757
labradorescencia 0486
laca 4230
lacolito 0734
lacuna 1723
lacune 1723
lacustre 2888
ladera, coluviones de – 2007
ladera, ruptura en la – 2277
lago 2330
lago anteglaciar 2393
lago de lava 1281
lago de playa 2343
lagoon 2458
lago residual 2344
lago salado 2893
lago semilunar 2232
lago sin desague 2338
lago termokárstica 2646
laguna 2458
laguna 3345
laguna, canal de – 3348
laguna de marea 2458
lahar caliente 1353
lahar de descarga 1354
lahar de fundición glacial 1357
lahar de lluvia 1355
lahar frío 1355
lama 4160
lamarckismo 1967
lamelar 0775
lamelas de presión 0894
lámina 2129
lámina 3101
laminación 1277
laminación 3101
laminación convoluta 3052
laminación cruzada 3133
laminación foreset 3140
laminación ígnea 0868
laminado 3101
laminador 2098
laminar 0775
laminar 0972
laminar 2728
laminar 3132
laminilla 3186
lamprofídica 0809
lapidario 0502
lapiez 2086
lapilli 1324
lapilli acrecional 1321
lapilli de cristales 1325
Larsen, diagrama de variación de – 0697
lastrabarrena 4270
lateral 4631
laterita 4124
lateritización 2681
latitud geocéntrica 0098
latitud geomagnética 0281
lava 1242
lava aa 1260
lava acrecional, bola de – 1296
lava afrolítica 1260
lava almohadillada 1253
lava, bola de – 1296
lava, colada de – 1263
lava, columna de – 1246
lava cordada 1250
lava cordada 1252
lava de barro 1397
lavadero 4149
lavadero 4158
lavadero 4659
lava de tobas aglomeradas 1346
lava, diluvio de – 1276
lavado 4873
lavado superficial 2031
lava en bloques 1259
lava en losas 1250
lava espumosa 1307
lava estriada 1270
lava farinácea 1263
lava, gota de – 1326
lava, inundación de – 1276
lava, lengua de – 1264
lava pahoehoe 1250
lava parásita, colada de – 1265
lava, retoños de – 1251
lava, revoque de – 1226
lava rugosa 1250
lava sin raíz, colada de – 1282
lava solidificada 1242
lava subterránea, colada de – 1267
lava, surtidor de – 1282
lava, torre de – 1299
laxitud 1179
lechada 4803
lecho 3097
lecho 3098
lecho anual 3127
lecho, forma del – 2246
lecho mayor 2242
lecho menor 2242
lecho ordinario 2242
lectoestratotipo 1745
lectotipo 1951
lengua 1752
lengua de tierra 2448
lengua de valle 2390
lengua glaciar 2390
lentejón 1752
lentejón 3102
lentícula 3102
lenticular 3102
lepidoblástica 0961
leptinolita 0940
leptita 0937
leptotermal 3984
leucocrático 0587
leucocrato 0587
leucosoma 1038
levantamiento 4830
levantamiento volcánico 1409
leyenda 4240
ley, promedio pesado de la – 4185
liberación, grado de – 4112
liberar 4150
licuación 0678
licuefacción 3084
licuefacción 4718
lidita 3369
lignito 3495
lignito friable 3556
lignito negro brillante 3558
lignito pardo 3495
lignito pardo 3555
lignito pardo terroso 3555
lignito xiloide 3495
limán 2457
límite elastico 4697
límnico 2888
limnilogía 2331
limnóbios 1802
limnoplancton 1801
limo 2715
limo 2720
limo 3204
limo 4281
limolita 3204
limonita 4126
limoso 3204
limpieza 3813
lindero 2062
lineación 0973
lineación 1607
lineación 2999
lineación de estiramiento 1613
línea de centros 4372
línea de flujo 0868
lineal 4631
línea neutra 0570
linear 0972
lipido 3723
lipotexita 1031
liquefacción del carbón 3609
liquidez, índice de – 4716
liquidez, límite de – 4716
liquido dominante, sistema de – 0393
líquido intercúmulo 0831
líquido madre 3377
líquido que moja 3705
líquido que no moja 3705
líquido residual 3377
líquido residual 0647
listón, en forma de – 0522
lithosfera 0089
lítico 3178
litificación 3421
litificación periférica 3411
litoestratigrafía 1709
litoestratigráfica, unidad – 1748
litoestratigráfica, zona – 1747
litoestratigraphique, horizon – 1749
litofacies 2861
litófago 1857
litofiso 0780
litogénesis 2853
litoidal 0835
litoideo 0835
litología 2853
litoral 2438
litoral 2836
litoral 2889
litoral 2906
litotipo 3579
litotopo 2862
lit-par-lit 1049
lixiviación 3929
lixiviación 4162
lodo 4160
lodo 4829
lodo de perforación 4281
loess, muñeca de – 3489
loma 2796
longitudinal 2289
longulito 0762
lopolito 0735
Love, onda de – 0189
lublinita 2153
lumaquélica 3309
luminescencia 0467
lunamoto 0061
lustre 0484
lustre 3407
lutita 3203
lutitica 3203

Ll

llama de soplete 1350
llanura abisal 2792
llanura aluvial fluvioglaciar 2424
llanura costera 2440
llanura de arena 2157
llanura de arena 2910
llanura de barro 2910
llanura de derrubios 2006
llanura de inondación 2239
llanura de lavado 2424
llanura litoral 2440
llegada 0172
llegada, primera – 4428
llegada violenta 0173
llenado, nivel de – 2251
lluvia de erupción 1373

lluvia de sangre 2159
lluvia, huella de – 3041
lluvia, impresión de – 3041

M

maar 1433
macadam 4843
macaluba 1395
macareo 2823
macarrón 2131
maceración 3594
maceral 3566
maceral activo 3578
maceral, grupo – 3566
maceral, perhidro– 3577
maceral, subhidro– 3577
maceral, variedad – 3567
machacadora de mandibulas 4148
macizo peneplanizado 2588
macizo truncado 2588
macla 0535
macla de contacto 0537
maclado polisintético 0538
macla, ley de la – 0535
macla por penetración 0537
macrocristal 0608
macrocristalina 0797
macroespora 1837
macroevolucién 1970
macropoliesquemático 0794
macrosférico 1868
macrosísmica, región – 0248
macrosismo 0138
macrospora 1837
madreperla 0503
maduración edáfica 2685
madurez 3725
madurez, índice de 3726
maduro 3176
maduro 3725
mafélsica 0598
máfico 0599
mafita 0599
magma 0639
magma 1198
magma latente 0643
magma original 0645
magma palingenético 0644
magma primario 0644
magma primario 0645
magma residual 0646
magma superficial 1244
magmática 0616
magmático, cuerpo – 0730
magmático, macizo – 0730
magmatogénico 3973
magnético, momento – 0296
magnético terrestre, campo – 0276
magnetisanda, fuerza – 0317
magnetismo, lito– 0293
magnetización, intensidad de – 4531
magnetohidrodinámica 0278
magnetómetro 0316
magnetómetro aeroportado 4522
magnetómetro a protón 4524
magnetómetro astático 0319
magnetómetro horizontal 4521
magnetómetro nuclear 4524
magnetómetro pájaro 4522
magnetómetro vertical 4521
magnetopausa 0277
magnetosfera 0277
magnitud 0220
magnitud, escala de – 0220
magnitud unificada 0220
maleable 0471
maleza 3554
malla 3157
malla, en – 0964
malla romboédrica 0516
mallosísmica, región – 0250
mamelar 4094
manantial 3630
manantial hipogénico 1408
manantial hirviente 1407
mancha de lava 1297
mancha rojiza 2692
manchas de dolomitización 3461
manganeso, nódulo de – 3477
manifold 3798
maniobra 4280
manómetro 4769
manto 1683
manto 3109
manto 3499
manto 4027
manto buzante 1688
manto de arena 3109
manto de ceniza 1320
manto de escoria 1330
manto de lava 1275
manto de recubrimiento 1682
manto, parte inferior del – 0091
manto, parte superior del – 0091
manto rocoso 2008
mantos cabalgantes 1691
mantos de corrimiento, tectónica de – 1679
mantos, paqueto de – 1691
mantos, pila de – 1691
mantos superpuestos, paqueto de – 1691
mantos superpuestos, pila de – 1691
manto terrestre 0091
mapa de correcciones isostáticas iguales 0373
mapa de curvas estructurales 4247
mapa de facies 4245
mapa de gravedad residual 4542
mapa de isocoras 4244
mapa de radar 4400
mapa de suelos 2671
mapa de suelos a nivel de reconocimiento 2671
mapa detallado de suelos 2671
mapa de zonas sísmicas 0244
mapa estructural 4247
mapa, foto– 4383
mapa geológico 4239
mapa geotécnico 4784
mapa isanómalo 0372
mapa isobárico 3854
mapa isogálico 0372
mapa isógamo 0372
mapa metalogenético 3905
mapa paleogeológico 4242
mapa regional de la gravedad 0371
mapa tectónico 4247
mar 2782
mar 2807
mar abierta 2840
mar adyacente 2783
mar ampollada 2824
marca 3018
marca falciforme 2398
marcas de colimación 4368
marcas fiduciales 4368
marchitez, índice de – 2773
mar de piedras 2057
Mar de Tethys 1127
marea 2818
marea, amplitud de – 2819
mareal, zona – 2504
marea roja 2843
marea terrestre 0199
maremoto 2816
marga 3221
marga arcillosa 3223
marga carbonatada 3224
margarita 0766
marino 2900
marisma 2514
marisma 2909
marisma 3532
marisma de manglares 2515
marisma litoral 2514
marismas, cinturón de – 2512
marjal 3537
mar lunar 0051
marmita de gigante 2028
marmita de gigante 2248
marmita de gigante 2389
marmita glaciar 2632
marmita torrenciál 2027
mármol 0920
marmorización 0920
mar, nivel medio del – 2817
masa 0715
masa 4032
masa carbonatada 3324
masa, déficit de – 0343
masa de mineral en sacos 4035
masa estacionaria 0125
masa, exceso de – 0343
masa extrusiva 1239
masa inerte 0125
masa mineralizada 3910
masa rocosa 4722
mascón 0052
masivo 0859
mástil 4267
material de partida 2657
material detrítico 2005
material rocoso 4722
materia mineral 3553
materia orgánica extraible 3724
matriz 0626
matriz 3080
máximo 1019
meandro 2100
meandro 2219
meandro abandonado 2232
meandro, arco del – 2232
meandro, cuello del – 2226
meandro de terraza 2223
meandro de valle 2221
meandro divagante 2220
meandro encajado 2222
meandro, escollo de – 2228
meandro, espolón del – 2227
meandro, estrangulación del – 2229
meandro heredado 2221
meandro libre 2220
meandro, núcleo del – 2227
meandros, zona de – 2224
mecánica de rocas 4664
mecánica de suelos 4664
mechinal 4811
medidor de desplazamiento 0121
medio 2859
medio ambiente 1900
mediogranular 0796
medio poroso 3680
medio primario 0402
medio secundario 0402
mediosilícea 0602
mediterránea, serie– 0690
megaciclotema 3120
megacristal 0608
megacristalina 0797
megaesporangio 1837
megagametófito 1838
megasférico 1868
megasísmica, región – 0249
megaspora 1837
MEH 4575
meiofauna 1853
melanocrático 0588
melanocrato 0588
melanosoma 1038
membrana semipermeable 3397
mena 3913
mena compleja 3914
mena cruda 4138
mena diseminada 4004
mena en bruto 4138
mena masiva 4003
mena mineral 3912
mena monomineral 3914
mena no explotable 3915
mena rica 3915
mercaptano 3716
merma, factor de – 3860
meroplancton 1801
mesa 2556
mesa concentradora 4157
mesa giratoria 4269
meseta basáltica 1257
meseta estructural 2555
meseta tectónica 1589
mesocrático 0587
mesocrato 0587
mesocristalina 0797
mesocumulado 0832
mesohalino 2832
mesonorma 0606
mesopelágico 1910
mesopelágico 2845
mesostasis 0626
mesotermal 3983
mesotermal, región – 0391
mesotrófica 3549
mesozona 0993
meta- 0883
meta-antracita 3562
metabasita 0883
metabentonita 3218
metablastesis 0884
metablástico 0884
metabolismo 1803
metacristal 0976
metales extraibles 4654
metales extraibles en caliente 4655
metales extraibles en frío 4655
metales pesados 4653
metálico 0484
metálico 3911
metálico, no– 3911
metalimnion 2333
metalogenético 3903
metalogenía 3903
metalotecto 3904
metaluminoso 0693
metamórfico 3997
metamórfico, de origen – 3974
metamórfico, grado – 0992
metamorfismo 0879
metamorfismo aloquímico 0881
metamorfismo cataclástico 0892
metamorfismo cáustico 0897
metamorfismo cinético 0892
metamorfismo de choque 0893
metamorfismo de contacto 0896
metamorfismo de dislocación 0892
metamorfismo de impacto 0893
metamorfismo de inyección 0889
metamorfismo dinámico 0892
metamorfismo, dinamo– 0892
metamorfismo, doble zona de – 1139
metamorfismo estático 0887
metamorfismo inicial 0902
metamorfismo isoquímico 0881
metamorfismo local 0891
metamorfismo mecánico 0892
metamorfismo múltple 0901
metamorfismo orgánico 3521
metamorfismo progresivo 0903

metamorfismo regional 0887
metamorfismo retrogrado 0904
metamorfismo superpuesto 0901
metamorfismo termal 0896
metamorfismo termico 0896
metamorfosis 1794
metasoma 1067
metasomatismo 3956
metasomatismo de contacto 0900
metasomatismo regional 1056
metastero 1041
metatectita 1031
metatecto 1040
metatexia 1030
metatexita 1030
meteorito 0063
meteorito férrico 0066
meteorito férrico-rocoso 0070
meteorito rocoso 0071
meteoritos, ciencia de los – 0062
meteorizacíon 1988
meteorización 2928
meteorización, costra de – 2010
meteorización esferoidal 2003
meteorizacíon física 1989
meteorización mecánica 2928
meteorización por helada 1996
meteorización por hidratación de sales 1997
meteorización química 1998
meteorización químiqua 2928
meteorización selectiva 2004
meteoro 0063
método de bucles horizontales 4575
método de bucles verticales 4575
método de líneas equipotenciales 4558
método de polarización inducida 4569
método de potencial espontáneo 4557
método de relación de caida de potencial 4560
método de resistividad 4561
método de tomo de tierra 4559
método eléctrico 4554
método electro-magnético 4572
método magneto-telúrico 4553
método telúrico 4553
MEV 4575
mezcla de suelos 4805
miarolítica 0873
micáceo 0775
micáceo 3177
micrita 3281
microacantilado 2470
microcoquina 3308
microcristalina 0797
microcristalino 3091
microespora 1836
microesporangio 1836
microevolución 1970
microfósil 1790
microgametófito 1838
microgranular 0797
microlítica 0836
microlito 0759
microlitotipo 3581
micrometeorito 0064
micromolinete de sondeo 3669
micropaleontología 1786
microplancton 1830
micropliegue 1643
microscopía de menas 3925
microscopio de polarización 0611
microscopio petrográfico 0611
microsférico 1868
microsísmica, región – 0251
microsismo 0097
microspora 1836
microtectónica 1479
microterremoto 0416
microtextura 0790
mictita 1023
miembro 1753
miembro inferior 1576
miembro mediano 1576
miembro superior 1576
migajosa 2727
migma 1022
migmatita 1022
migmatitización 1022
migración 3748
migración 4457
migración, camino de – 3751
migración capilar 3624
migración horizontal 3750
migración lateral 3750
migración primaria 3749
migración secondaria 3749
migración vertical 3750
migración, vía de 3751
miligal 0367
milonita 0915
milonita 1541
milonita 4729
milonítica 0969
milonitización 0915
mimética 1008
mineragrafía 3925
mineral 0460
mineral 3912
mineral accesorio 3174
mineral accesorio 0594
mineral aprovechable, bolsada de – 4178
mineral artificial 0461
mineral característico 0595
mineral característico 0689
mineral constituyente 0462
mineral crudo 4138
mineral de arcilla 3213
mineral de contacto 0978
mineral de embarque 4139
mineral en bruto 4138
mineral englobante 0844
minerales, beneficio de – 4137
mineral esencial 0594
mineral essentiel 0462
mineral estable 3175
mineral fundamental 0595
mineral guía 0988
mineral huésped 0844
mineral índice 0689
mineral índice 0988
mineralización 2682
mineralización 3936
mineral ligero 3173
mineral metalífero 0462
mineral normativo 0607
mineral oculto 0757
mineralogía 0459
mineraloide 0460
mineral pesado 3173
mineral posible 4210
mineral posterior 0844
mineral primario 0596
mineral primario 3913
mineral probable 4209
mineral probado 4209
mineral reconocido 4209
mineral relíctico – 0985
mineral secundario 0596
mineral técnico 0461
mineral tipomorfo 0988
miogeosinclinal 1096
miospora 1833
mirmequita 0785
mirmequítica 0852
mixoeuhalino 2834
mixohalino 2833
mixtos 4142
mobilismo 1113
mobilista, doctrina – 1113
modal 0593
modelado 4478
modelo 4775
modelo analógico 4782
modelo analógico de resistencia eléctrica 4783
modelo centrifugo 4781
modelo de estratos horizontales 4568
modelo, ensayo sobre – 4778
modelo equivalente 4778
modelo físico 4778
modelo fotoelástico 4783
modelo geomecanico 4779
modelo matematico 4776
modelo numérico 3866
modelo por rozamiento en la base 4780
moder 2744
modo 0166
modo 0593
modo fundamental 0167
modo superior 0167
modulación de frecuencía 4492
mofeta 1402
Mohorovičić, discontinudad de – 0086
Mohr, círculo de – 4690
Mohr-Coulomb, criterio de fractura – 4694
Mohr, curva intrínsica de – 4691
Mohr, diagrama de esfuerzos de – 4688
Mohs, escala de dureza de – 0473
Mohs, escala de dureza de – 4701
mol 3675
molasa 2926
molde 1824
molde de cristal de sal 3042
molde del tronco de un árbol por lava 1292
molde externo 1824
molde interno 1824
molienda autógena 4147
molienda, circuito de – 4146
molinete 4759
molinete, ensayo – 4759
molino 4149
molino de barras 4149
molino de bolas 4149
molino glaciar 2389
momento 1759
moneda actual, en – 3899
moneda constante, en – 3899
monística, hipótesis – 0020
monoclinal 1617
monoclínico 0515
monoesquemático 0794
monofilético 1960
monometálico 3911
monomicta 0589
monomicto 3265
monomineral 0589
monomineralógico 0589
monotípico 1945
mons 0037
montaje 0586
montaña alta 2575
montaña con estructura fallada 2539
montaña de circos 2618
montaña de plegamiento 2538
montaña de pliegues-falla 2539
montaña en bloques 2539
montaña en domo 2538
montaña en horst 2540
montaña, espolón de – 2605
montaña marginal, cadena de – 2542
montaña media 2575
montaña residual 2607
montañas, cadena de – 1102
montañas, nudo de – 2605
monte isla 2608
monte isla glaciar 2625
montera 3924
montera de oxidación 4125
monte submarino 2798
montmilch 2153
monzonítica 0813
mora de caverna 2149
morfología 1793
morfotectónica 1076
Morkill, fórmula de – 4217
morrena 2403
morrena central 2407
morrena de ablación 2412
morrena de borde 2416
morrena de empuje 2419
morrena de escoria 1295
morrena de flanco 2416
morrena de retroceso 2417
morrena estriada 2409
morrena frontal 2416
morrena inferior 2409
morrena intermedia 1295
morrena interna 2408
morrena lateral 2411
morrena longitudinal 2415
morrena móvil 2405
morrena subglaciar 2409
morrena superficial 2406
morrena terminal 2413
morro 2608
mortero, en – 0966
mosaico 4380
mosaico controlado 4381
mosaico de banda 4380
mosaico, en – 0960
mosaico no controlado 4381
mosaico semicontrolado 4381
mosqueada 0965
mota 2692
motado 2692
moteada 0965
moteado 3462
movilidad 3699
movilidad, relación de – 3699
movilización 1057
movilizado 1057
movil orogénico 1093
movimiento composicional 0999
movimiento en masa 2037
movimiento eustático 2858
movimiento, línea de – 2372
movimiento nagativo 2858
movimiento positivo 2858
muesca del estregamiento del hielo 2395
muestra 4741
muestra, bolsa para – 4229
muestra cincelada 4168
muestra contaminada 4742
muestra de fondo 3855
muestra de forma irregular 4744
muestra de mano 4229
muestra de referencia 0454
muestra de testigo 3593
muestra enriquecida 4179
muestra en roza 3593
muestra integral 4750
muestra no contaminada 4742
muestra original 4742
muestra, partir una – 4181
muestra por ranura 4168
muestra, reducción de – 4180
muestra salada 4179
muestras, mezcla de – 4176
muestra tomada al azar 4169
muestra triturada 4743
muestreador 4745
muestreador con pistón 4746
muestreador dirigido 4746

muestreador sumergible 4747
muestreo 4423
muestreo de control 4177
muestreo de frentes de trabajo 4167
muestreo de sondeo 4167
muestreo, distribución de – 4171
muestreo piloto 4175
muestreo por captura de una corriente 4170
muestreo, red de – 4171
multibanda, sistema – 4326
Munsell, código de colores – 2676
muro 0705
muro 3106
muro 3501
muro 3515
muro 4009
muro 4010
muro anular 1452
muro de contención 4797
muro de sostenimiento 4797
mutación 1982

N

nácar 0503
nacimiento 2193
nafta 3744
naftenos 3717
nanofósil 1790
nanoplancton 1799
neánico 1866
nebular, hipótesis – 0021
nebulita 1054
nebulosa solar 0021
necton 1798
nectónico 1798
neis 0936
nematoblástica 0962
neo-darwinismo 1966
neoestratotipo 1745
neo-lamarckismo 1967
neomineralización 0905
neomorfo 1983
neontología 1787
neosoma 1037
neotipo 1951
neovulcanismo 1150
nepiónico 1866
neptunismo 0004
neptunismo 3926
nerítico 2841
nerítico 2915
Neumann, líneas de – 0068
neumatolisis 3950
neumatolisis de contacto 0900
neumatolítico 3977
neumatolítico, estadio – 0653
nicho 1902
nicho de arranque 2052
nicho de partida 2052
nido 3105
nido 4032
nieve aptastada por el viento 2347
nieve blanda 2346
nieve de primavera 2350
nieve en polvo 2346
nieve granular 2350
nieve húmida 2349
nieve penitente 2365
nieve seca 2349
nieves perpetuas, nivel de las – 2354
nitrato de Chile 3384
nivación 2357
nivación, nicho de – 2612
nivelación 2589
nivelamiento 2589
nivelamiento parcial, nivel de – 2603
nivel cero 4577
nivel constante, ensayo de inyección a – 3711
nivel de agua 3777
nivel de crestas 2597
nivel de cumbres 2597
nivel de referencia 4249
nivel de ruido 0097
nivel de terrazas 2294
nivel estático 3666
nivel freático 3642
nivel guía 3500
nivel piezométrico 3646
nivel variable, ensayo de inyección a – 3711
niveo-eolico 2870
nodal, línea – 0234
nodal, plano – 0234
nodal, superficie 0234
nodular 3473
nódulo 0771
nódulo 3473
nódulo 4085
nódulo de manganeso 3477
nomenclatura binómica 1922
nomenclatura monómica 1922
nomenclatura trinómica 1922
norma 0606
novaculita 3366
nube ardiente 1366
nube de ceniza 1363
nube de ceniza electrizada 1365
nube volcánica 1360
nube volcánica en forma de coliflor 1362
nube volcánica en forma de pino 1362
nucleación 1360
núcleo 1070
núcleo 1628
núcleo 4861
núcleo de sal 1695
núcleo desprendido 1628
núcleo, parte exterior del – 0093
núcleo, parte interior del – 0093
núcleo separado 1628
núcleo terrestre 0093
nunatak 2618
nunatak volcánico 1289

O

obducción 1133
obsecuente 2285
observaciones, densidad de – 2675
observación vulcanológica 1190
obsidiana 1262
oceánico 2842
oceánico 2918
oceanización 1142
océano 2782
oceanografía 2781
oceanología 2781
octaedrito 0067
oficalcita 0930
ofiolita 0696
ofiolita 1098
ofítica 0845
oftalmita 1052
ojiva 2378
ojo de gato 0488
ojo de gato, efecto de – 0488
ojos 0952
ola, altura de – 2808
ola de arena 3007
ola de traslación 2815
ola, longitud de – 2808
ola, período de – 2808
oleaje 2807
olefinas 3717
oligohalino 2832
oligomícto 3265
oligotrófica 3549
olistolito 2977
olistostroma 2977
olla 1463
ombrógena 3550
omisión de capas 1557
omisión estratos 1557
oncolito 3300
onda acústica 4482
onda canalizado 0195
ondación 1122
ondación, hipótesis de la – 1121
onda cónica 0186
onda de acoplamiento 0193
onda de cizalla 0184
onda de compresión 0183
onda de choque 0274
onda de choque 4481
onda de dilatación 0183
onda de frontera 0196
onda de gravedad 0202
onda del manto 0194
onda del núcleo 0180
onda de presión 0182
onda de superficie 0188
onda de volumen 0179
onda, frente de – 4470
onda guiada 0195
onda larga 0188
onda, lentetud de – 0168
onda, longitud de – 0162
onda longitudinal 0182
onda, normal al plano de la – 0549
onda, número de – 0162
onda P 0182
onda, plano de la – 0549
onda primaria 0182
onda S 0184
onda secundaria 0184
onda sísmica 0161
onda sísmica 4481
onda sísmica interna 0179
onda superficial 0188
onda superficial 4439
onda, superficie de la – 0549
ondas W 0191
onda T 0201
onda transformada 0185
onda transversal 0184
onda, velocidad de – 0168
onda visible 0192
ondulación 4727
ontogénesis 1794
ontogenía 1794
ontología comparada 0007
ooespararenita 3285
ooesparita 3285
ooesparrudita 3285
oolítico 3479
oolito 3479
oomicrita 3286
oomicrudita 3286
opacita 0610
opaco 0545
opalescencia 0486
operación, convenio de – 3901
operación unificada, convenio de – 3901
operador 4447
Oppel-zona 1780
orbicular 0867
orden 0685
orden 1929
orden 2666
ordenada en el origen 4463
orgánica, materia – 2740
orientación cristalográfica 1013
orientación dimensional 1013
orientación, estudio de – 4639
orientación preferencial 1013
orientación preferente 1013
oriente 0486
origen 2943
origen, centro de – 1969
orla continental 2789
oroclino 1140
orogénesis 1100
orogénico, movimiento – 1100
orogénico, período – 1106
orogénico tardío 1108
orógeno 1101
orto 0986
ortoacumulado 0832
ortoconglomerado 3255
ortocuarcita 3230
ortodolomía 3320
ortoestereoscopio 4367
ortofídica 0826
ortogénesis 1975
ortogeosinclinal 1095
ortomagmático 3976
ortomagmático, estadio – 0651
ortoquímica 3268
ortoquímico 3271
ortorómbico 0515
ortoselección 1975
oscilación, hipótesis de la – 1119
oscilación libre 0198
oscilación toroidal 0198
otolito 1865
oulopholita 2151
oxidación 0416
oxidación-reducción, reacción de – 0416
ozokerita 3735

P

pacífica, serie– 0690
paisaje de morrena de fondo 2630
paisaje de morrenas terminales 2630
paisaje escalonado 2599
palagonita 1254
palagonitización 1254
paleobotánica 1827
paleocanal 4029
paleoclimatología 0011
paleocorriente 2864
paleoecología 1889
paleoespástica, reconstrucción – 0009
paleoetología 0007
paleogeografía 0011
paleoicnología 1872
paleoictiología 1848
paleomagnetismo 0294
paleontología 1786
paleosoma 1037
paleosuleo 3126
paleotemperatura 2864
paleotermometría 0456
paleotipo 0636
paleovulcanismo 1150
paleozoología 1847
palingénesis 1024
palingénesis 1968
palingenético 1024
palinología 1828
palinomorfo 1829
palúdico 2887
palustre 2887
palustre 3532
panal 2118
panalotriomórfica 0808
pandémico 1814
panel 4199
Pangea 1116
panidimórfico 0808
pantalla de impermeabilización 4861

pantalla drenante 4804
pantano 2330
pantano 3532
pantano de esfagno 3543
pantano de pendiente 3534
pantano de valle 3534
pantano, hierro de – 4116
pantano oscilante 3530
paquete de capas 3498
paquete productivo 3498
para 0986
para-apogéo 1985
paraconglomerado 3256
paracristalino 0997
paraestratotipo 1743
parafina 3717
paragénesis 0586
paragénesis 3907
paragénesis diagenética 3394
paragenético 0586
paralaje 4364
paralajes, cuña de – 4365
parálico 2898
paramagnetismo 0298
paramarginal 4212
parámetro focal 0151
paramorfismo 0465
paratipo 1950
parautóctono 0727
paravolcánicos, fenómenos – 1180
pared 0705
pared 1529
parentesco 0629
pareo 4852
par estereoscópico 4362
parrila 4144
parte alta 3539
parte baja 3539
parte central estirada 1680
parte encajada 2390
par termoeléctrico – 0379
parte terminal 0175
partición 2102
partición 2935
partícula 3078
partícula, diámetro de la – 3193
partículas, tamaño de – 2710
partícula, tamaño de – 3193
partidor 4182
pasada fotográfica 4358
pasa de difluencia glaciar 2621
pasadizo interdunal 2186
pasivo 0619
paso 2240
paso en tubo 2099
pasta 0626
patera 0037
pátina 3403
pátina de desierto 3405
patrón de calibre 4370
pavimento 4840
pavimento de basalto 1301
pavimento de piedras 2656
PDB 0454
pecilita 1262
pedernal 3366
pedernal 3491
pedillanura 2584
pedimento 2582
pedimento 2604
pedimentos coalescentes 2584
pedimentos soldades 2584
pedon 2689
pegmatítico 3978
pegmatítico, estadio – 0653
pegmatitización 1063
pegmatoidea 0814
pegostylita 2152
pelágico 1910
pelágico 2844
pelágico 2919
pelágico-abisal 1910
pelágico-abisal 2847
Pele, lágrima de – 1326
Pele, pelo de – 1326
pelesparita 3292
pelet 3094
pelet fecal 1863
pelet fecal 3094
pelicula 3100
película alocromática 4335
película calcítica 2142
película color-infrarroja 4337
película falso color 4337
película gaseosa 1370
película infrarroja 4336
película, método – 4230
película ortocromática 4334
película pancromática 4334
película registradora 4388
pelita 3208
pelítico 3208
pellet 4150
pellet fecal 1863
pelmicrita 3293
pendante 2130
pendantes 2117
pendiente de buzamiento 2559
pendiente de deslizamiento 2237
pendiente de origen glaciar, ruptura de – 2623
pendiente, ruptura de – 2199
pendiente socavada 2236
péndulo 0351
péndulo 4774
péndulo invertido 4774
péndulo, longitud equivalente del – 0126
péndulo reversible 0351
penecontemporanea 3474
peneplanización 2589
penesísmica, región – 0252
penetración 4513
penetración, dinámica de – 4711
penetración, estática de – 4711
penetración registrada 4513
penetración, resistancia a la – 4709
penetración standard, prueba de – 4712
penetración, velocidad de – 4298
penetrómetro 4758
penetrómetro estático 4758
penillanura 2589
penillanura embrionaria 2593
penillanura primaria 2593
penillanura relicto 2594
peñon 2608
peralcalino 0693
peraluminoso 0693
percolación 3622
percolación 4635
percolación, presión de corriente de – 4738
percusión, marca de – 3410
perfil ABC 2695
perfil AC 2695
perfil acústico 4480
perfil compuesto 4253
perfil de equilibrio 2198
perfil de equilibrio definitivo 2198
perfil de suelo 2686
perfil enterrado 2693
perfiles seriados 4251
perfil litoral 2441
perfil litoral en equilibrio 2441
perfil longitudinal de un valle 2271
perfil no equilibrado 2198
perfil sísmico poco profundo 4479
perfil transversal de un valle 2271
perforación 1882
perforación 3665
perforación 4261
perforación a percusión 4262
perforación a rotación 4263
perforación a rotación-percusión 4263
perforación con cable 4262
perforación con diamante 4265
perforación con escudo 4853
perforación con turbina 4264
perforación, equipo de – 4267
perforación por batido 4262
perforación rotativa 4263
perforador de balas 3799
periglacial 2633
periglaciar 2633
perimagmático 3985
perímetro mojado 2261
período 0163
período 1765
período libre 0130
peritaje 4216
perla cultivada 0503
perla de caverna 2149
perlado 0484
perlítica 0878
permafrost 2637
permeabilidad 2753
permeabilidad 3697
permeabilidad, barrera de – 3771
permeabilidad efectiva 3698
permeabilidad magnética 0304
permeabilidad primaria 3701
permeabilidad relativa 3698
permeabilidad secundaria 3701
permeámetro 4755
persistencia 4726
pertita 0785
pertítica 0851
pesca, herramientas de – 4287
pescar 4287
petrificación 3457
petroblastesis 1059
petrogénesis 0581
petrografía 0579
petrografía sedimentaria 2851
petróleo 3712
petróleo/agua, contacto – 3777
petróleo/agua, nivel – 3777
petróleo, altura de la columna de – 3775
petróleo, columna de – 3775
petróleo en tanque 3859
petróleo seco 3714
petróleo subsaturado 3857
petrología 0578
petrología de menas 3925
petrología estructural 0995
petrología sedimentaria 2851
petromorfo 2126
petroquímica 0580
petrotectónica 0996
pez 3743
pez, piedra – 1262
pF 2771
pH 0422
phi, escala – 3163
picnoclina 2830
pico y valle 0164
pié cúbico 3892
piedemonte abancalada 2604
piedemonte, glacis de – 2583
piedemonte, plataforma de – 2604
piedra 3242
piedra machacada 4876
piedra preciosa 0478
piedra sintética 0479
piezocristalización 0633
piezógeno 0977
piezómetro 3646
pila 4845
pilar de tierra 2029
pilar tectónico 1589
pilotaje 4838
pilotáxica 0825
pilote 4838
pilote columna 4839
pilote de fricción 4839
pilote flotante 4839
pilotes, equipo para hinca de – 4838
pináculo arrecifal 3329
pipa 4037
piperno 1342
pipkrake 2639
pirámide de tierra 2029
piritización 3458
pirobitumen 3737
piroclasto 1309
piroclasto 3170
piroclastos 1309
pirogénesis 0616
pirogénico 0655
pirólisis 3730
piromagma 1200
piromérido 0876
pirometamorfismo 0896
pirometasomatismo 0900
pirometasomatosis 0900
pirómetro 1195
pirómetro óptico 1195
pisada 1877
piso 0705
piso 1762
piso 3515
pisolita 1321
pisolita 3478
pisolítico 3478
pisolito 3478
pista 1877
pista 1878
pistas de solución 2087
pitch 4019
pitón 1424
pizarra 0942
pizarra 3220
pizarra carbonosa 3505
pizarra fuerte 3502
pizarra hojosa 3220
pizarra mosqueada 0941
pizarras cupríferas 4130
pizarra silícea 3367
pizarrosidad 1599
pizarroso 3220
placa 3132
placa para ensayo de carga 4708
placas, tectónica de – 1125
placer 3236
placer 3997
placer de playa 2912
planar 0972
planar 3183
plancha 0739
plancton 1799
planctónico 1799
planeta terrestre 0025
planetesimal, hipótesis – 0022
planetología 0023
planicie 0029
planicie 0033
planicie 2797
planicie anulare 0030
planicie anulare 0053
planización lateral, zona de – 2604
plano auxiliar 0238
plano axial 1631
plano de intersección 2557
plano esférico, proyección – 1021
plano geotécnico 4784
planta acumuladora 4647
planta convertidora 4647
planta de beneficio 4149

plantilla paraláctica 4373
plantilla perspectiva 4373
plasticidad 2747
plasticidad 4669
plasticidad, índice de – 4716
plasticidad, índices de – 2748
plasticidad, límite de – 4715
plasticidad, límite inferior de – 2749
plasticidad, límite superior de – 2749
plataforma 0705
plataforma carbonatada 3322
plataforma continental 2788
plataforma cratónica 1083
plataforma de abrasión 2479
plataforma de acumulación 2480
plataforma establizadora 4539
plataforma estructural 2555
plataforma insular 2791
platina universal 0614
playa 2485
playa 2580
playa 2892
playa alta 2487
playa baja 2487
playa costera 2892
playa, cresta de – 2490
playa de acreción 2492
playa, escarpe de – 2489
playa, microacantilado de – 2489
playa normal, cresta de – 2490
pleamar 2818
pleamar media 2820
plegamiento 0058
plegamiento embrional 1666
plegamiento en silla de montar 1635
plegamiento enterolítico 3076
plegamiento, fase de – 1107
plegamiento, fase de – 1669
plegamiento fluidal 1701
plegamiento incipiente 1666
plegamiento penecontemporáneo 1667
plegamiento precursor 1667
plegamiento sobreimpuesto 1668
plegamiento, tectónica de 1615
plegamiento, zona de – 1674
pleistosísmica, línea – 0246
pleistosista 0246
pleistosiste, región, – 0246
pleocroísmo 0562
plesiotipo 1952
plexo 1961
pliegue asimétrico 1644
pliegue buzante 1637
pliegue cabrío 1651
pliegue cilíndrico 1654
pliegue cobijante 1576
pliegue concéntrico 1652
pliegue cónico 1655
pliegue cruzado 1678
pliegue de arrastre 1570
pliegue de cizallamiento 1662
pliegue de estiramiento 1662
pliegue de fondo 1120
pliegue desenraizado 1659
pliegue diapírico 1698
pliegue disarmónico 1663
pliegue, eje del – 1630
pliegue en abanico 1650
pliegue en cofre 1649
pliegue en rodilla 1648
pliegue en zigzag 1651
pliegue, estadio embrional de un – 1666
pliegue-falla 1577
pliegue inclinado 1646
pliegue intrafoliar 1660
pliegue invertido 1647
pliegue no-cilíndrico 1654
pliegue paralelo 1652
pliegue parásito 1677
pliegue policlinal 1657
pliegue principal 1642
pliegue producido por desprendimiento 3055
pliegue ptigmático 1658
pliegue recumbente 1576
pliegue recumbente 1680
pliegue replegado 1668
pliegue secundario 1642
pliegues en bucle 1665
pliegues en cascada 1706
pliegues en relevo 1676
pliegues, fajo de – 1674
pliegue simétrico 1644
pliegue simétrico 1645
pliegue similar 1653
pliegues recumbentes apilados 1681
pliegues, sistema de – 1670
pliegue supratenuo 1664
plistosística, línea – 0246
plomada 0329
plomo anómalo 0441
plomo a un estadio, modelo de evolución del – 0439
plomo común 0437
plomo, curvas de evolución del – 0438
plomo, edad convencional modelo del – 0440
plomo en varios estadios, modelo de evolución del – 0439
plomo-plomo, datación – 0434
plomo primordial 0437
plomo radiogénico 0437
pluma del manto 1145
plurimetamorfismo 0901
plutón 0715
plutónica 0617
plutónico 0620
plutónico 3988
plutónico, complejo – 0620
plutonismo 0004
plutonismo 0615
plutonismo 3927
pneumatogénico 0655
pneumatolisis 3950
pneumatolítico 3977
pneumatolítico, estadio 0653
población 1941
pocillo de goteo 2139
poder calorífico inferior 3604
poder calorífico superior 3604
podsolización 2680
poeciloblástica 0974
poikilohalino 2831
poiquilítica 0844
poiquiloblástica 0974
poiquiloblasto 0981
poiquilofítica 0846
Poisson, coeficiente de – 4700
Poisson, número de – 4700
polaridad, época de – 0292
polaridad invertida 0290
polarización 0236
polarización 4557
polarización magnética 0296
polen 1839
poliesquemático 0794
polifilético 1960
poligénico 0590
polígono de piedras 2653
polígono de sal 3049
polígono de tundra 2655
polihalino 2832
polimetálico 3911
polimícto 3265
polimineral 0589
polimorfismo 0464
polimorfismo 1811
politípico 1945
politipismo 0463
politópico 1819
polje 2074
polo 1018
polo de la corona 1020
polo geomagnético 0280
polo magnético 0282
polvo 3206
polvo cósmico 0065
polvo de abrasión 3209
polvo meteorítico 0065
polvo volcánico 1316
populación 1941
porcelanita 3374
porcentaje de 15 atm 2773
porfídica 0817
pórfido 0817
porfidoblástica 0956
porfidoblasto 0981
porfidoclasto 0981
porfírica 0817
porfirítica 0817
porfirito 0817
pórfiro 0817
porfiroblástico 3470
porfiroida 0957
poro 3083
porosidad 2752
porosidad 3681
porosidad, coeficiente de – 3682
porosidad debida a cavidades 3692
porosidad debida a moldes 3691
porosidad de la matriz 3687
porosidad efectiva 3704
porosidad intergranular 3688
porosidad por fractura 3694
porosidad por interconexión 3695
porosidad por molde 3691
porosidad primaria 3684
porosidad relacionada con la textura – 3693
porosidad secundaria 3685
porosidad sin cemento 3442
porosidad total 3681
poros, relleno de – 3440
poros, volumen de los – 3681
posición estructural 1480
postcinemático 0906
postcinemático 1079
postcristalino 0997
postorogénico 0726
postorogénico 1108
posttectónico 0726
posttectónico 0906
postvolcánicos, fenómenos – 1374
potasio-argon, datación – 0449
potasio-calcio, datación – 0449
potencia 4449
potencial 3645
potencial de contacto 4615
potencial de membrana 4616
potencial electro-cinético 4617
potencial iónica 0419
Potsdamiano, sistema – 0334
Potsdamiano, valor – 0334
pozillo 4235
pozo 2105
pozo 3665
pozo 4261
pozo 4848
pozo bombeado 3791
pozo cerrado 3811
pozo con producción de agua 3794
pozo, controlar el – 4290
pozo de auxilio 4290
pozo de delimitación 3785
pozo de descubrimiento 3784
pozo de evaluación 3785
pozo de extrapolación 3785
pozo de inyección 3793
pozo del límite de permiso 3792
pozo de producción 3786
pozo descubierto 4277
pozo desnudo 4277
pozo desviado 4291
pozo desviado 4293
pozo dirigido 4291
pozo impermeable 3794
pozo inclinado 4293
pozo incontrolado 4288
pozo intermedio 3843
pozo marginal 3845
pozo, "matar" el – 4290
pozo muerto 3812
pozo perdido 4287
pozo piloto 4297
pozo productivo 3789
pozo, reconocimiento del – 4294
pozo seco 3794
pozo surgente 3790
pozo, tapar el – 4290
Pratt 0345
preadaptación 1980
precinemático 1079
precinemático 0906
precios al por mayor, índice de – 4223
precipitación 0411
precipitación 3614
precipitación, barrera de – 4642
preconsolidación 4703
precordillera 2605
precristalino 0997
precursor 0150
predominancia impar 3728
prefosa 1109
preponderante 1816
presa de arco gravedad 4865
presa de bóveda 4865
presa de contrafuertes 4864
presa de escollera 4859
presa de gravedad 4864
presa de tierra 4859
presión activa del terreno 4706
presión, bomba de – 3848
presión, caída de – 3853
presión de abandono 3849
presión de fondo 3847
presión de fondo, registrador de – 3848
presión de gas disuelto 3823
presión de gases, aumento retrógrado de – 1174
presión de gas libre 3823
presión del terreno en reposo 4705
presión del yacimiento 3819
presión, descarga de – 1173
presión de sobrecarga 3819
presión en erogación 3847
presión estatica 3847
presión, gradiente de – 3852
presión granular 3819
presión hidráulica 3824
presión hidráulica 4737
presión hidrostática 3820
presión hidrostática ascendente 4740
presión inicial 3849
presión lateral del suelo 4705
presión lateral del terreno 4705
presión magmática 0622
presión, mantenimiento de – 3832
presión natural del terreno 4705
presión pasiva del terreno 4706
presión, registro de subida de – 3850
presión total, celula de – 4771
prestamo 4871

presurómetro 4760
pretectónico 0906
prioridad, ley de – 1924
prismática 2730
prismático 0522
probabilista, combinación – 3877
proceso 2568
proceso discontinuo 4145
Proctor, ensayo de compactación– 4757
producción, capacidad de – 3806
producción, cruz de – 3796
producción cumulativa 3817
producción del pozo, capacidad de – 3806
producción final 3817
producción inicial 3805
producción, marcha de la – 3801
producción máxima 3807
producción, máxima de – 3807
producción, medir la – 3798
producción, puesta en – 3795
producción total 3817
producción, valor medio de – 3801
productividad, índice de – 3808
productividad, índice especifico de – 3808
profunda 0617
profundidad focal 0157
profundidad máxima 0214
profundidad, zona de – 0993
prolongación hacia abajo 4533
prolongación hacia abajo 4543
promontorio 2446
promontorio alado 2446
promontorio estructural 1639
propagación, curva de – 0171
propagación, tabla de tiempos de – 0171
propagación, tiempo de – 0171
propilita 0921
propilitización 0921
prorrata 3902
prospección en redes de drenaje 4641
prospección en roca 4644
prospección en sedimentos fluviales 4641
prospección en suelos 4644
prospección por gases 4648
protactinio-thorio, datación – 0445
protoclástica 0857
protodolomita 3321
protomilonta 0916
prototipo 4775
provinca geotérmica 0390
provincia faunística 1849
provincia geológica 4260
província geoquímica 0401
provincia metalogénica 3908
provincia petrográfica 0628
provincia petrológica sedimentaria 2852
proximal 2997
proyección de rocas 4818
proyecto antisísmico 0260
prueba 3787
prueba 3788
psammita 3226
psammítico 3225
psammítico 3226
psefita 3238
psefítico 3237
psefítico 3238
pseudoarrecifal 3334
pseudofósil 1789
pseudomórfico 1823
pseudomorfismo 0465
pseudomorfismo 0982
pseudomorfo 0982
pseudomorfo 1823
pseudomorfo según 0982
pseudonódula 3060
pseudopalagonita 1255
pseudoporfídica 0819
pseudotaquilita 0917
pseudotaquilita 1542
pseudovolcán 1189
pseudovolcán en escudo 1430
ptigmática 1051
pudinga 3253
puente 2102
puente kárstico 2079
puente natural 2079
pulido 3404
pulimentación 2399
pulimento 3404
pulimento glaciar, límite del – 2631
pulverización 1319
pulverización 3526
pumita 1308
pumítica 0875
pumítica 1308
punta 2449
punta caliente 1144
punta de inflexión 4577
punto 0494
punto de burbujeo 3857
punto de escape 3762
punto de toma 4348
punto de toma aérea 4348
punto principal 4369
punto principal conjugado 4372
putrefacción 3526
puy 1425
puzolana 1338
pyramidal 0522

Q

quebradizo 0471
queirónimo 1956
quelación 0409
quelación organo-metálico 0409
quelifítica, aureola – 0786
quilate 0482
quimogenético 1040
quiriosoma 1039
quiste 1841

R

radar 4384
radar coherente 4387
radar de apertura real 4385
radar de apertura sintética 4386
radar de onda contínua 4387
radar lateral 4385
radar, reflectividad de – 4392
radar, sombra de – 4391
radiación 0056
radiación calorífica 0381
radiación de calor 0381
radiación natural, sonda de – 4590
radiado 0776
radioactividad, hipótesis de – 1117
radiocarbono, datación – 0451
radio hidráulico 2262
radiolarios, barro de – 3368
radiolarios, tierra de – 3368
radiolarita 3368
radiolarita 3369
radiolítica 0843
radio medio 2262
radiómetro 4310
raices, región de – 1684
raices, zona de – 1684
raiz 0343
raíz 1104
rama 1582
ramificación 0723
ramificación 2210
ramificación 3826
ramificación 4071
ramificación de filones 4070
rango 0685
rango dinámico 0132
ranura 1538
rapaquivi, en – 0828
rápida 2202
rarefacción 0187
rasa 2479
rastro 1878
raya 0466
raya 1538
Rayleigh, onda de – 0189
rayo 0056
rayo de luz 0548
rayo, parámetro del – 0207
rayo sísmico 0206
raza 1939
reacción, borde de – 0786
reacción, zona de – 0786
realimentación 0133
rebosadero 2339
rebote elástico, teoría del – 0108
recalce 4796
recapitulación, ley de – 1968
recarga artificial 3671
receptor 4499
receptor programado 4499
rechazo horizontal 1551
rechazo horizontal aparente 1553
rechazo horizontal estratigráfico 1552
rechazo vertical 1551
rechazo vertical estratigráfico 1552
recirculación de gas 3835
reconocimiento del emplazamiento 4790
recorrido 0206
recorrido 1568
recorrido, tiempo de – 0171
recristalización 0905
recristalización 3452
recristalización dinámica 0967
rectificación 4376
rectificar 4296
recubrimiento 3501
recubrimiento 4232
recubrimiento 4788
recubrimiento, coeficiente de – 4202
recubrimiento estereoscópico 4361
recubrimiento rocoso 4788
recuesto 2603
recuperación, factor de – 3867
recuperación final estimada 3869
recuperación, herramientas de – 4287
recuperación primaria 3822
recuperación secundaria 3831
recuperación terciaria 3837
recuperación total 3867
recursos de posible interés económico 4211
recursos especulativos 4214
recursos hipotéticos 4214
recursos identificados 4206
recursos minerales 4205
recursos potenciales 4215
recursos sin identificación 4213
red dendritica, sistema en – 2292
red, densidad de la – 2255
redeposición 2993
red estereocristalina 0507
red fluvial 2205
red fluvial, falta de – 2257
red hidrográfica ausente 2257
red, línea de la – 0508
red, nudo de la – 0508
redondamiento 3187
redondamiento, índice de – 3187
redondeado 3188
redondeado, bien – 3188
red ortogonal, sistema en – 2292
redox, potencial – 0421
red radial, sistema en – 2291
reducción 0416
reducción bacteriana 0417
reducción de aire libre 0358
reducción de Bouguer 0359
reducción, esfera de – 3463
reducción isostática 0361
reducción topo-isostática 0364
reelaborado 2947
reemplazamiento 0931
reemplazamiento 3453
referencia geomagnética internacional, campo de – 4535
referencia, nivel de – 0354
referencia, nivel de – 3854
referencia, plano de – 0357
reflectancia 4316
reflectividad 0574
reflector difuso 4317
reflector especular 4318
reflexión 4436
reflexión, coeficiente de – 4473
reflexión difusa 4317
reflexión especular 4318
reflexión externa 0576
reflexión, horizonte de – 4461
reflexión interna 0576
reflexión, método de – 4405
reflexión, múltiple – 4441
reflexión primaria 4437
reflexión total 0213
reflujo 2822
refracción 2810
refracción, índice de – 0551
refracción, método de – 4405
refractario 0788
regalía 4224
regeneración, secuencia de – 3538
régimen helicoidal 2264
régimen hidráulico 2267
régimen hídrico, control del – 2775
régimen laminar 2264
régimen turbulento 2264
región 3908
regionalidad, grado de – 0340
regionalización 0255
regionalización 4194
registro 4420
registro 4500
registro analógico 4421
registro, carta de – 4500
registro de densidad variable 4421
registro de gradientes 4563
registro de superficie variable 4421
registro digital 4422
registro digital 4501
registro espectral de rayos gamma 4548
registro multicanal 4501
regolita 2008
regresión 1726
rejuvenecimiento 2572
rejuvenecimiento 3959
relajación de esfuerzo 1179
relámpago esférico 1372
relictica 0971
relicto 0985

relicto 1985
relicto armado 0787
relicto estable 0985
relicto inestable 0985
relieve 0558
relieve alpino 2576
relieve apalachiano 2564
relieve escarpado 2576
relieve, fortaleza del – 2575
relieve fuerte 2576
relieve geoquímico 0401
relieve inicial 2569
relieve jurasiano 2565
relieve, magnitud del – 2575
relieve multiconvexo 2600
rellano estructural 2307
rellano submarino 2801
relleno 3105
relleno 4825
relleno 4831
relleno de fisura 1515
relleno filoniano 1515
remolino 2269
rendimiento 0272
rendimiento 3898
reniforme 0776
renio-osmio, datación – 0446
rentabilidad, tasa de – 3898
reología 0102
reología 4671
repetición de capas 1557
repetición estratos 1557
repié 2602
réplica 0147
réplica 1825
réplica externa 1825
réplica interna 1825
reposo, período de – 1169
resaca 2826
resaca de fondo 2826
resalte 2470
resecuente 2284
resedimentación 2993
reserva razonable de mineral 4207
reserva reguladora de abastecimiento 4225
reservas 3868
reservas adicionales indicadas 3883
reservas comprobadas 3879
reservas de mineral 4208
reservas especulativas 3882
reservas, estimación de – 3870
reservas inferidas 3882
reservas posibles 3881
reservas probables 3880
residual 0971
residual 3971
residual, error – 0241
residual, material – 2855
residuo de denudación 2160
residuo insoluble 2930
resinaso 0484
resistencia 4682
resistencia a la cizalladura 4684
resistencia a la compresión 4683
resistencia a la compresión triaxial 4683
resistencia a la compresión uniaxial 4683
resistencia a la torsión, ensayo de – 4765
resistencia a la tracción 4684
resistencia al punto de carga 4733
resistencia eléctrica 4556
resistencia, línea de menor – 4818
resistencia punta 4685
resistencia residual 4685
resistividad aparente 4564
resistividad aparente 4609
resistividad de la formación 4607
resistividad del agua de formación 4608
resistividad específica 4556
resistividad verdadera de la formación 4607
resolución 4347
resolución espacial 4319
resolución horizontal, poder de – 4516
resolución, poder de – 4453
resolución vertical, poder de – 4516
resonancía, teoría de – 0046
resorción 0642
respuesta 4497
respuesta, curva de – 0133
respuesta de amplitud 4448
respuesta de fase 4448
respuesta de frecuencia 4448
respuesta en fase 0135
respuesta punctual 4446
restinga 2462
restita 1032
restita 1066
resurgencia 2090
retacado 4816
retardo 0563
retardo 4822
retención de humedad, curva de – 2772
retención efectiva 3704
retención inicial 3618
retención superficial 3618
retención, tiempo de – 3627
reticulado, método del – 4374
reticulado oblicuo 4375
reticular 0964
reticular, plano – 0508
reticulita 1331
retiro regresivo 1727
retiro regresivo 3112
retrabajado 2947
retracción, límite de – 4715
retrofosa 1110
retrometamorfismo 0904
retromorfismo 0904
retropais 1110
retroplegamiento 1689
retropliegue 1689
retumbo 1237
reverberación 4440
revesa de fondo 2786
revestido 3402
revestimiento 4858
revisión 3886
Reynolds 2265
rheomorfismo 1057
ría 2452
ribera abrigada 2336
ribera cóncava 2235
ribera convexa 2235
ribera escarpada 2236
ribera expuesta 2336
ribera exterior 2235
ribera externa 2461
ribera interior 2235
ribera interna 2461
Richter, escala de – 0220
rift concéntrico 1469
rift radial 1468
rift transversal 1470
rift valley 1590
rift, zona de – 1467
rimaya 2382
río 2191
río competente 2329
río confluente 2207
río de marea 2504
río, desarrollo del – 2197
río incompetente 2329
río, lecho del – 2242
río mareal 2504
río pirata 2325
ripabilidad 4828
ripar 4828
ripple 3000
ritmita 3121
rizadura 2814
rizadura 3000
rizadura, amplitud de la – 3002
rizadura de adhesión 3016
rizadura de corriente 3004
rizadura de interferencia 3013
rizadura de oscilación 3012
rizadura gigante 3007
rizadura incompleta 3015
rizadura, índice de la – 3002
rizadura linguoida 3006
rizadura, longitud de la – 3002
rizadura longitudinal 3011
rizadura lunada 3005
rizadura romboide 3010
rizadura transversal 3003
rizoconcreción 3490
roca 3389
roca 4722
roca aborregada 2400
roca ácida 0601
roca amigdalar 0872
roca amigdaloidea 0872
roca arrecifal 3342
roca asquística 0668
roca carbonatada 3266
roca compacta 3769
roca coerneánica 0935
roca cristalina 0583
roca de muro 3516
roca de playa 3412
roca diasquística 0668
roca efusiva 1240
roca encajante 0709
roca encajante 1530
roca encajante 3920
roca en equilibrio inestable 2163
roca en forma de hongo 2163
roca eolizada 2161
roca extrusiva 1240
roca facetada por el agua 2445
roca facetada por el viento 2162
roca ferruginosa 3464
roca filoniana 0741
roca foliácea 1604
roca formada por síntesis 0682
roca hojosa 1604
roca intrusiva 0715
roca laminar 3132
roca madre 0883
roca madre 2944
roca madre 3918
roca madre de gas 3722
roca madre de petróleo 3722
roca magmática 0616
roca metamórfica 0880
roca mosqueada 0941
roca moteada 0941
roca original 0883
roca piroclástica 1311
roca plutónica 0617
roca primaria 0631
roca redondeada 3242
rocas de caja 3501
roca sedimentaria 2850
roca sin discontinuidades 4722
roca sólida 2011
roca subyacente 4788
roca viva 2011
roca volcánica 1238
roción 2811
rocío, punto de – 3858
rodadura, capacidad de – 3191
rodadura, marca de – 3038
roentgen 4550
rombochasma 1143
rompiente 2811
rompiente, zona de – 2439
rompimiento, teoría del – 0046
rosa 3092
rosa del desierto 3406
Rosiwall, método micrométrico de 0613
rotación, ángulo de – 1126
rotación externa 1000
rotación interna 1000
rotación, polo de – 1126
rotación tectónica 1000
rotura 3431
rotura por calor 1993
rotura por insolación 1993
rozamiento lateral 4839
rozamiento negativo 4839
R, tectonita– 1003
rubidio-estroncio, datación – 0447
rudimentario 1978
rudita 3237
rudítico 3237
rugosidad 4727
rugoso 3190
ruido de fondo 4438
ruido del tiro 4438
ruido generado por la señal 0177
ruido sísmico 0097
ruido sísmico 4438
ruidos volcánicos 1237
rumbo 1483
rumbo 4016
ruptura cíclica 2199
ruptura, propagación de – 0239
ruptura, velocidad de – 0239
ruptura violenta de una roca 0115

S

sacaroidea 0811
sacaroideo 3090
sacudida 0138
sacudida local 0149
sacudida premonitoria 0147
saladar 2514
salar 2894
salbanda 0707
salbanda 4012
sal de roca 3382
sálico 0607
salina 2580
salina 2894
salina, tectónica – 1694
salinidad 2827
salinidad 3376
salinidad 3675
salino 3376
salitral 2909
salitre 2123
salmuera 3676
salobral 2154
salpicaduras, muro de – 1452
saltación 1982
saltación 2954
saltación, marca de – 3039
salto 1982
salvandas 3501
saprolita 2009
saprolita 4128
sapropel 3528
sapropelita 3721
satélite artificial 4302
saturación cromática 2677
saturación en agua de la formación 4611
saturación en agua, exponente de – 4606
saturación en hidrocarburos, coeficiente de – 4610
saturación en petróleo, coeficiente de – 4610

saturación, grado de – 3650
saturación, línia de – 0697
saturación, línea de – 4862
saturación residual en agua de la formación 4611
saturación, zona de – 3656
saturada, zona no – 3651
saturado 0605
saturado; no – 0605
sausurita 0927
sausuritización 0927
sazon 2757
schlieren 0665
Schmidt, fasilla de – 1015
Schmidt, red de – 1015
Schürmann, ley de – 3522
Schürmann, series de – 3935
Se 1004
seca 2816
sección 0685
sección mojada 2261
secreción lateral 3928
sectil 0471
secuencia 3110
secuencia de relieves 2573
secuencia positiva 3129
secuencia volcánica 1238
seda 0490
sedentario 1919
sedimentación 2989
sedimentación, balanza de – 3158
sedimentación, coeficiente de – 2989
sedimentación, grado de – 2989
sedimentación interna 3436
sedimentación rítmica 3121
sedimento 2850
sedimento carbonatado 3266
sedimento interno 3436
sedimentología 2849
sedimento relicto 2856
sedoso 0484
Seger, cono de – 1196
segregación 3395
segregación bandeada 0946
segregación cristalina 0671
segregación en bandas 0946
segregación magmática 3948
segregación magmático 3975
segregación tectónica 0947
seguridad, coeficiente de – 4777
selección natural 1966
sellada, formación – 3765
sellado 4800
sellar 4800
sello 3765
sello de alquitrán 3765
semi-antracita 3560
semicristalizada 0640
semidesintegración, período de – 0431
senderos de cabras 2060
seno 1636
seno, plano del – 1636
sensitividad – 4717
sensitometría 4341
señal 4435
señal 4517
señal acústica 4517
señal de entrada 4445
señal de salida 4445
señal-ruido, relación – 4443
señal sísmica 4517
señal transitoria 0176
separación de inclinación 1550
separación de rumbo 1554
separación en medio pesado 4153
separación, espiral de – 4156
separación magnética 4154
separación normal 1555
separación ortogonal 1555
separación, superficie de – 4461
separador 3797
septarias, nódulo de – 3487
seriada 0820
sericitización 0932
serie 1764
serie 1781
serie cabalgada 1576
serie cabalgante 1576
serie charnockítica 0990
serie de rocas ígneas 0630
serie de suelos 2663
serpentinita 0923
serpentinizanición 0923
sesgo 1614
sésil 1919
seudoarrecifal 3334
seudoestratificación 0869
seudofósil 1789
seudomórfico 1823
seudomorfismo 0465
seudomorfismo 0982
seudomorfo 0982
seudomorfo 1823
seudomorfo según 0982
seudopalagonita 1255
seudoporfídica 0819
seudoporfidoblástica 0959
seudotaquilita 0917
seudotaquilita 1542
seudovolcán 1189
seudovolcán en escudo 1430
sfenochasma 1143
Shore, escala de dureza de – 4701
siderita 0066
siderolito 0070
sifonamiento 4863
sigma-t 2827
signo óptico 0557
signos precursores 1191
silex 3366
silíceo 3365
silícica 0605
silicificación 0929
silificación 3458
sill 0740
sill mineralizado 4031
sima 2097
sima 2104
simbiosa 1898
simplectítica 0849
sincinemático 0906
sincinemático 1079
sinclinal 1619
sinclinal antiforme 1620
sinclinal doblemente buzante 1638
sinclinal periférico 1697
sinclinorio 1671
sincrónico 1729
sincrónico 1818
sineclise 1084
sinecología 1888
sinéresis, grieta de – 3044
sinforma 1619
singénesis 3398
singenética 3553
singenético 2854
singenético 3474
singenético 3966
sinneusis 0821
sinonimía 1955
sinónimo 1955
sinorogénico 1108
sinsedimentario 2854
sintectónico 0726
sintectónico 0906
sinter 3385
sintersilíceo 1406
síntesis, roca formada por – 0682
sintético 4478
sintipo 1951
sinuosidad 2216
sísmica, constante – 0257
sísmica, región – 0247
sismicidad específica 0242
sísmico, evento – 0095
sismo 0138
sismo de impacto 0141
sismo-eléctrico, efecto – 0113
sismógrafo 0116
sismógrafo de banda ancha 0116
sismograma 0116
sismo local 0149
sismología 0094
sismología legal 0264
sismología nuclear 0263
sismología planetaria 0094
sismométrico, tendido – 4417
sismómetro 0118
sismómetro 4414
sismómetros, intervalo entre – 4416
sismómetros, red de – 4416
sismo mundial 0138
sismo principal – 0142
sismo remoto 0150
sismoscopio 0117
sismos, enjambre de – 0144
sismo tectónico 0139
sismo volcánico 0140
sistema 1765
sistema activo 4301
sistema de filones 4063
Slingram 4575
SMOW 0454
sobreconsolidación 4703
sobrecrecimiento epitaxial 3456
sobrecrecimiento sintaxial 3456
sobreimpuesto 2290
sobreimpuesto 4630
sobrepresión tectónica 0886
sobrepressión 3820
sobrepuesto 0970
socavado 2471
socavamiento 2470
solapado 1556
solapada 2596
solapado aparente 1553
solapado estratigráfico 1552
solapado horizontal 1551
solape 3965
solape 4359
solape lateral 4359
solape longitudinal 4359
solar, sistema – 0018
solera 4852
solfatara 1391
solfatara ácida 1393
solidificación dermolítica 1200
solidificación, índice de – 0703
soliflucción 2041
solifluxión 2041
solígena 3550
solución 0411
solución intrastratal 3413
solución por presión 3445
solum 2694
solle 2632
sombra, zona de – 0216
somera 0617
somma 1461
sonda de cinta 4236
sonda lateral 4585
sondeabilidad 4298
sondeo 3665
sondeo 4261
sondeo 4710
sondeo de exploración 3782
sondeo delgado 3783
sondeo dinámico – 4711
sondeo estático 4711
sondeo estratigráfico 3783
sondeo por inyección 4751
sondeo, registro de – 4299
sondeo sísmico 4406
sondeo sísmico profundo 0094
sonido, fuente de – 4486
soplo 1385
sostenimiento 4858
sotavento 2964
sotavento, pendiente de – 2965
sparita 3289
sparker 4410
sparker 4493
Stefanescu, integral de – 4568
Steinmann, trilogía de – 0696
Steinmann, trilogía de – 1098
stenobentónico 1909
stockwork 4058
Stoneley, onda de – 0195
suave 1626
subacuático 2880
subaéreo 2866
subalcalino 0694
subaluminoso 0694
subangular 3189
subarcosa 3233
subarkosa 3233
subbase 4841
subducción, zona de – 1134
subedad 1763
subespecie 1939
subfacies metamórfica 0987
subfamilia 1930
subgénero 1933
subglacial 2873
subgrauvaca 3234
subhedral 0751
sublimación 0412
sublimación 3952
sublimación, producto de – 1394
sublimado 1394
sublitoral 2839
submareal 2908
submarginal 4212
submarino 2900
subofítica 0845
subpiso 1763
subproducto 4163
subredondeado 3188
subsaturado 0605
subsecuente 2286
subsidencia 2857
subsidencia 4704
substrato 0348
substrato 2696
substrato 3107
substrato 4788
subsuelo consolidado 4788
subterráneo 2876
subvolcán 1206
subvolcánica 0617
subvolcánico 3989
subyacente, masa – 0728
suelo 0705
suelo 2657
suelo 4636
suelo 4788
suelo ABC 2695
suelo AC 2695
suelo ácido 2737
suelo ácido de sulfatos 2739
suelo alcalino 2737
suelo azonal 2669
suelo, ciencia del – 2658
suelo de polígonos de tundra 2655
suelo enterrado 2693
suelo estriado 2653
suelo estructurado 2652
suelo expansivo 4855
suelo extruible 4855
suelo, factor de – 0259
suelo, factor de formación del – 2659
suelo, fracción del – 2712
suelo, génesis del – 2659

suelo geométrico 2652
suelo glacial 2875
suelo helado 2635
suelo, humedad del – 2764
suelo intrazonal 2668
suelo lunar 0060
suelo oceánico, formación de – 1128
suelo permanentemente helado 2637
suelo poligonal 2653
suelo rayado 2653
suelo residual 2009
suelo, textura del – 2709
suelo zonal 2667
sulfurización 3964
sumersión 1728
superespecie 1935
superestructura 1414
superfamilia 1930
superficie de envolvimiento 1672
superficie específica 2754
superficie específica 4109
superficie inicial 2569
superficie lístrica 1573
superficie mediana 1673
superficie pulimentada 1538
supergénico 3937
supergénico 3967
superimpuesta 4633
superindividual 1012
superparamagnetismo 0300
superpiso 1763
superposición 1714
superposición, ley de – 1714
superpuesto 0970
supraconsolidación 4703
supracortical 0584
supraestructura 1077
supralitoral 2837
supramareal 2908
supravolcán 1413
suprayacente 4631
supresión 4567
surco 1092
surco 2796
surco litoral 2460
surgencia 1358
surtidor de agua 1359
suspensión 0411
suspensión, corriente de – 2980
sustitución 0931
sustitución acoplada 0477
sustitución atómica 0477
sustrato 3107
sutura 1132
sutural 3450
synantético 0784
synorogénico 0726

T

tabla 0501
tablestacado 4799
tabular 0522
tabular 0755
tabular 3183
taconita 4114
taconomía 1920
tafocenosis 1896
tafonomía 1897
tafrogénesis 1091
talud 4860
talud arrecifal 3354
talud carbonatado 3322
talud continental 2790
talud continental, borde del – 2788
talud dunar 2966
talud frontal 2480
talud, inclinación de – 4827
talud insular 2792
talus de derrubios 2007
tamaño arena 3227
tamaño de bloques 4876
tamaño de grano, distribución de frecuencias de – 3197
tamaño de grano, grado de – 3162
tamizado 3157
tamiz de lodo 4282
tampón 0422
tanatocenosis 1896
tapón 3105
taquilita 1262
tautónimo 1957
taxita 1243
taxogénesis 1963
taxon 1926
taxon 2662
taxonomía de suelos 2662
taxón, zona de distribución de – 1783
tectita 0077
tectofacies 2861
tectogénesis 1100
tectónica 1073
tectónica del granito 0712
tectonita 1001
tectonita 1605
tectonita fundida 1001
tectonita primaria 1001
tectonita-S 1002
tectonita secundaria 1001
tectosfera 0080
tectotopo 2862
techo 3106
techo 3501
techo 4009
techo 4010
techo de pizarra fuerte 3502
techo, roca del – 0705
tefra 1310
tefracronología 1312
teledetección 4300
telemagmático 3986
telescoping 3965
telesismo 0150
teletermal 3984
telúrico 0002
temblor 0138
temblor 0146
temblor de tierra 0138
temperatura, sistema de baja – 0393
tenacidad 0470
tendido perpendicular 4419
tenor de cierre económico 4203
tensión 0103
tensión de humedad 2771
tensión desviatoria 0103
tensiones naturales en el fondo del taladro, ensayo de liberación de las – 4766
tension principal 0103
teñido 0492
terígeno 2902
termal normal, región – 0391
térmico, flujo – 0384
terminar en profundidad 1526
terminar hacia arriba 1526
termoclina 2830
termógeno 0977
termohalino 2829
termokarst 2645
termometría por isótopos 0456
termómetro geológico 0994
termopar 0379
terna estereoscópica 4362
terraplén 4841
terraplén morrénico 2414
terraza aluvial 2298
terraza artificial 2061
terraza, borde de – 2295
terraza buzante 2311
terraza cíclica 2299
terraza climática 2308
terraza continental 2790
terraza de armadura rocosa 2305
terraza de deposición 2337
terraza de deslizamiento 2302
terraza de erosión 2298
terraza de erosión 2337
terraza de espolón 2303
terraza de lava 1285
terraza del lago 2337
terraza de meandros 2304
terraza de ostrución 2310
terraza diastrófica 2307
terraza emergente 2311
terraza estructural 2307
terraza estructural 2555
terraza eustática 2308
terraza fluvial 2293
terraza fluvioglacial 2309
terraza formada por material de solifluxión 2309
terraza glaciar marginal 2420
terraza marina 2484
terraza no cíclica 2299
terraza poligénica 2302
terrazas emparejadas 2300
terrazas en escalón 2296
terrazas, entrecruzamiento de – 2311
terrazas policíclicas 2300
terraza sucesíva 2301
terraza, talud de – 2295
terremoto 0138
terremoto artificial 0265
terremoto continental 0148
terremoto, control de – 0262
terremoto cripto-volcánico 0140
terremoto de colapso 0141
terremoto de relé 0145
terremoto múltiple 0143
terremoto, predicción de – 0261
terremoto, prevención de – 0262
terremoto, riesgo de – 0255
terremoto, ruido del – 0203
terremotos, mecanismo de los – 0232
terremoto submarino 0148
terremoto terrestre 0148
terremoto volcánico 1193
terreno calcinado 0039
terreno caótico 0038
terreno, información en el – 4321
terreno laminado 0040
terrenos, unidades de – 4785
terrestre 2865
terrón 2725
terroso 3086
testificación 3787
testificación por cable 3787
testigo 3155
testigo 4233
testigo 4237
testigo continuo 4749
testigo integral 4750
testigo lateral 4238
testigo, partidor de – 4182
testigo, recuperación de – 4273
testigos, caja de – 4237
testiguero 3155
testiguero 4745
testiguero de gravedad 3156
testiguero de pistón 3156
testiguero de pistón vibratil 3156
Tethys 1127
tetragonal 0515
textura 0790
textura 1605
textura 3077
textura 4345
textura amasada 2649
textura celular 4089
textura en hoja de laurel 4082
textura en mallas 0856
textura en peine 4084
textura espinifex 0850
textura fina 2721
textura fluidal 0868
textura gráfica 0847
textura granular 4086
textura gruesa 2721
textura media 2721
textura ojosa 0952
textura pegmatítica 3470
textura reticular 0856
textura retorcida 2649
textura simplectítica 0783
textura superficial 3192
thalweg 2279
thixotropía 4670
tholeiita 0695
tholus 0037
thorio-plomo, datación – 0436
tiempo de origen 0159
tiempo-distancia, curva – 0171
tiempo, dominio de – 0223
tiempo perdido 3810
tiempo vertical 4428
tierra 3086
tierra armada 4798
tierra fina 2719
tierra firme 2518
tierra silícea 3367
tierras malas 2030
till 2404
tillita 2874
till marino 2903
tilloide 2978
tinte 2677
tipo, área – 1739
tipo, localidad – 1739
tipo, localidad – 1946
tipo, región – 1739
tiro central 4419
tiro, punto de – 4411
toba 1332
toba 3171
toba caótica 1333
tobáceo 1332
tobáceo 3171
toba cristalina 1333
toba de ceniza 1333
toba de escorias 1333
toba de explosión 1335
toba de lapilli 1333
toba de palagonita 1347
toba de polvo 1333
toba de pumita 1333
toba lítica 1333
toba pelítica 1333
toba pisolítica 1333
toba redepositada 1337
toba seca estratificada 1336
toba sedimentaria 1334
toba soldada 1341
toba subacuática 1335
toba vítrea 1333
toleita 0695
tómbolo 2467
tonstein 3505
topo 4854
topógena 3550
topomineral 3995
topotáctico 0543
topotaxia 0543
topotipo 1953
topozona 1784
torbellino 2269
torbellino de humo 1364
tormentas magnéticas 4529
tormenta volcánica 1373
torre 4267

torrencial 2884
torrente de fondo 2388
torrente glaciar 2392
torrente interno 2388
torrente intraglaciar 2388
torrente subglaciar 2388
torrente superficial 2388
torrente supraglaciar 2388
torsión, balanza de – 0350
tortuosidad 3707
tosca 4123
tostar 4151
T, plano – 1017
tracción 2953
trama 3079
trampa 3756
trampa anticlinal 3757
trampa estratigráfica 3758
trampa estructural 3757
trampa hidrodinámica 3761
trampa por cuña 3758
trampa por discordancia 3759
trampa por falla 3757
trampa por paleorelieve 3760
trampa por permeabilidad 3758
transevaporización 3931
transfluencia 2383
transformación 3426
transgresión 1726
transición, zona de – 3777
transición, zona de – 4620
translación 2370
translúcido 0545
transmisión, índice de – 4315
transmisividad 3702
transparente 0545
transporte en masa 2037
transporte eólico 2154
transporte tectónico 0998
transversal 2289
trapp 0638
traquítica 0838
traquitoidea 0839
traslación 1566
traslación continental 1115
tratado de limitación de pruebas nucleares 0266
tratado de valor umbral 0266
tratamiento artificial 0492
trayectoria 0206
trayectoria 2372
trayectoria 4471
trayectoria braquistocrónica 0209
trayectoria de vuelo 4303
traza 1492
traza 1876
traza 4420
traza de estación 1880
traza de pastos 1881
traza de reptación 1879
trépano 4271
triangulación radial 4324
tribu 1931
tributario 2207
triclínico 0515
trigonal 0515
trimorfismo 1869
trinchera 4235
tripartito, método – 0152
trípoli 3371
triquita 0524
triquito 0770
tritio, datación – 0452
trituración fina 4150
trituradora cónica 4148
trombolito 3330
tromómetro 0118
tubería agarrada 4286
tubería a presión 4867
tubería de producción 4275
tubería de revestimiento 4275
tubo 2099
tubo 4274
tubo ascendente 2106
tubo de lavado 4751
tubo de muestreo 4748
tubo drenante 4811
tubo filtro 3795
tubo portatestigos 4272
tubo sacatestigos 4272
tufisita 1339
tufita 1332
tufita 3171
tufítico 1332
túmulo 1272
tundra 2636
túnel 4847
túnel de lava 1290
turba 3494
turba de bosque 3536
turba de esfagno 3543
turba de matorral 3545
turba de pradera 3536
turba de tundra 3547
turba de turbera alta 3542
turba de turbera baja 3535
turba humo-sapropélica 3548
turba lacustre 3531
turba límnica 3531
turbera 3532
turbera alta 3542
turbera alta, de – 3550
turbera baja 3533
turbera baja, de – 3550
turbera, destrucción de la – 3541
turbera de transición 3536
turbera reticulada 2636
turbera terrestre 3546
turbidez, corriente de – 2981
turbidita 2983
turbogénesis 3524
turbonificación 3524
turmalinización 0932
T-X, gráfica – 4463

U

ultrabásica 0604
ultrabasita 0604
ultramáfica 0600
ultramafita 0600
ultrametamorfismo 0882
ultramilonita 0916
ultramilonita 1541
ultravolcánico 1166
umbral 2796
umbral del lago 2335
umbral glaciar 2625
uniaxial 0553
unidad cartográfica de suelos 2672
unificación 3901
uniformitarismo 0006
unión 0586
unión, plano de – 0536
uralitización 0928
uranio-plomo, datación – 0434
uvala 2073

V

vacio 3064
vacuola 0874
vagilidad 1917
validez 1923
valoración 4216
valor actual 3897
valor actual, en – 3899
valor a plazo 4220
valor constante, en – 3899
valor equivalente 3675
valores extremos 4186
valor medio que mueve 4198
valor umbral 4627
valuación 4216
valle abierto 2276
valle anticlinal 2544
valle asimétrico 2274
valle, cabecera de – 2276
valle ciego 2083
vallecito 2281
valle colapsado 2085
valle colgado 2483
valle con perfil en U 2273
valle con perfil en V 2272
valle decapitado 2324
valle de depresión axial 2545
valle de dos ciclos 2275
valle de falla 2547
valle de fondo 2278
valle de hundimiento 2085
valle de la muerte 1402
valle de línea de falla 2547
valle en forma de U 2273
valle en forma de V 2272
valle, fondo del – 2278
valle glaciar escalonado 2622
valle glaciar suspendido 2627
valle interdunal 2188
valle marginal anteglaciar 2393
valle monoclinal 2544
valle muerto 2084
valle muerto 2328
valle policíclico 2275
valles desmembrada, red de – 2482
valle seco 2084
valles, formación de los – 2270
valle sinclinal 2544
valles integrados, red de – 2283
valle submarino 2795
valle sumergido 2282
valle suspendido 2282
valle suspendido 2483
valle tectónico 1590
valle tectónico 2543
valle truncado 2481
valle, vertiente de un – 2277
vapor dominante, sistema de – 0393
variación, coeficiente de – 4190
variación, diagrama de – 0697
variación diurna 4528
variación secular 0289
varianza 4190
variedad 1940
varilla de perforación 4270
variograma 4196
variola 0781
variolita 0781
variolítica 0877
variómetro 0288
varva 3128
varvita 3128
vegetación, estudio de la – 4645
vehículo espacial 4302
vela 2140
velocidad aparente 0208
velocidad aparente 4468
velocidad, capa de baja – 0100
velocidad critical primaria 2265
velocidad critical secundaria 2266
velocidad cuadrática media 4466
velocidad de corrimiento 3703
velocidad de fase 0169
velocidad de fase, método de – 0230
velocidad de filtración 3703
velocidad de grupo 0170
velocidad de grupo, método de – 0229
velocidad de intervalo 4467
velocidad de las partículas 4414
velocidad, función de – 4469
velocidad, índice de – 4732
velocidad, limen de – 2959
velocidad media 4465
vena de agua subterránea 3638
venilla de mineral 4039
Vening Meinesz, eje de – 1137
venita 1045
ventana 4234
ventana kárstica 2078
ventisquero 2353
vermicular 0774
vertical 0329
vertical, desviación de la – 0329
vértice 0705
vértice 1635
vertientes, retroceso de – 2602
vesícula 0874
vesicular 0874
vesicular 1305
vestíbulo 2101
vestigio 4233
veta 3499
veta grande 4062
veta intrusiva 0748
veta madre 4062
viabilidad, estudio de – 4791
vibración 4809
vibración esferoidal 0198
vibrocompactación 4809
vibroflotación 4810
vibrógrafo 0123
vidrio 0627
vidrio volcánico 1262
viento, acción del – 0043
virgación 1675
viscosidad 4671
vitreno 3580
vitreo 0484
vitrificación 0708
vitrificación 0897
vitrofídica 0823
vitrófiro 0823
vitroporfídica 0823
voladura 4822
voladura controlada 4823
voladura, índice de – 4820
volar 4814
volátiles, proporción de carbon y – 3601
volátil, materia – 3597
volcán 1411
volcán aislado 1446
volcán apagado 1155
volcán compuesto 1429
volcán compuesto 1441
volcán con cono en cono 1445
volcán con cráter central 1434
volcán de arena 3071
volcán de barro 3071
volcán de barro 1395
volcán de estructura imbricada 1421
volcán de fango 1700
volcán de lava 1420
volcán embrionario 1222
volcán en actividad 1153
volcán encañado 1475
volcán en escudo 1420
volcán en ruinas 1476
volcán en subactividad 1153
volcanes en escudo, grupo de – 1450
volcánico 1148
volcánico 3989
volcán latente 1155
volcán maclado 1442
volcanoclástico 3172
volcán simple 1441
volcán subacuático 1415
volcán subaéreo 1415
volcán submarino 1416
volcán superficial 1413

volcán surcado 1474
voltaje 4556
volumen, control automático de – 4425
volumen, control automático de – 4508
volumen, porcentaje en – 4109
volumétrico, método – 3871
vuelco 2811
vuelco volcada 2811
vuelo, línea de – 4303
vuelo, línea de – 4358
vulcanismo 1147
vulcanismo epigenético 1152
vulcanismo extrusivo 1207
vulcanismo orogénico 1151
vulcanismo subreciente 1150
vulcanogenético 1148
vulcanología 1146

W

WDC 0082
Wentworth , escala de – 3163
Widmanstätten, figuras de – 0067
Wulff, fasilla de – 1015
Wulff, red de – 1015

X

xenoblástico 0980
xenoblasto 0980
xenocristal 0663
xenolito 0663
xenomórfica 0808
xenomórfico 0752
xenotermal 3984
xenotópico 3471
xilita 3495
xilófago 1857

Y

yacimiento 3754
yacimiento 3755
yacimiento 3773
yacimiento, comportamiento del – 3846
yacimiento filoniano 4038
yacimiento interestratificado 4021
yacimiento magmático de cristalización precoz 3975
yacimiento magmático de cristalización tardía 3975
yacimiento mineral 3910
yacimiento, modelo de – 3866
yacimientos, formación magmática de – 3946
yacimiento, volumen bruto del – 3872
yacimiento zonado 4015
yermo volcánico 1378
yesificación 3459
yeso 3381
Young, módulo de – 4695
yuxtaposición, plano de – 0536

Z

zanja de impermeabilización 4861
zapata 4835
zapata corrida 4837
zapata individual 4837
Zavaritsky, diagrama de – 0699
zócalo 1572
zócalo 2560
zona 0362
zona 0519
zona 1717
zona bioestratigráfica 1769
zona de asociación 1774
zona de distribución local 1784
zona de nivelamiento 2604
zonado 3907
zonado sferulítico 4102
zona filogenética 1779
zona fracturada 4821
zona invadida 4619
zona lavada por el filtrado 4620
zonalidad 3907
zona metamórfica 0993
zona petrolífera 3775
zona volcánica 1447
zonificación 0474
zonificación 0533
zonificación inversa 0475
zonificación normal 0475
zónula 1782
zoobentos 1852
zooclorelas 1845
zooplancton 1800
zooxantelas 1845